Technology Manual

Dorothy Wakefield • Kathleen McLaughlin
Beverly Dretzke • Brent Timothy

Statistics

Informed Decisions Using Data

SECOND EDITION

Michael Sullivan, III

Upper Saddle River, NJ 07458

Editor-in-Chief: Sally Yagan
Executive Acquisitions Editor: Petra Recter
Supplement Editor: Joanne Wendelken
Executive Managing Editor: Kathleen Schiaparelli
Assistant Managing Editor: Karen Bosch
Production Editor: Jennifer Zisa
Supplement Cover Manager: Paul Gourhan
Supplement Cover Designer: Christopher Kossa
Manufacturing Buyer: Ilene Kahn
Manufacturing Manager: Alexis Heydt-Long

Pearson Prentice Hall
Pearson Education, Inc.
Upper Saddle River, NJ 07458

Printed in the United States of America

10 9 8 7 6 5 4 3 2 1

ISBN 0-13-173151-3

Pearson Education Ltd., *London*
Pearson Education Australia Pty. Ltd., *Sydney*
Pearson Education Singapore, Pte. Ltd.
Pearson Education North Asia Ltd., *Hong Kong*
Pearson Education Canada, Inc., *Toronto*
Pearson Educación de Mexico, S.A. de C.V.
Pearson Education—Japan, *Tokyo*
Pearson Education Malaysia, Pte. Ltd.

MINITAB™ Manual

Dorothy Wakefield • Kathleen McLaughlin

Statistics

Informed Decisions Using Data

SECOND EDITION

Michael Sullivan, III

▸ Introduction

The MINITAB Manual is one of a series of companion technology manuals that provide hands-on technology assistance to users of Sullivan *Statistics: Informed Decisions Using Data, 2nd Edition.*

Detailed instructions for working selected examples and problems from *Statistics: Informed Decisions Using Data* are provided in this manual. To make the correlation with the text as seamless as possible, the table of contents includes page references for both the Sullivan text and this manual.

All of the data sets referenced in this manual are found on the data disk packaged in the back of every new copy of Sullivan *Statistics: Informed Decisions Using Data.* If needed, the MINITAB files (.mtp) may also be downloaded from the texts' companion website at www.prenhall.com/Sullivan.

Contents:

Getting Started with MINITAB

▸ Using MINITAB Files

MINITAB is a Windows-based Statistical software package. It is very easy to use, and can perform many statistical analyses. When you first open MINITAB, the screen is divided into two parts. The top half is called the Session Window. The results of the statistical analyses are often displayed in the Session Window. The bottom half of the screen is the Data Window. It is called a Worksheet and will contain the data.

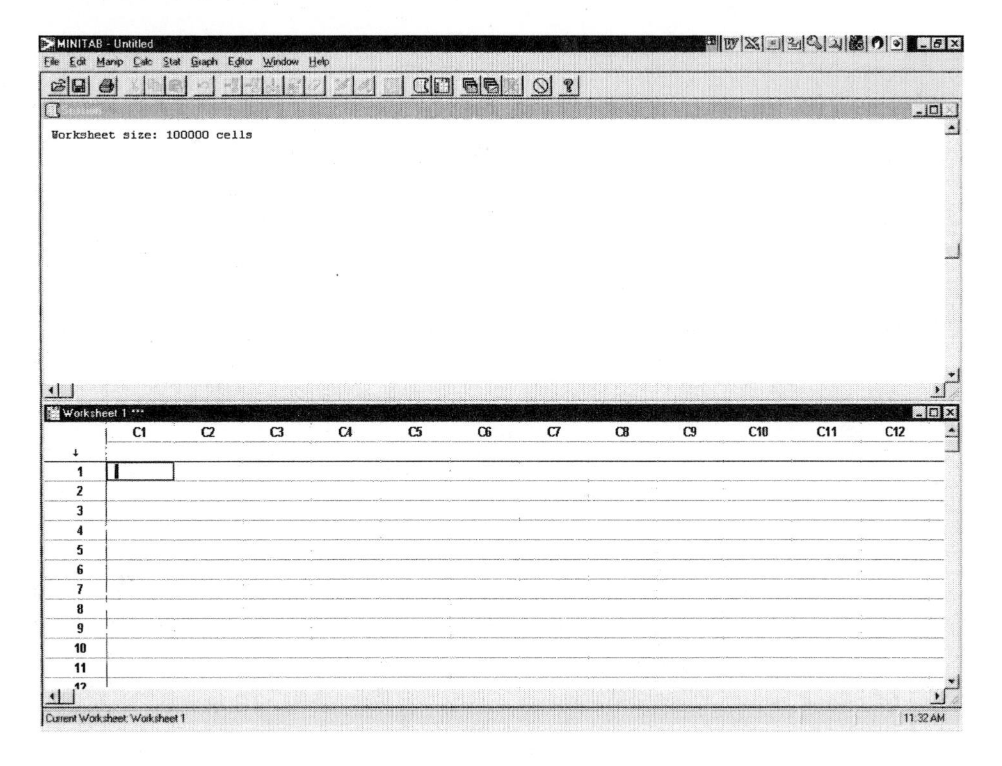

The data can either be entered directly into the Worksheet, or saved worksheets can be opened and used.

▸ Entering Data into the Data Window

To enter the data into the Data Window, you must first click on the bottom half of the screen to make the Data Window active. You can tell which half of the screen is active by the blue bar going across the screen. In the picture below, notice that the blue bar is in the middle of the screen, highlighting **Worksheet 1.** This indicates that the Data Window is active. The bar will be gray if the Window is not active. (Notice the Session Window bar is gray.)

In MINITAB, the columns are referred to as C1, C2, etc. Notice that there is an empty cell directly below each heading C1, C2, etc. This cell is for a column name. Column names are optional because you can refer to a column as C1 or C2, but a name helps to describe the data contained in a column. Enter the data beginning in cell 1. Notice that the cell numbers are located in the leftmost column of the worksheet.

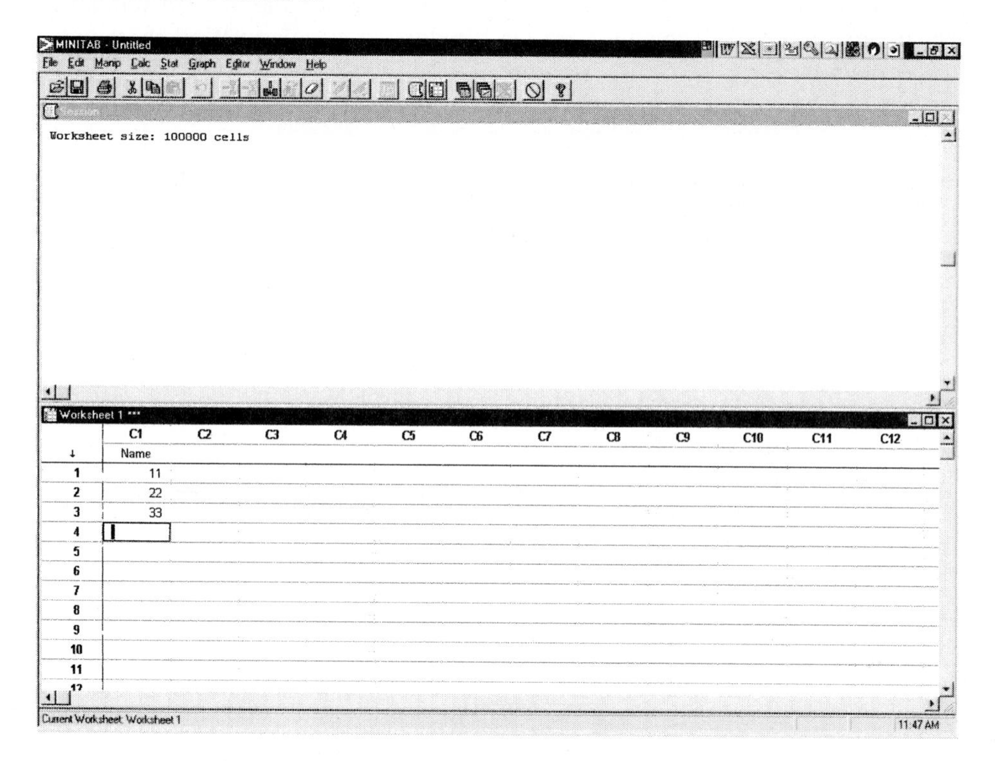

▸ Opening Saved Worksheets

Many of the worksheets that you will be using are saved on the enclosed data disk. To open a saved worksheet, click on **File → Open Worksheet.** The following screen will appear.

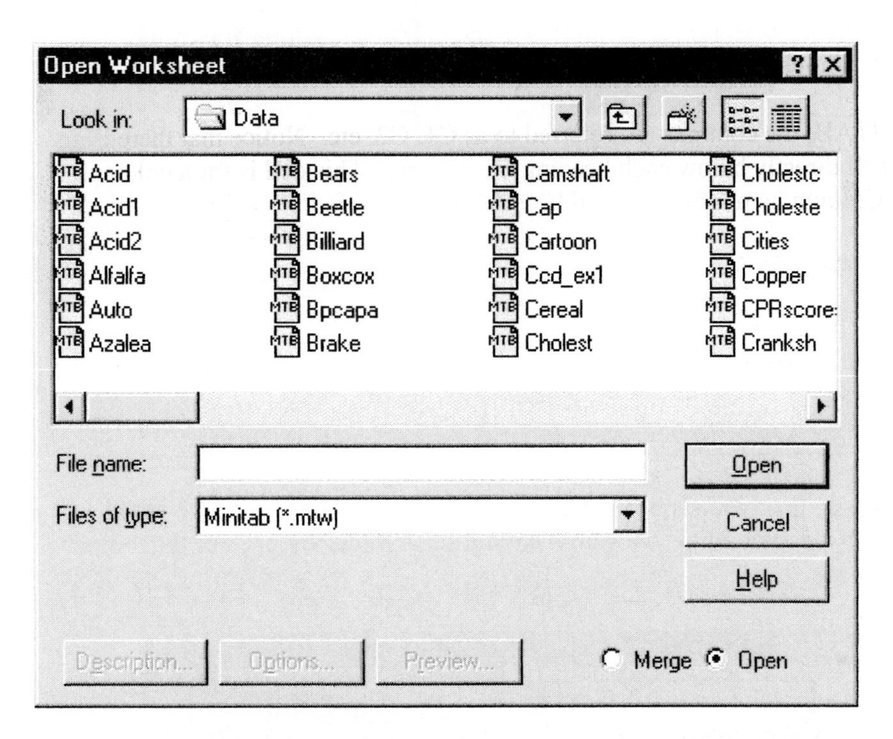

First, you must tell MINITAB where the data files are located. Since the data files are located on a CD data disk, you must tell MINITAB to **Look In** the CD drive. To do this, click on the down arrow to the right of the top input field and select the CD drive by double-clicking on it.

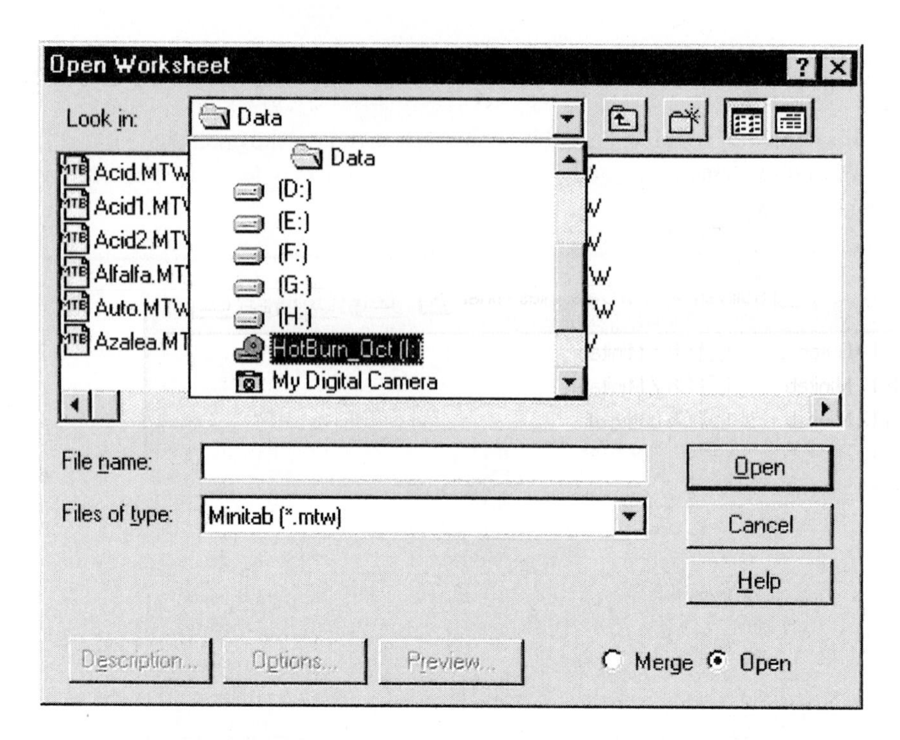

When you do this, you should see three folders listed. Select the MINITAB folder with a double-click. Now you should see a folder for each of the thirteen chapters of the book.

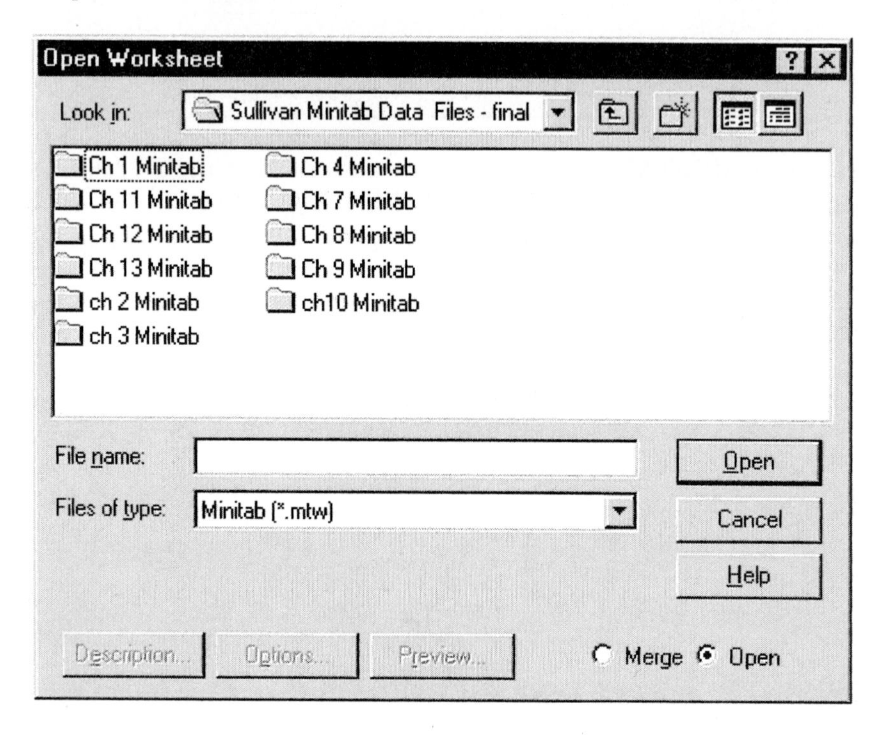

All data files are saved as MINITAB Portable worksheets and have the extension **.mtp.** Click on the down arrow for the field called **Files of type** and select **Minitab Portable (*.mtp).**

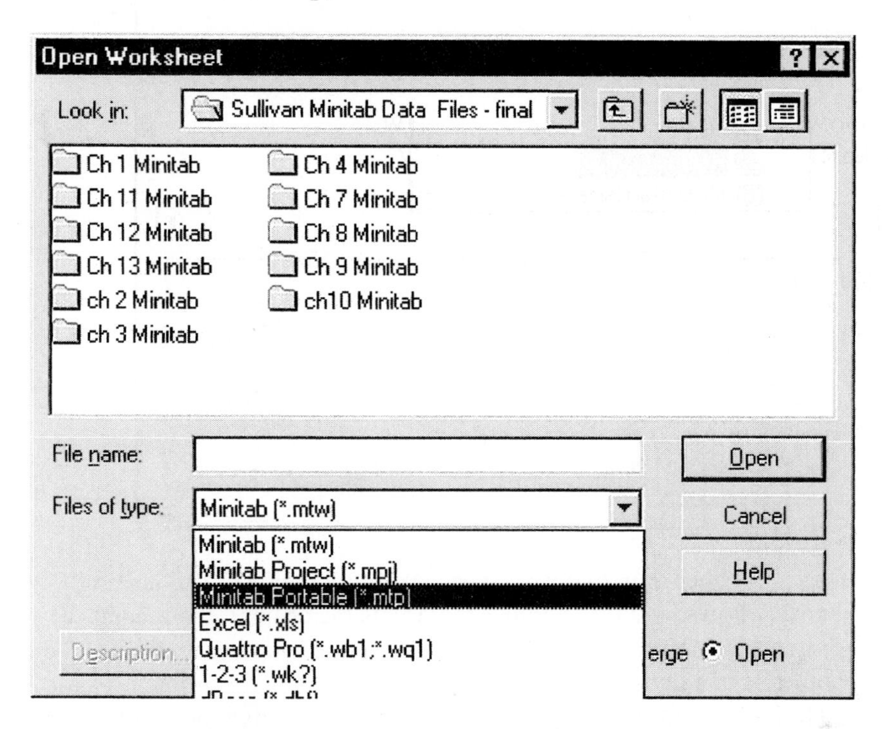

Now, select the folder called **Ch 1 Minitab** (by double-clicking) and you should see all the MINITAB worksheets for Chapter 1.

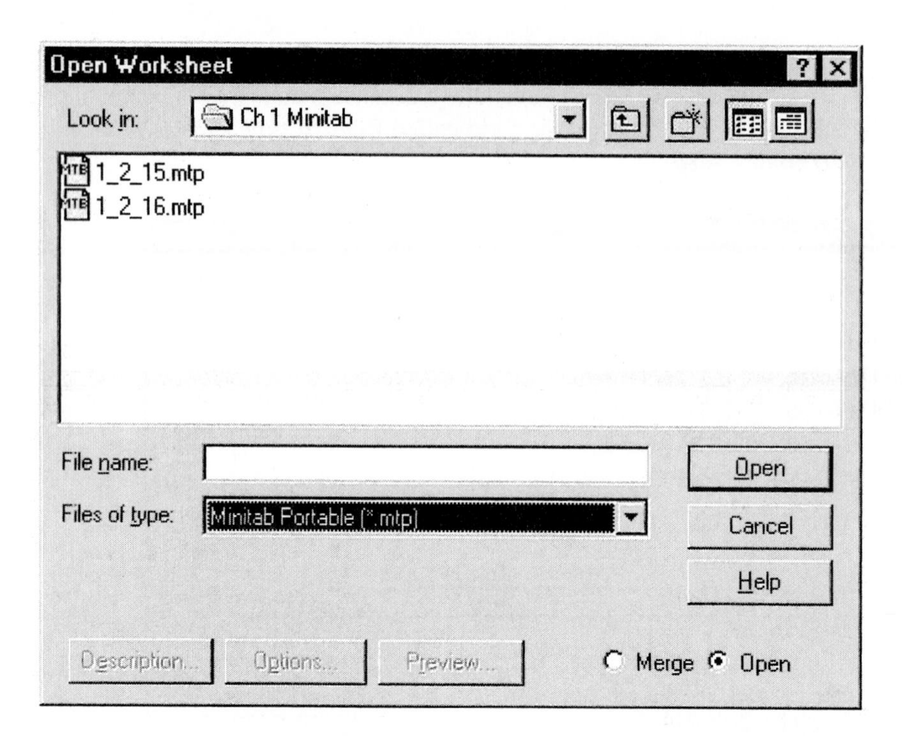

As you can see, **Ch 1** has only two worksheets saved on the CD. The naming convention is as follows: the first number represents the chapter, the second number represents the section of the chapter, and the third number represents the problem number. For example, file **1_2_15** contains the data for problem 15 in chapter 1 section 2. To open the worksheet **1_2_15**, double-click on it and the worksheet should appear in the Data Window.

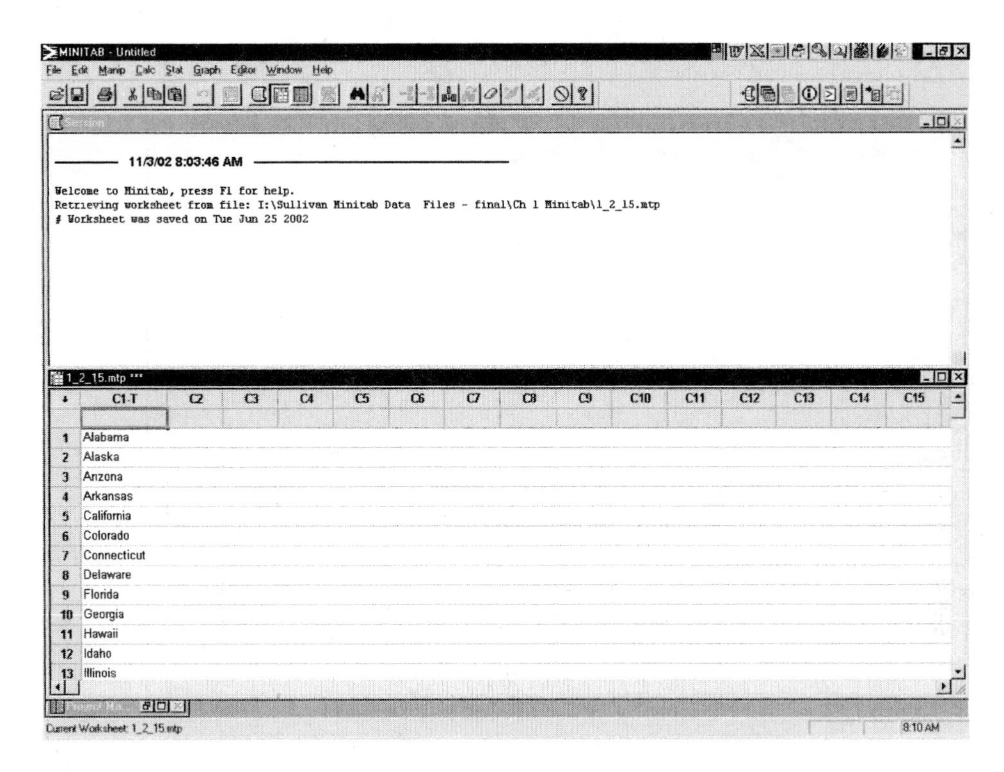

You are now ready to begin analyzing the data and learning more about MINITAB.

Data Collection

Section 1.2

Example 3 (pg. 18) Generating a Simple Random Sample

To select 5 clients randomly from a list of 30, first you must represent each client with a number. So, store the numbers 1 to 30 in C1. Click on **Calc → Make Patterned Data → Simple Set of Numbers.** You should **Store patterned data in** C1. The numbers will begin **From the first value** 1 and go **To last value** 30 **In steps of** 1.

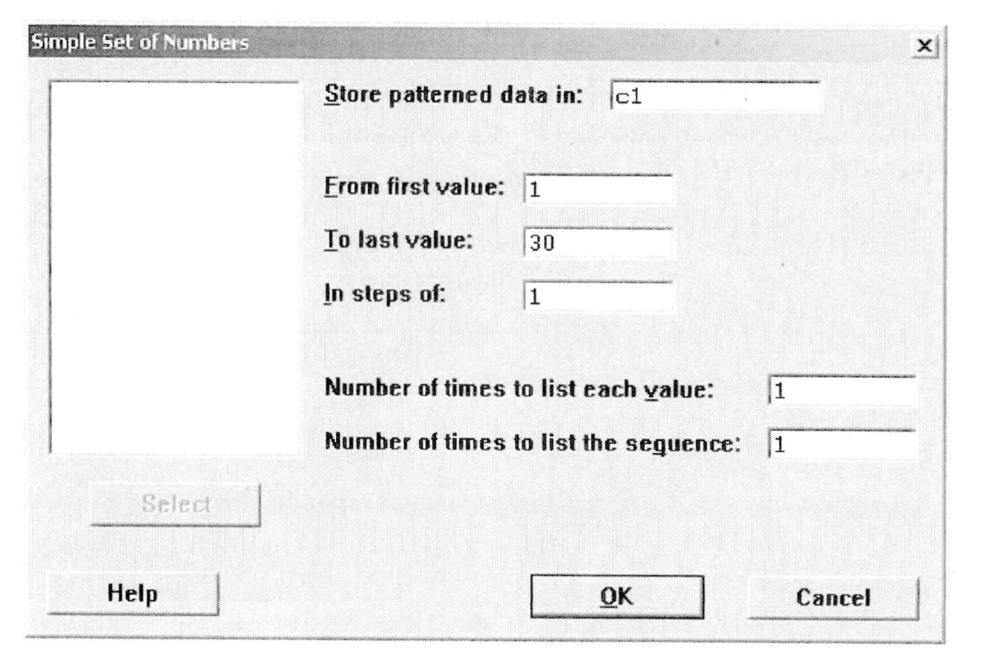

Click on **OK** and the numbers 1 to 30 should be in C1 of the Data Window.

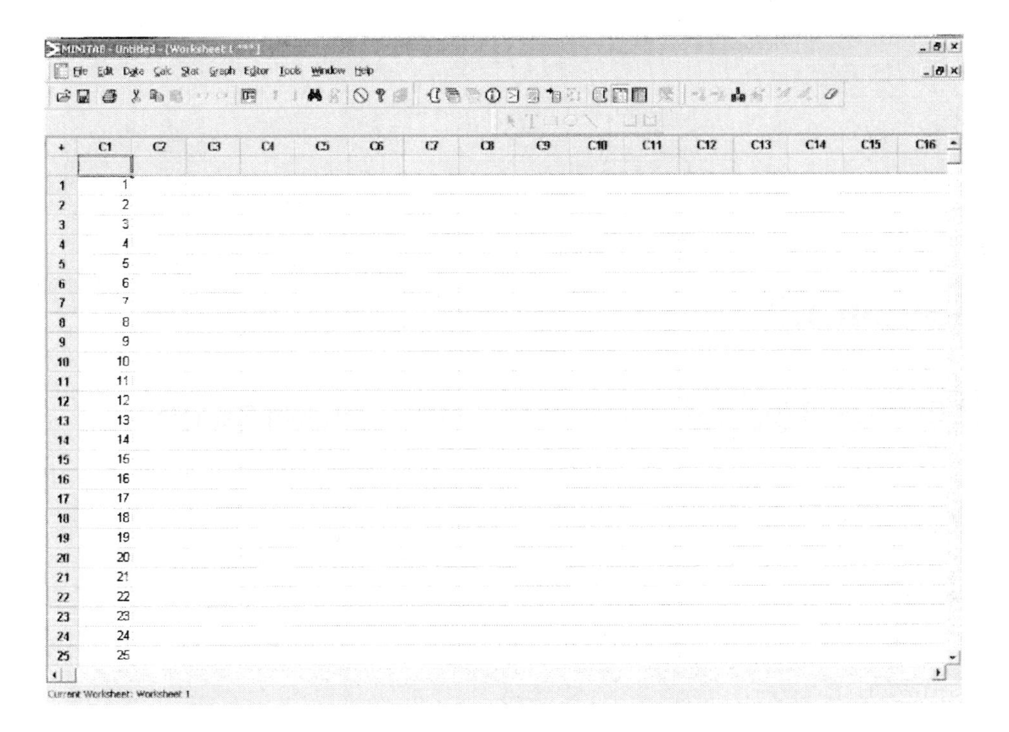

Next, you'd like to take a random sample of 5 clients. Since you do not want repeats, you will be sampling without replacement. This is the default type of sampling in MINITAB, so you won't have to do anything special for this sample. Click on **Calc → Random Data → Sample from columns.** You need to **Sample** 5 **rows from column** C1 and **Store samples in** C2.

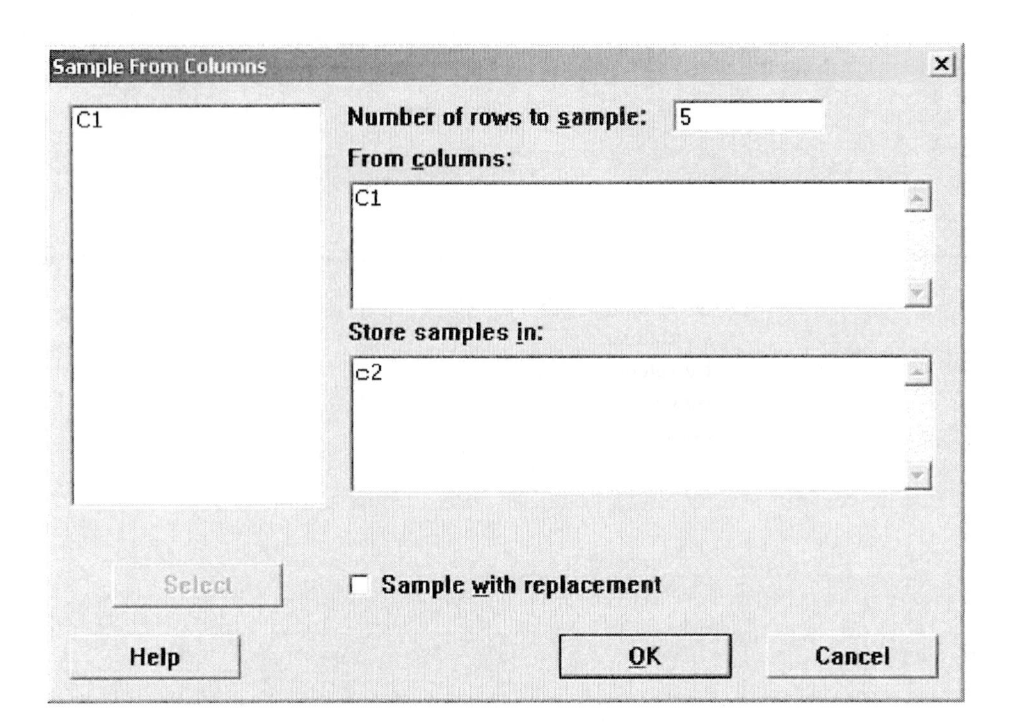

Click on **OK** and there should be a random sample of 5 client numbers in C2.

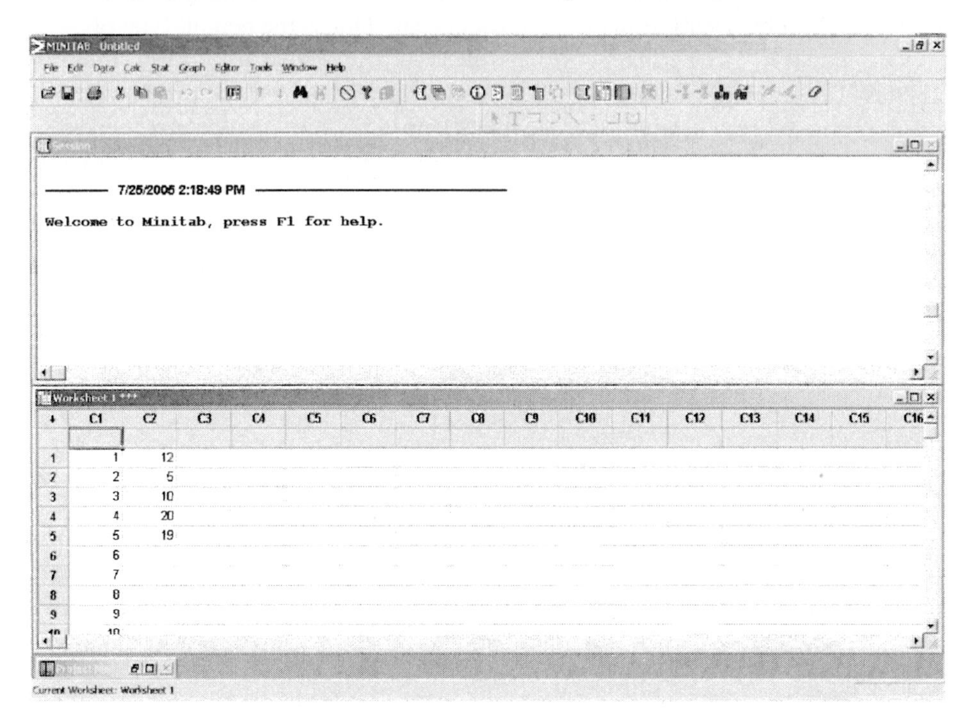

To order the sample list, click on **Data → Sort.** You should **Sort column** C2, **By column** C2, and *Store sorted data in* **Column of current worksheet** C3.

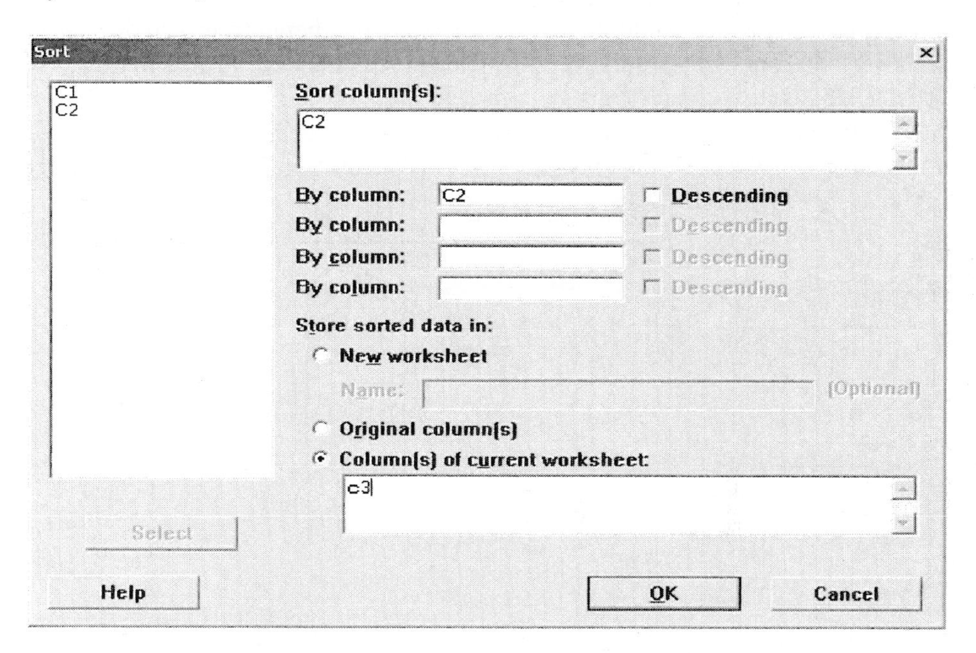

Click on **OK** and C3 should contain the sorted sample of 5 clients.

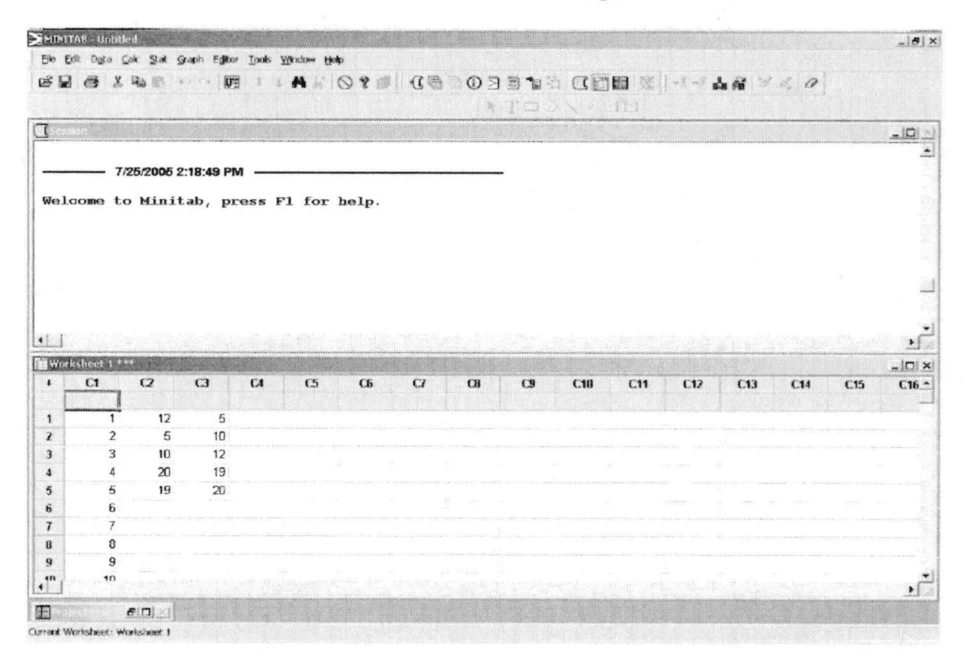

Since this is a *random* sample, each student will have different numbers in C2 and C3. ◀

▸ Problem 21 (pg. 21) Obtaining a Simple Random Sample

In this problem, notice that the 50 states are already numbered from 1 to 50. To select 10 states randomly from the 50, you need to store the numbers 1 to 50 in C1. Click on **Calc → Make Patterned Data → Simple Set of Numbers.** You should **Store patterned data in** C1. The numbers will begin **From the first value** 1 and go **To last value** 50 **In steps of** 1. Click on **OK** and the numbers 1 to 50 should be in C1 of the Data Window.

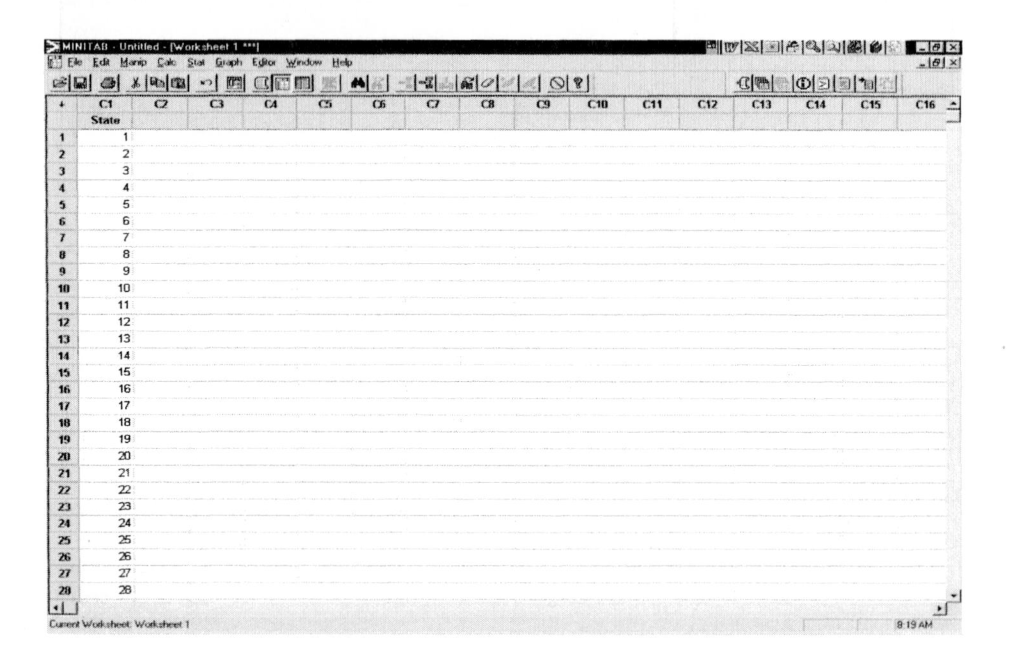

Next, to randomly sample 10 states, click on **Calc → Random Data → Sample from columns.** You need to **Sample** 10 **rows from column** C1 and **Store samples in** C2.

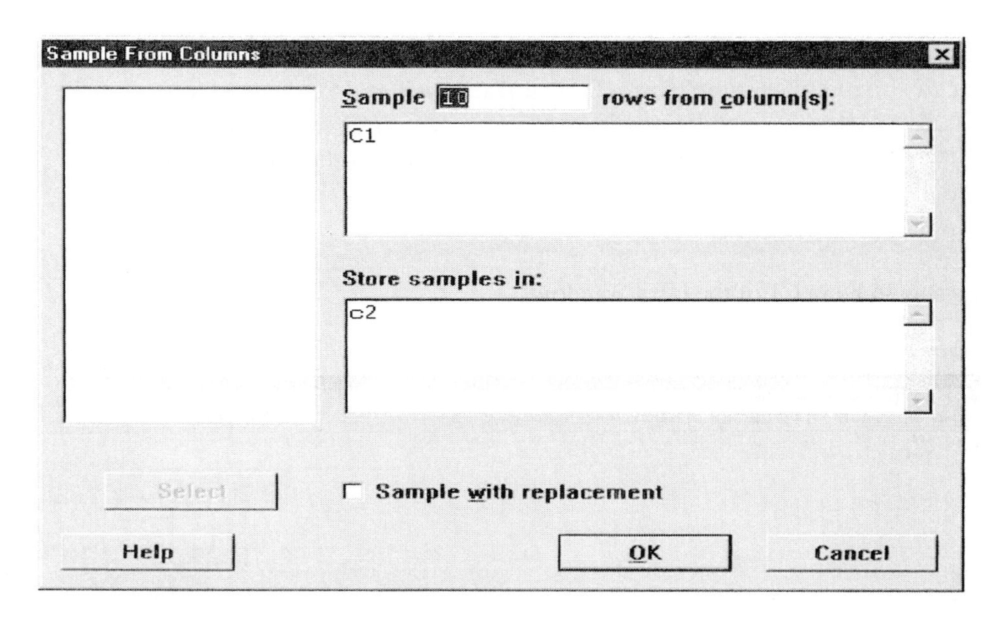

Click on **OK** and the sample of 10 state numbers should be in C2 of the Data Window.

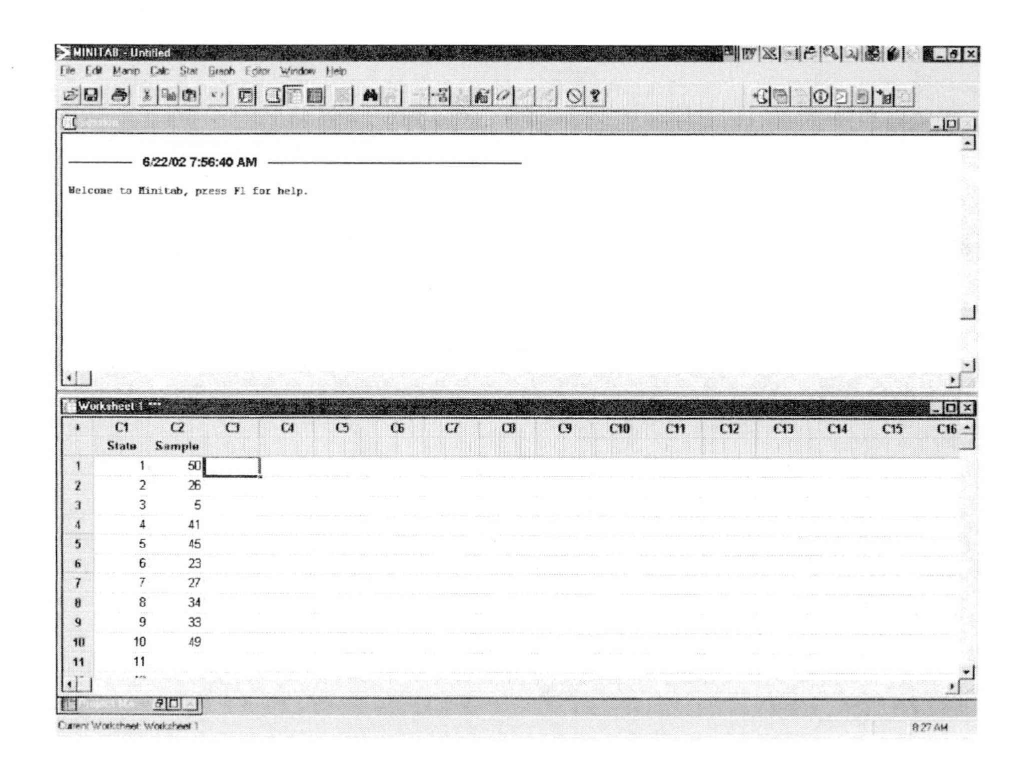

Finally, to order the sample list, click on **Data → Sort.** You should **Sort column** C2, **By column** C2, and *Store sorted data in* **Column of current worksheet** C3. Click on **OK** and the sorted sample of 10 state numbers should be in C3 of the Data Window.

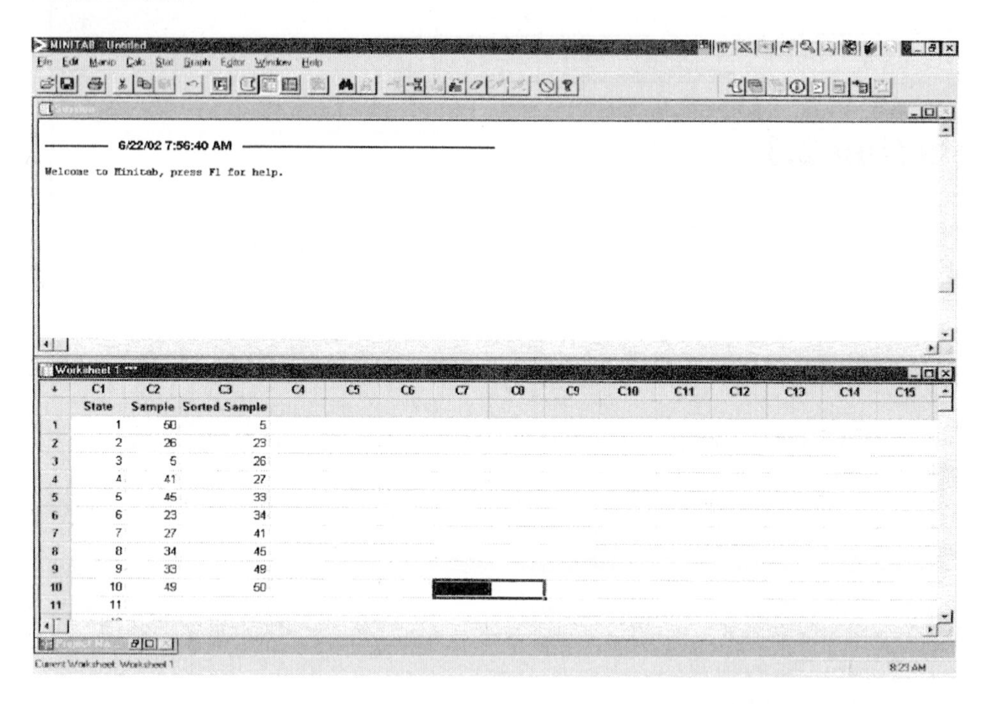

↓	C1 State	C2 Sample	C3 Sorted Sample
1	1	50	5
2	2	26	23
3	3	5	26
4	4	41	27
5	5	45	33
6	6	23	34
7	7	27	41
8	8	34	45
9	9	33	49
10	10	49	50
11	11		

Look at your random sample above. Using the list of states on page 21 of your textbook, see which states are in this sample. From the list, you can see that State 5 is California, State 23 is Minnesota, State 26 is Montana, etc.

For part b of this exercise, repeat these steps to obtain a second random sample of size 10. You will get a different set of 10 states since your sample is random.

Organizing and Summarizing Data

Section 2.1

Example 4 (pg. 63) Constructing Frequency and Relative-Frequency Bar Graphs, Pareto Charts

To create this bar chart, you will need to enter the data into the data worksheet. In column 1, type in the Body Part (as shown below). Label the column by typing "Body Part" in the gray cell in C1. Label column 2 "Frequency", and type in the frequencies.

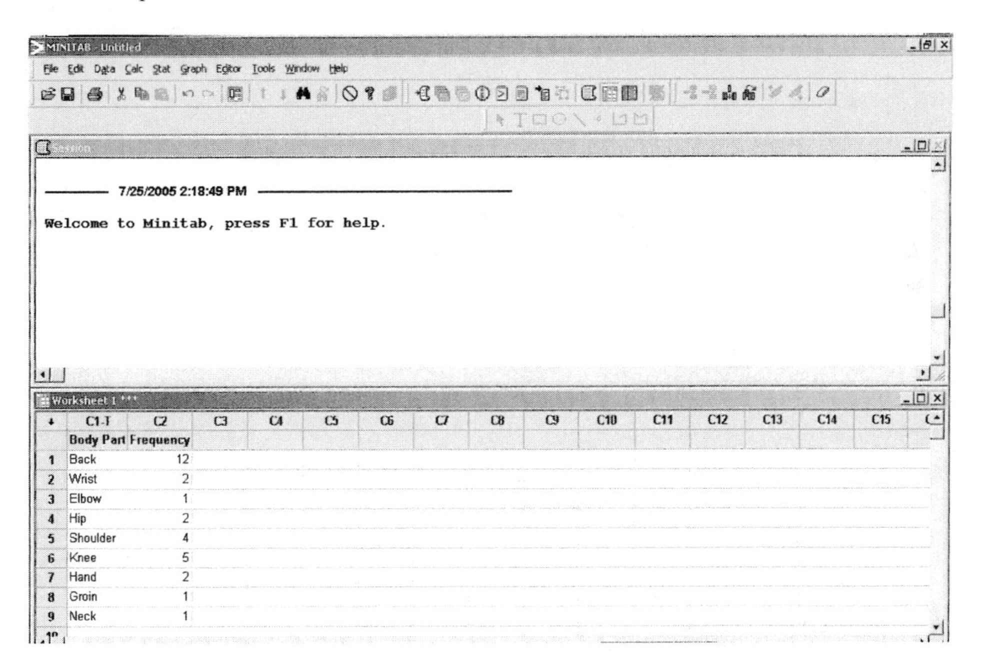

below. Next tell Minitab what you want to calculate. Click in the box beneath **Expression**, then type in **C2 / 30**. This tells Minitab to divide each value in C2 by 30.

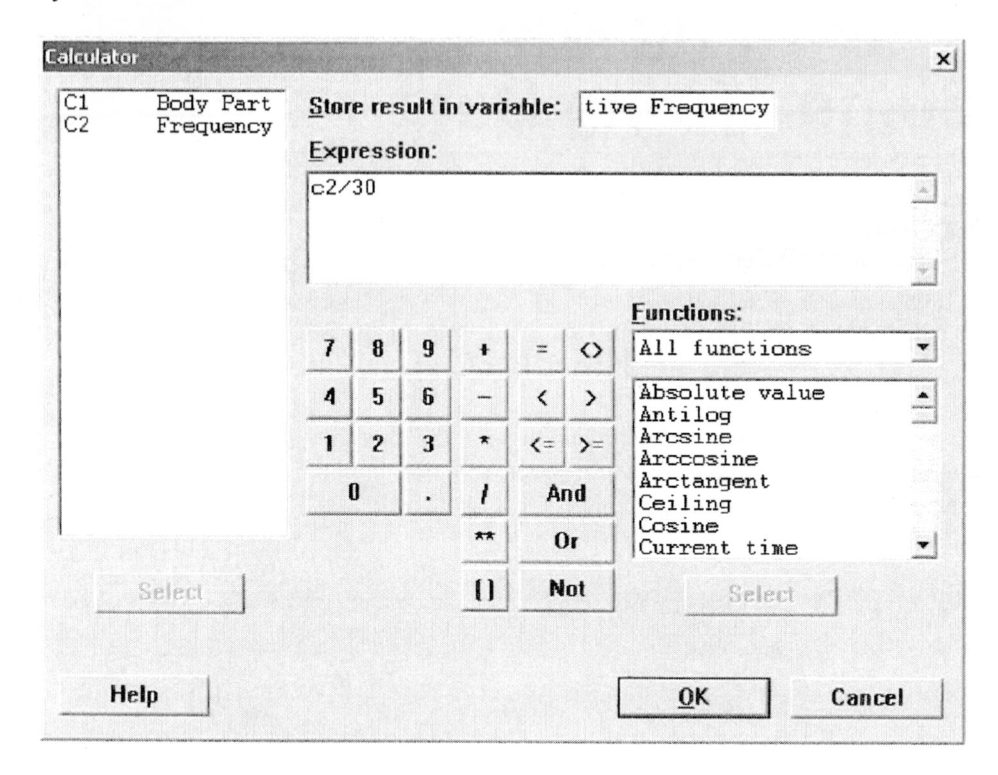

Click on **OK** and you should see the relative frequencies in C3.

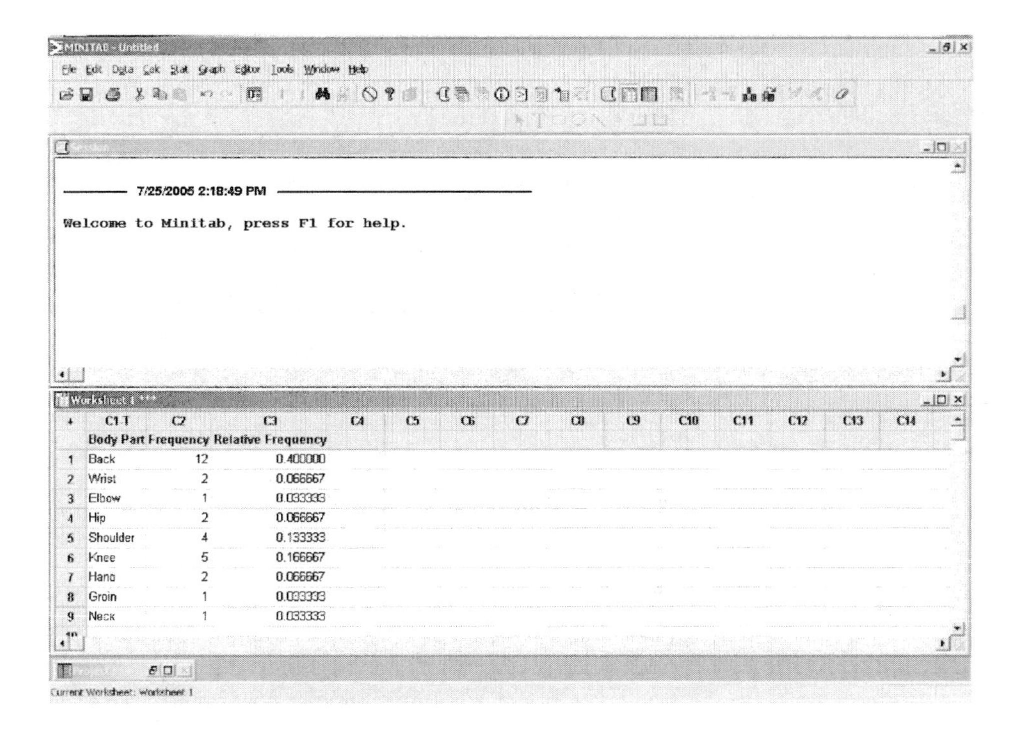

Now you are ready to make the bar graph. Click on: **Graph** → **Bar chart**. You will see a small pop-up that allows you to choose which type of bar chart you want. Since the information is in two columns (C1 and C2), the **Bars represent:** values from a table. Highlight a **Simple** bar chart (icon at the left of the top row), and click **OK**.

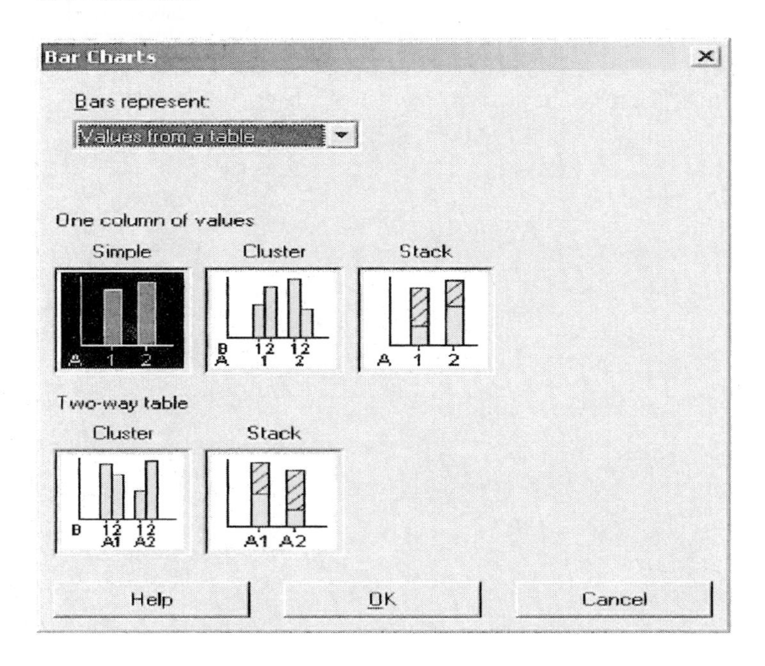

Now you must tell Minitab which variables to graph. Click in the field below **"Graph variables"**. This variable is the numerical measurement. In this case, it is the frequencies. Minitab is expecting a *column* here. When you click on this field, a list of numeric columns appears on the left side of the screen. You can select a column by double-clicking on it, or after a single click highlights the variable, click on **SELECT** at the bottom of the screen. Select C2, the frequencies. Next click on the field below **"Categorical variable"**. Select **C1**, the body parts.

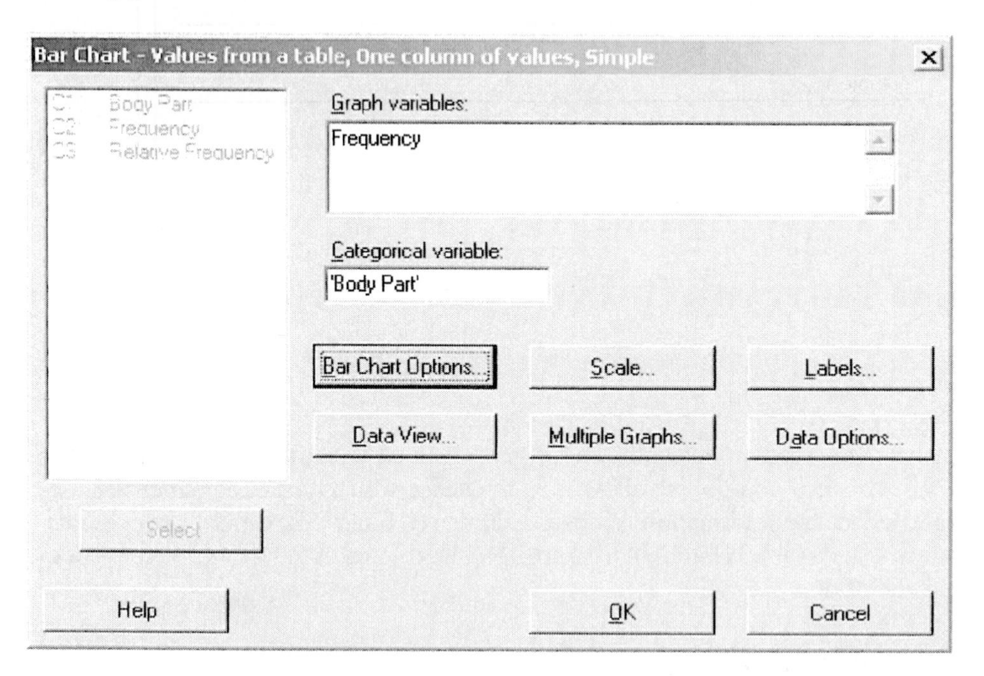

Next, click on the "**Labels**" button. In the top row, enter an appropriate **title** for the graph. If desired, you can also enter a sub-title and footnotes.

Bar Chart - Labels

Titles/Footnotes | Data Labels

Title:

Type of Rehabilitation

Subtitle 1:

Subtitle 2:

Footnote 1:

Footnote 2:

Help | OK | Cancel

Click on OK twice to view the chart.

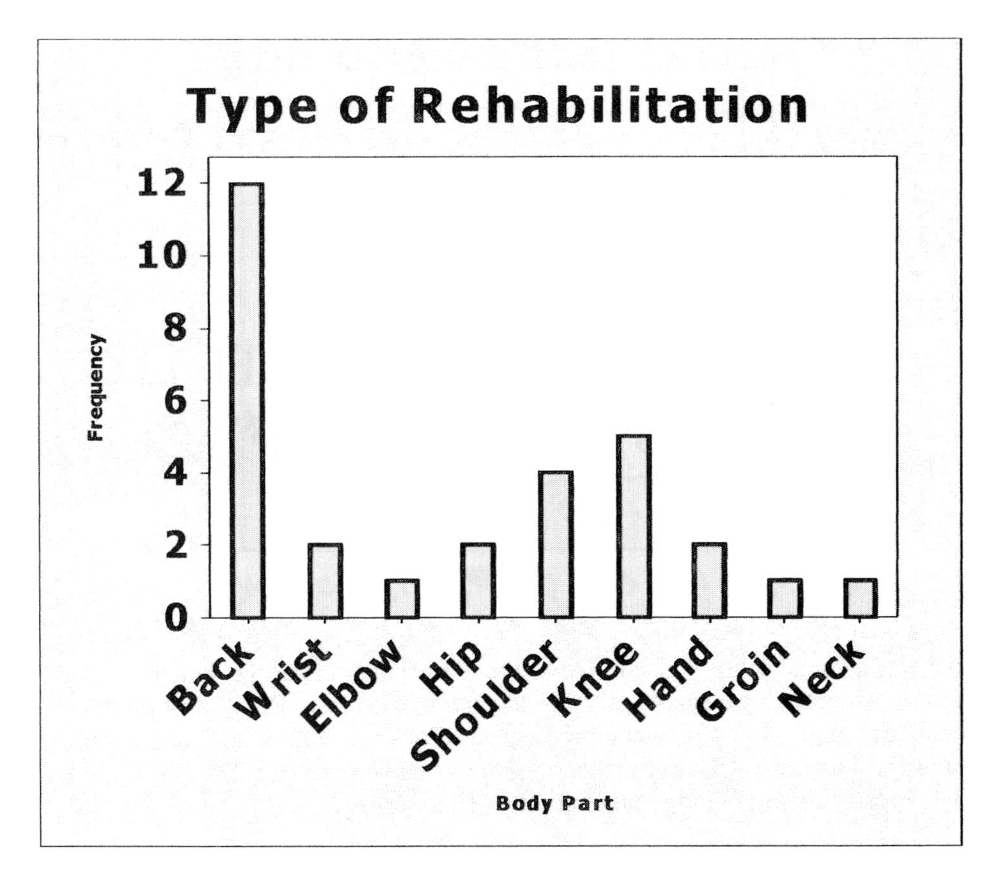

To add a title to the graph, click on the **Labels** button. Enter an appropriate title. Click on **OK** twice and you should see the clustered bar chart.

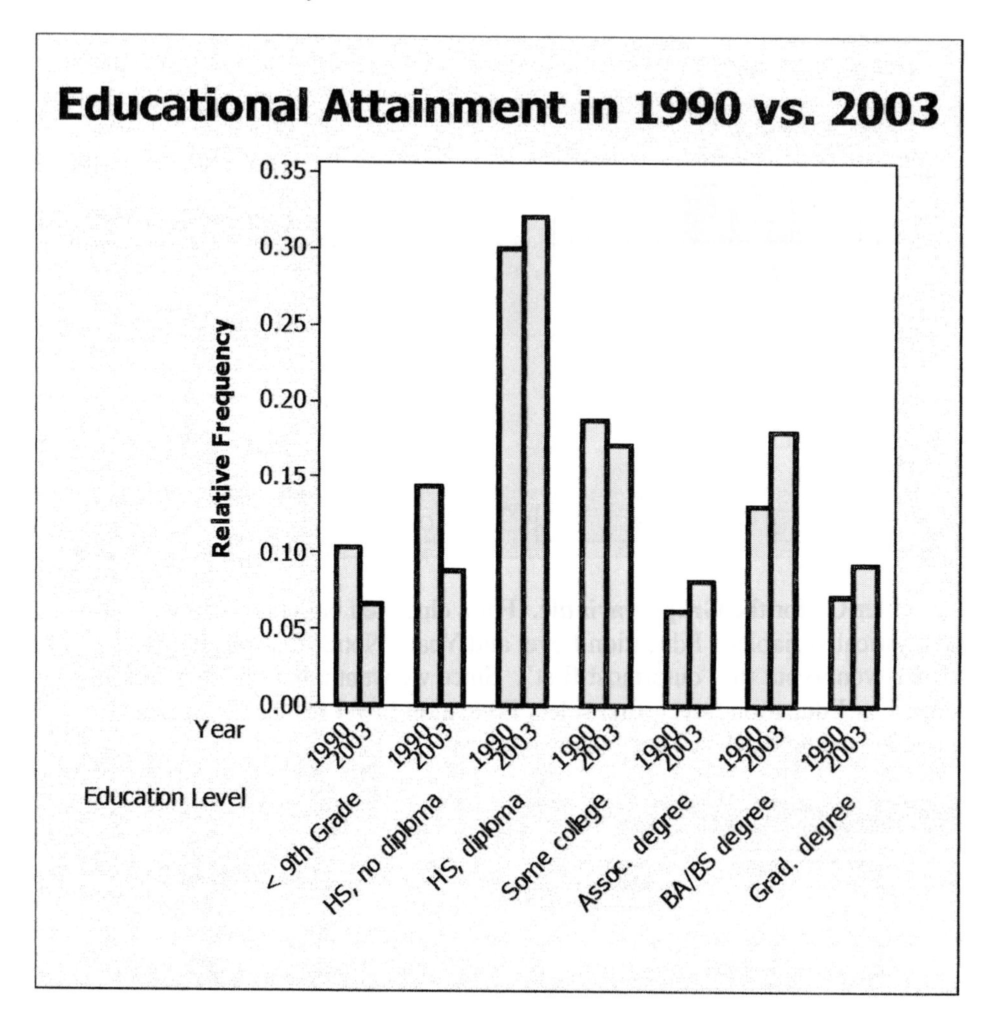

▸ Example 6 (pg. 66) Constructing a Pie Chart

This example uses the Educational Attainment data found in Table 6. It uses only the data for Year 2003. Enter the frequencies into a Minitab worksheet.

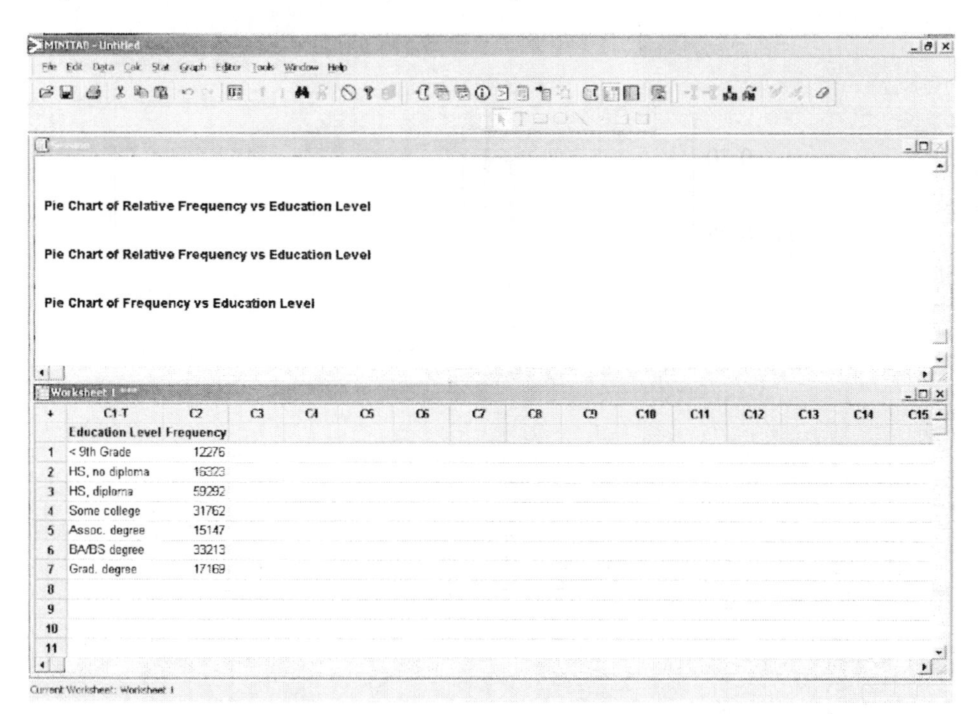

To create a Pie chart, click on **Graph → Pie chart.** On the screen that appears, select **Chart values from a table** by clicking on the small circle to the left of it. Select Education Level for **Categorical variable** and Frequency for **Summary variable**. Click on the **Labels** button, and type in an appropriate **Title.** Click on the tab **Slice labels** and select **Percent** and **Draw a line from label to slice.** Click on **OK** twice to view the pie chart.

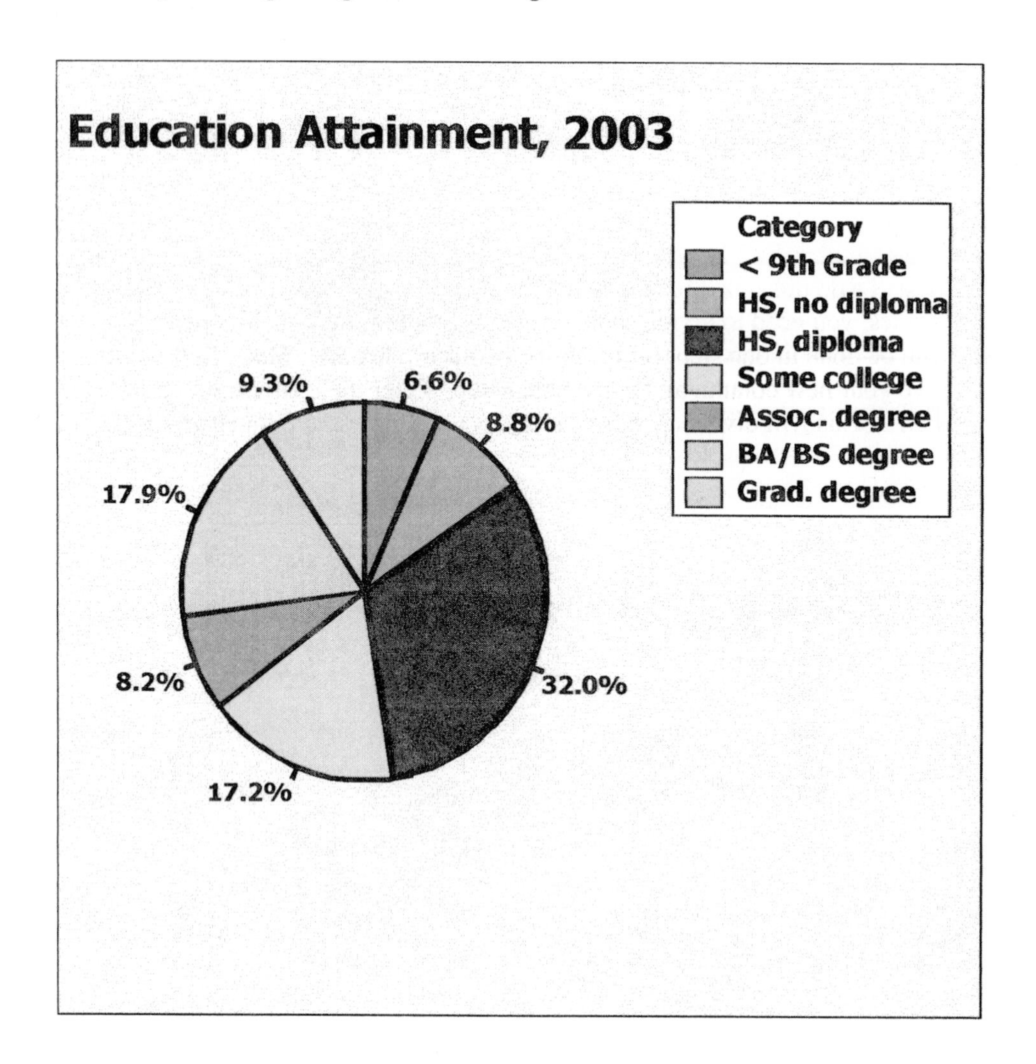
Education Attainment, 2003
Category
< 9th Grade
HS, no diploma
HS, diploma
Some college
Assoc. degree
BA/BS degree
Grad. degree
9.3%
6.6%
8.8%
17.9%
8.2%
32.0%
17.2%

▸ Problem 23 (pg. 71) Educational Attainment

Open worksheet **2_1_23.** **(Note:** You may want to shorten the Education level titles. One possibility is: HS, no diploma, HS diploma, Some college, Assoc degree, BA/BS degree, Grad degree. To change the titles, just type right over the ones in the worksheet.) First calculate relative frequencies for both Males and Females. To do this, click on **Calc → Calculator.** To calculate the relative frequencies, you need to divide each frequency by the sum of the frequencies. This can be done in one step. If you **Store result in variable:** Male_RelFreq, this will give your new column an appropriate name. Enter the **Expression** "C2 / sum(C2)". This will divide each frequency by the sum of the frequencies in C2.

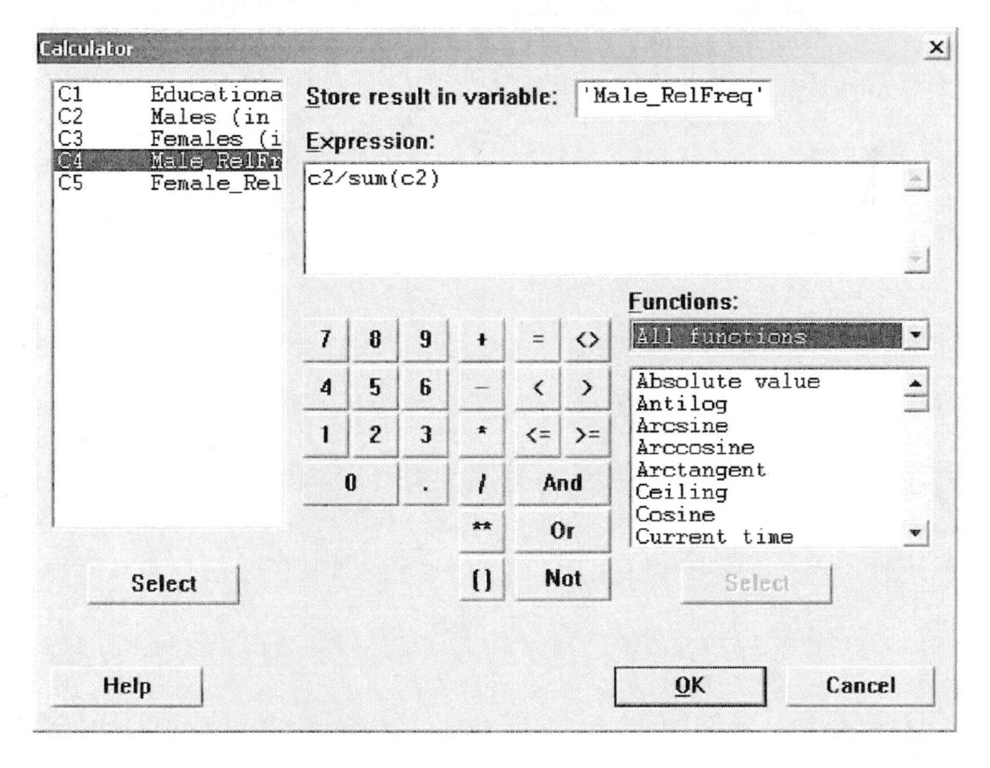

Click on **OK** and the relative frequencies should be in C4. Repeat this procedure to calculate the relative frequencies for Females. Name the new column "Female_RelFreq". The **Expression** will be "C3 / sum(C3)".

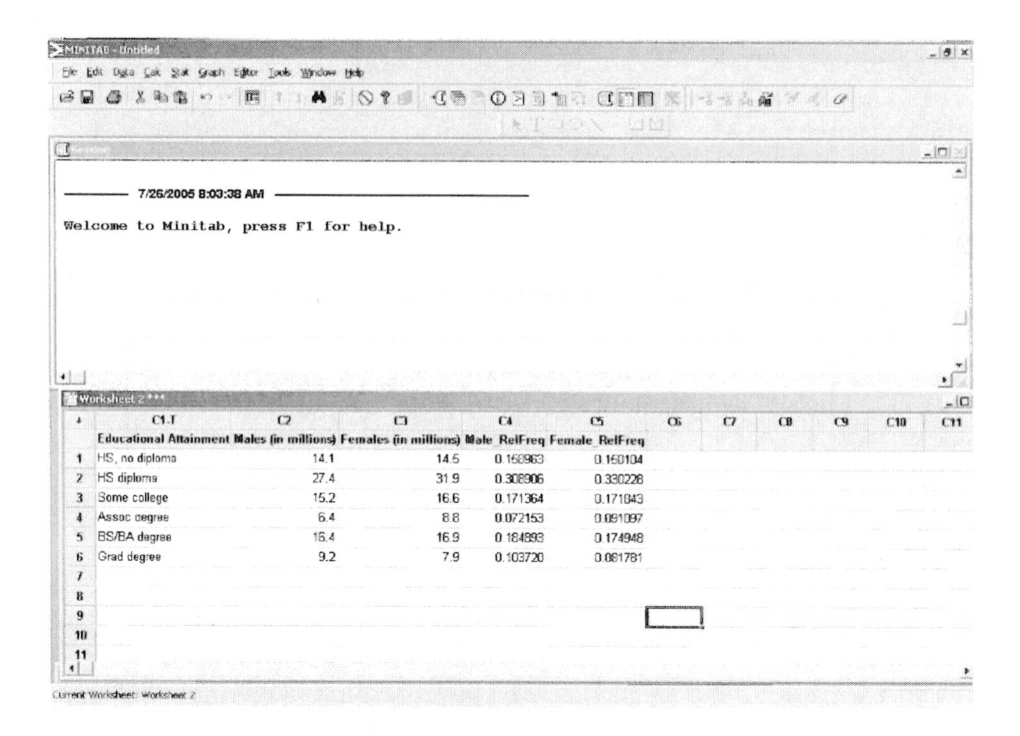

	C1-T	C2	C3	C4	C5
	Educational Attainment	Males (in millions)	Females (in millions)	Male_RelFreq	Female_RelFreq
1	HS, no diploma	14.1	14.5	0.158963	0.150104
2	HS diploma	27.4	31.9	0.308906	0.330228
3	Some college	15.2	16.6	0.171364	0.171843
4	Assoc degree	6.4	8.8	0.072153	0.091097
5	BS/BA degree	16.4	16.9	0.184893	0.174948
6	Grad degree	9.2	7.9	0.103720	0.081781

To make the bar graphs, click on **Graph** → **Bar Chart**. On the pop-up screen that appears, the **Bars represent "values from a table"**. Highlight a **Simple** bar chart (icon on the left of the top row), and click **OK**.

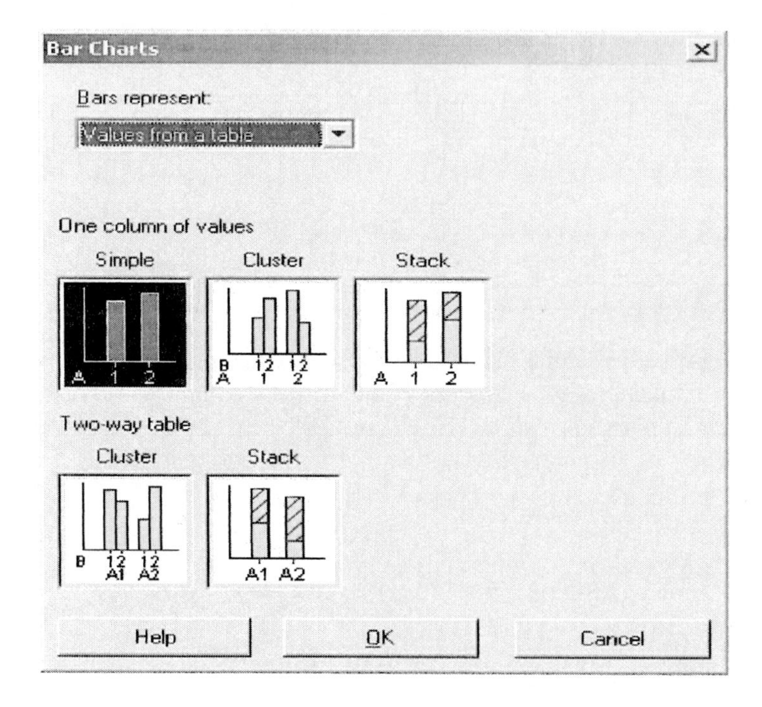

On the main bar chart screen, enter Male_RelFreq as the **Graph variable** and Education Level as the **Categorical variable.** Click on the **Labels** button, and enter an appropriate title (Educational Attainment for Males). Click on **OK** twice to view the bar graph.

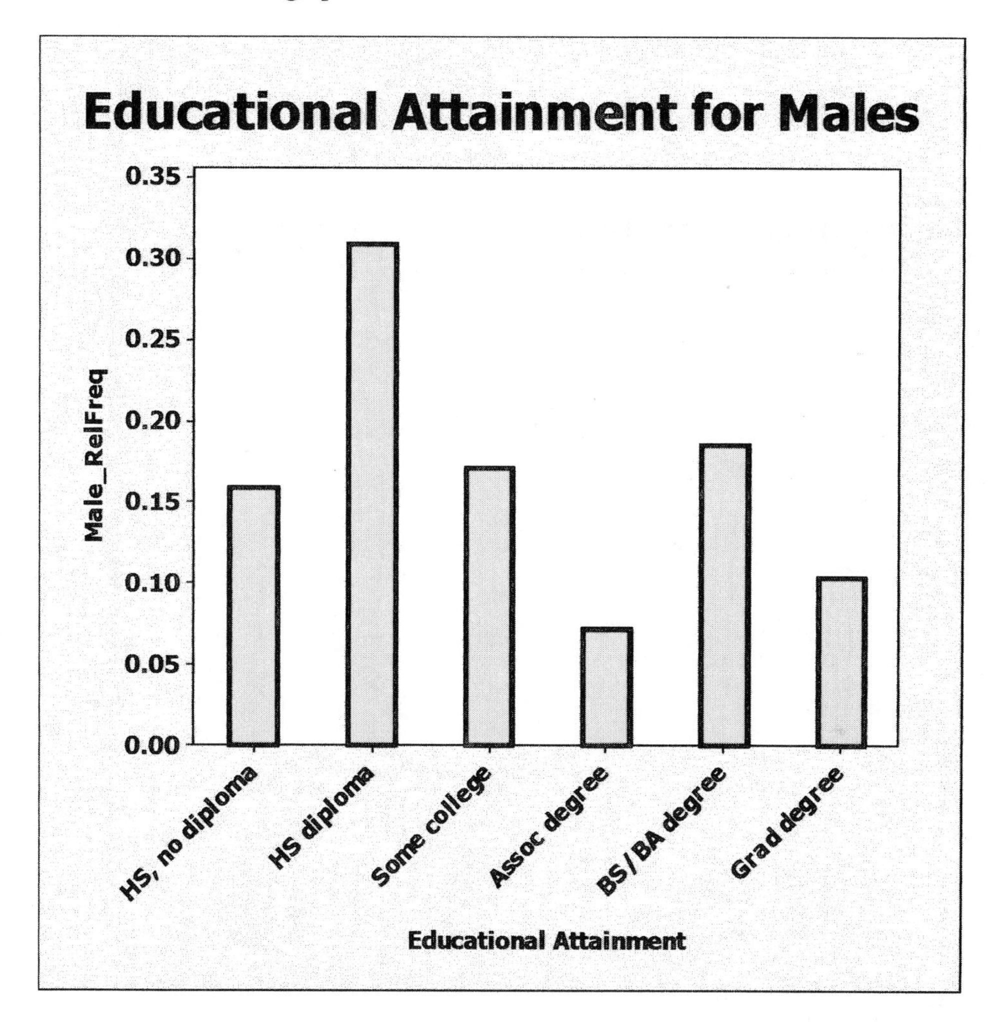

Repeat this procedure to create a bar for Females' educational attainment. Be sure to change the Y-variable to "Female_RelFreq" and to change the **Title.** The rest of the settings should still be set for you.

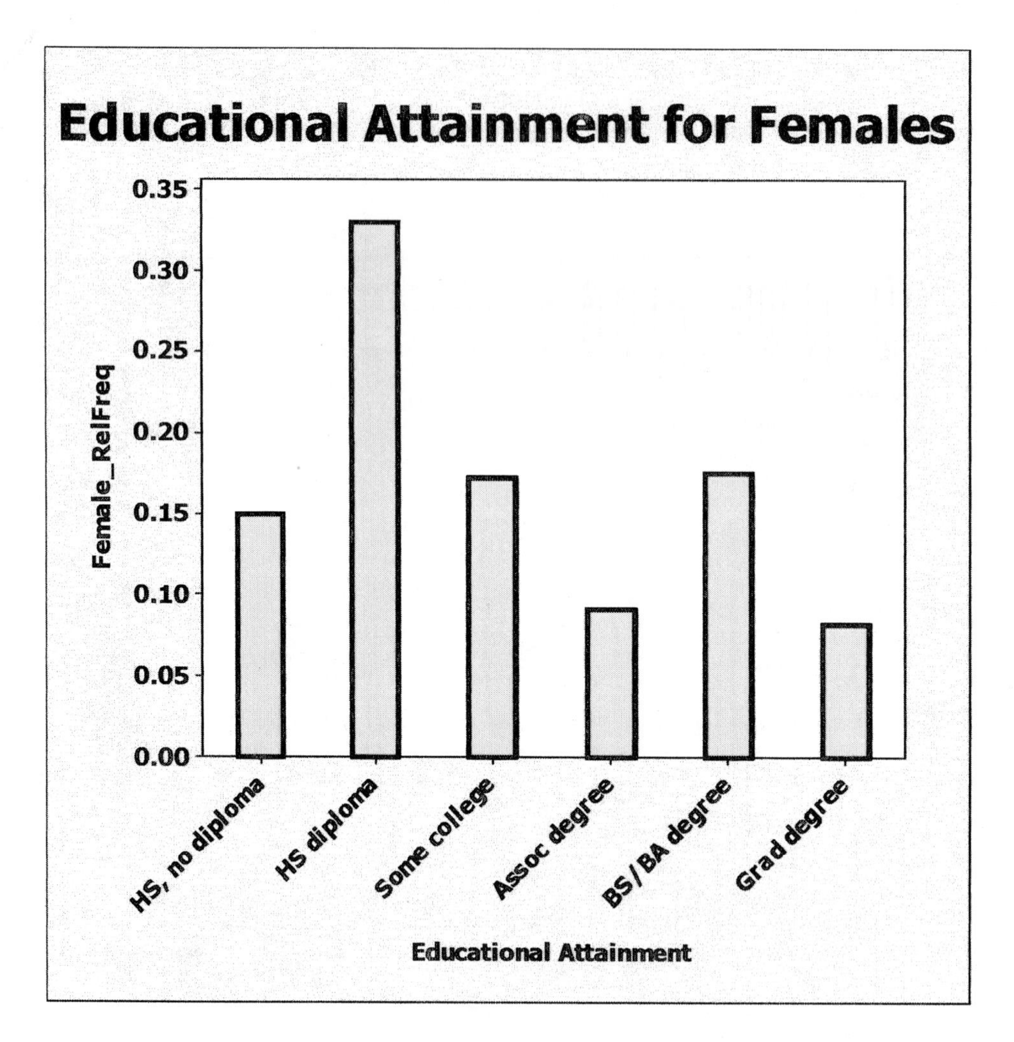

To create the cluster bar graph, click on **Graph** → **Bar Chart**. On the pop-up screen that appears, the **Bars represent "values from a table"**. Highlight a **Two-way table, Cluster** bar chart (icon on the left of the bottom row), and click **OK**.

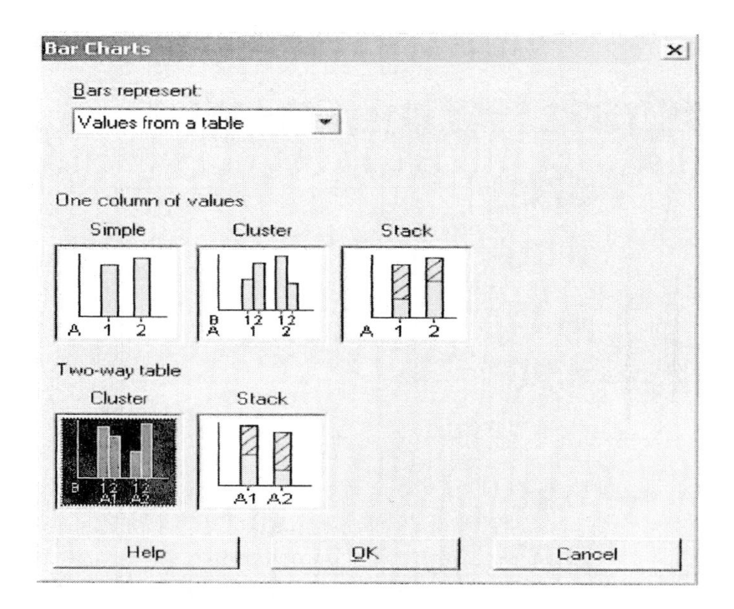

Select Male_RelFreq and Female_RelFreq for the **Graph variable** and Educational Attainment for the **Row labels**. Click on **Rows are outermost categories and columns are innermost**. Next click on the **Labels button**, and enter the title "Educational Attainment".

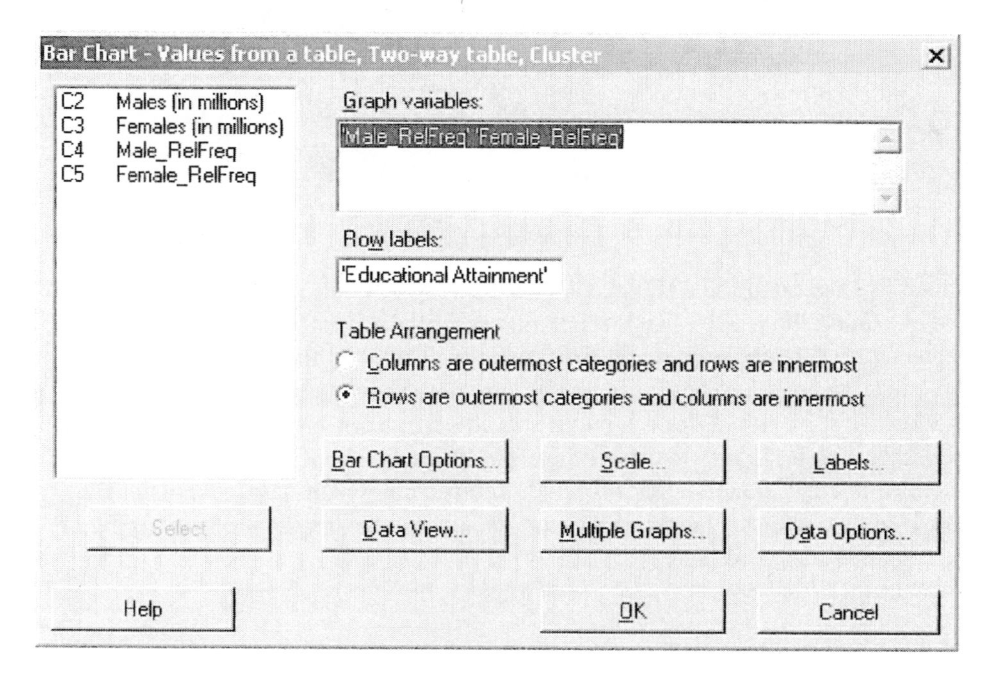

Click on **OK** twice to view the bar graph.

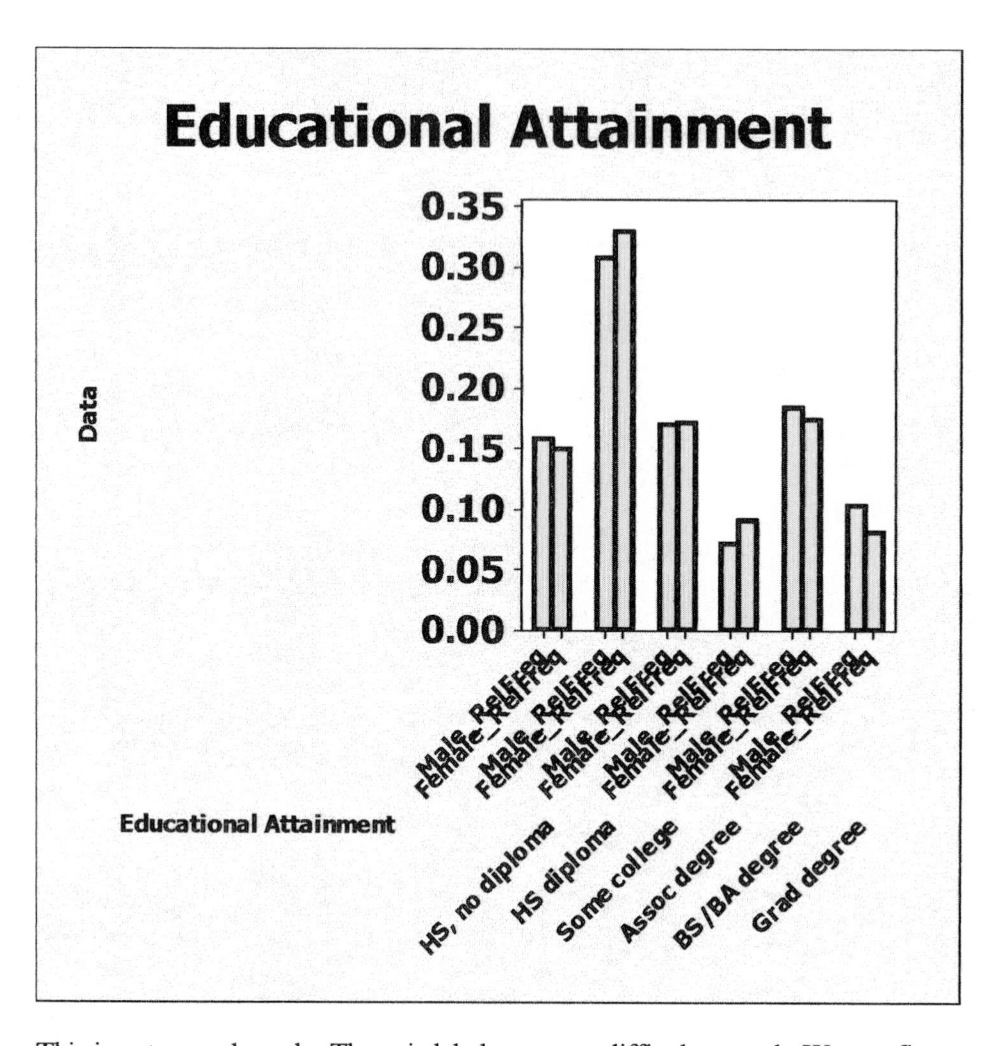

This is not a good graph. The axis labels are very difficult to read. We can fix this by editing the graph. Right-click on one of the labels "Male_RelFreq". On the menu list that pops up, select **Edit X scale**. This will bring up another pop-up screen. Click on the "Show" tab and be sure that **Show axis labels** is NOT checked. If it is, just click on it and it should become unselected. Next, click on the "Labels" tab. In the first row of the **Tick Labels** section, click on **Auto** to unselect it. Next, double-click on Male_RelFreq below **Custom.** You should now be able to edit the labels. Enter an "M" to replace "Male_RelFreq" and an F to replace "Female_RelFreq".

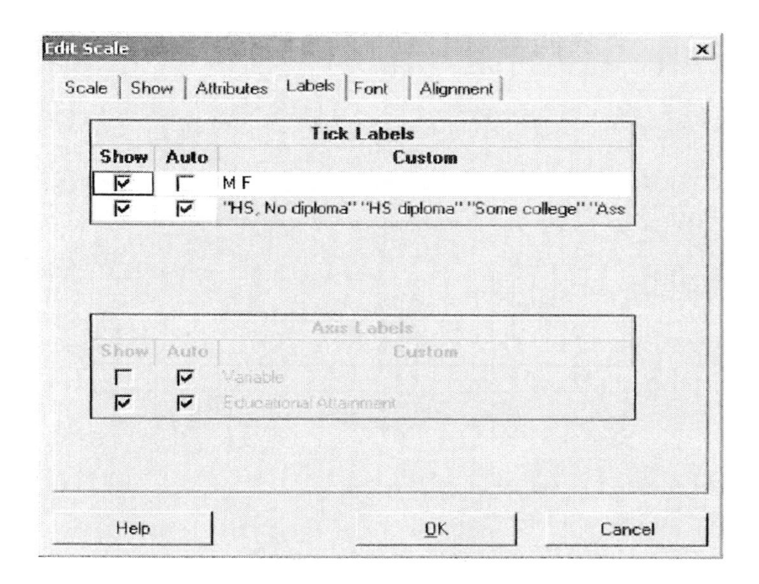

Click on **OK** and these changes should have been made to the graph.

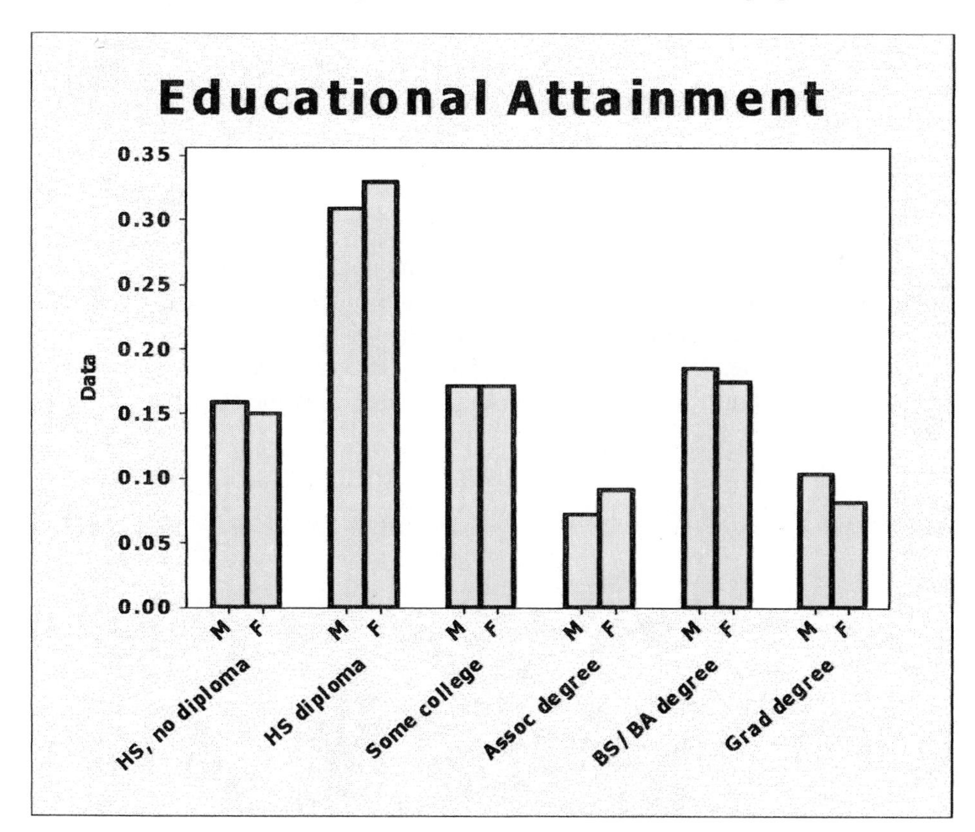

▸ Problem 27c,d,e (pg. 72) 2004 Presidential Election

Open Minitab worksheet **2_1_27.**

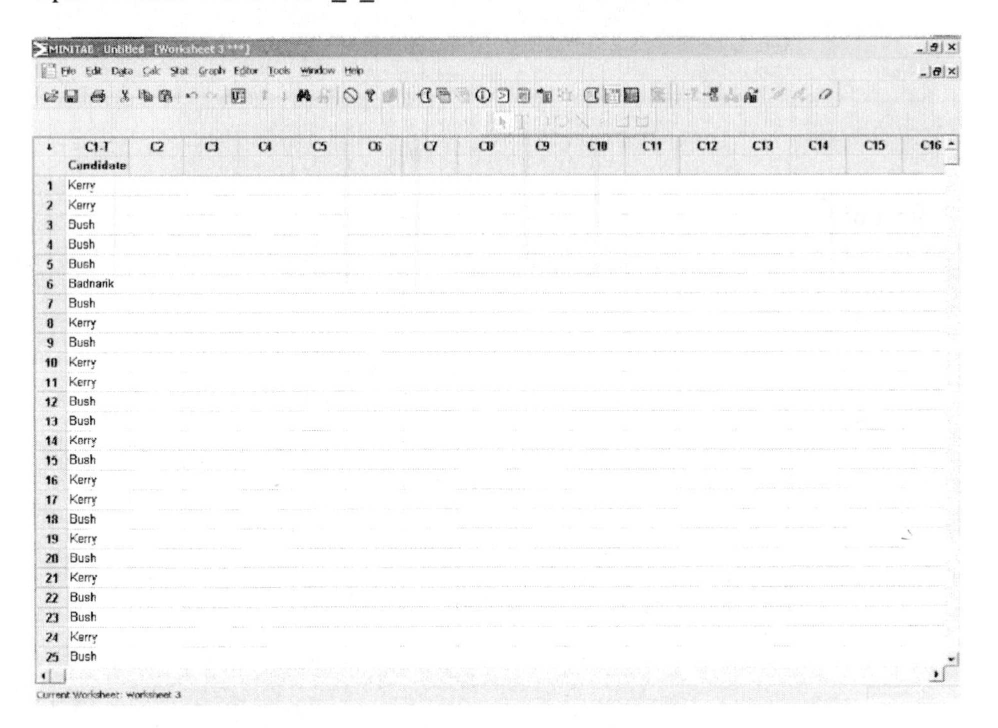

Click on **Graph → Bar Chart.** On the pop-up screen that appears, the **Bars represent "counts of unique values"**. Highlight a **Simple** bar chart (icon on the left of the top row), and click **OK**. Double-click on C1 to select it for the **Categorical variable**. Add a title to the graph by clicking on the **Labels** button and entering "Exit Poll Results". Click on **OK** twice to view the chart.

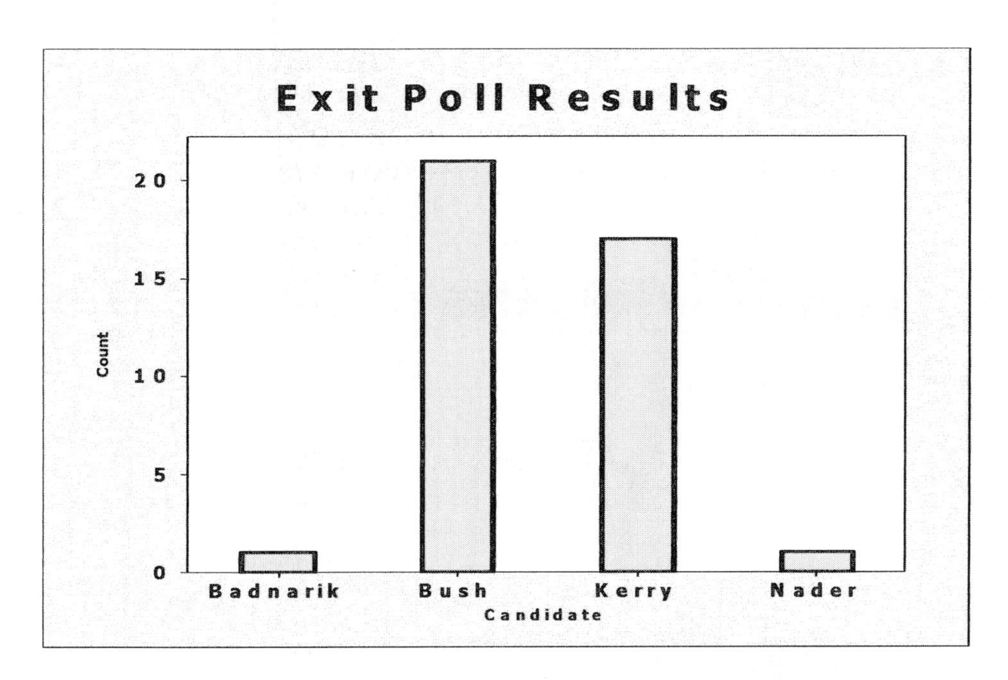

Notice the axis label is "Count" on the Y-axis. To give it a more appropriate label, right-click on "Counts" and select **Edit Y Axis Label.** Enter "Votes" below **Text** to replace "Counts". Click on **OK**. Your graph should now be labeled correctly.

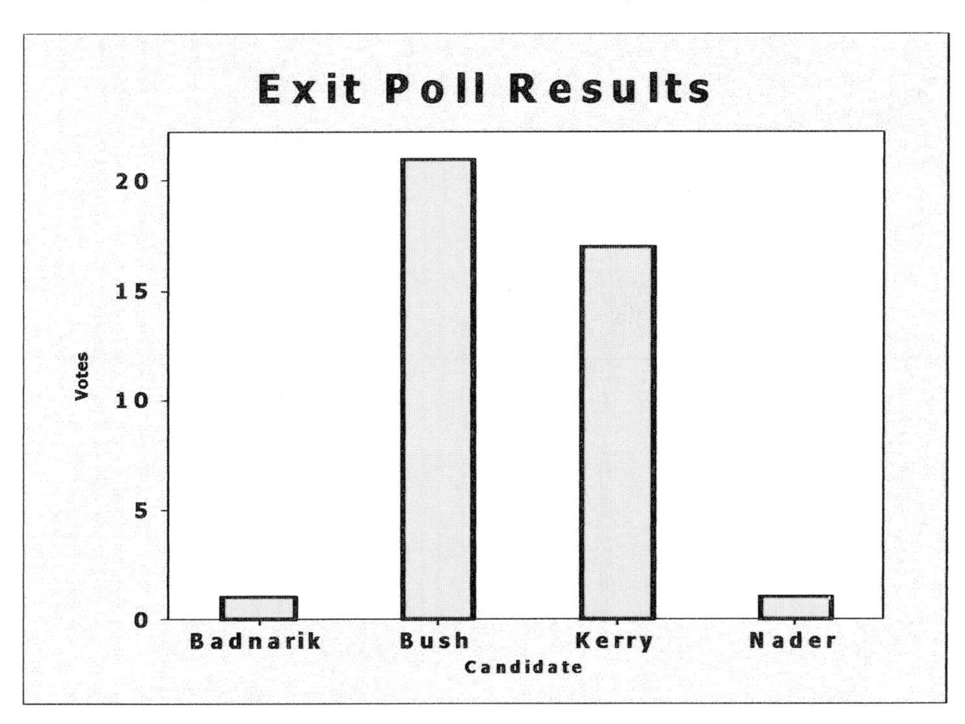

To create a Pie chart for this data, click on **Graph → Pie chart.** Select **Chart raw data** by clicking on the small circle to the left of it. Fill in C1 for the **Categorical variable**. Click on the **Labels** button and type in an appropriate title. Click on the **Slice labels** tab and select **Percent.** Click on **OK** to view the chart.

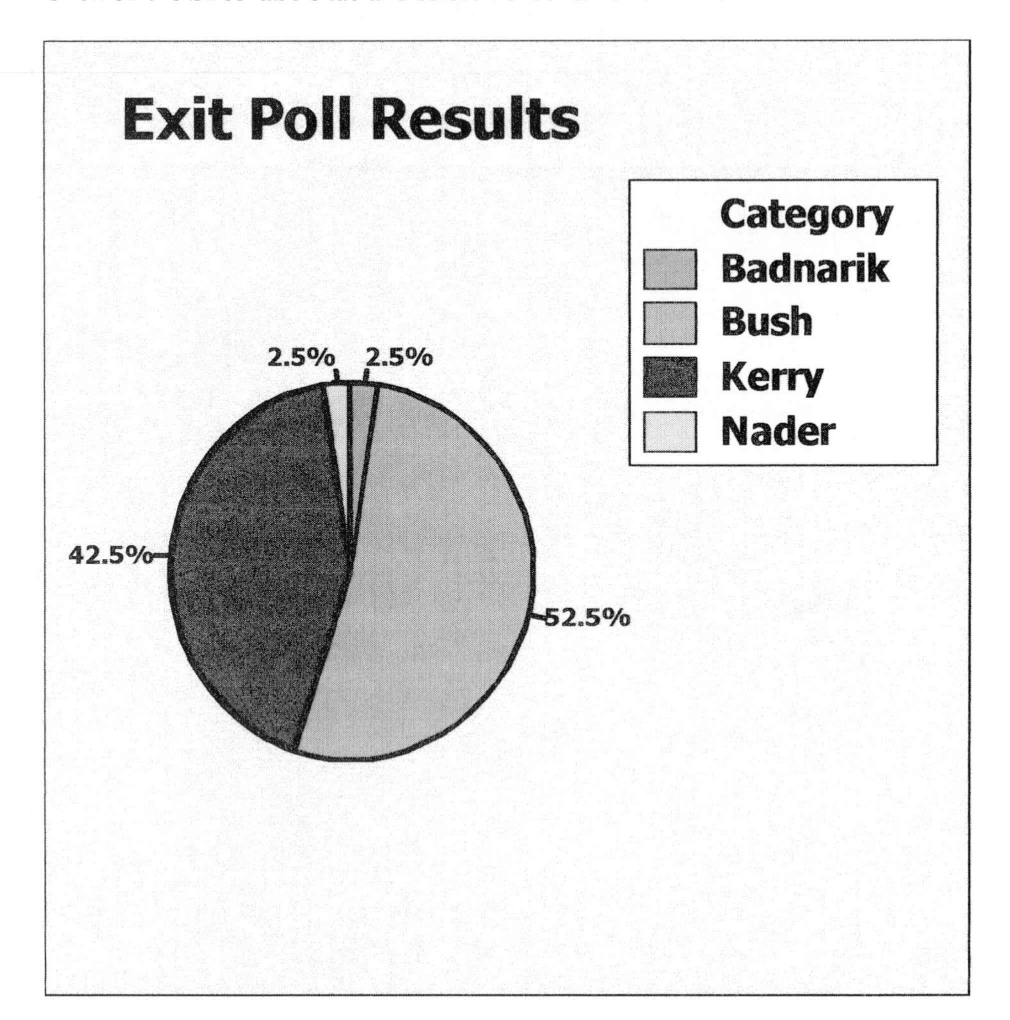

Section 2.2

Example 2 (pg. 78): Construct a Histogram for Discrete data

To create this histogram, you must enter the 40 data points (Table 8 on page 77) into column 1 of the Minitab worksheet. Name the column "Arrivals".

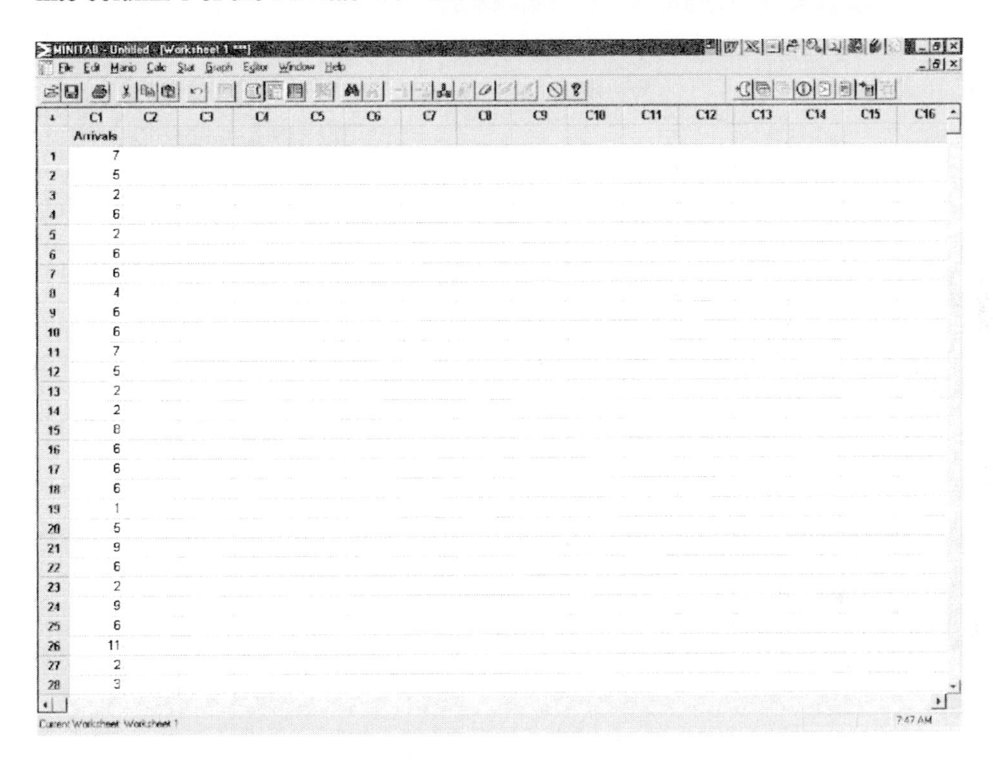

Now you are ready to make the histogram. Click on: **Graph** → **Histogram**. Select a **Simple** histogram and click **OK**. On the main Histogram screen, double-click on C1 in the large box at the left of the screen. "Arrivals" should now be filled in as the **Graph variable**.

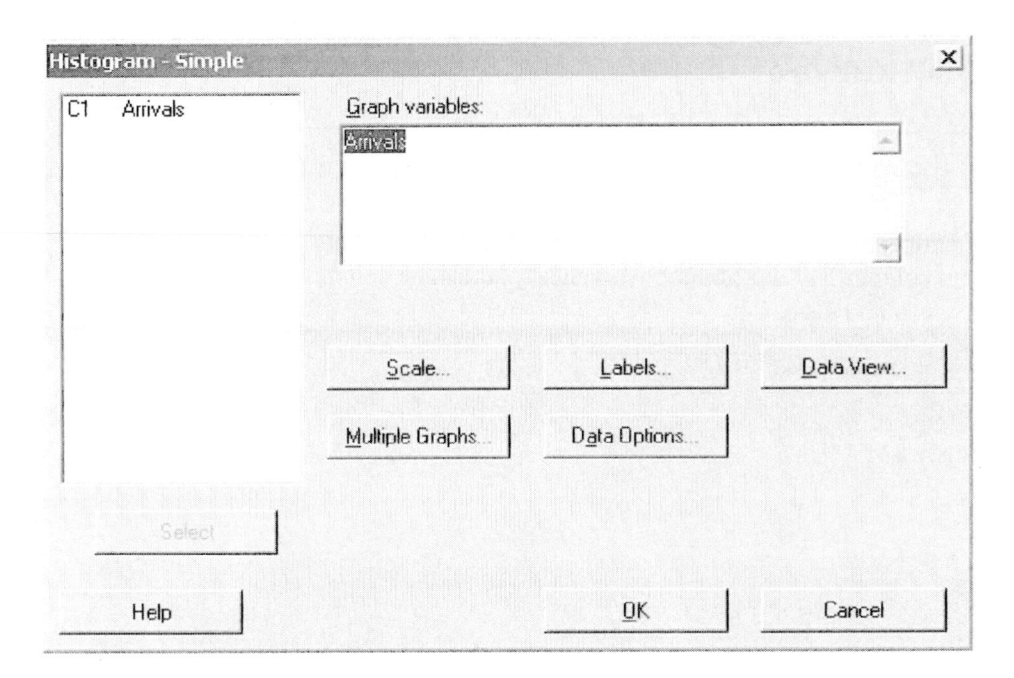

Click on the **Labels** button and enter a title for the graph. At this point, if you click on **OK**, MINITAB will draw a histogram using default settings. Your histogram will look like the one below.

If the font on the title is too large, just right-click on the title. Select **Edit Title**. You can change the font size from the pop-up screen that will appear.

Notice that the scaling on the X-axis is not as expected. We can instruct MINITAB to fix this by editing the graph. Right-click on the X-axis and select **Edit X Scale** from the menu that pops up. **On the Scale** tab, enter "1:11/1" for **Position of ticks**. This tells Minitab you want the numbering to go from 1 to 11 in steps of 1.

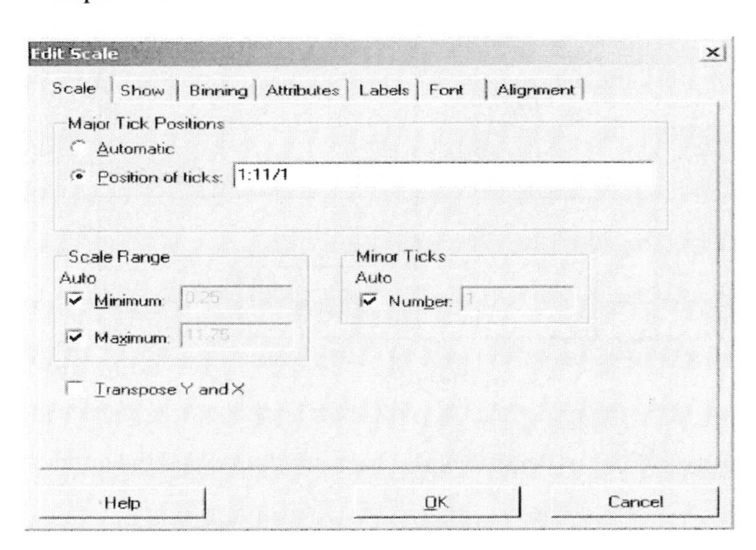

Next, click on the **Binning** tab. Select Midpoint for **Interval Type**. Enter "1:11/1" for **Midpoint/Cutpoint positions**.

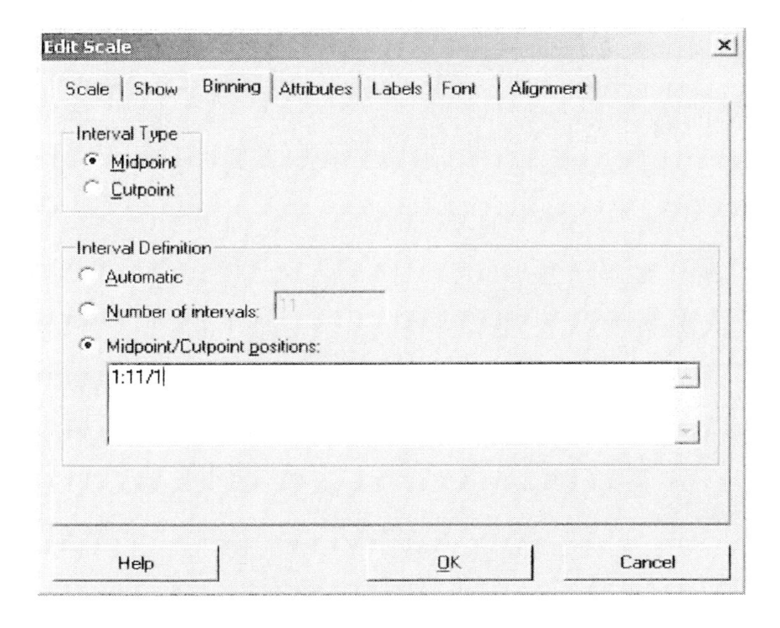

Click on **OK** to view the changes.

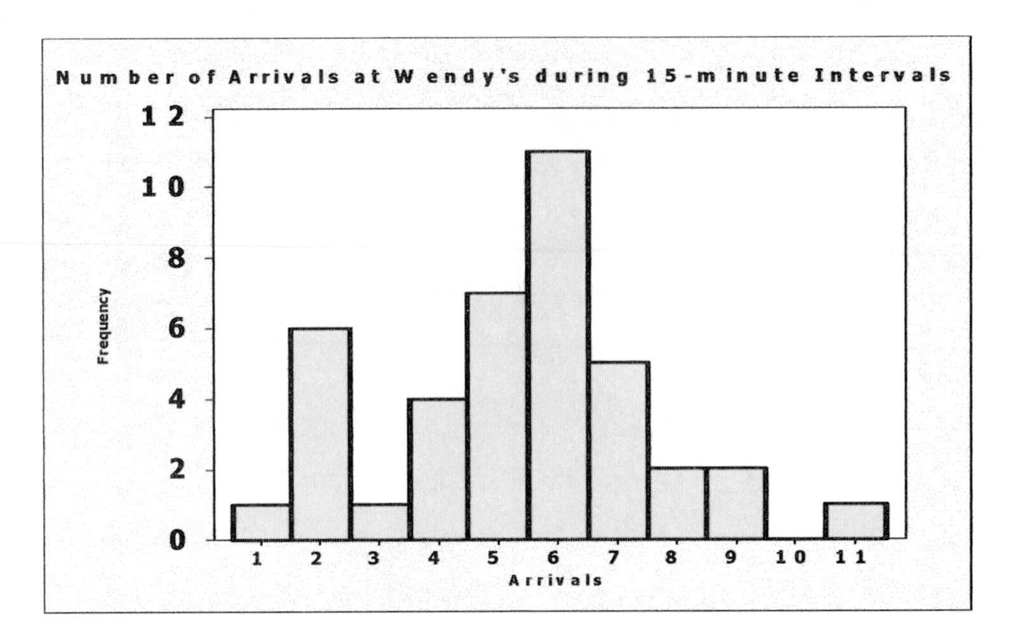

To print the graph, click on **File → Print Graph.** Next, click **OK** and the graph should print.

Example 5 (pg. 82) Construct a Histogram for Continuous data

To create this histogram, enter the data from Table 12 on page 80 of the textbook into a Minitab worksheet.

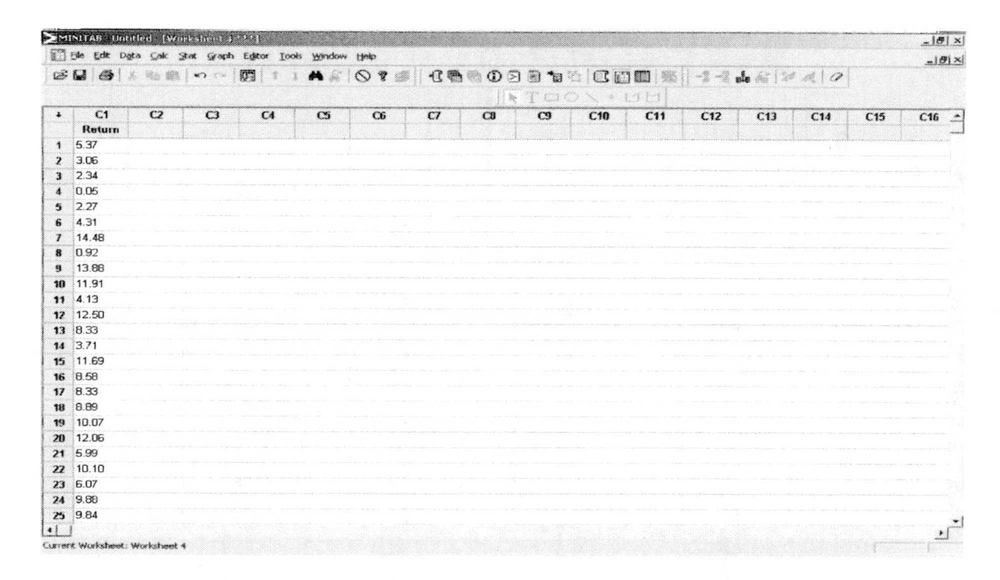

Now you are ready to make the histogram. Click on: **Graph → Histogram**. Select a **Simple** histogram and click OK. On the main Histogram screen, double-click on C1 in the large box at the left of the screen. "Return" should now be filled in as the **Graph variable**. Click on the **Labels** button and enter an appropriate title for the graph.

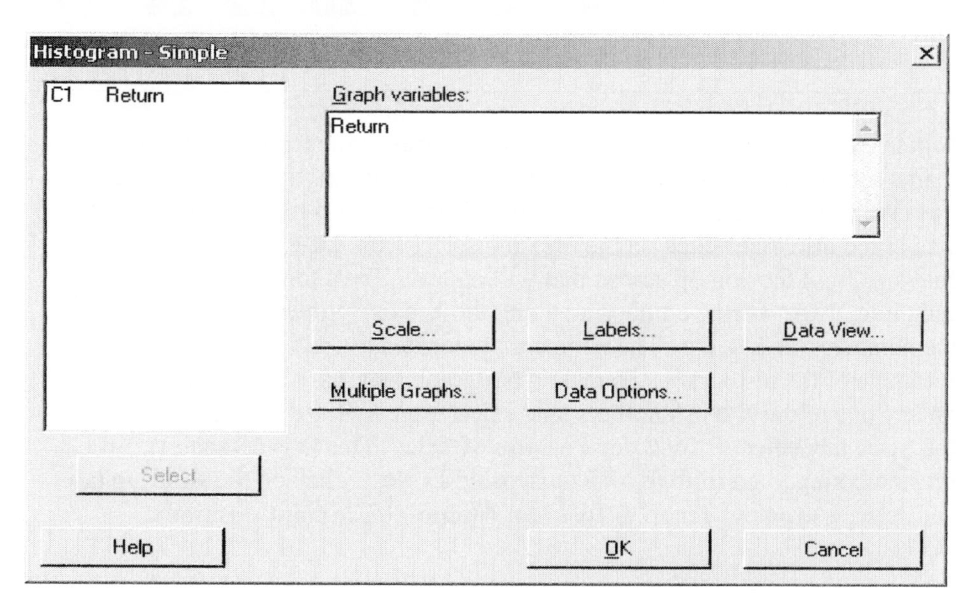

At this point, if you click on **OK**, MINITAB will draw a histogram using default settings. Your histogram will look like the one below.

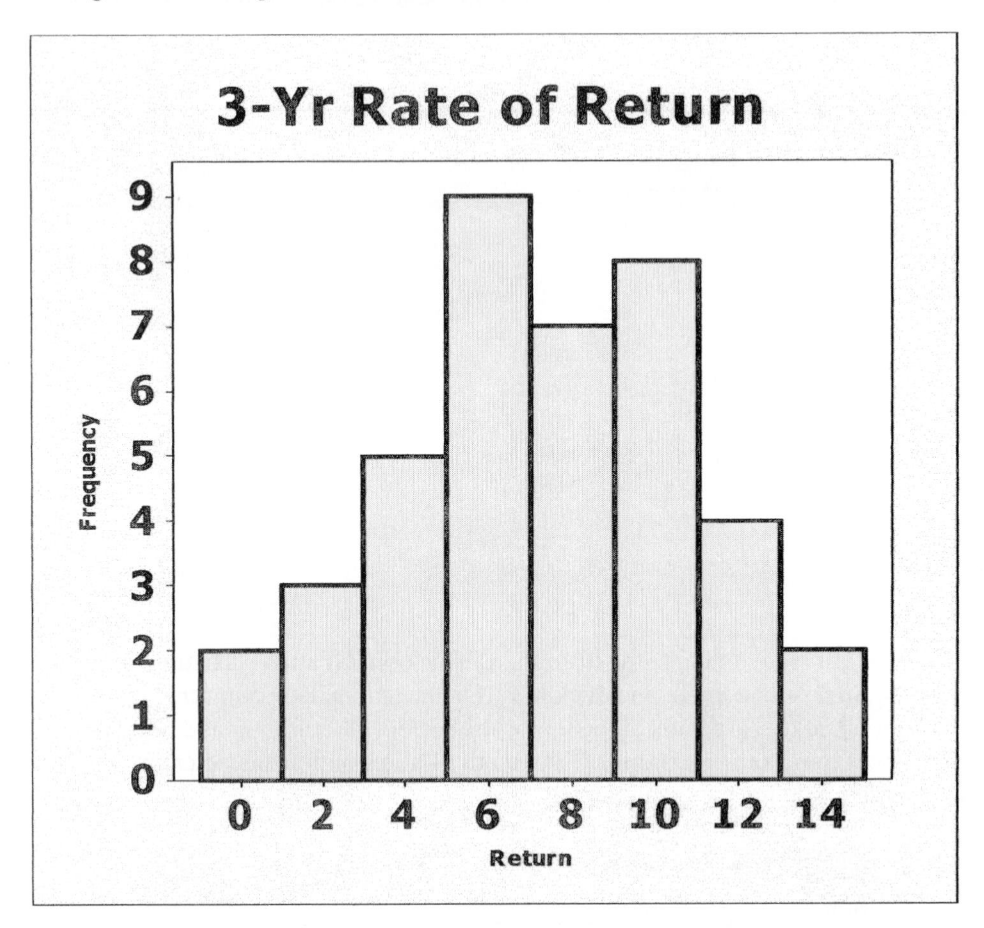

Notice that the X-axis classes are not like the ones in the textbook. The tick marks are in the middle of each rectangle so it is difficult to tell what the classes are. We can edit the graph to fix this, however. If you'd like to change the font size of the title, right-click on the title and select **Edit Title**. You can change the font size from the pop-up screen that will appear. Try a font size 10 for a very long title. Next, fix the numbering along the X-axis. Notice the textbook uses the numbers: 0, 2, 4, etc. The numbers are located at the beginning of each rectangle of the histogram. To make your graph look like this, simply right-click on any one of the X-axis numbers, and select **Edit X Scale** from the menu. On the **Scale** tab, enter "0:16/2" for **Position of ticks**. This tells Minitab you want the numbering to go from 0 to 16 in steps of 2. Next, click on the **Binning** tab, and select **Cutpoint.** Enter "0:16/2" for **Midpoint/Cutpoint positions**.

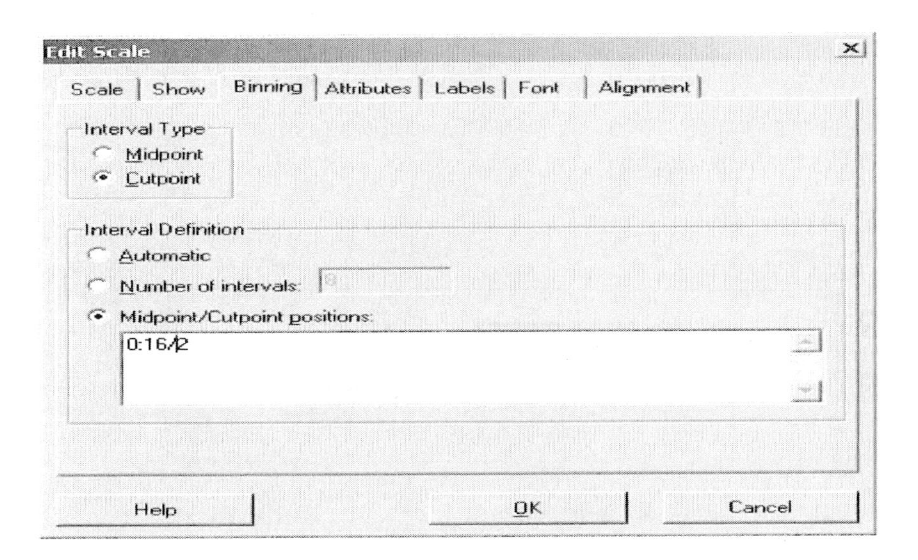

Click on **OK** to view the changes.

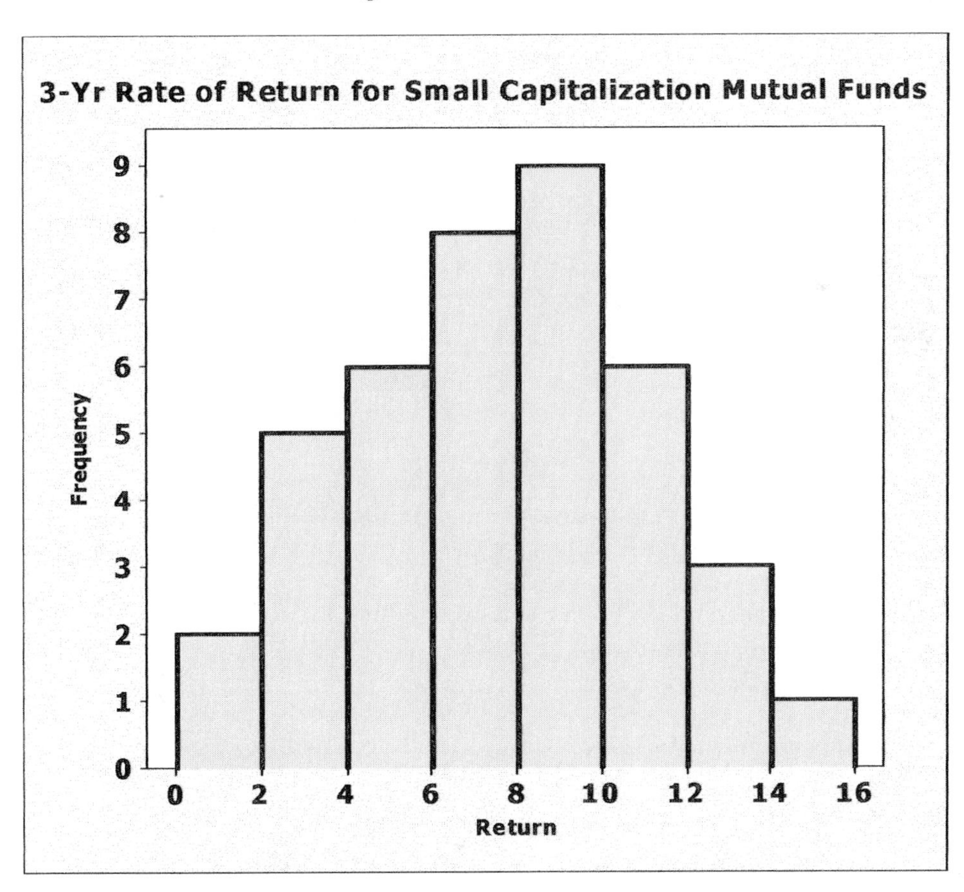

▸ Example 7 (pg. 83) Constructing a Stem-and-Leaf Plot

Open worksheet **2_2_T15.** To construct a Stem-and-leaf plot, click on **Graph → Stem-and-Leaf.** On the screen that appears, select C1 as your **Graph Variable** by doubling clicking on C1.

Stem-and-Leaf
C1 Percent
Graph variables:
Percent
By variable:
Trim outliers
Increment:
Select
Help
OK
Cancel

Click on **OK.** The stem and leaf plot will be displayed in the Session Window.

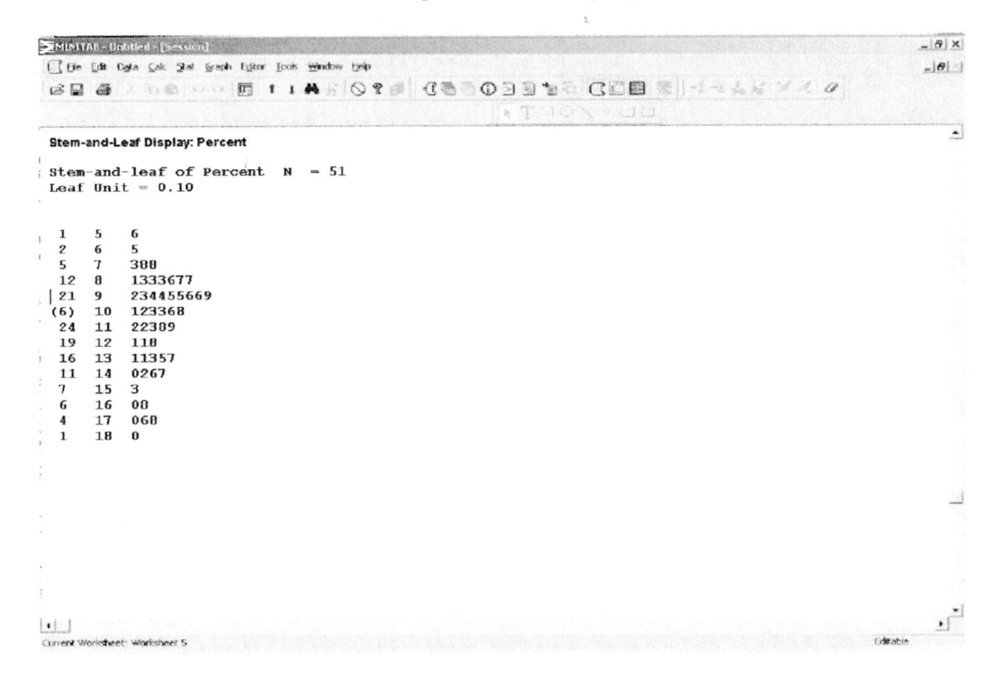

In this MINITAB display, the first column on the left is a counter. This column counts the number of data points starting from the smallest value (at the top of the plot) down to the median. It also counts from the largest data value (at the bottom of the plot) up to the median. Notice that there is one data point in the first row of the stem and leaf plot so the counter is "1". Row 2 has 1 data point so the counter increases to "2" (1+1=2). The row that contains the median has the number "6" in parentheses. This number counts the number of data points that are in the row that contains the median. The counter at the bottom starts at "1" because there is 1 data point in the bottom row.

The second column in the display is the **Stem**. In this example, the Stem values range from 5 to 18. Notice that MINITAB constructs an *ordered* stem and leaf.

The leaf values are shown to the right of the stem. The leaf values may be the actual data points or they may be the rounded data points. To find the actual values of the data points in the display, use the "Leaf Unit=" statement at the top of the display. The "Leaf Unit" gives you the place value of the leaves. In this stem and leaf plot, the first data point has a stem value of 1 and a leaf value of 0. Since the "Leaf Unit=0.10", the leaf value of 6 belongs in the "tenths" place and the stem value of 5 belongs in the "ones" places. Thus the data point is 5.6.

◄

Example 9 (pg. 86) Constructing a Dot Plot

Enter the 40 data points (Table 8 on page 77) into column 1 of the Minitab worksheet. Name the column "Arrivals".

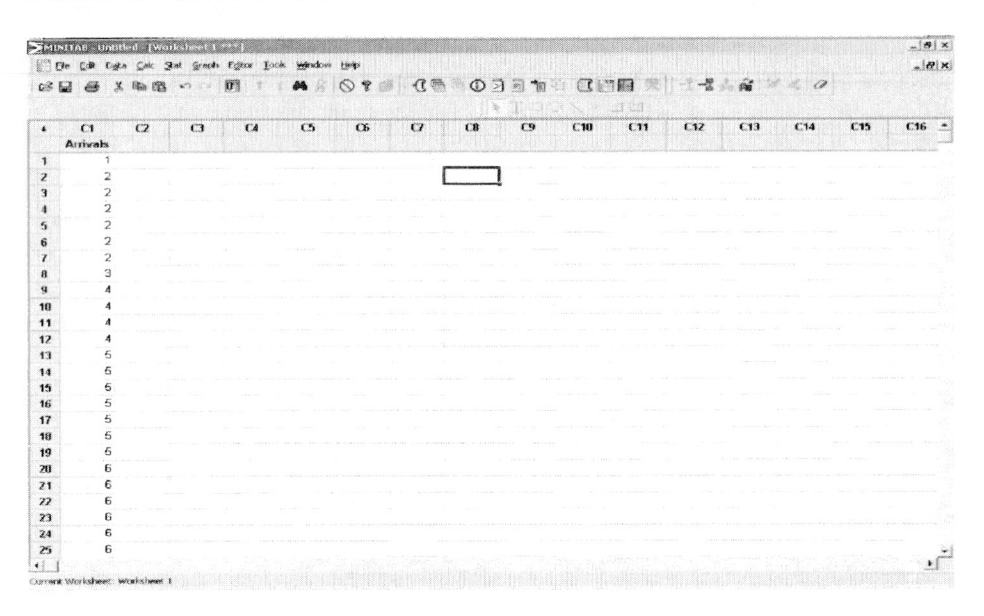

To construct a Dot Plot, click on **Graph** → **Dotplot** → **Simple.** On the screen that appears, select C1 as your **Graph Variable** by doubling clicking on C1.

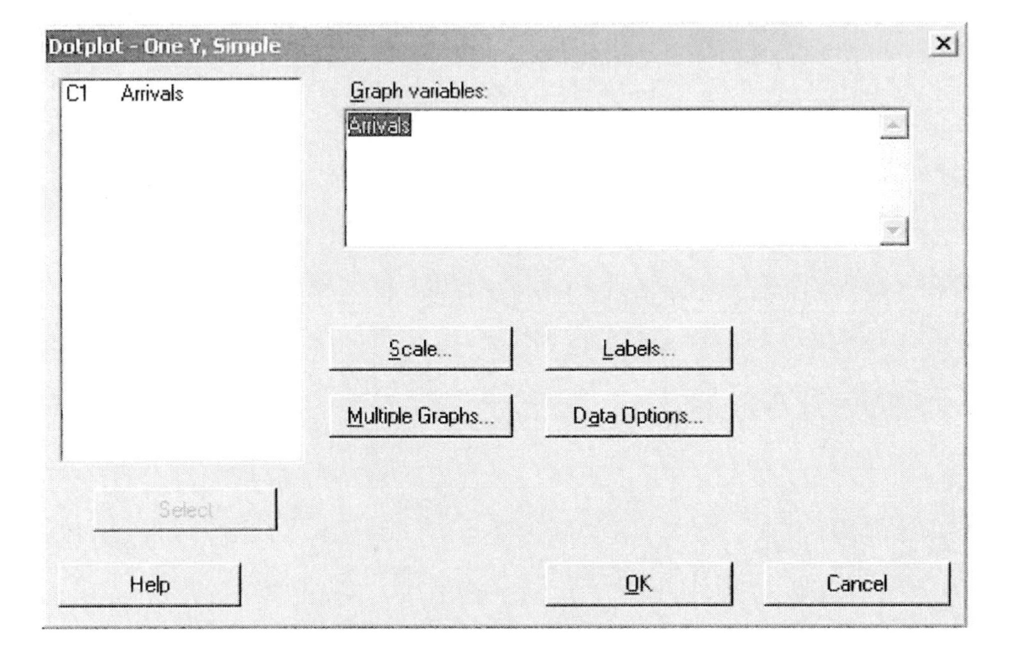

Click on the **Labels** button, and add an appropriate title. Click on **OK** to view the Dotplot.

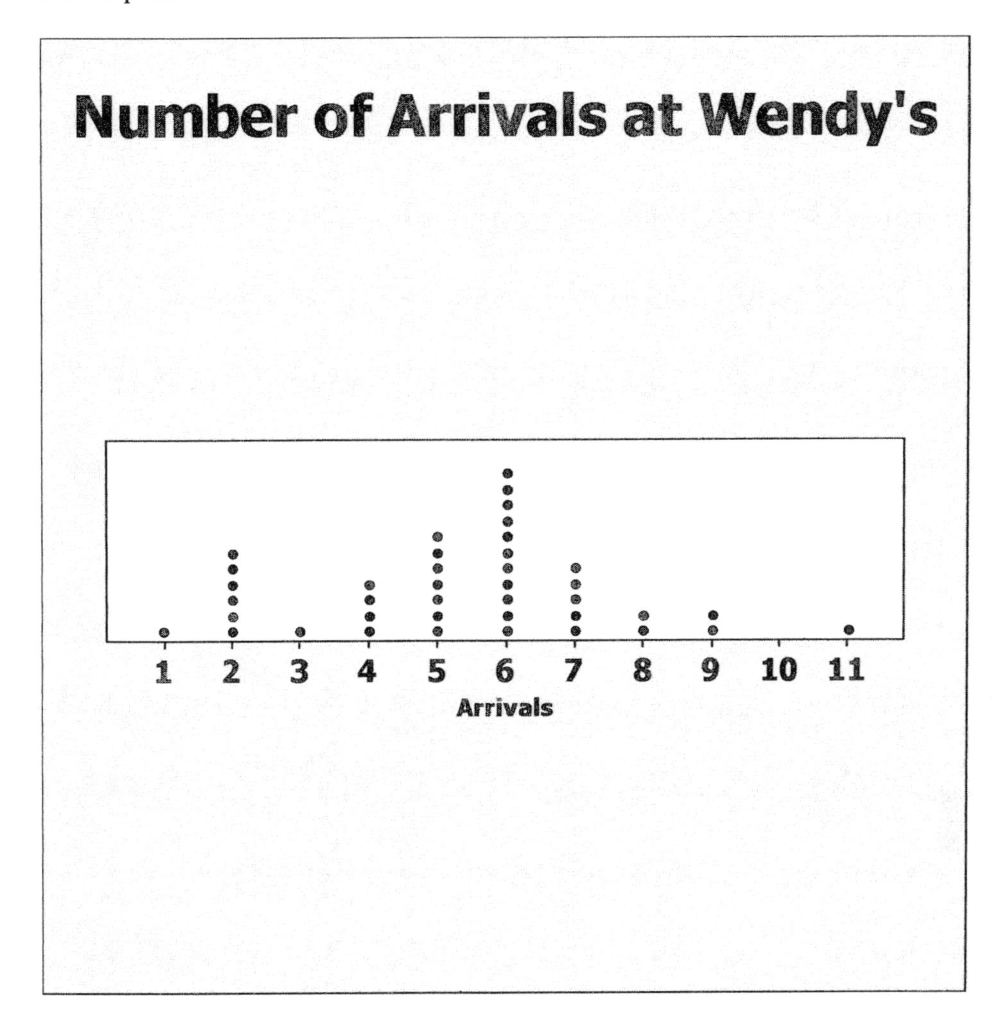

In the Dotplot shown, the X-scale has been edited so that the X-axis labels are from 1 to 11 in steps of 1.

▸ Problem 29c,d (pg. 91): Average Income

To create this histogram, open the Minitab worksheet **2_2_29.**

Click on: **Graph** → **Histogram** → **Simple**. First, double click on C1 in the large box at the left of the screen. "Income" should now be filled in as the **Graph variable**. Click on the **Labels** button, and add an appropriate title.
At this point, if you click on **OK**, MINITAB will draw a histogram using default settings. Do this and see what the default graph looks like. You can now edit it if necessary. You will have to fix the numbering along the X-axis. The instructions for this problem say to set the first class limit to 20,000 and to have a class width of 2500. To do this, simply right-click on the X-axis of the graph and select **Edit X Scale**. On the **Scale** tab, enter "20000:42500/2500" for **Position of ticks**. This tells Minitab you want the numbering to go from 20,000 to 42,500 in steps of 2500. Next, click on the **Binning** tab, and select **Cutpoint.** Enter "20000:42500/2500" for **Midpoint/Cutpoint positions**. Click on **OK** again to view your changes.

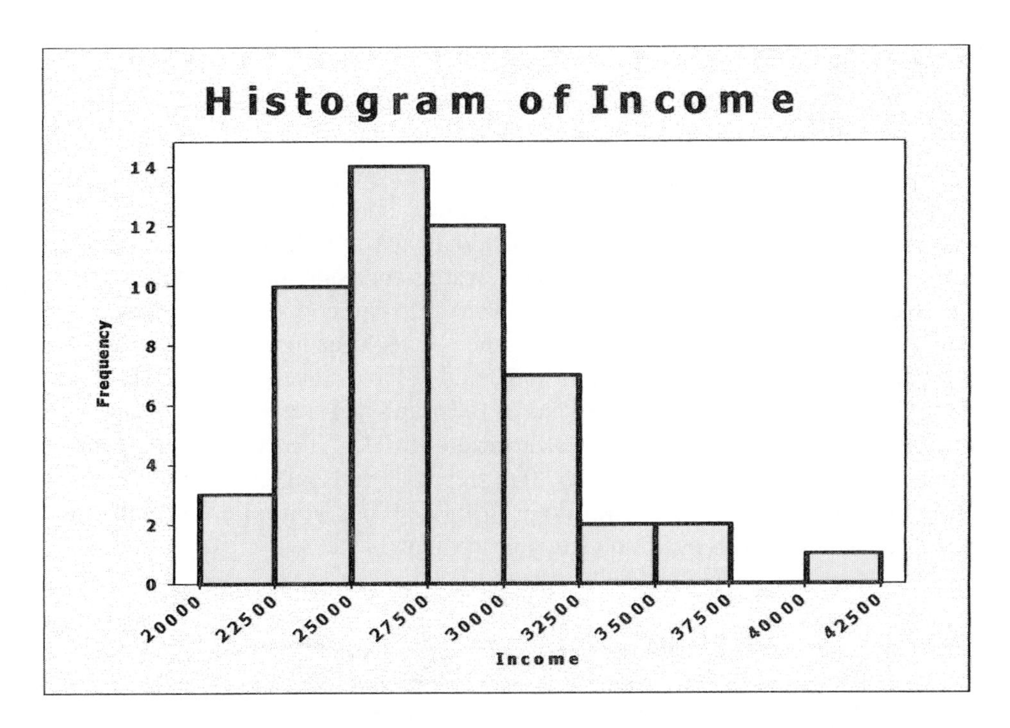

To create a relative frequency histogram, right-click on the Y-axis and select **Edit Y Scale.** Click on the **Type** tab and select **Percent.** All other settings remain the same. Click on **OK** to view the relative frequency histogram.

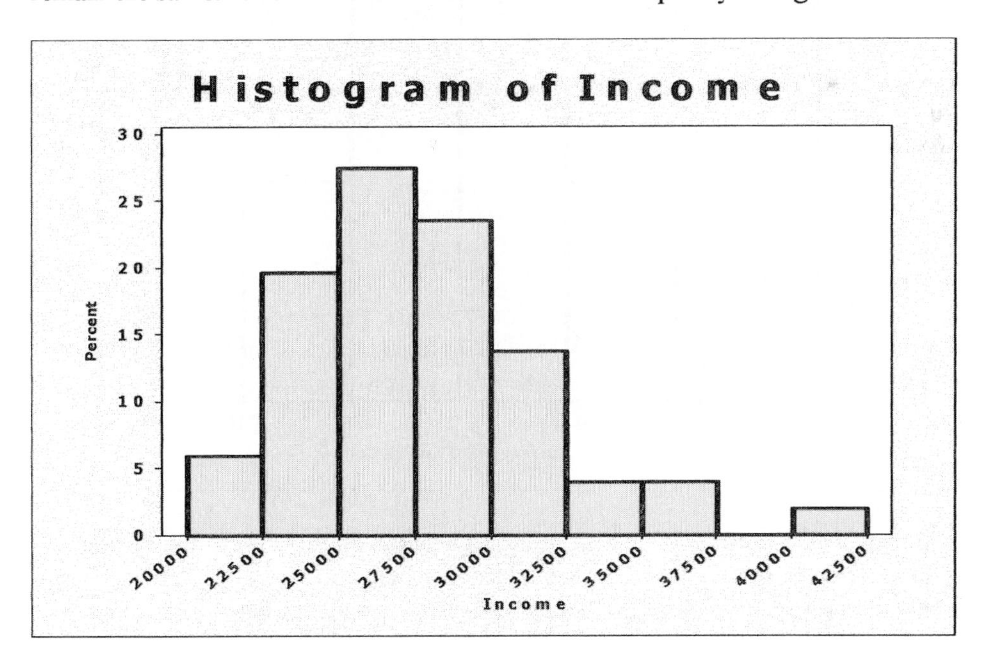

▸ Problem 31 (pg. 92): Serum HDL

To create this histogram, open the Minitab worksheet **2_2_31.** Click on **Graph** → **Histogram** → **Simple**. First, double click on C1 in the large box at the left of the screen. "Serum HDL Cholesterol" should now be filled in as the **Graph variable**. Click on the **Labels** button, and add an appropriate title. At this point, if you click on **OK**, MINITAB will draw a histogram using default settings. Do this and see what the default graph looks like. You can now edit it if necessary. You will have to fix the numbering along the X-axis. The instructions for this problem say to set the first class limit to 20 and to have a class width of 10. To do this, simply right-click on the X-axis of the graph and select **Edit X Scale**. On the **Scale** tab, enter "20:80/10" for **Position of ticks**. This tells Minitab you want the numbering to go from 20 to 80 in steps of 10. Next, click on the **Binning** tab, and select **Cutpoint.** Enter "20:80/10" for **Midpoint/Cutpoint positions**. Click on **OK** again to view your changes.

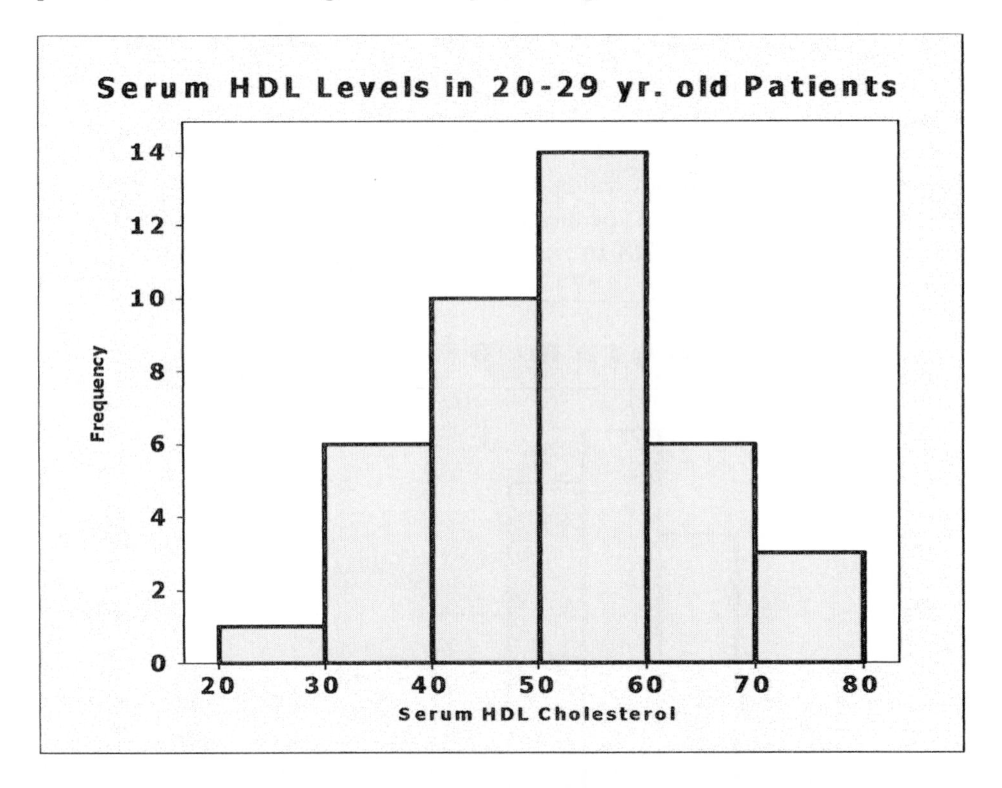

To create a relative frequency histogram, right-click on the Y-axis and select **Edit Y Scale.** Click on the **Type** tab and select **Percent.** All other settings remain the same. Click on **OK** to view the relative frequency histogram.

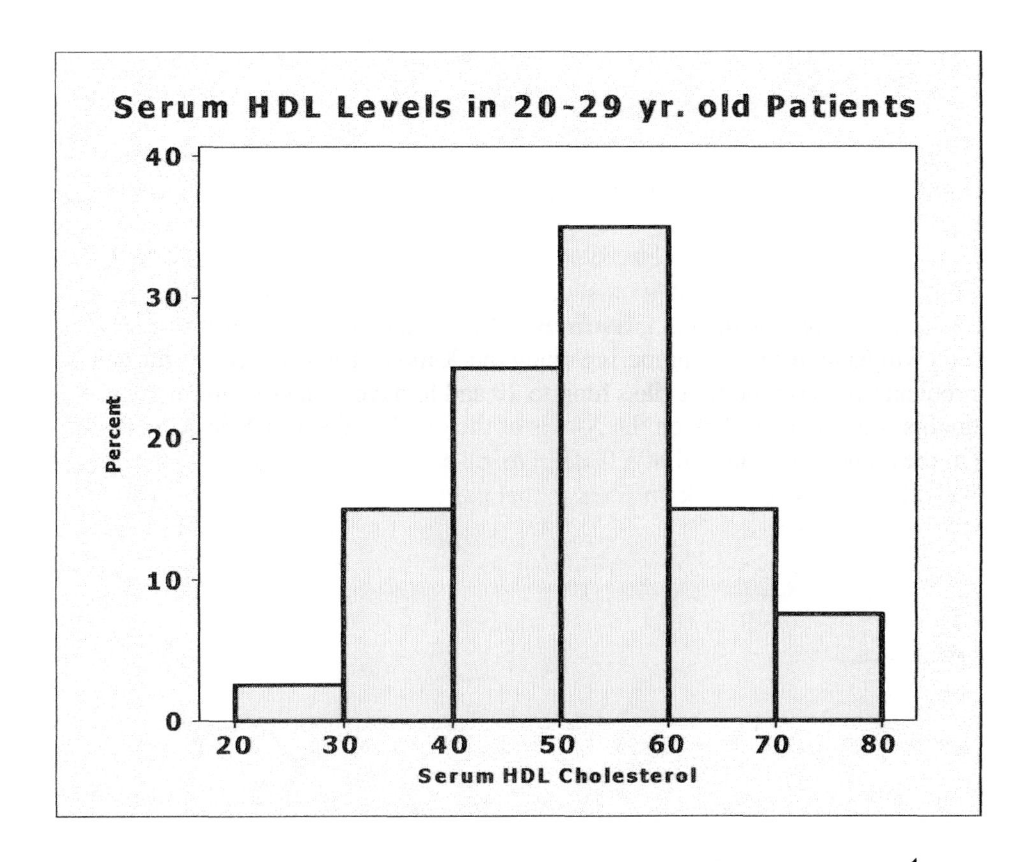
Serum HDL Levels in 20-29 yr. old Patients
Percent
40
30
20
10
0
20
30
40
50
60
70
80
Serum HDL Cholesterol

Section 2.3

‣ Polygon (pg. 97) Constructing Frequency Polygons

A frequency polygon is created the same way as a histogram in MINITAB. Enter the data into a Minitab worksheet. The Rate of Return data should be in column 1. Click on **Graph → Histogram → Simple.** First, double click on C1 in the large box at the left of the screen. "3-yr Rate of Return" should now be filled in as the **Graph variable**. Next, click on the **Labels** button and add an appropriate title.
To make a polygon, instead of a histogram, click on the **Data View** button. On the **Data Display** tab, click on **Bars** so that there is NO check mark, and select **Symbols** instead.

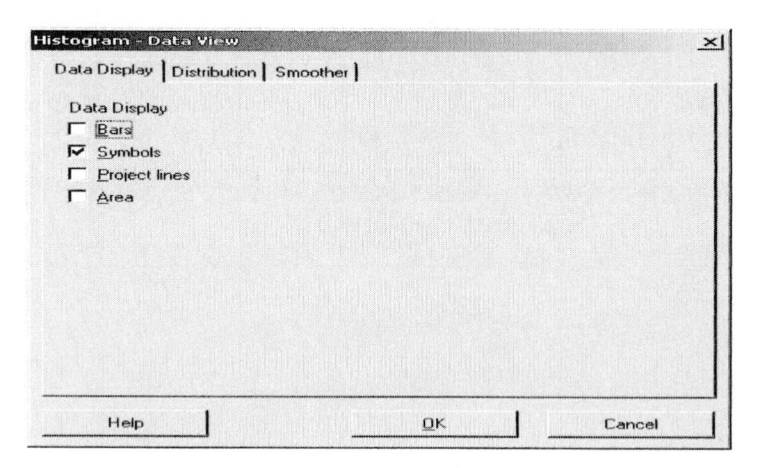

Next, click on the **Smoother** tab.

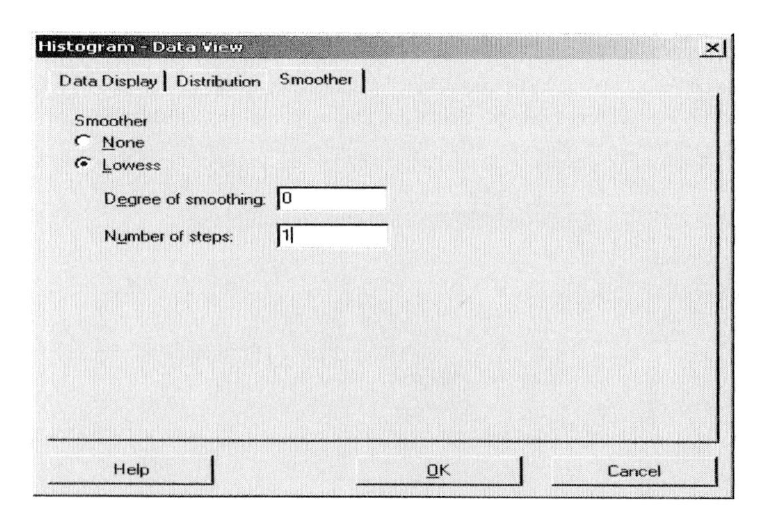

For the **Smoother,** select **Lowess**. The **Degree of smoothing** is 0 and the **Number of steps** is 1. Click on **OK**. These options will connect the points of the polygon. Click on **OK** again to view the graph.

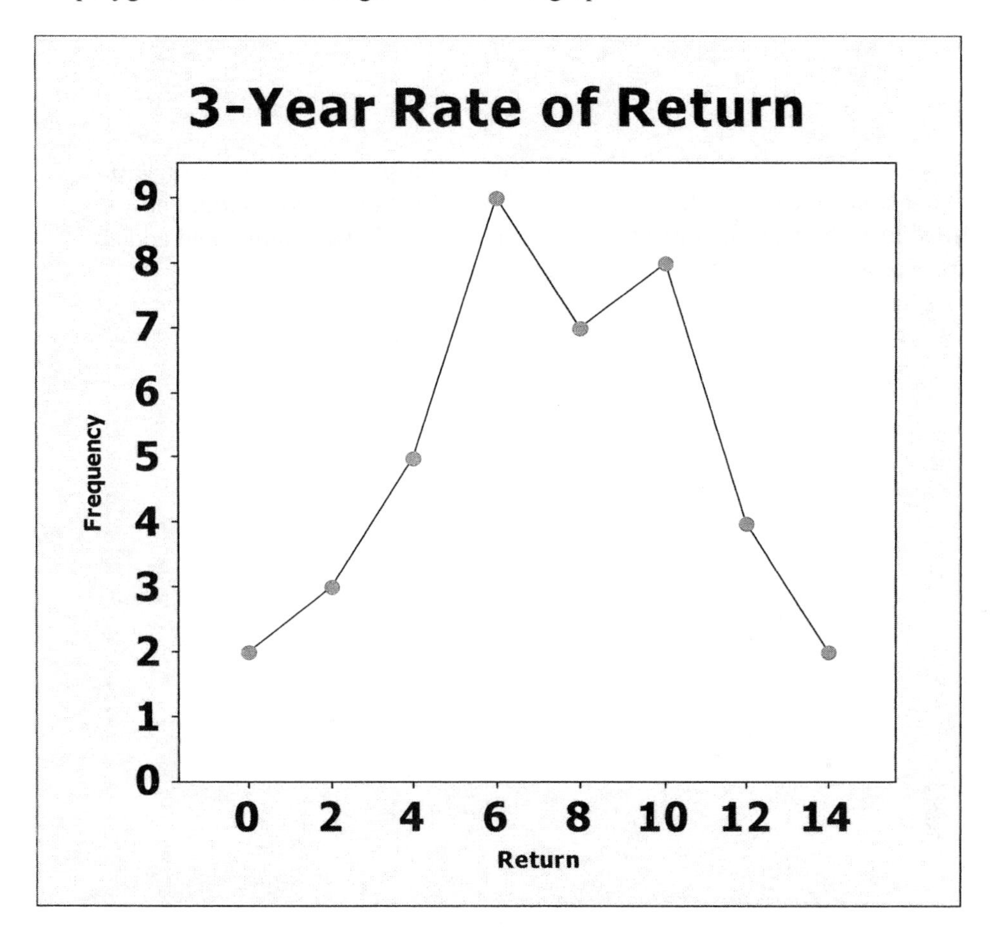

Next, fix the numbering along the X-axis. This time we want to use the midpoints of each class. To do this, simply right-click on the X-axis of the graph and select **Edit X Scale**. On the pop-up screen that appears, on the **Scale** tab, for **Position of ticks** enter 1 : 15 / 2. This tells MINITAB that the first tick is 1 and the last tick is 15, and the ticks are in steps of 2. Next, click on the **Binning** tab. The **Interval Type** is Midpoint. Enter **Midpoint/Cutpoint positions** of 1 : 15 / 2. This tells MINITAB that the first midpoint is 1 and the last midpoint is 15. The class width is 2. Click on **OK** to view the changes to your frequency Polygon.

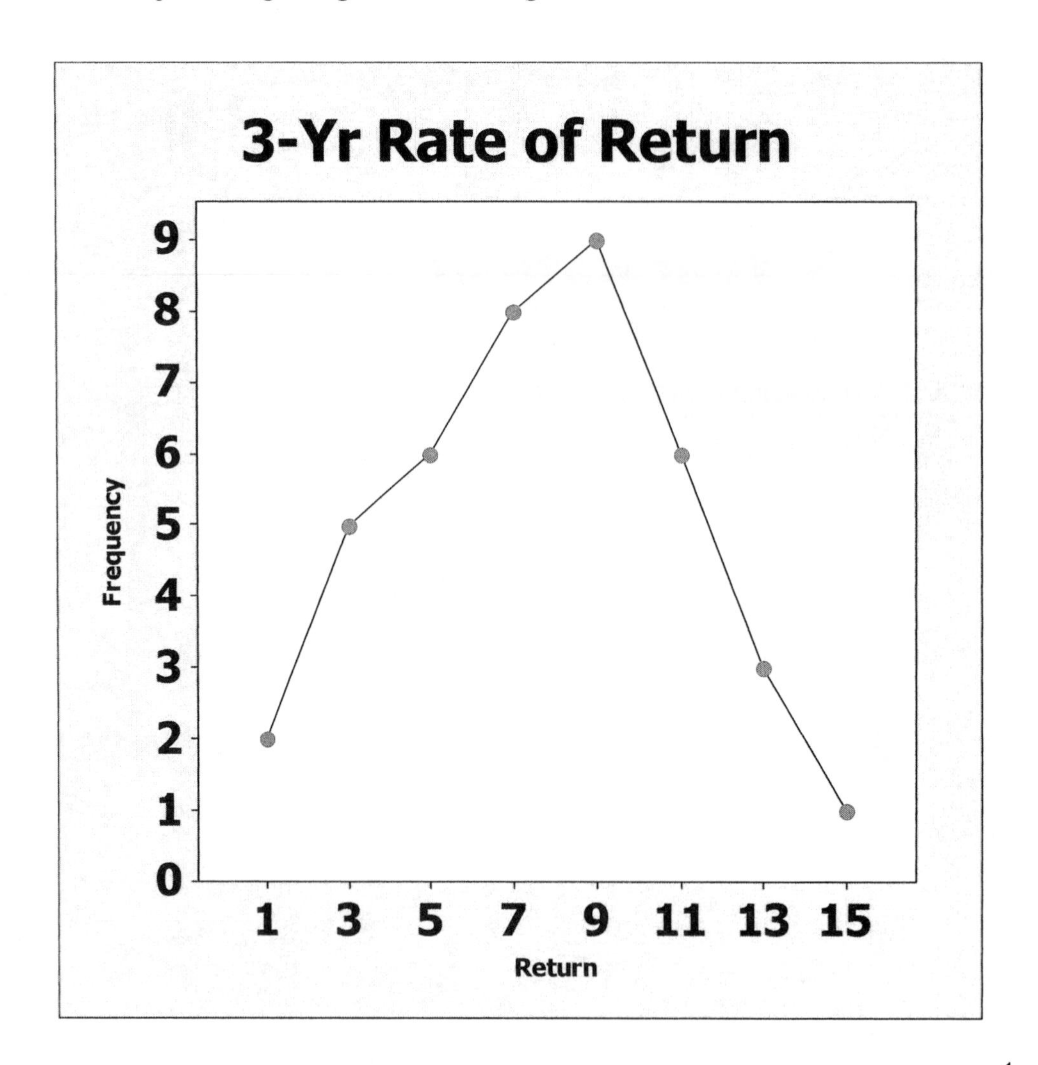
3-Yr Rate of Return
Frequency
0
1
2
3
4
5
6
7
8
9
1 3 5 7 9 11 13 15
Return

▸ Ogives (pg. 99) Constructing an Ogive

In this example, you must enter the data into the Data Window. Minitab does not have an automatic ogive function, but can plot the data for you after you enter the upper class limits and the cumulative relative frequencies. Begin with a clean worksheet. From Table 19 on page 99 of the text, enter the upper class limits into C1 and the cumulative relative frequencies into C2. Label each column appropriately as shown below.

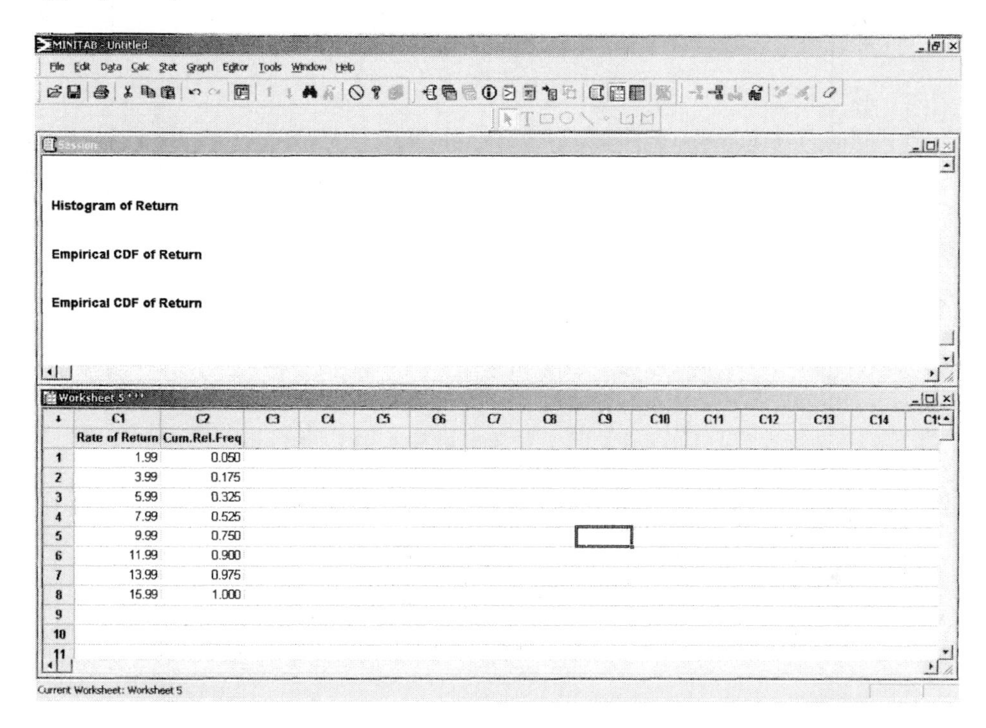

To plot the data, click on **Graph → Scatterplot → With Connect Line**. In the screen that appears, select C2 for the **Y-variable** and C1 for the **X-variable**. Click on the **Data View** button and be sure that both **Symbols** and **Connect line** are selected. By choosing both **Symbol** and **Connect line**, Minitab will connect the dots at each data point on the graph. Next, click on the **Labels** button and enter an appropriate title. Click on **OK** to view the ogive. You will have to edit the graph to number the X-axis to show each upper class limit. Right-click on the X-axis of the graph, and select **Edit X scale.** Enter the **Position of ticks** as 1.99: 15.99/ 2. This tells MINITAB that you want the tick marks to start at 1.99 and go up to 15.99 in steps of 2. Click on **OK** to view the changes to the ogive.

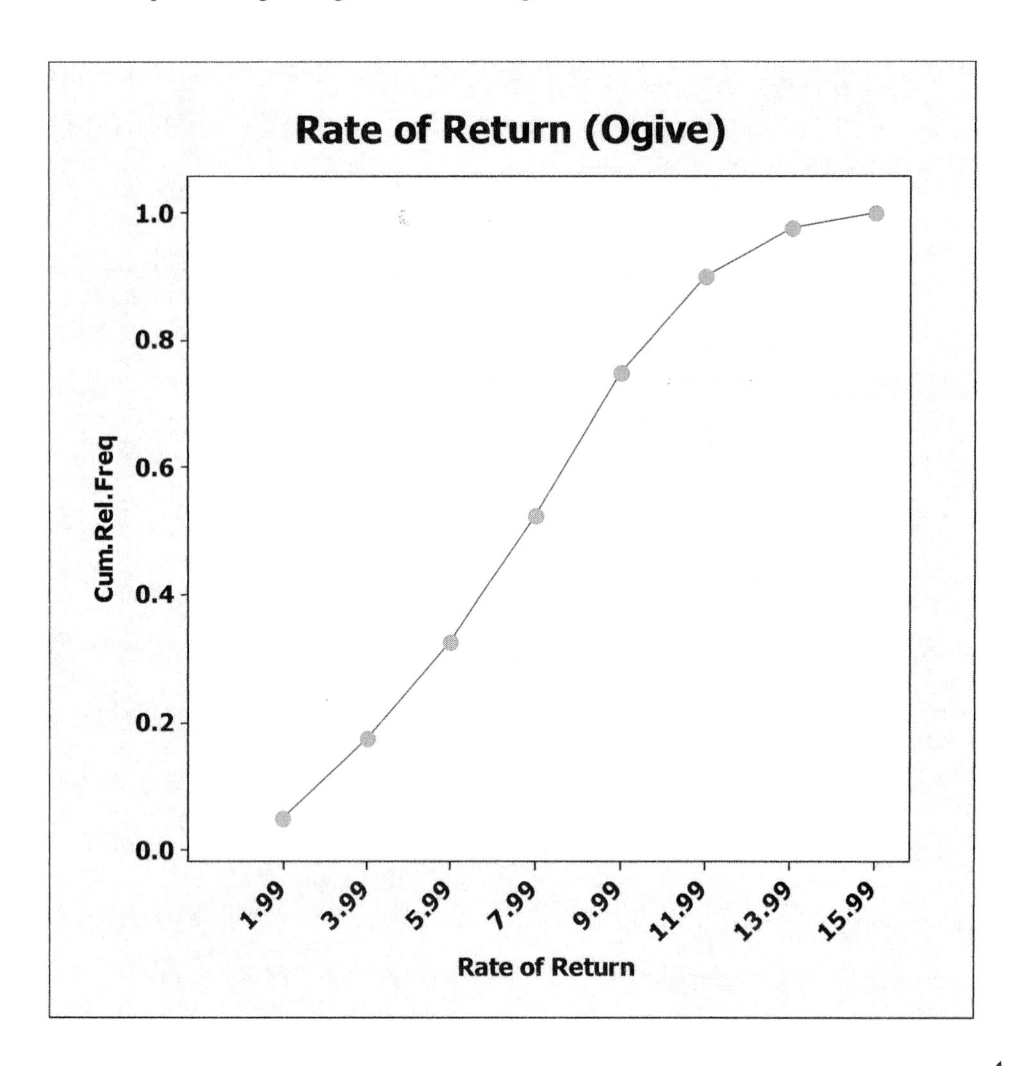
Rate of Return (Ogive)
Cum.Rel.Freq
1.0
0.8
0.6
0.4
0.2
0.0
1.99
3.99
5.99
7.99
9.99
11.99
13.99
15.99
Rate of Return

▸ Example 1 (pg. 100) Drawing a Time Series Plots

Open worksheet **2_3_T19.** Click on **Graph → Time Series Plot → Simple.** Select C2 (Closing Price) as the **Series**. Click on the **Time/Scale** button, and select **Stamp.** Select C1 (Date) for the **Stamp columns.** By choosing **Stamp**, you can select the column that contains the dates of your data. Click on **OK**.

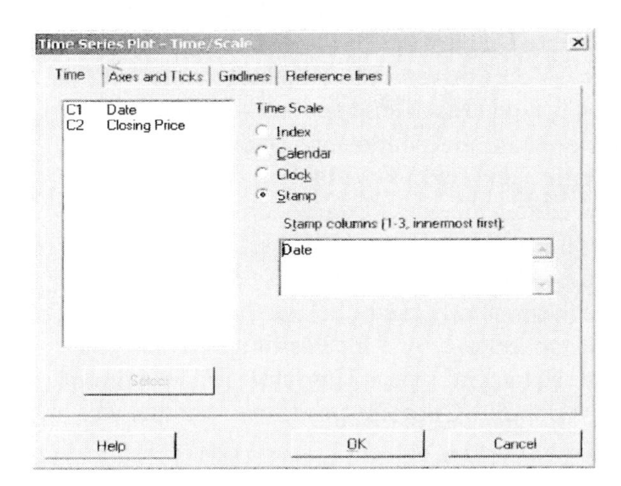

Next, click on the **Labels** button and enter an appropriate title for the plot. Click on **OK** twice, to view the plot.

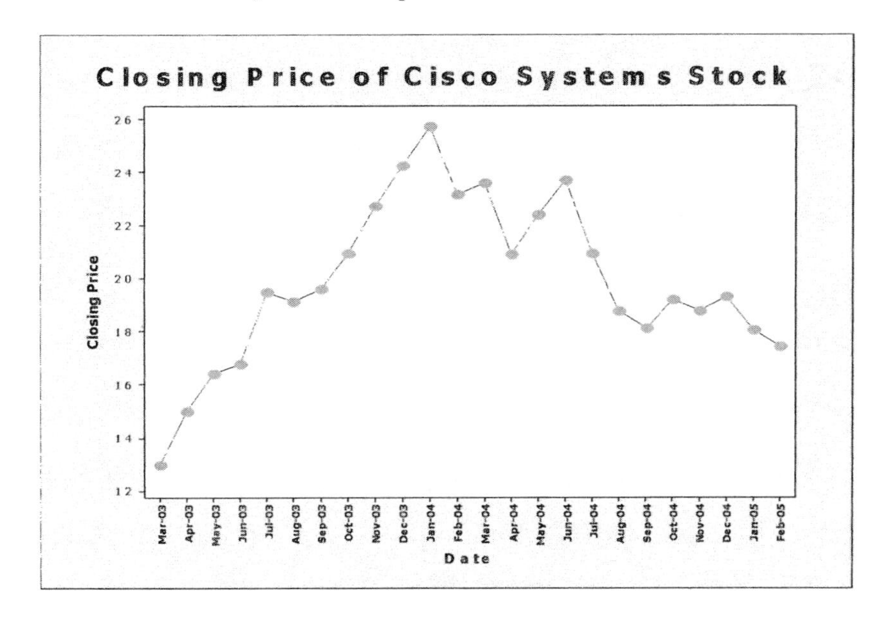

Note: If the dots on the graph are too large, right-click on a dot and select **Edit Symbols**. On the pop-up screen, select **Custom** and enter 1 for the **Size**.

◀

▸ Problem 22c-d (pg. 104) American League and National League

Open worksheet **2_3_22.** The American League data should be in C1 and the National League data in C3. Click on **Graph → Histogram → Simple.** First, double click on both C1 and C3 in the large box at the left of the screen. C1 and C3 should now be filled in as the **Graph variable**. Click on the **Data View** button and select **Symbols on the Date Display** tab. Next, click on the **Smoother** tab. For the **Smoother,** select **Lowess**. The **Degree of smoothing** is 0 and the **Number of steps** is 1. Click on **OK**. These options will connect the points of the polygon. Next, click on the **Labels** button and enter an appropriate title. Lastly, click on **Multiple Graphs** and select **Overlaid on the same graph**. All other changes will be made by editing the graph, so click on **OK** twice to view the polygon. Fix the numbering along the X-axis because for a polygon you should use the midpoints of each class. The text suggests the classes be: 2 to 6 with a class width of .5. To use these, simply right-click on the X-axis and select **Edit X Scale.** On the **Scale** tab, enter 2 : 6/.5 for **Position of ticks.** Click on the **Binning** tab and check that the interval type is **Cutpoint.** Under **Interval Definition,** enter 2 : 6 / .5 for the **Midpoint/Cutpoint positions.** Minitab will use the midpoints automatically.

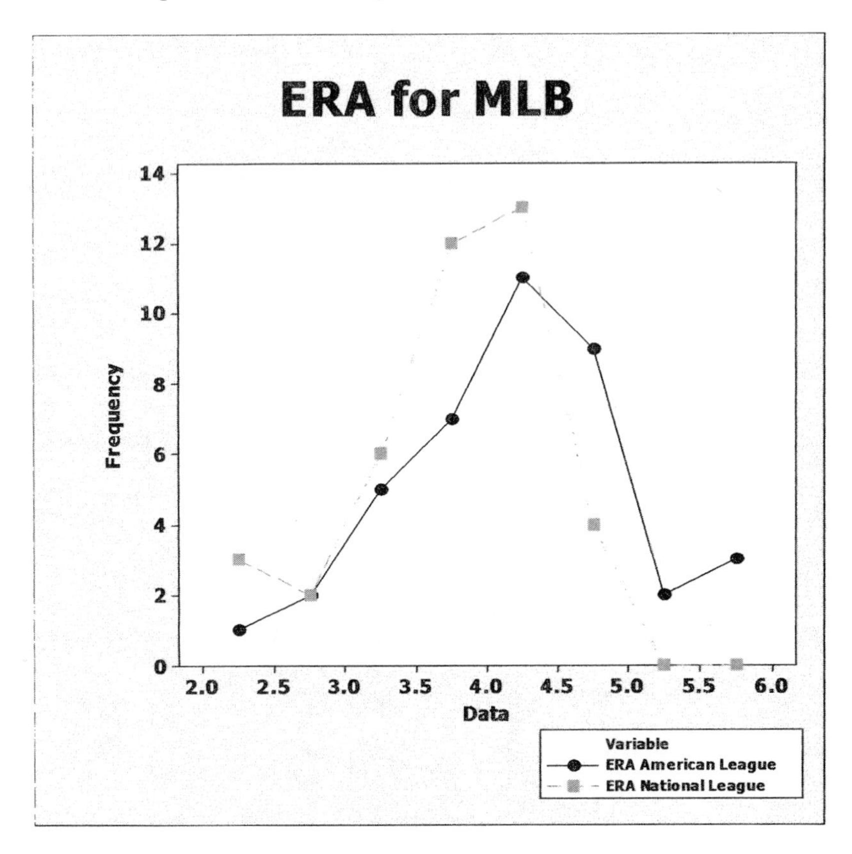

Note: Be sure that the Y-axis shows the frequencies rather than percents. To edit this, right-click on the Y-axis and select **Edit Y Scale.** On the **Type** tab, choose **Frequency.**

Part d of this problem is to draw an ogive for this data. The simplest way to do this is to edit the polygon that you just created. Right click on one of the numbers along the Y-axis, and select **Edit Y scale.** On the **Type** tab, choose **Percent** and **Accumulate values across bins.** Click on **OK** to view the ogive.

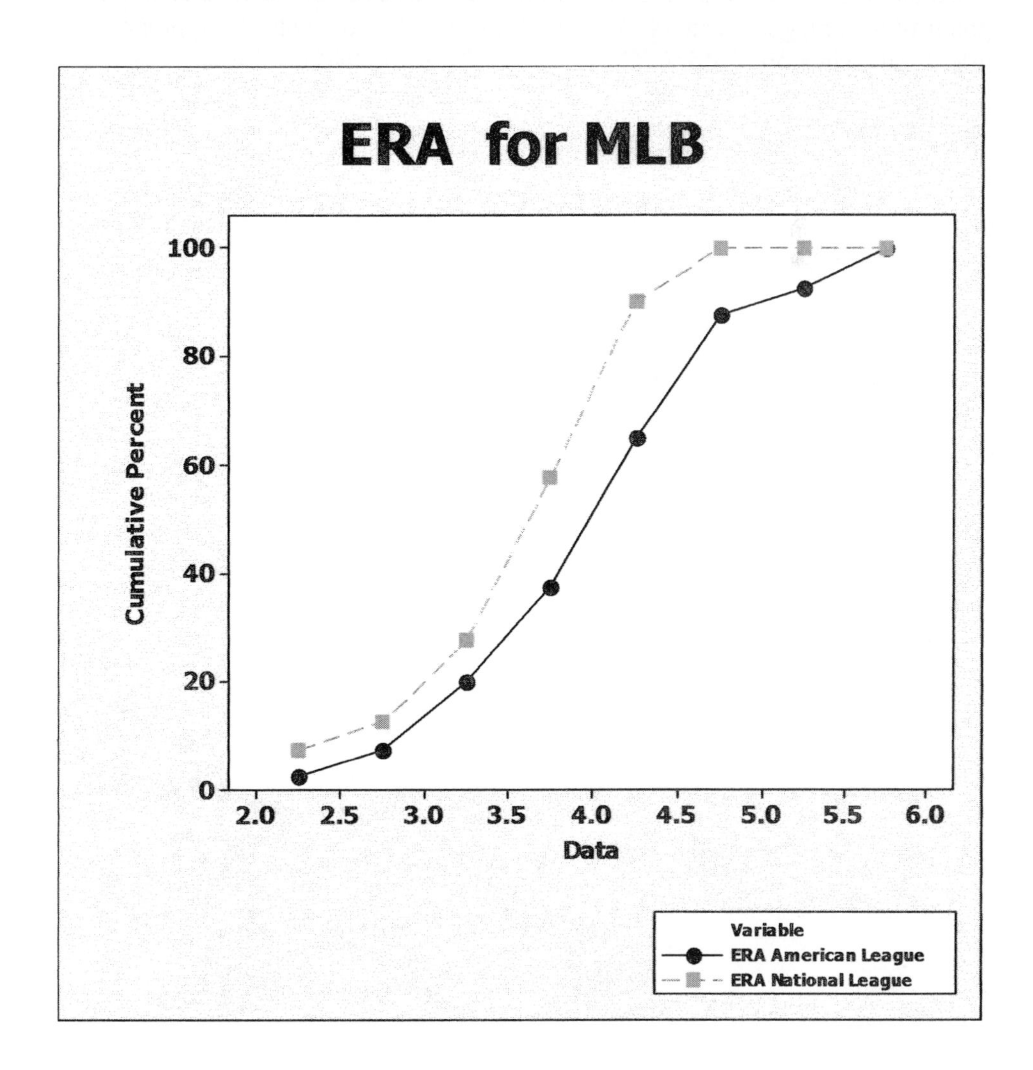

Numerically Summarizing Data

Section 3.1

Example 1 (pg. 122) Population mean and Sample mean

Finding the mean and standard deviation of a dataset is very easy using MINITAB. Enter the data into a Minitab worksheet. Column 1 can have the student name and Column 2 the test scores of the 10 students. To find the mean of this data, click on **Stat → Basic Statistics → Display Descriptive Statistics.** You should see the input screen below.

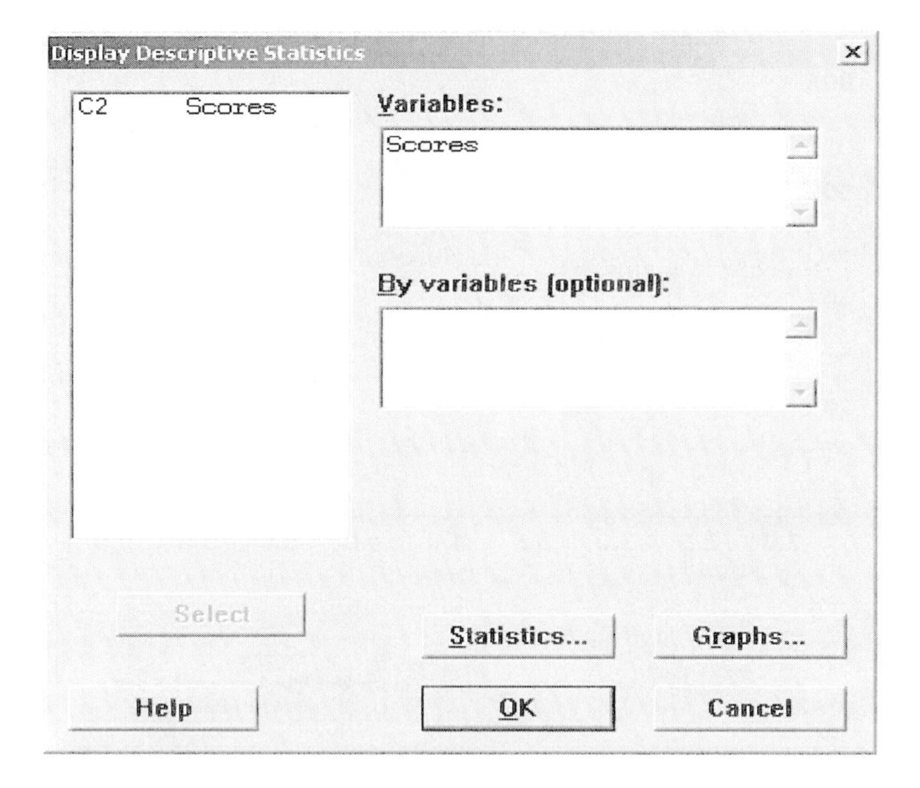

Double click on C2 to select the scores that are entered in C2. Click on **OK** and the descriptive statistics should appear in the Session Window.

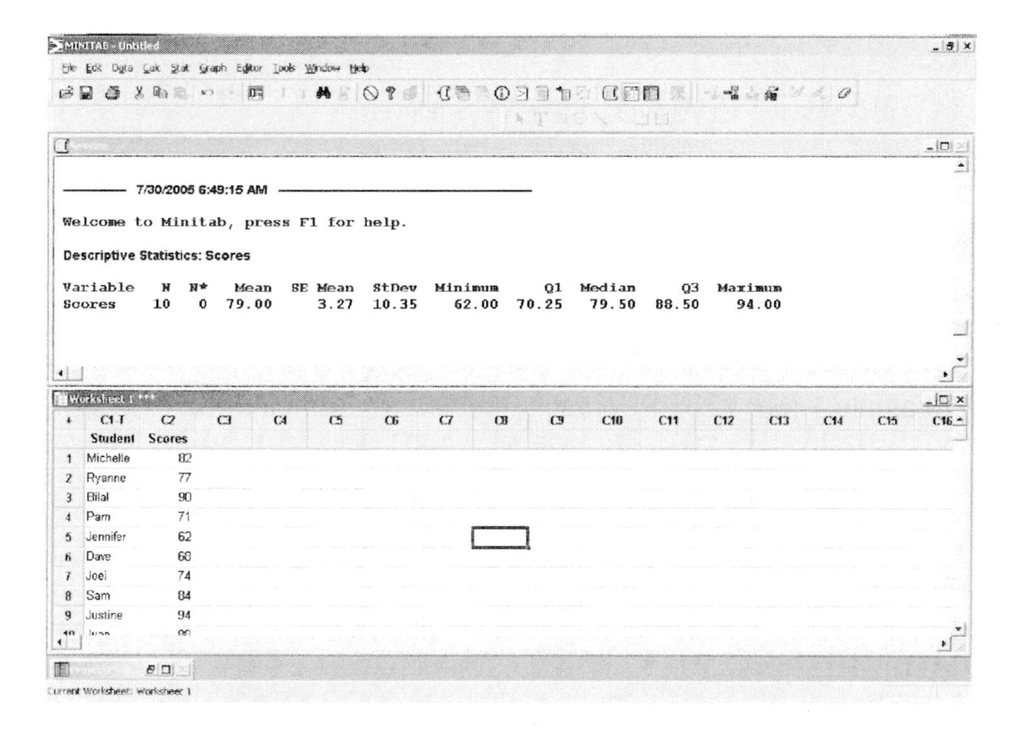

Notice that MINITAB displays several descriptive statistics: sample size (N), the number of missing values in the dataset (N*), mean, standard error of the mean, standard deviation, minimum value, first quartile, median, third quartile and maximum value. These statistics and more can be selected by clicking on the **Statistics** button from the input screen. So, the population mean test score is 79.

Now, take a random sample of 5 scores. Click on **Calc → Random Data → Sample from Columns.** You want to **Sample 4 rows from column C2** and **Store in C3.**

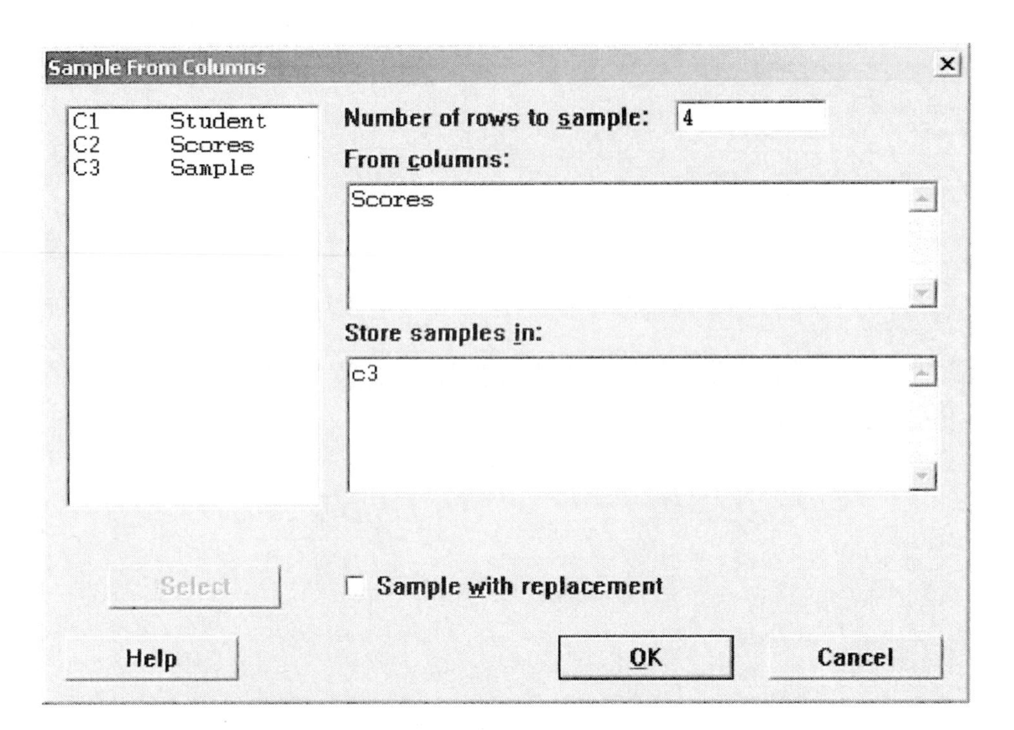

Click on **OK** to see the sample data. Since this is a ***random*** sample, your data will not look exactly like the data below.

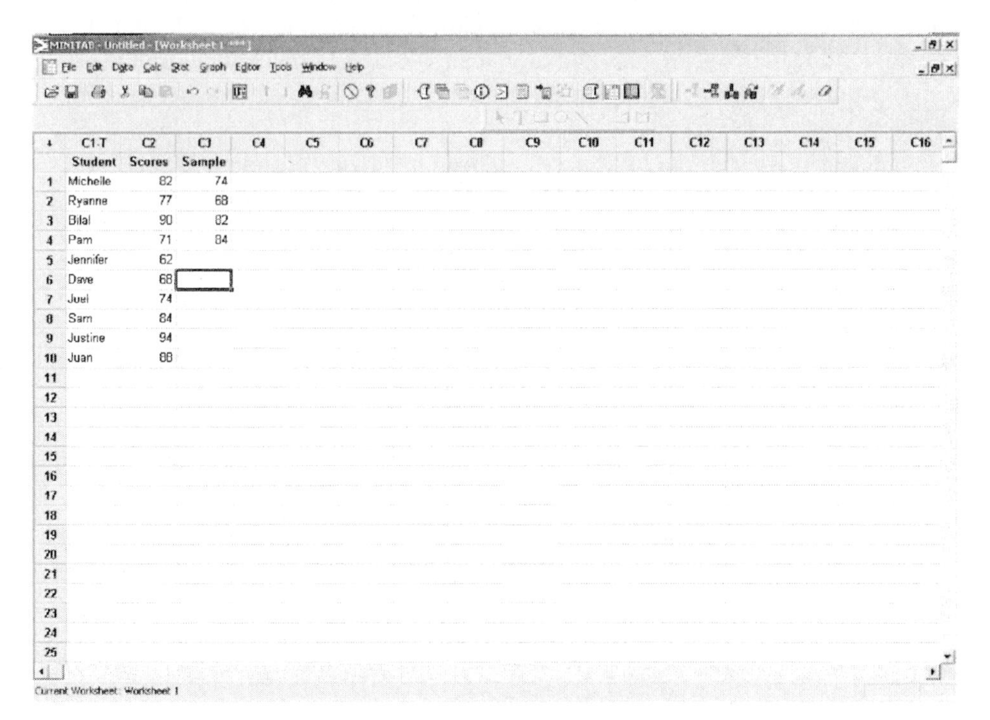

To find the sample mean of this data, click on **Stat → Basic Statistics → Display Descriptive Statistics.** Select C3 this time, click on **OK** and view the sample mean test score.

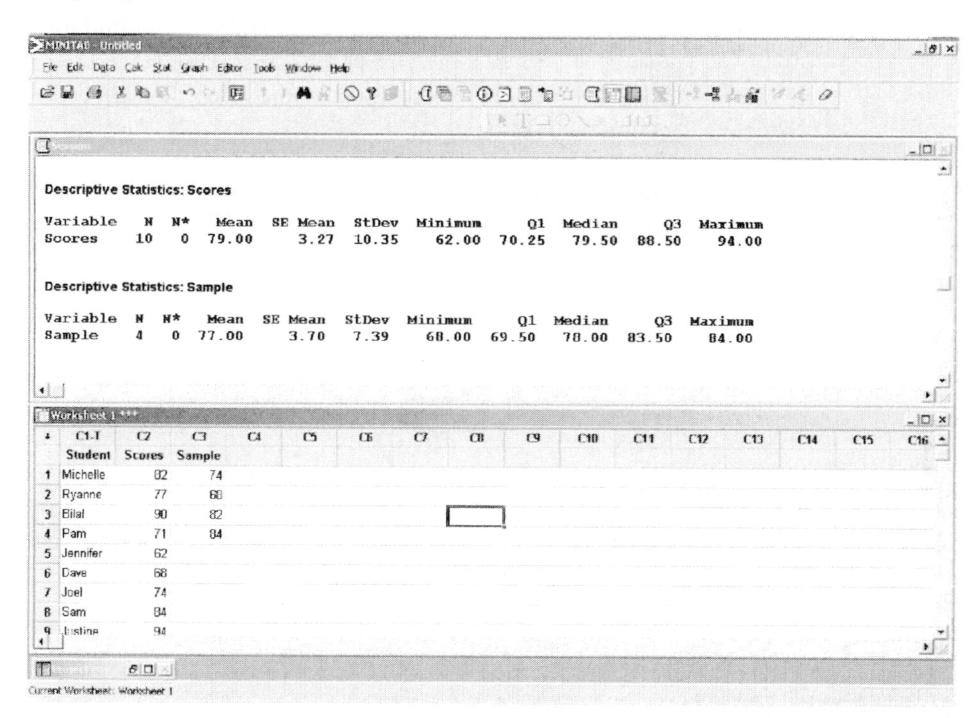

As you can see, the sample mean for this random sample is 77.0 and is very close to the population mean in this example.

◀

▸ Example 2 (pg. 124) Finding the median

Finding the median of a dataset is very easy using MINITAB. (Look at the Minitab output in Example 1 to see that the median test score is 79.5.) To find the median of Example 2, enter the song lengths into C1 of a Minitab worksheet. Click on **Stat → Basic Statistics → Display Descriptive Statistics.** Select C1 and, click on **OK** to see the summary statistics.

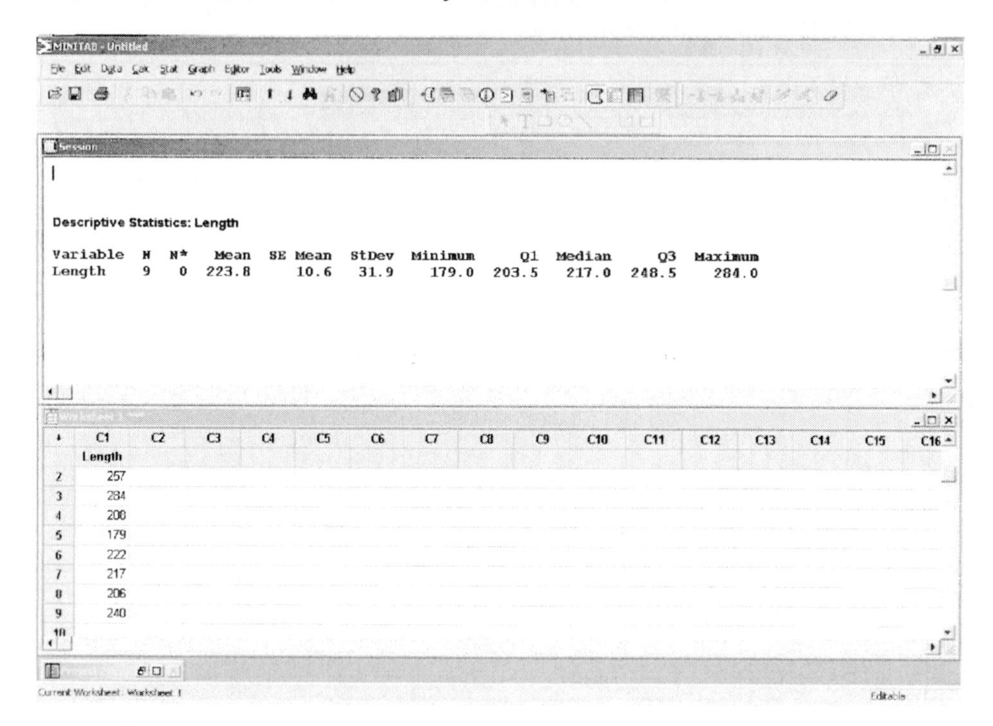

As you can see, the median length is 217.

◀

▸ Example 4 (pg. 126) Finding the mode

The mode is NOT produced automatically by the above procedure, however, it is quite simple to have MINITAB tally up the data values for you, and then you can select the one with the highest count. Enter the data for this example into C1 of a Minitab worksheet. Click on **Stat → Tables → Tally Individual Variables**. On the input screen, double-click on C1 to select it. Also, click on **Counts** to have MINITAB count up the frequencies for you.

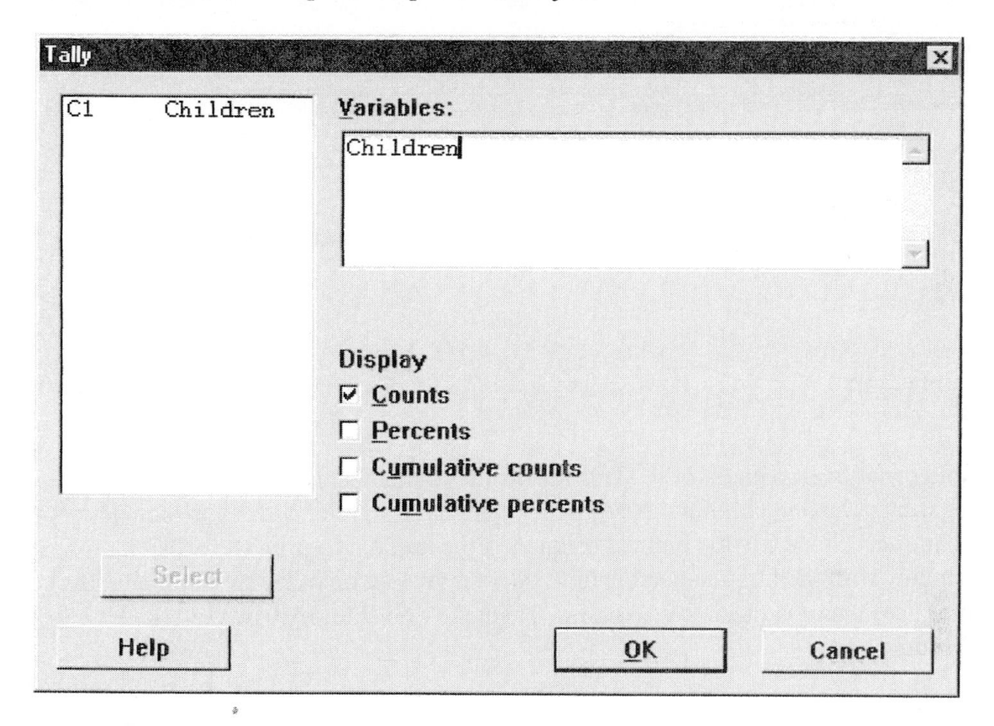

When you click on **OK**, a frequency table will appear in the Session Window.

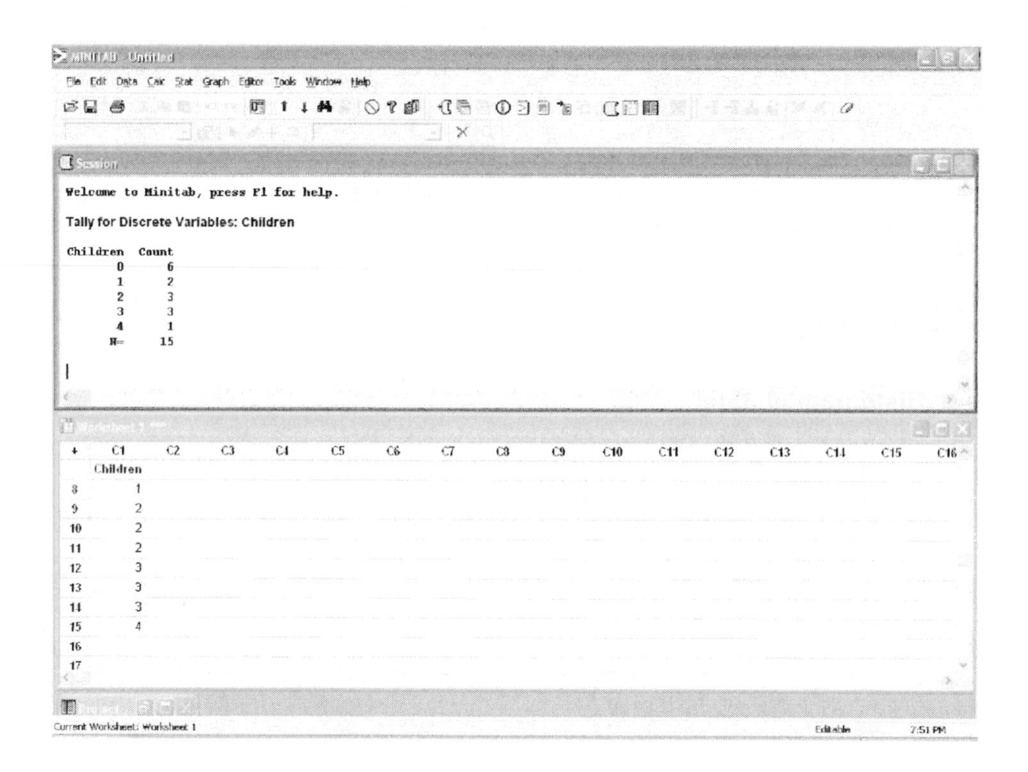

Notice that 0 has a count of 6. This means that 6 members of the math department have 0 children. Also, 2 members have 1 child, and 3 members have 2 children. Since 6 is the highest count, 0 is the mode. To print the Session Window with the frequency table in it, click anywhere up in the Session Window to be sure that it is the active window. Next click on **File → Print Session Window**.

◀

▸ Example 9 (pg. 129) Mean, median and shape of Birth weights

Enter the birth weights into C1. Click on **Stat → Basic Statistics → Display Descriptive Statistics.** Select C1 and, click on the **Graphs** button. A histogram is a good way to see the shape of the data. We will let Minitab help us by using its default settings. Click on **Histogram of data.**

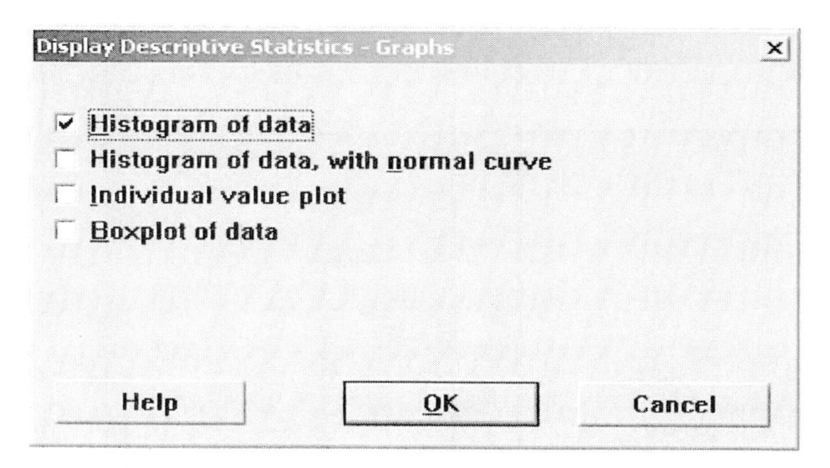

Click on OK twice to view the descriptive statistics in the Session Window, and the graph in the Graph Window.

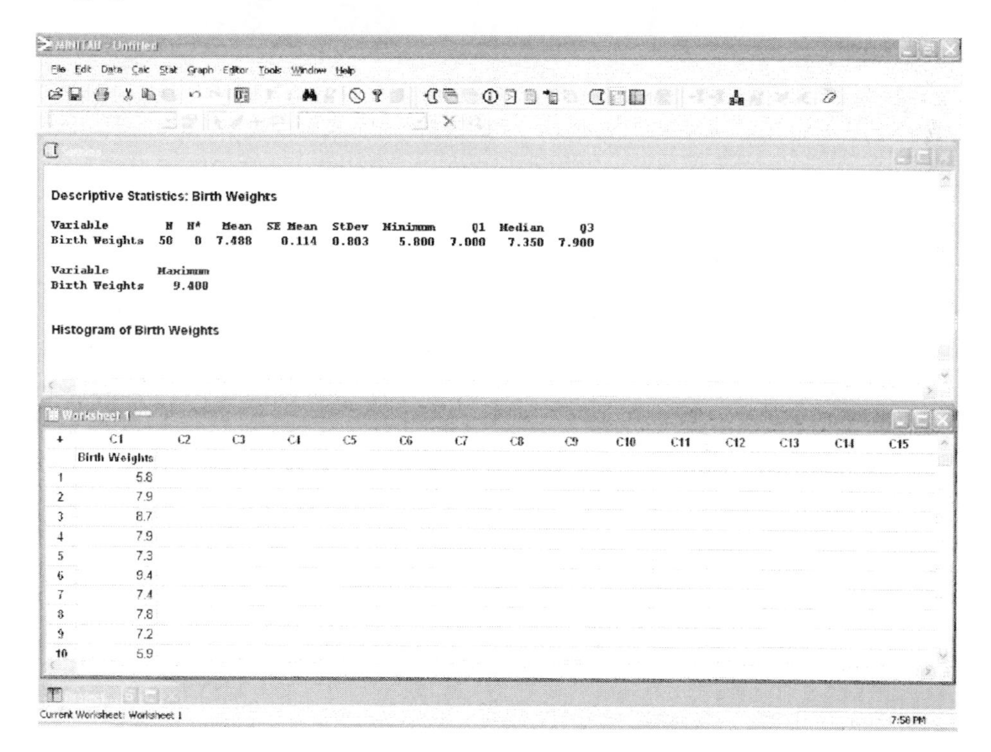

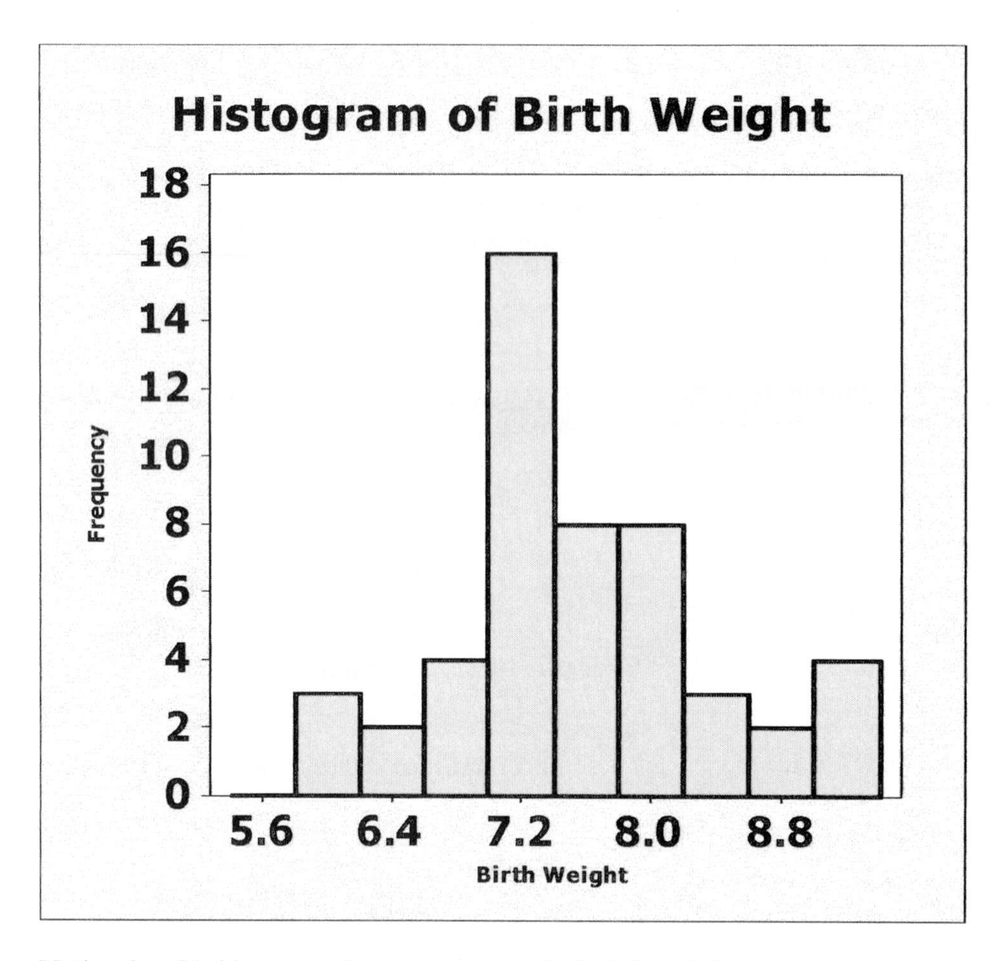

Notice that this histogram drawn automatically by Minitab is not exactly the same as the one in the textbook. This is not important since the shape of the distribution is of interest here. It is important to see that this histogram is fairly symmetric and bell-shaped.

◀

▸ Problem 25 (pg. 132) Pulse Rates

Enter the pulse rates into C1 of the Minitab worksheet. To take 2 random samples of size 3, click on **Calc → Random Data → Sample from Columns.** You want to **Sample 3 rows from column C1** and **Store in C2.** Repeat and this time **Store in C3.** Click on **Stat → Basic Statistics → Display Descriptive Statistics.** Select C1-C3 and click on **OK** to see the summary statistics.

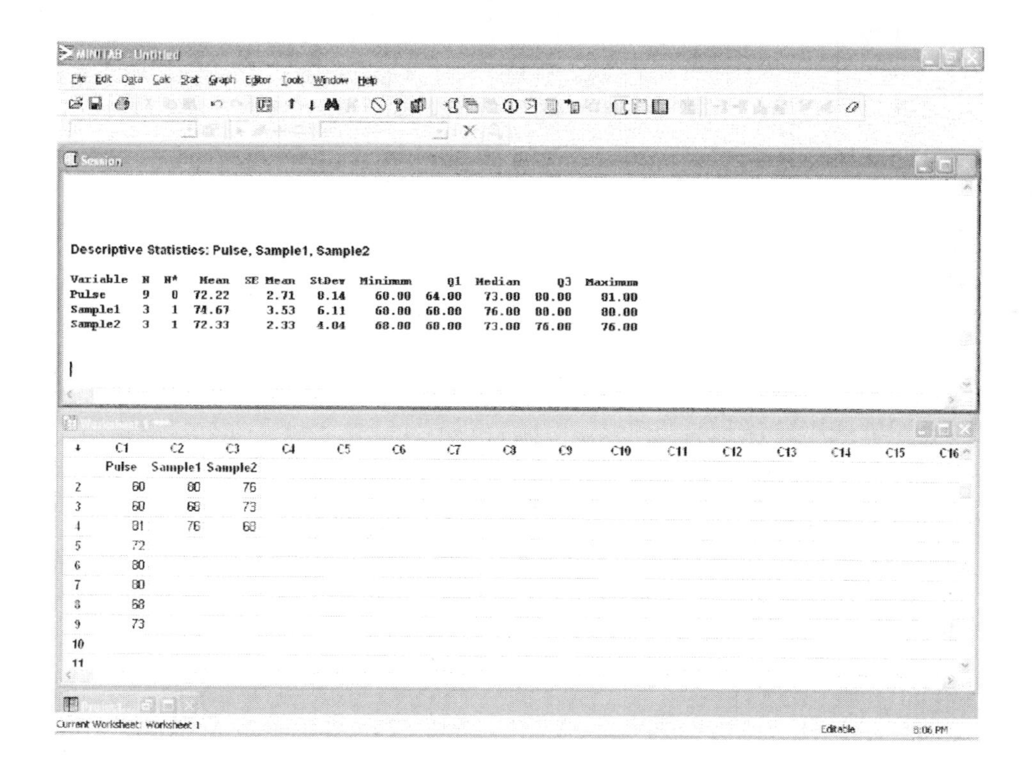

◂

▸ Problem 33 (pg. 133) Weights of M&Ms

Open worksheet **3_1_33**. The weights should be in C1. Click on **Stat → Basic Statistics → Display Descriptive Statistics.** Select C1, and click on the **Graphs** button. Click on **Histogram of data** and click on **OK** twice to see the summary statistics and the histogram.

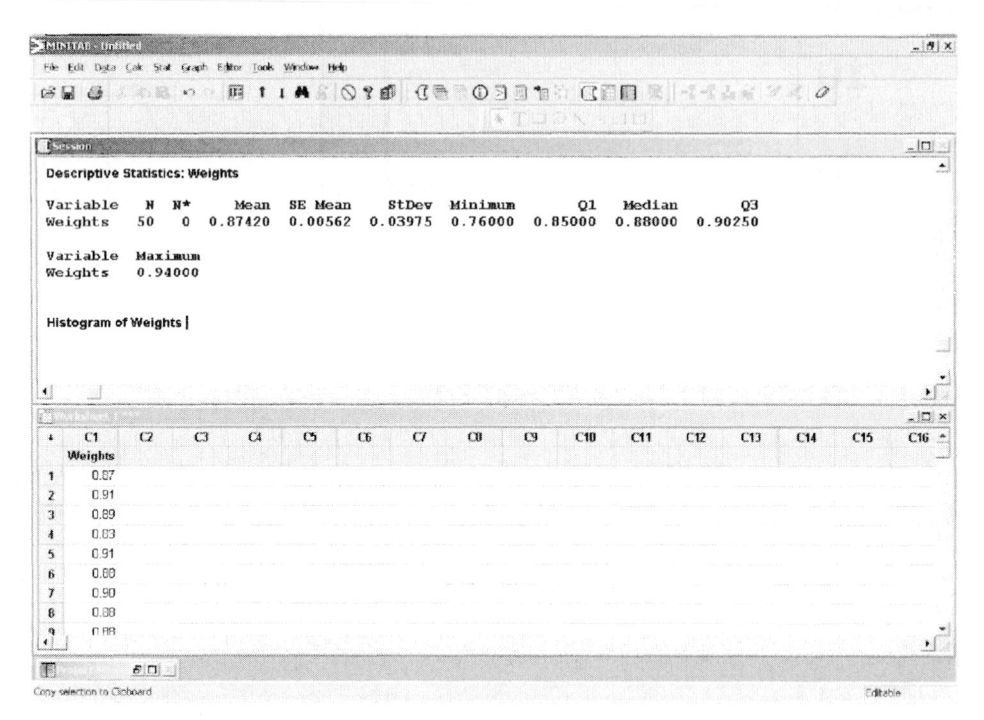

The mean is 0.87 and the median is 0.88.

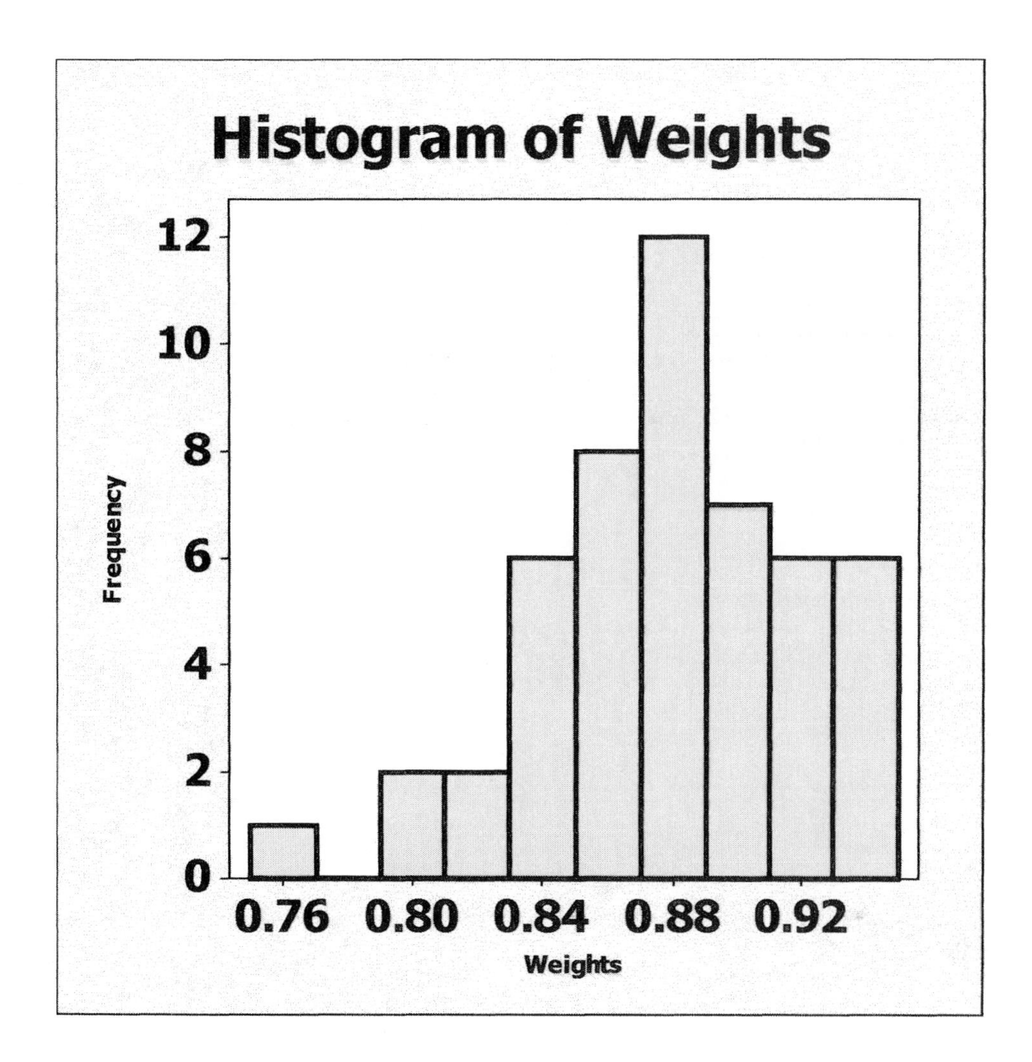

The shape of the histogram is fairly symmetric.

Section 3.2

Example 7 (pg. 144) Comparing the Variance and Standard Deviation of Two Datasets

Open Minitab worksheet **3_2_Ex1**. Click on **Stat → Basic Statistics → Display Descriptive Statistics.** Select both C1 and C2, and click on **OK** to see the summary statistics.

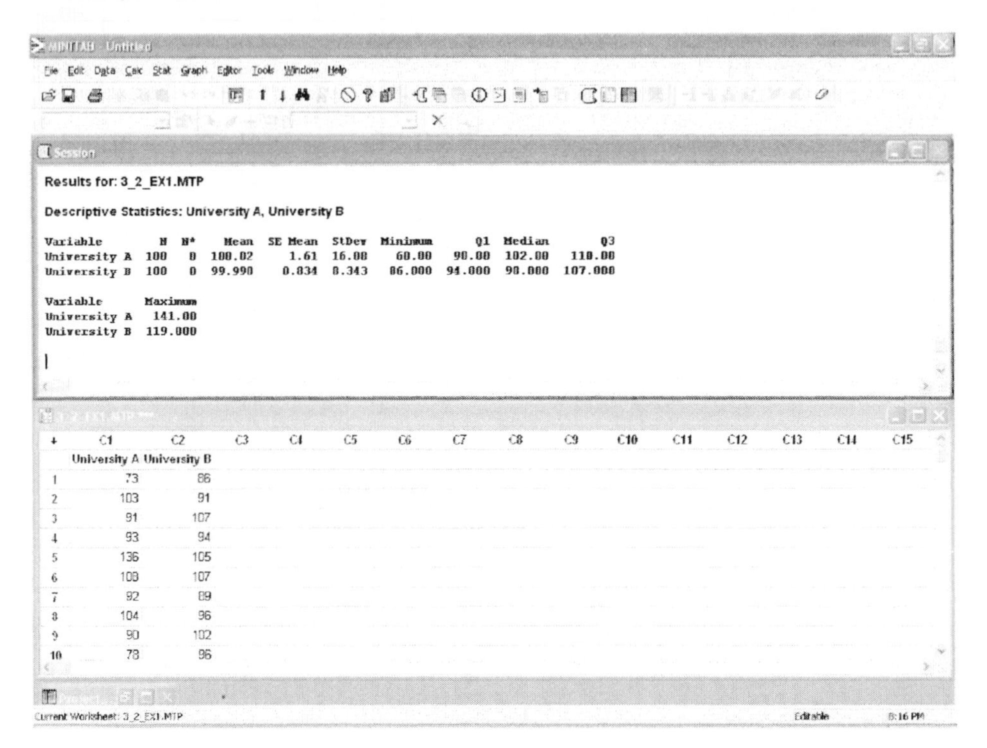

Problem 46 (pg. 154) Linear Transformations

Enter the 10 salaries into C1 of a Minitab worksheet. Click on **Stat → Basic Statistics → Display Descriptive Statistics.** Select C1 and click on **OK** to see the summary statistics.

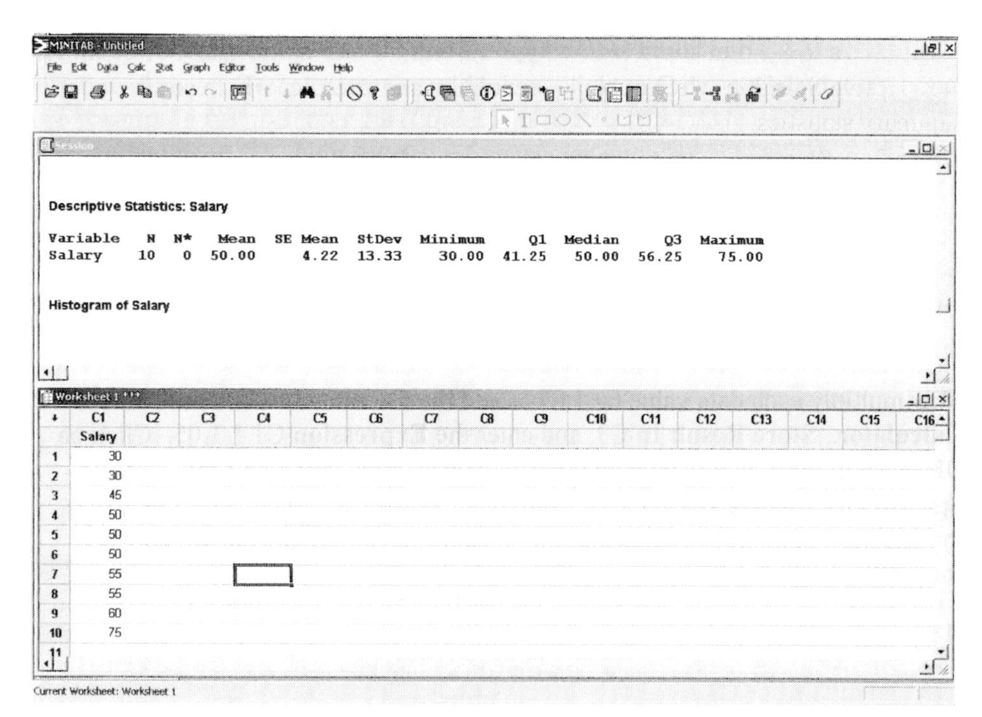

Now add $2500 to each data value. Click on **Calc → Calculator. Store Result In** C2, and enter the **Expression** C1 + 2.5 (since salary is in thousands of dollars). Click on OK and column 2 should contain the new salaries. Now, click on **Stat → Basic Statistics → Display Descriptive Statistics.** Select C2 and click on **OK** to see the summary statistics. Notice that the standard deviation is still 13.33.

Note: Minitab does not calculate the population standard deviation, so the results shown here use the sample standard deviation.

▸ Example 4 (pg. 179) Comparisons using Boxplots

Enter the data into a Minitab datasheet. Put the Flight data in C1 and the Control data in C2. To draw the boxplots, click on **Graph → Boxplot → Multiple Y's Simple.** Select C1 and C2 for the **Graph variables.** Click on the **Labels** button and enter an appropriate title. Click on the **Scale** button and select **Transpose value and category scales.** After clicking on **OK** to view the boxplot, edit the Y-axis label. Right-click on the label "Data" and select **Edit Y Axis Label.** Enter "Millimeters" for the **Text.**

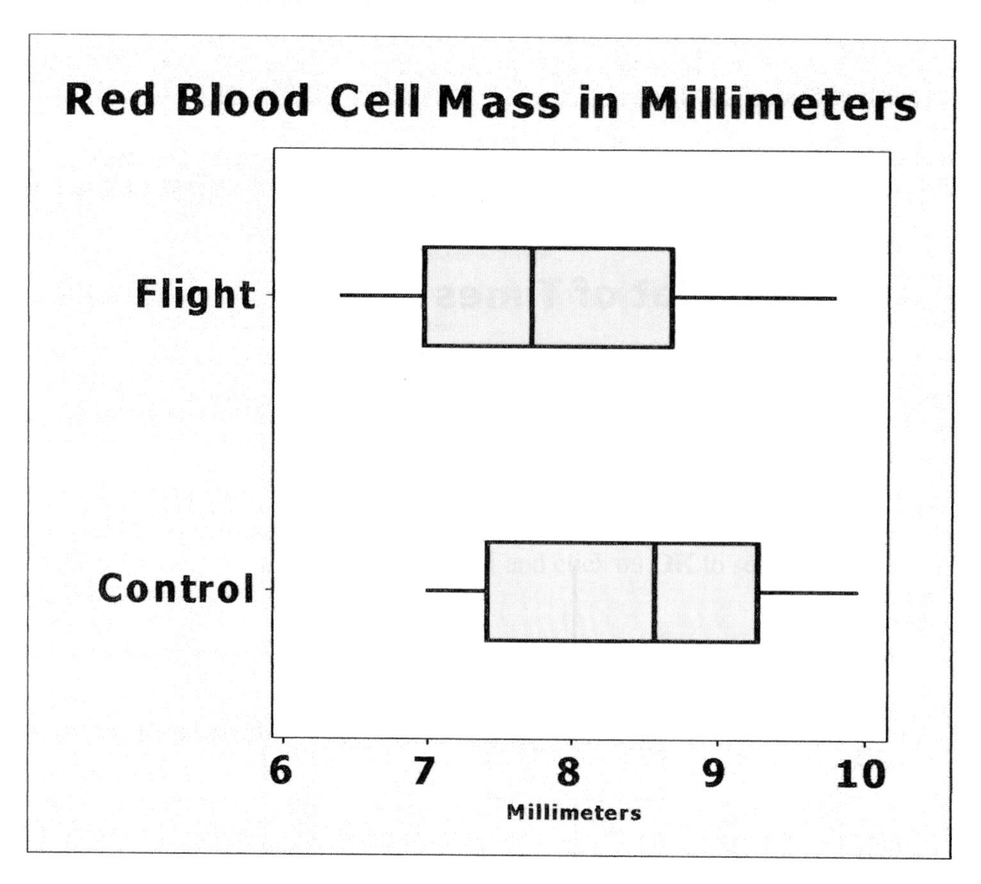

◂

Problem 5 (pg. 181) Age at Inauguration

Open Minitab worksheet 3_5_5. The Age data is in C1. To find the 5-number summary, click on **Stat → Basic Statistics → Display Descriptive Statistics.** Select C1, and click on **Graphs** to select **Boxplot of data**. Click on OK twice to view the summary statistics in the Session Window and the boxplot in the Graph Window.

Descriptive Statistics: Age of Presidents

Variable	N	N*	Mean	SE Mean	StDev
Age of President	43	0	54.814	0.951	6.235

Variable	Minimum	Q1	Median	Q3	Maximum
Age of President	42.000	51.000	58.000	55.000	69.000

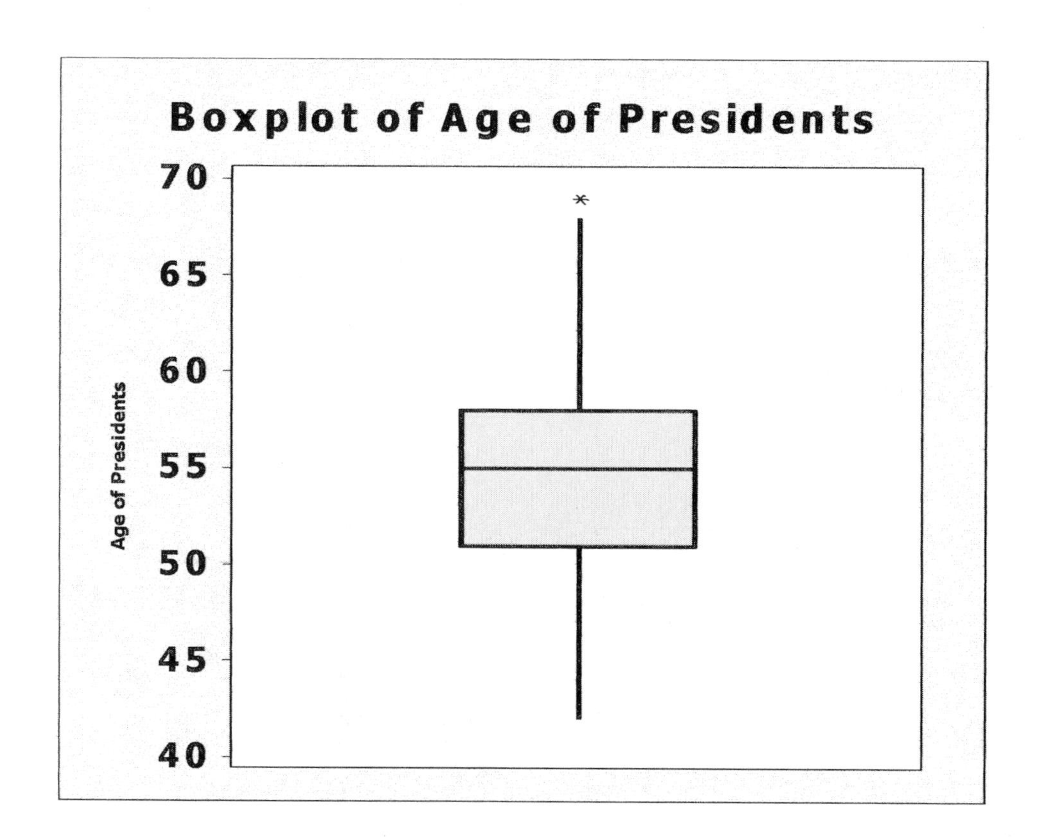

Describing the Relation between Two Variables

Section 4.1

Example 3 (pg. 201) Scatter Plots and Correlation Coefficients

Open Minitab worksheet **4_1_Ex1**. The speed should be in C1, and the distance should be in C2. Notice that speed is the x-variable and distance is the y-variable. To plot the data, click on **Graph → Scatterplot → Simple.** On the input screen, enter C2 for the **Y variable** and C1 for the **X variable.**

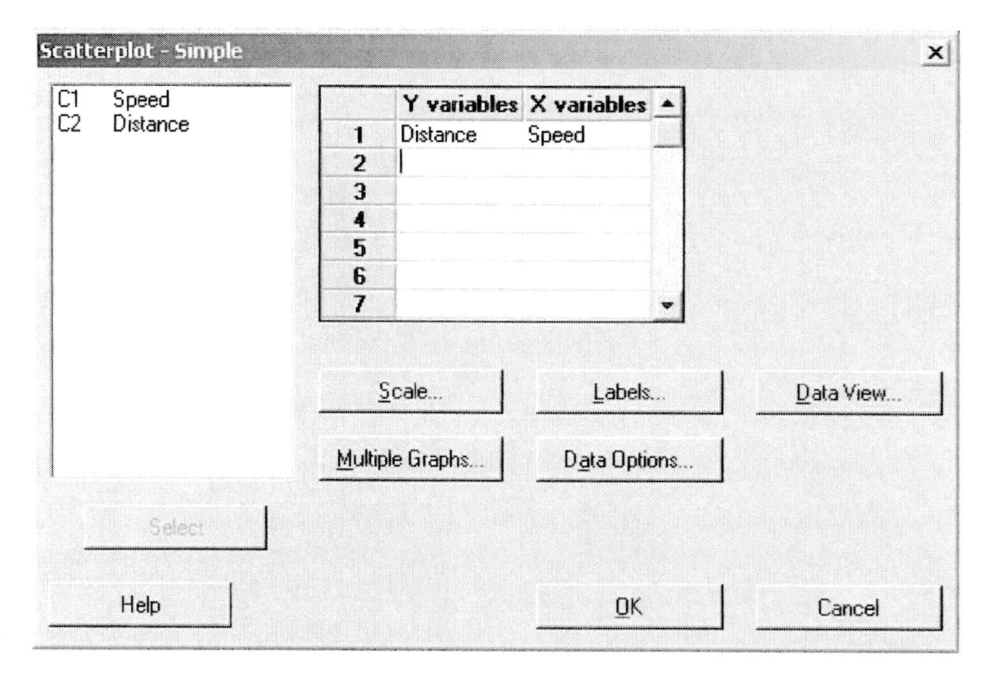

Next, click on the **Labels** button and enter an appropriate title. Click on **OK** twice to view the scatter plot that is created using Minitab default settings. If you need to change settings, you can edit the graph directly.

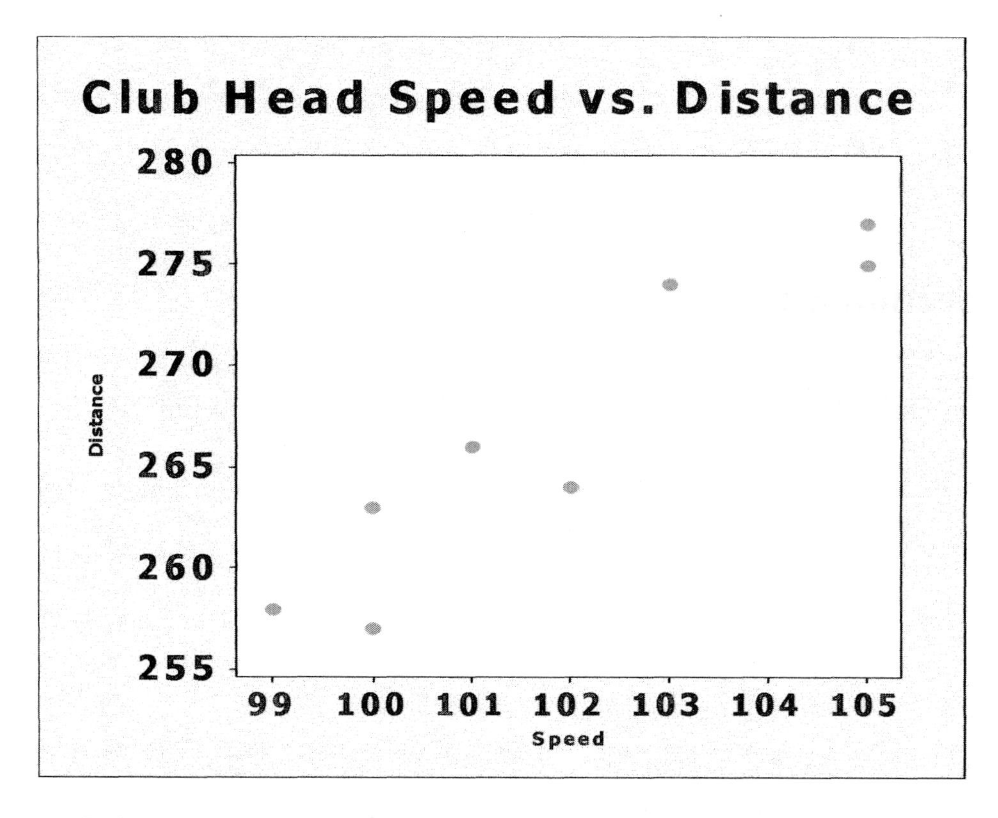

To find the correlation coefficient, click on **Stat** → **Basics Statistics** → **Correlation.** On the input screen, select both C2 and C3 for **Variables**, by double-clicking on each one.

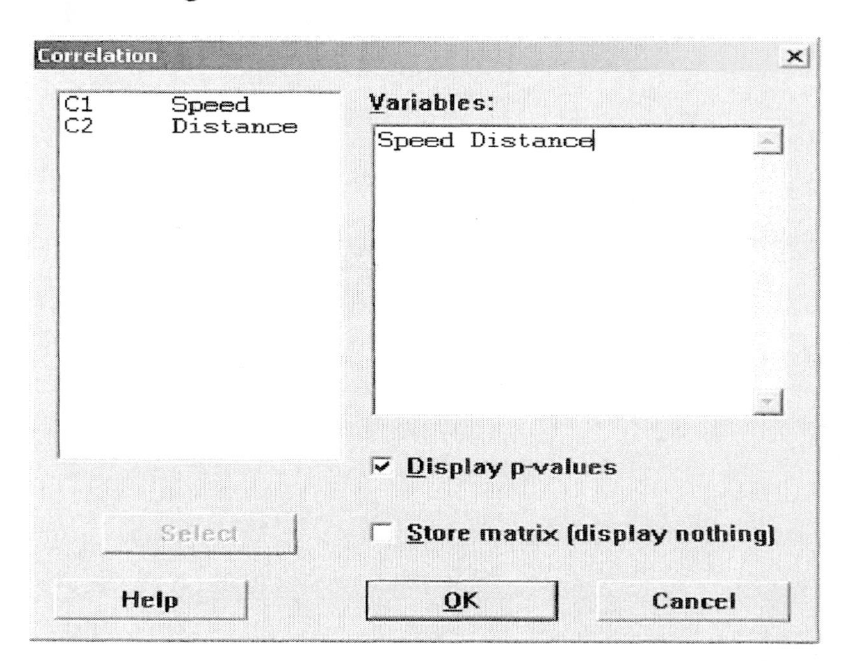

Click on **OK** and the Correlation Coefficient will be displayed in the Session Window.

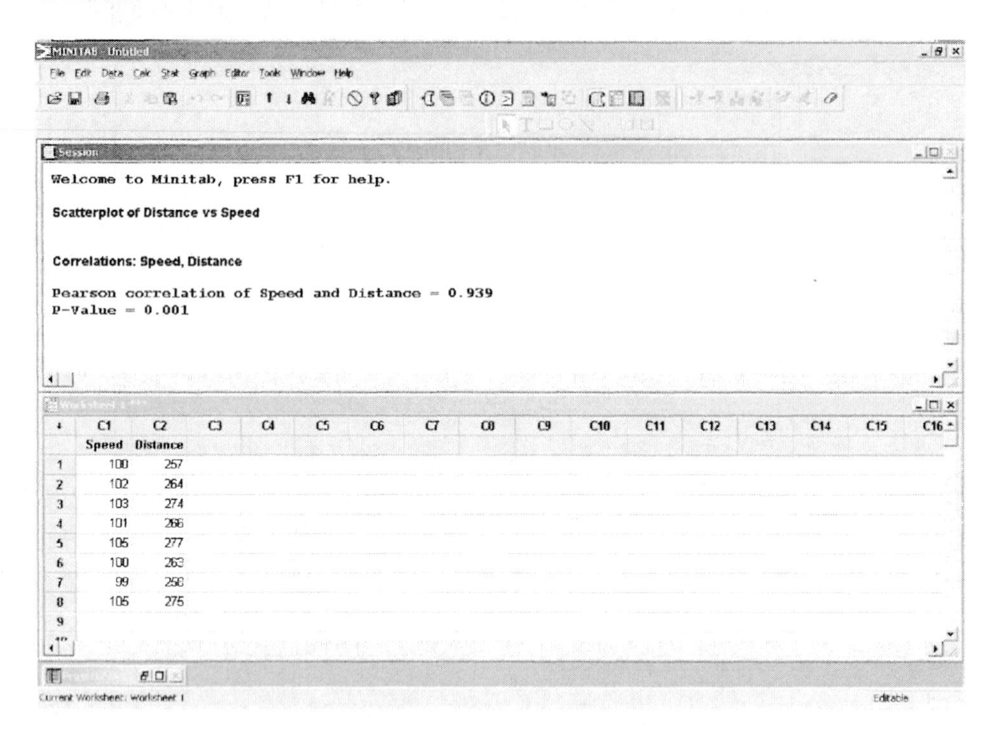

As you can see, the correlation coefficient is .939.

▸ Problem 23 (pg. 205) Height vs. Head Circumference

Open Minitab worksheet **4_1_23**. The height should be in C1, and head circumference should be in C2. Notice that "Height" is the x-variable and "Head Circumference" is the y-variable. To plot the data, click on **Graph → Scatterplot → Simple.** On the input screen, enter C2 for the **Y variable** and C1 for the **X variable.** Next, click on the **Labels** button and enter an appropriate title. Click on **OK** twice to view the scatter plot that is created using Minitab default settings. If you need to change settings, you can edit the graph directly.

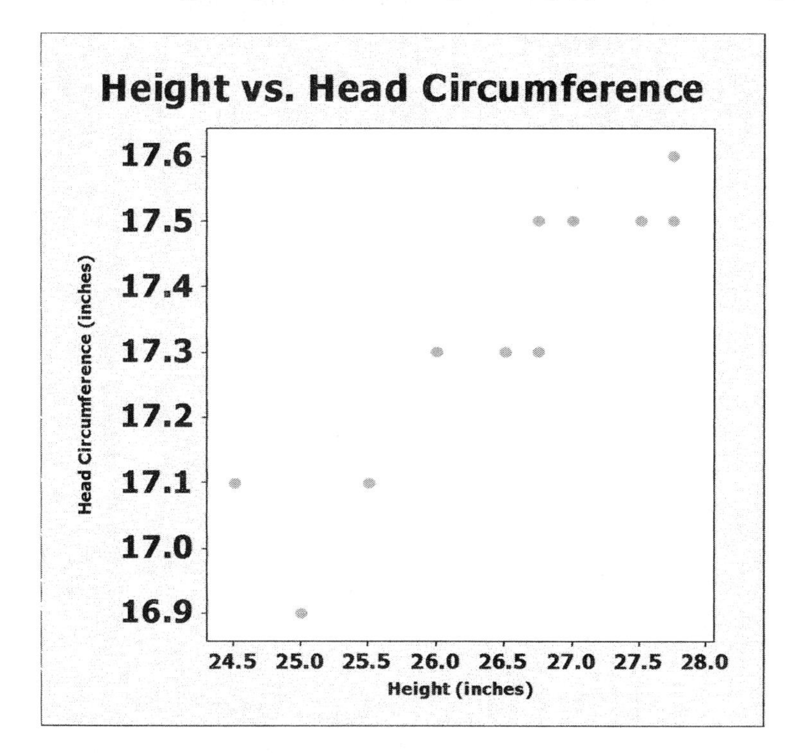

Now, to find the correlation coefficient, click on **Stat → Basics Statistics → Correlation.** On the input screen, select both C1 and C2 for **Variables**, by double-clicking on each one. Click on **OK** and the following output should be in the Session Window.

Correlations: Height, Head Cirumference

```
Pearson correlation of Height and Head Cirumference = 0.911
P-Value = 0.000
```

So, as you can see, the correlation coefficient is 0.911.

◀

Section 4.2

▸ Example 2 (pg. 216) Finding a Regression Equation

Open Minitab worksheet **4_1_Ex1**. The speed should be in C1, and the distance should be in C2. Notice that speed is the x-variable and distance is the y-variable. To find the least squares regression equation, click on **Stat → Regression → Regression.** Enter C2 for the **Response** variable, and C1 as the **Predictor.**

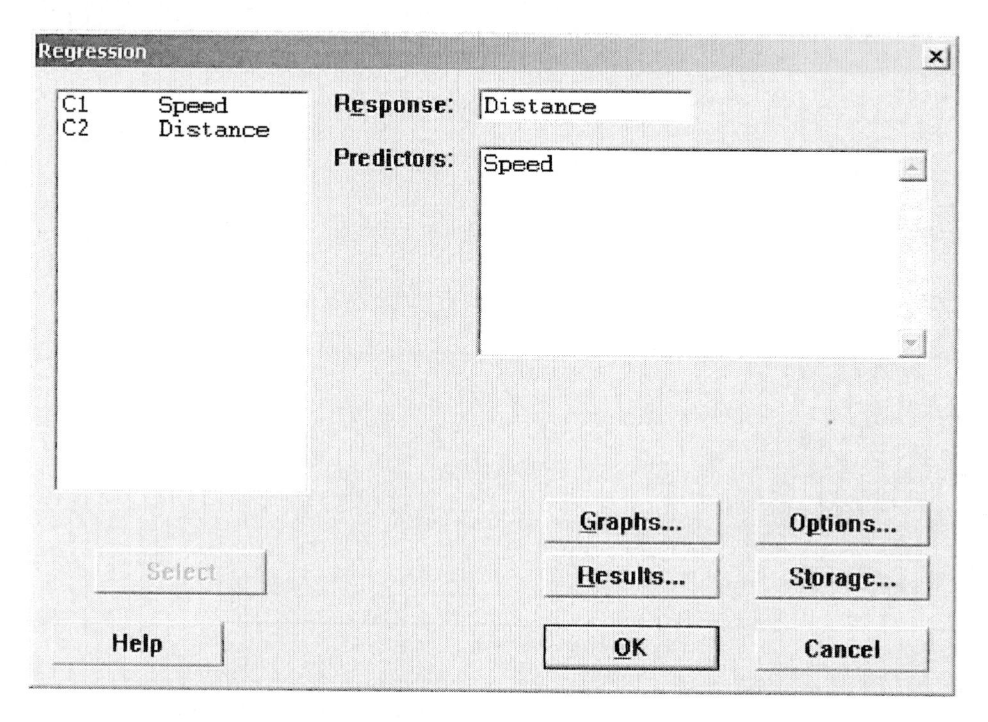

Click on **Results.** Select **Regression equation, table of coefficients, s, R-squared, and basic analysis of variance.**

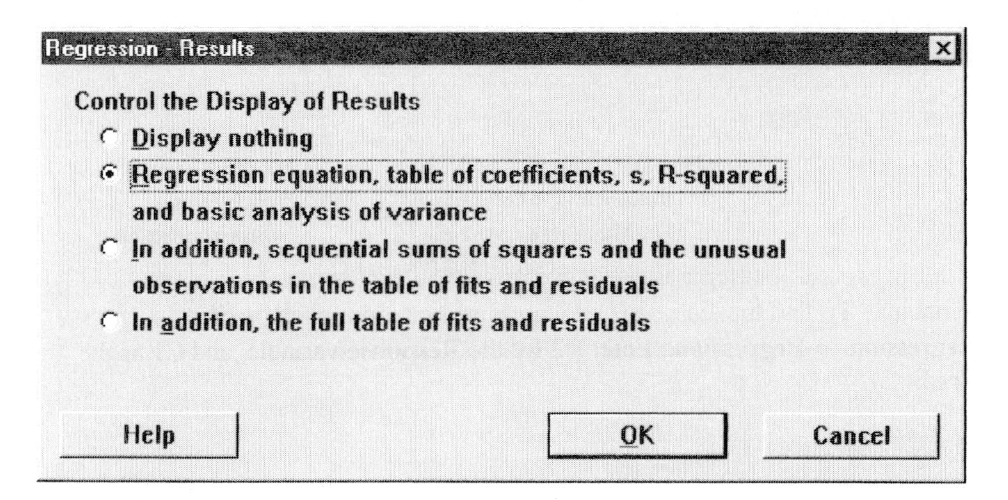

Next, click on **Storage** and select **Residuals.** Minitab will calculate the residuals for each data point and store them in an empty column.

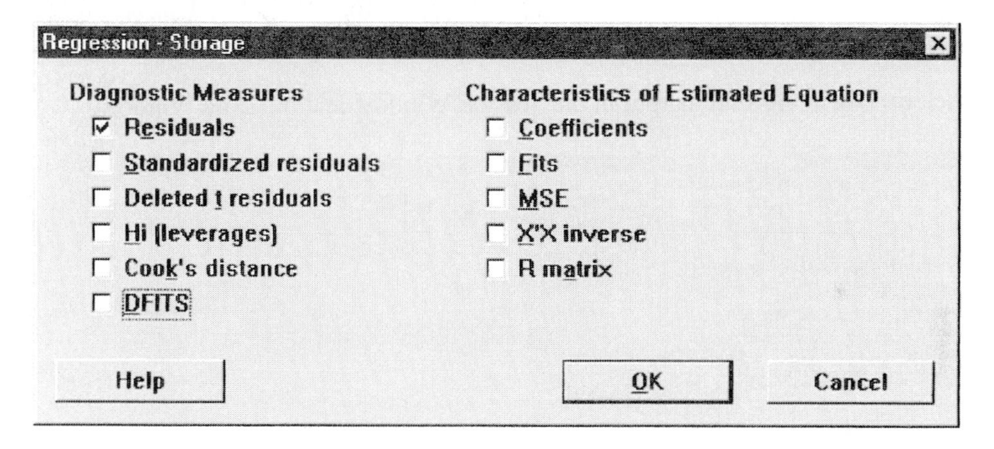

To tell Mintab to predict the distance a golf ball will travel if the club head speed is 103 mph, click on **Options** and enter 103 for **Prediction intervals for new observations.** Be sure to select **Fits.**

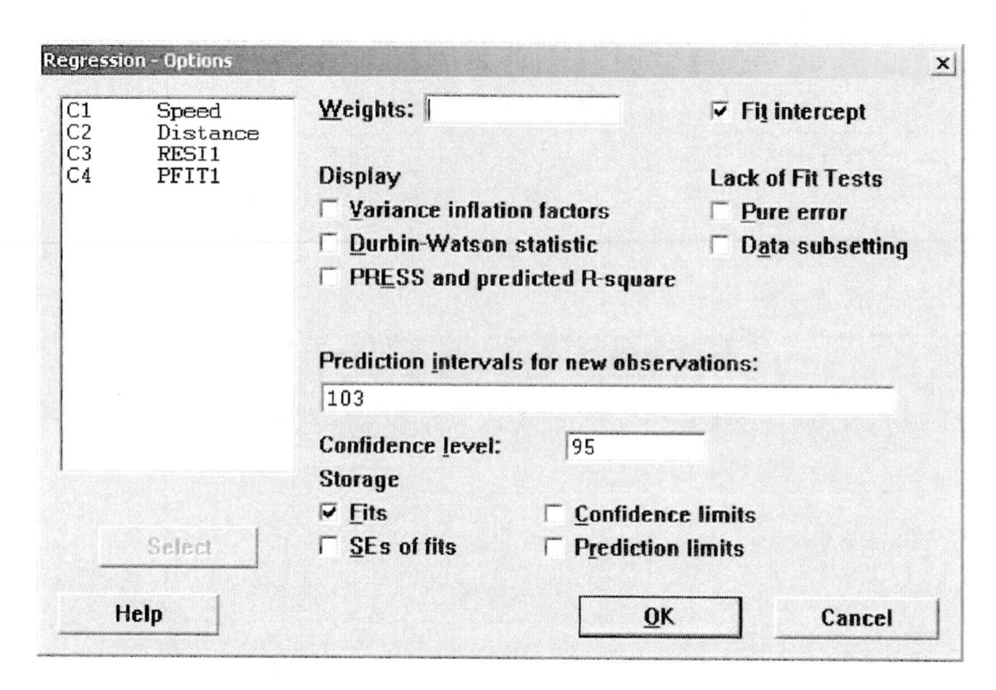

Click on **OK** to view the output in the Session Window and the Data Window.

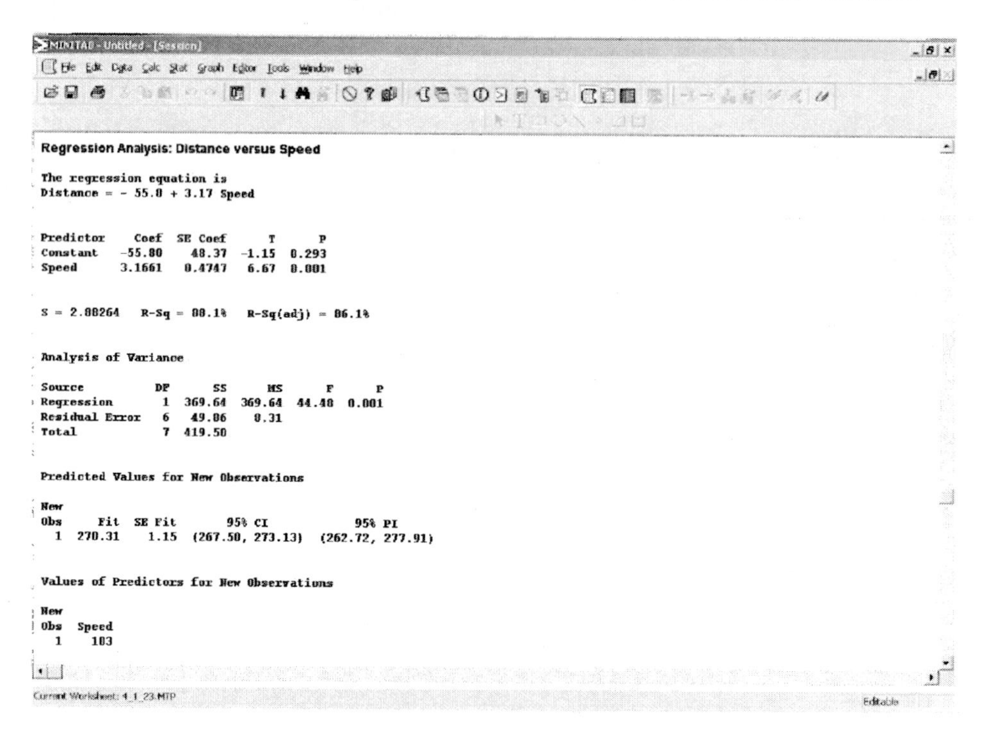

Notice that the regression equation is Distance = -55.8 + 3.17 * Speed. Also, in the data window there are now 2 extra columns. The column labeled **RESI1** contains the residuals. The column labeled **PFIT1** contains the predicted distance for a club head speed of 103, 270.31 yards. Notice that the actual distance for a speed of 103 mph is 274 yards, shown in row3 of the Minitab worksheet. Thus, the residual is 274 - 270.31 = 3.69. To draw the least-squares-regression line on the scatterplot, click on **Stat → Regression → Fitted Line plot.** Enter C2 for the **Response** variable, and C1 as the **Predictor.** Select **Linear** for the **Type of Regresssion Model** and click on **OK.**

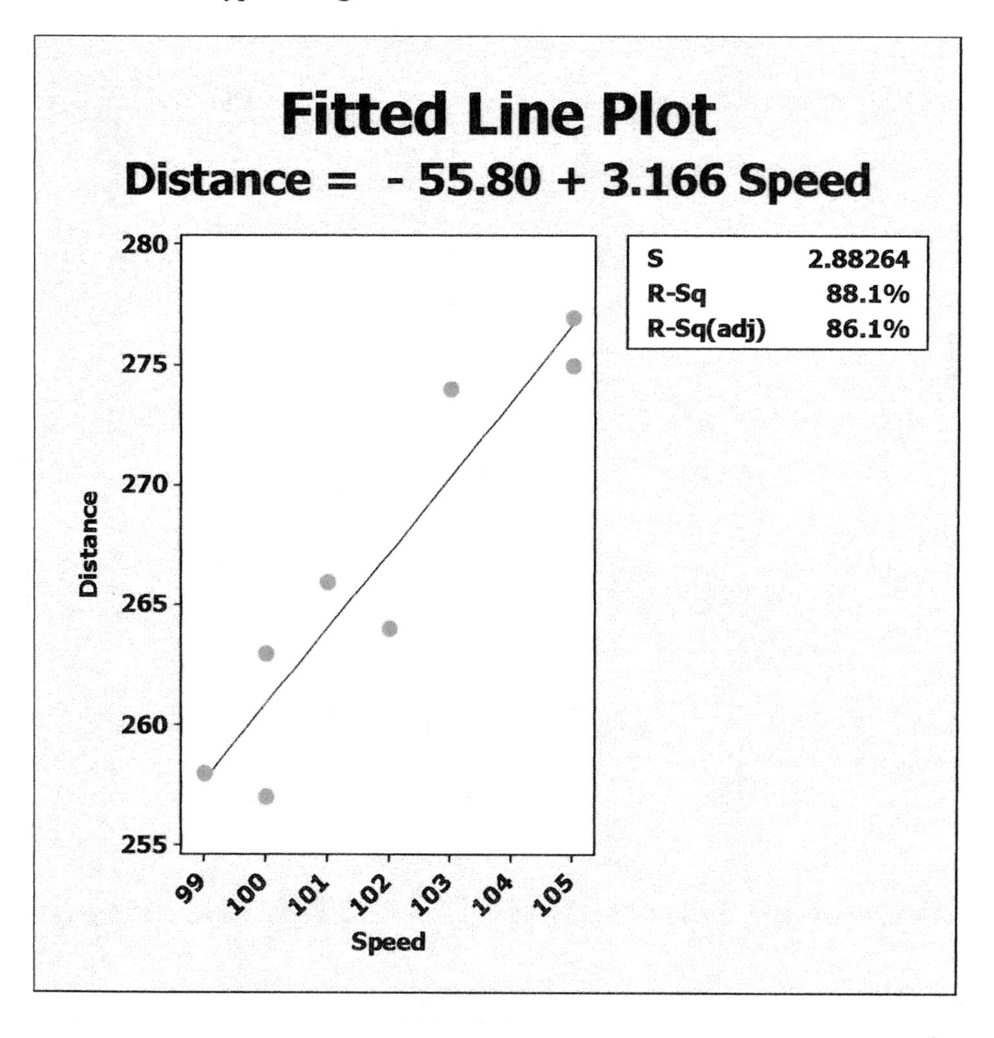

▸ Problem 18 (pg. 222) Gestation Period vs. Life Expectancy

Open worksheet **4_2_18**. Gestation period is in C2 and Life Expectancy is in C3. To plot the data, click on **Graph → Scatterplot → Simple.** On the input screen, enter C3 for the **Y variable** and C2 for the **X variable.** Next, click on the **Labels** button and enter an appropriate title. Click on **OK** twice to view the scatter plot that is created using Minitab default settings. If you need to change settings, you can edit the graph directly.

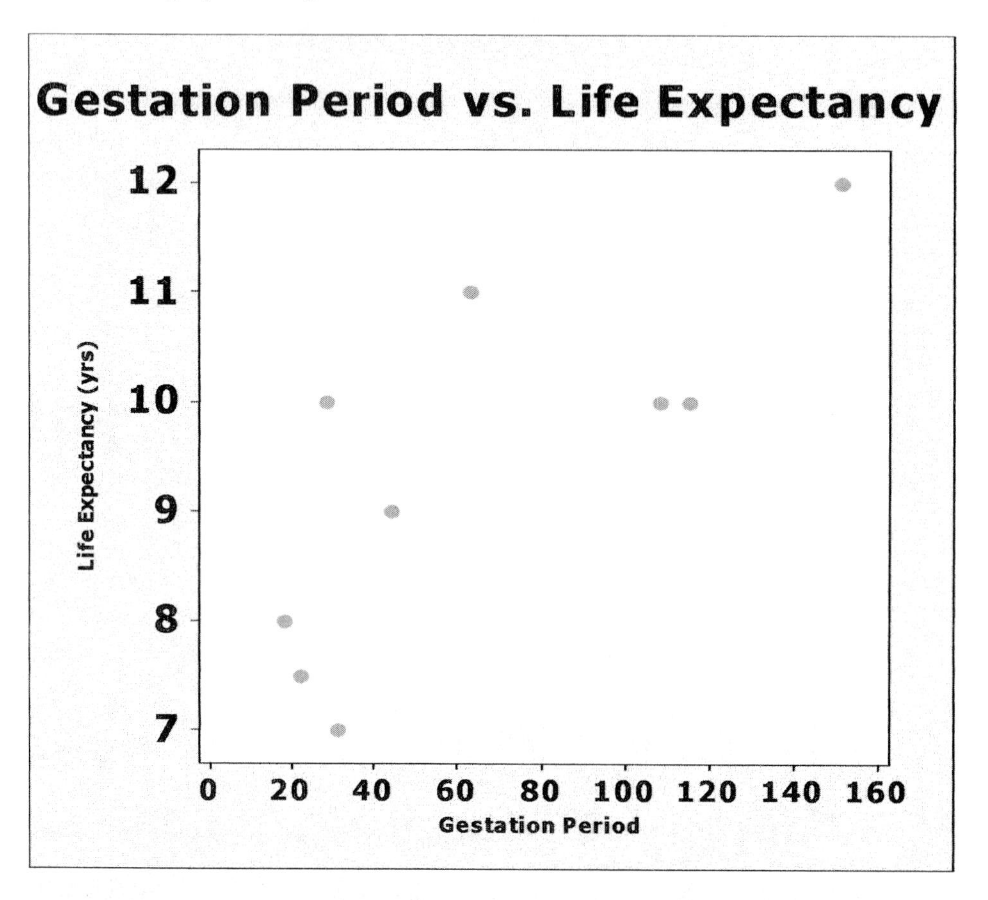

To find the regression equation, click on **Stat → Regression → Regression.** Enter C3 for the **Response** variable, and C2 as the **Predictor.** Click on **Results.** Select **Regression equation, table of coefficients, s, R-squared, and basic analysis of variance.** Next, click on **Storage** and select **Residuals.** Minitab will calculate the residuals for each data point and store them in an empty column. To tell Mintab to predict the life expectancy of a new species with a gestation period of 95 days, click on **Options** and enter 95 for **Prediction intervals for new observations.** Be sure to select **Fits.** Click on **OK**.

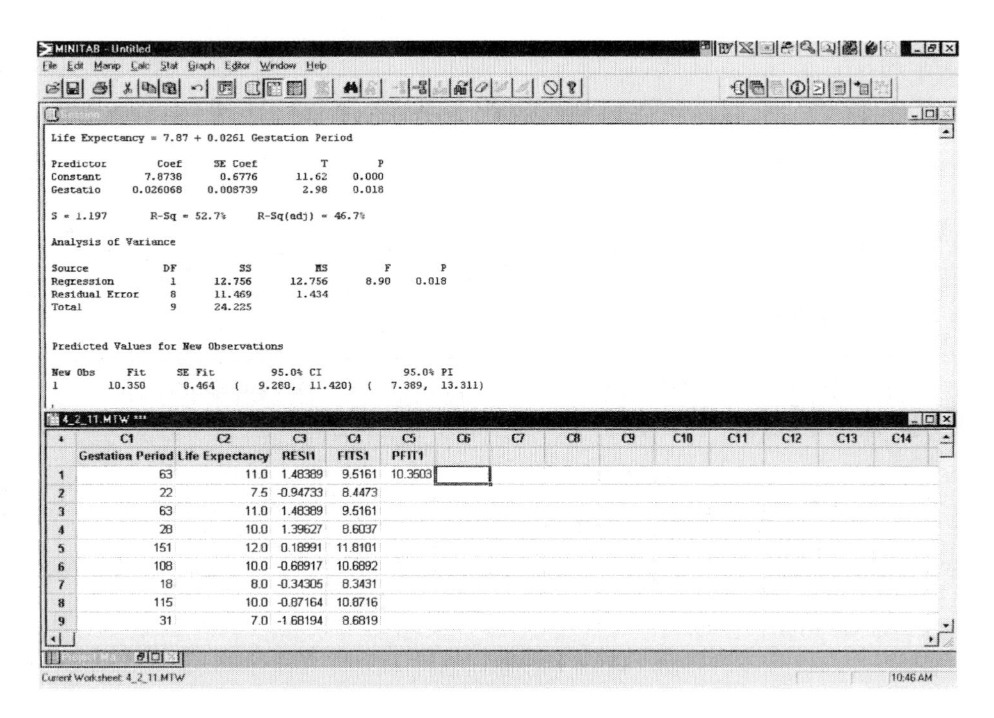

```
Life Expectancy = 7.87 + 0.0261 Gestation Period

Predictor      Coef      SE Coef        T       P
Constant     7.8738       0.6776    11.62   0.000
Gestatio   0.026068     0.008739     2.98   0.018

S = 1.197     R-Sq = 52.7%     R-Sq(adj) = 46.7%

Analysis of Variance

Source           DF        SS        MS       F       P
Regression        1    12.756    12.756    8.90   0.018
Residual Error    8    11.469     1.434
Total             9    24.225

Predicted Values for New Observations

New Obs      Fit     SE Fit         95.0% CI              95.0% PI
1         10.350      0.464   ( 9.280, 11.420)   ( 7.389, 13.311)
```

	C1 Gestation Period	C2 Life Expectancy	C3 RESI1	C4 FITS1	C5 PFIT1
1	63	11.0	1.48389	9.5161	10.3503
2	22	7.5	-0.94733	8.4473	
3	63	11.0	1.48389	9.5161	
4	28	10.0	1.39627	8.6037	
5	151	12.0	0.18991	11.8101	
6	108	10.0	-0.68917	10.6892	
7	18	8.0	-0.34305	8.3431	
8	115	10.0	-0.87164	10.8716	
9	31	7.0	-1.68194	8.6819	

The regression equation is Life Expectancy = 7.87 + .0261 * Gestation Period.
The predicted life expectancy for the new species is 10.35 years.
Look at the Data Window to see that the parakeet (row 7) has a gestation period of 18 days, and a predicted life expectancy of 8.3431 years.

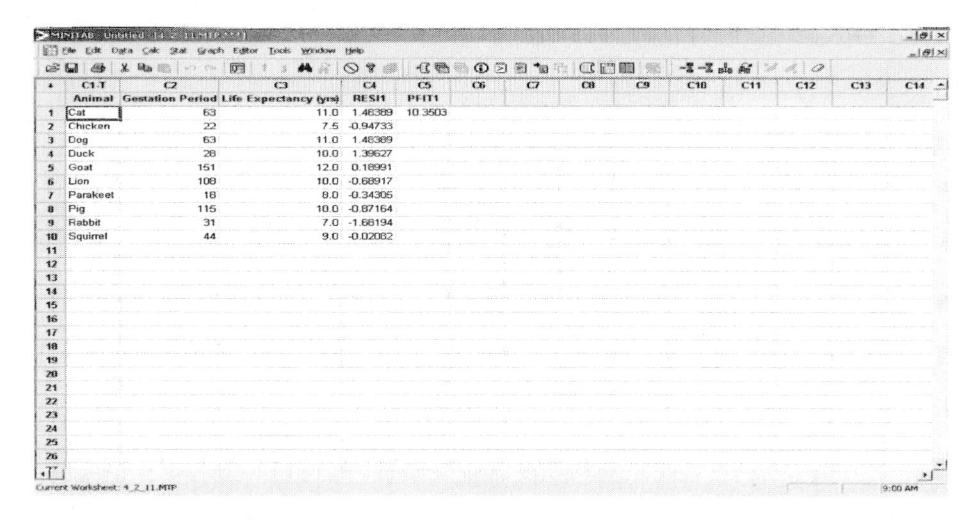

	C1-T Animal	C2 Gestation Period	C3 Life Expectancy (yrs)	C4 RESI1	C5 PFIT1
1	Cat	63	11.0	1.48389	10.3503
2	Chicken	22	7.5	-0.94733	
3	Dog	63	11.0	1.48389	
4	Duck	28	10.0	1.39627	
5	Goat	151	12.0	0.18991	
6	Lion	108	10.0	-0.68917	
7	Parakeet	18	8.0	-0.34305	
8	Pig	115	10.0	-0.87164	
9	Rabbit	31	7.0	-1.68194	
10	Squirrel	44	9.0	-0.02082	

Use the regression equation to find the predicted life expectancy of the rabbit, which has a gestation period of 31 days. The residuals are shown in C4. The residual for rabbit is –1.68.

◀

Section 4.3

▸ Example 1 (pg. 228) Finding the Coefficient of Determination

Open Minitab worksheet **4_1_Ex1**. The speed should be in C1, and the distance should be in C2. The coefficient of determination is part of the regression output. To find the regression equation, click on **Stat → Regression → Regression.** Enter C2 for the **Response** variable, and C1 as the **Predictor. Click on Results.** Select **Regression equation, table of coefficients, s, R-squared, and basic analysis of variance.** Click on **OK** twice.

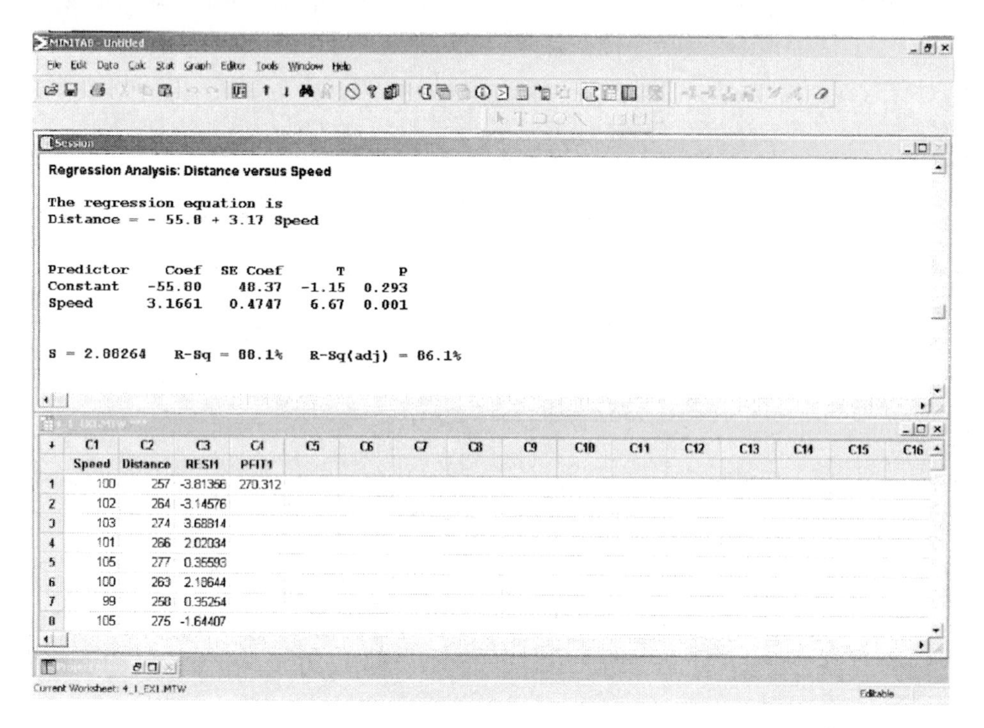

Notice that the coefficient of determination is R-Sq = 88.1%.

◀

Example 3 (pg. 230) Is a Linear Model Appropriate?

Enter the two columns of data in Table 6 (found on page 230 of the textbook) into the MINITAB Data Window. Enter Time into C1 and the Temperatures into C2. To find the regression equation, click on **Stat → Regression → Regression.** Enter C2 for the **Response** variable, and C1 as the **Predictor.** Click on **OK**. Click on **Results.** If you would like to see the other regression output, then select **Regression equation, table of coefficients, s, R-squared, and basic analysis of variance.** Finally, click on **Graphs** and beneath **Residuals versus the variables:** enter C1.

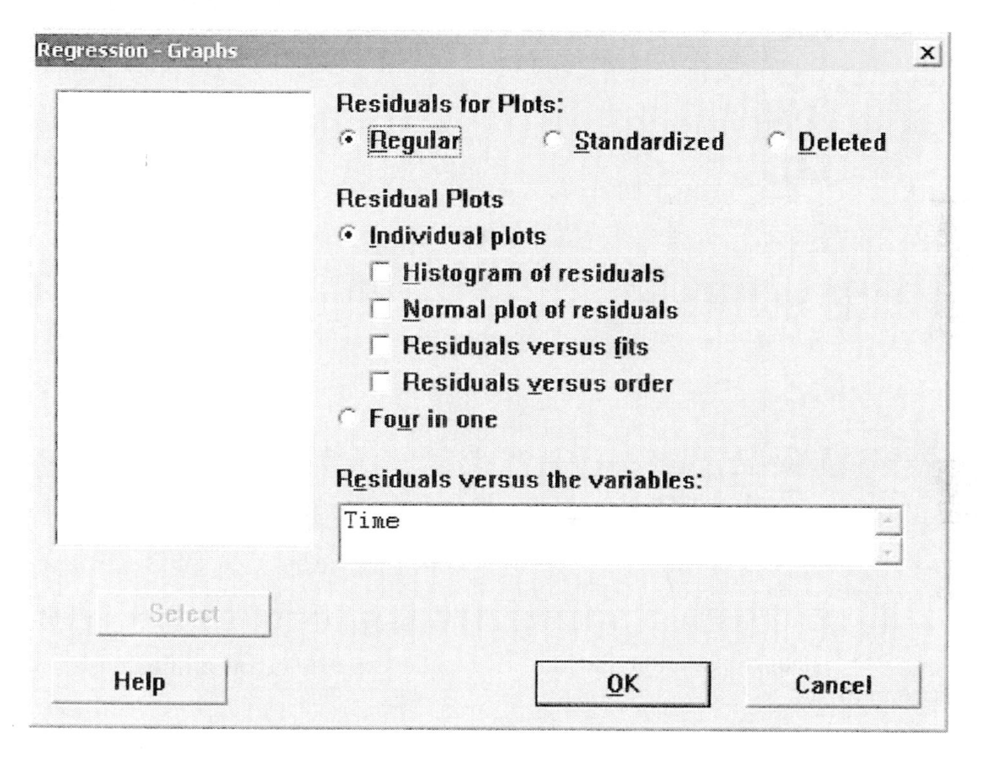

Click on **OK** twice and view the graph.

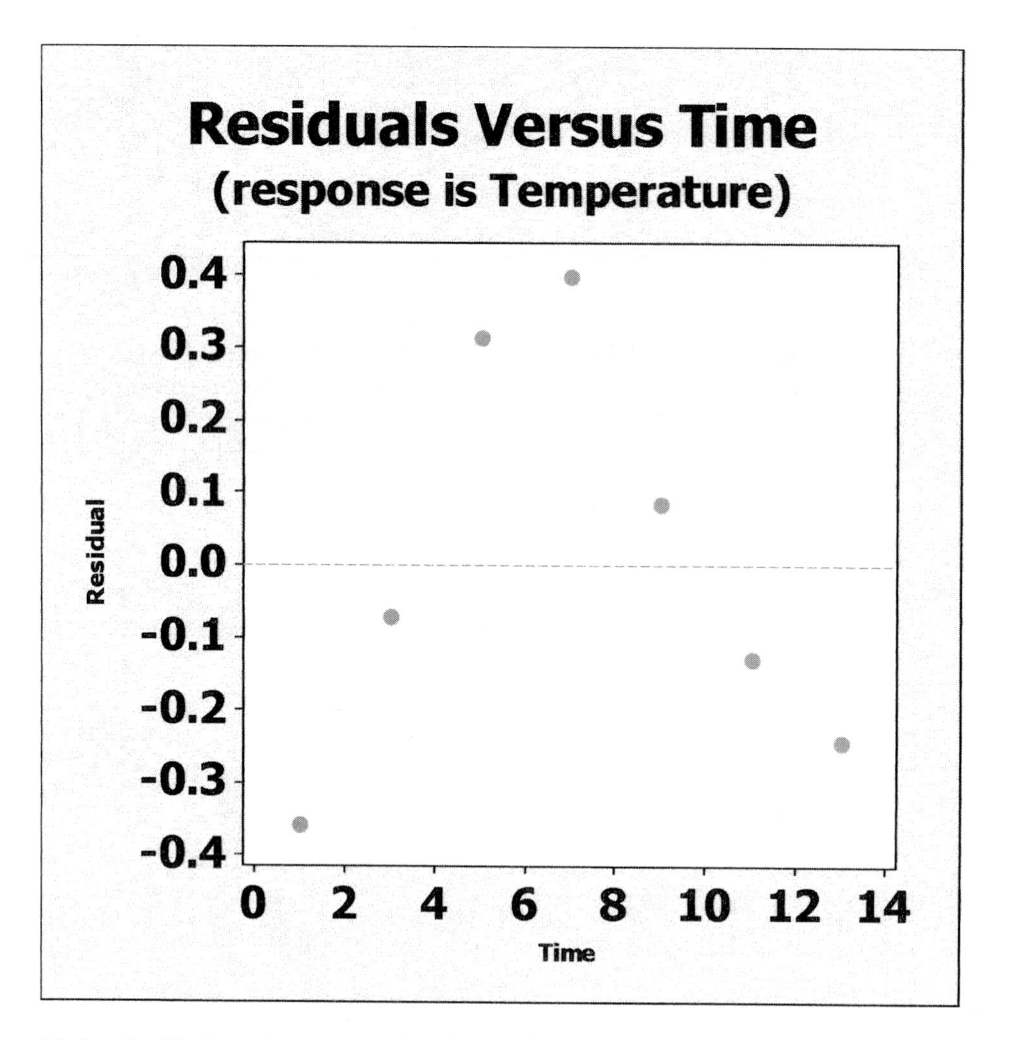

Notice the U-shaped pattern in the plot. This indicates that the linear model is not appropriate.

▸ Example 6 (pg. 232) Graphical Residual Analyses

Open Minitab worksheet **4_1_Ex1**. The speed should be in C1, and the distance should be in C2. Notice that speed is the x-variable and distance is the y-variable. To find the least squares regression equation, click on **Stat → Regression → Regression.** Enter C2 for the **Response** variable, and C1 as the **Predictor.** Next, click on **Storage** and select **Residuals.** Minitab will calculate the residuals for each data point and store them in an empty column. You will use this column to create the boxplot of the residuals. Click on **Results.** If you would like to see the other regression output, then select **Regression equation, table of coefficients, s, R-squared, and basic analysis of variance.** Finally, click on **Graphs** and enter C2 beneath **Residuals versus the variables**. Click on **OK** to view the residual plot.

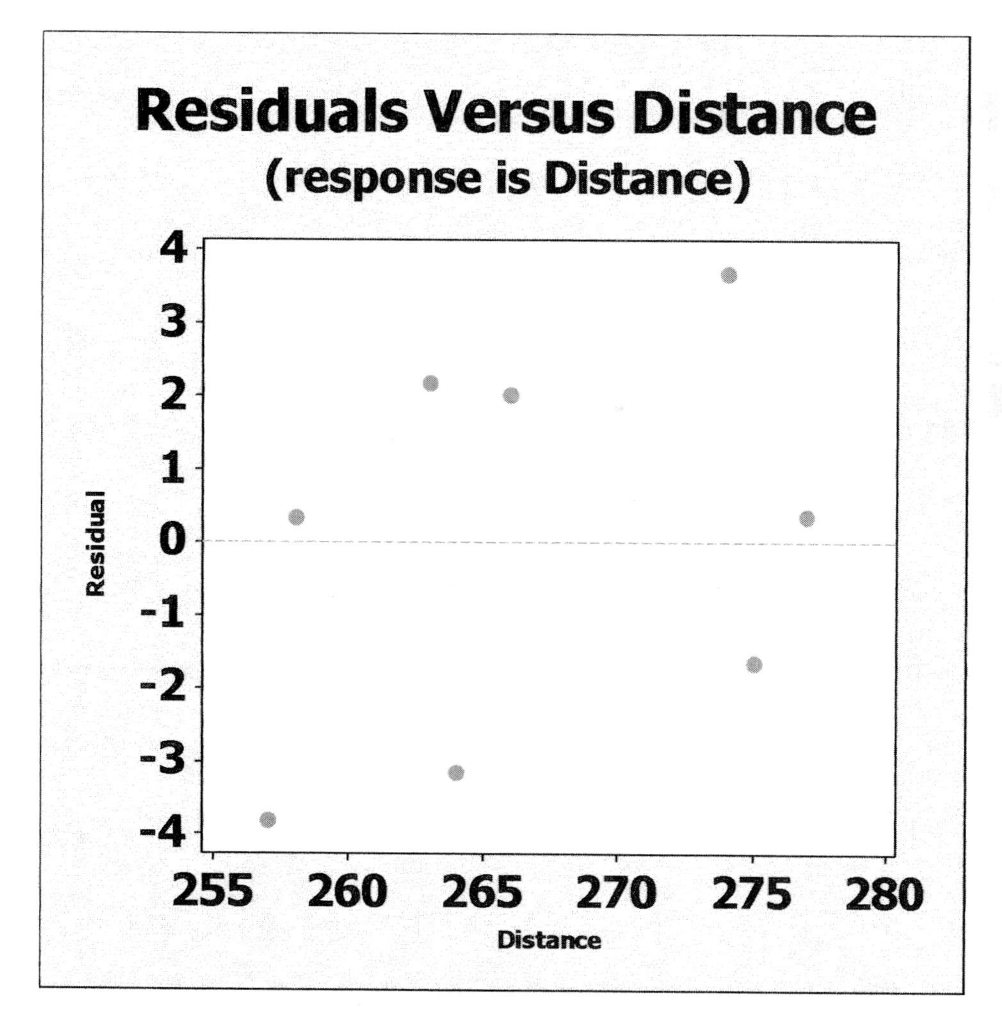

To check for possible outliers, create a boxplot of the residuals. The residuals should now be C3, labeled RESI1. Click on **Graph** → **Boxplot** → **Simple** and select C4 for the **Graph variable**. Click on the **Scale** button and select **Transpose value and category scales**. You may want to add an appropriate title **(Labels** button**).** Click on **OK** to view the boxplot of the residuals.

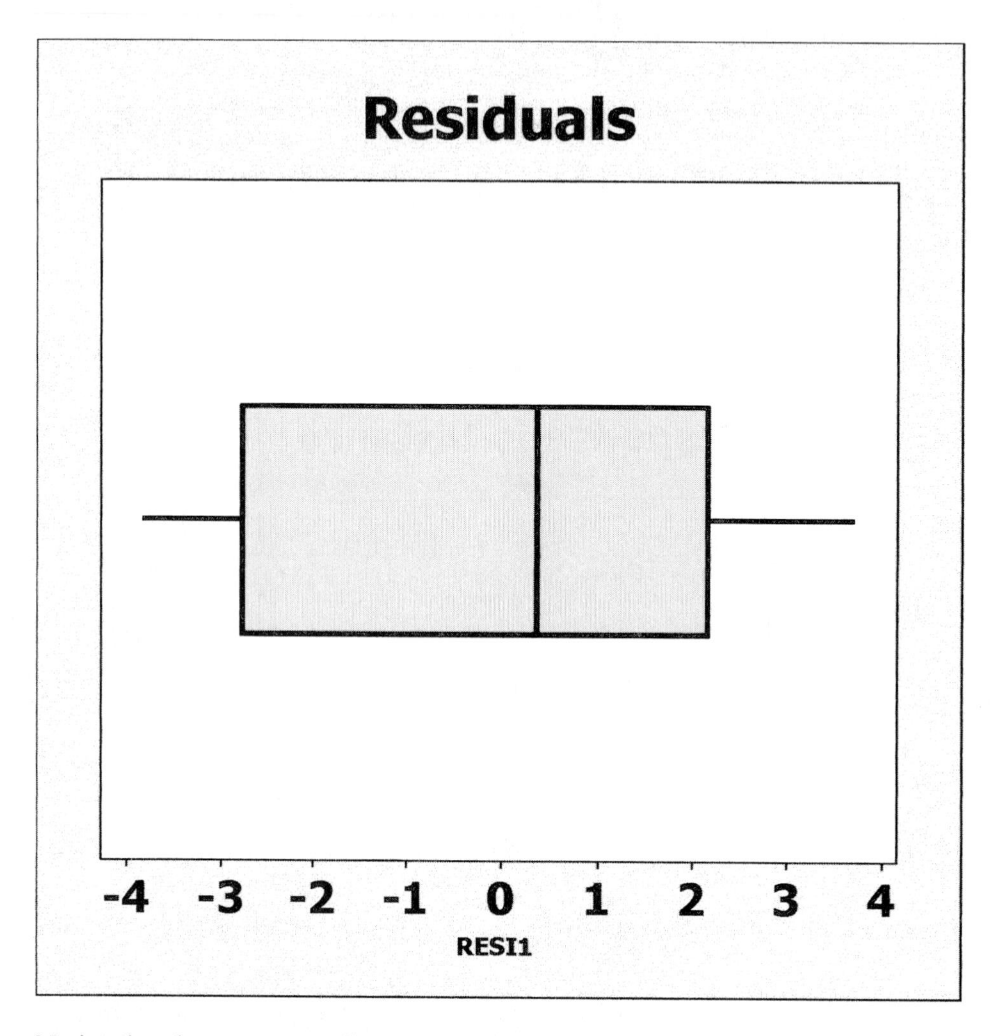

Notice that there are no outliers shown on the boxplot.

◀

▸ Example 7 (pg. 234) Identifying Influential Observations

Open Minitab worksheet **4_1_Ex1**. The speed should be in C1, and the distance should be in C2. Notice that speed is the x-variable and distance is the y-variable. Add the data for Tiger Woods to the bottom of the worksheet. Type 120 into C1 and 305 into C2. To find the least squares regression equation, click on **Stat → Regression → Regression.** Enter C2 for the **Response** variable, and C1 as the **Predictor.** Next, click on **Storage** and select **Residuals.** Minitab will calculate the residuals for each data point and store them in an empty column. You will use this column if you want to create a boxplot of the residuals. Click on **Results.** Since we want to see information on influential data points, select **In addition, sequential sums of squares and the unusual observations in the table of fits and residuals.** Finally, click on **Graphs** and enter C2 beneath **Residuals versus the variables**. Click on **OK** to view the residual plot.

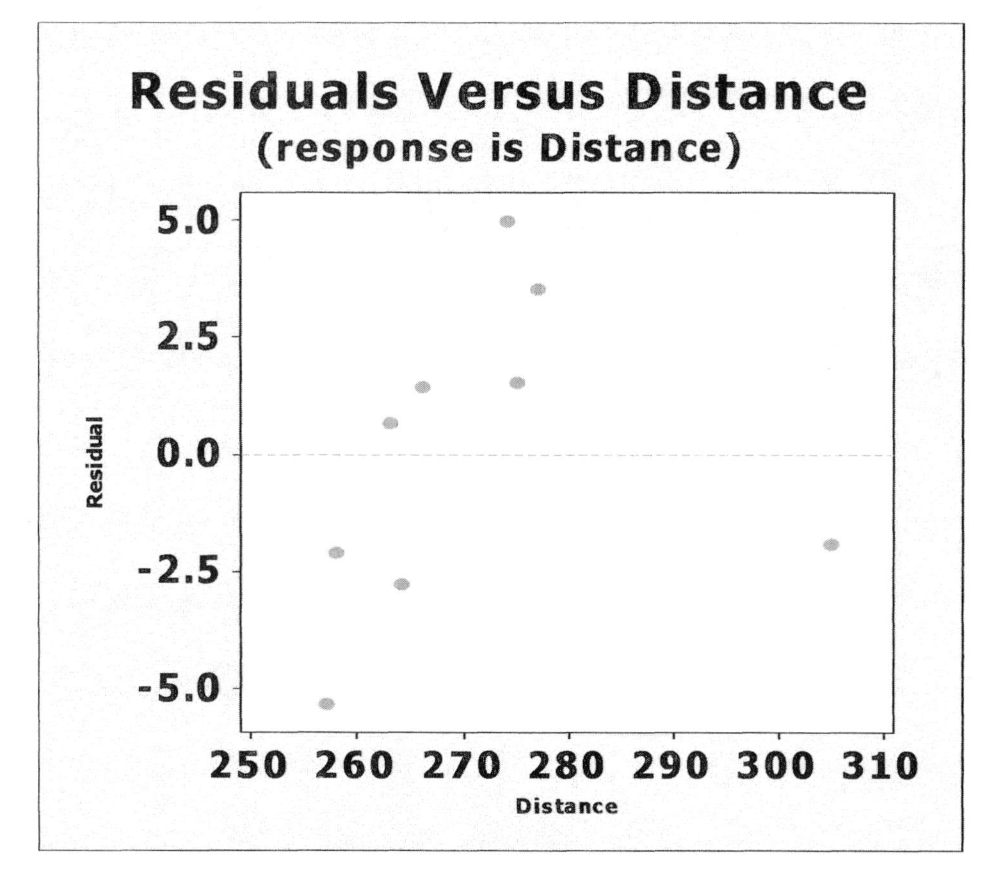

Notice that the residual corresponding to Tiger Woods (located above 305) is very different. Look at the output from the regression analysis.

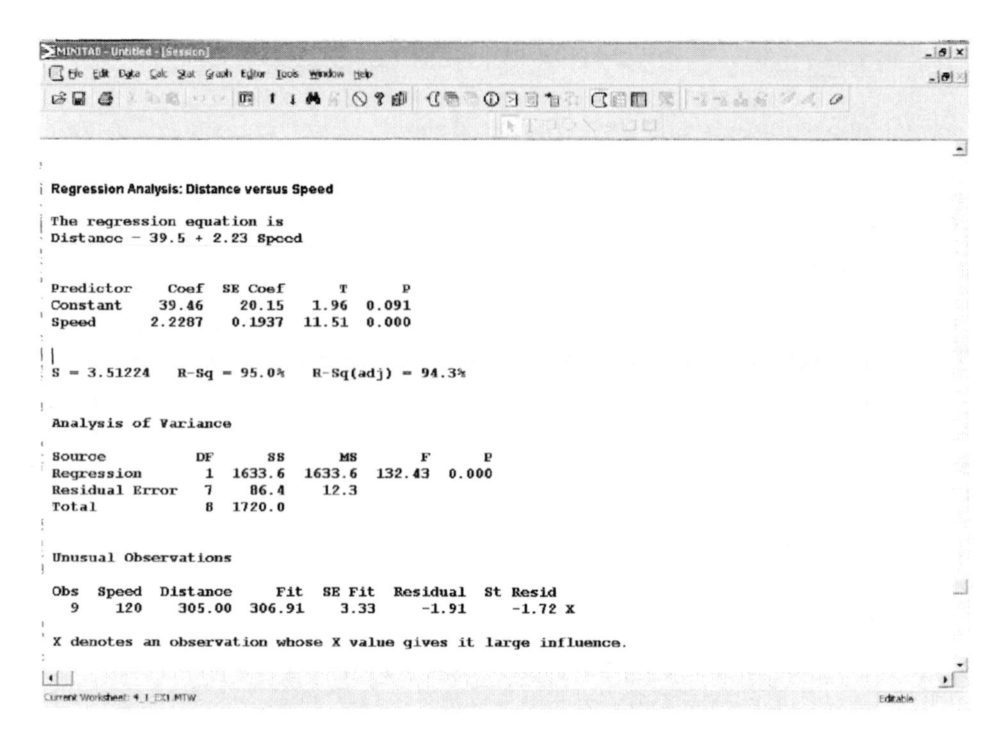
MINITAB - Untitled - [Session]

Regression Analysis: Distance versus Speed

The regression equation is
Distance = 39.5 + 2.23 Speed

Predictor	Coef	SE Coef	T	P
Constant	39.46	20.15	1.96	0.091
Speed	2.2287	0.1937	11.51	0.000

S = 3.51224 R-Sq = 95.0% R-Sq(adj) = 94.3%

Analysis of Variance

Source	DF	SS	MS	F	P
Regression	1	1633.6	1633.6	132.43	0.000
Residual Error	7	86.4	12.3		
Total	8	1720.0			

Unusual Observations

Obs	Speed	Distance	Fit	SE Fit	Residual	St Resid
9	120	305.00	306.91	3.33	-1.91	-1.72 X

X denotes an observation whose X value gives it large influence.

Near the bottom of the output, Obs 9 is identified as an unusual observation. Of course this refers to Tiger Woods since it is the 9th data point in the Minitab worksheet (row 9). The "X" at the end of the line indicates that Minitab has found that Tiger Woods is an "influential observation".

▸ Problem 29 (pg. 238) Kepler's Law of Planetary Motion

Open Minitab worksheet **4_3_29**. The distance from the sun should be in C2, and sidereal year should be in C3. First draw a scatter plot of the data. Click on **Graph→ Scatterplot → Simple.** Select "Year" for the **Y**-variable and "Distance" for the **X**-variable. Enter an appropriate title (**Labels** button) and click on **OK** to view the scatter plot.

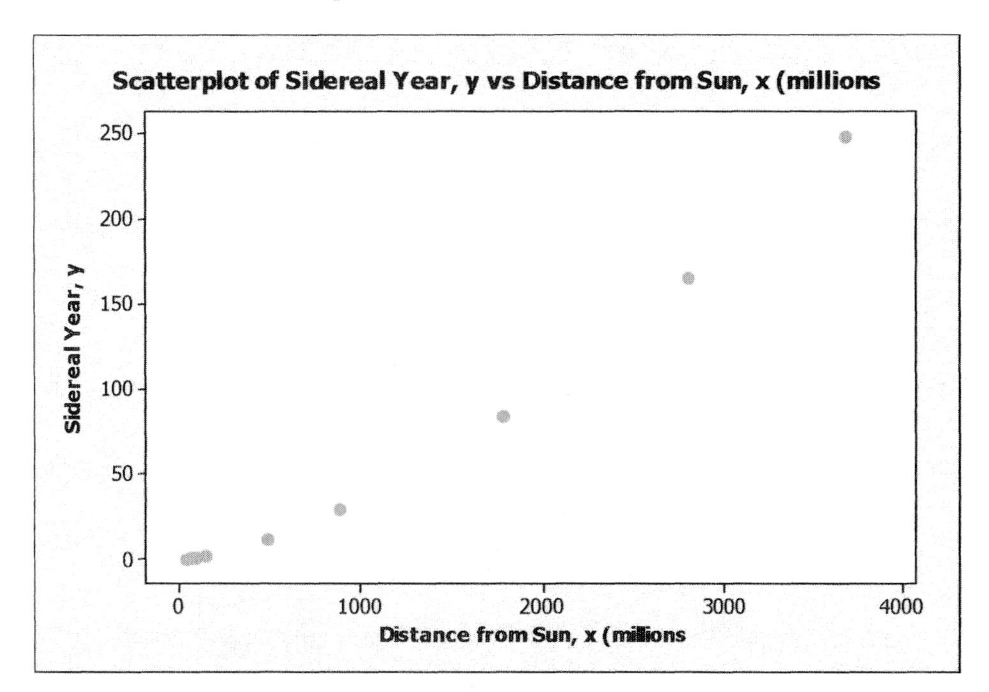

To do the regression analysis, first notice that "Distance" is the X-variable and "Year" is the Y-variable. To find the least squares regression equation, click on **Stat → Regression → Regression.** Enter C3 for the **Response** variable, and C2 as the **Predictor.** Click on **Results.** Since we want to see information on influential data points, select **In addition, sequential sums of squares and the unusual observations in the table of fits and residuals.** Finally, click on **Graphs** and enter C2 beneath **Residuals versus the variables**. Click on **OK** to view the regression analysis and the residual plot.

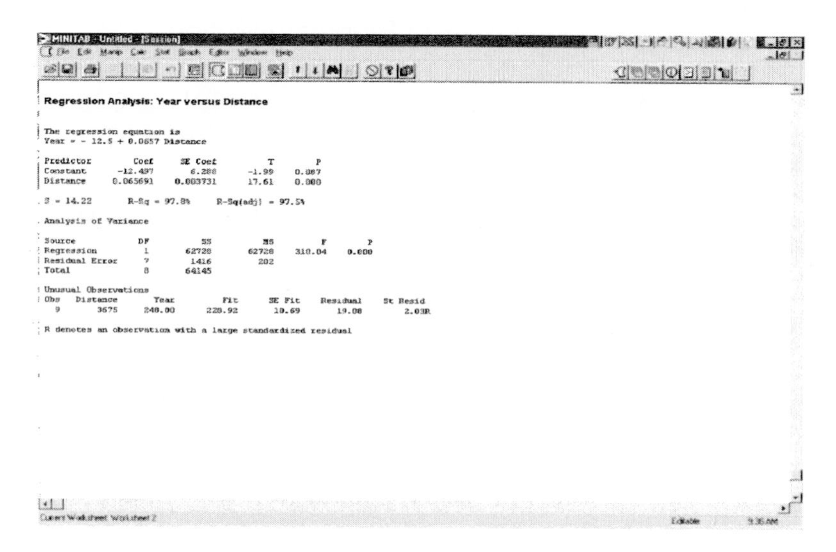

Notice that an Unusual Observation is reported. "Obs 9", which is Pluto, has a large standardized residual, which is indicated by the "R" at the end of the line. The regression equation is Year = -12.5 + 0.0657*Distance. Now look at the residual plot.

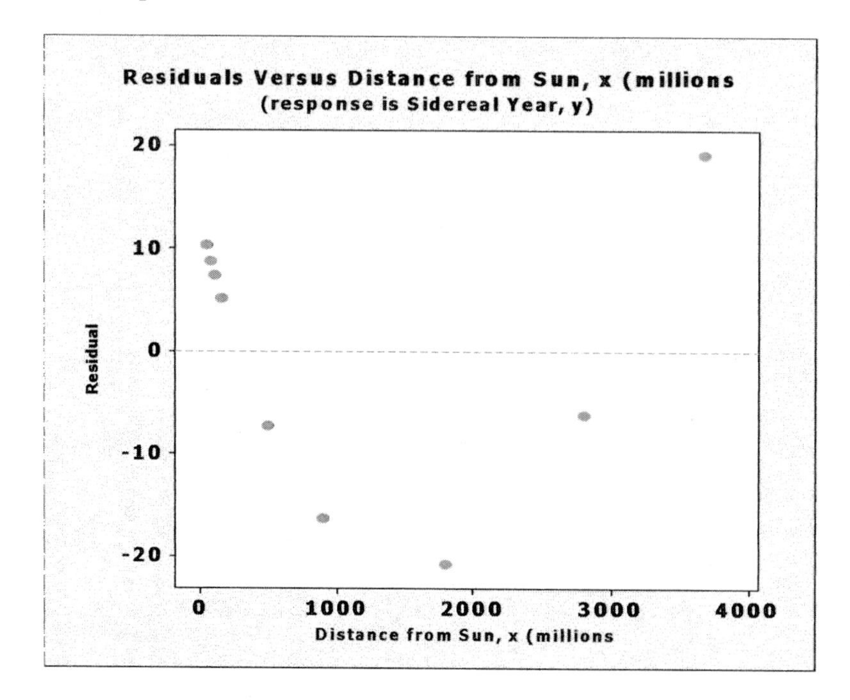

Notice that there is a distinct U-shape to the residuals. This indicates that the least squares regression line is not a good model for this data.

Problem 33 (pg. 238) Height vs. Weight

Open Minitab worksheet **4_3_33**. The Heights should be in C2, and Weight should be in C3. First draw a scatter plot of the data. Click on **Graph→ Scatterplot → Simple.** Select "Weight" for the **Y-variable** and "Height" for the **X-variable**. Enter an appropriate title (**Labels** button) and click on **OK** to view the scatter plot.

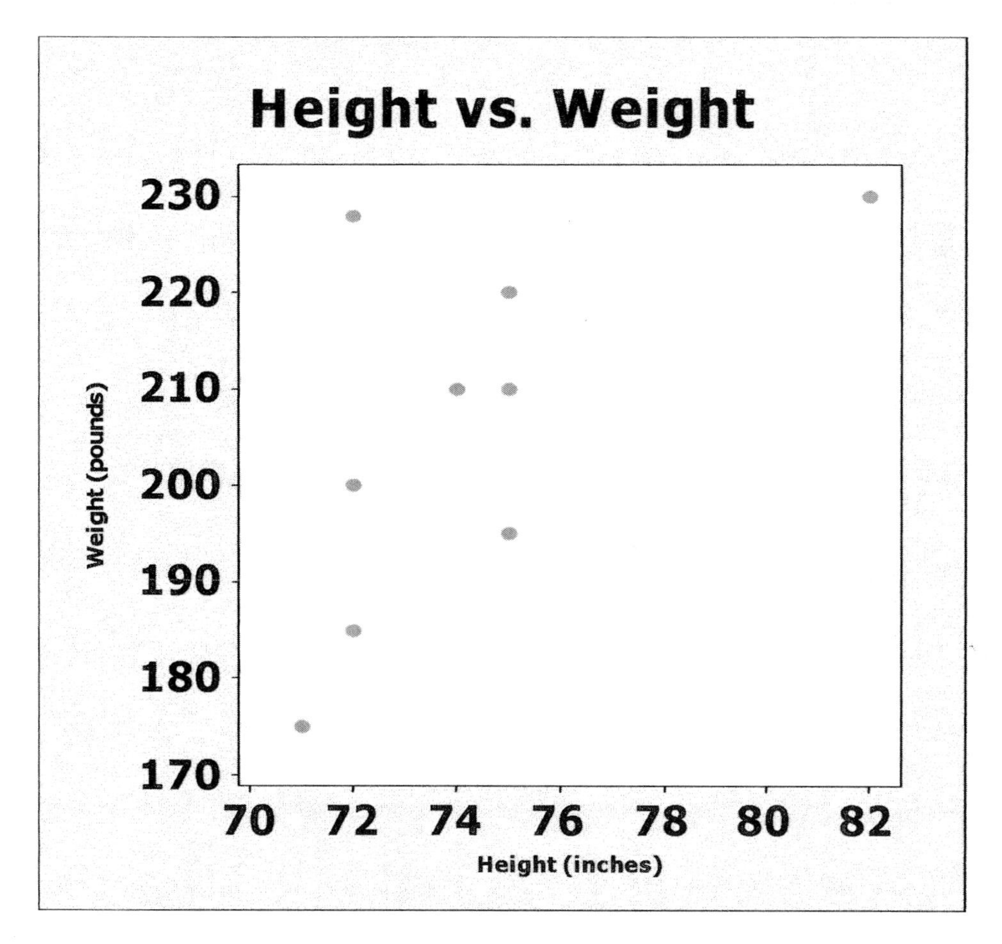

Notice that there is one data point that is far away from the other points. Run the regression to see if it is an "Unusual Observation". To find the least squares regression equation, click on **Stat → Regression → Regression.** Enter C3 for the **Response** variable, and C2 as the **Predictor.** Click on **Results.** Since we want to see information on influential data points, select **In addition, sequential sums of squares and the unusual observations in the table of fits and residuals.** Finally, click on **Graphs** and enter C2 beneath **Residuals versus the variables**. Click on **OK** to view the regression analysis and the residual plot.

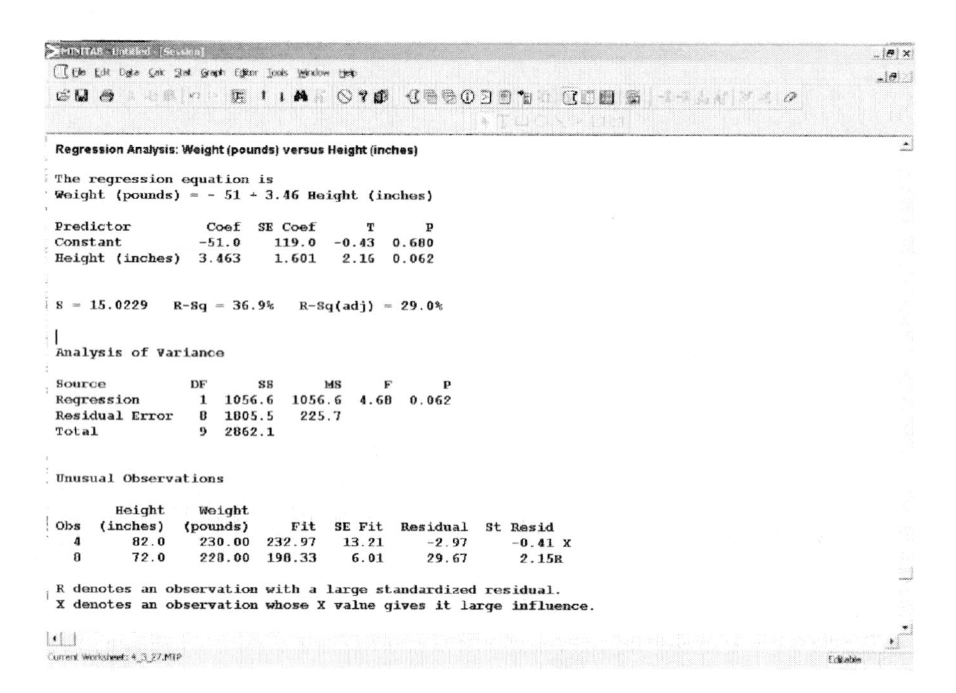

Regression Analysis: Weight (pounds) versus Height (inches)

The regression equation is
Weight (pounds) = - 51 + 3.46 Height (inches)

Predictor	Coef	SE Coef	T	P
Constant	-51.0	119.0	-0.43	0.680
Height (inches)	3.463	1.601	2.16	0.062

S = 15.0229 R-Sq = 36.9% R-Sq(adj) = 29.0%

Analysis of Variance

Source	DF	SS	MS	F	P
Regression	1	1056.6	1056.6	4.68	0.062
Residual Error	8	1805.5	225.7		
Total	9	2862.1			

Unusual Observations

Obs	Height (inches)	Weight (pounds)	Fit	SE Fit	Residual	St Resid
4	82.0	230.00	232.97	13.21	-2.97	-0.41 X
8	72.0	228.00	198.33	6.01	29.67	2.15R

R denotes an observation with a large standardized residual.
X denotes an observation whose X value gives it large influence.

The regression equation is Weight = -51 + 3.46 * Height. There are 2 unusual observations: Observation 4 (Randy Johnson) has an X at the end of the line which indicates that it is an influential observation, and observation 8 (Pete Harnisch) has an R at the end of the line which indicates that is has a large standardized residual.

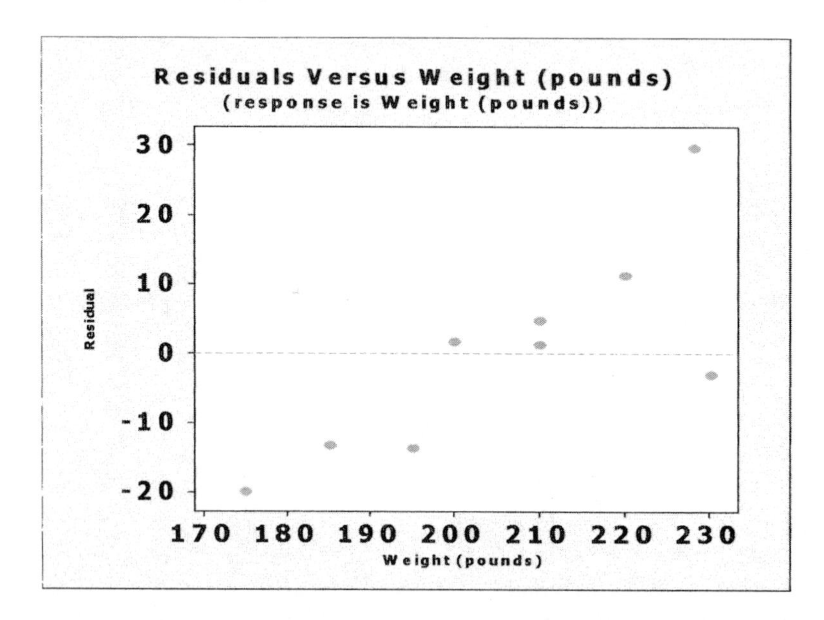

Notice that the residual corresponding to Randy Johnson is far to the right of the other residuals.

Repeat the above steps after removing the data for Randy Johnson from the Data Window. The regression analysis will look like the following.

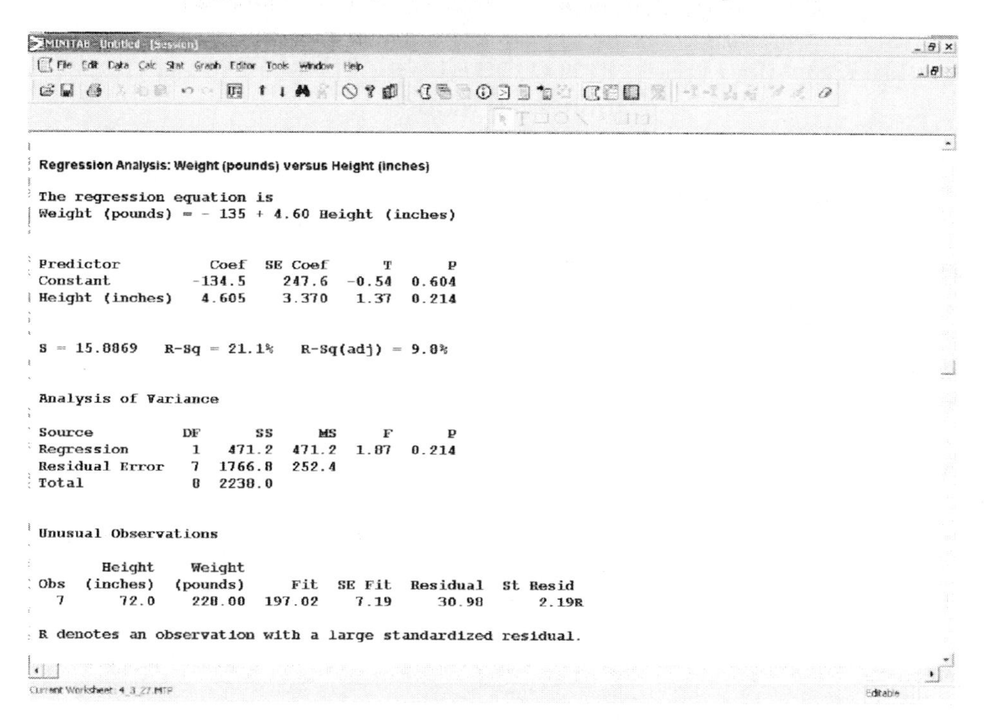

The regression equation changes to : Weight = -135 + 4.60 * Height. Pete Harnisch still shows up with a large standardized residual.

Section 4.4 (Note: available on CD)

▸ Example 4 (pg. 4-4) Finding the Curve of Best Fit to an Exponential Model

Enter the data into a Minitab worksheet. Year should be in C1, and Closing Price in C2. You will have to create a column called Index. Click on **Calc → Make Patterned Data → Simple Set of Numbers. Store patterned data in** "Index", **From first value** 1 **To last value** 15 **In steps of** 1. Click **OK** and C3 should be named "Index" and contain the numbers 1 to 15. First plot the closing price versus the index. Click on **Graph→ Scatterplot → Simple.** Select "Closing Price" for the **Y-variable** and "Index" for the **X-variable**. Enter an appropriate title (**Labels** button) and click on **OK** to view the scatter plot.

Notice that the data has a curved shape, indicating that it needs a transformation. Close the scatter plot, so that you are back to the Session Window. To try the log transformation, click on **Calc** → **Calculator. Store the result in** C4 and enter the **Expression** "LOGT(C2)". Click on **OK** and C4 will contain the transformed data. Name C4 "log(cp)" to indicate that it contains the log of the closing prices.

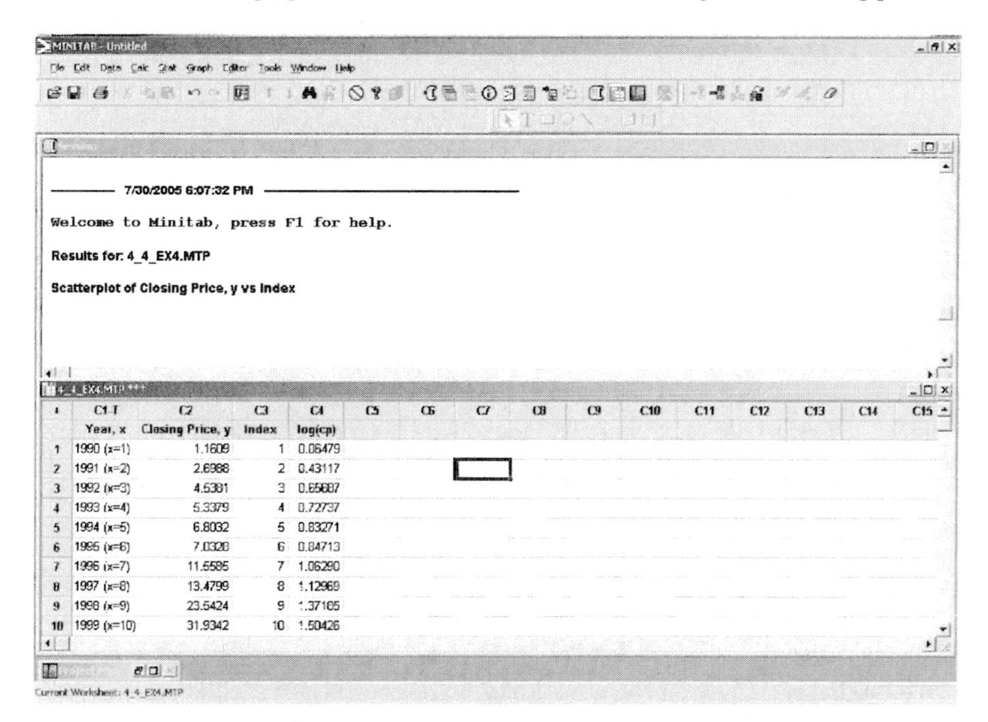

Create a new scatter plot. Click on **Graph**→ **Scatterplot** → **Simple.** Select "log(cp)" for the **Y-variable** and "Index" for the **X-variable**. Enter an appropriate title **(Labels button)** and click on **OK** to view the scatter plot.

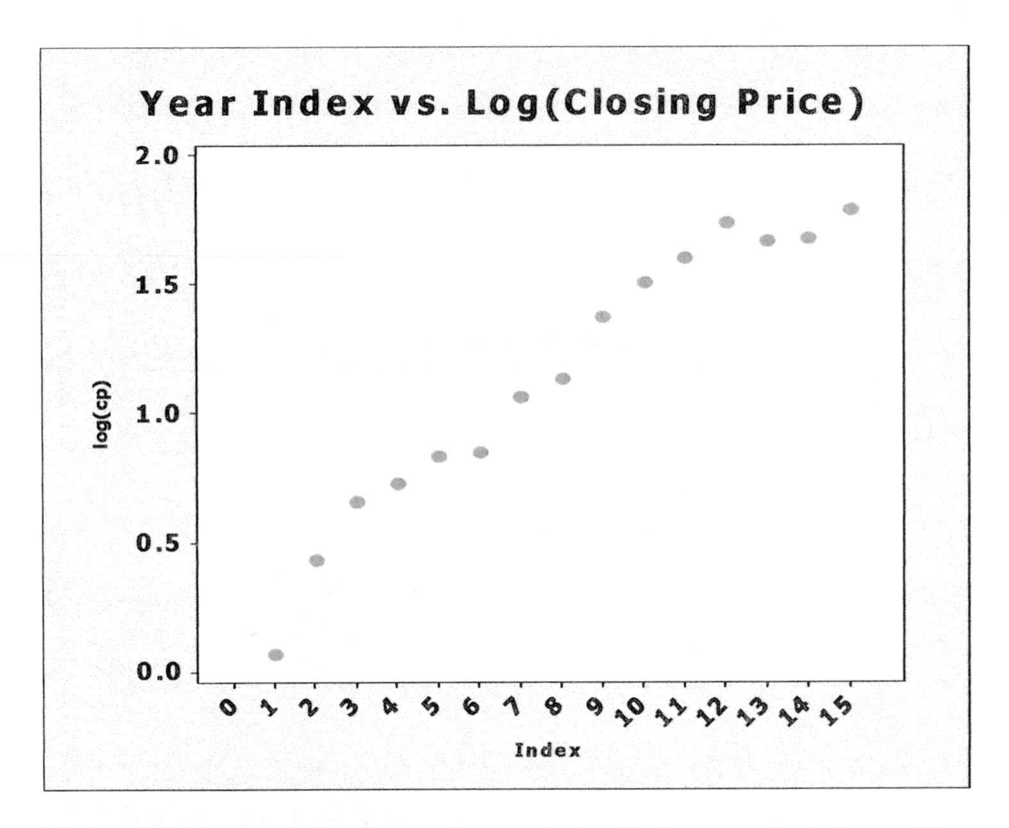

Notice that this scatter plot shows a linear relationship between the index and the transformed data. Now, find the regression equation. Click on **Stat** → **Regression** → **Regression.** Enter C4 for the **Response** variable, and enter C3 for the **Predictors.** Click on **Results.** Since you would like to see the other regression output, select **Regression equation, table of coefficients, s, R-squared, and basic analysis of variance.** Click on **OK** twice and view the output in the Session Window.

Regression Analysis: log(cp) versus Index

```
The regression equation is
log(cp) = 0.211 + 0.116 Index

Predictor      Coef   SE Coef      T      P
Constant    0.21070   0.06832   3.08  0.009
Index      0.116062  0.007515  15.44  0.000

S = 0.125745   R-Sq = 94.8%   R-Sq(adj) = 94.4%
```

The regression equation is log(cp) = 0.211 + .116*Index. You can now solve for a and b using a hand calculator following the step d on page 4-6 of the textbook.

▸ Example 5 (pg. 4-7) Curve of Best Fit to a Power Model

Enter the data from Table 9 into a Minitab worksheet. Time should be in C1 and Distance in C2. Click on **Graph→ Scatterplot → Simple.** Select "Distance" for the **Y-variable** and "Time" for the **X-variable**. Enter an appropriate title **(Labels** button) and click on **OK** to view the scatter plot.

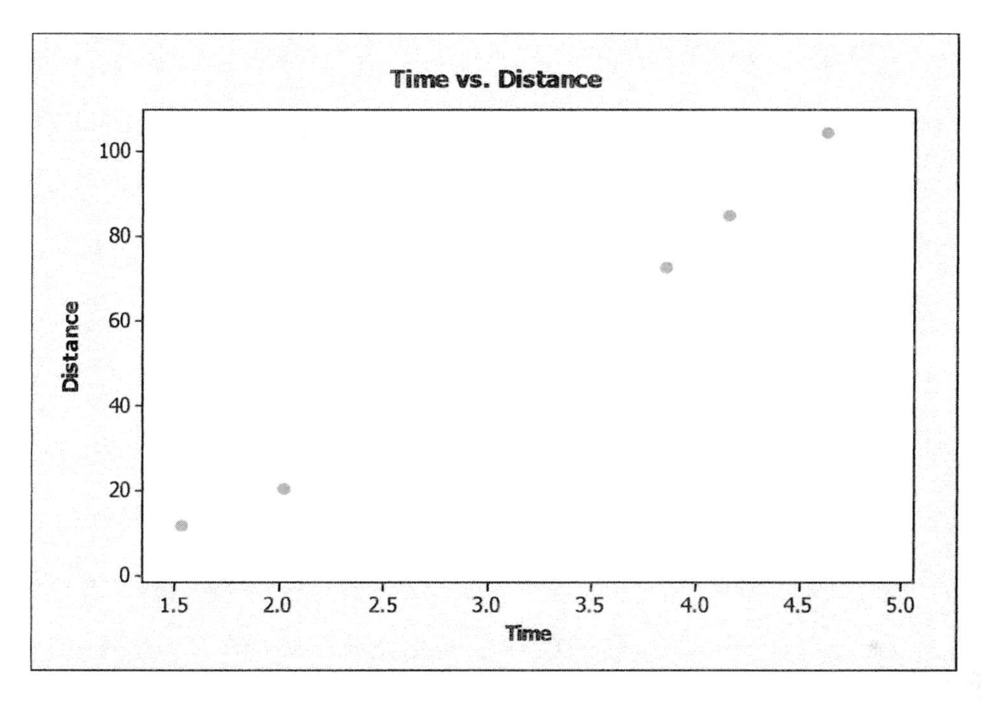

Now try a transformation. Close the scatter plot, so that you are back to the Session Window. To try the log transformation, click on **Calc → Calculator. Store the result in** C3 and enter the **Expression** "LOGT(c1)". Click on **OK** and C3 will contain the transformed data. Name C3 "log(time)" to indicate that it contains the log of the time. Repeat this step, so that you will have "log(distance)" in C4. Create a scatter plot of the transformed data. Click on **Graph→ Scatterplot → Simple.** Select "log(Distance)" for the **Y-variable** and "log(Time)" for the **X-variable**. Enter an appropriate title **(Labels button)** and click on **OK** to view the scatter plot.

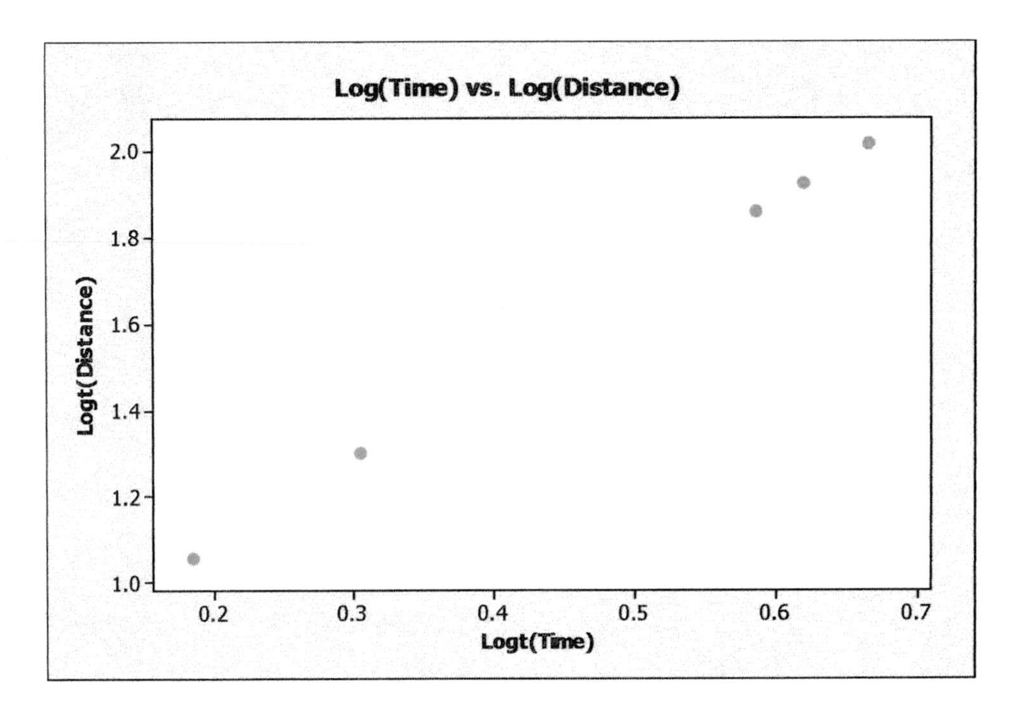

This scatter plot shows that the transformed data is linear. Now, do the regression analysis. Click on **Stat → Regression → Regression.** Enter C4 for the **Response** variable, and enter C3 as the **Predictors.** Click on **Results.** Since you would like to see the other regression output, select **Regression equation, table of coefficients, s, R-squared, and basic analysis of variance.** Click on **OK** twice and view the output in the Session Window.

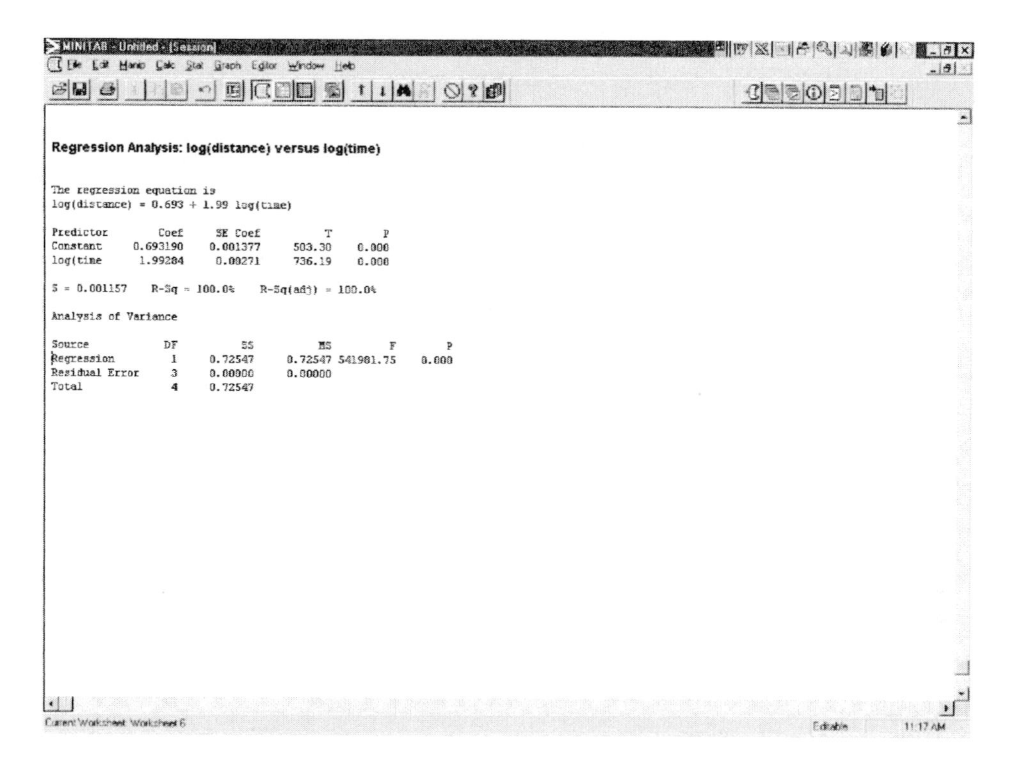

Regression Analysis: log(distance) versus log(time)

The regression equation is
log(distance) = 0.693 + 1.99 log(time)

Predictor	Coef	SE Coef	T	P
Constant	0.693190	0.001377	503.30	0.000
log(time	1.99284	0.00271	736.19	0.000

S = 0.001157   R-Sq = 100.0%   R-Sq(adj) = 100.0%

Analysis of Variance

Source	DF	SS	MS	F	P
Regression	1	0.72547	0.72547	541981.75	0.000
Residual Error	3	0.00000	0.00000		
Total	4	0.72547			

The regression equation is Log(distance) = .693 + 1.99 * Log(time). You can now solve for a and b using a hand calculator following the steps d and e on page 4-8 of the textbook.

◀

Problem 41 (pg. 4-10) Kepler's Law of Planetary Motion

Open Minitab worksheet **4_3_29** . The distance from the sun should be in C2, and sidereal year should be in C3. First draw a scatter plot of the data. Click on **Graph→ Scatterplot → Simple.** Select "Year" for the **Y-variable** and "Distance" for the **X-variable**. Enter an appropriate title (**Labels** button) and click on **OK** to view the scatter plot.

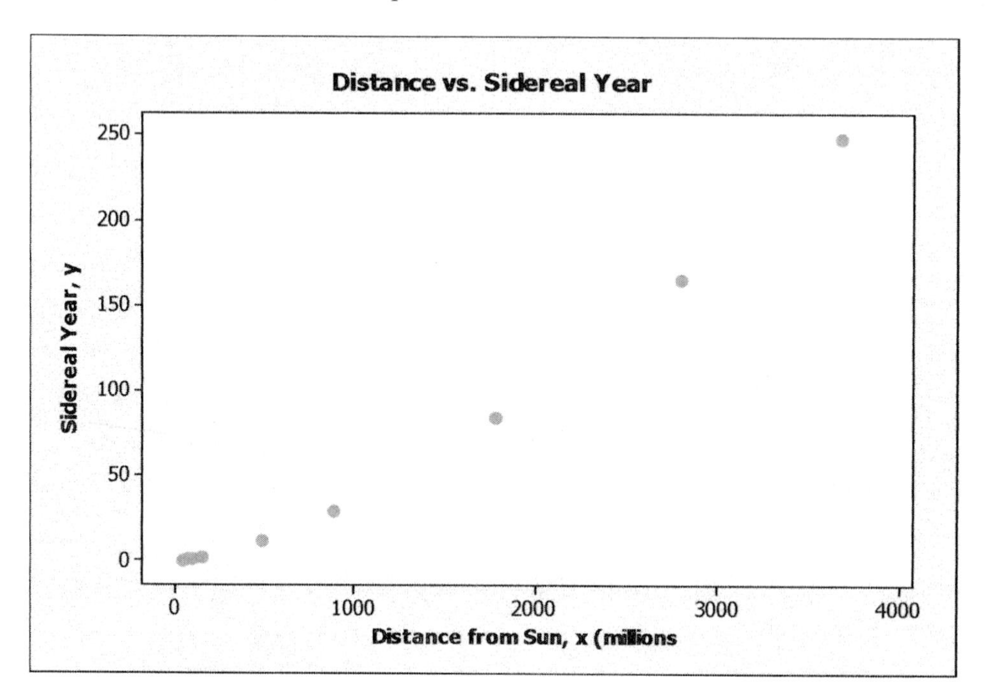

Notice the slight curve to the plot. Try a power transformation by calculating the log of both distance and year. Click on **Calc → Calculator. Store the result in** C4 and enter the **Expression** "LOGT(c2)". Click on **OK** and C4 will contain the transformed data. Name C4 "log(distance)" to indicate that it contains the log of the distance. Repeat this step, so that you will have "log(year)" in C5.

Now, plot the transformed data. Click on **Graph→ Scatterplot → Simple.** Select "log(Year)" for the **Y-variable** and "log(Distance)" for the **X-variable**. Enter an appropriate title (**Labels** button) and click on **OK** to view the scatter plot.

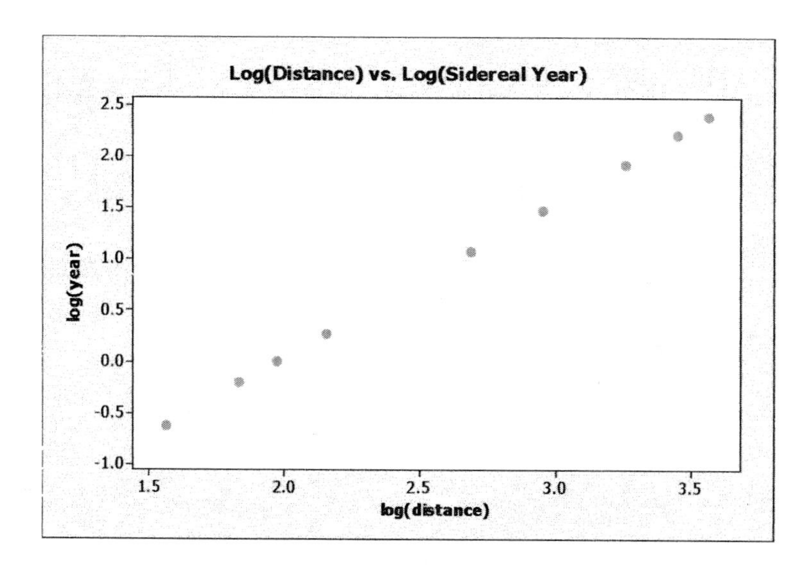

This scatter plot shows a linear relationship between the transformed variables. Now do the regression analysis on Log(distance) and Log(year). Click on **Stat → Regression → Regression.** Enter C5 for the **Response** variable, and enter C4 as the **Predictor.** Next, click on **Results.** Since you would like to see the other regression output, select **Regression equation, table of coefficients, s, R-squared, and basic analysis of variance.** Click on **OK** twice and view the output in the Session Window.

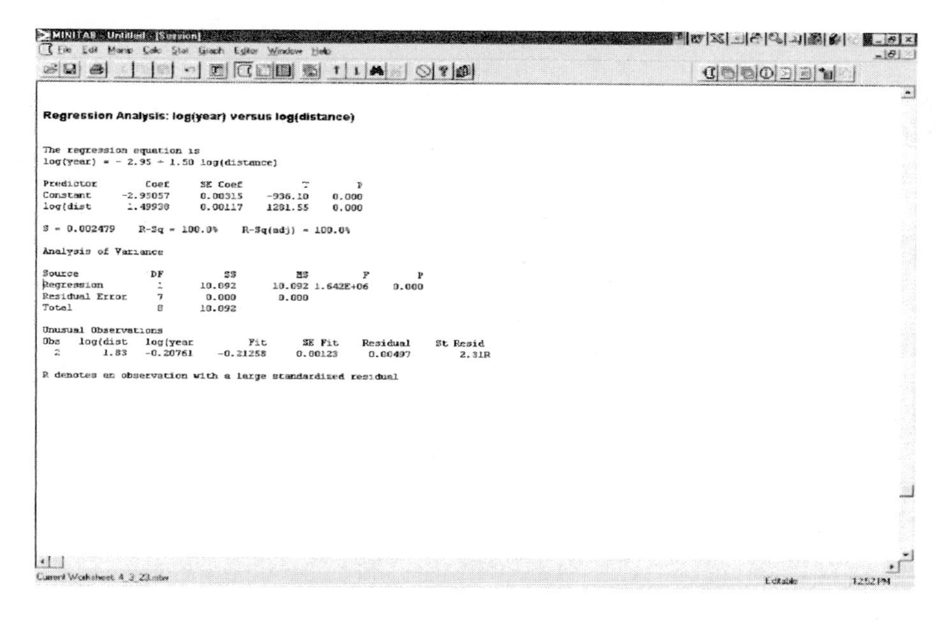

The regression equation is Log(Year) = -2.95 + 1.50 * Log(Distance). Use a hand calculator to find the power equation of best fit.

◀

CHAPTER 5

Probability

Section 5.1

Example 8 (pg. 259) Simulating Probabilities

To simulate the experiment of sampling 100 babies, let Minitab randomly generate 100 1's and 0's. To do this, click on **Calc → Random data → Integer.** You would like to **Generate** 100 **rows of data** and **Store in column** C1. Enter a **Minimum value** of 0 (to represent a boy) and a **Maximum value** of 1 (to represent a girl). Click on **OK** and you should see the random data in C1.

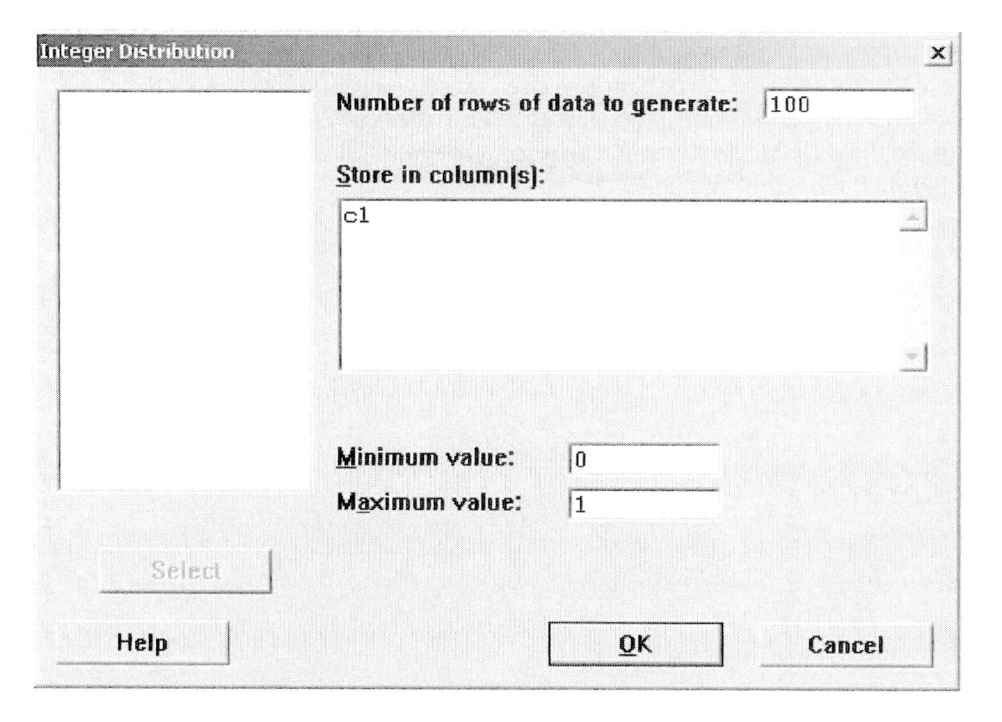

To count up the number of 1's and 0's, click on **Stat → Tables → Tally Individual Variables.** Select C1 as the **Variable** and click on both **Counts** and **Percents.**

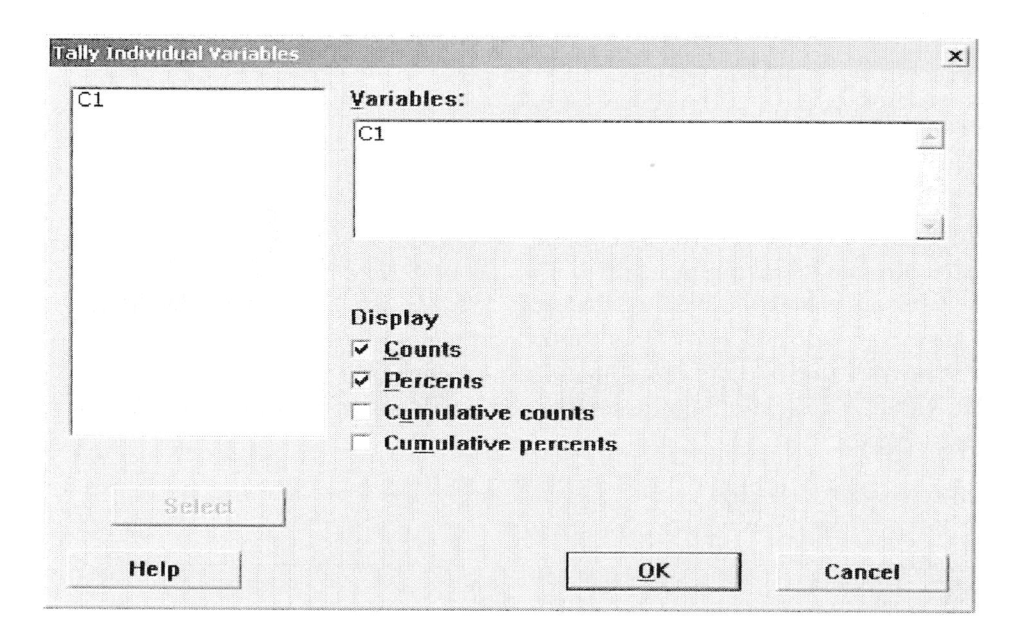

When you click on **OK,** the summary table will be in the Session Window.

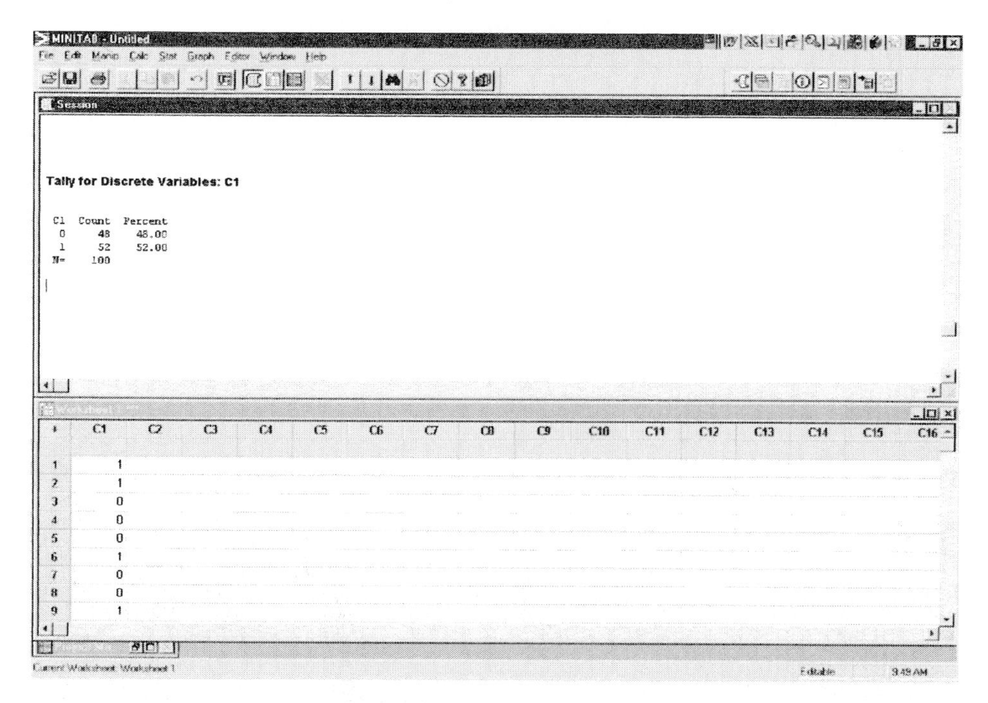

This random sample gave us 48% boys and 52% girls. Repeat these steps to generate a random sample of size 1000. Note: Since this is random data, every student's results will differ.

▸ Problem 49 (pg. 264) Simulation

To simulate the experiment of rolling a six-sided die 100 times, let Minitab randomly generate 100 integers with values 1 through 6. To do this, click on **Calc → Random data → Integer.** Since this simulation is repeated in part b, **Generate** 100 **rows of data** and **Store in column** C1-C2. Enter a **Minimum value** of 1 and a **Maximum value** of 6. Click on **OK** and you should see the random data in both C1 and C2. Repeat this step again, but this time **Generate** 500 **rows of data** and **Store in column** C3. Enter a **Minimum value** of 1 and a **Maximum value** of 6. To count up the results, click on **Stat → Tables → Tally Individual Variables.** Select C1-C3 as the **Variables** and click on both **Counts** and **Percents.**

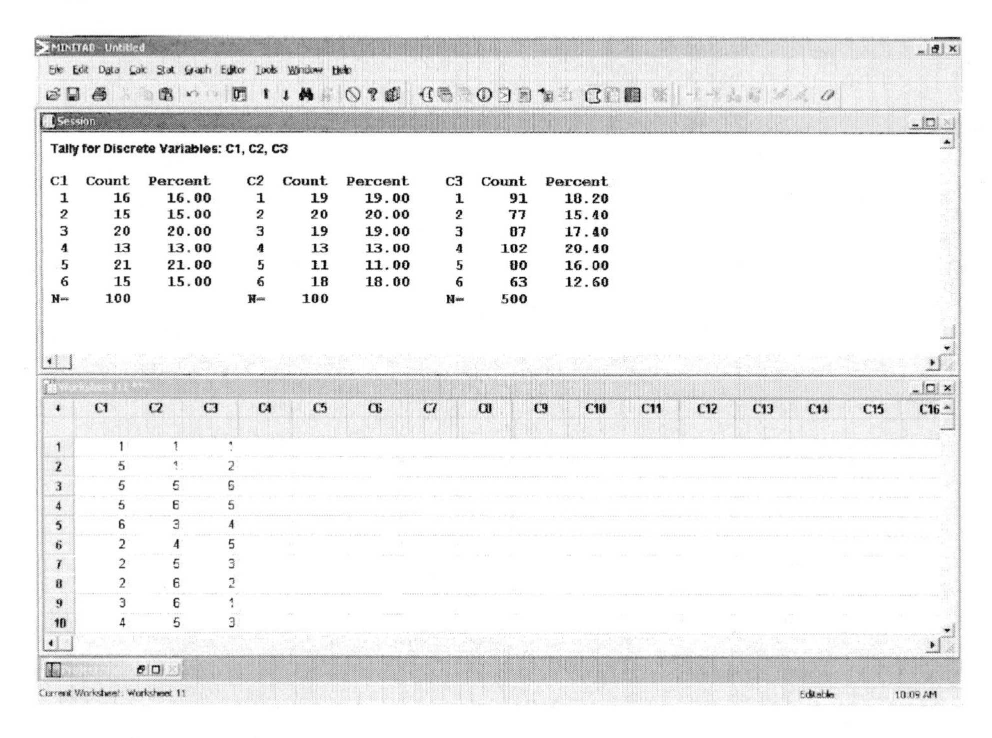

Notice that in the first sample there were 16% 1's, in the second sample there were 19% 1's, and in the third sample there were 18.2% 1's. Note: Since this is random data, every student's results will differ.

Discrete Probability Distributions

Section 6.2

▸ Example 2 (pg. 330) Constructing a Binomial Probability Distribution

In this example, 6% of the human population is blood type O-negative and a random sample of size 4 is selected. Thus, you want to find the binomial probability distribution for n = 4 and p = .06. First, enter the X values 0, 1, 2, 3, and 4 in C1. Next, click on **Calc → Probability Distributions → Binomial.** Since you want the probability for each value of X, select **Probability**. This tells MINITAB what type of calculation you want to do. The **Number of Trials** is 4 and the **Probability of Success** is .06. Enter C1 beside **Input Column.** Leave all other fields blank. Click on **OK.**

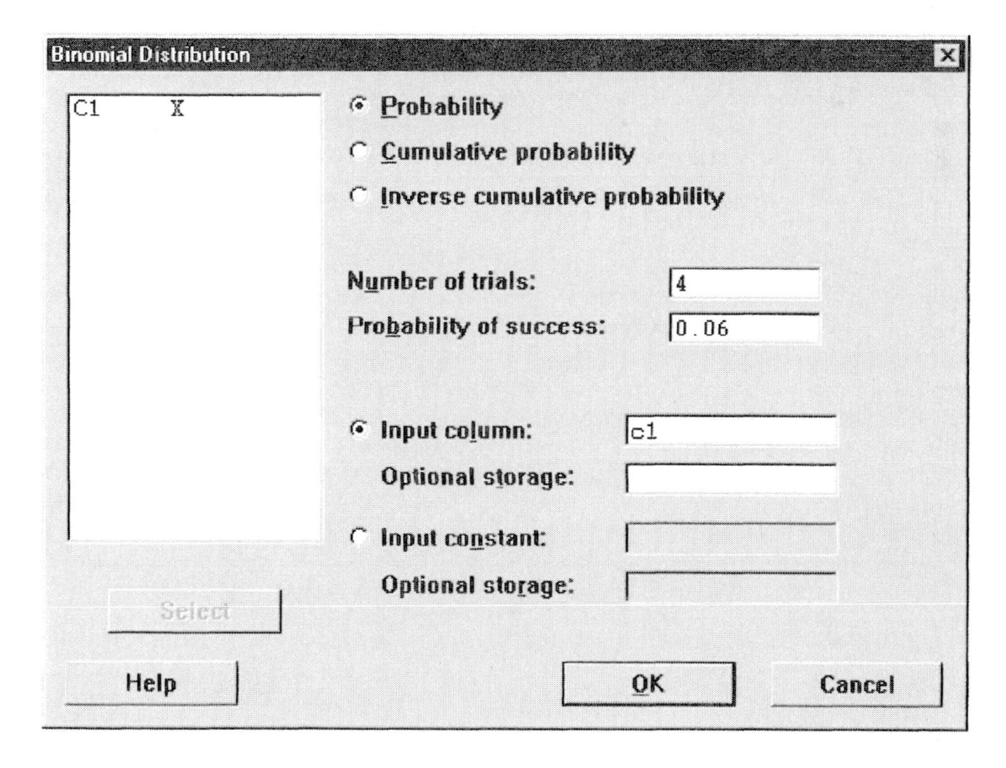

The binomial probability distribution for n=4 and p=.06 will be displayed in the Session Window. Notice that the probability that 2 people in a random sample of size 4 have blood type O-negative is .019086.

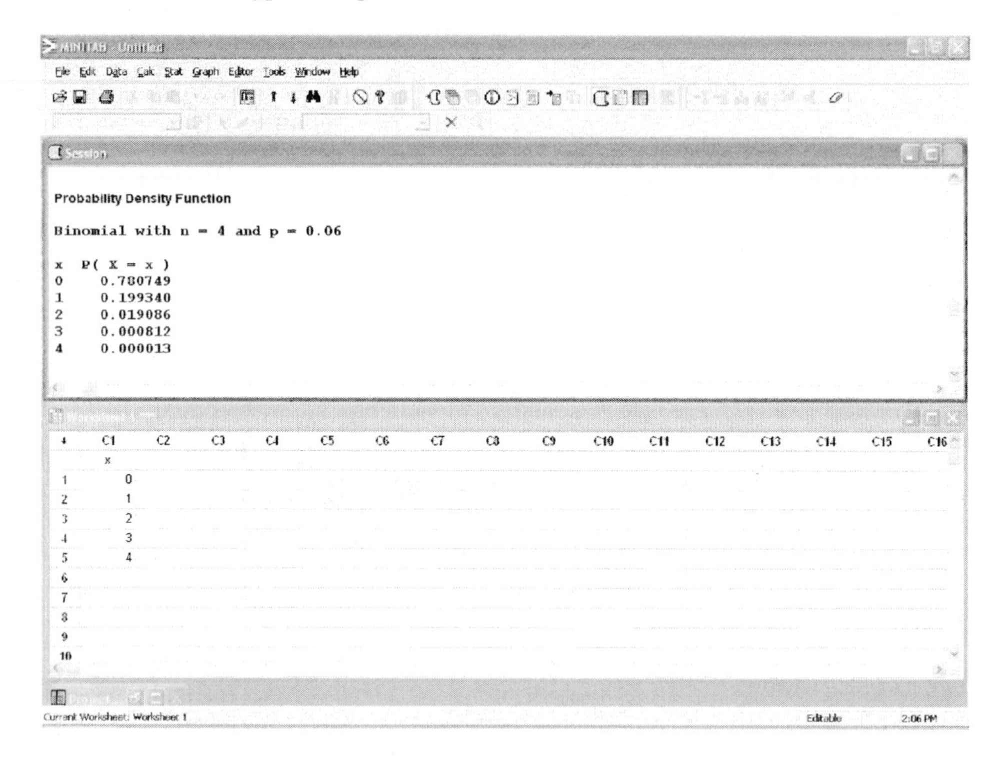

▸ Example 3 (pg. 332) Using the Binomial Distribution

In this example, 70% of American households have cable TV and a random sample of 15 American households is selected. Thus n = 15 and p = .75. Click on **Calc → Probability Distributions → Binomial.** To find the probability that exactly 10 of the 15 households have cable TV, select **Probability**. This tells MINITAB what type of calculation you want to do. The **Number of Trials** is 15 and the **Probability of Success** is .70. To find the probability of 10, enter 10 beside **Input Constant.** Leave all other fields blank. Click on **OK.**

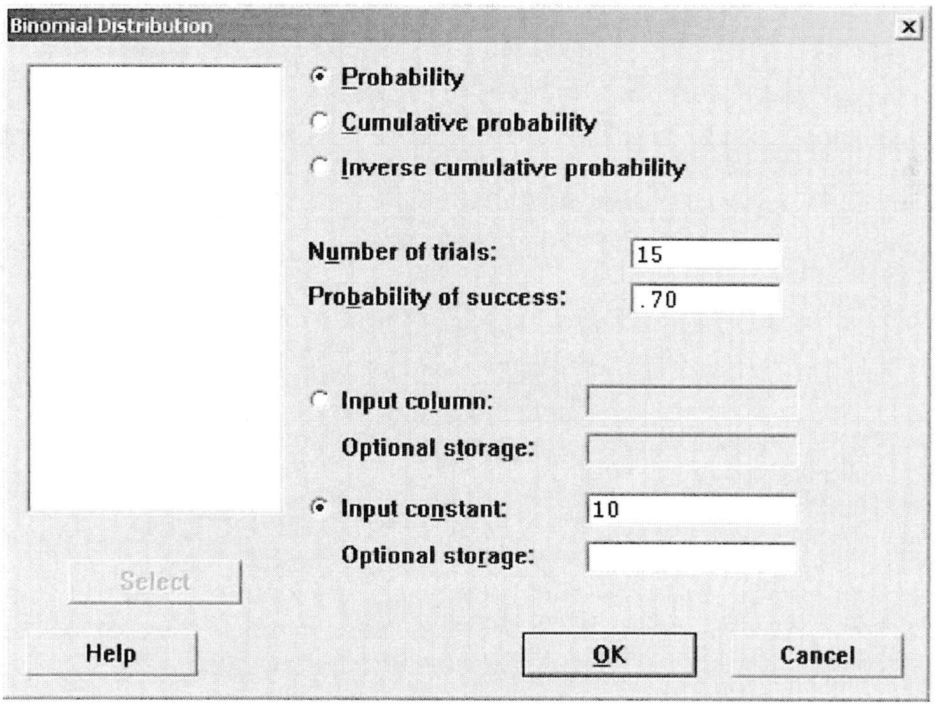

Binomial Distribution

Probability
Cumulative probability
Inverse cumulative probability

Number of trials: 15
Probability of success: .70

Input column:
Optional storage:
Input constant: 10
Optional storage:

Select
Help
OK
Cancel

The probability that 10 of the 15 households sampled have cable TV will be displayed in the Session Window. Notice that the probability is .206130.

Probability Density Function

```
Binomial with n = 15 and p = 0.7

 x  P( X = x )
10    0.206130
```

For part b, you want to find the probability that at least 13 of the 15 households have cable TV. One way to calculate this is to use the cumulative probability function. We will use this function to find the $P(X \leq 12)$, and then subtract the probability from 1 since we are interested in the complement of that probability. Click on **Calc → Probability Distributions → Binomial.** To find the probability that 12 or less of the 15 households have cable TV, select **Cumulative Probability**. This tells MINITAB what type of calculation you want to do. The **Number of Trials** is 15 and the **Probability of Success** is .75. To find the probability of 12 or less, enter 12 beside **Input Constant.** Leave all other fields blank. Click on **OK.**

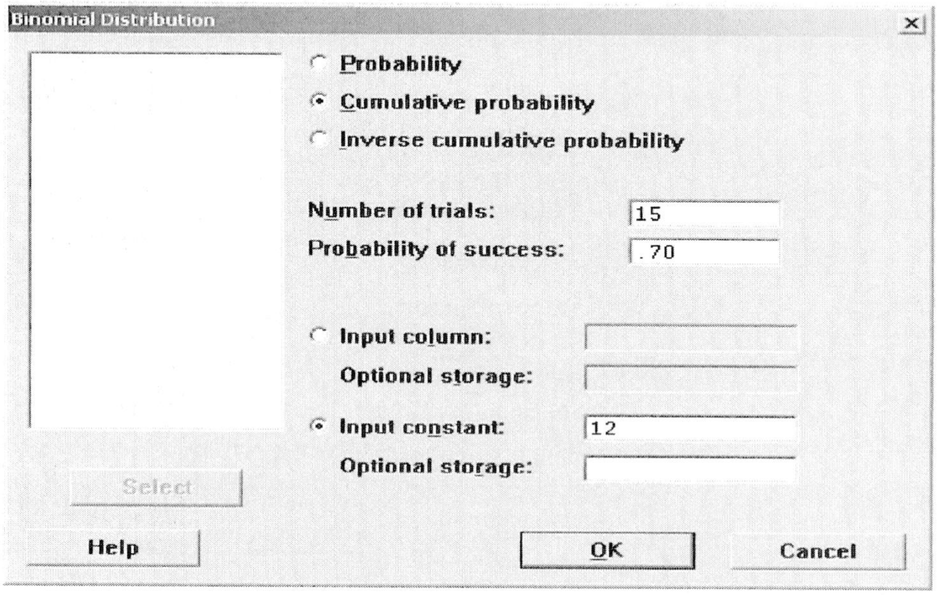

Binomial Distribution

Probability
Cumulative probability
Inverse cumulative probability

Number of trials: 15
Probability of success: .70

Input column:
Optional storage:

Input constant: 12
Optional storage:

Select

Help OK Cancel

The probability that 12 or less of the 15 households sampled have cable TV will be displayed in the Session Window. Notice that the probability is .873172. (Notice that this is the probability that you are looking for in part c.)

Cumulative Distribution Function

```
Binomial with n = 15 and p = 0.7
 x   P( X <= x )
12       0.873172
```

The probability that at least 13 households have cable TV is 1-.873172= .126828.

▸ Example 7 (pg. 336) Constructing Binomial Probability Histograms

In order to graph the binomial distribution, you must first create the distribution and save it in the Data Window. In C1, type in the values of X. Since n=10, the values of X are 0, 1, 2, 3, 4, 5, 6, 7, 8, 9 and 10.

Next, use MINITAB to generate the binomial probabilities for n=10 and p=0.20. Click on **Calc → Probability Distributions → Binomial.** Select **Probability**. The **Number of Trials** is 10 and the **Probability of Success** is .20. Now, tell MINITAB that the X values are in C1 and that you want the probabilities stored in C2. Enter C1 as the **Input Column** and enter C2 for **Optional Storage.**

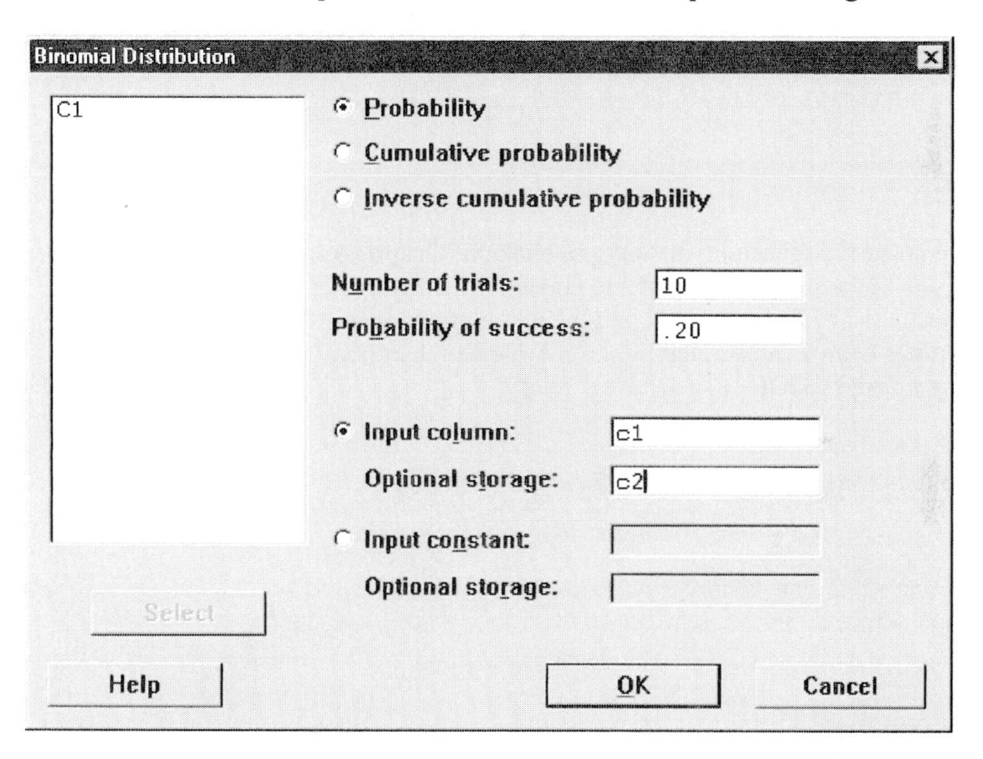

Click on **OK**. The probabilities should now be in C2. Label C1 as "X" and C2 as "P(X)". This will be helpful when you graph the distribution.

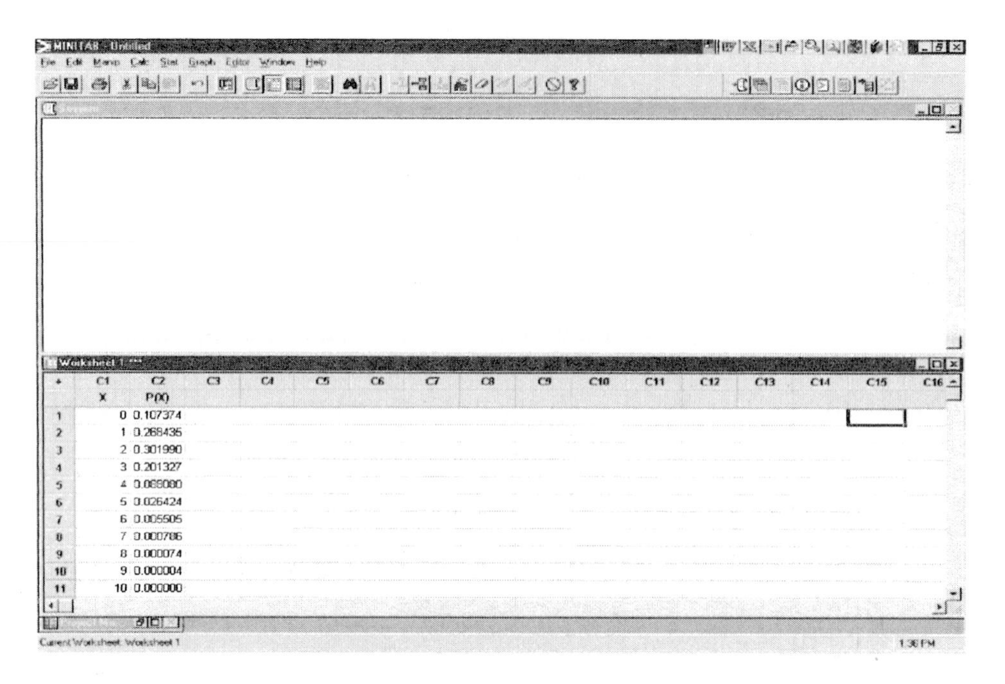

To create the probability histogram, click on: **Graph** → **Bar chart**. You will see a small pop-up that allows you to choose which type of bar chart you want. Since the X-values are in C1 and the probabilities are in C2, the **Bars represent Values from a table**. Highlight a **Simple** bar chart (icon at the left of the top row), and click **OK**.

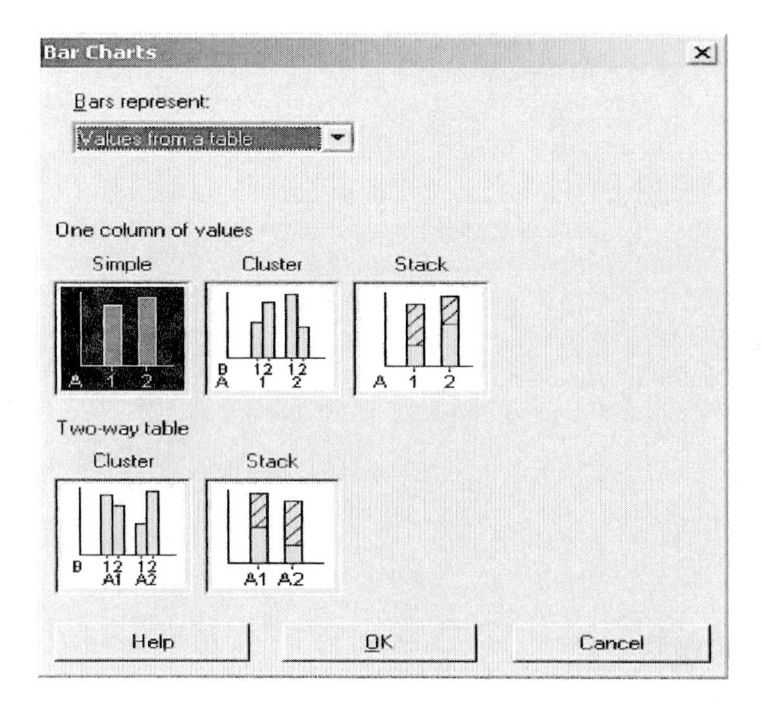

Now you must tell Minitab which variables to graph. Click in the field below "**Graph variables**". This variable is the set of probability values, which are in C2. Next click on the field below "**Categorical variable**". Select **C1**, which contains the X-values.

Bar Chart - Values from a table, One column of values, Simple

C1 X
C2 P(X)

Graph variables:
'P(X)'

Categorical variable:
X

Bar Chart Options... Scale... Labels...
Data View... Multiple Graphs... Data Options...

Select

Help OK Cancel

Next, click on the "**Labels**" button. In the top row, enter an appropriate **title** for the graph. If desired, you can also enter a sub-title and footnotes.

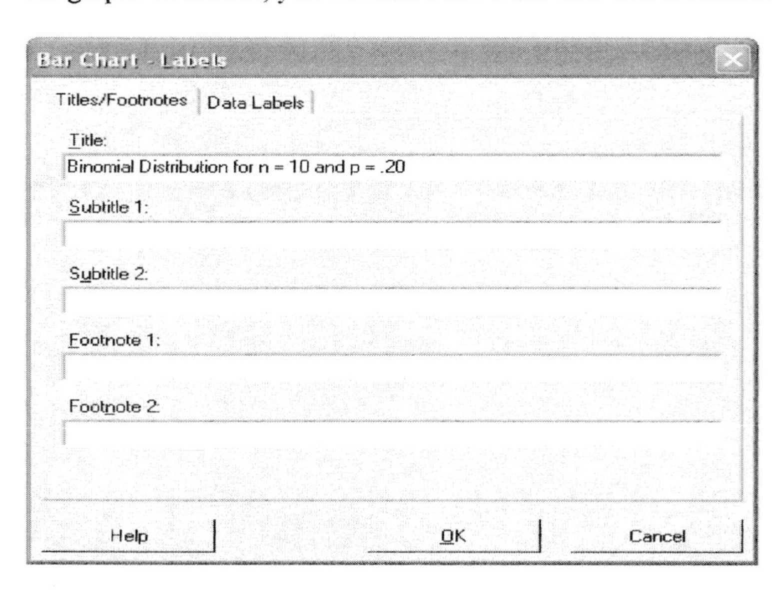

Click on OK twice to view the chart.

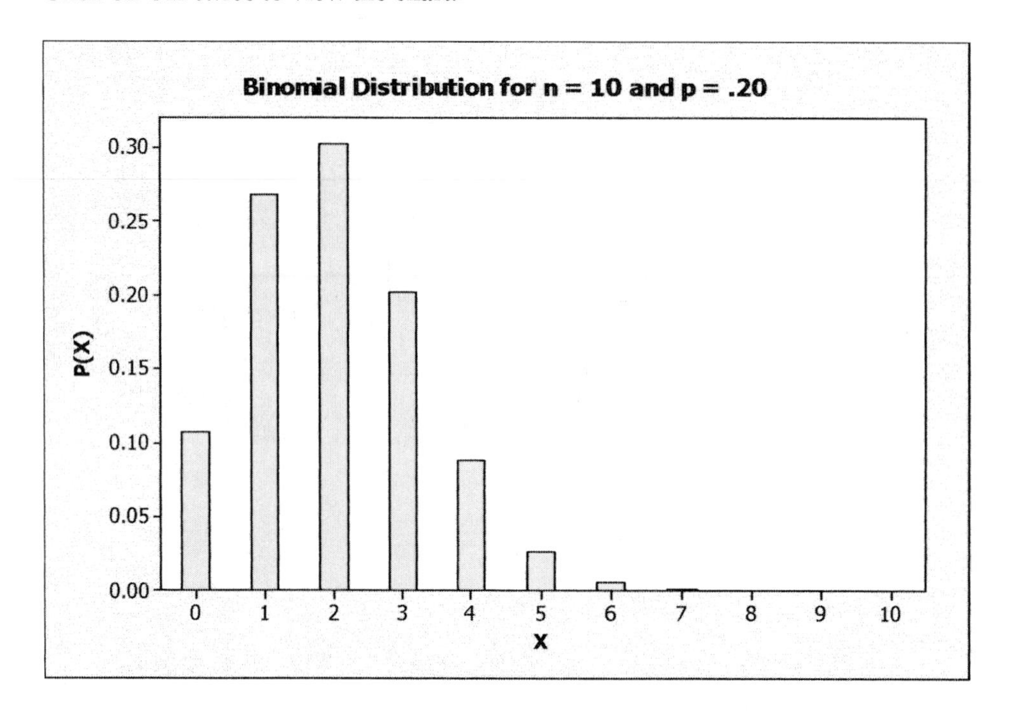

To adjust the graph so that the bars are connected, right click on any numerical value on the X-axis below the bars. (For example, right click on the number '5'). On the pull-down menu that appears, select **Edit X-scale**. Click on the checked box to the left of **Gap between clusters**. This will turn the 'check' off. Enter '0' in the box to the right of **Gap between clusters.**

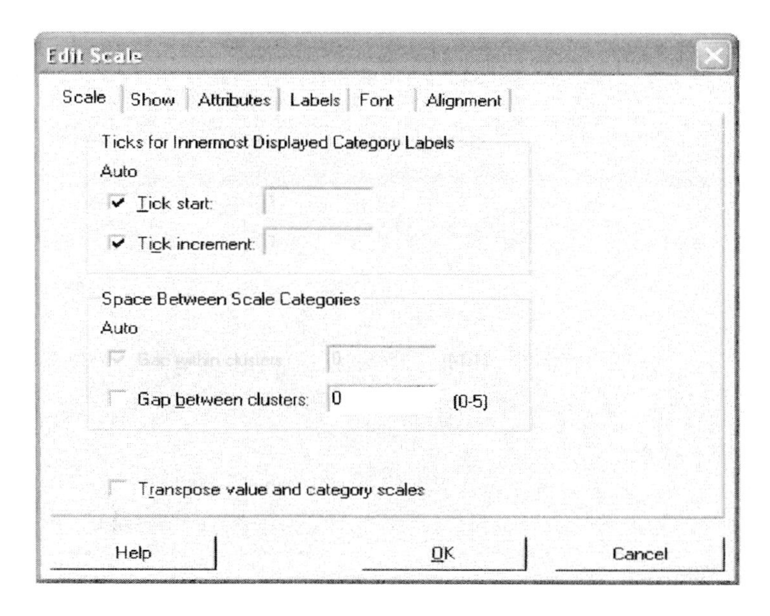

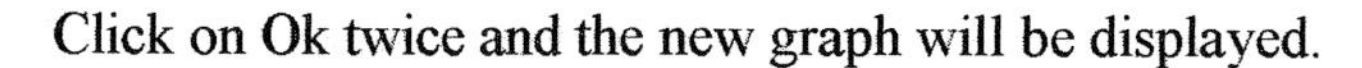

Click on Ok twice and the new graph will be displayed.

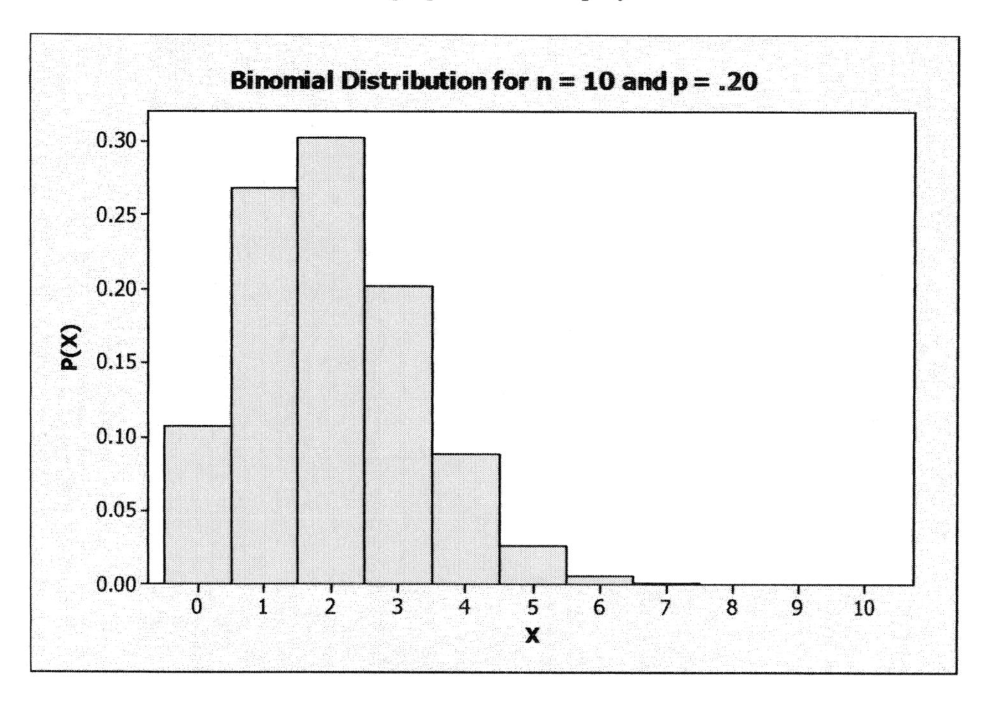

Repeat the steps above for parts b and c, changing the values of p.

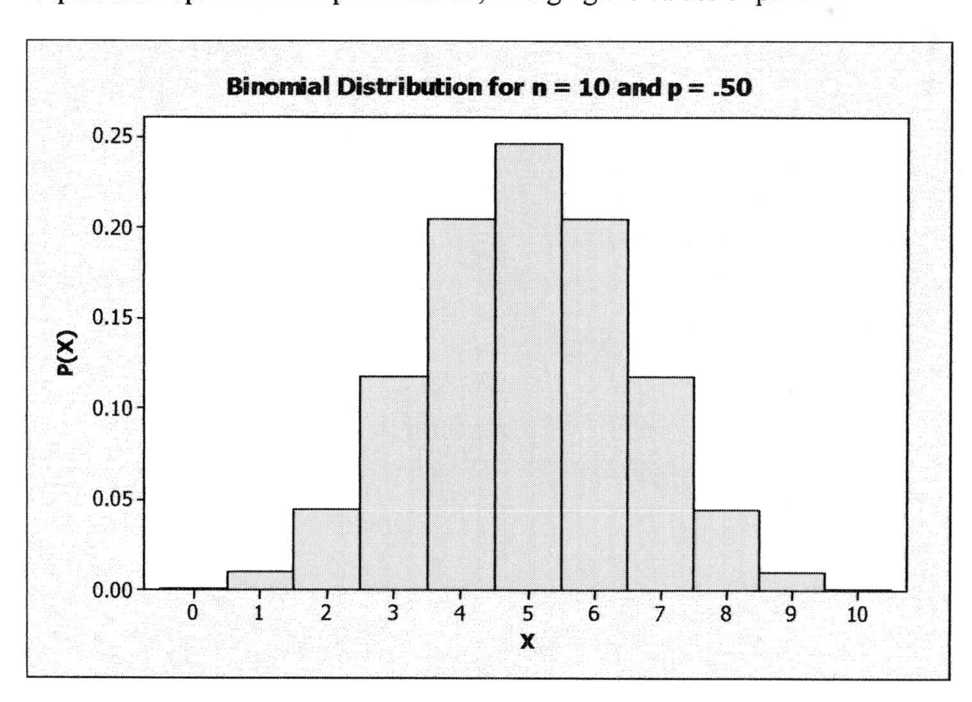

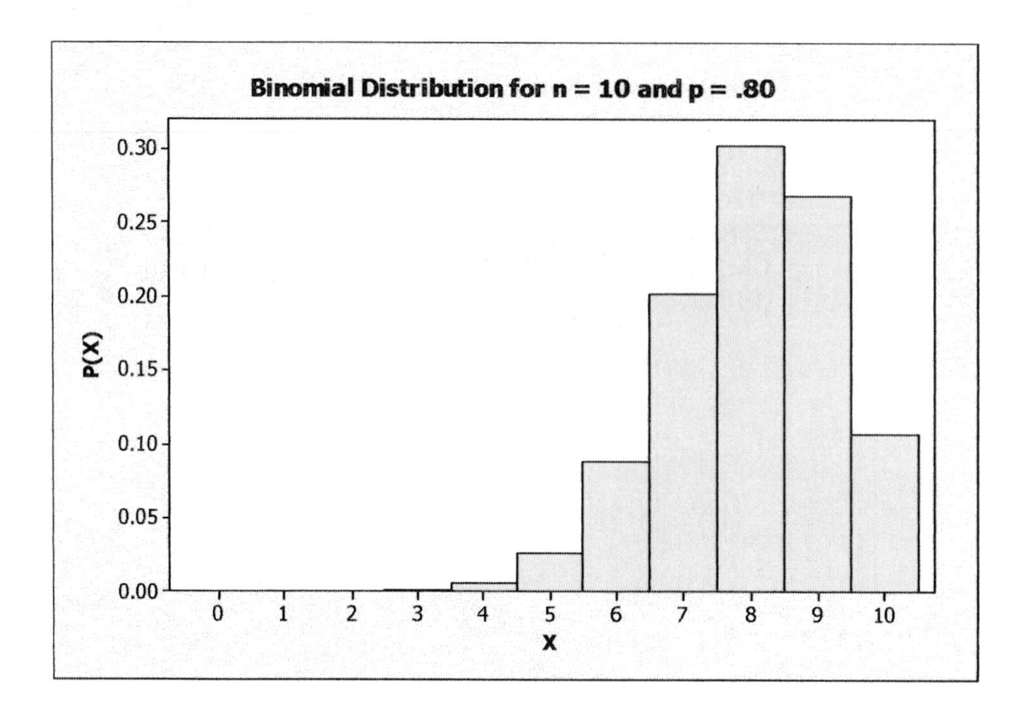
Binomial Distribution for n = 10 and p = .80
P(X)
0.30
0.25
0.20
0.15
0.10
0.05
0.00
0
1
2
3
4
5
6
7
8
9
10
X

▸ Problem 40 (pg. 341) Migraine Sufferers

In clinical trials, 2% of patients on Depakote experienced weight gain as a side effect. A random sample of 30 Depakote users is selected. Thus n = 30 and p = .02. Click on **Calc → Probability Distributions → Binomial.**

For part a, to find the probability that exactly 3 of the 30 users had a weight gain, select **Probability**. The **Number of Trials** is 30 and the **Probability of Success** is .02. To find the probability of 3, enter 3 beside **Input Constant.** Leave all other fields blank. Click on **OK** and the probability will be displayed in the Session Window. (.0188244)

For part b, you want to find the probability that 3 or fewer patients experienced weight gain as a side effect of using the drug. Repeat the steps above, but this time select **Cumulative Probability.** All other entries are the same. Click on **OK** and the probability will be displayed in the Session Window. (.997107)

For part c, you want to find the probability that 4 or more patients experienced this side effect. Since $P(X \geq 4) = 1 - P(X \leq 3)$, you can use the output from part b and subtract from 1. (1-.997107=.002893)

For part d, you want to find the probability that between 1 and 4 patients experienced this side effect. One way to calculate this is to find $P(X \leq 4)$ and subtract $P(X = 0)$. Click on **Calc → Probability Distributions → Binomial.** To find the probability that 4 or fewer of the 30 users had a weight gain, select **Cumulative Probability**. The **Number of Trials** is 30 and the **Probability of Success** is .02. To find the probability of 4 or less, enter 4 beside **Input Constant.** Leave all other fields blank. Click on **OK** and the probability will be displayed in the Session Window. Now, click on **Calc → Probability Distributions → Binomial.** To find the probability that exactly 0 of the 30 users had a weight gain, select **Probability**. The **Number of Trials** is 30 and the **Probability of Success** is .02. To find the probability of 0, enter 0 beside **Input Constant.** Leave all other fields blank. Click on **OK** and the probability will be displayed in the Session Window. (.999700-.545484=.45486)

Probability Density Function

```
Binomial with n = 30 and p = 0.02

x  P( X = x )
3   0.0188244
```

Cumulative Distribution Function

```
Binomial with n = 30 and p = 0.02

x  P( X <= x )
3     0.997107
```

Cumulative Distribution Function

```
Binomial with n = 30 and p = 0.02

x  P( X <= x )
4     0.999700
```

Probability Density Function

```
Binomial with n = 30 and p = 0.02

x  P( X = x )
0    0.545484
```

Problem 50 (pg. 342) Simulation

There is a 98% chance that a 20-year-old male will survive to age 30. To simulate 100 random samples of size 30 from this population, click on **Calc → Random Data → Binomial. Generate** 100 **rows of data,** and **store in column** C1. The **Number of trials** is 30 and the **Probability of success** is .98.

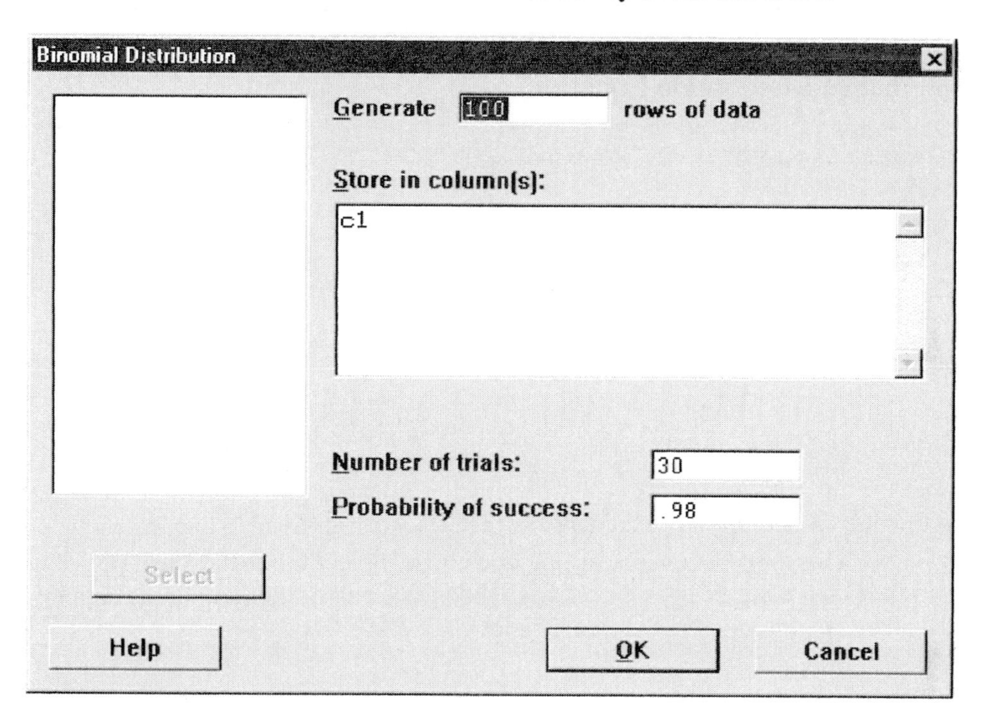

Click on OK and C1 should have the number of survivors for each of the 100 random samples in it. (Note: since this is random data, your results will be different from those shown below.)

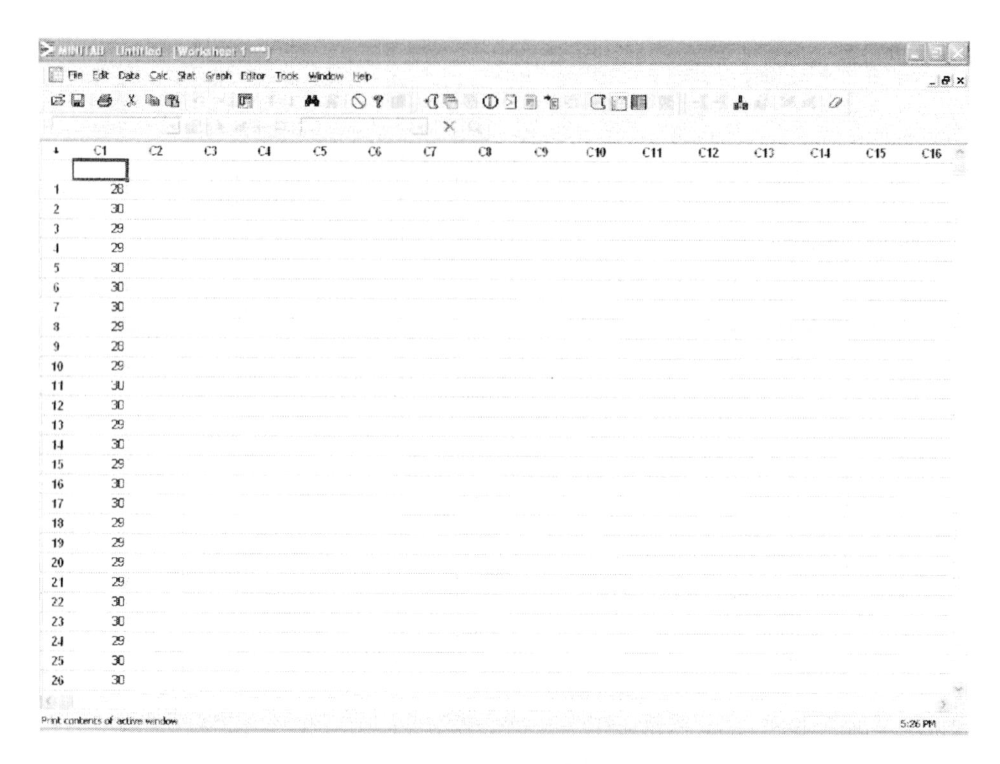

To find the probability that exactly 29 of the 30 males survived to age 30, click on **Stat→ Tables → Tally Individual Variables.** Select C1 as the **Variable**, and select both **Counts** and **Percents.** Click on OK and a summary of the data in C1 will be displayed in the Session Window.

Tally for Discrete Variables: C1

C1	Count	Percent
27	1	1.00
28	9	9.00
29	31	31.00
30	59	59.00
N=	100	

Notice that 31% of the time, 29 out of the 30 males survived. Thus, based on the simulation, the probability that 29 out of 30 males survive to age 30 is 0.31.

For part c, you want to find the exact probability based on the binomial distribution with n = 30 and p = .98. Click on **Calc → Probability Distributions → Binomial.** To find the probability that exactly 29 of the 30 males survive, select **Probability**. The **Number of Trials** is 30 and the **Probability of Success** is .98. Enter 29 beside **Input Constant** and leave all other fields blank. Click on **OK** and the probability will be displayed in the Session Window. As you can see below, the theoretical probability is .33397.

Probability Density Function

```
Binomial with n = 30 and p = 0.98

 x     P( X = x )
29     0.333970
```

For part d, you want to find the probability that at most 27 males survived. According to the summary table from the simulation, only 1% of the time did at most 27 survive.

To find the theoretical probability that at most 27 survive, repeat the steps for part c, but this time select **Cumulative Probability** and enter 27 beside **Input Constant.** All other entries are the same. Click on **OK** and the probability will be displayed in the Session Window.

Cumulative Distribution Function

```
Binomial with n = 30 and p = 0.98

 x     P( X <= x )
27      0.0217178
```

Finally, to find the mean number of survivors based on the simulations, click on **Stat→ Basic Statistics→ Display Descriptive Statistics.** Select C1 for the **Variable.**

Descriptive Statistics: C1

```
Variable        N   N*    Mean    StDev   Minimum      Q1
C1            100    0  29.480    0.703    27.000  29.000

Variable      Median      Q3    Maximum
C1                30  30.000     30.000
```

The mean number of survivors from the simulation is 29.480. The theoretical mean number of survivors is 30 * .98 = 29.4. Thus, this simulation gave results which are very close to the theoretical results.

◀

Section 6.3

▸ Example 2 (pg. 346) Finding Poisson Probabilities

Since cars arrive at McDonald's at a rate of 2 cars per minute, $\lambda = 2$ for this Poisson example. To find the probability that 6 cars arrive between 12:00 and 12:05, notice that this is a time interval of 5 minutes. This means that an average of (2)(5)=10 cars will arrive in a 5 minute interval. Click on **Calc → Probability Distributions → Poisson.** Since you want a simple probability, select **Probability** and enter 10 for the **Mean.** To find the probability that X=6, enter 6 for the **Input constant.**

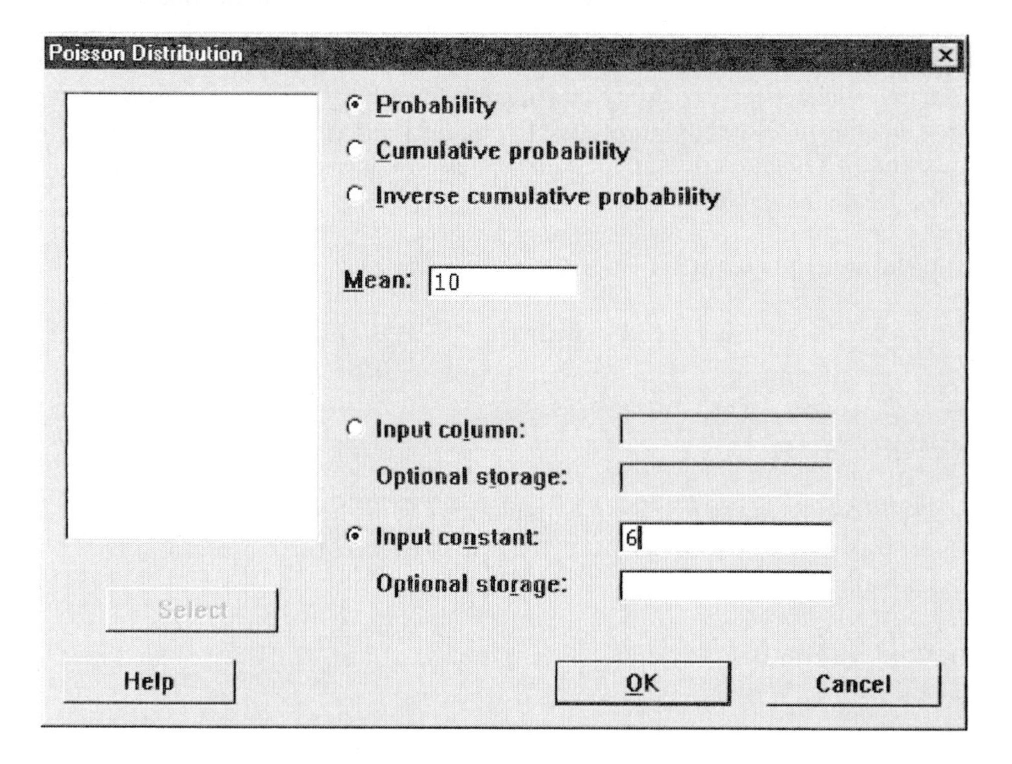

Click on **OK** and the probability will be in the Session Window.

Probability Density Function

```
Poisson with mean = 10

x  P( X = x )
6   0.0630555
```

Now, to find the probability that fewer than 6 cars arrive in the 5-minute interval, repeat the steps above. Click on **Calc → Probability Distributions → Poisson.** This time you want a cumulative probability, so select **Cumulative Probability** and enter 10 for the **Mean.** To find the probability that $X < 6$, enter 5 for the **Input constant.**

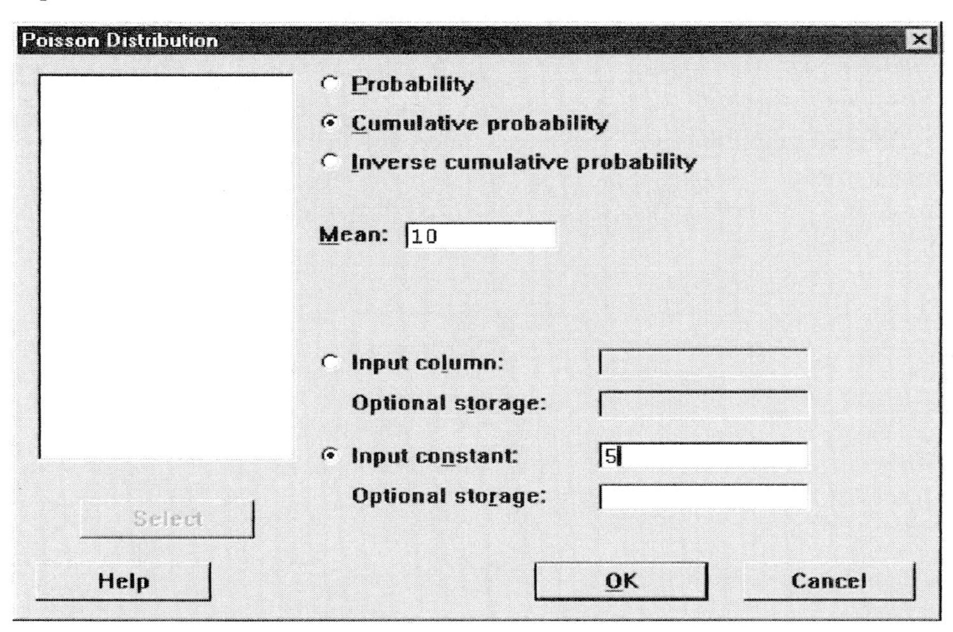

When you click on **OK**, the results will be in the Session Window.

Cumulative Distribution Function

```
Poisson with mean = 10

x  P( X <= x )
5    0.0670860
```

Finally, to find the probability that at least 6 cars arrive in the 5-minute interval, just use $1 - P(X < 6) = 1 - .067086 = .932914$.

◀

Example 3 (pg. 347) Beetles and the Poisson Distribution

In this example, 2000 beetles are spread out in 200 subsections. Assuming the beetles are evenly spread out, the mean is 2000/200 = 10. To find the probability that exactly 8 beetles are in a subsection, click on **Calc → Probability Distributions → Poisson.** Select **Probability**. The **Mean** is 10 and the **Input Constant** is 8.

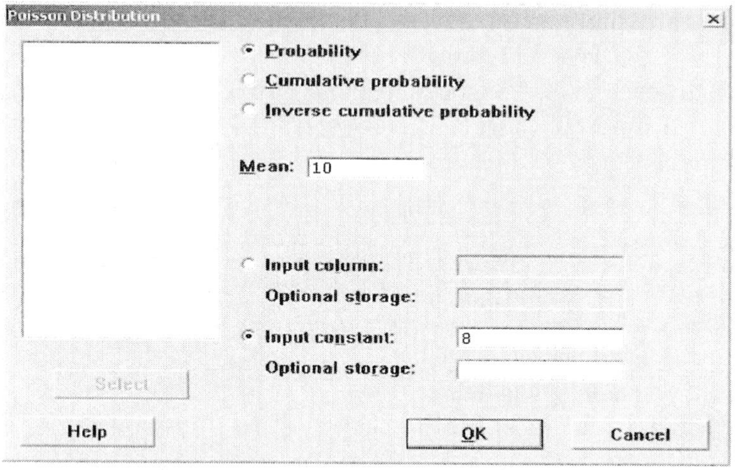

Poisson Distribution

Probability
Cumulative probability
Inverse cumulative probability

Mean: 10

Input column:
Optional storage:
Input constant: 8
Optional storage:

Select
Help
OK
Cancel

Click **OK.** The probability will be in the Session Window.

Probability Density Function
Poisson with mean = 10
x P(X = x)
8 0.112599

The probability of finding exactly 8 beetles in a subsection is 0.112599.

To find the probability of finding 17 or more beetles, Click on **Calc → Probability Distributions → Poisson.** Select **Cumulative Probability**. The **Mean** is 10. Enter 16 as the **Input Constant** and click **OK**. The result is in the Session Window.

Cumulative Distribution Function
Poisson with mean = 10
x P(X <= x)
16 0.972958

The probability of finding 17 or more beetles is 1 – 0.972758 = 0.027242.

▸ Problem 19 (pg. 349) Wendy's Drive-Through

Since cars arrive at Wendy's at a rate of 0.2 cars per minute, $\lambda = 0.2$ for this Poisson example. To find the probability distribution of the number that arrive between 12:00 and 12:30, notice that this is a time interval of 30 minutes. This means that an average of (.2)(30)=6 cars will arrive in a 30 minute interval.

First, create the Poisson distribution and save it in the Data Window. Enter the X values (number of cars arriving) into C1 and enter the frequencies into C2.

Next, use MINITAB to generate the Poisson probabilities for n=16 and μ=6. Click on **Calc → Probability Distributions → Poisson.** Select **Probability**. The **Mean** is 6. Now, tell MINITAB that the X values are in C1 and that you want the probabilities stored in C3. Enter **C1** as the **Input Column** and enter C3 for **Optional Storage.** Click on OK and the probability distribution will be in C3.
To find the *expected* number of arrivals, notice that there are 200 Wendy's restaurants. So multiply each probability in C3 by 200. Click on **Calc → Calculator**. **Store result in variable** C4 and type in the **Expression** 200*C3. Now compare the numbers in C2 to the numbers in C4.

MINITAB - Untitled - [6_3_15.MTP ***]

↓	C1	C2	C3	C4
	x (number of cars arriving)	Frequency	P(x)	Expected
1	1	4	0.014873	2.9745
2	2	5	0.044618	8.9235
3	3	13	0.089235	17.8470
4	4	23	0.133853	26.7705
5	5	25	0.160623	32.1246
6	6	28	0.160623	32.1246
7	7	25	0.137677	27.5354
8	8	27	0.103258	20.6515
9	9	21	0.068838	13.7677
10	10	15	0.041303	8.2606
11	11	5	0.022529	4.5058
12	12	3	0.011264	2.2529
13	13	2	0.005199	1.0398
14	14	2	0.002228	0.4456
15	15	0	0.000891	0.1783
16	16	2	0.000334	0.0668

Current Worksheet: 6_3_15.MTP 9:36 AM

◂

▸ Problem 23 (pg. 350) Simulation

Since the rate of colds is 23.8 per 100 18 to 24 year olds, $\lambda = 23.8$ for this Poisson example. To find the expected number of colds per 500 people in this age group, just multiply 23.8 by 5. This means that there is an average of (23.8)*(5)=119 colds per 500 people. To simulate taking 100 random samples of size 500, click on **Calc → Random Data → Poisson. Generate** 100 **rows of data** and **Store in column** C1. Enter 119 for the **Mean**.

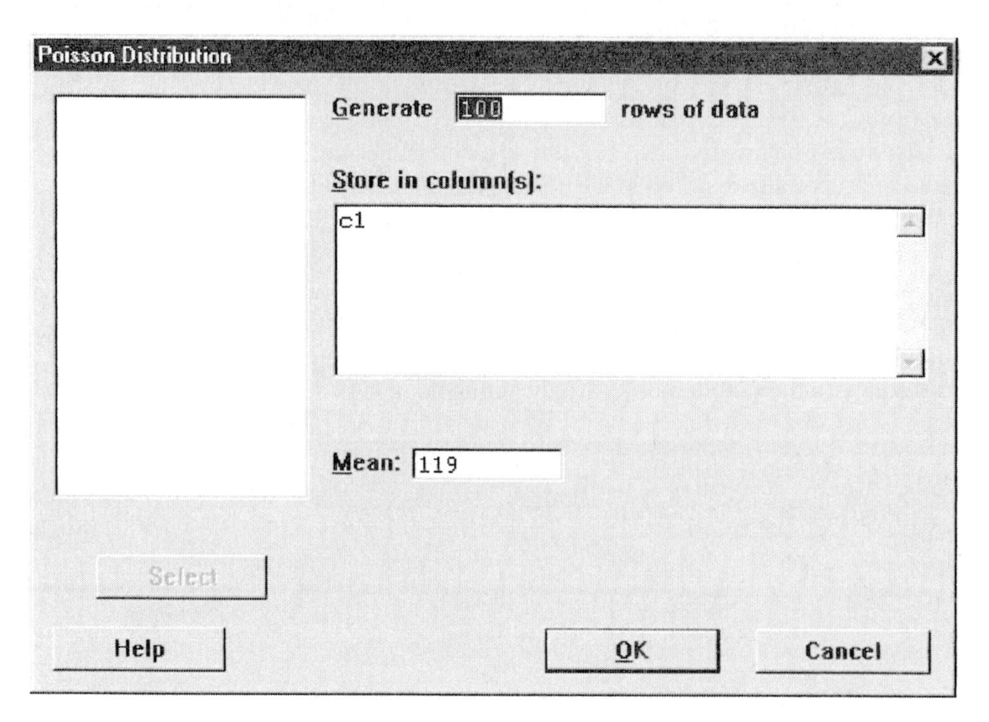

Click on OK and the random data will be in C1. Now summarize the random data by creating a table of the results. Click on **Stat → Tables → Tally Individual variables.** Enter C1 for the **Variable,** and select both **Percents** and **CumulativePercents.** The summary table will be in the Session Window. Since this is random data, everyone's summary table will be a little different.

Tally for Discrete Variables: C1

C1	Percent	CumPct
98	1.00	1.00
100	1.00	2.00
101	1.00	3.00
102	1.00	4.00
103	1.00	5.00
104	1.00	6.00
106	1.00	7.00
107	1.00	8.00
108	1.00	9.00
109	2.00	11.00
110	4.00	15.00
111	5.00	20.00
112	5.00	25.00
113	5.00	30.00
114	5.00	35.00
115	4.00	39.00
116	4.00	43.00
117	3.00	46.00
118	5.00	51.00
119	4.00	55.00
120	2.00	57.00
121	2.00	59.00
122	3.00	62.00
124	1.00	63.00
125	1.00	64.00
126	3.00	67.00
127	1.00	68.00
128	2.00	70.00
129	4.00	74.00
130	1.00	75.00
131	1.00	76.00
132	2.00	78.00
133	2.00	80.00
134	2.00	82.00
135	6.00	88.00
136	1.00	89.00
137	2.00	91.00
138	1.00	92.00
139	1.00	93.00
140	2.00	95.00
141	1.00	96.00
142	1.00	97.00
143	1.00	98.00
144	1.00	99.00
153	1.00	100.00

To answer parts c and d of this problem, you can use the table above:

$P(X \geq 150) = 1 - P(X < 150) = 99.00\%$ or .99.

$P(X < 100) = P(X \leq 99) = 1.00\%$ or .01.

Next, calculate the mean, standard deviation, and 5-number summary of the data. Click on **Stat → Basic Statistics → Display Descriptive Statistics.** Select C1 and click on **OK**. The results will be in the Session Window.

Descriptive Statistics: C1

Variable	N	N*	Mean	SE Mean	StDev	Minimum	Q1
C1	100	0	121.11	1.15	11.55	98.00	112.25

Variable	Median	Q3	Maximum
C1	118.00	130.75	153.00

Section 6.4 (Note: available on CD)

▸ Example 4 Finding Hypergeometric Probabilities

To determine whether to accept an order of 500 bolts, a random sample of 12 bolts is selected. If none of the 12 bolts are defective, the shipment is accepted. If 10% of the supplier's bolts are defective, what is the probability that the shipment is acceptable? To find this probability notice that N=500, n=12, and k=50 (0.10 * 500). Minitab uses "M" instead of k. Click on **Calc → Probability Distributions → Hypergeometric.** Since you want a simple probability, select **Probability** and enter values for N, M, and n. To find the probability that there are no defectives, enter 0 for the **Input constant.**

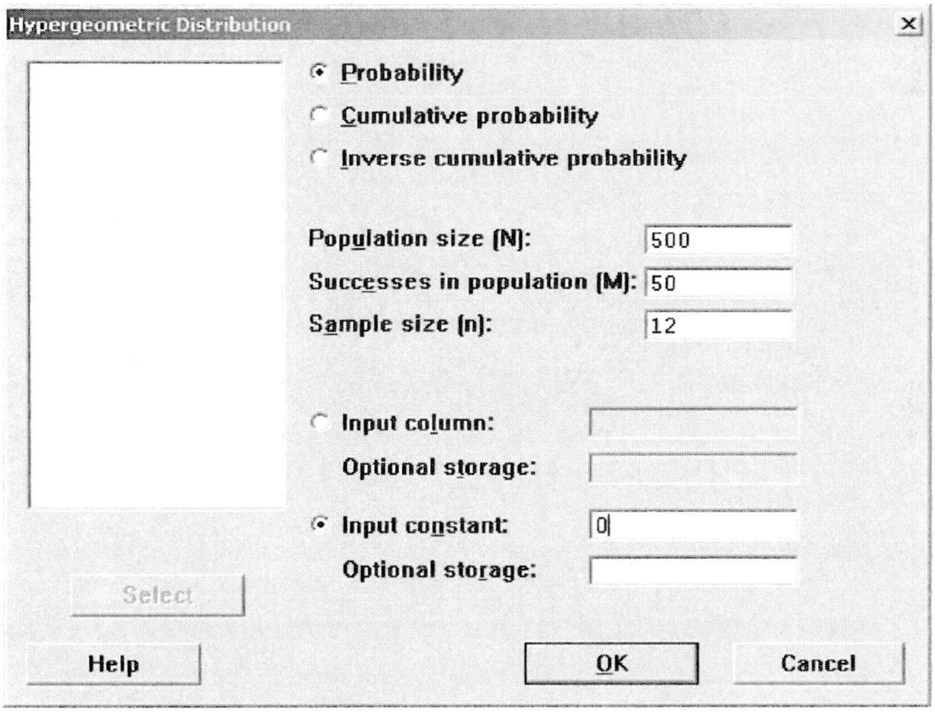

Hypergeometric Distribution

Probability
Cumulative probability
Inverse cumulative probability

Population size (N): 500
Successes in population (M): 50
Sample size (n): 12

Input column:
Optional storage:
Input constant: 0
Optional storage:

Select
Help OK Cancel

Click on **OK**. The result is in the Session Window. The probability that the shipment is acceptable is 0.278250.

Probability Density Function

```
Hypergeometric with N = 500, M = 50, and n = 12
x  P( X = x )
0    0.278250
```

◀

Normal Probability Distribution

Section 7.1

‣ Problem 37 (pg. 371) Hitting with a Pitching Wedge

Open Minitab worksheet **7_1_37**. The distances should be C1. To draw a histogram with a normal curve superimposed over it, click on **Stat→ Basic Statistics → Display Descriptive Statistics.** Select C1 for the variable, and then click on **Graphs.** Select **Histogram of data, with normal curve.**

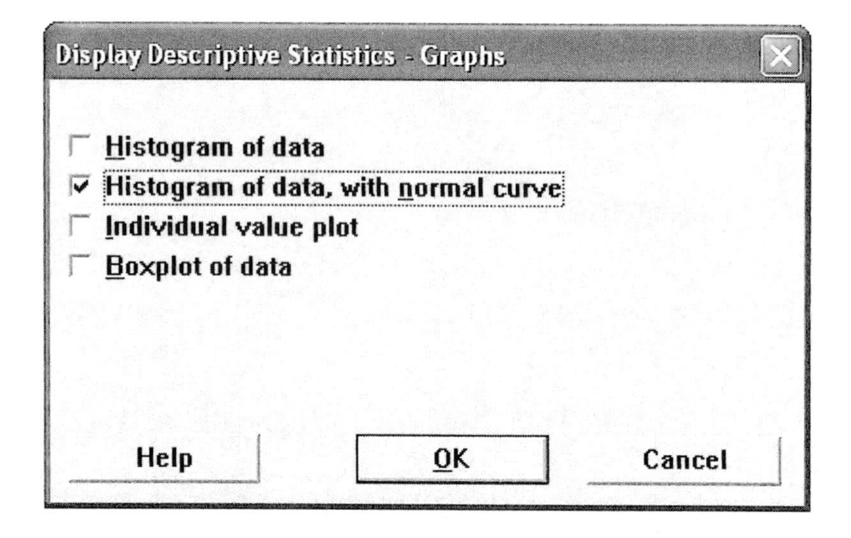

Click on **OK** twice to view the graph.

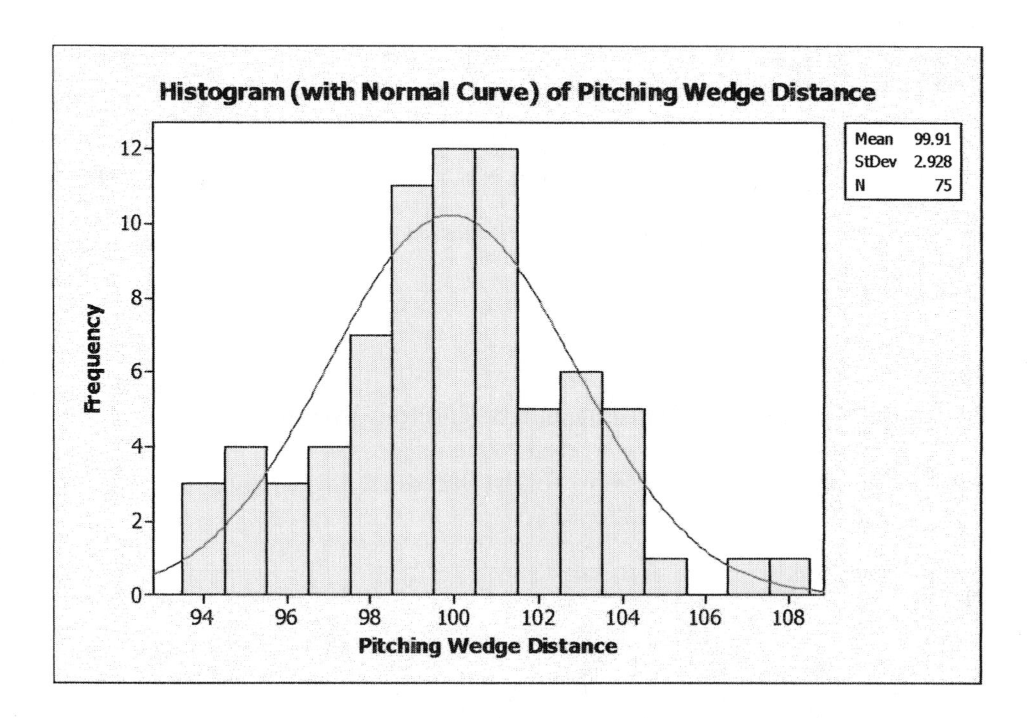

Use the graph to help you decide if the data follows a normal distribution.

Section 7.2
(Note: see Section 7.3 for more examples)

▸ Example 1 (pg. 372) Finding Area under a Standard Normal Curve

The standard normal distribution has μ=0 and σ=1. Find the area under the curve to the left of Z=1.68. To do this in MINITAB, click on **Calc → Probability Distributions → Normal.** On the input screen, select **Cumulative probability.** (Cumulative probability *'accumulates'* all probability to the left of the input constant.) Enter 0 for the **Mean** and 1 for the **Standard deviation.** Next select **Input Constant** and enter the value 1.68.

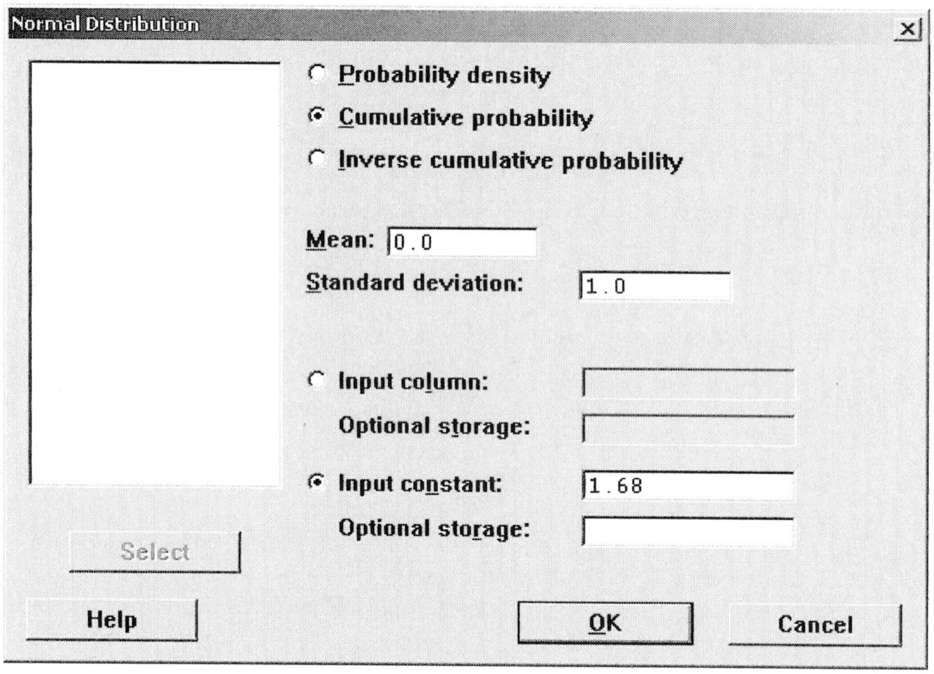

Normal Distribution

- ○ Probability density
- ● Cumulative probability
- ○ Inverse cumulative probability

Mean: 0.0

Standard deviation: 1.0

○ Input column:

Optional storage:

● Input constant: 1.68

Optional storage:

Select

Help OK Cancel

Click on **OK**. The result will be in the Session Window.

Cumulative Distribution Function

```
Normal with mean = 0 and standard deviation = 1
   x  P( X <= x )
1.68     0.953521
```

◂

Example 5 (pg. 376) Finding a Z-score from a Specified Area

Find the Z-score such that the area to the left of it is 0.32. To do this in MINITAB, click on **Calc → Probability Distributions → Normal.** On the input screen, select **Inverse cumulative probability.** Enter 0 for the **Mean** and 1 for the **Standard deviation.** Next select **Input Constant** and enter the value 0.32.

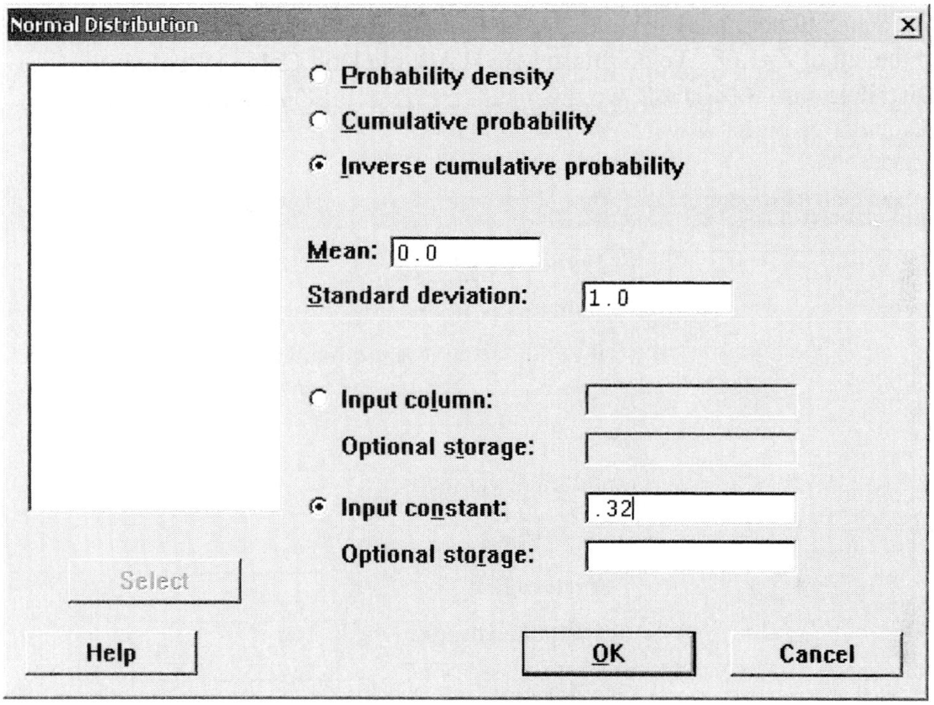

Normal Distribution

Probability density
Cumulative probability
Inverse cumulative probability

Mean: 0.0
Standard deviation: 1.0

Input column:
Optional storage:
Input constant: .32
Optional storage:

Select
Help OK Cancel

Click on **OK**. The result will be in the Session Window.

Inverse Cumulative Distribution Function

```
Normal with mean = 0 and standard deviation = 1

P( X <= x )          x
       0.32  -0.467699
```

Section 7.3

▸ Example 1 (pg. 385) Finding Area under a Normal Curve

The height of 3 year old girls is normally distributed with μ=38.72 and σ=3.17. What percent of 3 year old girls have a height less than 35 inches? To do this in MINITAB, click on **Calc → Probability Distributions → Normal.** On the input screen, select **Cumulative probability.** (Cumulative probability '*accumulates*' all probability to the left of the input constant.) Enter 38.72 for the **Mean** and 3.17 for the **Standard deviation.** Next select **Input Constant** and enter the value 35.

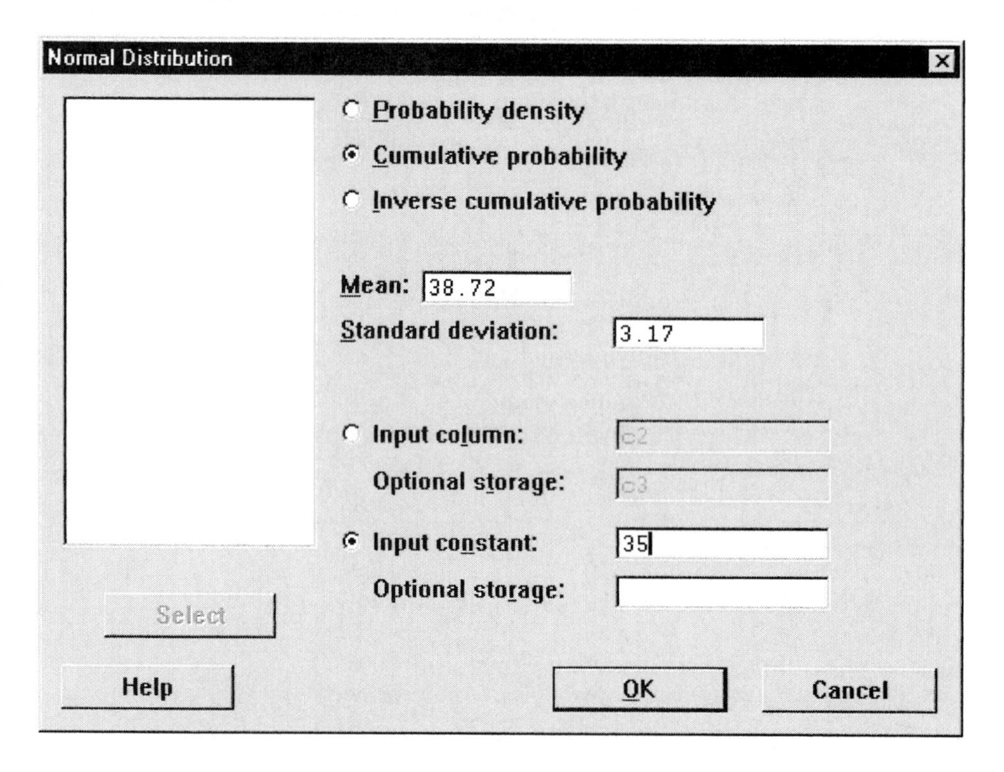

Click on **OK** and the probability should be displayed in the Session Window.

Cumulative Distribution Function

```
Normal with mean = 38.72 and standard deviation =
3.17

 x  P( X <= x )
35     0.120297
```

As you can see, the probability is .120297. Thus, about 12.03% of 3-year-old girls have a height of less than 35 inches.

◀

▸ Example 3 (pg. 387) Height of 3 year old girls

Heights are normally distributed with $\mu = 38.72$ inches and $\sigma = 3.17$ inches. To find $P(35 \le X \le 40)$, you will need MINITAB to give you two probabilities: one using X=40 and the other using X=35. Click on **Calc → Probability Distributions → Normal.** On the input screen, select **Cumulative probability.** Enter 38.72 for the **Mean** and 3.17 for the **Standard deviation.** Next select **Input Constant** and enter the value 40. Click on **OK.** Repeat the above steps using an **Input constant** of 35. Now the Session Window should have $P(X \le 40)$ and $P(X \le 35)$.

Cumulative Distribution Function

```
Normal with mean = 38.72 and standard deviation = 3.17

 x   P( X <= x )
40      0.656815
```

Cumulative Distribution Function

```
Normal with mean = 38.72 and standard deviation = 3.17

 x   P( X <= x )
35      0.120297
```

So, to find the $P(35 \le X \le 40)$, you must subtract the two probabilities. Thus, the $P(35 \le X \le 40) = .656815 - .120297 = .536518$.

(Note: Minitab uses more decimal places than the textbook does so these answers are slightly different.) ◂

Example 4 (pg. 388) Finding a specific data value

Heights are normally distributed with $\mu = 38.72$ inches and $\sigma = 3.17$ inches. Find the height of a 3 year old at the 20th percentile. To do this in MINITAB, click on **Calc → Probability Distributions → Normal.** On the input screen, select **Inverse Cumulative probability.** Enter 38.72 for the **Mean** and 3.17 for the **Standard deviation.** For this type of problem, the **Input constant** will be the area to the left of the X-value we are looking for. This input constant will be a decimal number between 0 and 1. For this example, select **Input Constant** and enter the value .20 since 20% of the heights are below this number. Click on **OK** and the X-value should be in the Session Window. Notice that the 20th percentile of heights of 3 year old girls is 36.0521 inches.

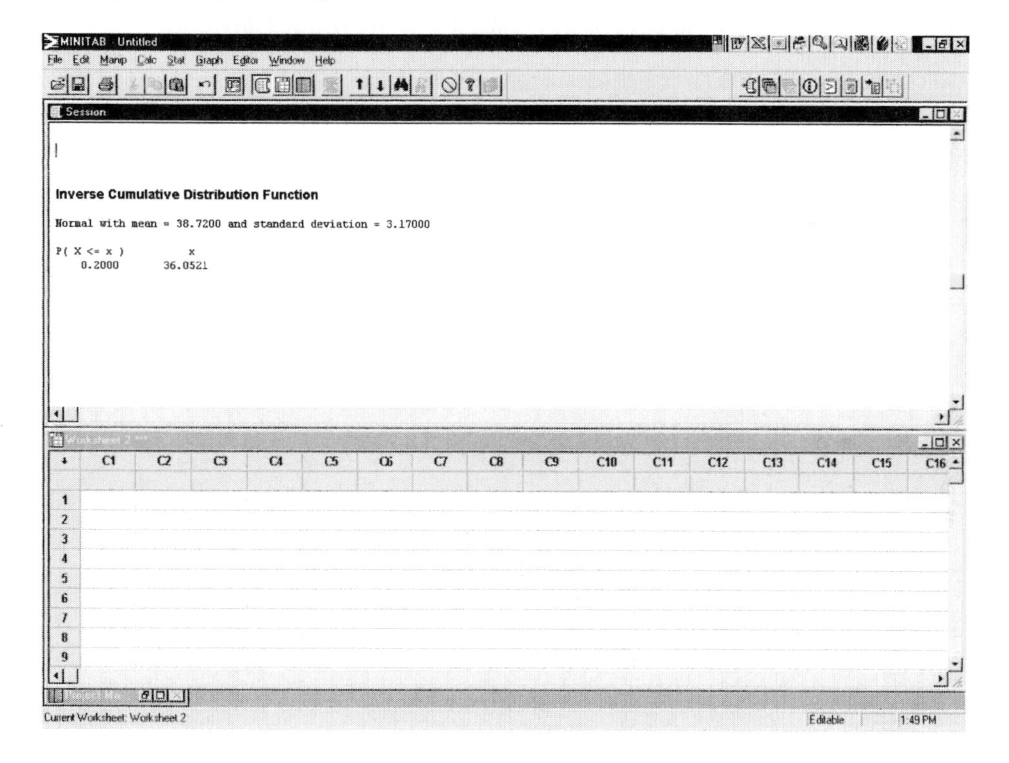

Example 6 (pg. 389) Finding a specific data value

Heights are normally distributed with μ=38.72 inches and σ=3.17 inches. Find the heights of 3 year old girls that separate the middle 98% of the distribution from the top and bottom 1%. To do this in MINITAB, click on **Calc → Probability Distributions → Normal.** On the input screen, select **Inverse Cumulative probability.** Enter 38.72 for the **Mean** and 3.17 for the **Standard deviation.** For this type of problem, the **Input constant** will be the area to the left of the X-value we are looking for. This input constant will be a decimal number between 0 and 1. For this example, select **Input Constant** and enter the value .99 since 99% of the heights are below the top 1%. Click on **OK** and the X-value should be in the Session Window. Repeat this step to find the bottom 1% cutoff point, but this time enter the value .01 as the **Input Constant.**

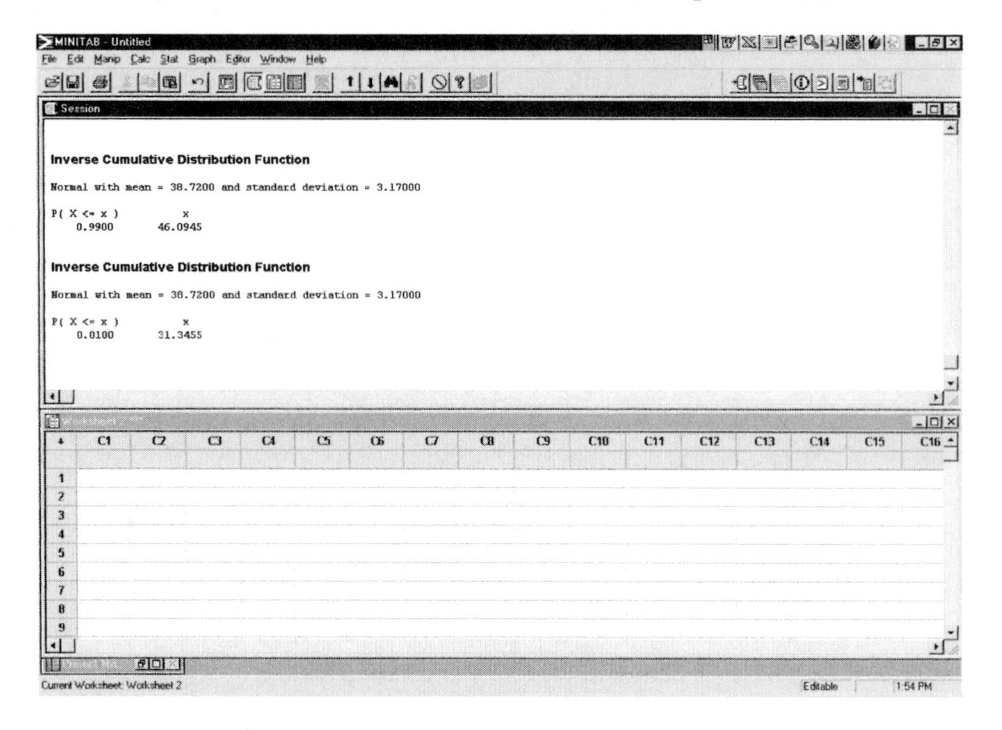

Notice that the 99th percentile of heights of 3 year old girls is 46.0945 inches and the 1st percentile of heights is 31.3455. These two values are the cut-off points for the middle 98%.

▸ Problem 21 (pg. 391) Hybrid Car

The Honda Insight has a gas mileage that is normally distributed with μ=56 and σ = 3.2. For each of the parts a – e of this question, click on **Calc → Probability Distributions → Normal.** On the input screen for each part, select **Cumulative probability.** Enter 56 for the **Mean** and 3.2 for the **Standard deviation.**

(a) Select **Input Constant** and enter the value 60. Click on **OK.** Subtract the result in the Session Window from 1. (1 - .894350 = 0.105650)

Cumulative Distribution Function

Normal with mean = 56 and standard deviation = 3.2

x P(X <= x)
60 0.894350

(b) Select **Input Constant** and enter the value 50. Click on **OK.**

Cumulative Distribution Function

Normal with mean = 56 and standard deviation = 3.2

x P(X <= x)
50 0.0303964

(c) Select **Input Constant** and enter the value 58. Click on **OK.** Repeat with an **Input Constant** of 62. Click on **OK.** Subtract the two probabilities. (.969604 - .734014 = .235590)

Cumulative Distribution Function

Normal with mean = 56 and standard deviation = 3.2

x P(X <= x)
58 0.734014

Cumulative Distribution Function

Normal with mean = 56 and standard deviation = 3.2

x P(X <= x)
62 0.969604

(d) Select **Input Constant** and enter the value 45. Click on **OK.** (.0002936)

Cumulative Distribution Function

Normal with mean = 56 and standard deviation = 3.2

x P(X <= x)
45 0.0002936

◂

▸ Problem 31 (pg. 392) Hybrid Car

The Honda Insight has a gas mileage that is normally distributed with μ=56 and σ = 3.2. Find the given percentiles of gas mileages. To do this in MINITAB, click on **Calc → Probability Distributions → Normal.** On the input screen, select **Inverse Cumulative probability.** Enter 56 for the **Mean** and 3.2 for the **Standard deviation.** For this type of problem, the **Input constant** will be the area to the left of the X-value we are looking for. This input constant will be a decimal number between 0 and 1.

(a) For this part, you are looking for the 97th percentile. Select **Input Constant** and enter the value .97 since 97% of the mileages are below the number we are looking for. Click on **OK** and the X-value should be in the Session Window. (62.0185 mpg)

(c) Find the mileages that make up the middle 86%. This area is between the 7th and 93rd percentiles. Select **Input Constant** and enter the value .93 since 93% of the mileages are below the number we are looking for. Click on **OK** and the X-value should be in the Session Window. Repeat this step, but enter the value .07 since there are 7% of the mileages below the number we are looking for. (51.2775, 60.7225)

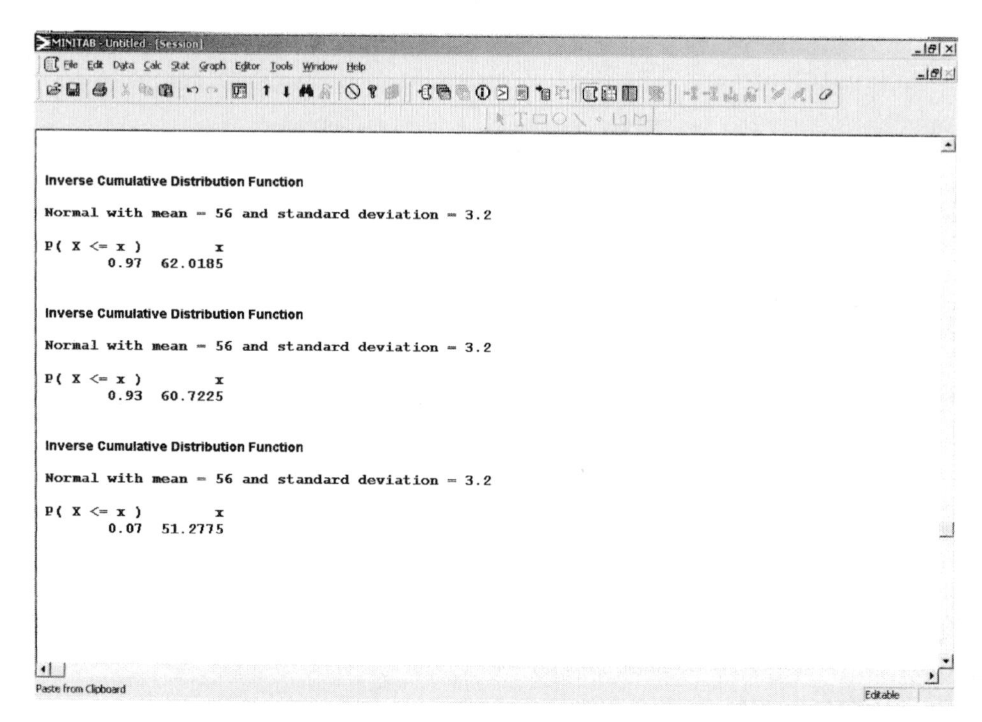

◀

Section 7.4

Example 3 (pg. 397) Assessing Normality

Open Minitab worksheet **7_4_EX3**. The waiting times are in C1. To draw a normal probability plot, click on **Graph → Probability Plot** and select the **Single** plot. Click on **OK.** Select C1 for the **Graph variable.**

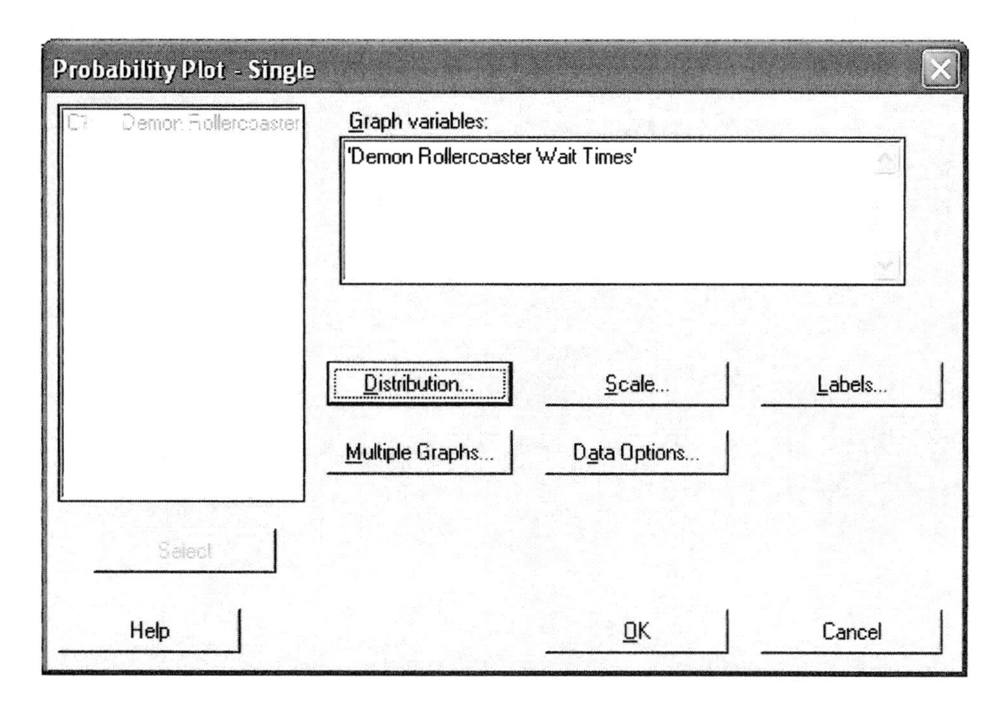

Click on **OK** and the normal probability plot should pop up.

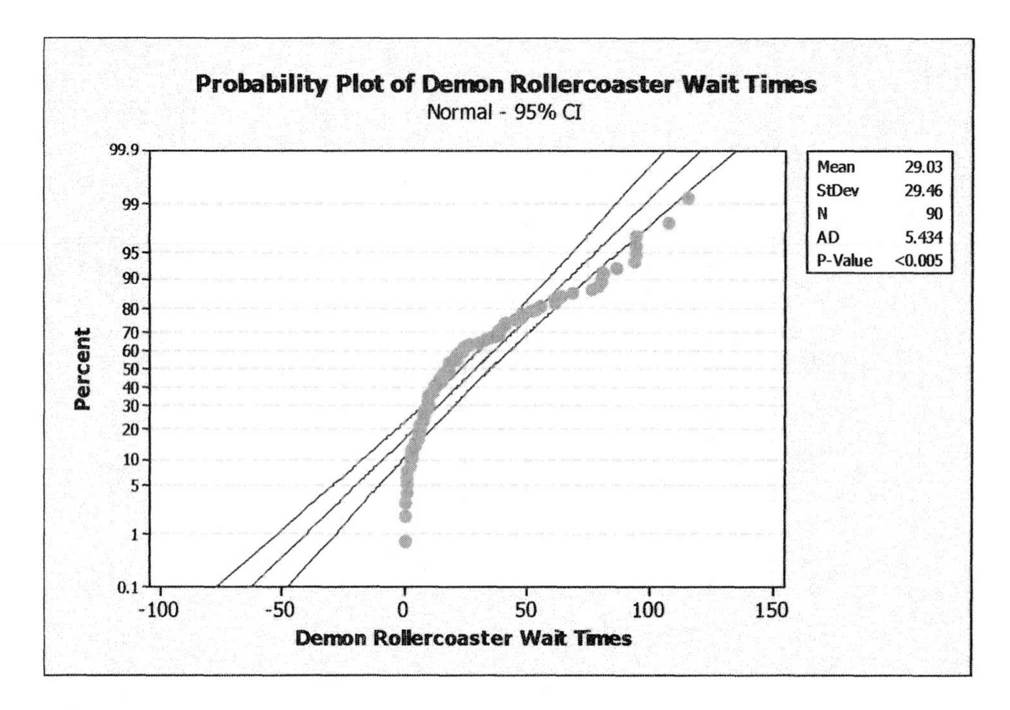

Clearly the normal probability plot is not linear. Thus, you can conclude that waiting times are not normally distributed.

◀

▸ Problem 12 (pg. 401) Customer Service

Open Minitab worksheet **7_4_12**. The number of customers is in C1. To draw a normal probability plot, click on **Graph → Probability Plot** and select the **Single** plot. Click on **OK.** Select C1 for the **Graph variable.** Click on **OK** and the normal probability plot should appear.

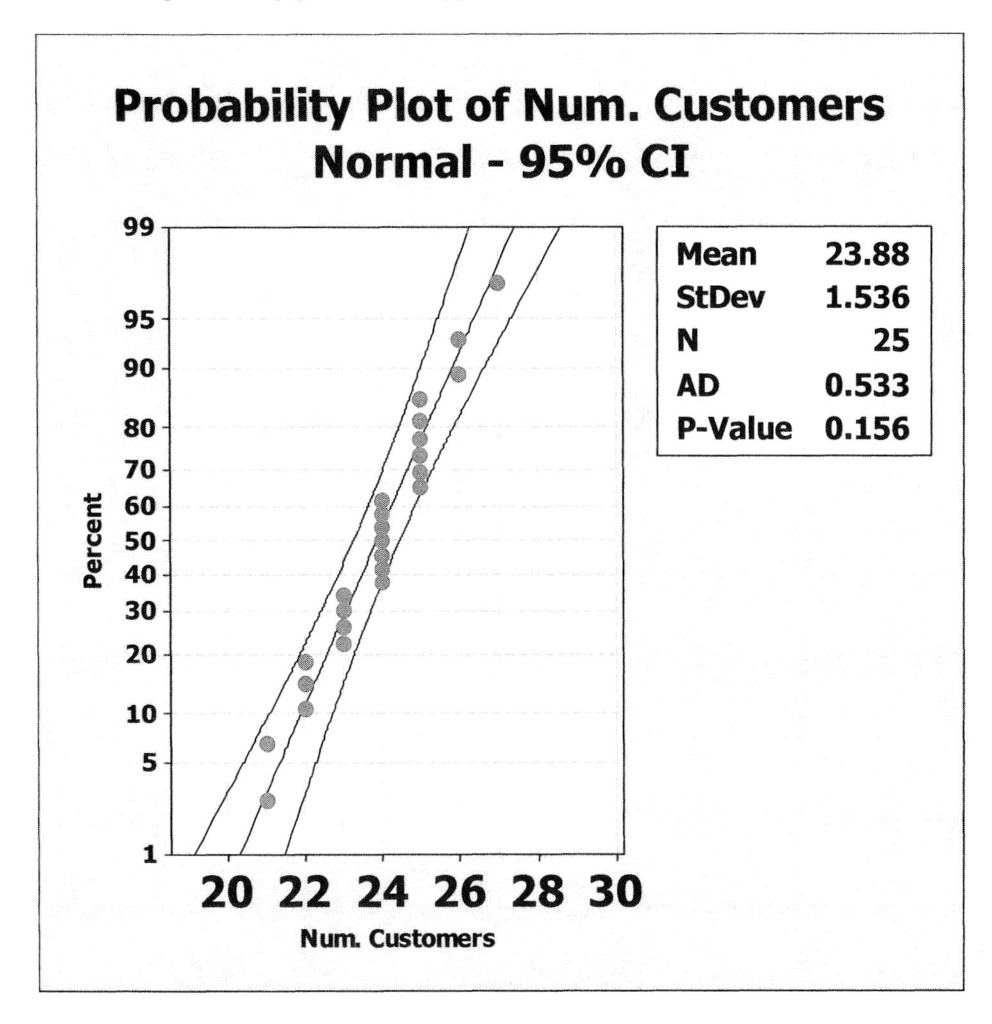

From this plot, you would conclude that the data is approximately normally distributed.

◀

Sampling Distributions

Section 8.1

Example 2 (pg. 420) Sampling Distribution of the Sample Mean

The heights of 3-year-old girls are normally distributed with μ=38.72 and σ=3.17. Approximate the sampling distribution of $\bar{x}$ by taking 100 simple random samples of size n = 5. To do this in MINITAB, click on **Calc → Random Data → Normal. Generate** 100 **rows of data** and **Store in columns** C1-C5. Enter the 38.72 for the **Mean** and 3.17 for the **Standard deviation.**

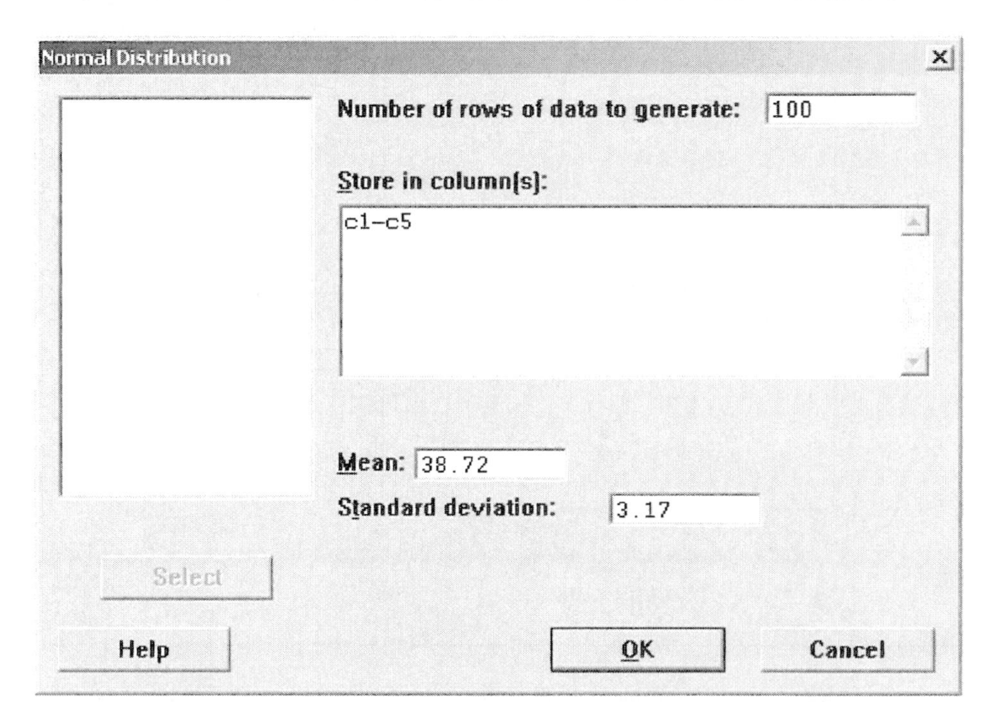

Click on **OK**. There will be 100 rows and 5 columns of random data in the Minitab worksheet. Each row represents a sample of size n=5. Since this is random data, everyone's data will be different.

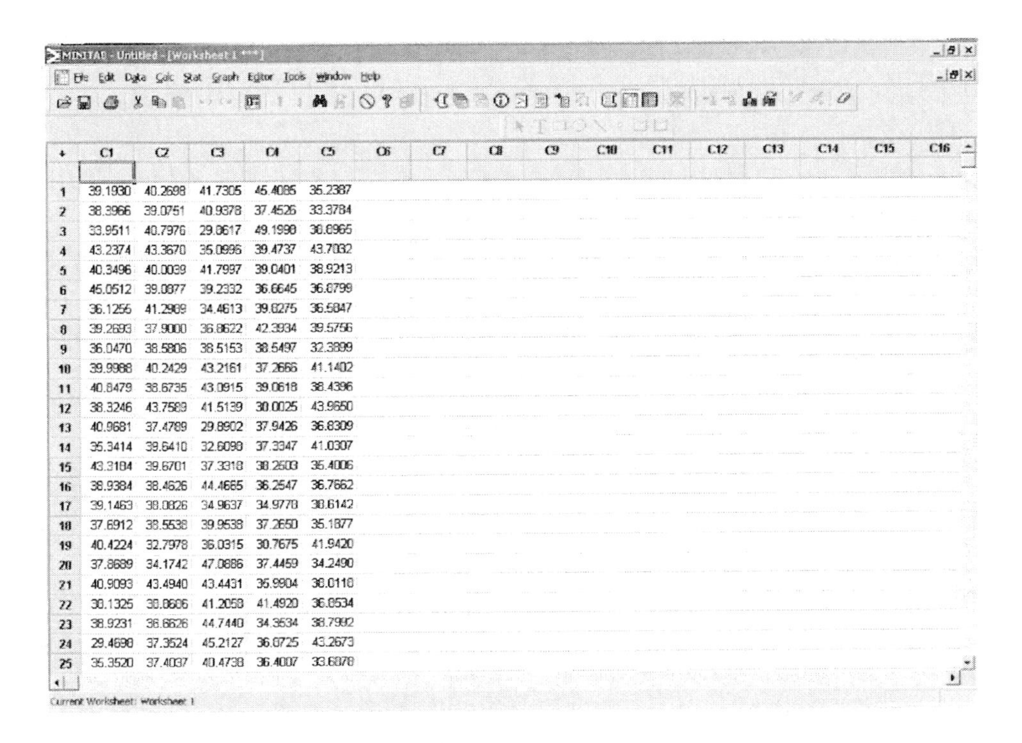

↓	C1	C2	C3	C4	C5
1	39.1930	40.2698	41.7305	45.4085	35.2387
2	38.3966	39.0751	40.9378	37.4526	33.3784
3	33.9511	40.7976	29.8617	49.1998	38.8965
4	43.2374	43.3670	35.0996	39.4737	43.7032
5	40.3496	40.0039	41.7997	39.0401	38.9213
6	45.0512	39.0877	39.2332	36.6645	36.8799
7	36.1256	41.2989	34.4613	39.8275	36.5847
8	39.2693	37.9000	36.8622	42.3934	39.5756
9	36.0470	38.5806	38.5153	38.5497	32.3899
10	39.9988	40.2429	43.2161	37.2666	41.1402
11	40.8479	38.6735	43.0915	39.0618	38.4396
12	38.3246	43.7589	41.5139	30.0025	43.9650
13	40.9681	37.4789	29.8902	37.9426	36.8309
14	35.3414	39.6410	32.6098	37.3347	41.0307
15	43.3184	39.6701	37.3318	38.2503	35.4006
16	38.9384	38.4626	44.4665	36.2547	36.7662
17	39.1463	38.0826	34.9637	34.9778	38.6142
18	37.6912	38.5538	39.9538	37.2650	35.1877
19	40.4224	32.7978	36.0315	30.7675	41.9420
20	37.8689	34.1742	47.0886	37.4459	34.2490
21	40.9093	43.4940	43.4431	35.9904	38.0118
22	38.1325	38.0606	41.2058	41.4920	36.8534
23	38.9231	36.6626	44.7440	34.3534	38.7992
24	29.4698	37.3524	45.2127	36.0725	43.2673
25	35.3520	37.4037	40.4738	36.4007	33.6878

Next, calculate the mean of each of the samples. Click on **Calc → Row Statistics.** Click on **Mean,** select **Input variables** C1-C5 and **Store result in** C6.

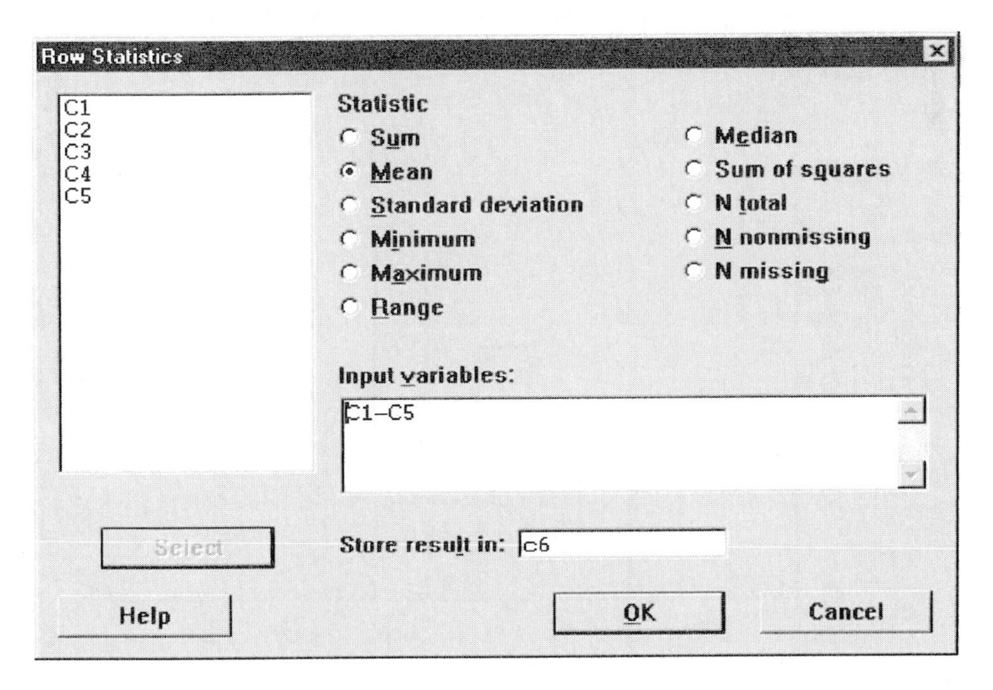

Click on **OK** and C6 will contain the averages for each row of 5 data points.

	C1	C2	C3	C4	C5	C6 Sample Mean
1	39.1930	40.2898	41.7305	45.4085	35.2387	40.3681
2	38.3966	39.0761	40.9378	37.4526	33.3784	37.8481
3	33.9511	40.7976	29.8617	49.1998	38.8965	38.5413
4	43.2374	43.3670	35.0996	39.4737	43.7032	40.9762
5	40.3496	40.0039	41.7997	39.0401	38.9213	40.0229
6	46.0512	39.0877	39.2332	36.6645	36.8799	39.3833
7	36.1255	41.2989	34.4613	39.8275	36.6847	37.6596
8	39.2693	37.9000	36.0622	42.3934	39.5756	39.2001
9	36.0470	38.5806	38.5153	38.6497	32.3809	36.8166
10	39.9988	40.2429	43.2161	37.2666	41.1402	40.3729
11	40.8479	38.6735	43.0915	39.0618	38.4396	40.0229
12	38.3246	43.7589	41.5139	30.0025	43.9650	39.5130
13	40.9681	37.4789	29.8902	37.9426	36.8309	36.6221
14	35.3414	39.6410	32.6098	37.3347	41.0307	37.1915
15	43.3184	39.6701	37.3318	38.2503	35.4006	38.7942
16	38.9384	38.4626	44.4665	36.2547	36.7662	38.9777
17	39.1463	38.0826	34.9637	34.9778	38.6142	37.1569
18	37.8912	38.5538	39.9638	37.2650	35.1877	37.7303
19	40.4224	32.7978	36.0315	30.7675	41.9420	36.3923
20	37.8689	34.1742	47.0886	37.4459	34.2490	38.1653
21	40.9093	43.4940	43.4431	35.9904	38.0118	40.3697
22	38.1325	38.8686	41.2058	41.4920	36.8534	39.3105
23	38.9231	36.6626	44.7440	34.3534	38.7992	38.6965
24	29.4698	37.3524	45.2127	36.8725	43.2673	38.4350
25	35.3520	37.4037	40.4738	36.4007	33.6878	36.6636

To draw a histogram of the sample means, click on **Stat** → **Basic Statistics** → **Display Descriptive Statistics.** Select C6 for the **Variable** and click on **Graphs**. Select **Histogram of Data** and click on **OK** twice to view the histogram.

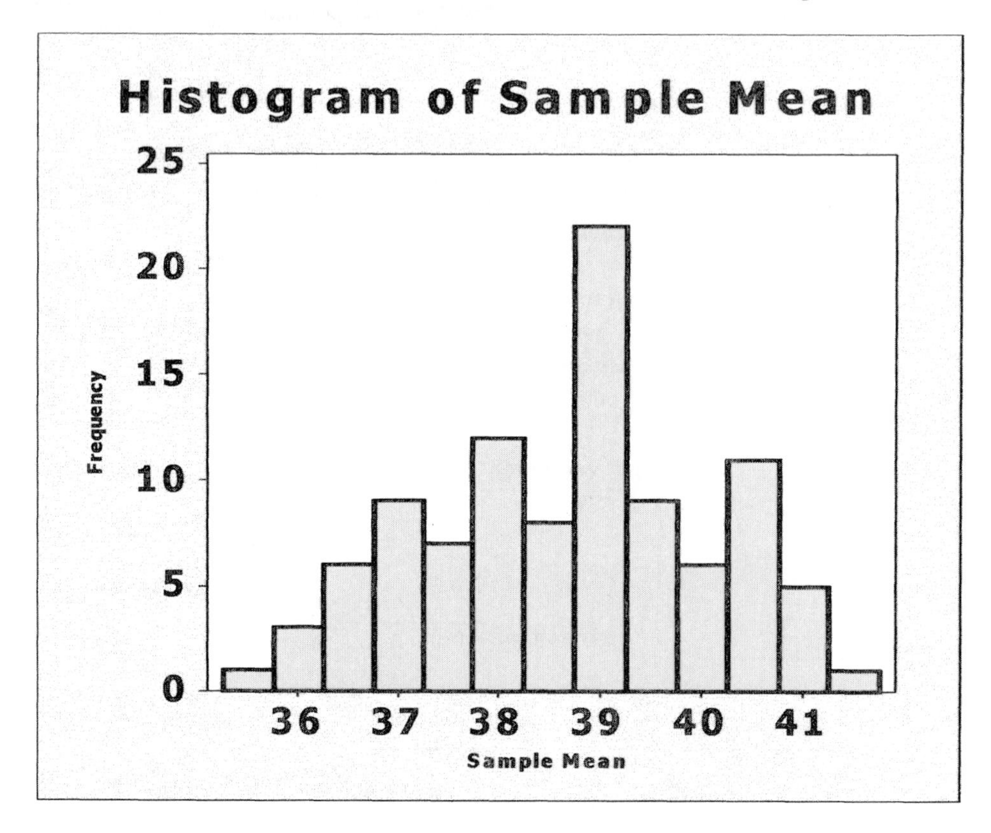

The descriptive statistics of the sample means are in the Session Window and can be seen after the Graph Window is closed.

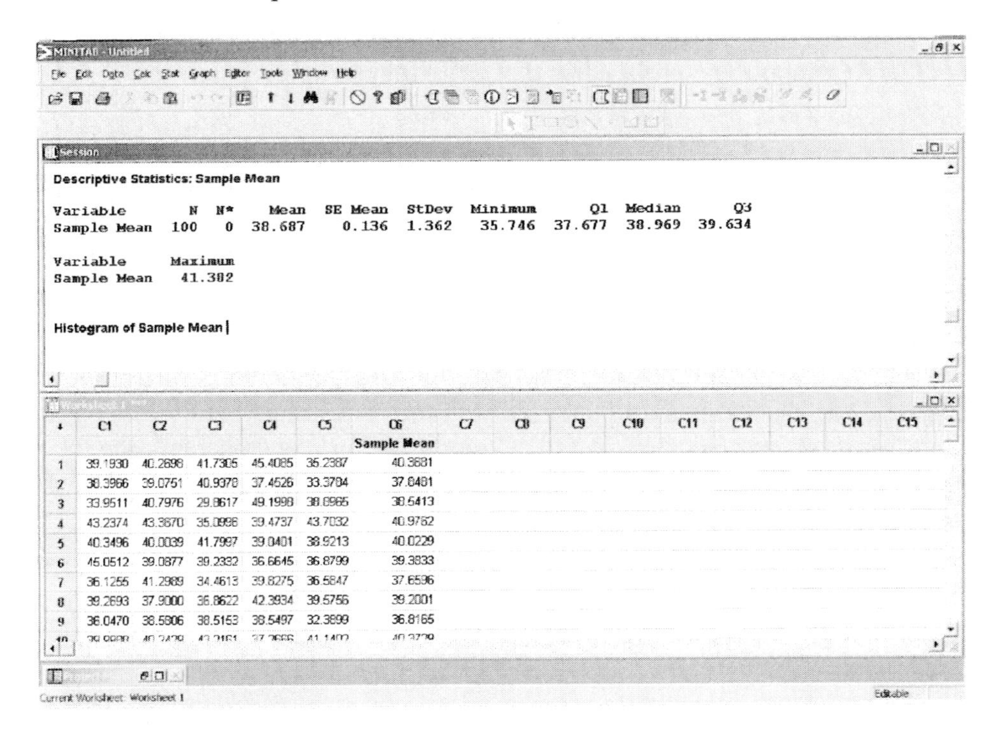

Notice that the mean of the 100 sample means is 38.687 and the standard deviation is 1.362.

▸ Example 4 (pg. 425) Describing the Sampling Distribution

The height of 3 year old girls is normally distributed with $\mu = 38.72$ and $\sigma = 3.17$. Compute the probability that a random sample of size n=10 results in a sample mean greater than 40 inches.

To find the probability that the *mean* height of 10 girls is more than 40 inches, you will need to calculate the standard deviation of $\bar{x}$ which is equal to 3.17 $/\sqrt{10} = 1.00$. (Use a hand calculator for this calculation.)

Now let MINITAB do the rest for you. Click on **Calc → Probability Distributions → Normal.** On the input screen, select **Cumulative probability.** Enter 38.72 for the **Mean** and 1.00 for the **Standard deviation.** Next select **Input Constant** and enter the value 40. Click on **OK** and the probability should appear in the Session Window.

Cumulative Distribution Function

Normal with mean = 38.72 and standard deviation = 1

```
 x  P( X <= x )
40    0.899727
```

Since you want to know the probability that the mean height is greater than 40, you should subtract this probability from 1. So 1 - .899727 = .100273.

◂

Example 5 (pg. 426) Sampling from a non-Normal Population

This time the population is the Exponential Distribution with mean and standard deviation equal to 10. Approximate the sampling distribution of $\bar{x}$ by taking 300 simple random samples of size (a) n = 3, (b) n=12, and (c) n=30.

To do this in MINITAB, you will repeat the following steps three times, once for each value of n. Click on **Calc → Random Data → Exponential. Generate** 300 **rows of data** and **Store in columns** C1-C3. Enter 10 for the **Scale**.

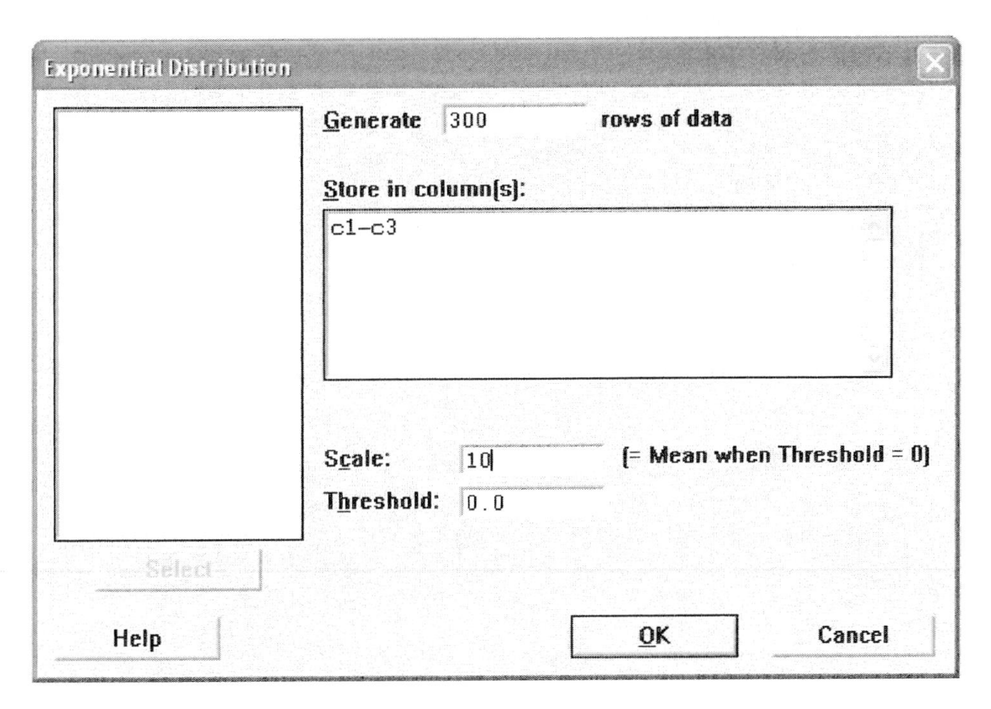

Click on **OK**. There will be 300 rows and 3 columns of random data in the Minitab worksheet. Each row represents a sample of size n=3. Since this is random data, everyone's data will be different. Next, calculate the mean of each of the samples. Click on **Calc → Row Statistics.** Click on **Mean,** select **Input variables** C1-C3 and **Store result in** C4. Click on **OK** and C4 will contain the averages for each row of 3 data points.

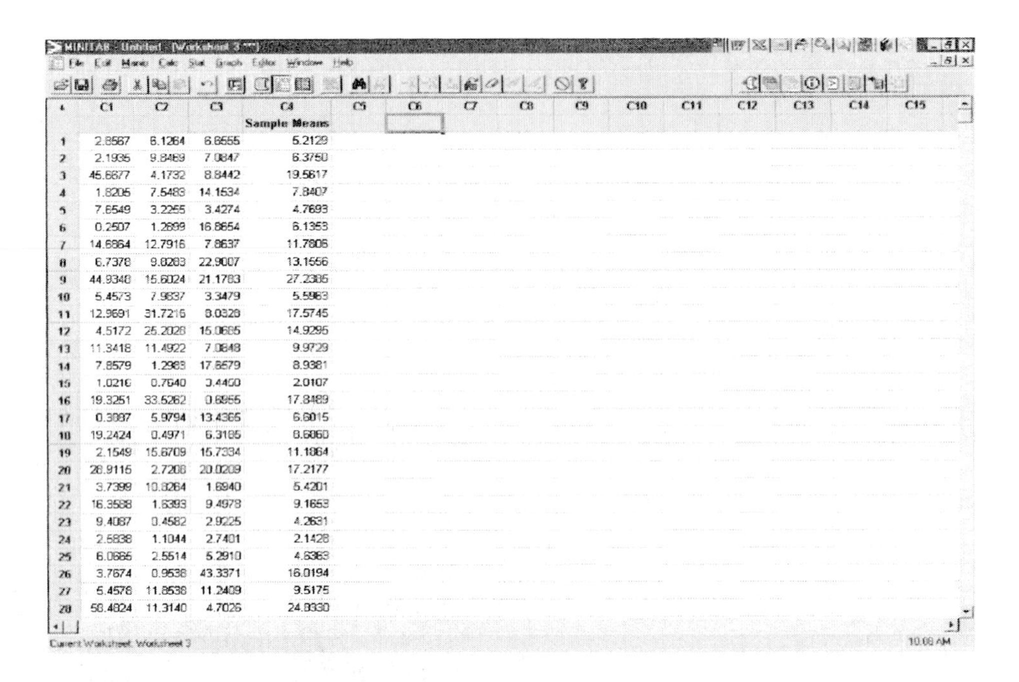

↓	C1	C2	C3	C4 Sample Means
1	2.8567	6.1264	6.6555	5.2129
2	2.1935	9.8469	7.0847	6.3750
3	45.6677	4.1732	8.8442	19.5617
4	1.8205	7.5483	14.1534	7.8407
5	7.6549	3.2255	3.4274	4.7693
6	0.2507	1.2899	16.8654	6.1353
7	14.6864	12.7916	7.8637	11.7806
8	6.7378	9.8283	22.9007	13.1556
9	44.9348	15.6024	21.1783	27.2385
10	5.4573	7.9837	3.3479	5.5963
11	12.9691	31.7216	8.0328	17.5745
12	4.5172	25.2028	15.0685	14.9295
13	11.3418	11.4922	7.0848	9.9729
14	7.8579	1.2983	17.6579	8.9381
15	1.0216	0.7640	3.4450	2.0107
16	19.3251	33.5262	0.6956	17.8489
17	0.3887	5.9794	13.4365	6.6015
18	19.2424	0.4971	6.3185	8.6860
19	2.1549	15.6709	15.7334	11.1864
20	28.9116	2.7208	20.0209	17.2177
21	3.7399	10.8264	1.6940	5.4201
22	16.3588	1.6393	9.4978	9.1653
23	9.4087	0.4582	2.9225	4.2631
24	2.5838	1.1044	2.7401	2.1428
25	6.0665	2.5514	5.2910	4.6363
26	3.7674	0.9538	43.3371	16.0194
27	5.4578	11.8538	11.2409	9.5175
28	58.4824	11.3140	4.7026	24.8330

To draw a histogram of the sample means, click on **Stat → Basic Statistics → Display Descriptive Statistics.** Select C4 for the **Variable** and click on **Graphs**. Select **Histogram of Data** and click on **OK** twice to view the histogram.

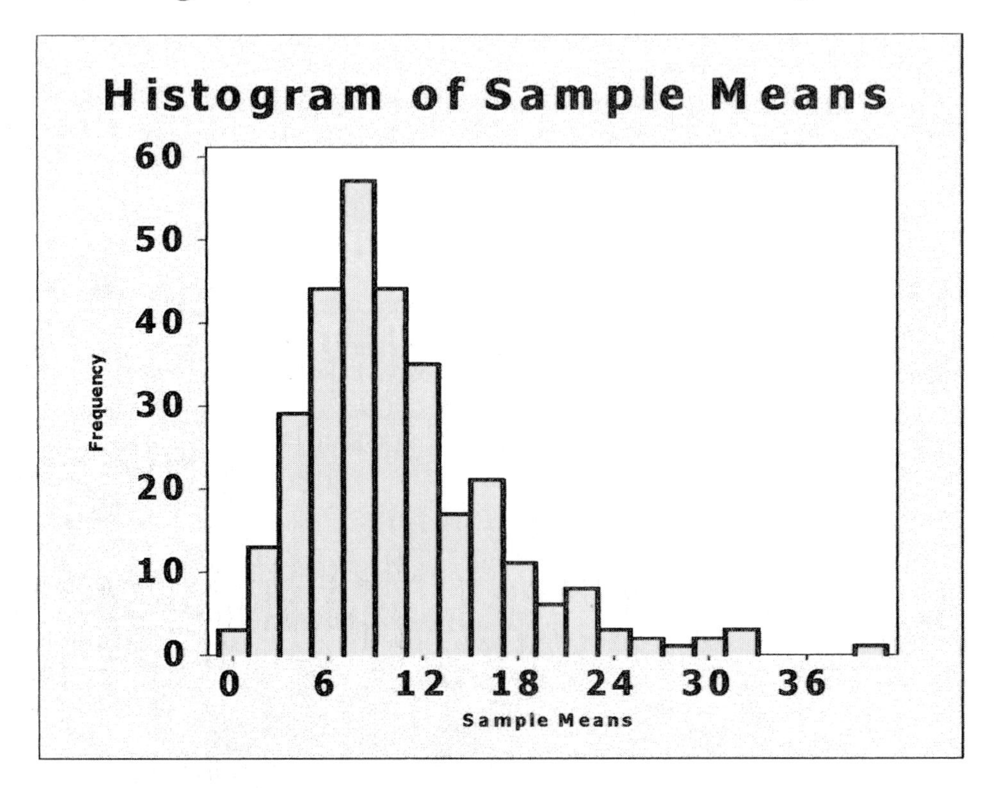

Notice that the histogram is still very skewed, just like the original population. The descriptive statistics of the sample means are in the Session Window and can be seen after the Graph Window is closed.

Descriptive Statistics: Sample Means

Variable	N	Mean	StDev
Sample Means	300	10.443	6.042

Notice that the mean of the 300 sample means is 10.443 and the standard deviation is 6.042.

Now repeat this for n=12. This time when you generate the random samples, you should **Store in columns** C1-C12.

Next, calculate the mean of each of the samples. Click on **Calc → Row Statistics.** Click on **Mean,** select **Input variables** C1-C12 and **Store result in** C13. Click on **OK** and C13 will contain the averages for each row of 12 data points.

To draw a histogram of the sample means, click on **Stat → Basic Statistics → Display Descriptive Statistics.** Select C13 for the **Variable** and click on **Graphs**. Select **Histogram of Data** and click on **OK** twice to view the histogram.

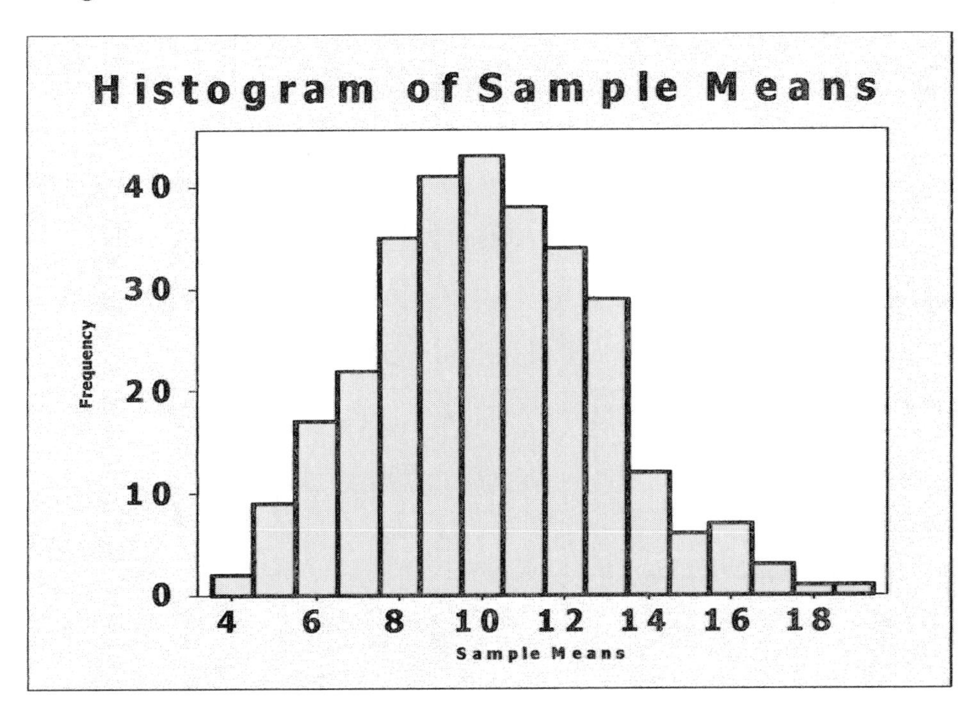

Descriptive Statistics: Sample Means

Variable	N	Mean	StDev
Sample Means	300	10.169	2.734

Notice that the histogram is not as skewed as before and the mean is 10.169. This time the standard deviation is 2.734 -- much smaller than the last time.

Now repeat this for n=30. This time when you generate the random samples, you should **Store in columns** C1-C30.
Next, calculate the mean of each of the samples. Click on **Calc → Row Statistics.** Click on **Mean,** select **Input variables** C1-C30 and **Store result in** C31. Click on **OK** and C31 will contain the averages for each row of 30 data points.
To draw a histogram of the sample means, click on **Stat → Basic Statistics → Display Descriptive Statistics.** Select C31 for the **Variable** and click on **Graphs**. Select **Histogram of Data** and click on **OK** twice to view the histogram.

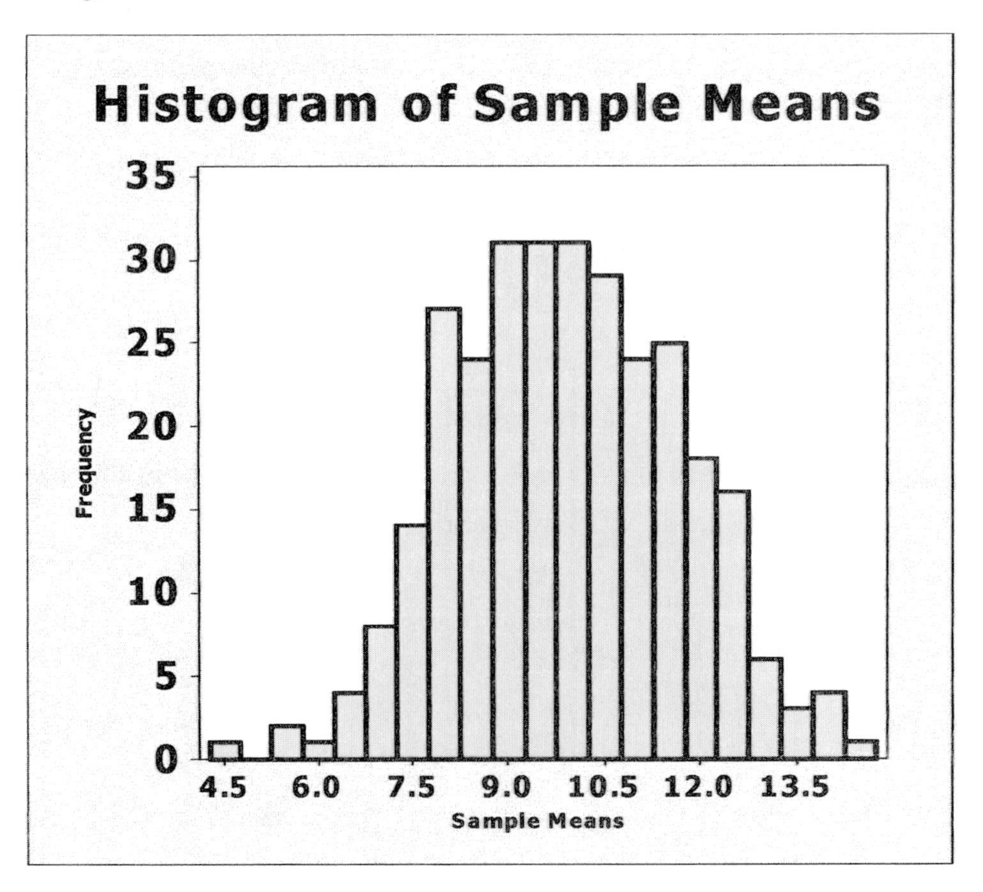

Descriptive Statistics: Sample Means

Variable	N	Mean	StDev
Sample Means	300	9.899	1.756

For n=30, the histogram has become fairly symmetric. The mean is 9.899 and notice that the standard deviation is now a very small 1.756.

◀

Example 6 (pg. 429) Applying the Central Limit Theorem

The mean calorie intake of males 20-39 years old is μ=2716 with σ=72.8. Compute the probability that a random sample of size n=35 results in a sample mean greater than 2750.

To find the probability that the *mean* calorie intake is more than 2750, you will need to calculate the standard deviation of $\bar{x}$, which is equal to $72.8 / \sqrt{35}$ = 12.3. (Use a hand calculator for this calculation.)

Now let MINITAB do the rest for you. Click on **Calc → Probability Distributions → Normal.** On the input screen, select **Cumulative probability.** Enter 2716 for the **Mean** and 12.3 for the **Standard deviation.** Next select **Input Constant** and enter the value 2750. Click on **OK** and the probability should appear in the Session Window.

Cumulative Distribution Function

Normal with mean = 2716 and standard deviation = 12.3

```
   x  P( X <= x )
2750     0.997147
```

Since you want to know the probability that the mean calorie intake is greater than 2750, you should subtract this probability from 1. So 1 - .997147 = .002853.

▸ Problem 20 (pg. 432) Serum Cholesterol

HDL cholesterol of females 20-29 years old is normally distributed with μ=53 and σ=13.4. For parts a – e of this probem, click on **Calc → Probability Distributions → Normal.** On the input screen, select **Cumulative probability.** Enter 53 for the **Mean.**

(a) Enter 13.4 for the **Standard deviation.** Next select **Input Constant** and enter the value 60. Click on **OK.** To find the probability that HDL is above 60, subtract the probability from 1. (1 - .6993 = .3007)

(b) Since you have a sample of n=15, use a hand calculator to calculate the standard deviation, $13.4/\sqrt{15} = 3.46$. Enter 3.46 for the **Standard deviation.** Next select **Input Constant** and enter the value 60. Click on **OK.** To find the probability that HDL is above 60, subtract the probability from 1. (1 - .97847 = .02153)

(c) Since you have a sample of n=20, use a hand calculator to calculate the standard deviation, $13.4/\sqrt{20} = 3.00$. Enter 3.00 for the **Standard deviation.** Next select **Input Constant** and enter the value 60. Click on **OK.** To find the probability that HDL is above 60, subtract the probability from 1. (1 - .990185 = .009815)

Cumulative Distribution Function

Normal with mean = 53 and standard deviation = 13.4

x	P(X <= x)
60	0.699300

Cumulative Distribution Function

Normal with mean = 53 and standard deviation = 3.46

x	P(X <= x)
60	0.978470

Cumulative Distribution Function

Normal with mean = 53 and standard deviation = 3

x	P(X <= x)
60	0.990185

◂

Problem 33 (pg. 433) Simulation

Scores on the Stanford-Binet IQ test are normally distributed with mean 100 and standard deviation 16.

Parts (a), (b), (c), and (e): Approximate the sampling distribution of $\bar{x}$ by taking 500 simple random samples of size n=20. Click on **Calc → Random Data → Normal. Generate** 500 **rows of data** and **Store in columns** C1-C20. Enter 100 for the **Mean** and 16 for the **Standard deviation**. Click on **OK**. There will be 500 rows and 20 columns of random data in the Minitab worksheet. Each row represents a sample of size n=20. Since this is random data, everyone's data will be different. Next, calculate the mean of each of the samples. Click on **Calc → Row Statistics.** Click on **Mean,** select **Input variables** C1-C20 and **Store result in** C21. Click on **OK** and C21 will contain the averages for each row of 20 data points. To draw a histogram of the sample means, click on **Stat → Basic Statistics → Display Descriptive Statistics.** Select C21 for the **Variable** and click on **Graphs**. Select **Histogram of Data** and click on **OK** twice to view the histogram.

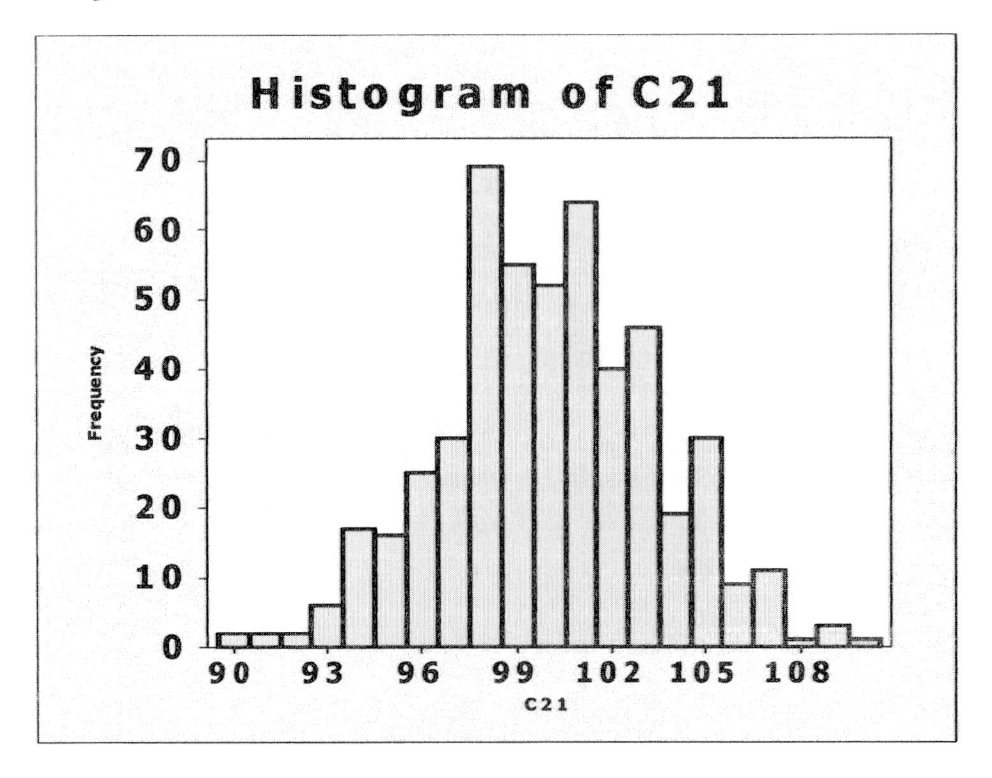

Notice that the histogram is normally distributed. The descriptive statistics of the sample means are in the Session Window and can be seen after the Graph Window is closed.

Descriptive Statistics: C21

Variable	N	Mean	StDev
C21	500	100.07	3.40

Notice that the mean of the 500 sample means is 100.07 and the standard deviation is 3.40. (Notice how close these are to the theoretical mean (100) and standard deviation ($16/\sqrt{20} = 3.58$) of the sampling distribution.)

Part (f): Click on **Calc → Probability Distributions → Normal.** On the input screen, select **Cumulative probability.** Enter 100 for the **Mean.** Since you have a sample of n=20, use a hand calculator to calculate the standard deviation, $16/\sqrt{20} = 3.58$. Enter 3.58 for the **Standard deviation.** Next select **Input Constant** and enter the value 108. Click on **OK.** To find the probability that the mean IQ is above 108, subtract the probability from 1. (1 - .987279 = .012721)

Part (g): To find the percent of the 500 random samples that had a sample mean IQ greater than 108, click on **Data → Sort.** For **Sort Column(s)** select C21and for **By column** also select C21. Click on the **Descending** option. For **Store sorted data in,** click on **Column(s) of current worksheet** and enter **C21.**

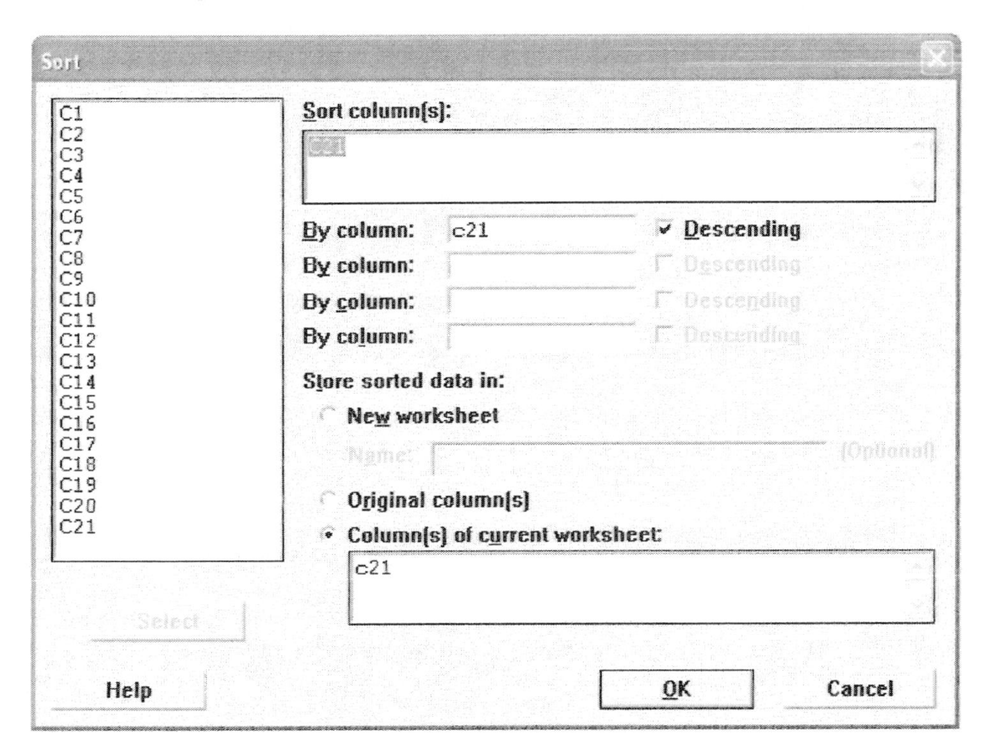

Click on **OK.** The data in C21 is now sorted in descending order. Count the number of data points that are greater than 108. In this example, there are 4. So the percent is 4/500 = .008. Notice that this is just a little smaller than the probability that was calculated in part (f) of this problem.

◀

Section 8.2

Example 2 (pg. 435) Sampling Distribution of the Sample Proportion

According to the CDC, 17% of Americans have high cholesterol. Approximate the sampling distribution of p by taking 100 simple random samples of size n = 10. To do this in MINITAB, click on **Calc → Random Data → Bernoulli. Generate** 100 **rows of data** and **Store in columns** C1-C10. Enter the 1.17 for the **Probability of success.**

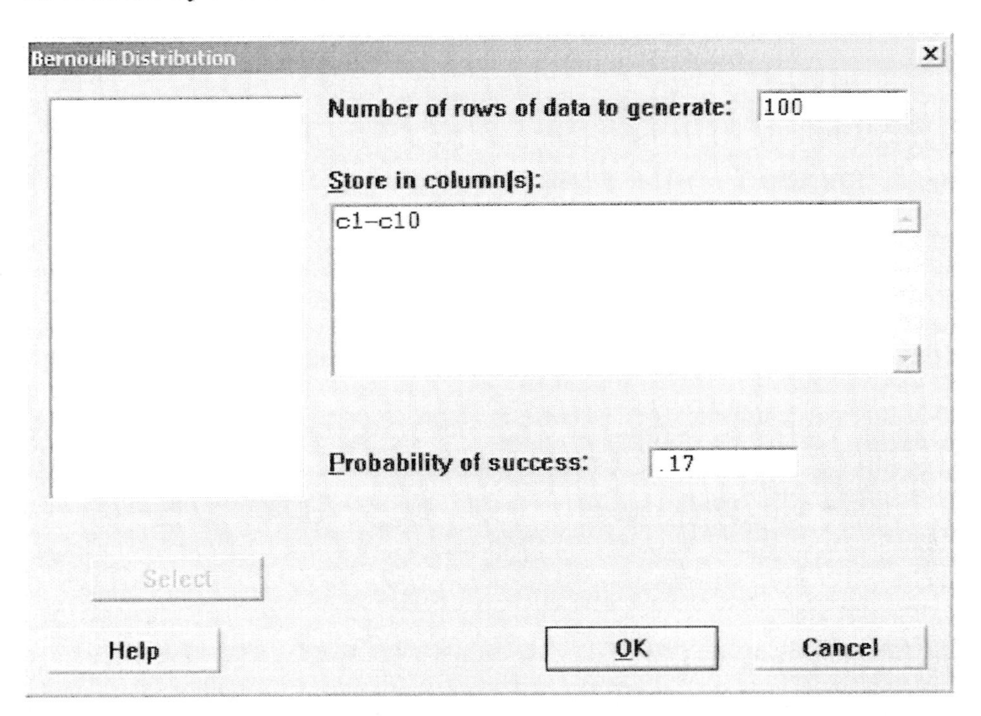

Click on **OK**. There will be 100 rows and 10 columns of random data in the Minitab worksheet. Each row represents a sample of size n=10. Since this is random data, everyone's data will be different. Next, calculate the proportion of successes in each of the samples. Click on **Calc → Row Statistics.** Click on **Mean,** select **Input variables** C1-C10 and **Store result in** C11. Finally, to draw a histogram of the sample means, click on **Stat → Basic Statistics → Display Descriptive Statistics.** Select C11 for the **Variable** and click on **Graphs**. Select **Histogram of Data** and click on **OK** twice to view the histogram.

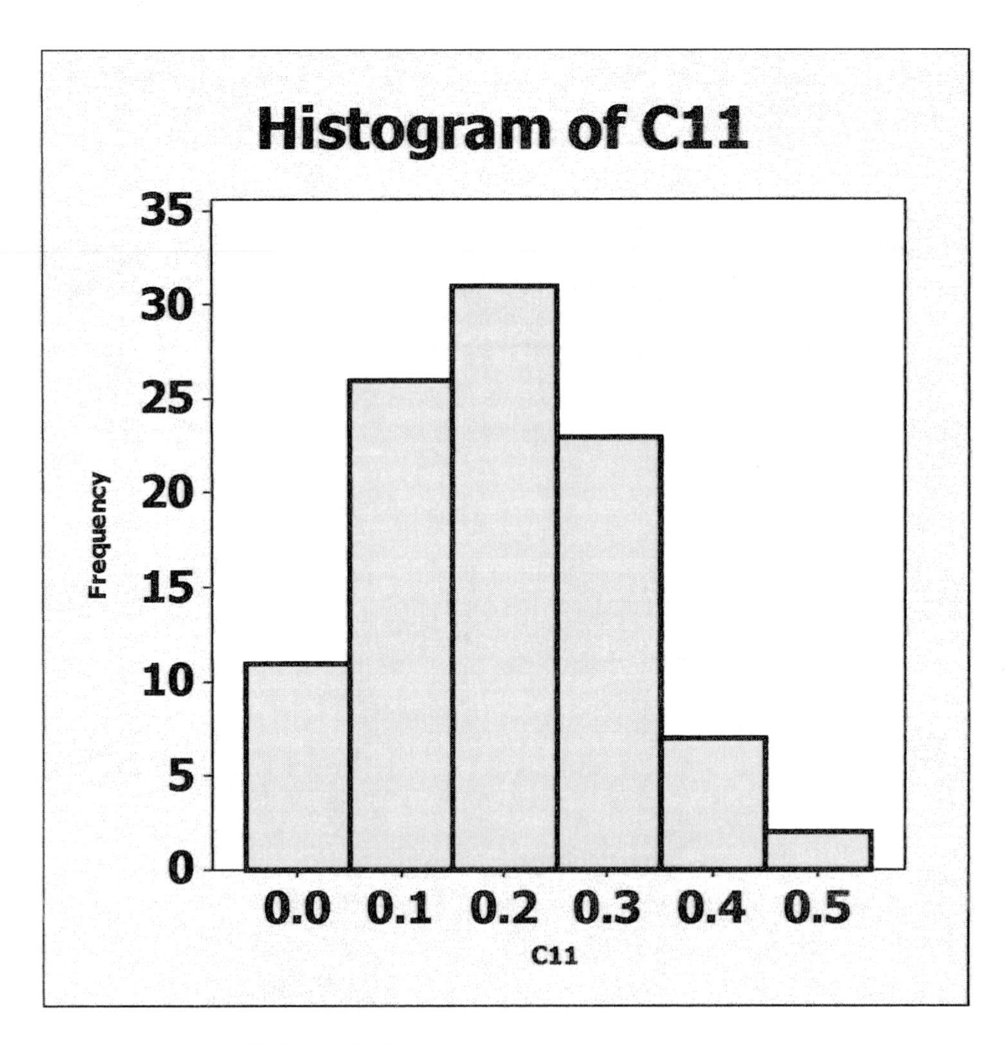

Descriptive Statistics: C11

Variable	N	Mean	StDev
C11	100	0.1950	0.1184

The mean of the sample proportions is 0.195 with a standard deviation of 0.1184.

Repeat the above steps for samples of size n=40 and n=80.

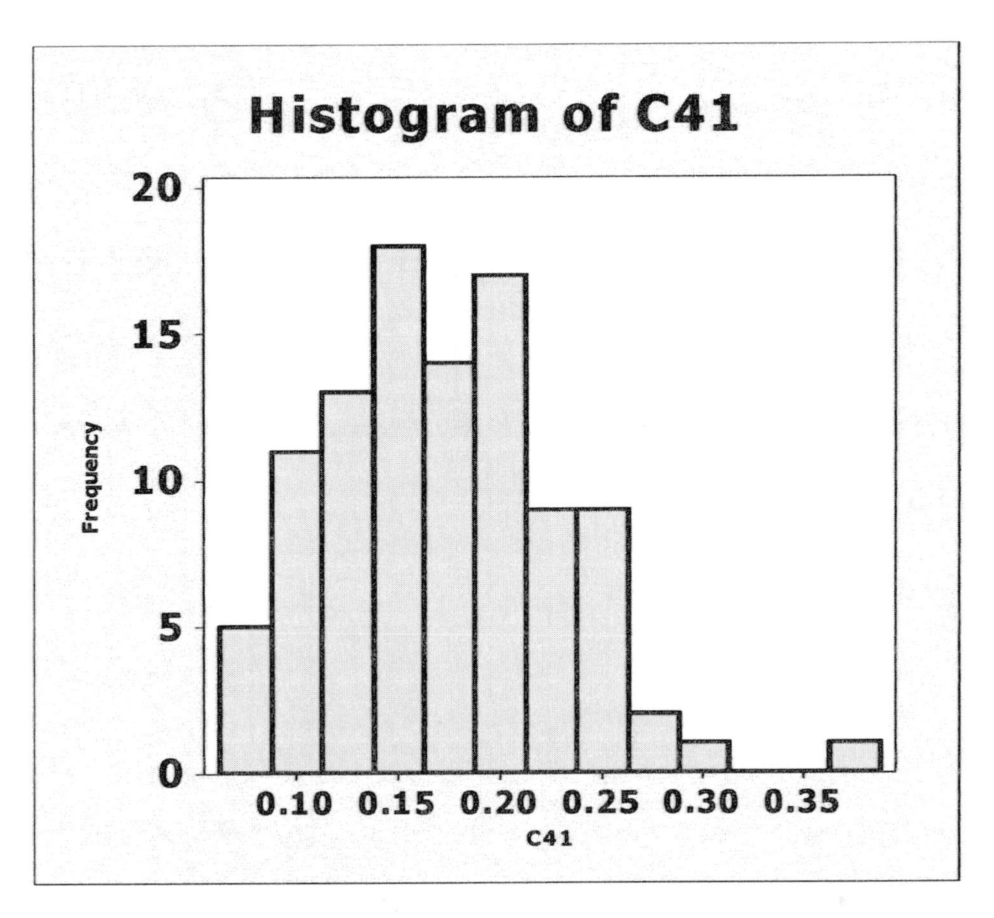

Descriptive Statistics: C41

Variable	N	Mean	StDev
C41	100	0.17150	0.05630

So, with a sample size of 40, the mean of the proportions is 0.1715 with a standard deviation of 0.0563.

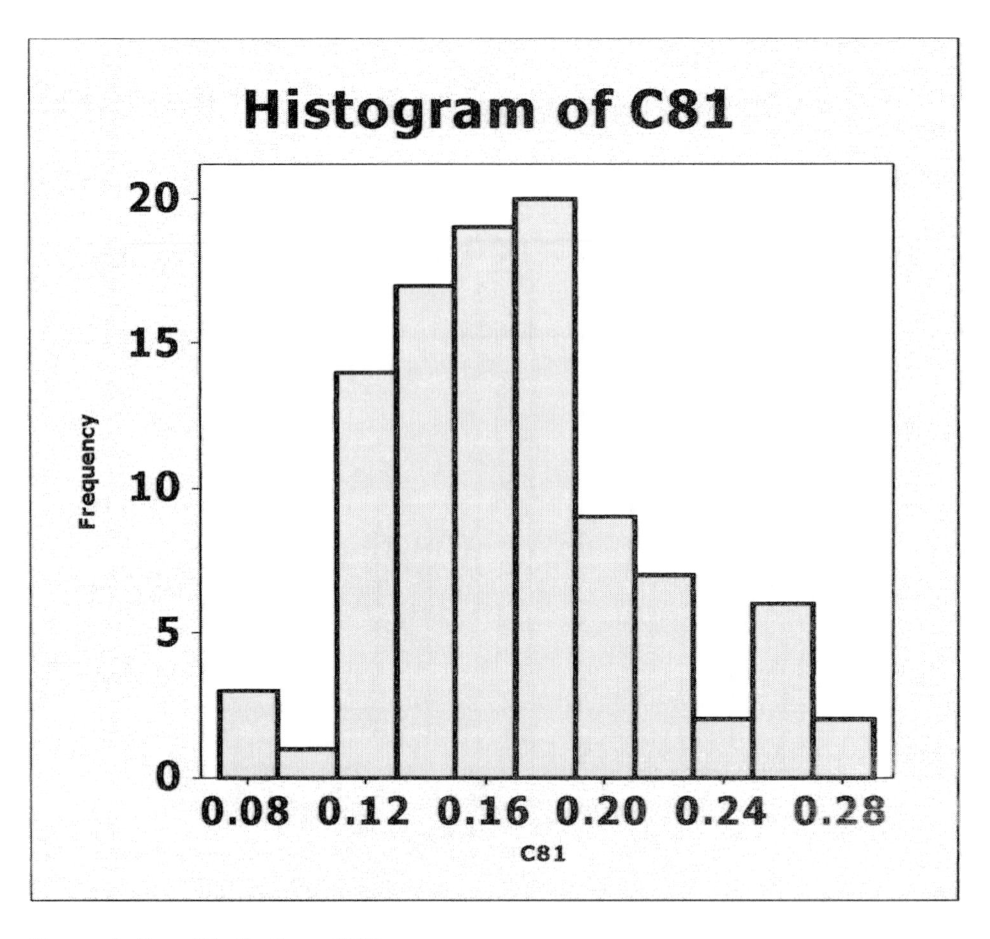

Descriptive Statistics: C81

Variable	N	Mean	StDev
C81	100	0.16800	0.04320

When the sample size is 80, the mean of the proportions is 0.168 with a standard deviation of 0.04320.

▸ Problem 19 (pg. 441) Phishing

43% of adults have received a 'phishing' contact. Suppose a random sample of 800 adults is obtained.

To find the probability that the *no more* than 40% have received a 'phishing' contact, you will need to calculate the mean and standard deviation of the sample proportion. The mean is 0.43 and the standard deviation is $\sqrt{(.43)(.57)/800} =$ 0.0175. (Use a hand calculator for this calculation.) The distribution is approximately normal.

Now let MINITAB do the rest for you. Click on **Calc → Probability Distributions → Normal.** On the input screen, select **Cumulative probability.** Enter 0.43 for the **Mean** and 0.0175 for the **Standard deviation.** Next select **Input Constant** and enter the value 0.40. Click on **OK** and the probability should appear in the Session Window.

Cumulative Distribution Function

Normal with mean = 0.43 and standard deviation = 0.0175

```
  x  P( X <= x )
0.4   0.0432381
```

To find the probability that 45% or more of the 800 adults received a phishing contact, repeat the above steps using 0.45 as the **Input Constant.**

Cumulative Distribution Function

Normal with mean = 0.43 and standard deviation = 0.0175

```
   x  P( X <= x )
0.45    0.873451
```

Since you want to know the probability that 45% or more were contacted, you should subtract this probability from 1. So 1 - .873451 = .126549.

◀

CHAPTER 9

Confidence Intervals

Section 9.1

▸ Example 2 (pg. 450) Constructing 95% Confidence Intervals Based on 20 Samples

Scores on the Stanford-Binet IQ test are normally distributed with mean 100 and standard deviation 16. Thus, you have a normally distributed population with a known mean and known standard deviation.

Simulate 20 simple random samples of size n=15. Click on **Calc → Random Data → Normal. Generate** 15 **rows of data** and **Store in columns** C1-C20. Enter 100 for the **Mean** and 16 for the **Standard deviation**. Click on **OK**. There will be 15 rows and 20 columns of random data in the Minitab worksheet. Each column represents a sample of size n=15 from the normal population with $\mu=100$ and $\sigma=16$. Since this is random data, everyone's data will be different. In this example, we already know that $\mu=100$, so we can actually see how well the confidence intervals do at estimating μ.

To construct the confidence intervals, click on **Stat → Basic Statistics → 1-Sample Z.** Select **Samples in Columns** and enter C1-C20. Enter 16 for the **Standard deviation.**

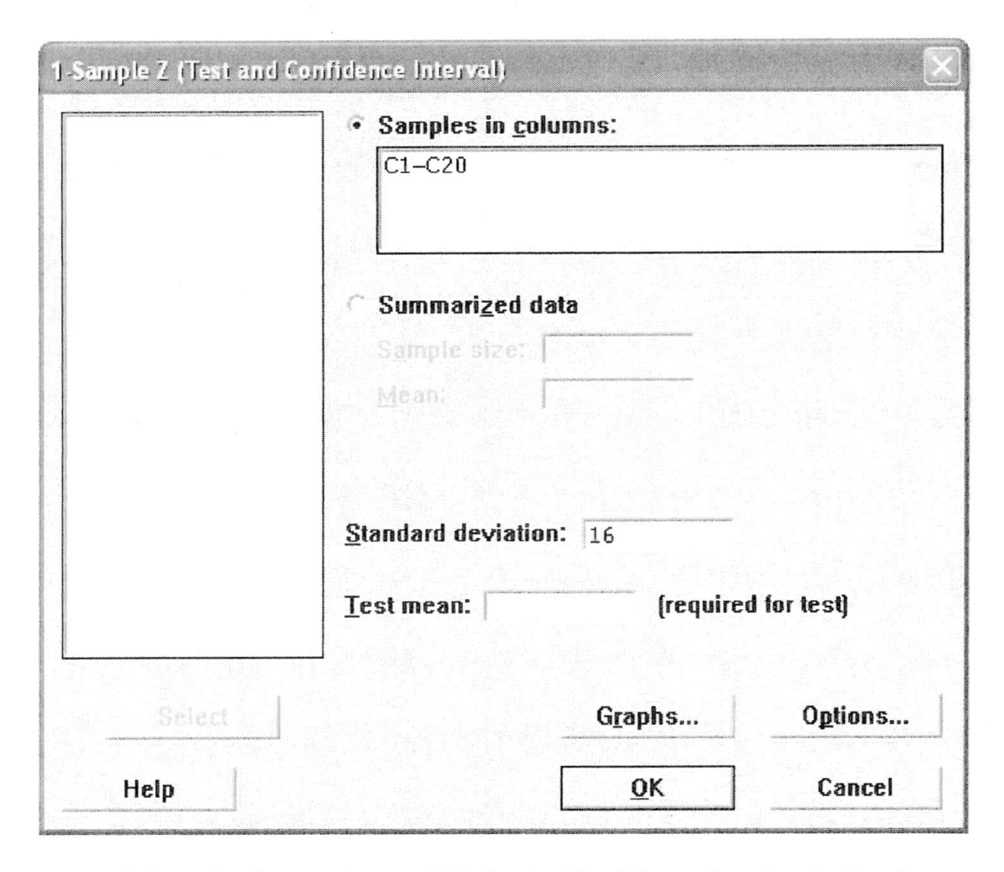

Next, click on **Options** and enter 95.0 for the **Confidence Level** and select 'not equal' for the **Alternative**.

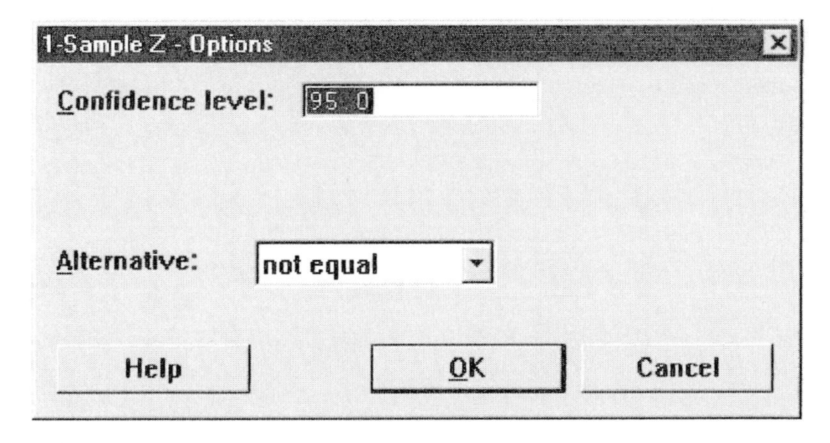

Click on **OK** twice and the results will be displayed in the Session Window.

One-Sample Z: C1, C2, C3, C4, C5, C6, C7, C8, ...

The assumed standard deviation = 16

Variable	N	Mean	StDev	SE Mean	95% CI
C1	15	103.886	9.427	4.131	(95.789, 111.983)
C2	15	101.887	13.680	4.131	(93.790, 109.984)
C3	15	94.2707	15.0754	4.1312	(86.1737, 102.3677)
C4	15	98.9203	19.0754	4.1312	(90.8233, 107.0172)
C5	15	103.282	14.020	4.131	(95.185, 111.379)
C6	15	96.0203	11.3153	4.1312	(87.9234, 104.1173)
C7	15	103.567	14.254	4.131	(95.470, 111.664)
C8	15	94.4526	17.6534	4.1312	(86.3556, 102.5495)
C9	15	96.6950	19.4770	4.1312	(88.5981, 104.7920)
C10	15	99.2813	10.4778	4.1312	(91.1843, 107.3783)
C11	15	98.9952	11.9066	4.1312	(90.8982, 107.0921)
C12	15	111.709	12.928	4.131	(103.612, 119.806)
C13	15	101.104	13.663	4.131	(93.007, 109.201)
C14	15	96.3307	13.2085	4.1312	(88.2338, 104.4277)
C15	15	101.958	11.956	4.131	(93.861, 110.055)
C16	15	105.277	18.585	4.131	(97.180, 113.374)
C17	15	97.9307	13.9386	4.1312	(89.8338, 106.0277)
C18	15	101.996	13.760	4.131	(93.899, 110.093)
C19	15	98.9821	21.0398	4.1312	(90.8852, 107.0791)
C20	15	94.0331	10.7566	4.1312	(85.9362, 102.1301)

Examine the confidence intervals and count how many do not contain 100, the true population mean. In these samples (above), only one confidence interval (#12) does not contain 100.

◀

Example 4 (pg. 455) Construct a 90% confidence interval

Enter the data from Table 1 on page 448 into C1.

To construct a 90% confidence interval for the speeds of 12 randomly selected cars, first verify that the data is approximately normal. To draw a normal probability plot, click on **Graph → Probability Plot** and select the **Single** plot. Click on **OK.** Select C1 for the **Graph variable.** Click on **OK** and the probability plot will be displayed.

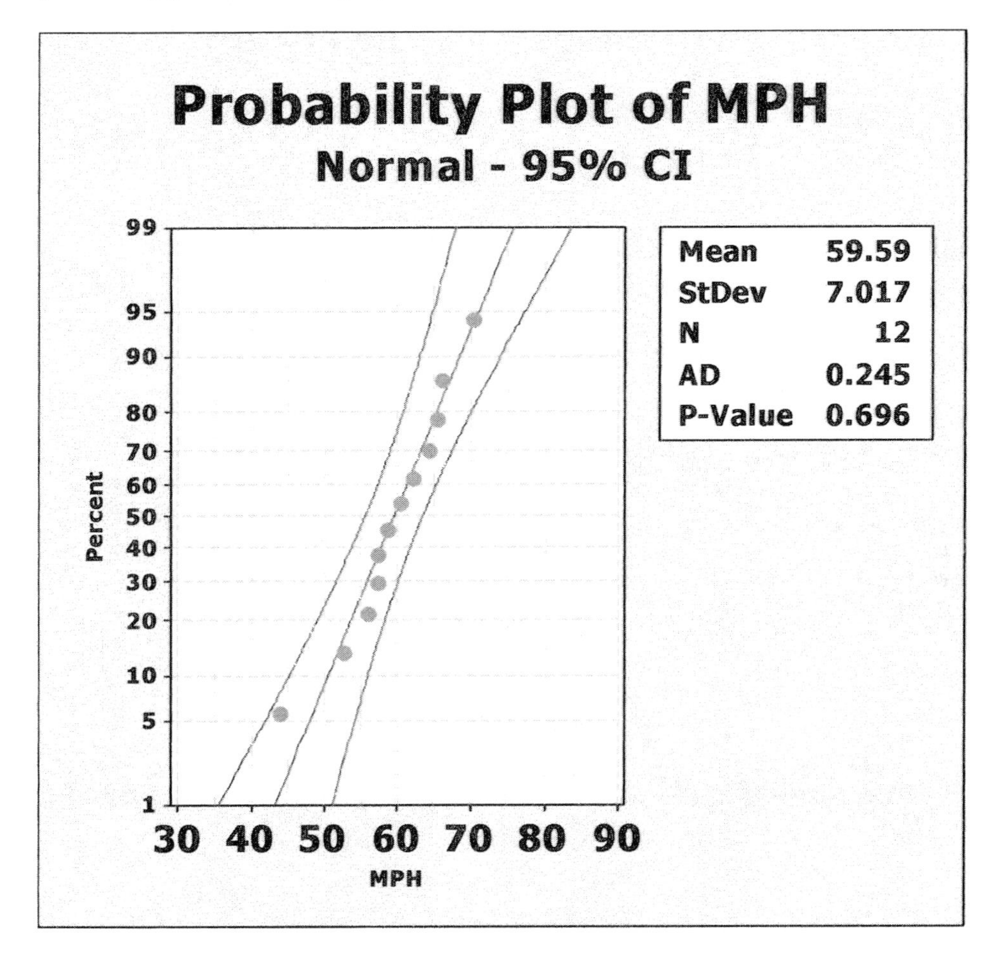

Notice that all data points are contained within the confidence bands of the plot.

Next, check for outliers using a boxplot. Click on **Graph → Boxplot** and select the **Simple** boxplot. Click on **OK**. Select C1 for the **Graph variable.** To view a horizontal boxplot (rather than a vertical one) click on **Scale** and select **Transpose value and category scales.** Click on **OK**.

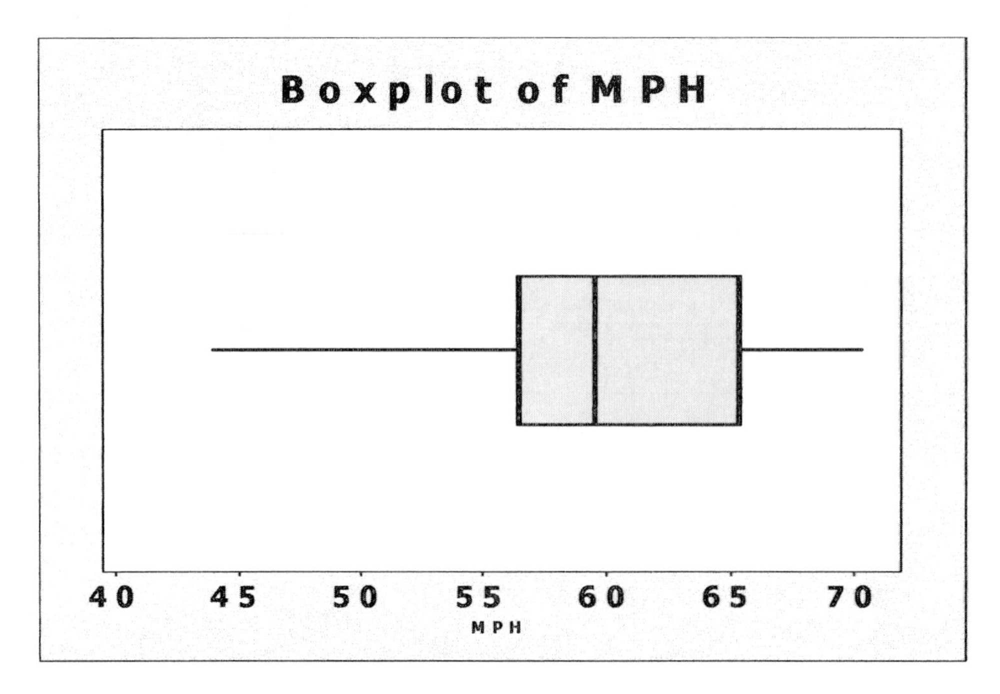

There are no outliers shown in the boxplot, so you may now proceed with the confidence interval.

Click on **Stat** → **Basic Statistics** → **1-Sample Z.** Select **Samples in Columns** and enter C1. Enter 8 for the **Standard deviation.** Next, select **Options** and enter 90.0 for the **Confidence Level** and select 'not equal' for the **Alternative**. Click on **OK** twice and the interval will be displayed in the Session Window.

One-Sample Z: MPH

The assumed standard deviation = 8

Variable	N	Mean	StDev	SE Mean	90% CI
MPH	12	59.5917	7.0171	2.3094	(55.7930, 63.3903)

The 90% confidence interval is (55.79, 63.39).

Problem 25 (pg. 460) Construct 90 and 95% confidence intervals for Repair Costs

Enter the data on cost of repairs into C1. To find the mean of the data and draw a normal probability plot at the same time, click on **Graph → Probability Plot** and select the **Single** plot. Click on **OK.** Select C1 for the **Graph variable.** Click on **OK** and the probability plot will be displayed.

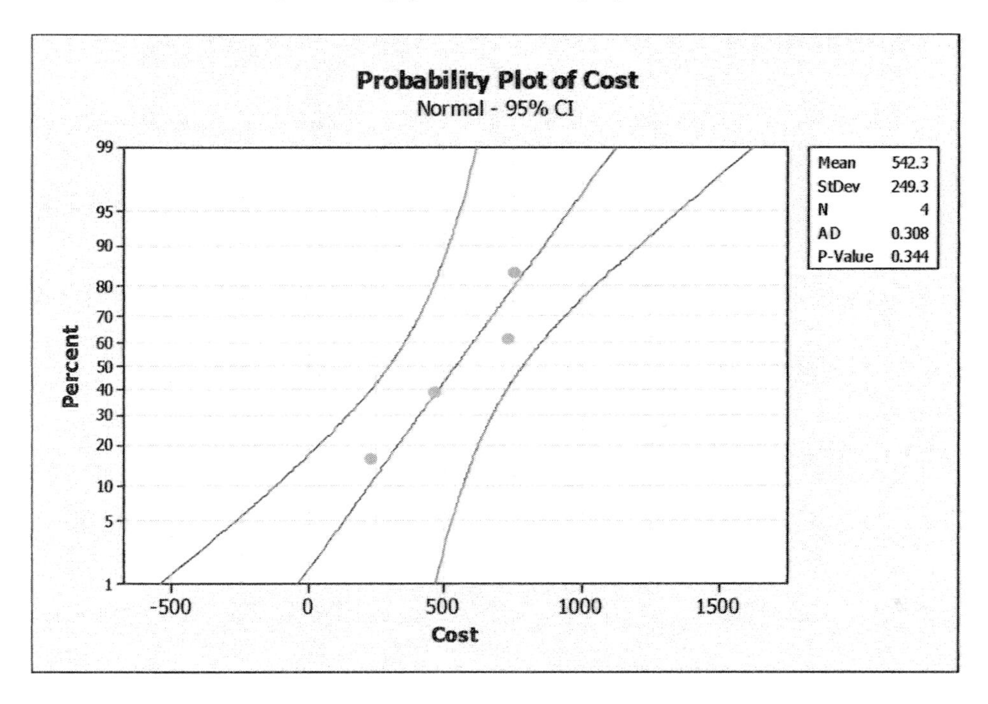

Notice that the mean of the data is listed on the plot, \$542.3. Also, notice that all data points are contained within the confidence bands of the plot.

Next, check for outliers using a boxplot. Click on **Graph → Boxplot** and select the **Simple** boxplot. Click on **OK**. Select C1 for the **Graph variable.** To view a horizontal boxplot (rather than a vertical one) click on **Scale** and select **Transpose value and category scales.** Click on **OK** twice.

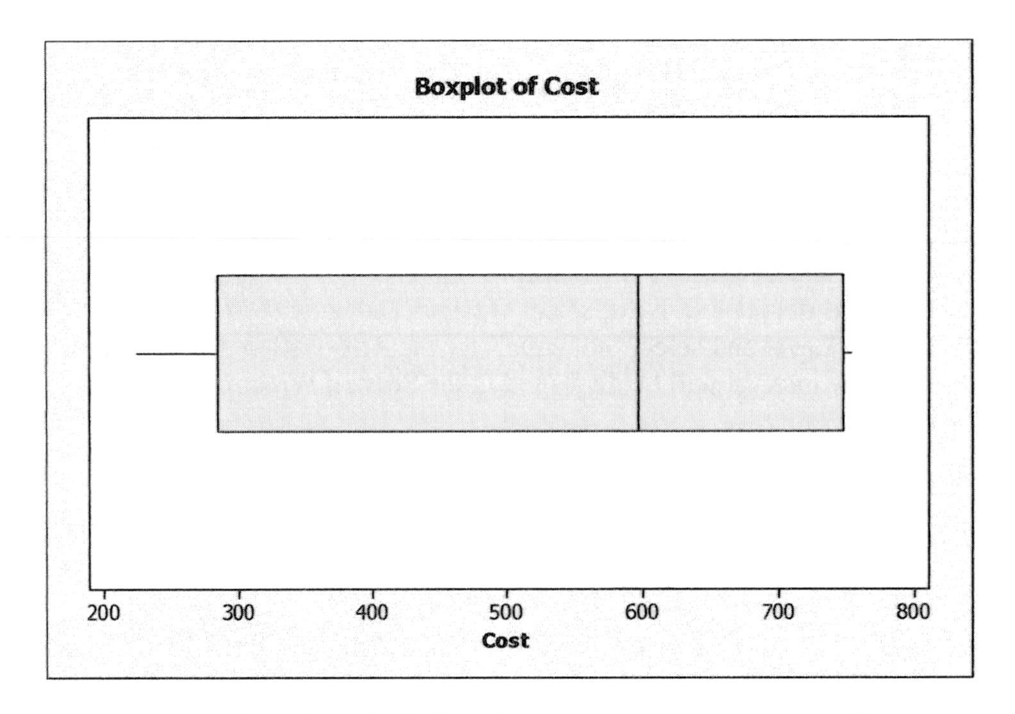

There are no outliers shown in the boxplot, so you may now construct the confidence interval.
Click on **Stat** → **Basic Statistics** → **1-Sample Z.** Select **Samples in Columns** and enter C1. Enter 220 for the **Standard deviation.** Next, select **Options** and enter 95.0 for the **Confidence Level** and select 'not equal' for the **Alternative**. Click on **OK** twice and the interval will be displayed in the Session Window. Repeat this last step to find a 90% confidence interval also.

One-Sample Z: Cost

```
The assumed standard deviation = 220


Variable  N     Mean    StDev  SE Mean          95% CI
Cost      4  542.250  249.251  110.000  (326.654, 757.846)
```

One-Sample Z: Cost

```
The assumed standard deviation = 220


Variable  N     Mean    StDev  SE Mean          90% CI
Cost      4  542.250  249.251  110.000  (361.316, 723.184)
```

Problem 35 (pg. 462) Miles on a Saturn

Open Minitab worksheet 9_1_35. The data values are in C1. The mean will be displayed in the output for the confidence intervals.

Construct the confidence interval. Click on **Stat → Basic Statistics → 1-Sample Z.** Select **Samples in Columns** and enter C1. Enter 19700 for the **Standard deviation.** Next, select **Options** and enter 99.0 for the **Confidence Level** and select 'not equal' for the **Alternative**. Click on **OK** twice and the interval will be displayed in the Session Window.

One-Sample Z: Miles

The assumed standard deviation = 19700

Variable	N	Mean	StDev	SE Mean	99% CI
Miles	33	49477.7	19715.6	3429.3	(40644.4, 58311.1)

Notice that the mean of the data is 49,477.7. Repeat for a 95% confidence interval.

One-Sample Z: Miles

The assumed standard deviation = 19700

Variable	N	Mean	StDev	SE Mean	95% CI
Miles	33	49477.7	19715.6	3429.3	(42756.4, 56199.1)

Section 9.2

▸ Example 2 (pg. 468) Finding t-values

Find the t-value such that the area under the t-distribution to right of the t-value is 0.10, assuming 15 degrees of freedom. Click on **Calc → Probability Distributions → t.** Select **Inverse cumulative probability** with a **Noncentrality parameter** of 0. Enter 15 **Degrees of freedom.** Recall that cumulative probability is calculated as the area to the left of a value. Since the area to the *right* of the t-value is 0.10, the area to the *left* of the t-value is 0.90. Select **Input constant** and enter .90. Click on **OK** and the t-value will be in the Session window.

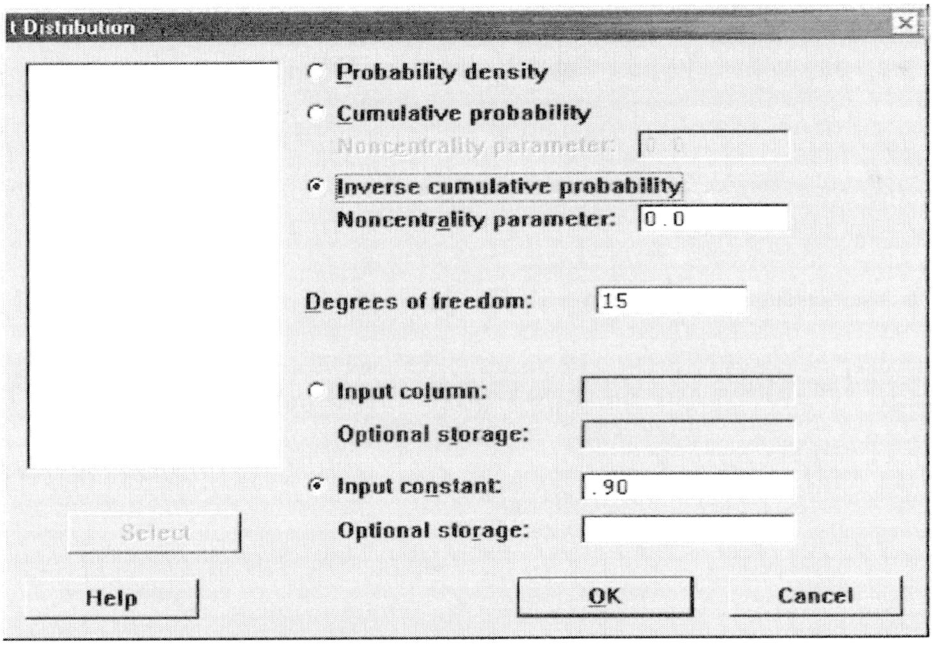

t Distribution

Probability density
Cumulative probability
Noncentrality parameter: 0.0
Inverse cumulative probability
Noncentrality parameter: 0.0
Degrees of freedom: 15
Input column:
Optional storage:
Input constant: .90
Optional storage:
Select
Help
OK
Cancel

Inverse Cumulative Distribution Function

```
Student's t distribution with 15 DF

P( X <= x )          x
     0.9000    1.34061
```

As you can see, the t-value is 1.34061.

Example 3 (pg. 470) Construct a 95% Confidence Interval for the mean diameter of matures trees

Construct a 95% confidence interval for the mean diameter of mature trees in the case in which σ, the population standard deviation is unknown. Enter the data from Table 5 on page 470 into C1.

First verify that the data is approximately normal. To draw a normal probability plot, click on **Graph → Probability Plot** and select the **Single** plot. Click on **OK.** Select C1 for the **Graph variable.** Click on **OK** and the probability plot will be displayed.

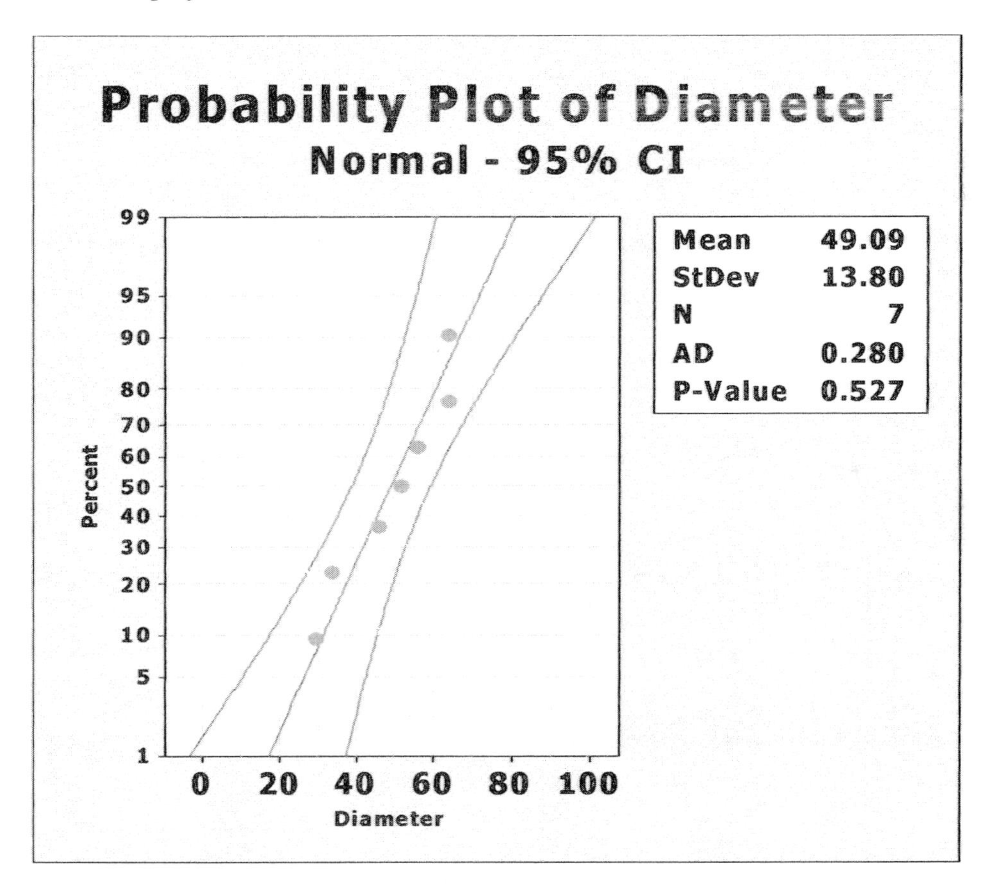

Notice that all data points are contained within the confidence bands of the plot.

Next, check for outliers using a boxplot. Click on **Graph → Boxplot** and select the **Simple** boxplot. Click on **OK**. Select C1 for the **Graph variable.** To view a horizontal boxplot (rather than a vertical one) click on **Scale** and select **Transpose value and category scales.** Click on **OK** twice.

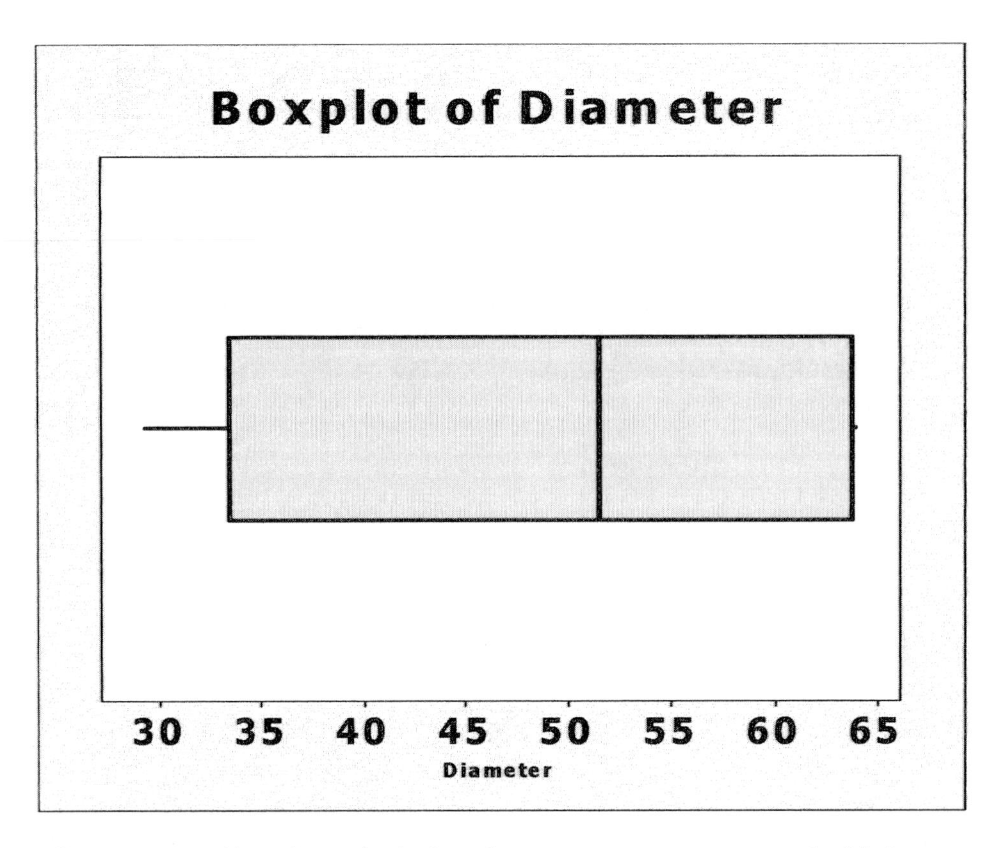

There are no outliers shown in the boxplot, so you may now proceed with the confidence interval.

Since n=7 and the population standard deviation is unknown, you should construct a t-interval for this problem. Click on **Stat → Basic Statistics → 1-Sample t.** Select **Samples in Columns** and enter C1. Next, select **Options** and enter 95.0 for the **Confidence Level** and select 'not equal' for the **Alternative**. Click on **OK** twice and the interval will be displayed in the Session Window.

One-Sample T: Diameter

Variable	N	Mean	StDev	SE Mean	95% CI
Diameter	7	49.0857	13.7957	5.2143	(36.3268, 61.8446)

Notice that the confidence interval is (36.3268, 61.8446).

◀

▸ Problem 7 (pg. 473) Finding t-values

For parts (a) - (d), click on **Calc → Probability Distributions → t.** Select **Inverse cumulative probability** with a **Noncentrality parameter** of 0. Then enter the appropriate **Degrees of freedom** and **Input constant**.

Part (a): Find the t-value such that the area under the t-distribution to right of the t-value is 0.10, assuming 25 degrees of freedom. Enter 25 **Degrees of freedom.** Since the area to the *right* of the t-value is 0.10, the area to the *left* of the t-value is 0.90. Select **Input constant** and enter .90. Click on **OK** and the t-value will be in the Session window.

Part (b): Find the t-value such that the area under the t-distribution to right of the t-value is 0.05, assuming 30 degrees of freedom. Enter 30 **Degrees of freedom.** Since the area to the *right* of the t-value is 0.05, the area to the *left* of the t-value is 0.90. Select **Input constant** and enter .95. Click on **OK** and the t-value will be in the Session window.

Part (c): Find the t-value such that the area under the t-distribution to left of the t-value is 0.01, assuming 18 degrees of freedom. Enter 18 **Degrees of freedom.** Select **Input constant** and enter .01. Click on **OK** and the t-value will be in the Session window.

Part (d): Find the critical t-value that corresponds to a 90% confidence level, assuming 20 degrees of freedom. Enter 20 **Degrees of freedom.** Since the center area between the t-values is 0.90, then the area to the *right* of the positive t-value is 0.05, and the area to the *left* of the t-value is 0.95. Select **Input constant** and enter .95. Click on **OK** and the t-value will be in the Session window.

Inverse Cumulative Distribution Function

```
Student's t distribution with 25 DF

P( X <= x )        x
        0.9  1.31635
```

Inverse Cumulative Distribution Function

```
Student's t distribution with 30 DF

P( X <= x )        x
       0.95  1.69726
```

Inverse Cumulative Distribution Function

```
Student's t distribution with 18 DF

P( X <= x )          x
       0.01   -2.55238
```

Inverse Cumulative Distribution Function

```
Student's t distribution with 20 DF

P( X <= x )         x
       0.95   1.72472
```

▸ Problem 28 (pg. 476) The Effect of Outliers

Enter the ages into C1. First construct a boxplot of the data to identify any outliers. Click on **Graph → Boxplot** and select the **Simple** boxplot. Click on OK. Select C1 for the **Graph variable.** To view a horizontal boxplot (rather than a vertical one) click on **Scale** and select **Transpose value and category scales.** Click on **OK** twice.

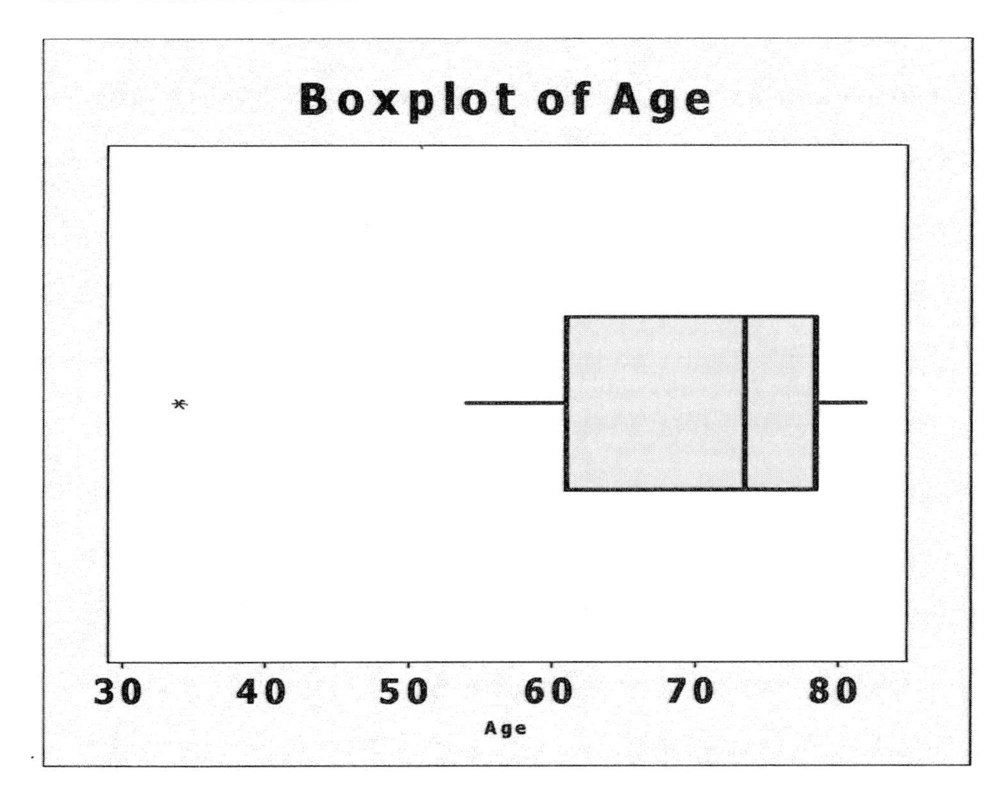

There is one outlier -- the person who died at age 34. Construct 90% confidence intervals, with and without the outlier. Since n=12 and the population standard deviation is unknown, you should construct a t-interval for this problem. Click on **Stat → Basic Statistics → 1-Sample t.** Select **Samples in Columns** and enter C1. Next, select **Options** and enter 99.0 for the **Confidence Level** and select 'not equal' for the **Alternative**. Click on **OK** twice and the interval will be displayed in the Session Window.

One-Sample T: Age

Variable	N	Mean	StDev	SE Mean	99% CI
Age	12	68.5833	13.9770	4.0348	(56.0520, 81.1147)

To construct the interval without the outlier, just delete age=34 from C1 and repeat the steps above. This time n=11.

One-Sample T: Age

Variable	N	Mean	StDev	SE Mean	99% CI
Age	11	71.7273	9.1879	2.7703	(62.9475, 80.5070)

Notice the difference between the two intervals. The first one, which included the outlier, is wider -- (56.1, 86.1). Compare that to the second interval, which did not include the outlier -- (62.9, 80.5).

◀

Section 9.3

▸ Example 3 (pg. 480) Construct a 95% confidence interval for p

A simple random sample of Americans were asked if they were in favor of tighter enforcement of government rules on TV content. Of the 1505 adults, 1129 said 'yes.' Construct a 95% confidence interval for the population proportion, p.

Use a hand-calculator to verify that the assumption of normality, required to perform the test, is met by evaluating the inequality: $(n\hat{p}(1-\hat{p}) \geq 10$. For this example, you will discover that this assumption is met.

To construct a 95% confidence interval, click on **Stat → Basic Statistics → 1 Proportion.** Select **Summarized Data.** The **Number of trials** is 1505 and the **Number of successes** is 1129.

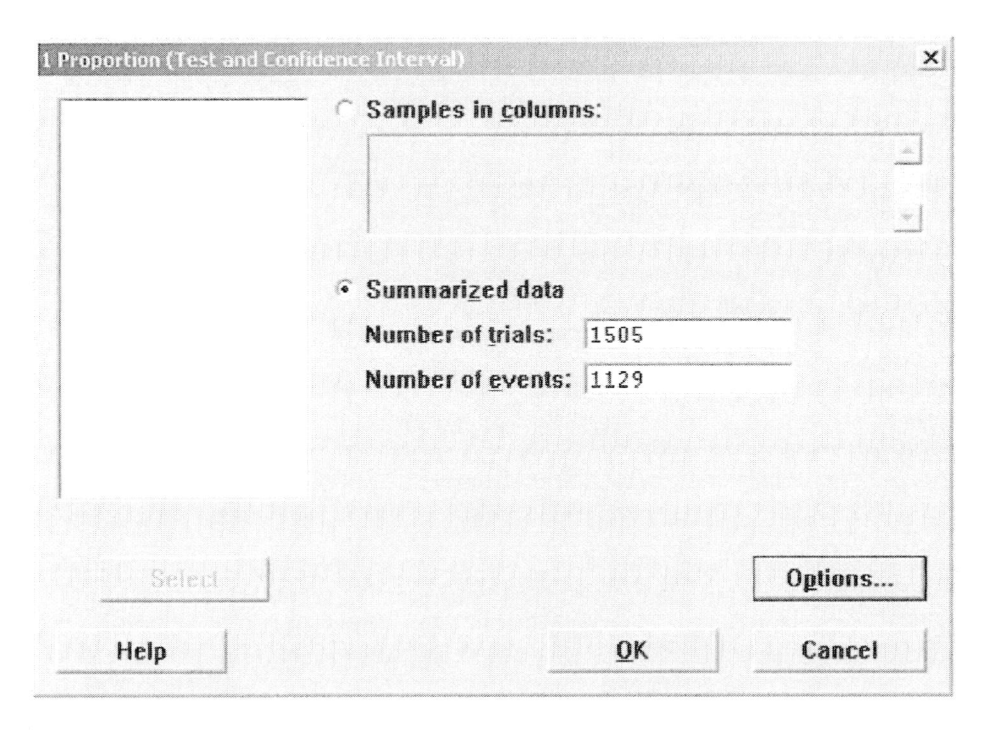

Next, to select the confidence level, click on **Options.** Enter 95.0 for the **Confidence Level** and select 'not equal' for the **Alternative**. Because the assumption of normality has been met, select **Use test and interval based on normal distribution.**

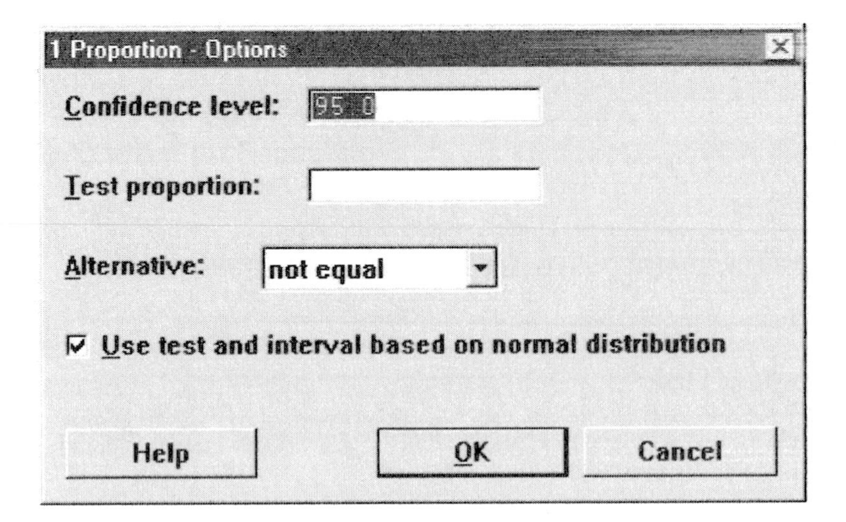

Click on **OK** twice and the output will be displayed in the Session Window.

Test and CI for One Proportion

```
Test of p = 0.5 vs p not = 0.5
```

Sample	X	N	Sample p	95% CI	Z-Value	P-Value
1	1129	1505	0.750166	(0.728294, 0.772038)	19.41	0.000

Notice the interval is (.728, .772).

▸ Problem 11 (pg. 484) Lipitor

A study of 863 patients who received 10 mg doses of Lipitor found that 47 reported headache as a side effect. Construct a 90% confidence interval for the true proportion of patients who reported headache as a side effect of Lipitor.

Use a hand-calculator to verify that the assumption of normality, required to perform the test, is met by evaluating the inequality: $(863)*(.054)*(.946) \geq 10$. For this example, you will discover that this assumption is met.

Click on **Stat → Basic Statistics → 1 Proportion.** Select **Summarized Data.** The **Number of trials** is 863 and the **Number of successes** is 47. Next, to select the confidence level, click on **Options.** Enter 90.0 for the **Confidence Level** and select 'not equal' for the **Alternative**. Because the assumption of normality has been met, select **Use test and interval based on normal distribution.** Click on **OK** twice and the results will be in the Session Window.

Test and CI for One Proportion

```
Test of p = 0.5 vs p not = 0.5

Sample  X     N  Sample p        90.0% CI              Z-Value  P-Value
1      47   863  0.054461  (0.041755, 0.067167)        -26.18    0.000
```

Notice that the sample proportion is displayed (.054461) as well as the confidence interval (.042, .067).

◀

Section 9.4

▸ Example 1 (pg. 488) Finding Critical Values for the Chi-Square

Find the critical values that separate the middle 90% of the chi-square distribution from the 5% area in each tail, assuming 15 degrees of freedom. Click on **Calc → Probability Distributions →Chi-square.** Select **Inverse cumulative probability** with a **Noncentrality parameter** of 0. Enter 15 **Degrees of freedom.** The calculation of the two critical values will have to be done in two steps. Since the center area between the chi-square values is 0.90, the area in both the upper and lower tail will be 5%. First find the upper chi-square value. The area to the *right* of the upper chi-square value is 0.05, so the area to the *left* of it is 0.95. Select **Input constant** and enter .95. Click on **OK** and the chi-square value will be in the Session window. Now find the lower chi-square value. The area to the *left* of the lower chi-square value is 0.05. Repeat the steps above, but this time select **Input constant** and enter .05.

Inverse Cumulative Distribution Function

```
Chi-Square with 15 DF

P( X <= x )          x
     0.9500    24.9958
```

Inverse Cumulative Distribution Function

```
Chi-Square with 15 DF

P( X <= x )          x
     0.0500     7.2609
```

As you can see, the lower chi-square value is 7.2609 and the upper chi-square value is 24.9958.

◀

▶ Example 2 (pg. 490) Constructing Confidence Intervals for the population Standard Deviation

Use the prices for the Corvettes (Table 7 on page 490) to construct a 90% confidence interval for the population standard deviation. In section 9.1, we verified that this data is normally distributed with a normal probability plot and we verified that there were no outliers with a boxplot. To calculate a 90% confidence for the population standard deviation, click on **Stat → Basic Statistics → Graphical Summary.** Select C1 as the **Variable** and enter 90 for the **Confidence level.** Click on **OK**.

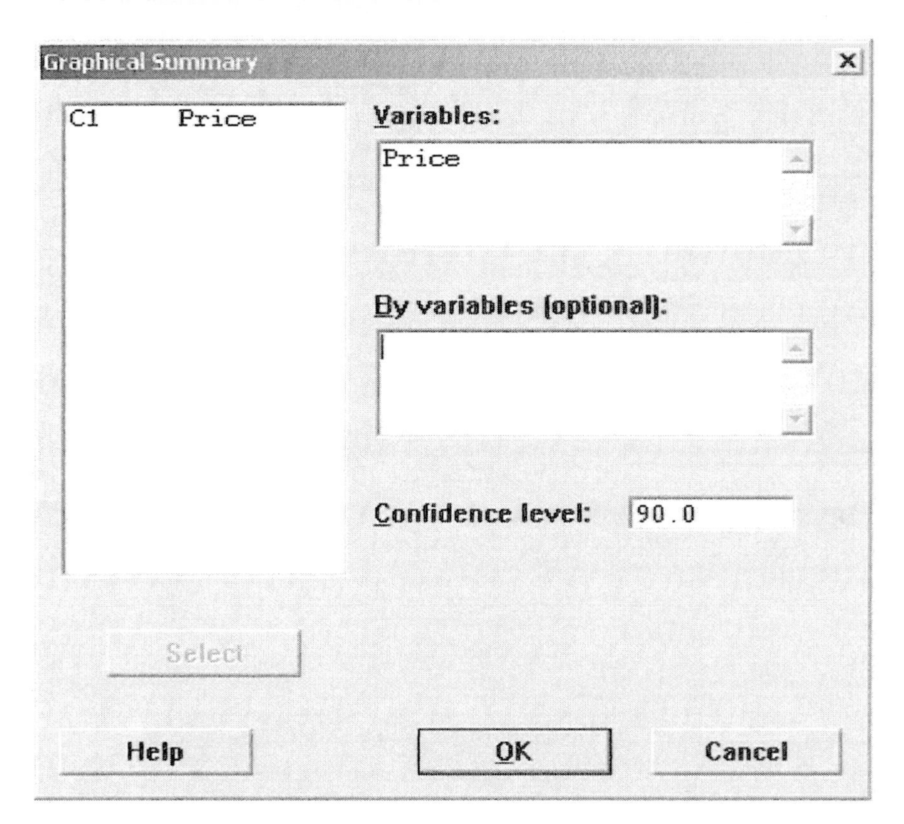

A Descriptive Statistics window will be displayed. The confidence interval for the standard deviation is displayed in the window.

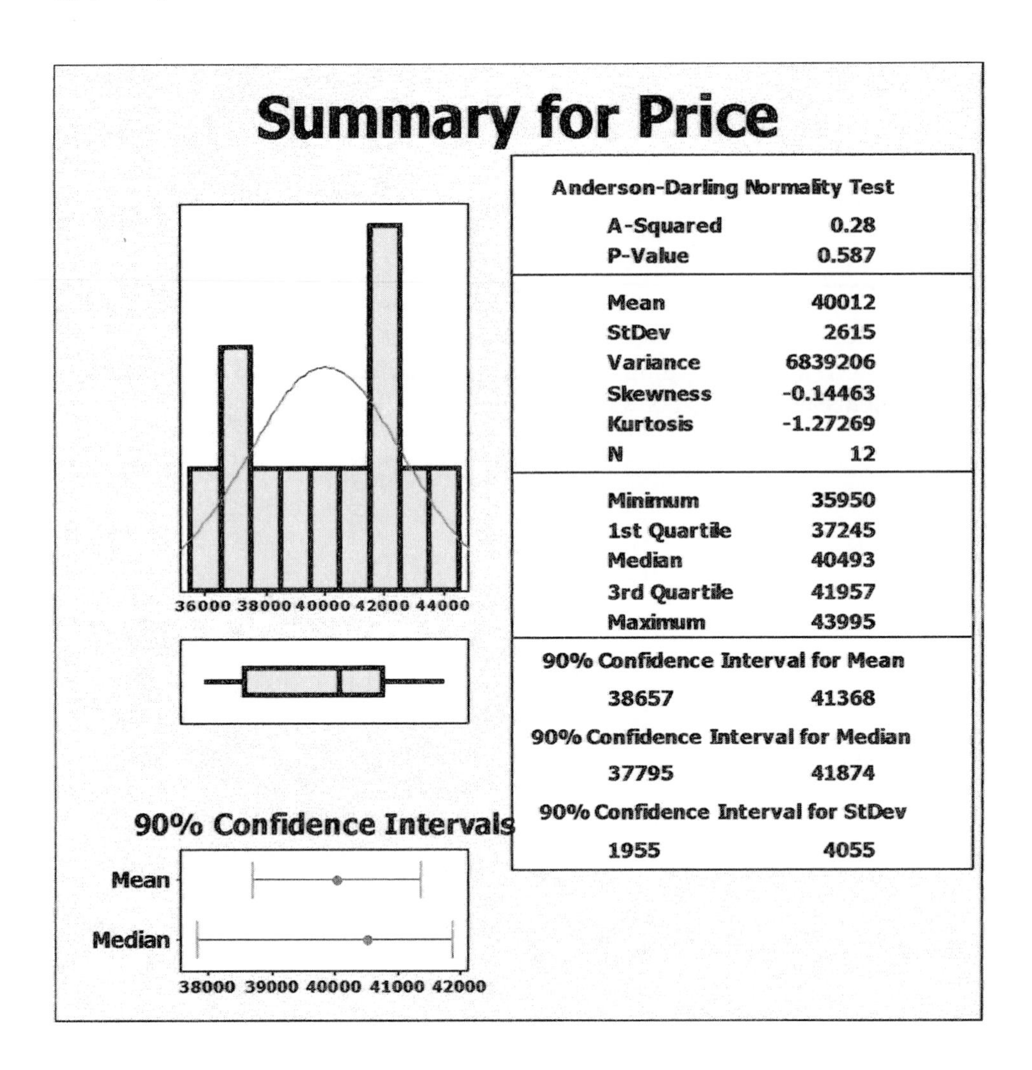

Notice on the right side of the window, near the bottom, the 90% Confidence Interval for the population Standard Deviation is (1955, 4055).

Problem 15 (pg. 492) Peanuts

Enter the data into C1 of a Minitab worksheet. First, verify that this data is normally distributed. (To draw a normal probability plot, click on **Graph** → **Probability Plot** and select the **Single** plot. Click on **OK**. Select C1 for the **Graph variable.** Click on **OK** and the probability plot will be displayed.)

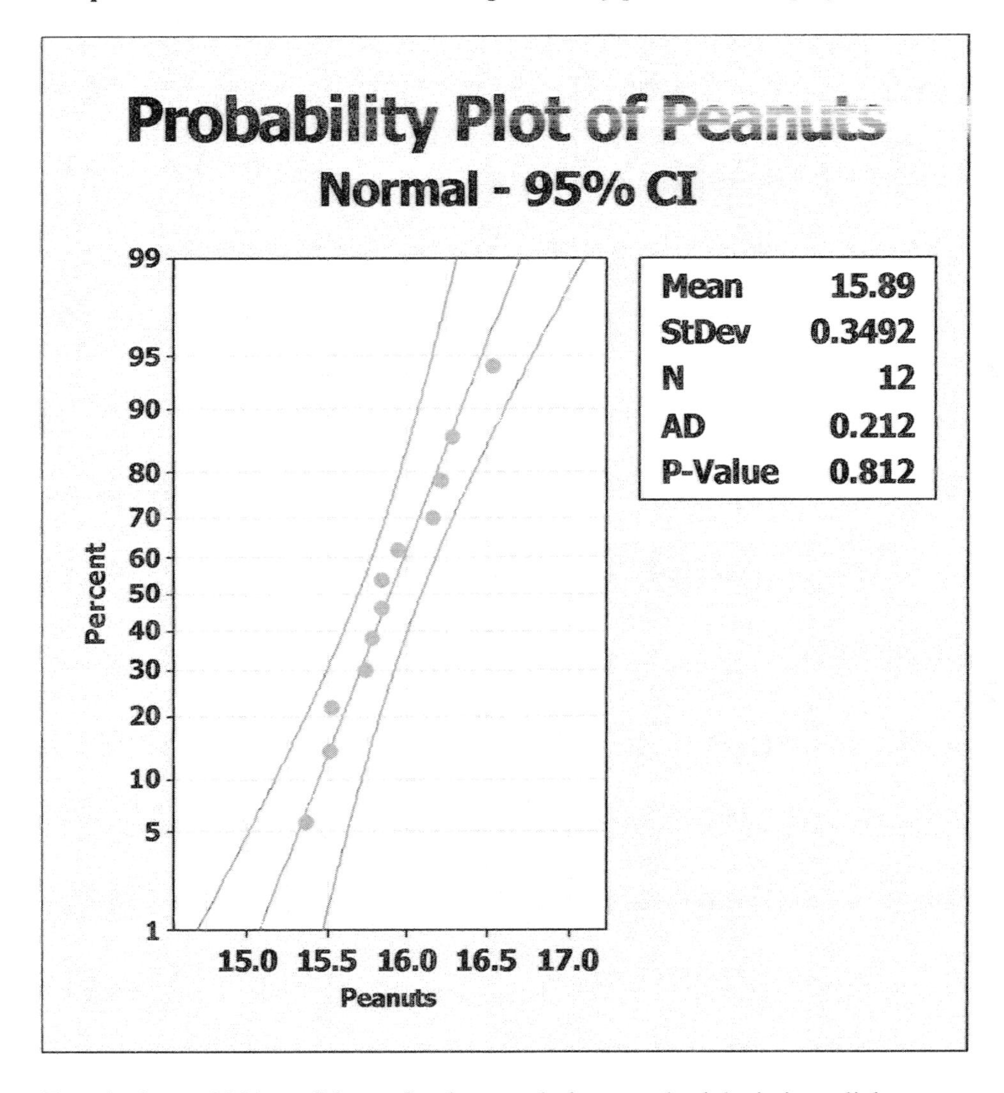

To calculate a 90% confidence for the population standard deviation, click on **Stat** → **Basic Statistics** → **Graphical Summary.** Select C1 as the **Variable** and enter 90 for the **Confidence level.** Click on **OK**.

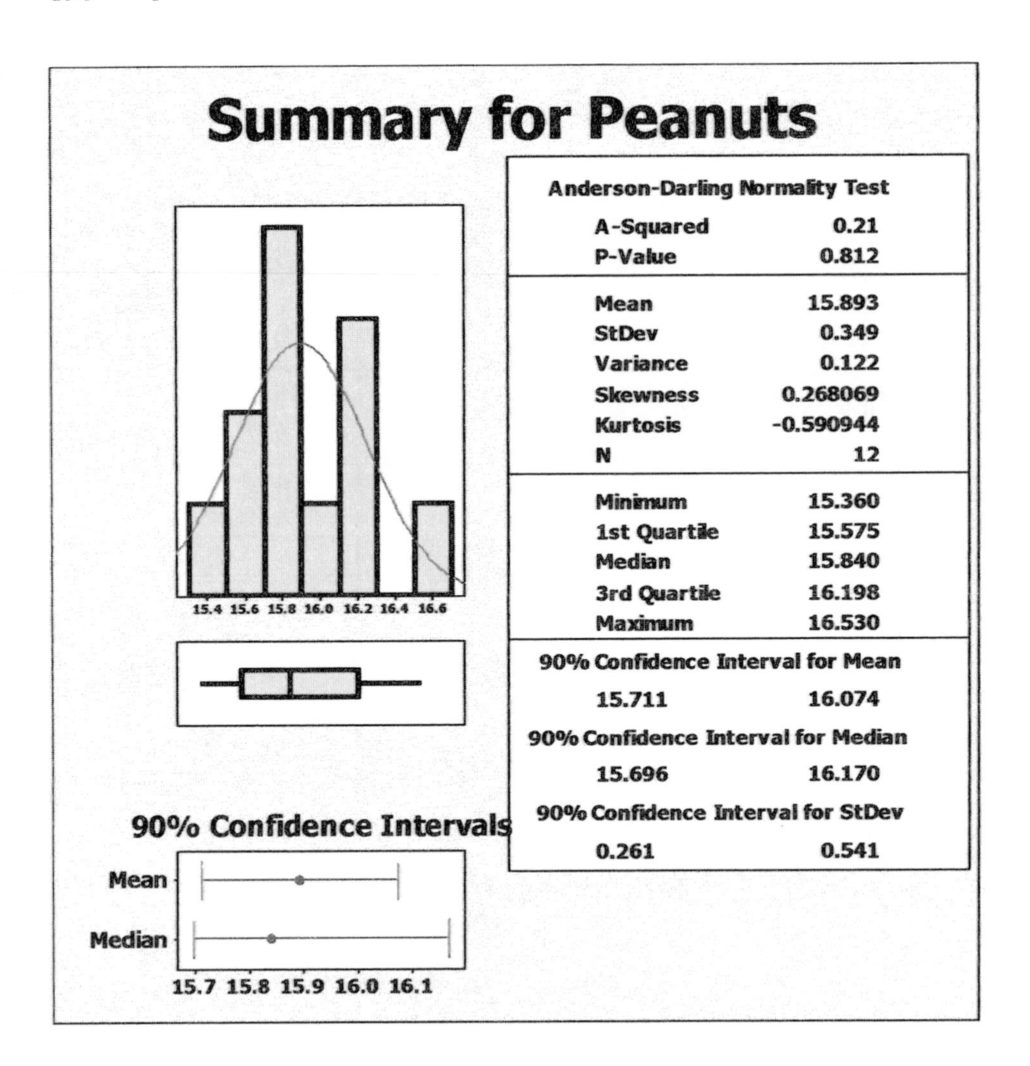

Notice on the right side of the window, near the bottom, the 90% Confidence Interval for the population standard deviation is (0.261, 0.541).

Testing Claims Regarding a Parameter

Section 10.2

▸ Example 5 (pg. 524) Hypothesis Testing

The mean monthly cell phone bill in 2004 was \$50.64. A market researcher believes that it is different today, so the researcher randomly selects 12 cell phone subscribers and obtains the data found in Table 2 on page 524. Is there enough evidence to support his claim that mean monthly bills are different from \$50.64 at $\alpha = .05$? Assume $\sigma = \$18.49$. Use the classical method to interpret.

Enter the data from Table 2 on page 524 of the text into C1 of a Minitab worksheet. **Do not enter any '\$' as part of the data.** Since $n < 30$, first test that the data is approximately normal with no outliers. Use a normal probability plot and a boxplot. To draw a normal probability plot, click on **Graph → Probability Plot** and select the **Single** plot. Click on **OK.** Select C1 for the **Graph variable.** Click on **OK** and the probability plot will be displayed. Notice that all data points are contained within the confidence bands of the plot.

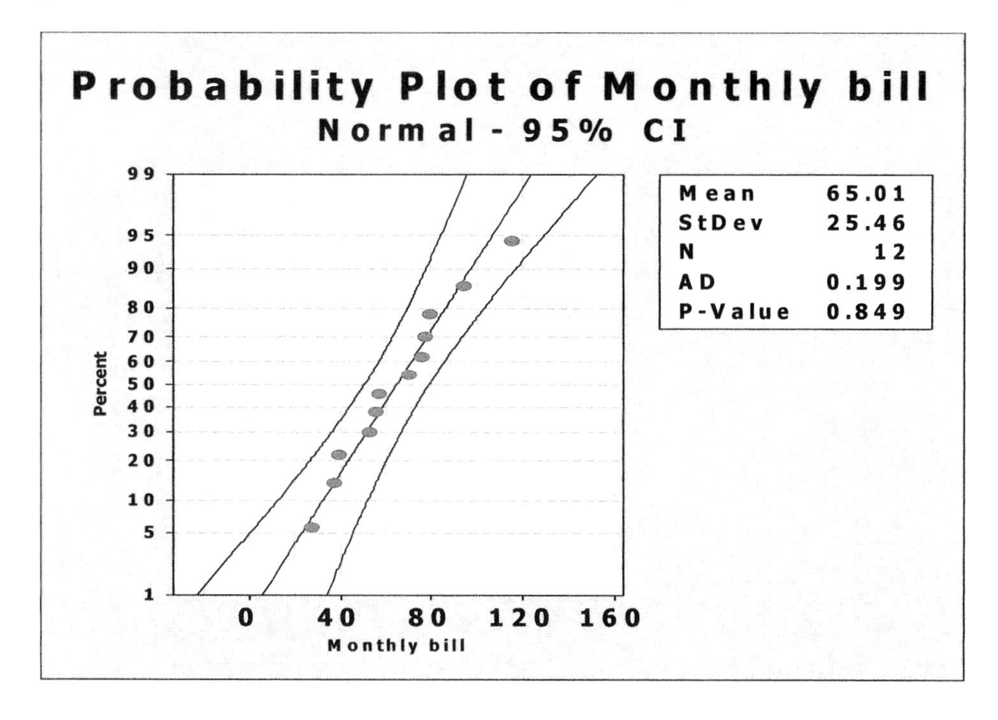

Next, check for outliers using a boxplot. Click on **Graph** → **Boxplot** and select the **Simple** boxplot. Click on **OK**. Select C1 for the **Graph variable.** To view a horizontal boxplot (rather than a vertical one) click on **Scale** and select **Transpose value and category scales.** Click on **OK**.

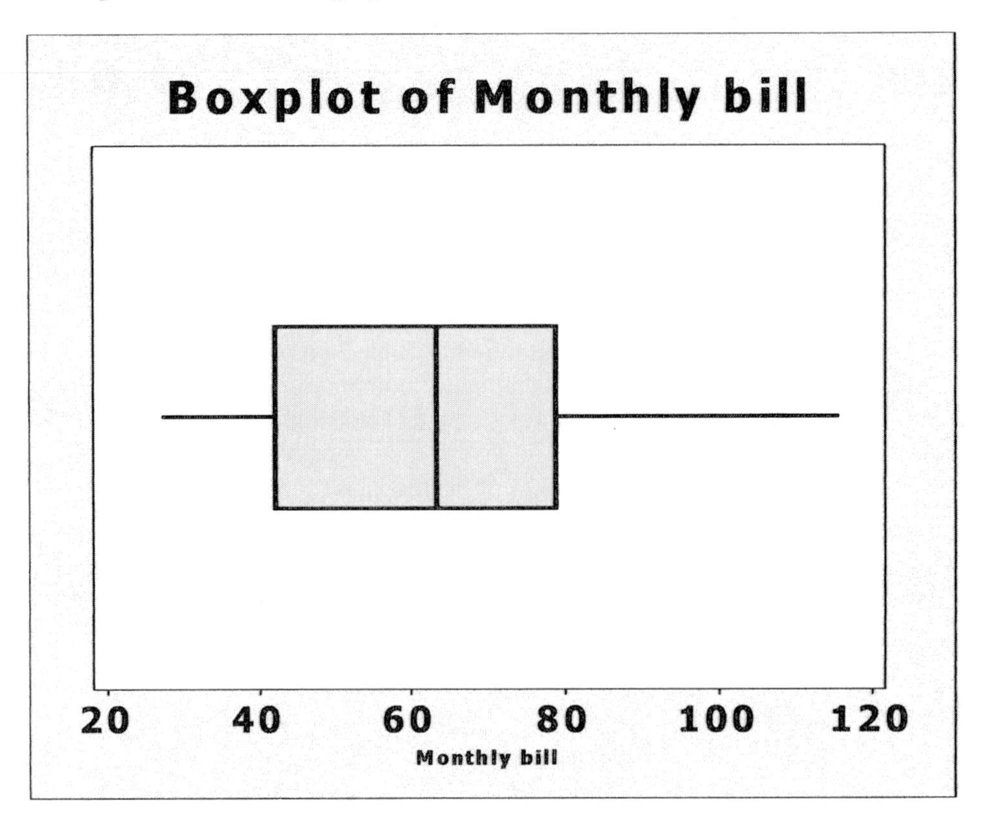

Since the data is approximately normal with no outliers, click on **Stat** → **Basic Statistics** → **1-Sample Z.** Select **Samples in Columns** and enter C1. Enter 18.49 for the **Standard deviation** and enter 50.64 for **Test mean.**

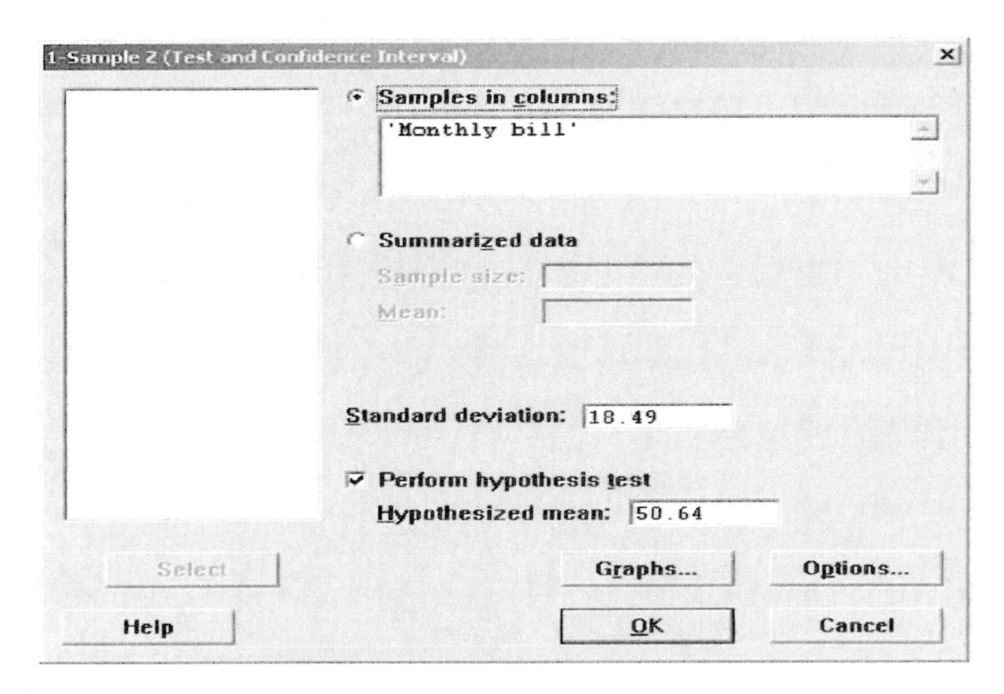

Since the claim is "the mean is different from $50.64", you will perform a two-tailed test. Click on **Options**, and set **Alternative** to "not equal".

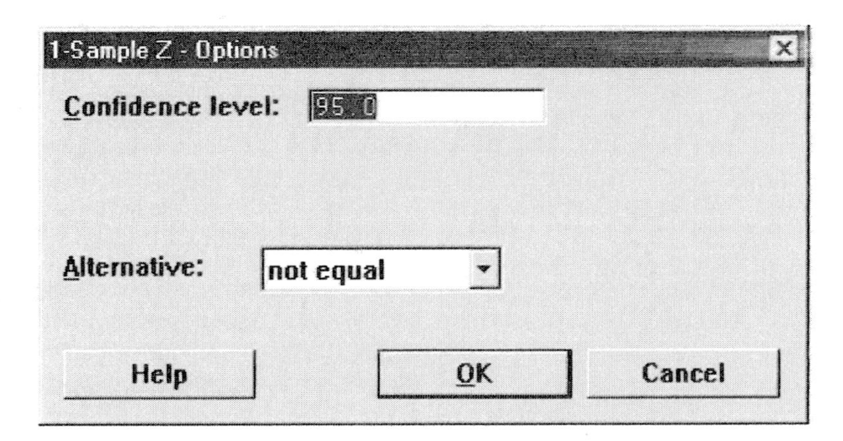

Click on **OK** twice and the results should be displayed in the Session Window.

One-Sample Z: Monthly bill

Test of mu = 50.64 vs not = 50.64
The assumed standard deviation = 18.49

Variable	N	Mean	StDev	SE Mean	95% CI	Z	P
Monthly Bill	12	65.0142	25.4587	5.3376	(54.5527, 75.4757)	2.69	0.007

Notice that MINITAB gives the test statistic and the P-value, so that you can make your conclusion using either value. Since the Z-value of 2.69 is greater than 1.96 you should Reject the null hypothesis. Also, since the P-value of 0.007 is less than α, you should Reject the null hypothesis.

◀

Problem 23 (pg. 528) Acid Rain

A biologist claims that the pH level of the rain in Pierce County, Washington has decreased since 1990. She obtains a random sample of 19 rain dates in the year 2004. Open worksheet **10_2_23.** The rain acidity for the 19 days is in C1. Because n=19, you must verify that pH level is normally distributed and does not contain any outliers.
Use a normal probability plot and a boxplot. To draw a normal probability plot, click on **Graph → Probability Plot** and select the **Single** plot. Click on **OK.** Select C1 for the **Graph variable.** Click on **OK** and the probability plot will be displayed. Notice that all data points are contained within the confidence bands of the plot.

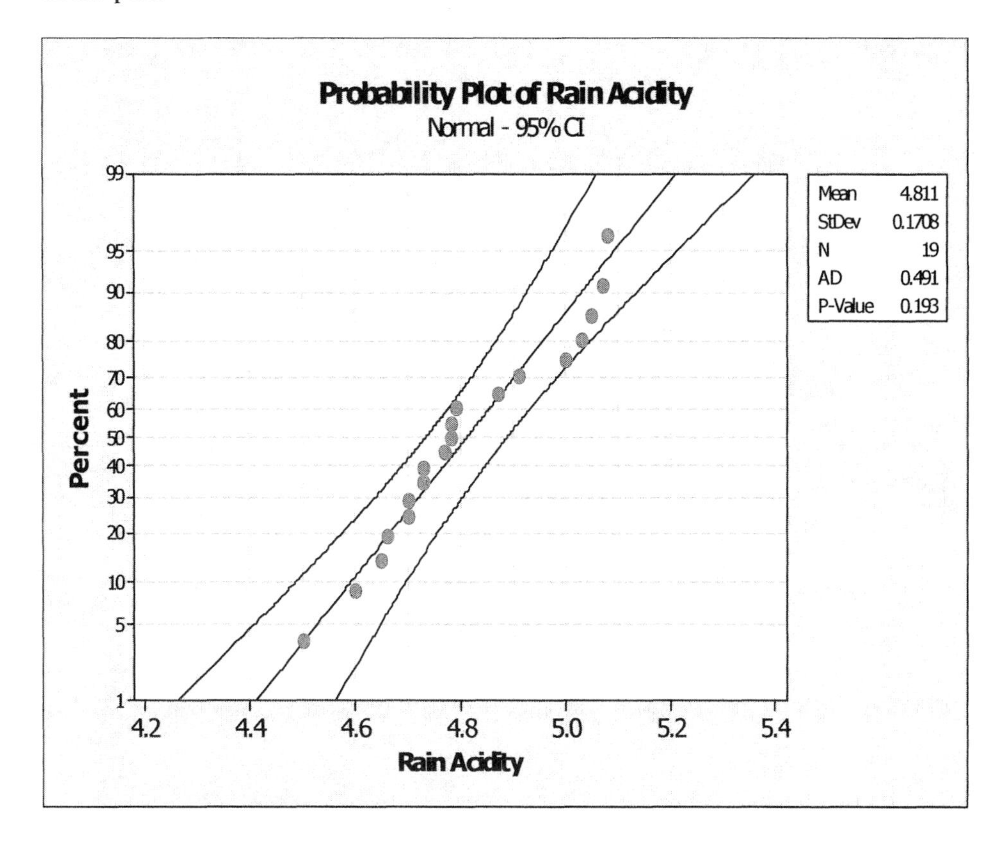

Next, check for outliers using a boxplot. Click on **Graph → Boxplot** and select the **Simple** boxplot. Click on **OK**. Select C1 for the **Graph variable.** To view a horizontal boxplot (rather than a vertical one) click on **Scale** and select **Transpose value and category scales.** Click on **OK**.

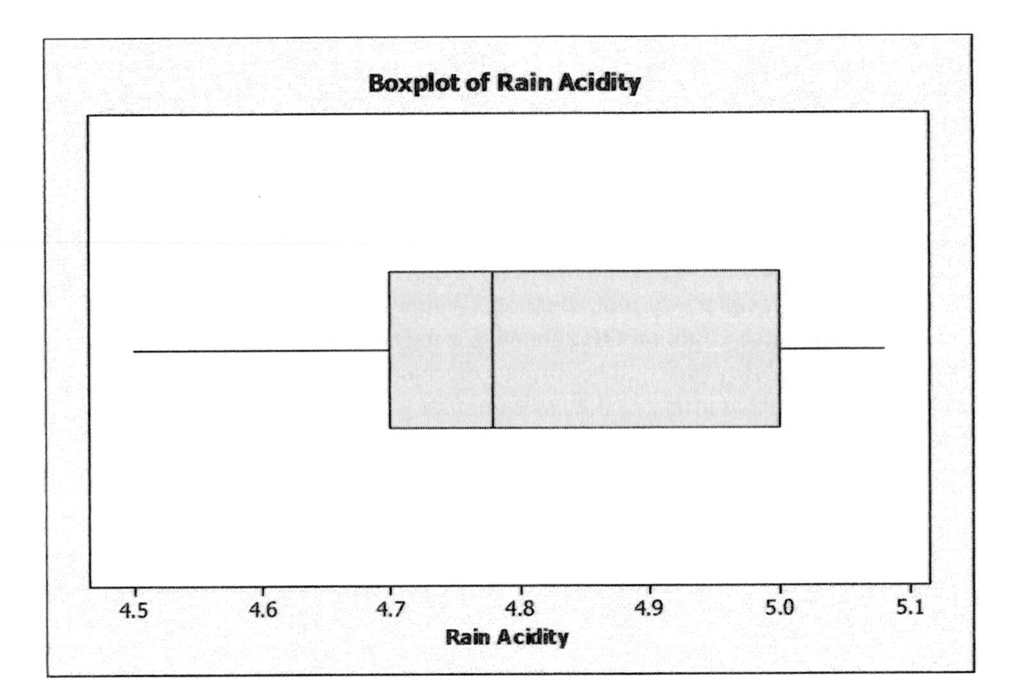

Since there are no outliers, close the Graph Window and perform the hypothesis test. Click on **Stat → Basic Statistics → 1-Sample Z.** Select **Samples in Columns** and enter C1. Beside **Standard deviation**, enter the assumed value, $\sigma = 0.2$. Click on **Test mean** and enter 5.03. Since the claim is "the rain acidity has decreased", you will perform a lower-tailed test. Click on **Options** and use the down arrow beside **Alternative** to select "less than". Click on **OK** and the results of the test should be displayed in the Session Window.

One-Sample Z: Rain Acidity

```
Test of mu = 5.03 vs mu < 5.03
The assumed sigma = 0.2

Variable          N      Mean    StDev   SE Mean
Rain Acidity     19   4.81053  0.17083   0.04588

Variable      95.0% Upper Bound        Z       P
Rain Acidity            4.88600    -4.78   0.000
```

Since P=0.00, reject the null hypothesis at any significance level. Therefore, the biologist is correct in thinking that the pH level of rain has decreased since 1990.

▸ Problem 25 (pg. 528) Filling Bottles

A quality control manager wishes to verify that the mean amount of juice in each bottle is 64.05 ounces. She obtains a random sample of 22 bottles and measures the content. Open worksheet **10_2_25.** The ounces of juice are in C1. Because n=22, we must verify that juice level is normally distributed and does not contain any outliers. Use a normal probability plot and a boxplot.
To draw a normal probability plot, click on **Graph → Probability Plot** and select the **Single** plot. Click on **OK.** Select C1 for the **Graph variable.** Click on **OK** and the probability plot will be displayed. Notice that all data points are contained within the confidence bands of the plot.

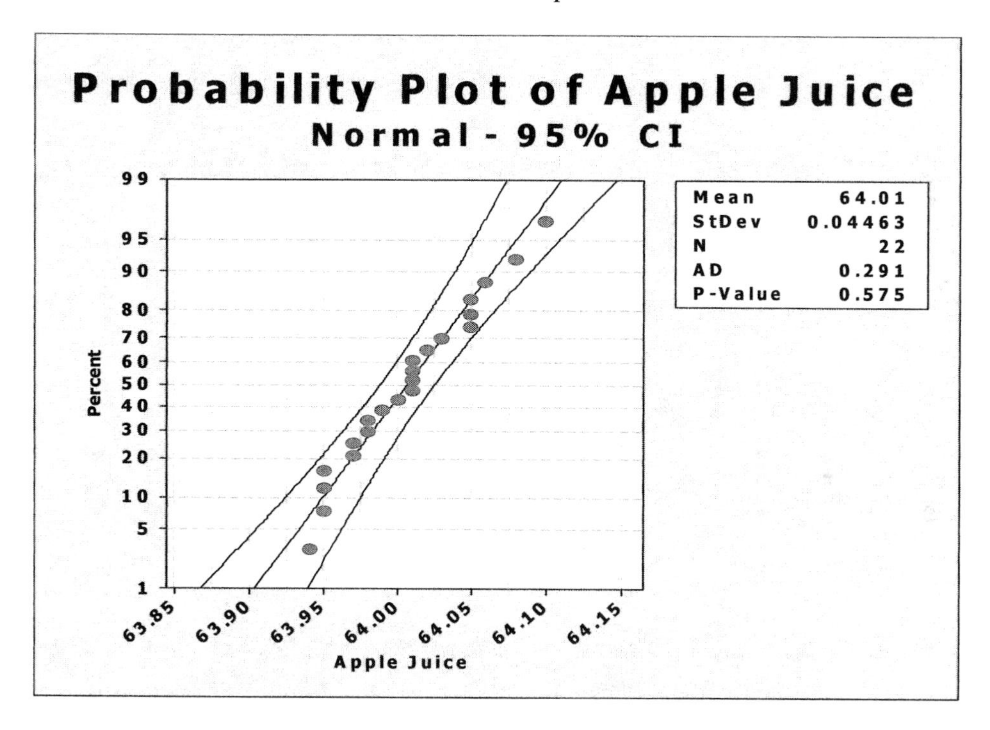

Next, check for outliers using a boxplot. Click on **Graph → Boxplot** and select the **Simple** boxplot. Click on **OK.** Select C1 for the **Graph variable.** To view a horizontal boxplot (rather than a vertical one) click on **Scale** and select **Transpose value and category scales.** Click on **OK.**

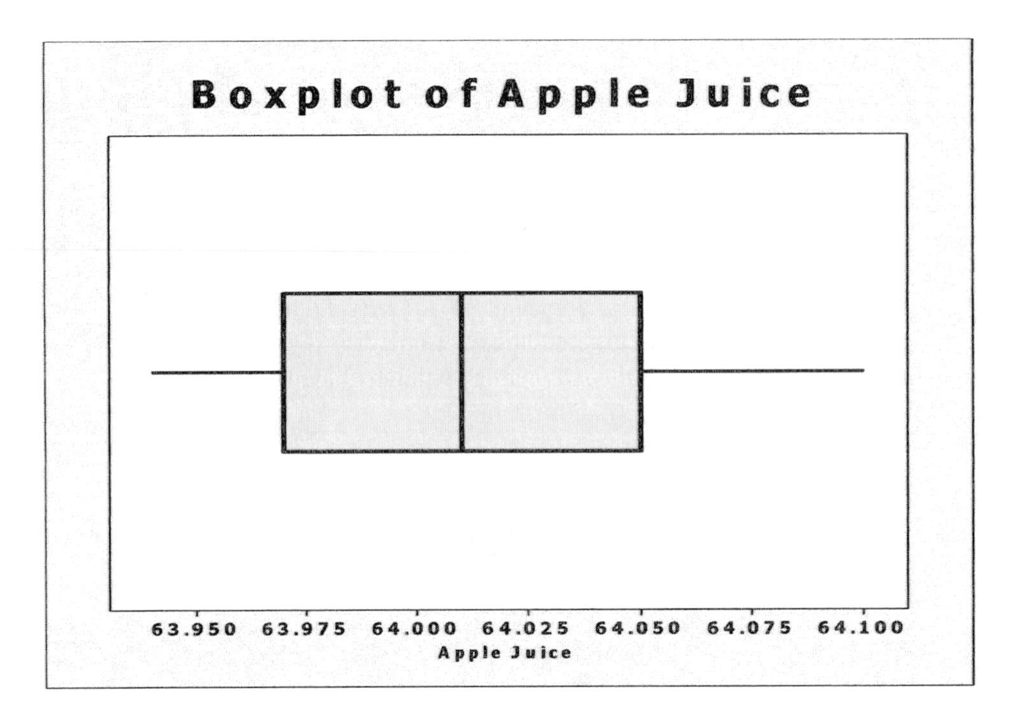

Since there are no outliers, proceed to perform the hypothesis test. Click on **Stat** → **Basic Statistics** → **1-Sample Z.** Select **Samples in Columns** and enter C1. Beside **Standard deviation**, enter the assumed value $\sigma = 0.06$. Click on **Test mean** and enter 64.05. Since the manager is interested in whether or not the bottles contain 64.05 oz., you will perform a two-tailed test. Click on **Options** and use the down arrow beside **Alternative** to select "not equal". Click on **OK** twice and the results of the test should be displayed in the Session Window.

One-Sample Z: Apple Juice

Test of mu = 64.05 vs not = 64.05
The assumed standard deviation = 0.06

Variable	N	Mean	StDev	SE Mean	95% CI	Z	P
Apple Juice	22	64.0073	0.0446	0.0128	(63.9822, 64.0323)	-3.34	0.001

Since P = .001 and is less than α, you should Reject the null hypothesis. Since you are rejecting the null hypothesis, the data provides evidence that the average amount of juice in the bottles is NOT 64.05 ounces, therefore the assembly line should be shut down so that the filling machine can be recalibrated.

◀

Section 10.3

▸ Example 1 (pg. 533) Testing μ with a Large Sample

According to the CDC, the mean number of cigarettes smoked per day by daily smokers is 18.1. Believing that retired adults smoke less than the general population, a researcher takes a random sample of 40 retired adult smokers and finds that their mean number of cigarettes smoked per day is 16.8 with a standard deviation of 4.7 cigarettes. At $\alpha = .10$, is there enough evidence to support the researcher's claim?

For this type of problem, the data is already summarized for you. Click on **Stat → Basic Statistics → 1-Sample t.** Select **Summarized data** and enter 40 for **Sample size**, 16.8 for **Mean**, 4.7 for **Standard deviation**. Click on **Perform hypothesis test** and enter 18.1 for the **Hypothesized mean**.

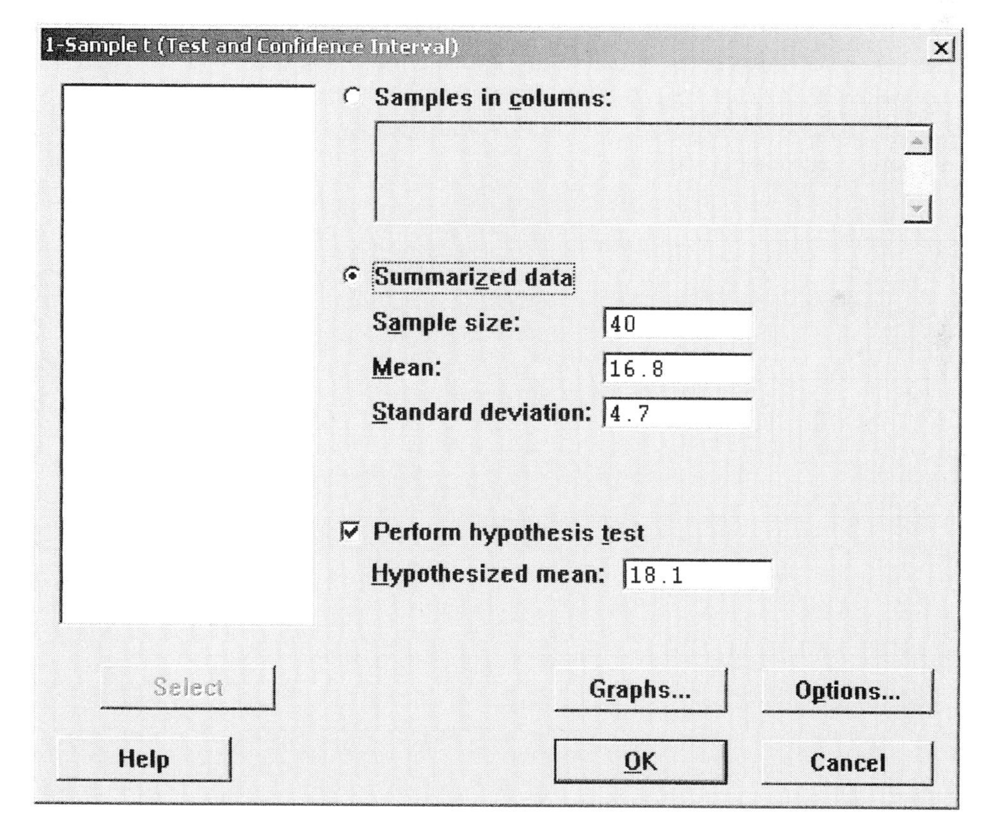

Since it is suspected that retired adults smoke less, you will perform a lower-tailed test. Click on **Options** and then on the down arrow beside **Alternative** to select "less than". Click on **OK** and the results of the test should be displayed in the Session Window.

One-Sample T

```
Test of mu = 18.1 vs < 18.1
```

N	Mean	StDev	SE Mean	95% Upper Bound	T	P
40	16.8000	4.7000	0.7431	18.0521	-1.75	0.044

Notice that MINITAB gives the test statistic and the P-value, so that you can make your conclusion using either value. Since the P-value is smaller than α, you would Reject the null hypothesis. There is enough evidence to support the researcher's claim that the mean number of cigarettes smoked by retired adults is less than the general population.

◀

▸ Problem 28 (pg. 542) Temperature of Surgical Patients

Open Minitab worksheet **10_3_28**. Click on **Stat → Basic Statistics →1-Sample t.** Select **Samples in Columns** and enter C1. Click on **Test mean** and enter 98.2. Since the claim is that the mean temperature of surgical patients is above the normal temperature, you will perform an upper-tailed test. Click on **Options** and then on the down arrow beside **Alternative** to select "greater than". Click on **OK** twice and the results of the test should be displayed in the Session Window.

One-Sample T: Patient Temperature

```
Test of mu = 98.2 vs mu > 98.2

Variable          N      Mean     StDev   SE Mean
Patient Temp     32   98.2906    0.6888    0.1218

Variable      95.0% Lower Bound        T       P
Patient Temp            98.0842     0.74   0.231
```

Since the P-value is .231 and is greater than α, you do not reject the null hypothesis. There is not enough evidence that surgical patients have a higher mean temperature than normal.

◀

▸ Problem 32 (pg. 542) Simulation

Create 40 random samples of size n=20. Click on **Calc → Random data → Normal.** **Generate** 20 **rows of data** and **Store in columns** C1-C40. Enter a **Mean** of 50 and a **Standard deviation** of 10. Click on **OK**. Each column contains a random sample of size n=20.

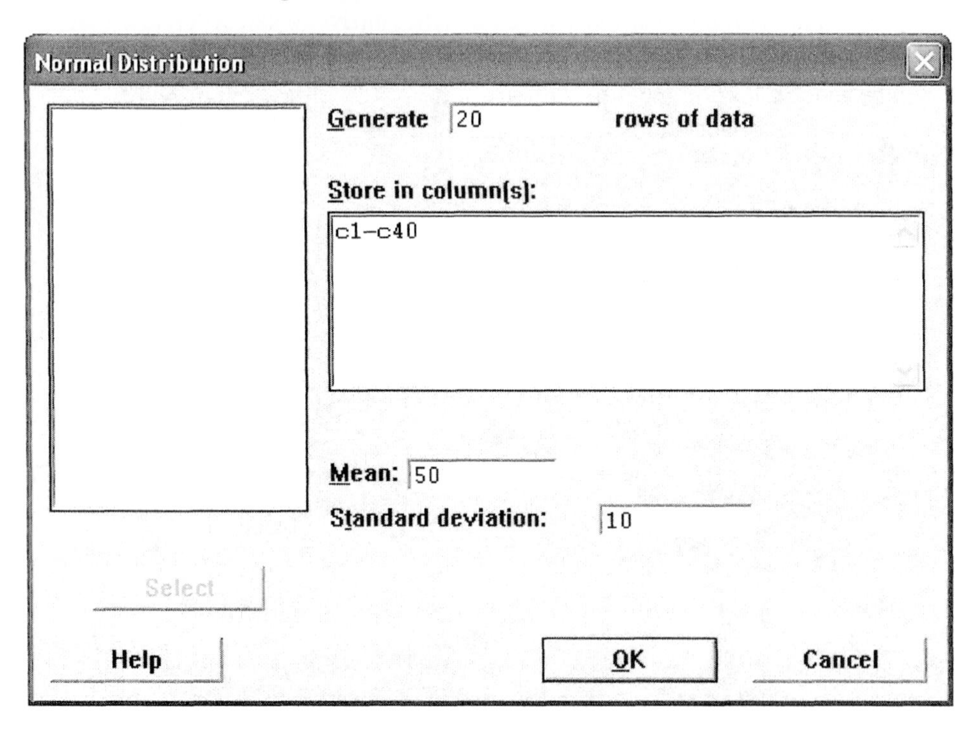

Now, perform t-tests on each column. Click on **Stat → Basic Statistics →1-Sample t.** Select **Samples in Columns** and enter C1- C40. Click on **Test mean** and enter 50. Click on **Options** and then on the down arrow beside **Alternative** to select "not equal". Click on **OK** and the results of the 40 tests should be displayed in the Session Window. To count the number of samples that would lead to a rejection of the null hypothesis, simply count the number of times that the P-value is less than .05. Divide this number by 40 for the result.

◂

Section 10.4

▸ Example 2 (pg. 547) Testing a Claim about a Proportion

The vaccine drug Prevnar was administered to 710 randomly selected infants. Of the 710 infants, 121 experienced a decrease in appetite. At $\alpha = .01$, is there enough evidence to conclude that the proportion of infants who receive Prevnar and experience a loss of appetite is different from 0.135, the proportion of children who experience a loss of appetite in competing medications?

Use a hand-calculator to verify that the assumption of normality, required to perform the test, is met by evaluating the inequality: $(n\hat{p}_o(1-\hat{p}_o) \geq 10$. For this example, you will discover that this assumption is met.

Click on **Stat → Basic Statistics → 1-Proportion.** The data is given in a summarized form, so select **Summarized data**. Enter 710 for the **Number of trials.** The **Number of Successes** is 121.

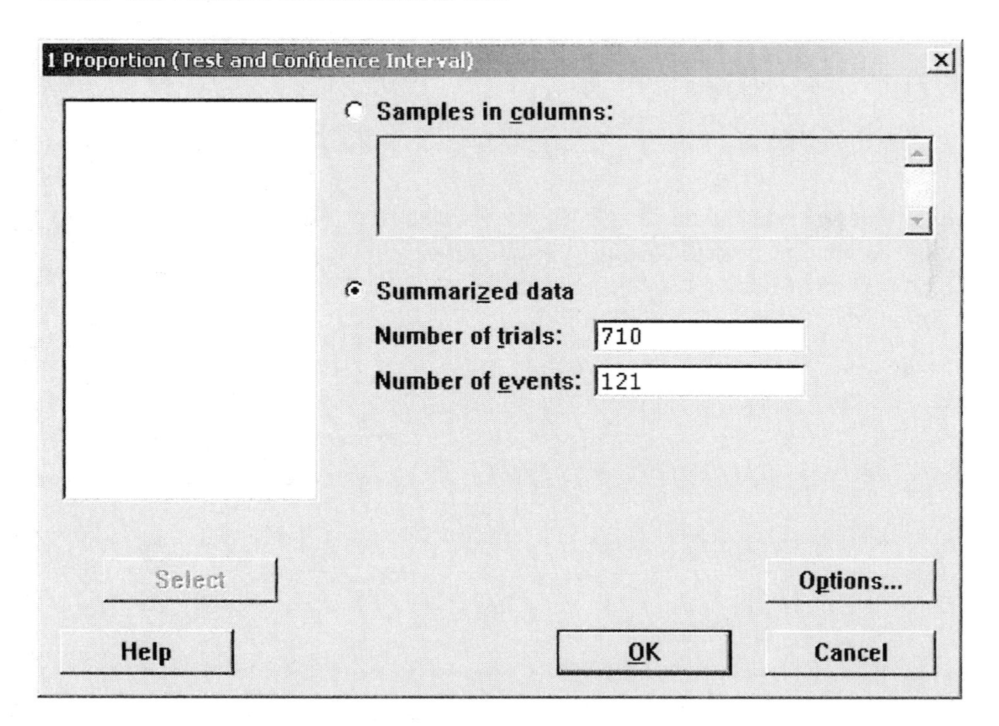

Click on **Options**. Enter .135 for the **Test Proportion** and select "not equal" for the **Alternative.** Because the assumption of normality has been met, select **Use test and interval based on normal distribution,** and then click on **OK** twice.

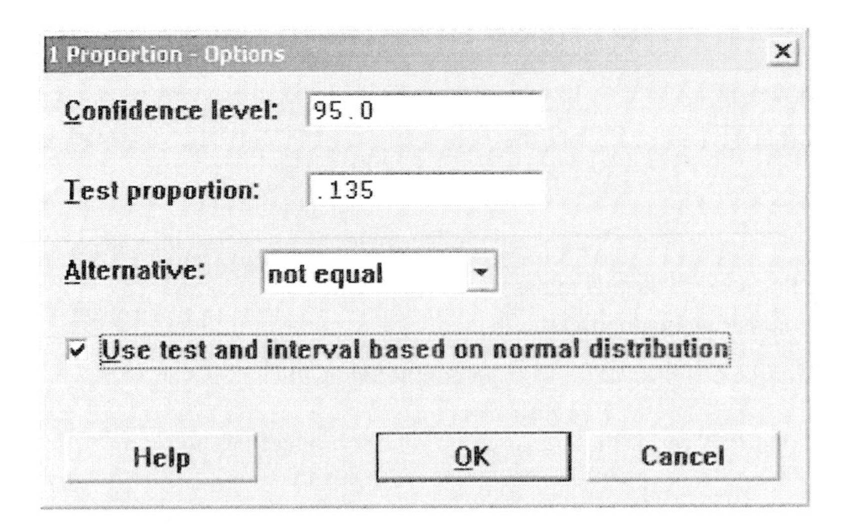

The results should be displayed in the Session Window.

Test and CI for One Proportion

Test of p = 0.135 vs p not = 0.135

Sample	X	N	Sample p	95% CI	Z-Value	P-Value
1	121	710	0.170423	(0.142765, 0.198080)	2.76	0.006

Notice that the test statistic (Z = 2.76), the P-value (P = .006) are displayed in the output. With such a small P-value, you should Reject the null hypothesis. Thus, the proportion of infants who experience loss of appetite while taking Prevnar is different from 0.135.

Example 4 (pg. 549) Hypothesis test for Population Proportion

Test whether an advertising campaign increased the proportion of males who consume the minimum daily requirement of calcium from the previous rate of 48.9% . A random sample of n=35 males taken after the campaign found that 21 consumed the RDA of calcium.

Use a hand-calculator to verify that the assumption of normality, required to perform the test, is met by evaluating the inequality: $(n\hat{p}_o(1-\hat{p}_o) \geq 10$.
For this example, you will discover that this assumption is NOT met.

Click on **Stat → Basic Statistics → 1-Proportion.** The data is given in a summarized form, so select **Summarized data**. Enter 35 for the **Number of trials.** The **Number of Successes** is 21. Click on **Options**. Enter .489 for the **Test Proportion** because it is was known that before the ad campaign, 48.9% consumed the RDA of calcium, and select "greater than" for the **Alternative**. Because the assumption of normality has not been met, do **NOT** select **Use test and interval based on normal distribution,** and then click on **OK** twice.

Test and CI for One Proportion

```
Test of p = 0.489 vs p > 0.489

                                                          Exact
Sample       X       N  Sample p  95.0% Lower Bound  P-Value
1           21      35  0.600000     0.447176            0.126
```

Since the P-value is .126 and is larger than $\alpha = .10$, you do not reject the null hypothesis. Thus, there is not enough evidence to prove that the advertising campaign was successful in increasing the daily intake of calcium among males in this age group.

▸ Problem 12 (pg. 551) Haunted Houses

Test whether there is significant evidence to support the claim that the proportion of adult Americans who believe in haunted houses has increased since 1996 when 33% of all adults believed. A random sample of n=1002 adults taken in 2005 found that 370 said they believed in haunted houses.

Use a hand-calculator to verify that the assumption of normality, required to perform the test, is met by evaluating the inequality: $(n\hat{p}_o(1-\hat{p}_o) \geq 10$.
For this example, you will discover that this assumption is met.

Click on **Stat → Basic Statistics → 1-Proportion.** The data is given in a summarized form, so select **Summarized data**. Enter 1002 for the **Number of trials.** The **Number of Successes** is 370. Click on **Options**. Enter .33 for the **Test Proportion** and select "greater than" for the **Alternative**. Because the assumption of normality has been met, select **Use test and interval based on normal distribution,** and then click on **OK** twice.

Test and CI for One Proportion

Test of p = 0.33 vs p > 0.33

Sample	X	N	Sample p	95% Lower Bound	Z-Value	P-Value
1	370	1002	0.369261	0.344184	2.64	0.004

Since the P-value is so small you would reject the null hypothesis. There is evidence to show that the proportion of adults Americans who believe in haunted houses has increased since 1996.

◂

▸ Problem 21 (pg. 552) Small Sample Hypothesis Test

Test whether there is significant evidence to support the claim that the percentage of mothers who smoke 21 cigarettes or more is less than 4%, the rate in 1997. A random sample of n=120 pregnant mothers found that 3 of them smoked 21 or more cigarettes per day.

Use a hand-calculator to verify that the assumption of normality, required to perform the test, is met by evaluating the inequality: $(n\hat{p}_o(1-\hat{p}_o) \geq 10$.
For this example, you will discover that this assumption is NOT met.

Click on **Stat → Basic Statistics → 1-Proportion.** The data is given in a summarized form, so select **Summarized data**. Enter 120 for the **Number of trials.** The **Number of Successes** is 3. Click on **Options**. Enter .04 for the **Test Proportion** and select "less than" for the **Alternative**. Because the assumption of normality has not been met, do NOT select **Use test and interval based on normal distribution,** and then click on **OK** twice.

Test and CI for One Proportion

```
Test of p = 0.04 vs p < 0.04

                                                           Exact
Sample        X      N  Sample p  95.0% Upper Bound  P-Value
1             3    120  0.025000           0.063344    0.289
```

Since the P-value is .289, which is larger than α = .05, you do not reject the null hypothesis. Thus, there is not enough evidence to prove that the proportion of pregnant mothers who smoke more than 21 cigarettes has decreased since 1997.

◀

Section 10.5

▸ Example 1 (pg. 555) Finding P-values for a Chi-Square Test

Once you have calculated the test statistic, χ^2, MINITAB can calculate the exact P-value of the test for you. In Example 1 on page 557 of the text, $\chi^2 = 7.291$ and there are 10 degrees of freedom. Click on **Calc → Probability distributions → Chi-square.** Select **Cumulative Probability** with **Noncentrality parameter = 0** and enter 19 **Degrees of Freedom.** Next, click on **Input Constant** and enter the test statistic, 25.996. Click on **OK** and the output will be displayed in the Session Window.

Cumulative Distribution Function

Chi-Square with 10 DF

x	P(X <= x)
7.291	0.302283

So the exact p-value is .302.

◂

▸ Problem 13 (pg. 557) Acid Rain

In this example, we are asked to test the hypothesis that $\sigma = 0.2$. The first step is to calculate the test statistic: $\chi^2 = (n-1) * s^2 / \sigma_o^2 = (19\text{-}1)*(.1708^2)/(.2^2) =$ 13.128.

Once you have calculated the test statistic, χ^2, MINITAB can calculate the exact P-value of the test for you. $\chi^2 = 13.128$ and there are 18 degrees of freedom. Click on **Calc → Probability distributions → Chi-square.** Select **Cumulative Probability** with **Noncentrality parameter = 0** and enter 18 **Degrees of Freedom.** Next, click on **Input Constant** and enter the test statistic, 13.128. Click on **OK** and the output will be displayed in the Session Window.

Cumulative Distribution Function

```
Chi-Square with 18 DF

        x    P( X <= x )
  13.1280         0.2161
```

Since the P-value = .2161, which is larger than $\alpha = .05$, you do not reject the null hypothesis that σ=.2.

◂

Inferences on Two Samples

Section 11.1

▸ Example 2 (pg. 577) Matched Pairs Data

Enter the data, found in Table 1 on page 577 of the textbook, into the MINITAB Data Worksheet. Put the Dominant Hand data in C1 and the Non-dominant Hand data in C2. Calculate the set of differences and check that they are approximately normal and there are no outliers. Click on **Calc → Calculator. Store result in** C3 and enter C1 - C2 for the **Expression.**
To draw a normal probability plot, click on **Graph → Probability Plot** and select the **Single** plot. Click on **OK.** Select C3 for the **Graph variable.** Click on **OK** and the probability plot will be displayed. Notice that all data points are contained within the confidence bands of the plot.
Next, check for outliers using a boxplot. Click on **Graph → Boxplot** and select the **Simple** boxplot. Click on **OK**. Select C3 for the **Graph variable.** To view a horizontal boxplot as shown in the textbook (rather than a vertical one) click on **Scale** and select **Transpose value and category scales.** Click on **OK**. Since there are no outliers, proceed to perform the hypothesis test.
Click on **Stat → Basic Statistics → Paired t.** Select **Samples in Columns** and enter C1 for the **First Sample** and C2 for the **Second Sample.**

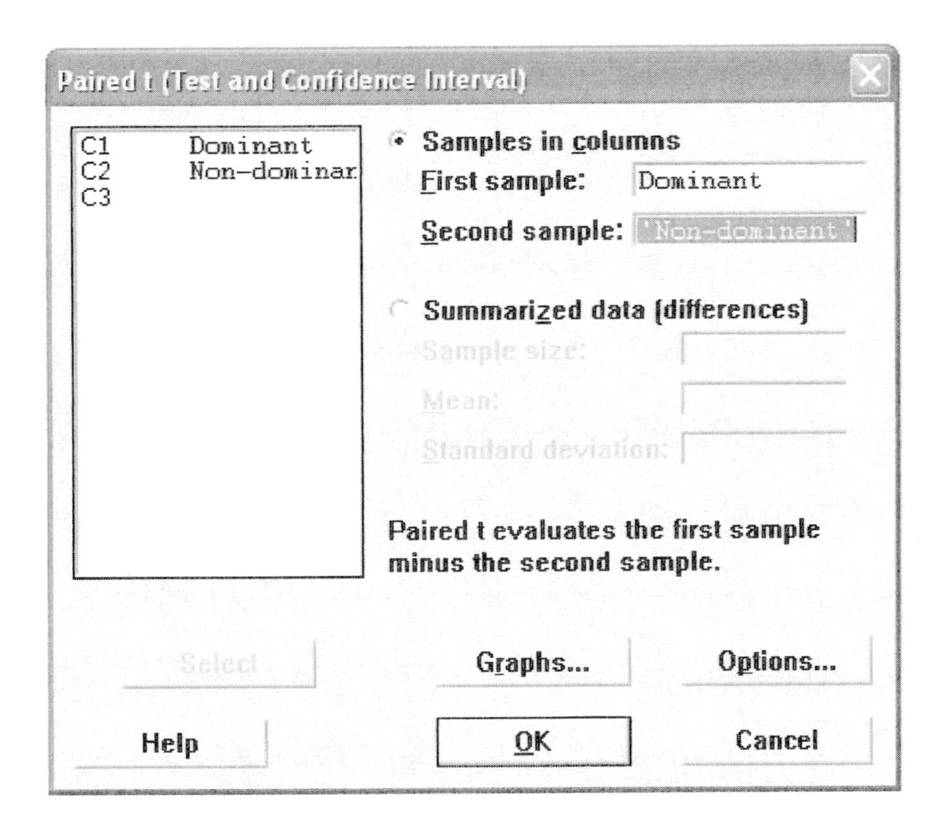

Click on **Options.** Enter 0 for **Test Mean** and select **less than** as the **Alternative.**

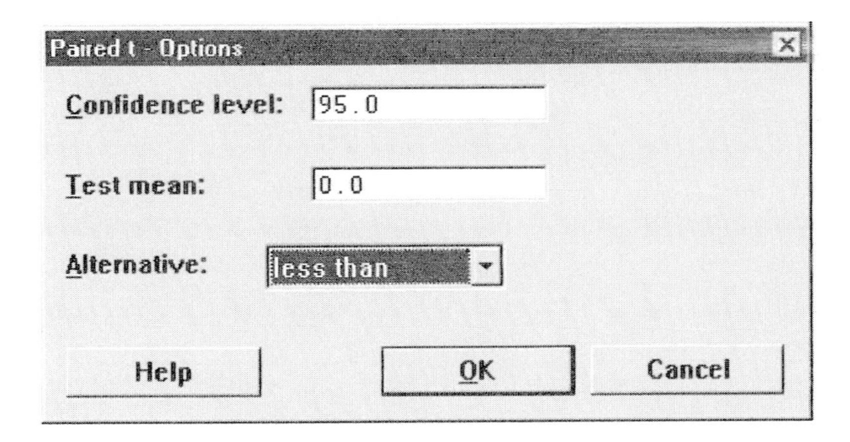

Click on **OK** twice to display the results in the Session Window.

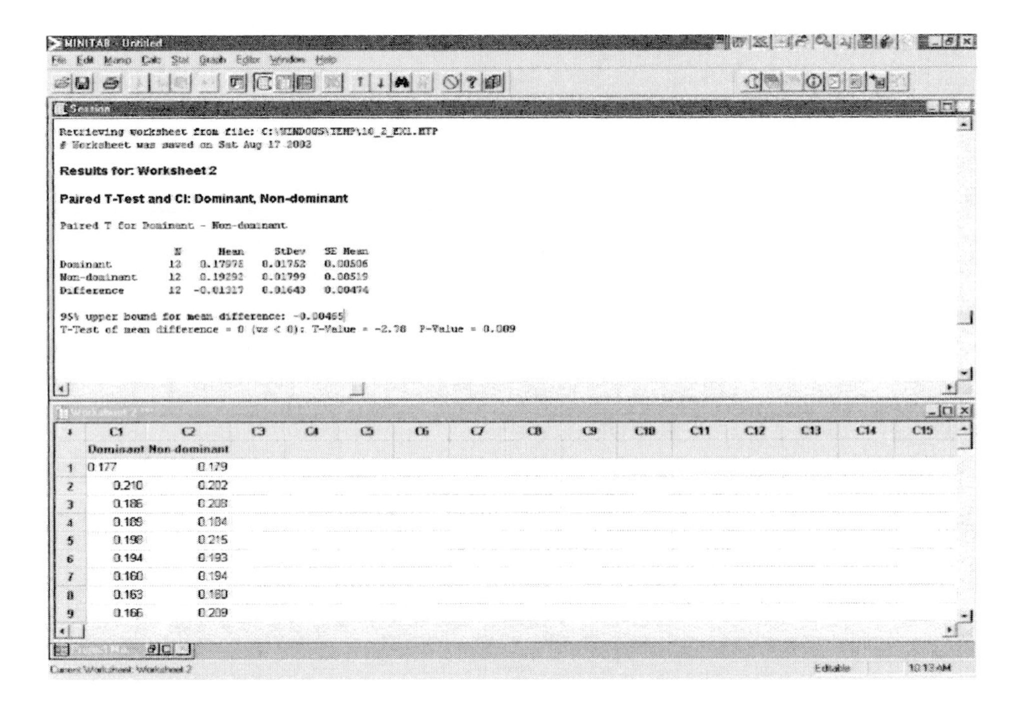

Notice that both the test statistic, t = -2.78, and the P-value = .009 are given. With such a small p-value, you Reject the null hypothesis. There is evidence to support Professor Neil's claim.

Note: Since you performed a left-tailed test, only the upper bound for the confidence interval is shown. To calculate the upper and lower bounds of a confidence interval, you must choose "not equal" as the alternative hypothesis.

◀

▸ Problem 13 (pg. 583) Muzzle Velocity

Open Minitab worksheet **11_1_13**. The Device A data is in C1 and the Device B data is in C2. Notice that for both samples, n = 12. The problem tells you that the differences are approximately normal and there are no outliers. Click on **Stat → Basic Statistics → Paired t.** Select **Samples in Columns**. Enter C1 for the **First Sample** and C2 for the **Second Sample.** Click on **Options.** Enter 99 for **Confidence level**, enter 0 for **Test Mean** and select **not equal** as the **Alternative** since you are testing if there is a difference in the measurements of Device A and Device B. Click on **OK** twice and the results of the test should be displayed in the Session Window.

Paired T for A - B

```
              N       Mean     StDev   SE Mean
A            12    792.458     1.407     0.406
B            12    792.342     1.603     0.463
Difference   12   0.116667  0.474501  0.136977


99% CI for mean difference: (-0.308757, 0.542090)
T-Test of mean difference = 0 (vs not = 0): T-Value = 0.85
P-Value = 0.413
```

Notice that the test statistic is T = .85 with a P-value = .413. Since this P-value is so large, you would Fail to Reject H_0 at any α level. Thus, there is not enough evidence to prove that Device A and Device B are different.

The 99% confidence interval is (-0.309, .542).

To draw the boxplot, create a column of differences. Click on **Calc → Calculator. Store result in** C3 and enter C1 - C2 for the **Expression.** Next, click on **Graph → Boxplot** and select the **Simple** boxplot. Click on **OK**. Select C3 for the **Graph variable.** To view a horizontal boxplot as shown in the textbook answer (rather than a vertical one) click on **Scale** and select **Transpose value and category scales.** Click on **OK**. ◂

Problem 15 (pg. 583) Seechi Disk

Open Minitab worksheet **11_1_15**. The Initial Depth data is in C2 and the Depth 5 years later is in C3. Notice that for both samples, n = 8. The problem tells you that the differences are approximately normal and there are no outliers. Click on **Stat → Basic Statistics → Paired t.** Select **Samples in Columns.** Enter C3 for the **First Sample** and C2 for the **Second Sample.** Click on **Options.** Enter 0 for **Test Mean** and select **greater than** as the **Alternative** since if the water is clearer 5 years later, the depth will be a larger number and thus the difference will be greater than 0. Click on **OK** twice and the results of the test should be displayed in the Session Window.

Paired T-Test and CI: Depth 5 years later, Initial Depth

```
Paired T for Depth 5 years later - Initial Depth

                    N     Mean    StDev  SE Mean
Depth 5 years la    8  59.5000   8.7342   3.0880
Initial Depth       8  54.3750  12.6935   4.4878
Difference          8  5.12500  6.08129  2.15006

95% lower bound for mean difference: 1.05154
T-Test of mean difference = 0 (vs > 0): T-Value = 2.38
P-Value = 0.024
```

Notice that the test statistic is T = 2.38 with a P-value = .024. Since this P-value is smaller than α=.05, you would Reject H_0. Thus, there is sufficient evidence to prove that the water is clearer 5 years later.

To construct a 95% confidence interval repeat the above steps, choosing **not equal** as the **Alternative** and 95 for the **Confidence level.** The 95% confidence interval for the mean difference is (0.04, 10.21).

```
Paired T for Depth 5 years later - Initial Depth

                    N     Mean    StDev  SE Mean
Initial Depth       8  54.3750  12.6935   4.4878
Depth 5 years la    8  59.5000   8.7342   3.0880
Difference          8  5.12500  6.08129  2.15006

95% CI for mean difference: (0.04091,10.20909)
T-Test of mean difference = 0 vs not = 0): T-Value = 2.38
P-Value = 0.049
```

To draw the boxplot, create a column of differences. Click on **Calc** → **Calculator. Store result in** C4 and enter C3 - C2 for the **Expression.** Next, click on **Graph** → **Boxplot** and select the **Simple** boxplot. Click on **OK**. Select C4 for the **Graph variable.** To view a horizontal boxplot as shown in the textbook answer (rather than a vertical one) click on **Scale** and select **Transpose value and category scales.** Click on **OK**.

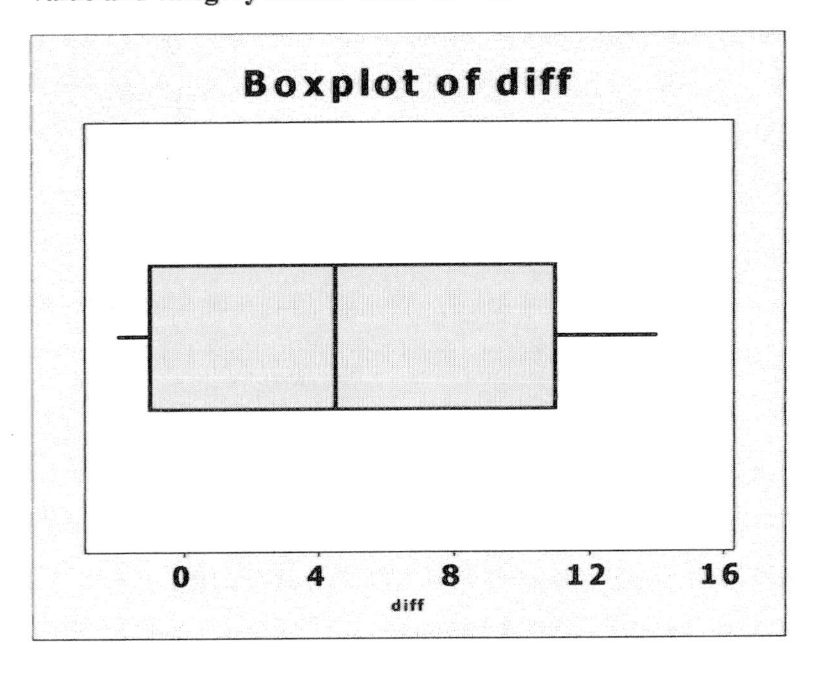

Section 11.2

▸ Example 1 (pg. 590) Testing a Claim Regarding Two Means

Enter the data from Table 3 on page 590 of the textbook into a Minitab worksheet. Enter the Flight measurements in C1 and the Control group measurements in C2. First check that the data is approximately normal and there are no outliers. To draw normal probability plots, click on **Graph → Probability Plot** and select the **Single** plot. Click on **OK.** Select C1 and C2 for the **Graph variables.** Click on **OK** and the probability plots will be displayed. Notice that all data points on each graph are contained within the confidence bands of the plot. Next, check for outliers using a boxplot. Click on **Graph → Boxplot** and select the **Multiple Y's Simple** boxplot. Click on **OK**. Select C1 and C2 for the **Graph variables.** To view horizontal boxplots as shown in the textbook (rather than vertical ones) click on **Scale** and select **Transpose value and category scales.** Click on **OK**. Since there are no outliers in either boxplot, proceed to perform the hypothesis test. Click on **Stat → Basic Statistics → 2-Sample t.** Select **Samples in different columns** and enter C1 for the **First** and C2 for the **Second** column. Click on **Options**, and then on the down arrow beside **Alternative** and select **not equal** since you want to test whether the mean red blood cell masses are different. Click on **OK** twice and the results of the test should be displayed in the Session Window.

Two-Sample T-Test and CI: Flight, Control
Two-sample T for Flight vs Control

	N	Mean	StDev	SE Mean
Flight	14	7.88	1.02	0.27
Control	14	8.43	1.01	0.27

Difference = mu (Flight) - mu (Control)
Estimate for difference: -0.549286
95% CI for difference: (-1.336653, 0.238082)
T-Test of difference = 0 (vs not =): T-Value = -1.44
P-Value = 0.163 DF = 25

Notice that the test statistic is T= -1.44. Since the P-value = .163 and is greater than the α-level of .05, there is not enough evidence to conclude that the mean red blood cell masses are different. The 95% confidence interval is also shown. It is (-1.337, 0.238). (Note that Minitab uses 25 degrees of freedom (a calculated value using Formula 2 on pg. 612), rather than 13 as is used in the textbook example. The p-values and the confidence intervals will be slightly different.) ◂

Problem 13 (pg. 597) Concrete strength

Open Minitab worksheet **11_2_13**. The Mixture 67-0-301 data is in C1 and the Mixture 67-0-400 data is in C2. The problem tells you that normal probability plots indicate that the data is approximately normal, and boxplots show no outliers.
Because the Confidence Interval in part c. of this problem asks for an interval for the difference: $\mu_{400} - \mu_{301}$, we will use C2 as our 'first' column and C1 as our 'second' column throughout the calculations.
For the hypothesis test in part b., click on **Stat → Basic Statistics → 2-Sample t.** Select **Samples in different columns** and enter C2 for the **First** and C1 for the **Second** column. Click on **Options**, and then on the down arrow beside **Alternative** and select **greater than** since you want to test if the 67-0-400 mixture is stronger than the 67-0-301 mixture. Click on **OK** twice and the results of the test should be displayed in the Session Window.

Two-Sample T-Test and CI: Mixture 67-0-400, Mixture 67-0-301

```
Two-sample T for Mixture 67-0-400 vs Mixture 67-0-301

                    N  Mean  StDev  SE Mean
Mixture 67-0-400   10  4483    474      150
Mixture 67-0301     9  3669    459      153


Difference = mu (Mixture 67-0-400) - mu (Mixture 67-0-301)
Estimate for difference:  814.111
95% lower bound for difference:  440.495
T-Test of difference = 0 (vs >): T-Value = 3.80  P-Value = 0.001
DF = 16
```

Since the p-value is so small, you would reject the null hypothesis. There is not enough evidence to prove that the 67-0-400 mix is stronger.

To find the 90% confidence interval, repeat the steps above, but this time click on **Options**, and select **not equal** as the **Alternative.** Be sure to enter 90 for the **Confidence Level.** Click on **OK** twice and the appropriate confidence interval should be displayed in the Session Window.

Two-Sample T-Test and CI: Mixture 67-0-400, Mixture 67-0-301

```
Two-sample T for Mixture 67-0-400 vs Mixture 67-0-301

                    N  Mean  StDev  SE Mean
Mixture 67-0-400   10  4483    474      150
Mixture 67-0301     9  3669    459      153


Difference = mu (Mixture 67-0-400) - mu (Mixture 67-0-301)
Estimate for difference:  814.111
90% CI for difference:   (440.495, 1187.727)
T-Test of difference = 0 (vs not =): T-Value = 3.80  P-Value =
0.002  DF = 16
```

The confidence interval is (440.5, 1187.7). This interval is slightly different than the one in the textbook answers because of the number of degrees of freedom used by Minitab.

To draw the boxplots, click on **Graph → Boxplot** and select the **Multiple Y's Simple** boxplot. Click on **OK**. Select C1 and C2 for the **Graph variables.** To view horizontal boxplots as shown in the textbook answer (rather than vertical ones) click on **Scale** and select **Transpose value and category scales.** Click on **OK**.

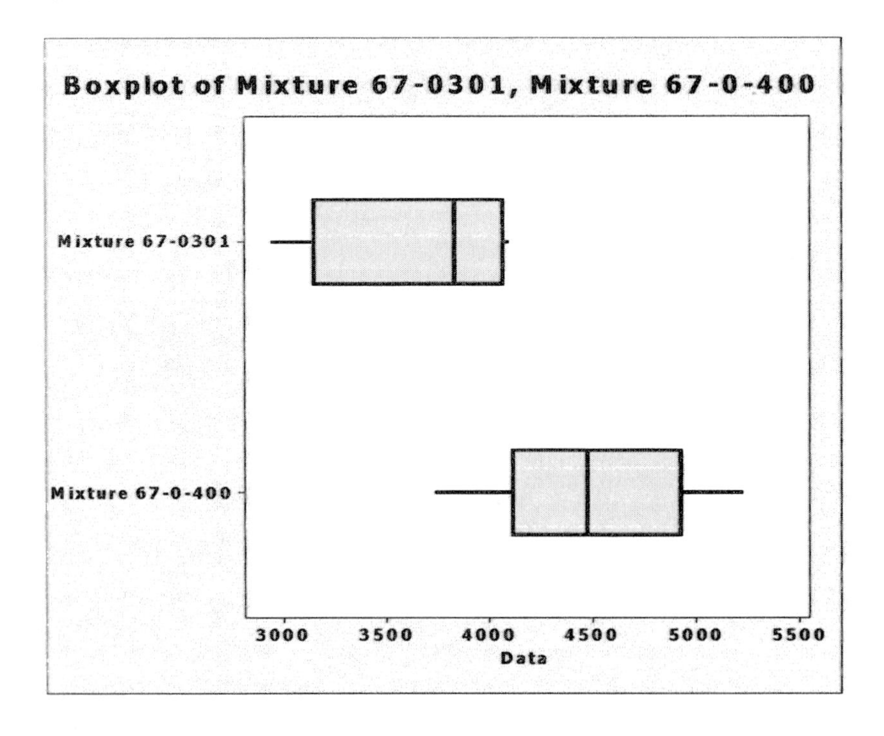

Section 11.3

Example 1 (pg. 604) Testing a claim regarding two proportions

In clinical trials of Nasonex, 3774 patients were randomly divided into an experimental group and a control group. Of the 2103 patients in the experimental group (Group 1), 547 reported headaches as a side effect. Of the 1671 patients in the control group (Group 2), 368 reported headaches as a side effect.

First, use a hand calculator to verify the requirements to perform the hypothesis test are met: ($n_1\hat{p}_1(1-\hat{p}_1)\geq 10$ and $n_2\hat{p}_2(1-\hat{p}_2)\geq 10$).

To test if the proportion of Nasonex users that experienced headaches is greater than the proportion in the control group, click on **Stat → Basic Statistics → 2 Proportions.** Select **Summarized Data** and use the data for Group 1 as the **First sample**. Enter 2103 **Trials** and 547 **Successes**. Use the data for Group 2 as the **Second sample**. Enter 1671 **Trials** and 368 **Successes.**

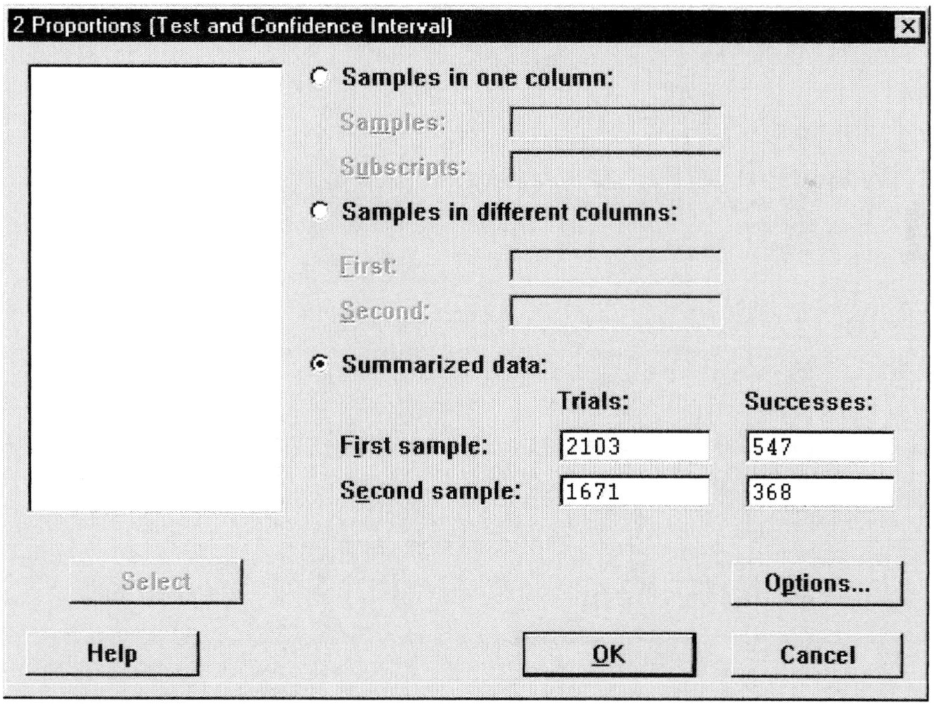

2 Proportions (Test and Confidence Interval)

Samples in one column:
Samples:
Subscripts:
Samples in different columns:
First:
Second:
Summarized data:

	Trials:	Successes:
First sample:	2103	547
Second sample:	1671	368

Select | Options... | Help | OK | Cancel

Click on **Options.** Enter 0 for **Test difference**, and select **greater than** as the **Alternative** since you want to test if proportion of Nasonex users whoexperienced headaches is greater than the proportion in the control group. Next click on **Use pooled estimate of p for test.**

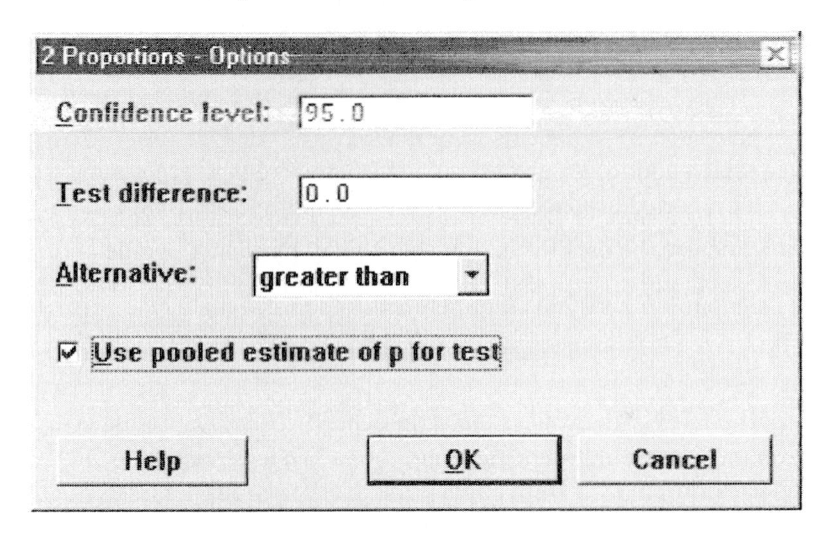

Click on **OK** twice to display the results in the Session Window.

Test and CI for Two Proportions

```
Sample      X      N  Sample p
1         547   2103  0.260105
2         368   1671  0.220227

Estimate for p(1) - p(2):  0.0398772
95% lower bound for p(1) - p(2):  0.0169504
Test for p(1) - p(2) = 0 (vs > 0):  Z = 2.84  P-Value = 0.002
```

Since the P-value is smaller than α, you should Reject the null hypothesis. The evidence suggests that Nasonex users reported more headaches than the control group.

(Note: to construct a confidence interval, the Alternative must be set to "not equal". You can enter any desired confidence level.)

▸ Problem 13 (pg. 610) Too Much Cholesterol in Your Diet?

Test the claim that a higher proportion of individuals with at most an 8th-grade education than individuals with at least some college consume too much cholesterol.

First, use a hand calculator to verify the requirements to perform the hypothesis test are met. Next, click on **Stat → Basic Statistics → 2 Proportions.** Select **Summarized Data** and use the data for individuals with at most an 8th-grade education (Group 1) as the **First sample**. Enter 320 **Trials** and 114 **Successes**. Use the data for individuals with some college (Group 2) as the **Second sample**. Enter 350 **Trials** and 112 **Successes**. Click on **Options.** Enter 0 for **Test mean,** and select **greater than** as the **Alternative** since you want to test if proportion of Group 1 individuals that consume too much cholesterol is greater than the proportion in Group 2. Next click on **Use pooled estimate of p for test.** Click on **OK** twice to display the results in the Session Window.

Test and CI for Two Proportions

```
Sample        X      N  Sample p
1           114    320  0.356250
2           112    350  0.320000

Estimate for p(1) - p(2):  0.03625
95% lower bound for p(1) - p(2):  -0.0239253
Test for p(1) - p(2) = 0 (vs > 0):  Z = 0.99  P-Value = 0.161
```

Since the P-value is larger than α, you should *Fail to Reject* the null hypothesis. The evidence does NOT suggest that Group 1 consumed more cholesterol than Group 2.

To construct a 95% confidence interval, repeat the steps above. This time the **Alternative** must be set to "**not equal**". Enter 95 for the **Confidence level.** Click on **OK** twice to display the results in the Session Window.

Test and CI for Two Proportions

```
Sample        X      N  Sample p
1           114    320  0.356250
2           112    350  0.320000

Estimate for p(1) - p(2):  0.03625
95% CI for p(1) - p(2):  (-0.0354533, 0.107953)
Test for p(1) - p(2) = 0 (vs not = 0):  Z = 0.99  P-Value = 0.322
```

The confidence interval is (-.035, 0.107). ◂

Section 11.4

Example 2 (pg. 616) Testing a claim regarding two Standard Deviations

Enter the data found in Table 5 on page 619 of the textbook into a Minitab worksheet. Enter the data for Cisco Systems is in C1 and the GE data is in C2. Test the investor's claim that Cisco Systems stock is more volatile than GE stock.

First, verify that both variables are approximately normal with normal probability plots. To draw normal probability plots, click on **Graph → Probability Plot** and select the **Single** plot. Click on **OK.** Select C1 and C2 for the **Graph variables.** Click on **OK** and the probability plots will be displayed. Notice that all data points on each graph are contained within the confidence bands of the plot.

Minitab performs a test of equal variances: $H_o: \sigma_1^2 = \sigma_2^2$ vs. $H_a: \sigma_1^2 \neq \sigma_2^2$. To perform the test of, click on **Stat → Basic Statistics → 2 Variances.** Select **Samples in different columns** and enter C1 for the **First** and C2 for the **Second.**

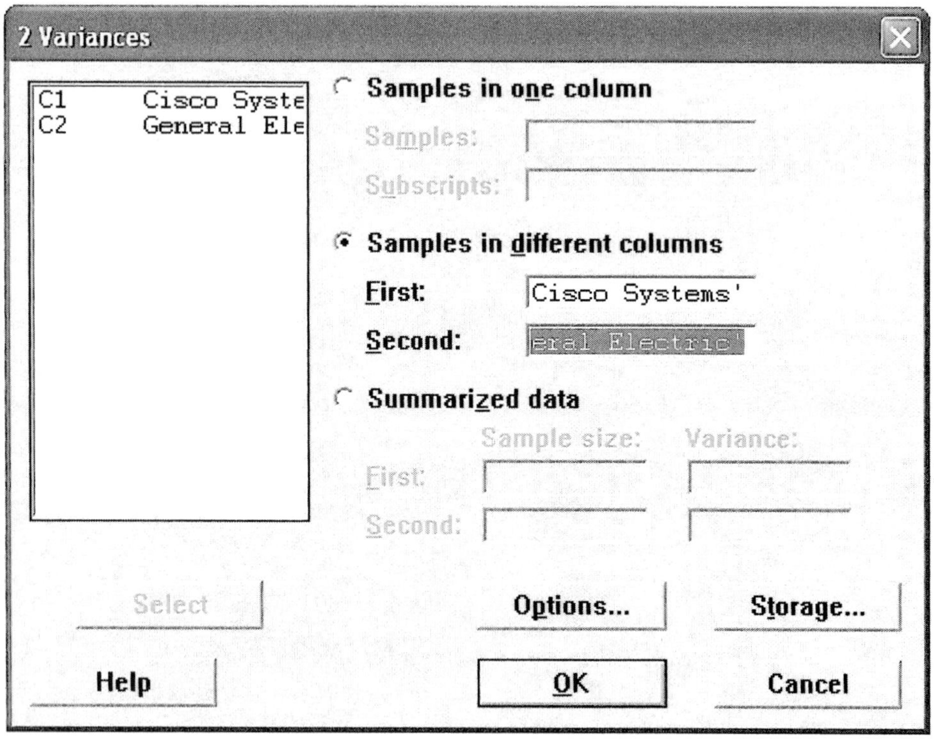

2 Variances

C1 Cisco Syste
C2 General Ele

Samples in one column
Samples:
Subscripts:
Samples in different columns
First: Cisco Systems'
Second: eral Electric'
Summarized data
Sample size: Variance:
First:
Second:

Select
Options...
Storage...
Help
OK
Cancel

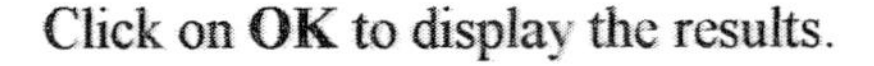

Click on **OK** to display the results.

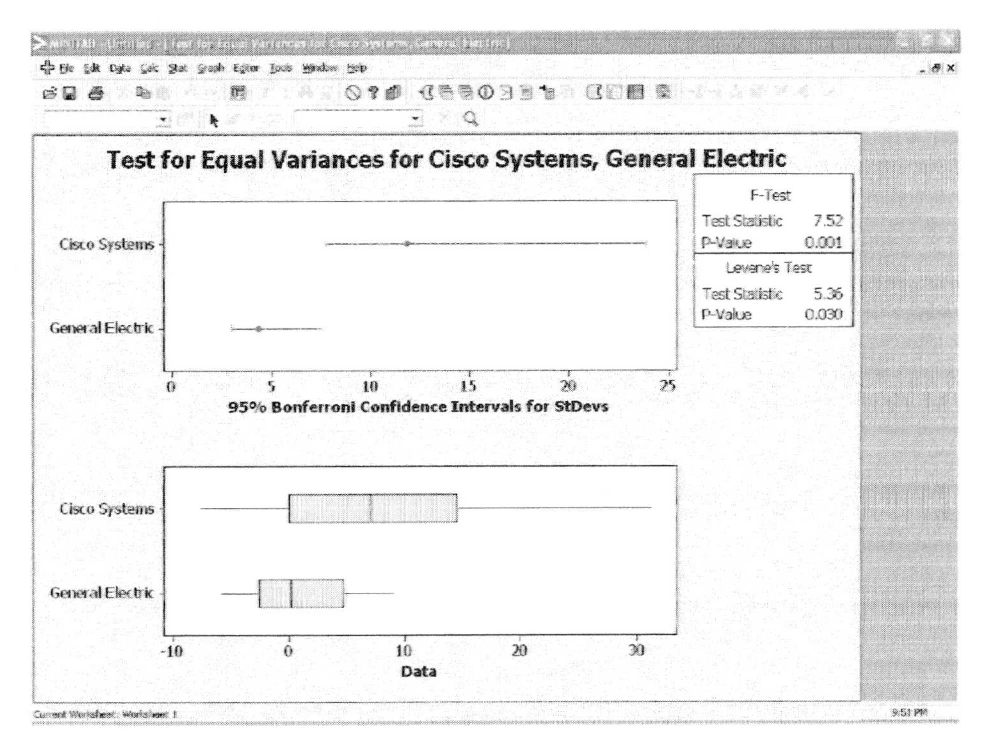

The results of the F-test are shown in the upper right corner of the window. F=7.52 and the p-value is .001. Thus, we would reject the null hypothesis. There is sufficient evidence to show that Cisco Systems stock and GE stock have different variances (and, therefore, different standard deviations).

(Note: Minitab performs the F-test as a two-tailed test. If you want to do a one-tailed test: $H_o: \sigma_1^2 = \sigma_2^2$ **vs.** $H_a: \sigma_1^2 > \sigma_2^2$**, divide the p-value in half. So, the appropriate p-value is .0005.)**

Problem 21 (pg. 621) Waiting Time in Line

Open worksheet **11_4_21.** The data for a Single Line is in C1 and the data for Multiple Lines is in C2. Test the nurse's claim that wait times have less variability when there is a single line. The problem states that the data is normally distributed.

Minitab performs a test of equal variances: $H_o: \sigma_1^2 = \sigma_2^2$ vs. $H_a: \sigma_1^2 \neq \sigma_2^2$. To perform the test, click on **Stat → Basic Statistics → 2 Variances.** Select **Samples in different columns** and enter C1 for the **First** and C2 for the **Second.**

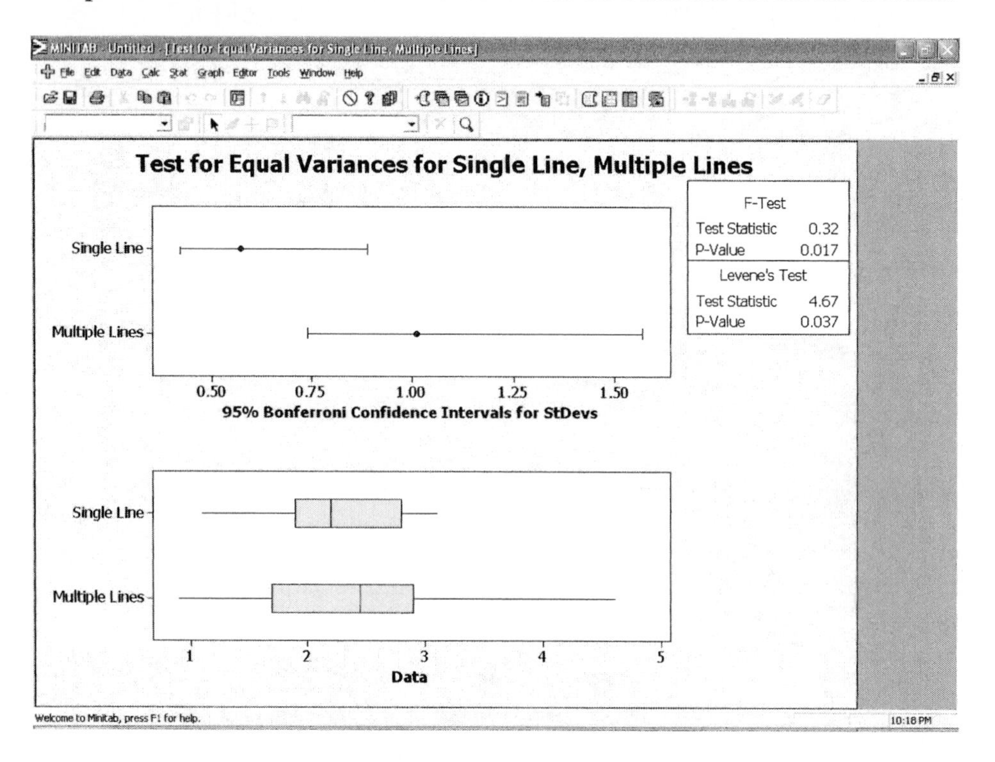

The F-statistic is 3.2 and the p-value is .017. Thus, we would reject the null hypothesis. There is sufficient evidence to show that the variances (and, therefore, the different standard deviations) are different.

(Note: Minitab performs the F-test as a two-tailed test. If you want to do the one-tailed test: $H_o: \sigma_1^2 = \sigma_2^2$ vs. $H_a: \sigma_1^2 < \sigma_2^2$, **divide the p-value in half. So, the appropriate p-value is .0085.)**

Inference on Categorical Data

Section 12.1

▸ Example 3 (pg. 635) The Chi-Square Goodness-of-Fit Test

Enter the data from Table 3 on page 636 of the textbook into a MINITAB worksheet. Enter the Day of the Week into C1 and name the column 'Day of Week.' Enter the frequencies into C2 and name it 'Observed.'

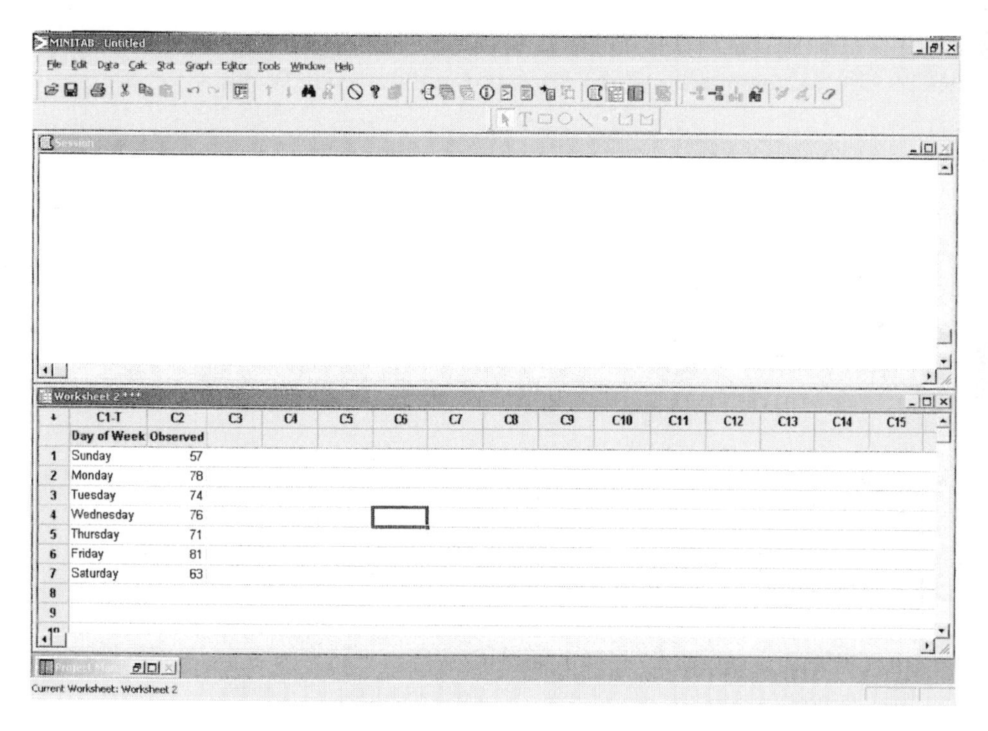

Click on **Stat → Tables → Chi-Square Goodness of Fit Test (one variable).** The **Observed counts** are in **C1** and the **Category names** are in **C2**. Select **Test Equal Proportions**.

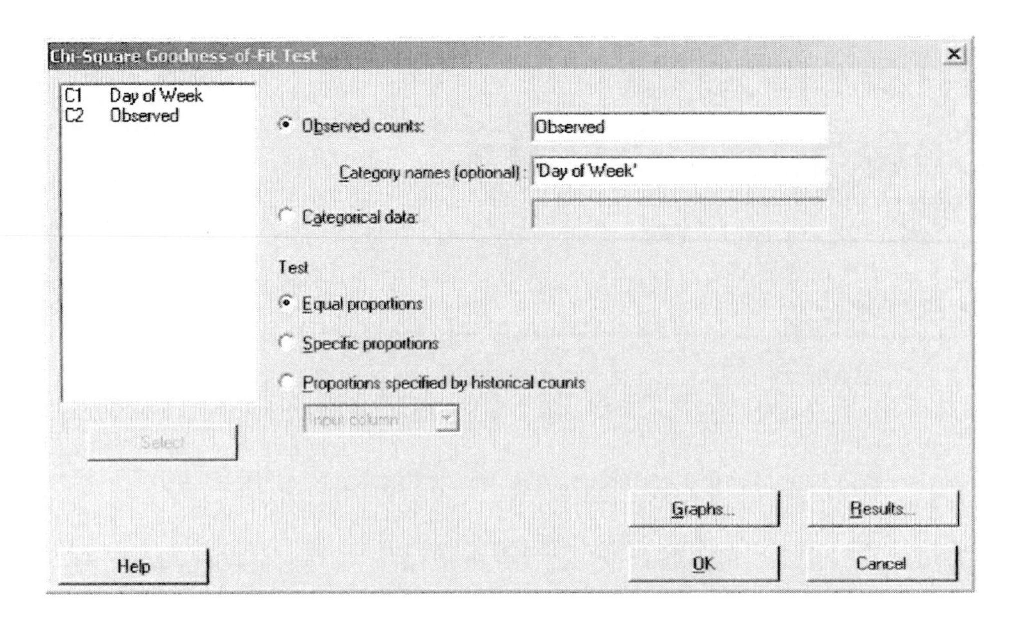

Click on **OK** and the results will be displayed in the Session Window.

Chi-Square Goodness-of-Fit Test for Observed Counts in Variable: Observed

Using category names in Day of Week

Category	Observed	Test Proportion	Expected	Contribution to Chi-Sq
Friday	81	0.142857	71.4286	1.28257
Monday	78	0.142857	71.4286	0.60457
Saturday	63	0.142857	71.4286	0.99457
Sunday	57	0.142857	71.4286	2.91457
Thursday	71	0.142857	71.4286	0.00257
Tuesday	74	0.142857	71.4286	0.09257
Wednesday	76	0.142857	71.4286	0.29257

N	**DF**	**Chi-Sq**	**P-Value**
500	6	6.184	0.403

Since the P-value is 0.403 and is larger than $\alpha = .05$, you should not reject the null hypothesis. You cannot reject the claim that the day on which a child is born occurs with equal frequency.

Problem 11 (pg. 639) Plain M&Ms

Enter the data into a MINITAB Data Window. Enter the Color into C1 and name the column 'Color.' Enter the frequencies into C2 and name it 'Observed.' Finally, enter the proportions stated by M&M/Mars into C3 and name it 'Proportions'.

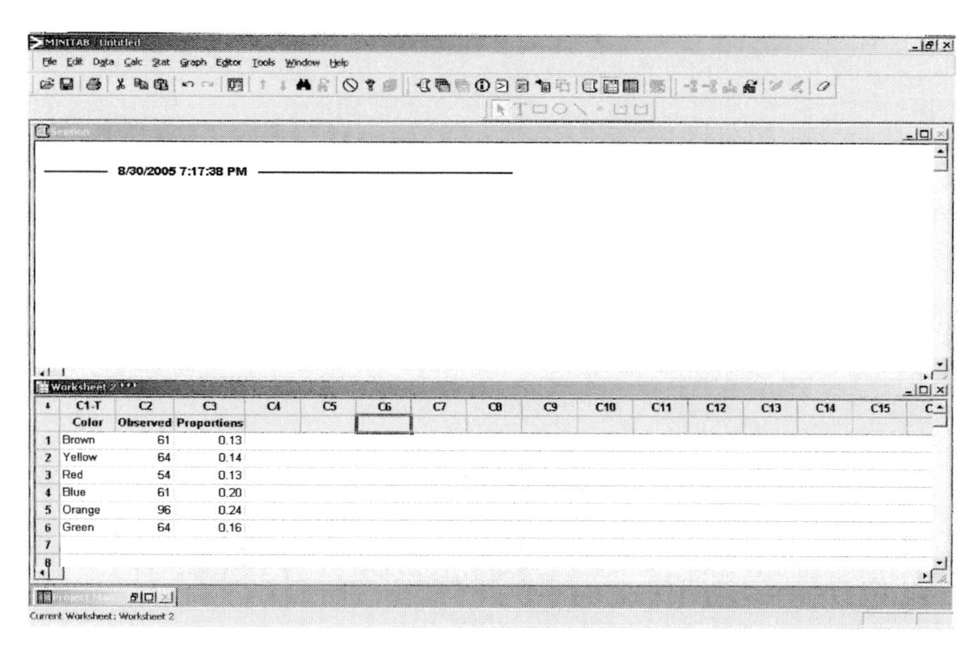

Click on **Stat → Tables → Chi-Square Goodness of Fit Test (one variable).** The **Observed counts** are in **C1** and the **Category names** are in **C2**. Select **Test Specific Proportions** and enter **C3**.

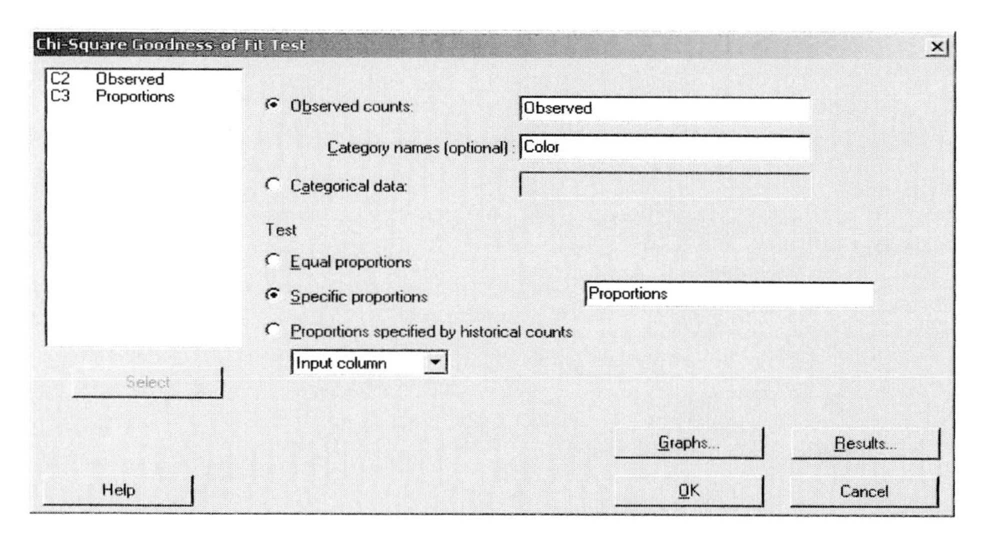

Click on **OK** and the Chi-Square test statistic will be displayed in the Session Window.

Chi-Square Goodness-of-Fit Test for Observed Counts in Variable: Observed

Using category names in Color

Category	Observed	Test Proportion	Expected	Contribution to Chi-Sq
Blue	61	0.20	80	4.51250
Brown	61	0.13	52	1.55769
Green	64	0.16	64	0.00000
Orange	96	0.24	96	0.00000
Red	54	0.13	52	0.07692
Yellow	64	0.14	56	1.14286

N	DF	Chi-Sq	P-Value
400	5	7.28997	0.200

The P-value is 0.200. Since this P-value is so large, you would ***fail to reject*** the null hypothesis at $\alpha = .05$. Thus, you cannot reject the manufacturer's claim regarding the distribution of colors of the M&Ms.

◀

Section 12.3

▸ Example 4 (pg. 658) Chi-Square Independence Test

Enter the data from Table 14 on page 657 of the textbook into the MINITAB Data Window. First label the columns: use Rh Level for C1, "A" for C2, "B" for C3, "AB" for C4, and "O" for C5. Now enter the data into the appropriate columns.

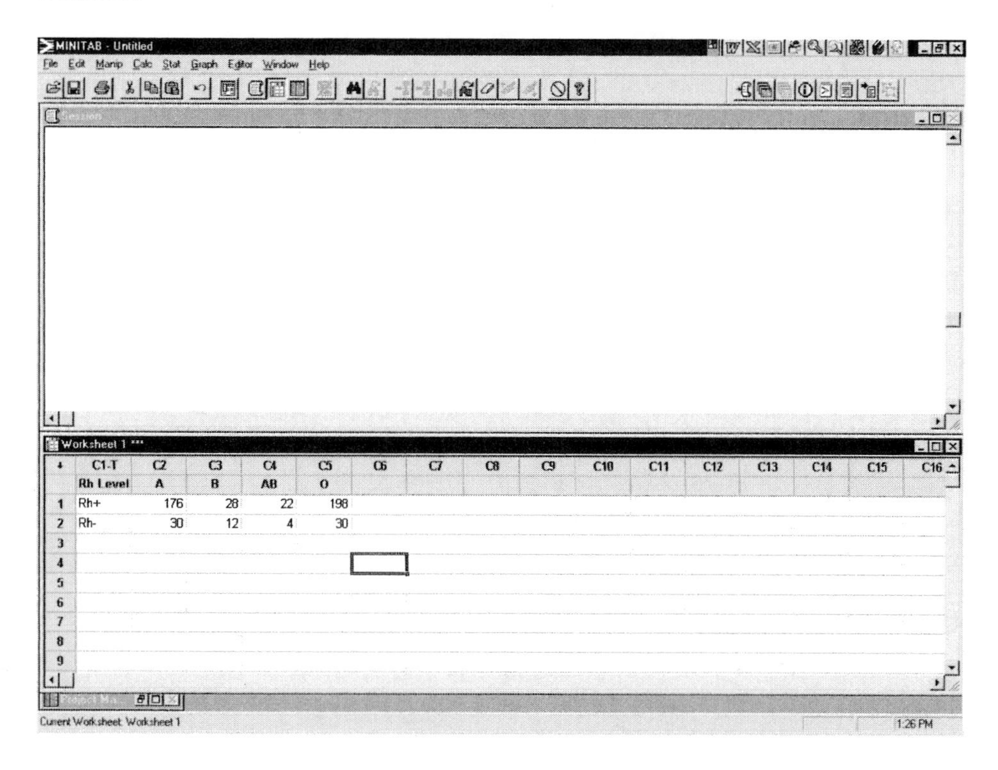

To perform the chi-square independence test, click on **Stat → Tables → Chi-square Test (Two way table in worksheet).** On the input screen, select C2 - C5 for the **Columns containing the table.** Click on **OK** and the test results will be displayed in the Session Window.

Chi-Square Test: A, B, AB, O

```
Expected counts are printed below observed counts
Chi-Square contributions are printed below expected counts

              A        B       AB        O  Total
    1       176       28       22      198    424
         174.69    33.92    22.05   193.34
          0.010    1.033    0.000    0.112

    2        30       12        4       30     76
          31.31     6.08     3.95    34.66
          0.055    5.764    0.001    0.626

Total       206       40       26      228    500

Chi-Sq = 7.601, DF = 3, P-Value = 0.055
1 cells with expected counts less than 5.
```

Notice that the Minitab output contains the observed counts, the expected counts and the Chi-Square contribution for each category in the table. The test statistic is Chi-Sq = 7.601 and the P-value = .055. Since this P-value is larger than α = .05, you should not reject the null hypothesis. Thus, there is not enough evidence to conclude that Rh-level and blood type are related.

◀

▸ Problem 9 (pg. 663) Education versus Area of Country

Open Minitab worksheet **12_3_9**. To perform the chi-square independence test, click on **Stat → Tables → Chi-square Test.** On the input screen, select C2 - C5 for the **Columns containing the table.** Click on **OK** and the test results will be displayed in the Session Window.

Chi-Square Test: Not a High Schoo, High School Grad, Some College, BA or higher

```
Expected counts are printed below observed counts
Chi-Square contributions are printed below expected counts

           Not a
            High      High
          School    School      Some    BA or
        Graduate  Graduate   College   higher  Total
    1         52       123        70       94    339
           66.90    108.80     80.99    82.31
           3.318     1.852     1.492    1.661

    2        123       146       102       96    467
           92.16    149.89    111.57   113.38
          10.322     0.101     0.821    2.666

    3        119       204       148      144    615
          121.36    197.39    146.93   149.32
           0.046     0.222     0.008    0.189

    4         62       106       111      104    383
           75.58    122.93     91.50    92.99
           2.440     2.330     4.154    1.304

Total        356       579       431      438   1804

Chi-Sq = 32.926, DF = 9, P-Value = 0.000
```

Notice that the test statistic is Chi-Sq = 32.926 and the P-value = .000. Since this P-value is smaller than $\alpha = .05$, you should reject the null hypothesis. Thus, there is evidence to conclude that the 'level of education' and 'region of the US' are related.

◂

Comparing Three or More Means

Section 13.1

Example 2 (pg. 680) Performing One-Way ANOVA

Enter the data in Table 1 (found on page 679 of your textbook) into C1 – C4 of a Minitab worksheet.

First verify that the data is approximately normal. Click on **Graph → Probability Plot** and select the **Single** plot. Click on **OK.** Select C1- C4 for the **Graph variables.** Click on **OK.** The four graphs will appear stacked on top of each other. Verify that the data is approximately normal for each dataset.

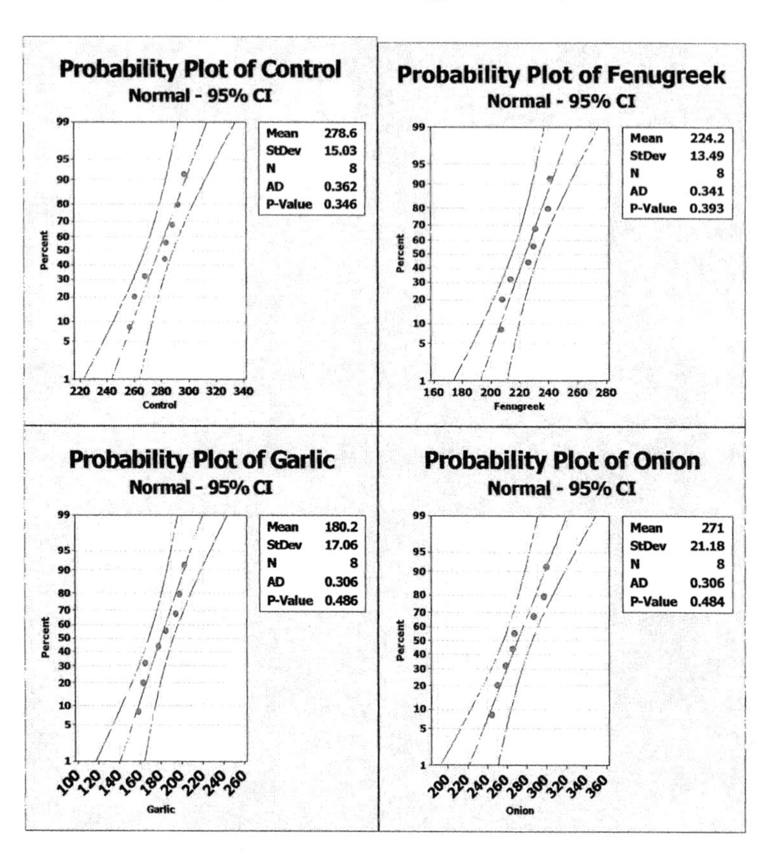

To test the claim that there is a difference in the mean glucose among the four treatment groups, perform a one-way analysis of variance. Click on **Stat →ANOVA → One Way (Unstacked).** Select all four columns. Click on **Graphs** and select **Boxplots of data.** Click on **OK** twice. The results of the test will be displayed in the Session Window and the boxplots will be displayed in the Graph Window.

One-way ANOVA: Control, Fenugreek, Garlic, Onion

Source	DF	SS	MS	F	P
Factor	3	50091	16697	58.21	0.000
Error	28	8032	287		
Total	31	58122			

S = 16.94 R-Sq = 86.18% R-Sq(adj) = 84.70%

```
                          Individual 95% CIs For Mean Based on
                          Pooled StDev
Level      N   Mean  StDev  --+---------+---------+---------+-------
Control    8  278.56  15.03                              (---*--)
Fenugreek  8  224.16  13.49                (--*---)
Garlic     8  180.24  17.06  (--*---)
Onion      8  271.00  21.18                           (--*---)
                            --+---------+---------+---------+-------
                             175       210       245       280
```

Pooled StDev = 16.94

Notice that F=58.21 with a P-value of 0.000. Since this is smaller than α=.05, you should reject the null hypothesis. There is sufficient evidence to conclude that the mean glucose levels are different.

The confidence intervals for the means are pictured in the Session Window. Notice that some of the intervals do not overlap, thus supporting the conclusion of the test -- that the means are not all equal.
Next, look at the boxplots of the four study groups.

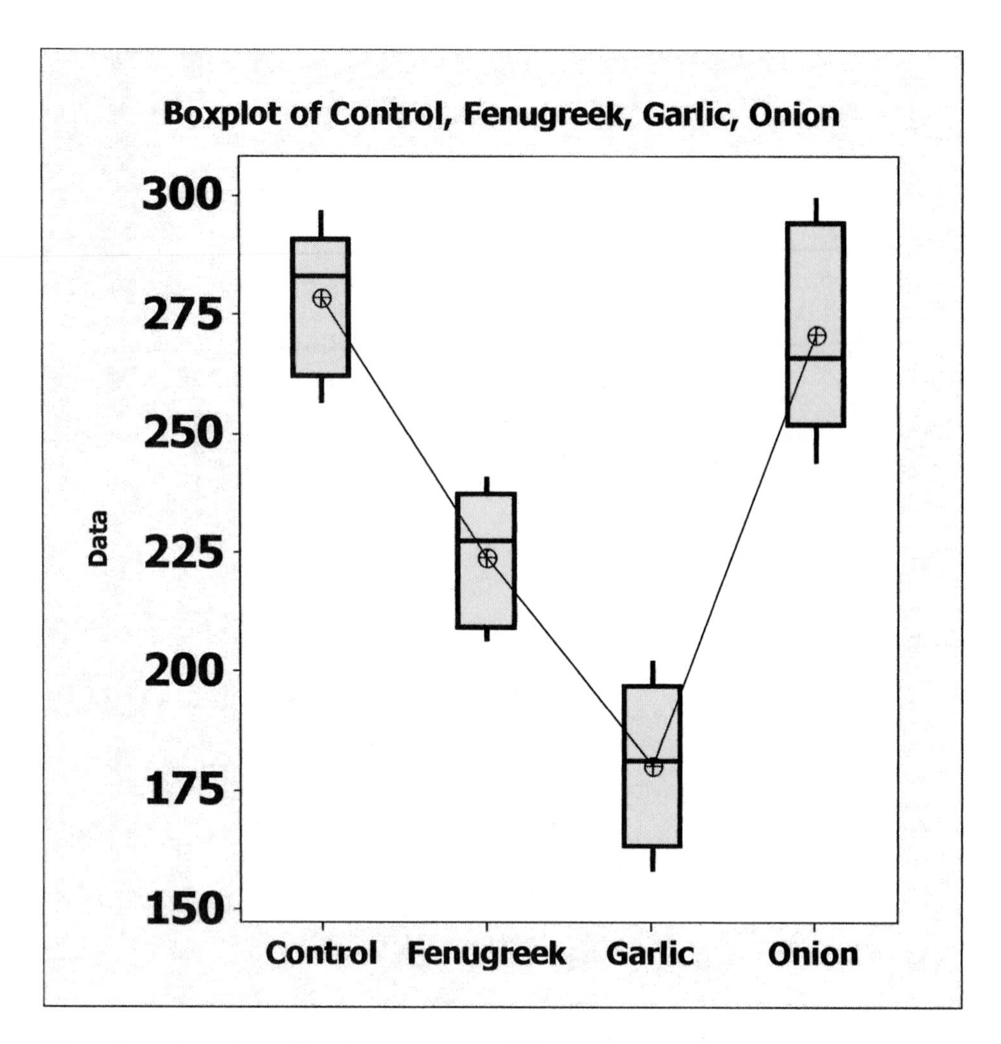

These boxplots show that the means are different, and results of the ANOVA show that they are also statistically significantly different.

Problem 17 (pg. 689) Crash Data

Open worksheet **13_1_17.** The chest compression data for the three different classes of vehicles should be in C2, C4, and C6. First verify that the data is approximately normal. Click on **Graph → Probability Plot.** Enter C2 for the **Variable**, be sure that the **Distribution** is Normal, and click on **OK**. Repeat this for C4 and C6. Notice that all are approximately normal. To perform a one-way analysis of variance, click on **Stat → ANOVA → One Way (Unstacked).** Select all three columns (C2, C4, and C6), and click on **Graphs.** Select **Boxplots of data.** Click on **OK.** The results of the test will be in the Session Window and the boxplots will be in the Graph Window.

One-way ANOVA: Large Family Cars, Passenger Vans, Midsize Utility Vehicles

Source	DF	SS	MS	F	P
Factor	2	70.6	35.3	2.94	0.079
Error	18	216.0	12.0		
Total	20	286.6			

S = 3.464 R-Sq = 24.63% R-Sq(adj) = 16.25%

Level	N	Mean	StDev
Large Family Car	7	30.571	3.259
Passenger Vans	7	27.714	3.729
Midsize Utility	7	32.143	3.388

Individual 95% CIs For Mean Based on Pooled StDev

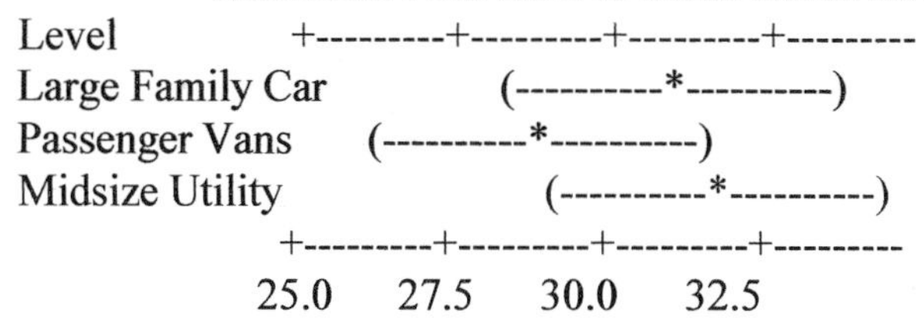

Pooled StDev = 3.464

The test statistic is F=2.94. Since the p-value is .079 and is larger than α=.01, do not reject the null hypothesis. The evidence supports theory that the mean chest compression for the three vehicle types is equal. The picture of the confidence intervals shows that the three intervals overlap, supporting the conclusion of the test. The boxplots also confirm the result.

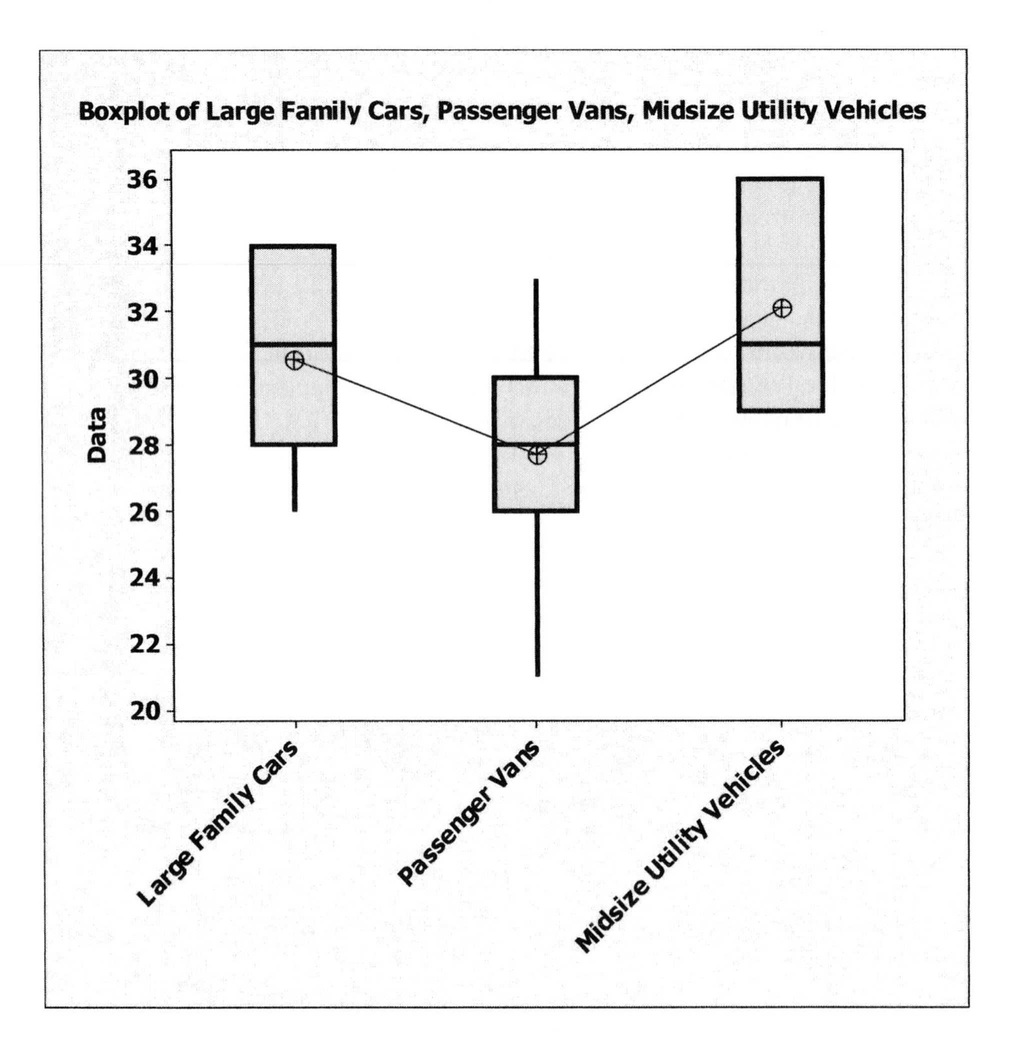

The mean chest compression for the three vehicle types is very similar.

Section 13.2

▸ Problem 16 (pg. 698) Price to Earnings Ratio

Open Minitab worksheet **13_2_16**. The data is in C1-C3 of the worksheet. To test whether the P/E ratio for the three companies differ significantly, perform an ANOVA. The Tukey Test is an option in Minitab's ANOVA. Click on **Stat → ANOVA → One Way (Unstacked).** Select all three columns. Click on **Comparisons** and select **Tukey's** and enter .05 for the **family error rate.** Next click on **Graphs.** Select **Boxplots of data.** Click on **OK.** The results of the test will be in the Session Window and the boxplots will be in the Graph Window.

One-way ANOVA: Financial, Food, Leisure

```
Source  DF      SS     MS     F      P
Factor   2  116.19  58.09  8.55  0.005
Error   12   81.58   6.80
Total   14  197.76

S = 2.607   R-Sq = 58.75%   R-Sq(adj) = 51.87%

                              Individual 95% CIs For Mean Based on
                              Pooled StDev
Level      N    Mean  StDev  ---------+---------+---------+---------+
Financial  5  11.908  2.365  (--------*-------)
Food       5  18.248  3.159                      (--------*-------)
Leisure    5  12.908  2.196    (-------*-------)
                             ---------+---------+---------+---------+
                                   12.0      15.0      18.0      21.0

Pooled StDev = 2.607
```

Notice that F=8.55 and the p-value is 0.005. Thus you would reject the null hypothesis that the three mean P/E ratios are the same. Next, look at the Tukey comparisons and boxplots.

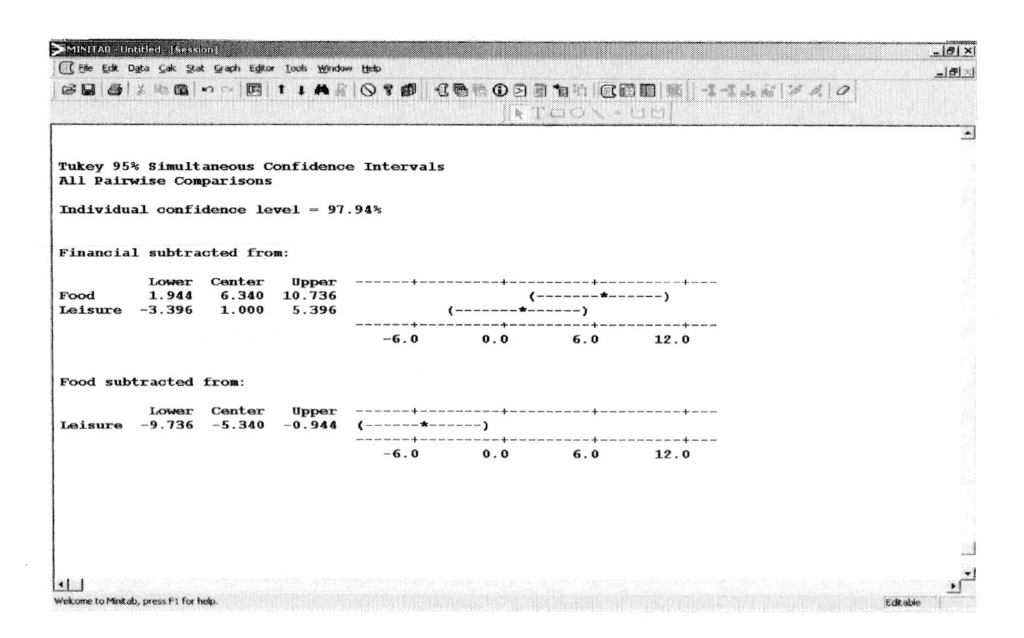

Notice that confidence intervals for the differences between Food and both Financial and Leisure do not include 0. This tells you that the P/E ratio for Food goods is different from that of both Financial and Leisure goods. The boxplots below also confirm the results.

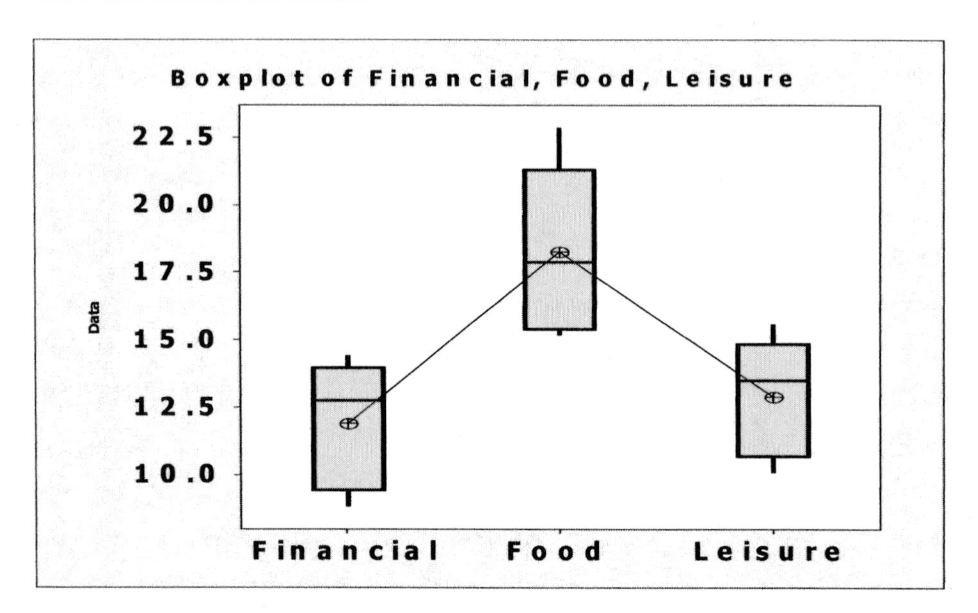

Section 13.3

Example 2 (pg. 704) Randomized Complete Block Design

Minitab is expecting the data to be in separate columns, one for the block, another for the treatment (diet), and a third for the response (weight gain). Enter the data into a Minitab worksheet.

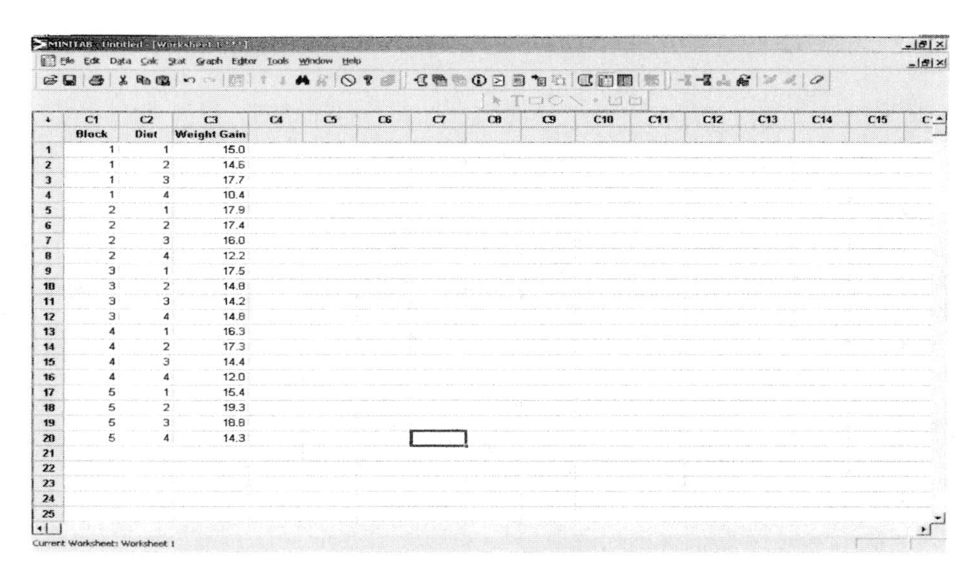

	C1 Block	C2 Diet	C3 Weight Gain
1	1	1	15.0
2	1	2	14.6
3	1	3	17.7
4	1	4	10.4
5	2	1	17.9
6	2	2	17.4
7	2	3	16.0
8	2	4	12.2
9	3	1	17.5
10	3	2	14.8
11	3	3	14.2
12	3	4	14.8
13	4	1	16.3
14	4	2	17.3
15	4	3	14.4
16	4	4	12.0
17	5	1	15.4
18	5	2	19.3
19	5	3	18.8
20	5	4	14.3

To test whether the mean weight gain for the four diets differ significantly, perform a two-way ANOVA. The Tukey Test is an option in Minitab's ANOVA. Click on **Stat → ANOVA → Two Way.** Enter the columns as shown below. Select **Display means** and also **Fit additive model.**

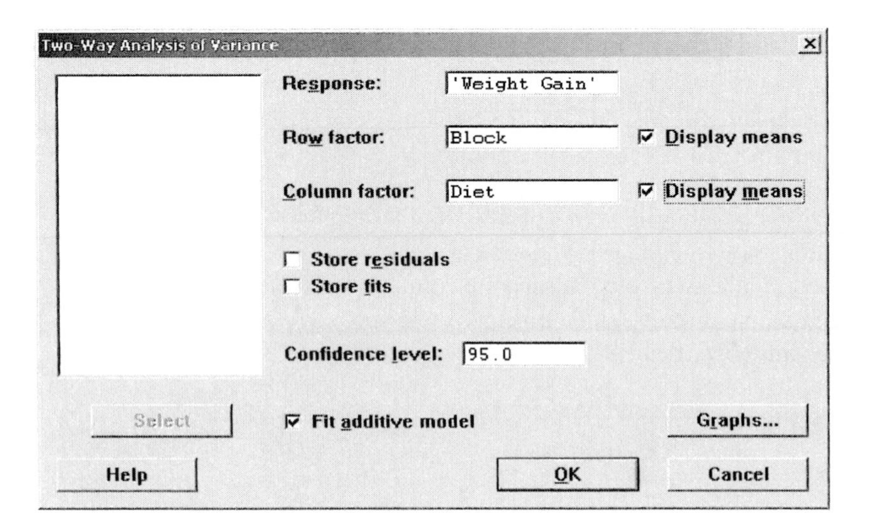

Click on **OK.** The results of the test will be in the Session Window.

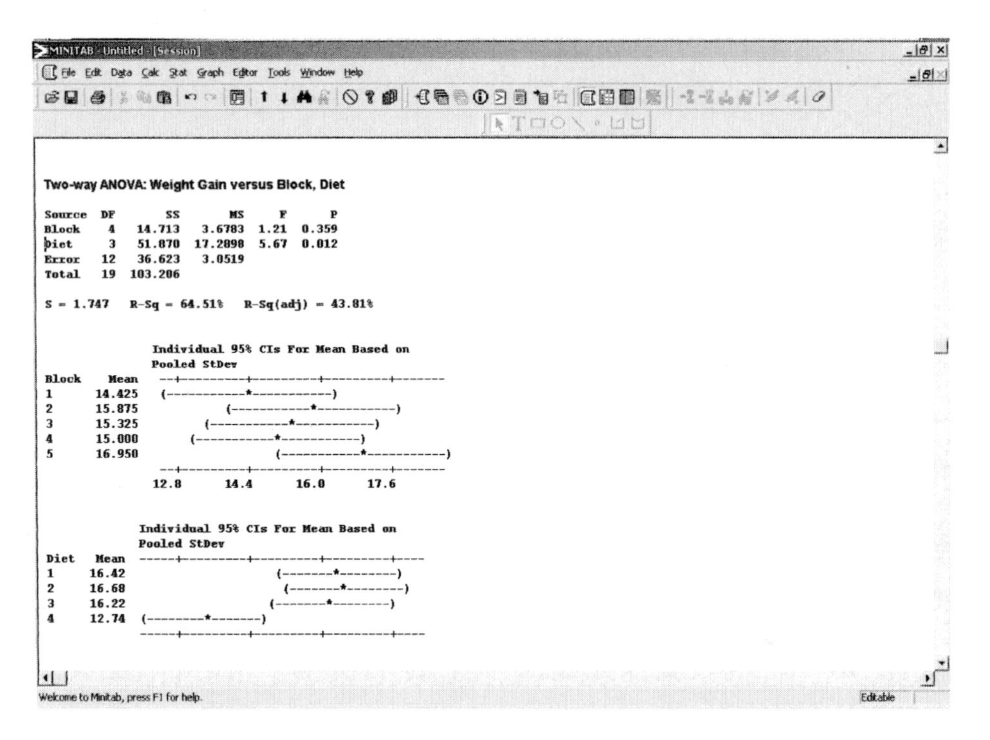

From the output, the p-value for Diet is 0.012. Since this is less than .05, reject the null hypothesis. Thus, there is evidence that the mean weight gain is not the same for all four diets.

◀

▸ Example 3 (pg. 705) Tukey's Test

To perform pairwise comparisons from a two-way ANOVA using Minitab, you must use the General Linear Model. This will give you the same output as the two-way ANOVA, but can also do the comparisons. Again, Minitab is expecting the data to be in separate columns, one for the block, another for the treatment (diet), and a third for the response (weight gain). Enter the data into a Minitab worksheet just as in Example 2 on the previous page. Click on **Stat → ANOVA → General Linear Model.** Enter Weight Gain for **Responses**, enter both Block and Diet for **Model**, and enter Block for **Random factors.**

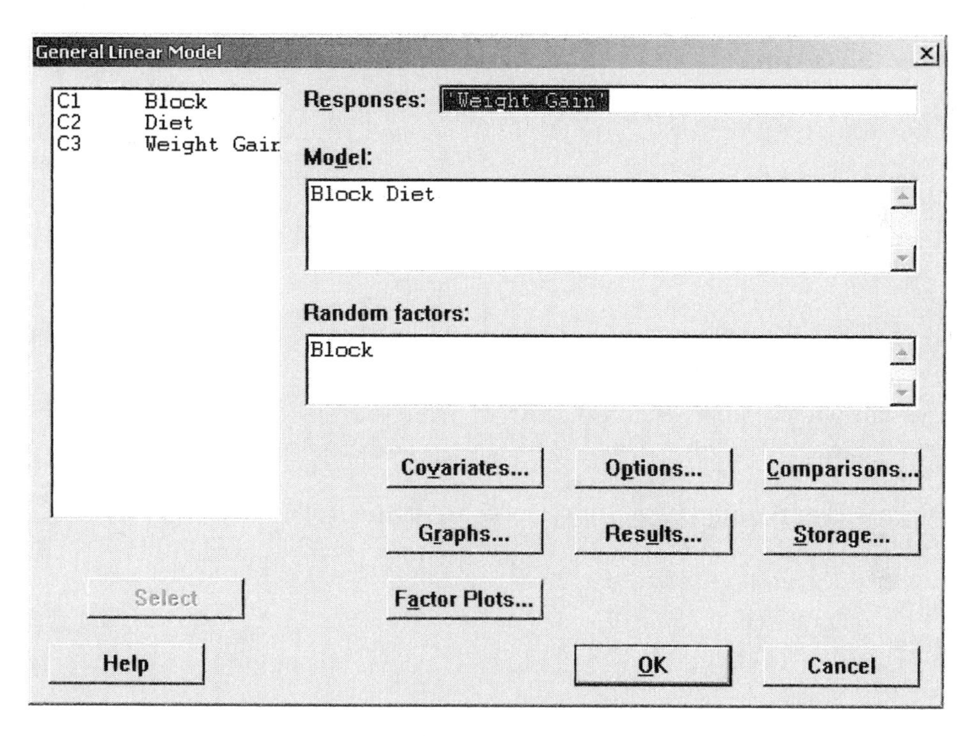

Next, click on **Comparisons.** Select **Pairwise comparisons** and enter Diet for **Terms.** This tells Minitab which variable you want to use for the pairwise comparisons. Select **Tukey**. If you want confidence intervals for the comparisons, select **Confidence interval**. Finally, if you want to test whether the pairs are different, select **Test.** Click **OK.**

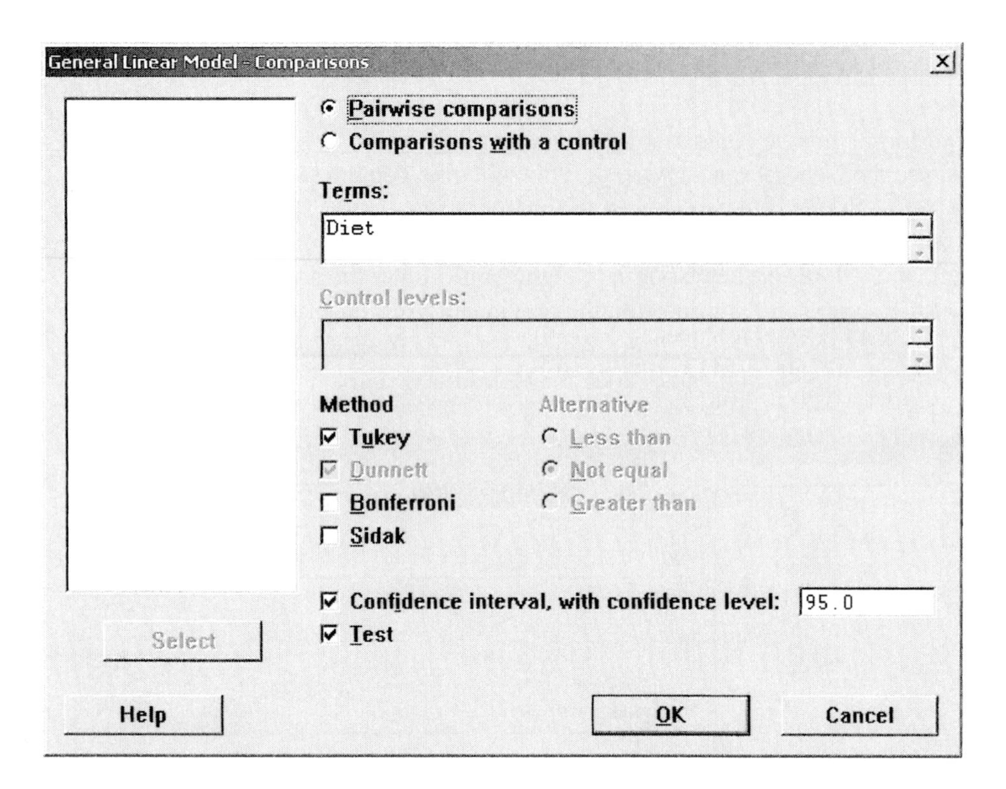

The results will be in the Session Window. First look at the results from the ANOVA. Notice that the p-value is 0.012 just as in Example 2.

General Linear Model: Weight Gain versus Block, Diet

```
Factor   Type     Levels  Values
Block    random        5  1, 2, 3, 4, 5
Diet     fixed         4  1, 2, 3, 4


Analysis of Variance for Weight Gain, using Adjusted
SS for Tests

Source   DF   Seq SS   Adj SS   Adj MS     F      P
Block     4   14.713   14.713    3.678  1.21  0.359
Diet      3   51.870   51.870   17.290  5.67  0.012
Error    12   36.623   36.623    3.052
Total    19  103.206


S = 1.74697   R-Sq = 64.51%   R-Sq(adj) = 43.81%
```

Next, look at the pairwise comparisons.

Tukey 95.0% Simultaneous Confidence Intervals
Response Variable Weight Gain
All Pairwise Comparisons among Levels of Diet

Diet = 1 subtracted from:

```
Diet  Lower  Center   Upper  ----+---------+---------+---------+--
2    -3.021   0.260  3.5413          (----------*----------)
3    -3.481  -0.200  3.0813         (----------*----------)
4    -6.961  -3.680 -0.3987  (----------*----------)
                             ----+---------+---------+---------+--
                               -6.0      -3.0       0.0       3.0
```

Diet = 2 subtracted from:

```
Diet  Lower  Center   Upper  ----+---------+---------+---------+--
3    -3.741  -0.460  2.8213              (---------*----------)
4    -7.221  -3.940 -0.6587  (----------*----------)
                             ----+---------+---------+---------+--
                               -6.0      -3.0       0.0       3.0
```

Diet = 3 subtracted from:

```
Diet  Lower  Center   Upper  ----+---------+---------+---------+--
4    -6.761  -3.480 -0.1987   (----------*----------)
                             ----+---------+---------+---------+--
                               -6.0      -3.0       0.0       3.0
```

From the confidence intervals, you can see that Diet 4 is different from all of the other 3 diets. The tests below confirm that.

Tukey Simultaneous Tests
Response Variable Weight Gain
All Pairwise Comparisons among Levels of Diet

Diet = 1 subtracted from:

Diet	Difference of Means	SE of Difference	T-Value	Adjusted P-Value
2	0.260	1.105	0.235	0.9952
3	-0.200	1.105	-0.181	0.9978
4	-3.680	1.105	-3.331	0.0266

The Levene's Test p-value is 0.993 indicating that the variances are equal. Next, perform an ANOVA test. Since you will also want pairwise comparisons for part c, use the General Linear Model function of Minitab. Click on **Stat → ANOVA → General Linear Model.** Enter Time for **Responses**, enter both Mouse and Drug for **Model**, and enter Mouse for **Random factors.** Next, click on **Comparisons.** Select **Pairwise comparisons** and enter Drug for **Terms.** This tells Minitab which variable you want to use for the pairwise comparisons. Select **Tukey**. If you want confidence intervals for the comparisons, select **Confidence interval**. Finally, if you want to test whether the pairs are different, select **Test.** Click **OK.**

General Linear Model: Time versus Mouse, Drug

Factor	Type	Levels	Values
Mouse	random	5	1, 2, 3, 4, 5
Drug	fixed	3	Drug 1, Drug 2, None

Analysis of Variance for Time, using Adjusted SS for Tests

Source	DF	Seq SS	Adj SS	Adj MS	F	P
Mouse	4	20.5173	20.5173	5.1293	36.16	0.000
Drug	2	1.6853	1.6853	0.8427	5.94	0.026
Error	8	1.1347	1.1347	0.1418		
Total	14	23.337				

S = 0.376608 R-Sq = 95.14% R-Sq(adj) = 91.49%

The p-value for Drug is 0.026 so you would reject the null hypothesis. There is evidence that the mean healing times are different among the three treatments. Now, examine the output from the Tukey comparisons to determine which treatment is different.

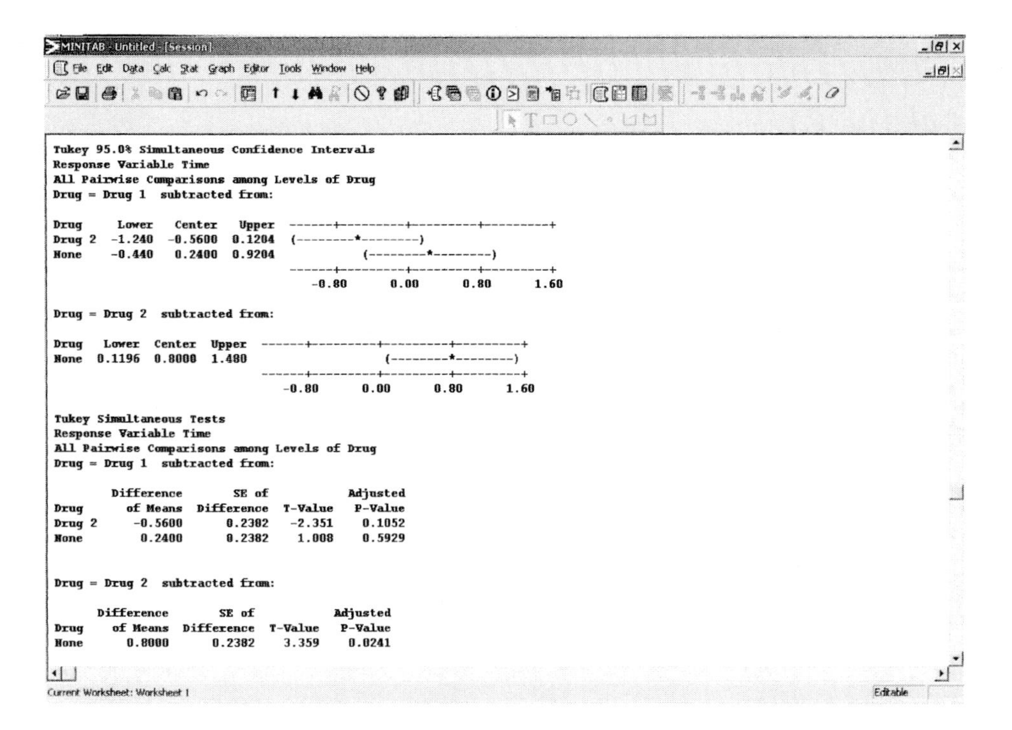

```
Tukey 95.0% Simultaneous Confidence Intervals
Response Variable Time
All Pairwise Comparisons among Levels of Drug
Drug = Drug 1  subtracted from:

Drug      Lower    Center    Upper   ------+---------+---------+---------+
Drug 2  -1.240  -0.5600  0.1204   (--------*--------)
None    -0.440   0.2400  0.9204            (--------*--------)
                                     ------+---------+---------+---------+
                                        -0.80      0.00      0.80      1.60

Drug = Drug 2  subtracted from:

Drug   Lower   Center   Upper   ------+---------+---------+---------+
None  0.1196  0.8000  1.480                (--------*--------)
                                ------+---------+---------+---------+
                                   -0.80      0.00      0.80      1.60

Tukey Simultaneous Tests
Response Variable Time
All Pairwise Comparisons among Levels of Drug
Drug = Drug 1  subtracted from:

        Difference        SE of              Adjusted
Drug      of Means   Difference   T-Value    P-Value
Drug 2     -0.5600       0.2382    -2.351     0.1052
None        0.2400       0.2382     1.008     0.5929


Drug = Drug 2  subtracted from:

      Difference        SE of              Adjusted
Drug    of Means   Difference   T-Value    P-Value
None      0.8000       0.2382     3.359     0.0241
```

Notice that the Tukey test comparing Drug #2 to No Drugs has a p-value 0.0241. Thus, Drug #2 is has a shorter mean healing time than No Drugs.

Section 13.4

Example 6 (pg. 720) Analyzing a 2 x 2 Factorial Design

An educational psychologist wanted to determine whether varying the conditions in which learning and testing take place affect test results. Enter the data found in Table 13 on page 720 of the text into a Minitab worksheet.

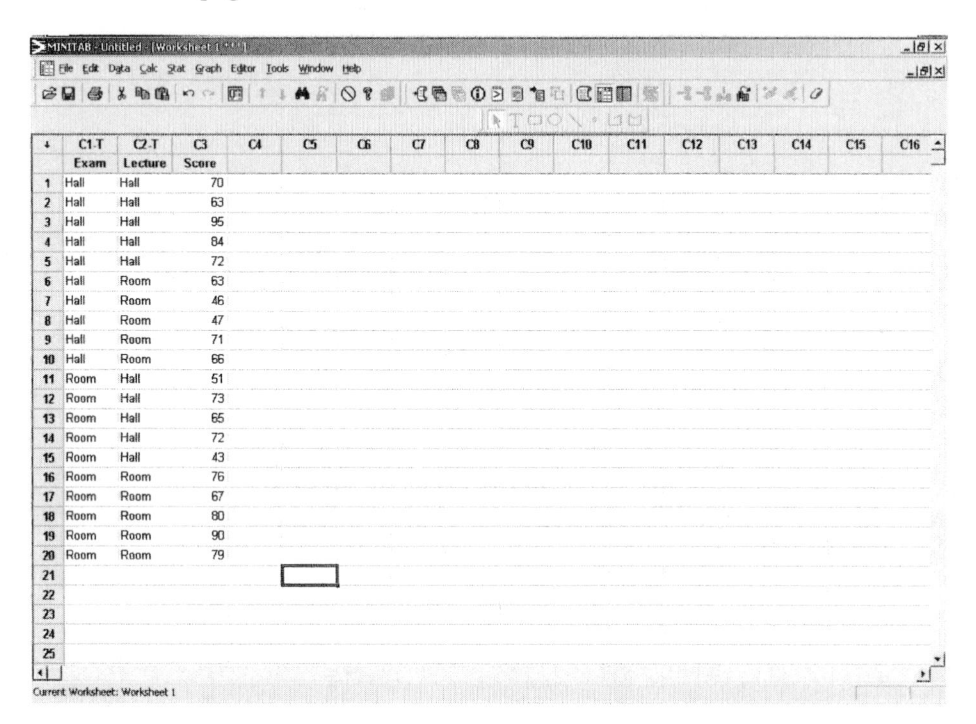

↓	C1-T Exam	C2-T Lecture	C3 Score
1	Hall	Hall	70
2	Hall	Hall	63
3	Hall	Hall	95
4	Hall	Hall	84
5	Hall	Hall	72
6	Hall	Room	63
7	Hall	Room	46
8	Hall	Room	47
9	Hall	Room	71
10	Hall	Room	66
11	Room	Hall	51
12	Room	Hall	73
13	Room	Hall	65
14	Room	Hall	72
15	Room	Hall	43
16	Room	Room	76
17	Room	Room	67
18	Room	Room	80
19	Room	Room	90
20	Room	Room	79

To test whether there is an interaction effect between exam location and lecture location, click on **Stat → ANOVA → Two Way.** Enter the columns as shown below. Select **Display means.** To see the interaction term, DO NOT select **Fit additive model.**

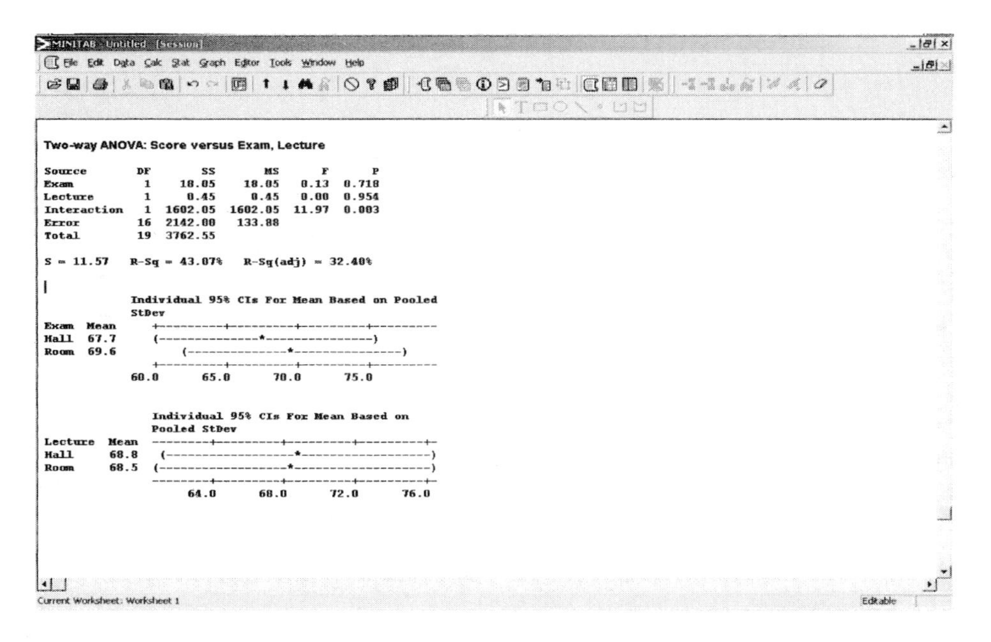

Two-way ANOVA: Score versus Exam, Lecture

Source	DF	SS	MS	F	P
Exam	1	18.05	18.05	0.13	0.718
Lecture	1	0.45	0.45	0.00	0.954
Interaction	1	1602.05	1602.05	11.97	0.003
Error	16	2142.00	133.88		
Total	19	3762.55			

```
S = 11.57   R-Sq = 43.07%   R-Sq(adj) = 32.40%

            Individual 95% CIs For Mean Based on Pooled
            StDev
Exam  Mean  ---------+---------+---------+---------+---------
Hall  67.7    (--------------*--------------)
Room  69.6         (--------------*--------------)
            ---------+---------+---------+---------+---------
            60.0      65.0      70.0      75.0

               Individual 95% CIs For Mean Based on
               Pooled StDev
Lecture  Mean  --------+---------+---------+---------+-
Hall     68.8    (------------------*------------------)
Room     68.5  (------------------*------------------)
               --------+---------+---------+---------+-
                    64.0      68.0      72.0      76.0
```

The p-value for the interaction term is 0.003 indicating that there is a significant interaction between the lecture location and the exam location. Therefore, do not test the main effects. To draw an interaction plot, click on **Stat** → **ANOVA** → **Interactions plot.** Enter Score for the **Response** and both Exam and Lecture for the **Factors**. Click on **OK**.

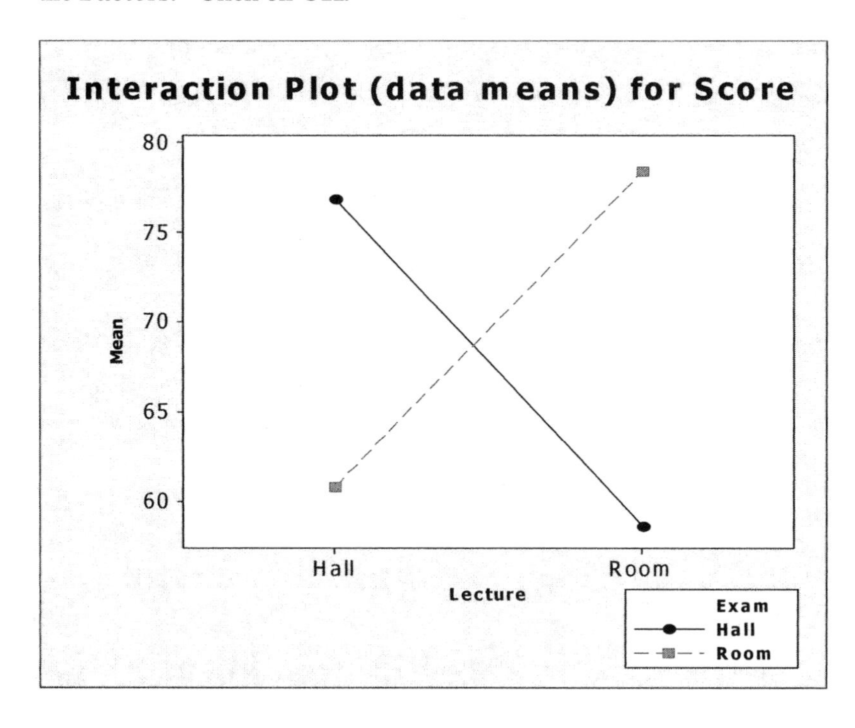

▸ Problem 17 (pg. 725) Cholesterol Levels

An physician wanted to know if age and gender were factors that explained cholesterol levels. Enter the data into a Minitab worksheet.

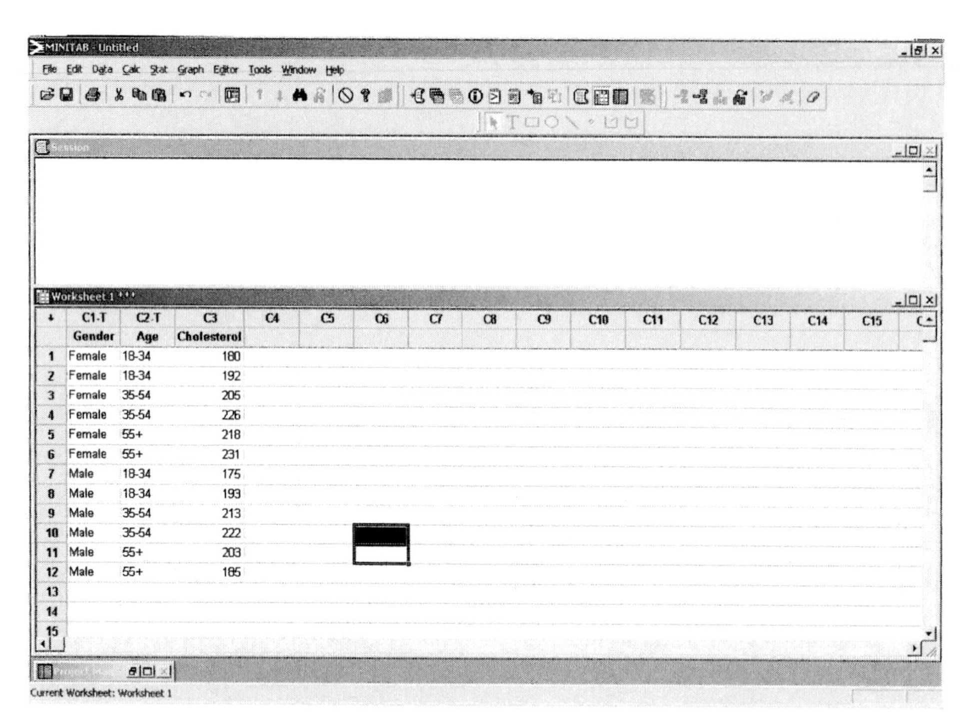

	C1-T Gender	C2-T Age	C3 Cholesterol
1	Female	18-34	180
2	Female	18-34	192
3	Female	35-54	205
4	Female	35-54	226
5	Female	55+	218
6	Female	55+	231
7	Male	18-34	175
8	Male	18-34	193
9	Male	35-54	213
10	Male	35-54	222
11	Male	55+	203
12	Male	55+	185

To determine if there is significant interaction between age and gender, click on **Stat → ANOVA → Two Way.** Enter the columns as shown below. Select **Display means.** To see the interaction term, DO NOT select **Fit additive model.**

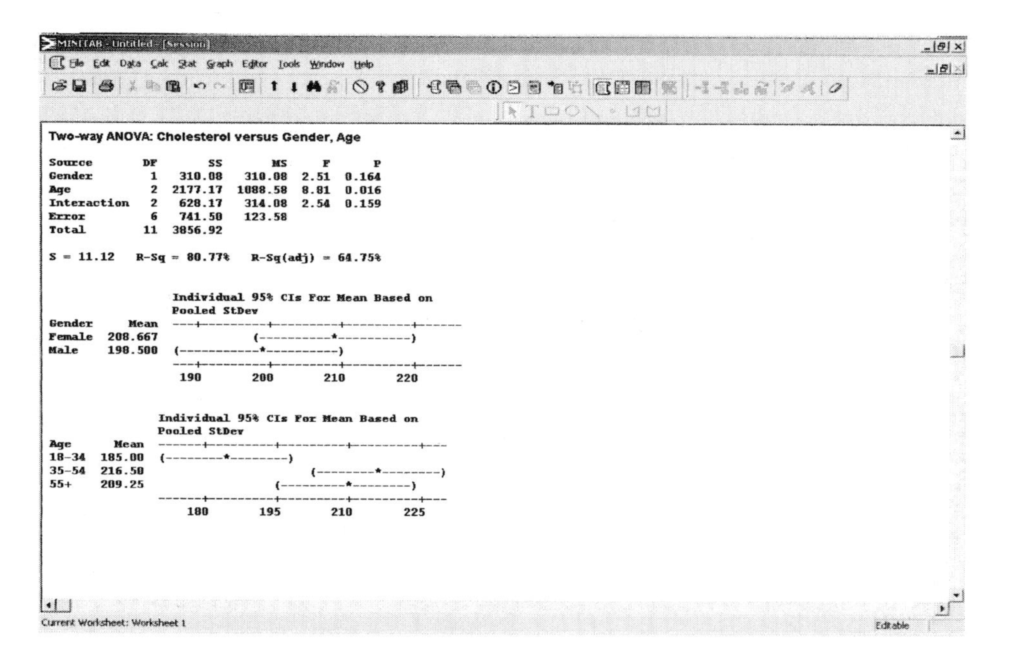

Two-way ANOVA: Cholesterol versus Gender, Age

Source	DF	SS	MS	F	P
Gender	1	310.08	310.08	2.51	0.164
Age	2	2177.17	1088.58	8.81	0.016
Interaction	2	628.17	314.08	2.54	0.159
Error	6	741.50	123.58		
Total	11	3856.92			

S = 11.12   R-Sq = 80.77%   R-Sq(adj) = 64.75%

```
                  Individual 95% CIs For Mean Based on
                  Pooled StDev
Gender    Mean    ---+---------+---------+---------+-----
Female  208.667            (----------*----------)
Male    198.500   (-----------*----------)
                  ---+---------+---------+---------+-----
                    190       200       210       220
```

```
                Individual 95% CIs For Mean Based on
                Pooled StDev
Age      Mean   ------+---------+---------+---------+---
18-34  185.00   (--------*--------)
35-54  216.50                     (---------*---------)
55+    209.25                (---------*--------)
                ------+---------+---------+---------+---
                     180       195       210       225
```

Since the p-value for the interaction term is 0.159, there is no significant interaction between age and gender. Re-run the ANOVA without the interaction term this time. Click on **Stat → ANOVA → Two Way.** Enter the columns as shown below. Select **Display means.** This time, select **Fit additive model.**

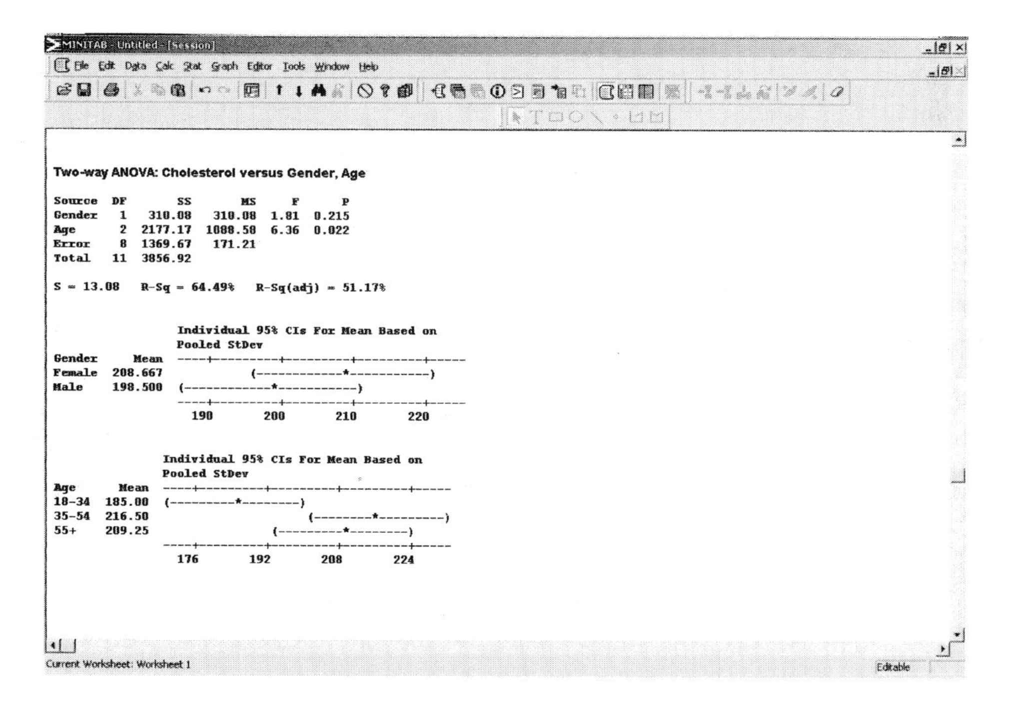

Two-way ANOVA: Cholesterol versus Gender, Age

Source	DF	SS	MS	F	P
Gender	1	310.08	310.08	1.81	0.215
Age	2	2177.17	1088.58	6.36	0.022
Error	8	1369.67	171.21		
Total	11	3856.92			

S = 13.08   R-Sq = 64.49%   R-Sq(adj) = 51.17%

```
                  Individual 95% CIs For Mean Based on
                  Pooled StDev
Gender    Mean    ----+---------+---------+---------+-----
Female  208.667           (------------*-----------)
Male    198.500   (------------*-----------)
                  ----+---------+---------+---------+-----
                     190       200       210       220
```

```
                Individual 95% CIs For Mean Based on
                Pooled StDev
Age      Mean   ----+---------+---------+---------+-----
18-34  185.00   (---------*--------)
35-54  216.50                   (---------*----------)
55+    209.25              (---------*--------)
                ----+---------+---------+---------+-----
                   176       192       208       224
```

Since the p-value for Age is 0.022, there is a significant difference in means for genders. However, there is no significant difference for age groups since the p-value is 0.215. To draw an interaction plot, click on **Stat** → **ANOVA** → **Interactions plot.** Enter Cholesterol for the **Response** and both Age and Gender for the **Factors**. Click on **OK**. Try reversing the order of the **Factors** to Gender and Age.

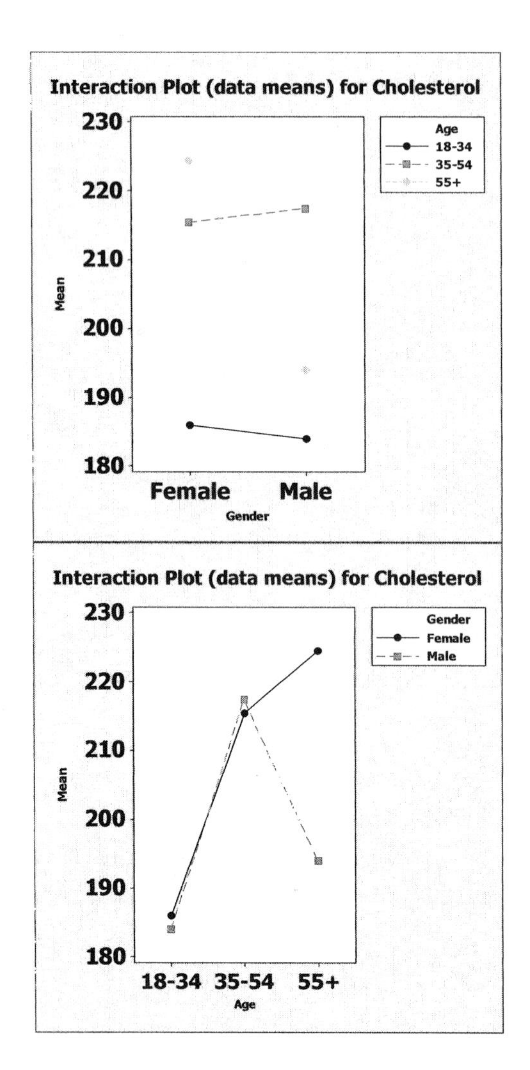

To perform the Tukey test on Age, click on **Stat** → **ANOVA** → **General Linear Model.** Enter Cholesterol for **Responses**, enter both Age and Gender for **Model.** Next, click on **Comparisons.** Select **Pairwise comparisons** and enter Age for **Terms.** This tells Minitab which variable you want to use for the pairwise comparisons. Select **Tukey**. If you want confidence intervals for the comparisons, select **Confidence interval**. Finally, if you want to test whether the pairs are different, select **Test.** Click **OK.**

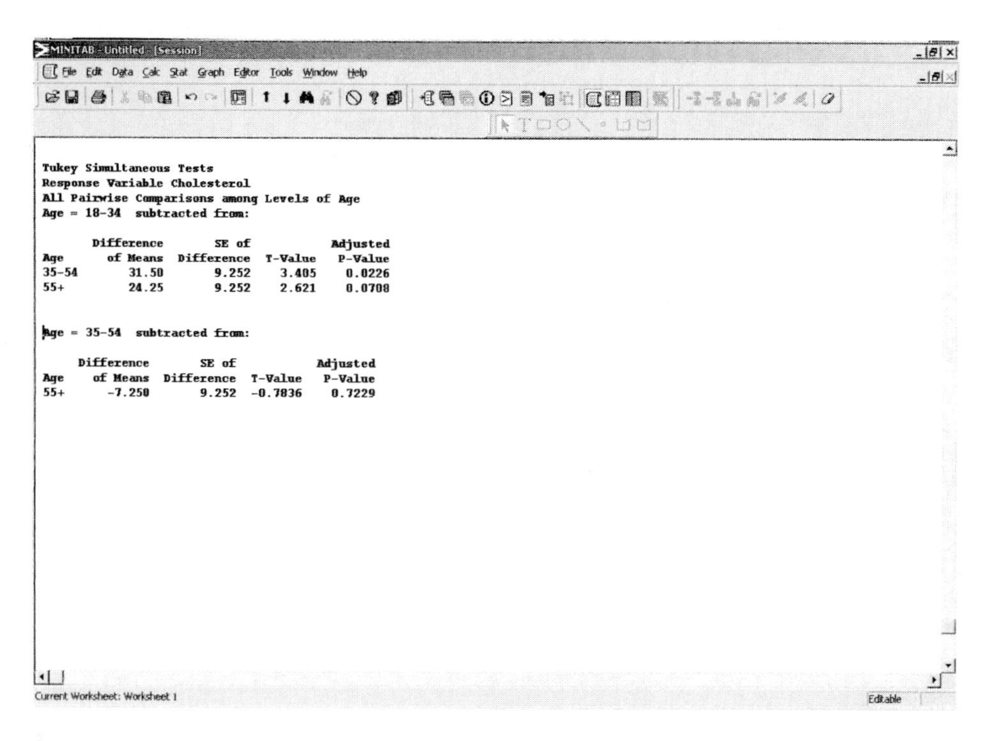

Tukey Simultaneous Tests
Response Variable Cholesterol
All Pairwise Comparisons among Levels of Age
Age = 18-34 subtracted from:

Age	Difference of Means	SE of Difference	T-Value	Adjusted P-Value
35-54	31.50	9.252	3.405	0.0226
55+	24.25	9.252	2.621	0.0708

Age = 35-54 subtracted from:

Age	Difference of Means	SE of Difference	T-Value	Adjusted P-Value
55+	-7.250	9.252	-0.7836	0.7229

Notice that Age group 18-24 is significantly different than age group 35-54 (p-value = 0.0276) and not quite significantly different than age group 55+ (p-value = 0.0708). There is no difference between age groups 35-54 and 55+.

Least-Squares Regression Model & Multiple Regression

Section 14.1

Examples 1-7 (pg. 737-747) Least-Squares Regression

Minitab's Regression procedure will give output for all of the first 5 examples in this section. Begin with the data on pg. 737. Enter the ages into C1 and the total cholesterol readings into C2. Age is the x-variable (Predictor) and Cholesterol is the y-variable (Response). To find the regression equation, click on **Stat → Regression → Regression.** Enter C2 for the **Response** variable, and C1 as the **Predictor.**

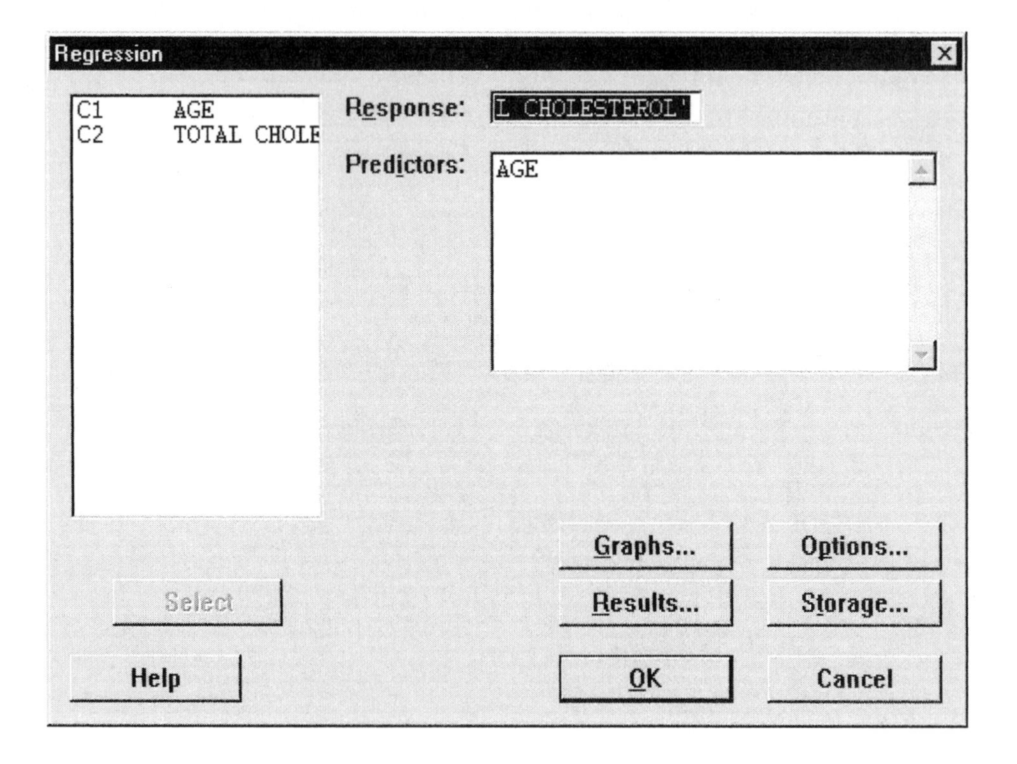

Click on **Results.** Select **Regression equation, table of coefficients, s, R-squared, and basic analysis of variance.**

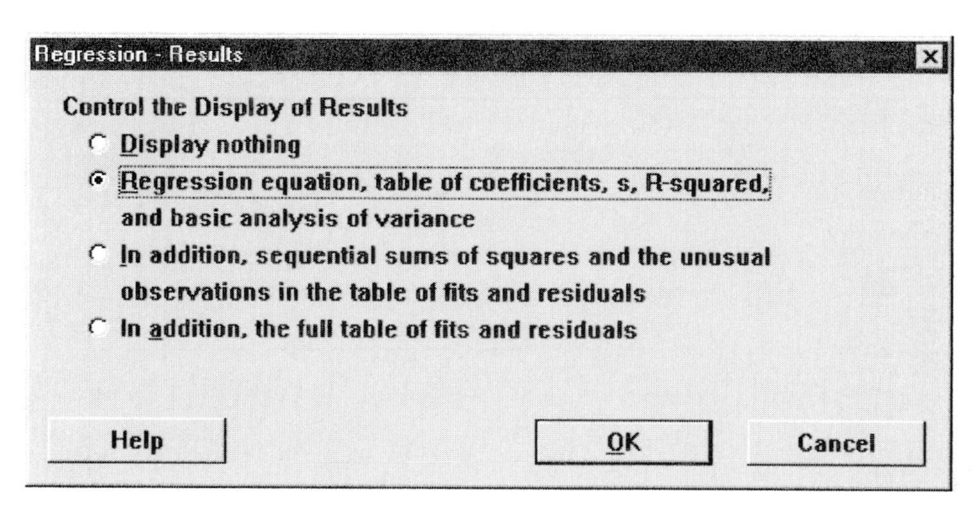

Click on **OK** to return to the main Regression Window, and now select **Graphs.** Select **Normal plot of residuals** and enter Age beneath **Residuals versus the variables.**

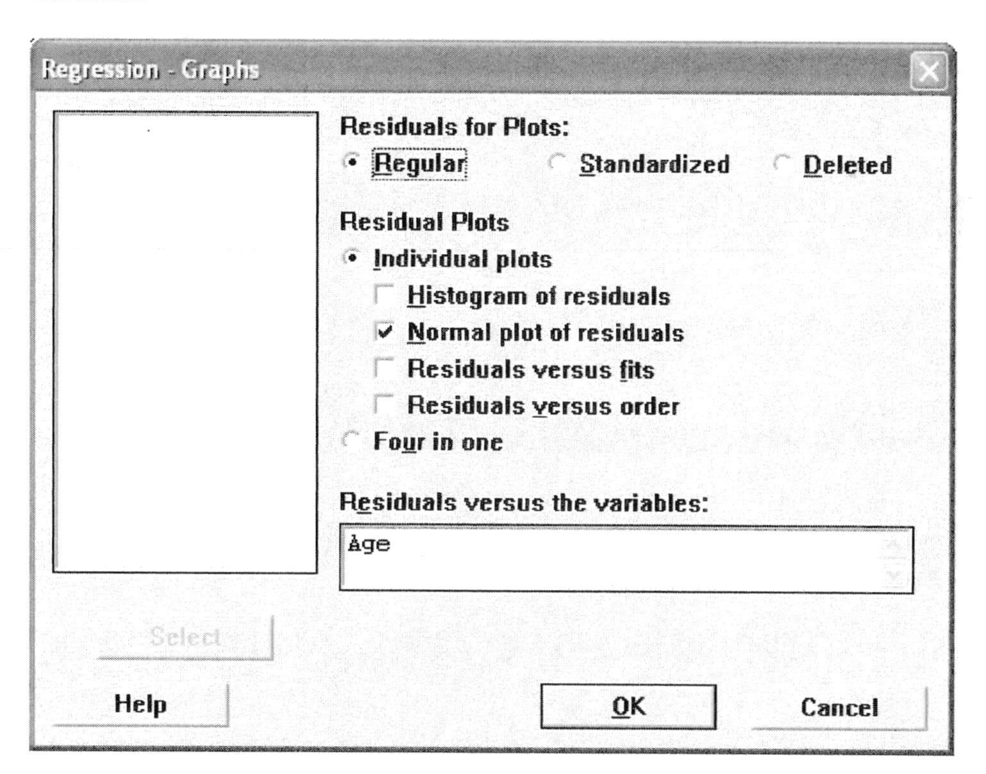

Click on **OK** twice to view the output in the Session Window.

First look at the two graphs created by Minitab.

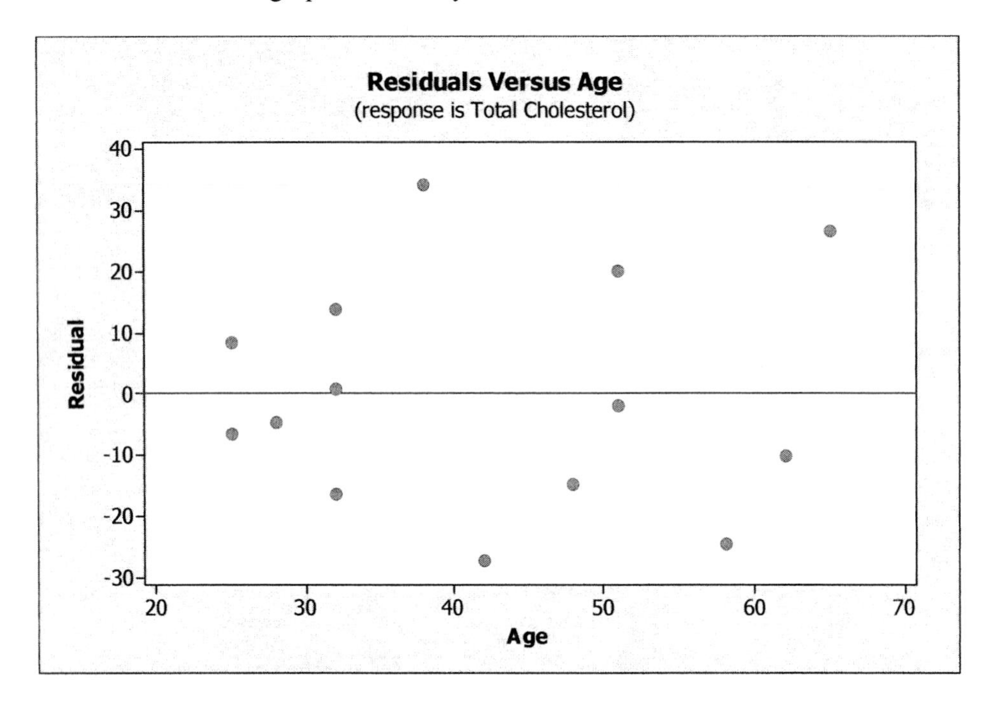

This is the plot showing the residuals plotted against the predictor variable, age. The errors are evenly spread out around the horizontal line at 0. Thus the assumption of constant error variance is satisfied. (Example 5, pg. 744)

Now look at the normal probability plot of the residuals. Notice that the points fall in a straight line. Thus the assumption of normality is satisfied. (Example 4, pg. 742)

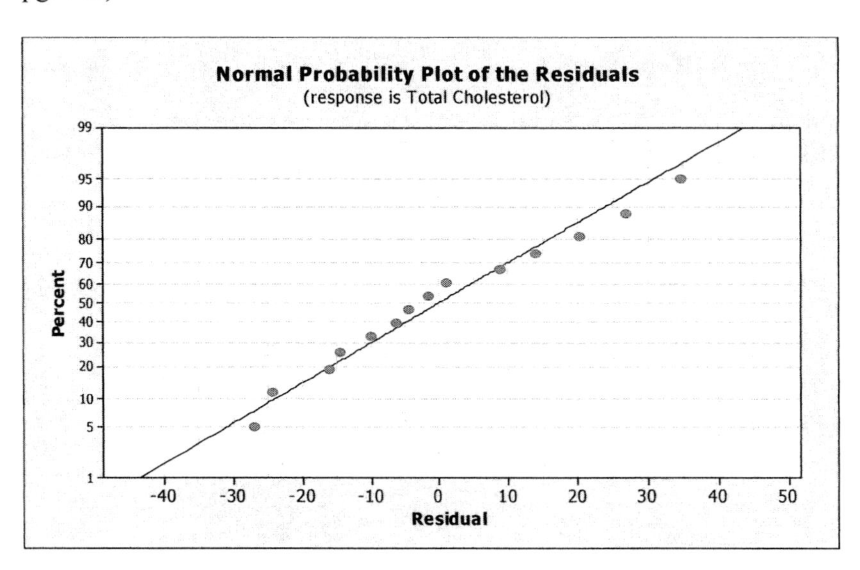

Finally, look at the results in the Session Window.

Regression Analysis: Total Cholesterol versus Age

```
The regression equation is
Total Cholesterol = 151 + 1.40 Age


Predictor    Coef  SE Coef     T      P
Constant   151.35    17.28  8.76  0.000
Age        1.3991   0.3917  3.57  0.004


S = 19.4805   R-Sq = 51.5%   R-Sq(adj) = 47.5%

Analysis of Variance

Source          DF      SS      MS      F      P
Regression       1  4840.5  4840.5  12.76  0.004
Residual Error  12  4553.9   379.5
Total           13  9394.4
```

Notice that the regression equation is Total Cholesterol = 151 + 1.40 * Age (Example 1, pg. 706). The standard error is also shown, S=19.48 (Example 2, pg. 710). Notice also that the t-value is shown to test the linear relationship between Age and Cholesterol, T=3.57 with a p-value of .004. Thus, we reject the null hypothesis and conclude that there is evidence to support the claim of a linear relationship between age and total cholesterol level (Example 5, pg. 746).

To construct the confidence interval for the slope of the regression line (pg. 749), use a hand-calculator and the Minitab output. From the output in the Session Window, you can see that b_1 is 1.3991 and its standard error is .3917. Simply look up the t-value in a table (2.179), and calculate 1.3991 ± 2.179 * .3917. Your confidence interval is (.545, 2.253).

◀

Problem 17 (pg. 749) United Technologies vs. S&P 500

Open worksheet **14_1_17**. The S&P rate of return should be in C2, and UTC's rate of return should be in C3. 'S&P rate of return' is the x-variable and 'UTC' is the y-variable. To find the regression equation, click on **Stat → Regression → Regression.** Enter C3 for the **Response** variable, and C2 as the **Predictor.** Click on **Results.** Select **Regression equation, table of coefficients, s, R-squared, and basic analysis of variance.** Click on **OK** to return to the main Regression Window, and now select **Graphs.** Select **Normal plot of residuals** and enter C2 beneath **Residuals versus the variables.** Click on **OK** to return to the main Regression Window. In this problem, you also want to find the mean rate of return for UTC's stock if the rate of return of the S&P 500 is 4.2 percent. Click on **Options** and enter 4.2 beneath **Prediction intervals for new observations.**

Regression - Options

C2 Rate of Ret
C3 Rate of Ret

Weights:

Fit intercept

Display
Variance inflation factors
Durbin-Watson statistic
PRESS and predicted R-square

Lack of Fit Tests
Pure error
Data subsetting

Prediction intervals for new observations:
4.2

Confidence level: 95

Storage
Fits
SEs of fits
Confidence limits
Prediction limits

Select

Help OK Cancel

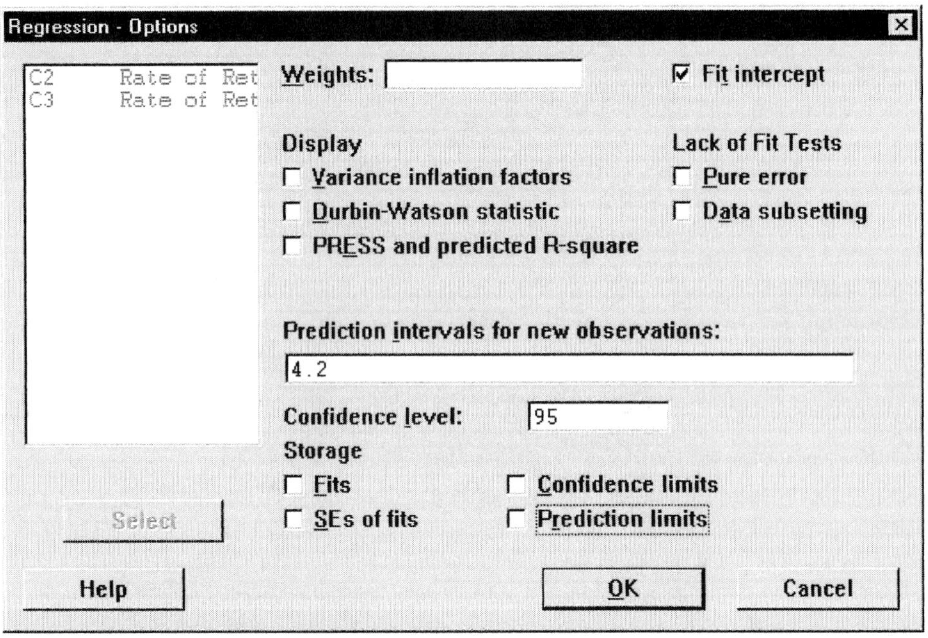

Click on **OK** twice to view the graphs and the output in the Session Window.

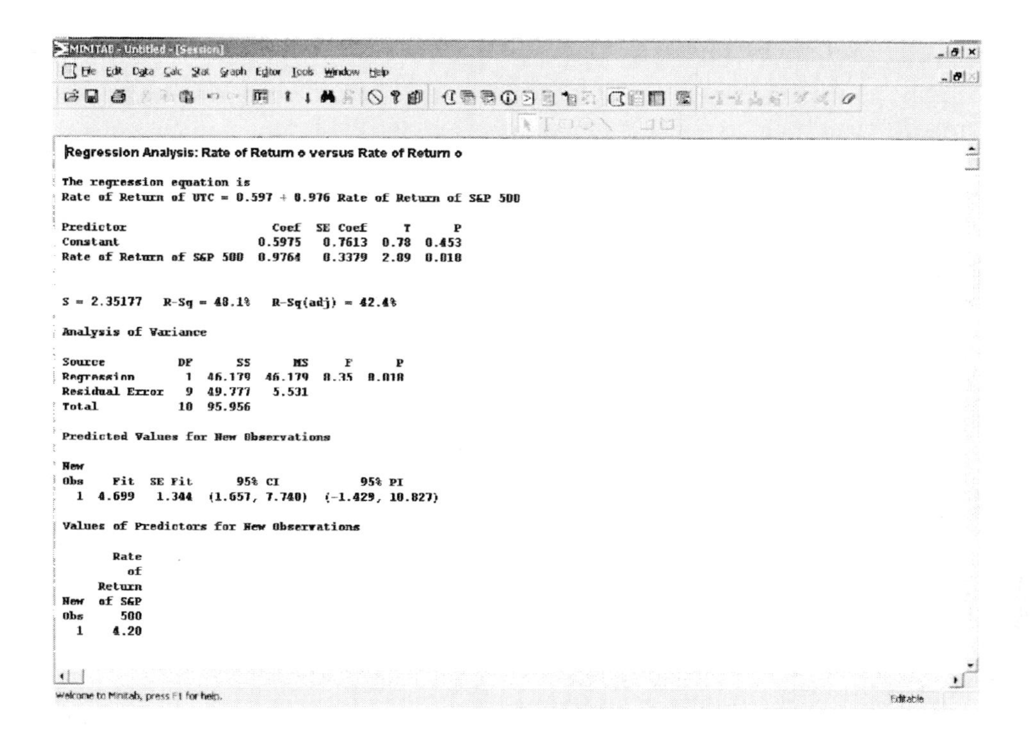

MINITAB - Untitled - [Session]

File Edit Data Calc Stat Graph Editor Tools Window Help

Regression Analysis: Rate of Return o versus Rate of Return o

```
The regression equation is
Rate of Return of UTC = 0.597 + 0.976 Rate of Return of S&P 500

Predictor                    Coef   SE Coef     T      P
Constant                   0.5975    0.7613  0.78  0.453
Rate of Return of S&P 500  0.9764    0.3379  2.89  0.018


S = 2.35177   R-Sq = 48.1%   R-Sq(adj) = 42.4%

Analysis of Variance

Source          DF      SS      MS     F      P
Regression       1  46.179  46.179  8.35  0.018
Residual Error   9  49.777   5.531
Total           10  95.956

Predicted Values for New Observations

New
Obs    Fit  SE Fit       95% CI            95% PI
  1  4.699   1.344  (1.657, 7.740)  (-1.429, 10.827)

Values of Predictors for New Observations

         Rate
           of
       Return
New    of S&P
Obs       500
  1      4.20
```

Welcome to Minitab, press F1 for help.

Editable

The regression equation is: (Rate of Return in UTC) = 0.597 + 0.976 * (Rate of Return of S&P 500). Thus, the unbiased estimates of β_0 and β_1 are 0.597 and 0.976. The standard error is S=2.35177.

To determine if the residuals are normally distributed with constant error variance, examine the following two plots. The first plot is a Normal Probability plot of the residuals.

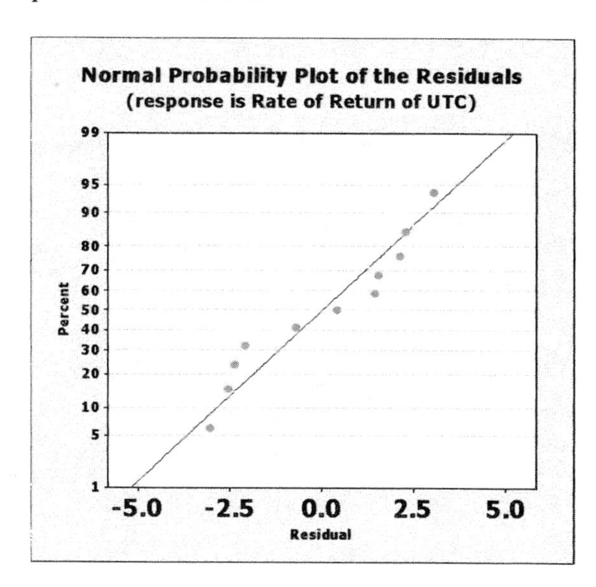

Since the normal probability plot is fairly linear, the residuals are approximately normal. The next plot is a plot of the Residuals vs. the Predictor.

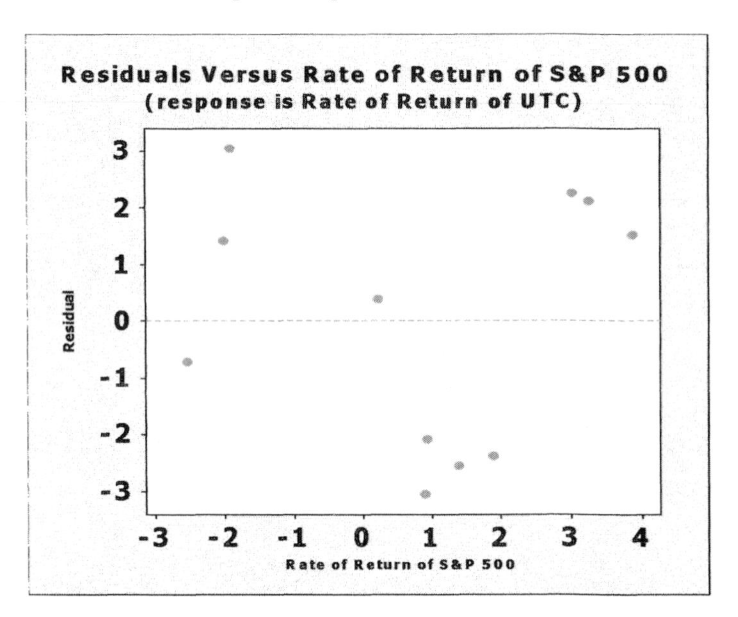

The errors are evenly spread out around the horizontal line at 0. Thus the assumption of constant error variance is satisfied.

The standard deviation of b_1, s_{b_1}, shown in the Session Window, is 0.3379.

The t-value to test if a linear relation exists between the S&P 6rate of return and the UTC rate of return is T=2.89 with a p-value of .018 (shown in the Session Window output). Since .018 is smaller than α=.10, you would reject the null hypothesis and conclude that there is a linear relationship.

To construct the confidence interval for the slope of the regression line, use a hand-calculator and the Minitab output. You know that b_1 is 0.9764 and you also know its standard error is .3379. Simply look up the t-value in a table ($t_{.005,9}$ = 3.25), and calculate .9764 $\pm$ 3.25 * .3379. Your confidence interval is (-.12, 2.07).

To find the mean rate of return for Cisco Systems stock if the rate of return of the S&P 500 is 4.2 percent, look at the output in the Session Window. Near the bottom, beneath **Predicted values for new observations**, the value for "Fit" is 4.699. This is the estimate of the mean rate of return for UTC when the rate of return for the S&P 500 is 4.2 percent.

◀

Section 14.2

▸ Examples 1&2 (pg. 754) Confidence and Prediction Intervals

Begin with the data on pg. 737 of the textbook. Enter the ages into C1 and the total cholesterol readings into C2. Age is the x-variable (Predictor) and Cholesterol is the y-variable (Response). To find the regression equation, click on **Stat → Regression → Regression.** Enter C2 for the **Response** variable, and C1 as the
Predictor. Click on **Options.** Enter '42' below **Prediction intervals for new observations**, and then enter 95 for **Confidence level.**

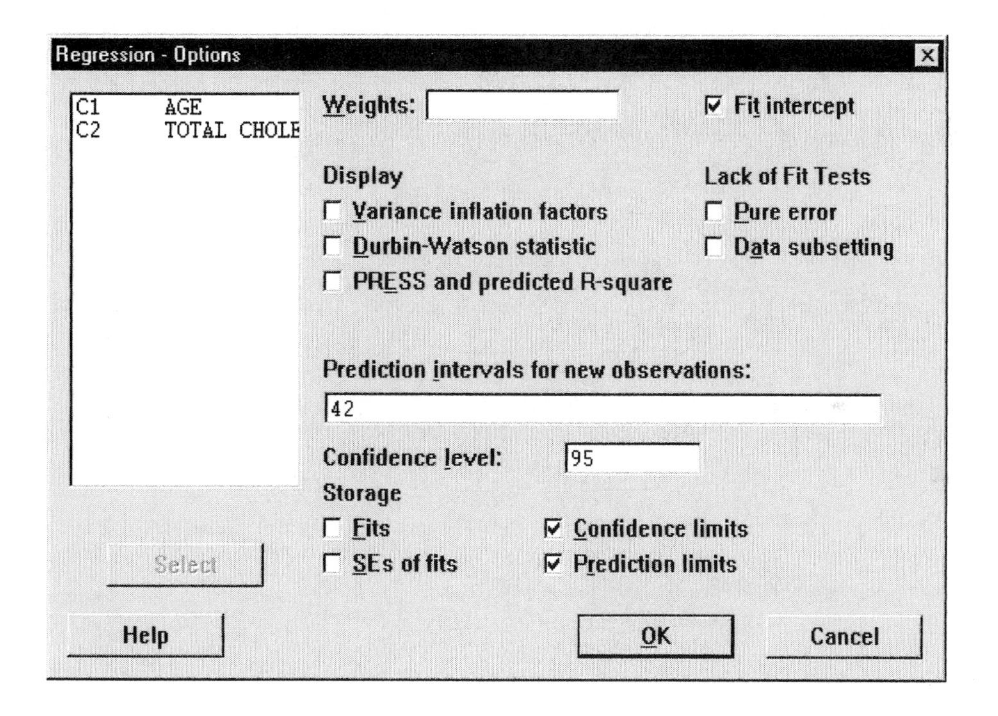

Click on **OK** twice to view the output in the Session Window.

At the bottom of the screen, below "Predicted Values for New Observations", there is the heading "Fit". The 210.11 is the predicted cholesterol level for a 42 year old. The confidence interval and prediction intervals are also shown.

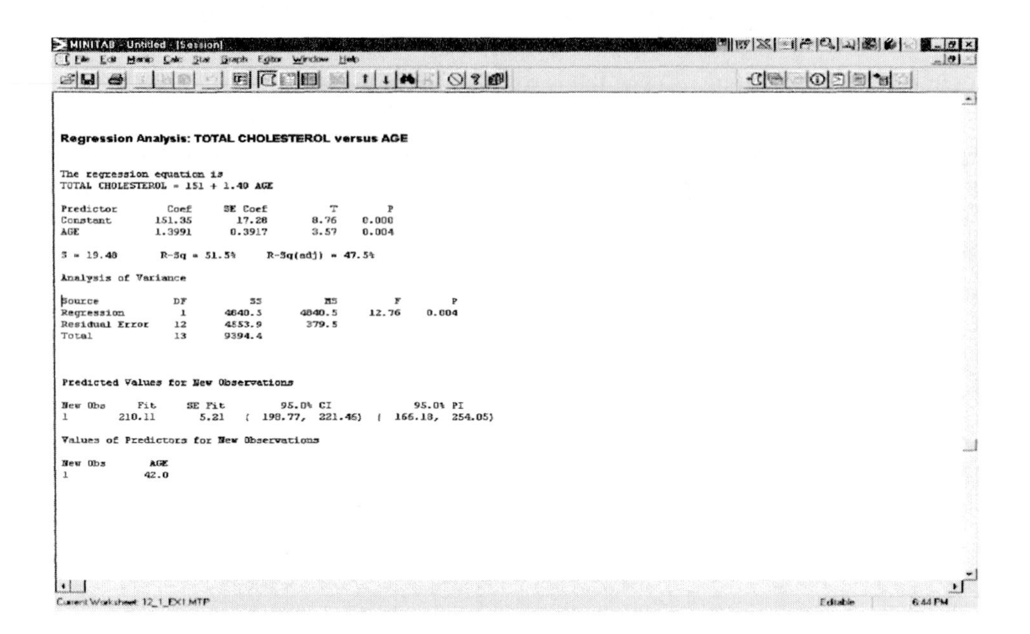

▸ Problem 7 (pg. 757) Height vs. Head Circumference

Open worksheet **14_1_7.** Height is in C1 and Head Circumference is in C2. Predict the mean head circumference of children who are 25.75 inches tall, and construct 95% confidence and prediction intervals. First, find the regression equation. Click on **Stat → Regression → Regression.** Enter C2 for the **Response** variable, and C1 as the **Predictor.** Click on **Results.** Select **Regression equation, table of coefficients, s, R-squared, and basic analysis of variance.** Click on **OK** to return to the main Regression Window, and now select **Options.** Enter 25.75 below **Prediction intervals for new observations**, and then enter 95 for **Confidence level.** Click on **OK** twice.

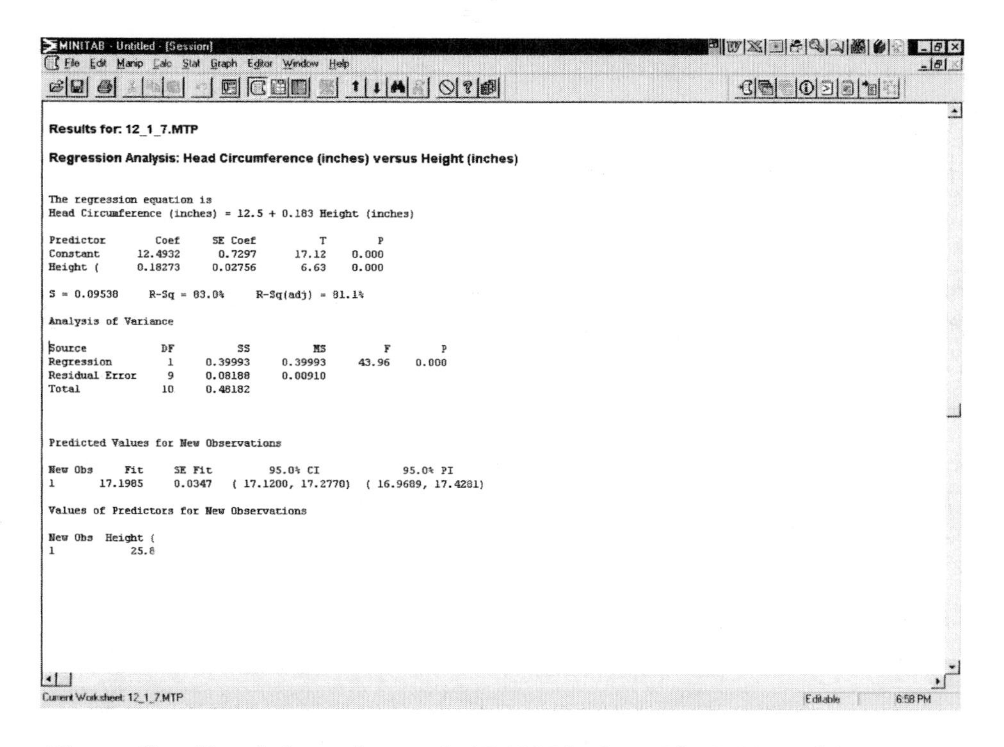

The predicted head circumference is 17.1985 inches. The 95% confidence interval is (17.12, 17.277) and the 95% prediction interval is (16.9689, 17.4281).

◀

Section 14.3

▸ Examples 1-7 (pg. 760-768) Multiple Linear Regression

Enter the data into a Minitab worksheet. The data for Age, Fat, and Cholesterol should be in C1, C2, and C3.

First create the correlation matrix. Click on **Stat → Basic statistics → Correlation.** Select C1- C3 for the **Variables** and click on **OK.** The correlation matrix will be displayed in the Session Window.

Correlations: Age, Fat, Cholesterol

	Age	Fat
Fat	0.324	
Cholesterol	0.718	0.778

Cell Contents: Pearson correlation

From the matrix, you can see that the linear correlation between age and cholesterol is .718, between fat consumption and cholesterol is .778, and since the correlation between age and fat consumption is only .324, multicollinearity is not a concern.

Second, find the least squares regression equation and draw the residual plots and boxplot of the residuals to assess the adequacy of the model. Click on **Stat → Regression → Regression.** The **Response** is Cholesterol, and the **Predictors** are both Age and Fat. Click on **Graphs.**

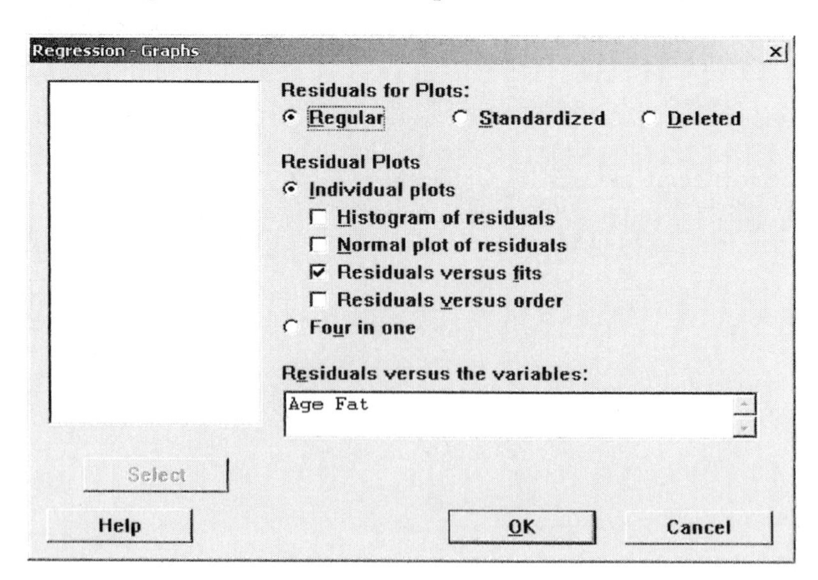

Select **Regular**, **Individual plots**, **Residuals versus fits**, and select both Age and Fat for **Residuals versus the variables**. Click **OK**. Click on **Storage** and select **Residuals**. Next, click on **Options**. Enter 32 23 below **Prediction intervals for new observations.**

Regression - Options

C1 Age
C2 Fat
C3 Cholesterol

Weights:

Fit intercept

Display

Variance inflation factors

Durbin-Watson statistic

PRESS and predicted R-square

Lack of Fit Tests

Pure error

Data subsetting

Prediction intervals for new observations:

32 23

Confidence level: 95

Storage

Fits

SEs of fits

Confidence limits

Prediction limits

Select

Help

OK

Cancel

Click **OK** twice to view the results. The plots are in Graph Windows.

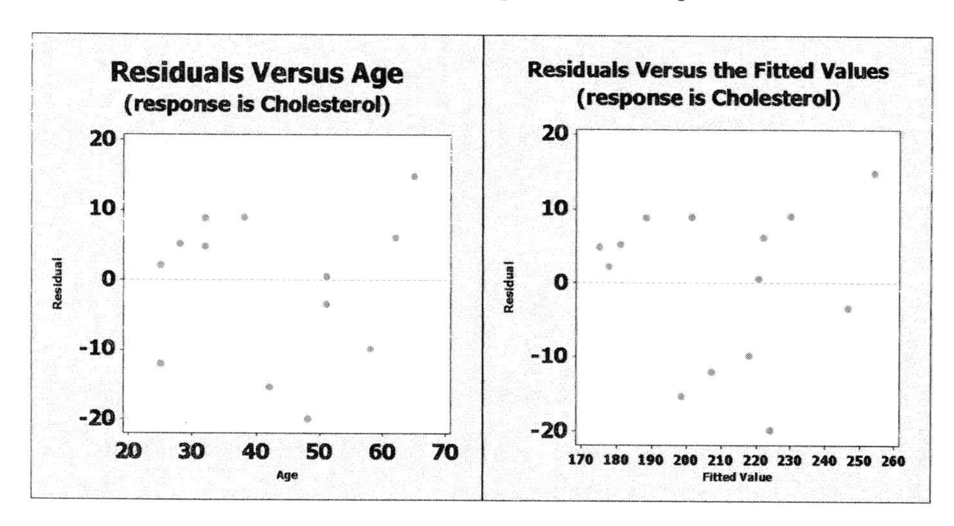

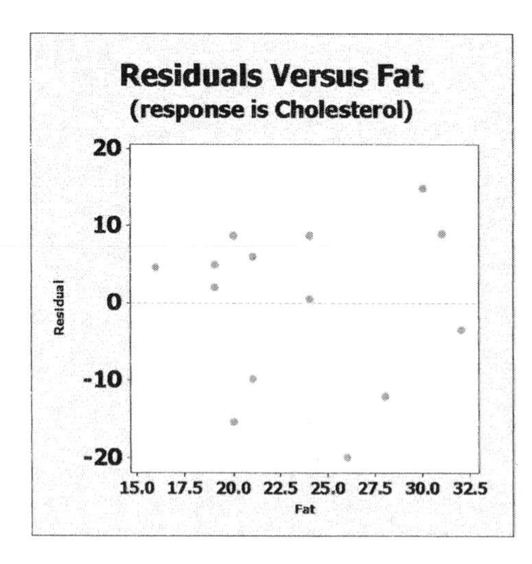

To create a boxplot of the stored residuals, click on **Graph** → **Boxplot** → **Simple.** Select RESI1 for **Graph variables**. Click on **Scale** and select **Transpose value and category scales**. Click on **OK** twice to view the boxplot.

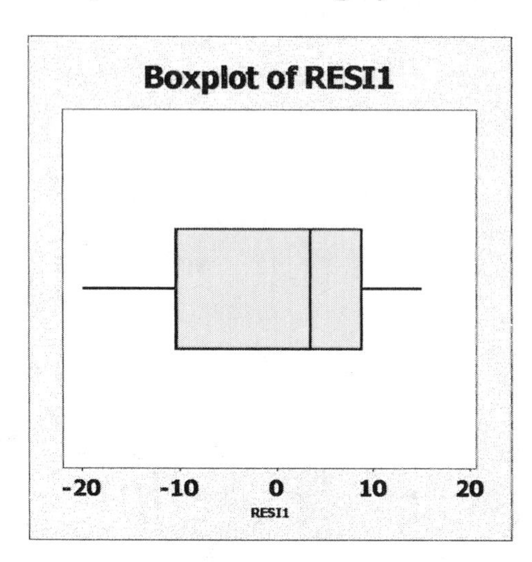

None of the residual plots show a pattern, and the boxplot has no outliers. The regression output is in the Session Window. The regression equation is Cholesterol = 90.8 + 1.01*Age + 3.24*Fat.

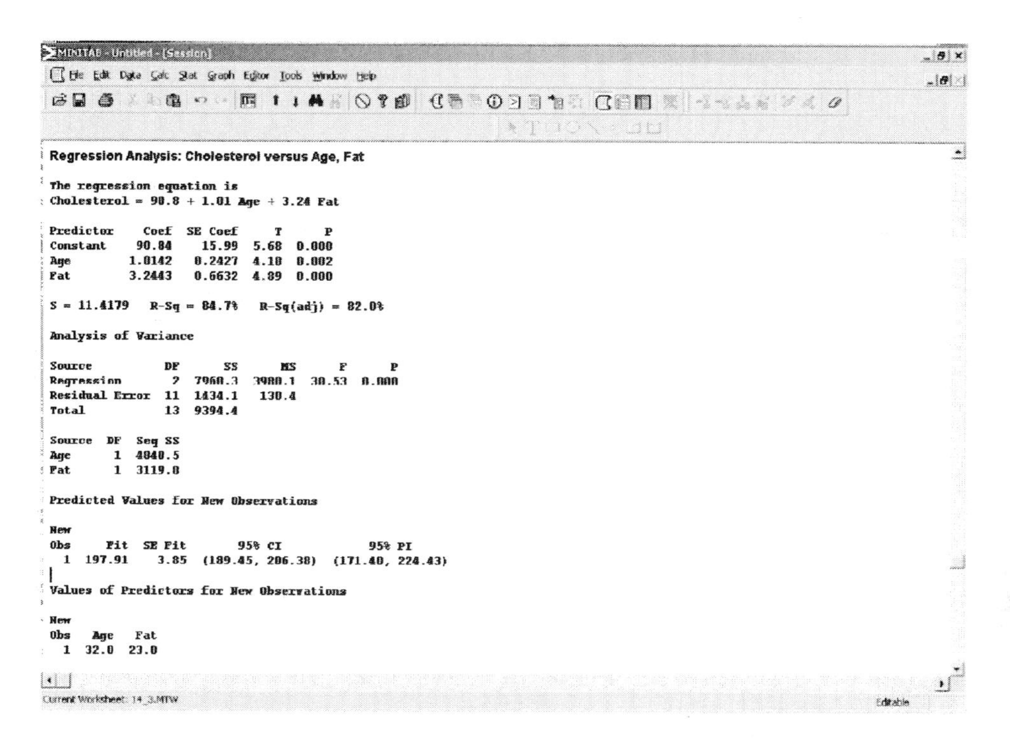

Notice that the coefficient of determination is listed in the output. $R^2 = 84.7\%$ and $R^2_{adj} = 82\%$. Next, before interpreting the F-test, draw a normal probability plot of the residuals. Click on **Graph → Probability Plot → Simple.** Select RESI1 for **Graph variables**. Click on **OK** to view the plot.

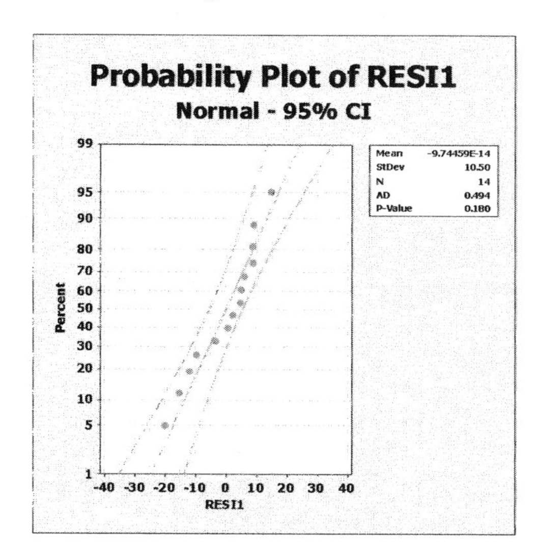

The residuals are approximately normal, so notice that the p-value for the F-test for the regression is 0.000 indicating that there is a significant linear relationship between at least one of the predictor variables (age, fat consumption) and the response variable (cholesterol).

Next, look at the tests of significance for each explanatory variable. The p-values are .002 for Age and .000 for Fat. Thus, both variables have a significant linear relationship with cholesterol.

Finally, look at the prediction interval. A 32 yr-old who consumes 23 grams of fat daily is predicted to have a cholesterol level of 197.91, and we are 95% confident that the total cholesterol level will be between 171.4 and 224.4.

◀

▸ Problem 21 (pg. 779) Miles per gallon

Open worksheet **14_3_21**. To develop a model that describes the MPG of sport utility vehicles possible predictors might be engine size, horsepower, and vehicle weight. First create the correlation matrix. Click on **Stat → Basic statistics → Correlation.** Select C1- C4 for the **Variables** and click on **OK.** The correlation matrix will be displayed in the Session Window.

Correlations: Engine, HP, Weight, MPG

	Engine	HP	Weight
HP	0.323		
Weight	0.092	0.370	
MPG	-0.812	-0.230	-0.477

Since the correlations between the 3 predictor variables are .323, .092, and .370, we do not need to worry about multicollinearity.

To find the regression equation, click on **Stat → Regression → Regression.** Enter C4 for the **Response** variable, and C1-C3 as the **Predictors.** Click on **Results.** Select **Regression equation, table of coefficients, s, R-squared, and basic analysis of variance.** Click on **OK** to return to the main Regression Window, and now select **Graphs.** Select **Regular**, **Individual plots**, and **Residuals versus fits**. Enter Engine, HP, and Weight for **Residuals versus the variables**. Click **OK**. Click on **Storage** and select **Residuals**. In this problem, you also want to find the mean MPG for a vehicle with a 4235 cc engine, 320 hp, and weight of 3950 lbs. Click on **Options** and enter 4235 320 39250 beneath **Prediction intervals for new observations.** Click **OK** twice to view the results in the Session Window.

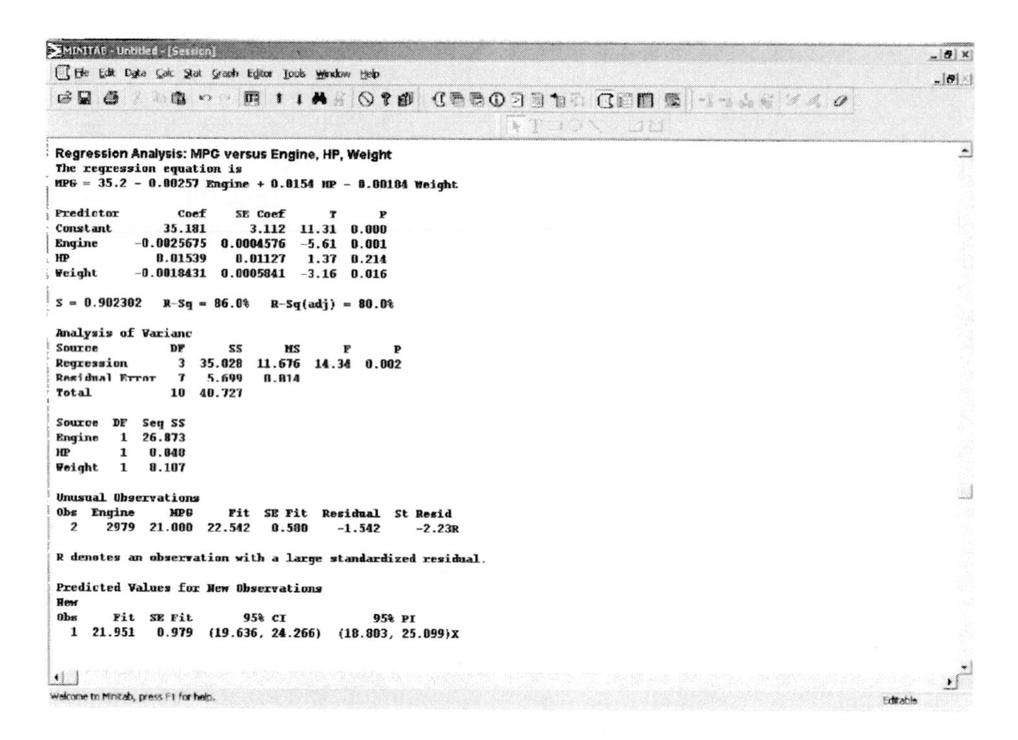

```
Regression Analysis: MPG versus Engine, HP, Weight
The regression equation is
MPG = 35.2 - 0.00257 Engine + 0.0154 HP - 0.00184 Weight

Predictor        Coef     SE Coef      T      P
Constant       35.181       3.112  11.31  0.000
Engine     -0.0025675   0.0004576  -5.61  0.001
HP            0.01539     0.01127   1.37  0.214
Weight     -0.0018431   0.0005841  -3.16  0.016

S = 0.902302   R-Sq = 86.0%   R-Sq(adj) = 80.0%

Analysis of Varianc
Source          DF      SS      MS      F      P
Regression       3  35.028  11.676  14.34  0.002
Residual Error   7   5.699   0.814
Total           10  40.727

Source  DF  Seq SS
Engine   1  26.873
HP       1   0.048
Weight   1   8.107

Unusual Observations
Obs  Engine     MPG     Fit  SE Fit  Residual  St Resid
  2    2979  21.000  22.542   0.580    -1.542    -2.23R

R denotes an observation with a large standardized residual.

Predicted Values for New Observations
New
Obs     Fit  SE Fit        95% CI              95% PI
  1  21.951   0.979  (19.636, 24.266)  (18.803, 25.099)X
```

From the Minitab output, the regression equation is:
MPG = 35.2 - .00257*Engine + .0154*HP - .00184*Weight.

The results of the F-test for the regression have a p-value of .002. Thus, you'd reject the null hypothesis. Next, check the p-values for each individual variable. Engine and weight have p-values of .001 and .016, but HP has a non-significant p-value of .214. So, re-run the regression model without HP as a predictor variable.

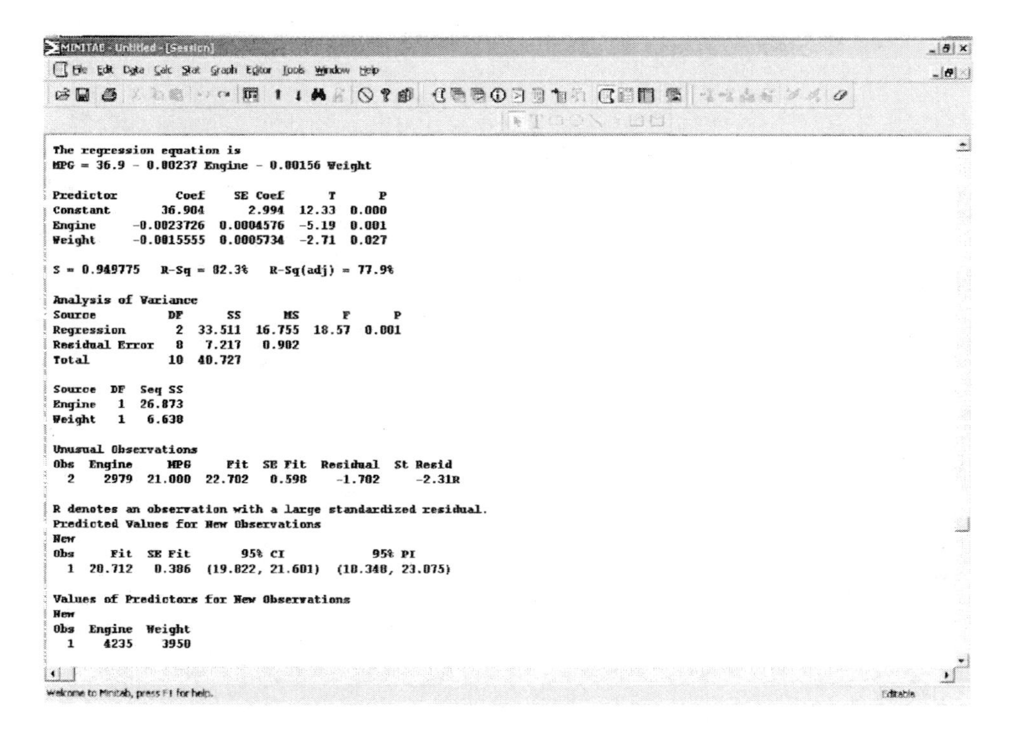

```
The regression equation is
MPG = 36.9 - 0.00237 Engine - 0.00156 Weight

Predictor        Coef     SE Coef      T      P
Constant       36.904       2.994  12.33  0.000
Engine     -0.0023726   0.0004576  -5.19  0.001
Weight     -0.0015555   0.0005734  -2.71  0.027

S = 0.949775   R-Sq = 82.3%   R-Sq(adj) = 77.9%

Analysis of Variance
Source          DF      SS      MS      F      P
Regression       2  33.511  16.755  18.57  0.001
Residual Error   8   7.217   0.902
Total           10  40.727

Source  DF  Seq SS
Engine   1  26.873
Weight   1   6.638

Unusual Observations
Obs  Engine     MPG     Fit  SE Fit  Residual  St Resid
  2    2979  21.000  22.702   0.598    -1.702    -2.31R

R denotes an observation with a large standardized residual.
Predicted Values for New Observations
New
Obs     Fit  SE Fit        95% CI              95% PI
  1  20.712   0.386  (19.822, 21.601)  (18.348, 23.075)

Values of Predictors for New Observations
New
Obs  Engine  Weight
  1    4235    3950
```

The new regression equation is: MPG = 36.9 - .00237*Engine - .00156*Weight. The p-values for both predictor variables are significant (.001 and .027). Thus, both variables have a significant linear relationship with MPG. Next, look at the residual plots.

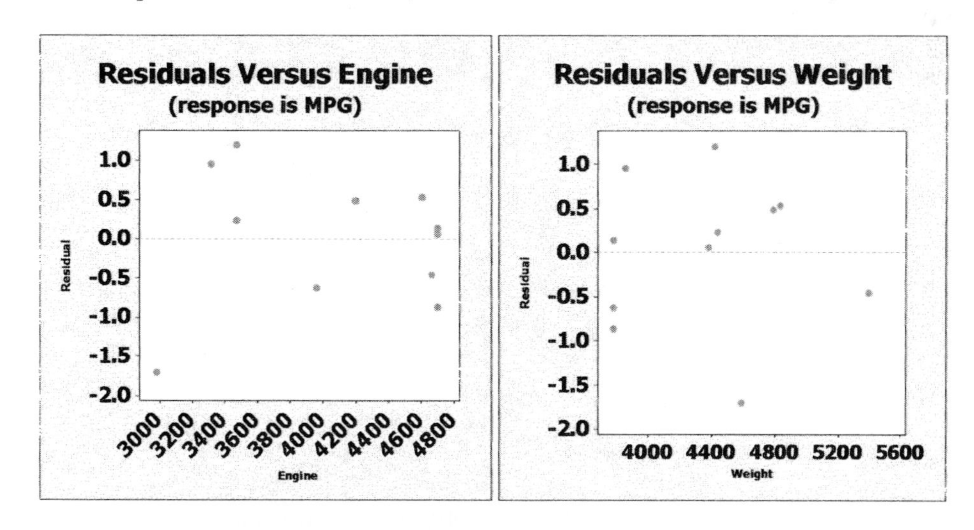

There is no pattern to the residuals.

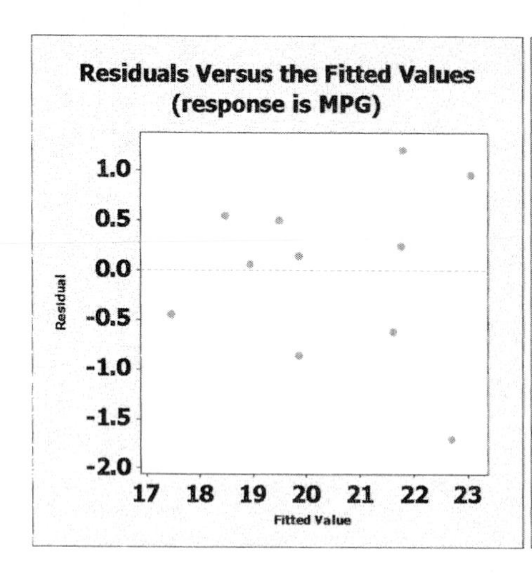

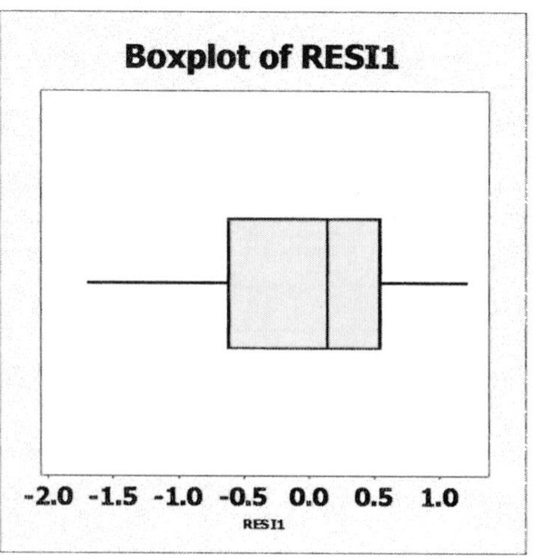

To create a boxplot of the stored residuals, click on **Graph → Boxplot → Simple.** Select RESI1 for **Graph variables**. Click on **Scale** and select **Transpose value and category scales**. Click on **OK** twice to view the boxplot. There are no outliers on the boxplot.

Notice that the coefficient of determination is listed in the output. $R^2 = 82.3\%$ and $R^2_{adj} = 77.9\%$. Next, before interpreting the F-test, draw a normal probability plot of the residuals. Click on **Graph → Probability Plot → Simple.** Select RESI1 for **Graph variables**. Click on **OK** to view the plot.

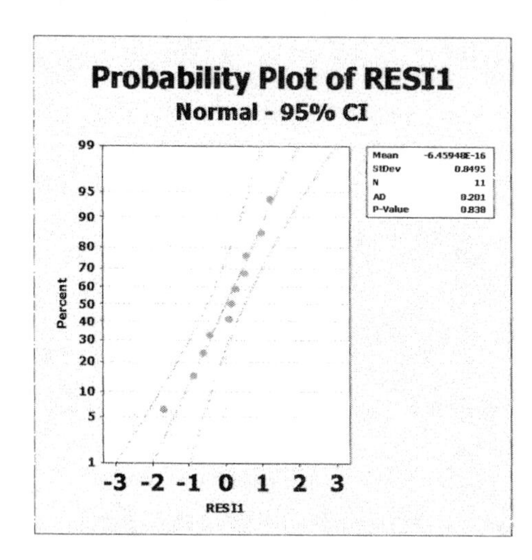

The residuals are approximately normal, so notice that the p-value for the F-test for the regression is 0.001 indicating that there is a significant linear relationship between at least one of the predictor variables (Engine and Weight) and the response variable (MPG).

Finally, look at the prediction interval. Since HP is no longer in the model, we will not use the fact that the vehicle is 320 horsepower. A vehicle with a 4235 cc engine that weighs 3950 lbs is predicted to get 20.712 mpg, and we are 95% confident that the total mpg will be between 18.348 and 23.075.

◀

Nonparametric Statistics

(Note: Textbook section available on CD)

Section 15.2

▸ Example 3 (pg. 15-7) Testing for Randomness (Small Sample)

Enter the data into C1 in a Minitab Worksheet. Minitab is expecting numerical data, so use a 0=Male and 1=Female (coded below in C2).

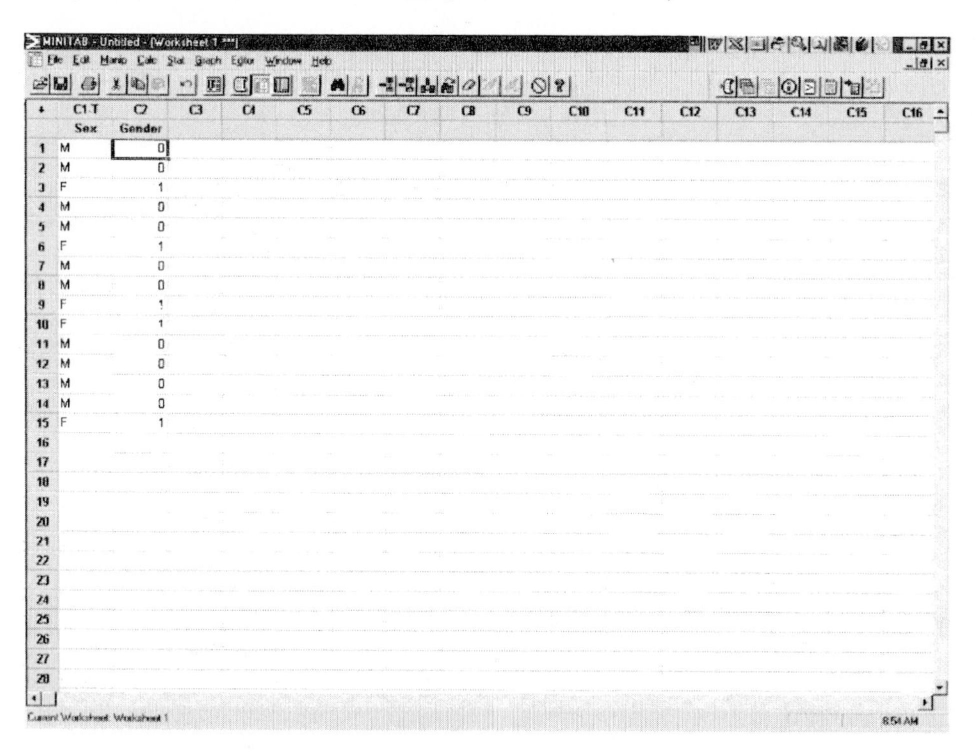

Click on **Stat** → **Nonparametrics** → **Runs Test.** Select C2 as the **Variable.** Since the data is 1s and 0s, it is binomial data. Select **Above and below** .5.

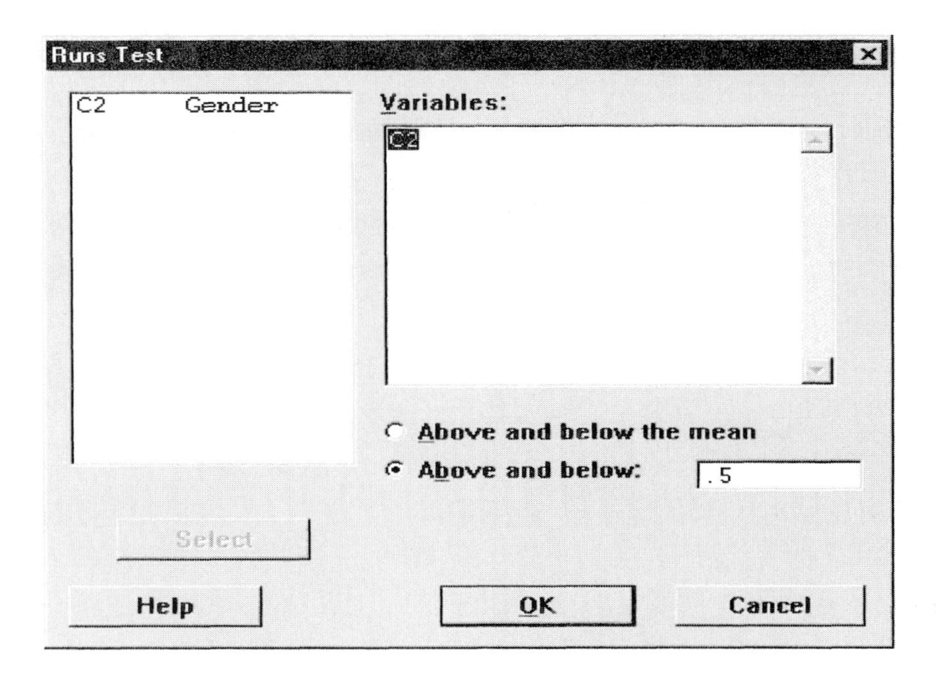

Click on **OK** and the results will appear in the Session Window.

Runs Test: Gender

```
Runs test for Gender

Runs above and below K = 0.5

The observed number of runs = 8
The expected number of runs = 7.66667
5 observations above K, 10 below
* N is small, so the following approximation may be
invalid.
P-value = 0.839
```

Notice that the P-value = .839. Since this value is larger than α=.05, you would fail to reject the null hypothesis. Thus, students enter the room in a *random* order as pertains to gender.

◀

▸ Problem 7 (pg. 15-9) Baseball

Enter the data into C1 in a Minitab Worksheet. Minitab is expecting numerical data, so use 0=Fastball and 1=Change-up.

Click on **Stat → Nonparametrics → Runs Test.** Select C1 as the **Variable.** Since the data is 1s and 0s, it is binomial data. Select **Above and below** .5. Click on **OK** and the results will appear in the Session Window.

Runs Test: Pitch

```
Pitch
K =      0.5000


The observed number of runs =    8
The expected number of runs =    9.5556
7 Observations above K    11 below
* N is small, so the following approximation may be
invalid.
P-value = 0.4250
```

Notice that the P-value = .4250. Since this value is larger than α=.05, you would not reject the null hypothesis. Thus, Keith Foulke chooses his pitches randomly.

◂

Section 15.3

▸ Example 1 (pg. 15-13) One-sample Sign Test

Enter the data into C1 in the MINITAB Data Window. (Do not type in the $ sign.) To perform the Sign Test, click on **Stat → Nonparametrics → 1-sample Sign.** Select C1 as the **Variable.** Since you would like to test if the median amount of credit card debt is $500, enter 500 for **Test median** and select **not equal** for the **Alternative.**

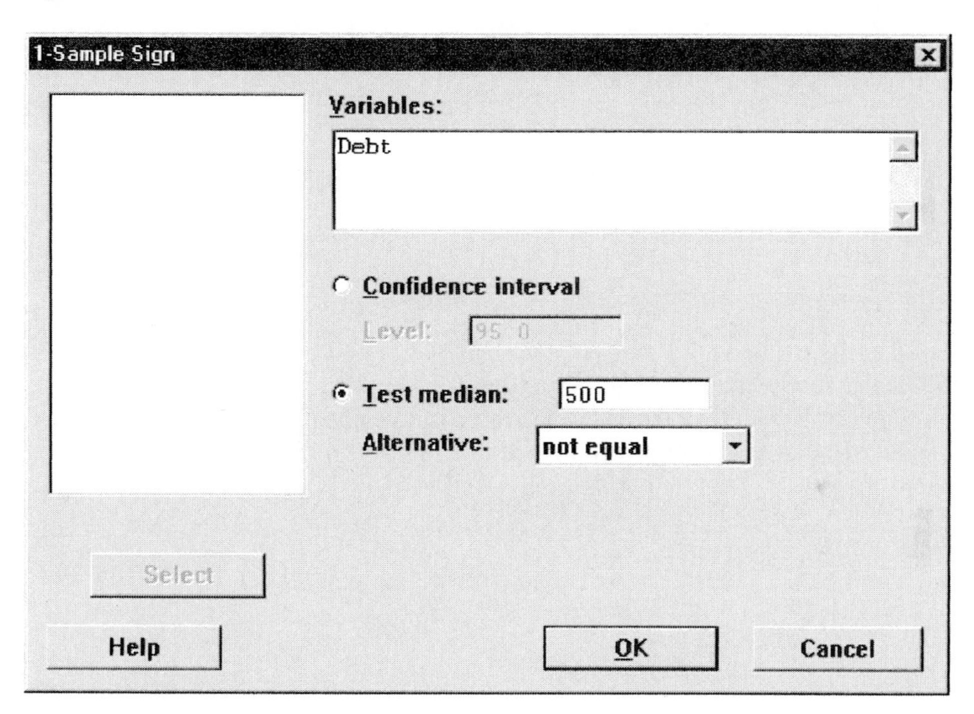

Click on **OK** and the results will be displayed in the Session Window.

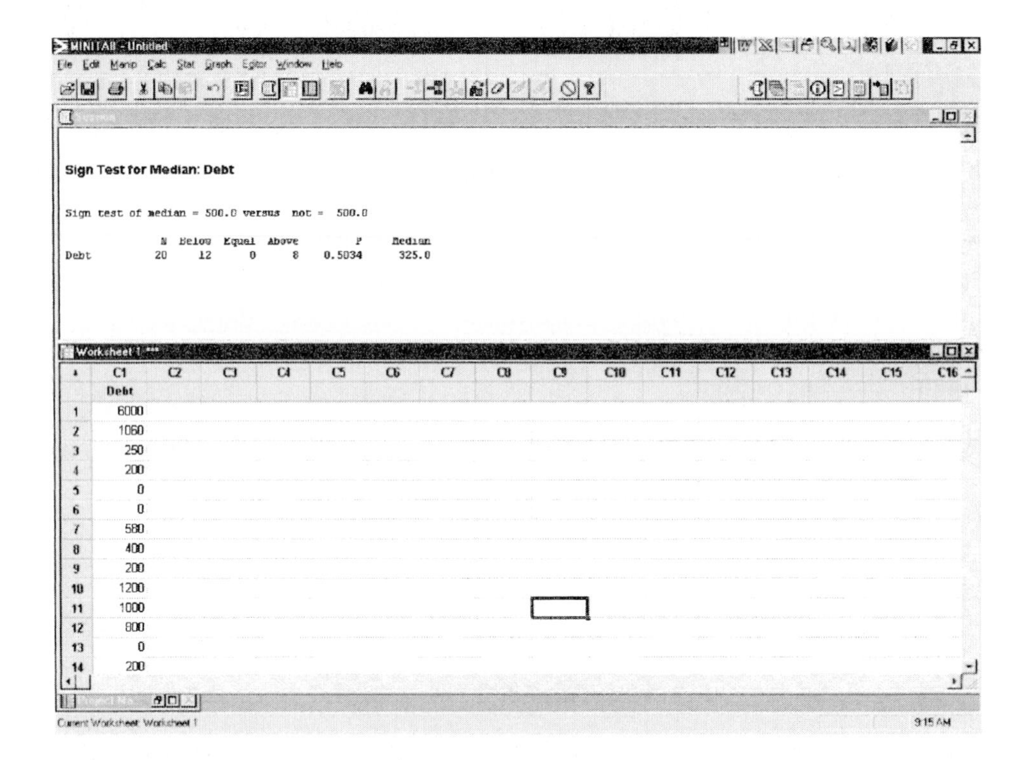

Notice that the P-value = .5034. Since this is such a large P-value, you would fail to reject the null hypothesis. Thus, there is no evidence that the median amount of credit card debt is different from $500.

◀

▸ Problem 11 (pg. 15-16) Acid Rain

Enter the pH levels into C1 of the MINITAB Data Window. Now perform a 1-sample Sign test. Click on **Stat → Nonparametrics → 1-sample Sign.** Select C1 as the **Variable.** Since you would like to test if the pH level is more than 4.90, enter 4.9 for **Test median** and select **greater than** for the **Alternative.** Click on **OK.**

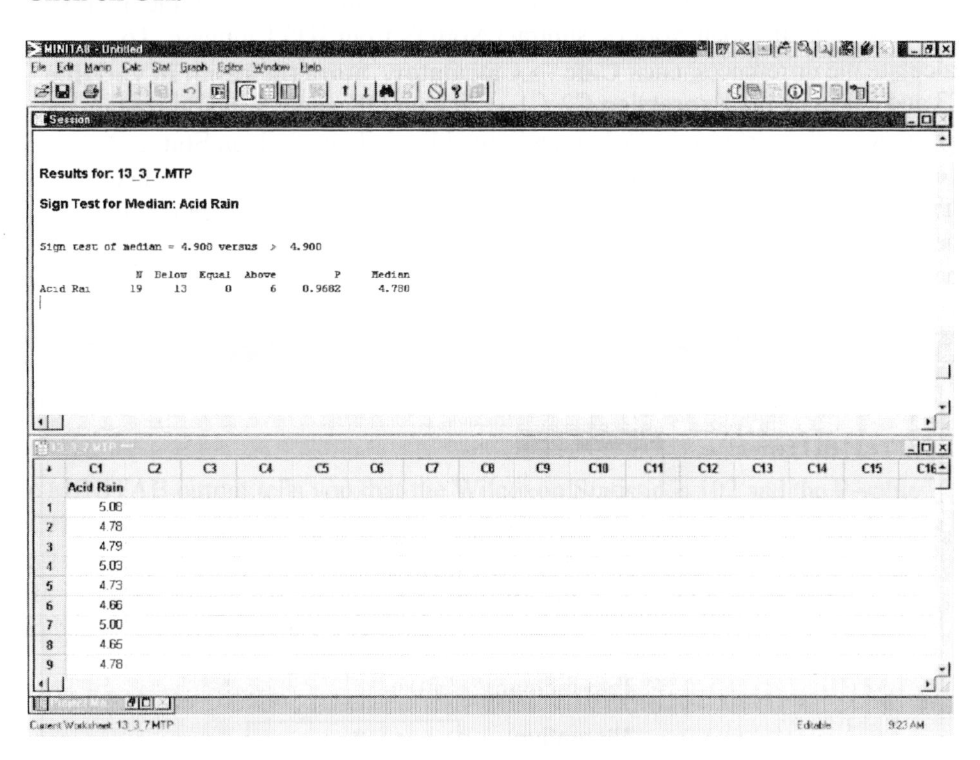

Notice that the P-value = .9682. Since this value is larger than α=.05, you would fail to reject the null hypothesis. Thus, there is not enough evidence to conclude that the pH level of rain has increased since 1990.

◀

Now, to do the Kruskal-Wallis test, click on **Stat → Nonparametrics → Kruskal-Wallis.** The **Response** variable is HDL level (C4) and the **Factor** is Age group (C5).

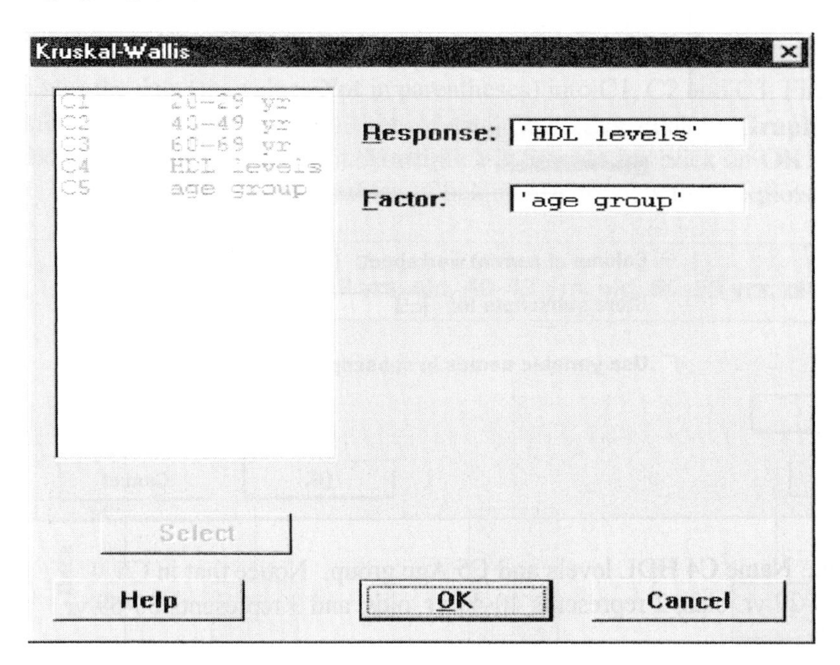

Click on **OK** and the results will be displayed in the Session Window.

Kruskal-Wallis Test: HDL levels versus age group

```
Kruskal-Wallis Test on HDL leve

age grou     N    Median    Ave Rank        Z
1           12     41.00        16.2    -0.92
2           12     45.50        18.8     0.12
3           12     46.50        20.5     0.81
Overall     36                  18.5

H = 1.01  DF = 2  P = 0.604
H = 1.01  DF = 2  P = 0.603 (adjusted for ties)
```

Notice that the test statistic is H=1.01 and the P-value=.604 (.603 when adjusted for ties). With such a large P-value, you do not reject the null hypothesis. So, there is a no difference in the distribution of HDL levels for the three age groups.

◀

▸ Problem 7 (pg. 15-49) Corn Production

Enter the data into C1 - C5 of a Minitab worksheet. Enter "Sludge Plot" in C1, "Spring disk" in C2, etc. First draw the boxplots of the data. To do this, click on **Graph → Boxplot** and select the Boxplot: **Multiple Y's Simple.** Click on **OK**. Select C1 - C5 for **Graph variables**. Click on **OK** to view the boxplots.
To perform a Kruskal-Wallis test, MINITAB requires that the data be stacked into one column with a second column identifying which sample each data value came from. To do this, click on **Data → Stack → Columns.** Select all five columns to be stacked on top of each other. Select **Column of current worksheet** and enter C6 and **Store subscripts in** C7. The subscripts will be numbers 1, 2, 3, 4 or 5 to indicate which column the data value came from. Be sure that **Use variable names in subscript column** is NOT selected. Click on **OK.** Name C6 Plants and C7 Plot Type. Notice that in C7, 1 represents Sludge Plot, 2 represents Spring Disk, etc. Now, to do the Kruskal-Wallis test, click on **Stat → Nonparametrics → Kruskal-Wallis.** The **Response** variable is Plants (C6) and the **Factor** is Plot Type (C7). Click on **OK** and the results will be displayed in the Session Window.

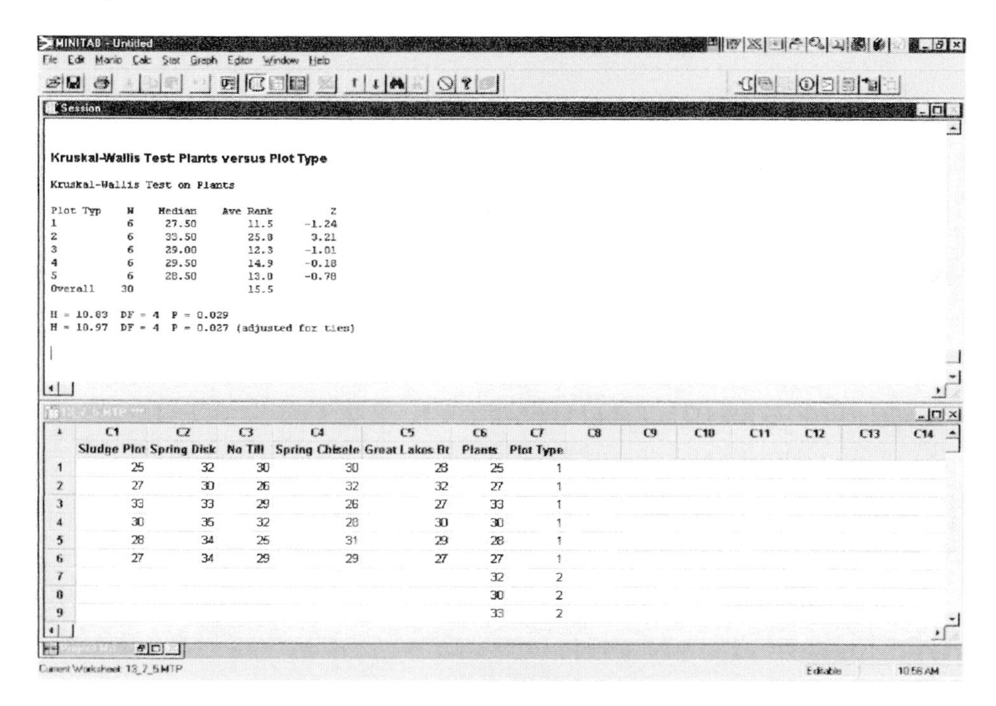

Notice that the test statistic is H=10.83 and the P-value=.029. (When adjusted for ties, H = 10.97 and the P-value = 0.027.) Since the P-value is smaller than α=.05, you should reject the null hypothesis. So, there is a difference in the distribution for each type of plot.

◀

TI-83 and TI-84 Series Graphing Calculators Manual

Dorothy Wakefield • Kathleen McLaughlin

Statistics

SECOND EDITION

Informed Decisions Using Data

Michael Sullivan, III

Introduction

A Manual for the TI-84 and TI-83 Graphing Calculators is one of a series of companion technology manuals that provide hands-on technology assistance to users of Sullivan *Statistics: Informed Decisions Using Data, 2nd edition.*

Detailed instructions for working selected examples and problems from *Statistics: Informed Decisions Using Data* are provided in this manual. To make the correlation with the text as seamless as possible, the table of contents includes page references for both the Sullivan text and this manual.

Contents:

Getting Started with the TI-84 and TI-83 Graphing Calculators

Overview

This manual is designed to be used with the TI-84 and TI-83 families of Graphing Calculators. These calculators have a variety of useful functions for doing statistical calculations and for creating statistical plots. The commands for using the statistical functions are basically the same for the TI-84's and TI-83's. All TI-84 calculators, the TI-83 Plus Calculator and the TI-83 Silver Edition can receive a variety of software applications that are available through the TI website (www.ti.com). TI also will provide downloadable updates to the operating systems of these calculators. These features are not available on the TI-83.

Your textbook comes with data files on the CD data disk that can be loaded onto any of the TI-84 or TI-83 calculators. The requirements for the transfer of data differ from one calculator to another. Some of these calculators (TI-84's and TI-83 Silver Edition) are sold with the necessary connection. For the TI-83 or TI-83 Plus, you can purchase a Graph Link manufactured by Texas Instruments which connects the calculator to the computer. (Note: In order to do examples in this manual, you can simply enter the data values for each example directly into your calculator. It is not necessary to use the graph link to download the data into your calculator. The download procedure using the computer link is an optional way of entering data.)

Throughout this manual all instructions and screen shots use the TI-84. These instructions and screen shots are also compatible with the TI-83 calculators.

Before you begin using the TI-84 or TI-83 calculator, spend a few minutes becoming familiar with its basic operations. First, notice the different colored keys on the calculator. On the TI-84's, the white keys are the number keys; the light gray keys on the right are the basic mathematical functions; the dark gray keys on the left are additional mathematical functions; the remaining dark gray keys are the advanced functions; the light gray keys just below the viewing screen are used to set up and display graphs, and the light gray arrow keys are used for moving the cursor around the viewing screen. On the TI-83's, the white keys are the number keys; the blue keys on the right are the basic mathematical functions; the dark gray keys on the left are additional mathematical functions;

the remaining dark gray keys are the advanced functions; the blue keys just below the viewing screen are used to set up and display graphs, and the blue arrow keys are used for moving the cursor around the viewing screen.

The primary function of each key is printed in white on the key. For example, when you press STAT, the STAT MENU is displayed.

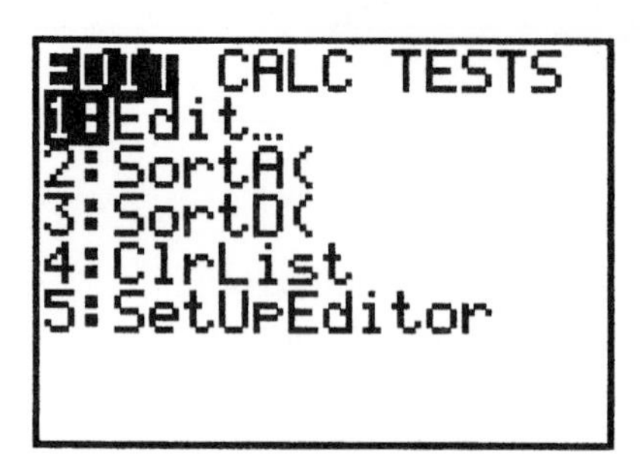

The secondary function of each key is printed in blue on the TI-84's (yellow on the TI-83's) above the key. When you press the **2nd** key (found in the upper left corner of the keys), the function printed above the key becomes active and the cursor changes from a solid rectangle to an ↑ (up-arrow). For example, when you press **2nd** and the x^2 key, the $\sqrt{\ }$ function is activated.
The notation used in this manual to indicate a secondary function is '**2nd**' followed by the name of the secondary function. For example, to use the LIST function, found above the STAT key, the notation used in this manual is **2nd** **[LIST]**. The LIST MENU will then be activated and displayed on the screen.

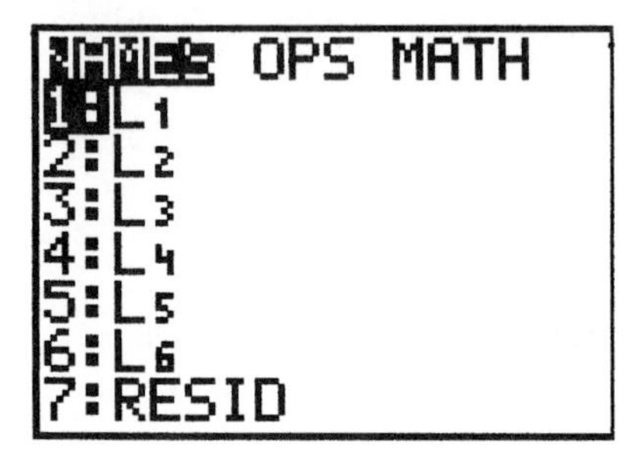

The alpha function of each key is printed in green above the key. When you press the green ALPHA key, the function printed in green above the key is activated and the cursor changes from a solid rectangle to A.

In this manual you will find detailed explanations of the different statistical functions that are programmed into the TI-84 and TI-83 graphing calculators. These explanations will accompany selected examples from your textbook. This will give you the opportunity to learn the various calculator functions as they apply to the specific statistical material in each chapter.

Getting Started

To operate the calculator, press **ON** in the lower left corner of the calculator. Begin each example with a blank screen, with a rectangular cursor flashing in the upper left corner. If you turn on your calculator and you do not have a blank screen, press the **CLEAR** key. You may have to press **CLEAR** a second time in order to clear the screen. If using the **CLEAR** key does not clear the screen, you can push **2nd** **[QUIT]** (Note: **QUIT** is found above the **MODE** key.)

Helpful Hints

To adjust the display contrast, push and release the 2nd key. Then push and hold the up arrow ▲ to darken or the down arrow ▼ to lighten.

The calculator has an automatic turn off that will turn the calculator off if it has been idle for several minutes. To restart, simply press the **ON** key.

There are several different graphing techniques available on the TI-84 and TI-83 calculators. If you inadvertently leave a graph on and attempt to use a different graphing function, your graph display may be cluttered with extraneous graphs, or you may get an ERROR message on the screen.

There are several items that you should check before graphing anything. First, press the **Y=** key, found in the upper left corner of the key pad, and clear all the Y-variables. The screen should look like the following display:

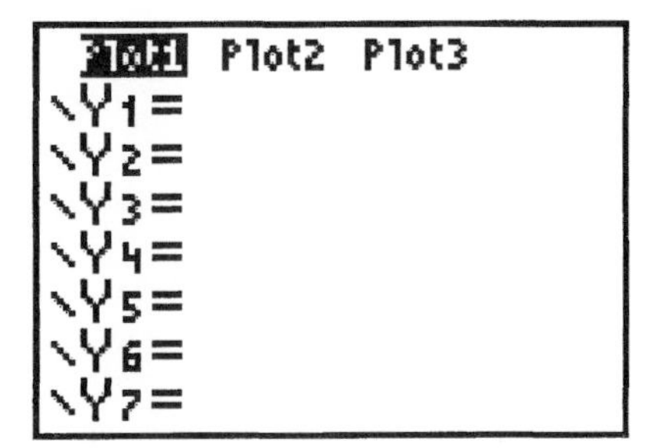

If there are any functions stored in the Y-variables, simply move the cursor to the line that contains a function and press CLEAR ENTER.

Next, press 2nd **[STAT PLOT]** (found on the Y= key) and check to make sure that all the STAT PLOTS are turned **OFF**.

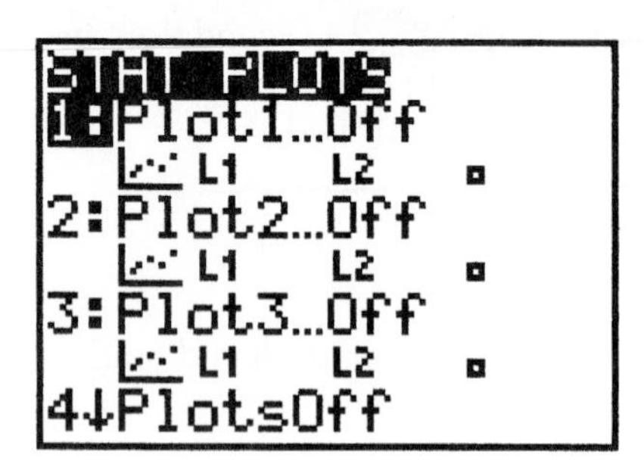

If you notice that a Plot is turned **ON**, select the Plot by using the down arrow key to highlight the number to the left of the Plot , press ENTER and move the cursor to **OFF** and press ENTER. Press **2nd** **[QUIT]** to return to the home screen.

Data Collection

Section 1.2

▸ Example 3 (pg.18) **Generating a Simple Random Sample**

The first step is to set the *seed* by selecting any 'starting number' and storing this number in **rand**. Suppose, for this example, that we select the number '34' as the starting number. Type **34** into your calculator and press the STO key found in the lower left section of the calculator keys. Next press the MATH key found in the upper left section of the calculator keys. The Math Menu will appear.

```
MATH NUM CPX PRB
1:▸Frac
2:▸Dec
3:³
4:³√(
5:ˣ√
6:fMin(
7↓fMax(
```

Use the right arrow key, ▸ found in the upper right section of the calculator keys, to move the cursor to highlight **PRB**. The Probability Menu will appear.

```
MATH NUM CPX PRB
1:rand
2:nPr
3:nCr
4:!
5:randInt(
6:randNorm(
7:randBin(
```

The first selection on the **PRB** menu is **rand,** which stands for 'random number'. Notice that this highlighted. Simply press ENTER twice and the starting value of '34' will be stored into **rand** and will be used as the *seed* for generating random numbers. (Note: This example uses '34' as the *seed,* but you can use any number as a seed for your random number generator.)

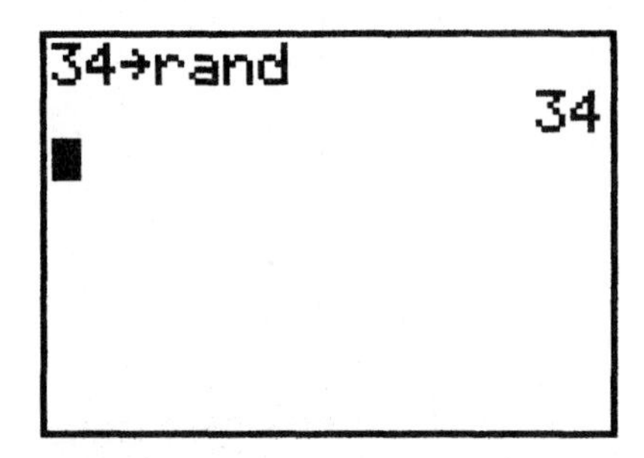

Now you are ready to generate a random integer. Press MATH again and the Math Menu will appear. Use the right arrow key, ▶, to move the cursor to highlight **PRB**. The Probability Menu will appear. Select **5:RandInt(** by using the down arrow key, ▼ , to highlight it and pressing ENTER or by pressing the 5 key. **RandInt(** should appear on the screen. This function requires two values: the starting integer, followed by a comma (the comma is found on the black key above the 7 key), and the ending integer. To complete the **RandInt** command, close the parentheses and press ENTER. (Note: Closing the parenthesis at the end of the command is optional.) This command will generate one random number.

For this example, the starting integer is **1** and the ending integer is **30**. The random number generated in this example is **11**.

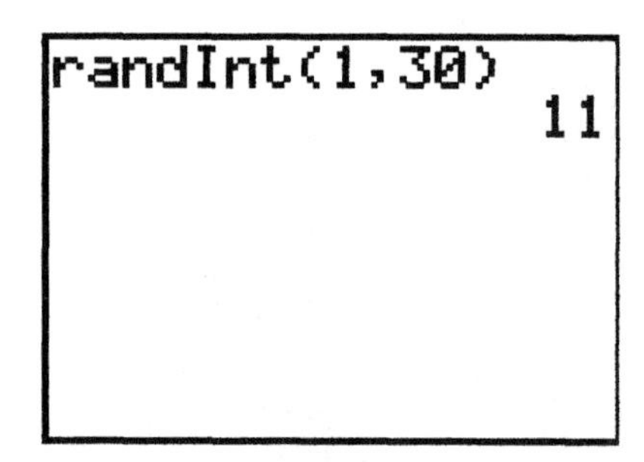

If you press ENTER again, a second random number between 1 and 30 will be generated. Continue pressing ENTER until you have generated 5 **distinct** random numbers. (Note: The TI-84 uses a method called "sampling with replacement" to generate random numbers. This means that it is possible to select the same integer twice.)

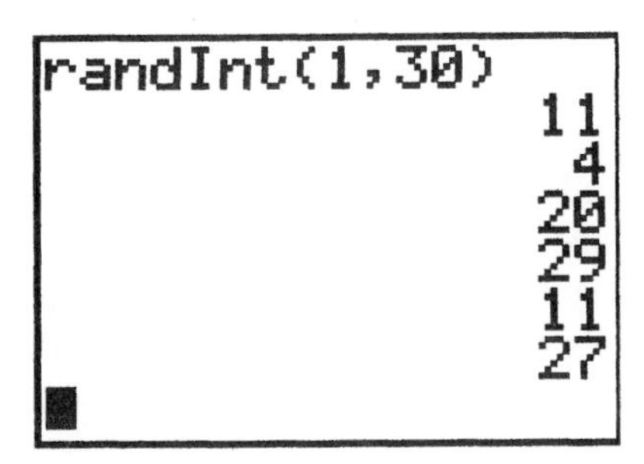

The five distinct random numbers generated are: 11, 4, 20, 29 and 27.

If you want to generate several random numbers with one command, you can change the **RandInt** command so that it contains three values: the starting value, the ending value and the number of values you want to generate. Since duplicates are possible, it is good practice to generate a few more numbers than are actually needed.

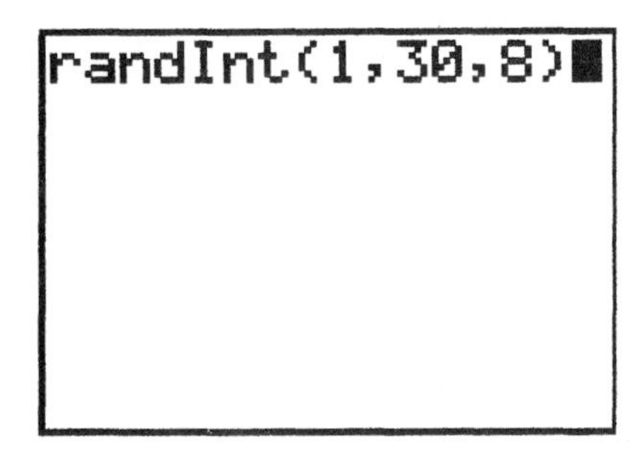

Press ENTER and a partial display of the 8 random integers should appear on your screen. (Note: your numbers will probably be different from the ones you see here. The numbers that are generated will depend on the *seed* that is initially selected.)

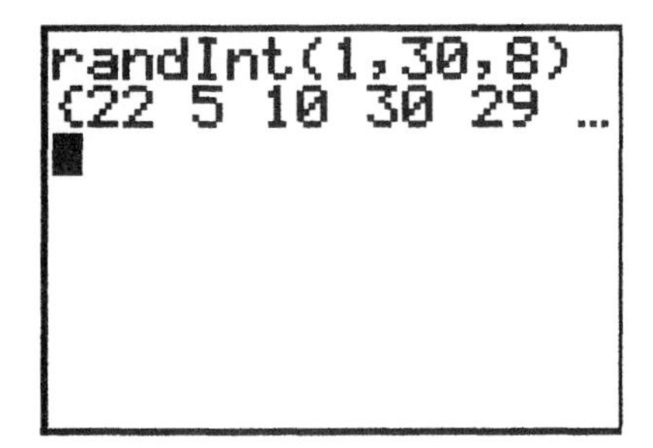

Use the right arrow to scroll through your 8 items and select the first five distinct random numbers.

Problem 21 (pg.21) Generating a Simple Random Sample

a. To obtain a simple random sample of size 10, press MATH, use the right arrow to highlight **PRB** and select **5: randInt**. Enter the starting value of **1**, the ending value of **50** and a sample size of **15**. We are generating more numbers than we actually need because of the possibility of getting duplicates in the sample. To obtain your sample of 10, select the first 10 distinct numbers in the sample that you generated. (Note: In this example, we did not set a new seed. Setting a new seed every time that you generate a random sample is optional. It is not required.)

b. Repeat the steps in part a. and generate another sample of integers.

Organizing and Summarizing Data

CHAPTER 2

Section 2.2

▸ Example 2 (pg. 78) A Histogram for Discrete Data

To create a histogram, you have two choices: 1): enter all the individual data points from Table 8 on pg. 77 into one column or 2): enter the data values into one column and the frequencies into another column using the summarized data in Table 9 on pg. 78. For this example, we will use the summarized data.

To create this histogram, you must enter information into List1 (**L1**) and List 2 (**L2**) on your calculator. You will enter the 'number of customers' into **L1** and the frequencies into **L2**. Press STAT and the Statistics Menu will appear.

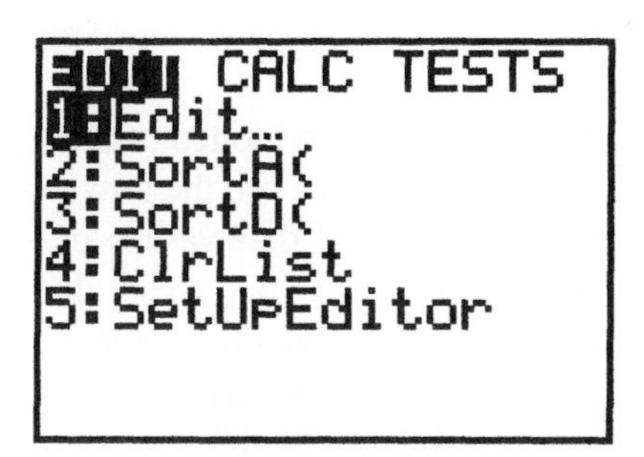

Press ENTER and lists **L1**, **L2** and **L3** will appear.

L1 L2 L3 2

L2(1)=

If the lists already contain data, you should clear them before beginning this example. Move your cursor so that the List name (**L1, L2**, or **L3**) of the list that contains data is highlighted.

L1	L2	L3 1
1	4	------
2	6	
3	8	
10		

L1 ={1,2,3,10}

Press CLEAR ENTER. Repeat this process until all three lists are empty.

L1	L2	L3 1
	4	------
	6	
	8	

L1(1)=

To enter the data values into **L1,** move your cursor so that it is positioned in the 1st position in **L1**. Type in the first value, **1**, and press ENTER or use the down arrow. Enter the next value, **2.** Continue this process until all 11 data values are entered into **L1**. Now use the up-arrow to scroll to the top of **L1**. As you scroll through the data, check it. If a data point is incorrect, simply move the cursor to highlight it and type in the correct value. When you have moved to the 1st value in **L1**, use the right arrow to move to the first position in **L2**. Enter the frequencies into **L2**.

L1	L2	L3 3
1	1	
2	6	
3	1	
4	4	
5	7	
6	11	
7	5	

L3(1)=

Before graphing the histogram, make sure that there are no functions in the Y-registers. To do this, press the Y= key. If there are any functions stored in any of the Y-values, simply move the cursor to the line that contains a function and press **CLEAR** . Now you are ready to graph the histogram. Press **2nd** **[STAT PLOT]** (located above the **Y=** key).

Select Plot1 by pressing **ENTER**.

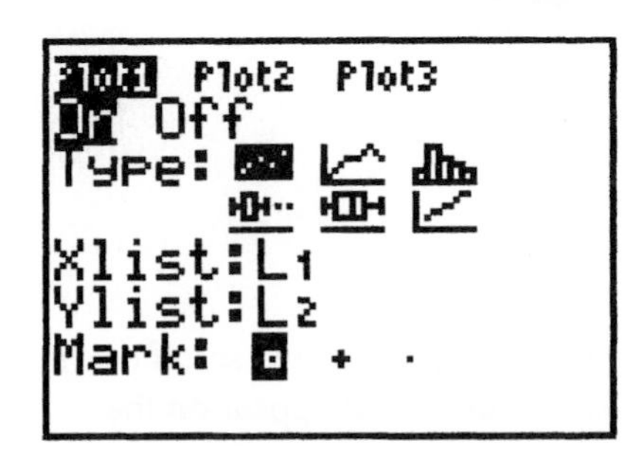

Notice that Plot1 is highlighted. On the next line, notice that the cursor is flashing on **ON** or **OFF**. Position the cursor to **ON** and press **ENTER** to select it. The next two lines on the screen show the different types of graphs. Move your cursor to the symbol for histogram (3rd item in the 1st line of **Type**) and press **ENTER**.

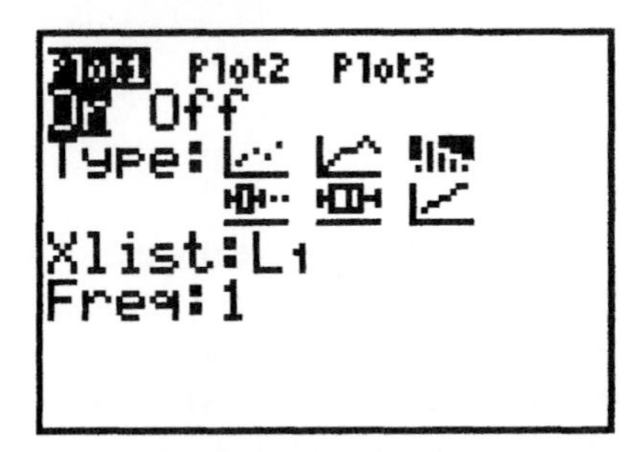

The next line is **Xlist**. Use the down arrow to move to this line. On this line, you indicate where the data values are stored. In most graphing situations, the data are entered into **L1,** so **L1** is the default option. Notice that the cursor is flashing on **L1**. Push **ENTER** to select **L1**. The last line is the frequency line. On this line '1' is the default. The cursor should be flashing on **1**. Change **1** to **L2** by pressing **2nd** **[L2]**. (Note: **L2** is found above the **2** key.)

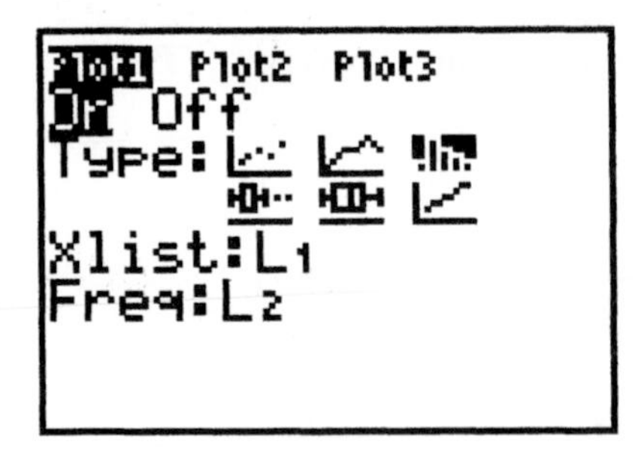

To view a histogram of the data, press **ZOOM**.

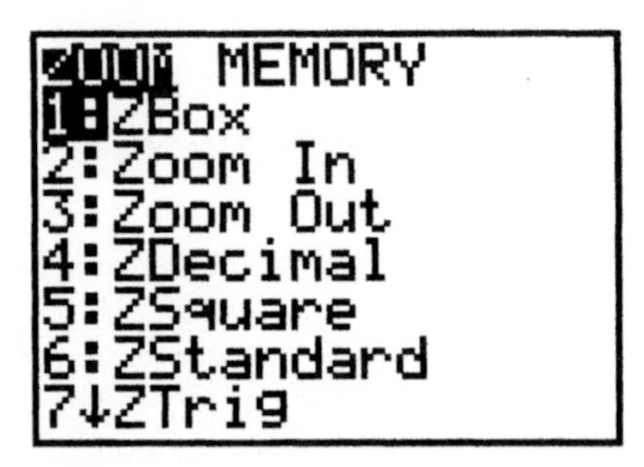

There are several options in the Zoom Menu. Using the down arrow, scroll down to option 9, **ZoomStat,** and press **ENTER**. A histogram should appear on the screen.

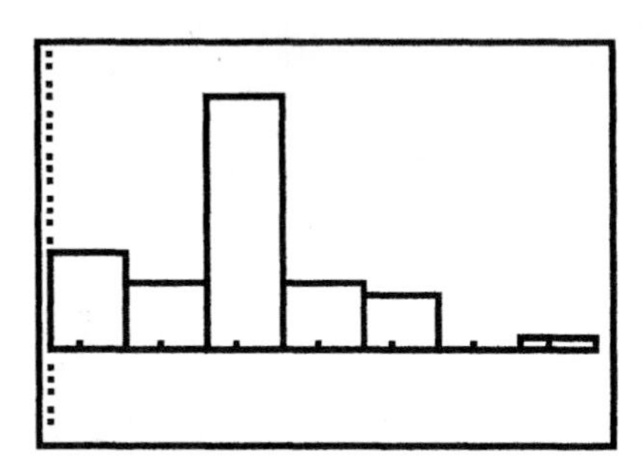

This histogram is not exactly the same as the ones on pg. 79 of your textbook. You can adjust the histogram so that it does look exactly like the one in your text. Press **Window** and set **Xmin** to 1, **Xmax** to 12 (this one extra data value is needed to complete the last bar of the histogram), and **Xscl** equal to 1, which is the difference between successive data values in the frequency distribution. Note: In many cases it is not necessary to change the values for **Ymin, Ymax** or **Yscl.** What you must do is to check these values and make sure that **Ymin** is a small negative value (a value between –6 and –1 would be good) and **Ymax** should be slightly larger than the largest frequency value in your dataset. You never need to adjust **Yscl.**

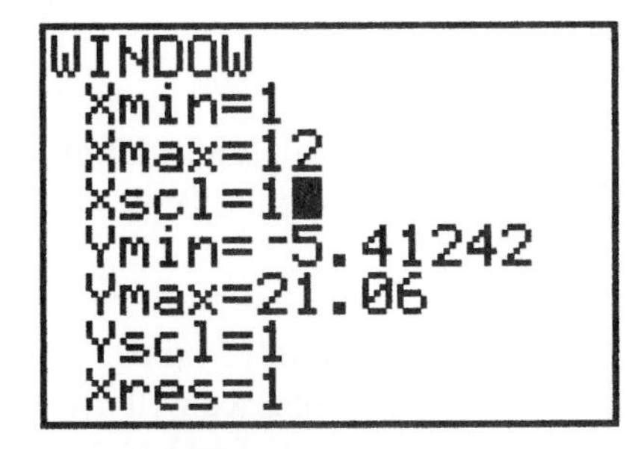

Press **GRAPH**.

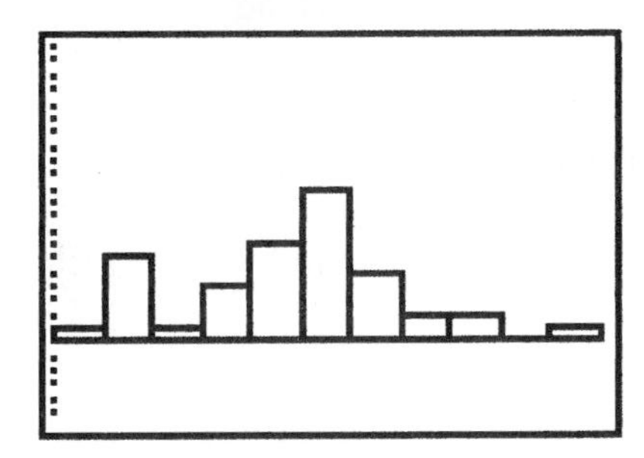

Notice the **TRACE** key. If you press it, a flashing cursor, ∗, will appear at the top of the 1st bar of the histogram.

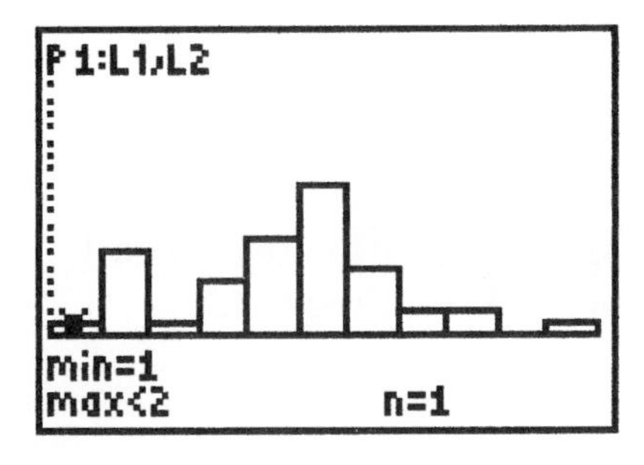

Notice the information at the bottom of the screen. **Min** is the actual data value for the first bar of the histogram. In this example, the first data value is **1**. We do not need to use the Max value in this example. "n=1" tells us that there is only one data point in the dataset that has a value of **1**. You can use the right arrow to move through each of the bars. For example, if you move to the 5th bar in the histogram, you will see that the data value for that bar is **5** and that there are 7 data points in the dataset that have a value of **5**.
Now that you have completed this example, turn Plot1 **OFF**. Using **2nd** **[STAT PLOT]**, select Plot1 by pressing **ENTER** and highlighting **OFF**. Press **ENTER** and **2nd** **[QUIT]** (located above the **MODE** key.) (Note: Turning Plot1 **OFF** is optional. You can leave it ON but leaving it ON will effect other graphing operations of the calculator.)

▸ Example 5 (pg. 82) A Histogram for Continuous Data

Press STAT and ENTER to select **1:Edit**. If there is data in **L1**, highlight **L1** at the top of the first list and press CLEAR and ENTER to clear the data. You should also clear **L2**.

To create this frequency histogram you can use the raw data in Table 12 on pg. 80 or the summarized data in Table 13 on page 80. For this example, we will use the summarized data. You must enter the midpoints of each class into List1 **(L1)** and the frequencies into List 2 (**L2**). To obtain the midpoints of each class, add two consecutive lower limits and divide by 2. For example, here is the calculation for the first class: (0+2)/2= 1.

To enter the midpoints in L1, you can do the calculation for the midpoints right on this screen. Simply type the calculation on the data entry line and push ENTER. The calculation will be automatically converted to the midpoint.

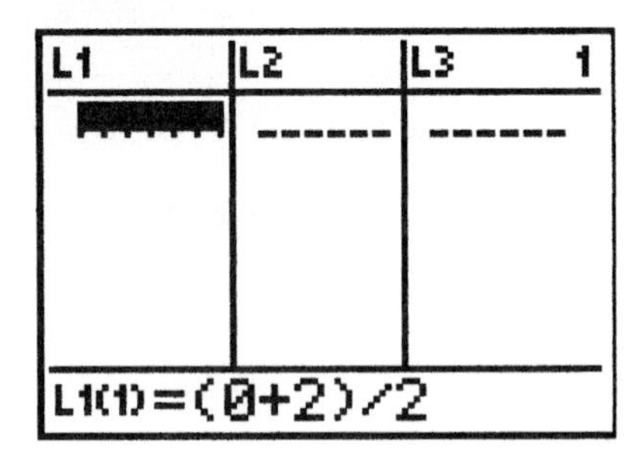

To set up the histogram, push 2nd [STAT PLOT] and ENTER to select **Plot 1**. Turn ON **Plot 1**, set **Type** to **Histogram**, set **Xlist** to **L1**,. set **Freq** to **L2.**

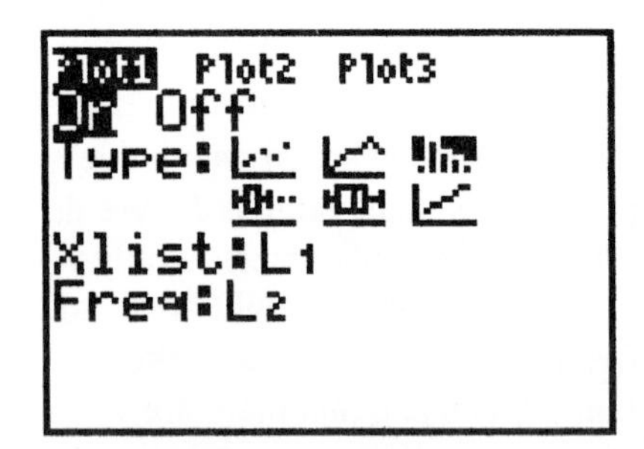

Press ZOOM, scroll down to **9:ZoomStat** and press ENTER to create a histogram. Press Window to adjust the Graph Window. Set **Xmin** equal to 0 (the lower limit of the first class) and **Xmax** equal to 16 (a value that would be the lower limit of an additional class at the end of the table. This extra value is needed to complete the last bar of the histogram). Set **Xscl** equal to 2, which is the class width. (Note: In many cases it is not necessary to change the values for **Ymin, Ymax** or **Yscl**. What you must do is to check these values and make sure that **Ymin** is a small negative value (a value between –6 and –1 would be good) and **Ymax** is slightly larger than the largest frequency value in your dataset. You never need to adjust **Yscl**.

Press **GRAPH** and the histogram should appear.

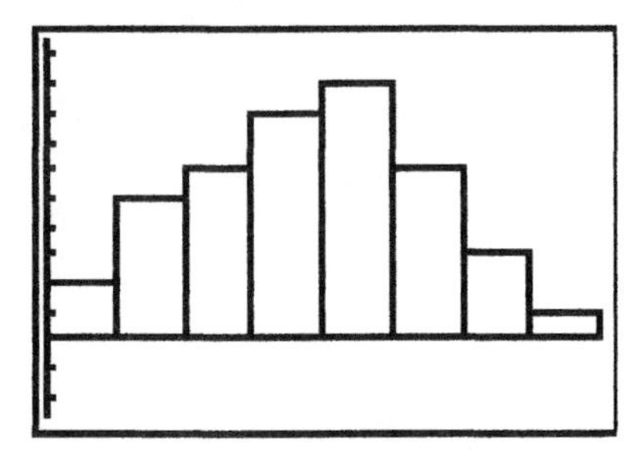

You can press **TRACE** and scroll through the bars of the histogram. The minimum value of the class will appear as **Min. Max** is written as an inequality that states that the maximum value in the class is *less than* the given value. **n** is the number of data points in the class.

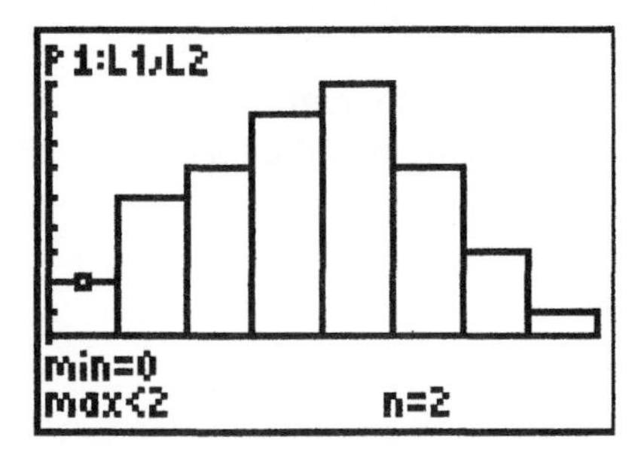

Notice, for example, with the cursor highlighting the first bar of the histogram, you will see that the first class contains values greater than or equal to 0 and less than 2 and that there are 2 data points in the this class.

To create the relative frequency histogram, replace the frequencies in L2 with relative frequencies. One way to do this is to **CLEAR** L2 and enter the relative frequencies from Table 13 on pg. 80. Alternately, without clearing the frequencies, move the cursor so that it highlights the label, L2, at the top of the column and press **ENTER**. On the data entry line, type in **2nd** **L2**/40 and press **ENTER**.

L1	L2	L3 2
1	2	------
3	5	
5	6	
7	8	
9	9	
11	6	
13	3	

L2 =L2/40

L1	L2	L3 3
1	.05	
3	.125	
5	.15	
7	.2	
9	.225	
11	.15	
13	.075	

L3(1)=

Press **Window** to adjust the Graph Window. Set **Xmin** equal to 0 and **Xmax** equal to 16. Set **Xscl** equal to 2. Set **Ymin** to -.1 and **Ymax** to .25.
Press **GRAPH** and the relative frequency histogram should appear.

You can press **TRACE** and scroll through the bars of the histogram.
The minimum and maximum values of the class will appear. **n** is the relative frequency of the class.

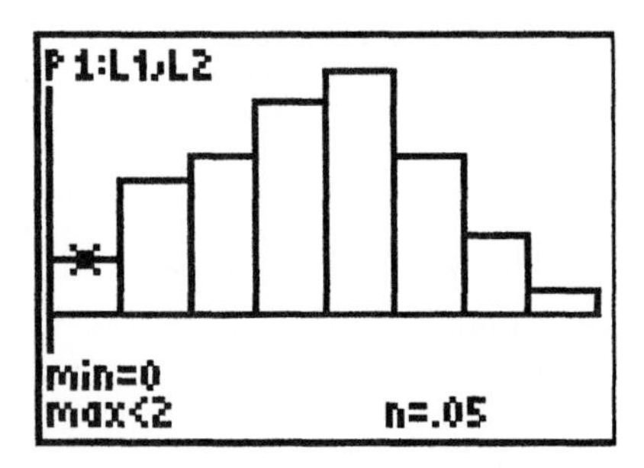

▸ Problem 30 (pg. 91)

For this example, we will construct the histogram first (part c.) and then use it to find the frequencies for the frequency distribution (part a.).

Press **STAT** and select **1:Edit** and press **ENTER**. Highlight the name "**L1**" and press **CLEAR** and **ENTER**. You can also clear **L2** but you will not be using **L2** in this example. Enter the data values into **L1**.
To set up the histogram, push **2nd** **STAT PLOT** and **ENTER** to select **Plot 1**. Turn ON **Plot 1**, set **Type** to **Histogram**, set **Xlist** to **L1.** Since you are using the raw data, you must set **Freq** to **1,** which indicates that you are entering individual data values. If the frequency is set on **L2** move the cursor so that it is flashing on **L2** and press **CLEAR**. The cursor is now in ALPHA mode (notice that there is an "A" flashing in the cursor). Push the **ALPHA** key and the cursor should return to a solid flashing square. Type in the number **1.**

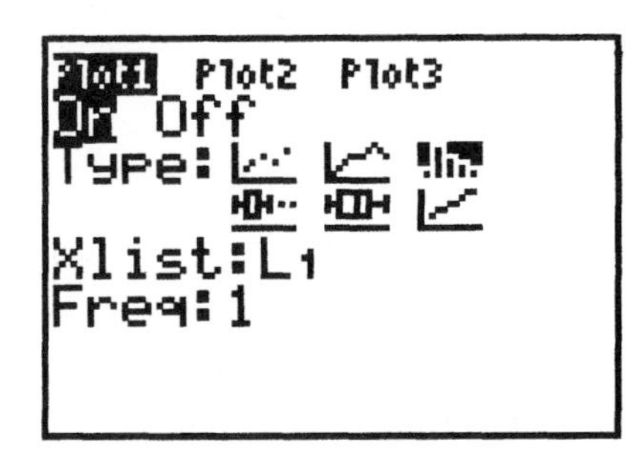

Press **ZOOM**, scroll down to **9:ZoomStat** and press **ENTER** to create a histogram. Press **Window** to set the Graph Window. The first value you must enter is the value for **Xmin.** This value will be the lower class limit of the first class which is **5**. The value for **Xmax** would be the lower class limit of the one extra class that would be needed to complete the last bar of the histogram. Look through the data in your textbook. Notice that the largest data point is 17.8, therefore, the last class would be 17 – 17.9. The lower class limit of the next class would be **18**. This is the value for **Xmax.** Set **Xscl** equal to 1, which is the class width. (Note: In many cases it is not necessary to change the values for **Ymin**, **Ymax** or **Yscl**. What you must do is to check these values and make sure that **Ymin** is a small negative value (a value between –6 and –1 would be good) and **Ymax** is slightly larger than the largest frequency value in your dataset. You never need to adjust **Yscl**.

Press **GRAPH** and the histogram should appear.

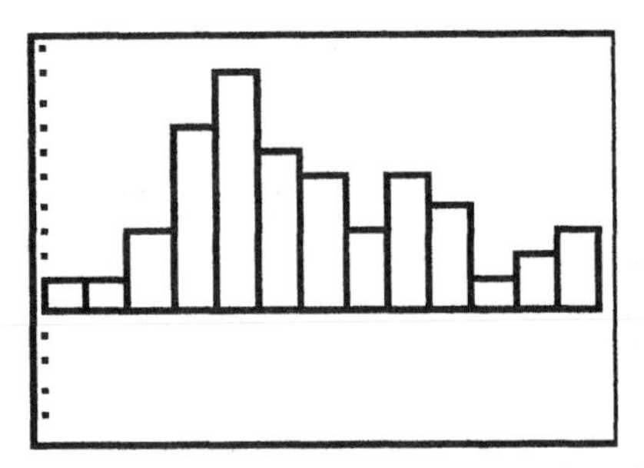

You can press TRACE and scroll through the bars of the histogram.
The minimum value of the class will appear as **Min.** **Max** is written as an inequality that states that the maximum value in the class is *less than* the given value. **n** is the number of data points in the class.

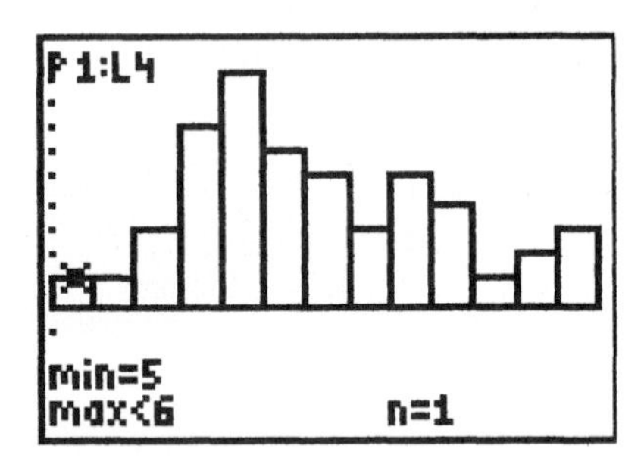

In the screen shown here, you see that the first class is 5-5.9 and there is one data point in this class. Use the right arrow to scroll through the bars of the histogram and use this information to construct the frequency distribution for part (a) of this problem. Here is a starting setup for the frequency distribution table:

Class	Frequency	Relative Frequency
5-5.9	1	
6-6.9		

To complete the relative frequency column (for part b.), simply divide each frequency by the total frequency. For example, the relative frequency for the first class would be 1/51.

To do part (f) of the problem, press **Window** to set the Graph Window. Set **Xscl** equal to 2, set **Xmax** to 19 and set **Ymax** to 20. Press **GRAPH** and the histogram should appear.

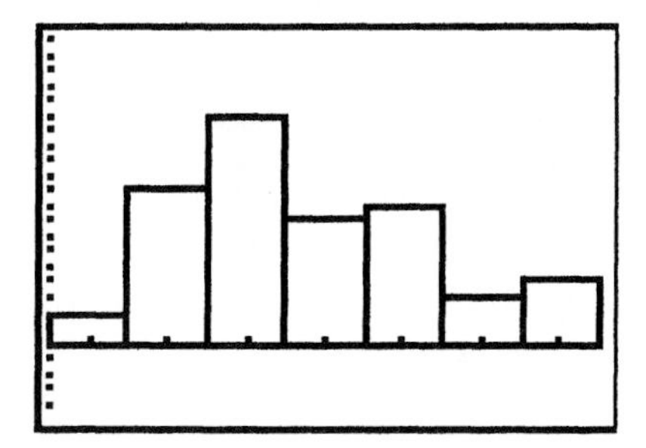

(Note: A relative frequency histogram (part d.) uses relative frequencies rather than frequencies. The actual picture on the TI-84 would be identical to the frequency histogram.)

Section 2.3

▸ Example (pg. 98) A Frequency Polygon

Press STAT and select **1:Edit** and press ENTER. Clear all data from **L1** and **L2**. Enter the midpoints from Table 18 on pg. 98 into **L1** and enter the frequencies into **L2**.

To set up the frequency polygon, press **2nd** STAT PLOT. Press ENTER to select **Plot 1**. Highlight **On** and press ENTER. Set **Type** to the frequency polygon which is the second selection and press ENTER. Set **Xlist** to **L1** and **Freq** to **L2**. Next, there are three different types of **Marks** that you can select for the graph. The first choice, a small square, is the best one to use.

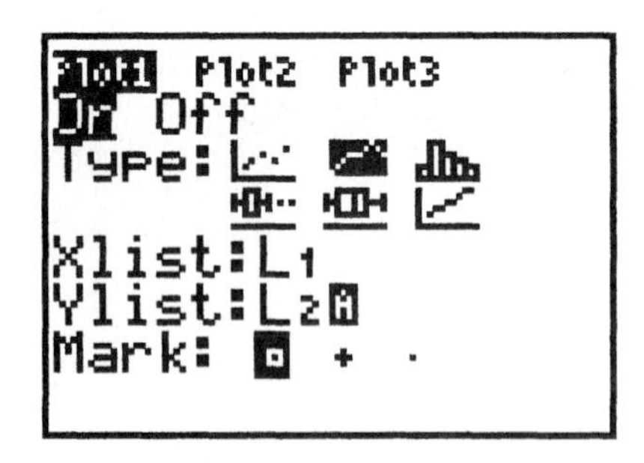

Press ZOOM and scroll down to **9:ZoomStat** and press ENTER or simply press 9 and **ZoomStat** will automatically be selected.

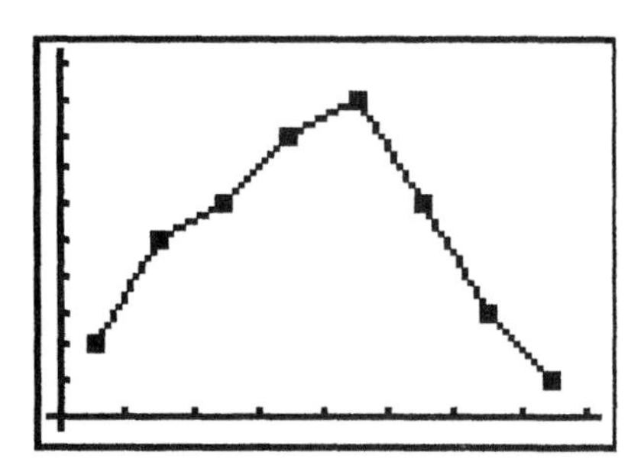

You can press TRACE and scroll through the points in the polygon. For example, the third data point represents class 3 which has a midpoint of 5 and a frequency of 6.

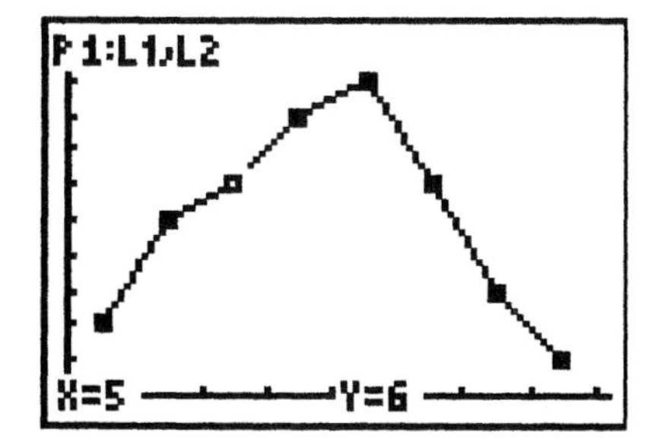
P1:L1,L2
X=5
Y=6

▸ Example (pg. 99) A Relative Frequency Ogive

Press **STAT** and select **1:Edit** and press **ENTER**. Clear all data from **L1** and **L2**. Using the data in Table 19 on pg. 99, enter the upper class limits into **L1** and enter the cumulative relative frequencies into **L2**.

L1	L2	L3 3
1.99	.05	
3.99	.175	
5.99	.325	
7.99	.525	
9.99	.75	
11.99	.9	
13.99	.975	

L3(1)=

To set up the ogive, press 2nd **STAT PLOT**. Press **ENTER** to select **Plot 1**. Highlight **On** and press **ENTER**. Set **Type** to the frequency polygon which is the second selection and press **ENTER**. Set **Xlist** to **L1** and **Freq** to **L2**.

Press **ZOOM** and scroll down to **9:ZoomStat** and press **ENTER** or simply press **9** and **ZoomStat** will automatically be selected.

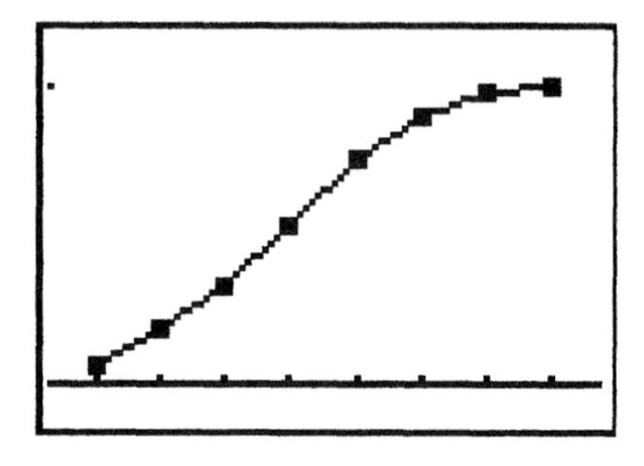

You can press **TRACE** and scroll through the points in the ogive. ◂

▸ Example 1 (pg. 100) A Time Series Plot

Press **STAT** and select **1:Edit**. Clear **L1** and **L2**. Notice that the dates in Table 20 on pg. 100 are "3/03" through "2/05." Rather than entering these actual dates into **L1**, you can simply number the months from "1" to "24" and enter these numbers into **L1**. Enter the closing prices into **L2**.

L1	L2	L3 2
1	12.98	------
2	15	
3	16.41	
4	16.79	
5	19.49	
6	19.14	
7	19.59	

L2 ={12.98,15,16...

To construct the time series chart, press **2nd** **STAT PLOT** and select **1:Plot 1** and **ENTER**. Turn ON **Plot 1**. Set the **Type** to **frequency polygon**. For **Xlist** select **L1** and for **Ylist** select **L2**. Press **ZOOM** and scroll down to **9:ZoomStat** and press **ENTER** or simply press **9** and **ZoomStat** will automatically be selected.

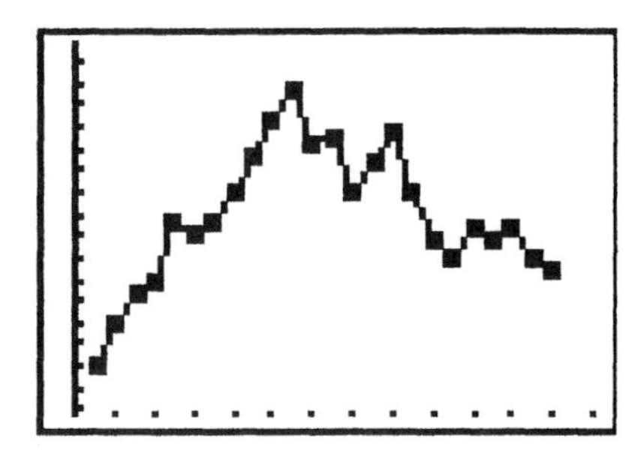

Use **TRACE** to scroll through the data values for each year. Notice for example, the closing price for the 4th month (which is June 2003) is 16.79.

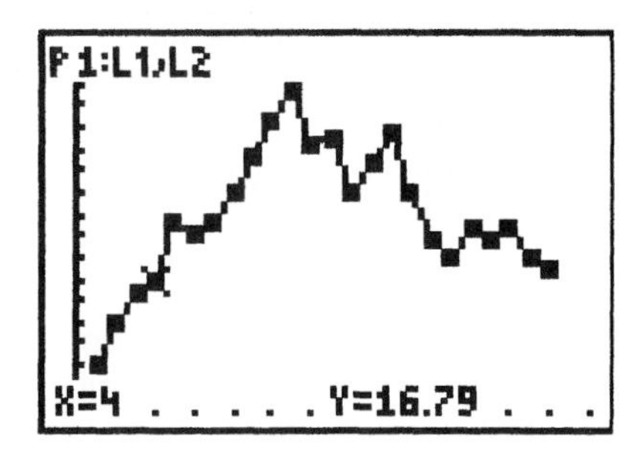

◂

▸ Problem 16 (pg. 102)

For this exercise, use the dataset in Problem 30 on page 91 and the frequency distribution table created on pg.18 of this manual.

Class	Midpoint	Frequency	Relative Frequency	Cumulative Frequency	Cumulative Relative Frequency
5-5.9	5.5				
6-6.9	6.5				

a.) Enter the remaining classes, the midpoints of each class and the frequencies and relative frequencies for each class. To complete the cumulative frequency column, simply accumulate the frequencies. For example, the cumulative frequency for the third class would be the sum of the frequencies for the 1st, 2nd and 3rd classes.

b.) To complete the cumulative relative frequency column, simply accumulate the relative frequencies.

c.) Press STAT and select **1:Edit**. Clear the lists and enter the midpoints into **L1** and the frequencies into **L2**. Press **2nd** [STAT PLOT] and select **Plot 1** and press ENTER. Set the **Type** to **frequency polygon**. Set **Xlist** to **L1** and Freq to **L2**. Press ZOOM and scroll down to **9:ZoomStat** and press ENTER or simply press 9 and **ZoomStat** will automatically be selected.

d.) Press STAT and select **1:Edit** and press ENTER. Clear all data from **L1and L2.** Enter the upper class limits into **L1** and enter the cumulative frequencies into **L2**. Press **2nd** [STAT PLOT] and select **Plot 1** and press ENTER. Set the **Type** to **frequency polygon**. Set **Xlist** to **L1** and Freq to **L2**. Press ZOOM and scroll down to **9:ZoomStat** and press ENTER or simply press 9 and **ZoomStat** will automatically be selected.

e.) A relative frequency ogive is constructed using the upper class limits and the relative frequencies (rather than the frequencies). The actual picture on the TI-84 is identical to the frequency ogive that you constructed in part (d).

Numerically Summarizing Data

Section 3.1

Example 1 (pg. 122) A Population Mean and a Sample Mean

The TI-84 has one method for calculating the mean of a dataset. This method is used for a population mean, μ, and a sample mean, $\bar{x}$. The symbol that the calculator uses for the mean of a dataset is always $\bar{x}$. If you are calculating the mean of a population, then the value for $\bar{x}$ that you obtain from the calculator is actually the value for μ.

(a). Press STAT and select **1:Edit**. Clear **L1** and enter the scores for the ten students into **L1**. Press STAT again and highlight **CALC** to view the Calc Menu.

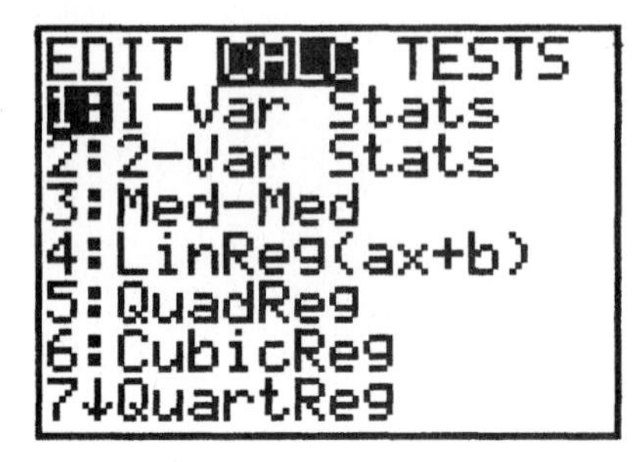

Select **1:1-Var Stats.** On this line, enter the name of the column that contains the data. Since you have stored the data in **L1,** simply enter **2nd** **[L1]** ENTER and the first page of the one variable statistics will appear. (Note: If you did not enter a column name, the default column, which is **L1,** would be automatically selected.)

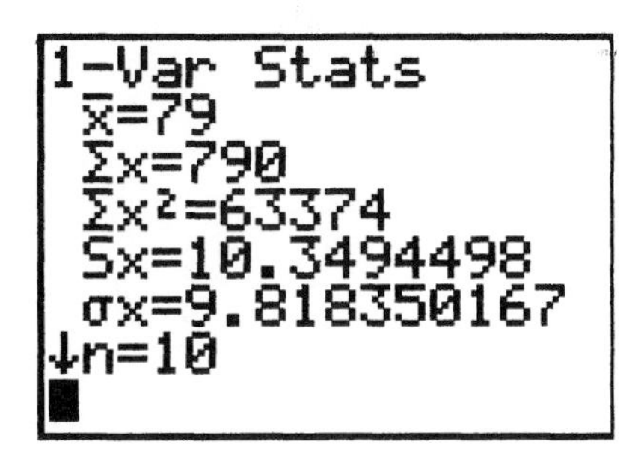

The first item is the mean of the dataset. The correct symbol for the mean of a population is μ. For this dataset, $\mu = 79$.

(b). To use a seed of '54', press **54** STO MATH and select **PRB**. Press ENTER to select **1:Rand** and press ENTER. (Note: Selecting a seed is an optional step which can be omitted when generating random data.)

To generate a random sample of 4 students from the 10 students, press MATH and select **PRB**. Select **5:RandInt** by pressing **'5'** or moving the cursor to **5:RandInt** and pressing ENTER. Enter a starting value of **1**, an ending value of **10** and a sample size of **8**. (Recall: The TI-84 samples with replacement. This method may result in duplicates in your sample. Selecting a few more values than you need is a good practice so that you will obtain the necessary number of distinct values for your sample.)

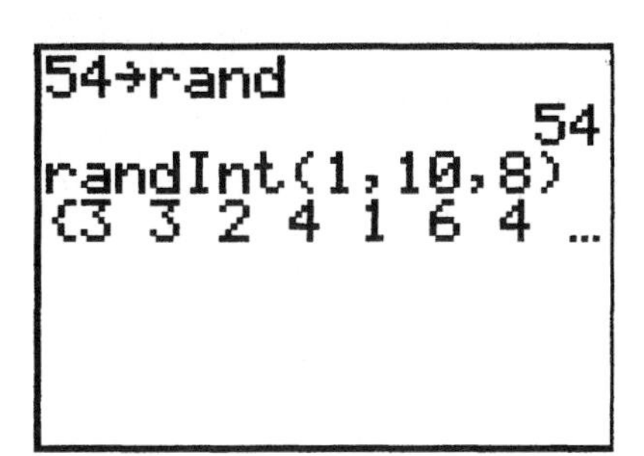

The random sample of 4 students is the first 4 distinct numbers: 3,2,4, and 1. (Note: If you had not selected '54' as the seed, you would have obtained a different set of random numbers.)

(c.) Press STAT and select **1:Edit**. Clear **L2** and enter the 4 data values into **L2**. Press STAT again and highlight **CALC** to view the Calc Menu. Select **1:1-Var Stats.** On this line, enter the name of the column that contains the data. Since you have stored the data in **L2,** simply enter **2nd** **[L2]** ENTER and the first page of the one variable statistics will appear.

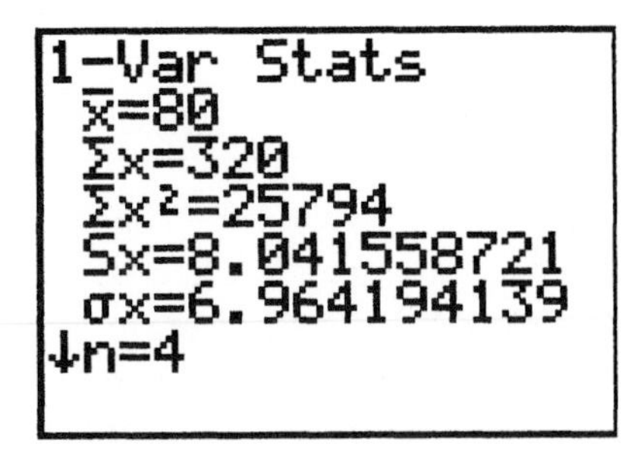

The sample mean, $\bar{x}$, is 80.

▸ Example 2 (pg. 124) The Median of a Dataset

Press **STAT** and select **1:Edit**. Clear **L1** and enter the data from Table 2 on pg. 125 into **L1**. Press **STAT** again and highlight **CALC** to view the Calc Menu. Select **1:1-Var Stats** and press **2nd** **[L1]** **ENTER**. Notice the small arrow in the bottom left corner of the screen. This indicates that more information follows this first page. Use the down arrow to scroll through this information. The third item you see on the second page is the median, Med = 217.

```
1-Var Stats
↑Sx=31.88172587
 σx=30.05837941
 n=9
 minX=179
 Q1=203.5
↓Med=217
```

(Note: It is not necessary to put the data in ascending order when calculating a median using the TI-84.)

◂

▸ Problem 34 (pg. 134)

Press STAT and select **1:Edit**. Clear **L1** and enter the data into **L1**. Press STAT again and highlight **CALC** to view the Calc Menu. Select **1:1-Var Stats** and press **2nd** [L1] ENTER.

(a.) The first item is the sample mean, 104.1 seconds. To find the median, scroll down to the next page of output. The median is 104 seconds.

(b.) Before graphing the histogram, make sure that there are no functions in the Y-registers. To do this, press the Y= key. If there are any functions stored in any of the Y-values, simply move the cursor to the line that contains a function and press CLEAR. Now you are ready to graph the histogram. Press **2nd** **[STAT PLOT]** (located above the Y= key).

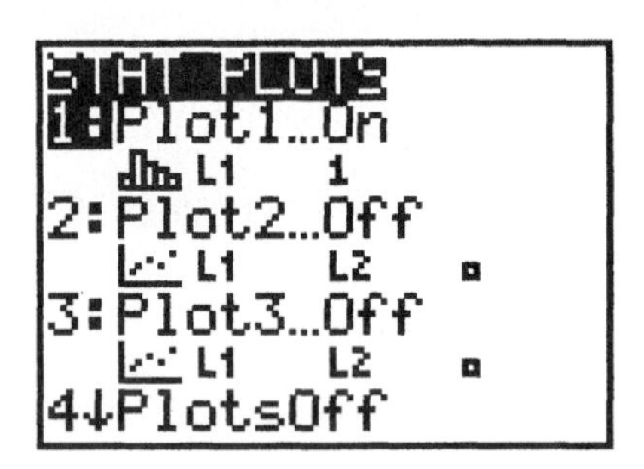

Select Plot1 by pressing ENTER. Position the cursor on **ON** and press ENTER to select it. On the next line, move your cursor to the symbol for histogram (3rd item in the 1st line of **Type**) and press ENTER. The next line is **Xlist**. Use the down arrow to move to this line. On this line, you tell the calculator where the data values are stored. In most graphing situations, the data are entered into **L1** so **L1** is the default option. Notice that the cursor is flashing on **L1**. Push ENTER to select **L1**. The last line is the frequency line. On this line, **1** is the default. The cursor should be flashing on **1** which indicates that you are entering the individual data values.

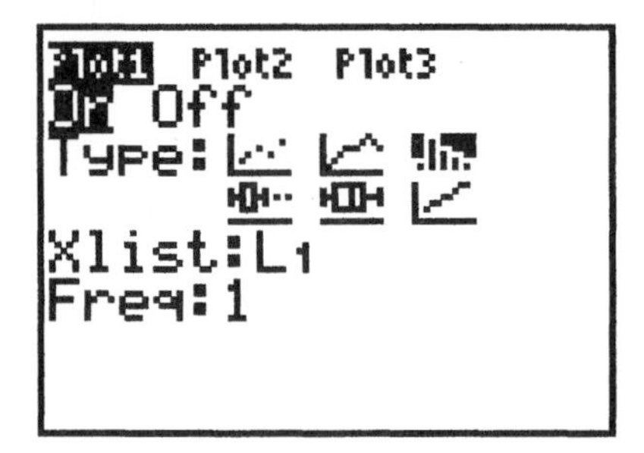

To view a histogram of the data, press ZOOM and scroll down to **9:ZoomStat** and press ENTER or simply press 9 and **ZoomStat** will automatically be selected.

Section 3.2

▸ Example 6 (pg.144) The Variance and Standard Deviation

Press STAT and select **1:Edit**. Clear **L1** and enter the population data from Table 9 on page 139 into **L1**. Press STAT and highlight **CALC** to display the Calc Menu. Select **1: 1-Var Stats** and press **2nd [L1]** ENTER.

```
1-Var Stats
 x̄=79
 Σx=790
 Σx²=63374
 Sx=10.3494498
 σx=9.818350167
↓n=10
```

The population standard deviation is σx, 9.818350167.

To calculate the population variance, $(\sigma x)^2$, type in the value of the standard deviation at the bottom of the screen and press the x^2 key and ENTER.

```
 Σx=790
 Σx²=63374
 Sx=10.3494498
 σx=9.818350167
↓n=10
9.818350167²
              96.4
```

The population variance is 96.4.

Next, obtain the sample standard deviation and sample variance for the sample of four students. Press STAT and select **1:Edit**. Clear **L2** and enter the sample data points (90, 77, 71 and 82) into **L2**. Press STAT and highlight **CALC** to display the Calc Menu. Select **1: 1-Var Stats** and press **2nd [L2]** ENTER.

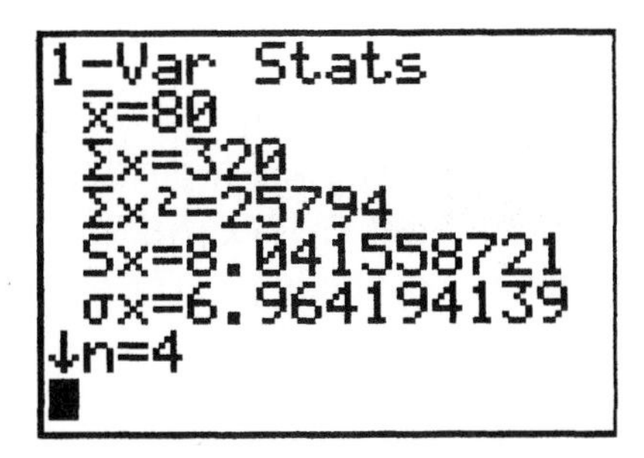

The sample standard deviation is Sx, 8.041558721.
To calculate the sample variance, $(Sx)^2$, type in the value of the standard deviation at the bottom of the screen and press the x^2 key and ENTER.

Problem 25 (pg. 150)

(a.) Press **STAT** and select **1:Edit**. Clear **L1** and enter the data into **L1**. Press **STAT** and highlight **CALC** to display the Calc Menu. Select **1: 1-Var Stats** and press **2nd** **[L1]** **ENTER**. The population standard deviation is σx, 7.67.

To calculate the population variance, $(\sigma x)^2$, type in the value of the standard deviation at the bottom of the screen and press the x^2 key and **ENTER**.

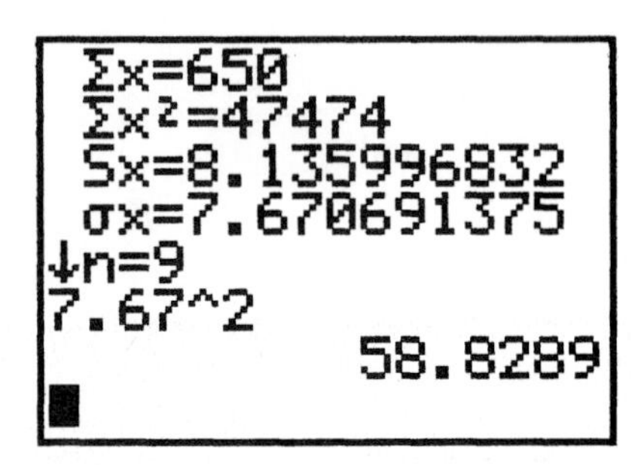

The population variance is 58.83.

(b.) To generate a random sample of size 3, first number the students from 1 to 9. The TI-84 samples with replacement so you will need to generate a sample that is larger than 3. Suppose you take a sample of size 5. Press **MATH** and **PRB**. Select **5:RandInt** and enter **1,9,5**. Here is one possible outcome.

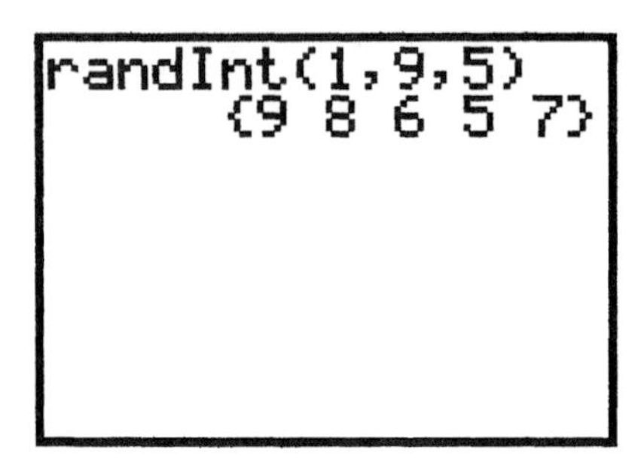

In this outcome we have selected 5 students: Student Numbers: 9,8,6,5 and 7. Since we need only need 3 students, simply select the first three: 9,8, and 6. (Note: Random samples vary so your results will be different.)

Enter the pulse rates of the 3 students into L2. Press **STAT** and highlight **CALC** to display the Calc Menu. Select **1: 1-Var Stats** and press **2nd** **[L2]** **ENTER**. The sample standard deviation is Sx, 6.0277.

To calculate the sample variance, $(Sx)^2$, type in the value of the standard deviation at the bottom of the screen and press the $\boxed{x^2}$ key and ENTER. The sample variance is 36.333.

Repeat this process to generate another random sample of size 3.

▸ Problem 46 (pg. 154)

(a.) Press STAT and select **1:Edit**. Clear **L1** and enter the salary data into **L1**. Press STAT and highlight **CALC** to display the Calc Menu. Select **1: 1-Var Stats** and press **2nd [L1]** ENTER. The population standard deviation, σx, is 12.6. The population variance, $(\sigma x)^2$, is 160. The range (75-30) is 45.

(b.) Press STAT and select **1:Edit**. Clear **L2**. Move the cursor to the top of **L2** again and press ENTER. The cursor should now be flashing on the bottom line of the screen. On this line, type in **2nd [L1] + 2.5.**

L1	L2	L3 2
30	------	------
30		
45		
50		
50		
50		
55		

L2 =L1+2.5

Press ENTER. Each value in **L2** should be 2.5 thousand dollars higher than the corresponding value in **L1**. Press STAT and highlight **CALC** to display the Calc Menu. Select **1: 1-Var Stats** and press **2nd [L2]** ENTER. The population standard deviation, σx, is 12.6, the same as the standard deviation for the original data. The range and variance also remain unchanged.

(c.) Press STAT and select **1:Edit**. Clear **L3**. Move the cursor to the top of **L3** again and press ENTER. The cursor should now be flashing on the bottom line of the screen. On this line, type in **2nd [L1] *1.05.** Press STAT and highlight **CALC** to display the Calc Menu. Select **1: 1-Var Stats** and press **2nd [L3]** ENTER. The population standard deviation, σx, is 13.3, which is 1.05 times the value of the standard deviation for the original data. The variance and range are also 1.05 times the variance and range of the original data.

(d.) Press STAT and select **1:Edit**. In **L1**, move the cursoar so that it is flashing on 75 (Benjamin's salary). Type in **100**. Press STAT and highlight **CALC** to display the Calc Menu. Select **1: 1-Var Stats** and press **2nd [L1]** ENTER. The population standard deviation, σx, has increased to 18.5. The population standard variance, $(\sigma x)^2$, has increased to 341.3. The range has increased to 70.

◂

Section 3.3

▸ Example 2 (pg. 159) The Weighted Mean

Press **STAT** and select **1:Edit**. Clear **L1** and **L2.** Enter the point values for each letter grade that Marissa earned into **L1**. Enter the corresponding credits earned into **L2**.

```
L1      L2      L3     2
4       4       ------
3       3
4       3
2       5
4       1
------  
L2(6) =
```

Press **STAT** and highlight **CALC** to display the Calc Menu. Select **1: 1-Var Stats** and press **2nd** [**L1**] [,] **2nd** [**L2**]. Press **ENTER**. (Note: You must place the comma between **L1** and **L2**).

```
1-Var Stats L1,L
2
```

```
1-Var Stats
 x̄=3.1875
 Σx=51
 Σx²=175
 Sx=.910585892
 σx=.8816709987
↓n=16
```

Her GPA (weighted average) is 3.1875.

◂

Example 4 (pg. 161) The Mean and Standard Deviation of a Frequency Distribution

Press **STAT** and select **1:Edit**. Clear **L1** and **L2.** Refer to Table 14 on pg. 158. Calculate the midpoints of each class by adding consecutive lower class limits and dividing by 2. You can do the calculations for the midpoints directly on this screen. For the first class, type in **(0+2)/2** and press **ENTER**.

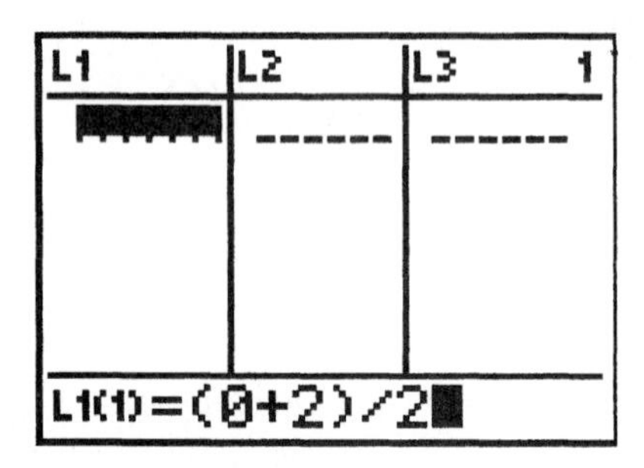

The value of the midpoint, 1, will appear as the first entry in **L1**. Continue this process to obtain the midpoints for each of the classes. Enter the frequencies into **L2**. Press **STAT** and highlight **CALC** to display the Calc Menu. Select **1: 1-Var Stats** and press **2nd [L1]** , **2nd [L2]**. Press **ENTER**. (Note: You must place the comma between **L1** and **L2**).

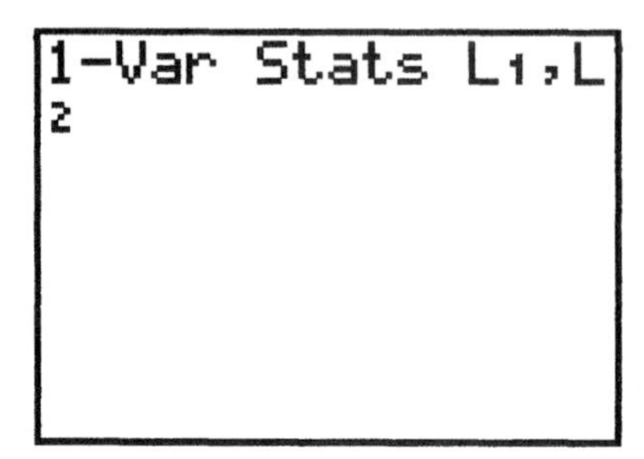

Using **L1** and **L2** in the **1:1-Var Stats** calculation is necessary when approximating a mean from a frequency distribution.. The calculator uses the data in **L1** and the associated frequencies in **L2** to approximate the sample statistics for the dataset. In this example, the approximate mean is 7.6 percent and the approximate standard deviation is 3.46 percent.

▸ Problem 9 (pg. 162)

(a.) Press **STAT** and select **1:Edit**. Clear **L1** and **L2**. Enter the midpoints for each of the temperature ranges into **L1** and the frequencies ('days') into **L2**. Press **STAT**, highlight **CALC**, select **1:1-Var Stats**, and press **2nd** **[L1]** **,** **2nd** **[L2]** **ENTER**.

L1	L2	L3 2
55	1	------
65	308	
75	1519	
85	1626	
95	503	
105	11	

L2(7) =

The population statistics will appear on the screen.

```
1-Var Stats
 x̄=80.93497984
 Σx=321150
 Σx²=26257400
 Sx=8.175217933
 σx=8.174187725
↓n=3968
```

The approximate value of the population mean is 80.93 and the approximate value of the population standard deviation (σx) is 8.174.

(b.) To set up the histogram, push **2nd** **[STAT PLOT]** and **ENTER** to select **Plot 1**. Turn ON **Plot 1**, set **Type** to **Histogram,** set **Xlist** to **L1,**. set **Freq** to **L2.**

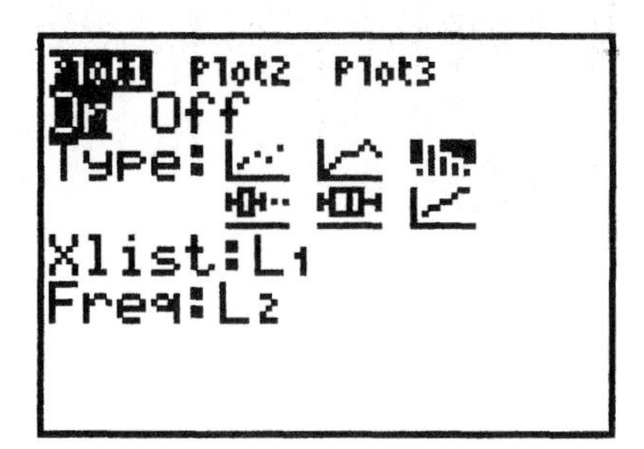

Press **Window** to adjust the Graph Window. Set **Xmin** equal to 55 (the midpoint of the first class) and **Xmax** equal to 115 (a value that would be the midpoint of an additional class at the end of the table. This extra value is needed to complete the last bar of the histogram). Set **Xscl** equal to 10, which is the class width. Set **Ymin = -5** and **Ymax = 1630**. You do not need to change **Yscl** or **Xres**. Press **GRAPH** and the histogram should appear.

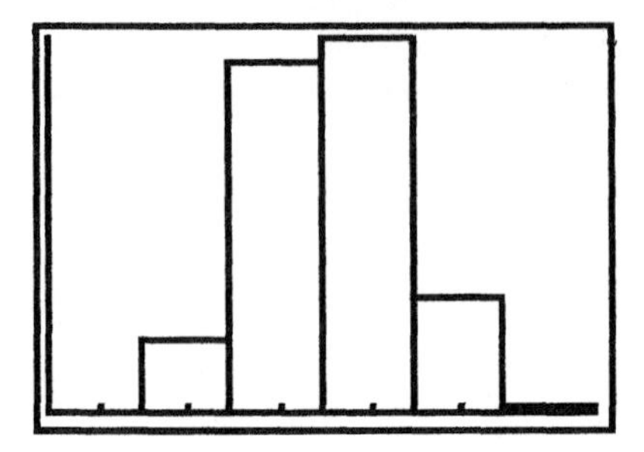

It may look as if there are only four bars in the histogram. There are actually 6 bars. The first and last bars have such small frequencies compared to the other bars that they are extremely small in the graph. Notice that the histogram is bell-shaped.

(c.) The Empirical Rule states that 95% of the data falls in the interval $(\mu \pm 2\sigma)$. To calculate the upper and lower limits of this interval, press **CLEAR** a few times until you get a blank screen. Enter **80.9-2*8.2** to get the lower limit. Press **2nd** **ENTER** and the calculation will appear again on the screen with the cursor flashing. Move the cursor so that it is positioned on the '-' sign and type in a '+' sign and press **ENTER** to get the upper limit.

Section 3.4

▸ Example 6 (pg. 170) Quartiles

Press STAT and select **1:Edit**. Clear **L1.** Enter the crime rate data from Table 17 on pg. 168 into **L1**. Press STAT, highlight **CALC**, select **1:1-Var Stats**, and press **2nd** **[L1]** ENTER. Scroll down to the 2nd screen to find the quartiles. **Q1** is the first quartile, **Med** is the 2nd quartile (or median) and **Q3** is the third quartile.

```
1-Var Stats
↑n=51
 minX=77.8
 Q1=270.4
 Med=365.8
 Q3=552.5
 maxX=1608.1
```

◂

Problem 15 (pg. 173)

Press **STAT** and select **1:Edit**. Clear **L1** and then enter the data. Press **STAT** and highlight **CALC**. Select **1:1-Var Stats** and press 2nd [**L1**] **ENTER**. The sample mean and sample standard deviation appear on the first screen. Scroll down to the 2nd screen to find the quartiles. **Q1** is the first quartile, **Med** is the 2nd quartile (or median) and **Q3** is the third quartile.

(a.) To calculate the Z-score for the data point 0.97 inches, do the following calculation: (0.97-sample mean)/sample standard deviation.

(c.) The interquartile range(**IQR**) is **Q3-Q1**.

(d.) The lower fence is **Q1-1.5*IQR.** The upper fence is **Q3+1.5*IQR.**

Section 3.5

▸ Example 2 (pg. 176) The Five Number Summary

Press **STAT** and select **1:Edit**. Clear **L1** and enter the data from Table 18 on pg. 176 into **L1.** Press **STAT** and highlight **CALC** to display the Calc Menu. Select **1: 1-Var Stats** and press 2nd **[L1]** **ENTER**. Scroll down to the 2nd screen to obtain the five values: **minX, Q1, med, Q3 and maxX.**

```
1-Var Stats
↑n=21
 minX=19.95
 Q1=26.055
 Med=30.95
 Q3=37.24
 maxX=64.63
```

Example 3 (pg. 177) A Boxplot

This is a continuation of Example 2. In Example 2 we entered the data from Table 18 on pg. 176 into **L1**.

Press **2nd** [STAT PLOT]. Select **1:Plot 1** and press ENTER. Turn On **Plot 1**. Move to the **Type** options. Using the right arrow (you can not use the down arrow to drop to the second line), scroll through the **Type** options and choose the first boxplot which is the first entry in row 2 of the **TYPE** options. Press ENTER. Move to **Xlist** and type in **L1**. Press ENTER and move to **Freq**. Set **Freq** to **1.** If **Freq** is set on **L2**, press CLEAR, and press ALPHA to return the cursor to a flashing solid rectangle and type in **1**. Press ZOOM and 9 to select **ZoomStat.** The Boxplot will appear on your screen.

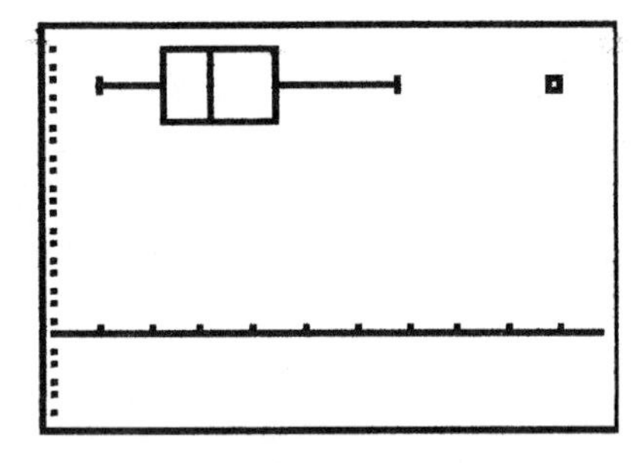

If you press TRACE and use the left and right arrow keys, you can display the following information: the smallest data point which is 19.95; Q1 (26.055); the median (30.95); Q3 (37.24); the largest data point which falls inside the upper fence which is 49.17 and the largest data point in the dataset, which is 64.63. This largest value is also an outlier because it lies outside the upper fence. Notice that this value is designated by a small box at the extreme right side of the diagram. (Note: The boxplot does not display the lower and upper fences.)

Describing the Relation between Two Variables

CHAPTER 4

Section 4.1

▸ Example 3 (pg. 201) A Scatter Diagram and Correlation Coefficient

Press STAT, highlight **1:Edit** and clear **L1** and **L2**. Refer to Table 1 on pg. 196. Enter the values of the predictor variable (Club Head Speed) into **L1** and the values of the response variable (Distance) into **L2**. Press **2nd [STAT PLOT]** , select **1:Plot1**, turn **ON** Plot 1 and press ENTER. For **Type** of graph, select the **scatter plot** which is the first selection. Press ENTER. Enter **L1** for **Xlist** and **L2** for **Ylist**. Highlight the first selection, the small square, for the type of **Mark**. Press ENTER. Press ZOOM and 9 to select **ZoomStat**.

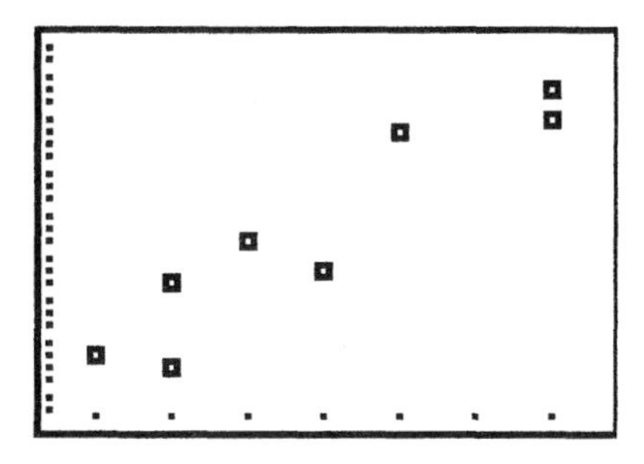

This graph shows a positive linear correlation: as 'Club Head Speed' increases, 'Distance' also increases.

In order to calculate r, the correlation coefficient, you must turn **On** the **Diagnostic** command. Press **2nd [CATALOG]** (Note: **CATALOG** is found above the 0 key). The CATALOG of functions will appear on the screen. Use the down arrow to scroll to the **DiagnosticOn** command.

Press ENTER ENTER.

Press STAT, highlight **CALC**, select **4:LinReg(ax+b)** and press ENTER ENTER. (Note: This command gives you the option of specifying which lists contain the X-values and Y-values. If you do not specify these lists, the defaults are used. The defaults are: **L1** for the X-values and **L2** for the Y-values.)

LinReg
y=ax+b
a=3.166101695
b=-55.79661017
r²=.8811498758
r=.9386958377

The correlation coefficient is r = .9386958377. This indicates a strong positive linear correlation between X and Y.

Section 4.2

‣ Example 3 (pg.219) Least Squares Regression Line

Press **STAT**, highlight **1:Edit** and clear **L1** and **L2**. Using Table 3 on pg.196, enter the values of the predictor variable (Club Head Speed) into **L1** and the values of the response variable (Distance) into **L2**. Press **STAT**, highlight **CALC** and select **4:LinReg(ax+b)**. This command has several options. One option allows you to store the regression equation into one of the Y-variables. To use this option, with the cursor flashing on the line **LinReg(ax+b)**, press **VARS**.

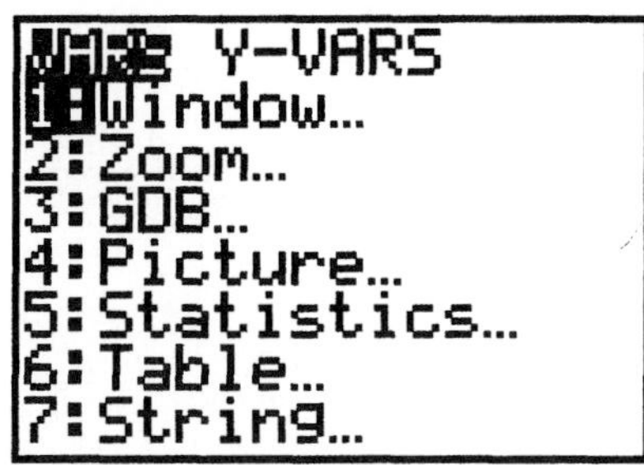

Highlight **Y-VARS.**

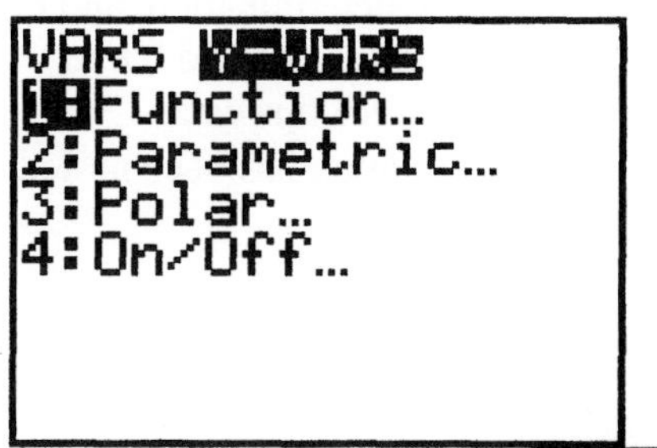

Select **1:Function** and press **ENTER**

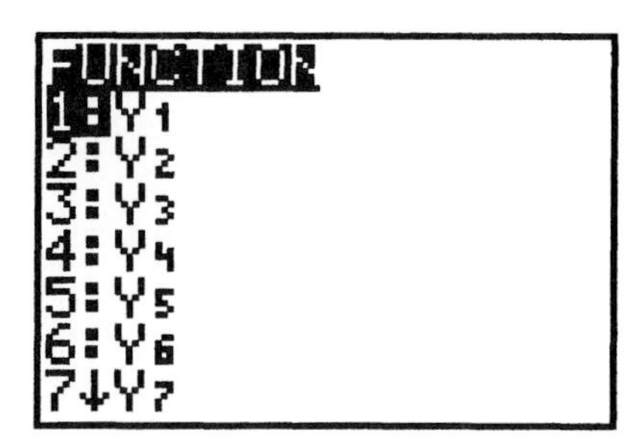

Notice that **1:Y1** is highlighted. Press **ENTER**.

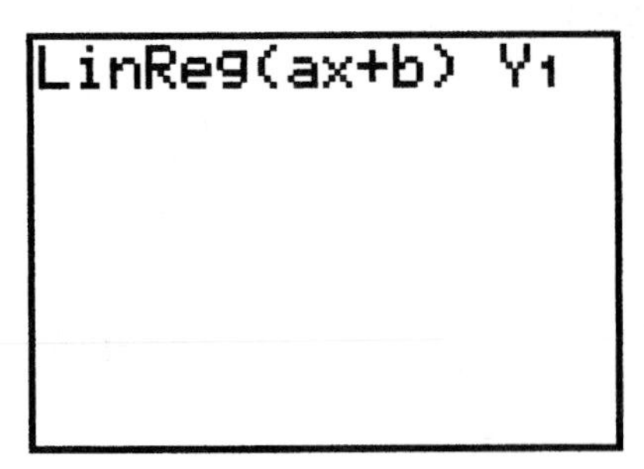

Press **ENTER**.

```
LinReg
 y=ax+b
 a=3.166101695
 b=-55.79661017
 r²=.8811498758
 r=.9386958377
```

The output displays the general form of the regression equation: y = ax+b followed by values for a and b. Next, r^2, the coefficient of determination, and r, the correlation coefficient , are displayed. If you put the values of a and b into the general equation, you obtain the specific linear equation for this data: $\hat{y}$ = 3.17x + -55.80. Press **Y=** and see that this specific equation has been pasted to **Y1**.

```
Plot1 Plot2 Plot3
\Y1=3.1661016949
153X+-55.7966101
695
\Y2=
\Y3=
\Y4=
\Y5=
```

Press 2nd **STAT PLOT]** , select **1:Plot1**, turn **ON** Plot1, select **scatter plot**, set **Xlist** to **L1** and **Ylist** to **L2**. Press **ZOOM** and **9**.

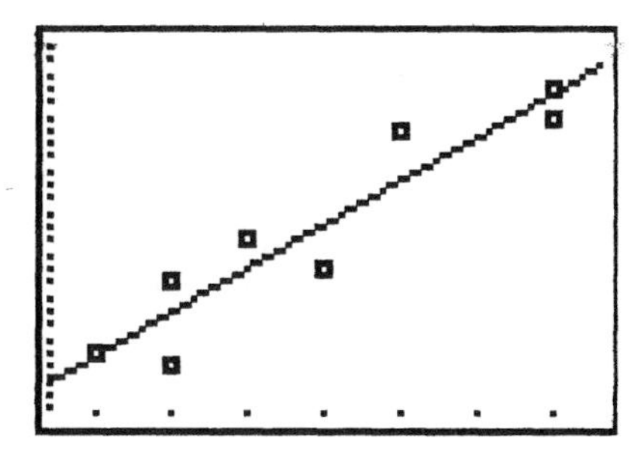

This picture displays a scatter plot of the data and the regression line. The picture indicates a strong positive linear correlation between X and Y, which is confirmed by the r-value of .939.

You can use the regression equation stored in **Y1** to predict Y-values for specific X-values. For example, suppose you would like to use the regression equation to predict the 'distance' a golf ball would travel when hit with a 'club head speed' of 103 mph. In other words, for X = 103, what does the regression equation predict for Y? To find this value for $\hat{y}$, press VARS, highlight **Y-VARS**, select **1:Function,** press ENTER, select **1:Y1** and press ENTER. Press (103) and press ENTER.

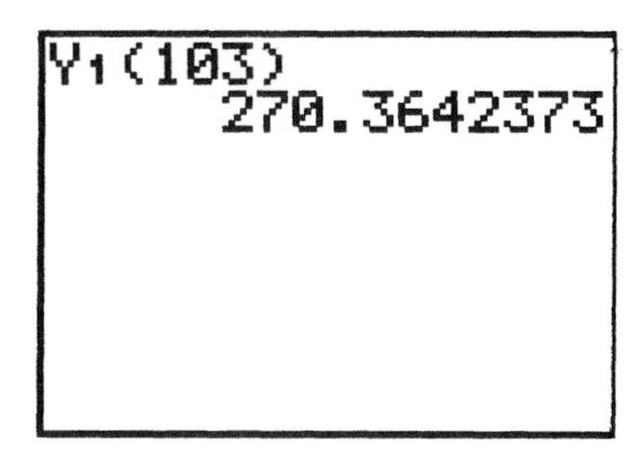

The output shows the predicted Y-value of 270.364 for the input X-value of 103.

The residual is: the actual Y-value (274) – the predicted Y-value (270.364).

Problem 18 (pg.222) The Equation of the Regression Line

Enter the predictor variable values into **L1** and the response variable values into **L2**. Press STAT. Highlight **CALC**, select **4:LinReg(ax+b)**, press ENTER. Press VARS, highlight **Y-VARS**, select **1:Function**, press ENTER and select **1:Y1** and press ENTER.

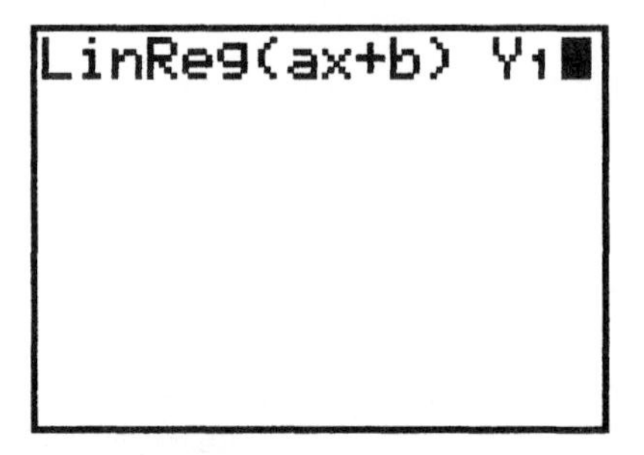

Press ENTER.

```
LinReg
 y=ax+b
 a=.0260679413
 b=7.873831377
 r²=.5265777937
 r=.7256568016
```

(a.) Using **a** and **b** from the output display, the resulting regression equation is $\hat{y}$ = .0261x + 7.8738. Press Y= to confirm that the regression equation has been stored in **Y1**. Press ZOOM and 9 for **ZoomStat** and a graph of the scatter plot with the regression line will be displayed.

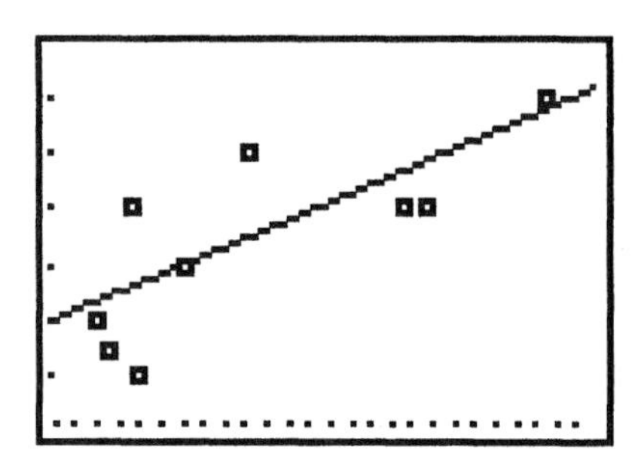

(c.) Next, you can use the regression equation to predict life expectancy for various animals. For example, to predict the life expectancy for an animal with a gestation period of 95 days, press VARS, highlight **Y-VARS**, select **1:Function** and press ENTER. Select **1:Y1** and press ENTER. Press (95) and ENTER.

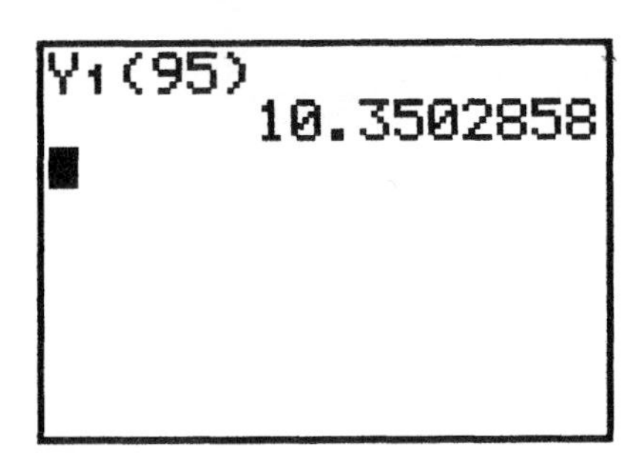

The predicted life expectancy for this species (with a gestation period of 95 days) is 10.35 years.

(d.) To predict the life expectancy of a parakeet, press **2nd** [ENTRY], (found above the ENTER key). Move the cursor so that it is flashing on **'9'** in the number **'95'** and type in **18**. Press ENTER.

Y1(95)
10.3502858
Y1(18)
8.34305432

The predicted life expectancy of a parakeet is 8.34 years.

(e.) Press **2nd** **[ENTRY]**. Move the cursor so that it is flashing on **'1'** in the number **'18'** and type in **31**. Press ENTER.

The predicted life expectancy of a rabbit is 8.7 years.

(f.) Residual = Actual Life Expectancy of a rabbit – predicted life expectancy of a rabbit: 7 – 8.7 = -1.7.

Section 4.3

▸ Example 2 (pg.229) **Coefficient of Determination, R^2**

Using Table 1 on pg. 196, enter the X-values into **L1** and the Y-values into **L2.** Press **STAT**. Highlight **CALC**, select **4:LinReg(ax+b)**, press **ENTER**. (For this example, we are not storing the regression equation in **Y1.**) .

```
LinReg
 y=ax+b
 a=3.166101695
 b=-55.79661017
 r²=.8811498758
 r=.9386958377
```

The value of r^2, .881, is displayed in the output.

◂

▸ Example 3 (pg. 230) Is a Linear Model Appropriate?

Using the data in Table 6, enter the X-values into **L1** and the Y-values into **L2**. Press STAT, highlight **CALC** and select **4:LinReg(ax+b)** and press ENTER

```
LinReg
 y=ax+b
 a=-.7928571429
 b=165.65
 r²=.9933086101
 r=-.9966486894
```

To plot the residuals, first make sure that there is nothing stored in the Y-registers. Press Y= and check the Y-registers. If any of them contain a function, move the cursor to that Y-register and press CLEAR.

Press **2nd** **[STAT PLOT]** , select **1:Plot1**, turn **ON** Plot 1 and press ENTER. For **Type** of graph, select the **scatter plot** which is the first selection. Press ENTER. Enter **L1** for **Xlist.** Move the cursor to **Ylist**. Press 2nd **[List]** and select 7:**Resid**. Highlight the first selection, the small square, for the type of **Mark**. Press ENTER. Press ZOOM and 9 to select **ZoomStat.**

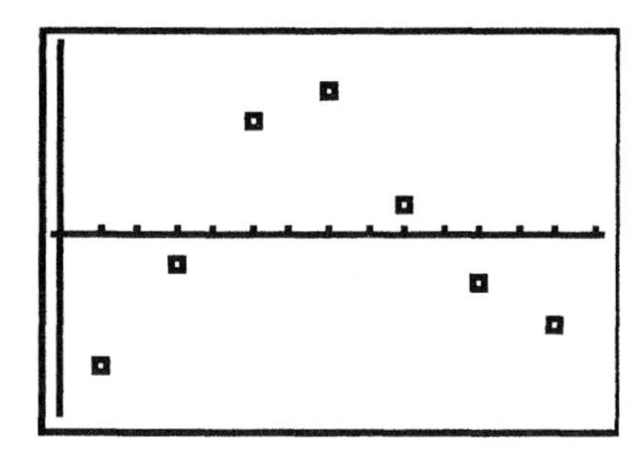

This graph of the residuals vs. the predictor variable (Time) shows a pattern (an upside-down U-shape), which indicates that the linear model is not appropriate.

◂

▸ Example 6 (pg. 232) Graphical Residual Analysis

Using the data in Table 4 on pg.220, enter the X-values, (Club Head Speed), into **L1** and the Y-values, (Distance), into **L2**. Press STAT, highlight **CALC** and select **4:LinReg(ax+b)** and press ENTER

To plot the residuals, first make sure that there is nothing stored in the Y-registers. Press Y= and check the Y-registers. If any of them contain a function, move the cursor to that Y-register and press CLEAR.

Press **2nd [STAT PLOT]** , select **1:Plot1**, turn **ON** Plot 1 and press ENTER. For **Type** of graph, select the **scatter plot** which is the first selection. Press ENTER. Enter **L1** for **Xlist.** Move the cursor to **Ylist**. Press 2nd **[List]** and select **7:Resid**. Highlight the first selection, the small square, for the type of **Mark**. Press ENTER. Press ZOOM and 9 to select **ZoomStat**.

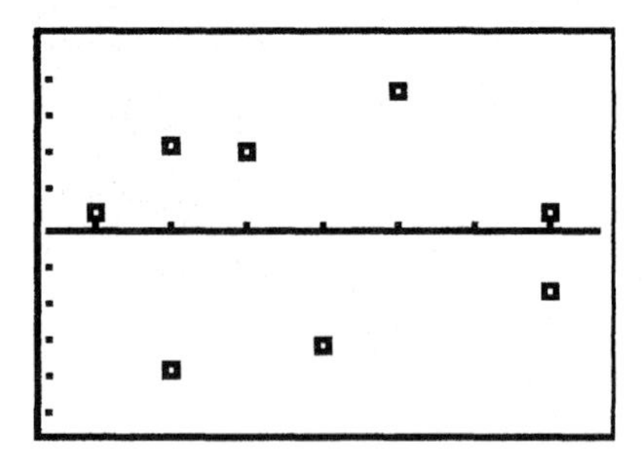

There is no discernable pattern in the plot of the residuals. This indicates support for the linear model that we used to calculate the regression equation.

The next step in analyzing the residuals is to construct a boxplot to determine if there are any unusual residuals values (called outliers.)

Press **2nd [STAT PLOT]**, select **1:Plot1**, turn **ON** Plot 1 and press ENTER. For **Type** of graph, select the **boxplot** with outliers, which is the first selection in the second row. Press ENTER. Move the cursor to **Xlist**. Press 2nd **[List]** and select **7:Resid**. Move the cursor to **Freq** and set this equal to **1**. Highlight the first selection, the small square, for the type of **Mark**. Press ZOOM and 9 to select **ZoomStat**.

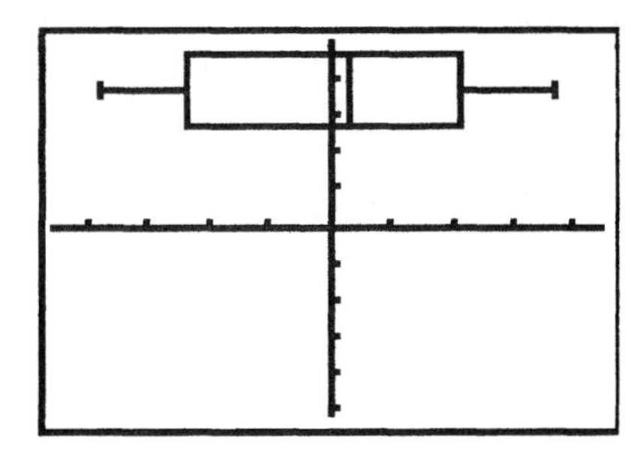

You can remove the X and Y axes from the boxplot by pressing **2nd [FORMAT]** (found above the **ZOOM** key). Scroll down to fourth line, use the right arrow to highlight **AxesOff** and press ENTER. Press ZOOM and 9 to select **ZoomStat**.

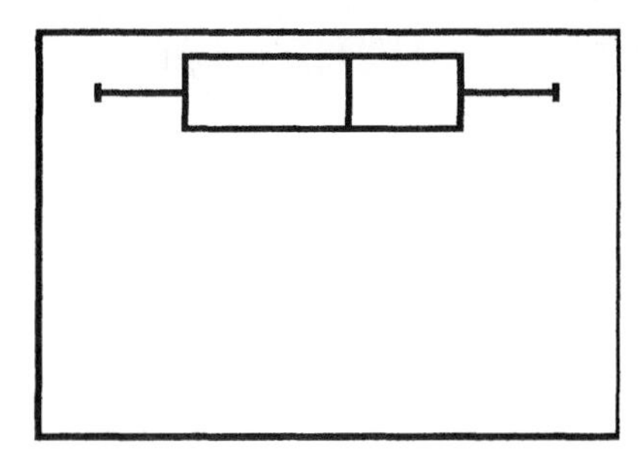

In this example, there are no outliers. This further supports the use of the linear model that was selected for this dataset.

▸ Problem 29 (pg. 238)

Enter the X-values (distance from the sun) into **L1** and the Y-values (sidereal year) into **L2**.

a.) First make sure that there is nothing stored in the Y-registers. Press Y= and check the Y-registers. If any of the Y-registers contain a function, move the cursor to that Y-register and press CLEAR.

Press **2nd** **[STAT PLOT]** , select **1:Plot1**, turn **ON** Plot 1 and press ENTER. For **Type** of graph, select the **scatter plot** which is the first selection. Press ENTER. Enter **L1** for **Xlist** and **L2** for **Ylist**. Highlight the first selection, the small square, for the type of **Mark**. Press ENTER. Press ZOOM and 9 to select **ZoomStat**.

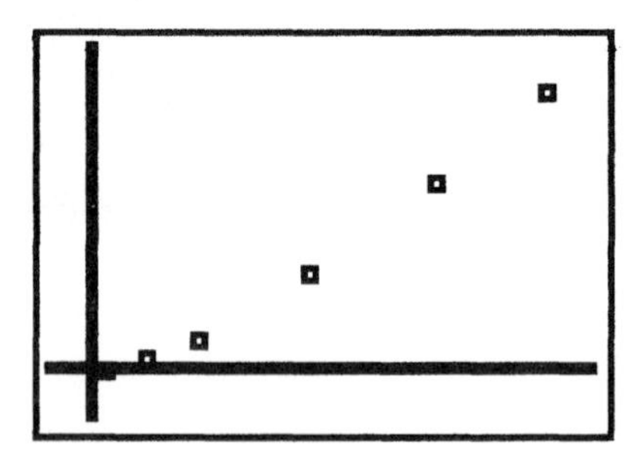

(Note: It is difficult to see all the nine points on this graph. It looks as if there are only 6 data points. In fact, there are nine points but, the first three points are so close together that they are indistinguishable from one another.)

b.) Press STAT, highlight **CALC** and select **4:LinReg(ax+b)** and press ENTER

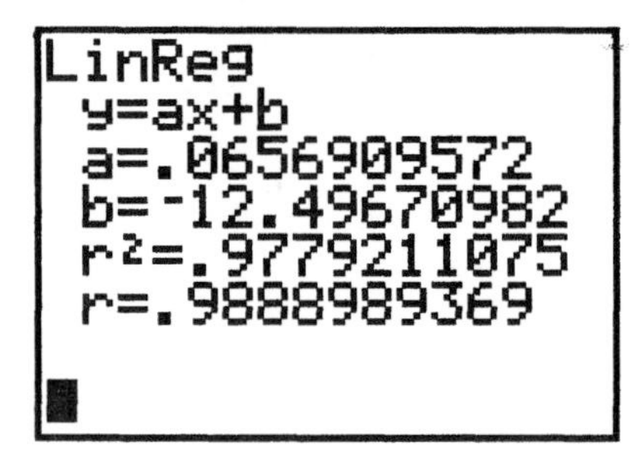

c.) Press **2nd** **[STAT PLOT]** , select **1:Plot1**, turn **ON** Plot 1 and press ENTER. For **Type** of graph, select the **scatter plot** which is the first selection. Press ENTER. Enter **L1** for **Xlist.** Move the cursor to **Ylist**. Press 2nd **[List]** and select **7:Resid**. Highlight the first selection, the small square, for the type of **Mark**. Press ENTER. Press ZOOM and 9 to select **ZoomStat**.

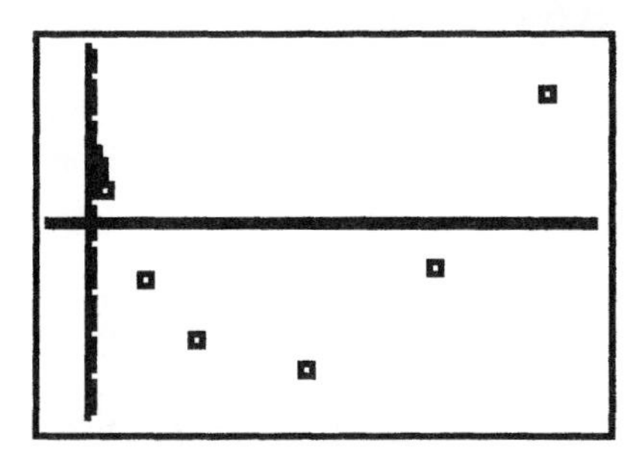

d.) This graph of the residuals vs. the x- variable shows a U-shaped pattern, which indicates that the linear model is not appropriate.

‣ Problem 33 (pg.238)

Enter the X-values (heights) into **L1** and the Y-values (weights) into **L2.**

a.) First make sure that there is nothing stored in the Y-registers. Press **Y=** and check the Y-registers. If any of them contain a function, move the cursor to that Y-register and press **CLEAR**.

Press 2nd **[STAT PLOT]** , select **1:Plot1**, turn **ON** Plot 1 and press **ENTER**. For **Type** of graph, select the **scatter plot** which is the first selection. Press **ENTER**. Enter **L1** for **Xlist** and **L2** for **Ylist**. Highlight the first selection, the small square, for the type of **Mark**. Press **ENTER**. Press **ZOOM** and **9** to select **ZoomStat**.

b.) Press **STAT**, highlight **CALC** and select **4:LinReg(ax+b)** and press **ENTER**

```
LinReg
 y=ax+b
 a=3.463110102
 b=-51.00908059
 r²=.3691677063
 r=.6075917266
```

c.) Press **STAT** and **Edit**. Move the cursor so that it is flashing on Randy Johnson's height of '82' in **L1** and press **DEL**. Move the cursor so that it is flashing on Randy Johnson's weight of '230' in **L2** and press **DEL**. Press **STAT**, highlight **CALC** and select **4:LinReg(ax+b)** and press **ENTER**

```
LinReg
 y=ax+b
 a=4.605
 b=-134.545
 r²=.2105652368
 r=.458873879
```

◂

Section 4.4 (Note: This Section is available on CD)

▸ Example 4 An Exponential Model

The TI-84 has the capability of creating an exponential model directly from the data. It is not necessary to transform the data by taking the log of the y-values.

a.) Enter the years (numbered 1 through 15) into **L1** and the Y-values into **L2**. Make sure that there is nothing stored in the Y-registers. Press Y= and check the Y-registers. If any of them contain a function, move the cursor to that Y-register and press CLEAR.
Press **2nd [STAT PLOT]** , select **1:Plot1**, turn **ON** Plot 1 and press ENTER. For **Type** of graph, select the **scatter plot** which is the first selection. Press ENTER. Enter **L1** for **Xlist** and **L2** for **Ylist**. Highlight the first selection, the small square, for the type of **Mark**. Press ENTER. Press ZOOM and 9 to select **ZoomStat.**

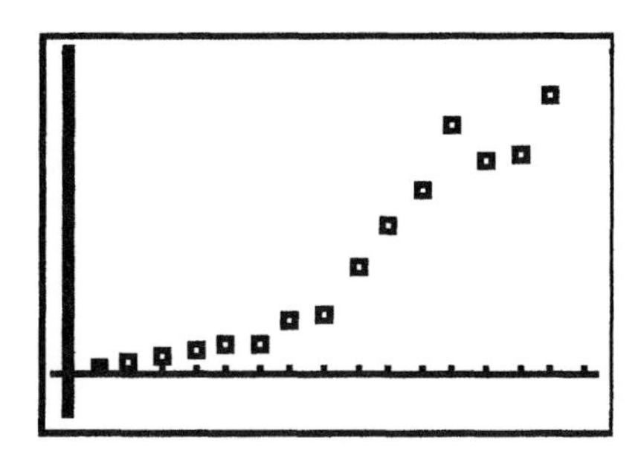

The shape of the scatter plot suggests that an exponential model would be appropriate for this dataset.

(d.) Press STAT, highlight **CALC** and select **0:ExpReg.** Press VARS, highlight **Y-VARS**, select **1:Function**, press ENTER and select **1:Y1** and press ENTER ENTER.

```
ExpReg
 y=a*b^x
 a=1.624420341
 b=1.306356354
 r²=.9483175405
 r=.9738159685
```

The exponential equation for this dataset is: $\hat{y} = 1.6244 * (1.3064)^x$.
If you press **GRAPH**, you can see a picture of the data along with the exponential model of best fit.

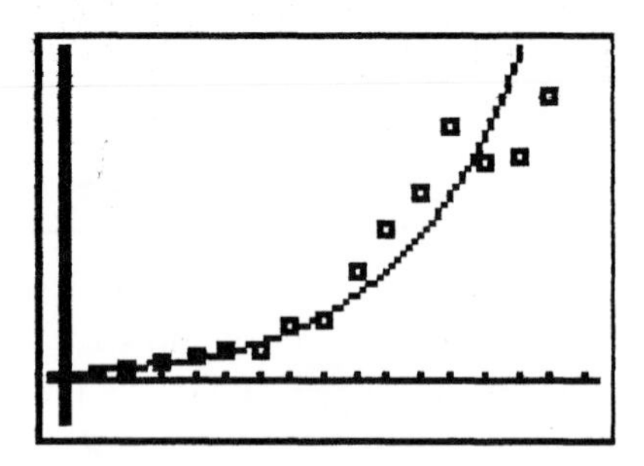

e.) To predict the closing price of Harley Davidson stock at the end of the year 2005, press **VARS**, highlight **Y-VARS**, select **1:Function** and press **ENTER**. Select **1:Y1** and press **ENTER**. Press **(** 16 **)** and **ENTER**. (Note: The year 2005 is represented by x =16.)

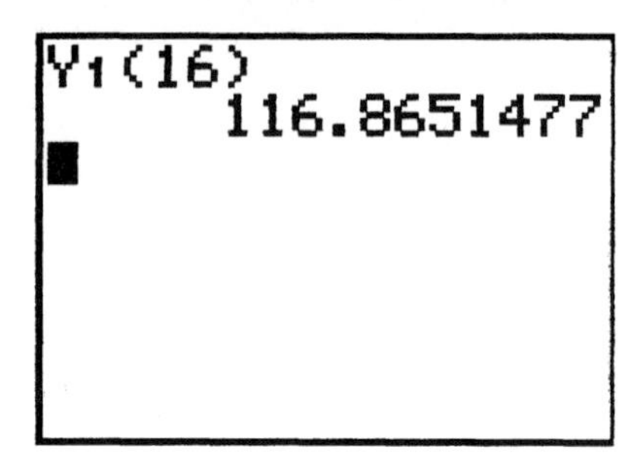

Example 5 A Power Model

The TI-84 has the capability of creating a power model directly from the data. It is not necessary to transform the equation by taking the log of the y-values. (Note: This model can used only in situations where $x > 0, y > 0$.)

a.) Enter the X-values into **L1** and the Y-values into **L2**. Make sure that there is nothing stored in the Y-registers. Press Y= and check the Y-registers. If any of them contain a function, move the cursor to that Y-register and press CLEAR. Press **2nd [STAT PLOT]** , select **1:Plot1**, turn **ON** Plot 1 and press ENTER. For **Type** of graph, select the **scatter plot** which is the first selection. Press ENTER. Enter **L1** for **Xlist** and **L2** for **Ylist**. Highlight the first selection, the small square, for the type of **Mark**. Press ENTER. Press ZOOM and 9 to select **ZoomStat.**

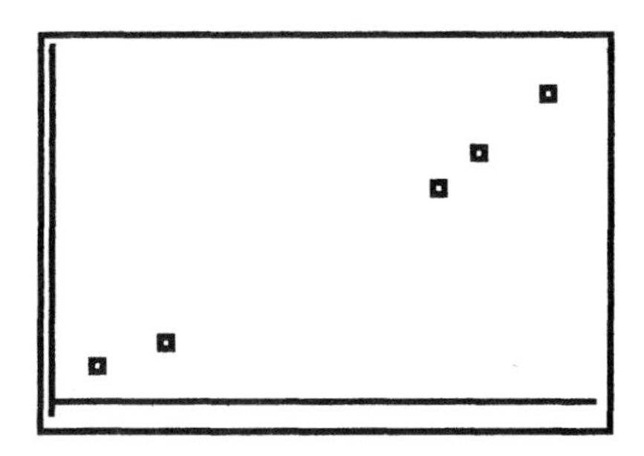

This is a very small dataset and it is difficult to determine which model would be the most appropriate one. For the purposes of this example, we will select the **power model.**

(d.) Press STAT, highlight **CALC** and scroll down to **A:PwrReg** and press ENTER. Press VARS, highlight **Y-VARS**, select **1:Function**, press ENTER and select **1:Y1** and press ENTER.

```
PwrReg
 y=a*x^b
 a=4.933897526
 b=1.992836932
 r²=.9999944648
 r=.9999972324
```

The power equation for this dataset is: $y = 4.934(x)^{1.99284}$.

If you press **GRAPH**, you can see a picture of the data along with the power model of best fit.

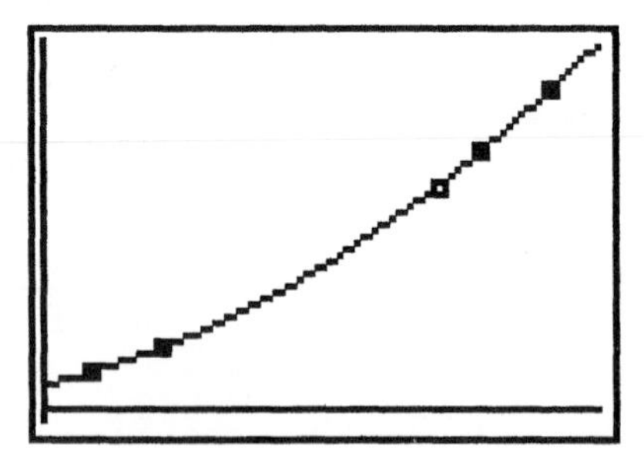

e.) To predict the distance a ball would have fallen if it took 4.2 seconds to hit the ground, press **VARS**, highlight **Y-VARS**, select **1:Function** and press **ENTER**. Select **1:Y1** and press **ENTER**. Press **(** 4.2 **)** and **ENTER**.

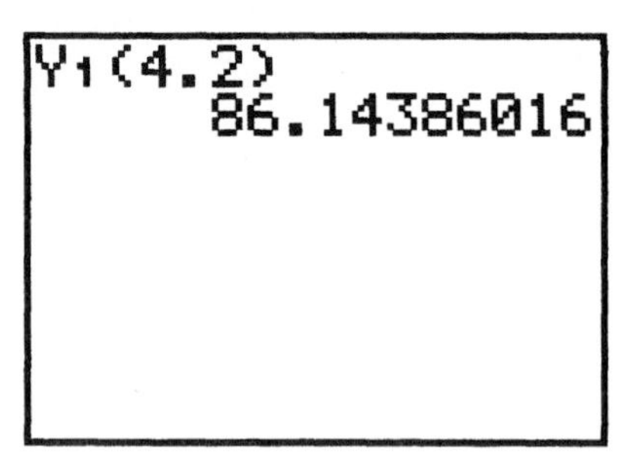

CHAPTER 5

Probability

Section 5.1

Example 8 (pg. 259) Simulating Probabilities

In this example, we will use simulation to estimate the probability of "having a boy." We assume the simple events, "having a boy," and "having a girl," are equally likely. In this simulation, we will designate "0" as a "boy," and "1" as a "girl."

a.) The first step is to set the *seed* by selecting a 'starting number' and storing this number in **rand**. Suppose, for this example, that we select the number '1204' as the starting number. Type **'1204'** into your calculator and press the STO key. Next press the MATH key, move the cursor to highlight **PRB**. Select **rand,** and press ENTER. The starting value of '1204' will be stored into **rand** and will be used as the *seed* for generating random numbers. (Note: This process of setting the *seed* is optional. You can omit it and simply go directly to the next step.)

(a.) For 100 births, press MATH, highlight **PRB**, and select **5:randInt(** and press ENTER. The **randInt(** command requires a minimum value, (which is 0 for this simulation), a maximum value (which is 1), and the number of trials (100). In the **randInt(** command type in **0** , **1** , **100.**

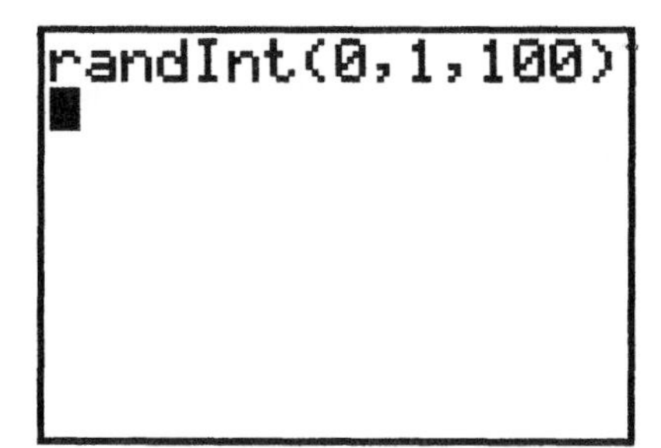

Press ENTER. It will take a few seconds for the calculator to generate 100 numbers. Notice, in the upper right hand corner a flashing | , indicating that the calculator is working. When the simulation has been completed, a string of

0's and **1's** will appear on the screen followed by **....**, indicating that there are more numbers in the string that are not shown.

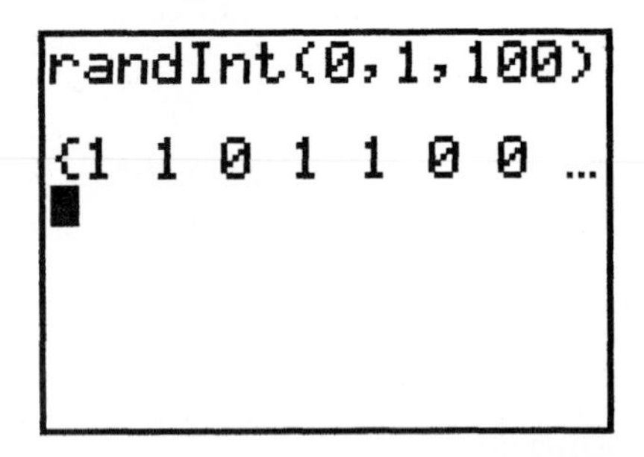

Press STO and 2nd [L1] ENTER. This will store the string of numbers in **L1**. Press **2nd [LIST]** and highlight MATH. Select **sum(** and type in **L1**.

```
randInt(0,1,100)
{1 1 0 1 1 0 0 ...
Ans→L1
{1 1 0 1 1 0 0 ...
sum(L1)
                62
```

The sum of **L1** equals the number of "1's" in the list. Since we have designated a "1" to be a "girl", we have 62 girls in the simulation. So, based on this simulation, we *estimate* the probability that a child is born a "girl" to be 62 out of 100 or 62%. And, the *estimated* probability that a child is a born a "boy" is 38% (100%-62%). These values are quite different from the theoretical probabilities. (The theoretical probabilities of both of these events are 50%). This difference is due to the fact that we did not do a very large number of repetitions.

(b.) Repeat the steps in part (a.) and increase the sample size to 999. This is the maximum sample size that the calculator will allow.

◀

▸ Problem 49 (pg 264)

In this simulation, we will use the integers 1,2,3,4,5 and 6 to represent the six possible outcomes on a six-sided die. Press **MATH**, highlight **PRB**, and select **5:randInt(** and press **ENTER**. The **randInt(** command requires a minimum value, (which is 1 for this simulation), a maximum value (which is 6), and the number of trials (100). In the **randInt(** command type in **1** , **6** , **100.**

```
randInt(1,6,100)
```

Press **ENTER**. It will take several seconds for the calculator to generate 100 rolls of the die. Notice, in the upper right hand corner a flashing |, indicating that the calculator is working. When the simulation has been completed, a string of **0's, 1's, 2's. etc.** will appear on the screen followed by **....**, indicating that there are more numbers in the string that are not shown.

Store the data in **L1** by pressing **STO** and 2nd **[L1]** **ENTER**.

(a.) One way to count the number of 1's in your simulation is to create a histogram of the results. First make sure that there is nothing stored in the Y-registers. Press **Y=** and check the Y-registers. If any register contains a function, move the cursor to that Y-register and press **CLEAR**.

To set up the histogram, push **2nd** [STAT PLOT] and **ENTER** to select **Plot 1**. Turn ON **Plot 1**, set **Type** to **Histogram**, set **Xlist** to **L1**,. set **Freq** to **1.** Press **Window** to adjust the Graph Window. Set **Xmin** equal to 1 (the minimum value in your simulation) and **Xmax** equal to 7 (a value that would be one integer larger than the maximum value on the roll of a die. This extra value is needed to complete the last bar of the histogram). Set **Xscl** equal to 1. (Note: In many cases it is not necessary to change the values for **Ymin, Ymax** or **Yscl**. What you must do is to check these values and make sure that **Ymin** is a small negative value (between –6 and –1 would be good) and **Ymax** must be larger than the largest frequency value in your dataset. A good value for **Ymax** would be 30. You never need to adjust **Yscl.**

Press **GRAPH** and the histogram should appear.

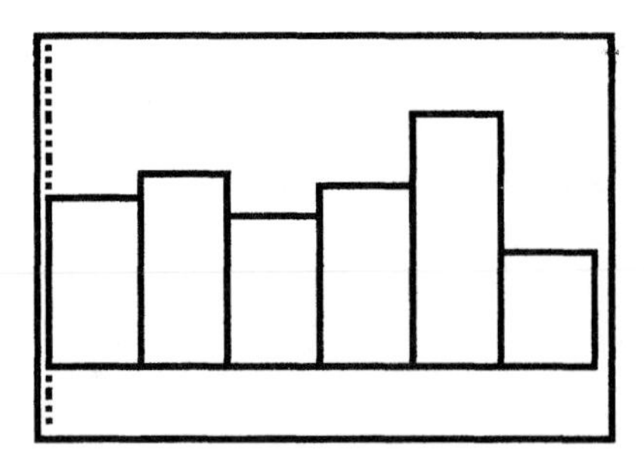

You can press **TRACE** and scroll through the bars of the histogram. The first bar of the histogram represents all the rolls that resulted in 1's.

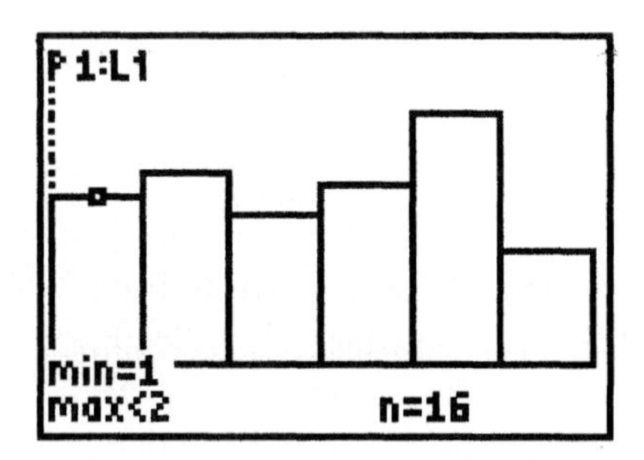

Notice, for this example, there were 16 rolls of 1's. So, based on this simulation, the estimated probability of rolling a '1' is 16 out of 100 or 16%.

(b.) Repeat the simulation and create a new histogram.

(c.) Repeat the simulation and increase the number of rolls of the die to 500 and create a new histogram. (Note: Increase the value of **Ymax** to 100 for this histogram.)

Section 5.5

▸ Example 4 (pg. 296) The Traveling Salesman - Factorials

The total number of different routes that are possible can be computed using the factorial function.

Press 7, MATH, highlight **PRB** and select **4:!** and ENTER.

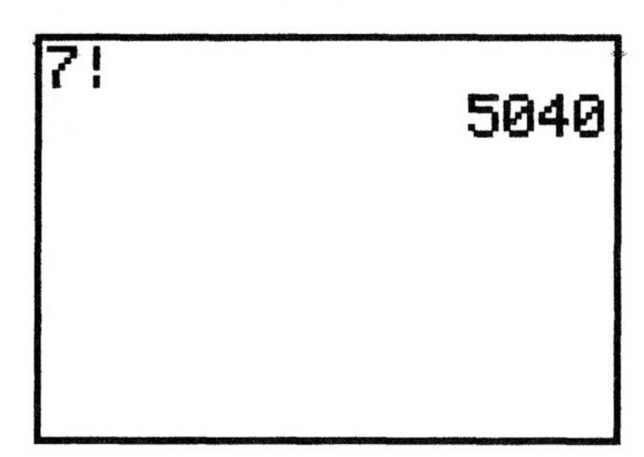

There are 5,040 different possible routes.

◂

▸ Example 6 (pg. 297) Permutations

(a.) In this example, there are 7 objects (n=7). From these 7 objects, 5 objects are selected (r=5). The permutation formula counts the number of different ways that these 5 objects can be selected and arranged from the total of 7 objects. The formula **nPr** is used with **n** = 7 and **r** =**5.** So, the formula is **7P5**.

Enter the first value, 7, and press MATH, highlight **PRB** and select **2:nPr** and ENTER.

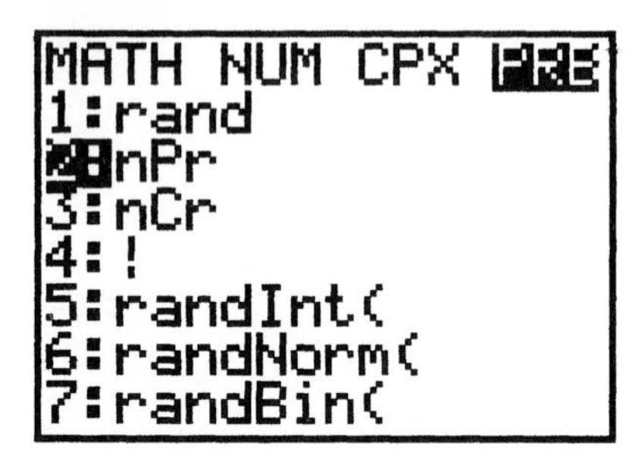

Now enter the second value, **5**, and ENTER. The answer, 2520, appears on the screen.

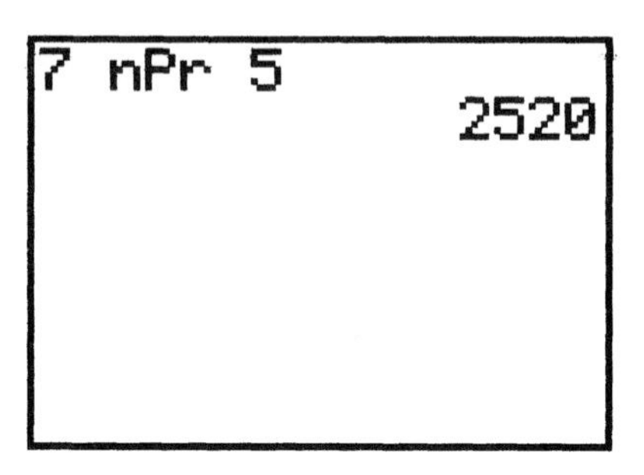

▸ Example10 (pg.300) Combinations

(b.) In this example, there are 6 objects (n=6). From these 6 objects, 4 objects are selected (r=4). The combination formula counts the number of different ways that these 4 objects can be selected from the total of 6 objects. The formula **nCr** is used with **n =6** and **r =4.** So, the formula is **6C4**.

Enter the first value, **6**, and press MATH, highlight **PRB** and select **3:nCr** and ENTER.

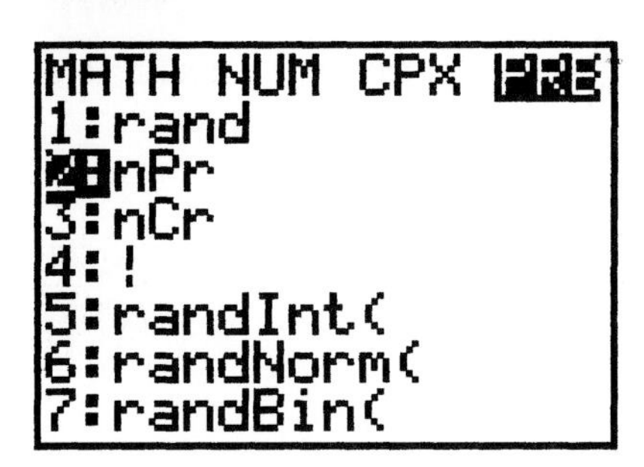

Enter the second value, **4**, and ENTER. The answer, 15, appears on the screen.

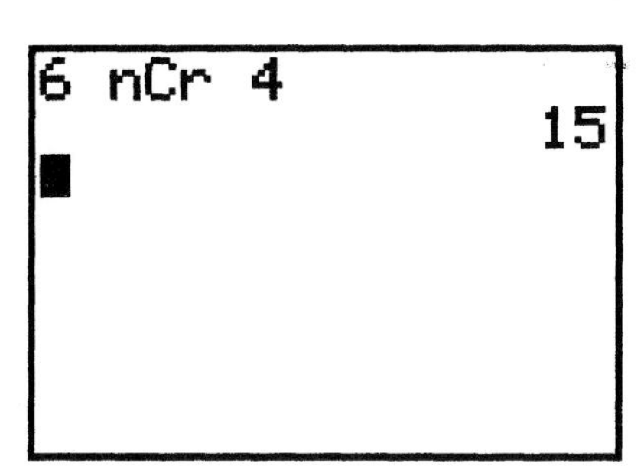

◀

Example 13 (pg. 302) Arranging Flags - Permutations with Non-distinct Items

To calculate $\frac{10!}{5!3!2!}$ you will use the factorial function (**!**). Enter the first value, **10**, press MATH, highlight **PRB** and select **4:!** . Then press ÷ . Open the parentheses by pressing (. Enter the next value, **5**, press MATH, **PRB**, and select **4:!** . To multiply by 3**!**, press x and enter the next value, **3**. Press MATH, **PRB**, and select **4:!** . To multiply by 2**!**, press x and enter the next value, **2**. Press MATH, **PRB**, and select **4:!** . Close the parentheses) and press ENTER.

```
10!/(5!*3!*2!)
             2520
```

Example 14 (pg. 303) Winning the Lottery - Probabilities involving Combinations

To calculate the probability of winning the Illinois Lottery, you must calculate $\frac{2}{{}_{54}C_6}$.

Enter the numerator, 2, into your calculator. Next press ÷ and enter the first value in the denominator, **54**, press **MATH**, highlight **PRB** and select **3:nCr**, enter the next value, **6.** Press **ENTER** and the answer will be displayed on your screen.

```
2/54 nCr 6
     7.7437845E-8
```

Notice that the answer appears in scientific notation. To convert to standard notation, move the decimal point 8 places to the left. The answer is .0000000774.

Example 15 (pg. 303) Probabilities involving Combinations

In this example, there are 120 fasteners in the shipment. Four fasteners in the shipment are defective. The remaining 116 fasteners are not defective. The quality-control manager randomly selects five fasteners.

To calculate the probability of selecting exactly one defective fastener, you must calculate: $\frac{{}_4C_1 * {}_{116}C_4}{{}_{120}C_5}$

To calculate the numerator, enter the first value, **4**, press **MATH**, highlight **PRB** and select **3:nCr** and enter the next value, **1**. Next press **x** and enter the next value, **116**, press **MATH**, highlight **PRB** and select **3:nCr**, enter the next value, **4.** Next press **÷** and enter the first value in the denominator, **120**, press **MATH**, highlight **PRB** and select **3:nCr**, enter the next value, **5.** Press **ENTER** and the answer will be displayed on your screen.

```
4 nCr 1*116 nCr
4/120 nCr 5
      .1502847988
```

‣ Problem 61 (pg. 305)

In this exercise, there are two groups made up of 8 students and 10 faculty. The combined number in the two groups is 18. Five individuals are to be selected from the total.

(a.) To select all students, you must choose 5 students from the group of 8 students and 0 faculty from the group of 10. Using the combination formula, you will do the following calculation: $\frac{{}_{8}C_{5} * {}_{10}C_{0}}{{}_{18}C_{5}}$.

To calculate the numerator, enter the first value, **8**, press **MATH**, highlight **PRB** and select **3:nCr** and enter the next value, **5**. Next press **x** and enter the next value, **10**, press **MATH**, highlight **PRB** and select **3:nCr**, enter the next value, **0**. Next press **÷** and enter the first value in the denominator, **18**, press **MATH**, highlight **PRB** and select **3:nCr**, enter the next value, **5**. Press **ENTER** and the answer will be displayed on your screen.

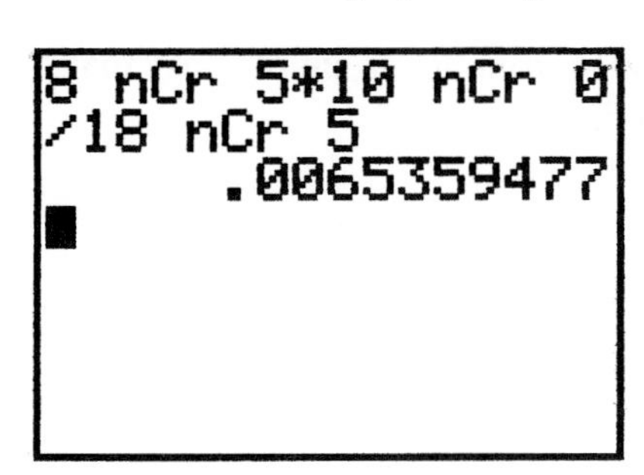

(b.) Repeat the steps in part (a.) but select 0 students from the group of 8 students and 5 faculty members from the group of 10 faculty members.

(c.) Repeat the steps in part (a.) but select 2 students from the group of 8 students and 3 faculty members from the group of 10 faculty members. ◂

▸ Problem 65 (pg. 306)

The compact disk has a total of 13 songs. Of the 13 songs, there are 5 songs that you like and 8 songs that you do not like. Suppose that four songs are randomly selected and played.

(a.) Calculate the probability that, among these first four songs selected, you like exactly two of them. Using the combination formula, you will do the following calculation: $\frac{{}_5C_2 * {}_8C_2}{{}_{13}C_4}$.

To calculate the numerator, enter the first value, **5**, press MATH, highlight **PRB** and select **3:nCr** and enter the next value, **2**. Next press × and enter the next value, **8**, press MATH, highlight **PRB** and select **3:nCr**, enter the next value, **2.** Next press ÷ and enter the first value in the denominator, **13**, press MATH, highlight **PRB** and select **3:nCr**, enter the next value, **4.** Press ENTER and the answer will be displayed on your screen.

(b.) Repeat the steps in part (a.) but select 3 songs from the group of songs that you like and 1 from the group of songs that you do not like.

(c.) Repeat the steps in part (a.) but select 4 songs from the group of songs that you like and 0 songs from the group that you do not like.

◂

Discrete Probability Distributions

CHAPTER 6

Section 6.1

▸ Example 4 (pg. 318) A Probability Histogram

Press **STAT** and select **1:EDIT**. Clear **L1** and **L2**. Using Table 1 on pg. 316, enter the X-values into **L1** and the P(x) values into **L2**.

To graph the probability distribution, first make sure that there is nothing stored in the Y-registers. Press **Y=** and check the Y-registers. If any of them contain a function, move the cursor to that Y-register and press **CLEAR**.

Press **2nd** **[STAT PLOT]** and press **ENTER**. Turn **ON** Plot 1, select **Histogram** for **Type**, type in **2nd** **[L1]** for **Xlist** and **2nd** **[L2]** for **Freq.** Press **WINDOW** and set **Xmin** = 0, **Xmax = 4, Xscl = 1, Ymin = 0** and **Ymax = .55.** Choosing 'Xmax=4' leaves some space at the right of the graph in order to complete the histogram. The Ymax value was selected by looking through the values in **L2** and then rounding the largest value UP to a convenient number. Press **GRAPH** to view the histogram.

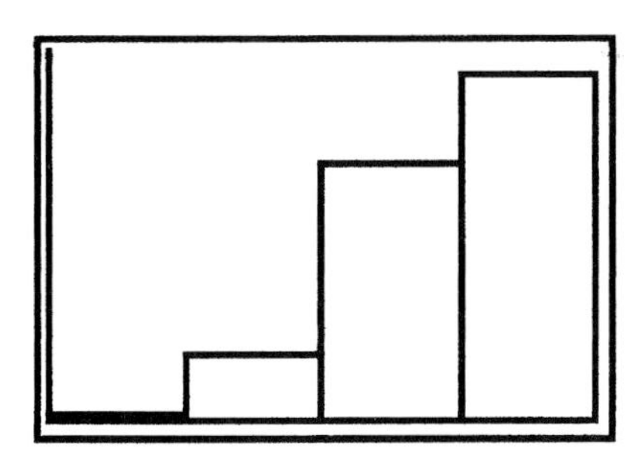

◀

▸ Example 7 (pg. 321) The Expected Value

Press **STAT** and select **1:EDIT**. Clear **L1** and **L2**. Enter the X-values from Table 5 into **L1** and the associated probabilities into **L2**.

L1	L2	L3 2
350	.99879	------
-2.5E5	.00121	

L2(3) =

Notice that the value of –249,650 appears as –2.5E5. This is a rounded value and it is written in scientific notation. The actual value is stored in the calculator; the rounded value is for display purposes only. Also, notice that the values in L2 are displayed as rounded values.

Press **STAT** and highlight **CALC**. Select **1:1-Var Stats,** press **ENTER** and press **2nd** **[L1]** **,** **2nd** **[L2]** **ENTER** to see the descriptive statistics.

```
1-Var Stats
 x̄=47.25
 Σx=47.25
 Σx²=75598075
 Sx=
 σx=8694.586962
↓n=1
```

The expected value of this discrete random variable is 47.25.

◂

▸ Example 9 (pg. 323) The Mean and Standard Deviation of a Discrete Random Variable

Press **STAT** and select **1:EDIT**. Clear **L1** and **L2**. Enter the X-values from Table 1 on pg. 316 into **L1** and the P(x) values into **L2**. Press **STAT** and highlight **CALC**. Select **1:1-Var Stats,** press **ENTER** and press **2nd [L1]** [,] **2nd [L2]** **ENTER** to see the descriptive statistics.

```
1-Var Stats L1,L
2
```

```
1-Var Stats
 x̄=2.39
 Σx=2.39
 Σx²=6.21
 Sx=
 σx=.7056202945
↓n=1
```

The mean of this discrete random variable is 2.39 and the standard deviation is .7056.

◀

▸ Problem 25 (pg. 325)

(a.) Enter the X-values into L1 and the frequencies into L2. Press 2nd [QUIT]. Press 2nd [LIST] and select **MATH**. Select **5:sum(** and type in **L2**. Press ENTER. The answer is the sum of the frequencies in L2. Press STAT and select **1:EDIT**. Move the cursor to highlight 'L2' at the top of the second list and press ENTER. With the cursor flashing at the bottom of the screen, type in **L2** ÷ (the sum of L2). This will convert the frequencies in L2 into probabilities. To confirm that you now have a probability distribution represented in L1 and L2, press 2nd [QUIT]. Press 2nd [LIST] and select **MATH**. Select **5:sum(** and type in **L2**. Press ENTER. The answer is the sum of the probabilities in L2. This sum should equal 1.

(b.) To draw the probability histogram, first make sure that there is nothing stored in the Y-registers. Press Y= and check the Y-registers. If any of them contain a function, move the cursor to that Y-register and press CLEAR. Press **2nd [STAT PLOT]** and press ENTER. Turn **ON** Plot 1, select **Histogram** for **Type**, type in **2nd [L1]** for **Xlist** and **2nd [L2]** for **Freq.** Press WINDOW and set **Xmin** = 1, **Xmax = 9, Xscl = 1, Ymin = 0 and Ymax = .14.** Choosing 'Xmax=9' leaves some space at the right of the graph in order to complete the histogram. The Ymax value was selected by looking through the values in **L2** and then rounding the largest value UP to a convenient number. Press GRAPH to view the histogram.

(c.-d.) Press STAT and highlight **CALC**. Select **1:1-Var Stats,** press ENTER and press 2nd [L1] [,] 2nd [**L2**] ENTER to see the descriptive statistics. ◂

Section 6.2

Example 5 (pg. 335) Binomial Probability Distribution Function

(a.) To find the probability that exactly 5 females in a random sample of 15 females attended a musical play in 2002, we will use the binomial probability density function, **binompdf(n,p,x).** For this example, n = 15, p = .20 and x = 5. Press **2nd [DISTR]**. Scroll down through the menu to select **A:binompdf(** and press **ENTER**. (Note: On the TI-83, the binomial probability distribution function is option **0:binompdf.**) Type in **15** , **.20** , **5**) and press **ENTER**. The answer, **.1032,** will appear on the screen.

```
binompdf(15,.2,5
)
      .1031822943
```

b. To calculate inequalities, such as the probability that *fewer than* 7 females in a group of 15 females attended a musical play in 2202, that is P(X < 7), you can use the cumulative probability command: **binomcdf (n,p,x)**. This command accumulates probability starting at X = 0 and ending at a specified X-value.

To calculate P(X < 7), we must accumulate probabilities for X = 0, 1, 2, …6, which is P(X ≤ 6). Press **2nd [DISTR]** scroll through the options and select **B:binomcdf(**. Type in **15** , **.20** , **6**) and press **ENTER**. The result, P(X ≤ 6) = P(X < 7) = .982.

Example 7 (pg. 336) Binomial Probability Histogram

a.) Construct a probability distribution for a binomial probability model with n = 10 and p = 0.2. Press **STAT**, select **1:EDIT** and clear **L1** and **L2**. Enter the values 0 through 10 into **L1**. Press **2nd [QUIT]**.

To calculate the probabilities for each X-value in **L1**, first change the display mode so that the probabilities displayed will be rounded to 3 decimal places. Press **MODE** and change from **FLOAT** to **3**. Press **ENTER**. This will round each of the probabilities to 3 decimal places. Press 2nd **[QUIT]**.

```
Normal Sci Eng
Float 0123456789
Radian Degree
Func Par Pol Seq
Connected Dot
Sequential Simul
Real a+bi re^θi
Full Horiz G-T
```

Press **STAT**, select **1:EDIT.** Move the cursor so that it highlights '**L2**' at the top of **L2**. Press **ENTER**. Next press **2nd [DISTR]** and select **A:binompdf(** and type in **10** , .2) and press **ENTER**.

L1	L2	L3 3
0.000	.107	
1.000	.268	
2.000	.302	
3.000	.201	
4.000	.088	
5.000	.026	
6.000	.006	

L3(1)=

To graph the binomial distribution, first make sure that there is nothing stored in the Y-registers. Press **Y=** and check the Y-registers. If any of them contain a function, move the cursor to that Y-register and press **CLEAR**.
Press **2nd [STAT PLOT]** and press **ENTER**. Turn **ON** Plot 1, select **Histogram** for **Type**, type in **2nd [L1]** for **Xlist** and **2nd [L2]** for **Freq.** Adjust the graph window by pressing **WINDOW** and setting **Xmin** = 0, **Xmax = 11, Xscl = 1, Ymin = 0** and **Ymax = .31.** Choosing 'Xmax=11' leaves some space at the right of the graph in order to complete the histogram. The Ymax value was selected by looking through the values in **L2** and then rounding the largest value UP to a convenient number. Press **GRAPH** to view the histogram.

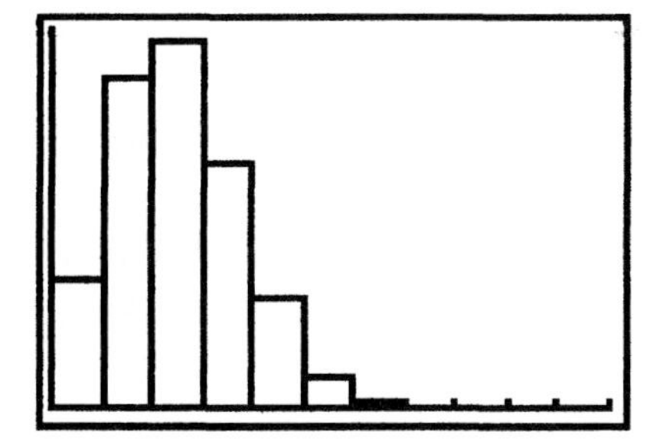

▸ Problem 31 (pg. 340)

(a.) Construct a probability distribution for a binomial probability distribution with n = 9 and p = .75. Press STAT, select **1:EDIT** and clear **L1** and **L2**. Enter the values 0 through 9 into **L1**. Press **2nd [QUIT]**. To calculate the probabilities for each X-value in **L1**, first change the display mode so that the probabilities displayed will be rounded to 3 decimal places. Press MODE and change from **FLOAT** to **3** and press **2nd [QUIT]**.

Press STAT, select **1:EDIT.** Move the cursor so that it highlights '**L2**' at the top of **L2**. Press ENTER. Next press **2nd [DISTR]** and select **A:binompdf(** and type in **9** , **.75**) and press ENTER.

(b.) Press STAT and highlight **CALC**. Select **1:1-Var Stats,** press ENTER and press **2nd [L1]** , **2nd [L2]** ENTER to see the descriptive statistics. The mean (μ) of the random variable is displayed as $\bar{x}$ on the calculator. The standard deviation is σx.

(c.) Use the formulas for the mean and standard deviation of a binomial random variable. The mean is: $\mu = n * p$; the standard deviation is: $\sigma = \sqrt{n * p * q}$.

(d.) To draw the probability histogram, first make sure that there is nothing stored in the Y-registers. Press Y= and check the Y-registers. If any of them contain a function, move the cursor to that Y-register and press CLEAR. Press **2nd [STAT PLOT]** and press ENTER. Turn **ON** Plot 1, select **Histogram** for **Type**, type in 2nd [L1] for **Xlist** and 2nd [L2] for **Freq.** Adjust the graph window by pressing WINDOW and setting **Xmin** = 0, **Xmax = 10, Xscl = 1, Ymin = 0** and **Ymax = .31.** Choosing 'Xmax=10' leaves some space at the right of the graph in order to complete the histogram. The Ymax value was selected by looking through the values in **L2** and then rounding the largest value UP to a convenient number. Press GRAPH to view the histogram.

◂

Problem 50 (pg. 342)

(a.) To generate random samples for this binomial model, press **MATH**, select **PRB** and select **7:randBin(**. This command requires three values: **n**, which is the sample size; **p**, the probability; and **x**, the number of samples. For this example type in **30** , **.98** , **100)**. Press ENTER It will take the calculator a few minutes to complete this simulation.

```
randBin(30,.98,1
00)
{30 29 27 30 30...
```

Store these probabilities in **L1** by pressing STO **2nd** **[L1]** .

(b.) To use the results of the simulation to compute the probability that exactly 29 of the 30 males survive to age 30, construct a histogram of the simulation. First make sure that there is nothing stored in the Y-registers. Press Y= and check the Y-registers. If any of them contain a function, move the cursor to that Y-register and press CLEAR.

Press **2nd** **[STAT PLOT]** and press ENTER. Turn **ON** Plot 1, select **Histogram** for **Type**, type in **2nd** **[L1]** for **Xlist** and **1** for **Freq.** Adjust the graph window by pressing WINDOW and setting **Xmin** = **26, Xmax** = **31, Xscl** = **1, Ymin** = **0** and **Ymax** = **60.** (Note: Choosing these values for Xmin and Xmax will display a small part of the probability histogram centered around the value of 29.) Press GRAPH to view the partial histogram. Press TRACE and scroll through the bars until you reach the bar for '29'. Take the frequency for that bar and divide it by 100 (the total number of simulations). Your result is the probability that exactly 29 males in a sample of 30 males will survive to age 30.

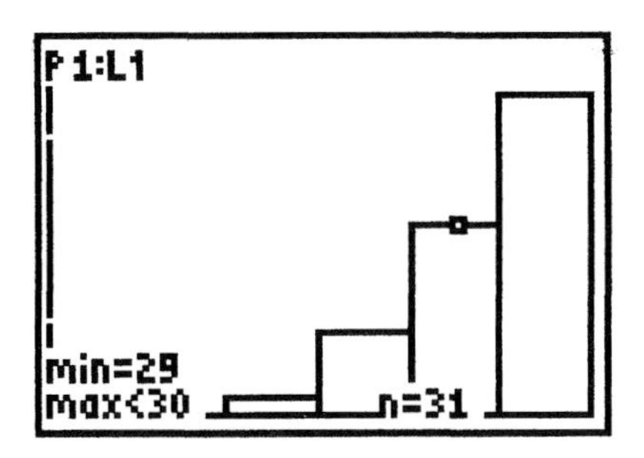

In this simulation, the probability is 31 out of 100 or 31%.

(c.) Press **2nd DISTR** and select **A:binompdf** and enter **30 , .98 , 29).**

(d.) Press **GRAPH** and the histogram of the simulation will appear. Press **TRACE** and scroll through the bars for '28', '29' and '30'. Sum the frequencies for these bars. Divide this sum by 100. This value is P(X≥28). The *complement* of this is P(X≤ 27). Subtract P(X≥28) from 1 to get P(X≤ 27).

(e.) Press **2nd DISTR** and select **B:binomcdf** and enter **30 , .98 , 27).** This value is P(X≤ 27).

(f-g.) First, calculate the mean and standard deviation of the 100 simulations. Press **STAT** and highlight **CALC**. Select **1:1-Var Stats,** press **ENTER** and press 2nd [L1] **ENTER** to see the descriptive statistics. The mean (μ) of the random variable is displayed as $\bar{x}$ on the calculator. The standard deviation is σx.

Then, use the formulas for the mean and standard deviation of a binomial random variable. The mean is: $\mu = n * p$; the standard deviation is: $\sigma = \sqrt{n * p * q}$.

▶ Problem 54 (pg. 343)

(a.) Suppose the probability that Shaquille O'Neal makes a *free throw* is .536. To find the probability that the *first* free throw he makes occurs on his third shot, press **2nd [DISTR]** and select **E:geometpdf(** and type in **.536** , **3**).

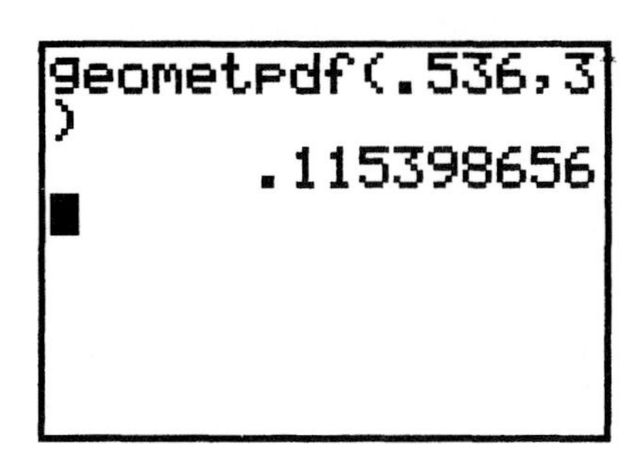

(b.) Construct a probability distribution for a geometric probability model with p = .563. Press STAT, select **1:EDIT** and clear **L1** and **L2**. Enter the values 1 through 10 into **L1**. Press **2nd [QUIT]**. To calculate the probabilities for each X-value in **L1**, first change the display mode so that the probabilities displayed will be rounded to 3 decimal places. Press MODE and change from **FLOAT** to **3** and press **2nd [QUIT]**.

Press STAT, select **1:EDIT.** Move the cursor so that it highlights '**L2**' at the top of **L2**. Press ENTER. Next press **2nd [DISTR]** and select **E:geometpdf(** and type in **.563** , **L1**) and press ENTER.

(c.) Press STAT and highlight **CALC**. Select **1:1-Var Stats,** press ENTER and press 2nd [L1] , 2nd [L2] ENTER to see the descriptive statistics. The mean (μ) of the random variable is displayed as $\bar{x}$ on the calculator.

(d.) Calculate the mean of a geometric probability model using the formula: $\mu = \frac{1}{p}$.

◀

Section 6.3

Example 2 (pg. 346) Probabilities of a Poisson Process

For problems that can be modeled with the poisson probability model, either the values of λ and t are given or the value of μ is given. These parameters (**λ, t and** μ) are related to each other in the following way: **μ=λ*t.** The **poissonpdf** command requires a value for **μ.**

(a.) In this example, **λ= 2** (cars per minute) and **t= 5** (minutes) and, therefore, **μ=2*5=10.** Use the command **poissonpdf (μ,x)** with μ = 10 and X = 6. Press 2nd [DISTR] and select **C:poissonpdf(** and type in **10** [,] **6** [)] and press [ENTER]. The answer will appear on the screen.

```
poissonpdf(10,6)
          .063055458
```

(b.) To calculate the probability that *less than 6* cars arrive in the 5 minute time period, use the command **poissoncdf (μ,x)** with μ = 10 and X = 5. Press **2nd [DISTR]** and select **D:poissoncdf(** and type in **10** [,] **5** [)] and press [ENTER]. The answer will appear on the screen.

```
poissoncdf(10,5)
         .0670859629
```

(c.) P(X≥ 6) = 1 – P(X≤5) = 1 - .0671 = .9329.

▸ Problem 7 (pg. 348)

This is an example of a Poisson process with **μ=5.**

(a.) Press **2nd [DISTR]** and select **C:poissonpdf(** and type in 5 [,] 6).

(b.) To calculate P(X < 6) press **2nd [DISTR]** and select **D:poissoncdf(** and type in 5 [,] 5).

(c.) P(X ≥ 6) = 1 – P(X < 6)

(d.) Press **2nd [DISTR]** and select **C:poissonpdf(** and type in 5 [,] **2nd** [{] 2 [,] **3** [,] **4** **2nd** [}] [)] and press **ENTER**.

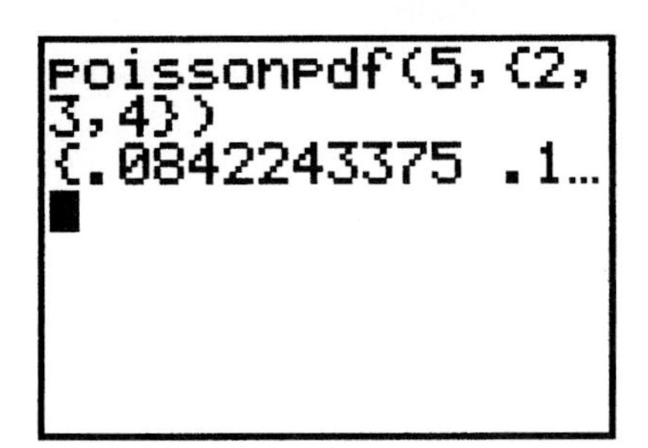

The first answer that appears in the output is P(2) which is .0842. Use the right arrow to scroll to the right to see P(3) and P(4).

◂

The Normal Probability Distribution

Section 7.2

▸ Example 2 (pg. 373) **Area Under the Standard Normal Curve to the left of a Z-score**

In this example of the standard normal curve, we will calculate the area to the *left* **of Z = 1.68.**

The TI-84 has two methods for calculating this area.

Method 1: **Normalcdf(**lowerbound, upperbound, μ, σ) computes the area between a lowerbound and an upperbound. In this example, you are computing the area from *negative infinity* to 1.68. Negative infinity is specified by (-) 1 2nd **[EE]** 9 9 (Note: **EE** is found above the comma [,]). Try entering –1 EE 99 into your calculator.

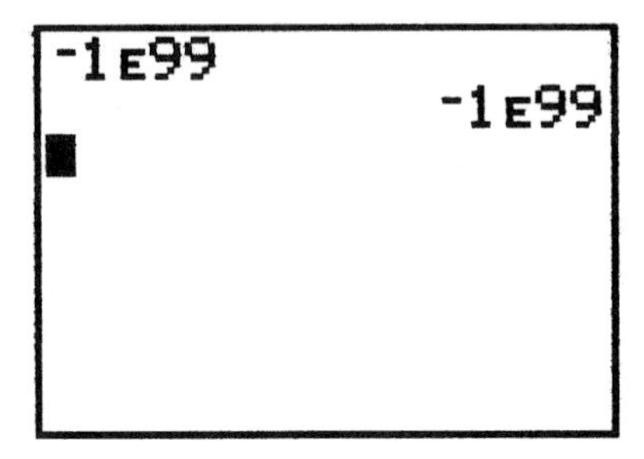

Now, to calculate the area to the left of 1.68, press 2nd [DISTR] and select 2:normalcdf(and type in -1E99 [,] 1.68 [,] 0 [,] 1 [)] and press P(X < 7) **ENTER. (Note: For the standard normal curve,** $\mu = 0$ **and** $\sigma = 1$**.)**

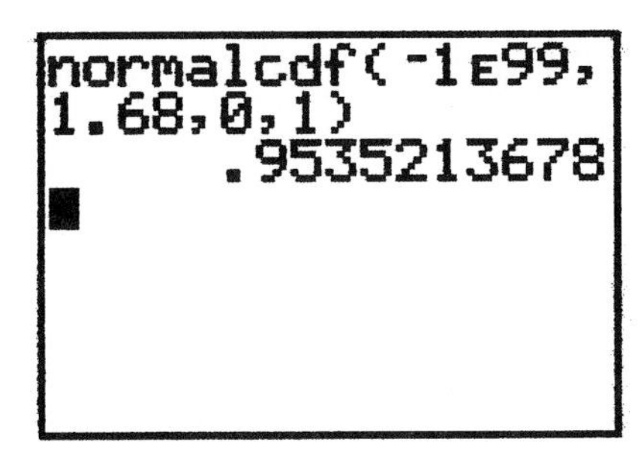

Method 2: This method calculates the area and also displays a graph of the probability distribution. You must first set up the WINDOW so that the graph will be displayed properly. Press WINDOW and set **Xmin** equal to -3 and set **Xmax** equal to 3. Set **Xscl** equal to 1.

Setting the Y-range is a little more difficult to do. A good "rule - of - thumb" is to set **Ymax** equal to .5 / σ . For this example, set **Ymax = .5.**

Use the up arrow to highlight **Ymin**. A good value for **Ymin** is **(-) Ymax / 4** so type in (-) .5 / 4.

Press 2nd [QUIT]. Clear all the previous drawings by pressing 2nd **[DRAW]** and selecting **1:ClrDraw** and pressing ENTER ENTER. Press 2nd **[STATPLOT]** and TURN OFF all PLOTS. Make sure that there is nothing stored in the Y-registers. Press Y= and check the Y-registers. If any of them contain a function, move the cursor to that Y-register and press CLEAR. Now you can draw the probability distribution. Press **2nd [DISTR]**. Highlight **DRAW** and select **1:ShadeNorm(** and type in **-1E99** , **1.68** , **0** , **1**) and press ENTER. The output displays a normal curve with the appropriate area shaded in and its value computed.

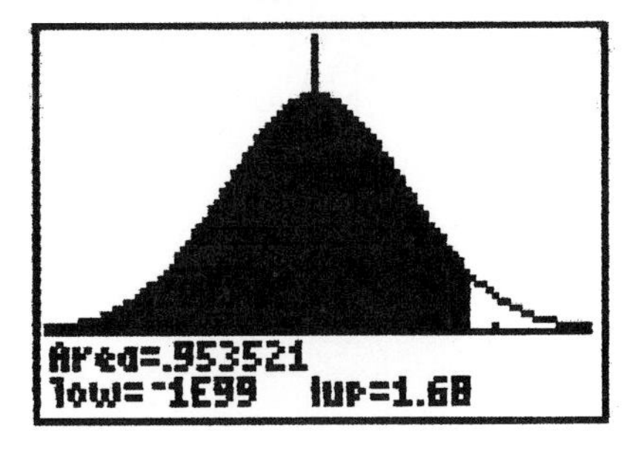

▸ Example 3 (pg. 374) Area Under the Standard Normal Curve to the right of a Z-score

In this example of the standard normal curve, we will calculate the area to the *right* of Z = -0.46.

Method 1: **Normalcdf(**lowerbound, upperbound, μ , σ **)** computes the area between a lowerbound and an upperbound. In this example, you are computing the area from –0.46 to *positive infinity*. Positive infinity is specified by 1 2nd [EE] 9 9 (Note: **EE** is found above the comma ,). To calculate the area to the right of –0.46, press **2nd [DISTR]** and select **2:normalcdf(** and type in **–0.46** , **1E99** , **0** , **1**) and press ENTER.

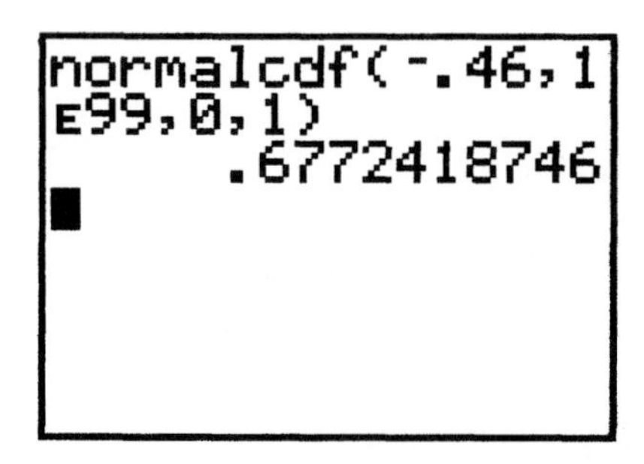

Method 2: This method calculates the area and also displays a graph of the probability distribution. Press WINDOW and set **Xmin** equal to -3 and set **Xmax** equal to 3. Set **Xscl** equal to 1. Set **Ymax = .5**. Set **Ymin = .5/4.** Press 2nd [QUIT]. Clear all the previous drawings by pressing 2nd **[DRAW]** and selecting **1:ClrDraw** and pressing ENTER ENTER. Press 2nd **[STATPLOT]** and **TURN OFF** all **PLOTS**. Make sure that there is nothing stored in the Y-registers. Press Y= and check the Y-registers. If any of them contain a function, move the cursor to that Y-register and press CLEAR. Now you can draw the probability distribution. Press 2nd [DISTR]. Highlight **DRAW** and select **1:ShadeNorm(** and type in **–0.46** , **1E99** , **0** , **1**) and press ENTER.

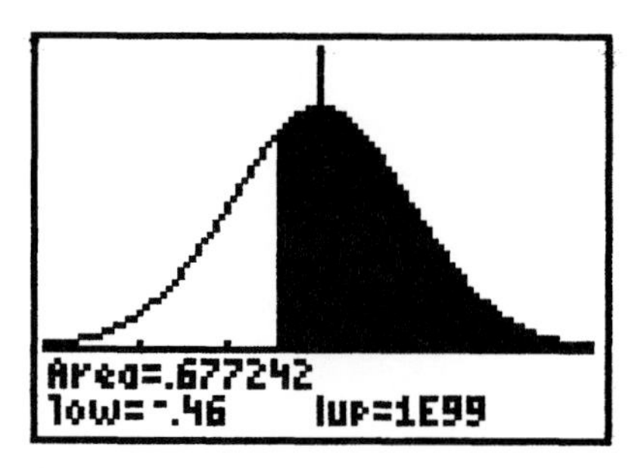

◂

▸ Example 4 (pg. 375) Area Under the Standard Normal Curve between two Z-scores

In this example of the standard normal curve, we will calculate the area between Z= -1.35 and Z= 2.01.

Method 1: **Normalcdf(**lowerbound, upperbound, μ, σ**)** computes the area between a lowerbound and an upperbound. In this example, you are computing the area from −1.35 to 2.01. To calculate the area between −1.35 and 2.01, press 2nd [DISTR] and select **2:normalcdf(** and type in **−1.35** [,] **2.01** [,] **0** [,] **1** [)] and press ENTER.

```
normalcdf(-1.35,
2.01,0,1)
       .8892764236
```

Method 2: This method calculates the area, and also displays a graph of the probability distribution. Press WINDOW and set **Xmin** equal to -3 and set **Xmax** equal to 3. Set **Xscl** equal to 1. Set **Ymax = .5.** Set **Ymin = .5/4.**

Press 2nd [QUIT]. Clear all the previous drawings by pressing 2nd **[DRAW]** and selecting **1:ClrDraw** and pressing ENTER ENTER. Press 2nd **[STATPLOT]** and **TURN OFF** all **PLOTS**. Make sure that there is nothing stored in the Y-registers. Press Y= and check the Y-registers. If any of them contain a function, move the cursor to that Y-register and press CLEAR. Now you can draw the probability distribution. Press 2nd [DISTR]. Highlight **DRAW** and select **1:ShadeNorm(** and type in **−1.35** [,] **2.01** [,] **0** [,] **1** [)] and press ENTER.

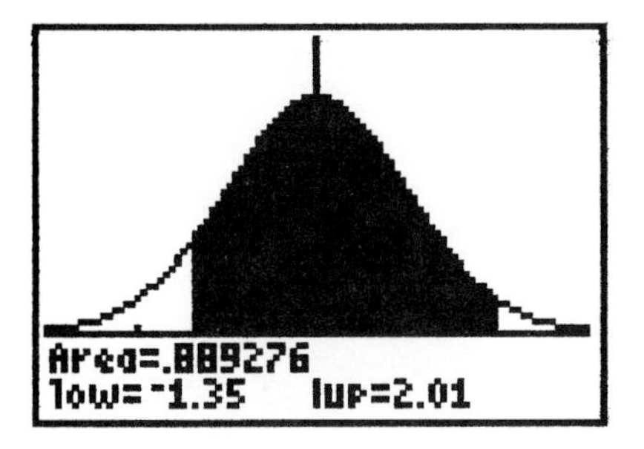

◂

Example 6 (pg. 377) Finding a Z-Score from a Specified Area to the Left

This is called an inverse normal problem and the command **invNorm(area,** μ, σ**)** is used. In this type of problem, an area under the normal curve is given and you are asked to find the corresponding Z-score. In this example, the area given is the area to the *left* of a Z-score. The area is 0.32 . (The area value that you **ENTER** into the TI-84must always be area to the left of a Z-score.) .

To find the Z-score corresponding to *left area* of 0.32, press 2nd [DISTR] and select **3:invNorm(** and type in **.32** , **0** , **1**) and press ENTER.

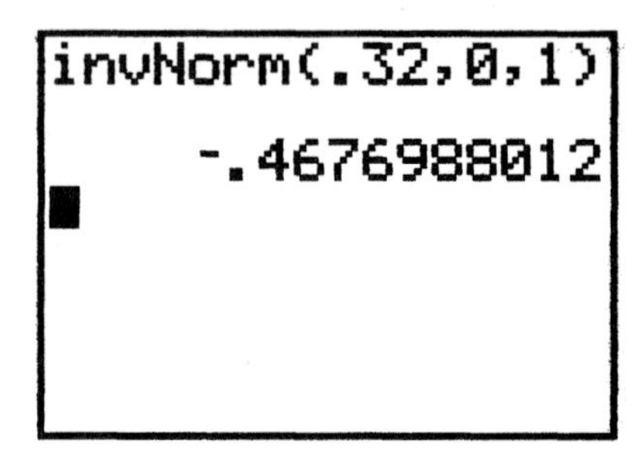

The Z-score of -.47 has an area of 0.32 to the *left*.

▸ Example 7 (pg. 378) Finding a Z-Score from a Specified Area to the Right

This is an inverse normal problem and the command **invNorm(area,** μ **,** σ **)** is used. In this type of problem, an area under the normal curve is given and you are asked to find the corresponding Z-score. In this example, the area given is the area to the *right* of a Z-score. The area is 0.4332 . (The area value that you enter into the TI-84 must always be area to the left of a Z-score.) .

To find the Z-score corresponding to *right area* of 0.4332, subtract 0.4332 from 1 to obtain the area to the *left* of the Z-score. Press 2nd [DISTR] and select **3:invNorm(** and type in **.5668** [,] **0** [,] **1**) and press **ENTER**.

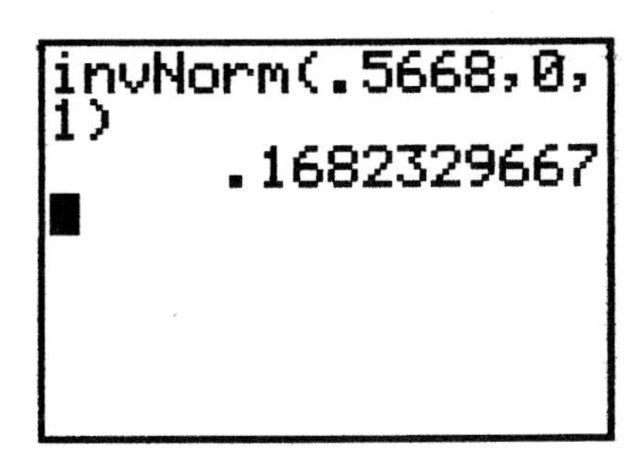

The Z-score that has an area to the *right* equal to 0.4332 is Z= .17.

◂

▸ Example 8 (pg. 379) Finding Z-Scores for an Area in the Middle

This is an inverse normal problem and the command **invNorm(area,** μ, σ**)** is used. In this problem the *middle area* is .90. That leaves an area of .10 to be equally divided between the *left* and *right* tail areas. Each of these areas are, therefore, equal to .05. The Z-score that marks the lower edge of the middle area is the Z-score that corresponds to a *left area* of .05.

To find the Z-score corresponding to a *left area* of 0.05, press 2nd [DISTR] and select **3:invNorm(** and type in **.05** [,] **0** [,] **1**) and press ENTER.

```
invNorm(.05,0,1)
        -1.644853626
```

The Z-score that marks the lower edge of the middle area of .90 is **–1.645.**

To find the Z-score corresponding to a *left area* of 0.95, press 2nd [DISTR] and select **3:invNorm(** and type in **.95** [,] **0** [,] **1**) and press ENTER. The Z-score that marks the upper edge is **1.645.**

◂

Section 7.3

Example 2 (pg. 386) Finding Area Under a Normal Curve

In this exercise, use a normal distribution with $\mu = 38.72$ and $\sigma = 3.17$.

Method 1:To find the percentile rank of a three-year-old female whose height is 43 inches, we calculate P(X < 43). Press **2nd** **[DISTR]**, select **2:normalcdf(** and type in **-1E99** , **43** , **38.72** , **3.17**) and press ENTER.

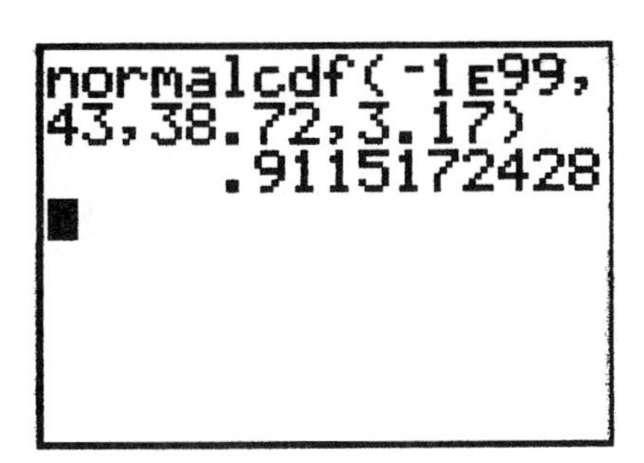

Method 2: To find P(X < 43) and include a graph, you must first set up the WINDOW so that the graph will be displayed properly. You will need to set Xmin equal to (μ - 3 σ) and Xmax equal to (μ + 3 σ). Press WINDOW and set **Xmin** equal to (μ - 3 σ) by typing in **38.72 - 3 * 3.17.** Press ENTER and set **Xmax** equal to (μ + 3 σ) by typing in **38.72 + 3 * 3.17.** Set **Xscl** equal to σ, which is 3.17.

Setting the Y-range is a little more difficult to do. A good "rule - of - thumb" is to set **Ymax** equal to .5 / σ. For this example, set **Ymax = .5/3.17.**

Use the up arrow to highlight **Ymin**. A good value for **Ymin** is **(-) Ymax / 4** so type in (-) **.158 / 4.**

Press **2nd** **[DRAW]** and select **1:ClrDraw** and press ENTER ENTER. Press 2nd **[STATPLOT]** and **TURN OFF** all **PLOTS**. Make sure that there is nothing stored in the Y-registers. Press Y= and check the Y-registers. If any of them contain a function, move the cursor to that Y-register and press CLEAR. Press **2nd** **[DISTR]**, highlight **DRAW** and select **1:ShadeNorm(** and type in **-1E99** , **43** , **38.72** , **3.17**) and press ENTER.

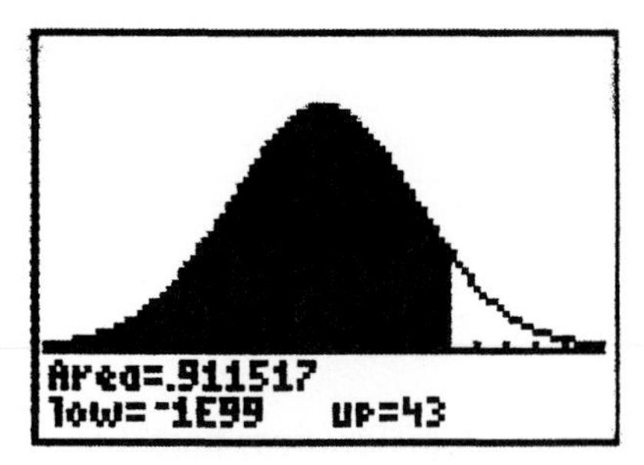

Conclusion: 91 % of all three-year-old females are less than 43 inches tall.

Note: **When using the TI-84 (or any other technology tool), the answers you** obtain may vary slightly from the answers that you would obtain using the **standard normal table. Consequently, your answers may not be exactly the same** as the answers found in your textbook. The differences are simply due to rounding.

▸ Example 3 (pg. 387) Finding the Probability of a Normal Random Variable

In this exercise, use a normal distribution with $\mu = 38.72$ and $\sigma = 3.17$.

Method 1:To find P(35 ≤ X ≤ 40) press **2nd [DISTR]**, select **2:normalcdf(** and type in **35** , **40** , **38.72** , **3.17**) and press ENTER.

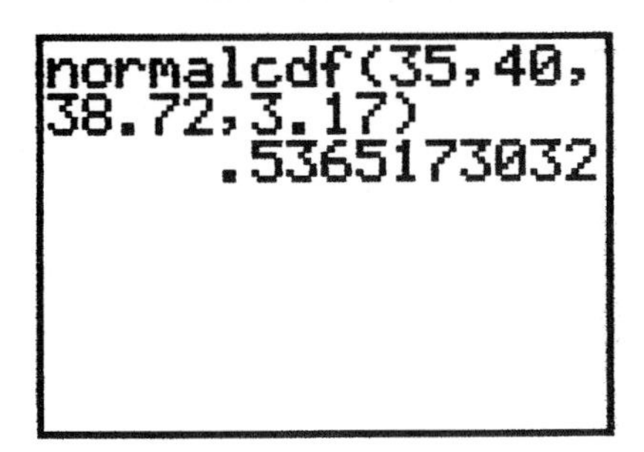

Method 2: To find the probability and include a graph, you must first set up the **WINDOW so that the graph will be displayed properly. You will need to set Xmin equal to (μ - 3 σ) and Xmax equal to (μ + 3 σ).** Press WINDOW and set **Xmin** equal to (μ - 3 σ) by typing in **38.72 - 3 * 3.17.** Press ENTER and set **Xmax** equal to (μ + 3 σ) by typing in **38.72 + 3 * 3.17.** Set **Xscl** equal to σ, which is 3.17.

Setting the Y-range is a little more difficult to do. A good "rule - of - thumb" is to set **Ymax** equal to .5 / σ . For this example, set **Ymax = .5/3.17.**

Use the up arrow to highlight Ymin. A good value for Ymin is (-) Ymax / 4 so type in (-) **.158 / 4.**

Press 2nd [DRAW] and select 1:ClrDraw and press ENTER ENTER. **Press 2nd [STATPLOT] and TURN OFF all PLOTS. Make sure that there is nothing stored in the Y-registers. Press** Y= **and check the Y-registers. If any of them contain a function, move the cursor to that Y-register and press** CLEAR. **Press 2nd [DISTR], highlight DRAW and select 1:ShadeNorm(** and type in **35** , **40** , **38.72** , **3.17**) and press ENTER.

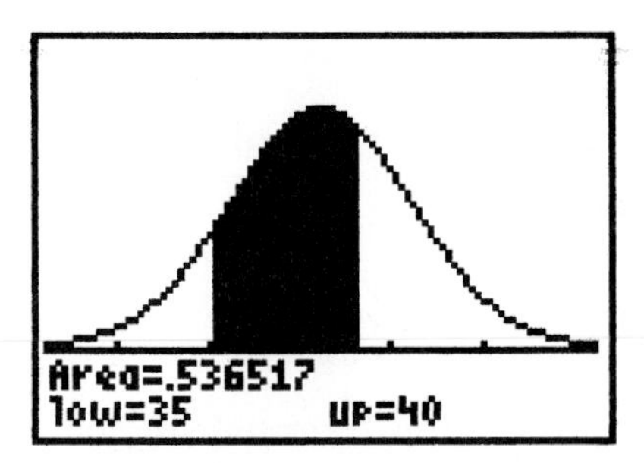

Conclusion: The probability that a randomly selected three-year-old female is between 35 and 40 inches tall is .5365 or 53.65%.

◀

▸ Example 5 (pg. 388) Finding the Value of a Normal Random Variable

This is an inverse normal problem and the command **invNorm(area,** μ **,** σ **)** is used. In this type of problem, a percentage of the area under the normal curve is given and you are asked to find the corresponding X-value. In this example, the percentage given is the bottom 20 %, (the 20th percentile). Press **2nd** **[DISTR]** and select **3:invNorm(** and type in **.20** [,] **38.72** [,] **3.17**) and press **ENTER**.

```
invNorm(.20,38.7
2,3.17)
      36.05206069
```

Conclusion: The height that separates the bottom 20% of three-year-old females from the top 80% is 36.05 inches.

◂

▸ Example 6 (pg.389) Finding the Value of a Normal Random Variable

This is an inverse normal problem and the command **invNorm(area, μ, σ)** is used. In this problem the *middle area* is .98. That leaves an area of .02 to be equally divided between the *left* and *right* tail areas. Each of these areas are, therefore, equal to .01.

The height that marks the lower edge of the middle 98% corresponds to a left area of .01. To find the X-value corresponding to a *left area* of 0.01, press 2nd [DISTR] and select **3:invNorm(** and type in **.01** [,] **38.72** [,] **3.17**) and press ENTER.

The height that separates the middle 98% from the top 1% is actually the height that separates the bottom 99% (the middle 98% plus the 1% in the left tail) from the top 1%. To find the X-value corresponding to a *left area* of 0.99, press 2nd [DISTR] and select **3:invNorm(** and type in **.99** [,] **38.72** [,] **3.17**) and press ENTER.

```
invNorm(.01,38.7
2,3.17)
     31.34547723
invNorm(.99,38.7
2,3.17)
     46.09452277
```

The middle 98% of the distribution lies between 31.34 inches and 46.09 inches.

◀

Section 7.4

Example 2 (pg. 397) A Normal Probability Plot

Press **STAT** and select **1:Edit** and press **ENTER**. Clear all data from **L1.** **ENTER** the data from Table 4 on pg. 395 into **L1**.

To set up the normal probability plot, first make sure that there is nothing stored in the Y-registers. Press **Y=** and check the Y-registers. If any of them contain a function, move the cursor to that Y-register and press **CLEAR**.
Press **2nd [STAT PLOT]**. Press **ENTER** to select **Plot 1**. Highlight **On** and press **ENTER**. Set **Type** to the normal probability plot which is the third selection in the second row. Press **ENTER**. Set **Data List** to **L1** and **Data Axis** to **X**. Next, there are three different types of **Marks** that you can select for the graph. The first choice, a small square, is the best one to use.

Press **ZOOM** and select **9:ZoomStat** and **ENTER**.

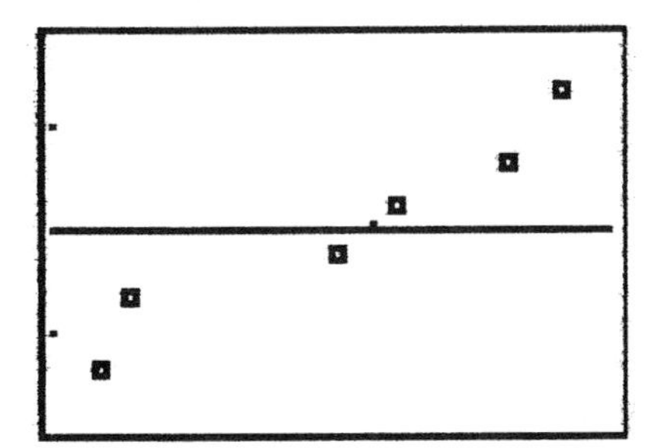

The calculator draws the normal probability plot with a horizontal line at the X-axis. This plot is *fairly* linear, indicating that the data generally follows a normal distribution.

Section 7.5

Example 1 (pg. 404) Normal Approximation to the Binomial

In this binomial experiment, the random variable, X, is the number of individuals with blood type O-negative, the probability, p, that an individual has type O-negative blood is .06 and the sample size, n, is 500. We will approximate the probability that fewer than 25 individuals in the sample have type O-negative blood, that is, $P(X < 25)$ using the normal approximation to the binomial. In order to use the normal approximation to the binomial, the first step is to verify that $n*p*(1-p) \geq 10$. In this example, $500 * .06 * .94 = 28.2$, so the requirement is satisfied.

Next, calculate the mean and standard deviation of this binomial random variable. The mean, μ, equals $n*p = 500 * .06 = 30$. The standard deviation, σ, equals
$$\sqrt{n*p*(1-p)} = \sqrt{500*.06*.94} = 5.31.$$

To approximate $P(X < 25)$ with a normal probability we calculate $P(X \leq 24.5)$. (Note: This adjustment from 25 to 24.5 is called a *continuity correction*).

Press 2nd [DISTR], select **2:normalcdf(** and type in **-1E99** [,] **24.5** [,] **30** [,] **5.31** [)] and press **ENTER**.

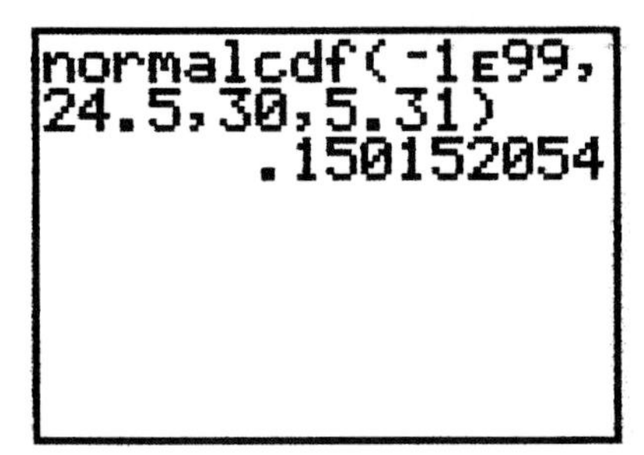

You can compare this probability (.1505) that you obtained through a normal approximation to the actual probability obtained from the binomial distribution.

Press 2nd [DISTR], select **B:binomcdf(** and type in **500** [,] **.06** [,] **24** [)] and press **ENTER**.

```
binomcdf(500,.06
,24)
     .1493809338
```

The actual $P(X < 25)$ is .1494.

Example 2 (pg. 405) Normal Approximation to the Binomial

In this binomial experiment, the random variable, X, is the number of households with cable TV, the probability, p, that a household has cable TV is .70 and the sample size, n, is 1000. We will approximate the probability that at least 734 households in the sample have cable TV, that is, $P(X \geq 734)$ using the normal approximation to the binomial. In order to use the normal approximation to the binomial, the first step is to verify that $n*p*(1-p) \geq 10$. In this example, $1000 * .70 * .30 = 210$, so the requirement is satisfied.

First, calculate the mean and standard deviation of this binomial random variable. The mean, μ, equals $n*p = 1000 * .70 = 700$. The standard deviation, σ, equals

$$\sqrt{n*p*(1-p)} = \sqrt{1000*.70*.30} = 14.491.$$

To approximate $P(X \geq 734)$ with a normal probability we calculate $P(X \geq 733.5)$ (Note: This adjustment from 734 to 733.5 is called a *continuity correction*).

Press 2nd [DISTR], select **2:normalcdf(** and type in **733.5** [,] **1E99** [,] **700** [,] **14.491** [)] and press **ENTER**.

```
normalcdf(733.5,
1E99,700,14.491)
         .0103948862
```

The approximate $P(X \geq 734)$ is .0104, which is small probability. This suggests that the percentage of households in DuPage County with cable TV is actually higher than 70%.

Sampling Distributions

CHAPTER 8

Section 8.1

▸ Example 6 (pg. 429) Applying the Central Limit Theorem

The calorie intake of 20-39 year old males is described as a population with a mean, μ, = 2716 and a standard deviation, σ, = 72.8. A sample of 35 males is selected from the population and the calorie intake of each male is recorded. What is the probability that the sample average, $\bar{x}$, is 2750 calories or higher?

Since n > 30, you can conclude that the sampling distribution of the sample mean is approximately normal with $u_{\bar{x}}$ =2716 and $\sigma_{\bar{x}} = 72.8/\sqrt{35}$.

To calculate P($\bar{x} \geq 2750$), press **2nd** **[DISTR]**, select **2:normalcdf(** and type in **2750** , **1E99** , **2716** , **72.8/**$\sqrt{35}$) and press **ENTER**.

```
normalcdf(2750,1
E99,2716,72.8/√(
35))
      .0028636574
```

◂

▶ Problem 21 (pg. 432) Old Faithful

The times between eruptions of the geyser Old Faithful are normally distributed with $\mu = 85$ minutes and $\sigma = 21.25$ minutes.

a.) To find P(X > 95) press **2nd** **[DISTR]**, select **2:normalcdf(** and type in **95** [,] **1 EE 99** [,] **85** [,] **21.25** [)] and press **ENTER**.

b.) To calculate $P(\bar{x} > 95)$, press **2nd** **[DISTR]**, select **2:normalcdf(** and type in **95** [,] **1E99** [,] **85** [,] **21.25/**$\sqrt{20}$ [)] and press **ENTER**.

c.) To calculate $P(\bar{x} > 95)$, press **2nd** **[DISTR]**, select **2:normalcdf(** and type in **95** [,] **1E99** [,] **85** [,] **21.25/**$\sqrt{30}$ [)] and press **ENTER**.

```
normalcdf(95,1E9
9,85,21.25)
      .3189674148
normalcdf(95,1E9
9,85,21.25/√(20)

      .0176658586
```

```
normalcdf(95,1E9
9,85,21.25/√(30)

      .0049756658
```

Section 8.2

Example 4 (pg. 438) Probabilities of a Sample Proportion

(a.) According to the National Center for Health Statistics, 15 % of all Americans have hearing trouble. To calculate the probability that at least 18 % of Americans in a sample of 120 have hearing trouble, we need to consider the distribution of the sample proportion, $\hat{p}$. This distribution is approximately normal (provided n*p*(1 - p) ≥ 10) with a mean, $\mu_{\hat{p}} = p$, and with standard deviation, $\sigma_{\hat{p}} = \sqrt{\frac{p(1-p)}{n}}$.

To calculate P($\hat{p} \geq .18$), press **2nd [DISTR]**, select **2:normalcdf(** and type in **.18** [,] **1E99** [,] **.15** [,] $\sqrt{\frac{.15(1-.85)}{120}}$ [)] and press ENTER.

```
normalcdf(.18,1E
99,.15,√(.15(1-.
15)/120)
        .178692832
```

(b.) To determine if it would be unusual to find only 10 Americans with hearing trouble in a sample of 120 Americans, we first calculate $\hat{p}$, the sample **proportion. In this case,** $\hat{p} = \frac{10}{120} = 0.083$. **To calculate P(** $\hat{p} \leq .083$**), press 2nd [DISTR], select 2:normalcdf(** **and type in -1E99** [,] **.083** [,] **.15** [,] $\sqrt{\frac{.15(1-.85)}{120}}$ [)] and press ENTER.

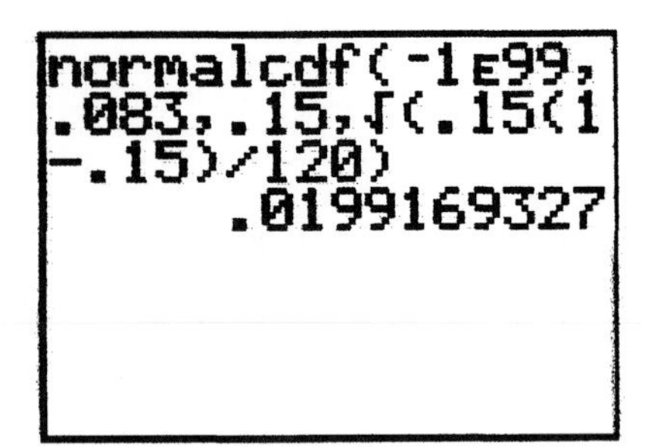

P($\hat{p} \leq .083$) = .0199. Since this probability is very small, this indicates that it **would be unusual to find only 10 Americans with hearing trouble in a sample of** 120.

◀

Estimating the Value Of a Parameter Using Confidence Intervals

Section 9.1

▸ Example 4 (pg. 455) A Confidence Interval for μ (σ known)

Enter the data from Table 1 on pg. 448 into **L1**. Since the sample size is less than 30, we will check for normality using a normal probability plot and we will check for outliers using a Boxplot.

To set up the normal probability plot, press **2nd [STAT PLOT]**. Press ENTER to select **Plot 1**. Highlight **On** and press ENTER. Set **Type** to the normal probability plot which is the third selection in the second row. Press ENTER. Set **Data List** to **L1** and **Data Axis** to **X**. Next, there are three different types of **Marks** that you can select for the graph. The first choice, a small square, is the best one to use.

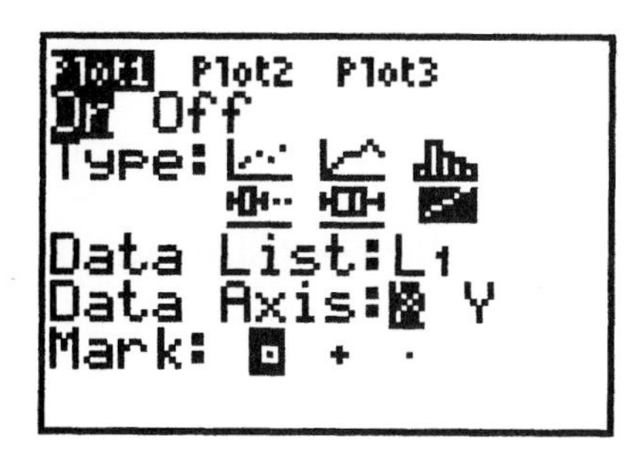

Press ZOOM and select **9:ZoomStat** and ENTER.

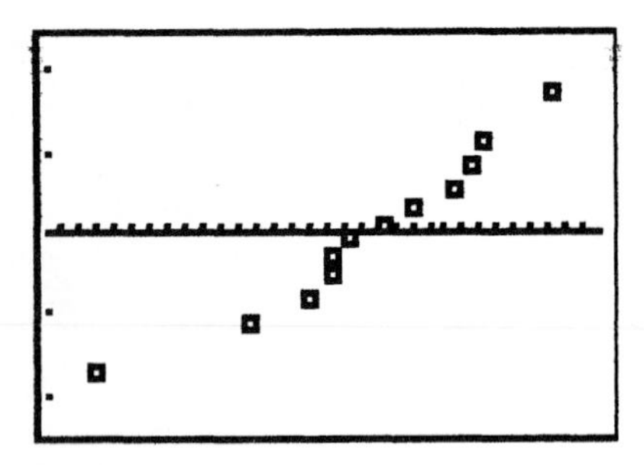

(Note: The calculator draws the normal probability plot along with a horizontal line at the X-axis.) This plot is *fairly* linear, indicating that the data generally follow a normal distribution.

To set up the boxplot, press 2nd [**STAT PLOT**]. Press ENTER to select **Plot 1**. Highlight **On** and press ENTER. Set **Type** to the boxplot with outliers which is the first selection in the second row. Press ENTER. Set **XList** to **L1** and **Freq** to **1**. Next, there are three different types of **Marks** that you can select for the graph. The first choice, a small square, is the best one to use.

Press ZOOM and select **9:ZoomStat** and ENTER.

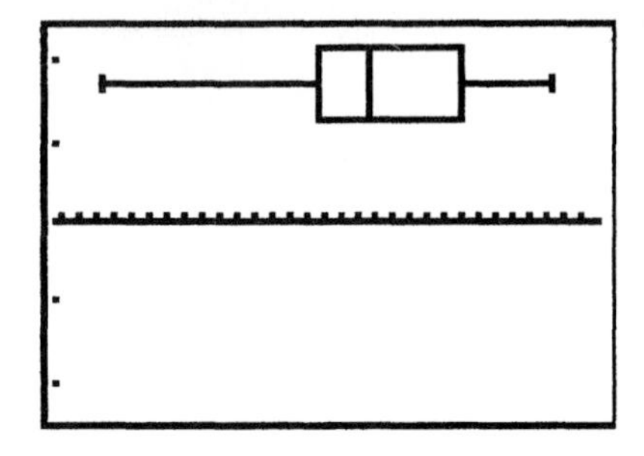

There are no outliers indicated in the boxplot. (Note: Outliers would appear as *'s at the extreme left or right ends of the boxplot.)

Since the data appear to be normally distributed with no outliers, and the population standard deviation is given, the criteria for a Z-interval have been met.

To estimate μ, the population mean, using a 90% confidence interval, press STAT, highlight **TESTS** and select **7:Zinterval.**

```
EDIT CALC TESTS
1:Z-Test...
2:T-Test...
3:2-SampZTest...
4:2-SampTTest...
5:1-PropZTest...
6:2-PropZTest...
7:ZInterval...
```

On the first line of the display, you can select **Data** or **Stats.** For this example, select **Data** because you want to use the actual data which is in **L1**. Press ENTER. Move to the next line and enter 8, the assumed value of σ. On the next line, enter **L1** for **LIST**. For **Freq,** enter **1.** For **C-Level**, enter **.90** for a 90% confidence interval. Move the cursor to **Calculate.**

```
ZInterval
 Inpt:Data Stats
 σ:8
 List:L1
 Freq:1
 C-Level:.90
 Calculate
```

Press ENTER.

```
ZInterval
 (55.818,63.415)
 x̄=59.61666667
 Sx=6.956335678
 n=12
```

A 90% confidence interval estimate of μ, the population mean, is (55.818, 63.415). The output display includes the sample mean (59.617), the sample standard deviation (6.956), and the sample size (12).

▸ Problem 19 (pg. 459)

(a.) A random sample of size n = 25 is selected from a population that is normally distributed with a standard deviation, σ, equal to 13. The sample mean, $\bar{x}$, is equal to 108. To estimate μ, the population mean, using a 96% confidence interval, press **STAT**, highlight **TESTS** and select **7:Zinterval.**

```
EDIT CALC TESTS
1:Z-Test...
2:T-Test...
3:2-SampZTest...
4:2-SampTTest...
5:1-PropZTest...
6:2-PropZTest...
7:ZInterval...
```

On the first line of the display, you can select **Data** or **Stats.** For this example, select **Stats** because you have the sample mean but not the actual data. Press **ENTER**. Move to the next line and enter **13**, the value of σ. On the next line, enter **108**, the value for $\bar{x}$, the sample mean. On the next line, enter the sample size, **25**. For **C-Level**, enter **.96** for a 96% confidence interval. Move the cursor to **Calculate.**

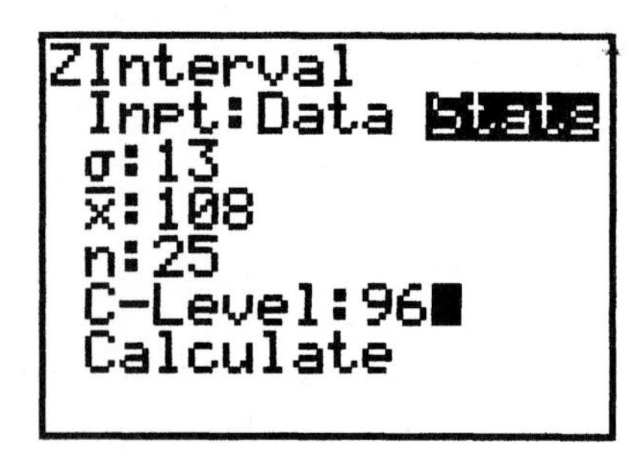

Press **ENTER**.

```
ZInterval
 (102.66,113.34)
 x̄=108
 n=25
```

Section 9.2

Example 4 (pg. 471) A Confidence Interval for μ (σ Unknown)

Enter the data from Table 5 on pg. 470 into **L1**. Since the sample size is less than 30, the first step is to check for normality using a normal probability plot and then to check for outliers using a Boxplot.

To set up the normal probability plot, press **2nd** **[STAT PLOT]**. Press ENTER to select **Plot 1**. Highlight **On** and press ENTER. Set **Type** to the normal probability plot which is the third selection in the second row. Press ENTER. Set **Data List** to **L1** and **Data Axis** to **X**. Next, there are three different types of **Marks** that you can select for the graph. The first choice, a small square, is the best one to use.

Press ZOOM and select **9:ZoomStat** and ENTER.

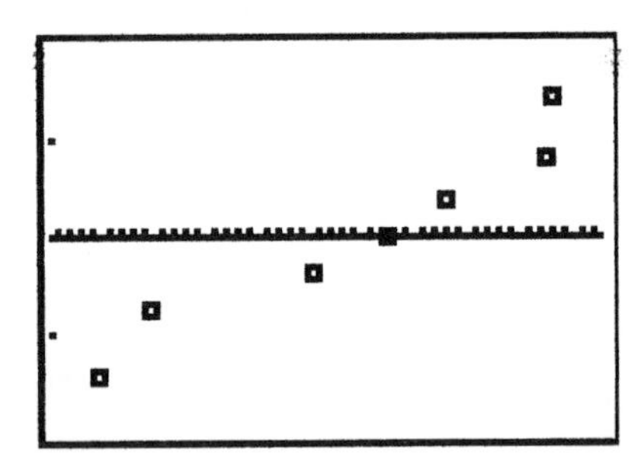

The calculator draws the plot with a horizontal line at the X-axis. This plot is *fairly* linear, indicating that the data generally follow a normal distribution.

To set up the boxplot, press **2nd** **[STAT PLOT]**. Press ENTER to select **Plot 1**. Highlight **On** and press ENTER. Set **Type** to the boxplot with outliers which is the first selection in the second row. Press ENTER. Set **XList** to **L1** and **Freq** to **1**. Next, there are three different types of **Marks** that you can select for the graph. The first choice, a small square, is the best one to use.

Press **ZOOM** and select **9:ZoomStat** and **ENTER**.

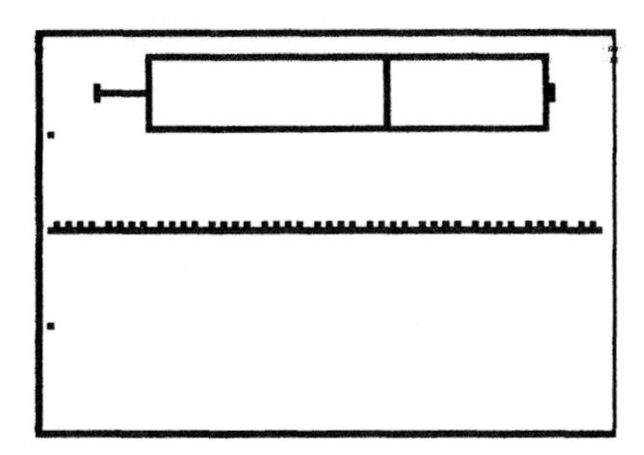

There are no outliers indicated in the boxplot. (Note: Outliers would appear as *'s at the extreme left or right ends of the boxplot.)

In this example, notice that σ is unknown. To construct the confidence interval for μ, the correct procedure under these circumstances ($n < 30$, σ unknown, the population approximately normally distributed and no outliers) is to use a T-Interval.

Press **STAT**, highlight **TESTS**, scroll through the options and select **8:TInterval** and press **ENTER** . Select **Data** for **Inpt** and press **ENTER**. For **List**, enter **L1** and for **Freq**, enter **1**. Set **C-level** to **.95**. Highlight **Calculate**.

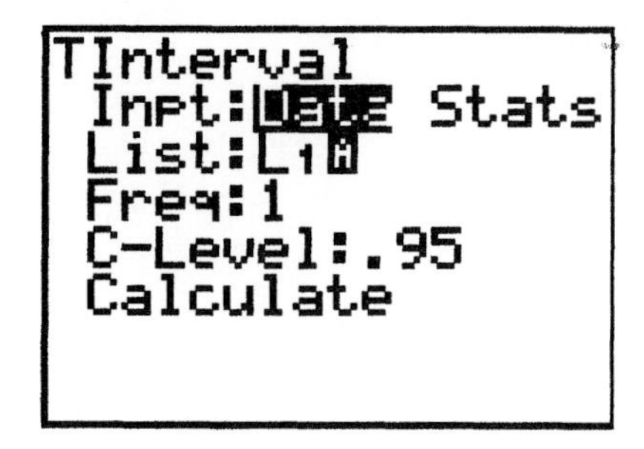

Press **ENTER**.

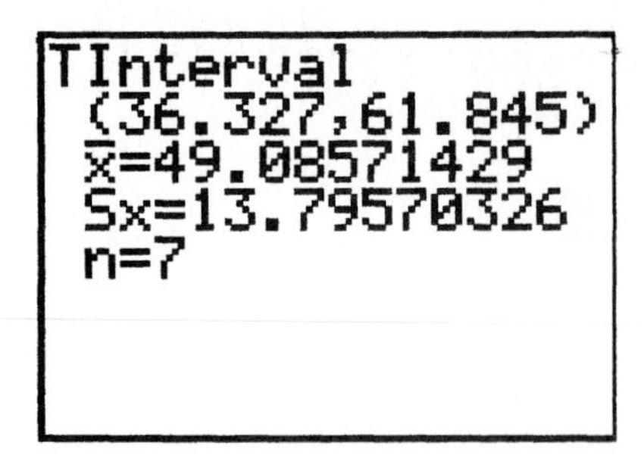

A 90% confidence interval for μ is (36.327, 61.845). The sample statistics (mean, standard deviation and sample size) are also given in the output screen.

▸ Problem 11 (pg. 473)

(a.) In this example, $\bar{x} = 18.4$, s= 4.5 and the sample size, n, = 35.
Since σ, the population standard deviation is unknown, the correct procedure for constructing a confidence interval for μ is the T-procedure.

Press STAT, highlight **TESTS**, scroll through the options and select **8:TInterval** and press ENTER. In this example, you do not have the actual data. What you have are the summary statistics of the data, so select **Stats** and press ENTER. Enter the values for $\bar{x}$, **Sx** and **n.** Enter **.95** for **C-level.** Highlight **Calculate.**

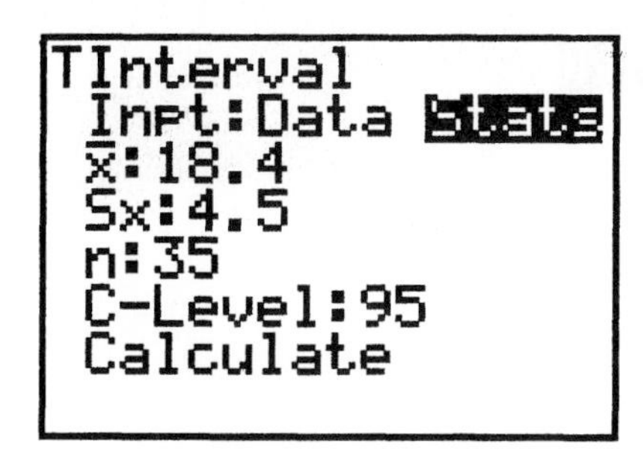

Press ENTER.

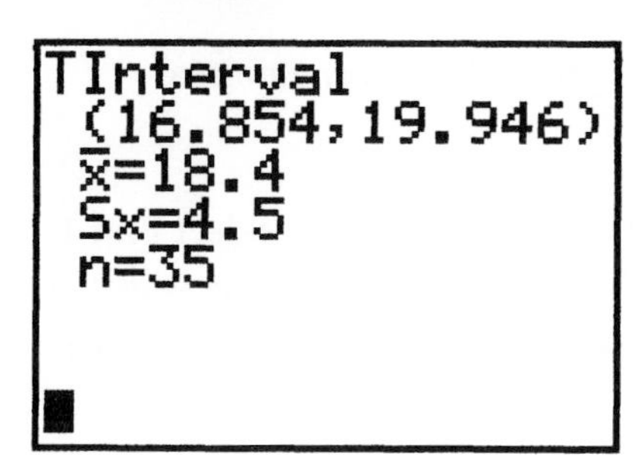

A 95% confidence interval estimate for μ is (16.854, 19.946).

◂

▸ Problem 25 (pg.475)

Enter the data into **L1**. Since the sample size is less than 30, the first step is to check for normality using a normal probability plot and then to check for outliers using a Boxplot.

To set up the normal probability plot, press **2nd** [STAT PLOT] . Press ENTER to select **Plot 1**. Highlight **On** and press ENTER. Set **Type** to the normal probability plot which is the third selection in the second row. Press ENTER. Set **Data List** to **L1** and **Data Axis** to **X**. For **Marks** select the small square.

Press ZOOM and select **9:ZoomStat** and ENTER.

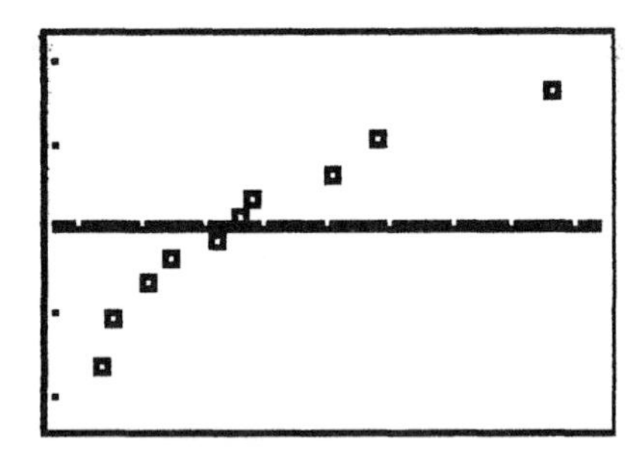

This plot is *fairly* linear, indicating that the data generally follow a normal distribution.

To set up the boxplot, press **2nd** [STAT PLOT] . Press ENTER to select **Plot 1**. Highlight **On** and press ENTER. Set **Type** to the boxplot with outliers which is the first selection in the second row. Press ENTER. Set **XList** to **L1** and **Freq** to **1**. For **Marks** select the small square.

Press ZOOM and select **9:ZoomStat** and ENTER.

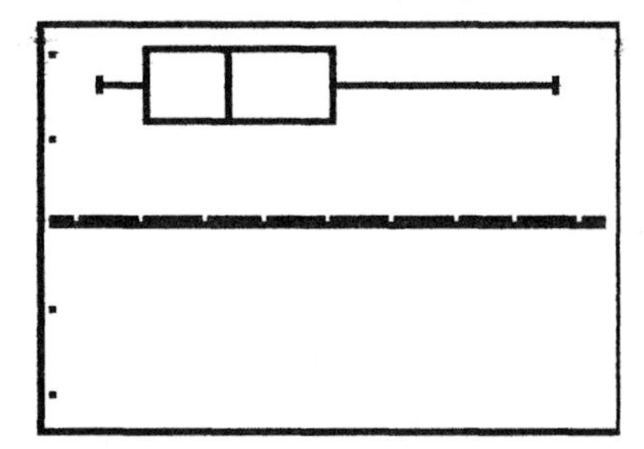

There are no outliers indicated in the boxplot. (Note: Outliers would appear as *'s at the extreme left or right ends of the boxplot.)

Since the data appear to be normally distributed with no outliers, and the population standard deviation is unknown, the criteria for the T-interval have been met.

Press STAT, highlight **TESTS** and select **8:T-interval.** Select **Data** for **Inpt** and press ENTER. For **List**, enter **L1** and for **Freq**, enter **1**. Set **C-level** to **.95**. Highlight **Calculate.**

```
TInterval
 Inpt:Data Stats
 List:L1
 Freq:1
 C-Level:95
 Calculate
```

Press ENTER.

```
TInterval
 (151.85,183.15)
 x̄=167.5
 Sx=21.87972171
 n=10
```

Section 9.3

Example 3 (pg. 480) A Confidence Interval for a Population Proportion

In this example, 1505 Americans were asked whether they were in favor of tighter enforcement of government rules on TV content during hours when children are most likely to be watching. 1129 individuals responded 'Yes.' Construct a 95 % confidence interval for p, the true proportion of all Americans who favor tighter enforcement.

Press **STAT**, highlight **TESTS**, scroll through the options and select **A:1-PropZInt**. The value for X is the number of Americans in the group of 1505 who responded 'Yes', so **X = 1129**. The number who were surveyed is n, so **n = 1505**. Enter **.95** for **C-level.**

```
1-PropZInt
 x:1129
 n:1505
 C-Level:.95
 Calculate
```

Highlight **Calculate** and press **ENTER**.

In the output display the confidence interval for p is (.72829, .77204). The sample proportion, $\hat{p}$, is .75017 and the number surveyed is 1505.

Note: You should calculate $n * \hat{p} * (1 - \hat{p})$. This value must be greater than or equal to 10 in order to use this confidence interval procedure. (It is actually easier to do this calculation after you have calculated the confidence interval because the calculator displays the value of $\hat{p}$ as part of the output.). For this example, the calculation is 1505*.09*.91. This value is greater than 10, so this supports the use of the confidence interval procedure. Also, make sure that the sample size is no more than 5% of the population size. In this example, because the population is so large (all Americans), the sample size criteria is certainly satisfied.

◀

Testing Claims Regarding a Parameter

Section 10.2

Example 3 (pg. 521) Testing a Claim about μ (σ known): Large Sample

In this example, we are given the sample statistics: $\bar{x} = 12895.9$ and n = 35. Since n > 30, the Central Limit Theorem applies so there is no need to test for normality or for the presence of outliers.

The hypothesis test, $H_o: \mu = 12200$ vs. $H_a: \mu > 12200$, is a right-tailed test. Since, the population standard deviation, σ, is given, the **Z-Test** is the appropriate test. To run the test, press **STAT**, highlight **TESTS** and select **1:Z-Test**. Since we are using the sample statistics for the analysis, select **Stats** for **Inpt** and press **ENTER**. For μ_0 enter 12200, the value for μ in the null hypothesis. For σ enter 3800, for $\bar{x}$ enter 12895.9 and for n enter 35. On the next line, choose the appropriate alternative hypothesis and press **ENTER**. For this example, it is > μ_0, a right-tailed test.

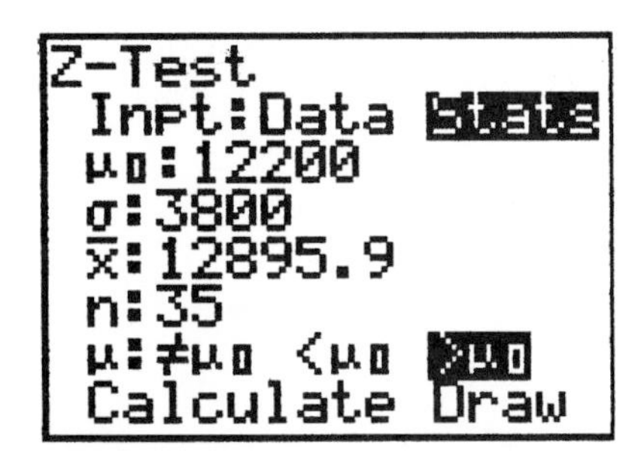

There are two choices for the output of this test. The first choice is **Calculate**. The output displays the alternative hypothesis, the calculated z-value, the P-value, $\bar{x}$ and n.

```
Z-Test
 μ>12200
 z=1.083421032
 p=.1393108277
 x̄=12895.9
 n=35
```

Since p = .1393, which is greater than α, the correct conclusion is to **Fail to Reject** H_o. (Note: The P-value calculated using the TI-84 is slightly different from the P-value obtained using the Z-table. That difference is simply due to rounding.)

To view the second output option, clear all the previous drawings by pressing 2nd **[DRAW]** and selecting **1:ClrDraw** and pressing ENTER ENTER. Press 2nd **[STATPLOT]** and TURN OFF all PLOTS. Make sure that there is nothing stored in the Y-registers. Press Y= and check the Y-registers. If any of them contain a function, move the cursor to that Y-register and press CLEAR. Now press STAT, highlight **TESTS**, and select **1:Z-Test**. All the necessary information for this example is still stored in the calculator. Scroll down to the bottom line and select **DRAW**. A normal curve is displayed with the right-tail area of .1393 shaded. This shaded area is the area to the right of the calculated Z-value. The Z-value and the P-value are also displayed.

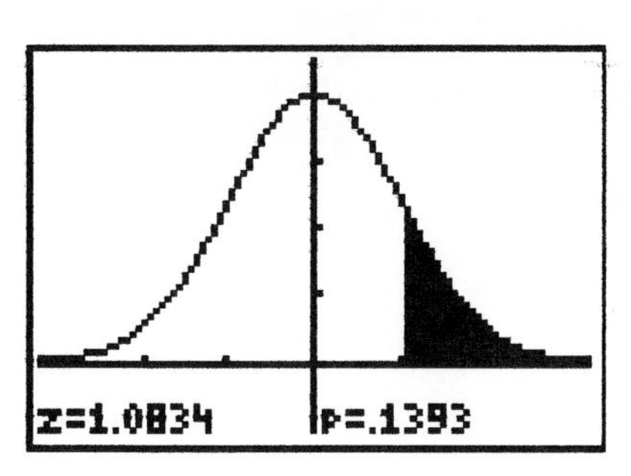

Example 5 (pg. 524) Testing a Claim about μ (σ known): Small Sample

Enter the data from Table 2 on pg. 522 into L1. Because the sample size is less than 30, the data must be tested for normality and checked for outliers.

To set up the normal probability plot, press 2nd [STAT PLOT] . Press ENTER to select **Plot 1**. Highlight **On** and press ENTER. Set **Type** to the normal probability plot which is the third selection in the second row. Press ENTER. Set **Data List** to **L1** and **Data Axis** to **X**. For **Marks** select the small square.

Press ZOOM and select **9:ZoomStat** and ENTER.

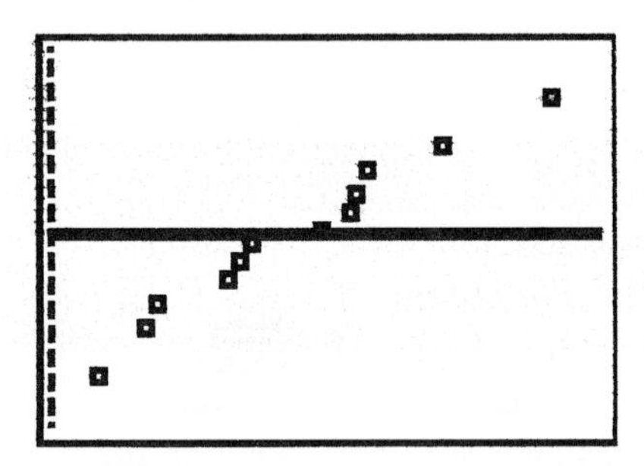

This plot is *fairly* linear, indicating that the data generally follow a normal distribution.

To set up the boxplot, press 2nd [STAT PLOT]. Press ENTER to select **Plot 1**. Highlight **On** and press ENTER. Set **Type** to the boxplot with outliers which is the first selection in the second row. Press ENTER. Set **XList** to **L1** and **Freq** to **1**. For **Marks** select the small square.

Press ZOOM and select **9:ZoomStat** and ENTER.

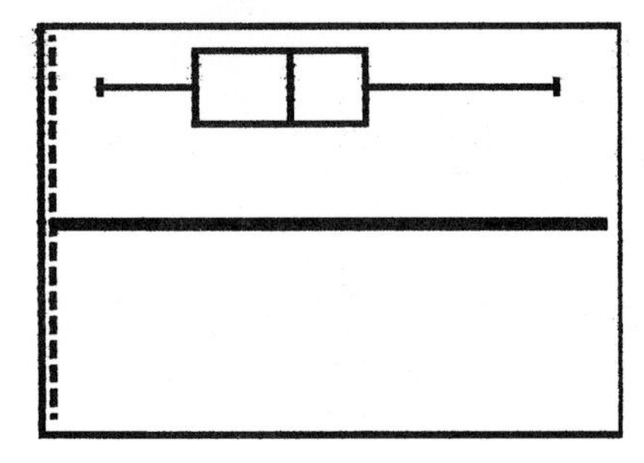

There are no outliers indicated in the boxplot. (Note: Outliers would appear as *'s at the extreme left or right ends of the boxplot.)

The hypothesis test, $H_o: \mu = 50.64$ vs. $H_a: \mu \neq 50.64$, is a two-tailed test. Since the population standard deviation, σ, is given, the **Z-Test** is the appropriate test. To run the test, press STAT, highlight **TESTS** and select **1:Z-Test**. Since you are using the actual data, which is stored in L1, for the analysis, select **Data** for **Inpt** and press ENTER. For μ_0 enter 50.64, the value for μ in the null hypothesis. For σ enter 18.49. Enter **L1** for **List, and 1** for **Freq**. On the next line, choose the appropriate alternative hypothesis and press ENTER. For this example, it is $\neq \mu_0$, a two-tailed test.

```
Z-Test
 Inpt:Data Stats
 μ0:50.64
 σ:18.49
 List:L1
 Freq:1
 μ:≠μ0 <μ0 >μ0
 Calculate Draw
```

There are two choices for the output of this test. The first choice is Calculate. The output displays the alternative hypothesis, the calculated z-value, the P-value, $\bar{x}$ and n.

```
Z-Test
 μ≠50.64
 z=2.693000215
 p=.007081319
 x̄=65.01416667
 Sx=25.45873362
 n=12
```

Since p = .007, which is less than α, the correct conclusion is to **Reject** H_o.

To view the second output option, clear all the previous drawings by pressing 2nd **[DRAW]** and selecting **1:ClrDraw** and pressing ENTER ENTER. Press 2nd **[STATPLOT]** and TURN OFF all PLOTS. Make sure that there is nothing stored in the Y-registers. Press Y= and check the Y-registers. If any of them contain a function, move the cursor to that Y-register and press CLEAR. Now press STAT, highlight **TESTS**, and select **1:Z-Test**. All the necessary information for this example is still stored in the calculator. Scroll down to the bottom line and select **DRAW. A normal curve is displayed with each tail area** of .0035 shaded. (Note: Because the areas are so small in this example, they are

not really visible in the curve.) The shaded areas total to 0.007 which is the P-
value. The Z-value and the P-value are also displayed.

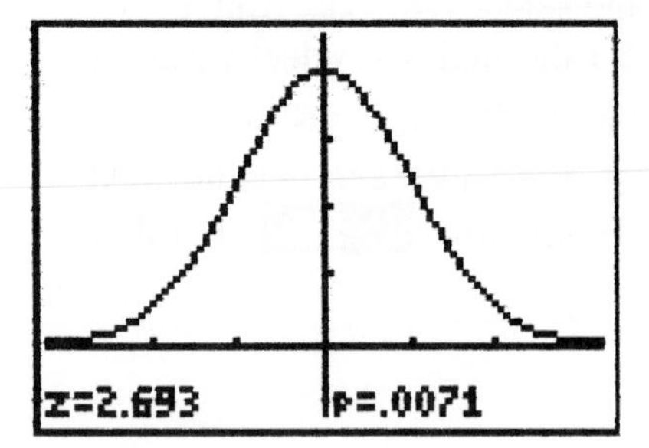

▸ Example 6 (pg. 525) Using a Confidence Interval to Test a Claim

Enter the data from Table 2 on pg. 522 into L1. Because the sample size is less than 30, the data must be tested for normality and checked for outliers. (Note: **These tests were done with the previous Example and the results indicated that** the data were normally distributed with no outliers.)

To estimate μ, the population mean, using a 95% confidence interval, press STAT, highlight **TESTS** and select **7:Zinterval.**

```
EDIT CALC TESTS
1:Z-Test...
2:T-Test...
3:2-SampZTest...
4:2-SampTTest...
5:1-PropZTest...
6:2-PropZTest...
7:ZInterval...
```

On the first line of the display select **Data.** Press ENTER . Move to the next line and enter 18.49, the assumed value of σ . On the next line, enter **L1** for **LIST**. For **Freq**, enter **1**. For **C-Level** , enter **.95** for a 95% confidence interval. Move the cursor to **Calculate.**

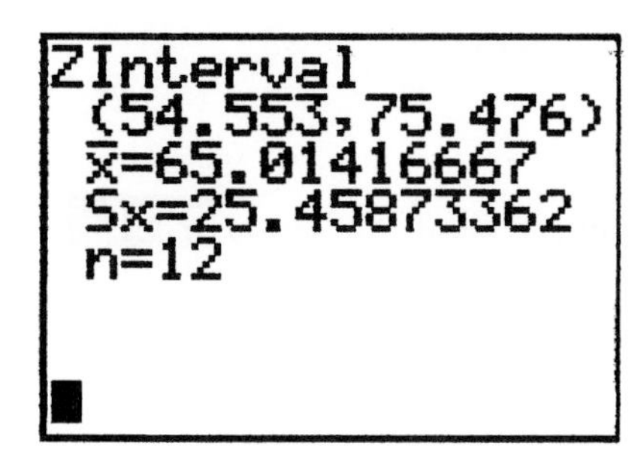

The 95% confidence interval for μ is (54.553, 75.476). Notice that this **confidence interval does not contain the hypothesized value for μ (50.64). Since** the hypothesized value is not contained in the confidence interval, the correct decision is: **Reject the null hypothesis.** There is sufficient evidence at the .05 significance level to support the claim that the mean monthly cell phone bill is different from $50.64.

◀

Problem 15 (pg. 527)

Test the hypotheses: $H_o: \mu = 20$ vs. $H_a: \mu < 20$. The underlying population is assumed to be normally distributed with $\sigma = 3$. The sample mean, $\bar{x}$, = 18.3, and n = 18. Press STAT, highlight **TESTS** and select **1:Z-Test**. For **Inpt**, choose **Stats** and press ENTER. Fill the input screen with the appropriate information. Choose < μ_0 for the alternative hypothesis and press ENTER.

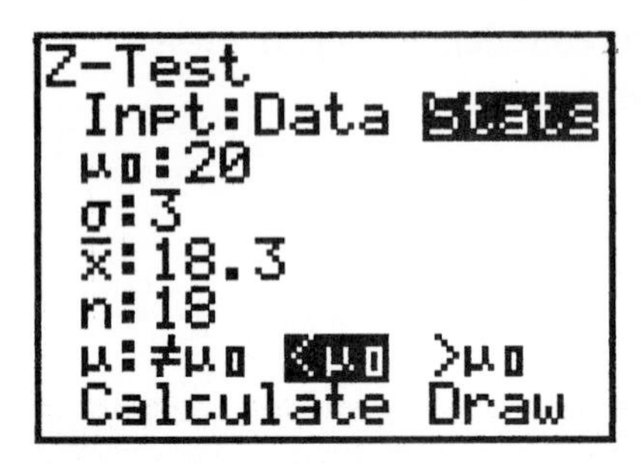

Highlight **Calculate** and press ENTER.

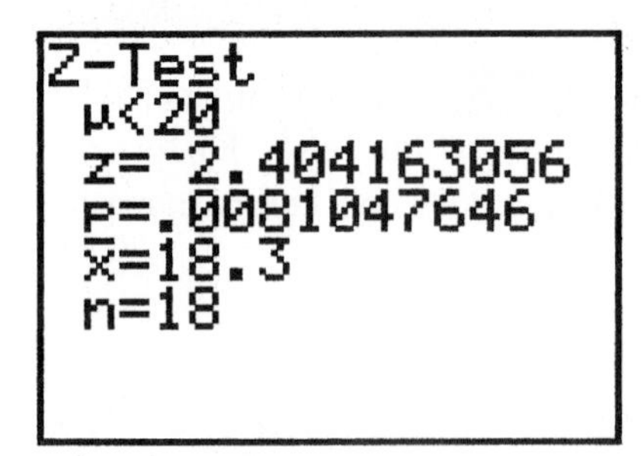

Or, (after clearing all graphs and Y-registers) highlight **Draw** and press ENTER.

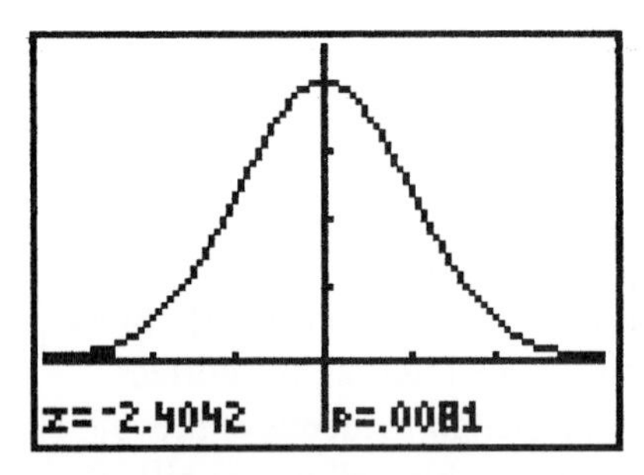

The P-value is .008. Since the P-value is less than α, the correct conclusion is to **Reject** H_o.

▸ Problem 17 (pg. 527)

Test the hypotheses: $H_o: \mu = 105$ vs. $H_a: \mu \neq 105$. In this example, the sample size is greater than 30. Since the sample size is *large,* the Central Limit Theorem applies and we can assume that the sampling distribution of $\bar{x}$ is approximately normal. The population standard deviation, σ, is equal to 12.

To run the test, press STAT, highlight **TESTS** and select **1:Z-Test**. For **Inpt**, choose **Stats** and press ENTER. Fill the input screen with the appropriate information. Choose $\neq \mu_0$ for the alternative hypothesis and press ENTER.

Highlight **Calculate** and press ENTER.

```
Z-Test
 μ≠105
 z=-1.873425265
 p=.0610095551
 x̄=101.2
 n=35
```

Or, (after clearing all graphs and Y-registers) highlight **Draw** and press ENTER.

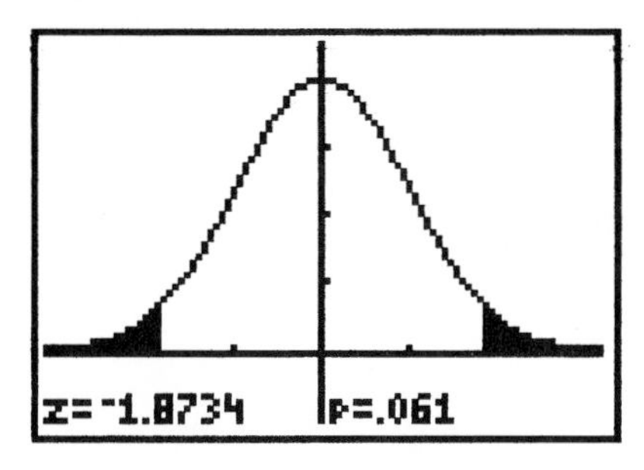

The P=value is .061. Since the P-value is greater than α, the correct conclusion is to **Fail to Reject** H_o.

◂

Section 10.3

▶ Example 2 (pg. 535) Testing a Claim about μ (σ unknown): Large Sample

Refer to Example1 on pg. 533. Test the hypotheses: $H_o: \mu = 18.1$ vs. $H_a: \mu <$ 18.1. In this example, the sample size is greater than 30. Since the sample size is *large*, the Central Limit Theorem applies and we can assume that the sampling distribution of $\bar{x}$ is approximately normal. Also, notice that σ, the population standard deviation is not given. Instead the sample standard deviation is given; therefore, the appropriate test is the T-test.

To run the test, press **STAT**, highlight **TESTS** and select **2:T-Test**. For **Inpt**, choose **Stats** and press **ENTER**. Fill the input screen with the appropriate information. (μ_0=18.1, $\bar{x}$=16.8, s = 4.7 and n = 40.) Choose < μ_0 for the alternative hypothesis and press **ENTER**. Highlight **Calculate** and press **ENTER**.

```
T-Test
 μ<18.1
 t=-1.749345089
 p=.0440489391
 x̄=16.8
 Sx=4.7
 n=40
```

Or, (after clearing all graphs and the Y-registers) highlight **Draw** and press **ENTER**.

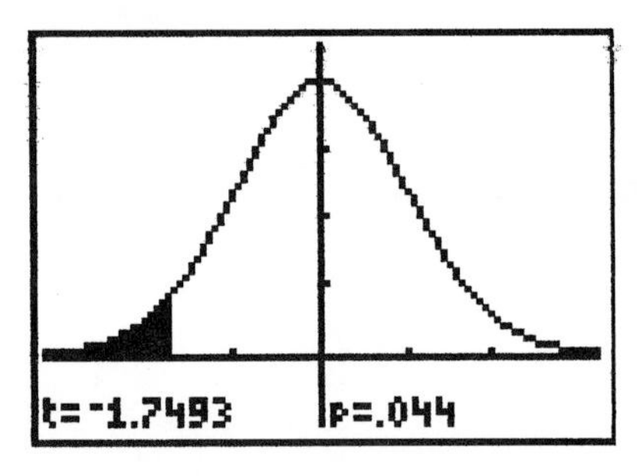

The P=value is .044. Since the P-value is less than α. the correct conclusion is to **Reject** H_o. ◀

▸ Problem 9 (pg. 538)

Test the hypotheses: $H_o: \mu = 20$ vs. $H_a: \mu < 20$. The underlying population is known to be normally distributed. The sample statistics are $\bar{x} = 18.3$, s = 4.3 and n = 18. Press **STAT**, highlight **TESTS** and select **1:T-Test**. For **Inpt**, choose **Stats** and press **ENTER**. Fill the input screen with the appropriate information. Choose < μ_0 for the alternative hypothesis and press **ENTER**.

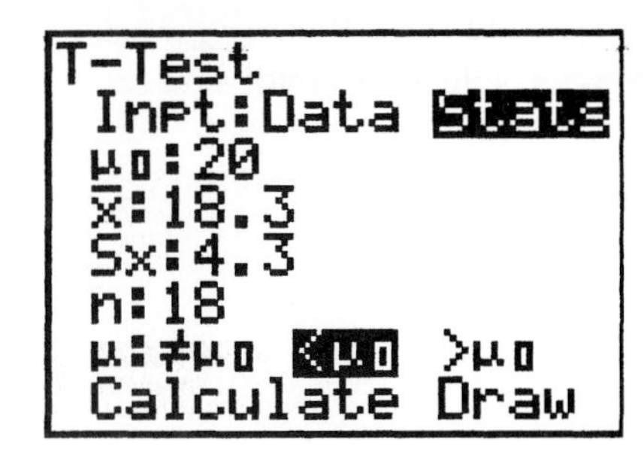

Highlight **Calculate** and press **ENTER**.

```
T-Test
 μ<20
 t=-1.677323062
 p=.0558841671
 x̄=18.3
 Sx=4.3
 n=18
```

Or, (after clearing all graphs and the Y-registers) highlight **Draw** and press **ENTER**.

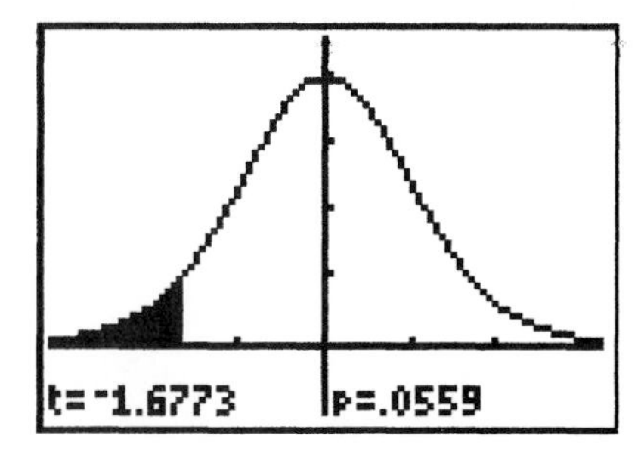

The P=value is .0559. Since the P-value is greater than α, the correct conclusion is to **Fail to Reject** H_o.

▸ Problem 25 (pg. 541)

Enter the data into L1. Because the sample size is less than 30, the data must be tested for normality and checked for outliers.

To set up the normal probability plot, press **2nd** [STAT PLOT]. Press ENTER to select **Plot 1**. Highlight **On** and press ENTER. Set **Type** to the normal probability plot which is the third selection in the second row. Press ENTER. Set **Data List** to **L1** and **Data Axis** to **X**. For **Marks** select the small square.

Press ZOOM and select **9:ZoomStat** and ENTER.

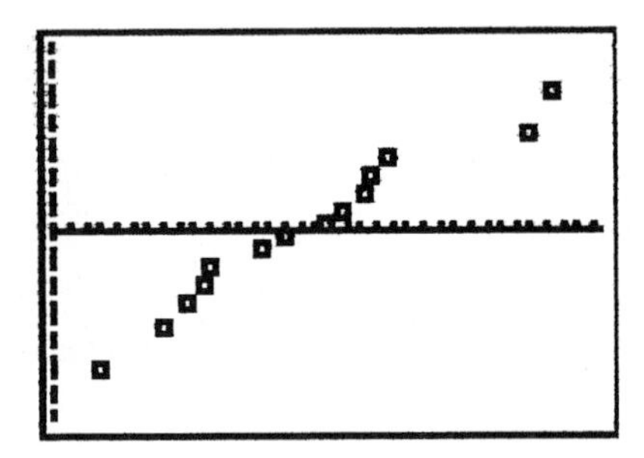

This plot is *fairly* linear, indicating that the data generally follow a normal distribution.

To set up the boxplot, press **2nd** [STAT PLOT]. Press ENTER to select **Plot 1**. Highlight **On** and press ENTER. Set **Type** to the boxplot with outliers which is the first selection in the second row. Press ENTER. Set **XList** to **L1** and **Freq** to **1**. For **Marks** select the small square.

Press ZOOM and select **9:ZoomStat** and ENTER.

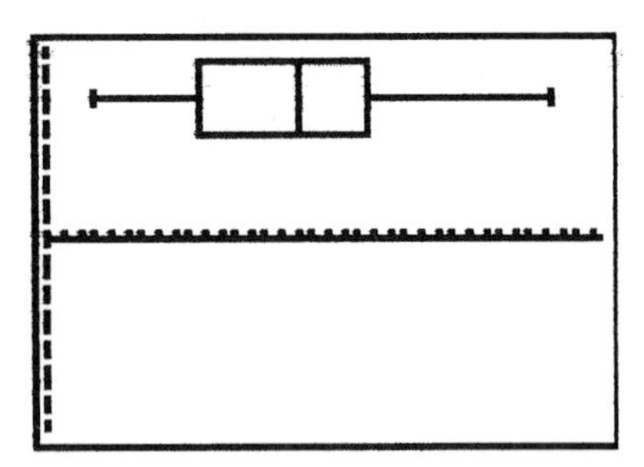

There are no outliers indicated in the boxplot. (Note: Outliers would appear as *'s at the extreme left or right ends of the boxplot.)

This test is a left-tailed test of $H_o: \mu = 22.0$ vs. $H_a: \mu < 22.0$. Since n < 30, and the population standard deviation, σ, is unknown, the T-Test is the appropriate test. This test requires the underlying population to be approximately normally distributed with no outliers, as was verified in the plots.

Press STAT, highlight **TESTS** and select **2:T-Test**. Choose **Data** for **Inpt** and press ENTER. Fill in the following information: μ_0 **= 22.0, List = L1,** and **Freq =1**. Choose the right-tailed alternative hypothesis, $< \mu_0$, and press ENTER. Highlight **Calculate** and press ENTER

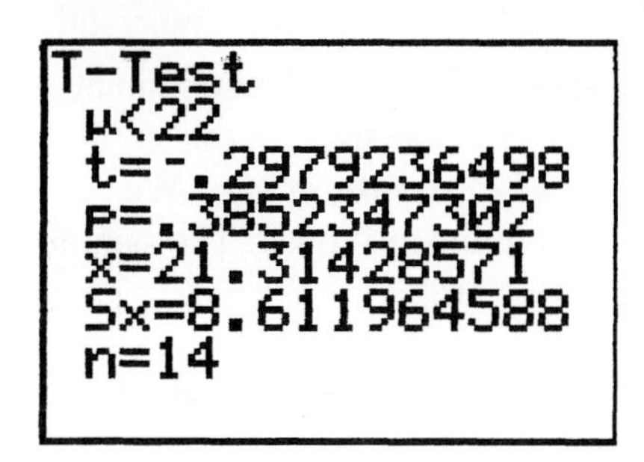

Or, highlight **Draw** and press ENTER.

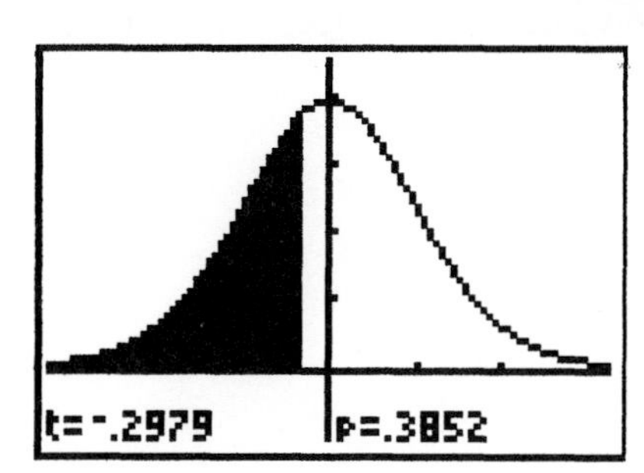

Since the P-value is greater than α, the correct conclusion is to **Fail to Reject** H_o.

Section 10.4

Example 3 (pg. 548) Testing a Claim about a Population Proportion: Large Sample

Refer to Example 2 on pg. 547. This hypothesis test is a two-tailed test of: $H_o: p = .135$ vs. $H_a: p \neq .135$. The procedure that is used for this test is the **1-Proportion Test**. This test has two requirements. The first requirement is: $n * p_0 * (1 - p_0) \geq 10$. To verify this, calculate 710*.135*(1-.135). The result is greater than 10, so the first requirement is satisfied. The second requirement is that the sample size is not more than 5% of the *population* size. In this example, the population is *all babies between 12 and 15 months of age.* We don't know the exact size of the population, but it is in the millions. The sample size of 710 is definitely less than 5% of the population size.

To run the test, press **STAT**, highlight **TESTS** and select **5:1-PropZTest**. This test requires a value for p_0, which is the value for p in the null hypothesis. Enter **.135** for p_0. Next, a value for X is required. X is the number of "successes" in the sample. In this example, a success is " experiencing a loss of appetite", so **X** is equal to **121**. Next, enter the value for **n**. Select $\neq p_0$ for the alternative hypothesis and press **ENTER**.

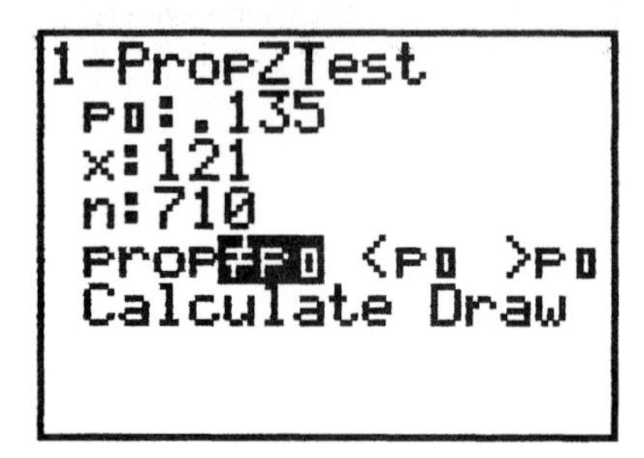

Highlight **Calculate** and press **ENTER**.

```
1-PropZTest
 prop≠.135
 z=2.762064888
 p=.0057438195
 p̂=.1704225352
 n=710
```

The output displays the alternate hypothesis that was selected, the calculated Z-value, the P-value, the sample proportion, $\hat{p}$, and the sample size. (Note: $\hat{p}$ = 121/710.)

Or, (after clearing all graphs and Y-registers) highlight **Draw** and press ENTER.

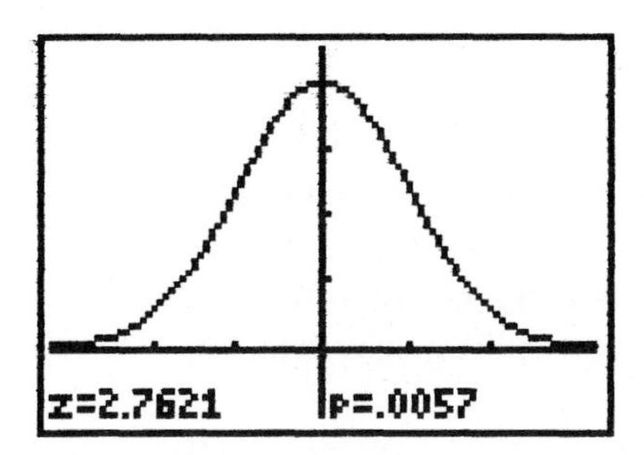

Since the P-value is less than α, the correct conclusion is to **Reject** H_o.

Example 4 (pg. 549) Testing a Claim about a Population Proportion: Small Sample

In this test of a population proportion, the requirement $n * p_0 * (1 - p_0) \geq 10$ is **not satisfied. (The calculation 35*.489*(1-.489) is equal to 8.75.) (Note: In** cases in which the sample size is relatively small this requirement is often not satisfied.)

An alternative method of testing a hypothesis about a population proportion is to use the binomial probability formula to calculate the likelihood of the sample result. If the sample result is *unusual* then we will **reject the null hypothesis.** We define *unusual* events as events that have a probability less than .05.

The hypothesis test is: $H_o: p = .489$ vs. $H_a: p > .489$. The sample statistics are n = 35 and X = 21. Using the binomial probability formula, we calculate the likelihood of obtaining 21 or more males who consume the recommended daily allowance of calcium in the sample of 35. We assume that the proportion of males in the population who consume the recommended daily allowance of calcium is .489.

Press **1 – 2nd DISTR** and select **B:binomcdf(**. Type in **35 , .489 , 20)**.

```
1-binomcdf(35,.4
89,20)
      .1261068497
```

(Note: The command **binomcdf** calculates the probability that $X \leq 20$, which is the *complement* of the probability that $X \geq 21$. To obtain $P(X \geq 21)$, we calculate $P(X \leq 20)$ and subtract this value from 1.)

The result is: $P(X \geq 21) = .126$. Since this probability is greater than .05, the correct conclusion is to **Fail to Reject** H_o.

▸ Problem 20 (pg. 551)

This hypothesis test is a right-tailed test of: $H_o: p = .85$ vs. $H_a: p > .85$. To use the **1-Proportion Test,** first you must determine whether the requirements for this test have been satisfied. The first requirement is: $n * p_0 * (1 - p_0) \geq 10$. To verify this, calculate 200*.85*(1-.85). The result is greater than 10, so the first requirement is satisfied. The second requirement is that the sample size is not more than 5% of the *population* size. In this example, the population is *all American adults.* We don't know the exact size of the population, but it is in the millions. The sample size of 150 is definitely less than 5% of the population size.

To run the test, press **STAT**, highlight **TESTS** and select **5:1-PropZTest**. Enter **.85** for p_0. For **X,** enter **171** and for n, enter **200**. Select $> p_0$ for the alternative hypothesis and press **ENTER**.

Highlight **Calculate** and press **ENTER**.

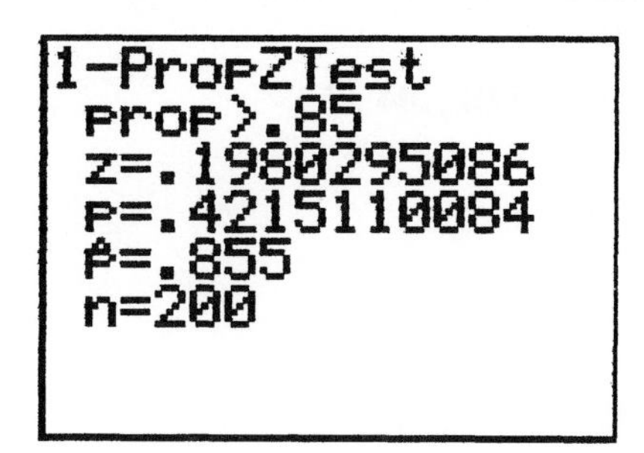

Or highlight **Draw** and press **ENTER**.

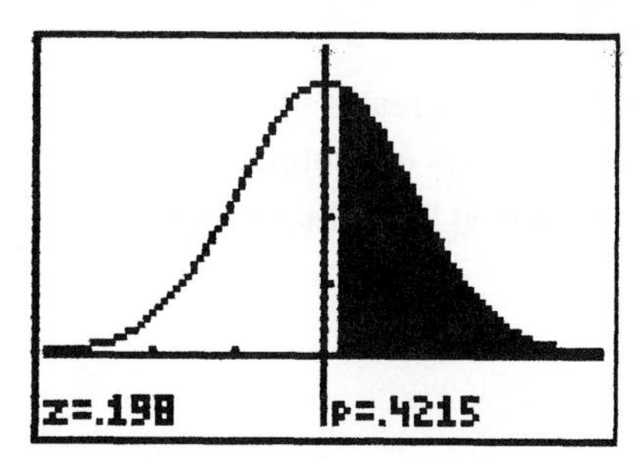

Since the P-value is greater than α, the correct conclusion is to **Fail to Reject** H_o. The data does not support the nutritionist's claim that the percentage of American adults who eat salad at least once a week is higher than 85%.

◂

▸ Problem 21 (pg. 552)

In this test of a population proportion, the requirement $n * p_0 * (1 - p_0) \geq 10$ is **not satisfied. (The calculation 120*.04*(1-.04) is equal to 4.608.) (Note: In** cases in which the sample size is relatively small this requirement is often not satisfied.)

An alternative method of testing a hypothesis about a population proportion is to use the binomial probability formula to calculate the likelihood of the sample result. **If the sample result is *unusual* then we will reject the null hypothesis.** We define *unusual* events as events that have a probability less than .05.

The hypothesis test is: $H_o: p = .04$ vs. $H_a: p < .04$. The sample statistics are n = 120 **and** X = 3. **Using the binomial probability formula, we calculate the** likelihood of obtaining 3 or fewer mothers who smoked 21 or more cigarettes **during pregnancy. We assume that the proportion of mothers who smoked 21** or more cigarettes during pregnancy is .04.

Press 2nd **DISTR** and select **A:binomcdf(**. Type in **120 , .04 , 3)**.

```
binomcdf(120,.04
,3)
          .288658855
```

The result is: P(X ≤ 3) = .2887. Since this probability is greater than .05, the correct conclusion is to **Fail to Reject** H_o. The data does not support the **obstetrician's belief that less than 4% of mothers smoked 21 or more cigarettes** during pregnancy.

◂

Section 10.5

▸ Example 1 (pg. 555) Testing a Claim About a Population Standard Deviation

This is a left-tailed hypothesis test about σ. The hypotheses are: H_o: $\sigma = 0.75$ vs. $H_a: \sigma < 0.75$.

Enter the data from Table 4 into L1. Because the sample size is less than 30, the data must be tested for normality.

To set up the normal probability plot, press **2nd [STAT PLOT]**. Press **ENTER** to select **Plot 1**. Highlight **On** and press **ENTER**. Set **Type** to the normal probability plot which is the third selection in the second row. Press **ENTER**. Set **Data List** to **L1** and **Data Axis** to **X**. For **Marks** select the small square.

Press **ZOOM** and select **9:ZoomStat** and **ENTER**.

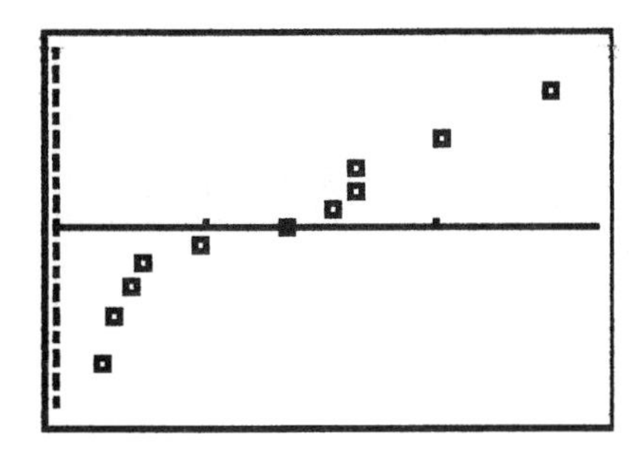

This plot is *fairly* linear, indicating that the data generally follows a normal distribution.

The next step is to find the value of the sample standard deviation. Press **STAT**, highlight **CALC** and select **1:1-Var Stats** and press **ENTER**. Type in **L1** and press **ENTER**. The sample statistics are displayed on the screen. The sample standard deviation, **Sx = 0.6404**.

Next, calculate the test statistic: $\chi^2 = (n-1)s^2 / \sigma_0^2$. The calculation, $10(.6404)^2/(.75)^2 = 7.291$. Because this is a left-tailed test, to find the P-value associated with the test statistic value of 7.291, we calculate the area under the χ^2-curve to the *left* of 7.291. Press **2nd DISTR**, and select **7: χ^2 cdf(**. This calculation requires a *lowerbound*, an *upperbound* and the *degrees of freedom.*

The lowerbound is negative infinity (-1E99 on the calculator), the upperbound is 7.291 and the degrees of freedom value is 10 (which is equal to n-1).

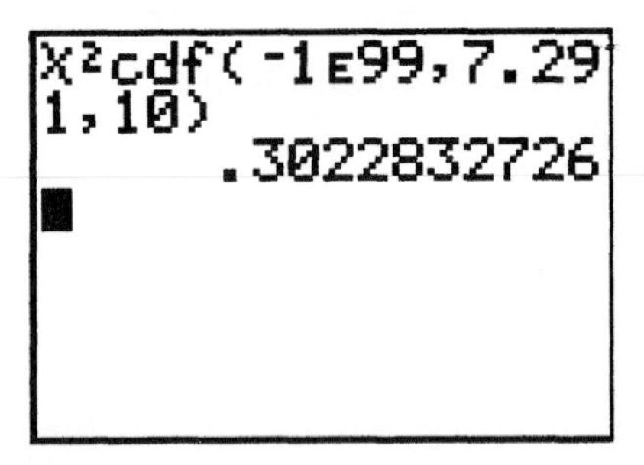

Since the P-value of .3023 is greater than α, the correct conclusion is to **Fail to Reject** H_o. There is not sufficient evidence to support that claim that σ < 0.75.

▸ Problem 7 (pg. 556)

This is a two-tailed hypothesis test about σ. The hypotheses are: H_o: $\sigma = 4.3$ vs. $H_a: \sigma \neq 4.3$. The population is normally distributed, the sample size is 12 and the sample standard deviation is 4.8.

The first step is to calculate the test statistic: $\chi^2 = (n-1)s^2 / \sigma_0^2$. The calculation, $11(4.8)^2/4.3^2 = 13.71$. To find the P-value associated with the test statistic value of 13.71, we calculate the area under the χ^2-curve to the *right* of 13.71. The reason that we are finding the area *to the right* of 13.71 is because the sample standard deviation, 4.8, is *greater than* 4.3. Press **2nd DISTR**, and select **7: χ^2 cdf(**. This calculation requires a *lowerbound*, an *upperbound* and the *degrees of freedom.* The lowerbound is 13.71, the upperbound is positive infinity (1E99 on the calculator) and the degrees of freedom value is 11 (which is equal to n-1).

```
X²cdf(13.71,1E99
,11)
      .2494577162
```

The p-value is .2495. Because this is a two-tailed test, you must compare this P-value to α/2 which is .025. Since the P-value is greater than .025, the correct decision is to **Fail to reject** H_o. There is not sufficient evidence to support that claim that $\sigma \neq 4.3$.

◂

Inferences on Two Samples

Section 11.1

Example 3 (pg. 580) Testing a Claim Regarding Matched Pairs Data

In this example, the data (found in Table 2 on pg. 578) is paired data, with two reaction times for each of the students. Enter the reaction times for the individual's dominant hand in L1 and enter the reaction times for the individual's non-dominant hand into L2. Next, you must create a set of differences, d = reaction time of dominant hand - reaction time of non-dominant hand. To create this set, move the cursor to highlight the label **L3,** found at the top of the third column, and press **ENTER**. Notice that the cursor is flashing on the bottom line of the display. Press **2nd [L1] - 2nd [L2]**

L1	L2	L3 3
.177	.179	------
.21	.202	
.186	.208	
.189	.184	
.198	.215	
.194	.193	
.16	.194	

L3 =L1-L2

and press **ENTER**.

L1	L2	L3 3
.177	.179	-.002
.21	.202	.008
.186	.208	-.022
.189	.184	.005
.198	.215	-.017
.194	.193	.001
.16	.194	-.034

L3(1)= -.002

Each value in **L3** is the difference **L1 - L2**.

To test the claim that the reaction time in an individual's dominant hand is less than the reaction time in his/her non-dominant hand, the hypothesis test is:
$H_o: \mu_d = 0$ vs. $H_a: \mu_d < 0$.

Because the sample size is less than 30, the set of differences must be tested for normality and checked for outliers.

To set up the normal probability plot, press **2nd** [STAT PLOT] . Press ENTER to select **Plot 1**. Highlight **On** and press ENTER. Set **Type** to the normal probability plot which is the third selection in the second row. Press ENTER. Set **Data List** to **L3** and **Data Axis** to **X**. For **Marks,** select the small square.

Press ZOOM and select **9:ZoomStat** and ENTER.

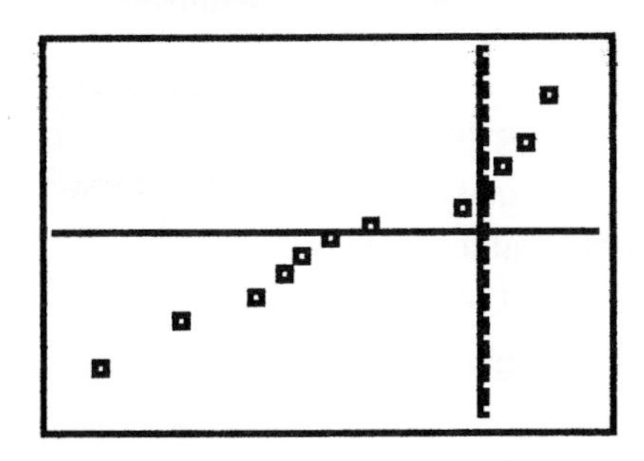

This plot is *fairly* linear, indicating that the data generally follow a normal distribution.

To set up the boxplot, press **2nd** [STAT PLOT]. Press ENTER to select **Plot 1**. Highlight **On** and press ENTER. Set **Type** to the boxplot with outliers which is the first selection in the second row. Press ENTER. Set **XList** to **L3** and **Freq** to **1**. For **Marks** select the small square.

Press ZOOM and select **9:ZoomStat** and ENTER.

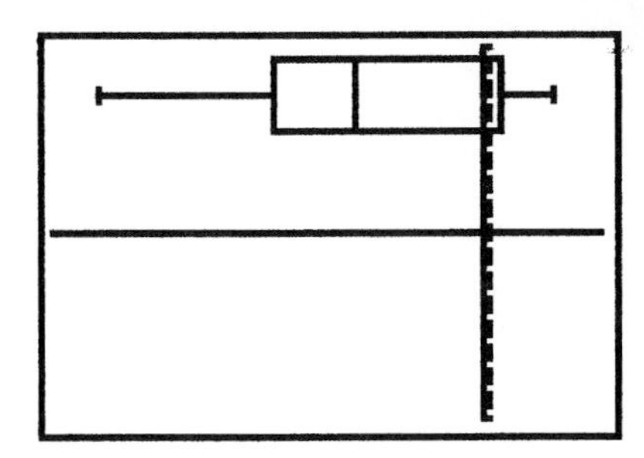

Note: Both of these graphs include the X-axis and Y-axis with the plot. To turn the axes off, press GRAPH . Next press **2nd [FORMAT]** and scroll down to

Axes On. Move the cursor to **Axes Off** and press **ENTER**. If you redo the graphs, the axes will no longer appear on the screen. You may prefer the way the graphs look without the axes. Here is the graph of the Boxplot:

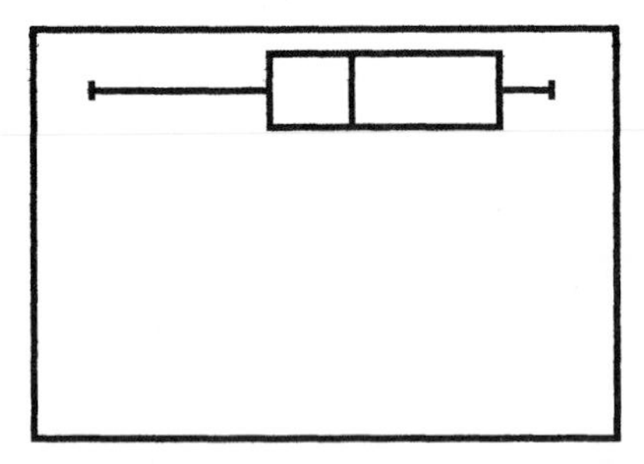

There are no outliers indicated in the boxplot. (Note: Outliers would appear as *'s at the extreme left or right ends of the boxplot.)

To run the hypothesis test, press **STAT**, highlight **TESTS** and select **2:T-Test**. In this example, you are using the actual data to do the analysis, so select **Data** for **Inpt** and press **ENTER**. The value for μ_o is **0**, the value in the null hypothesis. The set of differences is found in **L3**, so set **List** to **L3**. Set **Freq** equal to **1**. Choose < μ_o as the alternative hypothesis and highlight **Calculate** and press **ENTER**.

```
T-Test
 μ<0
 t=-2.775932838
 p=.0090173095
 x̄=-.0131666667
 Sx=.0164307546
 n=12
```

You can also highlight **DRAW** and press **ENTER**.

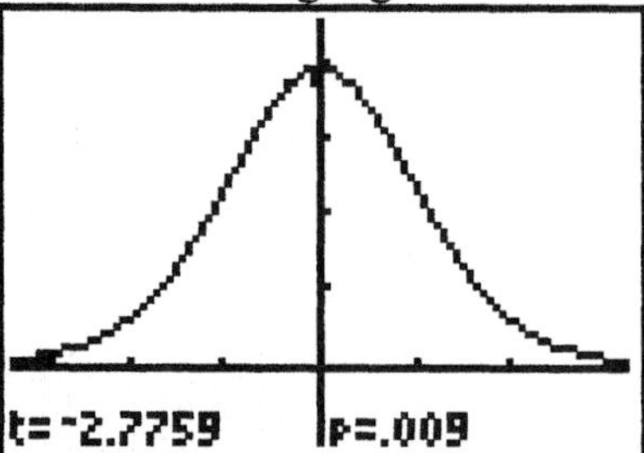

Since the P-value is less than α, the correct decision is to **Reject** H_o. We conclude that an individual's dominant hand has a faster reaction time.

▸ Example 4 (pg. 581) Constructing a Confidence Interval for Matched-Pairs Data

In Example 3 we created a column of differences in L3. We also confirmed that the data was normally distributed with no outliers. To create a 95% confidence interval for the mean difference, press STAT, highlight **TESTS** and select **8:T-Interval**. In this example, you are using the actual data to do the analysis, so select **Data** for **Inpt** and press ENTER. The set of differences is found in **L3**, so set **List** to **L3**. Set **Freq** equal to **1**. Set **C-Level** to **.95**. Highlight **Calculate** and press ENTER.

```
TInterval
 (-.0236,-.0027)
 x̄=-.0131666667
 Sx=.0164307546
 n=12
```

The 95% confidence interval for μ_d, the mean difference in reaction time, is (-.0236, -.0027)seconds. We interpret this interval to mean that, on average, the reaction time of a person's dominant hand is between .0027 and .0236 seconds *less* than the reaction time of a person's non-dominant hand.

◂

▸ Problem 15 (pg. 583)

In this example the data is paired data, with two measurements of water clarity at the same location in the lake at Joliet Junior College. The first reading is taken at a specific location on the lake at a particular time during a given year. The second reading is taken at the same location 5 years later. Enter the **initial depth** readings into **L1** and the readings **5 years later** into **L2**. Next, you must create a set of differences, d = intial depth reading - reading 5 years later. To create this set, move the cursor to highlight the label **L3,** found at the top of the third column, and press ENTER. Notice that the cursor is flashing on the bottom line of the display. Press **2nd [L1]** - **2nd [L2]** and press ENTER.

L1	L2	L3 3
38	52	-14
58	60	-2
65	72	-7
74	72	2
56	54	2
36	48	-12
56	58	-2

L3(1)= -14

Each value in **L3** is the difference **L1 - L2**.

We are interesting in testing the claim that the clarity of the lake is improving. If the clarity of the lake *is improving*, then the depth at which the disk is no longer visible should be getting deeper. If that is the case, then the difference, **L1 – L2**, should be *negative*. In other words, the original depths that were measured should be "less deep" than the measurements 5 years later. So the appropriate alternate hypothesis is: $\mu_d < 0$.

Note: A normal probability plot and boxplot of the data indicate that the differences are approximately normal with no outliers.

(b.) To run the hypothesis test, press STAT, highlight **TESTS** and select **2:T-Test**. In this example, you are using the actual data to do the analysis, so select **Data** for **Inpt** and press ENTER. The value for μ_o is **0**, the value in the null hypothesis. The set of differences is found in **L3**, so set **List** to **L3**. Set **Freq** equal to **1**. Choose < μ_o as the alternative hypothesis and highlight **Calculate** and press ENTER.

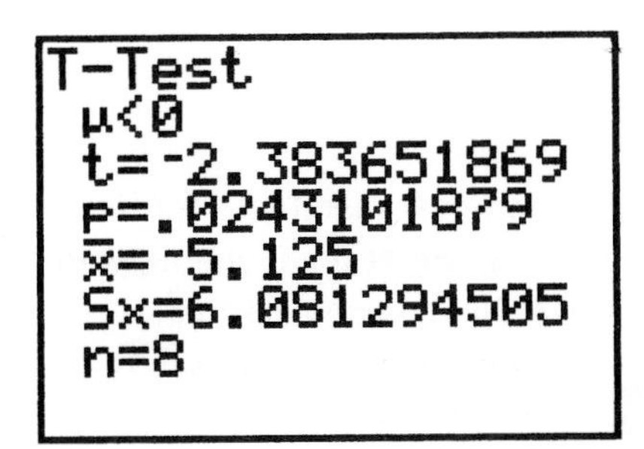

Or, highlight **Draw** and press ENTER.

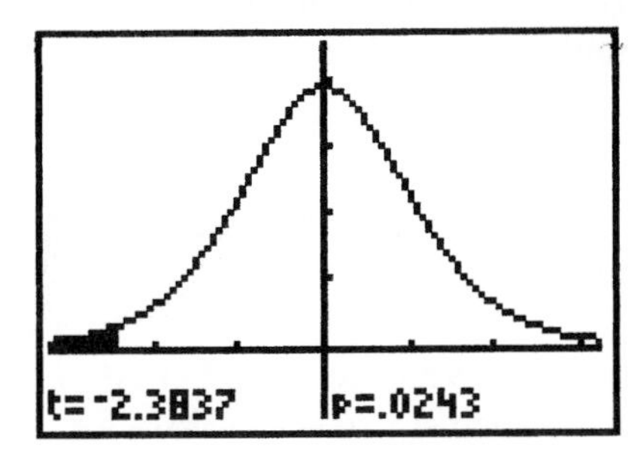

Since the P-value is less than α, the correct decision is to **Reject** H_o. We conclude that the clarity of the lake is improving.

(c.) Press STAT, highlight **TESTS** and select **8:T-Interval**. Select **Data** for **Inpt**, set **List** to **L3**, set **Freq** equal to **1** and set **C-Level** to **.99**. Highlight **Calculate** and press ENTER.

```
TInterval
 (-10.21,-.0409)
 x̄=-5.125
 Sx=6.081294505
 n=8
```

The average difference in the depth at which the disk is visible is (-.04, -10.21) feet. This means that the average depth at which the disk is visible has increased by an amount between .04 and 10.21 feet.

(d) To set up the boxplot, press **2nd [STAT PLOT]**. Press ENTER to select **Plot 1**. Highlight **On** and press ENTER. Set **Type** to the boxplot with outliers which is the first selection in the second row. Press ENTER. Set **XList** to **L3** and **Freq** to **1**. For **Marks** select the small square. Press ZOOM and select **9:ZoomStat** and ENTER.

▸ Problem 20 (pg. 585)

In this example the data is paired data consisting of daily car rental fees for two different companies at 10 locations in the United States. Enter the rental rates for *Thrifty* into **L1** and the rental rates for *Hertz* into **L2**. Next, create a set of differences, d = *Thrifty* - *Hertz*. Move the cursor to highlight the label **L3,** found at the top of the third column, and press ENTER. Press 2nd [L1] - 2nd [L2] and press ENTER. Each value in **L3** is the difference **L1 - L2**.

We are interested in testing the claim that *Thrifty* is less expensive than *Hertz*. If that is the case, then the difference, **L1 – L2,** should be *negative*. So the appropriate alternate hypothesis is: $\mu_d < 0$.

Note: A normal probability plot and boxplot of the data indicate that the differences are approximately normal with no outliers.

To run the hypothesis test, press STAT, highlight **TESTS** and select **2:T-Test**. In this example, you are using the actual data to do the analysis, so select **Data** for **Inpt** and press ENTER. The value for μ_o is **0**, the value in the null hypothesis. The set of differences is found in **L3**, so set **List** to **L3**. Set **Freq** equal to **1**. Choose < μ_o as the alternative hypothesis and highlight **Calculate** and press ENTER.

```
T-Test
 μ<0
 t=.0890092826
 p=.5344881153
 x̄=.259
 Sx=9.201623589
 n=10
```

Or, highlight **Draw** and press ENTER.

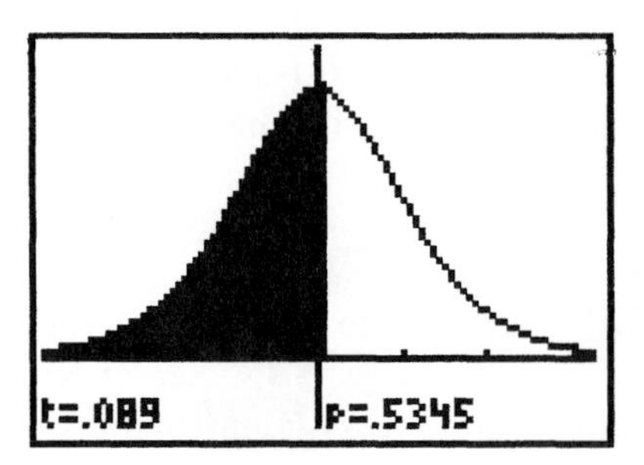

Since the P-value is greater than α, the correct decision is to **Fail to Reject** H_o.
We cannot conclude that *Thrifty* is less expensive than *Hertz*.

◀

Section 11.2

▸ Example 2 (pg. 592) Testing a Claim Regarding Two Means

To test the claim that the flight animals have a different red blood cell mass than the control animals, use a two-tailed test: H_o: $\mu_1 = \mu_2$ vs H_a: $\mu_1 \neq \mu_2$. Refer to Table 3 on pg. 590. Enter the data from the 14 flight rats into **L1** and the data from the control rats into **L2**. Because the data sets are small, the first step is to use normal probability plots and boxplots to verify that both datasets are approximately normal and contain no outliers.

To set up the normal probability plot for the **Flight** data, press **2nd** [STAT PLOT]. Press ENTER to select **Plot 1**. Highlight **On** and press ENTER. Set **Type** to the normal probability plot which is the third selection in the second row. Press ENTER. Set **Data List** to **L1** and **Data Axis** to **X**. For **Marks,** select the small square. Press ZOOM and select **9:ZoomStat** and ENTER.

To set up the normal probability plot for the **Control** data, repeat the above steps changing the **Data List** to **L2**.

To set up both boxplots on the same graph, press **2nd** [STAT PLOT] . Press ENTER to select **Plot 1**. Highlight **On** and press ENTER. Set **Type** to the boxplot with outliers which is the first selection in the second row. Press ENTER. Set **XList** to **L1** and **Freq** to **1**. For **Marks** select the small square. Use the up arrow to move to the top of the screen and highlight **Plot 2**. Highlight **On** and press ENTER. Set **Type** to the boxplot with outliers which is the first selection in the second row. Press ENTER. Set **XList** to **L2** and **Freq** to **1**. For **Marks** select the small square.

Press ZOOM and select **9:ZoomStat** and ENTER.

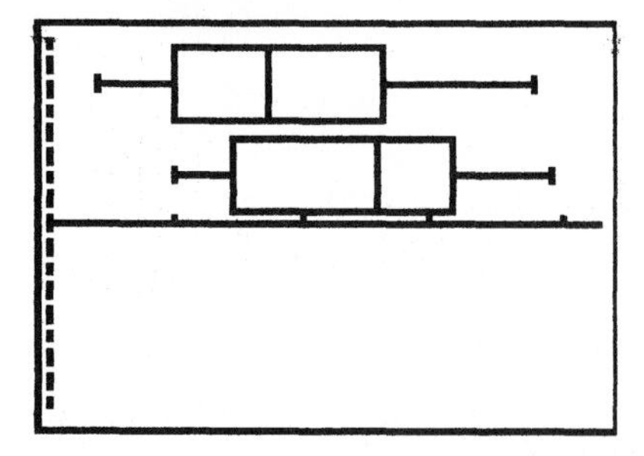

By drawing both boxplots on the same graph, you can visually compare the two datasets. Notice that the Flight data graph appears shifted slightly to the left of the Control data graph. The next step is to determine whether this shift is statistically significant.

To run the hypothesis test, press STAT, highlight **TESTS**, and select **4:2-SampTTest**. Since you are inputting the sample data, select **Data** and press ENTER. Enter **L1** for **List1** and **L2** for **List 2**. Set **Freq1** and **Freq2** to **1**. Select $\neq \mu_2$ as the alternative hypothesis and press ENTER. Scroll down to the next line. On this line, there are two options. Select **NO** because, in the procedure we are using (called Welch's approximate t-test), the variances are NOT assumed to be equal and therefore, we do not want a pooled variance. Press ENTER.

```
2-SampTTest
 Inpt:Data Stats
 List1:L1
 List2:L2
 Freq1:1
 Freq2:1
 μ1:≠μ2 <μ2 >μ2
↓Pooled:No Yes
```

Scroll down to the next line, highlight **Calculate** and press ENTER.

```
2-SampTTest
 μ1≠μ2
 t=-1.436781704
 p=.1627070766
 df=25.99635232
 x̄1=7.880714286
↓x̄2=8.43
```

The output (shown above) for **Calculate** displays the alternative hypothesis, the test statistic, the P-value, the degrees of freedom and the sample statistics. Notice the degrees of freedom = 25.996. In cases, such as this one, in which the population variances are not assumed to be equal, the calculator calculates an adjusted degrees of freedom, (see the formula on pg. 592 of your textbook.)

If you choose **Draw**, the output includes a graph with the area associated with the P-value shaded.

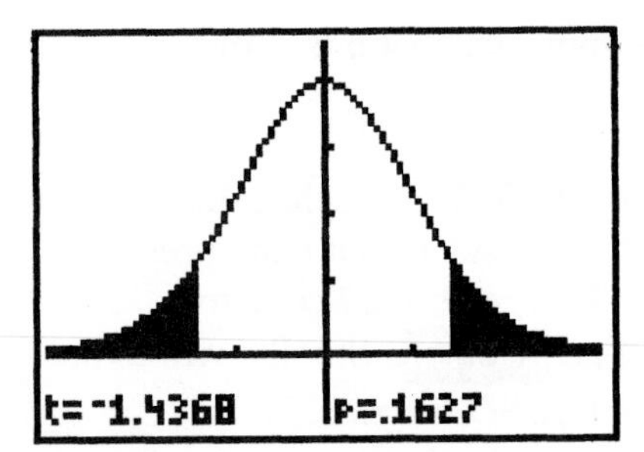

Since the P-value is greater than α, we fail to reject H_0. There is not sufficient evidence to support the claim that there is a significant difference in the red blood cell mass of the flight animals and the control animals.

▸ Example 3 (pg.594) Constructing a Confidence Interval about the Difference of Two Means

For this example, we will use the summary data from Table 4 on pg. 591. In Example 2, we tested for normality using normal probability plots and for outliers using Boxplots. Both data sets appeared to be approximately normal with no outliers.

To construct a 95% confidence interval for (μ_1-μ_2), press STAT, highlight **TESTS** and select **0:2-SampTint**. For **Inpt,** select **Stats** and press ENTER. Enter the mean of the first sample, 7.881, the standard deviation, 1.017, and the sample size, 14. Then enter the mean (8.43), standard deviation (1.005), and sample size (14) of the second sample. Scroll down to the next line and enter .95 for the **C-level**. On the next line, select **No,** because we are not using a pooled variance.

```
2-SampTInt
 Inpt:Data Stats
 x̄1:7.881
 Sx1:1.017
 n1:14
 x̄2:8.43
 Sx2:1.005
↓n2:14
```

Scroll down to the next line, **Calculate,** and press ENTER

```
2-SampTInt
 (-1.334,.23648)
 df=25.9963378
 x̄1=7.881
 x̄2=8.43
 Sx1=1.017
↓Sx2=1.005
```

A 95 % confidence interval for the difference in the population means is (-1.334, .23648). Since this interval contains 0, the correct conclusion is that there is no difference in the red blood cell mass of the two groups. (Note: This interval differs slightly from the textbook interval because the calculator uses the adjusted degrees of freedom in the calculations.)

◂

▸ Problem 9 (pg. 596)

To test the claim that the treatment group experienced a larger mean improvement than the control group, use a one-tailed test: H_o: $\mu_1 = \mu_2$ vs H_a: $\mu_1 > \mu_2$. Because the data sets are large (n_1 and $n_2 > 30$), we know that the sampling distribution of ($\bar{x}_1 - \bar{x}_2$) is approximately normal and we can safely use the Two-sample T-test.

To run the hypothesis test, press **STAT**, highlight **TESTS**, and select **4:2-SampTTest.** Since you are inputting the sample statistics, select **Stats** and press **ENTER**. Enter the sample statistics for the two samples.

```
2-SampTTest
 Inpt:Data Stats
 x̄1:14.8
 Sx1:12.5
 n1:55
 x̄2:8.1
 Sx2:12.7
↓n2:60
```

Scroll down to the next line, select $>\mu_2$ as the alternative hypothesis and press **ENTER**. Scroll down to the next line. Select **NO** because we are not using a pooled variance. Press **ENTER**.

Scroll down to the next line, highlight **Calculate** and press **ENTER**.

```
2-SampTTest
 μ1≠μ2
 t=-1.436781704
 p=.1627070766
 df=25.99635232
 x̄1=7.880714286
↓x̄2=8.43
```

The output (shown above) for **Calculate** displays the alternative hypothesis, the test statistic, the P-value, the adjusted degrees of freedom and the sample statistics.

Or, select **Draw** and press **ENTER**.

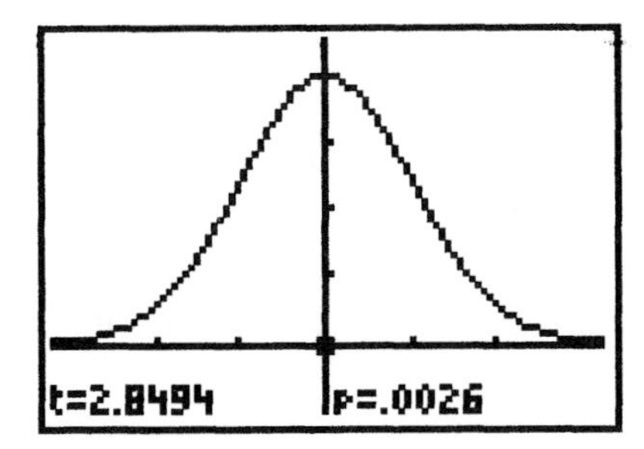

Since the P-value is less than α, we reject H_0. The data supports the claim that the treatment group has a larger mean improvement than the control group.

To construct a 95% confidence interval for $(\mu_1 - \mu_2)$, press STAT, highlight **TESTS** and select **0:2-SampTint**. For **Inpt,** select **Stats** and press ENTER. Enter the sample statistics for each group. Scroll down to the next line and enter .95 for the **C-level**. On the next line, select **No,** because we are not using a pooled variance.

Scroll down to the next line, **Calculate,** and press ENTER

```
2-SampTInt
 (2.0412,11.359)
 df=112.4181377
 x̄1=14.8
 x̄2=8.1
 Sx1=12.5
↓Sx2=12.7
```

A 95% confidence interval for the difference in the population means is (2.04, 11.36). We can interpret this in the following way: the mean improvement in the total score for the treatment group is 2.04 to 11.36 points more than the mean improvement for the control group. (Note: This interval differs slightly from the textbook interval because the calculator uses the adjusted degrees of freedom in the calculations.)

Problem 15 (pg. 598)

To test the claim that carpeted rooms contain more bacteria than uncarpeted rooms, use a one-tailed test: H_o: $\mu_1 = \mu_2$ vs H_a: $\mu_1 > \mu_2$. Enter the data from the 8 carpeted rooms into **L1** and the data from the uncarpeted rooms into **L2**. A normal probability plot and boxplot indicate that the data are normally distributed with no outliers.

To run the hypothesis test, press **STAT**, highlight **TESTS**, and select **4:2-SampTTest**. Since you are inputting the sample data, select **Data** and press **ENTER**. Enter **L1** for **List1** and **L2** for **List 2**. Set **Freq1** and **Freq2** to **1**. Select $>\mu_2$ as the alternative hypothesis and press **ENTER**. Scroll down to the next line. Select **NO** because, in the procedure we are using, we do not want a pooled variance. Press **ENTER**. Scroll down to the next line, highlight **Calculate** and press **ENTER**.

```
2-SampTTest
 μ1>μ2
 t=.9557706856
 p=.1779571293
 df=13.56321553
 x̄1=11.2
↓x̄2=9.7875
```

The output (shown above) for **Calculate** displays the alternative hypothesis, the test statistic, the P-value, the adjusted degrees of freedom and the sample statistics.

If you choose **Draw**, the output includes a graph with the area associated with the P-value shaded.

Since the P-value (.178) is greater than α, we fail to reject H_0. There is not sufficient evidence to support the claim that there is more bacteria in carpeted rooms than in uncarpeted rooms.

To construct a 95% confidence interval for (μ_1-μ_2), press **STAT**, highlight **TESTS** and select **0:2-SampTint**. For **Inpt**, select **Data** and press **ENTER**. Enter **L1** and **L2** for **List1** and **List2**. Scroll down and enter .95 for the **C-level**. On the next line, select **No**, because we are not using a pooled variance.

Scroll down to the next line, **Calculate,** and press **ENTER**

```
2-SampTInt
 (-1.767,4.5918)
 df=13.56321553
 x̄1=11.2
 x̄2=9.7875
 Sx1=2.6774188
↓Sx2=3.21000111
```

A 95% confidence interval for the difference in the population means is (-1.8, 4.8) Since this interval contains 0, the correct conclusion is that there is no difference in the amount of bacteria in carpeted rooms vs. uncarpeted rooms. (Note: This interval differs slightly from the textbook interval because the calculator uses the adjusted degrees of freedom in the calculations.)

◀

Section 11.3

▸ Example 2 (pg. 606) Testing a Claim Regarding Two Population Proportions

Refer to Example 1 on pg. 604. To test the claim that the proportion of Nasonex users who experienced headaches as a side effect is greater than the proportion in the control group who experienced headaches, the correct hypothesis test is: $H_o: p_1 = p_2$ vs $H_a: p_1 > p_2$. Designate the Nasonex users as Group 1 and the Control Group as Group 2. The sample statistics are $n_1 = 2103$, $x_1 = 547$, $n_2 = 1671$, and $x_2 = 368$.

First, verify that the requirements for the hypothesis test are satisfied. The problem states that the individuals were randomly divided into two groups so the first requirement (independent random samples) is satisfied. The second requirement is: $n\hat{p}(1-\hat{p}) \geq 10$ for each of the groups. For the first group, $\hat{p} = \frac{x}{n} = \frac{547}{2103} = 0.26$. The calculation, $n\hat{p}(1-\hat{p}) = 2103 * 0.26 * (1 - 0.26)$ is greater than 10. Repeat this calculation for the second group. Both calculations are greater than 10 so the second requirement is satisfied. The final requirement is that the sample sizes are not more than 5% of the population sizes. The population of Americans 12 years of age or older is in the millions so this requirement is easily satisfied.

Next, to run the hypothesis test, press **STAT**, highlight **TESTS** and select **6:2-PropZTest** and fill in the appropriate information. Highlight **Calculate** and press **ENTER**.

```
2-PropZTest
 p1>p2
 z=2.839330068
 p=.0022604818
 p̂1=.2601046125
 p̂2=.2202274087
↓p̂=.2424483307
```

The output displays the alternative hypothesis, the test statistic, the P-value, the sample proportions, the weighted estimate of the population proportion, $\hat{p}$, and the sample sizes.

Or, highlight **Draw** and press ENTER.

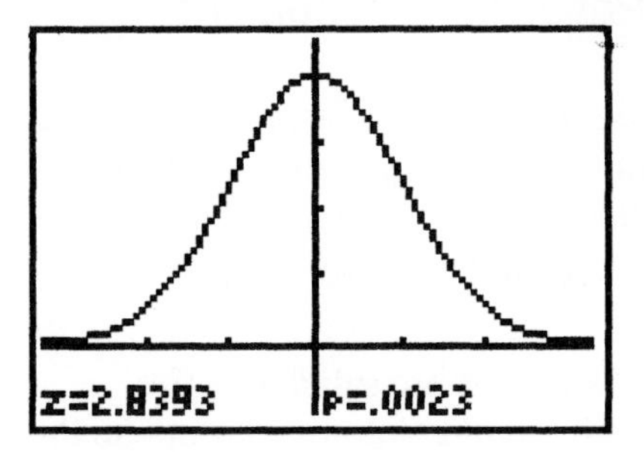

Since the P-value is less than α, the correct decision is to **Reject** H_o. There is sufficient evidence to support the claim that there is a higher incidence of headaches among the Nasonex users than among the individuals taking a placebo.

Example 3 (pg. 607) A Confidence Interval for the Difference between Two Population Proportions

Construct a confidence interval to compare the proportion of Nasonex users who experience headaches to the proportion of nonusers who experience headaches. Designate the Nasonex users as Group 1 and the Control group as Group 2. The sample statistics are $n_1 = 374$, $x_1 = 64$, $n_2 = 376$, and $x_2 = 68$.

First, verify that the requirements for the hypothesis test are satisfied. The problem states that the individuals were randomly divided into two groups so the first requirement (independent random samples) is satisfied. The second requirement is: $n\hat{p}(1-\hat{p}) \geq 10$ for each of the groups. For the first group, $\hat{p} = \frac{x}{n} = \frac{64}{374} = 0.17$. The calculation, $n\hat{p}(1-\hat{p}) = 374 * 0.17 * (1 - 0.17)$ is greater than 10. Repeat this calculation for the second group. Both calculations are greater than 10 so the second requirement is satisified. The final requirement is that the sample sizes are not more than 5% of the population sizes. The population of Americans between the ages of 3 and 11 is in the millions so this requirement is easily satisfied.

Press STAT, highlight **TESTS** and select **B:2-PropZInt** and fill in the appropriate information. Highlight **Calculate** and press ENTER.

```
2-PropZInt
 (-.0555,.03601)
 p̂1=.1711229947
 p̂2=.1808510638
 n1=374
 n2=376
```

The confidence interval (-.056, .036) contains 0. This means that there is no evidence to support the claim that the proportion of Nasonex patients complaining of headaches is different from those individuals who do not take Nasonex.

Section 11.4

Example 2 (pg. 616) Testing a Claim Regarding Two Population Standard Deviations

To test the claim that Cisco Systems is a more volatile stock than General Electric, use a one-tailed test: H_o: $\sigma_1 = \sigma_2$ vs H_a: $\sigma_1 > \sigma_2$. Enter the data from the Cisco Systems stock into **L1** and the data from the General Electric stock into **L2**.

The hypothesis test we are using requires that the data be normally distributed. Even minor deviations from normality will affect the validity of the test. (Note: In other hypothesis tests, minor deviations from normality did not seriously affect the test validity.) So, an important first step is to use normal probability plots to verify that both datasets are normally distributed. Once this has been confirmed, we can run the hypothesis test.

To run the hypothesis test, press STAT, highlight **TESTS**, and select **D:2-SampFTest**. Since you are inputting the sample data, select **Data** and press ENTER. Enter **L1** for **List1** and **L2** for **List 2**. Set **Freq1** and **Freq2** to **1**. On the next line, select $>\sigma_2$ as the alternative hypothesis. Highlight **Calculate** and press ENTER.

```
2-SampFTest
 σ1>σ2
 F=7.519296727
 p=6.9363271E-4
 Sx1=11.8356524
 Sx2=4.31622017
↓x̄1=8.333
```

The output displays the alternative hypothesis, the test statistic (F), the P-value, the sample standard deviations, the sample means and the sample sizes.

Or, highlight **Draw** and press ENTER.

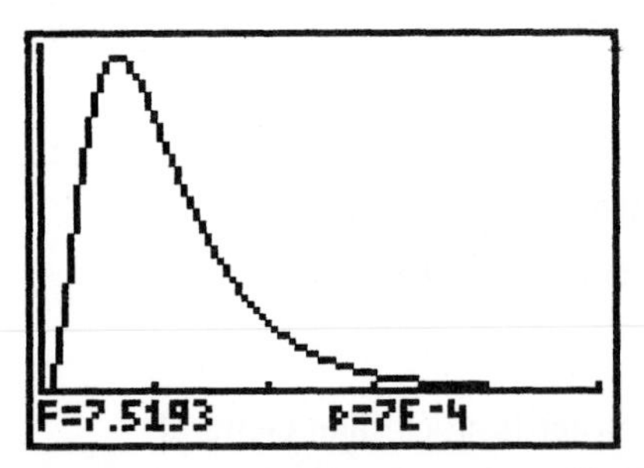

The output shows the F-distribution curve, the value of the F-statistic and the P-value. Since the P-value (.0007) is less than α, the correct decision is to **Reject** H_o. There is sufficient evidence to support the claim that Cisco Systems stock is more volatile than General Electric stock.

Example 3 (pg. 618) Testing a Claim Regarding Two Population Standard Deviations

To test the claim that the standard deviation of the red blood cell count in the flight animals is different from the standard deviation of the red blood cell count in the control animals , use a two-tailed test: H_o: $\sigma_1 = \sigma_2$ vs. H_a: $\sigma_1 \neq \sigma_2$. Enter the data from the flight rats into **L1** and the data from the control rats into **L2**. Use normal probability plots to verify that each data set is normally distributed.

To run the hypothesis test, press **STAT**, highlight **TESTS**, and select **D:2-SampFTest**. Since you are inputting the sample data, select **Data** and press **ENTER**. Enter **L1** for **List1** and **L2** for **List 2**. Set **Freq1** and **Freq2** to **1**. On the next line, select $\neq \sigma_2$ as the alternative hypothesis. Highlight **Calculate** and press **ENTER**.

```
2-SampFTest
 σ1≠σ2
 F=1.023974926
 p=.9665798398
 Sx1=1.0174513
 Sx2=1.00546966
↓x̄1=7.880714286
```

The output displays the alternative hypothesis, the test statistic (F), the P-value, the sample standard deviations, the sample means and the sample sizes.

Or, highlight **Draw** and press **ENTER**.

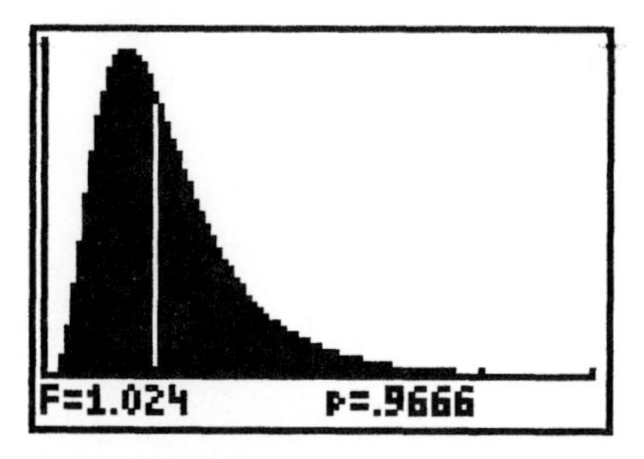

The output shows the F-distribution curve, the value of the F-statistic and the P-value. Since the P-value (.9666) is greater than α, the correct decision is to **Fail to Reject** H_o. There is not sufficient evidence to support the claim that the standard deviation of the red blood cell count for the flight rats is different from the standard deviation of the red blood cell count for the control group.

Inference on Categorical Data

Section 12.1

▸ Example 2 (pg. 634) Testing a Claim using the Goodness-of-Fit Test

In this example, we test the claim that the population distribution of the United States is different now than it was in 2000. The procedure that we use is the Chi-Square (χ^2) Goodness-of-Fit test.

The χ^2 test has 3 requirements: (1) the data are randomly selected; (2) all *expected frequencies* are greater than or equal to 1 and (3) no more than 20% of the *expected frequencies* are less than 5. For the first requirement, we *assume* that the data was randomly selected. For requirements (2) and (3), we set up a table using the data and then check to see if these requirements are satisfied.

Enter the percentages (in decimal form) given in the problem for each of the four regions into **L1**. Move the cursor so that it is flashing on '**L2**' at the top of the second column and press ENTER. The cursor will move to the bottom of the screen and will be flashing next to '**L2=**'. Type in **L1*1500**. (Note: 1500 is the sample size.) Press ENTER . **L2** will contain the *expected frequencies.*

Enter the *observed frequencies* found in Table 1 into **L3**.

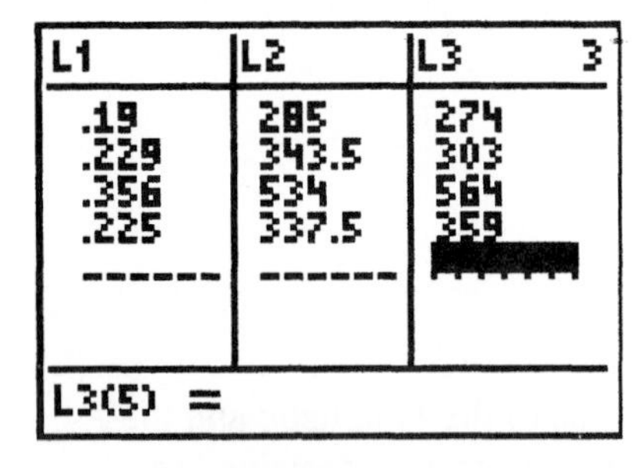

Notice the *expected frequencies* in L2. All the values are greater than 5, so the requirements of the test have been satisfied.

This test is a test of the hypotheses: H_o: The distribution of residents in the U.S. is the same today as it was in 2000, vs. H_a: The distribution of residents in the U.S. is different today than in 2000.

The test statistic is $\chi^2 = \sum \frac{(O_i - E_i)^2}{E_i}$. To calculate the χ^2-value, move the cursor so that it is flashing on '**L4**' at the top of the fourth column. (Note: To view the fourth column, use the right arrow.) Press ENTER. The cursor will move to the bottom of the screen and will be flashing next to '**L4=**'. Type in **(L3-L2)²/L2**. Press ENTER. Press 2nd **[QUIT]**. Next, press 2nd **[LIST]**. Select **Math** and **5:sum(**. Type in **L4** and press ENTER. This is the value of the χ^2 statistic.

To calculate the P-value associated with this test statistic of 8.255, we must find the area to the *right* of 8.255 in the χ^2 curve. Press 2nd **[DISTR]** and select **8:** χ^2 **cdf**. The calculation requires *a lower bound, an upper bound and the degrees of freedom.)* The lower bound is the test statistic, the upper bound is positive infinity (1E99) and *degrees of freedom* is equal to the 'number of categories minus 1'. In this example: df = 3 which is: (regions in the U.S.) – 1.

```
χ²cdf(8.255,1E99
,3)
          .0410253664
```

Since the P-value (.041) is less than α (.05), the correct decision is to **reject** the null hypothesis. There is sufficient evidence to support the claim that the distribution of residents in the U.S. is different today than it was in 2000.

◀

▸ Example 3 (pg. 635) Testing a Claim Using the Goodness Of Fit Test

In this example, we test the claim that the day (Sunday, Monday, Tuesday, etc.) on which a child is born occurs with equal frequency. This is a test of the hypotheses: H_o: $p_1 = p_2 = p_3 = p_4 = p_5 = p_6 = p_7 = \frac{1}{7}$ vs. H_a: At least one of the proportions is different than the others.

Enter the fraction $\frac{1}{7}$ (by pressing 1 ÷ 7) into **L1** seven times (for the seven days of the week). Move the cursor so that it is flashing on '**L2**' at the top of the second column and press ENTER. The cursor will move to the bottom of the screen and will be flashing next to '**L2**='. Type in **L1*500**. (Note: 500 is the sample size.) Press ENTER. **L2** will contain the *expected frequencies.*

Enter the *observed frequencies* into **L3**.

L1	L2	L3 3
.14286	71.429	78
.14286	71.429	74
.14286	71.429	76
.14286	71.429	71
.14286	71.429	81
.14286	71.429	63
------	------	
L3(8) =		

Notice the *expected frequencies* in L2. All the values are greater than 5, so the requirements of the test have been satisfied.

The test statistic is $\chi^2 = \sum \frac{(O_i - E_i)^2}{E_i}$. To calculate the χ^2-value, move the cursor so that it is flashing on '**L4**' at the top of the fourth column. (Note: To view the fourth column, use the right arrow.) Press ENTER. The cursor will move to the bottom of the screen and will be flashing next to '**L4**='. Type in **(L3-L2)²/L2**. Press ENTER. Press 2nd **[QUIT]**. Next, press 2nd **[LIST]**. Select **Math** and **5:sum(**. Type in **L4** and press ENTER. This is the value of the χ^2 statistic.

To calculate the P-value associated with this test statistic of 6.184, we must find the area to the *right* of 6.184 in the χ^2 curve. Press 2nd **[DISTR]** and select **8: χ^2cdf.** The calculation requires *a lower bound, an upper bound and the degrees of freedom.)* The lower bound is the test statistic, the upper bound is positive infinity (1E99) and *degrees of freedom* is equal to the 'number of categories minus 1'. In this example: df = (days of the week) –1 = 6.

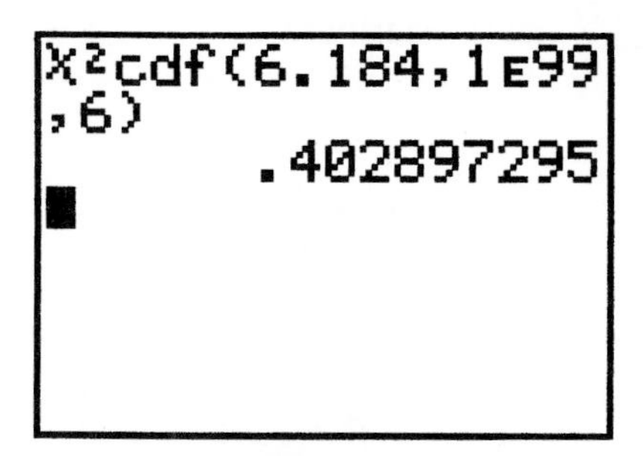

Since the P-value (.403) is greater than α (.01), the correct decision is to **fail to reject** the null hypothesis. There is not sufficient evidence to reject the claim that the day of the week on which a child is born occurs with equal frequency.

◀

▸ Problem 21 (pg.641)

(a.) In this problem, we will use the TI-84 to generate 500 random integers numbered 1 through 5. The first step is to set the *seed* by selecting a 'starting number' and storing this number in **rand**. Suppose, for this example, that we select the number '22' as the starting number. Type **22** into your calculator and press the STO key. Next press the MATH key and select **PRB** and select **rand,** which stands for 'random number'. Press ENTER and the starting value of '22' will be stored into **rand** and will be used as the *seed* for generating random numbers.

Now you are ready to generate the random integers. Press MATH and select **PRB**. Select **5:RandInt(.** This function requires three values: the starting value, the ending value and the number of values you want to generate. For this example, you want to generate 500 values from the integers ranging from 1 to 5. The command is **randInt(1,5,500)**. Press STO **L1** and press ENTER . The values will be stored into **L1**.

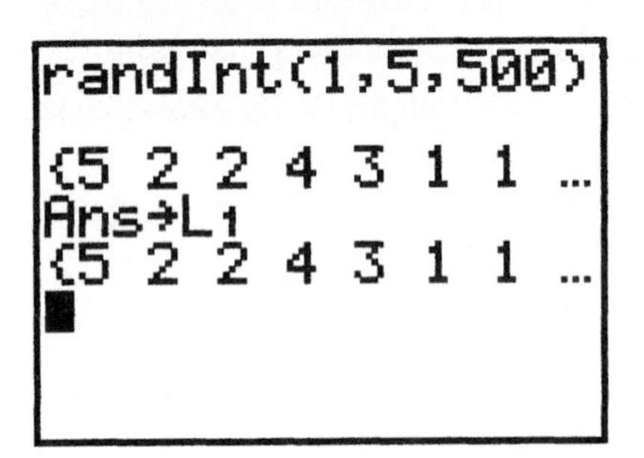

Next, press 2nd **[LIST]**. Select **OPS** and **1:SortA(** . Type in **L1** and press ENTER. This will put all the values in **L1** in numerical order. Press 2nd **[QUIT]**. Press STAT , select **EDIT.** Move the cursor to the first entry in L1 and scroll through the list using the down arrow. Hold the 'down arrow' button down and scroll through the 1's. Record the number of 1's that you have on a piece of paper.

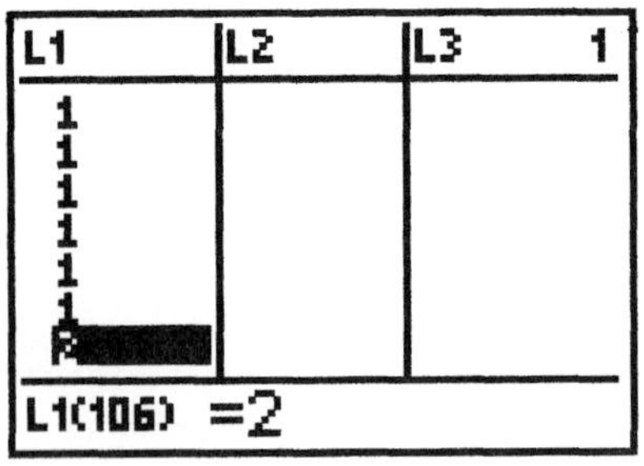

```
L1      L2      L3      1
1
1
1
1
1
1
2
L1(106) =2
```

In this illustration, there are 105 1's. (Note: If you started with a different seed, you would get a different result).

Continue scrolling through the column. Stop at the final '2' in the list.

L1	L2	L3 1
2		
2		
2		
2		
2		
2		
2		
2		
3		

L1(205) =3

The number of 2's in this illustration is 204 – 105 = 99. Record this number on your paper.

Continue this process until you have recorded the counts for all the data values in L1.

(b.) Each of the numbers, 1 through 5, should occur with equal frequency. So, the proportion of 1's, 2's, 3's, 4's and 5's should equal .20.

(c.) In this example, we are testing the claim that the random number generator is generating random numbers between 1 and 5. This is a test of the hypotheses: H_o: $p_1 = p_2 = p_3 = p_4 = p_5 = .2$, vs. H_a: At least one of the proportions is different than the others.

Press **STAT**, select **EDIT. L1** contains the 500 randomly selected integers. Enter the values 1 through 5 into **L2** and their corresponding frequencies into **L3**. L3 contains the *observed frequencies*

L1	L2	L3 3
1	1	105
1	2	98
1	3	105
1	4	88
1	5	103
1	------	
1		

L3(6) =

Move to **L4** and enter the expected proportion, .20, five times. Move the cursor so that it is flashing on 'L5' at the top of the fifth column and press **ENTER**. The cursor will move to the bottom of the screen and will be flashing next to '**L5=**'. Type in **L4*500.** (Note: 500 is the sample size.) Press **ENTER**. **L5** will contain the *expected frequencies.* All the *expected frequencies* in L5 are greater than 5, so the requirements of the test have been satisfied.

The test statistic is $\chi^2 = \sum \frac{(O_i - E_i)^2}{E_i}$. To calculate the χ^2-value, move the cursor so that it is flashing on '**L6**' at the top of the sixth column. Press **ENTER**. The cursor will move to the bottom of the screen and be flashing next to '**L6=**'. Type in **(L3-L5)²/L5.** Press **ENTER**. Press 2nd **[QUIT].** Next, press 2nd **[LIST].** Select **Math** and **5:sum(**. Type in **L6** and press **ENTER**. This is the value of the χ^2 statistic.

To calculate the P-value associated with this test statistic of 2.07 (in this illustration), we must find the area to the *right* of 2.07 in the χ^2 curve. Press 2nd **[DISTR]** and select **8:** χ^2 **cdf.** The calculation requires *a lower bound, an upper bound and the degrees of freedom.* The lower bound is the test statistic, the upper bound is positive infinity (1E99) and *degrees of freedom* is equal to the 'number of categories minus 1'. In this example: df = 4 which is (the number of different possible outcomes of the random variable) – 1.

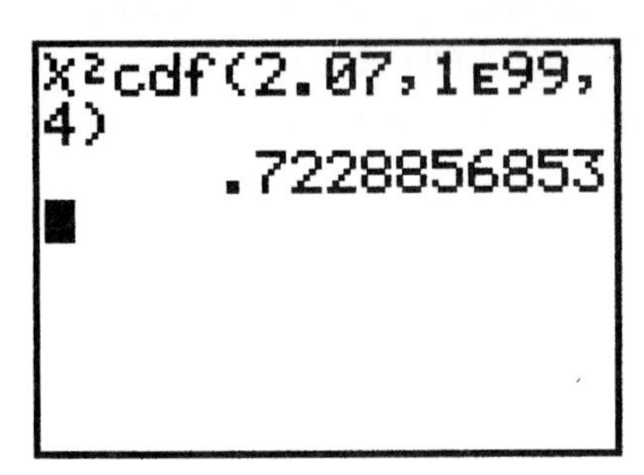

Since the P-value (.723) is greater than α (.01), the correct decision is to **fail to reject** the null hypothesis. There is not sufficient evidence to reject the claim that the random number generator is generating random numbers between 1 and 5. In other words, we can conclude that the random number generator is working correctly.

◀

Section 12.3

▸ Example 4 (pg. 658) Performing a Chi-Square Independence Test

Use the data in Table 14 on pg. 655 to test the claim that blood type and Rh level are independent. The correct procedure is the Chi-Square (χ^2) Independence test.

The hypotheses are: H_o: Blood type and Rh status are independent (or not related), vs. H_a: Blood type and Rh status are dependent (or somehow related).

The χ^2 test has 3 requirements: (1) the data are randomly selected; (2) all *expected frequencies* are greater than or equal to 1 and (3) no more than 20% of the *expected frequencies* are less than 5. For the first requirement, we *assume* that the data was randomly selected. We will verify requirements (2) and (3) at the end of our analysis.

The first step is to enter the data in the table into **Matrix A**. On the TI-84 Plus and TI-83 Plus, press **2nd [MATRIX]**. **MATRIX** is found above the $\mathbf{x^{-1}}$ key. (On the TI-83, press **MATRX**). Highlight **EDIT** and press **ENTER**.

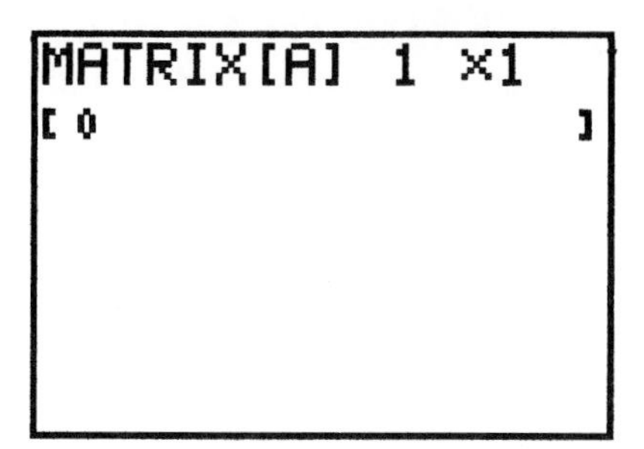

On the top row of the display, enter the size of the matrix. The matrix has 2 rows and 4 columns, so press **2**, press the right arrow key, and press **4**. Press **ENTER**. Enter the first value, **176**, and press **ENTER**. Enter the second value, **28**, and press **ENTER**. Continue this process and fill the matrix.

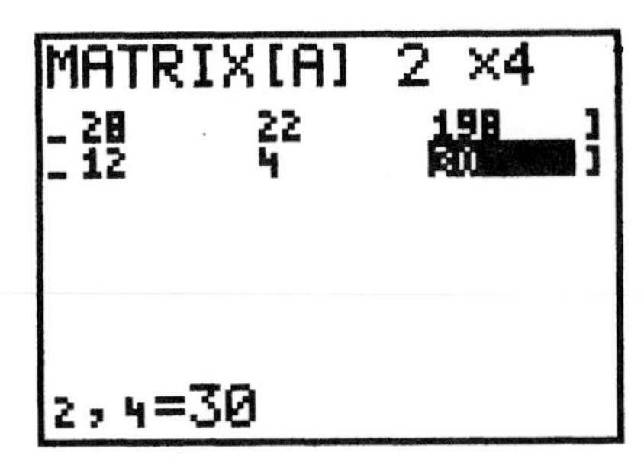

Press **2nd [Quit]** . To perform the test of independence, press **STAT**, highlight **TESTS**, and select **C: χ^2-Test** and press **ENTER**.

```
χ²-Test
 Observed:[A]
 Expected:[B]
 Calculate Draw
```

For **Observed**, **[A]** should be selected. If **[A]** is not already selected, press **2nd [MATRX]**, highlight **NAMES**, select **1:[A]** and press **ENTER**. For, **Expected**, **[B]** should be selected. Move the cursor to the next line and select **Calculate** and press **ENTER**.

```
χ²-Test
 χ²=7.600585552
 p=.0550295317
 df=3
```

The output displays the test statistic and the P-value. Since the P-value is greater than α, the correct decision is to **Fail to Reject** the null hypothesis. This means that blood type is *independent* of Rh status.

Or (after clearing all graphs and Y-registers), you could highlight **Draw** and press **ENTER**.

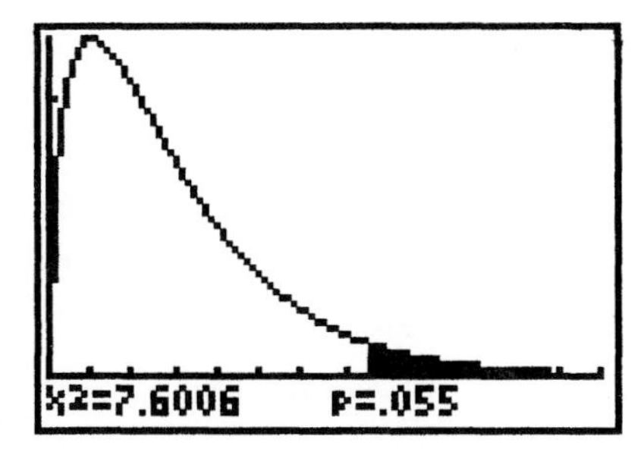

This output displays the χ^2 –**curve** with the area associated with the P-value shaded in. The test statistic and the P-value are also displayed.

The final step in this procedure is to confirm that the test requirements have been satisfied. The two requirements that we need to verify are: (1) all *expected frequencies* are greater than or equal to 1 and (2) no more than 20% of the *expected frequencies* are less than 5. Both requirements involve the *expected frequencies* which are stored in **Matrix B**. To view **Matrix B**, press **2nd [MATRX]** highlight **NAMES**, select **2:[B]** and press ENTER ENTER.

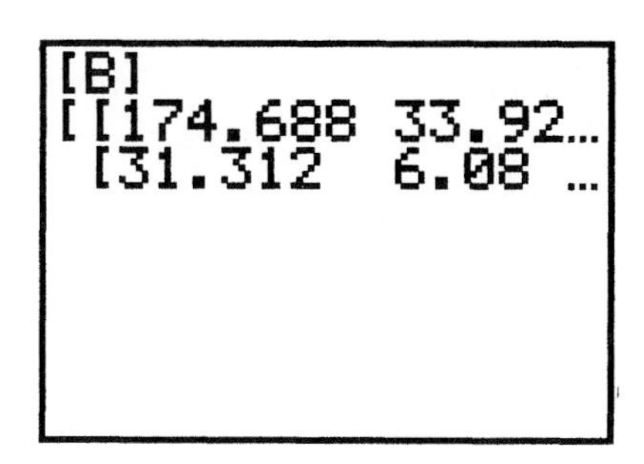

Scroll through the **8** entries in **Matrix B** and confirm that all entries are greater than 5. This confirms that the test requirements have been satisfied.

◀

▸ Example 5 (pg. 659) A Test of Homogeneity of Proportions

In this example, we are testing the claim that the proportions of subjects who experience abdominal pain are equal among all three groups (those taking Zocor, those taking a placebo and those taking Cholestyramine). The correct procedure is the Chi-Square (χ^2) Independence test. The hypotheses are: H_o: $p_1 = p_2 = p_3$ vs. H_a: At least one of the proportions is different than the others.

The χ^2 test has 3 requirements: (1) the data are randomly selected; (2) all *expected frequencies* are greater than or equal to 1 and (3) no more than 20% of the *expected frequencies* are less than 5. For the first requirement, we *assume* that the data was randomly selected. We will verify requirements (2) and (3) at the end of our analysis.

The first step is to enter the data in the table into **Matrix A**. On the TI-84 Plus and TI-83 Plus, press **2nd [MATRIX]**. **MATRIX** is found above the $\mathbf{x^{-1}}$ key. (On the TI-83, press **MATRX**). Highlight **EDIT** and press **ENTER**.

On the top row of the display, enter the size of the matrix. The matrix has 2 rows and 3 columns, so press **2**, press the right arrow key, and press **3**. Press **ENTER**. Enter the first value, **51**, and press **ENTER**. Enter the second value, **5**, and press **ENTER**. Continue this process and fill the matrix.

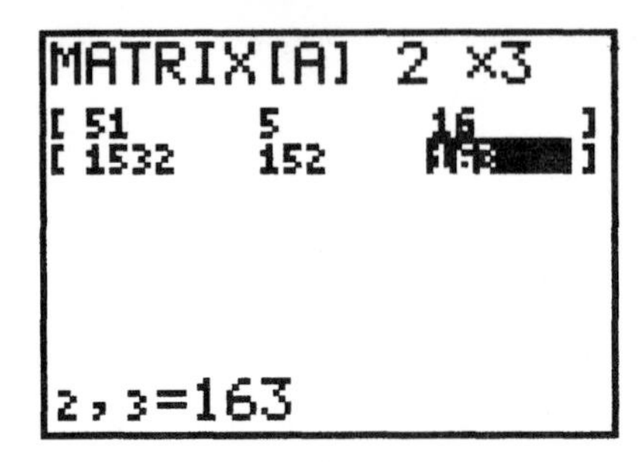

Press **2nd [Quit]**. To perform the test of independence, press **STAT**, highlight **TESTS**, and select **C: χ^2-Test** and press **ENTER**.

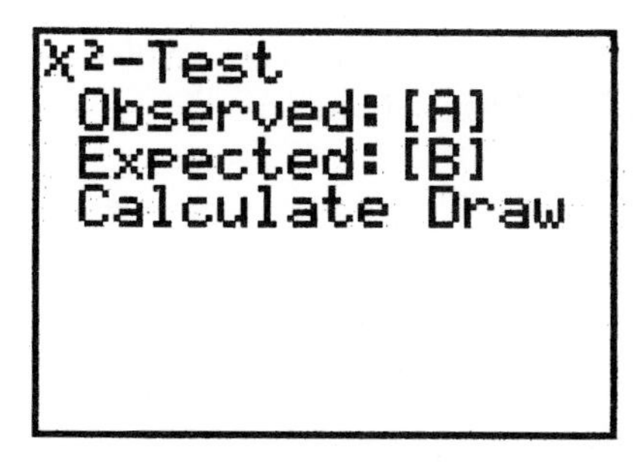

For Observed, [A] should be selected. If **[A]** is not already selected, press 2nd **[MATRX]**, highlight **NAMES**, select **1:[A]** and press **ENTER**. For, **Expected**, **[B]** should be selected. Move the cursor to the next line and select **Calculate** and press **ENTER**.

```
χ²-Test
 χ²=14.70651321
 p=6.4050309E-4
 df=2
```

The output displays the test statistic and the P-value. Since the P-value is less than α, the correct decision is to **Reject** the null hypothesis. We conclude that at least one of the three groups experiences abdominal pain at a rate different from the other two groups.

The final step in this procedure is to confirm that the test requirements have been satisfied. The two requirements that we need to verify are: (1) all *expected frequencies* are greater than or equal to 1 and (2) no more than 20% of the *expected frequencies* are less than 5. Both requirements involve the *expected frequencies* which are stored in **Matrix B**. To view **Matrix B, press 2nd [MATRX]** highlight **NAMES**, select **2:[B]** and press **ENTER** **ENTER**.

```
 df=2

[B]
[[59.39343408 5...
 [1523.606566 1...
```

Scroll through the 6 entries in **Matrix B** and confirm that all entries are greater than 5. This confirms that the test requirements have been satisfied.

◀

▸ Problem 9 (pg. 663)

In this example, we are testing the claim that education level is independent of region of the United States. The correct procedure is the Chi-Square (χ^2) Independence test. The hypotheses are: H_o : *education level* is independent of *region of the United States.* vs. H_a: *education level* is not independent of *region of the United States.*

Enter the data in the table into **Matrix A**.. Highlight **EDIT** and press ENTER. On the top row of the display, enter the size of the matrix. The matrix has 4 rows and 4 columns. Press ENTER. Fill the matrix with the data values.

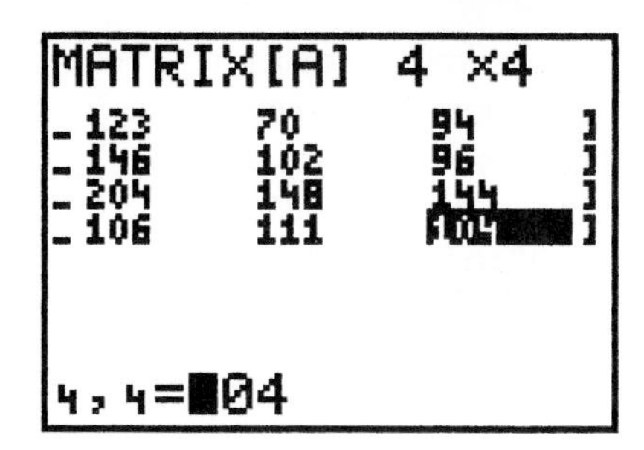

Press **2nd** **[Quit]**. To perform the test of independence, press STAT, highlight **TESTS**, and select **C: χ^2-Test** and press ENTER.

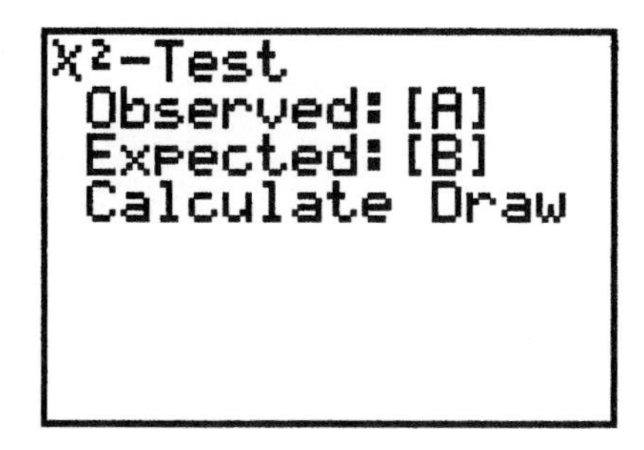

For **Observed, [A]** should be selected. If **[A]** is not already selected, press **2nd** **[MATRX]**, highlight **NAMES**, select **1:[A]** and press ENTER. For, **Expected**, **[B]** should be selected. Move the cursor to the next line and select **Calculate** and press ENTER.

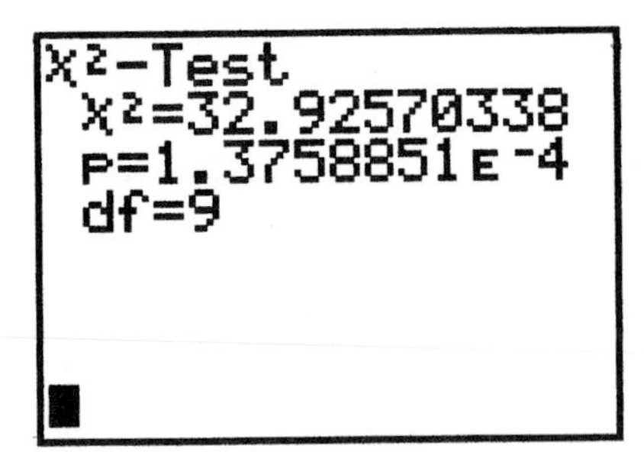

The output displays the test statistic and the P-value. Since the P-value is less than α, the correct decision is to **Reject** the null hypothesis. We conclude that *education level* and *region of the country* are **not** independent.

The final step in this procedure is to confirm that the test requirements have been satisfied. The two requirements that we need to verify are: (1) all *expected frequencies* are greater than or equal to 1, and (2) no more than 20% of the *expected frequencies* are less than 5. Both requirements involve the *expected frequencies* which are stored in **Matrix B**. To view **Matrix B,** press **2nd** **[MATRX]** highlight **NAMES**, select **2:[B]** and press **ENTER** **ENTER**.

```
[B]
[[66.89800443 1...
 [92.15742794 1...
 [121.3636364 1...
 [75.58093126 1...
```

Scroll through the 16 entries in **Matrix B** (as you scroll through the entries, record these entries (rounded to the nearest whole number) on paper to use in part (b). of these problem) and confirm that all entries are greater than 5. This confirms that the test requirements have been satisfied.

(b). To determine which cell contributed most to the test statistic, we will store the two matrices into Lists.

Press **STAT**, highlight **EDIT.** Enter the *observed values* given in the Table into L1 (enter the 1st column, followed by the 2nd column, etc.) Enter the *expected values* from Matrix B (enter the 1st column, followed by the 2nd column, etc.) into L2. To calculate each cell's contribution to the χ^2-value, move the cursor so that it is flashing on '**L3**' at the top of the third column. Press **ENTER** . The cursor will move to the bottom of the screen and will be flashing next to '**L3=**'. Type in **(L1-L2)^2/L2**. Press **ENTER**. Scroll through the values in L3 and find the largest value, 3.9. This is the entry for 'West' region and 'Some College.' The observed value for this cell is '111' and the expected value is '93'. This tells us that, for this cell, **more** residents had 'some college' than is expected.

CHAPTER 13

Comparing Three Or More Means

Section 13.1

Example 1 (pg. 679) Testing the Requirements of a One-Way ANOVA

The four requirements for the One-way ANOVA test are listed in your textbook on pg. 678. The first two requirements state that the data must be obtained using simple random sampling techniques and that all samples must be independent. We will assume that these requirements have been satisfied. The third requirement states that the populations must be normally distributed. This requirement can be validated through normal probability plots of each of the samples. The final requirement, that population variances are equal, can be tested using the sample standard deviations. The criteria that we will use is the following: the largest standard deviation must be no more than two times larger than the smallest standard deviation.

Enter the sample data from Table 1 on pg. 679 into **L1, L2, L3** and **L4**. We will check for normality using a normal probability plot. To set up the normal probability plot, press **2nd** [**STAT PLOT**] . Press ENTER to select **Plot 1**. Highlight **On** and press ENTER. Set **Type** to the normal probability plot which is the third selection in the second row. Press ENTER. Set **Data List** to **L1** and **Data Axis** to **X**. Next, there are three different types of **Marks** that you can select for the graph. The first choice, a small square, is the best one to use.

Press ZOOM and select **9:ZoomStat** and ENTER.

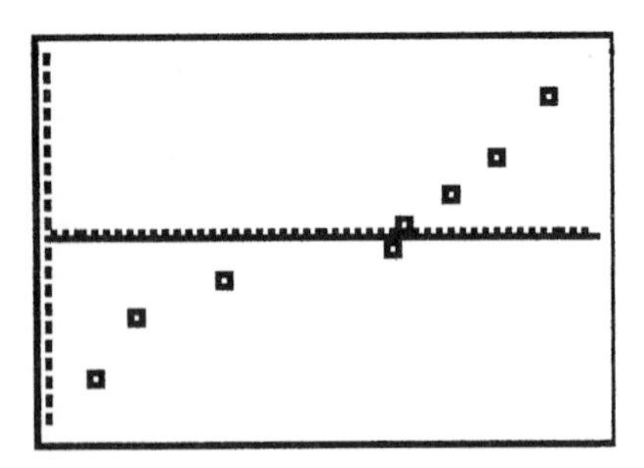

This plot is *fairly* linear, indicating that the data generally follow a normal distribution.

Repeat this process for the sample data in **L2, L3** and **L4**. All four plots are fairly normal so the requirement of normality has been satisfied.

The next requirement involves the sample standard deviations. To obtain the standard deviation of the data in **L1**, press STAT, select **CALC** and **1-Var Stats** and type in 2nd [L1. Press ENTER. The standard deviation for the data in **L1** will appear on the screen (sx = 15.026). Repeat this process to obtain the standard deviation of the data in **L2** (13.490), **L3** (17.060) and **L4** (21.180). Calculate the ratio of the largest standard deviation to the smallest standard deviation: 21.180/13.490 = 1.570. Since this ratio is less than 2, the requirement of equal population variances is satisfied.

◀

Example 2 (pg. 680) One-Way ANOVA Test

This example is a continuation of Example1.

The researcher wishes to determine if there is a difference in the mean glucose among the four treatment groups. The test of hypothesis is:
$H_o : \mu_1 = \mu_2 = \mu_3 = \mu_4$ vs. H_a :at least one mean is different from the others.

In Example 1, we showed that the requirements of the One-Way ANOVA test have been satisfied. We also entered the data into **L1, L2**, and **L3**.

To run the hypothesis test, press STAT, highlight **TESTS** and select **F:ANOVA(** and type in **2nd [L1]** , **2nd [L2]** , **2nd [L3]** , **2nd [L4]**.

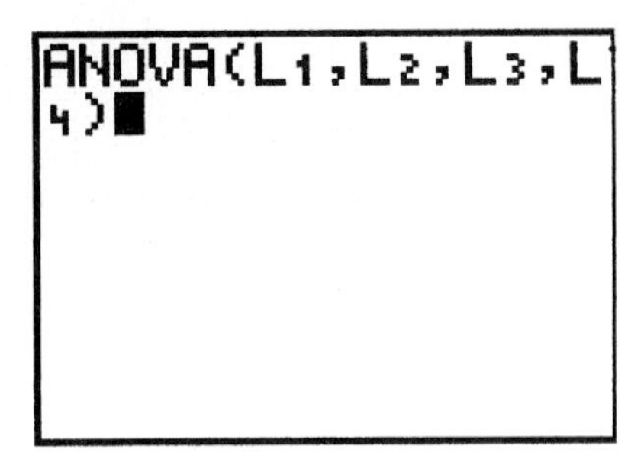

Press ENTER and the results will be displayed on the screen.

```
One-way ANOVA
 F=58.20909526
 p=3.743987E-12
 Factor
  df=3
  SS=50090.6909
↓ MS=16696.897
```

```
One-way ANOVA
↑ MS=16696.897
 Error
  df=28
  SS=8031.61625
  MS=286.843437
 Sxp=16.9364529
```

The output displays the test statistic, F = 58.21, and the P-value, p = 3.744E-12 (written in scientific notation). Since the P-value is less than α, the correct decision is to **Reject** the null hypothesis. There is sufficient evidence to support the claim that at least one of the population means for glucose levels among the four treatment groups is different.

The output also displays several other pieces of information. This information can be used to set up an Analysis of Variance Table (similar to the tables given as output in Minitab and Excel).

Note: The TI-84 can only do three boxplots on the same graph so it is not possible to get a graphical representation of the four data sets together.

Inferences on the Least Squares Regression Model

CHAPTER 14

Section 14.1

▸ Example 1 (pg. 737) Least-Squares Regression

Press **STAT**, highlight **1:Edit** and clear **L1** and **L2**. For each of the fourteen patients, enter the age into **L1** and the total cholesterol into **L2**. Press **2nd** **[STAT PLOT]** , select **1:Plot1**, turn **ON** Plot 1 and press **ENTER**. For **Type** of graph, select the **scatter plot** which is the first selection. Press **ENTER**. Enter **L1** for **Xlist** and **L2** for **Ylist**. Highlight the first selection, the small square, for the type of **Mark**. Press **ENTER**. Press **ZOOM** and **9** to select **ZoomStat**.

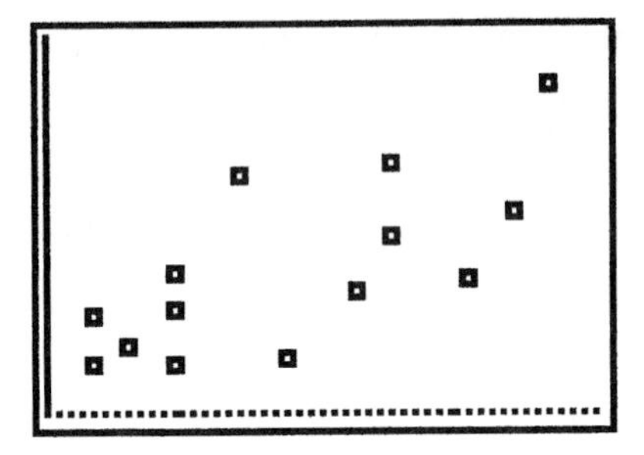

This graph shows a positive linear correlation, with quite a bit of scatter.

In order to calculate r, the correlation coefficient, and, r^2, the coefficient of determination, you must turn **On** the **Diagnostic** command. Press **2nd** **[CATALOG]** (Note: **CATALOG** is found above the **0** key). The CATALOG of functions will appear on the screen. Use the down arrow to scroll to the **DiagnosticOn** command.

Press ENTER ENTER.

To calculate the correlation coefficient, the coefficient of determination and the regression equation, press STAT, highlight **CALC**, select **4:LinReg(ax+b).** Press VARS, highlight **Y-VARS**, select **1:Function** by pressing ENTER, select **1:Y1** by pressing ENTER and press ENTER. This stores the regression equation in **Y1.** (Note: This command gives you the option of specifying which lists contain the X-values and Y-values. If you do not specify these lists, the defaults are used. The defaults are: **L1** for the X-values and **L2** for the Y-values.)

```
LinReg
 y=ax+b
 a=1.399064152
 b=151.3536582
 r²=.5152520915
 r=.7178106237
```

The correlation coefficient is r = .718. This suggests a positive linear correlation between X and Y, but not a very strong one. The coefficient of determination is .515. This tells us that 51.5% of the variation in cholesterol levels can be explained by the predictor variable, age. The regression equation is $\hat{y} = 151.35 + 1.399x$.

To see a scatterplot of the data along with the regression equation, press GRAPH

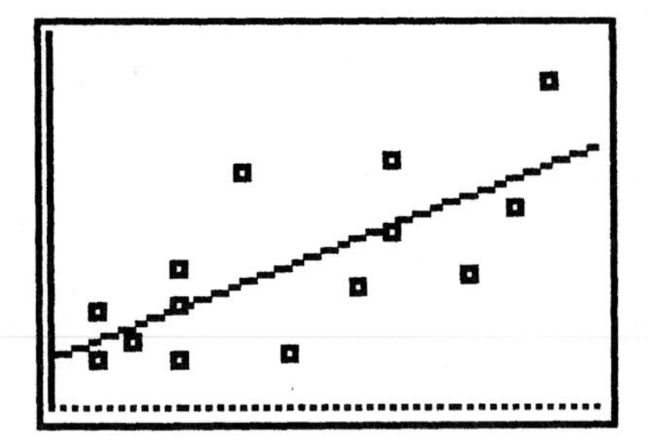

▸ Example 3 (pg. 741) Computing the Standard Error

This example is a continuation of the previous one. (Note: If you have not already done so, enter the data from Table 1 on pg.737 into **L1** and **L2**. Press STAT, highlight **CALC**, select **4:LinReg(ax+b)**. Press VARS, highlight **Y-VARS**, select **1:Function** by pressing ENTER, select **1:Y1** by pressing ENTER and press ENTER. (This stores the regression equation in **Y1.)**

Press STAT, highlight **TESTS** and select **E:LinRegTTest**. Enter **L1** for **Xlist**, **L2** for **Ylist**, and **1** for **Freq**. On the next line, β and ρ, select $\neq 0$ and press ENTER. Leave the next line, RegEQ, blank. Highlight **Calculate**.

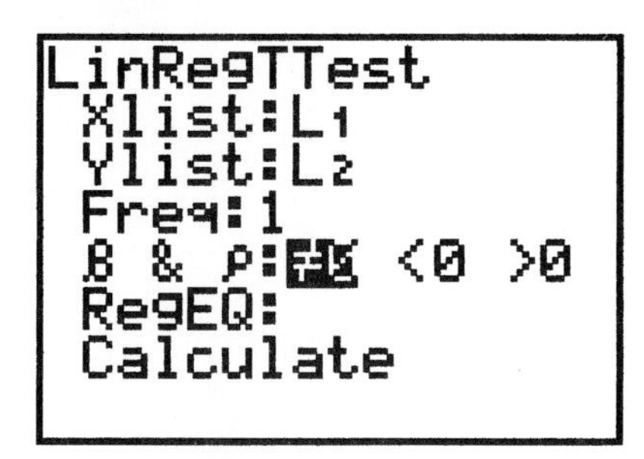

Press ENTER.

```
LinRegTTest
 y=a+bx
 β>0 and ρ>0
 t=3.57143321
 p=.0019211307
 df=12
↓a=151.3536582
```

The output displays several pieces of information describing the relationship between X and Y. What you are interested in for this example is the standard error. Scroll down to the next page of output and you will see s = 19.48. This is the standard error.

◀

▸ Example 4 (pg. 742) Verifying that the Residuals are Normally Distributed

This example is a continuation of the previous ones. (Note: If you have not already done so, enter the data from Table 1 on pg.737 into **L1** and **L2**. Press STAT, highlight **CALC**, select **4:LinReg(ax+b)**. Press VARS, highlight **Y-VARS**, select **1:Function** by pressing ENTER, select **1:Y1** by pressing ENTER and press ENTER. (This stores the regression equation in **Y1.**)

The values for $(y_i - \hat{y}_i)$, called Residuals, are automatically stored to a list called **RESID**. Press **2nd [LIST]**, select **7:RESID**. Press STO, **2nd [L3]** and ENTER. This stores the residuals to **L3**.

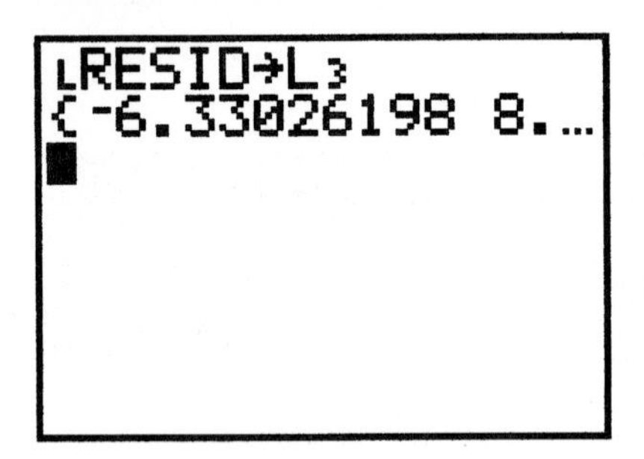

To set up the normal probability plot, press **2nd [STAT PLOT]**. Press ENTER to select **Plot 1**. Highlight **On** and press ENTER. Set **Type** to the normal probability plot which is the third selection in the second row. Press ENTER. Set **Data List** to **L3** and **Data Axis** to **X**. For **Marks** select the small square.

Press ZOOM and select **9:ZoomStat** and ENTER.

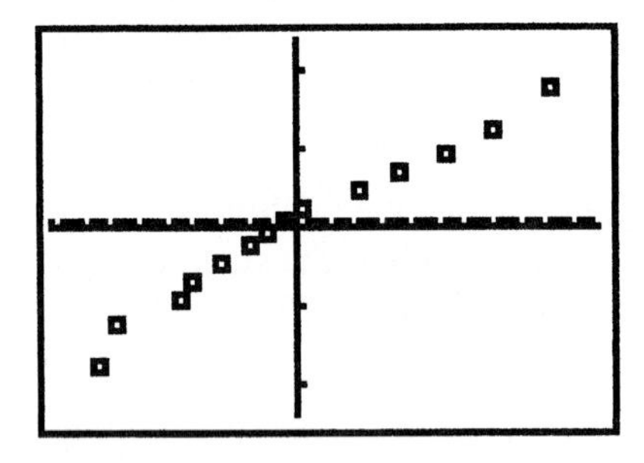

This plot is *fairly* linear, indicating that the Residuals follow a normal distribution.

Example 6 (pg. 745) Testing for a Linear Relation

This example is a continuation of the previous examples. (Note: If you have not already done so, enter the data from Table 1 on pg.737 into **L1** and **L2**. Press STAT, highlight **CALC**, select **4:LinReg(ax+b)**. Press VARS, highlight **Y-VARS**, select **1:Function** by pressing ENTER, select **1:Y1** by pressing ENTER and press ENTER. This stores the regression equation in **Y1.)**

To test the claim that there is a linear relationship between *age* and *total cholesterol,* the appropriate hypothesis test is: $\beta_1 = 0$ vs. $\beta_1 \neq 0$.

The first step is to verify that the assumptions required to perform the test are satisfied. The first assumption is that the sample has been obtained using random sampling. This has been confirmed and is stated in Example 1. The next assumption is that the residuals are normally distributed and this has been confirmed by the normal probability plot in Example 4. The last assumption is that the residuals have constant error variance. This assumption can be validated using a graph of the Residuals vs. the predictor variable, Age. If the Residuals do, in fact, have constant error variance, then the Residuals will appear as a scatter of points about the horizontal line at 0.

To construct the graph of the Residuals vs. Age, press **2nd [STAT PLOT]**. Press ENTER to select **Plot 1**. Highlight **On** and press ENTER. Set **Type** to the scatter plot which is the first selection in the first row. Press ENTER. Set **Xlist** to **L1** and **YList** to **L3**. For **Marks,** select the small square.

Press WINDOW and use the data to set the Window. First, look at the *Age* variable in Table 1 on pg. 739. Notice the minimum and maximum values (25 and 65). Choose **Xmin** and **Xmax** to encompass these values. For example, choose **Xmin** = 20 and **Xmax** = 70. To set the Y-values, you must look through the Residuals that are stored in **L3**. Press **2nd [QUIT]**. Press STAT, highlight **EDIT** and scroll through the values in **L3**. The minimum value is –24.5 and the maximum value is 34.482. To make the graph symmetric about the X-axis, use **Ymin** = -35 and **Ymax** = 35. Press WINDOW and enter these values for **Ymin** and **Ymax**. Press GRAPH

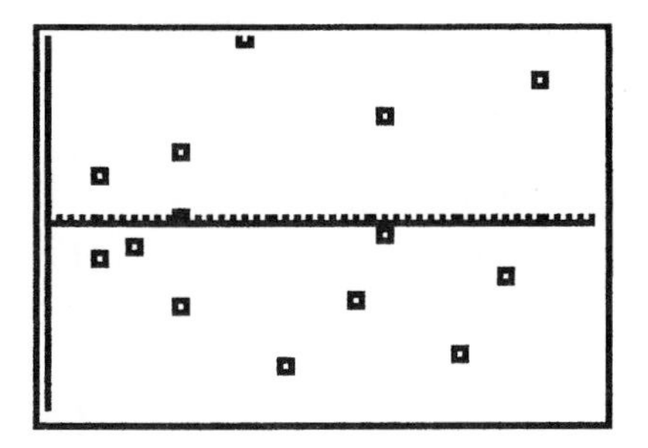

The errors are evenly spread about the horizontal line at 0, so the assumption of constant error variance is satisfied.

To run the test, press STAT, highlight **TESTS** and select **E:LinRegTTest**. Enter **L1** for **Xlist**, **L2** for **Ylist**, and **1** for **Freq**. On the next line, β and ρ, select $\neq 0$ and press ENTER. Leave the next line, RegEQ, blank. Highlight **Calculate**.

```
LinRegTTest
 Xlist:L1
 Ylist:L2
 Freq:1
 β & ρ:≠0 <0 >0
 RegEQ:
 Calculate
```

Press ENTER.

```
LinRegTTest
 y=a+bx
 β≠0 and ρ≠0
 t=3.57143321
 p=.0038422614
 df=12
↓a=151.3536582
```

```
LinRegTTest
 y=a+bx
 β≠0 and ρ≠0
↑b=1.399064152
 s=19.48053511
 r²=.5152520915
 r=.7178106237
```

The output displays several pieces of information describing the relationship between X and Y. What you are interested in for this example is the P-value (p = .0038). Since the P-value is less than α, the correct decision is to **Reject** the null hypothesis. This indicates that there is a linear relationship between X and Y.

◀

▸ Example 7 (pg. 747) Constructing a Confidence Interval about the Slope of the True Regression Line

This example is a continuation of the previous examples. (Note: If you have not already done so, enter the data from Table 1 on pg.737 into **L1** and **L2**. Press STAT, highlight **CALC**, select **4:LinReg(ax+b)**. Press VARS, highlight **Y-VARS**, select **1:Function** by pressing ENTER, select **1:Y1** by pressing ENTER and press ENTER. This stores the regression equation in **Y1.**)

The 95% confidence interval for β_1, the slope of the true regression line is given by the following formula:

$$b_1 \pm t_{\frac{\alpha}{2}} \cdot \frac{s_e}{\sqrt{\sum (x_i - \bar{x})^2}}.$$

From the previous examples in this Section, we have already obtained the following values: b_1 and s_e. b_1 is the coefficient of x in the regression equation and s_e is the standard error. The value for $t_{\frac{\alpha}{2}}$ can be found using the TI-84.

(Note: The TI-83 does not have this option.) To find the t-value on the TI-84, you will need a value for $\frac{\alpha}{2}$ and also, a value for the degrees of freedom. Since we are constructing a 95% confidence interval for β_1, α = 5% or .05. So, $\frac{\alpha}{2}$ = .025. The degrees of freedom value = (n – 2) = (14 – 2) = 12. To find the t-value, press **2nd DISTR** and select **4:invT**. This command requires two values, $\frac{\alpha}{2}$ and degrees of freedom. Type in **.025** , **12**) and press ENTER.

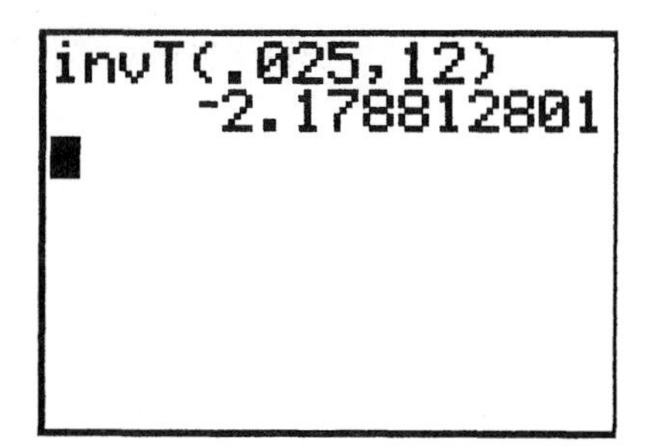

The t-value for the lower bound of the confidence interval is – 2.179. The corresponding t-value for the upper bound of the confidence interval is 2.179.

The only part of the formula that we now need to calculate is the denominator and we can use an equivalent form of this expression that can be easily evaluated on the calculator. The denominator, $\sqrt{\sum(x_i - \bar{x})^2}$, can be written equivalently as: $\sqrt{(\sum x_i^2 - \frac{(\sum x_i)^2}{n}}$. The values for $\sum x_i and \sum x_i^2$ are stored in your calculator as part of the regression procedure. To access them, press VARS and select **5:Statistics.** On the Statistics Menu, select Σ. Then select the first entry, Σx by pressing ENTER ENTER. The result, 589, will appear on the screen. To obtain $\sum x_i^2$, press VARS and select **5:Statistics.** On the Statistics Menu, select Σ. Then select the second entry, Σx^2, by pressing ENTER ENTER. The result is 27253. Enter these values into the equation:

$$\sqrt{(27253 - \frac{(589)^2}{14}} = 49.73.$$

Now fill the values into the formula for the confidence interval.
The lower bound of the 95% confidence interval is:

$$1.399 - 2.179.\frac{19.48}{49.73} = 0.545.$$

The upper bound is:

$$1.399 + 2.179.\frac{19.48}{49.73} = 2.253.$$

Problem 16 (pg. 749)

In this exercise, we will go directly to the **LinRegTTest**. This procedure combines all the steps for analyzing the model. It has more output for analyzing the regression model then the command **LinReg(ax+b).**

To begin the analysis, enter the data into L1 and L2. Next, press **STAT**, highlight **TESTS** and select **E:LinRegTTest**. Enter **L1** for **Xlist**, **L2** for **Ylist**, and **1** for **Freq**. On the next line, β and ρ, select $\neq 0$ and press **ENTER**. Move the cursor to the next line, RegEQ. On this line, tell the calculator where to store the regression equation. We will store it in **Y1.** Press **VARS**, highlight **Y-VARS**, select **1:Function** by pressing **ENTER**, select **1:Y1** by pressing **ENTER**. Highlight **Calculate**.

```
LinRegTTest
 Xlist:L1
 Ylist:L2
 Freq:1
 β & ρ:≠0 <0 >0
 RegEQ:Y1
 Calculate
```

Press **ENTER**.

```
LinRegTTest
 y=a+bx
 β≠0 and ρ≠0
 t=12.50191637
 p=7.6259241E-8
 df=11
↓a=.2088364861
```

```
LinRegTTest
 y=a+bx
 β≠0 and ρ≠0
↑b=.057527857
 s=.1121198008
 r²=.9342490303
 r=.9665655851
```

(a.) First, notice the regression equation: y= a+bx. The unbiased estimator of β_0 is 'a' which is equal to 0.2088. The unbiased estimator of β_1 is 'b' which is equal to 0.0575.
(b.) The standard error is 's' which is equal to 0.1121.
(c.) The values for $(y_i - \hat{y}_i)$, called Residuals, are automatically stored to a list called **RESID**. Press **2nd** **[LIST]**, select **7:RESID**. Press STO, **2nd** **[L3]** and ENTER. This stores the residuals to **L3**.

To set up the normal probability plot, you must first 'deselect' Y1. Press Y= and move the cursor so that it highlights = in the Y1 equation and press ENTER. This will deselect Y1 so that it will not appear on the graph with the Residual plot. Next, press **2nd** **[STAT PLOT]**, select **1:Plot1**, turn **ON** Plot 1 and press ENTER. Set **Type** to the normal probability plot which is the third selection in the second row. Press ENTER. Set **Data List** to **L3** and **Data Axis** to **X**. For **Marks** select the small square.

Press ZOOM and select **9:ZoomStat** and ENTER.

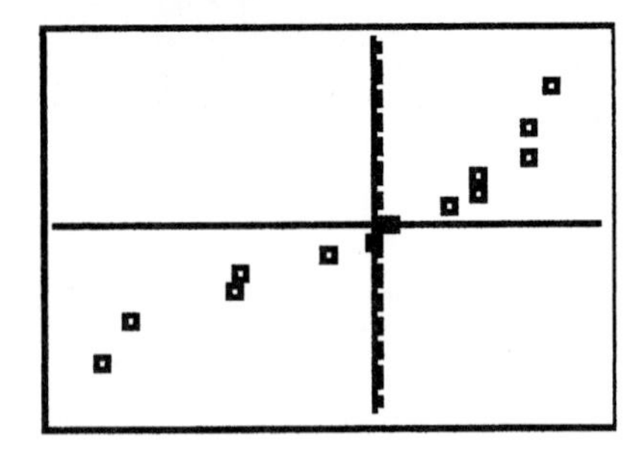

This plot is *fairly* linear, indicating that the Residuals follow a normal distribution.

(d.) To calculate s_{b_1}, we will use the formula $\frac{s_e}{\sqrt{\sum (x_i - \bar{x})^2}}$. The denominator, $\sqrt{\sum (x_i - \bar{x})^2}$, can be written equivalently as:

$\sqrt{(\sum x_i^2 - \frac{(\sum x_i)^2}{n}}$. The values for $\sum x_i and \sum x_i^2$ are stored in your calculator as part of the regression procedure. To access them, press VARS and select **5:Statistics.** On the Statistics Menu, select Σ. Then select the first entry, Σx by pressing ENTER and press ENTER The result, 184, will appear on the screen. To obtain $\sum x_i^2$, press VARS and select **5:Statistics.** On the Statistics Menu, select Σ. Then select the second entry, Σx^2, by pressing ENTER ENTER. The result is 3198. Enter these values into the equation:

$$\frac{.1121}{\sqrt{3198-\frac{(184)^2}{13}}}=.0046$$

(e.) To test the claim that there is a linear relationship between the predictor variable, 'Tar' and the response variable, 'Nicotine', the appropriate hypothesis test is: $\beta_1 = 0$ vs. $\beta_1 \neq 0$. This is the test that we set up in the **LinRegTTest.** Notice that a p-value is displayed in the output. Since the p-value of 7.626E-8 (written in scientific notation) is less than the α-value of .10, the correct decision is to **reject H_0.** The data supports the claim that there is a linear relationship between the variables.

(f.) The 90% confidence interval for β_1 is: $b_1 \pm t_{\frac{\alpha}{2}} s_{b_1}$. To calculate the confidence interval, replace all the variables with their appropriate values (b_1 = .0575, $s_{b_1} = .0046$). To find the t-value on the TI-84, you will need a value for $\frac{\alpha}{2}$ and also, a value for the degrees of freedom. Since we are constructing a 90% confidence interval for β_1, α = 10% or .10. So, $\frac{\alpha}{2} = .05$. The degrees of freedom value = (n – 2) = (13 – 2) = 11. To find the t-value, press **2nd DISTR** and select **4:invT**. This command requires two values, $\frac{\alpha}{2}$ and degrees of freedom. Type in **.05 , 11)** and press **ENTER** to obtain the t-value. The confidence interval for β_1 is: .0575 ± 1.796(.0046).

(g.) To obtain the mean amount of nicotine in a cigarette that has 12 mg of tar, press **VARS**, highlight **Y-VARS**, select **1:Function** by pressing **ENTER**, select **1:Y1** by pressing **ENTER** Type in (12) and press **ENTER** **Y1(12) = 0.899.**

Section 14.2

Example 3 (pg. 756) Constructing a Confidence Interval and a Prediction Interval

This example is a continuation of the Examples in Section 14.1

To construct a confidence interval for the mean y-value at a specific x-value, we must obtain the value for $\hat{y}$ from the regression equation and then calculate the margin of error. The confidence interval is then created using the formula $\hat{y} \pm$ margin of error. The formula that we will use for the margin of error is equivalent to the one in the textbook. The formula we will use is:

$$E = t_{\frac{\alpha}{2}} s_e \sqrt{\left(\frac{1}{n} + \frac{n(x^* - \bar{x})^2}{n(\sum x_i^2) - (\sum x_i)^2}\right.}$$

In the following steps we will find all the values needed to construct the confidence interval.

First, start with the regression equation that we obtained in Example 1 is: $\hat{y} = 151.35 + 1.399x$. Using this equation we calculate $\hat{y}$ for x = 42 and obtain the value, 210.1.

Next, we will find the t-value on the TI-84. We will need a value for $\frac{\alpha}{2}$ and also, a value for the degrees of freedom. Since we are constructing a 95% confidence interval, α = 5% or .05. So, $\frac{\alpha}{2}$ = .025. The degrees of freedom value = (n – 2) = (14 – 2) = 12. To find the t-value, press **2nd DISTR** and select **4:invT**. This command requires two values, $\frac{\alpha}{2}$ and degrees of freedom. Type in **.025** [,] **12** [)] and press ENTER.

Next, we need the standard error, s_e, which we obtained in Example 3 in Sec. 14.1. The value is 19.48.

Next, we need to calculate $\bar{x}$, $\sum x^2$, and $(\sum x)^2$. Press VARS, select **5:Statistics.** Highlight **2:** $\bar{x}$ and press ENTER ENTER. Notice that $\bar{x} = 42.07$.

Press VARS again, select **5:Statistics,** highlight $\sum$, select **1:** $\sum x$ and press ENTER ENTER. So, $\sum x = 589$. Press VARS again, select **5:Statistics,** highlight $\sum$, select **2:** $\sum x^2$, and press ENTER ENTER. Notice that $\sum x^2 =$ 27253.

Now, calculate the margin of error, E, when $x^* = 42$ using the formula for E shown above.

```
2.179*19.48*√(1/
14+(14(42-42.07)
²/(14((27253)-(5
89)^2))
        11.34441534
```

The confidence interval for the mean y-value at x = 42 is $\hat{y} \pm 11.34$. Since $\hat{y}$ =210.1, we calculate 210.1± 11.34 to obtain the lower bound of 198.76 and the upper bound of 221.44. These are the lower and upper bounds of the 95% confidence interval for the mean cholesterol level of all 42-year-old females.

To construct a prediction interval for the y-value at a specific x-value, we will follow the same type of procedure as we did for the confidence interval: $\hat{y} \pm$ margin of error. The only change is in the formula for the margin of error. The formula that we will use for the margin of error is equivalent to the one in the textbook. The formula we will use is:

$$E = t_{\frac{\alpha}{2}} s_e \sqrt{(1 + \frac{1}{n} + \frac{n(x^* - \bar{x})^2}{n(\sum x_i^2) - (\sum x_i)^2}}$$

Calculate the margin of error, E, when $x^* = 42$ using the formula for E shown above.

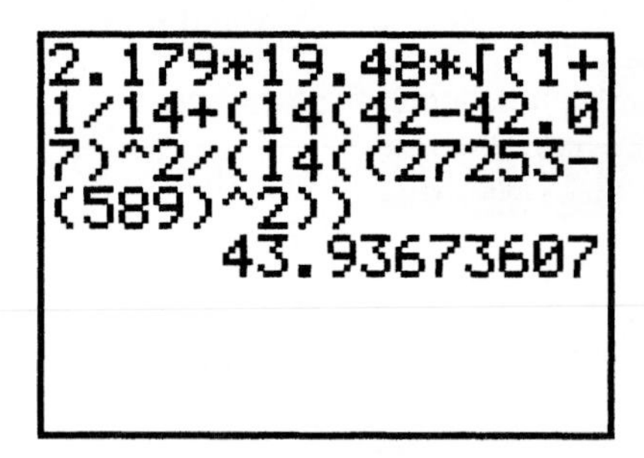

The prediction interval for the y-value at x = 42 is $\hat{y} \pm 43.94$. Since $\hat{y} = 210.1$, we calculate 210.1± 43.94 to obtain the lower bound of 166.16 and the upper bound of 254.04. These are the lower and upper bounds of the 95% prediction interval for the cholesterol level of a randomly selected 42-year-old female.

◄

Excel Manual

Beverly Dretzke

Statistics

SECOND EDITION

Informed Decisions Using Data

Michael Sullivan, III

▸ Contents:

Getting Started with Microsoft Excel

Overview

This manual is intended as a companion to Sullivan's *Statistics: Informed Decisions Using Data, 2nd edition*. The manual presents instructions on how to use Microsoft Excel to carry out selected examples and exercises from *Statistics: Informed Decisions Using Data*.

The first section of the manual contains an introduction to Microsoft Excel and how to perform basic operations such as entering data, using formulas, saving worksheets, retrieving worksheets, and printing. The screens pictured in this manual were obtained using the Office 2003 version of Microsoft Excel on a PC. You may notice slight differences if you are using a different version or a different computer.

Getting Started with the User Interface

GS 1.1 The Mouse

The mouse is a pointer device that allows you to move around the Excel worksheet and to select specific locations and objects. There are four main mouse operations: Select, click, double-click, and right-click.

1. To **select** generally means to move the mouse pointer so that the white arrow is pointing at or is positioned directly over an object. You will often **select** commands in the standard toolbar located near the top of the screen. Some of the more familiar of these commands are open, save, and print.

2. To **click** means to press down on the left button of the mouse. You will frequently select cells of the worksheet and commands by "clicking" the left button.

3. To **double-click** means to press the left mouse button twice in rapid succession.

4. To **right-click** means to press down on the right button of the mouse. A right-click is often used to display special shortcut menus.

GS 1.2 The Excel Worksheet

The figure shown below presents a blank Excel worksheet. Important parts of the worksheet are labeled.

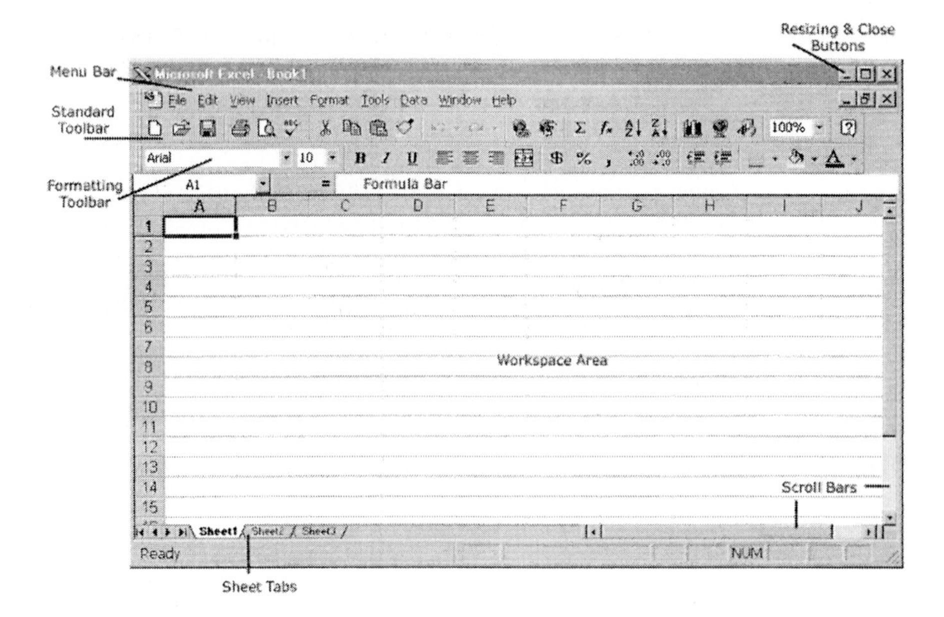

GS 1.3 Menu Conventions

Excel uses standard conventions for all menus. For example, the Menu bar contains the commands File, Edit, View, etc. Selecting one of these commands will "drop down" a menu. The Edit menu is displayed at the top of the next page.

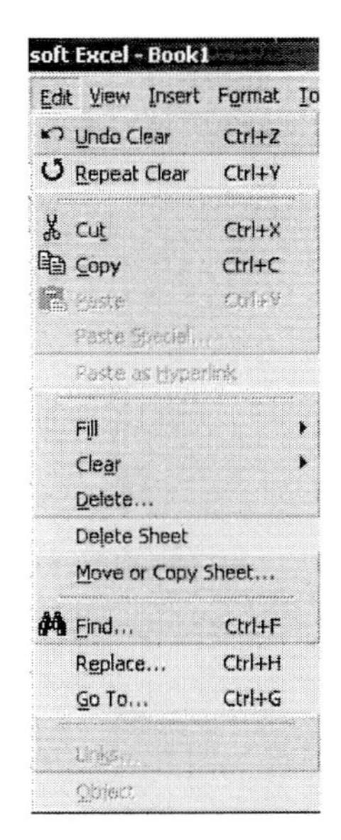

Icons to the left of the Cut, Copy, Paste, and Find commands indicate toolbar buttons that are equivalent to the menu choices.

Keyboard shortcuts are displayed to the right of the commands. For example, Ctrl+X is a keyboard shortcut for Cut.

The triangular markers to the right of Fill and Clear indicate that selection of these commands will result in a second menu of choices.

Selection of commands that are followed by an ellipsis (e.g., Paste Special… and Delete…) will result in the display of a dialog box that usually must be responded to in some way in order for the command to be executed.

The menus found in other locations of the Excel worksheet will operate in the same way.

GS 1.4 Dialog Boxes

Many of the statistical analysis procedures that are presented in this manual are associated with commands that are followed by dialog boxes. Dialog boxes usually require that you select from alternatives that are presented or that you enter your choices.

For example, if you click **Insert** and select **Function**, a dialog box like the one shown at the top of the next page will appear. You may search for a function by entering a brief description and then clicking Go, or you may select a specific category (e.g., Math & Trig, Statistical) and/or a specific function (e.g., STEVP, CHIDIST). You make your selections by clicking on them.

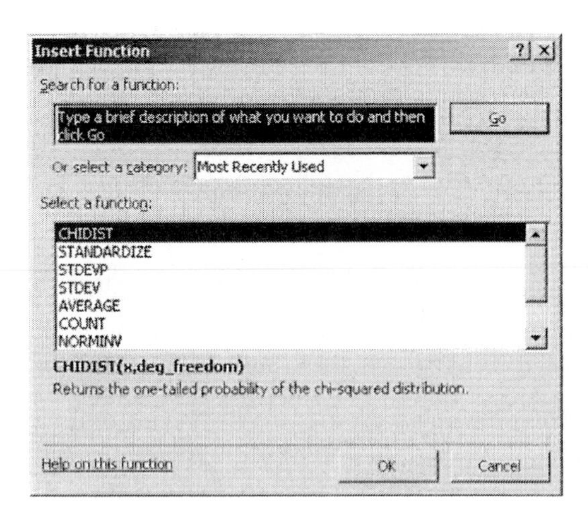

When you click the OK button at the bottom of the dialog box, another dialog box will often be displayed that asks you to provide information regarding location of the data in the Excel worksheet

Getting Started with Opening Files

GS 2.1 Opening a New Workbook

When you start Excel, the screen will open to **Sheet 1** of **Book 1**. Sheet names appear on tabs at the bottom of the screen. The name "Book 1" will appear in the top left corner.

If you are already working in Excel and have finished the analyses for one problem and would like to open a new book for another problem, follow these steps: First, at the top of the screen, click **File** and select **New**. Next, click **OK** in the New dialog box. If you were previously working in Book 1, the new worksheet will be given the default name Book 2.

The names of books opened during an Excel work session will be displayed at the bottom of the Window menu. To return to one of these books, click **Window** and then click the book name.

GS 2.2 Opening a File That Has Already Been Created

To open a file that you or someone else has already created, click on **File** and select **Open**. A list of file locations will appear. Select the location by clicking on it. Many of the data files that are presented in your statistics textbook are available a CD-ROM that accompanies this manual. To open any of these files, you will select the appropriate CD drive on your computer. This drive will be designed with a symbol that looks like .

After you select the CD drive, a list of folders and files available on the CD will appear. You will need to select the folder or file you want by clicking on it. If you have selected a folder, another screen will appear with a list of files contained in the folder. Click on the name of the file that you would like to open.

Getting Started with Entering Information

GS 3.1 Cell Addresses

Columns of the worksheet are identified by letters of the alphabet and rows are identified by numbers. The cell address A1 refers to the cell located in column A row 1. The dark outline around a cell means that it is "active" and is ready to receive information. In the figure shown below, cell C1 is ready to receive information. You can also see C1 in the **Name Box** to the left of the **Formula Bar**. You can move to different cells of the worksheet by using the mouse pointer and clicking on a cell. You can also press [**Tab**] to move to the right or left, or you can use the arrow keys on the keyboard.

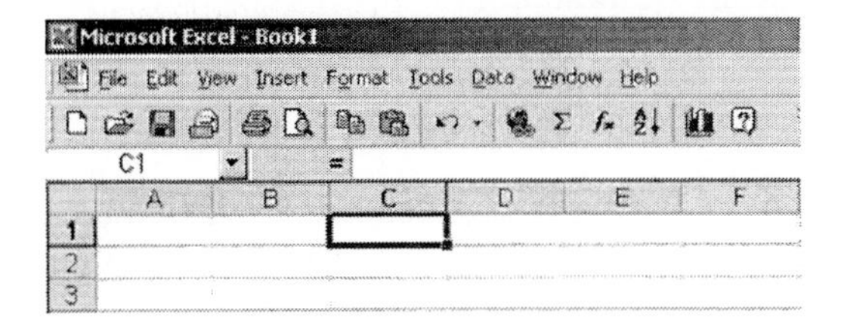

You can also activate a **range** of cells. To activate a range of cells, first click in the top cell and drag down and across (or click in the bottom cell and drag up and across). The range of cells highlighted in the figure below is designated B2:D6.

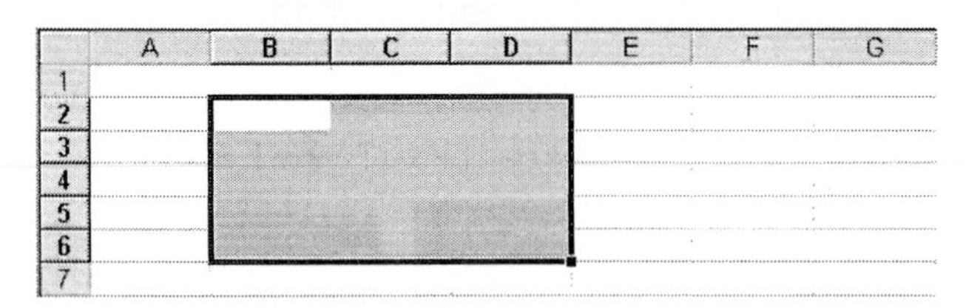

GS 3.2 Types of Information

Three types of information may be entered into an Excel worksheet.

1. **Text.** The term "text" refers to alphabetic characters or a combination of alphabetic characters and numbers, sometimes called "alphanumeric." The figure provides an example of an entry comprised solely of alphabetic characters (cell A1) and an entry comprised of a combination of alphabetic characters and numbers (cell B1).

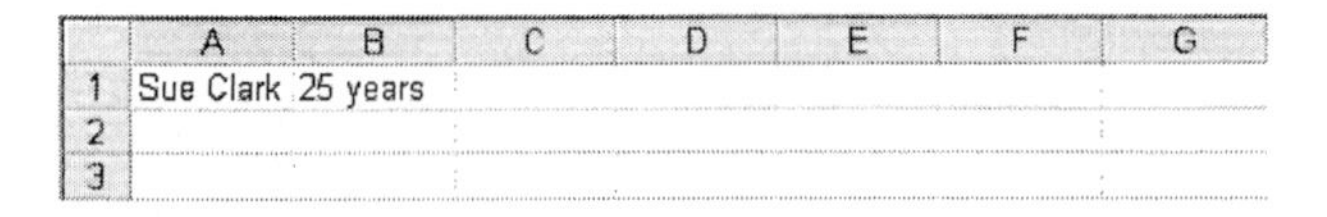

2. **Numeric.** Any cell entry comprised completely of numbers falls into the "numeric" category.

3. **Formulas.** Formulas are a convenient way to perform mathematical operations on numbers already entered into the worksheet. Specific instructions are provided in this manual for problems that require the use of formulas.

GS 3.3 Entering Information

To enter information into a cell of the worksheet, first activate the cell. Then key in the desired information and press [**Enter**]. Pressing the [Enter] key moves you down to the next cell in that column. The information shown at the top of the next page was entered as follows:

1. Click in cell **A1**. Key in **1**. Press **[Enter]**.

2. Key in **2**. Press **[Enter]**.

3. Key in **3**. Press **[Enter]**.

	A	B	C	D	E	F	G
1	1						
2	2						
3	3						
4							
5							

GS 3.4 Using Formulas

When you want to enter a formula, begin the cell entry with an equal sign (=). The arithmetic operators are displayed below.

Arithmetic operator	Meaning	Example
+	Addition	=3+2
-	Subtraction	=3-2
*	Multiplication	=3*2
/	Division	=3/2
^	Exponentiation	=3^2

Numbers, cell addresses, and functions can be used in formulas. For example, to sum the contents of cells A1 and B1, you can use the formula =A1+B1. To divide this sum by 2, you can use the formula =(A1+B1)/2. Note that Excel carries out expressions in parentheses first and then uses the results to complete the calculations of the formula. Formulas will sometimes not produce the desired results because parentheses were necessary but were not used.

Getting Started with Changing Information

GS 4.1 Editing Information in the Cells

There are several ways that you can edit information that has already been entered into a cell.

- If you have not completed the entry, you can simply backspace and start over. Clicking on the red X to the left of the Formula Bar will also delete an incomplete cell entry.

- If you have already completed the entry and another cell is activated, you can click on the cell you want to edit and then press either [**Delete**] or [**Backspace**] to clear the contents of the cell.

- If you want to edit part of the information in a cell instead of deleting all of it, follow the instructions provided in the example.

1. Let's say that you wanted to enter 1234 in cell A1 but instead entered 124. Return to cell **A1** to make it the active cell by either clicking on it with the mouse or by using the arrow keys.

2. You will see A1 in the Name Box and 124 in the Formula Bar. Click between 2 and 4 in the Formula Bar so that the **I-beam** is positioned there.

3. Enter the number 3 and press [**Enter**].

= 1234
A B C D E F G
1 1234

GS 4.2 Copying Information

To copy the information in one cell to another cell, follow these steps:

1. First click on the source cell. Then, at the top of the screen, click **Edit** and select **Copy**.

2. Click on the target cell where you want the information to be placed. Then, at the top of the screen, click **Edit** and select **Paste**.

To copy a range of cells to another location in the worksheet, follow these steps:

1. First click and drag over the range of cells that you want to copy so that they are highlighted. Then, at the top of the screen, click **Edit** and select **Copy**.

2. Click in the topmost cell of the target location. Then, at the top of the screen, click **Edit** and select **Paste**.

To copy the contents of one cell to a range of cells follow these steps:

1. Let's say that you have entered a formula in cell C1 that adds the contents of cells A1 and B1 and you would like to copy this formula to cells C2 and C3 so that C2 will contain the sum of A2 and B2 and cell C3 will contain the sum of A3 and B3.

2. First click in cell **C1** to make it the active cell. You will see =A1+B1 in the Formula Bar.

C1 = =A1+B1

	A	B	C	D	E	F	G
1	1	1	2				
2	2	2					
3	3	3					

3. At the top of the screen, click **Edit** and select **Copy**.

4. Highlight cells C2 and C3 by clicking and dragging over them.

	A	B	C	D	E	F	G
1	1	1	2				
2	2	2					
3	3	3					
4							

5. At the top of the screen, click **Edit** and select **Paste**. The sums should now be displayed in cells C2 and C3.

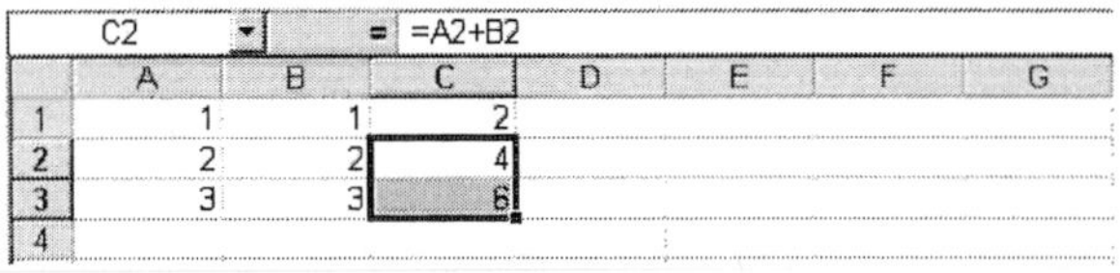

C2 = =A2+B2

	A	B	C	D	E	F	G
1	1	1	2				
2	2	2	4				
3	3	3	6				
4							

GS 4.3 Moving Information

If you would like to move the contents of one cell from one location to another in the worksheet, follow these steps:

1. Click on the cell containing the information that you would like to move.

2. At the top of the screen, click **Edit** and select **Cut**.

3. Click on the target cell where you want the information to be placed.

4. At the top of the screen, click **Edit** and select **Paste**.

If you would like to move the contents of a range of cells to a different location in the worksheet, follow these steps:

1. Click and drag over the range of cells that you would like to move so that it is highlighted.

2. At the top of the screen, click **Edit** and select **Cut**.

3. Click the topmost cell of the new location. (It is not necessary to click and drag over the entire range of the new location.)

4. At the top of the screen, click **Edit** and select **Paste**.

---

*If you make a mistake, just click **Edit** and select **Undo**.*

GS 4.4 Changing the Column Width

There are a couple of different ways that you can use to change the column width. Only one way will be described here. Output from the Descriptive Statistics data analysis tool will be used as an example. As you can see in the output displayed below, many of the labels in column A can only be partially viewed because the column width is too narrow.

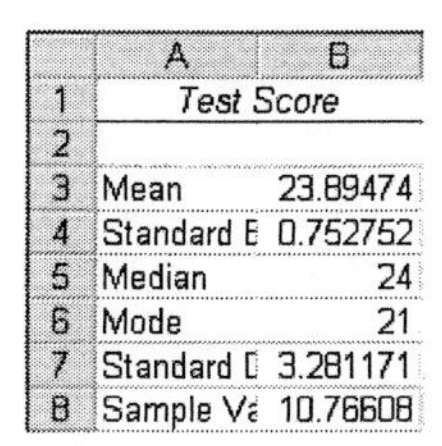

	A	B
1	*Test Score*	
2		
3	Mean	23.89474
4	Standard E	0.752752
5	Median	24
6	Mode	21
7	Standard D	3.281171
8	Sample Va	10.76608

Position the mouse pointer directly on the vertical line between A and B in the letter row at the top of the worksheet — A | B —so that it turns into a black plus sign.

Click and drag to the right until you can read all the output labels. (You can also click and drag to the left to make columns narrower.) After adjusting the column width, your output should appear similar to the output shown below.

	A	B
1	*Test Score*	
2		
3	Mean	23.89474
4	Standard Error	0.752752
5	Median	24
6	Mode	21
7	Standard Deviation	3.281171
8	Sample Variance	10.76608

Getting Started with Sorting Information

GS 5.1 Sorting a Single Column of Information

Let's say that you have entered "Score" in cell A1 and four numbers directly below it and that you would like to sort the numbers in ascending order.

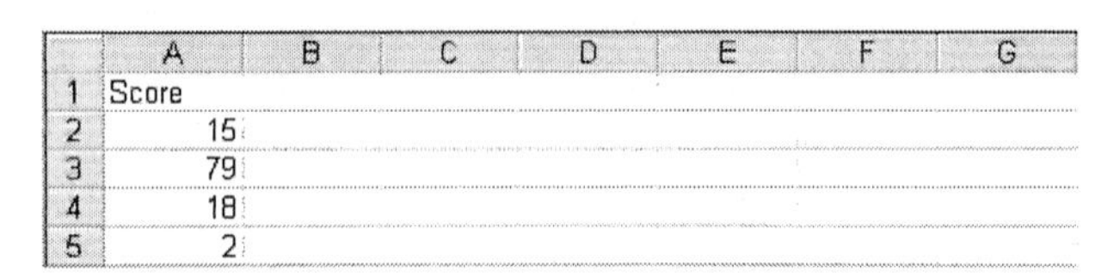

	A	B	C	D	E	F	G
1	Score						
2	15						
3	79						
4	18						
5	2						

1. Click and drag from cell A1 to cell A5 so that the range of cells is highlighted.

You could also click directly on A *in the letter row at the top of the worksheet. This will result in all cells of column A being highlighted.*

	A	B	C	D	E	F	G
1	Score						
2	15						
3	79						
4	18						
5	2						
6							

2. At the top of the screen, click **Data** and select **Sort**.

3. In the Sort dialog box that appears, you are given the choice of sorting the information in column A in either ascending or descending order. The ascending order has already been selected. Header row has also been selected. This means that the "Score" header will stay in cell A1 and will not be included in the sort. Click **OK** at the bottom of the dialog box.

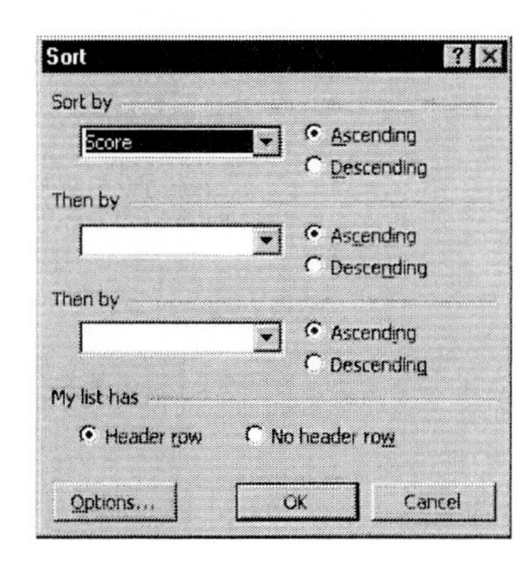

The cells in column A should now be sorted in ascending order as shown below.

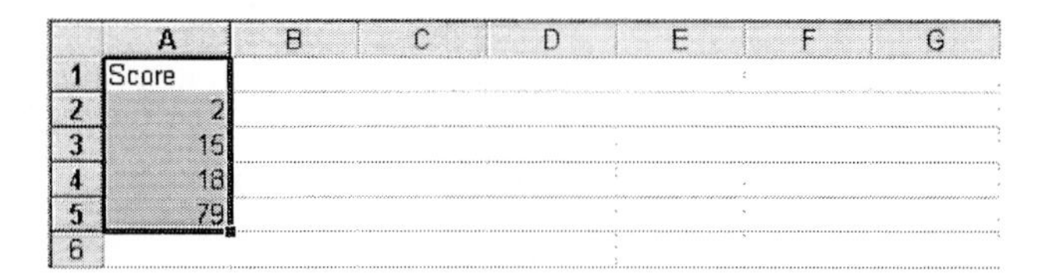

GS 5.2 Sorting Multiple Columns of Information

Your Excel data files will frequently contain multiple columns of information. When you sort multiple columns at the same time, Excel provides a number of options.

Let's say that you have a data file that contains the information shown below and that you would like to sort the file by GPA in descending order.

	A	B	C	D	E	F	G
1	Score	Age	Major	GPA			
2	2	19	Music	3.1			
3	15	19	History	2.4			
4	18	22	English	2.7			
5	79	20	English	3.7			

1. Click and drag from A1 down and across to D5 so that the entire range of cells is highlighted.

	A	B	C	D	E	F	G
1	Score	Age	Major	GPA			
2	2	19	Music	3.1			
3	15	19	History	2.4			
4	18	22	English	2.7			
5	79	20	English	3.7			

2. At the top of the screen, click **Data** and select **Sort**.

3. In the Sort dialog box that appears, you are given the option of sorting the data by three different variables. You want to sort only by GPA in descending order. Click the down arrow to the right of the Sort by window until you see GPA and click on **GPA** to select it. Then click the button to the left of **Descending** so that a black dot appears there. You want the variable labels to stay in row 1, so **Header row** should be selected. Click **OK**.

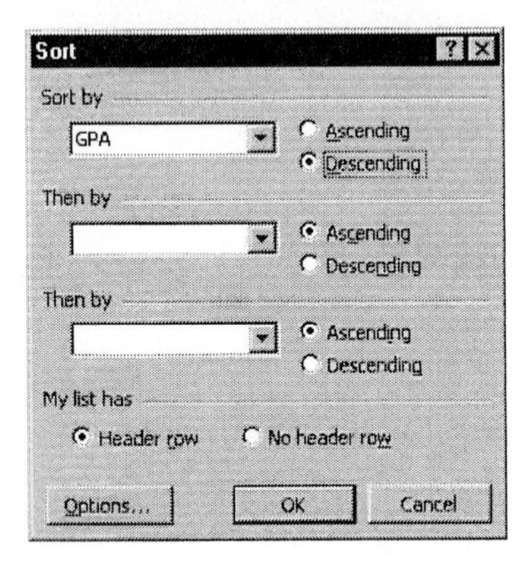

The sorted data file is shown below.

	A	B	C	D	E	F	G
1	Score	Age	Major	GPA			
2	79	20	English	3.7			
3	2	19	Music	3.1			
4	18	22	English	2.7			
5	15	19	History	2.4			

Getting Started with Saving Information

GS 6.1 Saving Files

To save a newly created file for the first time, click **File** and select **Save** at the top of the screen. A Save As dialog box will appear. You will need to select the location for saving the file by clicking on it. In the dialog box shown below, the 3½ Floppy has been selected.

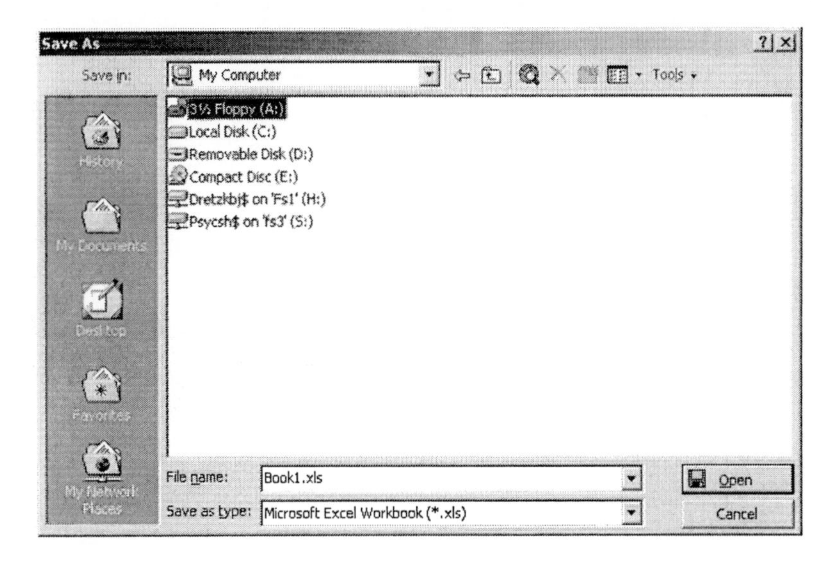

The default file name, displayed in the File name window, is **Book1.xls**. It is recommended you replace the default name with a name that is more descriptive. It is also recommended that you use the **xls** extension for all your Excel files.

Once you have saved a file, clicking **File** and selecting **Save** will result in the file being saved in the same location under the same file name. No dialog box will appear. If you would like to save the file in a different location, you will need to click **File** and select **Save As**.

GS 6.2 Naming Files

Recent Windows and Mac versions of Excel will allow file names to have around 200 characters. The extension can have up to three characters. You will find that long, descriptive names will be easier to work with than really short names. For example, if a file contains data that was collected in a survey of Milwaukee residents, you may want to name the file **Milwaukee residents survey.xls**.

Several symbols cannot be used in file names. These include: forward slash (/), backslash (\), greater-than sign (>), less-than sign (<), asterisk (*), question mark (?), quotation mark (“), pipe symbol (|), color (:), and semicolon (;).

Getting Started with Printing Information

GS 7.1 Printing Worksheets

To print a worksheet, click **File** and select **Print**. The Print dialog box, displayed at the top of the next page, will appear.

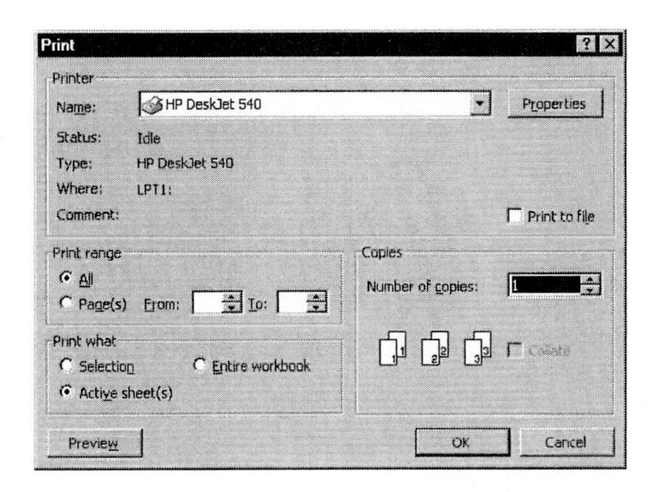

Under Print range, you will usually select **All**, and under Print what, you will usually select **Active sheet(s)**. The default number of copies is 1, but you can increase this if you need more copies. When the Print dialog box has been completed as you would like, click **OK**.

GS 7.2 Page Setup

Excel provides a number of page setup options for printing worksheets. To access these options, click **File** and select **Page Setup**. Under **Page**, you may want to select the **Landscape** orientation for worksheets that have several columns of data. Under **Sheet**, you may want to select **Gridlines**.

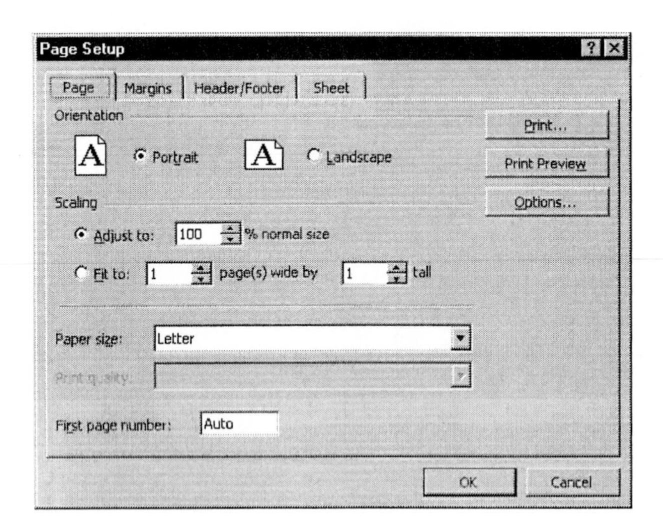

Getting Started with Add-ins

GS 8.1 Loading Excel's Analysis Toolpak

The Analysis Tookpak is an Excel Add-In that may not necessarily be loaded. If it does not appear at the bottom of the Tools menu, then click on **Add-Ins** in the Tools menu to get the dialog box shown below. Click in the box to the left of **Analysis ToolPak** to place a checkmark there. Then click **OK**. The ToolPak will load and will be listed at the bottom of the Tools menu as shown at the top of the next page.

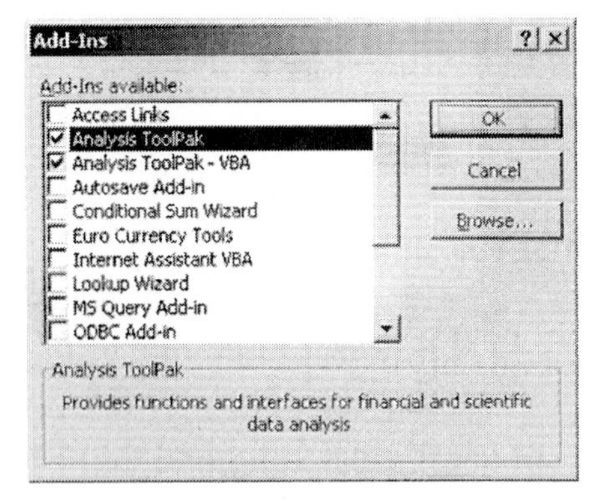

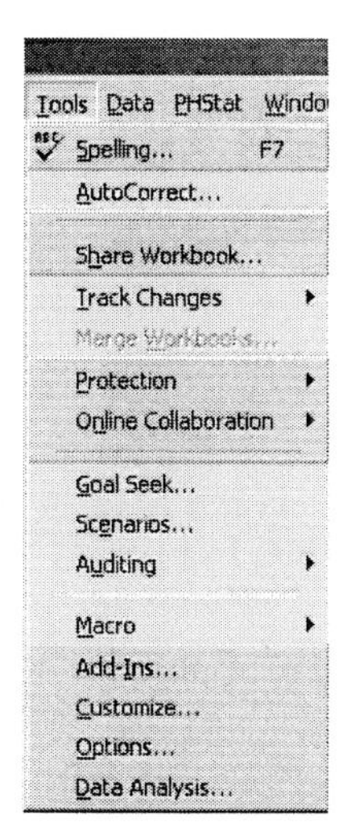

GS 8.2 Loading the PHStat2 Add-In

PHStat2 is a Prentice Hall statistical add-in that is included on the CD-ROM that accompanies your statistics textbook. PHStat2 is Windows software. The instructions that are given here also appear in the PHStat2 readme file. Read through that file completely to make sure that you are aware of all the technical requirements.

To use the Prentice Hall PHStat2 Microsoft Excel add-in, you first need to run the setup program (Setup.exe) located in the PHStat2 folder on your textbook CD-ROM. The setup program will install the PHStat2 program files to your system and add icons on your Desktop and Start Menu for PHStat2. Depending on the age of your Windows system files, some Windows system files may be updated during the setup process as well. Note that PHStat2 is compatible with Microsoft Excel 97, Microsoft Excel 2000, and Microsoft Excel 2003. PHStat2 is not compatible with Microsoft Excel 95.

During the Setup program you will have the opportunity to specify the directory into which to copy the PHStat2 files (default is \Program Files\Prentice Hall\PHStat2).

To begin using the Prentice Hall PHStat2 Microsoft Excel add-in, click the appropriate Start Menu or Desktop icon for PHStat2 that was added to your system during the setup process.

When a new, blank Excel worksheet appears, check the Tools menu to make sure that both the **Analysis ToolPak** and **Analysis ToolPak–VBA** have checkmarks next to them.

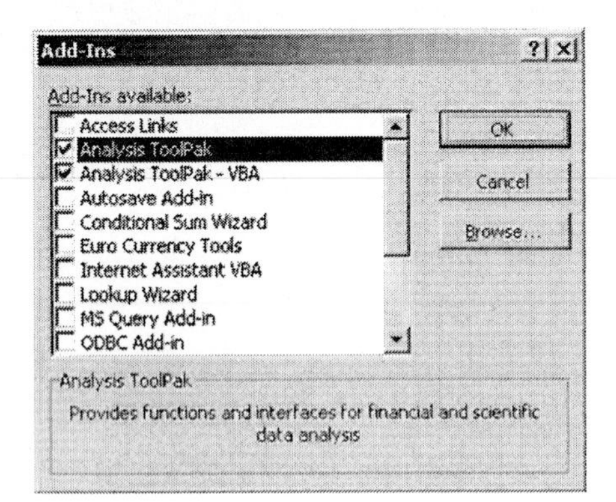

Data Collection

CHAPTER 1

1.2 Observational Studies, Experiments, and Simple Random Sampling

► Example 3 (pg. 18) | Obtaining a Simple Random Sample Using Technology

You will be generating a sample of size 5 from the list of 30 clients. To obtain the random sample, you will be using the PHStat add-in.

If the PHStat add-in has not been loaded, you will need to load it before continuing. Follow the instructions in Section GS 8.2.

---

1. Open a new, blank Excel worksheet.

2. At the top of the screen, click **PHStat**. Select **Sampling** → **Random Sample Generation**.

3. Complete the Random Sample Generator dialog box as shown below. A sample of five clients will be randomly selected from a population of 30. The topmost cell of the output will contain the label "Client #." Click **OK**.

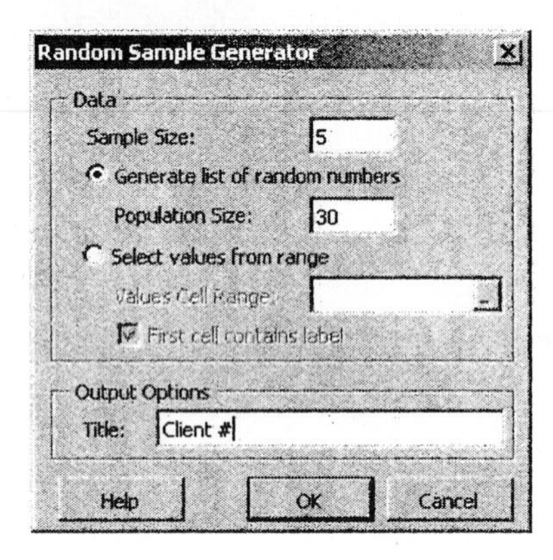

The output is displayed in a new worksheet named "RandomNumbers." Because the numbers were generated randomly, it is not likely that your output will be exactly the same.

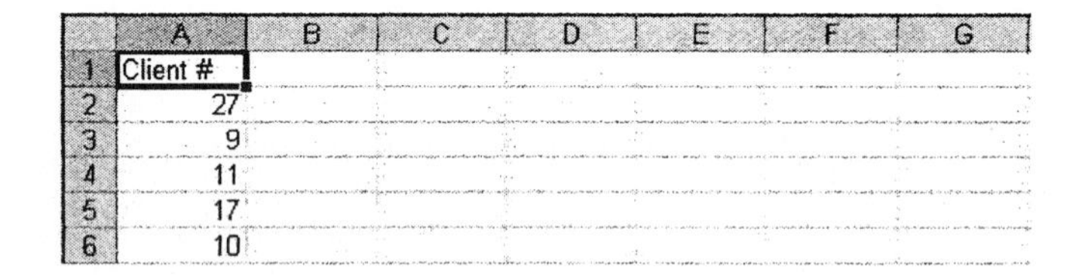

	A	B	C	D	E	F	G
1	Client #						
2	27						
3	9						
4	11						
5	17						
6	10						

◄

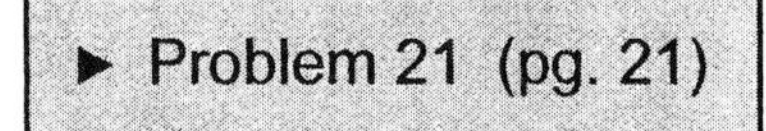

Obtaining Two Simple Random Samples of States

You will be generating two lists of 10 random numbers between 1 and 50 to use in selecting random samples of states.

If the PHStat add-in has not been loaded, you will need to load it before continuing. Follow the instructions in Section GS 8.2.

1. Open a new, blank Excel worksheet.

2. At the top of the screen, click **PHStat**. Select **Sampling → Random Sample Generation**.

3. Complete the Random Sample Generator dialog box as shown below. A sample of 10 states will be randomly selected from a population of 50. "State #" will appear in the top cell of the output. Click **OK**.

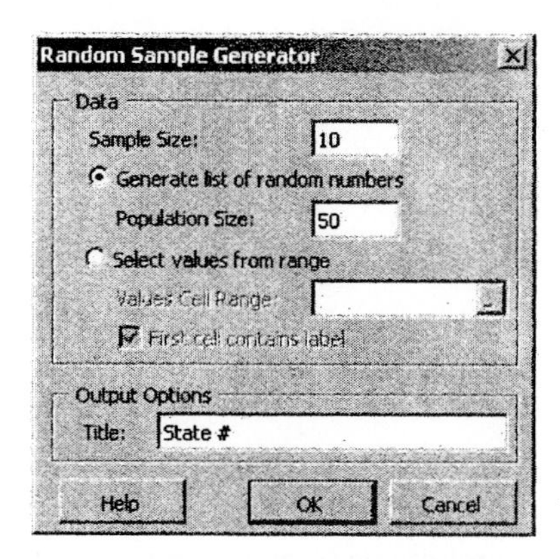

The ten randomly selected states are displayed below. Because the numbers were generated randomly, it is not likely that your output will be exactly the same.

	A	B	C	D	E	F	G
1	State #						
2	2						
3	23						
4	4						
5	46						
6	37						
7	31						
8	3						
9	13						
10	12						
11	10						

4. Repeat the procedure to obtain a second sample of 10 states. At the top of the screen, click **PHStat**. Select **Sampling → Random Sample Generation**.

5. Complete the Random Sample Generator dialog box as shown below. A second sample of 10 states will be randomly selected. Click **OK**.

The output is placed in a worksheet named "RandomNumbers2." The ten randomly selected states are displayed below. Again, because the numbers were generated randomly, it is not likely that your output will be exactly the same.

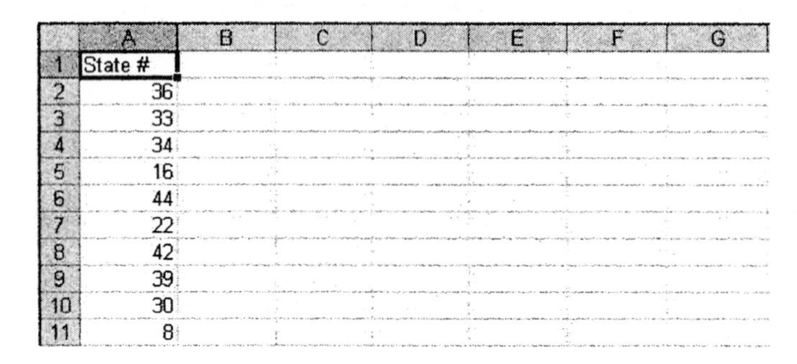

	A	B	C	D	E	F	G
1	State #						
2	36						
3	33						
4	34						
5	16						
6	44						
7	22						
8	42						
9	39						
10	30						
11	8						

◀

▶ Problem 22 (pg. 21) Obtaining a Simple Random Sample of Presidents

If the PHStat add-in has not been loaded, you will need to load it before continuing. Follow the instructions in Section GS 8.2.

1. Open a new Excel worksheet.

2. At the top of the screen, click **PHStat**. Select **Sampling → Random Sample Generation**.

3. Complete the Random Sample Generator dialog box as shown below. A sample of eight presidents will be randomly selected from a population of 43. "President #" will appear in the top cell of the generated output. Click **OK**.

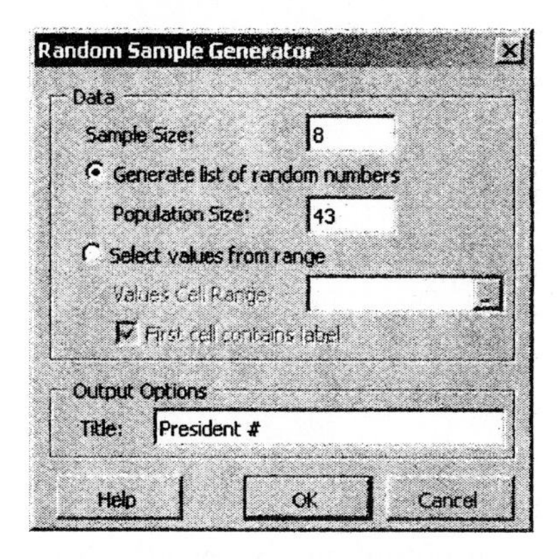

The set of eight random numbers is displayed below. Because the numbers were generated randomly, it is unlikely that your output will be exactly the same.

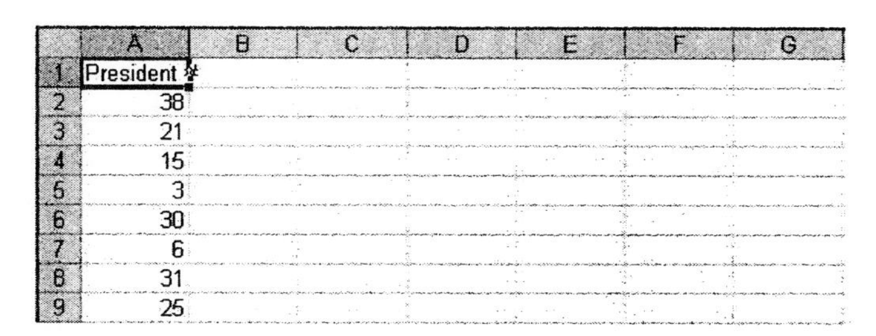

	A	B	C	D	E	F	G
1	President #						
2	38						
3	21						
4	15						
5	3						
6	30						
7	6						
8	31						
9	25						

4. Repeat the procedure to obtain a second sample of eight presidents. At the top of the screen, click **PHStat**. Select **Sampling → Random Sample Generation**.

5. Complete the Random Sample Generator dialog box as shown below. A second random sample of eight presidents will be selected. Click **OK**.

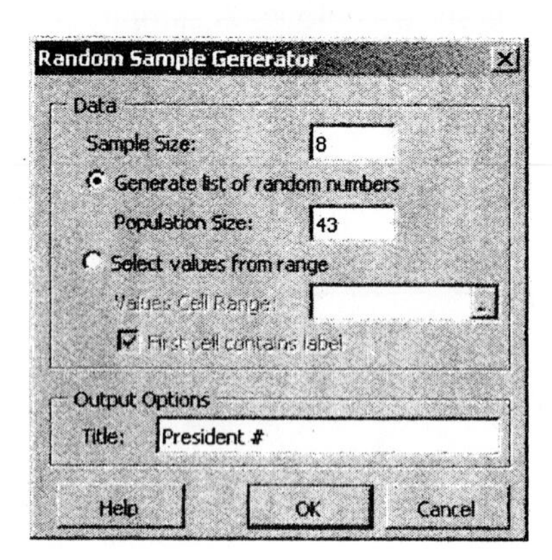

The second set of eight randomly selected presidents is displayed below. Again, because the numbers were generated randomly, it is unlikely that your output will be exactly the same.

	A	B	C	D	E	F	G
1	President #						
2	11						
3	23						
4	26						
5	38						
6	2						
7	17						
8	14						
9	41						

◀

Section 1.3 Other Effective Sampling Methods

Obtaining a Stratified Sample

You will be generating a sample of size 100 comprised of 28 resident students, 61 non-resident students, and 11 staff.

If the PHStat add-in has not been loaded, you will need to load it before continuing. Follow the instructions in Section GS 8.2.

---

1. Open a new, blank Excel worksheet.

2. You will first select the resident students. At the top of the screen, click **PHStat**. Select **Sampling → Random Sample Generation**.

3. Complete the Random Sample Generator dialog box as shown below. A sample of 28 resident students will be randomly selected from a population of 6,204. "Resident Student #" will appear in the top cell of the generated output. Click **OK**.

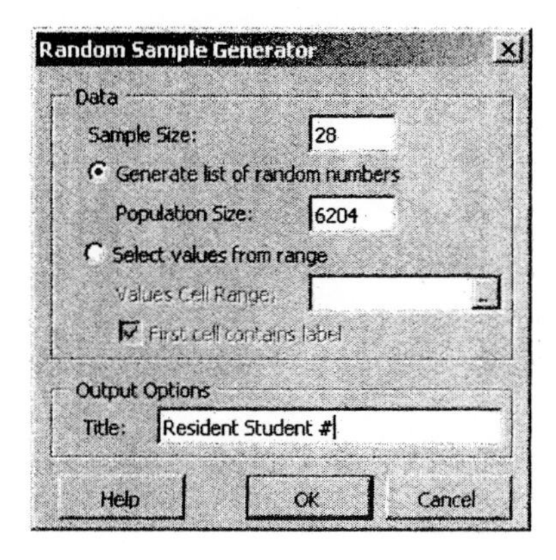

A partial listing of the 28 randomly selected resident students is displayed below. The output appears in a sheet labeled "RandomNumbers." Because the numbers were generated randomly, it is unlikely that your output will be exactly the same.

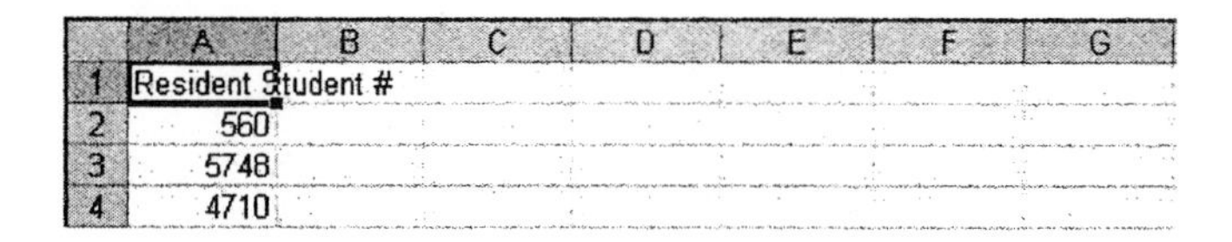

4. Repeat the procedure to randomly select 61 non-resident students. At the top of the screen, click **PHStat**. Select **Sampling → Random Sample Generation**.

5. Complete the Random Sample Generator dialog box as shown below. A sample of 61 nonresident students will be selected from a population of 13,304. "Nonresident Student #" will appear in the top cell of the generated output. Click **OK**.

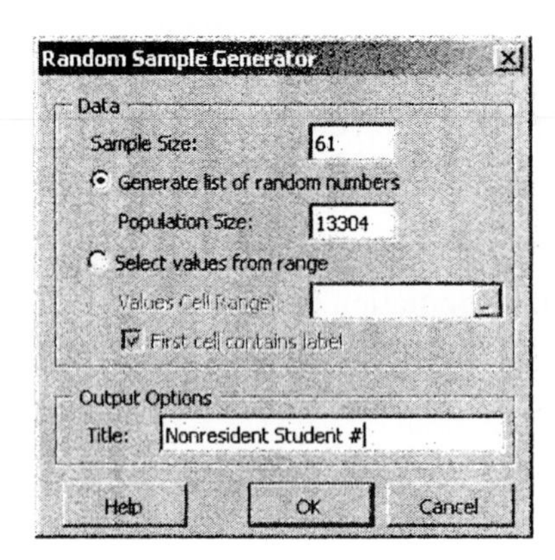

A partial listing of the 61 randomly selected nonresident students is displayed below. The output appears in a sheet labeled "RandomNumbers2." Again, because the numbers were generated randomly, it is unlikely that your output will be exactly the same.

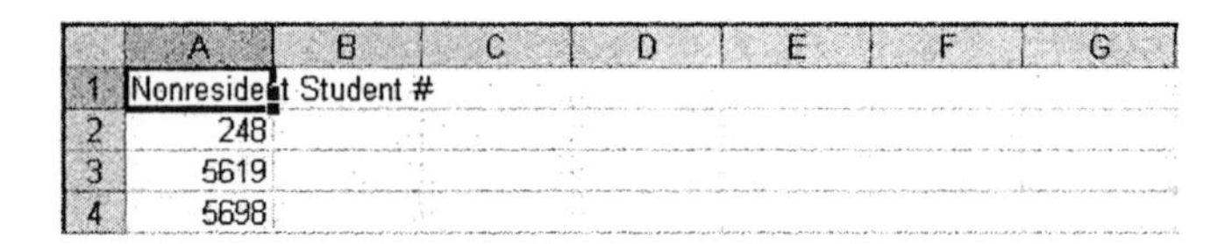

6. Repeat the procedure to randomly select the 11 staff members. At the top of the screen, click **PHStat**. Select **Sampling → Random Sample Generation**.

7. Complete the Random Sample Generator dialog box as shown below. A sample of 11 staff will be randomly selected from a population of 2,401. "Staff #" will appear in the top cell of the generated output. Click **OK**.

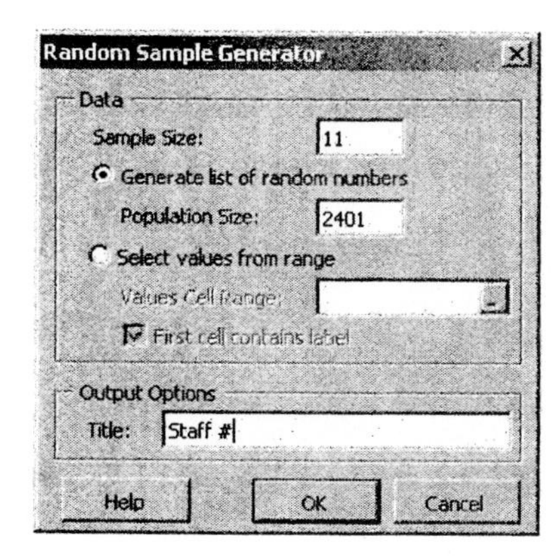

A partial listing of the 11 randomly selected numbers are displayed below. The output appears in a sheet labeled "RandomNumbers3." Again, because the numbers were generated randomly, it is unlikely that your output will be exactly the same.

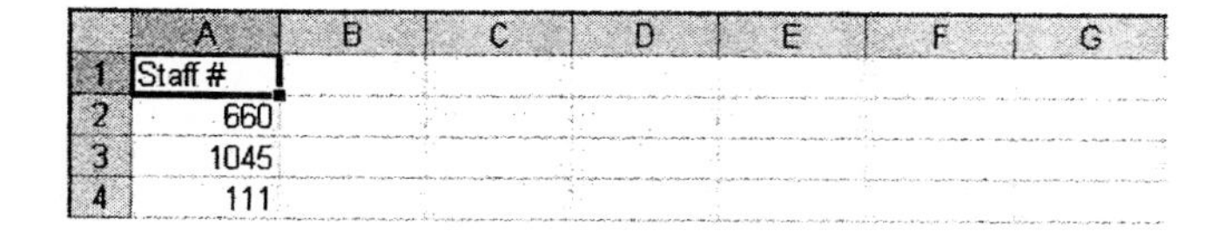

	A	B	C	D	E	F	G
1	Staff #						
2	660						
3	1045						
4	111						

◀

Organizing and Summarizing Data

CHAPTER 2

Section 2.1 Organizing Qualitative Data

▶ Problem 23 (pg. 71) **Constructing a Side-by-Side Relative Frequency Bar Graph**

1. Open worksheet "2_1_23" in the Chapter 2 folder. The first two rows are shown below.

	A	B	C	D	E	F	G
1	Education	Males (in r	Females (in millions)				
2	Not a high	14.1	14.5				

2. The table displays frequencies (in millions). You will be graphing relative frequencies, so the first step is to calculate relative frequencies. Begin by summing the numbers in the column labeled "Males." Click in cell **B8**. At the top of the screen, click on AutoSum. It looks like Σ.

3. Cell B8 should now show =SUM(B2:B7) indicating that Excel will sum the numbers in cells B2 through B7. Press [**Enter**].

	A	B	C	D	E	F	G
1	Education	Males (in r	Females (in millions)				
2	Not a high	14.1	14.5				
3	High scho	27.4	31.9				
4	Some coll	15.2	16.6				
5	Associate	6.4	8.8				
6	Bachelor's	16.4	16.9				
7	Advanced	9.2	7.9				
8		=SUM(B2:B7)					
9		SUM(number1, [number2], ...)					
10							

4. Next, sum the numbers in the column labeled "Females." Click in cell **C8**. Click Σ at the top of the screen. You should now see =SUM(C2:C7) in cell C8. Press [**Enter**].

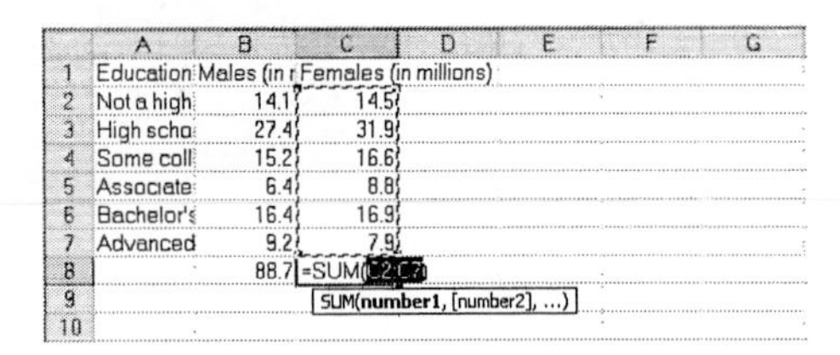

	A	B	C	D	E	F	G
1	Education	Males (in r	Females (in millions)				
2	Not a high	14.1	14.5				
3	High scho	27.4	31.9				
4	Some coll	15.2	16.6				
5	Associate	6.4	8.8				
6	Bachelor's	16.4	16.9				
7	Advanced	9.2	7.9				
8		88.7	=SUM(C2:C7)				
9			SUM(number1, [number2], ...)				
10							

5. Copy the original table without the sums to a space a little below the original table so that upper left cell of the copy is A10 as shown below.

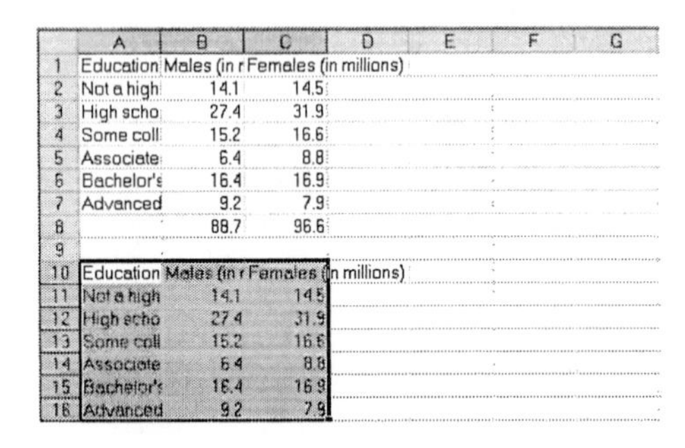

	A	B	C	D	E	F	G
1	Education	Males (in r	Females (in millions)				
2	Not a high	14.1	14.5				
3	High scho	27.4	31.9				
4	Some coll	15.2	16.6				
5	Associate	6.4	8.8				
6	Bachelor's	16.4	16.9				
7	Advanced	9.2	7.9				
8		88.7	96.6				
9							
10	Education	Males (in r	Females (in millions)				
11	Not a high	14.1	14.5				
12	High scho	27.4	31.9				
13	Some coll	15.2	16.6				
14	Associate	6.4	8.8				
15	Bachelor's	16.4	16.9				
16	Advanced	9.2	7.9				

6. To calculate the relative frequencies, you will divide each frequency by its column total. Start by clicking in cell **B11**. Key in the formula **=B2/$B$8**. Press [**Enter**].

The dollar signs are necessary for the cell B8 address because you will be copying the formula to calculate all the relative frequencies in the "Males" column.

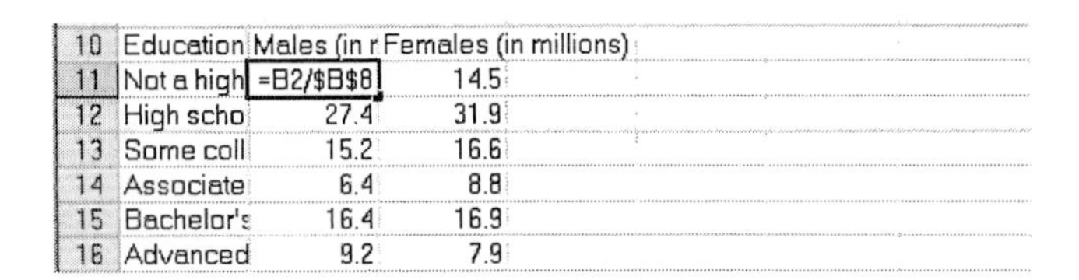

	A	B	C
10	Education	Males (in r	Females (in millions)
11	Not a high	=B2/$B$8	14.5
12	High scho	27.4	31.9
13	Some coll	15.2	16.6
14	Associate	6.4	8.8
15	Bachelor's	16.4	16.9
16	Advanced	9.2	7.9

7. Copy the formula in cell **B11** to cells **B12** to **B16**.

8. Click in cell **C11** and enter the formula =**C2/$C$8**. Press [**Enter**].

9. Copy the formula in cell **C11** to cells **C12** to **C16**.

10	Education	Males (in r	Females (in millions)
11	Not a high	0.158963	0.150104
12	High scho	0.308906	0.330228
13	Some coll	0.171364	0.171843
14	Associate	0.072153	0.091097
15	Bachelor's	0.184893	0.174948
16	Advanced	0.10372	0.081781

10. Next, remove "(in millions)" from the column headings because "(in millions)" refers to frequencies, not relative frequencies. The column headings in cells **B10** and **C10** should be **Males** and **Females**, respectively.

10	Education	Males	Females
11	Not a high	0.158963	0.150104

11. Click in any cell of the relative frequency table. Then, at the top of the screen, click **Insert** and select **Chart**.

If you activate a cell containing data prior to selecting Chart, the data range will be automatically entered in the chart dialog box.

---

12. In the Chart Type dialog box, click on the **Column** Chart type to select it. Under Chart sub-type, select the leftmost diagram in the top row. Click **Next>**.

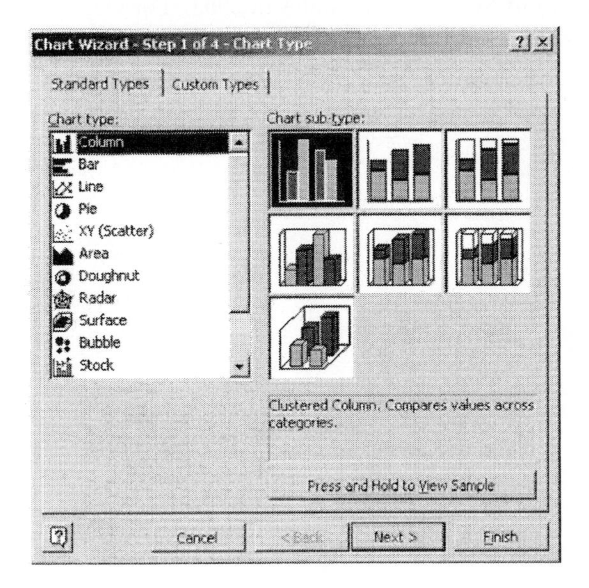

13. Check the data range window in the lower portion of the Step 2 of 4 dialog box. It should be $A$10:$C$16. Click **Next>**.

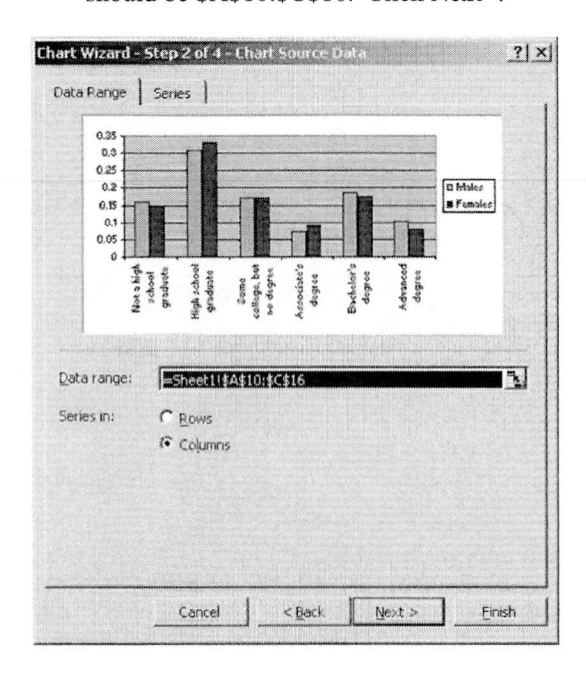

14. Click on the **Titles** tab at the top of the Chart Options dialog box. For the Chart title, enter **Educational Attainment of Males and Females in 2003**. For the Category (X) axis, enter **Educational Attainment**. For the Value (Y) axis, enter **Relative Frequency**. Click **Next>**.

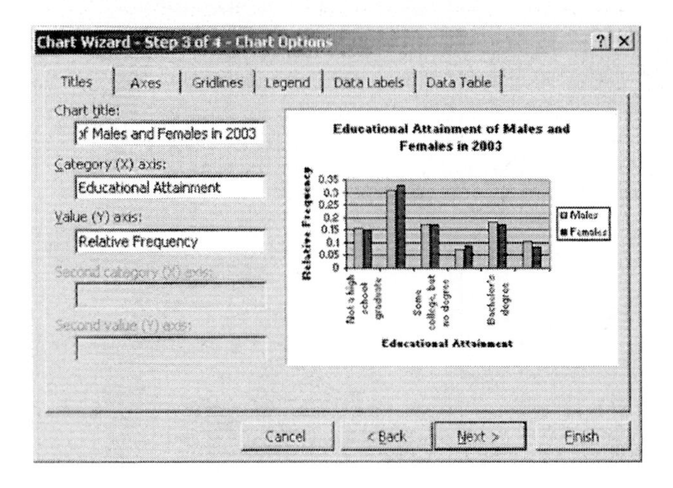

15. The Chart Location dialog box presents two options for placement of the chart. For this example, select **As new sheet**. Click **Finish**.

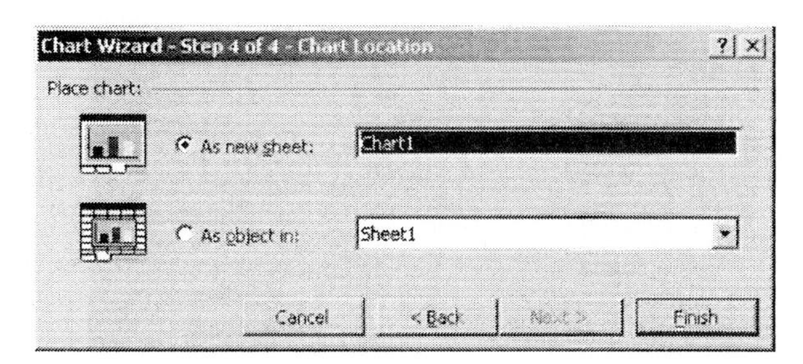

16. Your completed graph should look similar to the one shown below. Let's display two decimal places in all relative frequency values. **Right-click** on any relative frequency value. I right-clicked on 0.25.

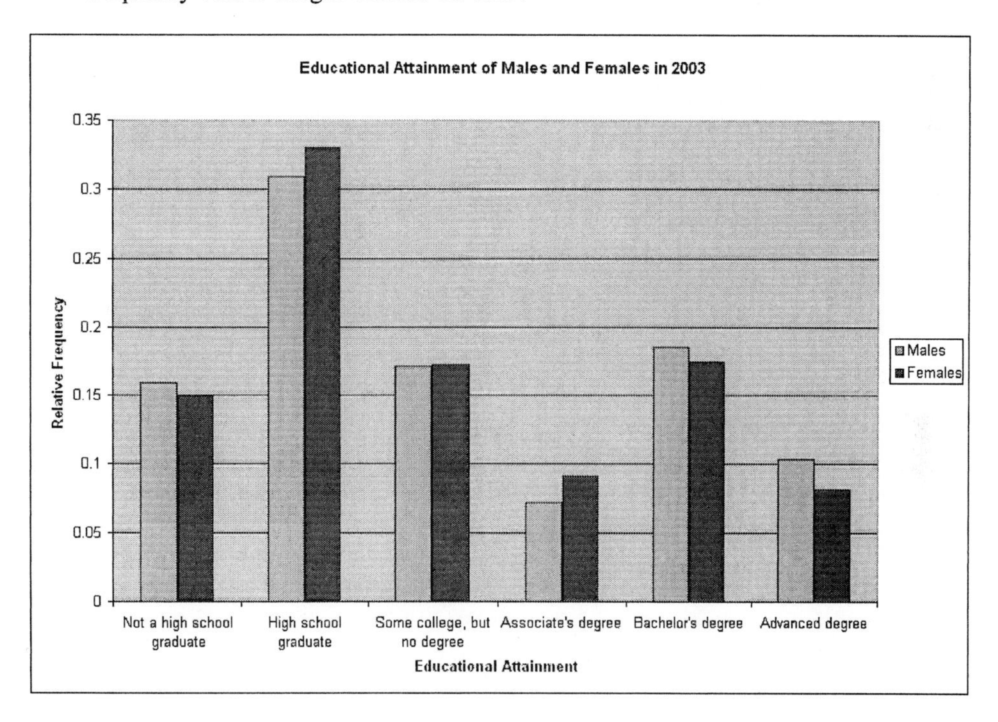

17. Select **Format Axis** from the shortcut menu.

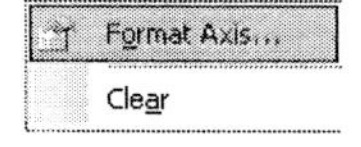

18. Click the **Number** tab at the top of the Format Axis dialog box. Select the **Number** category, and select **2** decimal places. Click **OK**.

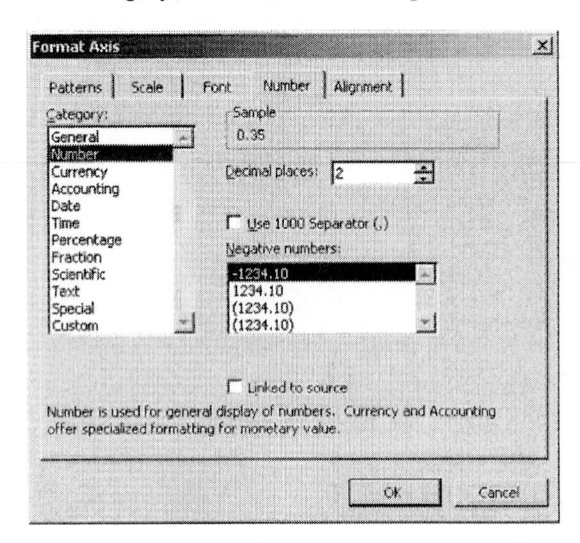

The completed graph with the number scale modification is displayed below.

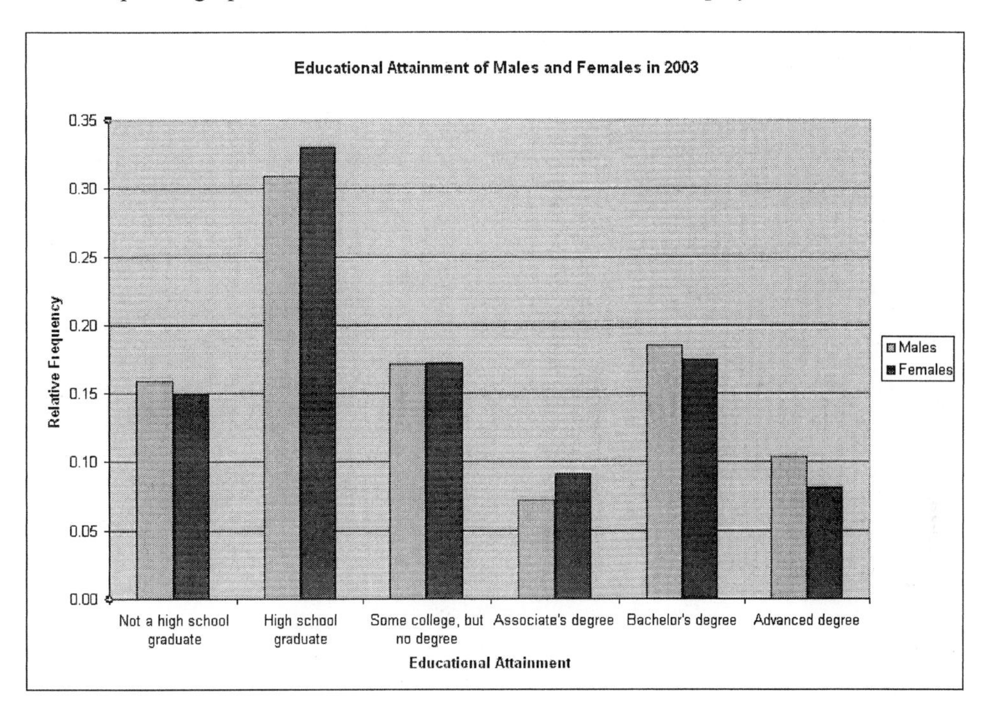

◀

▶ Problem 27 (pg. 72)

Constructing a Frequency Distribution, a Frequency Bar Graph, and a Pie Chart

1. Open worksheet "2_1_27" in the Chapter 2 folder. The first two rows are shown below.

	A	B	C	D	E	F	G
1	C1-T						
2	Kerry						
3	Bush						

2. Change the entry in the top cell from **C1-T** to **Candidate**.

3. Click in any worksheet cell that contains a candidate's name. Then, at the top of the screen, click **Data** and select **Pivot Table and Pivot Chart Report**.

4. Select **Microsoft Office Excel list or database** and **Pivot Chart report (with Pivot Table Report)** in the Step 1 of 3 dialog box. Click **Next>**.

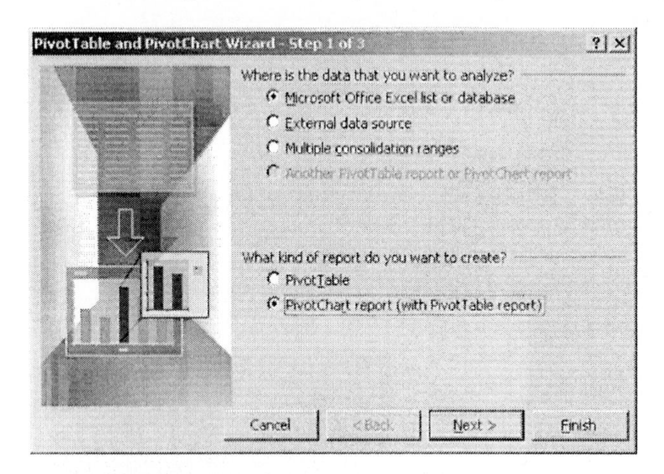

5. Check the range in the Step 2 of 3 dialog box. It should be $A$1:$A$41. Click **Next>**.

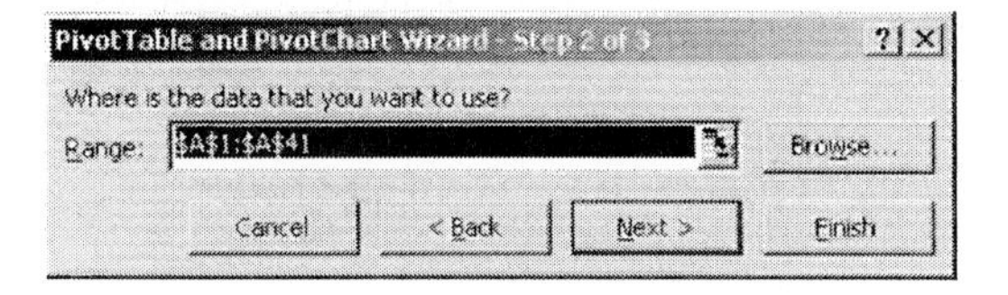

6. Click **Layout** in the Step 3 of 3 dialog box.

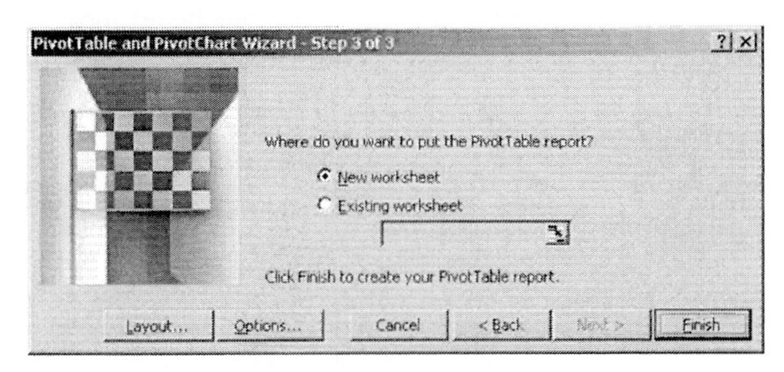

7. Drag the **Candidate** field button to ROW. Drag the **Candidate** field button to DATA. Click **OK**.

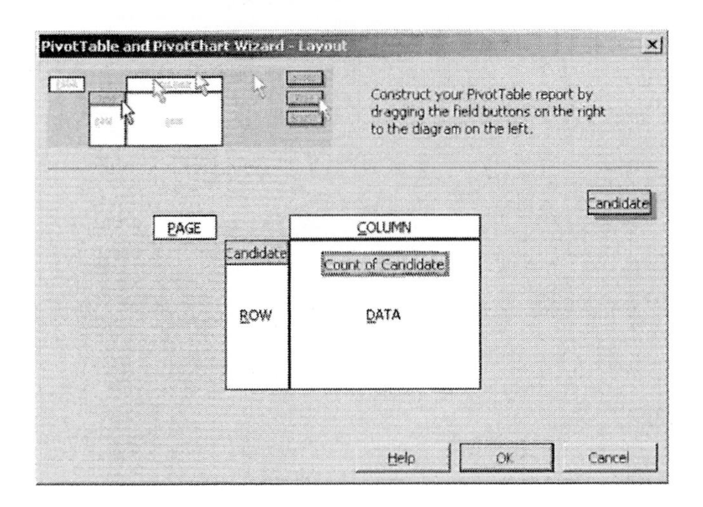

8. Select **New worksheet** and click **Finish**.

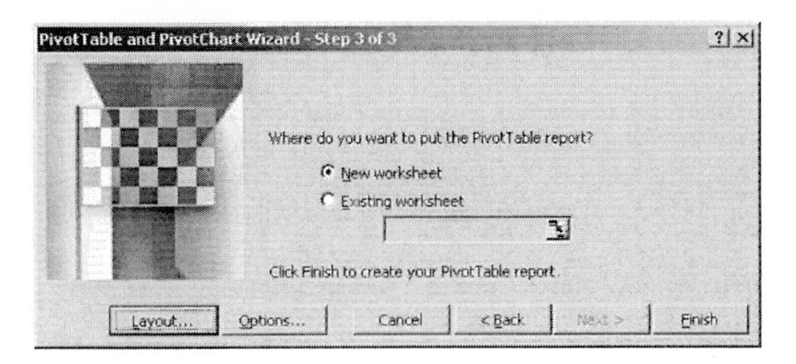

9. The frequency distribution table is displayed in Sheet 2, and the frequency bar graph is displayed in a worksheet labeled Chart1. You will next follow steps to modify the graph. Click **PivotChart** on the Pivot Table toolbar and select **Hide PivotChart Field Buttons**.

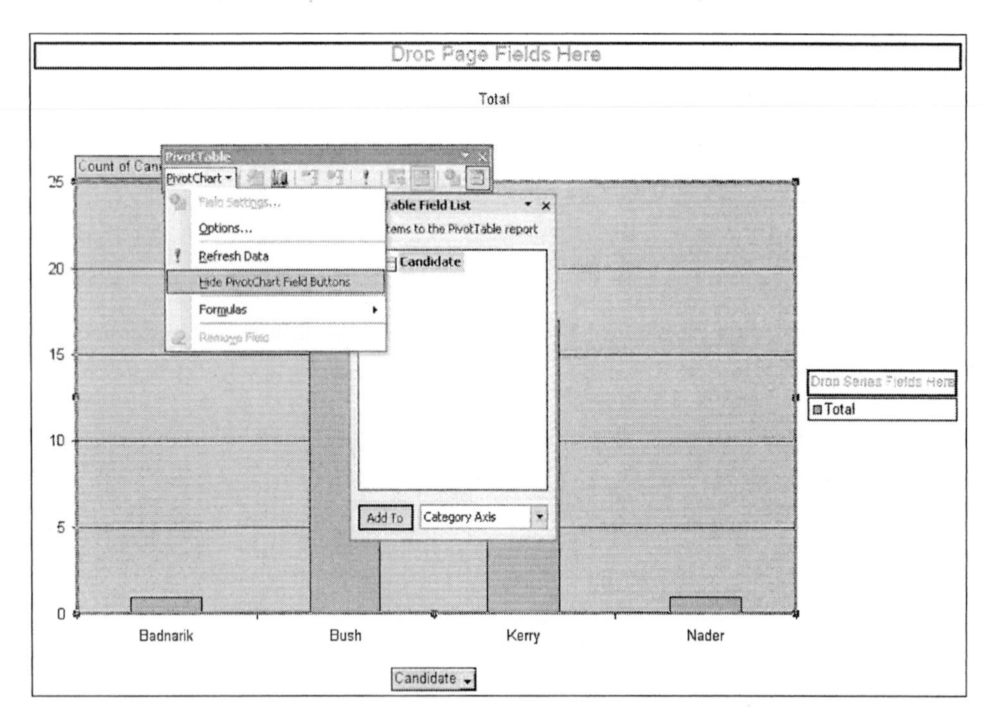

10. Click the **Chart Wizard** icon on the Pivot Table toolbar and click **Next>** in the Step 1 of 4 Chart Wizard dialog box.

11. Click on the **Titles** tab at the top of the Step 3 of 4 dialog box. For the Chart title, enter **2004 Presidential Election Exit Poll in Los Alamos County, New Mexico**. For the Category (X) axis, enter **Candidate**. For the Value (Y) axis, enter **Frequency**.

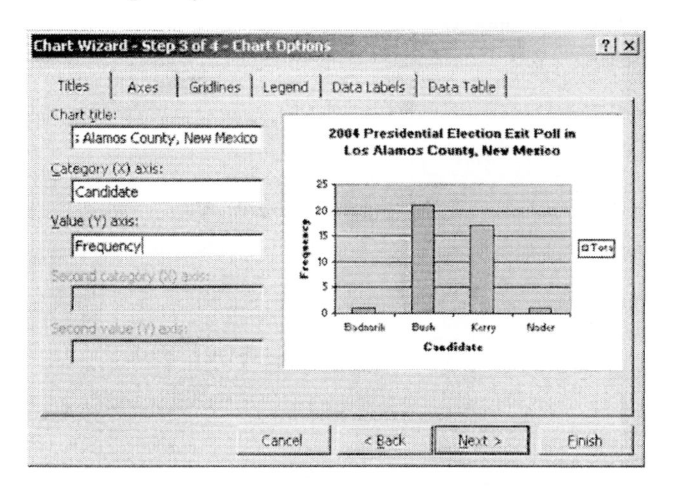

12. Click the **Legend** tab at the top of the dialog box. Click in the **Show legend** box to remove the checkmark that appears there. Click **Next>**.

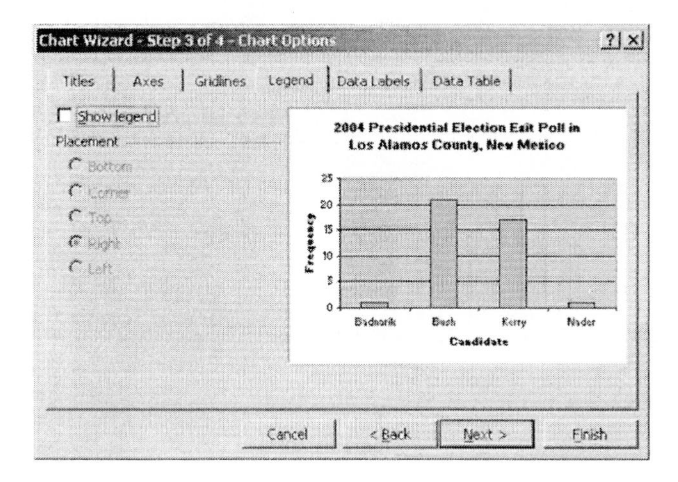

13. Select **As new Sheet** in the Step 4 Chart Location dialog box. Click **Finish**.

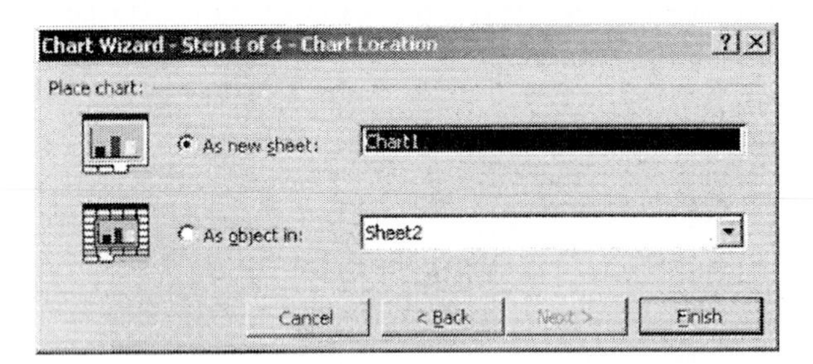

14. Your frequency bar graph should look similar to the one displayed below. You will now construct a pie chart. Click the **Chart Wizard** icon on the Pivot Table

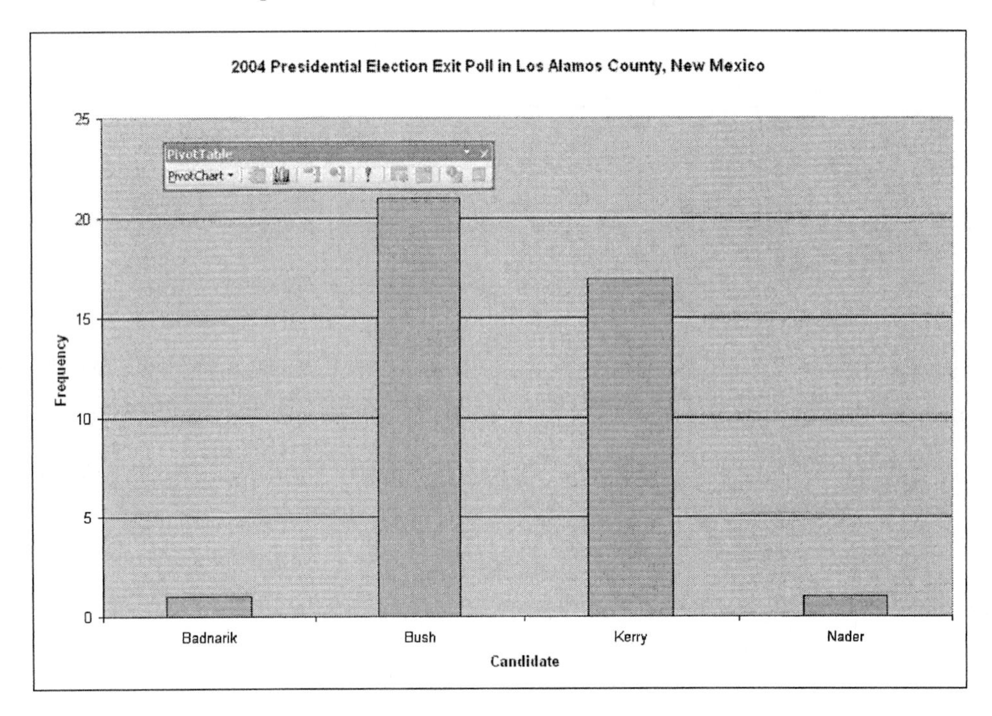

toolbar. Under Chart type at the left of the dialog box, click **Pie** to select it. Then, at the right of the dialog box, under Chart sub-type, click on the first diagram at the far left of the top row to select it. Click **Next>**.

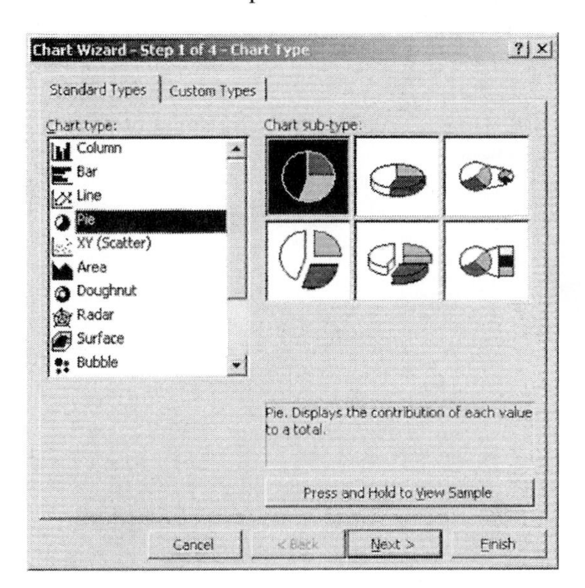

15. Click in the box to the left of **Show legend** to place a checkmark here.

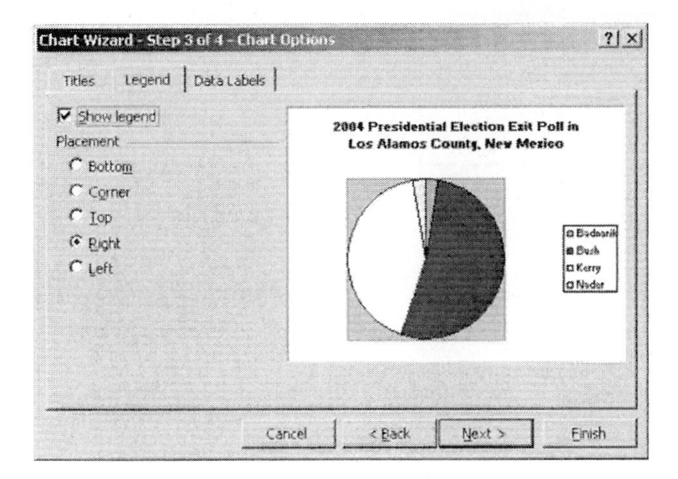

16. Click the **Data Labels** tab at the top of the dialog box. Under Label Contains, select **Category** name and **Percentage**. **Show leader lines** should already be selected. Click **Next>**.

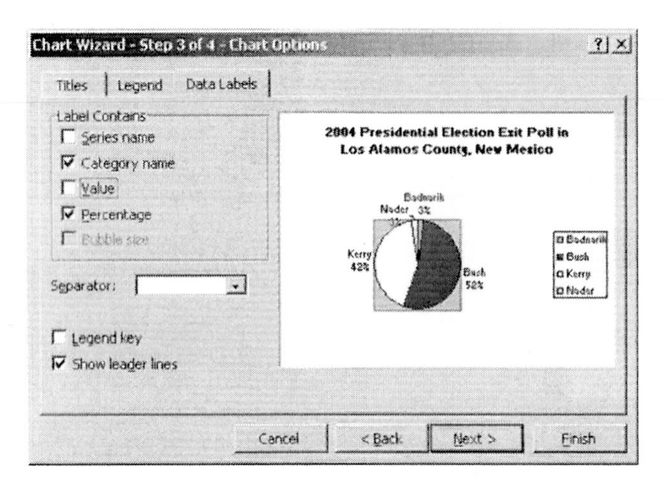

17. Select **As new sheet** in the Step 4 of 4 Chart Location dialog box. Click **Finish**.

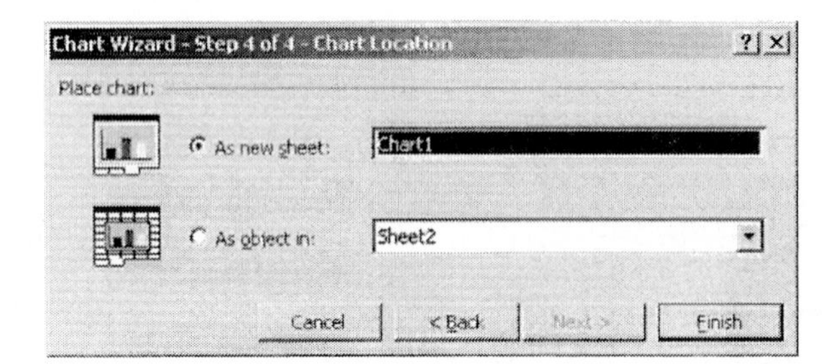

Your completed pie chart should look similar to the one show below.

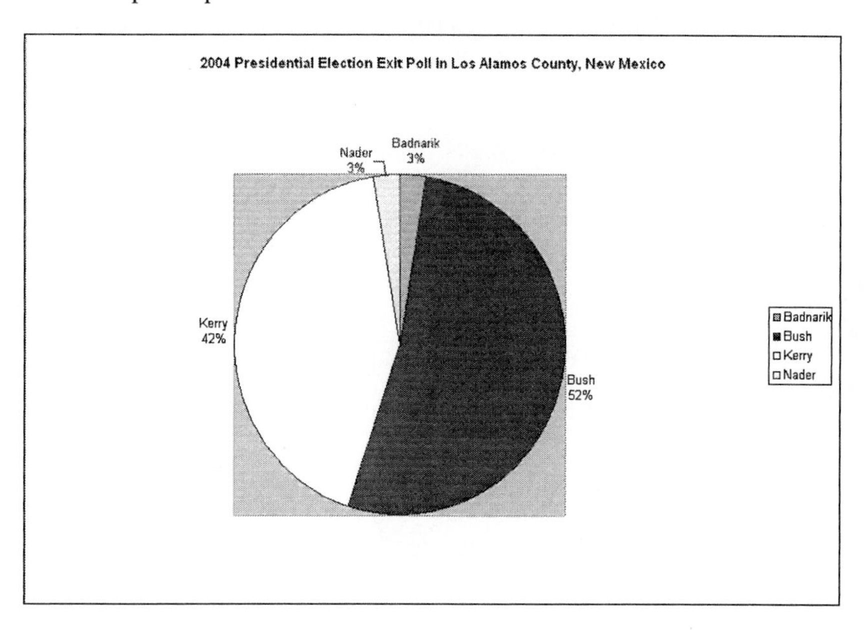

◀

▶ Problem 28 (pg. 73)

Constructing a Frequency Distribution and a Frequency Bar Graph

1. Open worksheet "2_1_28" in the Chapter 2 folder. The first three rows are shown below.

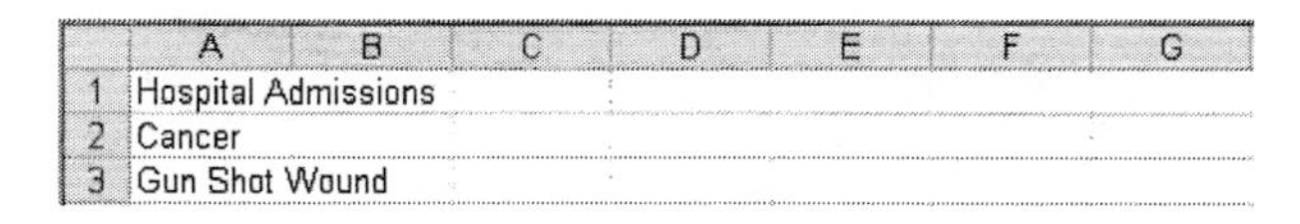

	A	B	C	D	E	F	G
1	Hospital Admissions						
2	Cancer						
3	Gun Shot Wound						

2. Click in any cell in column A that contains an entry. Then, at the top of the screen, click **Data** and select **Pivot Table and Pivot Chart Report**.

3. At the top of the dialog box, select **Microsoft Excel list or database**. At the bottom of the dialog box, select **Pivot Chart (with Pivot Table)**. Excel will prepare both a frequency distribution and a frequency bar graph of the diagnoses data. Click **Next>**.

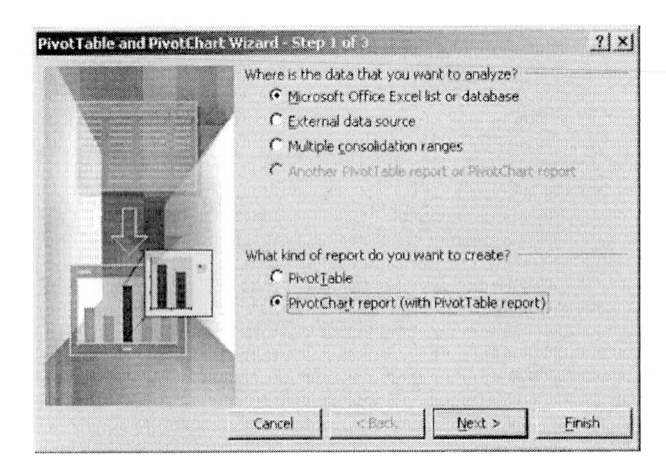

4. The data range is automatically placed in the Range window of the dialog box. It should read $A$1:$A$21. Make any necessary revisions to the range, and then click **Next>**.

5. You can place the table and chart in a new worksheet or in the existing worksheet. For this problem, select **New worksheet**. Click **Finish**.

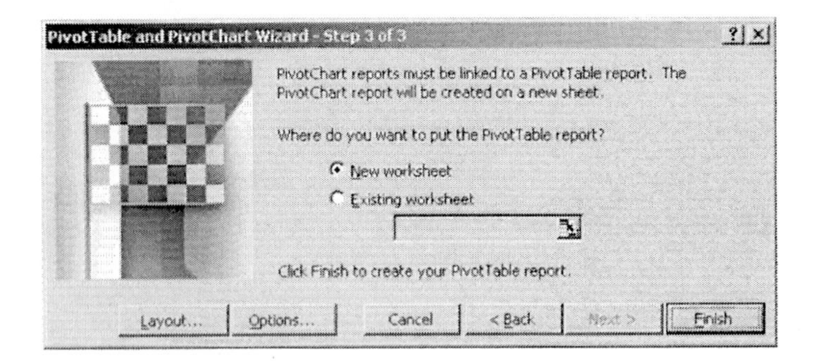

6. In the Pivot Table Field List, click **Hospital Admissions** and then click **Add To** Category Axis. Next, click on **Hospital Admissions** and drag it to the chart area that says Drop Data Items Here. The diagnoses will now be displayed on the X-axis in alphabetical order, and the frequency bars for the diagnoses will be displayed in the chart. Close the Pivot Table Field list by clicking X in the top right corner.

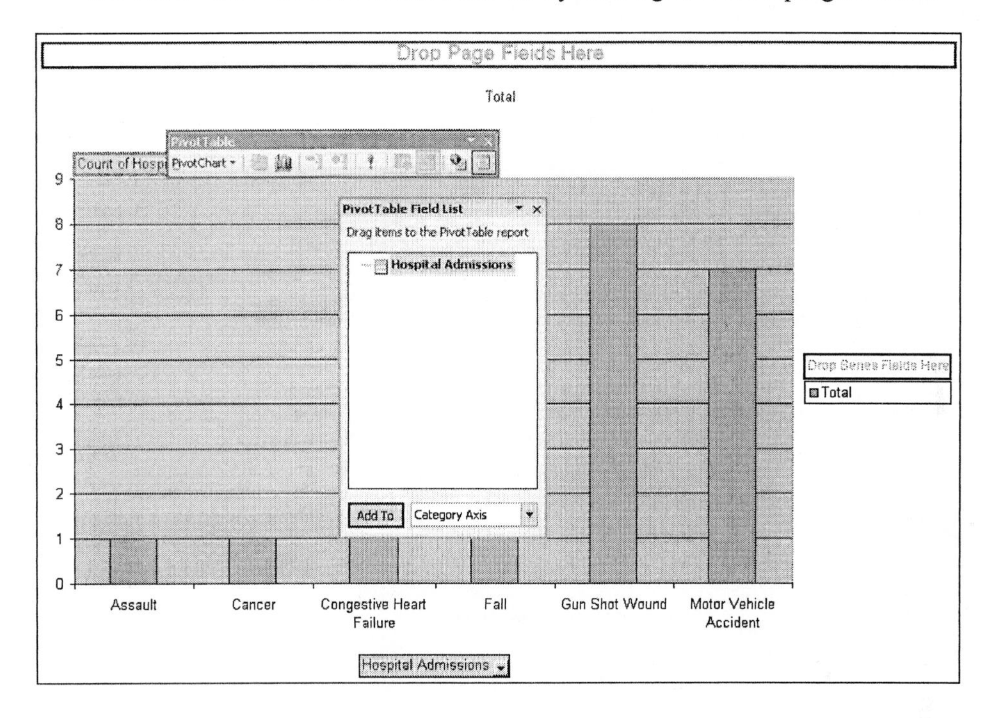

7. Click **Pivot Chart** and select **Hide Pivot Chart Field Buttons**.

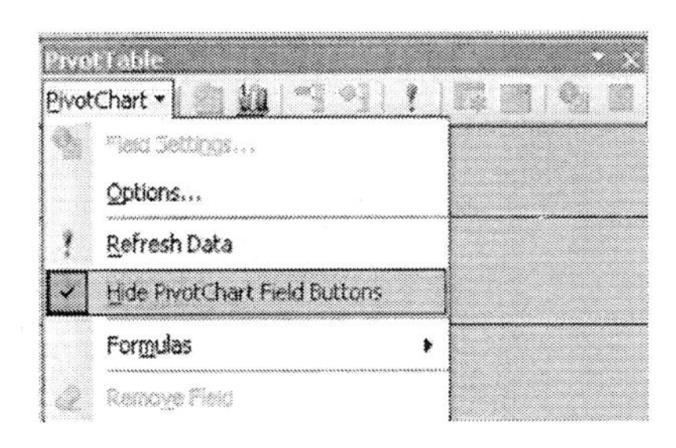

8. You will now add a title and labels for the X- and Y-axis. Click the **Chart Wizard** icon on the Pivot Table toolbar and click **Next>** in the Step 1 of 4 Chart Wizard dialog box.

9. Click the **Titles** tab at the top of the Chart Options dialog box. For the Chart title, enter **Diagnoses of Patients Admitted to a Hospital**. For the Category (X) axis, enter **Diagnosis**. For the Value (Y) axis, enter **Frequency**.

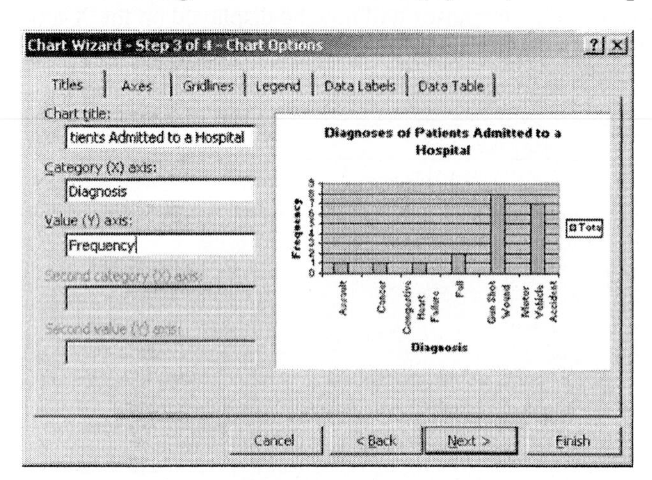

10. Click the **Legend** tab at the top of the dialog box. Then click in the box to the left of **Show legend** to remove the checkmark that appears there. Click **Next>**.

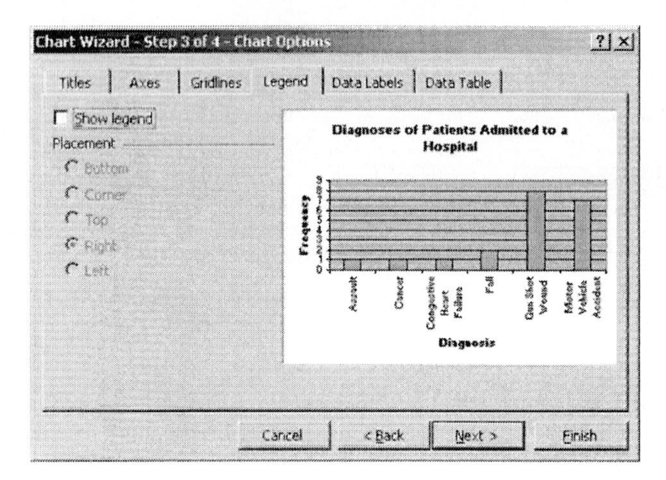

11. Select **As object** in Sheet2 so that the graph and frequency distribution table will be in the same worksheet. Click **Finish**.

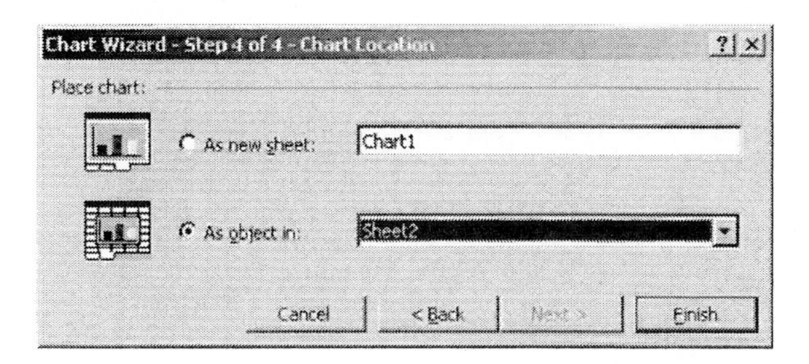

12. Close the Pivot Table toolbar. Your worksheet should appear similar to the one shown below. If you would like, you can drag the graph to a different location in the worksheet.

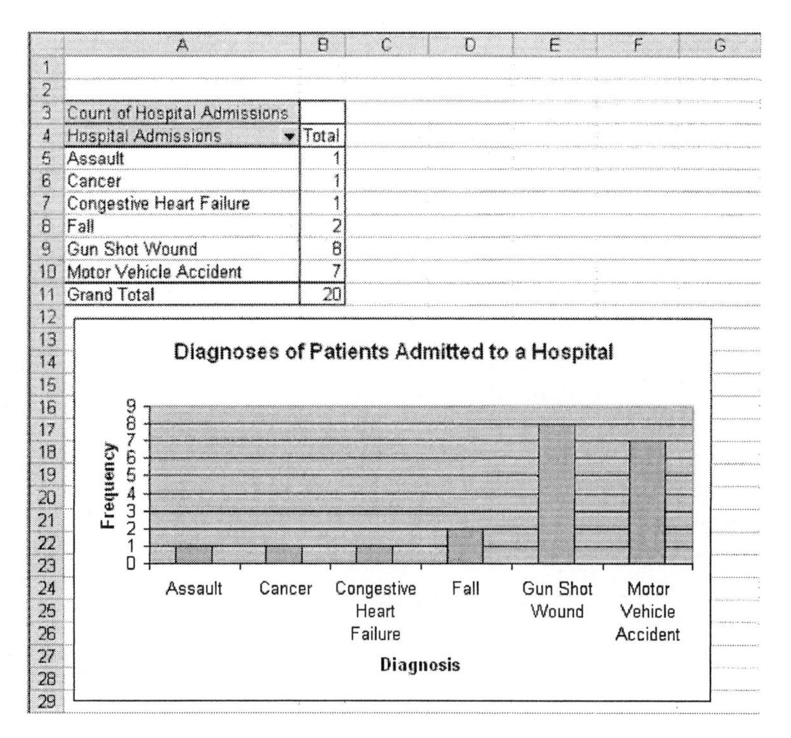

◀

Section 2.2 Organizing Quantitative Data: The Popular Displays

► Problem 27 (pg. 90) Constructing a Frequency Histogram

You will use the Wait Time data to learn how to construct a frequency histogram.

1. Open worksheet "2_2_27" in the Chapter 2 folder. The first few lines are shown below.

	A	B	C	D	E	F	G
1	Wait Time						
2	11						
3	4						
4	13						

2. Sort the data in ascending order. In the sorted data set, you can see that the minimum number of customers is three. Scroll down to find the maximum number of customers. The maximum number, displayed in cell A41, is 14.

For instructions on how to sort, refer to Sections GS 5.1 and GS 5.2.

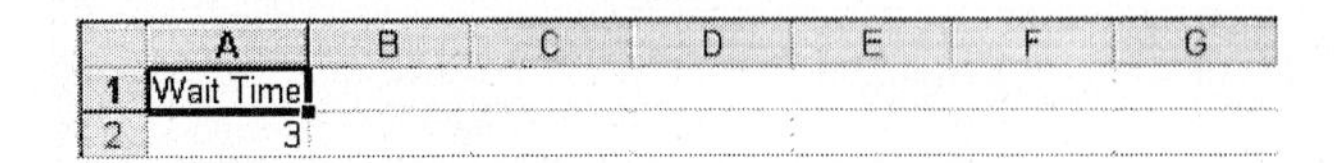

	A	B	C	D	E	F	G
1	Wait Time						
2	3						

3. Enter the label **Number of Customers** in cell B1 of the worksheet. You will use the minimum value as the first "bin" for tallying values, so enter **3** in cell B2. Because you will be using a class width of 1, continue by entering the values **4** through **14** in column B.

	A	B	C	D	E	F	G
1	Wait Time	Number of Customers					
2	11	3					
3	4	4					
4	13	5					
5	10	6					
6	7	7					
7	6	8					
8	6	9					
9	8	10					
10	8	11					
11	8	12					
12	5	13					
13	5	14					

4. At the top of the screen, click **Tools** and select **Data Analysis**. In the Data Analysis dialog box, select **Histogram** and click **OK**.

If Data Analysis does not appear as a choice in the Tools menu, you will need to load the Microsoft Excel Analysis ToolPak add-in. Follow the procedure in Section GS 8.1 before continuing.

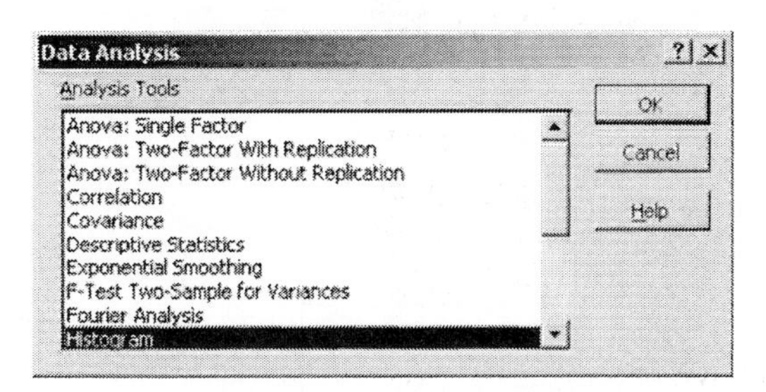

5. Complete the fields in the histogram dialog box as shown below. Be sure to select **Labels** and **Chart Output**. Click **OK**.

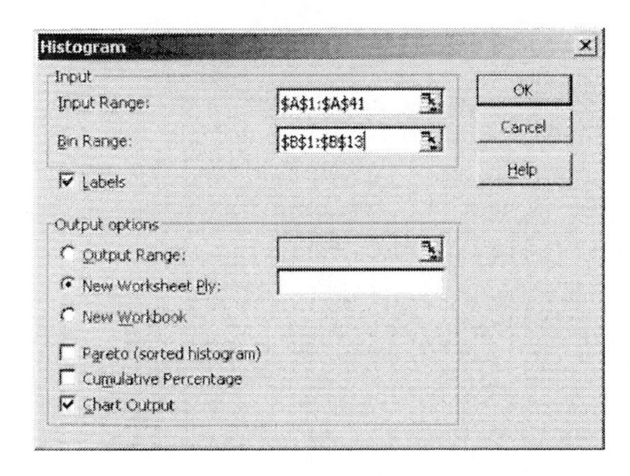

6. The histogram is placed in a new worksheet. You will now follow steps to modify this histogram so that it is presented in a more informative manner. Begin by making the chart taller so that it is easier to read. To do this, first click within the figure near a border. Black square handles appear. Click on the center handle at the bottom border of the figure and drag it down a few rows.

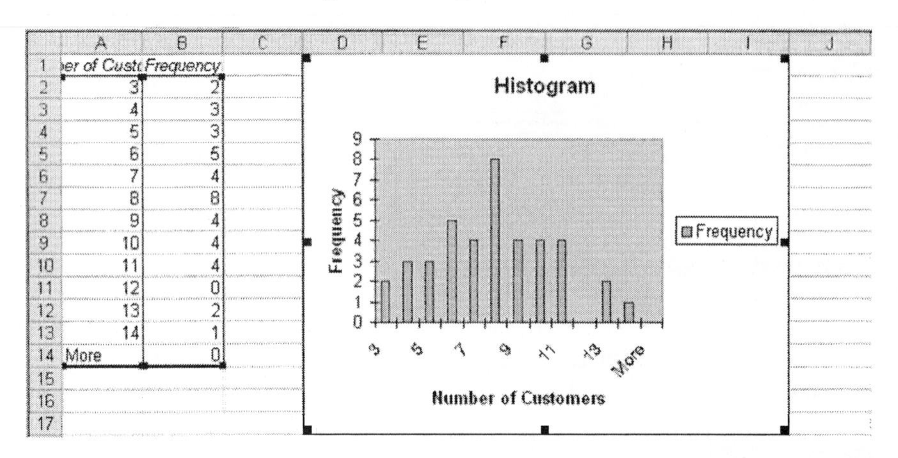

7. Remove the "More" category from the X axis. To do this, **right-click** in the gray plot area of the histogram and select **Source Data** from the shortcut menu that appears.

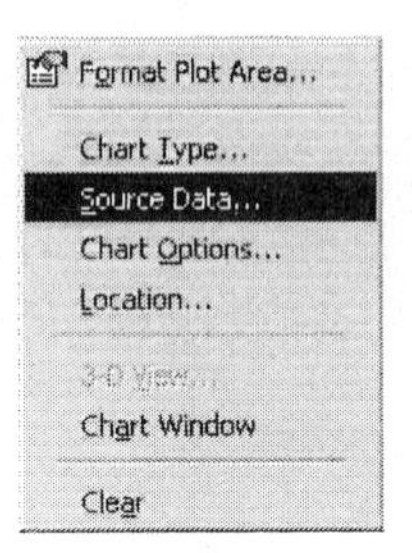

8. Click the Series tab at the top of the Source Data dialog box. "More" appears in cell A14 of the worksheet and its zero frequency appears in cell B14. To exclude that information, edit the entry in the Values window and the entry in the Category (X) axis labels window. The entry in the Values window should read **=Sheet1!$B$2:$B$13**. The entry in the Category (X) axis labels window should read **=Sheet1!$A$2:$A$13**. Click **OK**.

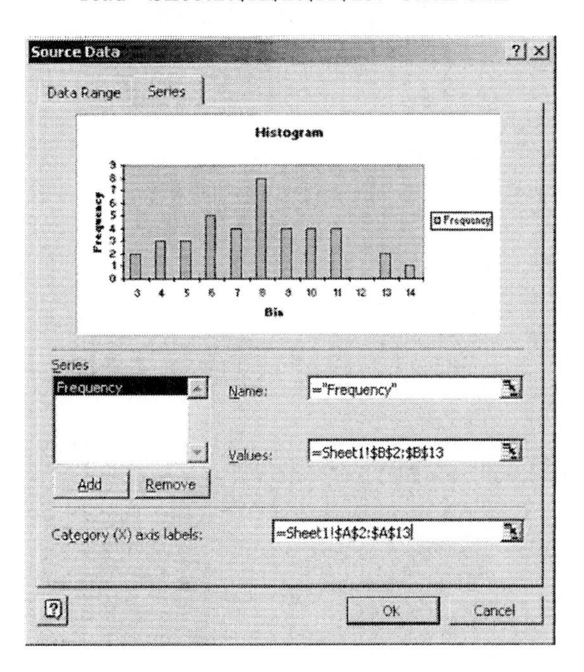

9. **Right-click** in the gray plot area of the histogram and select **Chart Options** from the menu that appears.

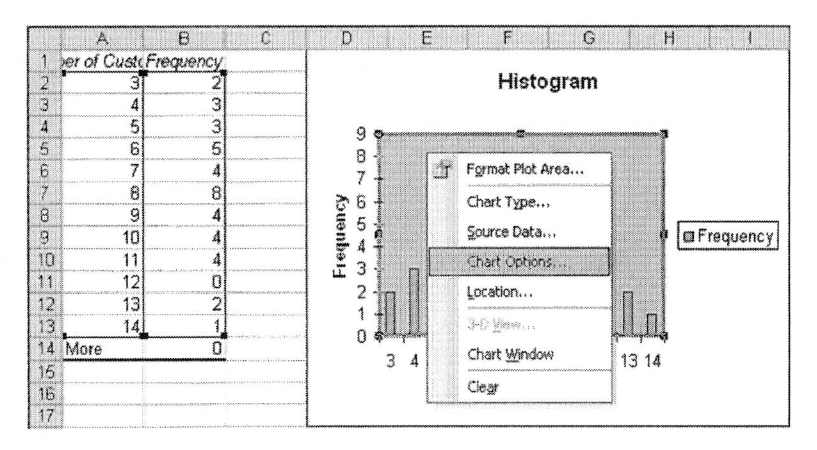

10. Click the **Titles** tab at the top of the Chart Options dialog box. In the Chart title field, replace "Histogram" with **Customers Waiting for a Table**. You should see **Number of Customers** in the Category (X) axis window and **Frequency** in the Value (Y) axis window.

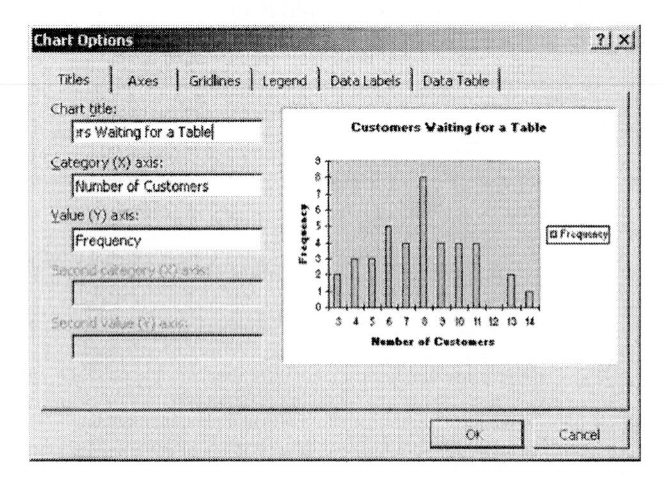

11. Click the **Gridlines** tab at the top of the dialog box. Under Value (Y) axis, click in the box next to **Major gridlines** so that a checkmark appears there.

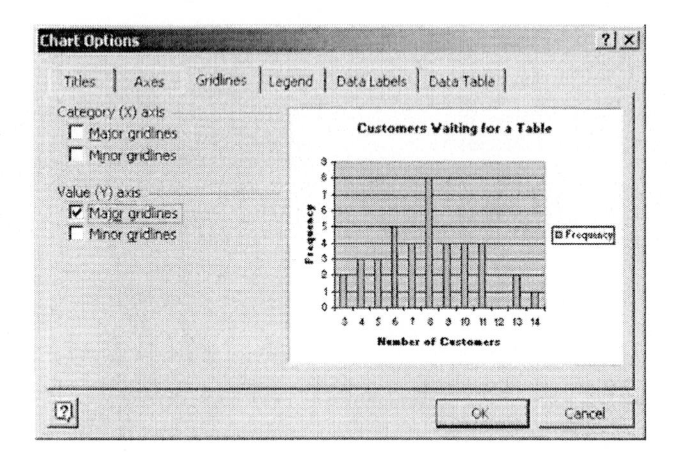

12. Click the **Legend** tab at the top of the dialog box. Click in the box to the left of **Show legend** to remove the checkmark. The removal of the checkmark will delete the frequency legend from the right side of the histogram chart. Click **OK**.

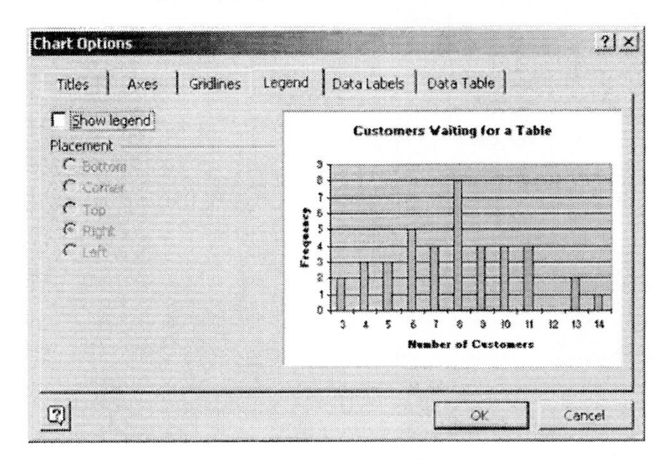

13. Remove the space between the vertical bars. **Right-click** directly on one of the vertical bars. Select **Format Data Series** from the shortcut menu that appears.

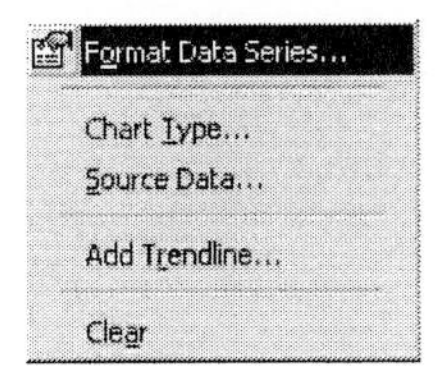

14. Click the **Options** tab at the top of the Format Data Series dialog box. Change the value in the Gap width box to 0. Click **OK**.

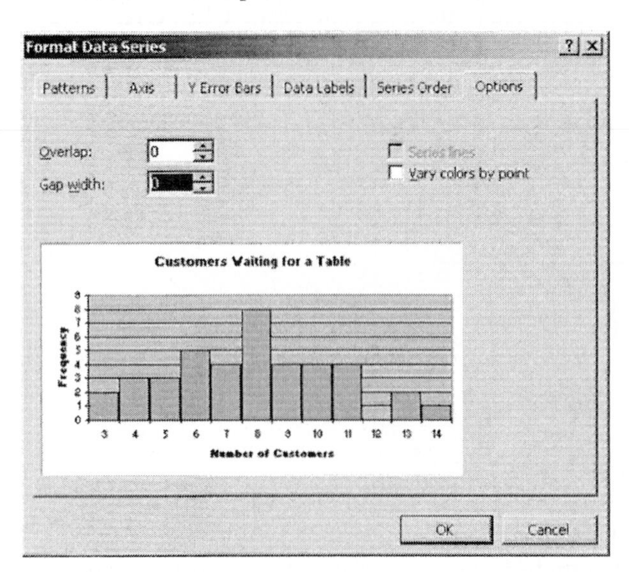

Your completed histogram should look similar to the one displayed below.

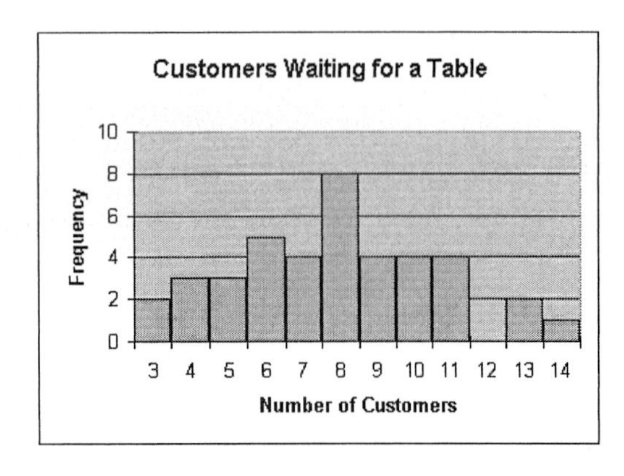

▶ Problem 29 (pg. 91)

Constructing a Frequency Histogram with a Class Limit of 2500

1. Open worksheet "2_2_29" in the Chapter 2 folder. The first two lines are shown at the top of the next page.

	A	B	C	D	E	F	G
1	Income						
2	24028						

2. Sort the data in ascending order. In the sorted data set, you can see that the minimum income is 21,677. Scroll down to find the maximum income. The maximum income, displayed in cell A52 is 42345.

For instructions on how to sort, refer to Sections GS5.1 and GS 5.2.

3. Enter the label **Lower Limit** in cell B1 of the worksheet. The textbook instructions tell you to use 20000 and as the lower limit of the first class and to use a class width of 2500. Enter **20000** in cell B2. Calculate the remaining lower limits by adding the class width of 2500 to the lower limit of each previous class. You will use a formula to do these computations in the Excel worksheet. Click in cell **B3** and key in **=B2+2500** as shown below. Press [**Enter**].

	A	B	C	D	E	F	G
1	Income	Lower Limit					
2	21677	20000					
3	22123	=B2+2500					
4	22252						

4. Click in cell **B3** (where 22500 now appears) and copy the contents of cell B3 to cells B4 through B11. Because the maximum income is 42345, you have calculated one more lower limit than is needed for the histogram. The value of 42500, however, will be used when calculating the upper limit of the last class.

	A	B	C	D	E	F	G
1	Income	Lower Limit					
2	21677	20000					
3	22123	22500					
4	22252	25000					
5	22581	27500					
6	23301	30000					
7	23528	32500					
8	23567	35000					
9	23584	37500					
10	23753	40000					
11	23889	42500					

5. Enter the label **Upper Limit** in cell C1. The upper limit is equal to one less than the lower limit of the next higher class. To do these calculations, you will enter a formula in the Excel worksheet. Click in cell **C2** and enter the formula **=B3-1** as shown in the worksheet below. Press [**Enter**].

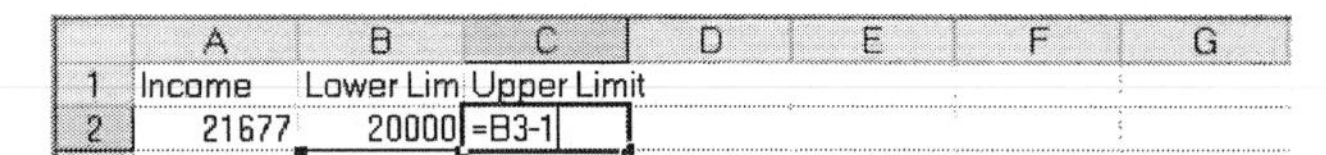

	A	B	C	D	E	F	G
1	Income	Lower Lim	Upper Limit				
2	21677	20000	=B3-1				

6. Copy the formula in C2 (where 22499 now appears) to cells C3 through C10. You will use these upper limits for the bins when you construct the histogram chart.

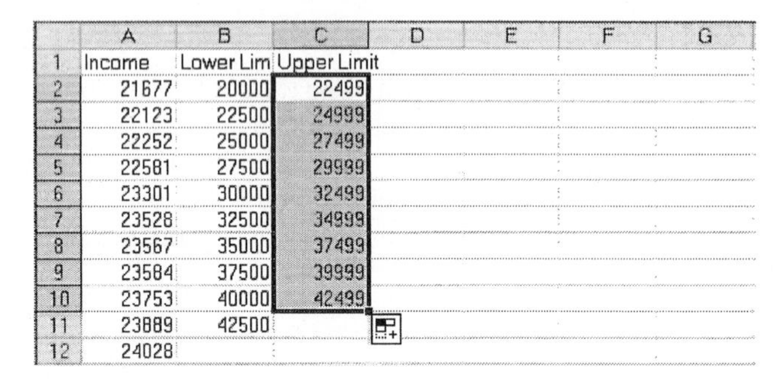

	A	B	C	D	E	F	G
1	Income	Lower Lim	Upper Limit				
2	21677	20000	22499				
3	22123	22500	24999				
4	22252	25000	27499				
5	22581	27500	29999				
6	23301	30000	32499				
7	23528	32500	34999				
8	23567	35000	37499				
9	23584	37500	39999				
10	23753	40000	42499				
11	23889	42500					
12	24028						

7. At the top of the screen, click **Tools** and select **Data Analysis**. In the Data Analysis dialog box, select **Histogram** and click **OK**.

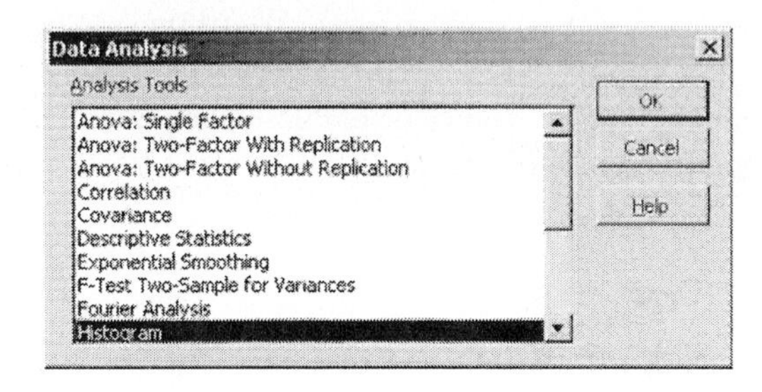

If Data Analysis does not appear as a choice in the Tools menu, you will need to load the Microsoft Excel Analysis ToolPak add-in. Follow the procedure in Section GS 8.1 before continuing.

---

8. Complete the fields in the histogram dialog box as shown below. The Input Range is $A$1:$A$52 and the Bin Range is $C$1:$C$10. The bin is actually the upper limit of the class. Be sure to select **Labels** and **Chart Output**. Click **OK**.

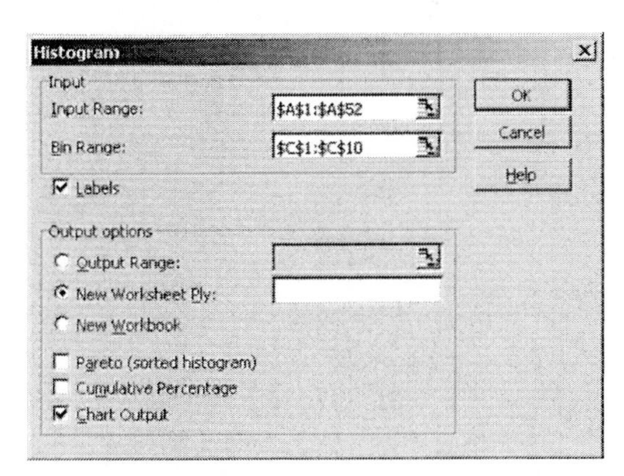

9. The histogram is placed in a new worksheet. You will now follow steps to modify this histogram so that it is presented in a more informative manner. Begin by making the chart taller so that it is easier to read. To do this, first click within the figure near a border. Black square handles appear. Click on the center handle at the bottom border of the figure and drag it down a few rows.

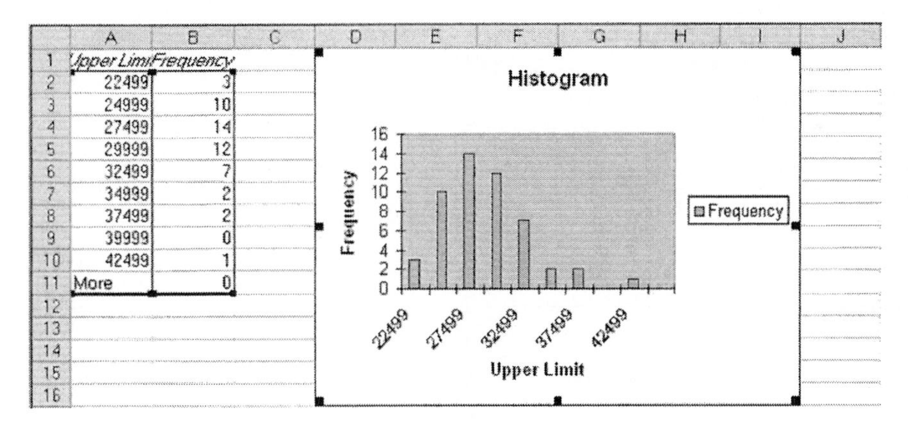

10. Remove the "More" category from the X axis. To do this, **right-click** in the gray plot area of the histogram and select **Source Data** from the shortcut menu that appears.

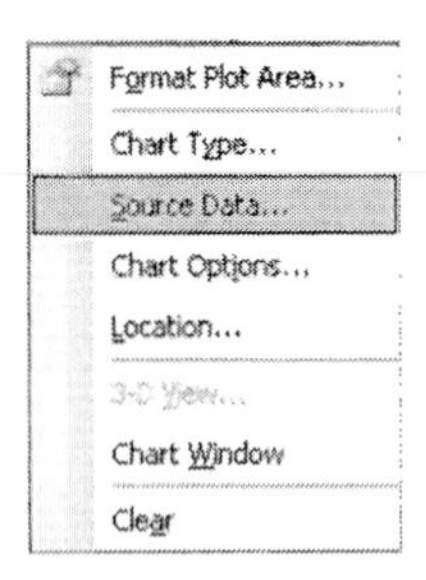

11. Click the **Series** tab at the top of the Source Data dialog box. "More" appears in cell A11 of the worksheet and its zero frequency appears in cell B11. To exclude that information, edit the entry in the Values window and the entry in the Category (X) axis labels window. The entry in the Values window should read =Sheet2!$B$2:$B$10 (assuming that your chart was placed in Sheet 2), and the entry in the Category (X) axis labels window should read =Sheet2!$A$2:$A$10. Click **OK**.

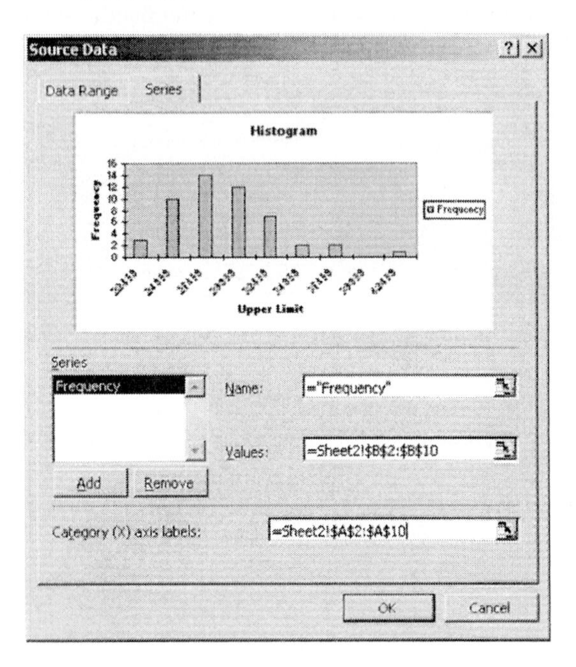

12. **Right-click** in the gray plot area of the histogram and select **Chart Options** from the menu that appears.

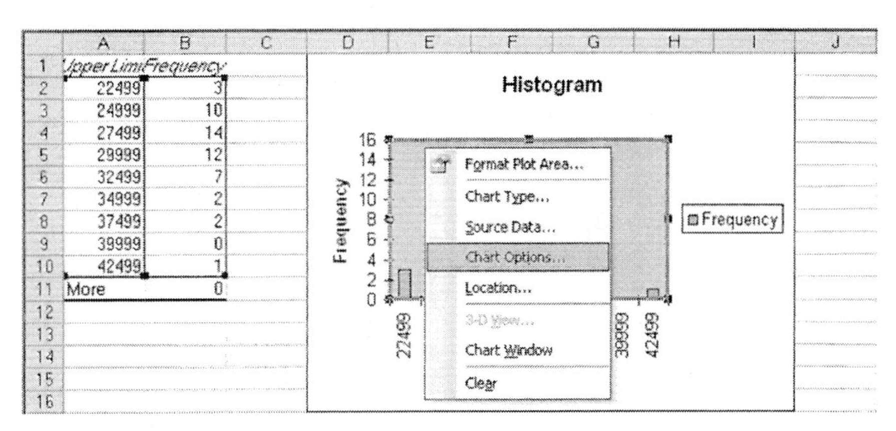

13. Click the **Titles** tab at the top of the Chart Options dialog box. In the Chart title field, replace "Histogram" with **Per Capita Disposable Income for the 50 States and District of Columbia in 2003**. In the Category(X) axis field, enter **Disposable Income**.

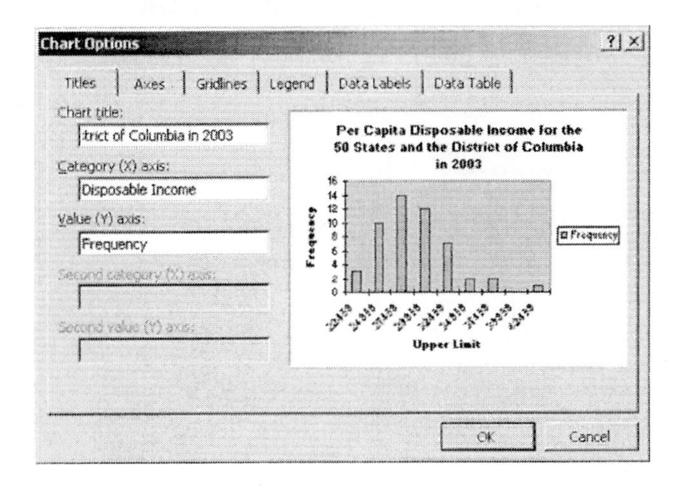

14. Click the **Gridlines** tab at the top of the dialog box. Under Value (Y) axis, click in the box next to Major gridlines so that a checkmark appears there.

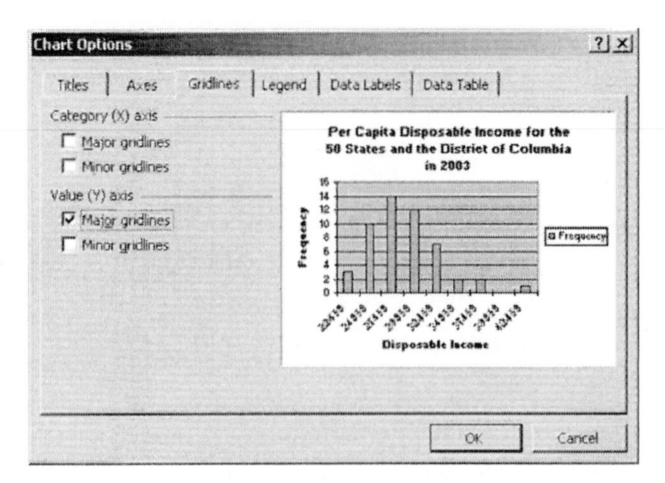

15. Click the **Legend** tab at the top of the dialog box. Click in the box to the left of **Show legend** to remove the checkmark. The removal of the checkmark will delete the frequency legend from the right side of the histogram chart. Click **OK**.

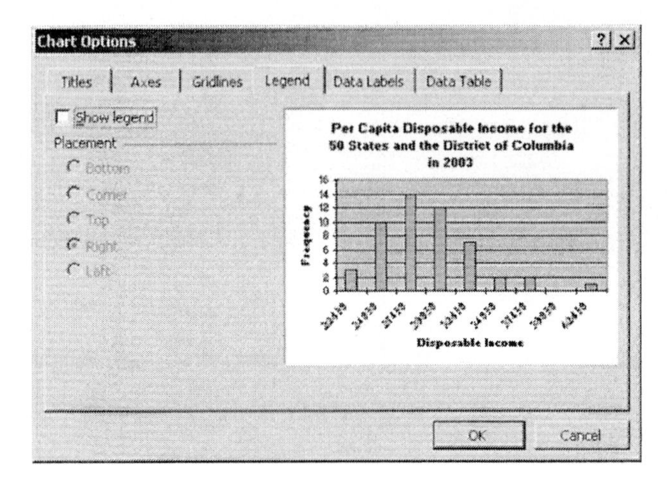

16. Remove the space between the vertical bars. **Right-click** directly on one of the vertical bars. Select **Format Data Series** from the shortcut menu that appears.

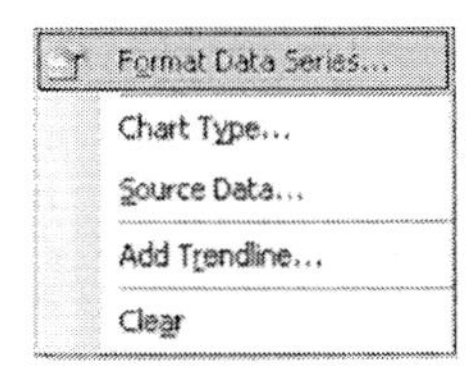

17. Click the **Options** tab at the top of the Format Series dialog box. Change the value in the Gap width box to 0. Click **OK**.

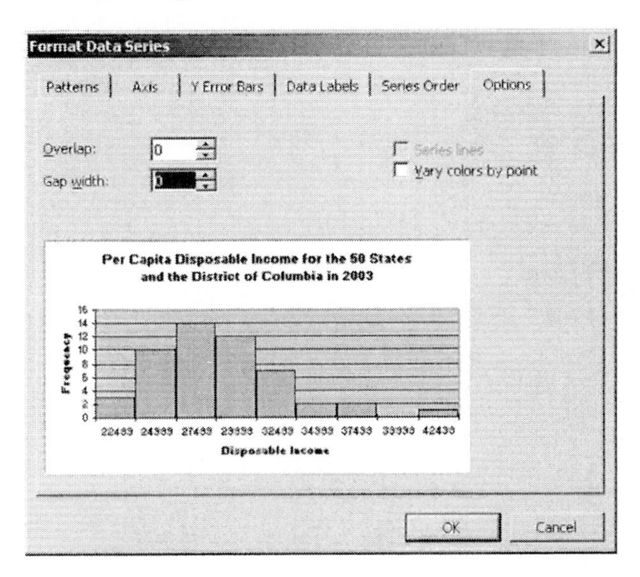

18. If you would like to include both the upper and lower interval for each class in the X-axis labels, follow these steps. Enter the upper and lower intervals in column A of the worksheet as shown below.

13	20000-22499
14	22500-24999
15	25000-27499
16	27500-29999
17	30000-32499
18	32500-34999
19	35000-37499
20	37500-39999
21	40000-42499

19. **Right-click** in the gray plot area of the chart and select **Source Data** from the menu that appears.

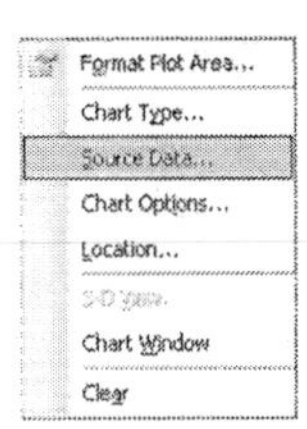

20. Click the **Series** tab at the top of the Source Data dialog box. Change the entry in the Category (X) labels so that it contains the range \$A\$13:\$A\$21 as shown below. Click **OK**.

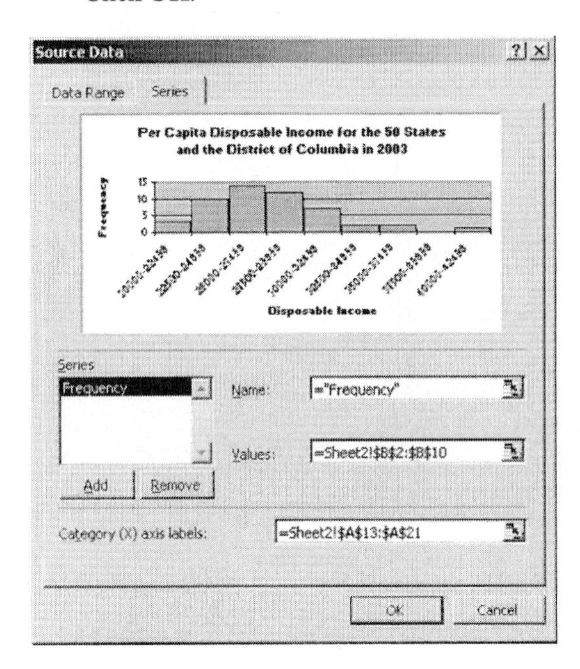

21. **Right-click** on any number on the X axis. I right-clicked on 20000. Select **Format Axis** from the shortcut menu.

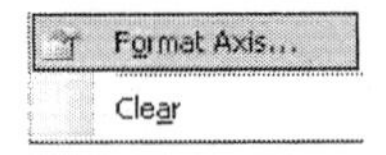

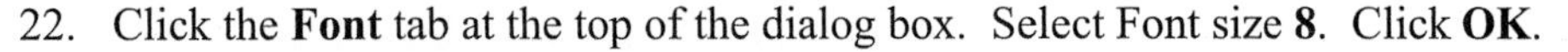

22. Click the **Font** tab at the top of the dialog box. Select Font size **8**. Click **OK**.

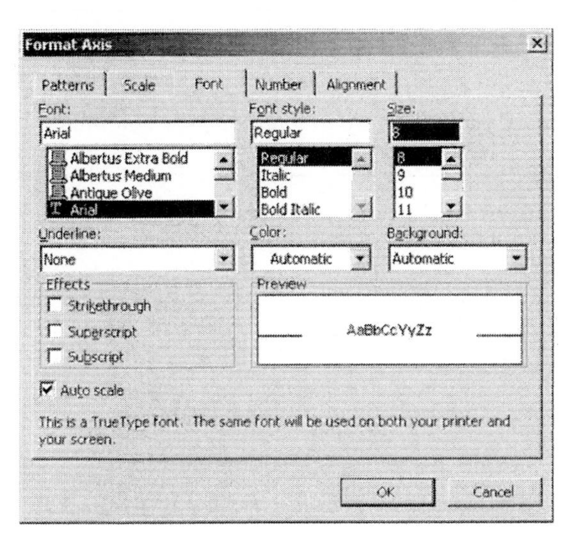

23. Click within the outer white area of the chart so that black handles appear. Click and drag on the handles to make the figure larger. Your completed chart should look similar to the one shown below. You can do further adjusting to font sizes and styles if you would like.

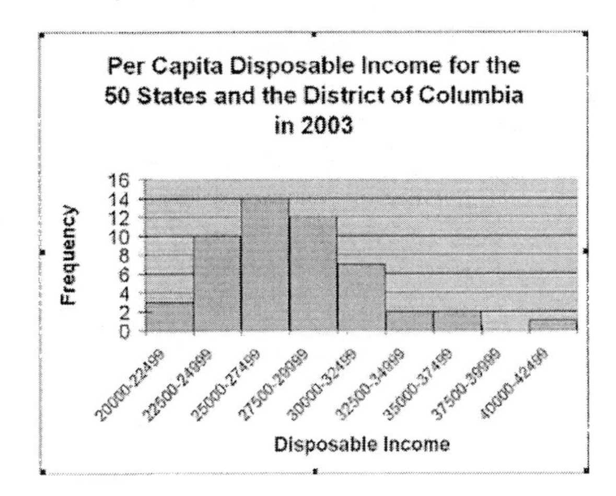

◀

► Problem 35 (pg. 93) Constructing a Stem-and-Leaf Plot

If the PHStat add-in has not been loaded, you will need to load it before continuing. Follow the instructions in Section GS 8.2.

1. Open worksheet "2_2_35" in the Chapter 2 folder. The first few rows are shown below.

	A	B	C	D	E	F	G
1	Pres. Age at Inaguration						
2	57						
3	54						
4	65						

2. At the top of the screen, click **PHStat**. Select **Descriptive Statistics →Stem-and-Leaf Display**.

3. In the Stem-and-Leaf Display dialog box, click in the **Variable Cell Range Window** so that the flashing I-beam appears there. Then click and drag over the Age at Inauguration data (cell A1 through cell A44) to enter the range in the Variable Cell Range window. Set the stem unit as **10**. Click **OK**.

*If you prefer to key in the range rather than clicking and dragging, you can key in **A1:A44**.*

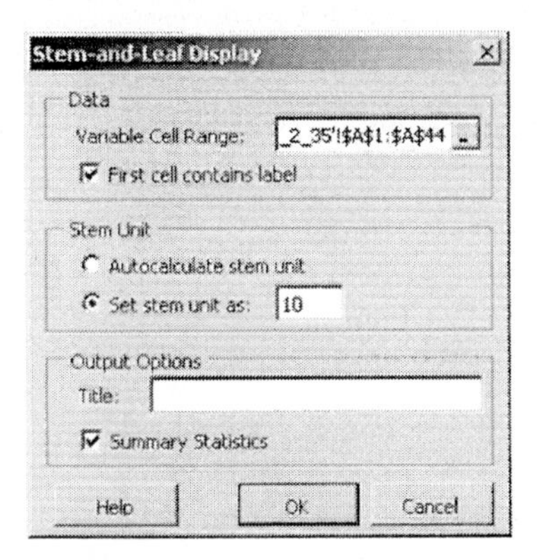

Your completed stem-and-leaf display output should look similar to the display shown below.

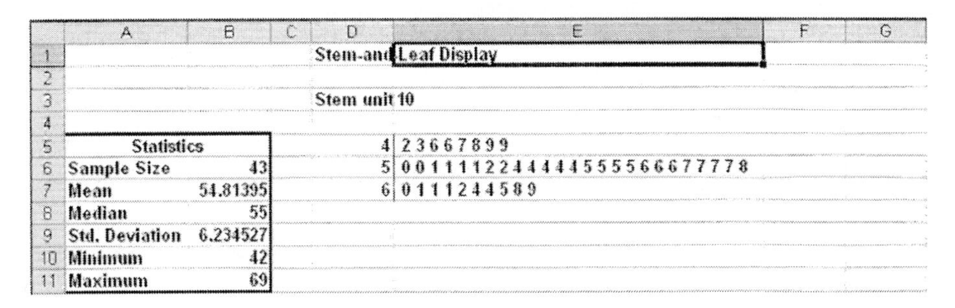

	A	B	C	D	E
1				Stem-and-Leaf Display	
2					
3				Stem unit 10	
4					
5	Statistics			4	2 3 6 6 7 8 9 9
6	Sample Size	43		5	0 0 1 1 1 1 2 2 4 4 4 4 4 5 5 5 5 6 6 6 7 7 7 7 8
7	Mean	54.81395		6	0 1 1 1 2 4 4 5 8 9
8	Median	55			
9	Std. Deviation	6.234527			
10	Minimum	42			
11	Maximum	69			

◀

Section 2.3 Additional Displays of Quantitative Data

▶ Problem 15 (pg. 102)

Constructing a Histogram, Frequency Polygon, Percentage Polygon, and Cumulative Percentage Polygon (Ogive)

If the PHStat add-in has not been loaded, you will need to load it before continuing. Follow the instructions in Section GS 8.2.

1. Open worksheet "2_2_29" in the Chapter 2 folder. Instructions for constructing a histogram of these data were given in the preceding section of this manual. In this section, you will learn how to use the PHStat add-in. If you have not already done so, complete steps 2 through 6 on pages 57-58 of this manual. The worksheet with those steps completed is shown below.

	A	B	C	D	E	F	G
1	Income	Lower Lim	Upper Limit				
2	21677	20000	22499				
3	22123	22500	24999				
4	22252	25000	27499				
5	22581	27500	29999				
6	23301	30000	32499				
7	23528	32500	34999				
8	23567	35000	37499				
9	23584	37500	39999				
10	23753	40000	42499				
11	23889	42500					

2. PHStat requires that you enter the midpoints. The first class is 20000 to 22499. To find the midpoint, first subtract 20000 from 22400 and then divide by 2. This is equal to 1249.5. So, the midpoint of the first class equals 20000 plus 1249.5 or 21249.5. The remaining midpoints increase in increments of 2500 up to 41249.5. You will be using Excel's Fill feature to obtain these midpoints. Enter **Midpoint** in cell **D1**. Enter **21249.5** in cell **D2**.

	A	B	C	D	E	F	G
1	Income	Lower Lim	Upper Lim	Midpoint			
2	21677	20000	22499	21249.5			

3. Click in cell **D2** where 21249.5 now appears. At the top of the screen, click **Edit**. Select **Fill** → **Series**.

4. Complete the Series dialog box as shown below. Be sure to select **Columns**. The step value is **2500**, and you want to stop at **41249.5**. Click **OK**.

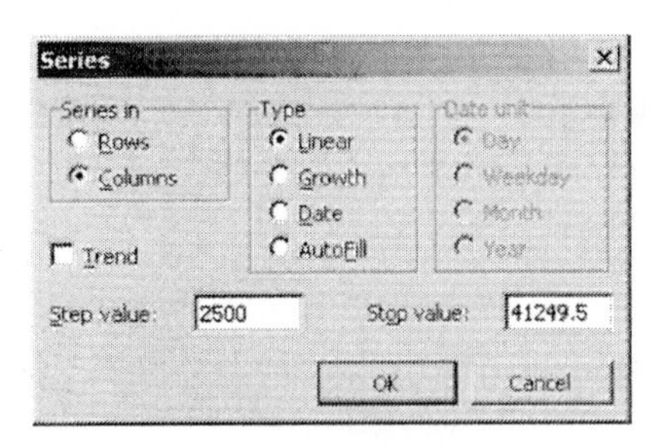

5. Your worksheet should now look like the one displayed below. At the top of the screen, click **PHStat**. Select **Descriptive Statistics** → **Histograms & Polygons**.

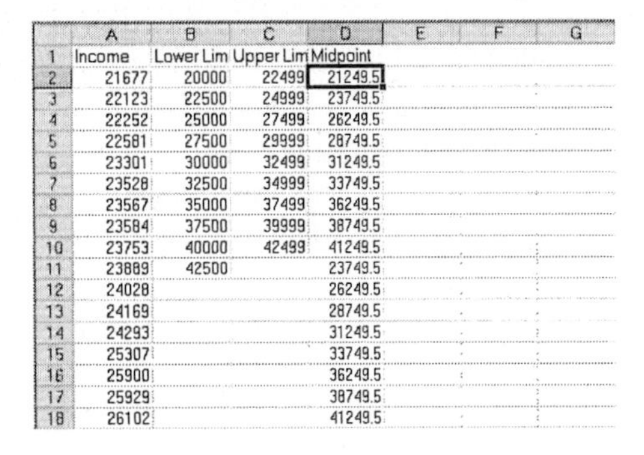

	A	B	C	D	E	F	G
1	Income	Lower Lim	Upper Lim	Midpoint			
2	21677	20000	22499	21249.5			
3	22123	22500	24999	23749.5			
4	22252	25000	27499	26249.5			
5	22581	27500	29999	28749.5			
6	23301	30000	32499	31249.5			
7	23528	32500	34999	33749.5			
8	23567	35000	37499	36249.5			
9	23584	37500	39999	38749.5			
10	23753	40000	42499	41249.5			
11	23889	42500		23749.5			
12	24028			26249.5			
13	24169			28749.5			
14	24293			31249.5			
15	25307			33749.5			
16	25900			36249.5			
17	25929			38749.5			
18	26102			41249.5			

6. Complete the dialog box as shown at the top of the next page. Click in the **Variable Cell Range** window. Then click and drag over the data range in column A of the worksheet to enter the variable cell range. Click in the **Bins Cell Range** window. Then click and drag over the Upper Limit range in column C of the worksheet to enter the bins. (Note: Bins are the upper limits of the classes.) Click

in the **Midpoints Cell Range** window and then click and drag over the Midpoint range in column D to enter the midpoints. Select **Single Group Variable**. Under Output Options, select all four: **Histogram**, **Frequency Polygon**, **Percentage Polygon**, and **Cumulative Percentage Polygon (Ogive)**. Click **OK**.

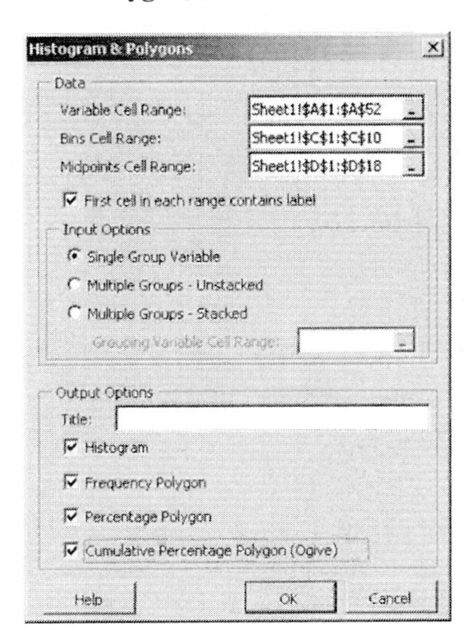

The four requested graphs are constructed and placed in four different worksheets. The sheet labeled Polygon 1 contains the frequency polygon.

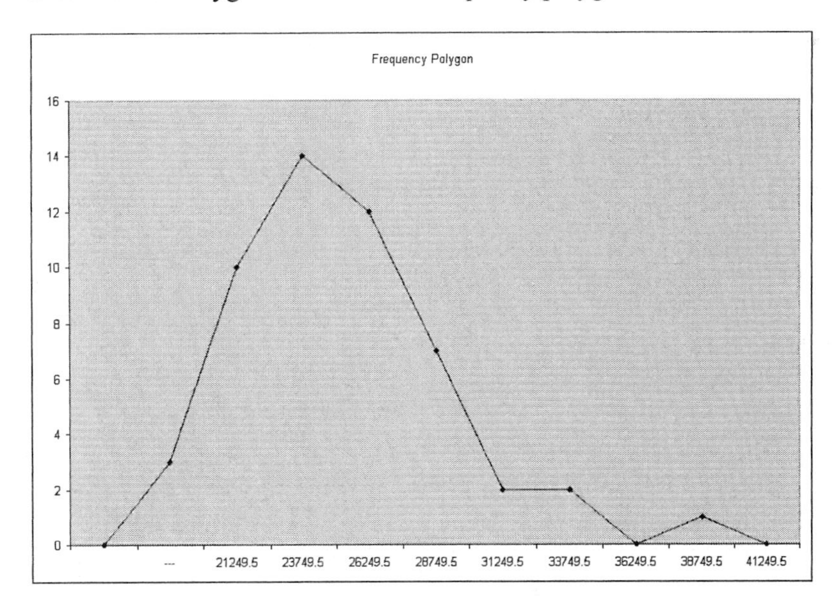

The sheet labeled Polygon 2 contains the percentage polygon.

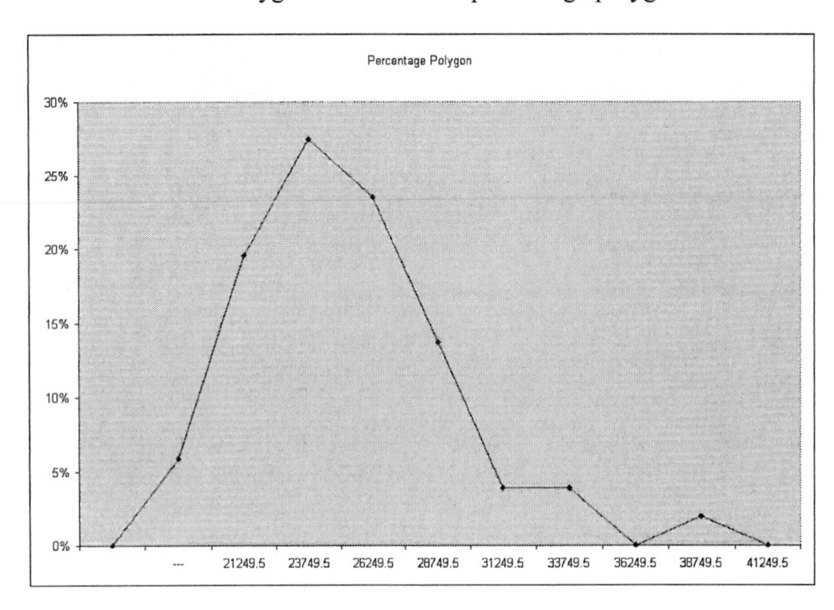

The sheet labeled Polygon 3 contains the cumulative percentage polygon (ogive).

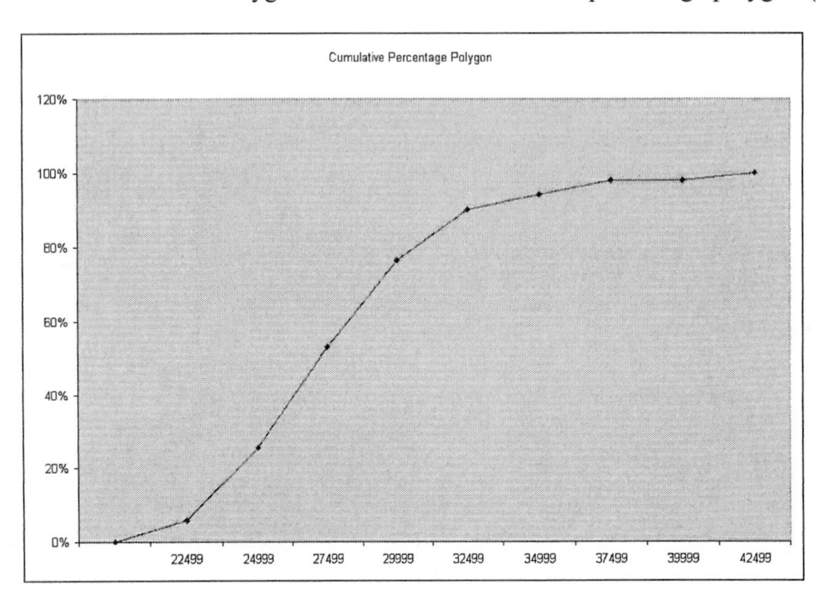

The sheet labeled Frequencies contains the histogram and summary data.

	A	B	C	D	E
1	Frequencies (Income)				
2	Upper Limit	Frequency	Percentage	Cumulative %	Midpts
3		0	0	0	
4	22499	3	5.88%	5.88%	—
5	24999	10	19.61%	25.49%	21249.5
6	27499	14	27.45%	52.94%	23749.5
7	29999	12	23.53%	76.47%	26249.5
8	32499	7	13.73%	90.20%	28749.5
9	34999	2	3.92%	94.12%	31249.5
10	37499	2	3.92%	98.04%	33749.5
11	39999	0	0.00%	98.04%	36249.5
12	42499	1	1.96%	100.00%	38749.5
13		0	0		41249.5
14					23749.5
15					26249.5
16					28749.5
17					31249.5
18					33749.5
19					36249.5
20					38749.5
21					41249.5

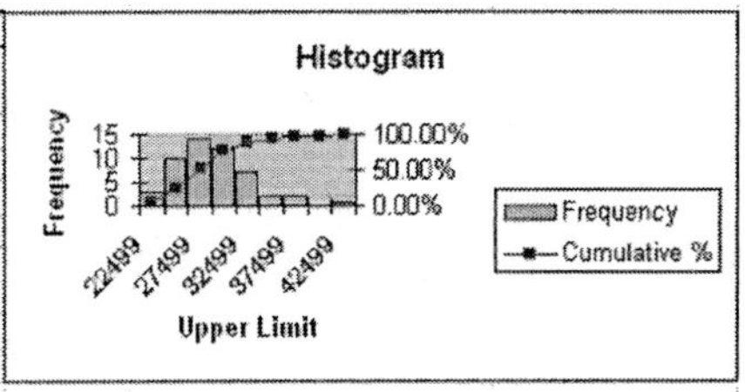

Histogram

Frequency: 0, 5, 10, 15

100.00%, 50.00%, 0.00%

22499, 27499, 32499, 37499, 42499

Upper Limit

Frequency

Cumulative %

Numerically Summarizing Data

CHAPTER 3

Section 3.1 Measures of Central Tendency

► Problem 25 (pg. 132)

Computing a Population Mean and a Sample Mean

If the PHStat add-in has not been loaded, you will need to load it before continuing. Follow the instructions in Section GS 8.2.

1. Open worksheet "3_1_25" in the Chapter 3 folder. The first two rows are shown below.

	A	B	C	D	E	F	G
1	Student	Pulse					
2	Perpectua	76					

2. Click in cell **B11** below the pulse rate data.

3. At the top of the screen, click **Insert** and select **Function**.

4. Select the **Statistical** category and the **AVERAGE** function. Click **OK**.

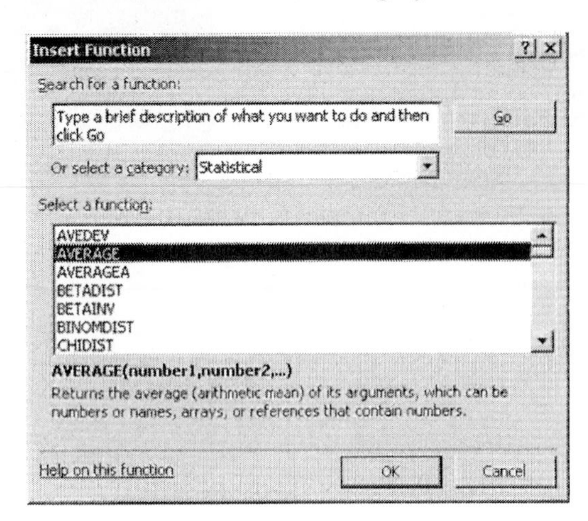

5. You should see the range B2:B10 in the Number 1 window of the dialog box. If this range does not appear in the window, you will need to enter it. Click **OK**.

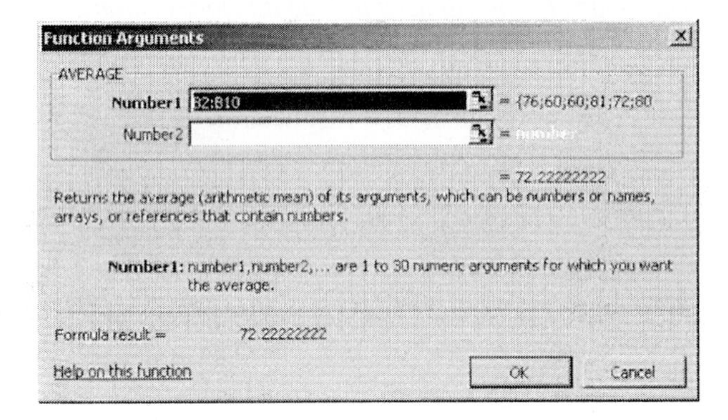

6. You will now see 72.22222 in cell B11. You will want to enter the label **Mean** in cell A11 for reference purposes.

7. You will now generate a simple random sample of size n = 3 from this population. At the top of the screen, click **PHStat**. Select **Sampling → Random Sample Generation**.

8. Complete the Random Sample Generator dialog box as shown below. A sample of three values will be selected from the range B1:B10. The first cell of this range contains a label (Pulse). The topmost cell of the output will contain the label "Home Run Sample." Click **OK**.

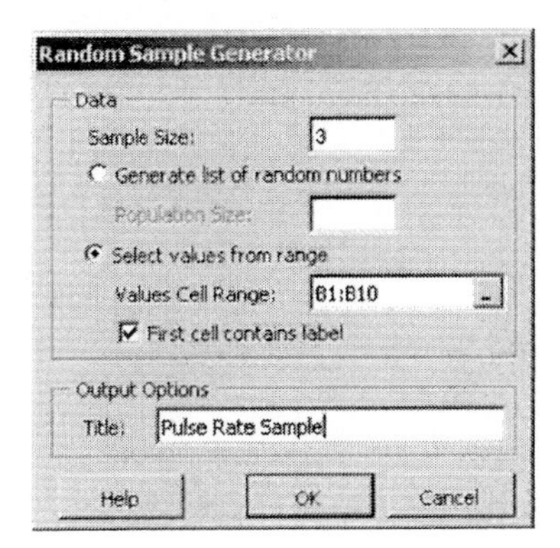

9. The output is displayed in a new worksheet named "RandomNumbers." Because the numbers were generated randomly, it is not likely that your output will be exactly the same. Click in cell **A5** below the sample data.

	A	B	C	D	E	F	G
1	Pulse Rate Sample						
2	72						
3	68						
4	76						

10. At the top of the screen, click **Insert** and select **Function**.

11. Select the **Statistical** category and the **AVERAGE** function. Click **OK**.

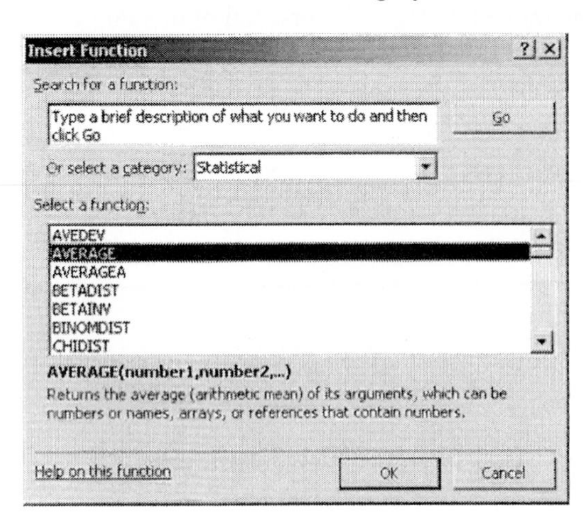

12. You should see the range A2:A4 in the Number 1 window of the dialog box. If this range does not appear, you will need to enter it. Click **OK**.

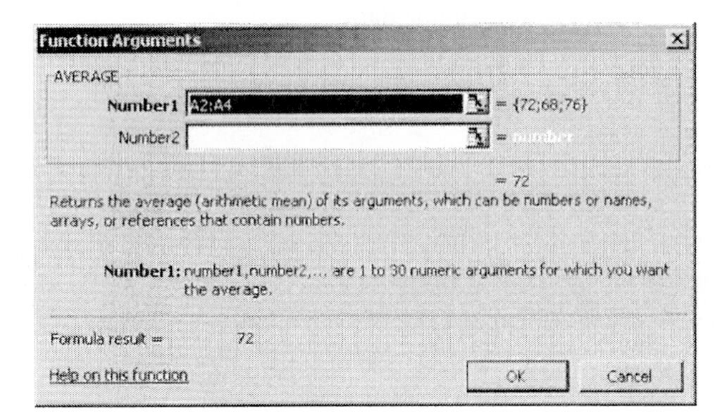

The sample mean will appear in cell A5. The mean for the sample that I generated is 72.

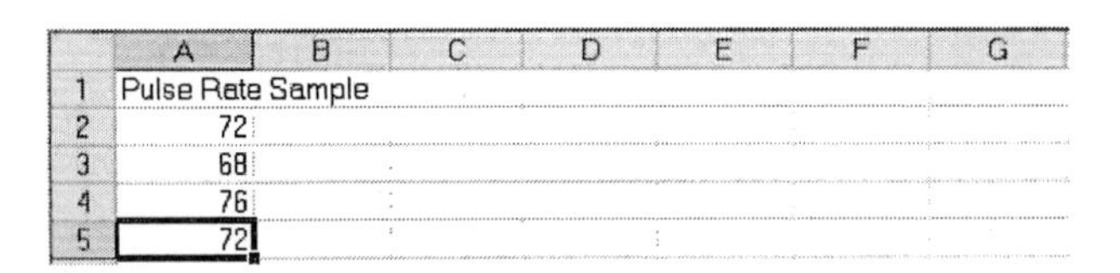

	A	B	C	D	E	F	G
1	Pulse Rate Sample						
2	72						
3	68						
4	76						
5	72						

13. Repeat steps 7-12 to generate another random sample and calculate its mean.

◀

► Problem 31 (pg. 133)

Computing the Mean and Median

1. Open worksheet "3_1_31" in the Chapter 3 folder. The first two rows are shown below.

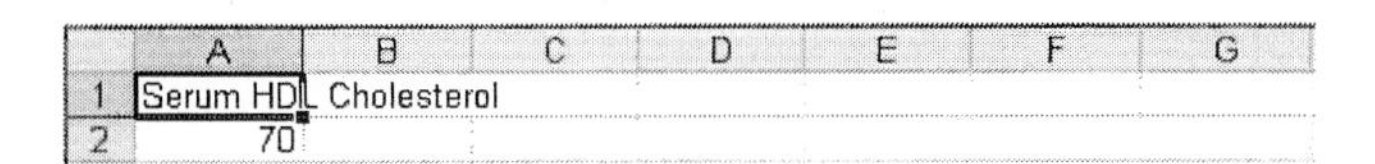

	A	B	C	D	E	F	G
1	Serum HDL Cholesterol						
2	70						

2. At the top of the screen, click **Tools** and select **Data Analysis**. In the Data Analysis dialog box, select **Descriptive Statistics** and click **OK**.

If Data Analysis does not appear as a choice in the Tools menu, you will need to load the Microsoft Excel Analysis ToolPak add-in. Follow the procedure in Section GS 8.1 before continuing.

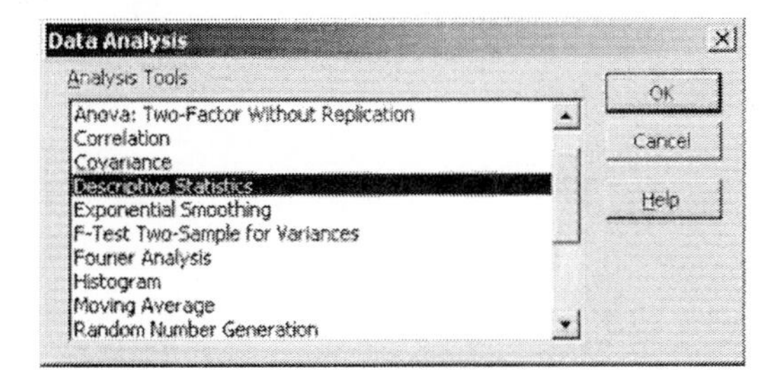

3. Complete the fields in the Descriptive Statistics dialog box as shown below. Be sure to click in the boxes to the left of **Labels in First Row** and **Summary Statistics**. Click **OK**.

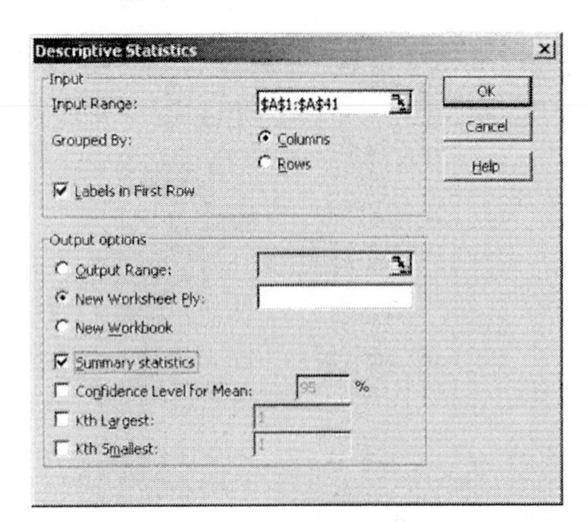

4. Make the first column wider so that you can read all the output labels. The mean is 51.125 and the median is 51.

	A	B	C	D	E	F	G
1	*Serum HDL Cholesterol*						
2							
3	Mean	51.125					
4	Standard Error	1.717253					
5	Median	51					
6	Mode	56					
7	Standard Deviation	10.86086					
8	Sample Variance	117.9583					
9	Kurtosis	-0.25119					
10	Skewness	-0.01096					
11	Range	45					
12	Minimum	28					
13	Maximum	73					
14	Sum	2045					
15	Count	40					

◄

▶ Problem 39 (pg. 134) Determining the Mode of Qualitative Data

1. Open worksheet "3_1_39" in the Chapter 3 folder. The first two rows are shown below.

2. The mode of qualitative data is the value with the highest frequency. You will need to construct a frequency distribution table in order to identify the mode. At the top of the screen, click **Data** and select **Pivot Table and Pivot Chart Report**.

3. Select **Microsoft Office Excel list or database** and **Pivot Table** in the Step 1 of 3 dialog box. Click **Next>**.

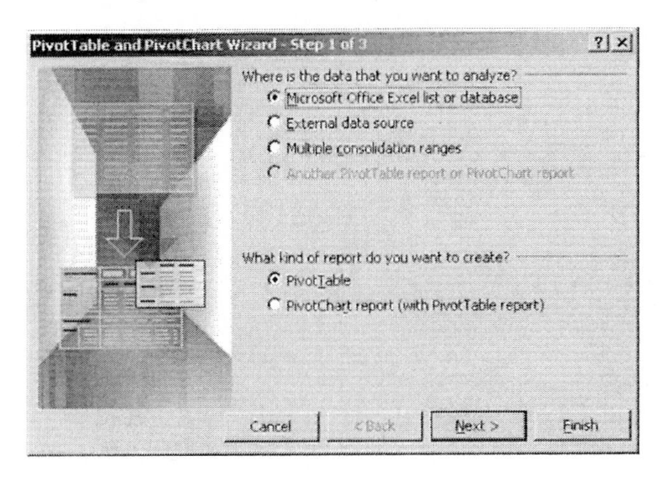

4. Check the range in the Step 2 of 3 dialog box. It should be $A$1:$A$41. Click **Next>**.

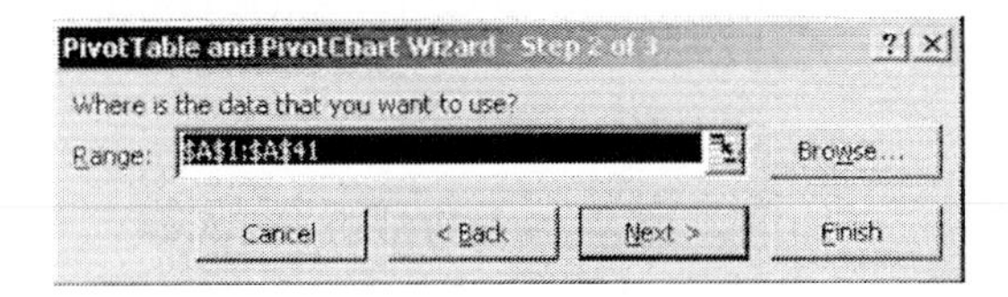

5. Click **Layout** in the Step 3 of 3 dialog box.

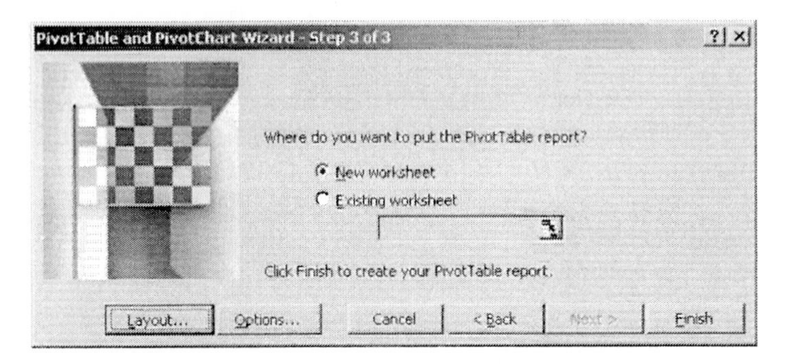

6. Drag the **C1-T** field button to ROW. Drag the **C1-T** field button to DATA. Click **OK**.

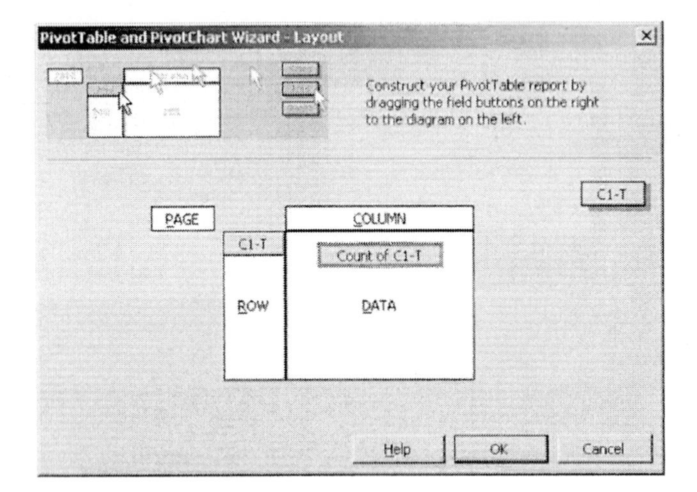

7. Select **New worksheet** and click **Finish**.

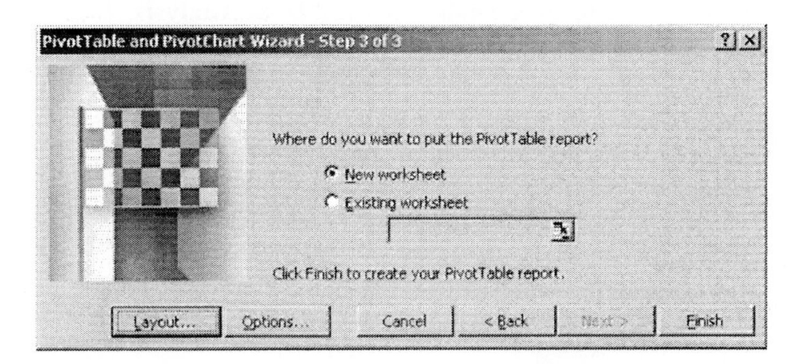

The mode is Bush with a frequency of 21.

	A	B	C	D	E	F	G
1							
2							
3	Count of C1-T						
4	C1-T	Total					
5	Badnarik	1					
6	Bush	21					
7	Kerry	17					
8	Nader	1					
9	Grand Total	40					

◀

▶ Problem 42 (pg. 135) Determining the Mean, Median, and Mode of Three Quantitative Variables

1. Open worksheet "3_1_42" in the Chapter 3 folder. The first two lines are shown below.

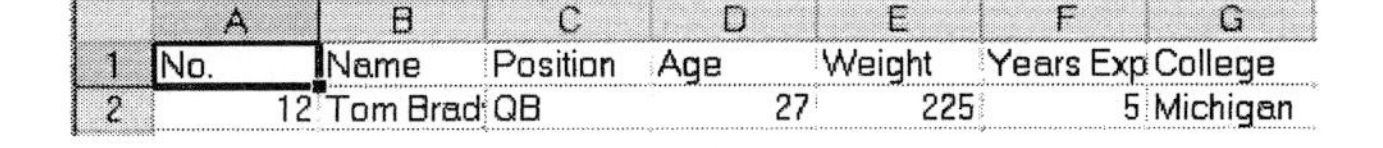

	A	B	C	D	E	F	G
1	No.	Name	Position	Age	Weight	Years Exp	College
2	12	Tom Brad	QB	27	225	5	Michigan

2. You will be finding the mean, median and mode of age, weight, and years of experience. At the top of the screen, click **Tools** and select **Data Analysis**. In the Data analysis dialog box, select **Descriptive Statistics** and click **OK**.

If Data Analysis does not appear as a choice in the Tools menu, you will need to load the Microsoft Excel Analysis ToolPak add-in. Follow the procedure in Section GS 8.1 before continuing.

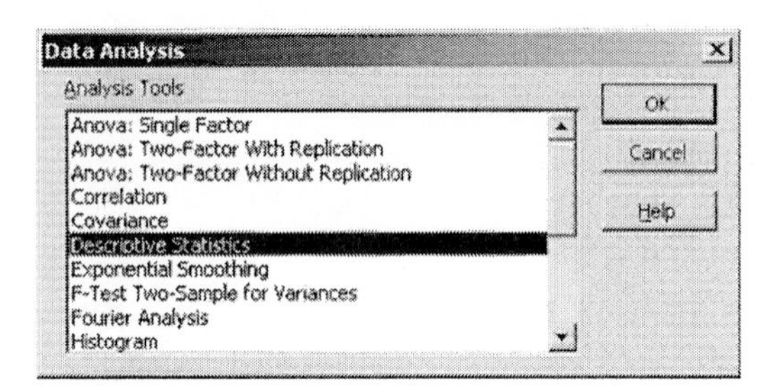

3. Complete the fields in the Descriptive Statistics dialog box as shown below. You are entering the range of all three quantitative variables. Be sure to click in the boxes to the left of **Labels in First Row** and **Summary Statistics**. Click **OK**.

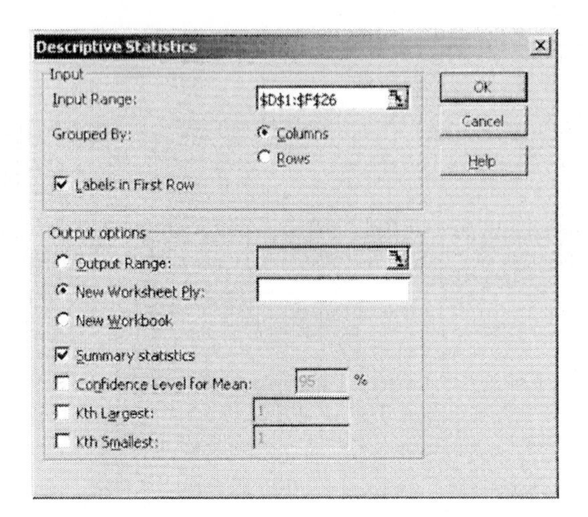

4. Make the columns wider so that you can read all the output labels.

	A	B	C	D	E	F	G
1	*Age*		*Weight*		*Years Exp.*		
2							
3	Mean	27.08	Mean	249.8	Mean	4.64	
4	Standard Error	0.50636	Standard Error	8.646965	Standard Error	0.528772	
5	Median	27	Median	245	Median	4	
6	Mode	26	Mode	305	Mode	3	
7	Standard Deviation	2.531798	Standard Deviation	43.23482	Standard Deviation	2.643861	
8	Sample Variance	6.41	Sample Variance	1869.25	Sample Variance	6.99	
9	Kurtosis	0.563541	Kurtosis	-1.50475	Kurtosis	-0.37753	
10	Skewness	0.923906	Skewness	0.245078	Skewness	0.586561	
11	Range	9	Range	122	Range	10	
12	Minimum	24	Minimum	190	Minimum	0	
13	Maximum	33	Maximum	312	Maximum	10	
14	Sum	677	Sum	6245	Sum	116	
15	Count	25	Count	25	Count	25	

A word of caution is in order regarding the value reported for the mode. Three situations are possible: l) If all values occur only once in a distribution, Excel will return #N/A. 2) If a variable has only one mode, the MODE function will return that value. 3) If a variable has more than one mode, however, the MODE function will still return only one value. The value that is returned will be the one associated with the modal value that occurs first in the data set. To check for multiple modes, I suggest that you construct a frequency distribution.

◄

Section 3.2 Measures of Dispersion

► Example 1 (pg. 137) Computing the Mean, Range, Variance, and Standard Deviation for a Sample

1. Open worksheet "3_2_Ex1" in the Chapter 3 folder. The first few rows are shown below.

	A	B	C	D	E	F	G
1	University	University B					
2	73	86					
3	103	91					
4	91	107					

2. At the top of the screen, click **Tools** and select **Data Analysis**.

If Data Analysis does not appear as a choice in the Tools menu, you will need to load the Microsoft Excel Analysis ToolPak add-in. Follow the procedure in Section GS 8.1 before continuing.

---

3. Select **Descriptive Statistics** and click **OK**.

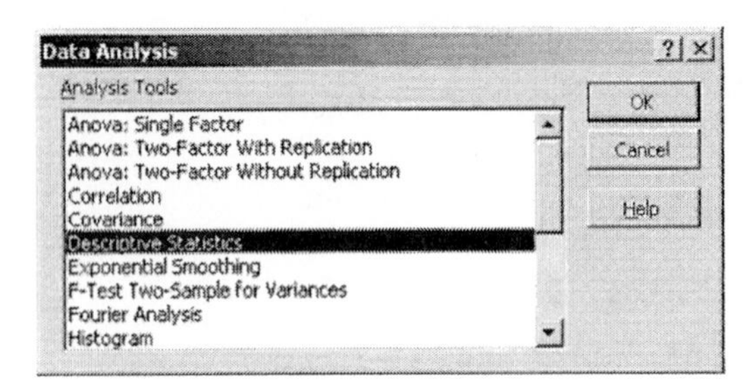

4. Complete the Descriptive Statistics dialog box as shown below. You will be obtaining descriptive statistics for the data in the worksheet range A1:B101. There are labels ("University A" and "University B") in the top cell of each column. The output will be placed in a new worksheet. Be sure to click in the **Summary statistics** box to place a checkmark there. The output will include summary statistics for both universities. Click **OK**.

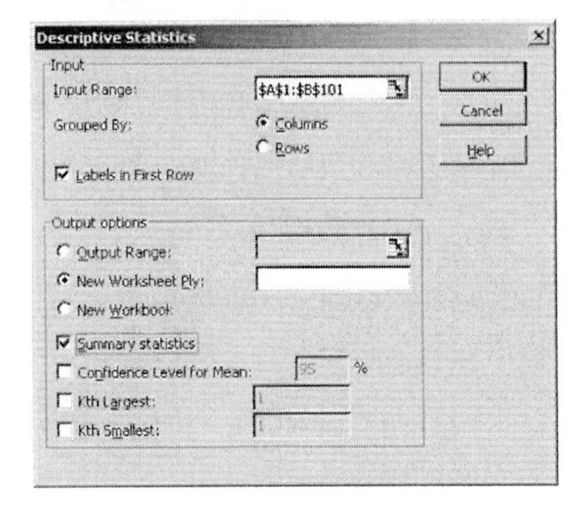

Your output will look similar to the output displayed below. I recommend that you make columns A and D wider so that you can read the complete output label for each summary value. The descriptive statistics output includes the mean, standard deviation, sample variance, and the range. Note that the variance and standard deviation are computed using the unbiased estimate formula that has n – 1 in the denominator.

	A	B	C	D	E	F	G
1	University A		University B				
2							
3	Mean	100.02	Mean	99.99			
4	Standard E	1.607996	Standard E	0.834302			
5	Median	102	Median	98			
6	Mode	103	Mode	107			
7	Standard D	16.07996	Standard D	8.343019			
8	Sample Va	258.5653	Sample Va	69.60596			
9	Kurtosis	-0.03652	Kurtosis	-0.65097			
10	Skewness	0.0468	Skewness	0.340169			
11	Range	81	Range	33			
12	Minimum	60	Minimum	86			
13	Maximum	141	Maximum	119			
14	Sum	10002	Sum	9999			
15	Count	100	Count	100			

◀

► Example 1 (pg. 137)

Constructing a Histogram

You are instructed to construct a histogram for each university's IQ score data. I will provide directions for University A only. You can then repeat the steps on your own to construct the histogram for University B.

1. Open worksheet "3_2_Ex1" in the Chapter 3 folder. The first few rows are shown below.

	A	B	C	D	E	F	G
1	University	University B					
2	73	86					
3	103	91					
4	91	107					

2. Excel's histogram procedure uses grouped data to generate a frequency distribution and a frequency histogram. The procedure requires that you indicate a "bin" for each class. The number that you specify for each bin is actually the upper limit of the class. You are instructed to use a lower class limit of 50 for the first class and a class width of 10. The upper limit is equal to one less than the lower limit of the next higher class. For this problem, the upper limits will be 59, 69, 79, etc. By

referring to the output of the previous example in this manual, you will see that the maximum value in University A is 141. So, the highest upper limit you will need 149. Enter **Bin** in cell C1 and key in the upper limits in column C as shown below.

	A	B	C	D	E	F	G
1	University	University	Bin				
2	73	86	59				
3	103	91	69				
4	91	107	79				
5	93	94	89				
6	136	105	99				
7	108	107	109				
8	92	89	119				
9	104	96	129				
10	90	102	139				
11	78	96	149				

3. Click **Tools** and select **Data Analysis**. Select **Histogram** and click **OK**.

If Data Analysis does not appear as a choice in the Tools menu, you will need to load the Microsoft Excel Analysis ToolPak add-in. Follow the procedure in Section GS 8.1 before continuing.

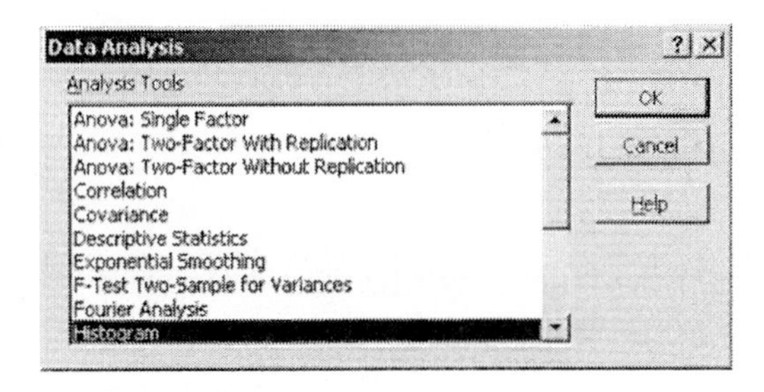

4. Complete the Histogram dialog box as shown below and click **OK**.

Note the checkmark in the box to the left of Labels. In your worksheet, "University A" appears in cell A1, and "Bin" appears in cell C1. Because you included these cells in the Input Range and Bin Range, respectively, you need to let Excel know that these cells contain labels rather than data. Otherwise Excel will attempt to use the information in these cells when constructing the frequency distribution and histogram.

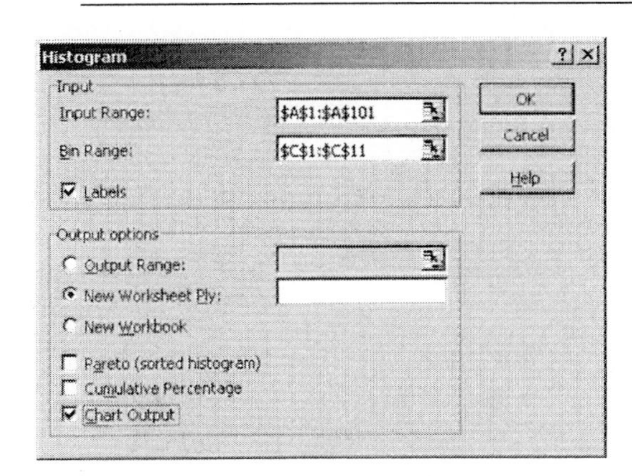

You should see output similar to the output displayed below.

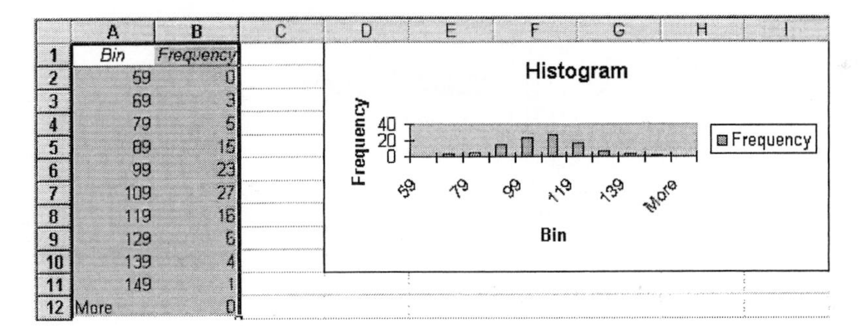

Bin	Frequency
59	0
69	3
79	5
89	15
99	23
109	27
119	16
129	6
139	4
149	1
More	0

5. You will now modify this histogram so that it is displayed in a more informative manner. First, make the chart taller so that it is easier to read. To do this, click within the figure near a border. Black square handles appear. Click on the center handle on the bottom border of the figure and drag it down a few rows.

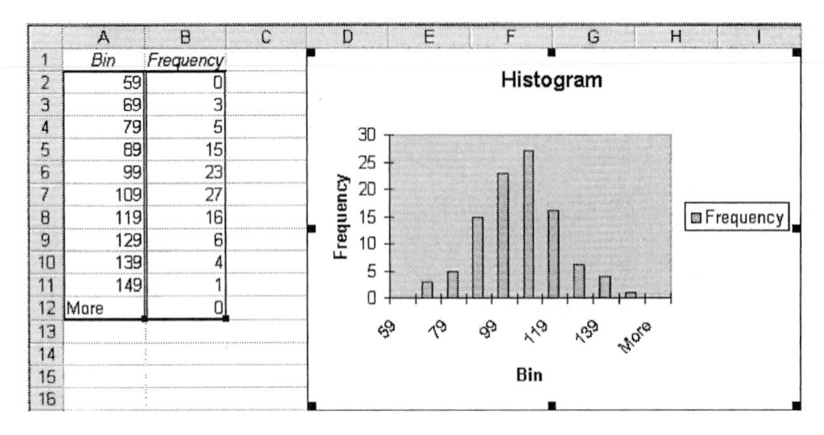

Bin	Frequency
59	0
69	3
79	5
89	15
99	23
109	27
119	16
129	6
139	4
149	1
More	0

6. Next, remove the space between the vertical bars. **Right-click** on one of the vertical bars. Select **Format Data Series** from the shortcut menu that appears.

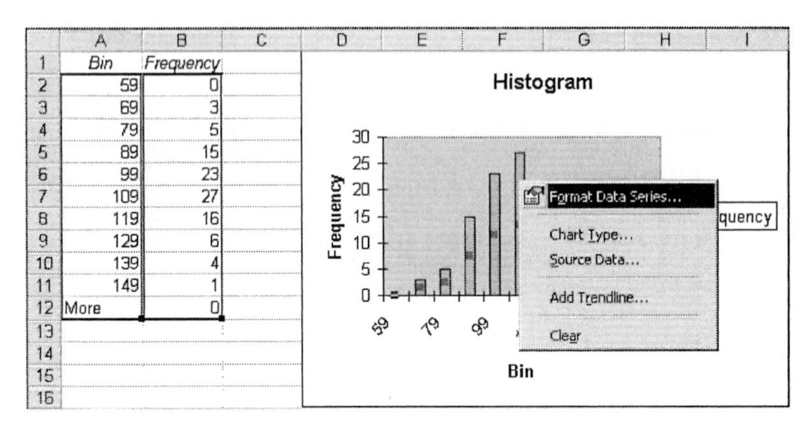

Bin	Frequency
59	0
69	3
79	5
89	15
99	23
109	27
119	16
129	6
139	4
149	1
More	0

7. Click the **Options** tab at the top of the Format Data Series dialog box. Change the value in the Gap width box to **0**. Click **OK**.

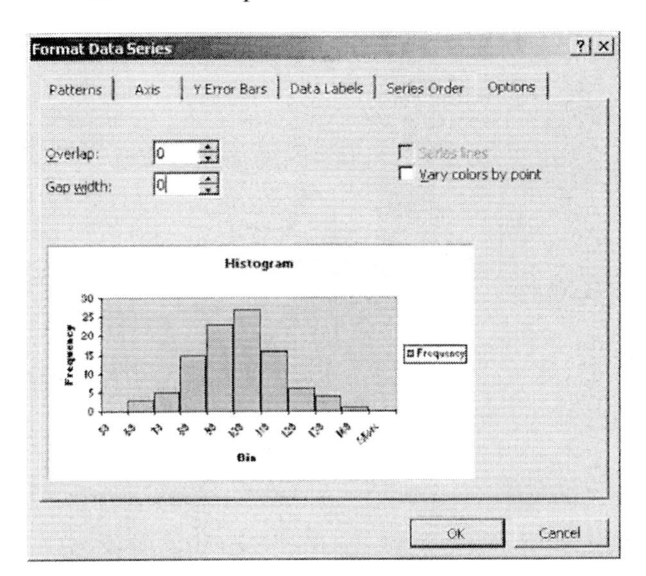

8. Change the X-axis values from upper limits to lower limits. To do this, first enter the lower limits in column C of the Excel worksheet as shown below.

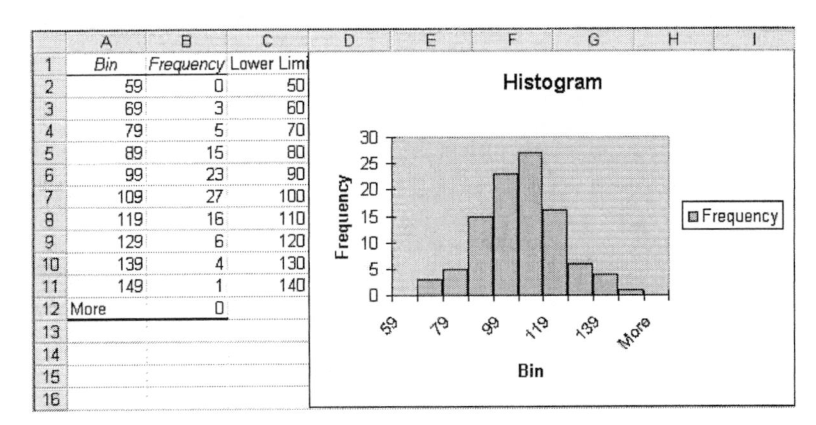

	A	B	C
1	Bin	Frequency	Lower Limi
2	59	0	50
3	69	3	60
4	79	5	70
5	89	15	80
6	99	23	90
7	109	27	100
8	119	16	110
9	129	6	120
10	139	4	130
11	149	1	140
12	More	0	
13			
14			
15			
16			

9. **Right-click** on a vertical bar in the chart. Select **Source Data** from the shortcut menu that appears.

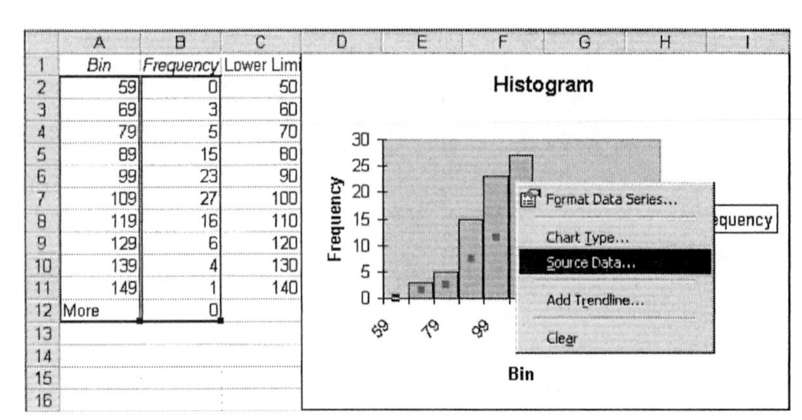

10. Click the **Series** tab at the top of the Source Data dialog box. The ranges displayed in the Values field and the Category (X) axis label field refer to the frequency distribution table in the top left of the worksheet. You do not want to include row 12, because that is the row containing information related to the "More" category. You also want the lower limit values in column C to be displayed on the X axis rather than the column A Bin values. First, change the 12 to 11 in the **Values** field. Assuming that your histogram is displayed in sheet 2, the entry should read **=Sheet2!$B$2:$B$11**. Next, edit the **Category (X) axis labels** field so that the entry reads **=Sheet2!$C$2:$C$11**. Click **OK**.

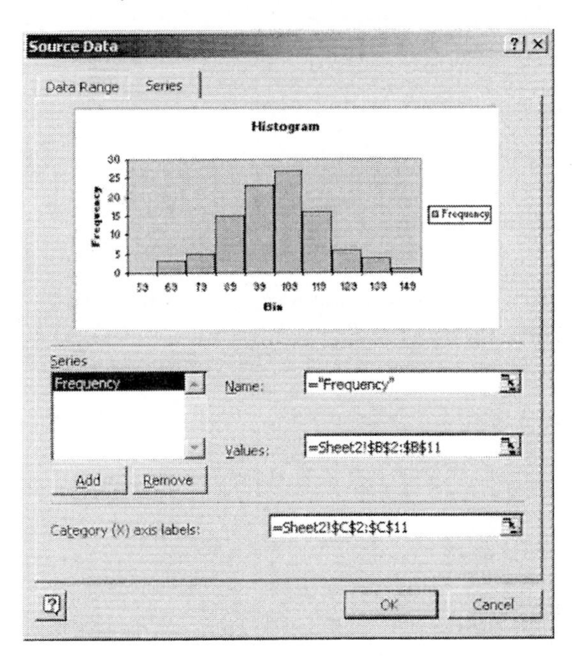

11. You will use Chart Options to modify three aspects of the histogram: Titles, gridlines, and legend. **Right-click** in the gray plot area of the chart and select **Chart Options** from the shortcut menu that appears.

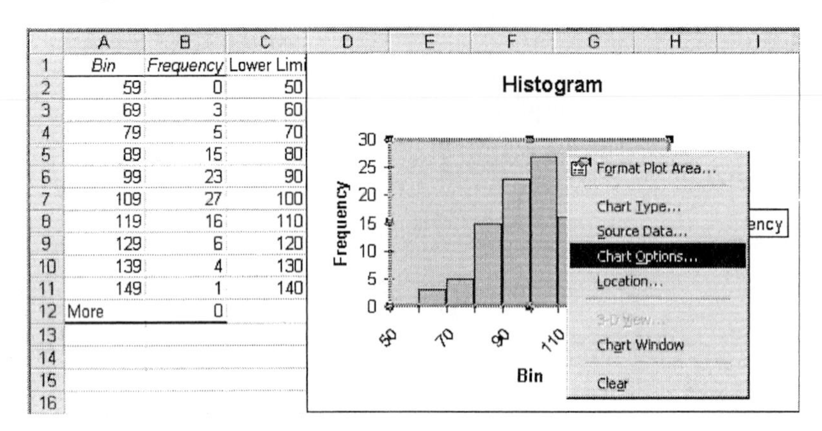

	A	B	C
1	Bin	Frequency	Lower Limi
2	59	0	50
3	69	3	60
4	79	5	70
5	89	15	80
6	99	23	90
7	109	27	100
8	119	16	110
9	129	6	120
10	139	4	130
11	149	1	140
12	More	0	

12. Click the **Titles** tab at the top of the Chart Options dialog box. Change the Chart title from "Histogram" to **University A IQ Scores**. Change the Category (X) axis label from "Bin" to **IQ Scores**.

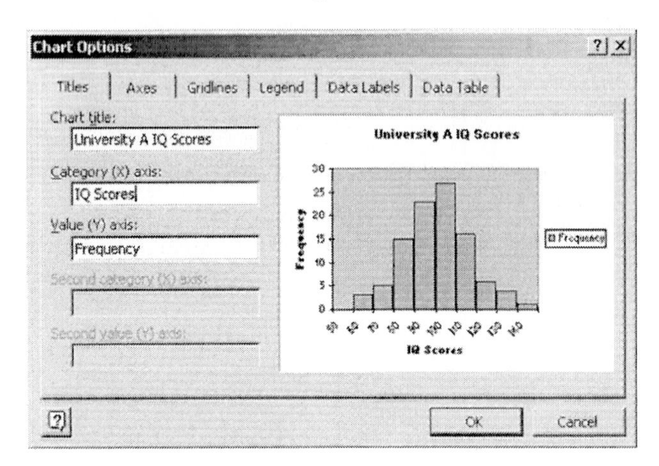

13. Click the **Gridlines** tab at the top of the Chart Options dialog box. Under Value (Y) axis, click in the **Major gridlines** box so that a checkmark appears there.

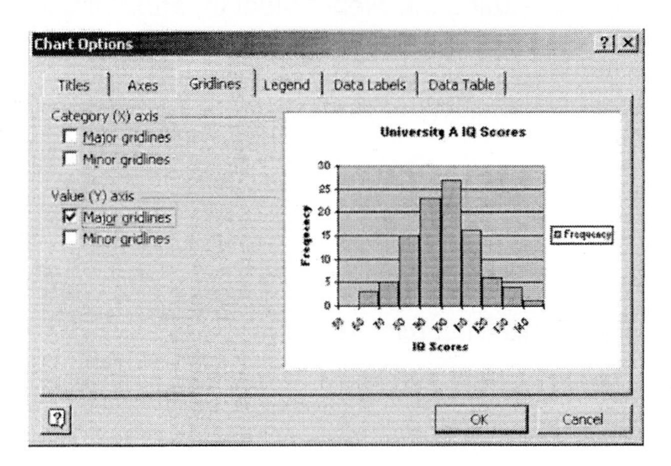

14. Click the **Legend** tab. To remove the frequency legend displayed at the right of the histogram chart, click in the **Show legend** box to remove the checkmark. Click **OK**. Your histogram chart should now appear similar to the one displayed below.

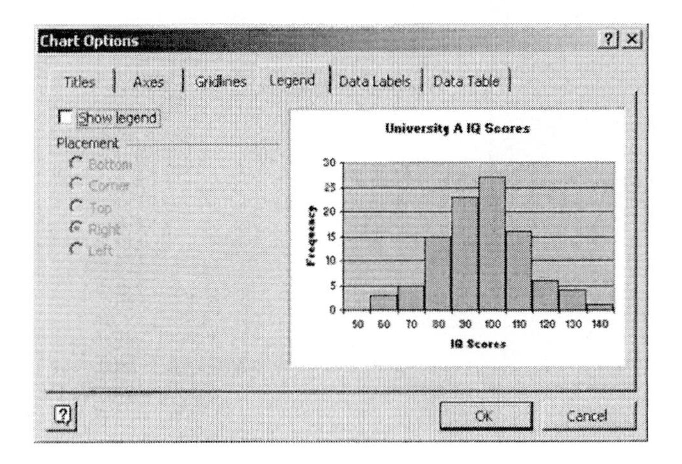

◀

Problem 37 (pg. 153) Computing the Mean, Median, and Sample Standard Deviation

1. Open worksheet "3_2_37" in the Chapter 3 folder. The first few rows are shown below.

	A	B	C	D	E	F	G
1	Financial S	Energy Stocks					
2	17.1	11.43					

2. At the top of the screen, click **Tools** and select **Data Analysis**.

If Data Analysis does not appear as a choice in the Tools menu, you will need to load the Microsoft Excel Analysis ToolPak add-in. Follow the procedure in Section GS 8.1 before continuing.

3. Select **Descriptive Statistics** and click **OK**.

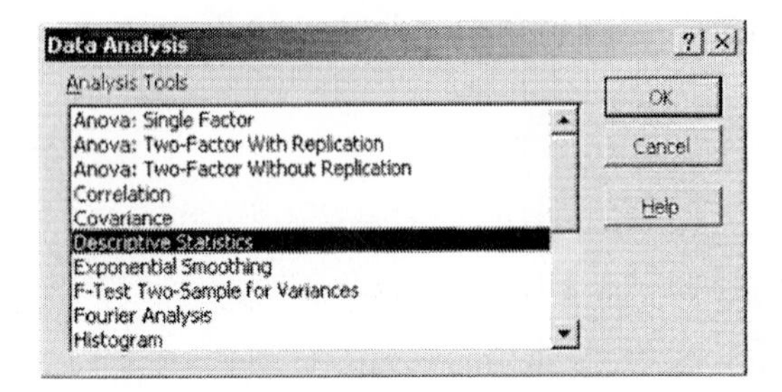

4. Complete the Descriptive Statistics dialog box as shown below. You will be obtaining descriptive statistics for the data in the worksheet range A1:B33. There are labels ("Financial Stocks" and "Energy Stocks") in the top cell of each column. The output will be placed in a new worksheet. Be sure to click in the **Summary statistics** box to place a checkmark there. The output will include summary statistics for both types of stocks. Click **OK**.

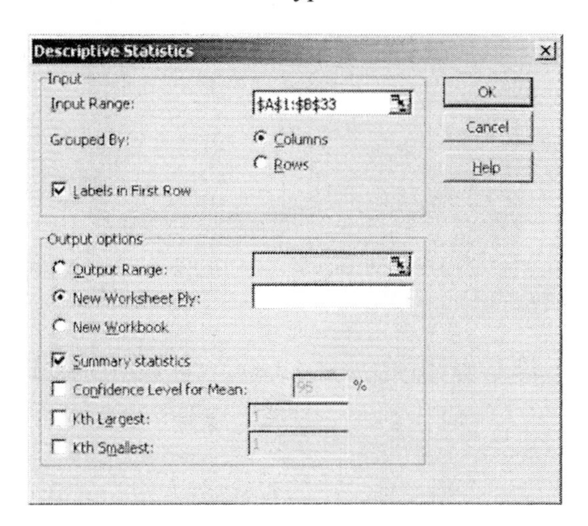

Your output will look similar to the output displayed below. I recommend that you make columns A and D wider so that you can read the complete label for each summary value.

	A	B	C	D	E	F	G
1	*Financial Stocks*		*Energy Stocks*				
2							
3	Mean	15.71563	Mean	22.48125			
4	Standard Error	1.30433	Standard Error	2.254096			
5	Median	16.09	Median	19.585			
6	Mode	#N/A	Mode	#N/A			
7	Standard Deviation	7.378407	Standard Deviation	12.75109			
8	Sample Variance	54.4409	Sample Variance	162.5903			
9	Kurtosis	-0.31416	Kurtosis	-0.11948			
10	Skewness	0.485813	Skewness	0.825906			
11	Range	30.36	Range	47.5			
12	Minimum	3.27	Minimum	6.11			
13	Maximum	33.63	Maximum	53.61			
14	Sum	502.9	Sum	719.4			
15	Count	32	Count	32			

◀

Section 3.4 Measures of Position

▶ Example 2 (pg. 167)

Determining the Percentile of a Data Value

1. Open worksheet "3_4_EX2" in the Chapter 3 folder. The first two lines are shown below.

	A	B	C	D	E	F	G
1	State	Crime Rate					
2	North Dak	77.8					

2. At the top of the screen, click **Tools** and select **Data Analysis**. In the Data Analysis dialog box, select **Rank and Percentile** and click **OK**.

If Data Analysis does not appear as a choice in the Tools menu, you will need to load the Microsoft Excel Analysis ToolPak add-in. Follow the procedure in Section GS 8.1 before continuing.

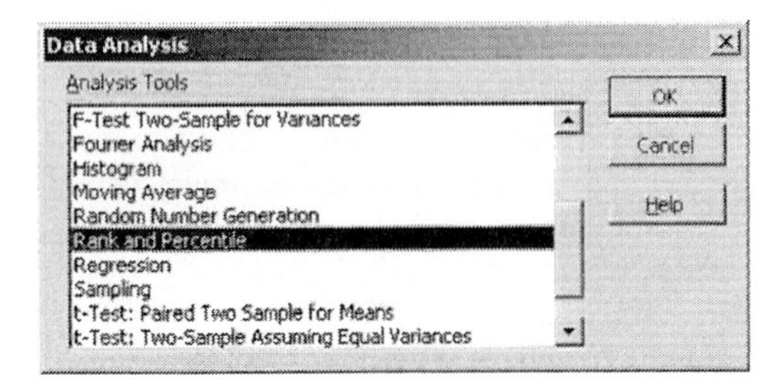

3. Complete the Rank and Percentile dialog box as shown below. Click **OK**.

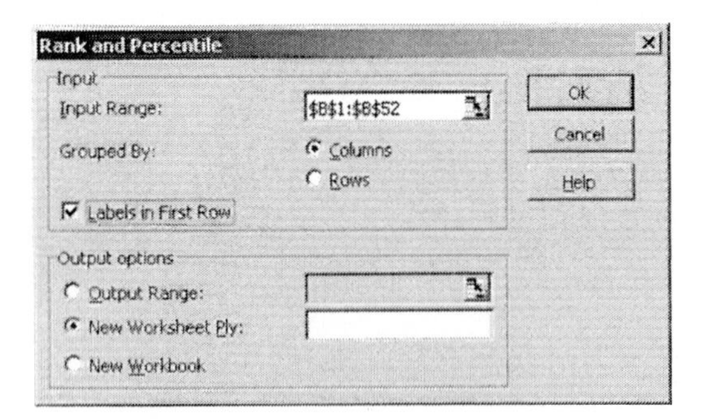

The first 15 entries in the output are shown below. The 75th percentile corresponds to the 39th observation in the data set. Texas is the 39th observation in the data set. It has a crime rate of 552.5.

	A	B	C	D	E	F	G
1	*Point*	*Crime Rate*	*Rank*	*Percent*			
2	51	1608.1	1	100.00%			
3	50	793.5	2	98.00%			
4	49	730.2	3	96.00%			
5	48	703.9	4	94.00%			
6	47	687.8	5	92.00%			
7	46	665.2	6	90.00%			
8	45	658	7	88.00%			
9	44	646.3	8	86.00%			
10	43	614.2	9	84.00%			
11	42	593.4	10	82.00%			
12	41	579.3	11	80.00%			
13	40	556.8	12	78.00%			
14	39	552.5	13	76.00%			
15	38	513.2	14	74.00%			
16	37	511.2	15	72.00%			

◀

▶ Problem 15 (pg. 173)

Computing z-Scores and Determining the Quartiles

1. Open worksheet "3_4_15" in the Chapter 3 folder. The first two rows of the worksheet are displayed below.

	A	B	C	D	E	F	G
1	Inches of Rain						
2	0.97						

2. In order to calculate z-scores, we need the mean and standard deviation. The easiest way to get these summary measures is to use Excel's Descriptive Statistics tool. At the top of the screen, click **Tools** and select **Data Analysis**. In the Data Analysis dialog box, select **Descriptive Statistics** and click **OK**.

If Data Analysis does not appear as a choice in the Tools menu, you will need to load the Microsoft Excel Analysis ToolPak add-in. Follow the procedure in Section GS 8.1 before continuing.

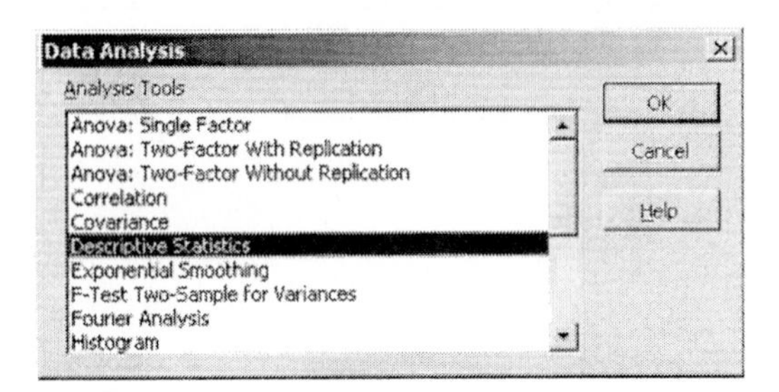

3. Complete the Descriptive Statistics dialog box as shown below. The data are located in the range $A$1:$A$21. The data are grouped by columns. There is a label in the first row. The output will placed in the current worksheet with A23 as the upper leftmost cell. The output will display summary statistics. Click **OK**.

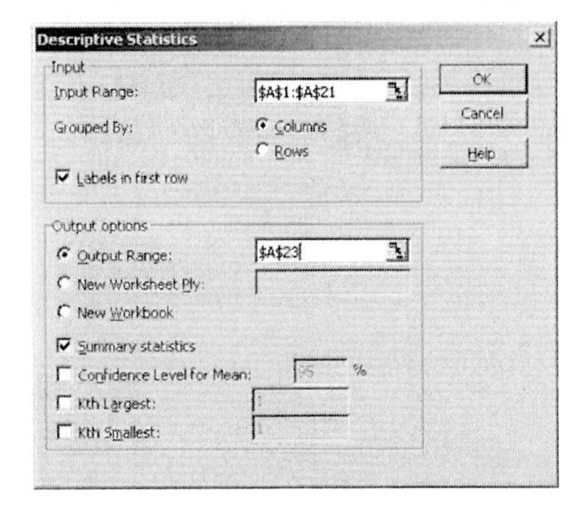

4. The mean is 3.9935 and the standard deviation is 1.7790. Type the label **z-Score** in cell **C1** of the worksheet.

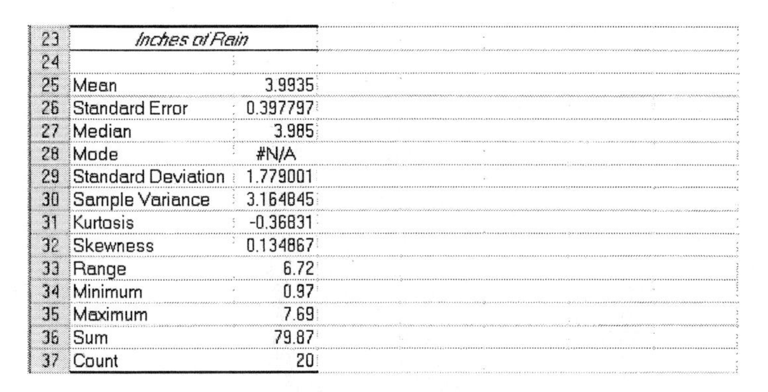

	Inches of Rain	
23	*Inches of Rain*	
24		
25	Mean	3.9935
26	Standard Error	0.397797
27	Median	3.985
28	Mode	#N/A
29	Standard Deviation	1.779001
30	Sample Variance	3.164845
31	Kurtosis	-0.36831
32	Skewness	0.134867
33	Range	6.72
34	Minimum	0.97
35	Maximum	7.69
36	Sum	79.87
37	Count	20

5. You will use the STANDARDIZE function to calculate z-scores. Click in cell **C2** where the first z-score will be placed. At the top of the screen, click **Insert** and select **Function.**

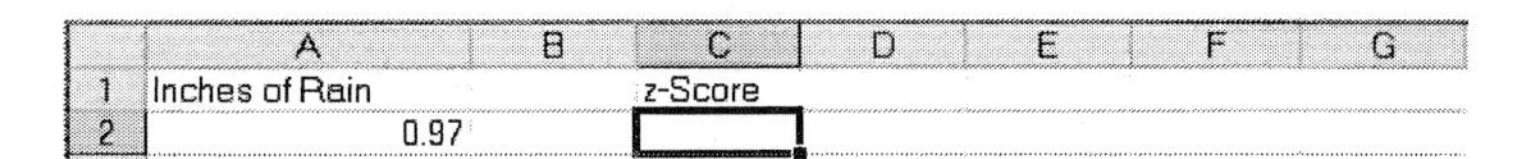

	A	B	C	D	E	F	G
1	Inches of Rain		z-Score				
2	0.97						

6. Select the **Statistical** category and the **STANDARDIZE** function. Click **OK.**

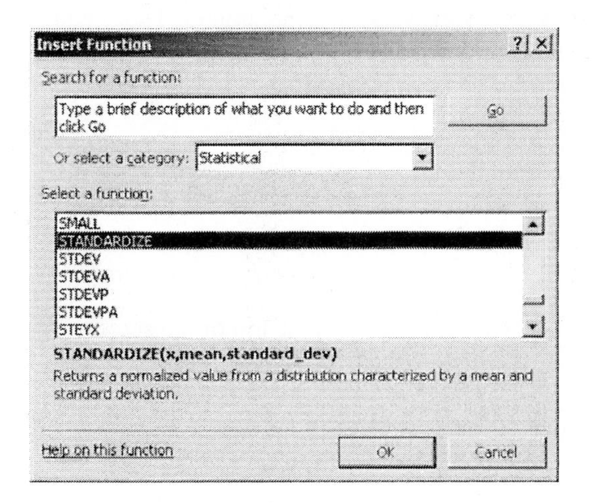

7. The cell address of the first observation is **A2**, the mean of the distribution is **3.9935**, and the standard deviation is **1.7790**. Click **OK**.

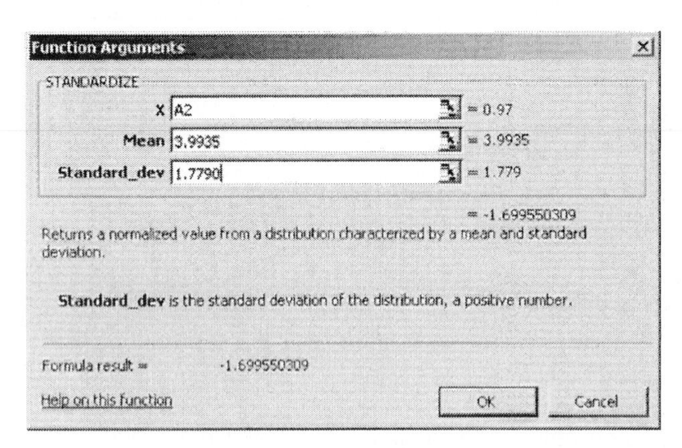

8. Copy the contents of cell C2 to cells **C3** through **C21**.

	A	B	C	D	E	F	G
1	Inches of Rain		z-Score				
2	0.97		-1.69955				

9. The z-scores for all the rainfall values are now displayed in column C. Next, you will determine the quartiles. Type the label **Quartile** in cell E1 followed by the numbers **1**, **2**, and **3** as shown below.

	A	B	C	D	E	F	G
1	Inches of Rain		z-Score		Quartile		
2	0.97		-1.69955		1		
3	1.14		-1.60399		2		
4	1.85		-1.20489		3		

10. Click in cell **F2** where the first quartile will be placed. At the top of the screen, click **Insert** and select **Function**.

	A	B	C	D	E	F	G
1	Inches of Rain		z-Score		Quartile		
2	0.97		-1.69955		1		
3	1.14		-1.60399		2		
4	1.85		-1.20489		3		

11. Select the **Statistical** category and the **QUARTILE** function. Click **OK**.

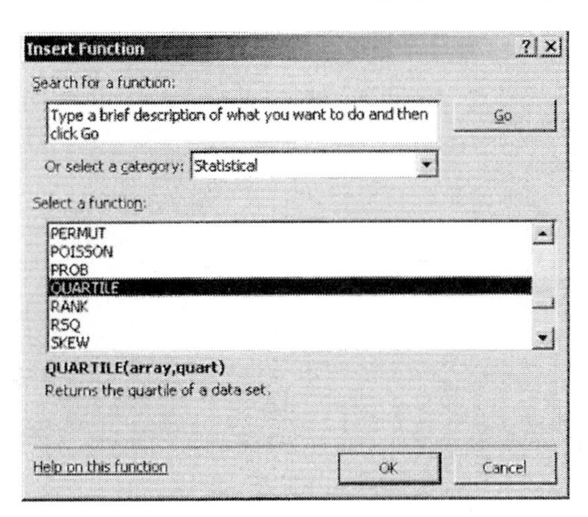

12. Complete the QUARTILE dialog box as shown below. The array refers to the worksheet location of the data values. You want dollar signs in the Array cell addresses because you will be copying the contents of cell F2 to cells F3 and F4. Click **OK**.

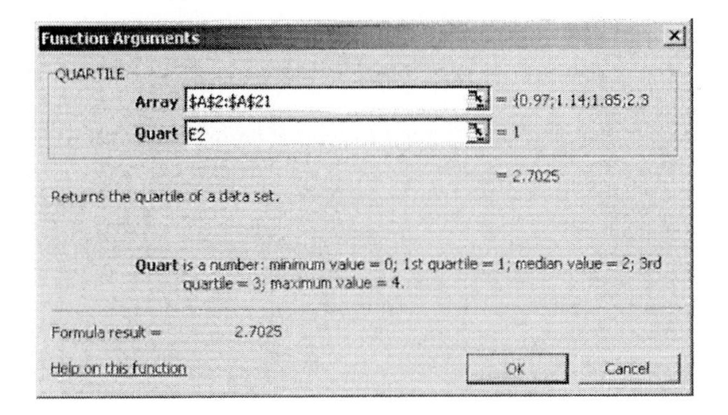

13. Now, copy the contents of cell F2 to cells F3 and F4. The first, second, and third quartiles are 2.7025, 3.985, and 5.29, respectively.

	A	B	C	D	E	F	G
1	Inches of Rain		z-Score		Quartile		
2	0.97		-1.69955		1	2.7025	
3	1.14		-1.60399		2	3.985	
4	1.85		-1.20489		3	5.29	

◀

Section 3.5 The Five-Number Summary and Boxplots

▶ Problem 5 (pg. 181) — Finding the Five-Number Summary and Constructing a Boxplot

You will use the inauguration data to learn how to find the five-number summary and construct a boxplot.

If the PHStat add-in has not been loaded, you will need to load it before continuing. Follow the instructions in Section GS 8.2.

1. Open worksheet "3_5_5" in the Chapter 3 folder. The first few lines of the worksheet are shown below.

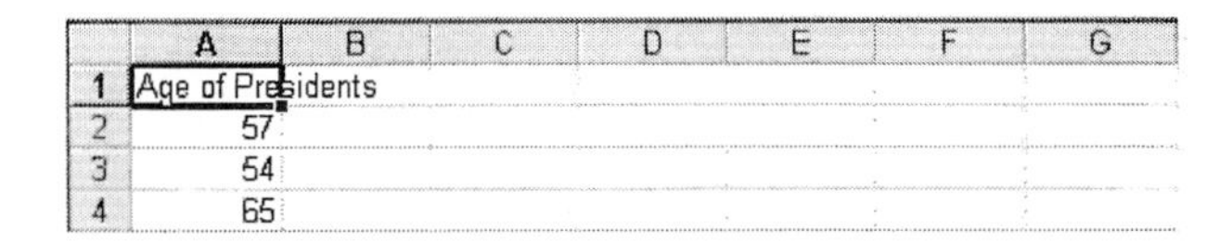

	A	B	C	D	E	F	G
1	Age of Presidents						
2	57						
3	54						
4	65						

2. At the top of the screen, click **PHStat** and select **Descriptive Statistics → Box-and-Whisker Plot**.

3. Complete the Box-and-Whisker Plot dialog box as shown at the top of the next page. To enter the cell range, first click in the Raw Data Cell Range Window of the dialog box. Then click and drag over the range A1 through A45 of the worksheet to enter the range. Or you could simply key in A1:A45. Click **OK**.

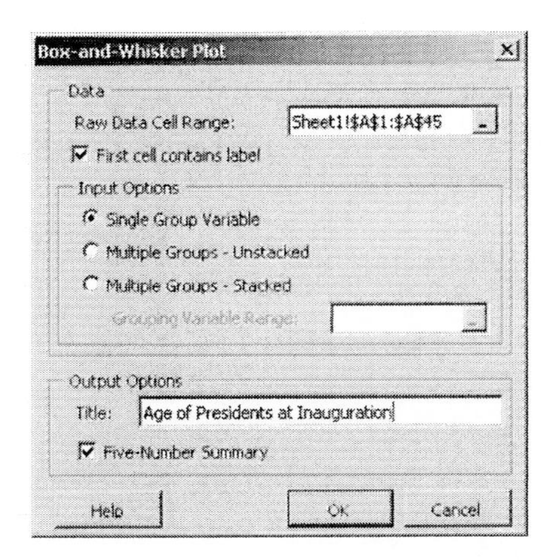

The Five-Number Summary is displayed in a worksheet named "Five Numbers."

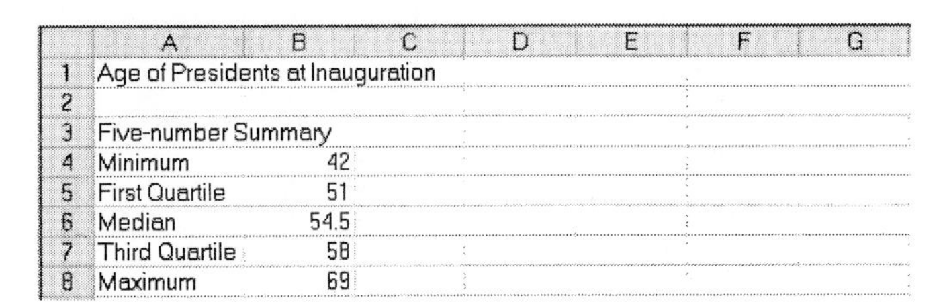

	A	B	C	D	E	F	G
1	Age of Presidents at Inauguration						
2							
3	Five-number Summary						
4	Minimum	42					
5	First Quartile	51					
6	Median	54.5					
7	Third Quartile	58					
8	Maximum	69					

The box-and-whisker plot is displayed in a worksheet named "BoxWhiskerPlot."

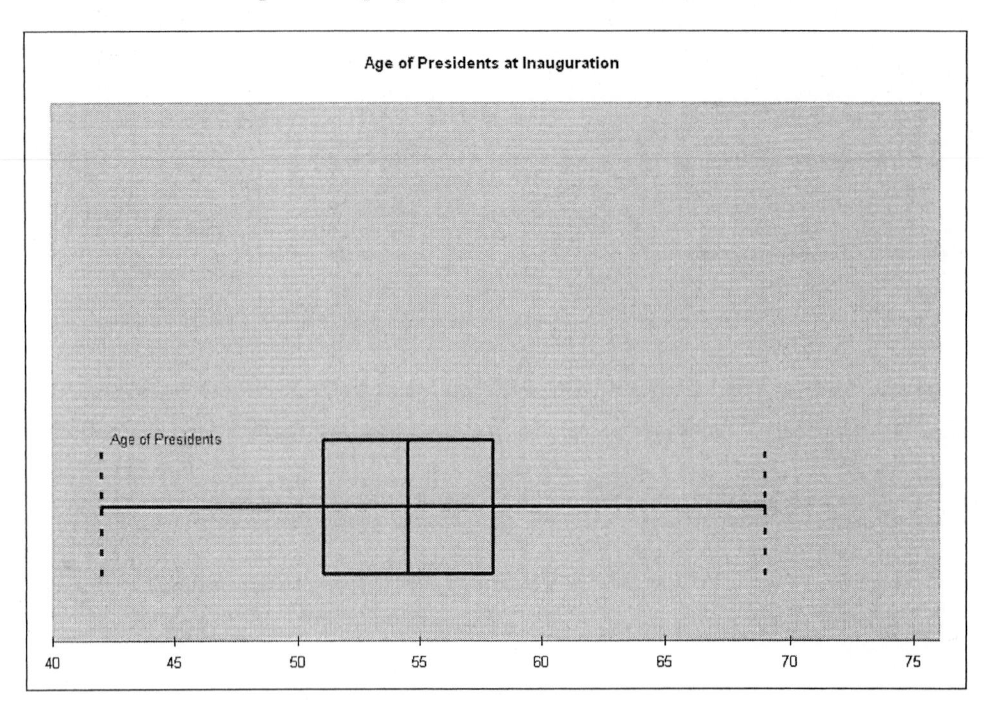

◀

Describing the Relation between Two Variables

CHAPTER 4

Section 4.1 Scatter Diagrams and Correlation

► Problem 23 (pg. 205)

Drawing a Scatter Diagram

1. Open worksheet "4_1_23" in the Chapter 4 folder. The first few rows are shown below.

	A	B	C	D	E	F	G
1	Height (inc	Head Circumference (inches)					
2	27.75	17.5					
3	24.5	17.1					

2. Click in any cell of the data table. Then, at the top of the screen, click **Insert** and select **Chart**.

3. Under Chart type, select **XY (Scatter)**. Under Chart sub-type, select the topmost diagram. Click **Next>**.

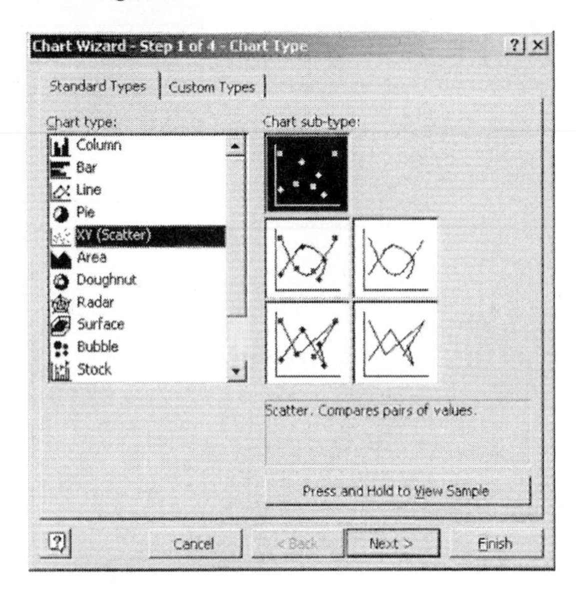

4. The data range should be A1:B12. Click **Next>**.

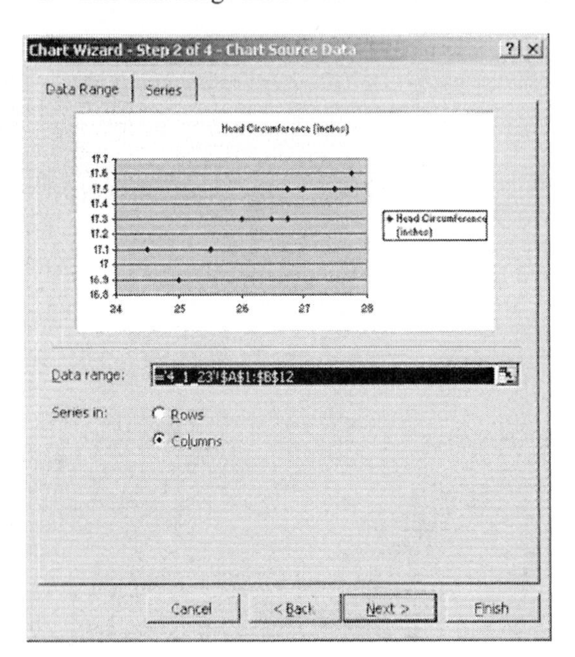

5. Click the **Titles** tab at the top of the Chart Options dialog box. Enter a title for the chart and labels for the X and Y axes. Chart title: **Height and Head Circumference**. Value (X) axis: **Height (inches)**. Value (Y) axis: **Head Circumference (inches)**.

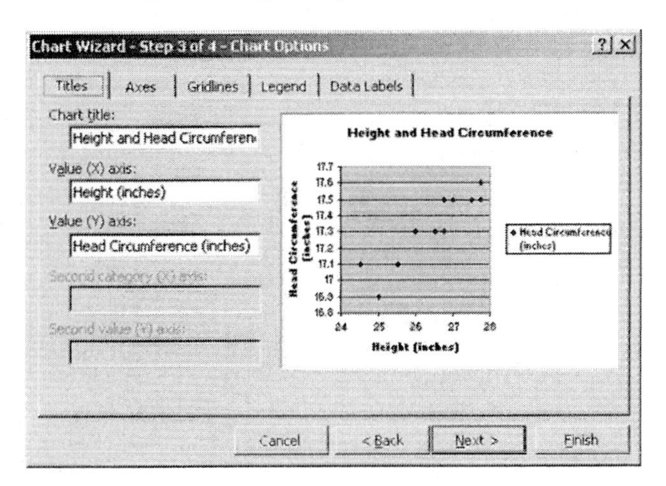

6. Click the **Gridlines** tab at the top of the Chart Options dialog box. Under Value (X) axis, click in the box to the left of **Major gridlines** to place a checkmark there.

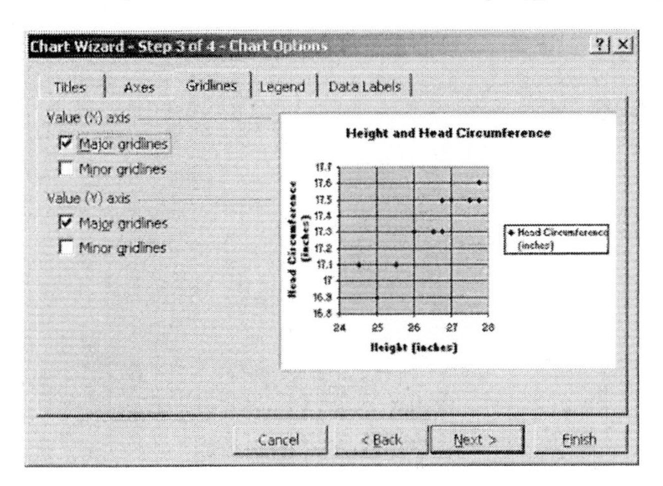

7. Click the **Legend** tab at the top of the Chart Options dialog box. Click in the box to the left of **Show legend** to remove the checkmark. Click **Next>**.

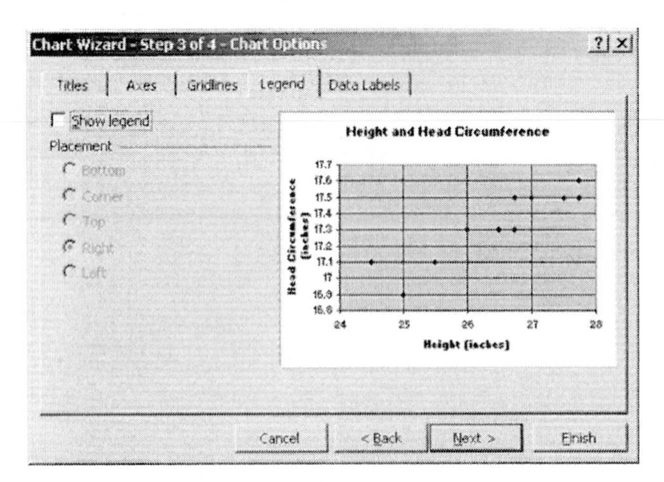

8. In the Chart Location dialog box, select **As new sheet**. Click **Finish**.

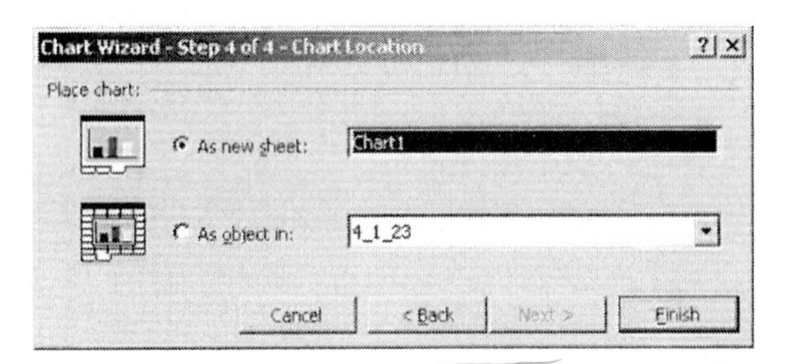

9. The dots in the scatter diagram are very small and hard to see. If you would like to make them larger, first **right-click** directly on one of the dots. Then select **Format Data Series** from the shortcut menu that appears.

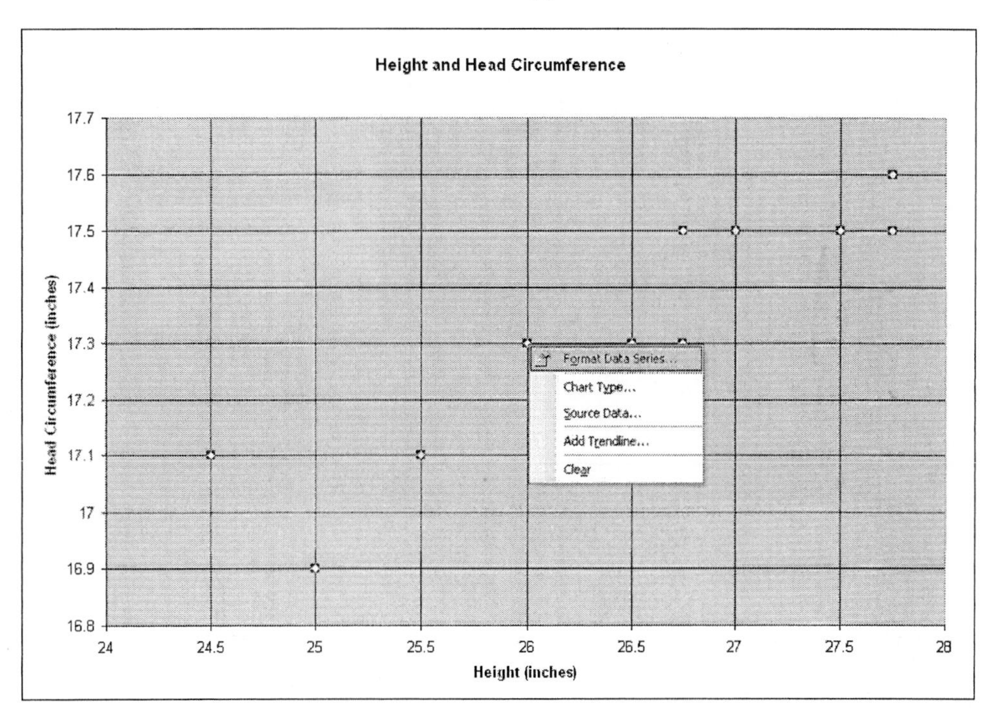

10. Click the **Patterns** tab at the top of the dialog box. You can make a size adjustment in the lower right of the Format Data Series dialog box. Change the size from 5 to **10** pts. Click **OK**.

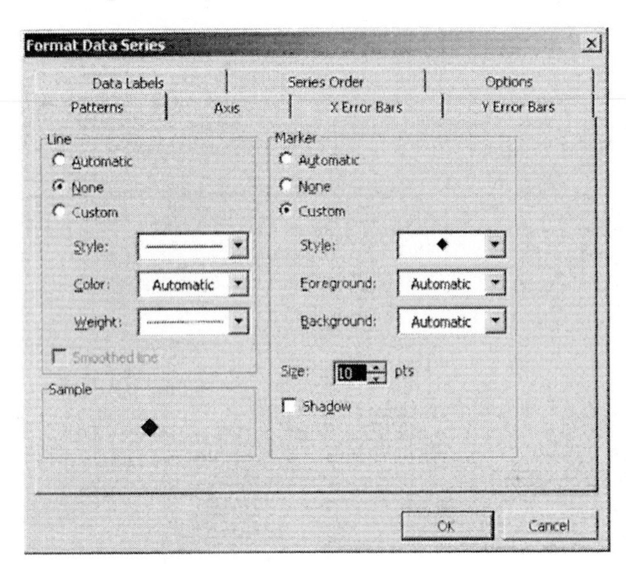

The completed scatter diagram should look similar to the one displayed below.

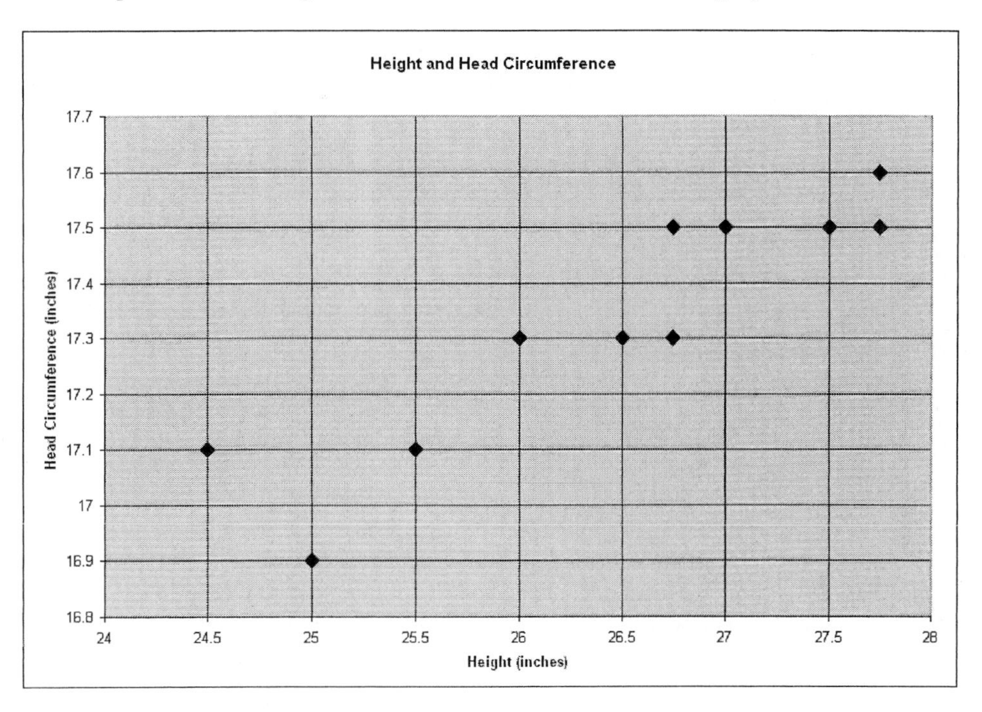

▶ Problem 23 (pg. 205) Computing the Correlation Coefficient

1. Open worksheet "4_1_23" in the Chapter 4 folder. The first few rows are shown below.

	A	B	C	D	E	F	G
1	Height (inc	Head Circumference (inches)					
2	27.75	17.5					
3	24.5	17.1					

2. At the top of the screen, click **Tools** and select **Data Analysis**.

If Data Analysis does not appear as a choice in the Tools menu, you will need to load the Microsoft Excel Analysis ToolPak add-in. Follow the procedure in Section GS 8.1 before continuing.

---

3. In the Data Analysis dialog box, select **Correlation** and click **OK**.

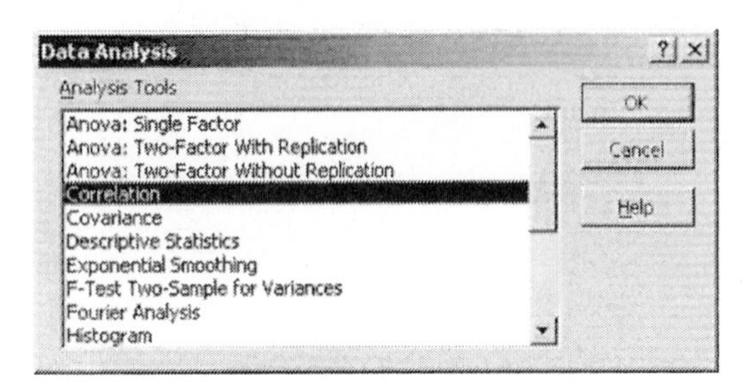

4. Complete the fields in the Correlation dialog box as shown below. The input range is A1:B12, the first row contains labels, and the output will be placed in the same worksheet as the data with A14 as the uppermost left cell. Click **OK**.

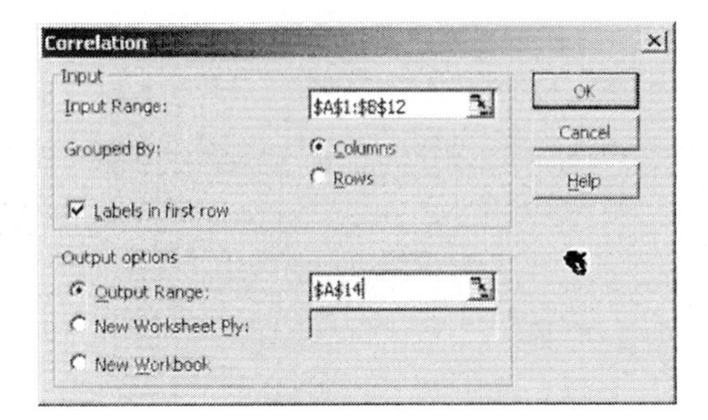

The output is a correlation matrix. I recommend that you make the columns wider so that you can easily read all the labels. The correlation between height and head circumference is 0.9111.

14		Height (inches)	Head Circumference (inches)
15	Height (inches)	1	
16	Head Circumference (inches)	0.911072733	1

◀

Section 4.2 Least-Squares Regression

▶ Problem 17 (pg. 222)

Finding the Least-Squares Regression Line

1. Open worksheet "4_1_23" in the Chapter 4 folder. The first few rows are shown below.

	A	B	C	D	E	F	G
1	Height (inc	Head Circumference (inches)					
2	27.75	17.5					
3	24.5	17.1					
4	25.5	17.1					

2. At the top of the screen, click **Tools** and select **Data Analysis.**

If Data Analysis does not appear as a choice in the Tools menu, you will need to load the Microsoft Excel Analysis ToolPak add-in. Follow the procedure in Section GS 8.1 before continuing.

3. In the Data Analysis dialog box, select **Regression** and click **OK.**

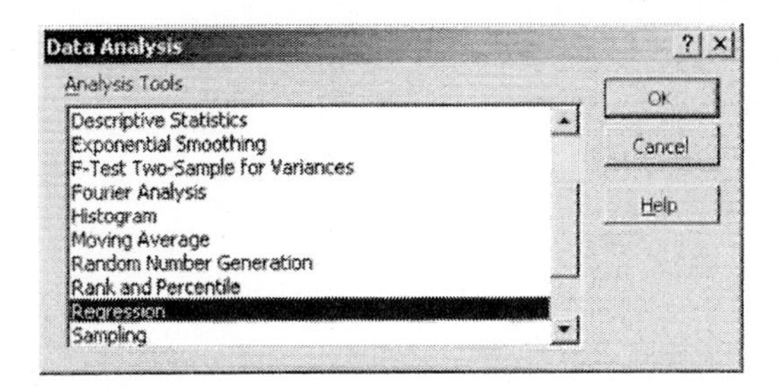

4. Complete the Regression dialog box as shown below. Click **OK**.

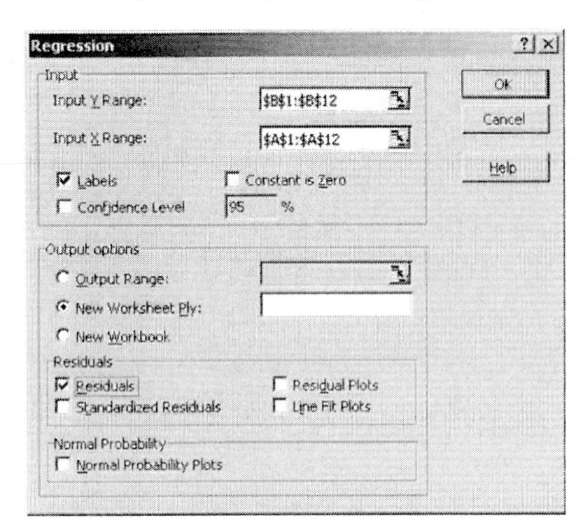

Because the output is lengthy, I will display it in two parts. The Multiple R, 0.9111, near the top of the output is the same as r for bivariate regression analysis. The intercept, shown near the bottom, is 12.4932. The slope, immediately below the intercept, is 0.1827.

	A	B	C	D	E	F	G
1	SUMMARY OUTPUT						
2							
3	*Regression Statistics*						
4	Multiple R	0.911073					
5	R Square	0.830054					
6	Adjusted R	0.811171					
7	Standard E	0.095384					
8	Observatio	11					
9							
10	ANOVA						
11		*df*	*SS*	*MS*	*F*	*ignificance F*	
12	Regressior	1	0.399935	0.399935	43.95785	9.59E-05	
13	Residual	9	0.081883	0.009098			
14	Total	10	0.481818				
15							
16		*Coefficients*	*tandard Err*	*t Stat*	*P-value*	*Lower 95%*	*Upper 95%*
17	Intercept	12.49317	0.729685	17.12132	3.56E-08	10.84251	14.14383
18	Height (inc	0.182732	0.027561	6.630072	9.59E-05	0.120385	0.24508

The output section entitled RESIDUAL OUTPUT presents predicted circumference for each height value in the data set. For example, the first height value is 27.75. The predicted circumference for that height is 17.5640. The Residuals column presents the difference between the observed height and predicted height. The observed height for the first observation is 17.5. The difference between 17.5 and 17.5640 is –0.0640.

22	RESIDUAL OUTPUT		
23			
24	*Observation*	*Circumfe*	*Residuals*
25	1	17.56399	-0.06399
26	2	16.97011	0.129886
27	3	17.15285	-0.05285
28	4	17.24421	0.055787
29	5	17.06148	-0.16148
30	6	17.56399	0.036006
31	7	17.33558	-0.03558
32	8	17.42694	0.073055
33	9	17.38126	-0.08126
34	10	17.38126	0.118738
35	11	17.51831	-0.01831

◀

Section 4.3 Diagnostics on the Least-Squares Regression Line

▶ Problem 21 (pg. 236)

Drawing a Scatter Diagram and a Residual Plot

You will draw a scatter diagram and a residual plot using the Other Old Faithful data.

1. Open worksheet "4_3_21" in the Chapter 4 folder. The first few rows are shown below.

	A	B	C	D	E	F	G
1	Time Betw	Length					
2	12.17	1.88					
3	11.63	1.77					
4	12.03	1.83					

2. You will begin by drawing the scatter diagram. Click in any cell of the data table. Then, at the top of the screen, click **Insert** and select **Chart**.

3. Under Chart type, select **XY (Scatter)**. Under Chart sub-type, select the topmost diagram. Click **Next>**.

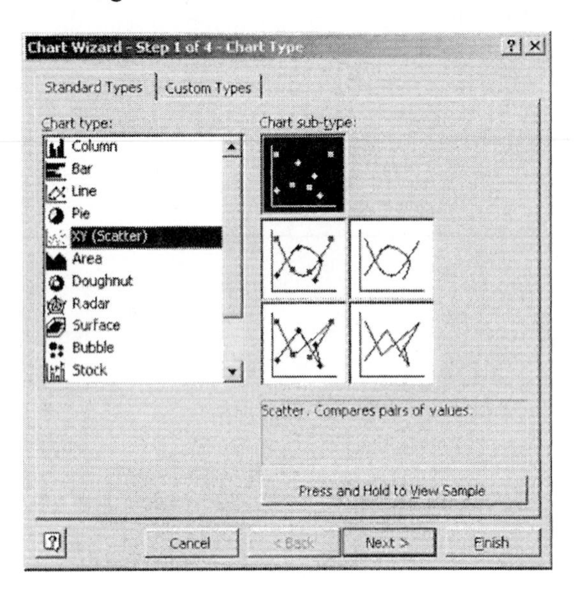

4. The data range should be the same as shown below, A1:B10. Make any necessary corrections. Then click **Next>**.

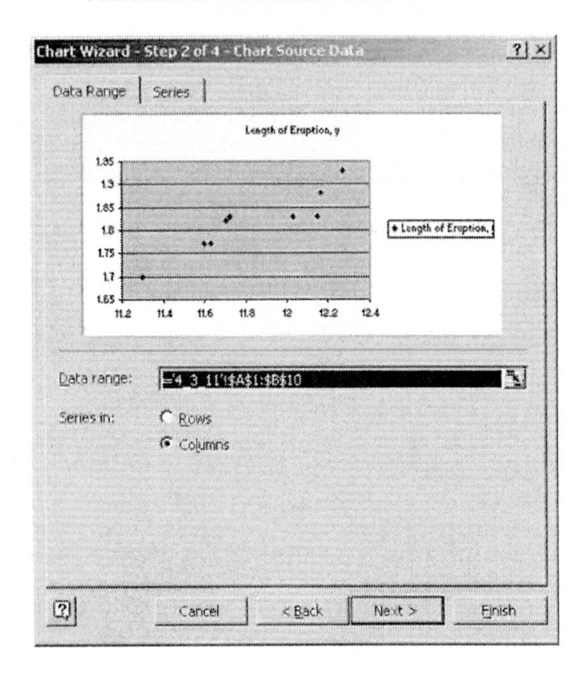

5. Enter a title for the chart and labels for the X and Y axes. Chart title: **Scatter Diagram of Time between Eruptions and Length of Eruptions**. Value (X) axis: **Time between Eruptions**. Value (Y) axis: **Length of Eruptions**.

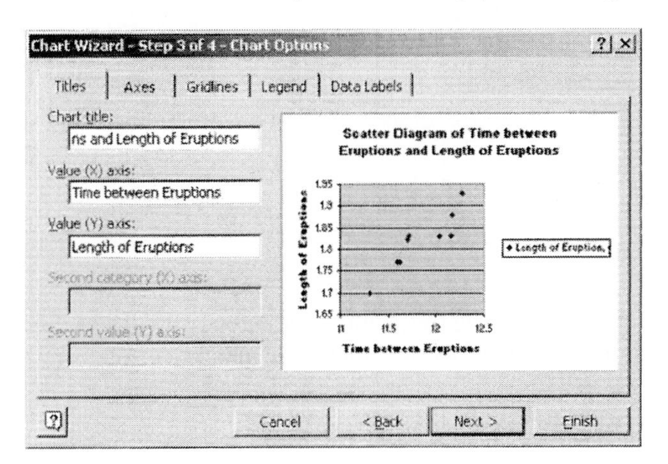

6. At the top of the Chart Options dialog box, click the **Legend** tab.

7. Click in the box to the left of **Show legend** to remove the checkmark. Click **Next>**.

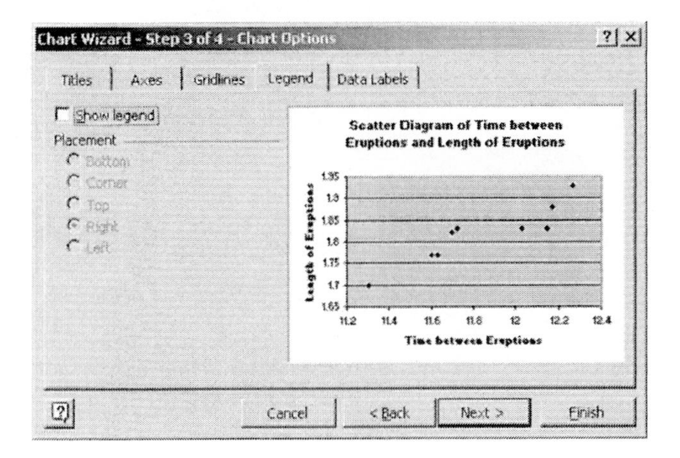

8. In the Chart Location dialog box, select **As new sheet**. Click **Finish**.

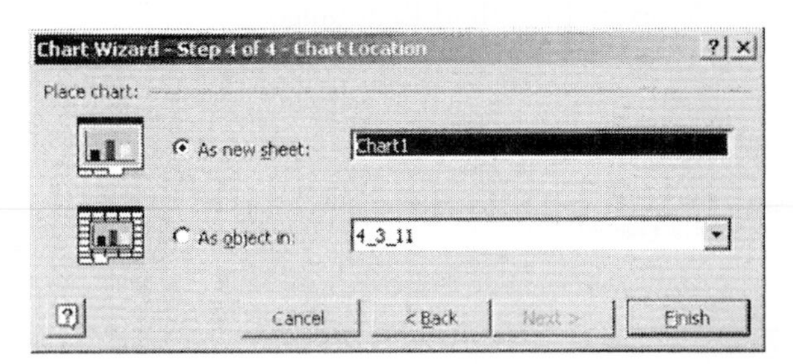

9. Place the regression line on the chart. To do this, **right-click** directly on one of the dots. Then select **Add Trendline** from the shortcut menu that appears.

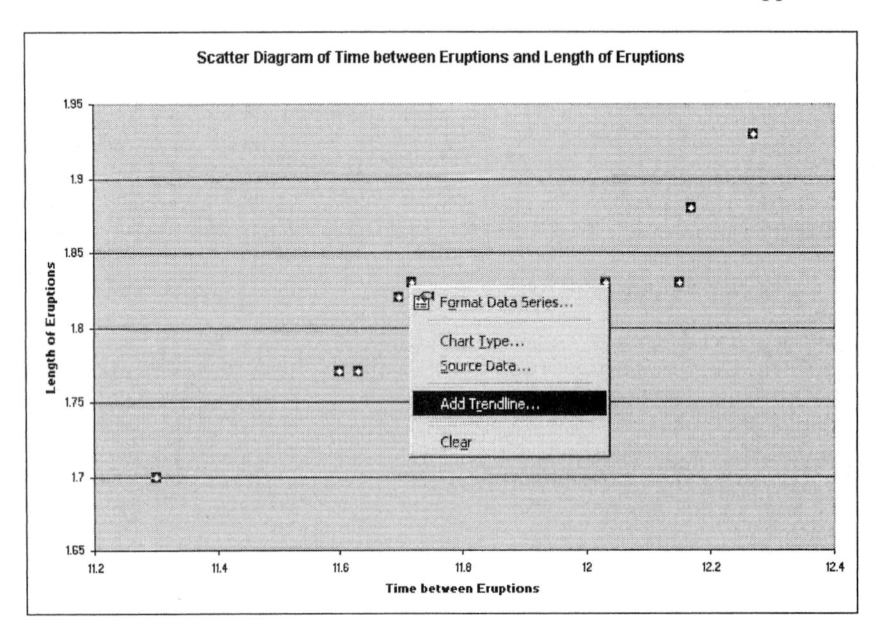

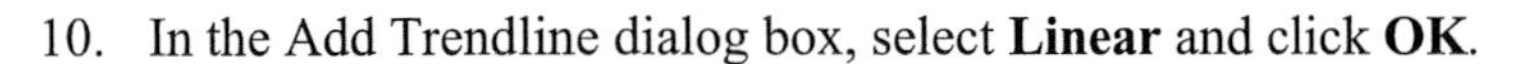

10. In the Add Trendline dialog box, select **Linear** and click **OK**.

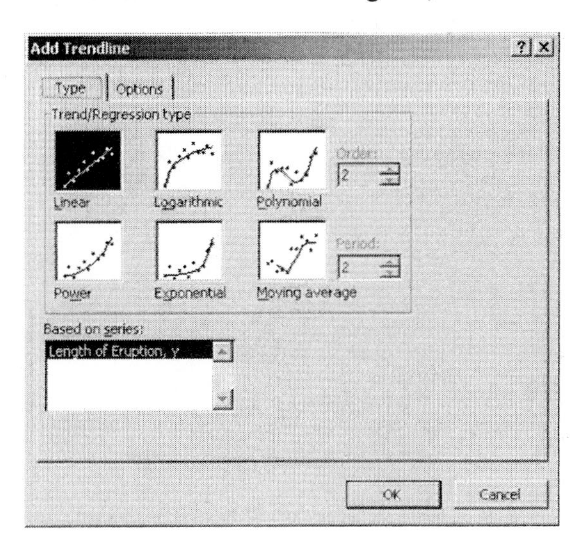

The completed scatter diagram should look similar to the diagram shown below.

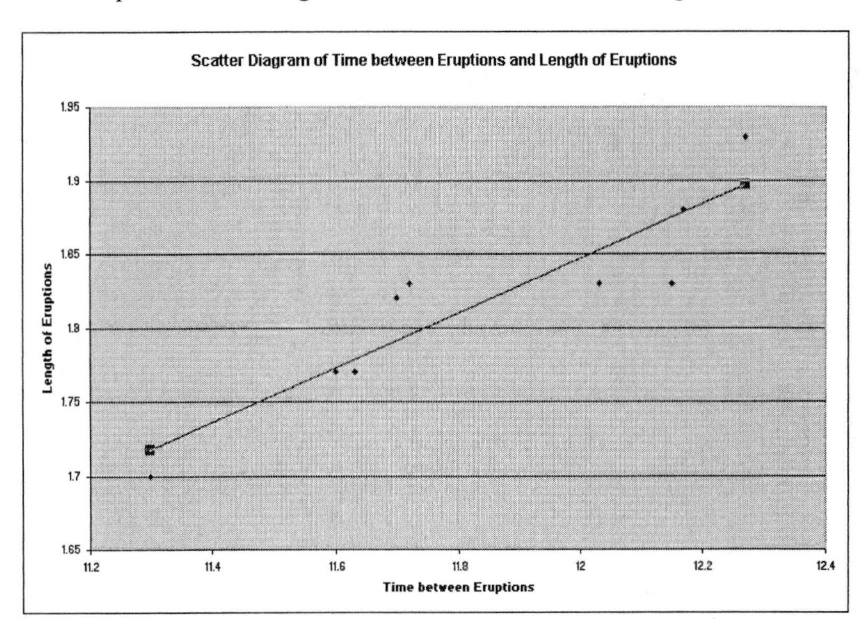

11. Next, you will carry out a regression analysis in order to obtain the residuals. Return to the worksheet that contains the data by clicking on the **Sheet 1** tab at the bottom of the screen.

12. At the top of the screen, click **Tools** and select **Data Analysis**.

If Data Analysis does not appear as a choice in the Tools menu, you will need to load the Microsoft Excel Analysis ToolPak add-in. Follow the procedure in Section GS 8.1 before continuing.

---

13. In the Data Analysis dialog box, select **Regression** and click **OK**.

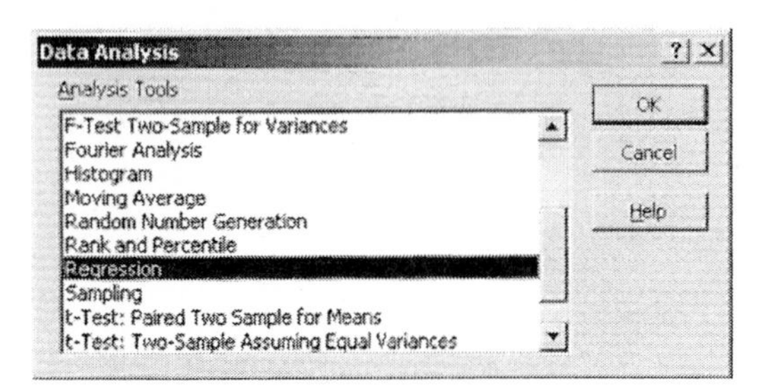

14. Complete the Regression dialog box as shown below. Be sure to select **Residuals** in the lower part of the dialog box. Click **OK**.

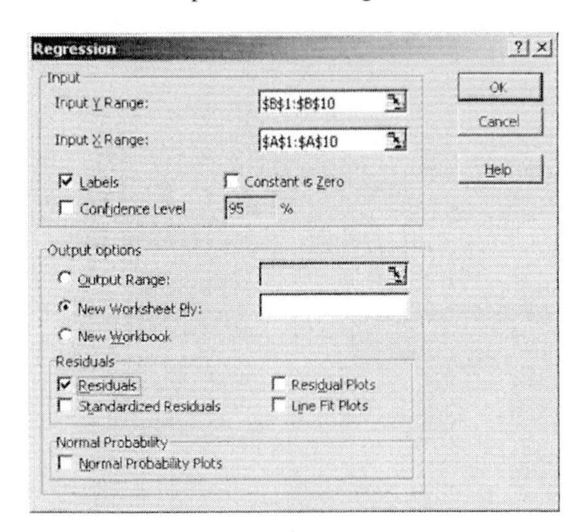

15. Next, you will copy the residuals to the worksheet containing the data so that you can use them in the residual plot. Click and drag over the residuals in the lower part

of the output so that the range **C24:C33** is highlighted. Then, at the top of the screen, click the copy button.

22	RESIDUAL OUTPUT		
23			
24	Observation	Length of E	Residuals
25	1	1.878762	0.001238
26	2	1.778632	-0.00863
27	3	1.852803	-0.0228
28	4	1.875054	-0.04505
29	5	1.717442	-0.01744
30	6	1.791612	0.028388
31	7	1.897305	0.032695
32	8	1.773069	-0.00307
33	9	1.795321	0.034679
34			

16. Return to the worksheet containing the data by clicking on the **Sheet 1** tab at the bottom of the screen.

17. Click in cell **D1**and then, at the top of the screen, click the paste button.

	A	B	C	D	E	F	G
1	Time Betw	Length of Eruption, y		Residuals			
2	12.17	1.88		0.001238			
3	11.63	1.77		-0.00863			
4	12.03	1.83		-0.0228			
5	12.15	1.83		-0.04505			
6	11.3	1.7		-0.01744			
7	11.7	1.82		0.028388			
8	12.27	1.93		0.032695			
9	11.6	1.77		-0.00307			
10	11.72	1.83		0.034679			

18. To construct the residual plot, you will be using Time between Eruptions as the X variable and Residuals as the Y variable. These two variables need to be in adjacent columns of the worksheet. Copy the Time between Eruptions values to column C of the worksheet.

	A	B	C	D	E	F	G
1	Time Betw	Length of E	Time Betw	Residuals			
2	12.17	1.88	12.17	0.001238			
3	11.63	1.77	11.63	-0.00863			
4	12.03	1.83	12.03	-0.0228			
5	12.15	1.83	12.15	-0.04505			
6	11.3	1.7	11.3	-0.01744			
7	11.7	1.82	11.7	0.028388			
8	12.27	1.93	12.27	0.032695			
9	11.6	1.77	11.6	-0.00307			
10	11.72	1.83	11.72	0.034679			

19. Highlight the contents of both column C and column D by clicking and dragging over the range **C1:D10**. Then, at the top of the screen, click **Insert** and select **Chart**.

20. Under Chart type, select **XY (Scatter)**. Under Chart sub-type, select the topmost diagram. Click **Next>**.

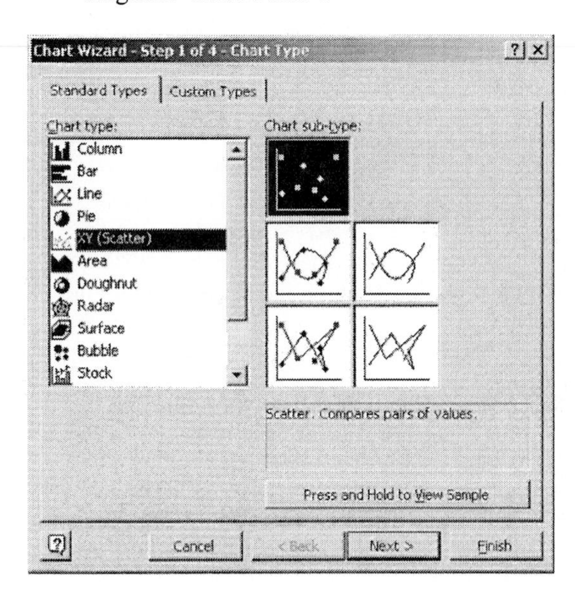

21. The data range should be the same as shown below, C1:D10. Make any necessary corrections. Then click **Next>**.

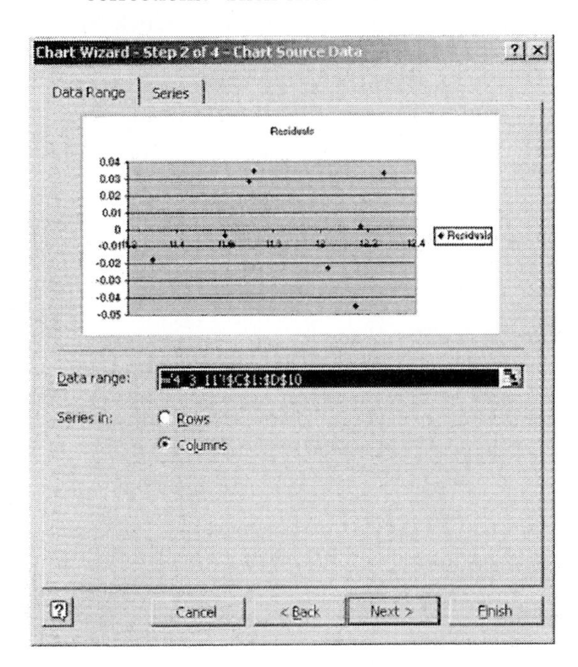

22. Click the **Titles** tab at the top of the Chart Options dialog box. Enter a title for the chart and labels for the X and Y axes. Chart title: **Residual Plot**. Value (X) axis: **Time between Eruptions**. Value (Y) axis: **Residuals**.

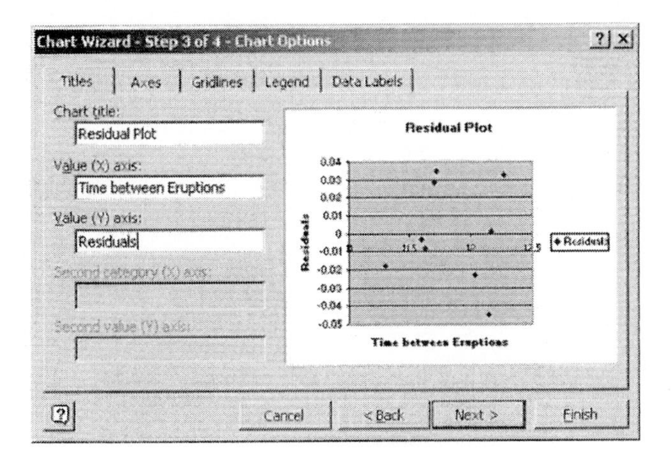

23. Click the **Legend** tab at the top of the Chart Options dialog box.

24. Click in the box to the left of **Show legend** to remove the checkmark. Click **Next>**.

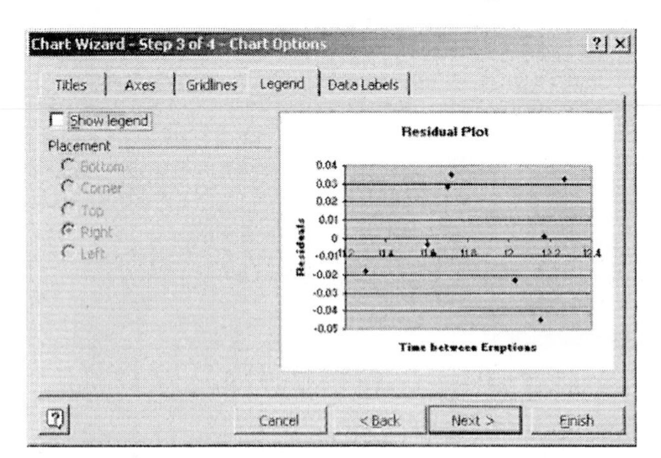

25. In the Chart Location dialog box, select **As new sheet**. Click **Finish**.

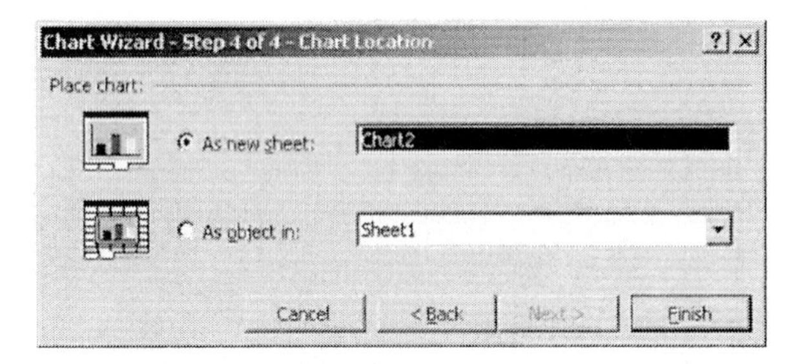

Your completed residual plot should look similar to the one shown below.

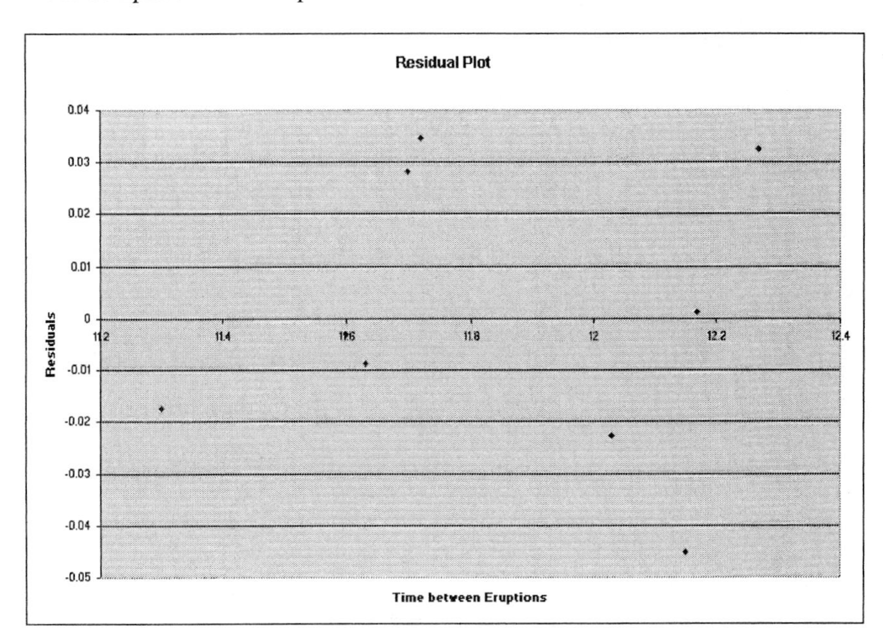

◀

Section 4.4 Nonlinear Regression: Transformations (on CD)

► Example 4 (on CD)

Drawing a Scatter Diagram and Finding Regression Line of Transformed Data

1. The data below are taken from Table 7 in Example 4. Enter these data in an Excel worksheet.

	A	B	C	D	E	F	G
1	Year, x	Closing price, y					
2	1990 (x = 1)	1.1609					
3	1991 (x - 2)	2.6988					
4	1992 (x = 3)	4.5381					
5	1993 (x = 4)	5.3379					
6	1994 (x = 5)	6.8032					
7	1995 (x = 6)	7.0328					
8	1996 (x = 7)	11.5585					
9	1997 (x = 8)	13.4799					
10	1998 (x = 9)	23.5424					
11	1999 (x = 10)	31.9342					
12	2000 (x = 11)	36.7277					
13	2001 (x = 12)	54.31					
14	2002 (x = 13)	46.2					
15	2003 (x = 14)	47.53					
16	2004 (x = 15)	60.75					

2. Click in any cell of the data table. Then, at the top of the screen, click **Insert** and select **Chart**.

3. Under chart type, select **XY (Scatter)**. Under Chart sub-type, select the topmost diagram. Click **Next>**.

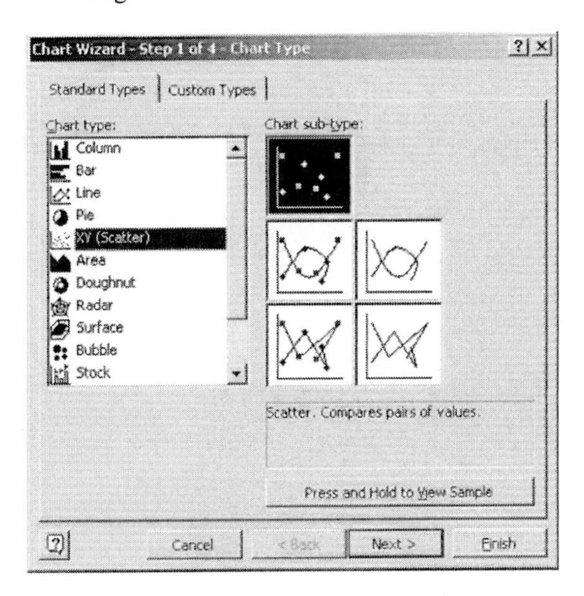

4. The data range should be the same as shown below, A1:B16. Make any necessary corrections. Then click **Next>**.

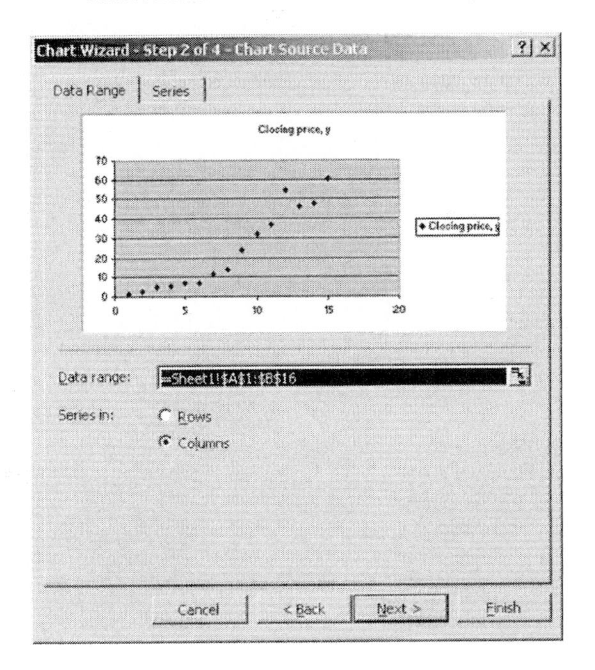

5. Enter a title for the chart and labels for the X and Y axes. Chart title: **Scatter Diagram of Year and Closing Price**. Value (X) axis: **Year**. Value (Y) axis: **Closing Price**.

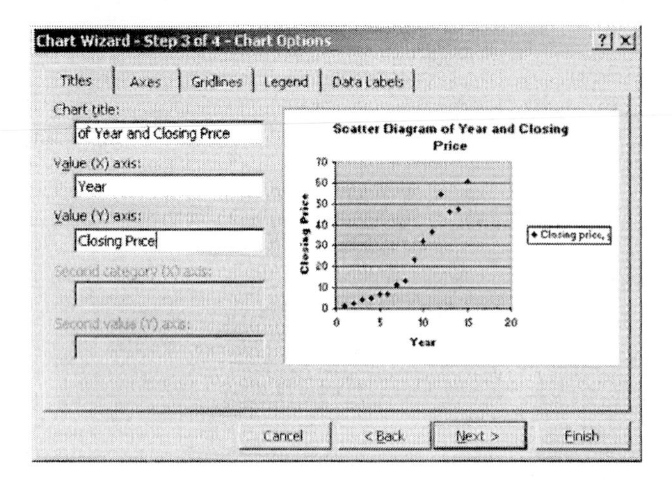

6. At the top of the Chart Options dialog box, click the **Legend** tab.

7. Click in the box to the left of **Show legend** to remove the checkmark. Click **Next>**.

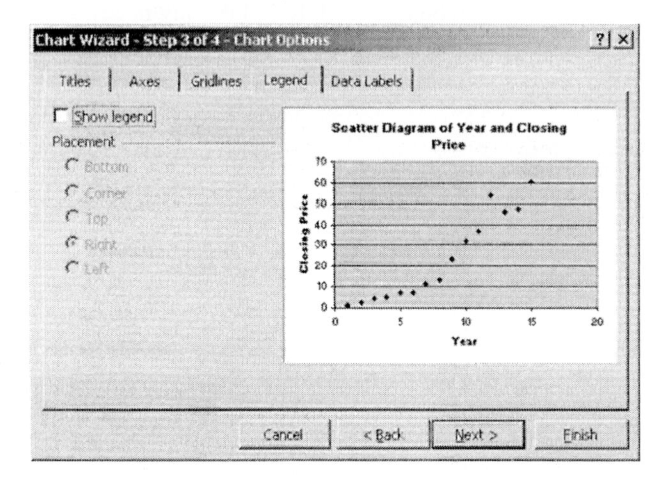

8. In the Chart Location dialog box, select **As new sheet**. Click **Finish**.

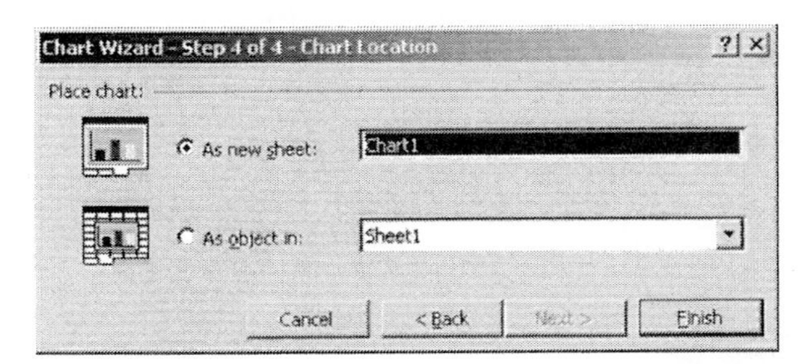

The completed scatter diagram should look similar to the one shown below.

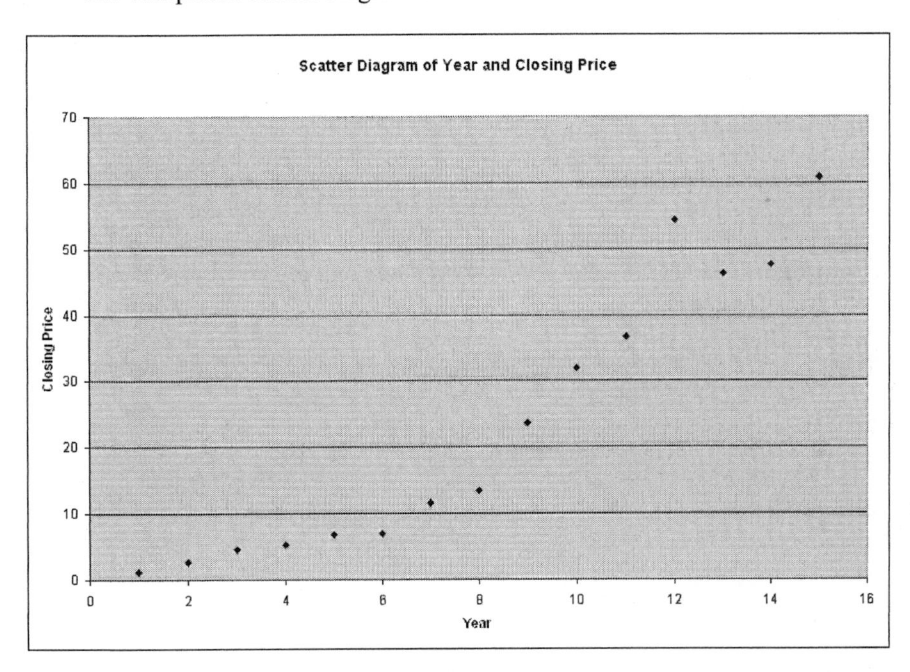

9. Next, you will determine the logarithm of the Y values. Return to the worksheet containing the data by clicking on the **Sheet 1** tab at the bottom of the screen. Enter the label **Y=log y** in cell D1.

	A	B	C	D	E	F	G
1	Year, x	Closing price, y		Y=log y			
2	1990 (x = `	1.1609					

10. Click in cell **D2** where you will place the log of 1.1609.

11. At the top of the screen, click **Insert** and select **Function**.

12. Select the **Math & Trig** category and the **LOG** function. Click **OK**.

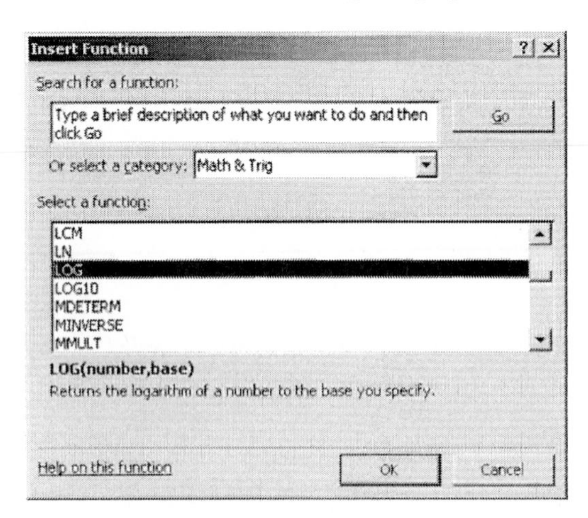

13. Complete the LOG dialog box as shown below. Rather than entering 0.392 in the Number window, you are entering the cell reference **B2**. By using the cell reference, you will be able to copy the LOG function rather than filling it out for each Y value. Click **OK**.

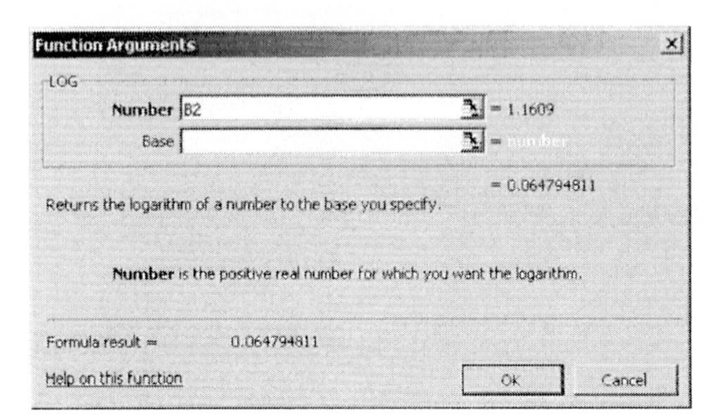

14. Copy the contents of cell D2 to cells D3 through D16.

	A	B	C	D	E	F	G
1	Year, x	Closing price, y		Y=log y			
2	1990 (x =	1.1609		0.064795			
3	1991 (x - 2	2.6988		0.431171			
4	1992 (x =	4.5381		0.656874			
5	1993 (x =	5.3379		0.72737			
6	1994 (x =	6.8032		0.832713			
7	1995 (x =	7.0328		0.847128			
8	1996 (x =	11.5585		1.062901			
9	1997 (x =	13.4799		1.129687			
10	1998 (x =	23.5424		1.371851			
11	1999 (x =	31.9342		1.504256			
12	2000 (x =	36.7277		1.564994			
13	2001 (x =	54.31		1.73488			
14	2002 (x =	46.2		1.664642			
15	2003 (x =	47.53		1.676968			
16	2004 (x =	60.75		1.783546			

15. Next, you will need to enter the X values of 1, 2, 3, … 15 in a separate column of the worksheet. In their present form in column A, they will be treated as non-numeric data. Enter the label **X** in C1 and the values 1 through 15 in the cells immediately below.

	A	B	C	D	E	F	G
1	Year, x	Closing pri	X	Y=log y			
2	1990 (x =	1.1609	1	0.064795			
3	1991 (x - 2	2.6988	2	0.431171			
4	1992 (x =	4.5381	3	0.656874			
5	1993 (x =	5.3379	4	0.72737			
6	1994 (x =	6.8032	5	0.832713			
7	1995 (x =	7.0328	6	0.847128			
8	1996 (x =	11.5585	7	1.062901			
9	1997 (x =	13.4799	8	1.129687			
10	1998 (x =	23.5424	9	1.371851			
11	1999 (x =	31.9342	10	1.504256			
12	2000 (x =	36.7277	11	1.564994			
13	2001 (x =	54.31	12	1.73488			
14	2002 (x =	46.2	13	1.664642			
15	2003 (x =	47.53	14	1.676968			
16	2004 (x =	60.75	15	1.783546			

16. At the top of the screen, click **Tools** and select **Data Analysis.**

If Data Analysis does not appear as a choice in the Tools menu, you will need to load the Microsoft Excel Analysis ToolPak add-in. Follow the procedure in Section GS 8.1 before continuing.

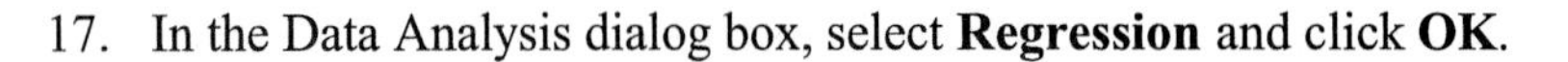

17. In the Data Analysis dialog box, select **Regression** and click **OK**.

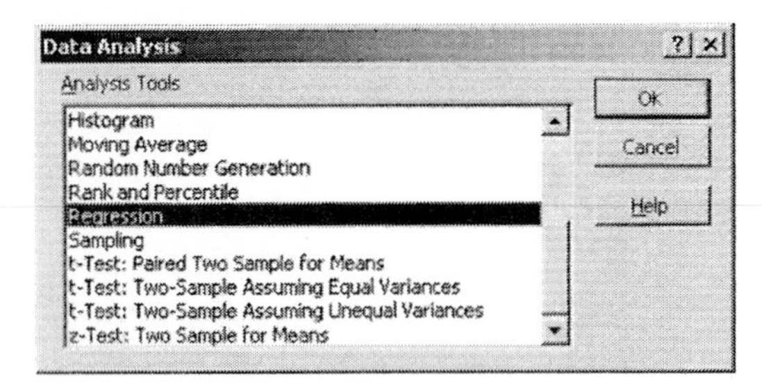

18. Complete the Regression dialog box as shown below. Click **OK**.

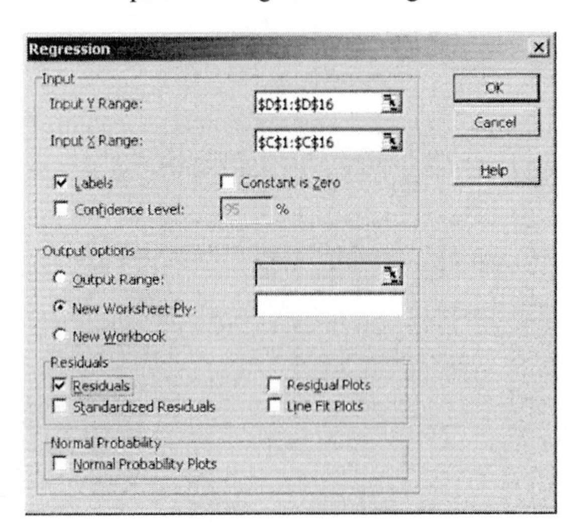

The output is displayed below. The intercept is 0.2113 and the slope is 0.1157.

	A	B	C	D	E	F	G
1	SUMMARY OUTPUT						
2							
3	*Regression Statistics*						
4	Multiple R	0.974467					
5	R Square	0.949586					
6	Adjusted F	0.945708					
7	Standard E	0.123719					
8	Observatio	15					
9							
10	ANOVA						
11		*df*	*SS*	*MS*	*F*	*ignificance F*	
12	Regressior	1	3.747978	3.747978	244.8644	8.18E-10	
13	Residual	13	0.198982	0.015306			
14	Total	14	3.94696				
15							
16		*Coefficients*	*tandard Err*	*t Stat*	*P-value*	*Lower 95%*	*Upper 95%*
17	Intercept	0.211348	0.067224	3.143954	0.007761	0.06612	0.356576
18	X	0.115696	0.007394	15.64814	8.18E-10	0.099723	0.131669

◀

Probability

CHAPTER 5

Section 5.1 Probability Rules

▶ Example 8 (pg. 259) — Simulating Probabilities of Having a Baby Boy or a Baby Girl

You will be generating random numbers selecting from 0 and 1 where 0 represents a boy and 1 represents a girl.

If the PHStat add-in has not been loaded, you will need to load it before continuing. Follow the instructions in Section GS 8.2

1. Open a new Excel worksheet.

2. Enter the numbers 0 and 1 in column A of the worksheet as shown below.

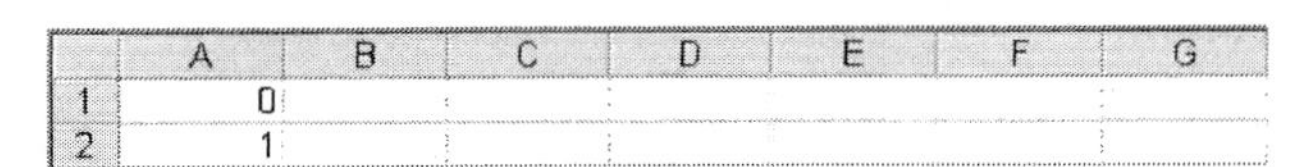

	A	B	C	D	E	F	G
1	0						
2	1						

3. At the top of the screen, click **Tools** and select **Data Analysis**.

If Data Analysis does not appear as a choice in the Tools menu, you will need to load the Microsoft Excel Analysis ToolPak add-in. Follow the procedure in Section GS 8.1 before continuing.

4. Select **Sampling** and click **OK**.

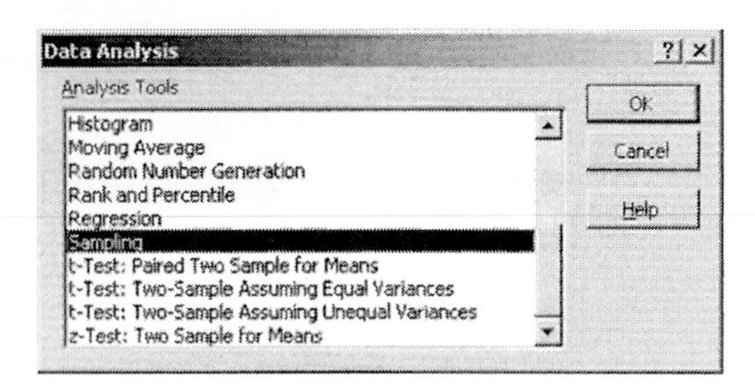

5. Complete the Sampling dialog box as shown below. You will be generating a sample of 100 numbers. The output will be placed in column B. Click **OK**.

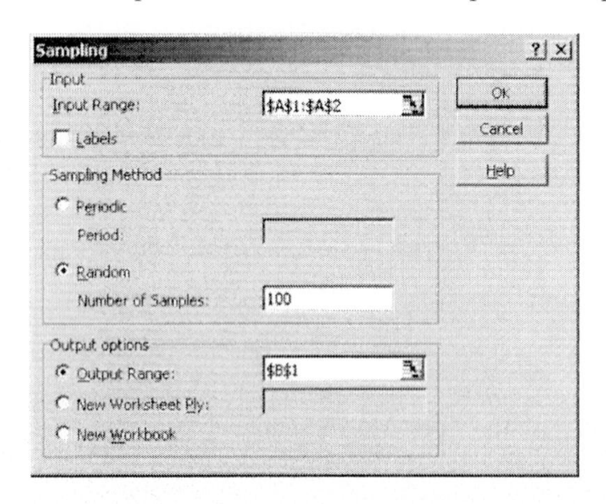

6. The first few rows of the output are displayed below. Because the numbers were generated randomly, it is not likely that your output will be exactly the same. Construct a frequency distribution to obtain a count of boys (0's) and girls (1's). At the top of the screen, click **PHStat** and select **Descriptive Statistics → One-Way Tables & Charts**.

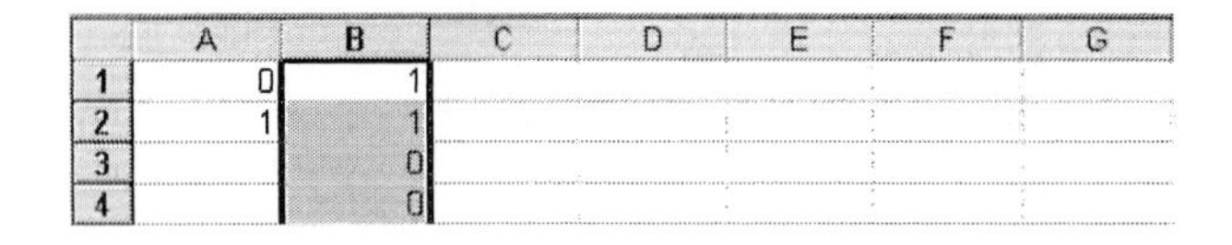

	A	B	C	D	E	F	G
1	0	1					
2	1	1					
3		0					
4		0					

7. Complete the One-Way Tables & Charts dialog box as shown at the top of the next page. Click **OK**.

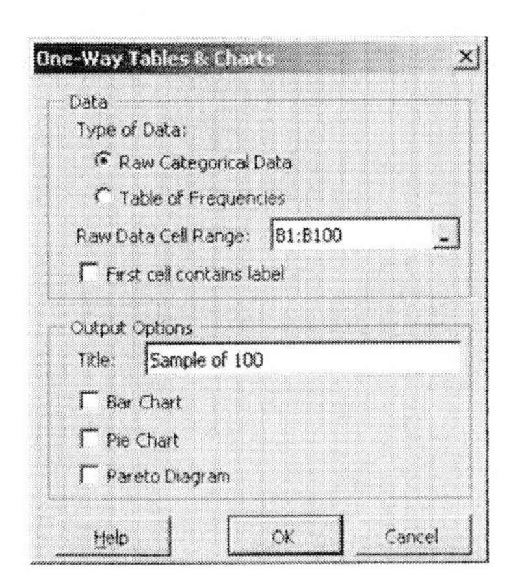

8. The frequency distribution table is displayed in a worksheet named "OneWayTable." In this example, you can see that there are 50 boys (0's) and 50 girls (1's).

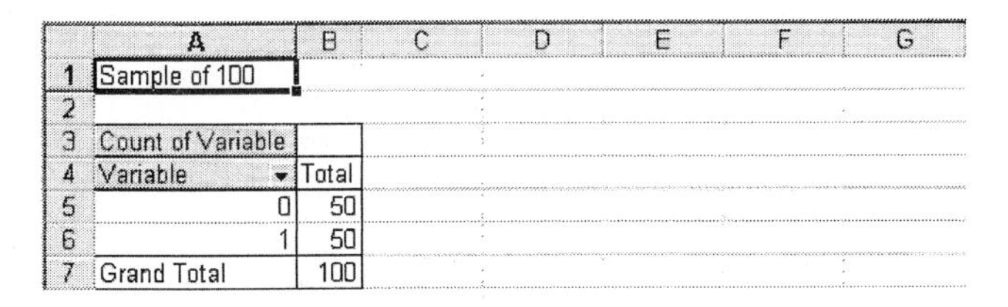

	A	B	C	D	E	F	G
1	Sample of 100						
2							
3	Count of Variable						
4	Variable	Total					
5	0	50					
6	1	50					
7	Grand Total	100					

9. Generate a second sample with 1,000 random numbers. Go back to the worksheet containing the numbers 0 and 1 by clicking on the **Sheet1** tab near the bottom of the screen. Then click **Tools** and select **Data Analysis**. In the Data Analysis dialog box, select **Sampling** and click **OK**. Complete the Sampling dialog box as shown below. The 1,000 random numbers will be placed in column C. Click **OK**.

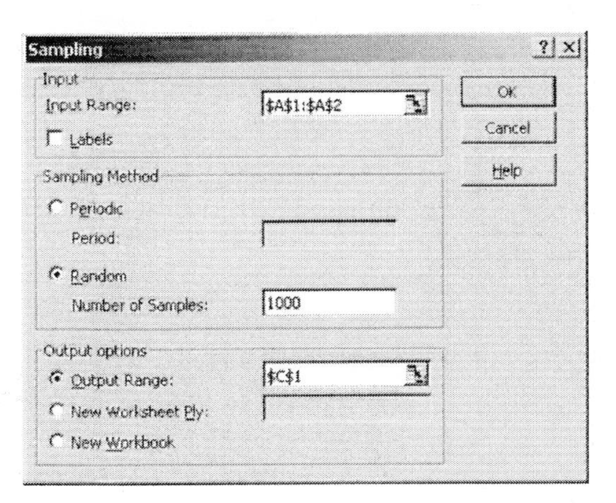

10. Construct a frequency distribution to obtain a count of the boys (0's) and girls (1's). At the top of the screen, click **PHStat** and select **Descriptive Statistics → One-Way Tables & Charts**.

11. Complete the One-Way Tables & Charts dialog box as shown below. Click **OK**.

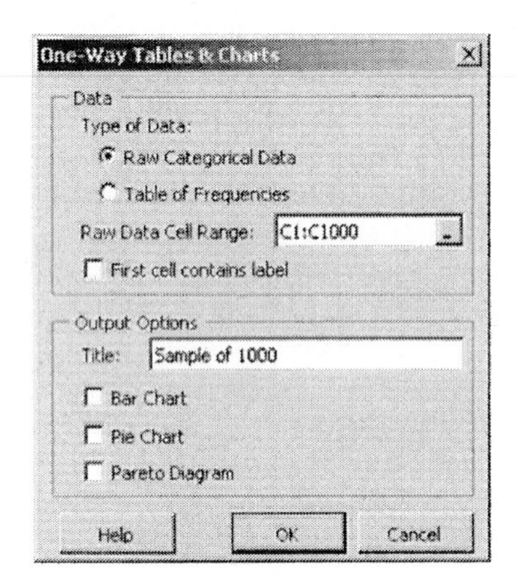

The frequency distribution is displayed below for the random sample that I generated. Your frequencies will likely be somewhat different. There are 509 boys and 491 girls in my sample.

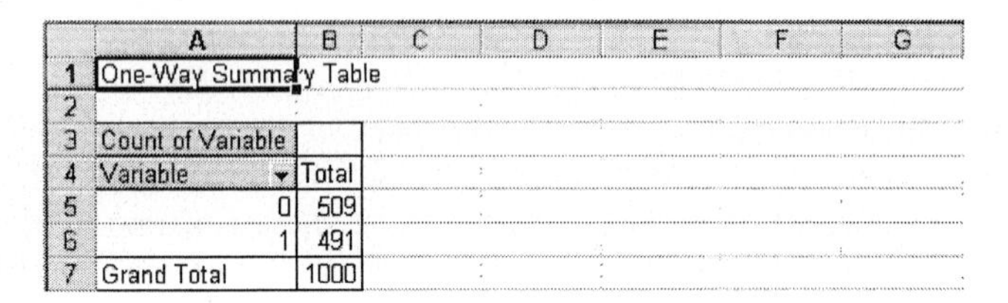

	A	B
1	One-Way Summary Table	
2		
3	Count of Variable	
4	Variable	Total
5	0	509
6	1	491
7	Grand Total	1000

Computing Probabilities

You will be computing probabilities associated with health-risk behaviors.

1. Open a new Excel worksheet.

2. Enter the information shown below. This is the college survey data displayed on page 263 of the textbook. You will be placing probabilities associated with the responses in column C.

	A	B	C	D	E	F	G
1	Response	Frequency	Probability				
2	Never	125					
3	Rarely	324					
4	Sometimes	552					
5	Most of the time	1257					
6	Always	2518					

3. Calculate the sum of the frequencies. To do this, first click in cell **B7** where you will place the sum. Then at the top of the screen, click the AutoSum button **Σ** .

	A	B	C	D	E	F	G
1	Response	Frequency	Probability				
2	Never	125					
3	Rarely	324					
4	Sometimes	552					
5	Most of the time	1257					
6	Always	2518					
7		=SUM(B2:B6)					

4. You should see =SUM(B2:B6). Make any necessary corrections, and then press [**Enter**]. Click in cell **C1** where you will place the probability associated with Never.

	A	B	C	D	E	F	G
1	Response	Frequency	Probability				
2	Never	125					
3	Rarely	324					
4	Sometimes	552					
5	Most of the time	1257					
6	Always	2518					
7		4776					

5. To compute the probability of Never, you will divide 125 by 4776. Carry out this division by entering a formula that uses cell references, **=B2/$B$7**, as shown below. The dollar signs are necessary to make the sum in B7 an absolute reference that will not change when you copy the formula. Press [**Enter**].

	A	B	C	D	E	F	G
1	Response	Frequency	Probability				
2	Never	125	=B2/$B$7				
3	Rarely	324					
4	Sometimes	552					
5	Most of the time	1257					
6	Always	2518					
7		4776					

6. Copy the contents of cell C2 to cells C3 through C7.

	A	B	C	D	E	F	G
1	Response	Frequency	Probability				
2	Never	125	0.026173				
3	Rarely	324	0.067839				
4	Sometimes	552	0.115578				
5	Most of the time	1257	0.263191				
6	Always	2518	0.527219				
7		4776	1				

Discrete Probability Distributions

CHAPTER 6

Section 6.1 Discrete Random Variables

▶ Problem 19 (pg. 324)

Drawing a Probability Distribution, Computing the Mean and Variance

1. Open a new Excel worksheet.

2. Enter the information shown below. This information is displayed in problem 19 on page 324.

	A	B	C	D	E	F	G
1	x	P(x)					
2	0	0.035					
3	1	0.074					
4	2	0.197					
5	3	0.32					
6	4	0.374					

3. Highlight the range B1:B6. Then, at the top of the screen, click **Insert** and select **Chart**.

	A	B	C	D	E	F	G
1	x	P(x)					
2	0	0.035					
3	1	0.074					
4	2	0.197					
5	3	0.32					
6	4	0.374					

4. Under Chart type, select **Column**. Under Chart sub-type, select the leftmost diagram in the top row. Click **Next>**.

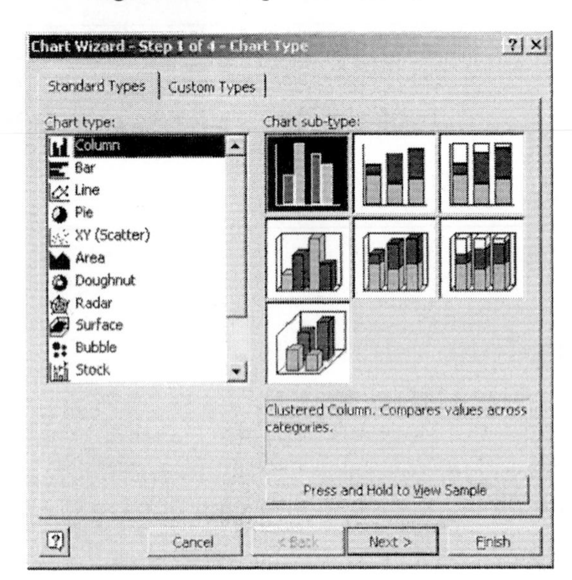

5. At the top of the Source Data dialog box, click the **Series** tab.

6. You want the values in column A to be the Category (X) axis labels. Click in the Category (X) axis labels window at the bottom of the dialog box. Then click and drag in the worksheet over the range A2:A6 to enter that information. Click **Next>**.

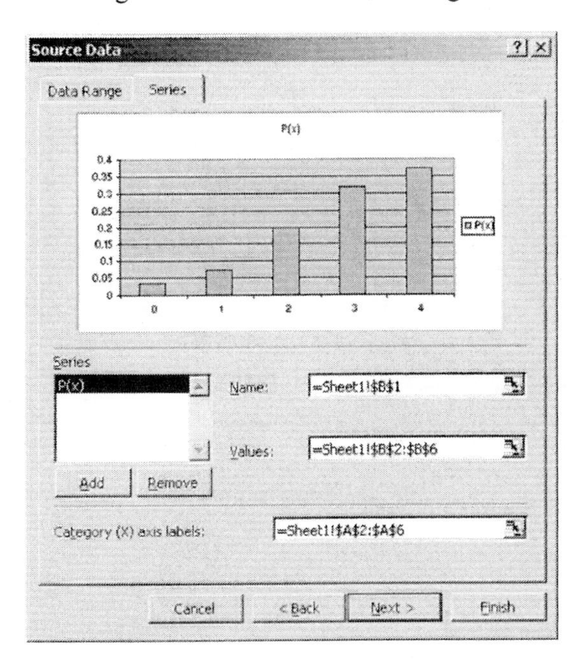

7. Enter a title for the chart and labels for the X and Y axes. Chart title: **Probability Distribution of Parental Involvement**. Category (X) axis: **Number of Activities**. Value (Y) axis: **Probability**.

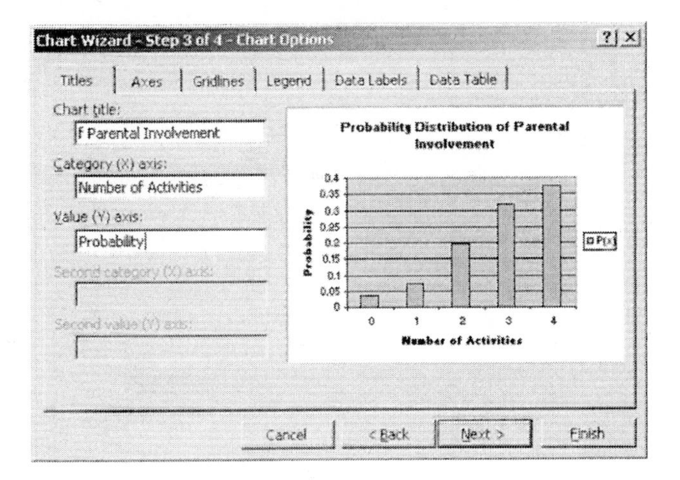

8. Click the **Legend** tab at the top of the Chart Options dialog box.

9. Click in the box to the left of **Show legend** to remove the checkmark. Click **Next>**.

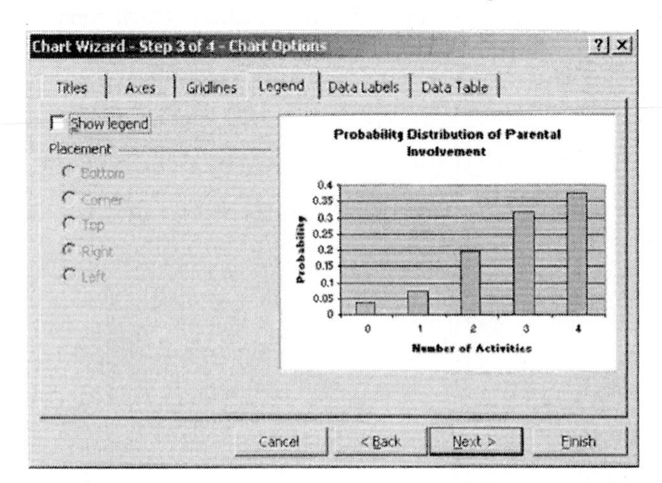

10. In the Chart Location dialog box, select **As new sheet**. Click **Finish**.

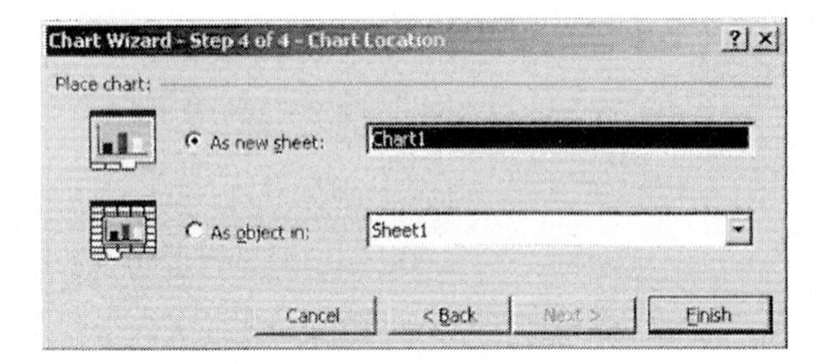

11. Remove the space between the vertical bars. **Right-click** on one of the vertical bars. Select **Format Data Series** from the shortcut menu that appears.

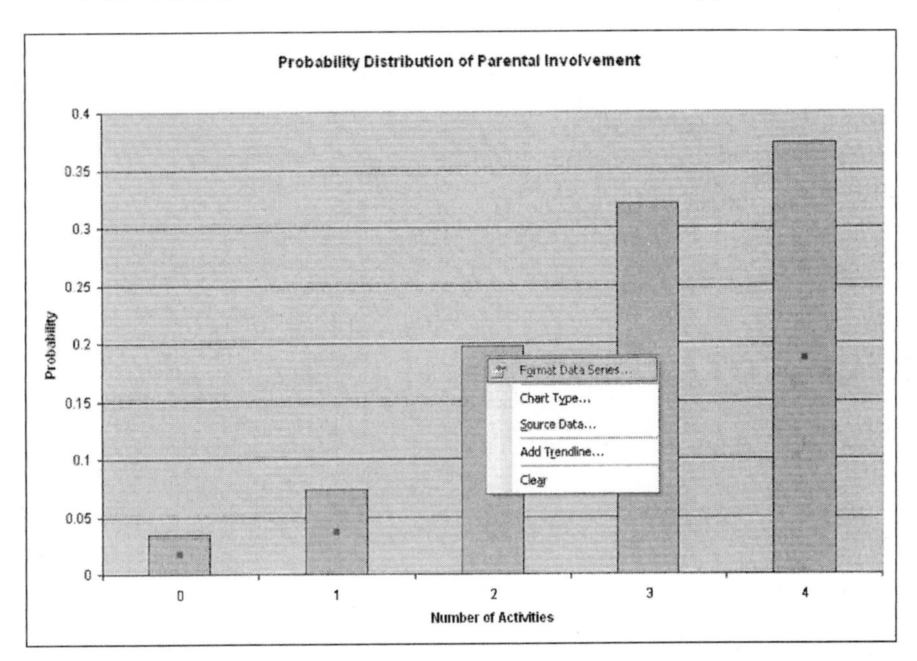

12. Click the **Options** tab at the top of the Format Data Series dialog box. Change the value in the Gap width box to **0**. Click **OK**.

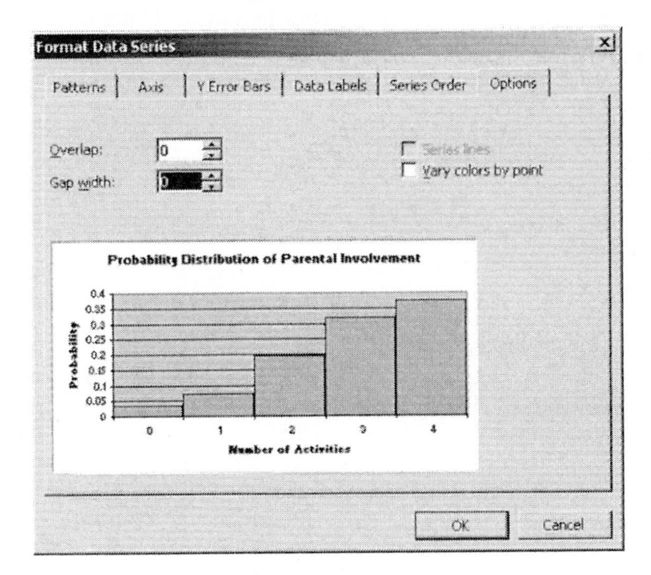

Your completed probability distribution should look similar to the one displayed below.

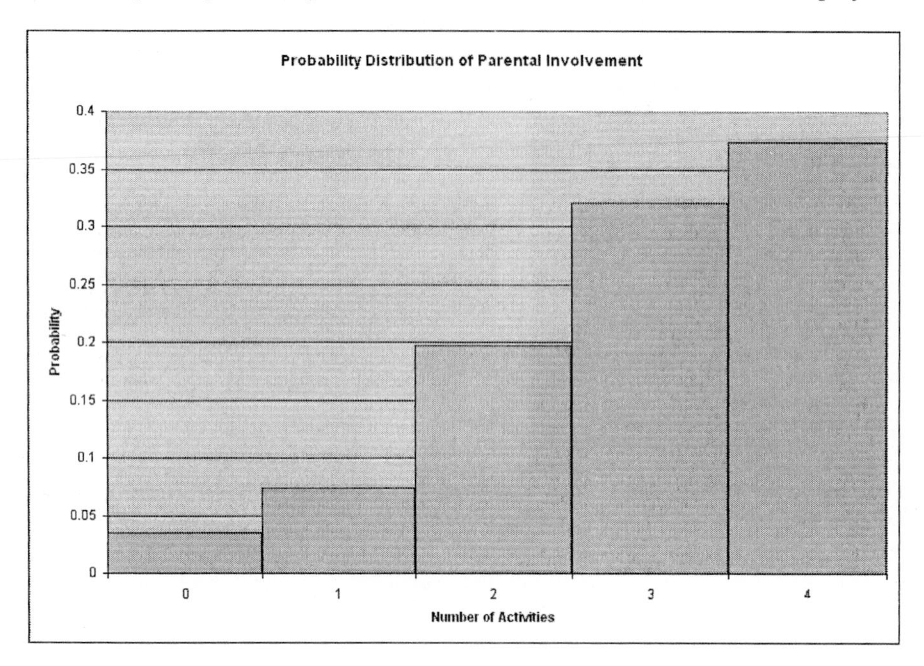

13. You will now compute the mean of this variable. Click the **Sheet1** tab at the bottom of the screen to return to the worksheet containing the data.

14. You will be applying the formula shown on page 319 in your textbook. Click in cell **C2** of the worksheet and enter the formula shown below, **=A2*B2**, to multiply the x value in column A by the probability in column B. Press [**Enter**].

	A	B	C	D	E	F	G
1	x	P(x)					
2	0	0.035	=A2*B2				
3	1	0.074					
4	2	0.197					
5	3	0.32					
6	4	0.374					

15. Copy the formula in cell C2 to cells C3 through C6.

	A	B	C	D	E	F	G
1	x	P(x)					
2	0	0.035	0				
3	1	0.074	0.074				
4	2	0.197	0.394				
5	3	0.32	0.96				
6	4	0.374	1.496				

16. To add the products in column C, first click in cell **C7**. Then, at the top of the screen, click the AutoSum button **Σ**. You should see =SUM(C2:C6). Press **[Enter]**.

	A	B	C	D	E	F	G
1	x	P(x)					
2	0	0.035	0				
3	1	0.074	0.074				
4	2	0.197	0.394				
5	3	0.32	0.96				
6	4	0.374	1.496				
7			=SUM(C2:C6)				

The mean is 2.924 activities.

	A	B	C	D	E	F	G
1	x	P(x)					
2	0	0.035	0				
3	1	0.074	0.074				
4	2	0.197	0.394				
5	3	0.32	0.96				
6	4	0.374	1.496				
7			2.924				

17. You will now compute the variance of this variable. The formula is given on page 322 of your textbook. Click in cell **D2** of the worksheet. Enter the formula shown below, **=(A2-$C$7)^2*B2**, to multiply the squared deviation score by the probability. Note that the dollar signs are necessary for the mean's cell reference. That reference must be an absolute reference that does not change when the formula is copied. Press **[Enter]**.

	A	B	C	D	E	F	G
1	x	P(x)					
2	0	0.035	0	=(A2-$C$7)^2*B2			
3	1	0.074	0.074				
4	2	0.197	0.394				
5	3	0.32	0.96				
6	4	0.374	1.496				
7			2.924				

18. Copy the contents of cell D2 to cells D3 through D6.

	A	B	C	D	E	F	G
1	x	P(x)					
2	0	0.035	0	0.299242			
3	1	0.074	0.074	0.273931			
4	2	0.197	0.394	0.168194			
5	3	0.32	0.96	0.001848			
6	4	0.374	1.496	0.433008			

19. Click in cell **D7** to place the sum there. At the top of the screen, click the AutoSum button Σ. In cell D7, you should see =SUM(D2:D6). Press [**Enter**].

	A	B	C	D	E	F	G
1	x	P(x)					
2	0	0.035	0	0.299242			
3	1	0.074	0.074	0.273931			
4	2	0.197	0.394	0.168194			
5	3	0.32	0.96	0.001848			
6	4	0.374	1.496	0.433008			
7			2.924	=SUM(D2:D6)			

20. The variance of this variable is 1.1762. Click in cell **D8** to place the standard deviation there. Enter the formula shown below, **=SQRT(D7)**, to take the square root of the variance. Press [**Enter**].

	A	B	C	D	E	F	G
1	x	P(x)					
2	0	0.035	0	0.299242			
3	1	0.074	0.074	0.273931			
4	2	0.197	0.394	0.168194			
5	3	0.32	0.96	0.001848			
6	4	0.374	1.496	0.433008			
7			2.924	1.176224			
8				=SQRT(D7)			

The standard deviation is 1.0845.

	A	B	C	D	E	F	G
1	x	P(x)					
2	0	0.035	0	0.299242			
3	1	0.074	0.074	0.273931			
4	2	0.197	0.394	0.168194			
5	3	0.32	0.96	0.001848			
6	4	0.374	1.496	0.433008			
7			2.924	1.176224			
8				1.084539			

◄

▶ Problem 37 (pg. 327)

Simulation

1. Open a new Excel worksheet.

2. Enter the probability distribution from Problem 21, p. 325, as shown below.

	A	B	C	D	E	F	G
1	x	P(X=x)					
2	0	0.1677					
3	1	0.3354					
4	2	0.2857					
5	3	0.1491					
6	4	0.0373					
7	5	0.0248					

3. At the top of the screen, click **Tools** and select **Data Analysis.**

If Data Analysis does not appear as a choice in the Tools menu, you will need to load the Microsoft Excel Analysis ToolPak add-in. Follow the procedure in Section GS 8.1 before continuing.

---

4. Select **Random Number Generation** and click **OK.**

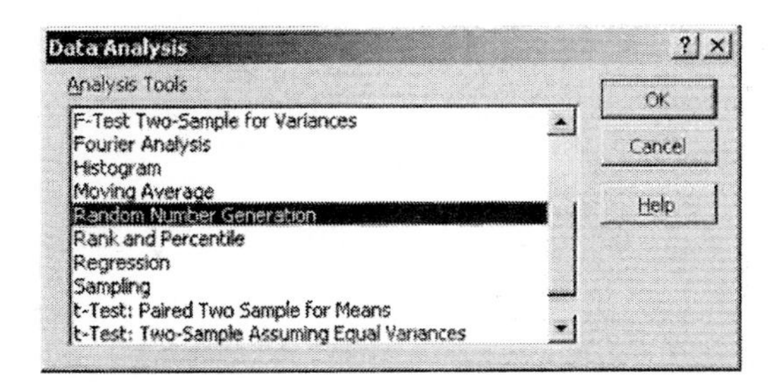

5. Complete the Random Number Generation dialog box as shown below. Click **OK.**

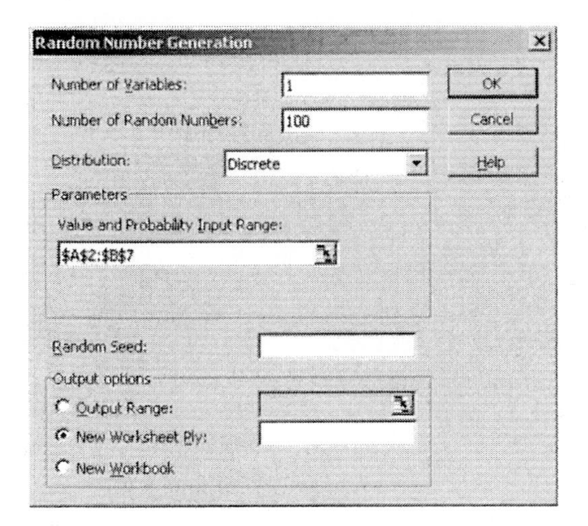

6. The first few lines of my output are displayed below. Because I did not input a Random Seed in the previous dialog box, Excel used a default number to start the random generation process. If you also left the Random Seed window blank, it is likely that your output will be the same as mine.

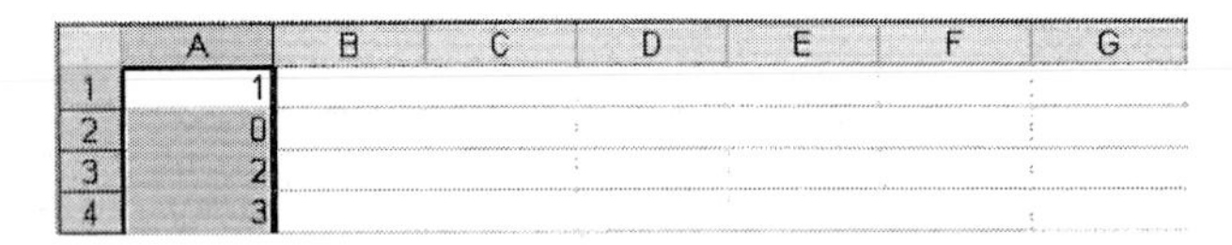

	A	B	C	D	E	F	G
1	1						
2	0						
3	2						
4	3						

7. To construct a frequency distribution using Excel's Pivot Table, you need a label in the topmost cell. Click in cell **A1**. At the top of the screen, click **Insert** and select **Rows**. Then type **X** in cell A1.

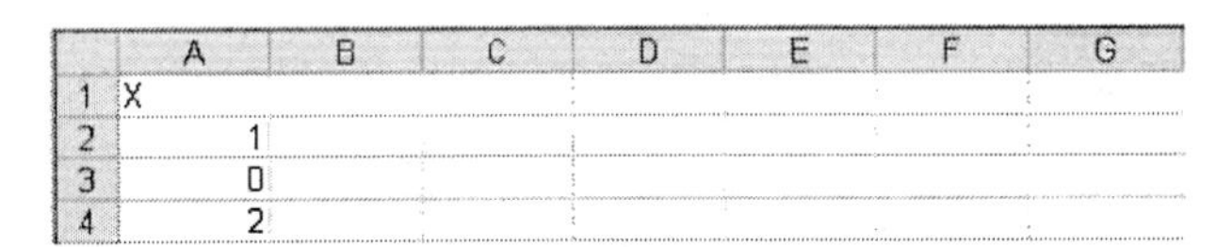

	A	B	C	D	E	F	G
1	X						
2	1						
3	0						
4	2						

8. You are now ready to construct a frequency distribution of these data. At the top of the screen, click **Data** and select **Pivot Table and Pivot Chart Report**.

9. Select **Microsoft Excel list** at the top of the dialog box. Select **Pivot Table** at the bottom of the dialog box. Click **Next>**.

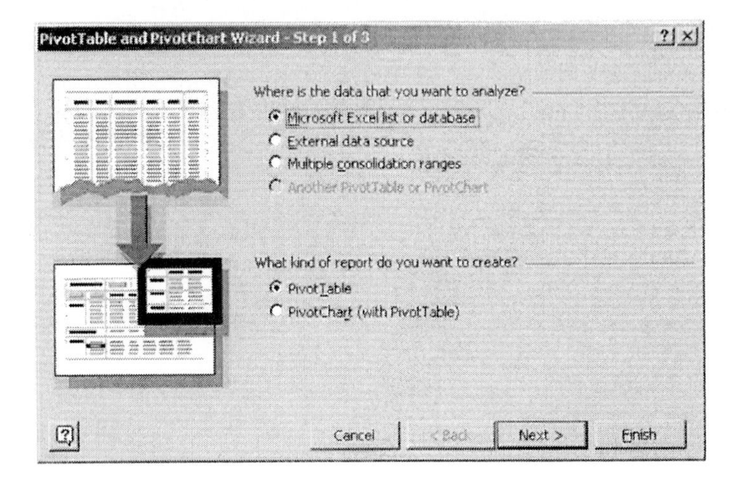

10. Enter the data range A1:A101 in the Step 2 dialog box as shown below. Click **Next>**.

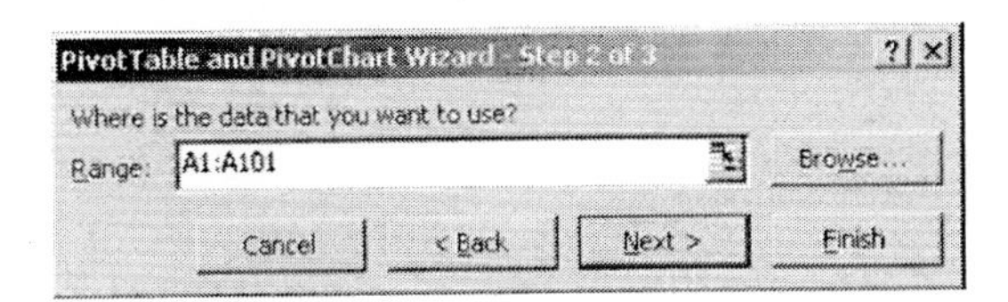

11. In the Step 3 dialog box, select **New worksheet**. At the bottom of the dialog box, click **Layout**.

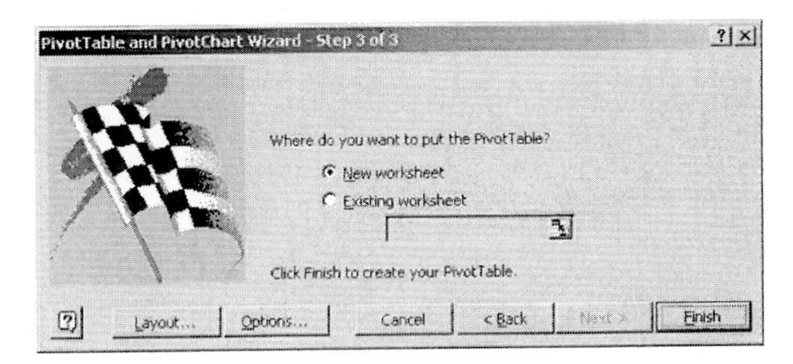

12. Drag the X field button at the right to ROW. Then drag the X field at the right to DATA.

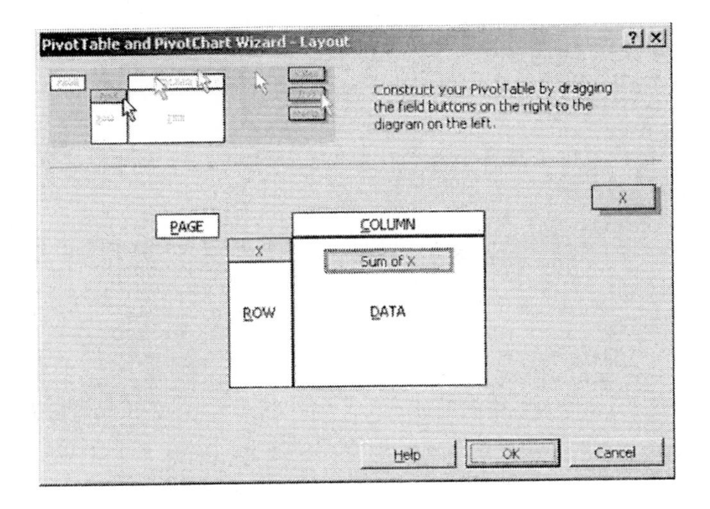

13. The default is Sum of X. You need to change this to count. Double-click on the **Sum of X** button.

14. Under Summarize by, select **Count**. Click **OK**. Also click **OK** in the Layout dialog box .

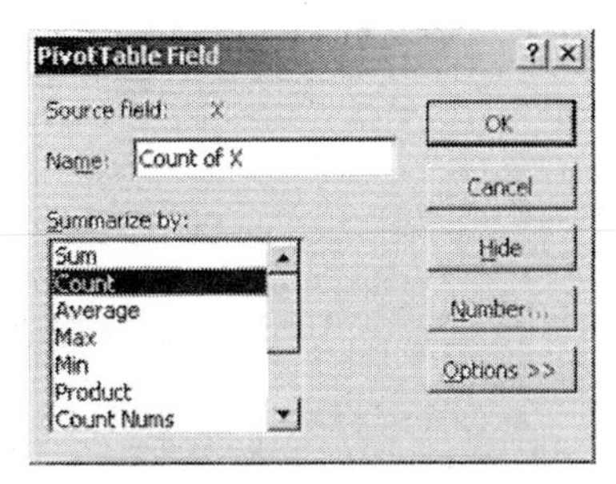

15. In the Step 3 dialog box, click **Finish**.

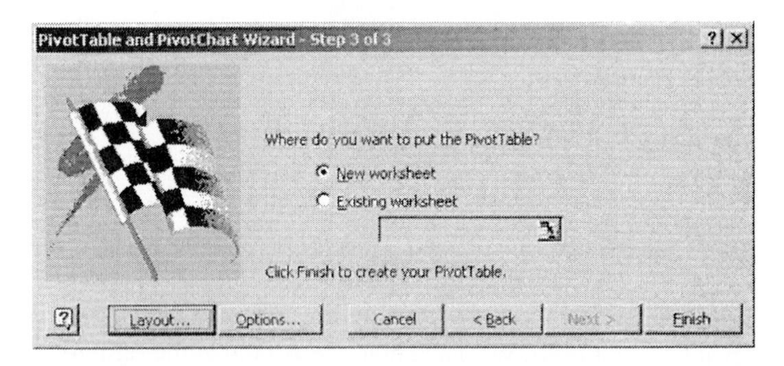

16. Enter the label **P(X=x)** in cell C4.

	A	B	C	D	E	F	G
1							
2							
3	Count of X						
4	X	Total	P(X=x)				
5	0	20					
6	1	34					
7	2	23					
8	3	15					
9	4	4					
10	5	4					
11	Grand Total	100					

17. You will now compute the probability of observing x = 1 using a formula. Click in cell **C5**. Enter the formula **=B5/$B$11** as shown below. The dollar signs are necessary for the cell B11 address to make it an absolute reference that will not change when it is copied. Press [**Enter**].

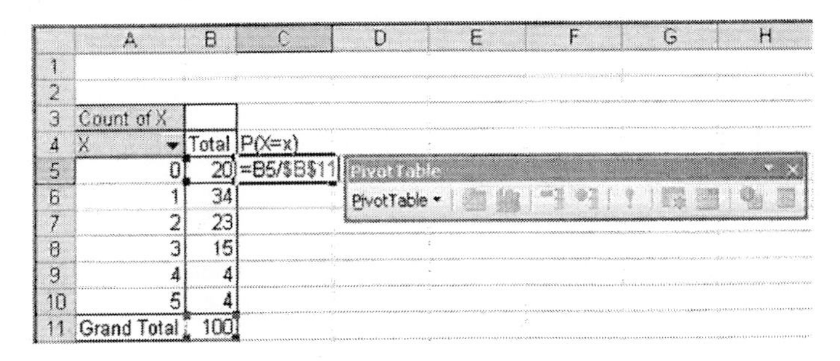

18. Copy the contents of cell C5 to cells C6 through C11.

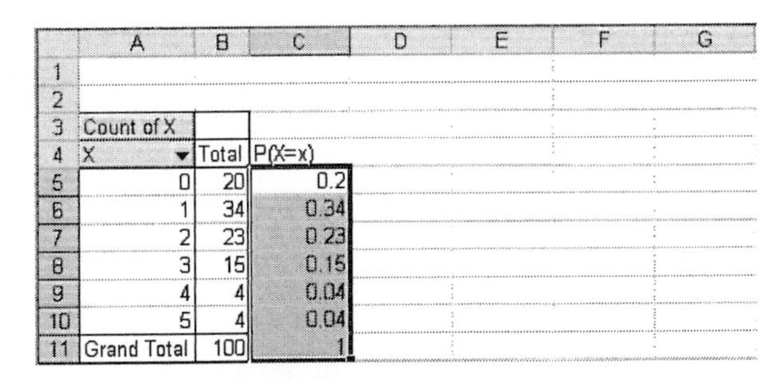

19. You will now compute the mean of this distribution. You will be applying the formula shown on page 319 of your textbook. Click in cell **D5** of the worksheet and enter the formula **=A5*C5** to multiply the x value in column A by the probability in column C. Press [**Enter**].

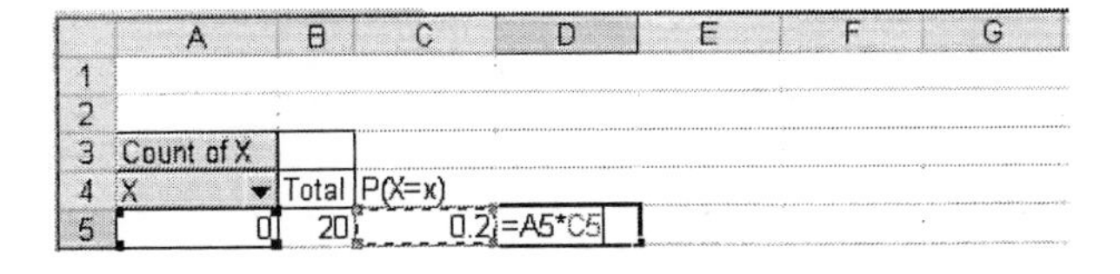

20. Copy the formula in cell D5 to cells D6 through D10.

	A	B	C	D	E	F	G
1							
2							
3	Count of X						
4	X	Total	P(X=x)				
5	0	20	0.2	0			
6	1	34	0.34	0.34			
7	2	23	0.23	0.46			
8	3	15	0.15	0.45			
9	4	4	0.04	0.16			
10	5	4	0.04	0.2			

21. Add the products in column D. To do this, first click in cell **D11**. Then, at the top of the screen, click the AutoSum button Σ. You should see =SUM(D5:D10). Press [**Enter**]. The mean is 1.61.

	A	B	C	D	E	F	G
1							
2							
3	Count of X						
4	X	Total	P(X=x)				
5	0	20	0.2	0			
6	1	34	0.34	0.34			
7	2	23	0.23	0.46			
8	3	15	0.15	0.45			
9	4	4	0.04	0.16			
10	5	4	0.04	0.2			
11	Grand Total	100	1	=SUM(D5:D10)			

22. You will now compute the variance of this distribution. The formula is given on page 322 of your textbook. Click in cell **E5** of the worksheet. Enter the formula shown below to multiply the squared deviation score by the probability. Note that the dollar signs are necessary for the cell address of the mean. The address must be an absolute reference that does not change when the formula is copied. Press [**Enter**].

	A	B	C	D	E	F	G
1							
2							
3	Count of X						
4	X	Total	P(X=x)				
5	0	20	0.2	0	=(A5-$D$11)^2*C5		

23. Copy the contents of cell E5 to cells E6 through E10.

	A	B	C	D	E	F	G
1							
2							
3	Count of X						
4	X	Total	P(X=x)				
5	0	20	0.2	0	0.51842		
6	1	34	0.34	0.34	0.126514		
7	2	23	0.23	0.46	0.034983		
8	3	15	0.15	0.45	0.289815		
9	4	4	0.04	0.16	0.228484		
10	5	4	0.04	0.2	0.459684		

24. Click in cell **E11** to place the sum there. At the top of the screen, click the AutoSum button Σ. In cell E11 you should see =SUM(E5:E10). Press [**Enter**]. The variance is 1.6579.

	A	B	C	D	E	F	G
1							
2							
3	Count of X						
4	X	Total	P(X=x)				
5	0	20	0.2	0	0.51842		
6	1	34	0.34	0.34	0.126514		
7	2	23	0.23	0.46	0.034983		
8	3	15	0.15	0.45	0.289815		
9	4	4	0.04	0.16	0.228484		
10	5	4	0.04	0.2	0.459684		
11	Grand Total	100	1	1.61	=SUM(E5:E10)		

25. Click in cell **E12** to place the standard deviation there. Enter the formula shown below to take the square root of the variance. Press [**Enter**].

	A	B	C	D	E	F	G
1							
2							
3	Count of X						
4	X	Total	P(X=x)				
5	0	20	0.2	0	0.51842		
6	1	34	0.34	0.34	0.126514		
7	2	23	0.23	0.46	0.034983		
8	3	15	0.15	0.45	0.289815		
9	4	4	0.04	0.16	0.228484		
10	5	4	0.04	0.2	0.459684		
11	Grand Total	100	1	1.61	1.6579		
12					=SQRT(E11)		

26. The standard deviation of this distribution is 1.2876. If you would like to repeat the simulation by performing 500 repetitions, start with step 1 of these instructions. In step 5, be sure to request 500 random numbers. To prevent Excel from starting the generation process with the default as used previously, you should also enter a number in the Random Seed window of the Random Number Generation dialog box.

	A	B	C	D	E	F	G
1							
2							
3	Count of X						
4	X	Total	P(X=x)				
5	0	20	0.2	0	0.51842		
6	1	34	0.34	0.34	0.126514		
7	2	23	0.23	0.46	0.034983		
8	3	15	0.15	0.45	0.289815		
9	4	4	0.04	0.16	0.228484		
10	5	4	0.04	0.2	0.459684		
11	Grand Total	100		1.61	1.6579		
12					1.287595		

◀

Section 6.2 The Binomial Probability Distribution

▸ Example 5 (pg. 335)

Constructing Binomial Probability Histograms; Finding Mean and Standard Deviation

If the PHStat add-in has not been loaded, you will need to load it before continuing. Follow the instructions in Section GS 8.2.

1. The problem states that the probability of attending a musical play is 0.20 and that the sample size is 15. You are asked to find that probability that exactly 5 women have attended a musical play. Open a new Excel worksheet. At the top of the screen, click **PHStat** and select **Probability and Prob. Distributions → Binomial**.

2. Complete the Binomial Probability Distribution dialog box as shown below. Click **OK**.

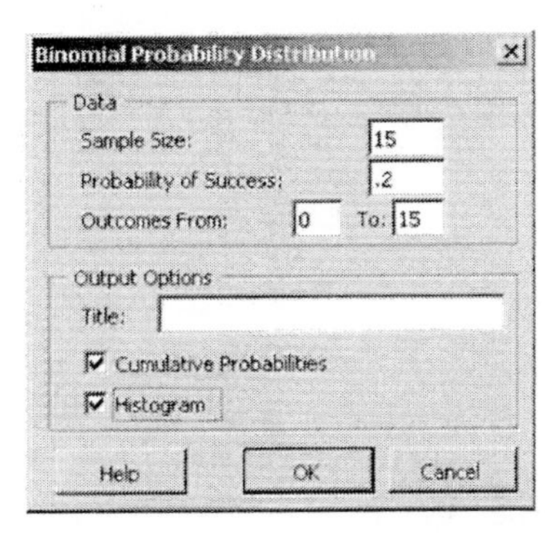

The worksheet labeled "Histogram" contains a histogram of this binomial distribution.

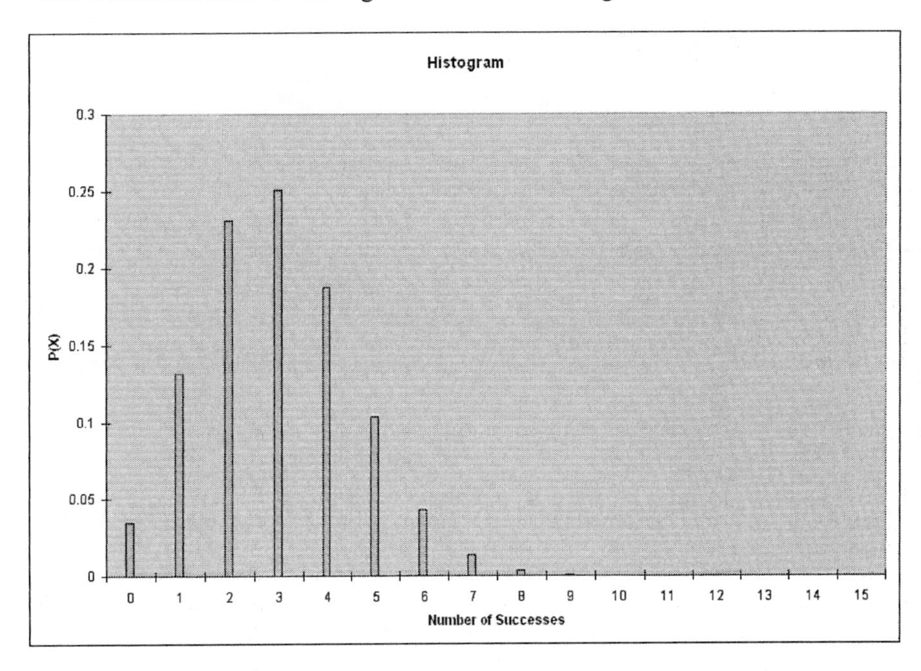

The worksheet labeled "Binomial" contains summary statistics and a binomial probabilities table.

	A	B	C	D	E	F	G
1	Binomial Probabilities						
2							
3	Data						
4	Sample size	15					
5	Probability of success	0.2					
6							
7	Statistics						
8	Mean	3					
9	Variance	2.4					
10	Standard deviation	1.549193					
11							
12	Binomial Probabilities Table						
13		X	P(X)	P(<=X)	P(<X)	P(>X)	P(>=X)
14		0	0.035184	0.035184	0	0.964816	1
15		1	0.131941	0.167126	0.035184	0.832874	0.964816
16		2	0.230897	0.398023	0.167126	0.601977	0.832874
17		3	0.250139	0.648162	0.398023	0.351838	0.601977
18		4	0.187604	0.835766	0.648162	0.164234	0.351838
19		5	0.103182	0.938949	0.835766	0.061051	0.164234
20		6	0.042993	0.981941	0.938949	0.018059	0.061051
21		7	0.013819	0.99576	0.981941	0.00424	0.018059
22		8	0.003455	0.999215	0.99576	0.000785	0.00424
23		9	0.000672	0.999887	0.999215	0.000113	0.000785
24		10	0.000101	0.999988	0.999887	1.25E-05	0.000113
25		11	1.15E-05	0.999999	0.999988	1.01E-06	1.25E-05
26		12	9.54E-07	1	0.999999	5.7E-08	1.01E-06
27		13	5.51E-08	1	1	2E-09	5.7E-08
28		14	1.97E-09	1	1	3.28E-11	2E-09
29		15	3.28E-11	1	1	0	3.28E-11

◀

Section 6.3 The Poisson Probability Distribution

Problem 19 (pg. 349)	Constructing a Poisson Probability Distribution

If the PHStat add-in has not been loaded, you will need to load it before continuing. Follow the instructions in Section GS 8.2.

1. Open a new Excel worksheet. At the top of the screen, click **PHStat** and select **Probability and Prob. Distributions → Poisson**.

2. The formula for the mean of a Poisson distribution is λt. For this exercise, $\lambda = .2$ and $t = 30$. So, $\lambda t = 6$. Complete the Poisson Probability Distribution dialog box as shown below. Click **OK**.

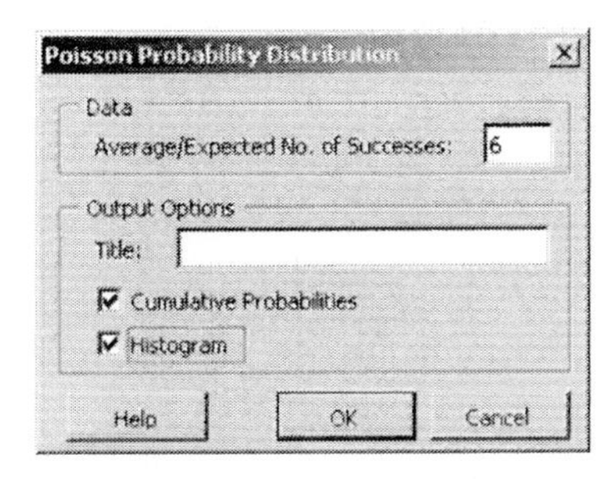

3. The probability distribution of x before the advertising is displayed in a worksheet labeled Poisson. Open worksheet 6_3_19 in the chapter 6 folder.

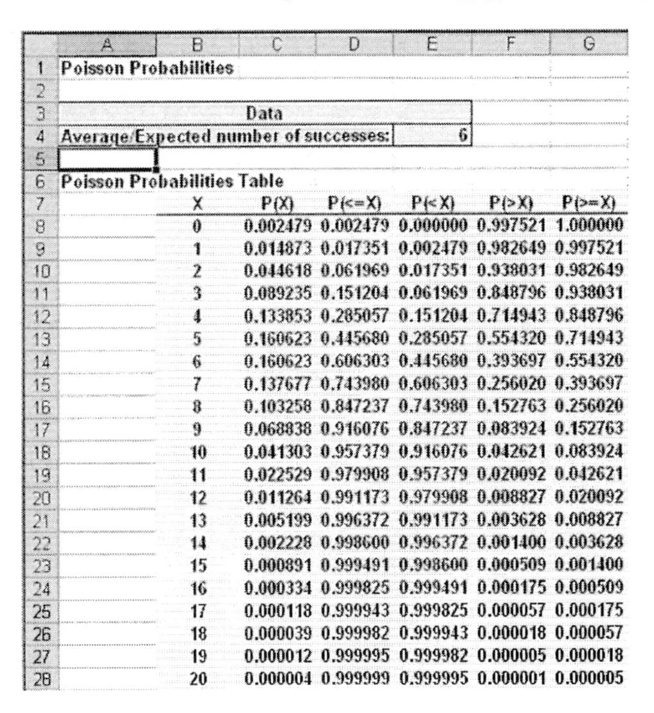

Poisson Probabilities

Data	
Average/Expected number of successes:	6

Poisson Probabilities Table

X	P(X)	P(<=X)	P(<X)	P(>X)	P(>=X)
0	0.002479	0.002479	0.000000	0.997521	1.000000
1	0.014873	0.017351	0.002479	0.982649	0.997521
2	0.044618	0.061969	0.017351	0.938031	0.982649
3	0.089235	0.151204	0.061969	0.848796	0.938031
4	0.133853	0.285057	0.151204	0.714943	0.848796
5	0.160623	0.445680	0.285057	0.554320	0.714943
6	0.160623	0.606303	0.445680	0.393697	0.554320
7	0.137677	0.743980	0.606303	0.256020	0.393697
8	0.103258	0.847237	0.743980	0.152763	0.256020
9	0.068838	0.916076	0.847237	0.083924	0.152763
10	0.041303	0.957379	0.916076	0.042621	0.083924
11	0.022529	0.979908	0.957379	0.020092	0.042621
12	0.011264	0.991173	0.979908	0.008827	0.020092
13	0.005199	0.996372	0.991173	0.003628	0.008827
14	0.002228	0.998600	0.996372	0.001400	0.003628
15	0.000891	0.999491	0.998600	0.000509	0.001400
16	0.000334	0.999825	0.999491	0.000175	0.000509
17	0.000118	0.999943	0.999825	0.000057	0.000175
18	0.000039	0.999982	0.999943	0.000018	0.000057
19	0.000012	0.999995	0.999982	0.000005	0.000018
20	0.000004	0.999999	0.999995	0.000001	0.000005

4. Click in cell **B18** where you will place the sum of the frequencies. Then, at the top of the screen, click the AutoSum button Σ. You should see =SUM(B2:B17). Press [**Enter**]. The sum is 200.

	A	B	C	D	E	F	G
1	x (Num. of	Frequency					
2	1	4					
3	2	5					
4	3	13					
5	4	23					
6	5	25					
7	6	28					
8	7	25					
9	8	27					
10	9	21					
11	10	15					
12	11	5					
13	12	3					
14	13	2					
15	14	2					
16	15	0					
17	16	2					
18		=SUM(B2:B17)					

5. You will compute probabilities and place them in column C. Click in **C1** and enter the label **P(X=x)**.

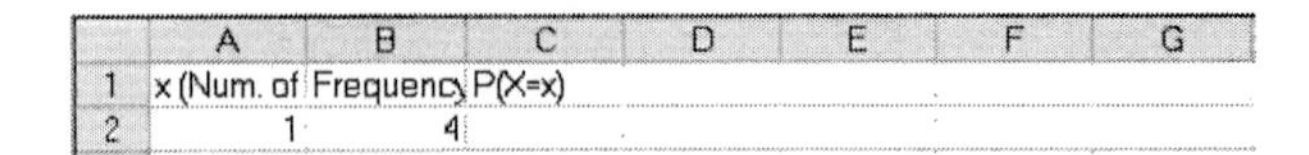

	A	B	C	D	E	F	G
1	x (Num. of	Frequency	P(X=x)				
2	1	4					

6. Click in cell **C2** and enter the formula **=B2/$B$18** as shown below. The dollar signs are necessary for the B18 address because the sum must have an absolute reference that does not change when it is copied. Press [**Enter**].

	A	B	C	D	E	F	G
1	x (Num. of	Frequency	P(X=x)				
2	1	4	=B2/$B$18				

7. Copy the contents of cell C2 to cells C3 through C17. Column C of the worksheet displays the probability distribution of x after the advertising.

	A	B	C	D	E	F	G
1	x (Num. of	Frequency	P(X=x)				
2	1	4	0.02				
3	2	5	0.025				
4	3	13	0.065				
5	4	23	0.115				
6	5	25	0.125				
7	6	28	0.14				
8	7	25	0.125				
9	8	27	0.135				
10	9	21	0.105				
11	10	15	0.075				
12	11	5	0.025				
13	12	3	0.015				
14	13	2	0.01				
15	14	2	0.01				
16	15	0	0				
17	16	2	0.01				
18		200					

◄

The Normal Probability Distribution

CHAPTER 7

Section 7.1 Properties of the Normal Distribution

▶ Problem 37 (pg. 371)

Constructing a Relative Frequency Histogram

1. Open worksheet "7_1_37" in the Chapter 7 folder. The first few rows are shown below.

	A	B	C	D	E	F	G
1	Pitching Wedge Distance						
2	100						
3	104						
4	104						

2. Click in any cell of column A that has an entry and then, at the top of the screen, click **Data** and select **Pivot Table and Pivot Chart Report**.

3. At the top of the Step 1 dialog box, select **Microsoft Excel list**. At the bottom of the dialog box, select **Pivot Table**. Click **Next>**.

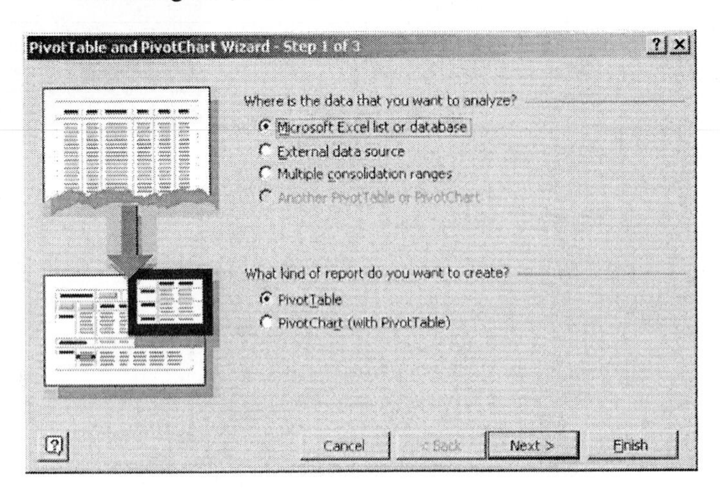

4. The data range $A$1:$A$76 should automatically appear in the Range window. Make any necessary corrections. Then click **Next>**.

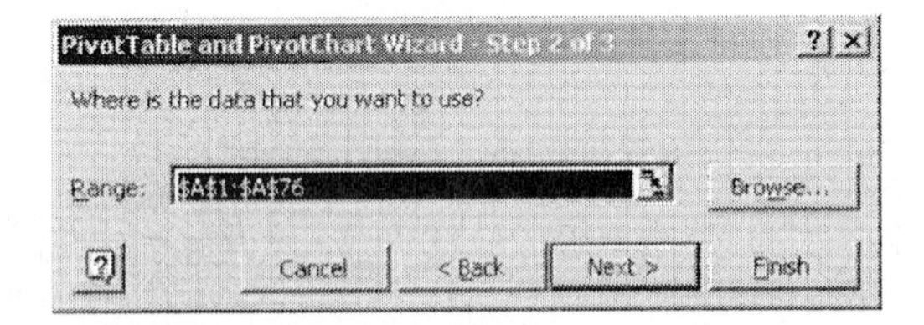

5. Select **New worksheet**. Click the **Layout** button at the bottom of the dialog box.

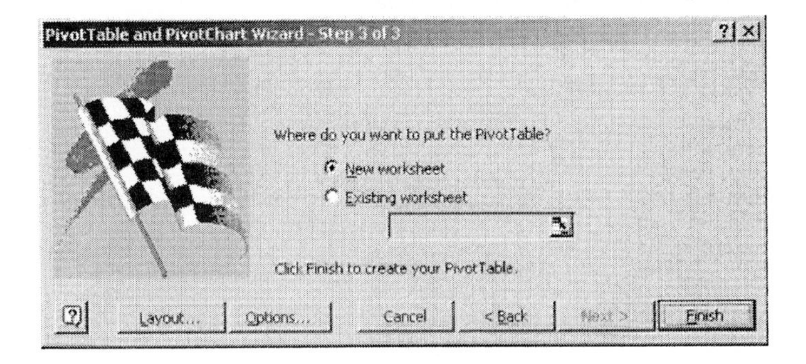

6. Drag the Pitching field button to ROW, and drag the Pitching field button to DATA.

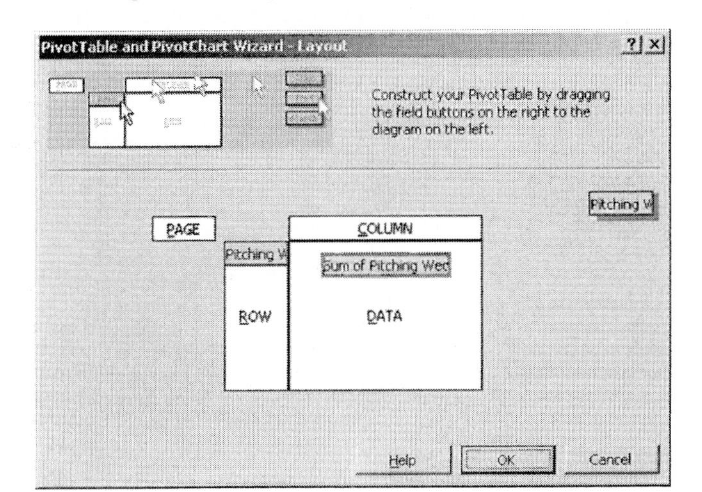

7. The default summary measure is Sum. You want Count. To make this change, double-click on the **Sum of Pitching** button.

8. Under Summarize by, select **Count**. Click **OK**. Also click **OK** in the Layout dialog box. Click **Finish** in the Step 3 dialog box.

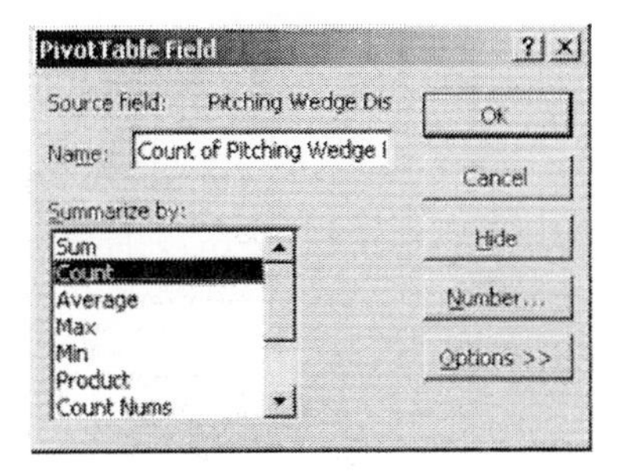

9. The output consists of frequencies and you want relative frequencies. You will need to compute the relative frequencies. Enter the label **Relative Frequency** in cell C4. Then click in cell **C5** where you will place the relative frequency associated with a distance of 94.

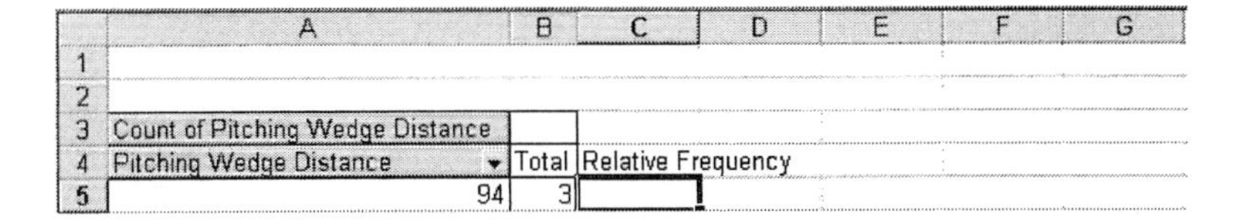

	A	B	C	D	E	F	G
1							
2							
3	Count of Pitching Wedge Distance						
4	Pitching Wedge Distance	Total	Relative Frequency				
5	94	3					

10. Relative frequency is equal to frequency divided by the sum of the frequencies. In cell C5, enter the formula **=B5/$B$19** as shown below. The dollar signs are necessary for the B19 address so that the sum has an absolute reference that will not change when the formula is copied. Press [**Enter**].

	A	B	C	D	E	F	G
1							
2							
3	Count of Pitching Wedge Distance						
4	Pitching Wedge Distance	Total	Relative Frequency				
5	94	3	=B5/$B$19				

11. Copy the contents of cell C5 to cells C6 through C18

	A	B	C	D
1				
2				
3	Count of Pitching Wedge Distance			
4	Pitching Wedge Distance	Total	Relative Frequency	
5	94	3	0.04	
6	95	4	0.053333	
7	96	3	0.04	
8	97	4	0.053333	
9	98	7	0.093333	
10	99	11	0.146667	
11	100	12	0.16	
12	101	12	0.16	
13	102	5	0.066667	
14	103	6	0.08	
15	104	5	0.066667	
16	105	1	0.013333	
17	107	1	0.013333	
18	108	1	0.013333	
19	Grand Total	75		

12. You will now construct a relative frequency line graph. Click and drag over the range **C4:C18** so that these cells are highlighted.

13. At the top of the screen, click **Insert** and select **Chart**.

14. Under Chart type, select **Column**. Under Chart sub-type, select the leftmost diagram in the first row. Click **Next>**.

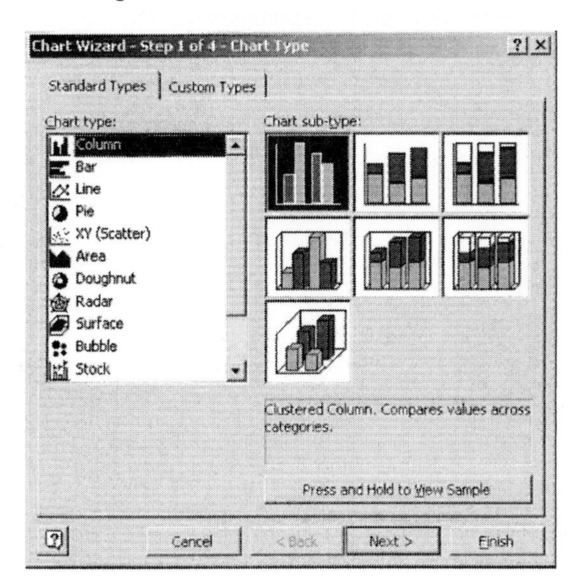

15. You should see =Sheet1!$C$4:$C$18 in the Data range window. Click the **Series** tab at the top of the Chart Source Data dialog box. At the bottom of the dialog box, enter **=Sheet1!$A$5:$A$18** in the Category (X) axis labels window so that the values, 94, 95, 96, etc. appear below the X-axis. Click **Next>**.

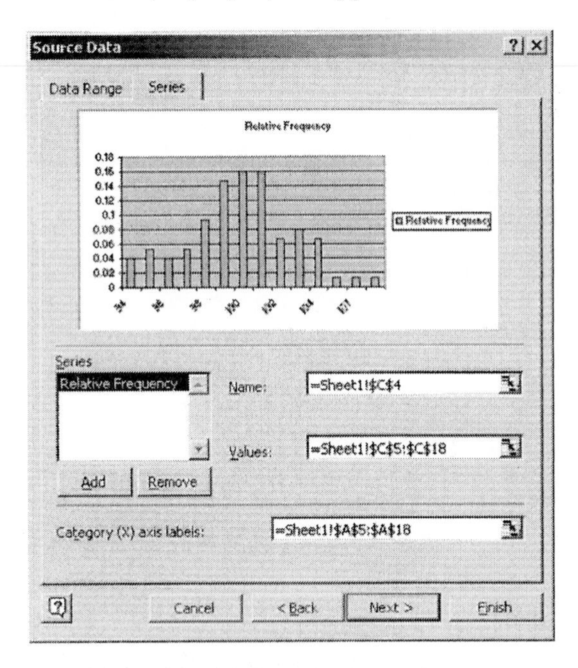

16. Enter a chart title and labels for the X and Y axes. Chart title: **Relative Frequencies of Pitching Wedge Distances**. Category (X) axis: **Pitching Wedge Distance**. Value (Y) axis: **Relative Frequency**. Click **Next>**.

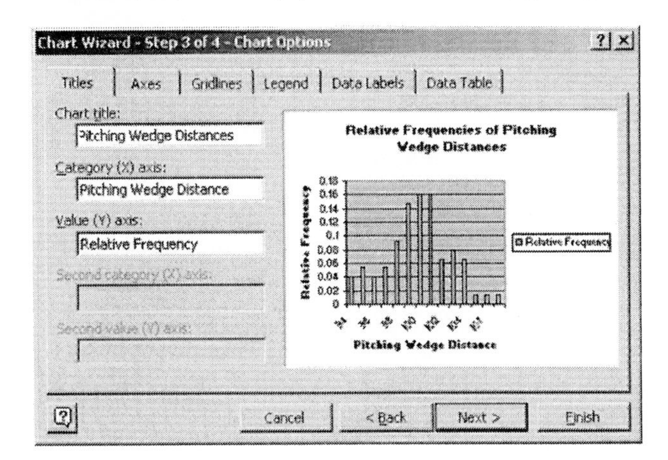

17. Click the **Legend** tab at the top of the dialog box. Click in the box to the left of Show legend to remove the checkmark that appears there. Click **Next>**.

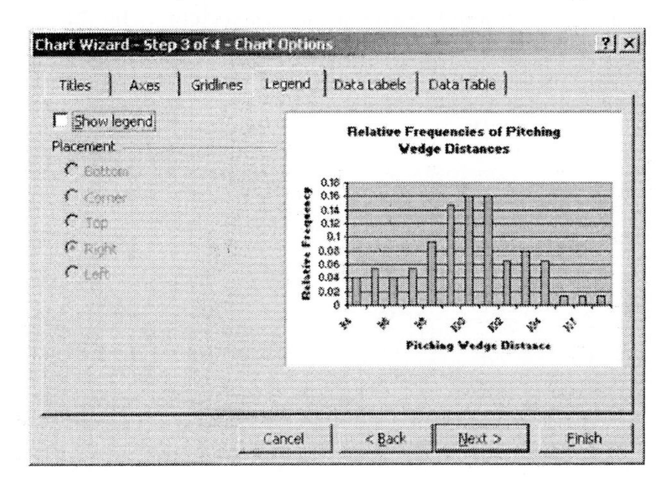

18. In the Chart Location dialog box, select **As new sheet**. Click **Finish**.

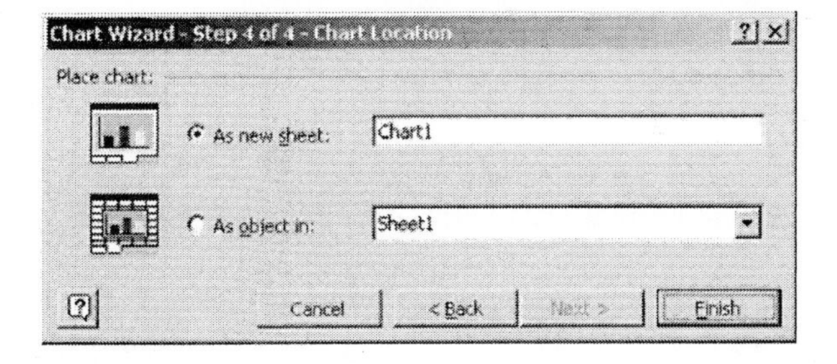

19. Remove the space between the vertical bars. To do this, **right-click** directly on one of the vertical bars. Then select **Format Data Series** from the shortcut menu that appears.

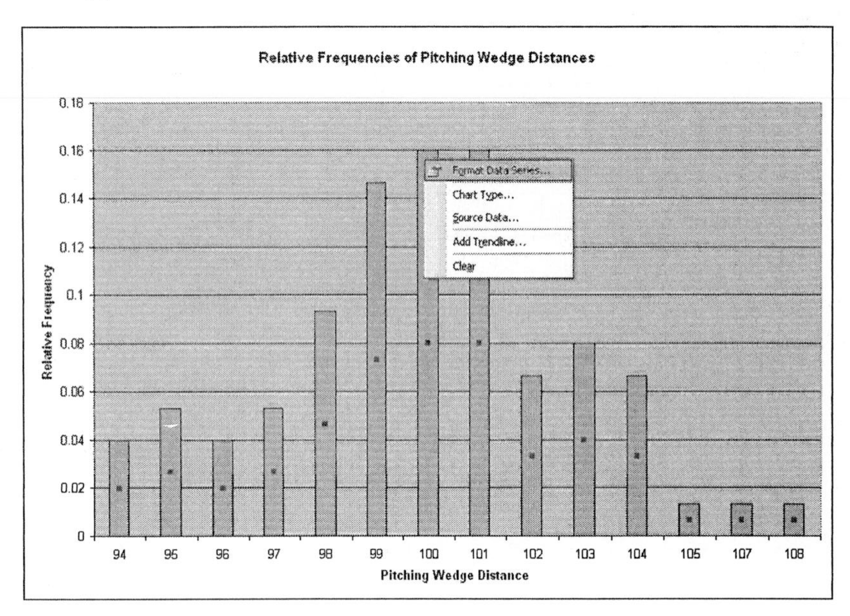

20. Click the **Options** tab at the top of the Format Data Series dialog box. Change the gap width to **0**. Click **OK**.

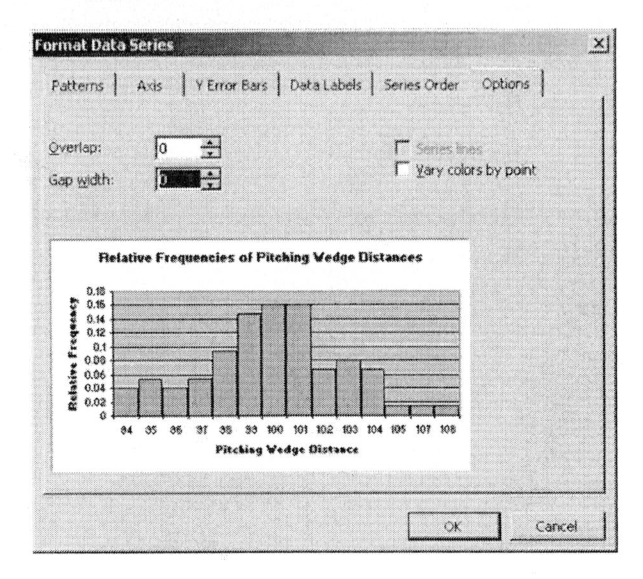

Your completed histogram should look similar to the one shown below.

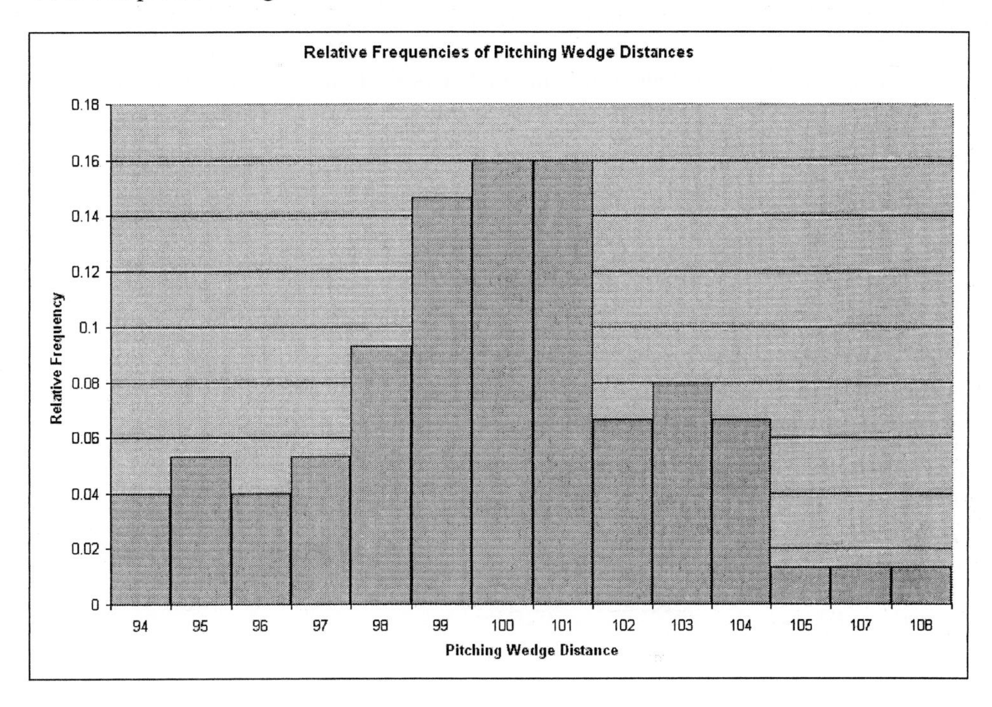

◄

Section 7.2 The Standard Normal Distribution

► Problem 5 (pg. 381) — **Finding Area Under the Standard Normal Curve**

1. Open a new Excel worksheet. Enter the information shown below. You will be using the NORMSDIST function to find the area to the left of specified z values.

	A	B	C	D	E	F	G
1	Z	Area to Left					
2	-2.45						
3	-0.43						
4	1.35						
5	3.49						

2. Click in cell **B2** where you will place the area to the left of z = -2.45. At the top of the screen, click **Insert** and select **Function**.

3. Select the **Statistical** category and the **NORMSDIST** function. Note that the NORMSDIST function returns the standard normal cumulative distribution. Click **OK**.

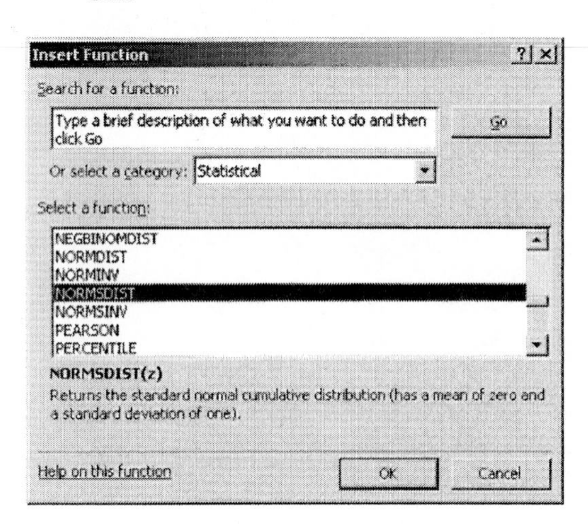

4. Complete the NORMSDIST dialog box as shown below. You are entering the cell address of z = -2.45 rather than the numerical value so that you will be able to copy the function. Click **OK**.

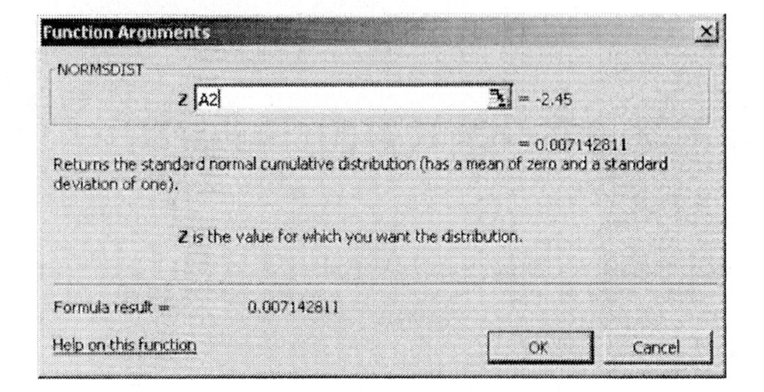

5. Copy the contents of cell B2 to cells B3 through B5. The areas are displayed below.

	A	B	C	D	E	F	G
1	Z	Area to Left					
2	-2.45	0.007143					
3	-0.43	0.333598					
4	1.35	0.911492					
5	3.49	0.999758					

◄

► Problem 7 (pg. 381) Finding Area Under the Standard Normal Curve

If the PHStat add-in has not been loaded, you will need to load it before continuing. Follow the instructions in Section GS 8.2.

1. Part a asks you to find the area under the standard normal curve that lies to the right of Z = -3.01. At the top of the screen, click **PHStat** and select **Probability and Prob. Distributions → Normal**.

2. Complete the Normal Probability Distribution dialog box as shown below. The normal standard curve as a mean of 0 and a standard deviation of 1. The area to the right of -3.01 would be obtained by requesting the probability of X > -3.01. Click **OK**.

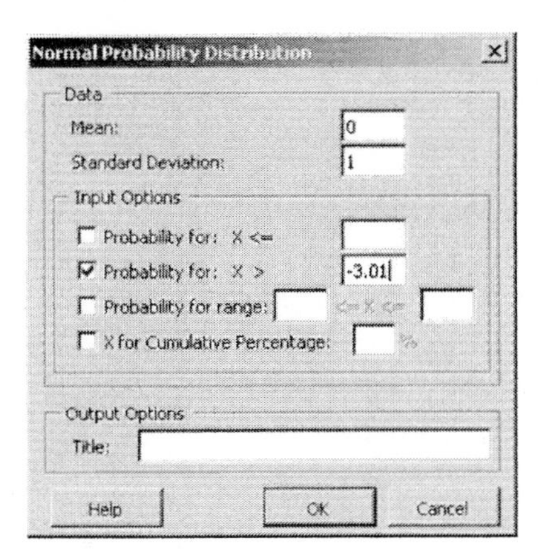

3. The output is shown below. Repeat these steps to obtain probabilities for the other z-values given in the problem.

	A	B	C
1	Normal Probabilities		
2			
3	Common Data		
4	Mean	0	
5	Standard Deviation	1	
6			
7	Probability for X >		
8	X Value	-3.01	
9	Z Value	-3.01	
10	P(X>-3.01)	0.9987	

◀

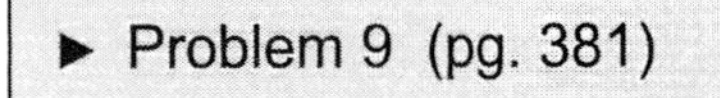

Finding Area Under the Standard Normal Curve

If the PHStat add-in has not been loaded, you will need to load it before continuing. Follow the instructions in Section GS 8.2.

1. Open a new Excel worksheet. At the top of the screen, click **PHStat** and select **Probability and Prob. Distributions → Normal**.

2. Complete the Normal Probability Distribution dialog box as shown below. The normal standard curve as a mean of 0 and a standard deviation of 1. You want to find the area between Z = -2.04 and z = 2.04. Click **OK**.

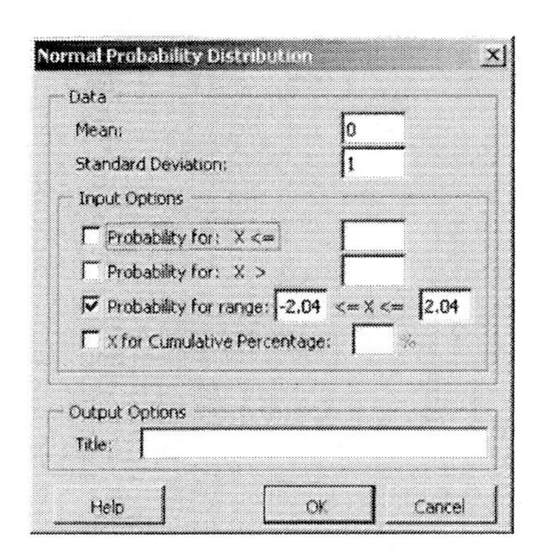

The area is equal to 0.9586.

	A	B	C	D	E	F	G
1	Normal Probabilities						
2							
3	Common Data						
4	Mean	0					
5	Standard Deviation	1					
6				Probability for a Range			
7				From X Value	-2.04		
8				To X Value	2.04		
9				Z Value for -2.04	-2.04		
10				Z Value for 2.04	2.04		
11				P(X<=-2.04)	0.0207		
12				P(X<=2.04)	0.9793		
13				P(-2.04<=X<=2.04)	0.9586		

◀

► Problem 15 (pg. 382)

Finding the Z-score that Corresponds to a Specified Area

1. Open a new Excel worksheet. You will be using the NORMSINV function to find the Z-score such that the area under the standard normal curve to the left is 0.1.

2. Click in the cell of the worksheet where you would like to place the output. I clicked in cell **A1**.

3. At the top of the screen, click **Insert** and select **Function**.

4. Select the **Statistical** category and the **NORMSINV** function Click **OK**.

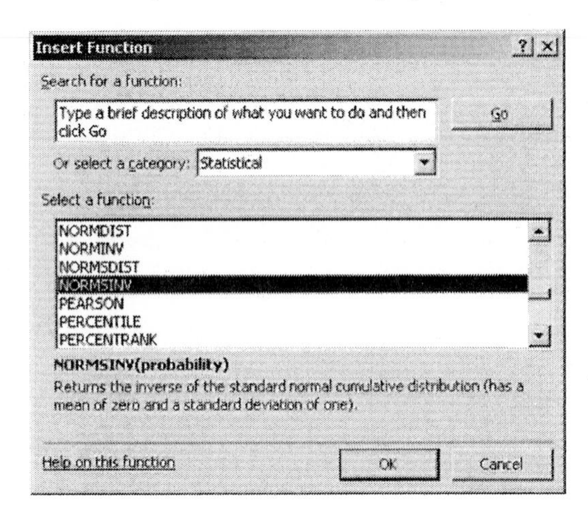

5. Enter the area **.1** in the Probability window. Click **OK**.

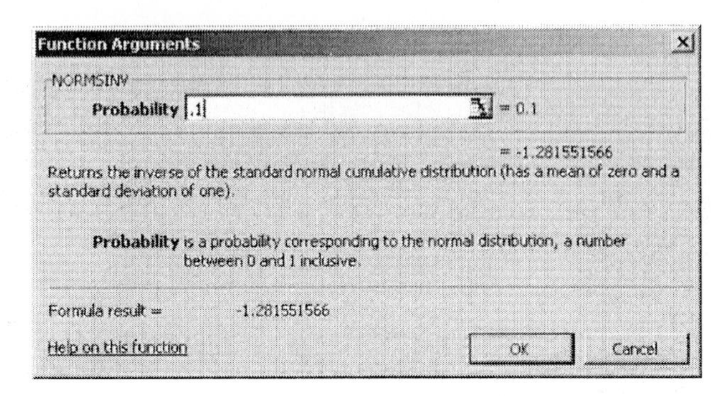

The NORMSINV function returns a Z-score of –1.2816.

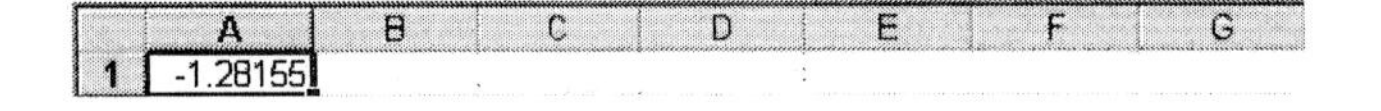

Section 7.3 Applications of the Normal Distribution

▶ Problem 17 (pg. 390) | Computing Probabilities of a Normally Distributed Variable

If the PHStat add-in has not been loaded, you will need to load it before continuing. Follow the instructions in Section GS 8.2.

1. Open a new Excel worksheet. At the top of the screen, click **PHStat** and select **Probability and Prob. Distributions → Normal.**

2. Complete the Normal Probability Distribution dialog box as shown below. The mean incubation time is 21 days and the standard deviation is 1 day. For part a, you want to find the probability that a randomly selected egg hatches in less than 20 days. Click **OK**.

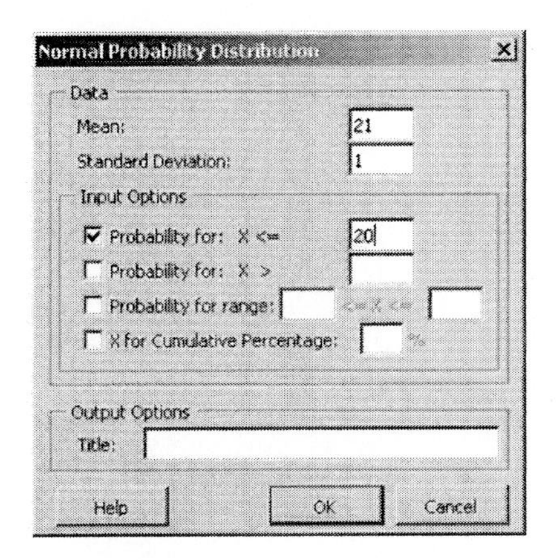

The probability is 0.1587.

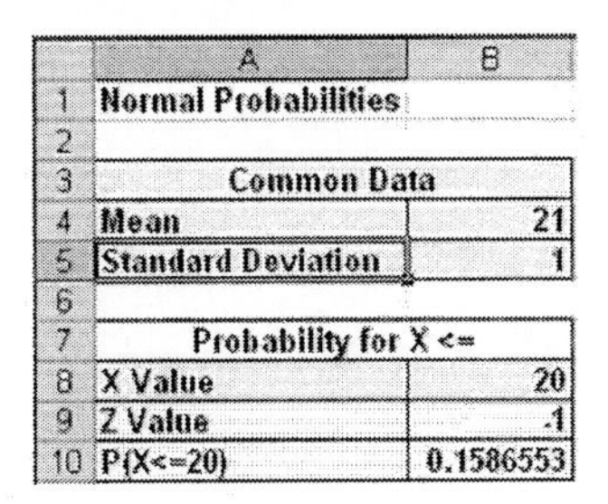

	A	B
1	Normal Probabilities	
2		
3	Common Data	
4	Mean	21
5	Standard Deviation	1
6		
7	Probability for X <=	
8	X Value	20
9	Z Value	-1
10	P(X<=20)	0.1586553

3. For part b, you want to find the probability that a randomly selected egg takes over 22 days to hatch. At the top of the screen, click **PHStat** and select **Probability and Prob. Distributions → Normal**. Complete the Normal Probability Distribution dialog box as shown below. Click **OK**.

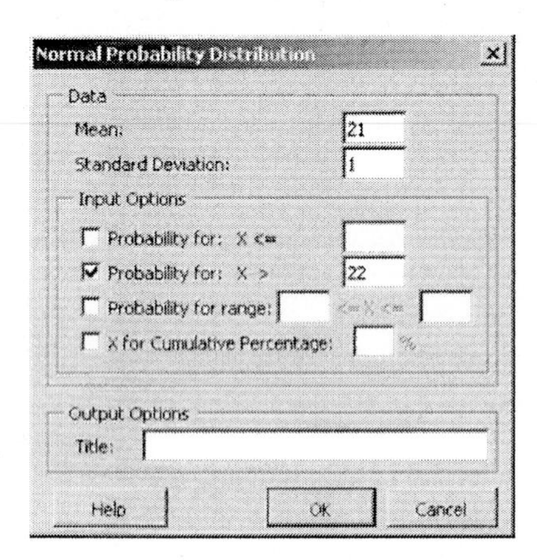

The probability is 0.1587.

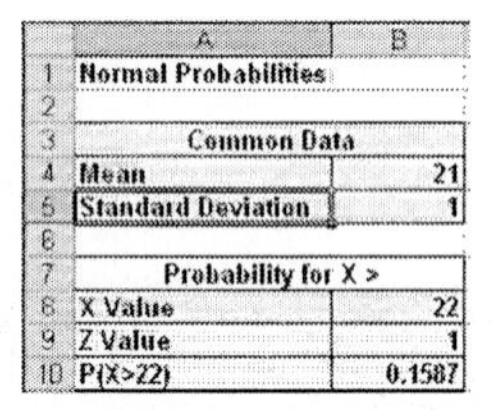

	A	B
1	Normal Probabilities	
2		
3	Common Data	
4	Mean	21
5	Standard Deviation	1
6		
7	Probability for X >	
8	X Value	22
9	Z Value	1
10	P(X>22)	0.1587

4. For part c, you want to find the probability that a randomly selected egg hatches between 19 and 21 days. At the top of the screen, click **PHStat** and select **Probability and Prob. Distributions → Normal**. Complete the Normal Probability Distribution dialog box as shown below. Click **OK**.

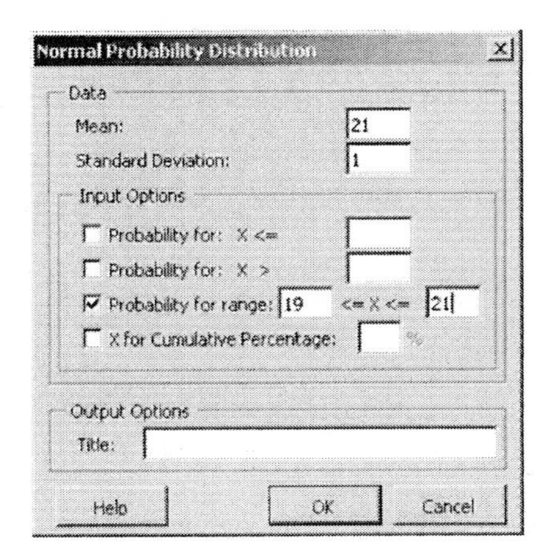

The probability is 0.4772.

	A	B	C	D	E
1	**Normal Probabilities**				
2					
3	**Common Data**				
4	**Mean**	21			
5	**Standard Deviation**	1			
6				**Probability for a Range**	
7				From X Value	19
8				To X Value	21
9				Z Value for 19	-2
10				Z Value for 21	0
11				P(X<=19)	0.0228
12				P(X<=21)	0.5000
13				P(19<=X<=21)	0.4772

5. For part d, you want to find the probability that an egg hatches in less than 18 days. At the top of the screen, click **PHStat** and select **Probability and Prob. Distributions** → **Normal**. Complete the Normal Probability Distribution dialog box as shown below. Click **OK**.

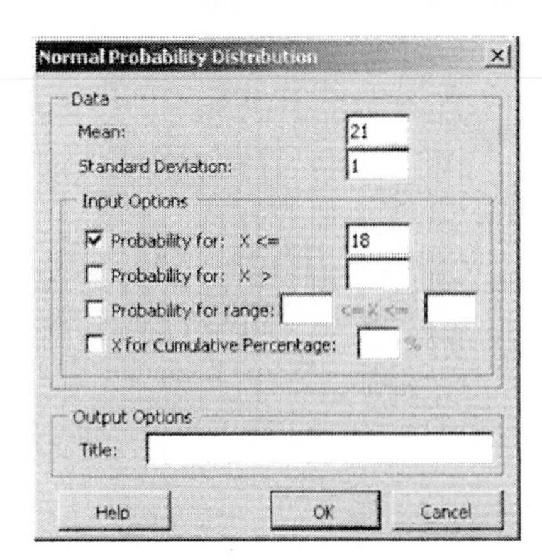

The probability is 0.0013.

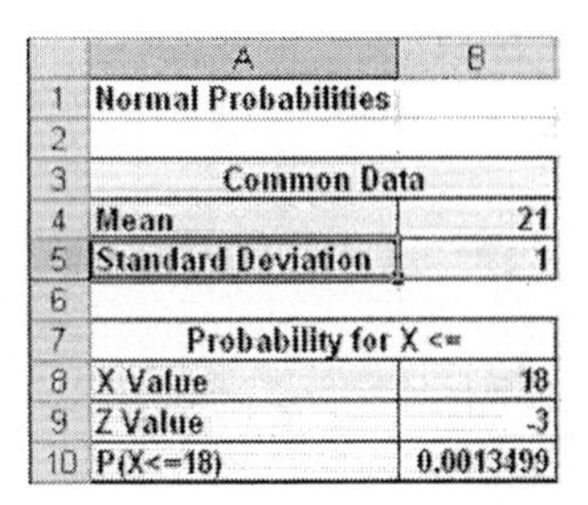

	A	B
1	**Normal Probabilities**	
2		
3	**Common Data**	
4	**Mean**	21
5	**Standard Deviation**	1
6		
7	**Probability for X <=**	
8	X Value	18
9	Z Value	-3
10	P(X<=18)	0.0013499

Section 7.4 Assessing Normality

► Example 3 (pg. 397) Drawing a Normal Probability Plot and Constructing a Histogram

1. Open worksheet "7_4_Ex3" in the Chapter 7 folder. The first few lines are shown below.

If the PHStat add-in has not been loaded, you will need to load it before continuing. Follow the instructions in Section GS 8.2

	A	B	C	D	E	F	G
1	Demon Rollercoaster Wait Times						
2	7						
3	33						
4	30						

2. At the top of the screen, click **PHStat**. Then select **Probability and Prob. Distributions → Normal Probability Plot**.

3. Complete the Normal Probability Plot dialog box as shown below. Click **OK**.

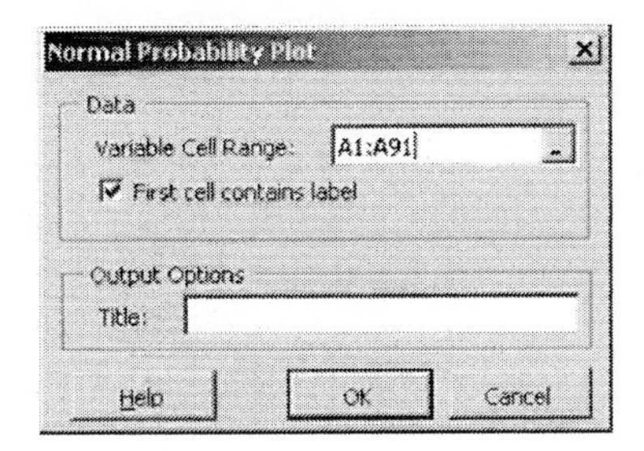

The plot is displayed in a worksheet named NormalPlot. Additional output is provided in a worksheet named Plot.

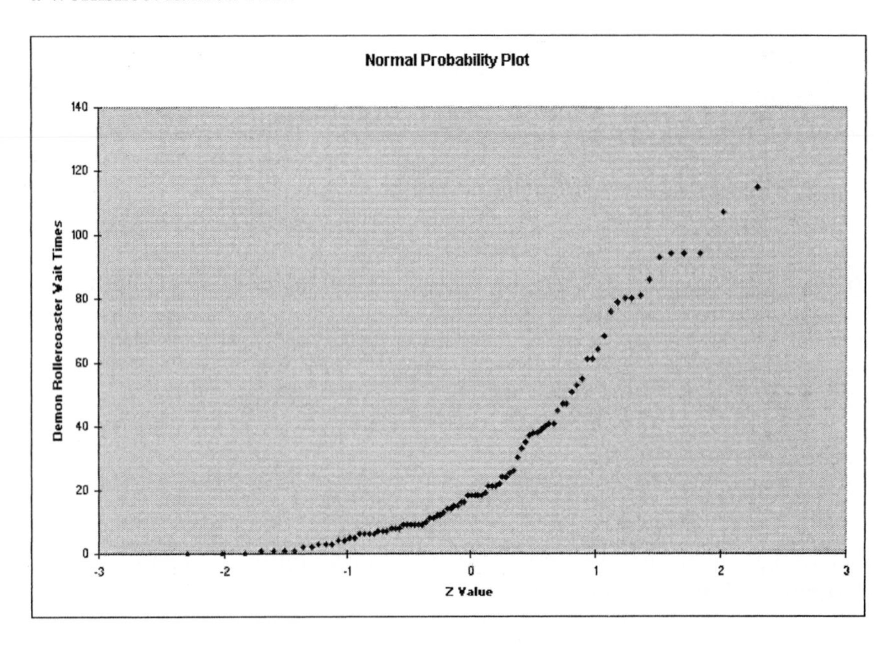

4. To construct a quick histogram of the data, use Pivot Chart. First, click in any cell in the column of data. Then, at the top of the screen, click **Data** and select **Pivot Table and Pivot Chart Report**.

5. At the top of the dialog box, select **Microsoft Excel list or database**. At the bottom of the dialog box, select **Pivot Chart (with Pivot Table)**. Click **Next>**.

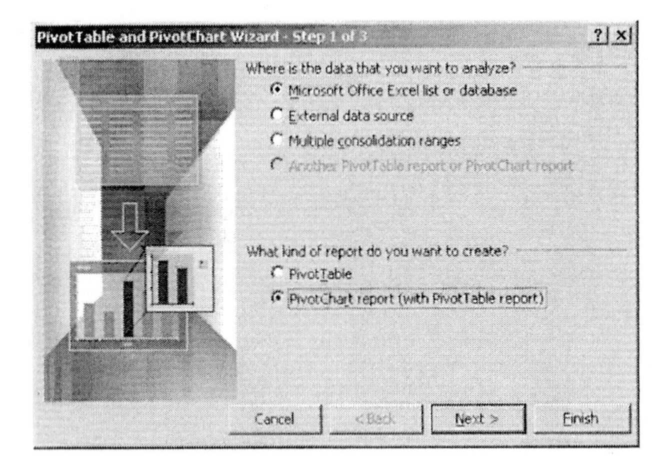

6. The data range is automatically placed in the Range window of the dialog box. It should read $A$1:$A$91. Make any necessary revisions to the range, and then click **Next>**.

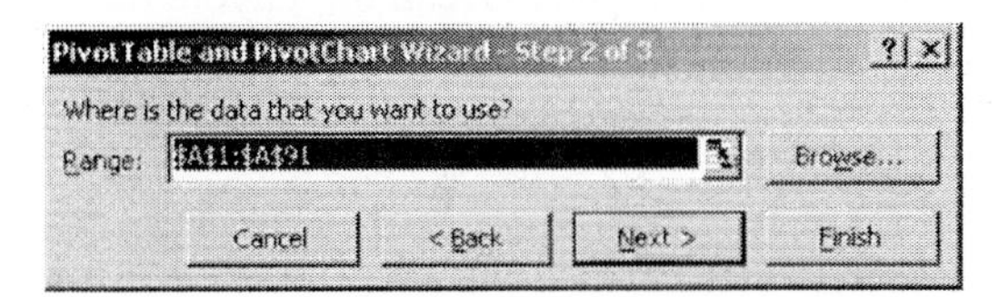

7. You can place the table and chart in new worksheets or in the existing worksheet. For this example, select **New worksheet**. Click **Layout** at the bottom of the dialog box.

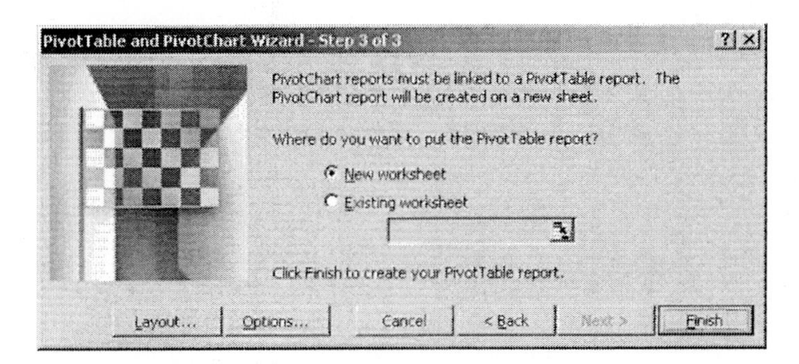

8. Drag the **Demon** field button to **ROW**. Drag the **Demon** field button to **DATA**.

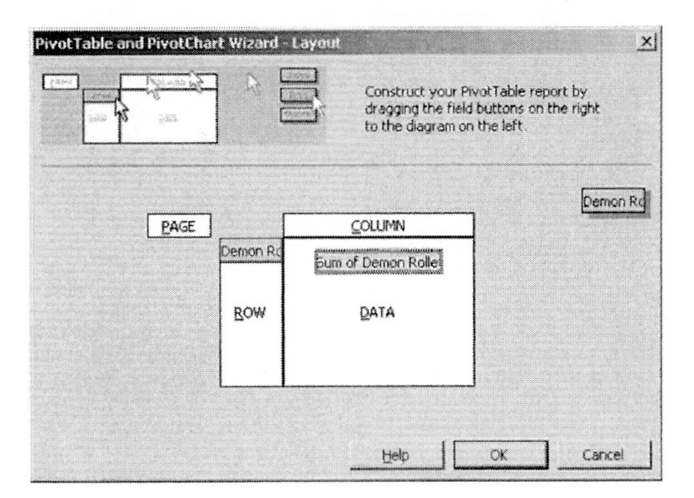

9. The default is Sum of Demon. You need to change this to count. Double-click on the Sum of Demon button.

10. Under Summarize by, select **Count**. Click **OK**. Also click **OK** in the Layout dialog box.

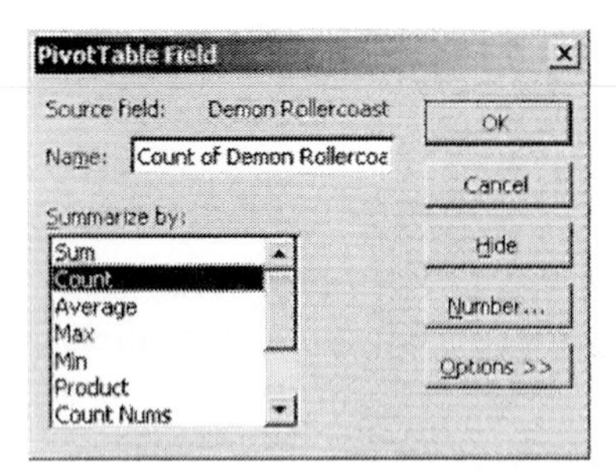

11. In the Step 3 dialog box, click **Finish**.

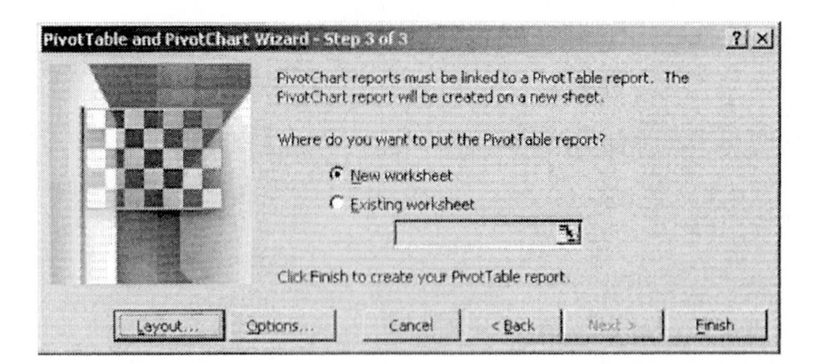

12. Close Pivot Table and Field List so that you can view the entire graph. Now you can easily see that the distribution is not normally distributed.

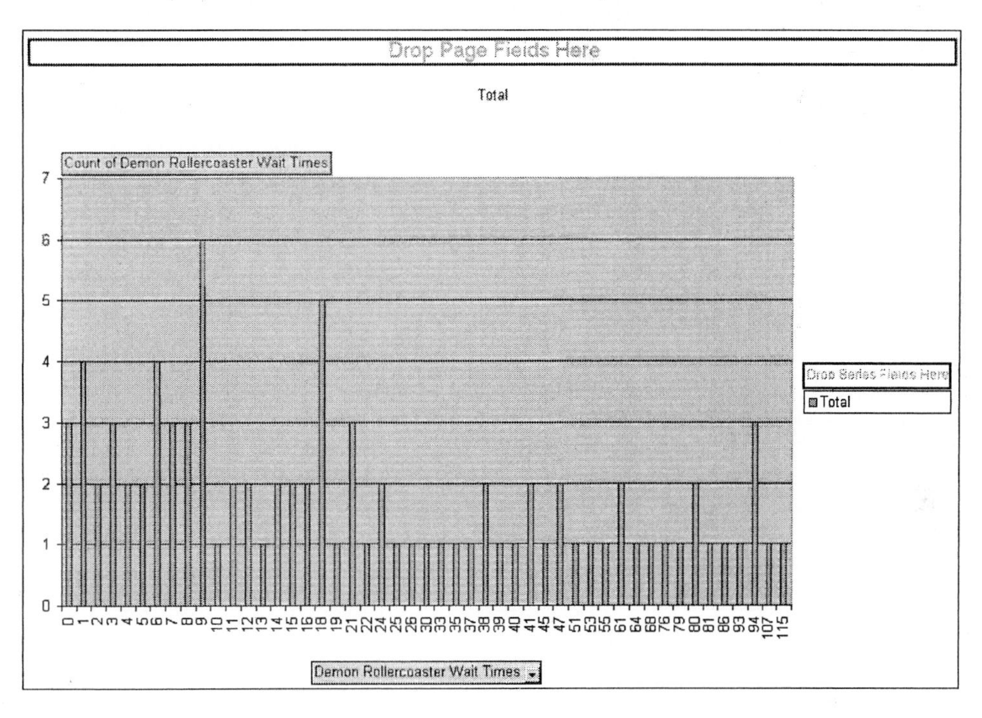

Section 7.5 The Normal Approximation to the Binomial Probability Distribution

▶ Problem 22 (pg. 406) | Using the Normal Approximation to the Binomial

1. Open a new Excel worksheet.

If the PHStat add-in has not been loaded, you will need to load it before continuing. Follow the instructions in Section GS 8.2

2. At the top of the screen, click **PHStat**. Select **Probability and Prob. Distributions → Normal**.

3. At the top of page 403 in your textbook, you see that the mean is equal to np and that the standard deviation is equal to $\sqrt{np(1-p)}$. In this problem, n = 100 and p = 0.80. So the mean = (100)(0.80) = 80 and the standard deviation = $\sqrt{100(0.80)(0.20)} = 4$. These values were used for the mean and standard deviation in the dialog box shown below. To find the answer to part a, we apply the correction for continuity to the value of 80 and ask for the probability that the value falls between 79.5 and 80.5. Click **OK**.

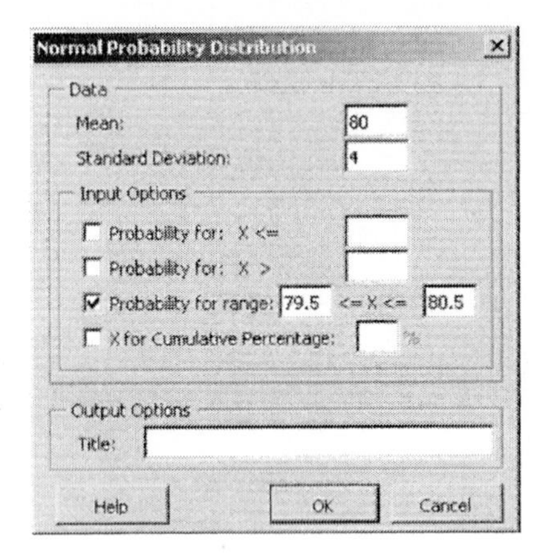

The probability is 0.0995.

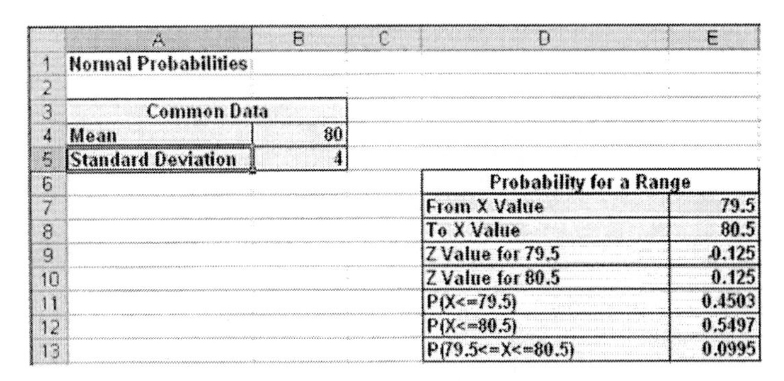

	A	B	C	D	E
1	Normal Probabilities				
2					
3	Common Data				
4	Mean	80			
5	Standard Deviation	4			
6				Probability for a Range	
7				From X Value	79.5
8				To X Value	80.5
9				Z Value for 79.5	-0.125
10				Z Value for 80.5	0.125
11				P(X<=79.5)	0.4503
12				P(X<=80.5)	0.5497
13				P(79.5<=X<=80.5)	0.0995

4. At the top of the screen, click **PHStat**. Select **Probability and Prob. Distributions → Normal**.

5. Part b asks for the probability that at least 80 of them start smoking before they were 18 years old. We again apply the correction for continuity and ask for the

probability that at least 79.5 of them started smoking before they were 18 years old. Click **OK**.

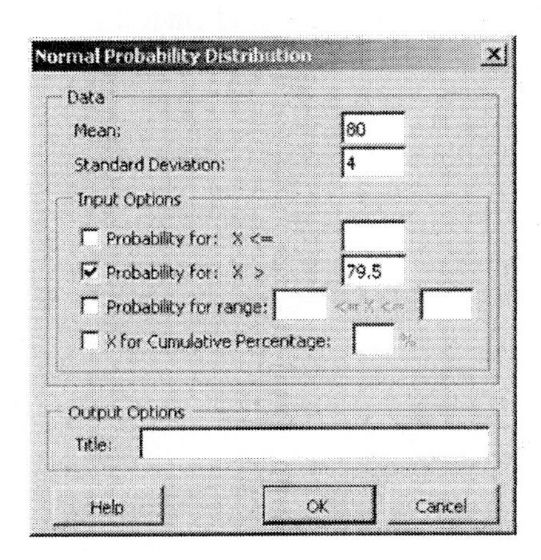

The probability is 0.5497.

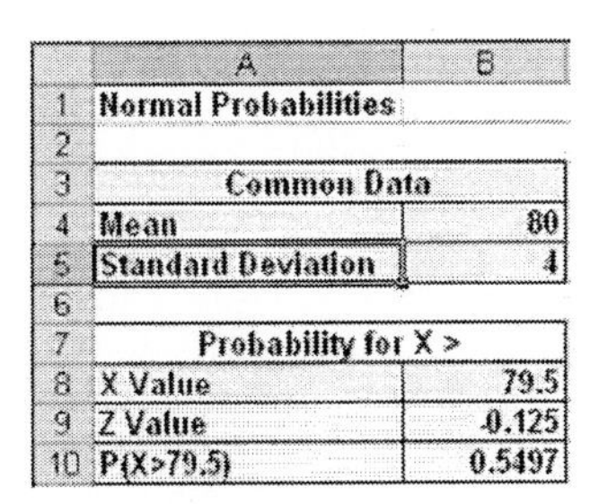

	A	B
1	**Normal Probabilities**	
2		
3	**Common Data**	
4	**Mean**	80
5	**Standard Deviation**	4
6		
7	**Probability for X >**	
8	X Value	79.5
9	Z Value	-0.125
10	P(X>79.5)	0.5497

6. At the top of the screen, click **PHStat**. Select **Probability and Prob. Distributions → Normal**.

7. Part c asks for the probability that fewer than 70 of them started smoking before they were 18 years old. Again, apply the correction for continuity and enter the value 69.5. Click **OK**.

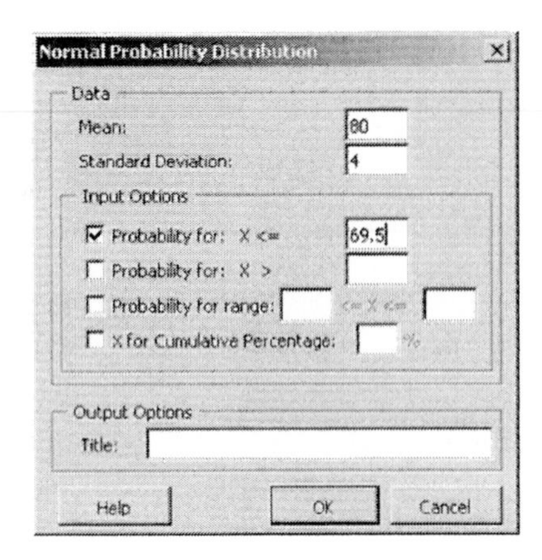

The probability is 0.0043.

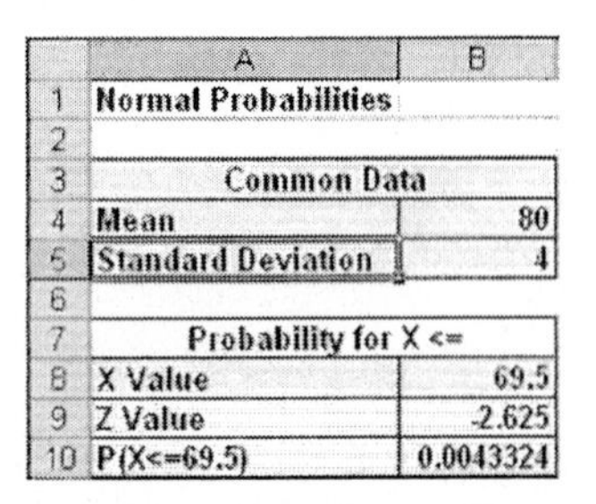

	A	B
1	**Normal Probabilities**	
2		
3	**Common Data**	
4	Mean	80
5	Standard Deviation	4
6		
7	**Probability for X <=**	
8	X Value	69.5
9	Z Value	-2.625
10	P(X<=69.5)	0.0043324

8. At the top of the screen, click **PHStat**. Select **Probability and Prob. Distributions → Normal**.

9. Part d asks for the probability that between 70 and 90 of them, inclusive, started smoking before they were 18 years old. Apply the correction for continuity and enter 69.5 and 90.5. Click **OK**.

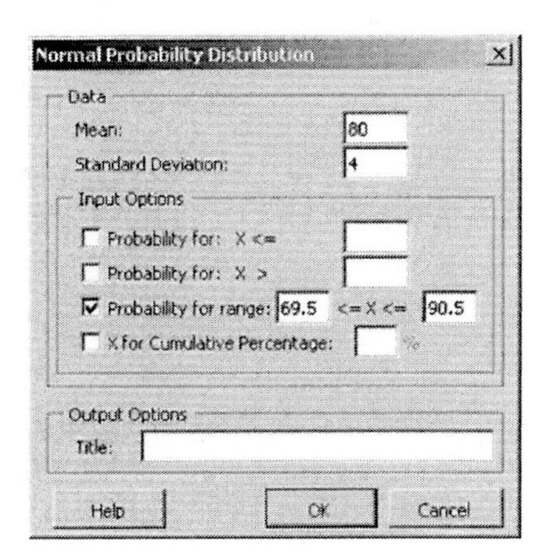

The probability is 0.9913.

	A	B	C	D	E
1	Normal Probabilities				
2					
3	Common Data				
4	Mean	80			
5	Standard Deviation	4			
6				Probability for a Range	
7				From X Value	69.5
8				To X Value	90.5
9				Z Value for 69.5	-2.625
10				Z Value for 90.5	2.625
11				P(X<=69.5)	0.0043
12				P(X<=90.5)	0.9957
13				P(69.5<=X<=90.5)	0.9913

◀

Sampling Distributions

CHAPTER 8

Section 8.1 Distribution of the Sample Mean

► Example 2 (pg. 420)

Sampling from a Normal Population

1. Open a new Excel worksheet.

If the PHStat add-in has not been loaded, you will need to load it before continuing. Follow the instructions in Section GS 8.2

2. At the top of the screen, click **PHStat**. Select **Sampling → Sampling Distributions Simulation**.

3. Complete the Sampling Distributions Simulation dialog box as shown below. You are asking for 100 samples of size n = 5 drawn from a standardized normal distribution. A histogram will be included in the output. Click **OK**.

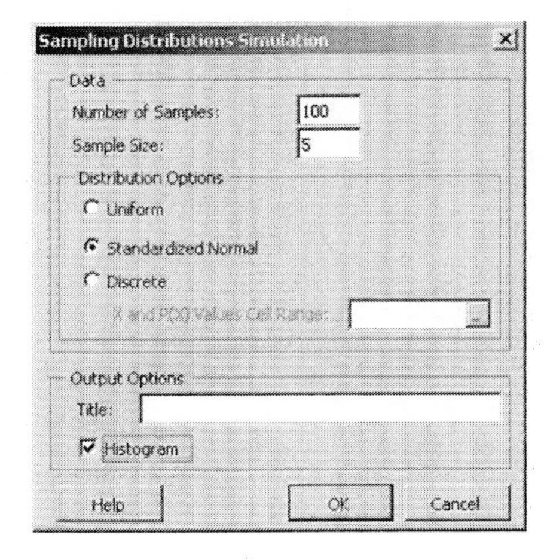

4. The PHStat procedure samples from a standardized normal distribution and generates z-scores. If you wish to transform the z-scores to height values, apply the following formula: $X = \mu + z\sigma$ where X refers to the height value, $\mu = 38.72$ inches, $\sigma = 3.17$ inches, and z refers to the generated z score.

   The worksheet labeled Histogram contains a histogram of the sample means and a frequency distribution of the sample means.

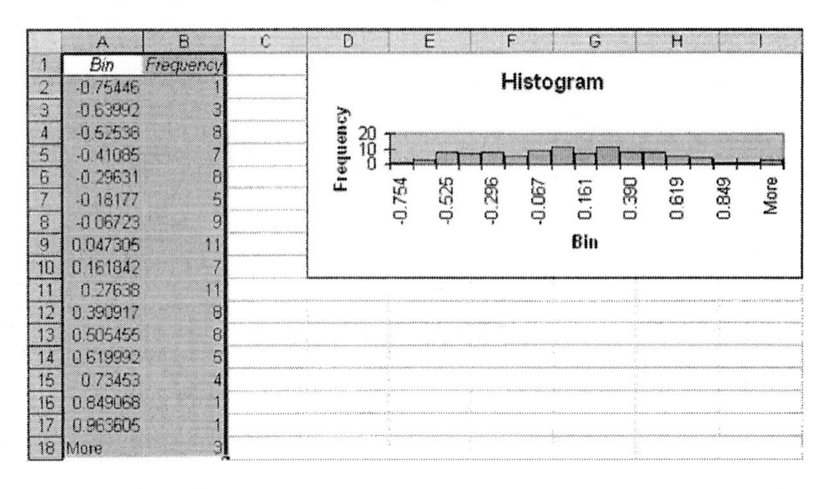

Bin	Frequency
-0.75446	1
-0.63992	3
-0.52538	8
-0.41085	7
-0.29631	8
-0.18177	5
-0.06723	9
0.047305	11
0.161842	7
0.27638	11
0.390917	8
0.505455	8
0.619992	5
0.73453	4
0.849068	1
0.963605	1
More	3

5. To make the chart taller so that it is more informative, click within the figure near a border so that black square handles appear. Then, click on the center handle on the bottom border of the figure and drag it down a few rows.

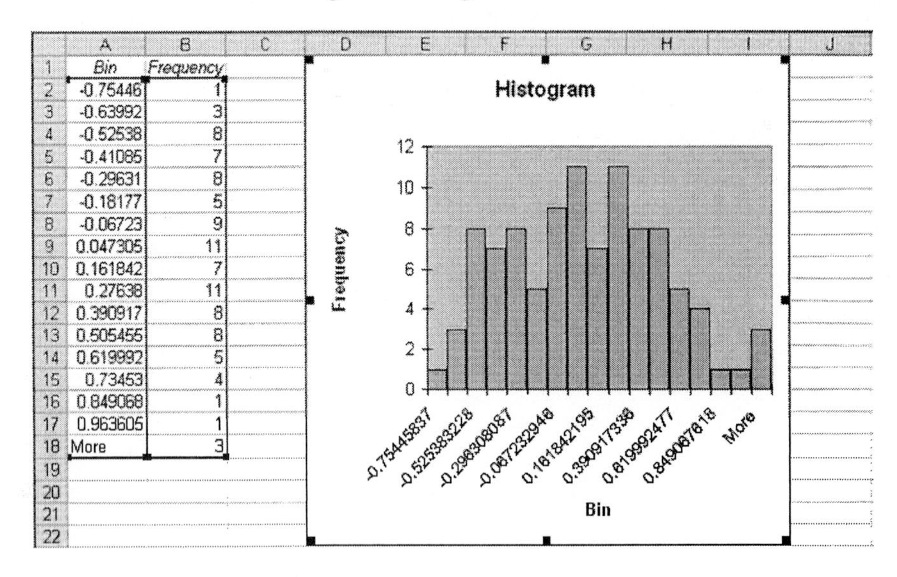

Bin	Frequency
-0.75446	1
-0.63992	3
-0.52538	8
-0.41085	7
-0.29631	8
-0.18177	5
-0.06723	9
0.047305	11
0.161842	7
0.27638	11
0.390917	8
0.505455	8
0.619992	5
0.73453	4
0.849068	1
0.963605	1
More	3

6. Click on the **SDS** sheet tab at the bottom of the screen. This worksheet presents the z-scores in each of the 100 samples of size n = 5, the Sample Means, the Overall Average, and the Standard Error of the Mean. Let's transform the means from z-values to height (in inches) values. First, enter labels for your output as shown below.

	A	B	C	D	E	F	G
1	Sampling Distribution Simulation						
2							
3	-0.30023	-1.27768	0.244257	1.276474	1.19835	1.733133	-2.18359
4	-0.5238	0.849429	0.513207	-0.6083	1.304979	-1.76094	0.550572
5	0.226954	0.214491	-0.00746	-1.22383	0.235361	0.148221	0.752424
6	-0.62756	0.87899	-0.83829	-0.11489	-0.23214	1.518415	-1.31745
7	-0.25863	0.639438	-1.27406	0.51915	0.97652	-0.86389	0.371469
8	Sample Means:						
9	-0.29665	0.260933	-0.27247	-0.03028	0.696615	0.154988	-0.36531
10	Overall Average:						
11	0.044289						
12	Standard Error of the Mean:						
13	0.437819						
14	Sample Means:						
15							
16	Overall Average:						
17							
18	Standard Error of the Mean:						
19							

7. Click in cell **A15** where you will place the height value corresponding to the z-value placed in A9 (-0.29665). Enter the formula **=38.72+A9*3.17** as shown below. Press [**Enter**].

14	Sample Means:				
15	=38.72+A9*3.17				
16	Overall Average:				
17					
18	Standard Error of the Mean:				

8. Copy the formula in A15 all the way across from cell B15 to the last column containing a sample mean value, CV15. Row 15 of the worksheet now contains the means expressed as height values. Next, click in cell **A17** where you will place the overall average of these means. At the top of the screen, click **Insert** and select **Function**.

14	Sample Means:						
15	37.77961	39.54716	37.85627	38.62401	40.92827	39.21131	37.56195
16	Overall Average:						
17							
18	Standard Error of the Mean:						

9. Select the **Statistical** category and the **AVERAGE** function. Click **OK**.

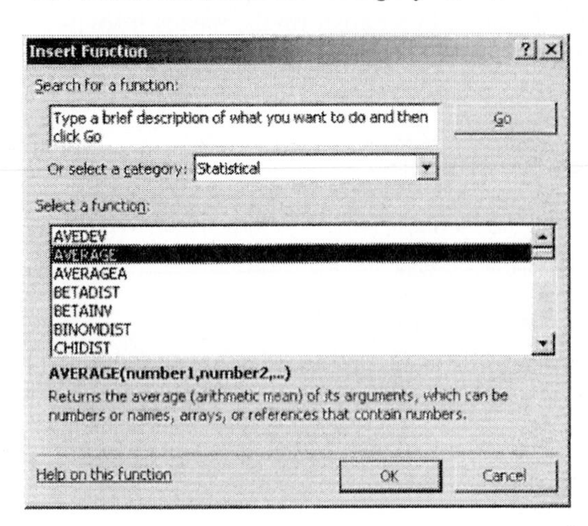

10. Complete the Function Arguments dialog box as shown below. You are asking for the average of the values in the range A15:CV15. Click **OK**.

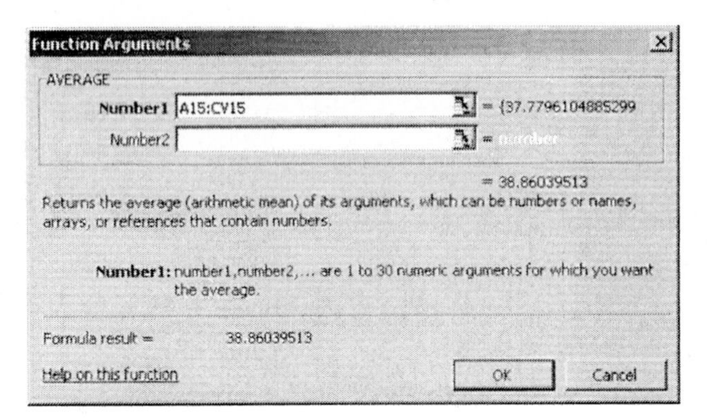

11. The function returns a value of 38.8604. Click in cell **A19** where you will place the standard deviation of the means (i.e., the standard error of the mean). At the top of the screen, click **Insert** and select **Function**.

12. Select the **Statistical** category and the **STDEVP** function. STDEVP refers to the standard deviation of a population. Click **OK**.

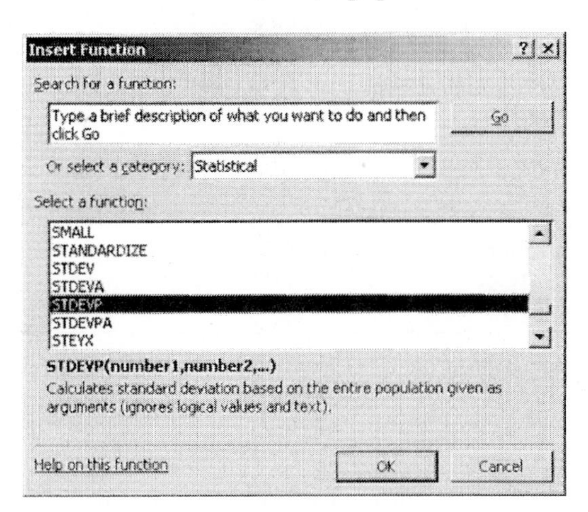

13. Complete the Function Arguments dialog box as shown below. Click **OK**.

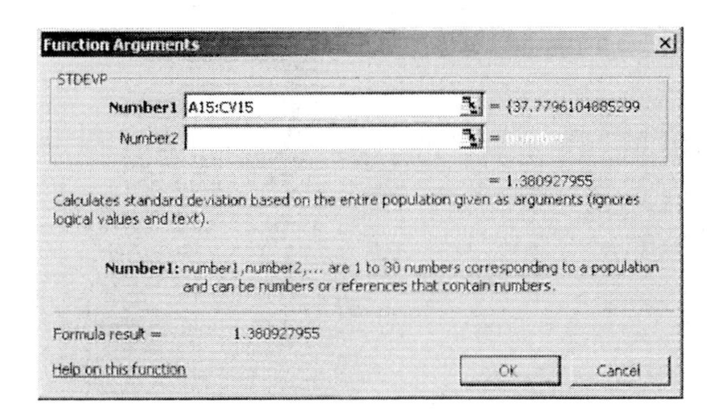

14. The function returns a value of 1.3809. Because the samples were generated randomly, it is not likely that your result will be exactly the same.

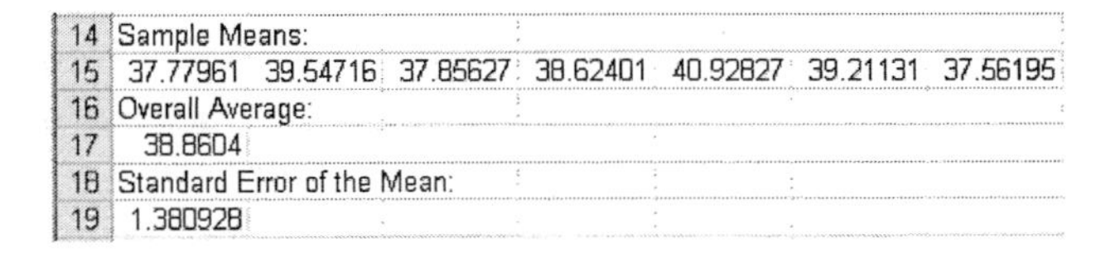

14	Sample Means:						
15	37.77961	39.54716	37.85627	38.62401	40.92827	39.21131	37.56195
16	Overall Average:						
17	38.8604						
18	Standard Error of the Mean:						
19	1.380928						

Section 8.2 Distribution of the Sample Proportion

► Example 2 (pg. 435) Sampling from the Binomial Distribution

1. Open a new Excel worksheet.

2. You will be using Excel's Random Number Generation tool. At the top of the screen, click **Tools** and select **Data Analysis**.

If Data Analysis does not appear as a choice in the Tools menu, you will need to load the Microsoft Excel Analysis ToolPak add-in. Follow the procedure in Section GS 8.1 before continuing.

3. In the Data Analysis dialog box, select **Random Number Generation** and click **OK**.

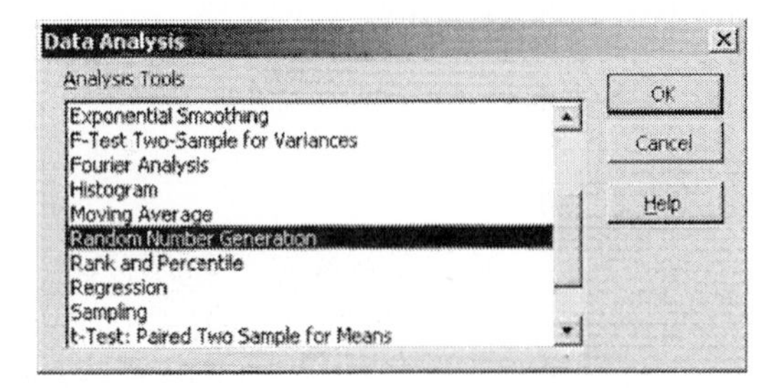

4. Complete the Random Number Generation dialog box as shown below. You have one variable which is the number in the sample with high cholesterol. You are generating a total of 100 samples. The distribution is binomial. The p value refers to the proportion of Americans that have high cholesterol. In this problem, p = 0.17. Sample size is n = 10. Input a number for the seed to begin the random generation. I input 34. Finally, the output will be played in a new worksheet labeled "n = 10." Click **OK**.

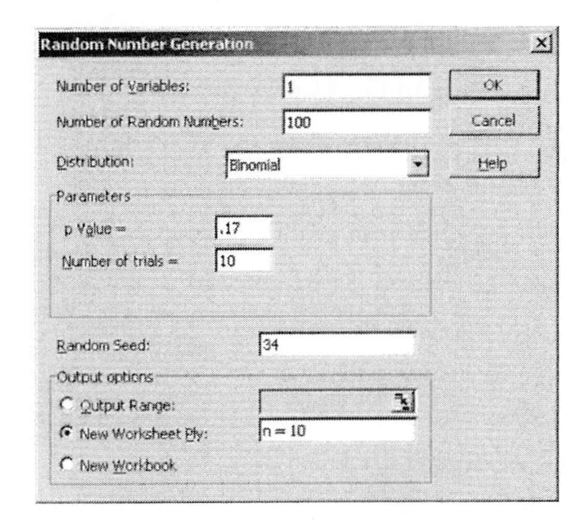

5. The second step of the problem asks you to compute the sample proportion for each of the 100 samples. First, you will want to enter a label for the proportions. Click in cell **B1**. Then at the top of the screen, click **Insert** and select **Row**.

6. Type **Proportion** in cell B1 and press **[Enter]**.

7. In cell B2, the type formula **=A2/10** and press **[Enter]**.

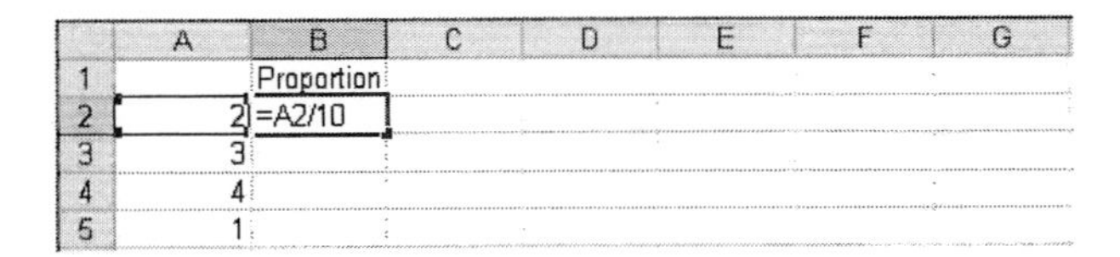

8. Copy the formula in cell B2 all the way down to cell B101. Only the first five rows are shown below.

	A	B	C	D	E	F	G
1		Proportion					
2	2	0.2					
3	3	0.3					
4	4	0.4					
5	1	0.1					

9. The third step asks you to draw a histogram of the sample proportions. At the top of the screen, click **Data** and select **Pivot Table and Pivot Chart Report**.

10. Near the top of the dialog box, select **Microsoft Office Excel list or database**. Near the bottom, select **Pivot Chart report (with Pivot Table report)**. Click **Next>**.

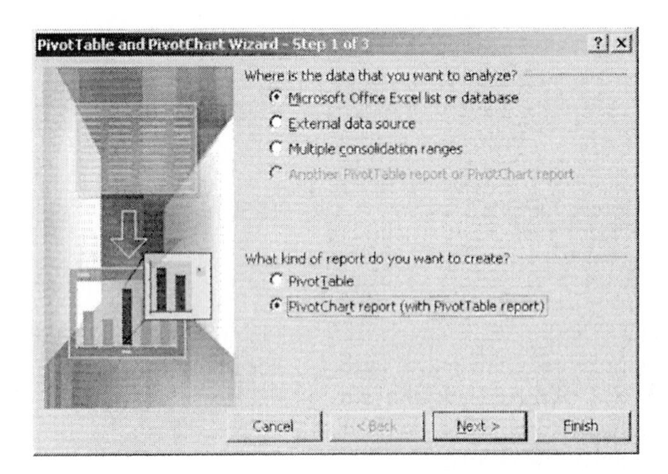

11. The data range should be B1:B101. Make any necessary corrections, and click **Next>**.

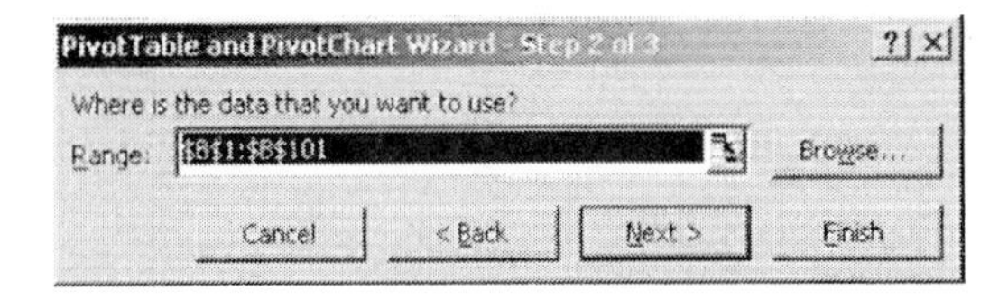

12. Select **New worksheet**. Then click **Layout** near the bottom of the Step 3 of 3 dialog box.

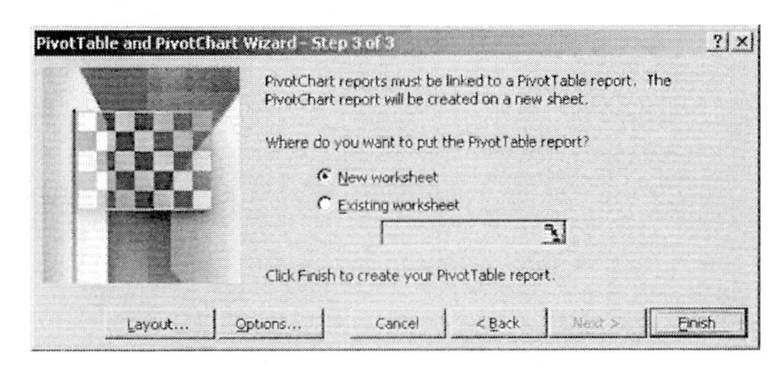

13. Drag the **Proportion** field button to ROW. Draw the **Proportion** field button to DATA.

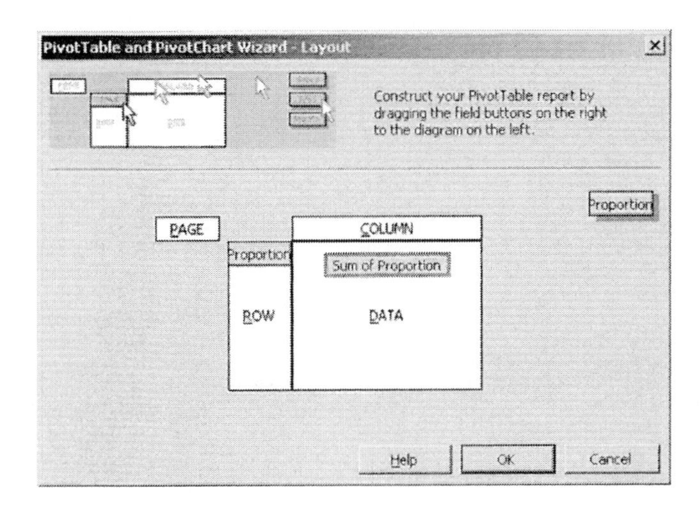

14. The default DATA selection for a numeric variable is Sum. You will need to change this to Count. Double-click on **Sum of Proportion**. Under Summarize by, select **Count**. Click **OK**.

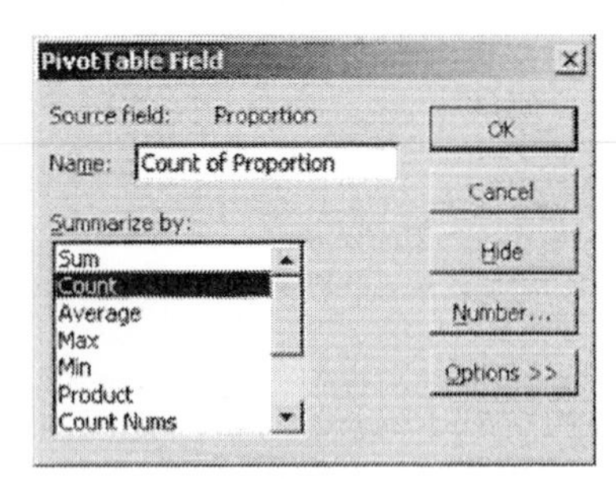

15. Click **OK** in the Layout dialog box.

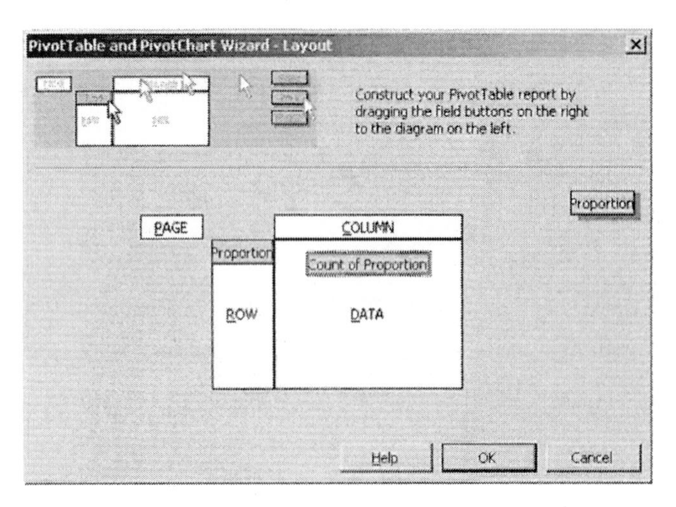

16. Click **Finish** in the Step 3 of 3 dialog box.

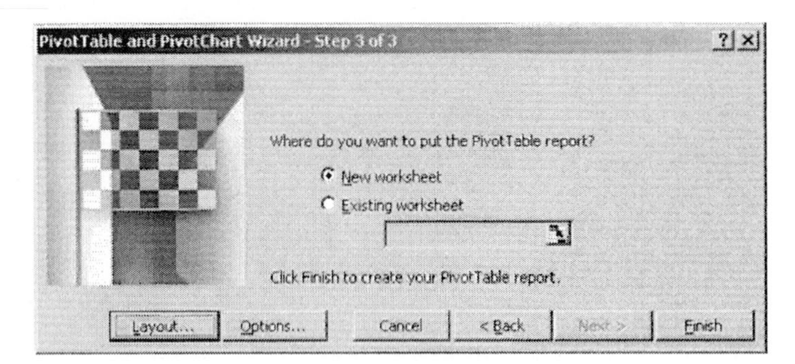

17. Click Chart Wizard in the Pivot Table toolbar.

18. Hide the Pivot Chart Field Buttons.

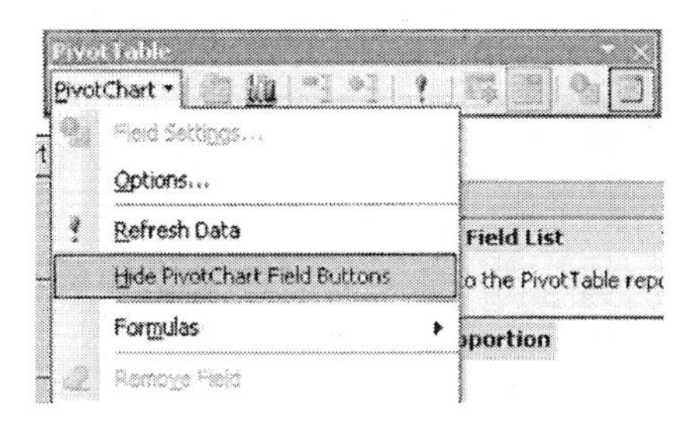

19. Close the Pivot Table toolbar.

20. Click on the **Total** legend to the right of the chart. Then right-click and select **Clear** from the menu.

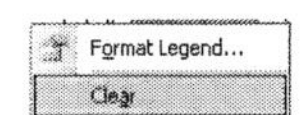

21. Next, you will remove the gap between the vertical bars. Right-click on one of the vertical bars and select **Format Data Series** from the menu.

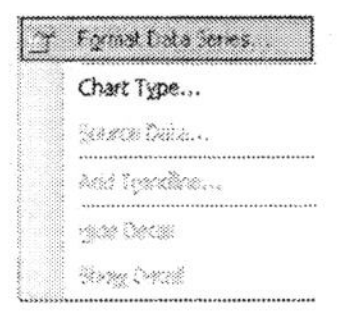

22. Click the Options tab at the top of the dialog box. Change the Gap width to 0. Click **OK**.

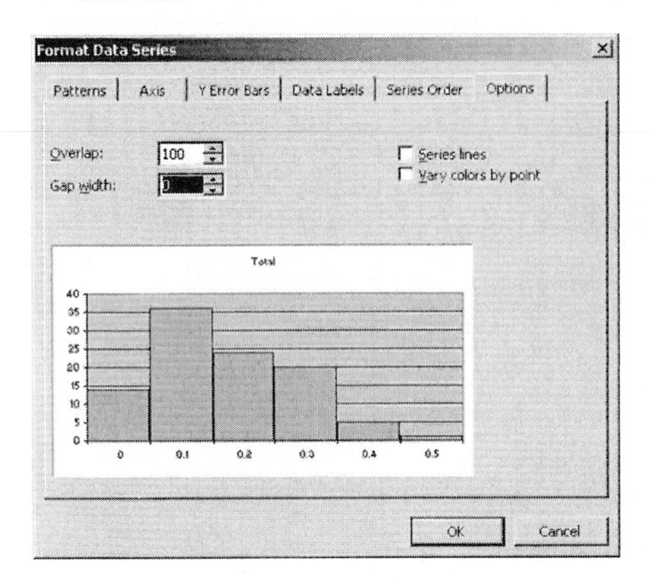

23. Click on the word "Total" at the top of the chart. A grey box will appear around the word indicating that it is ready to be edited. Type **Distribution of p with n = 10**. Press [**Enter**].

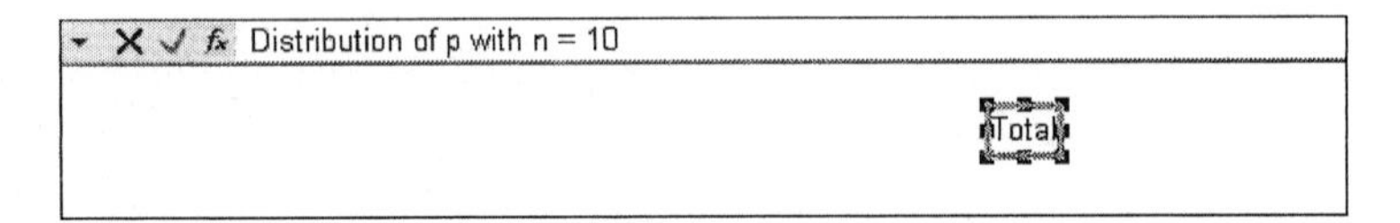

24. The completed histogram is shown below. The fourth step asks you to compute the mean and standard deviation of the sample proportions. To do this, you will first need to return to the worksheet that displays the proportions. Click on the **n = 10** sheet tab at the bottom of the screen.

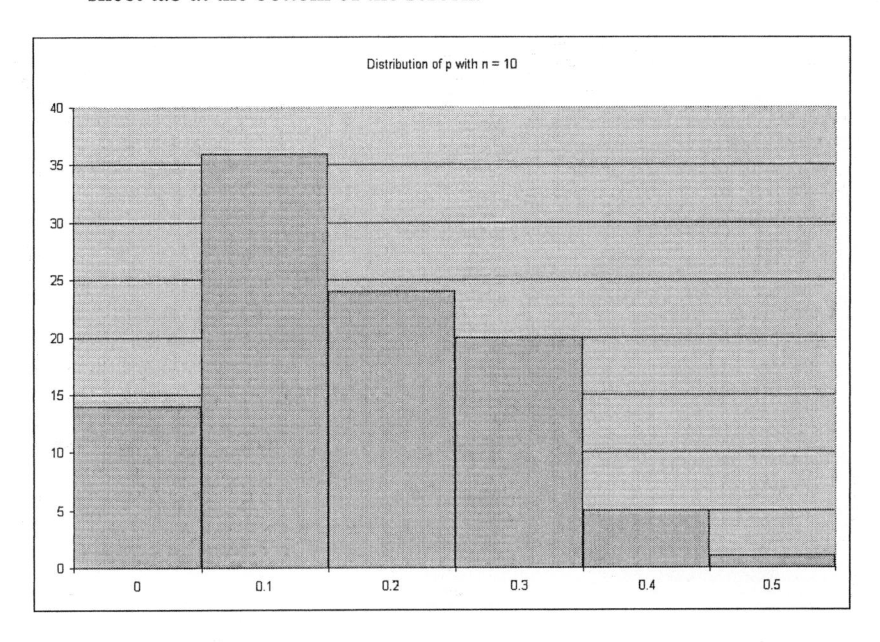

25. Type **Mean** in cell **A102** and then click in cell **B102** where you place the numeric value of the mean.

99	2	0.2
100	0	0
101	1	0.1
102	Mean	
103		

26. At the top of the screen, click **Insert** and select **Function**.

27. Select the **Statistical** category and the **AVERAGE** function. Click **OK**.

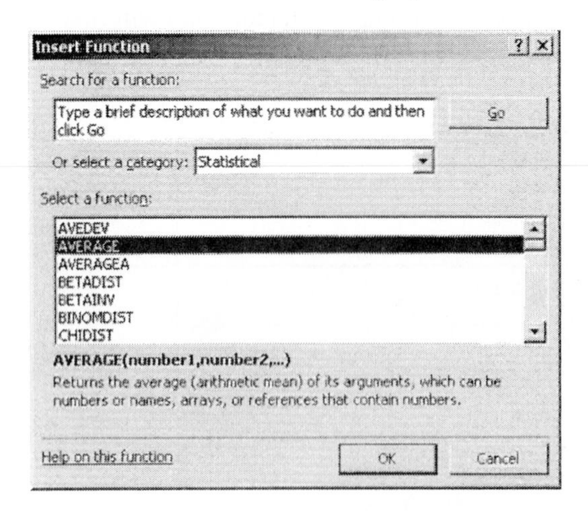

28. The range should show **B2:B101**. Click **OK**.

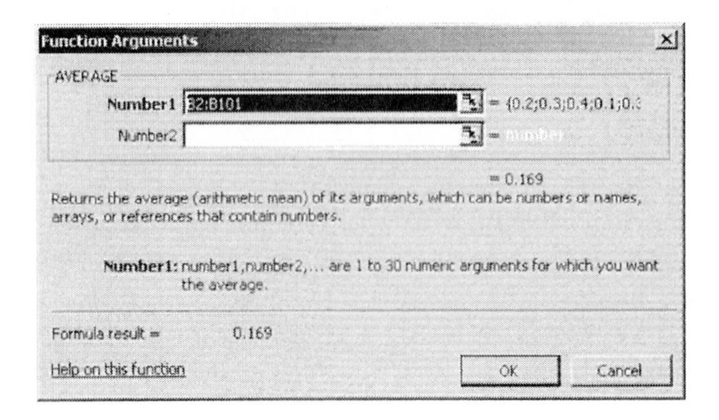

29. The mean of this distribution of proportions is equal to 0.169. Because the samples were generated randomly, it is not likely that your mean will be exactly the same. Type **St. Dev.** in cell **A103**, and click in cell **B103** to place the numerical value of the standard deviation there.

98	2	0.2
99	2	0.2
100	0	0
101	1	0.1
102	Mean	0.169
103	St. Dev.	

30. At the top of the screen, click **Function** and select **Insert**.

31. Select the **Statistical** category and the **STDEV** function. Click **OK**.

32. The range should be **B2:B101**. Make any necessary corrections and then click **OK**.

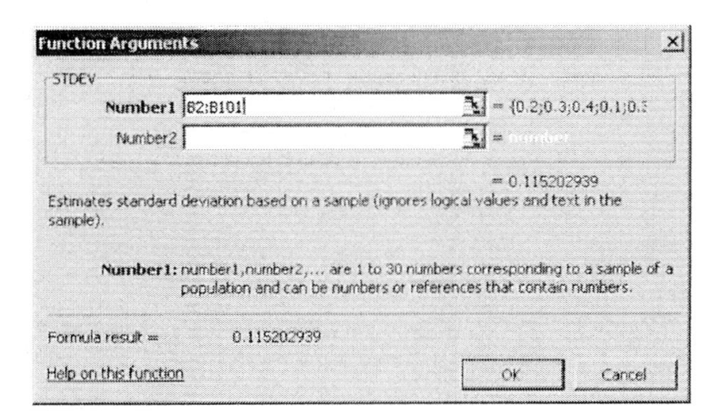

33. The standard deviation of this distribution of proportions is equal to 0.1152. Because the samples were generated randomly, it is likely that your standard deviation will not be exactly the same.

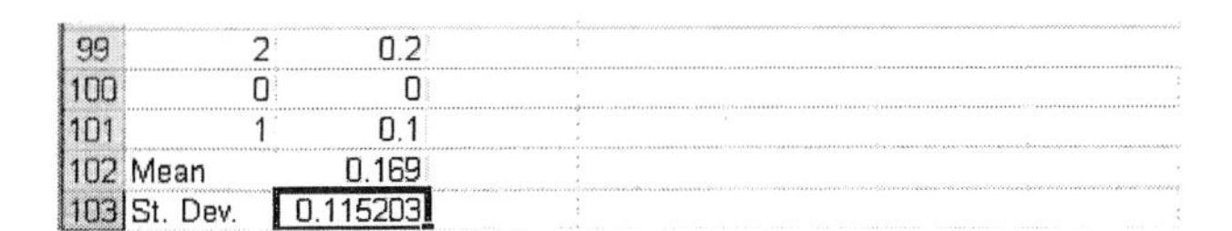

99		2	0.2
100		0	0
101		1	0.1
102	Mean		0.169
103	St. Dev.		0.115203

34. Repeat these steps to generate random samples of size n = 40 and n = 80.

◄

Estimating the Value of a Parameter Using Confidence Intervals

Section 9.1 The Logic in Constructing Confidence Intervals about a Population Mean Where the Population Standard Deviation Is Known

► Problem 25 (pg. 460) | Constructing 95% and 90% Confidence Intervals

1. Open a new Excel worksheet and enter the costs as shown below.

	A	B	C	D	E	F	G
1	Cost of Repairs						
2	225						
3	462						
4	729						
5	753						

If the PHStat add-in has not been loaded, you will need to load it before continuing. Follow the instructions in Section GS 8.2.

2. At the top of the screen, click **PHStat** and select **Confidence Intervals → Estimate for the Mean, sigma known**.

3. First, you will request the 95% confidence level. Complete the Estimate for the Mean dialog box as shown below. Click **OK**.

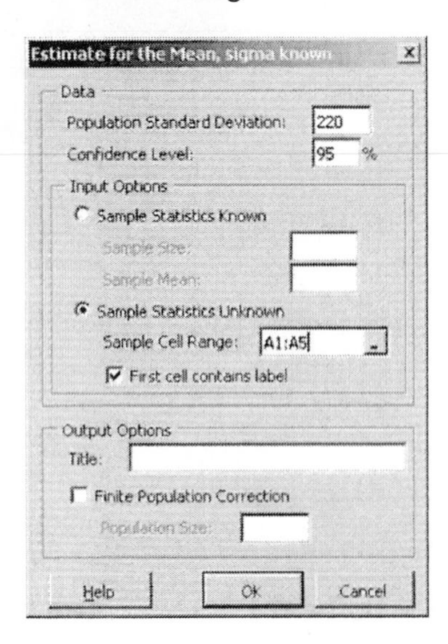

4. The output is placed in a worksheet named Confidence. At the bottom of the screen, click on the **Sheet1** tab to return to the data.

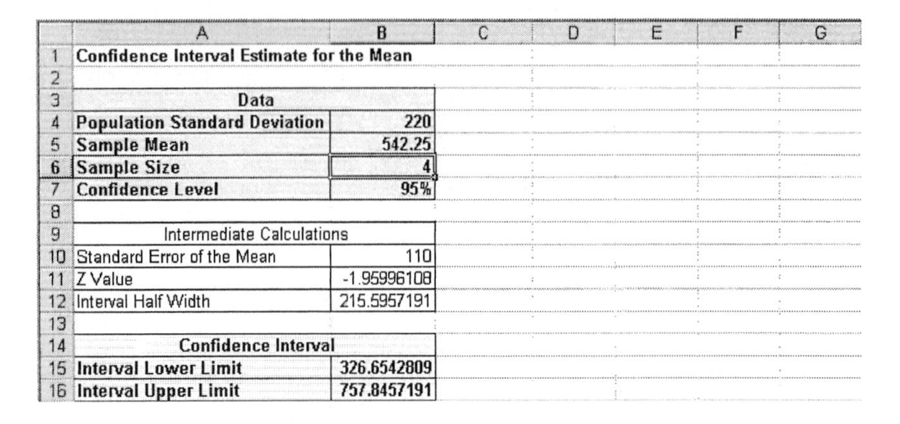

	A	B	C	D	E	F	G
1	**Confidence Interval Estimate for the Mean**						
2							
3	**Data**						
4	**Population Standard Deviation**	220					
5	**Sample Mean**	542.25					
6	**Sample Size**	4					
7	**Confidence Level**	95%					
8							
9	Intermediate Calculations						
10	Standard Error of the Mean	110					
11	Z Value	-1.95996108					
12	Interval Half Width	215.5957191					
13							
14	**Confidence Interval**						
15	**Interval Lower Limit**	326.6542809					
16	**Interval Upper Limit**	757.8457191					

5. Repeat these steps to obtain the 90% confidence level. At the top of the screen, click **PHStat**. Select **Confidence Intervals → Estimate for the mean, sigma known**.

6. Complete the Estimate for the Mean dialog box as shown below. The Sample Cell Range is A1:A5. Click **OK**.

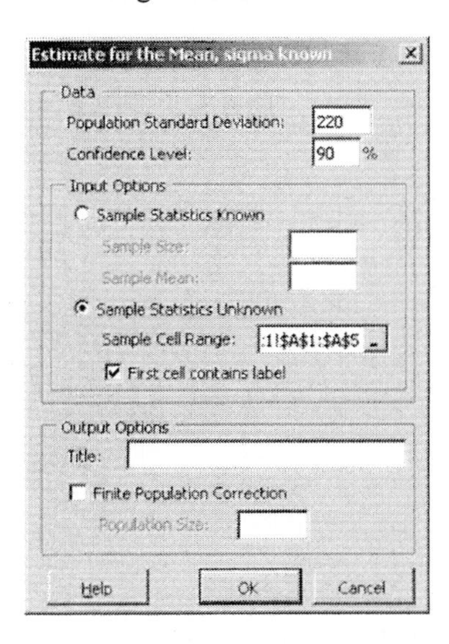

The output is placed in a worksheet named Confidence 2.

	A	B	C	D	E	F	G
1	**Confidence Interval Estimate for the Mean**						
2							
3	**Data**						
4	**Population Standard Deviation**	**220**					
5	**Sample Mean**	**542.25**					
6	**Sample Size**	**4**					
7	**Confidence Level**	**90%**					
8							
9	Intermediate Calculations						
10	Standard Error of the Mean	110					
11	Z Value	-1.644853					
12	Interval Half Width	180.9338301					
13							
14	**Confidence Interval**						
15	**Interval Lower Limit**	**361.3161699**					
16	**Interval Upper Limit**	**723.1838301**					

◀

Section 9.2 Confidence Intervals about a Population Mean in Practice Where the Population Standard Deviation is Unknown

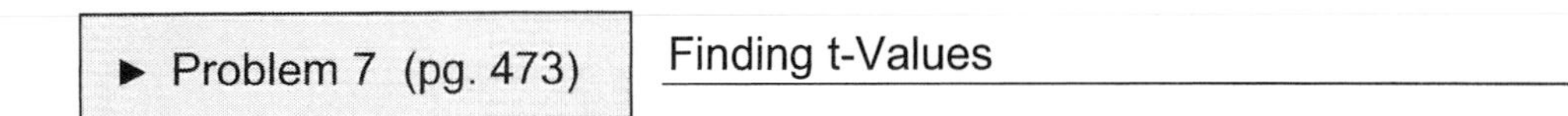

1. Open a new Excel worksheet and enter labels for items a, b, c, and d as shown below.

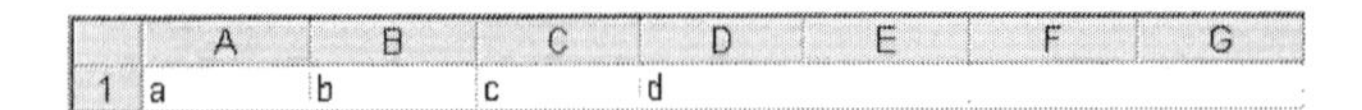

	A	B	C	D	E	F	G
1	a	b	c	d			

2. Click in cell **A2** where you will place the answer to part a. At the top of the screen, click **Insert** and select **Function**.

3. Under Function category, select **Statistical**. Under Function name, select **TINV**. Click **OK**.

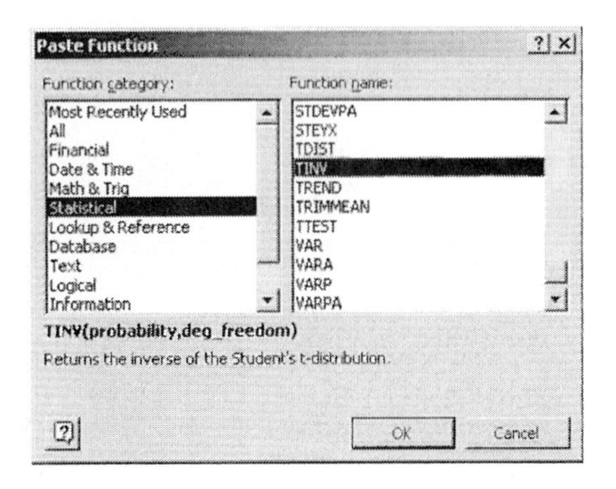

4. Complete the TINV dialog box as shown below. The probability refers to a two-tailed probability. So, to get obtain the t-value for area of 0.10 in the right tail alone, request probability of **.20** (i.e., 0.10 in each tail). Click **OK**.

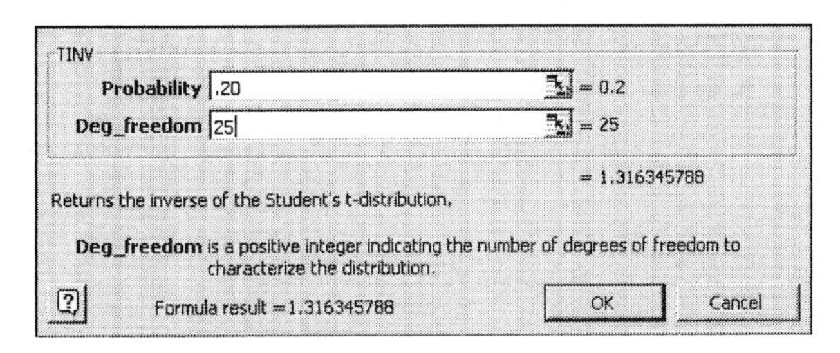

5. The function returns a value of 1.3163. Click in cell **B2** where you will place the answer to part b. At the top of the screen, click **Insert** and select **Function**.

6. Select **Statistical** and select **TINV**. Click **OK**.

7. Complete the TINV dialog box as shown below. The probability refers to a two-tailed probability. So, to get obtain the t-value for area of 0.05 in the right tail alone, request probability of **.10** (i.e., 0.05 in each tail). Click **OK**.

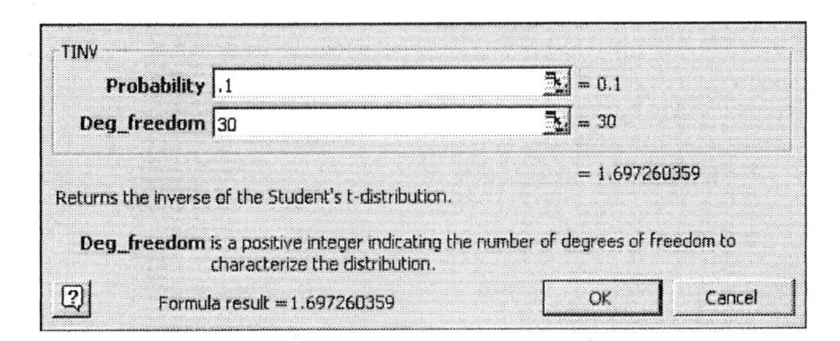

8. The function returns a value of 1.6973. Click in cell **C2** where you will place the answer to part c.

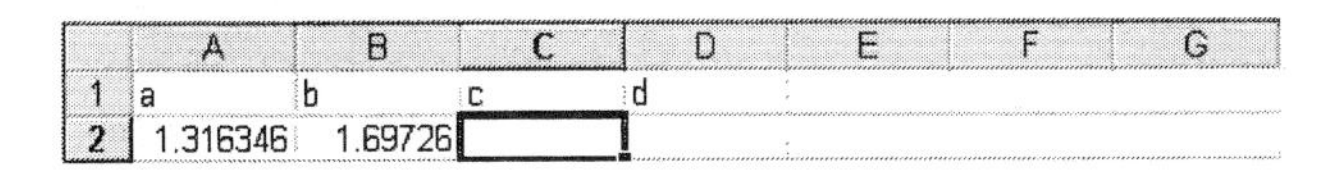

	A	B	C	D	E	F	G
1	a	b	c	d			
2	1.316346	1.69726					

9. At the top of the screen, click **Insert** and select **Function**.

10. Under Function category, select **Statistical**. Under Function name, select **TINV**. Click **OK**.

11. Complete the TINV dialog box as shown below. The probability refers to a two-tailed probability. So, to get obtain the t-value for area of 0.01 in the left tail, request probability of **0.02** (i.e., 0.01 in each tail). Click **OK**.

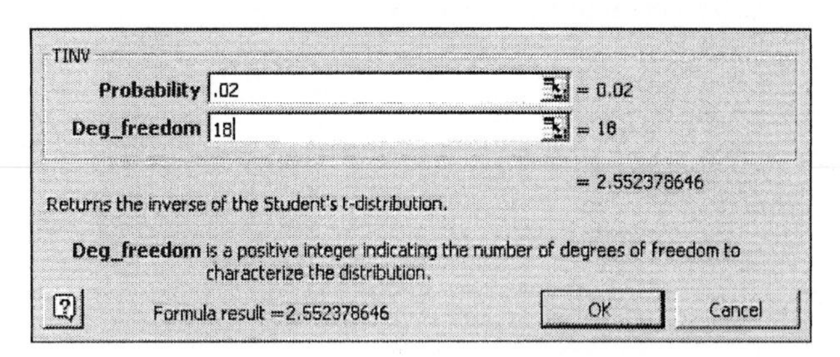

12. The function returns a value of 2.5524. Because the function returns the absolute value and the t-value you need is in the left tail, the correct result is negative, -2.5524. Click in cell **D2** where you will place the answer to part d.

13. At the top of the screen, click **Insert** and select **Function**.

14. Under Function category, select **Statistical**. Under Function name, select **TINV**. Click **OK**.

15. Complete the TINV dialog box as shown below. To obtain the t-value corresponding to 90% confidence, you input a two-tailed probability of **.10** (i.e., 0.05 in each tail). Click **OK**.

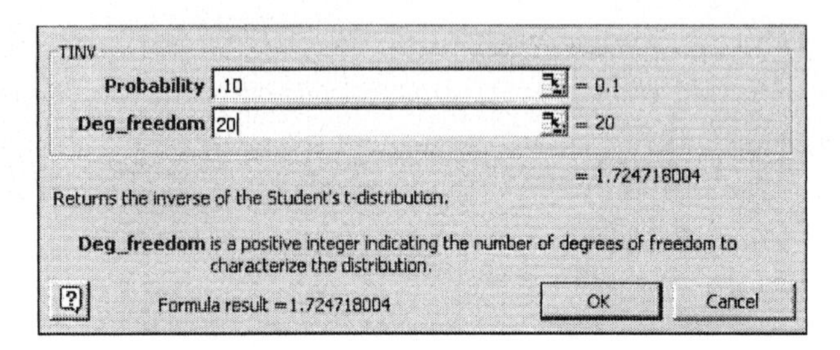

The function returns a value of 1.7247. The completed worksheet is displayed below.

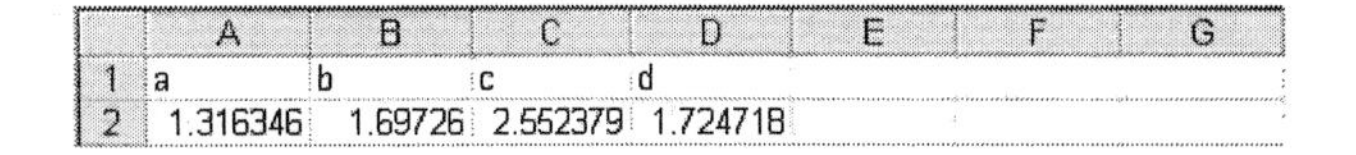

	A	B	C	D	E	F	G
1	a	b	c	d			
2	1.316346	1.69726	2.552379	1.724718			

◀

▶ Problem 9 (pg. 473) Constructing 96% and 90% Confidence Intervals

1. Open a new Excel worksheet.

If the PHStat add-in has not been loaded, you will need to load it before continuing. Follow the instructions in Section GS 8.2.

2. At the top of the screen, click **PHStat**. Select **Confidence Intervals → Estimate for the Mean, sigma unknown**.

3. To construct the 96% confidence interval for n = 25, complete the Estimate for the Mean dialog box as shown below. Click **OK**.

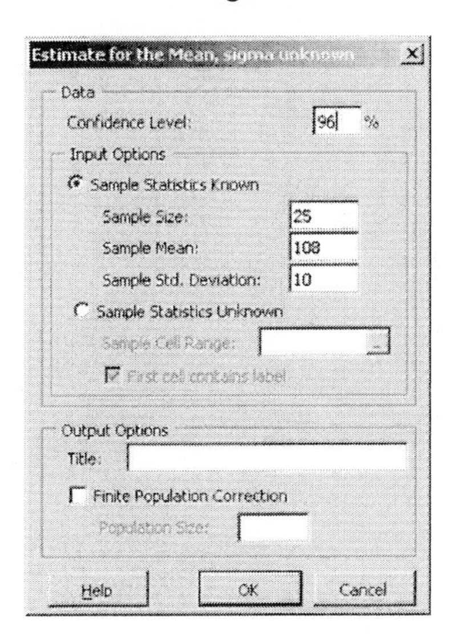

4. The output is displayed in a worksheet named Confidence. To construct the 96% confidence interval for n = 10, begin by clicking **PHStat** at the top of the screen. Select **Confidence Intervals → Estimate for the Mean, sigma unknown**.

	A	B	C	D	E	F	G
1	**Confidence Interval Estimate for the Mean**						
2							
3	**Data**						
4	**Sample Standard Deviation**	10					
5	**Sample Mean**	108					
6	**Sample Size**	25					
7	**Confidence Level**	96%					
8							
9	Intermediate Calculations						
10	Standard Error of the Mean	2					
11	Degrees of Freedom	24					
12	*t* Value	2.17154593					
13	Interval Half Width	4.343091859					
14							
15	**Confidence Interval**						
16	**Interval Lower Limit**	103.66					
17	**Interval Upper Limit**	112.34					

5. Complete the Estimate for the Mean dialog box as shown below. Click **OK**.

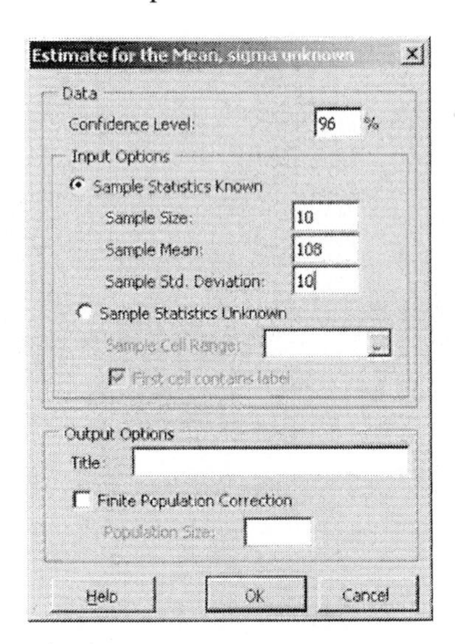

6. The output is displayed in a worksheet named Confidence2. To construct the 90% confidence interval with n = 25, begin by clicking **PHStat** at the top of the screen. Select **Confidence Intervals → Estimate for the Mean, sigma unknown**.

7. Complete the Estimate for the Mean dialog box as shown below. Click **OK**.

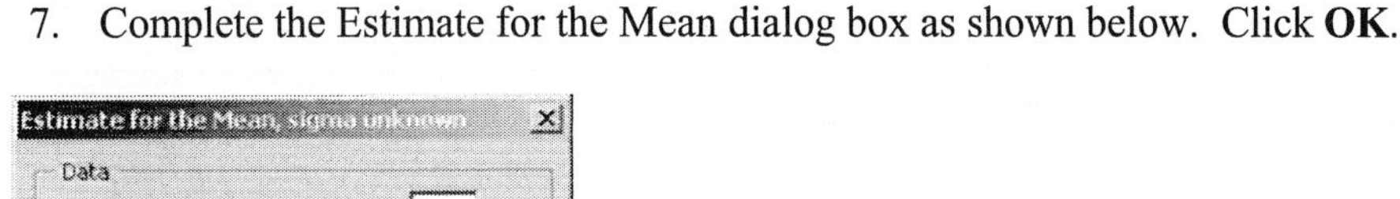

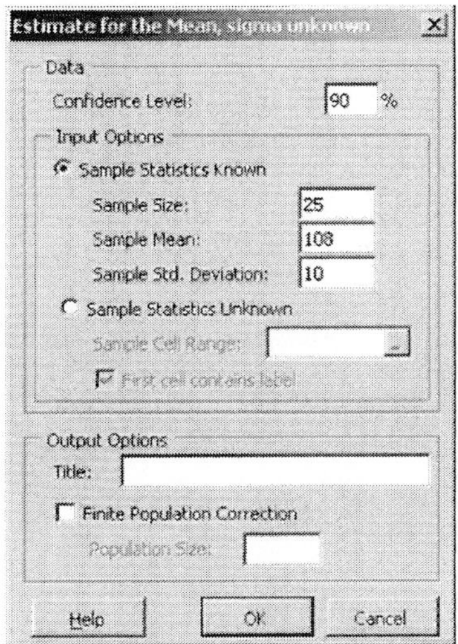

The output is displayed in a worksheet named Confidence3.

	A	B	C	D	E	F	G
1	**Confidence Interval Estimate for the Mean**						
2							
3	**Data**						
4	**Sample Standard Deviation**	**10**					
5	**Sample Mean**	**108**					
6	**Sample Size**	**25**					
7	**Confidence Level**	**90%**					
8							
9	Intermediate Calculations						
10	Standard Error of the Mean	2					
11	Degrees of Freedom	24					
12	*t* Value	1.710882316					
13	Interval Half Width	3.421764632					
14							
15	**Confidence Interval**						
16	**Interval Lower Limit**	**104.58**					
17	**Interval Upper Limit**	**111.42**					

◀

▶ Problem 27 (pg. 476) Constructing a Boxplot and Confidence Intervals

1. Open a new Excel worksheet and enter the price data as shown at the top of the next page.

If the PHStat add-in has not been loaded, you will need to load it before continuing. Follow the instructions in Section GS 8.2.

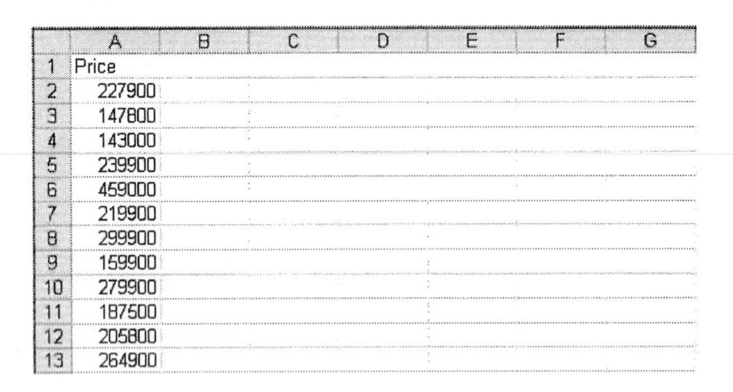

	A	B	C	D	E	F	G
1	Price						
2	227900						
3	147800						
4	143000						
5	239900						
6	459000						
7	219900						
8	299900						
9	159900						
10	279900						
11	187500						
12	205800						
13	264900						

2. At the top of the screen, click **PHStat** and select **Descriptive Statistics → Box-and-Whisker Plot**.

3. Complete the Box-and-Whisker Plot dialog box as shown below. Click **OK**.

Instead of Sheet1!$A$1:$A$13, you could just enter A1:A13.

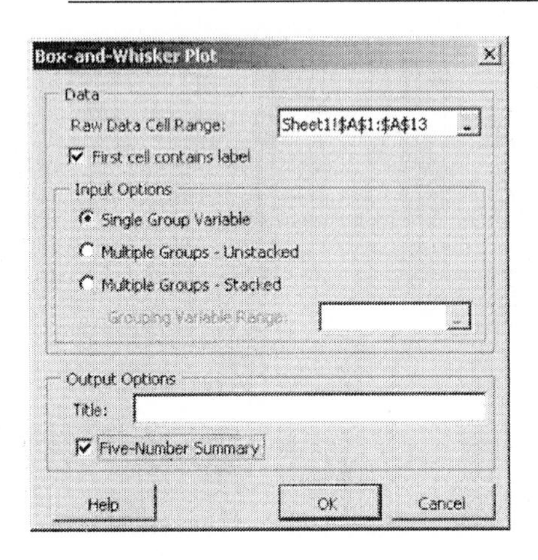

4. The box-and-whisker plot is displayed in a worksheet named BoxWhiskerPlot. Return to the worksheet containing the data. To do this, click on the **Sheet1** tab near the bottom of the screen.

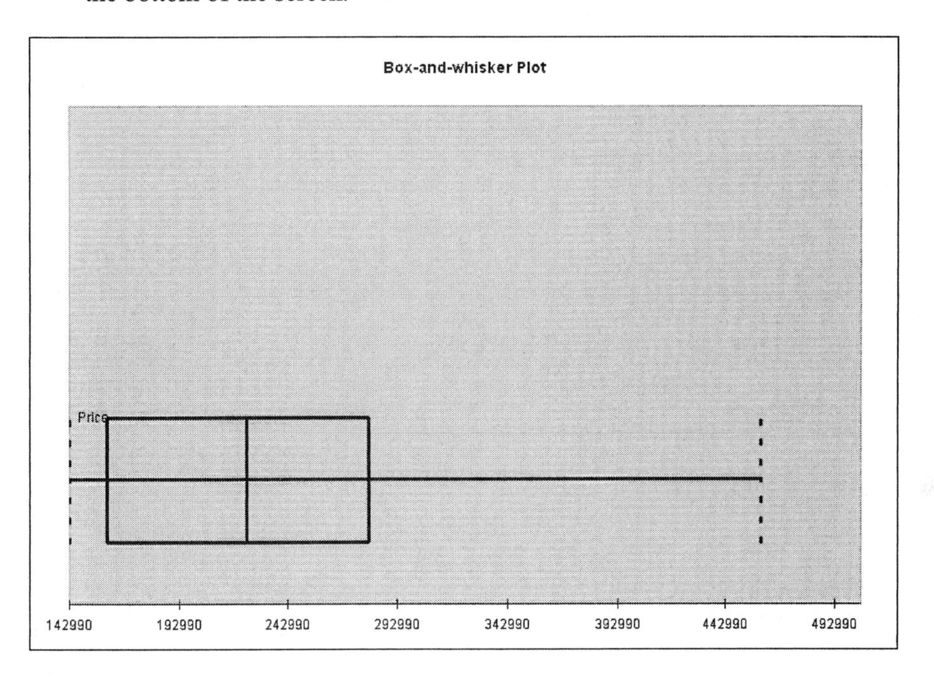

5. The outlier is 459,000. Next you will construct a 99% confidence interval that includes this outlier. At the top of the screen, click **PHStat**. Select **Confidence Intervals → Estimate for the Mean, sigma unknown**.

6. Complete the Estimate for the Mean dialog box as shown below. The Sample Cell Range is A1:A13. Click **OK**.

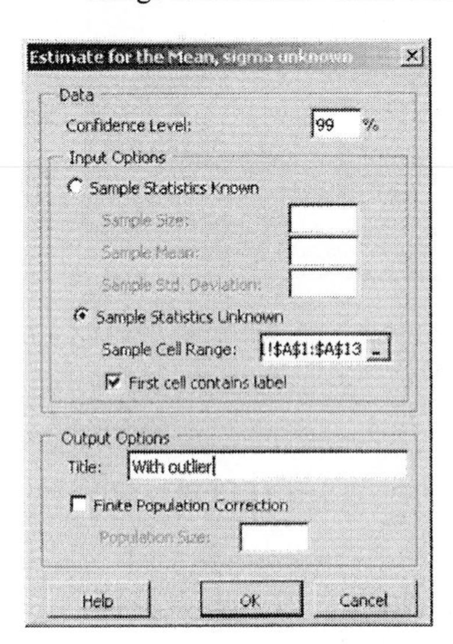

7. The output is placed in a worksheet named Confidence. You now will construct a 99% confidence interval that does not include the outlier. Return to the sheet that contains the data by clicking on the **Sheet1** tab at the bottom of the screen.

	A	B
1	**With outlier**	
2		
3	**Data**	
4	**Sample Standard Deviation**	86548.69185
5	**Sample Mean**	236283.3333
6	**Sample Size**	12
7	**Confidence Level**	99%
8		
9	Intermediate Calculations	
10	Standard Error of the Mean	24984.45527
11	Degrees of Freedom	11
12	*t* Value	3.105806514
13	Interval Half Width	77596.88391
14		
15	**Confidence Interval**	
16	**Interval Lower Limit**	158686.45
17	**Interval Upper Limit**	313880.22

8. Sort the data in ascending order so that the outlier (459,000) is at the bottom of the data set.

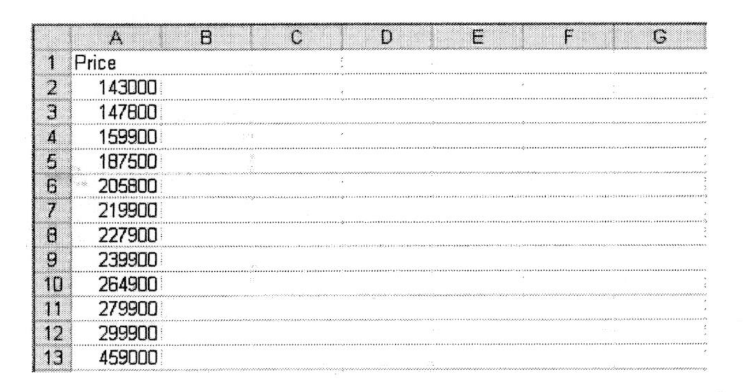

	A	B	C	D	E	F	G
1	Price						
2	143000						
3	147800						
4	159900						
5	187500						
6	205800						
7	219900						
8	227900						
9	239900						
10	264900						
11	279900						
12	299900						
13	459000						

9. At the top of the screen, click PHStat. Select **Confidence Intervals → Estimate for the Mean, sigma unknown**.

10. Complete the Estimate for the Mean dialog box as shown below. Note that the range excludes 459,000. Click **OK**.

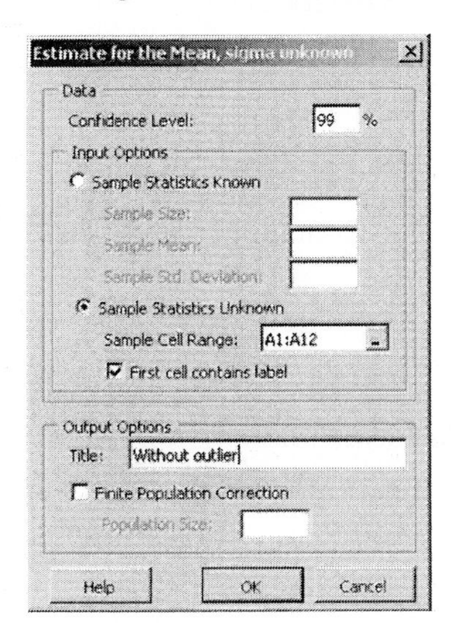

The output is placed in a worksheet named Confidence2.

	A	B
1	**Without outlier**	
2		
3	**Data**	
4	**Sample Standard Deviation**	53184.00648
5	**Sample Mean**	216036.3636
6	**Sample Size**	11
7	**Confidence Level**	99%
8		
9	Intermediate Calculations	
10	Standard Error of the Mean	16035.5813
11	Degrees of Freedom	10
12	*t* Value	3.169272672
13	Interval Half Width	50821.1296
14		
15	**Confidence Interval**	
16	**Interval Lower Limit**	165215.23
17	**Interval Upper Limit**	266857.49

◀

Section 9.3 Confidence Intervals about a Population Proportion

▶ Problem 11 (pg. 484) — Estimating a Population Proportion and Constructing a 90% Confidence Interval

If the PHStat add-in has not been loaded, you will need to load it before continuing. Follow the instructions in Section GS 8.2.

1. At the top of the screen, click **PHStat**. Select **Confidence Intervals → Estimate for the Proportion**.

2. Complete the Estimate for the Proportion dialog box as shown below. Click **OK**.

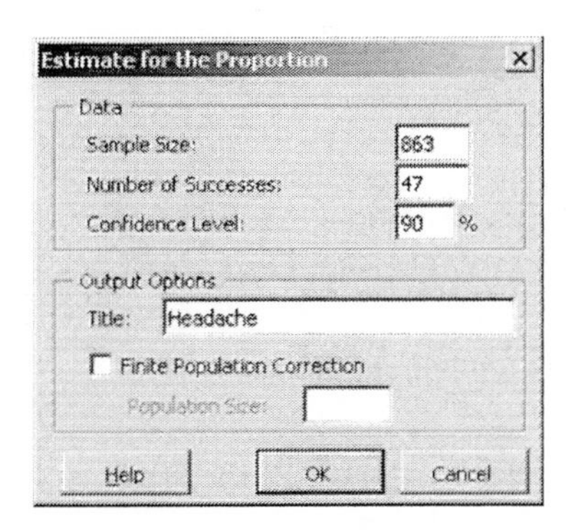

The output is placed in a worksheet named Confidence.

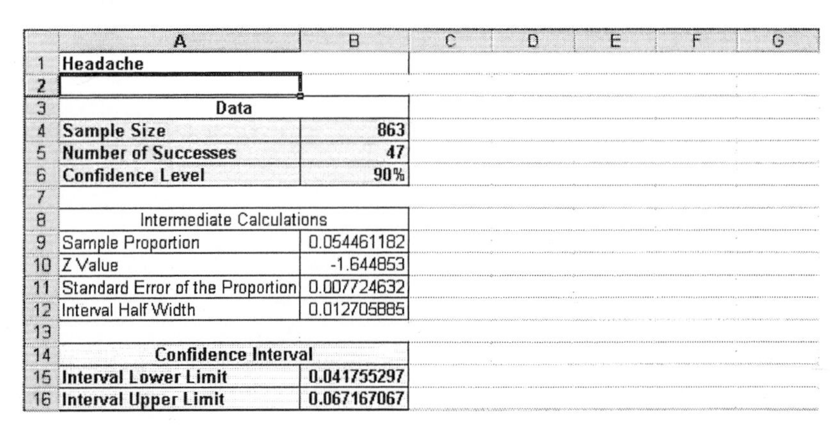

	A	B
1	**Headache**	
2		
3	**Data**	
4	**Sample Size**	**863**
5	**Number of Successes**	**47**
6	**Confidence Level**	**90%**
7		
8	Intermediate Calculations	
9	Sample Proportion	0.054461182
10	Z Value	-1.644853
11	Standard Error of the Proportion	0.007724632
12	Interval Half Width	0.012705885
13		
14	**Confidence Interval**	
15	**Interval Lower Limit**	**0.041755297**
16	**Interval Upper Limit**	**0.067167067**

◀

Section 9.4 Confidence Intervals about a Population Standard Deviation

▶ Problem 11 (pg. 491) **Constructing a 95% Confidence Interval about a Population Standard Deviation**

If the PHStat add-in has not been loaded, you will need to load it before continuing. Follow the instructions in Section GS 8.2.

1. Open a new Excel worksheet and enter the data as shown below.

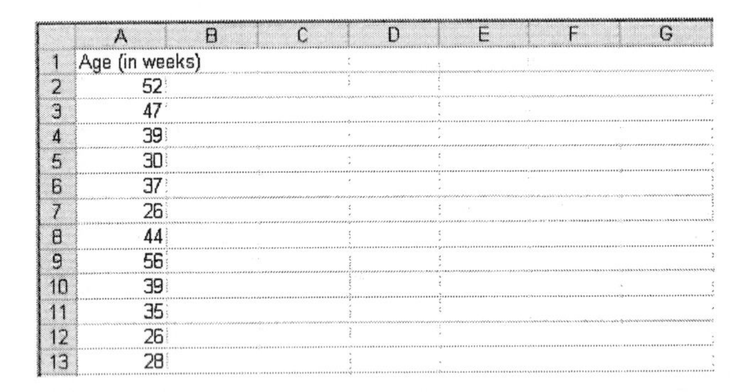

	A	B	C	D	E	F	G
1	Age (in weeks)						
2	52						
3	47						
4	39						
5	30						
6	37						
7	26						
8	44						
9	56						
10	39						
11	35						
12	26						
13	28						

2. At the top of the screen, click **PHStat**. Select **Confidence Intervals → Estimate for the Population Variance.**

3. Complete the Estimate for the Population Variance dialog box with a sample size of **12**, sample standard deviation of **10.00**, a confidence interval of **95**, and a title of **Crawling Babies**. Click **OK**.

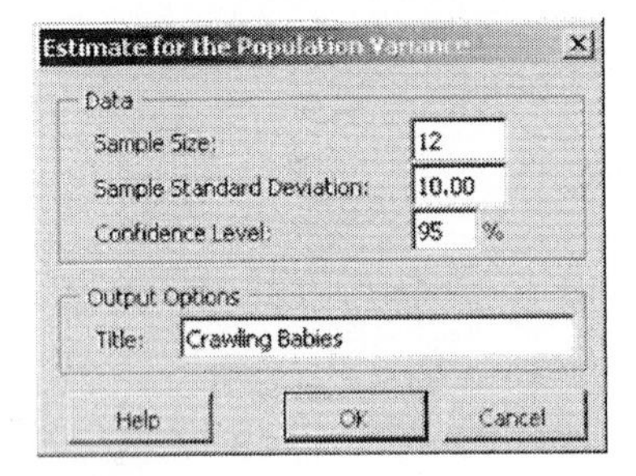

The output is displayed in a worksheet named PVInterval.

	A	B	C	D	E
1	**Crawling Babies**				
2					
3	**Data**				
4	**Sample Size**	12			
5	**Sample Standard Deviation**	10			
6	**Confidence Level**	95%			
7					
8	Intermediate Calculations				
9	Degrees of Freedom	11			
10	Sum of Squares	1100			
11	Single Tail Area	0.025			
12	Lower Chi-Square Value	3.815748			
13	Upper Chi-Square Value	21.92005			
14					
15	**Results**				
16	**Interval Lower Limit for Variance**	50.18237			
17	**Interval Upper Limit for Variance**	288.279			
18					
19	**Interval Lower Limit for Standard Deviation**	7.083952			
20	**Interval Upper Limit for Standard Deviation**	16.97878			
21					
22	*Assumption:*				
23	**Population from which sample was drawn has an approximate normal distribution.**				

Testing Claims Regarding a Parameter

CHAPTER 10

Section 10.2 Testing Claims about a Population Mean Assuming the Population Standard Deviation Is Known

► Problem 17 (pg. 527) — Computing the P-Value

If the PHStat add-in has not been loaded, you will need to load it before continuing. Follow the instructions in Section GS 8.2.

1. Open a new Excel worksheet.

2. At the top of the screen, click **PHStat**. Select **One-Sample Tests → Z Test for the Mean, sigma known**.

3. Complete the Z Test for the Mean dialog box as shown below. Click **OK**.

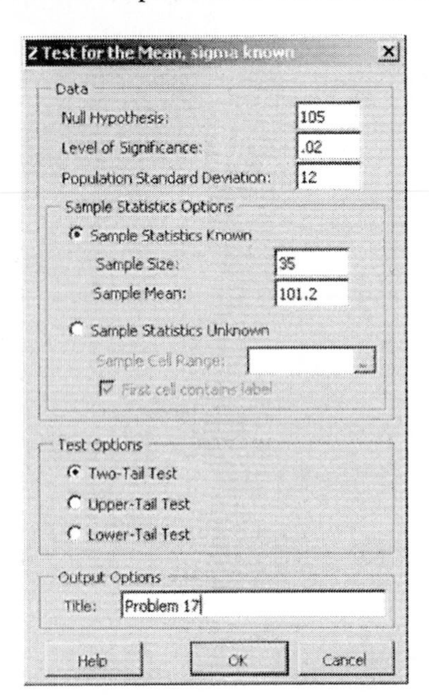

The output is placed in a worksheet named Hypothesis.

	A	B
1	**Problem 17**	
2		
3	**Data**	
4	**Null Hypothesis** μ=	105
5	**Level of Significance**	0.02
6	**Population Standard Deviation**	12
7	**Sample Size**	35
8	**Sample Mean**	101.2
9		
10	Intermediate Calculations	
11	Standard Error of the Mean	2.028370211
12	**Z Test Statistic**	-1.873425265
13		
14	**Two-Tail Test**	
15	**Lower Critical Value**	-2.326347874
16	**Upper Critical Value**	2.326347874
17	***p*-Value**	0.061009684
18	**Do not reject the null hypothesis**	

◀

▶ Problem 29 (pg. 529) — Testing the Researcher's Claim, $\alpha = .05$

If the PHStat add-in has not been loaded, you will need to load it before continuing. Follow the instructions in Section GS 8.2.

1. Open a new Excel worksheet.

2. At the top of the screen, click PHStat. Select **One-Sample Tests → Z Test for the Mean, sigma known**.

3. Complete the Z Test for the Mean dialog box as shown below. Click **OK**.

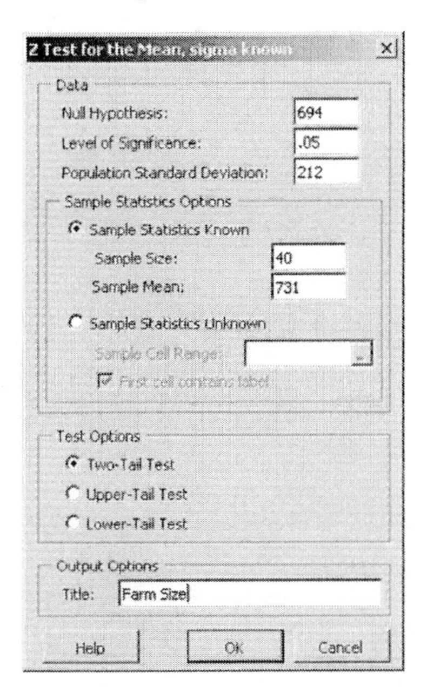

The output is placed in a worksheet named Hypothesis.

	A		B
1	Farm Size		
2			
3	Data		
4	Null Hypothesis	μ=	694
5	Level of Significance		0.05
6	Population Standard Deviation		212
7	Sample Size		40
8	Sample Mean		731
9			
10	Intermediate Calculations		
11	Standard Error of the Mean		33.5201432
12	Z Test Statistic		1.1038139
13			
14	Two-Tail Test		
15	Lower Critical Value		-1.959963985
16	Upper Critical Value		1.959963985
17	*p*-Value		0.269673874
18	Do not reject the null hypothesis		

◄

Section 10.3 Testing Claims about a Population Mean in Practice

► Problem 9 (pg. 538) Testing the Hypothesis, $\alpha = 0.05$

If the PHStat add-in has not been loaded, you will need to load it before continuing. Follow the instructions in Section GS 8.2.

1. Open a new Excel worksheet.
2. At the top of the screen, click **PHStat**. Select **One-Sample Tests → t Test for the Mean, sigma unknown**.

3. Complete the t Test for the Mean dialog box as shown below. Be sure to select Lower-Tail Test. Click **OK**.

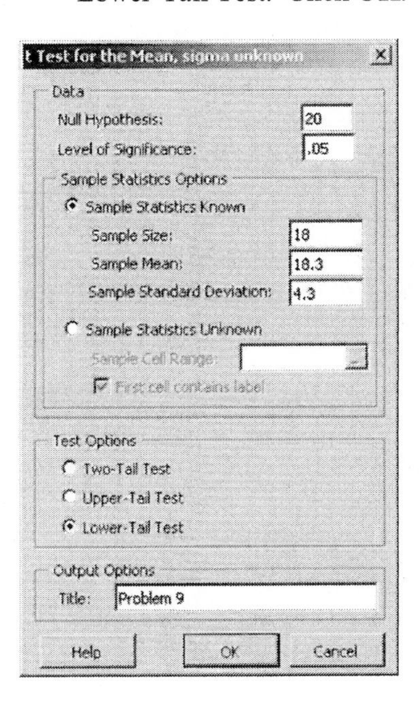

The output is placed in a worksheet named Hypothesis.

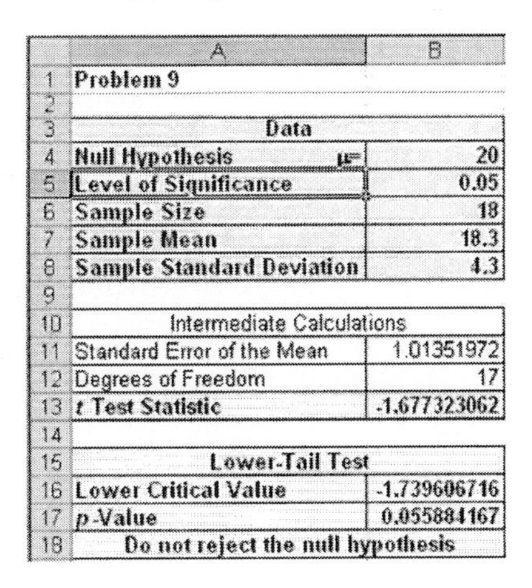

	A	B
1	**Problem 9**	
2		
3	**Data**	
4	**Null Hypothesis** μ=	20
5	**Level of Significance**	0.05
6	**Sample Size**	18
7	**Sample Mean**	18.3
8	**Sample Standard Deviation**	4.3
9		
10	Intermediate Calculations	
11	Standard Error of the Mean	1.01351972
12	Degrees of Freedom	17
13	***t*** **Test Statistic**	-1.677323062
14		
15	**Lower-Tail Test**	
16	**Lower Critical Value**	-1.739606716
17	***p*-Value**	0.055884167
18	**Do not reject the null hypothesis**	

Section 10.4 Testing Claims about a Population Proportion

▶ Problem 3 (pg. 550) | Testing the Hypothesis p = 0.3

If the PHStat add-in has not been loaded, you will need to load it before continuing. Follow the instructions in Section GS 8.2.

1. Open a new Excel worksheet.

2. At the top of the screen, click **PHStat**. Select **One-Sample Tests → Z Test for the Proportion**.

3. Complete the Z Test for the Proportion dialog box as shown below. Be sure to select Upper-Tail Test. Click **OK**.

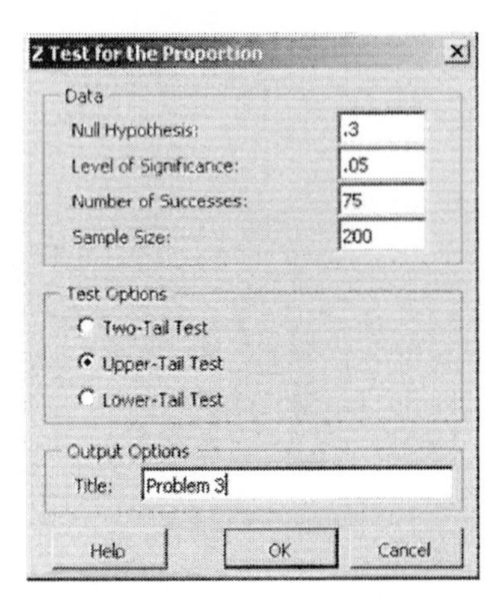

The output is placed in a worksheet named Hypothesis.

	A	B
1	Problem 3	
2		
3	Data	
4	Null Hypothesis $p=$	0.3
5	Level of Significance	0.05
6	Number of Successes	75
7	Sample Size	200
8		
9	Intermediate Calculations	
10	Sample Proportion	0.375
11	Standard Error	0.032403703
12	Z Test Statistic	2.314550249
13		
14	Upper-Tail Test	
15	Upper Critical Value	1.644853627
16	p-Value	0.010318779
17	Reject the null hypothesis	

◀

Section 10.5 Testing Claims about a Population Standard Deviation

► Problem 3 (pg. 556) | Testing the Hypothesis $\sigma = 50$

If the PHStat add-in has not been loaded, you will need to load it before continuing. Follow the instructions in Section GS 8.2.

1. Open a new Excel worksheet.

2. At the top of the screen, click **PHStat**. Select **One-Sample Tests → Chi-Square Test for the Variance**.

3. Complete the Chi-Square Test for the Variance dialog box as shown below. Click **OK**.

The Null Hypothesis value of 2500 shown in the dialog box is the variance. The variance was obtained by squaring the numerical value of σ (i.e., 50) that appears in the null hypothesis.

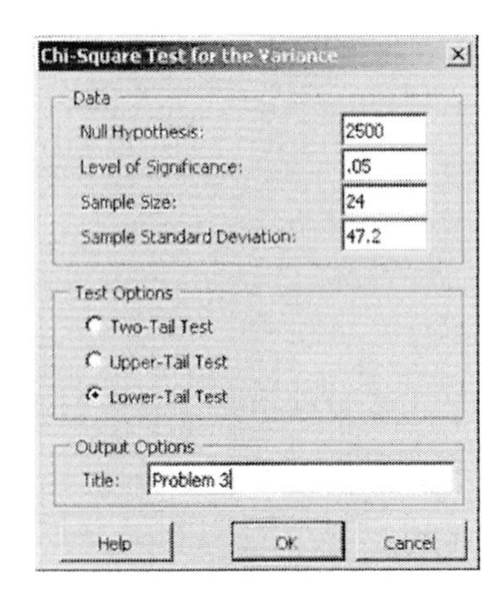

The output is placed in a worksheet named Hypothesis.

	A		B
1	**Problem 3**		
2			
3	**Data**		
4	**Null Hypothesis**	σ^2=	2500
5	**Level of Significance**		0.05
6	**Sample Size**		24
7	**Sample Standard Deviation**		47.2
8			
9	Intermediate Calculations		
10	Degrees of Freedom		23
11	Half Area		0.025
12	**Chi-Square Statistic**		20.496128
13			
14	Lower-Tail Test		
15	Lower Critical Value		13.09051436
16	*p*-Value		0.388171516
17	**Do not reject the null hypothesis**		

◀

Inferences on Two Samples

CHAPTER 11

Section 11.1 Inference about Two Means: Dependent Samples

▸ Problem 13 (pg. 583) Testing the Difference in the Measurement of Muzzle Velocity

1. Open worksheet "11_1_13" in the Chapter 11 folder. The first few rows are shown below.

	A	B	C	D	E	F	G
1	A	B					
2	793.8	793.2					
3	793.1	793.3					
4	792.4	792.6					

2. At the top of the screen, click **Tools** and select **Data Analysis**.

If Data Analysis does not appear as a choice in the Tools menu, you will need to load the Microsoft Excel Analysis ToolPak add-in. Follow the procedure in Section GS 8.1 before continuing.

3. Select **t-Test: Paired Two Samples for Means**. Click **OK**.

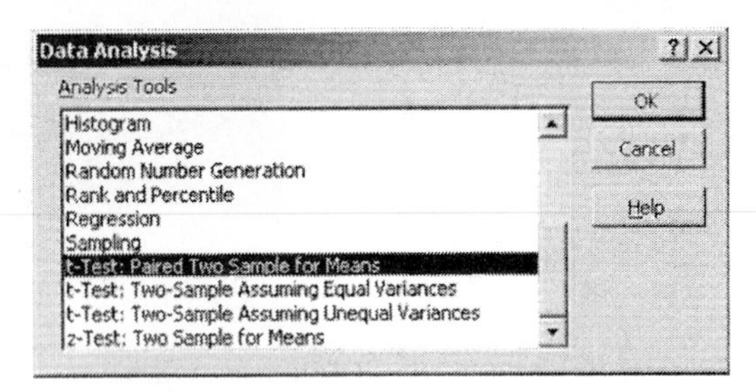

4. Complete the t-Test: Paired Two Sample for Means dialog box as shown below. Be sure to change alpha to 0.01. Click **OK**.

The default value for the Hypothesized Mean Difference is zero. So you can leave this window blank if your hypothesized difference is zero. A checkmark is necessary in the box to the left of Labels if the first row in the variable range is a label rather than a data value.

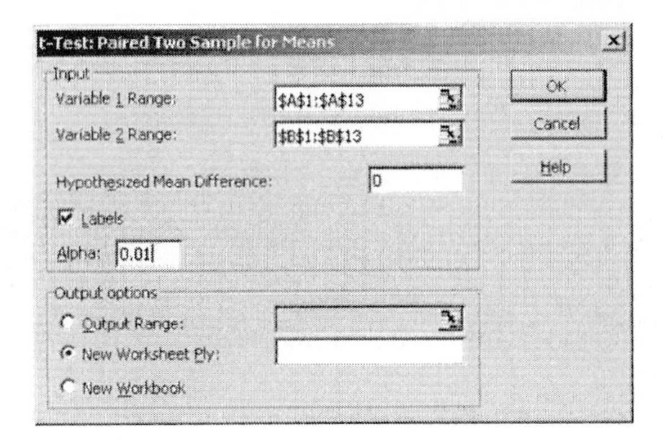

The output is displayed in a new worksheet. Make column A wider so that you can read the labels.

	A	B	C	D	E	F	G
1	t-Test: Paired Two Sample for Means						
2							
3		A	B				
4	Mean	792.4583	792.3417				
5	Variance	1.979015	2.568106				
6	Observations	12	12				
7	Pearson Correlation	0.958563					
8	Hypothesized Mean Difference	0					
9	df	11					
10	t Stat	0.851726					
11	P(T<=t) one-tail	0.206262					
12	t Critical one-tail	2.718079					
13	P(T<=t) two-tail	0.412525					
14	t Critical two-tail	3.105807					

◀

Section 11.2 Inference about Two Means: Independent Samples

▸ Problem 13 (pg. 597)

Testing a Hypothesis Regarding Two Different Concrete Mix Designs

If the PHStat add-in has not been loaded, you will need to load it before continuing. Follow the instructions in Section GS 8.2.

1. Open worksheet "11_2_13" in the Chapter 11 folder. The first few rows are shown below.

	A	B	C	D	E	F	G
1	Mixture 67	Mixture 67-0-400					
2	3960	4070					
3	3830	4640					
4	2940	5020					

2. At the top of the screen, click **Tools** and select **Data Analysis**.

3. Select **t-Test: Two-Sample Assuming Equal Variances** and click **OK**.

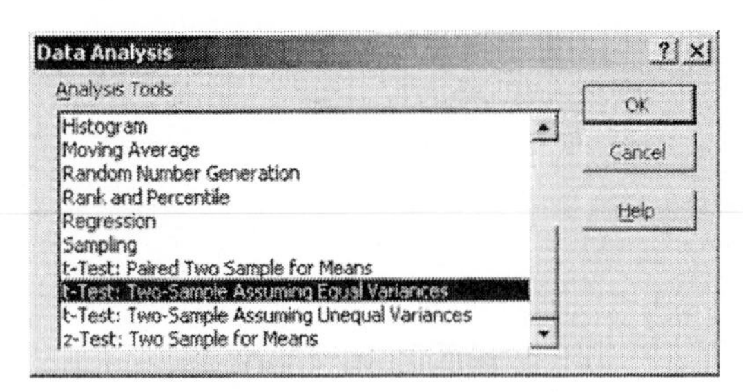

4. Complete the t-Test: Two-Sample Assuming Equal Variances dialog box as shown below. Click **OK**.

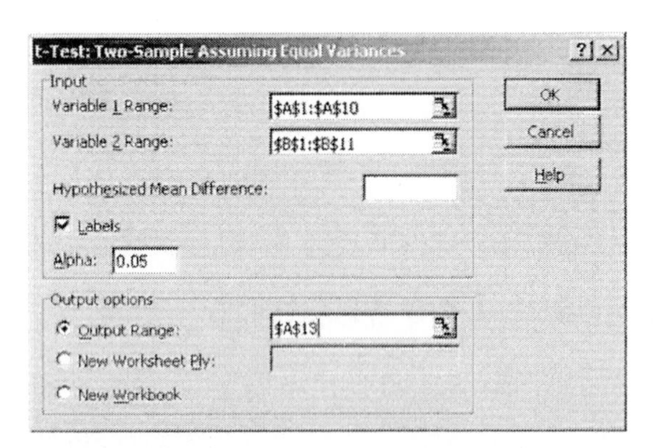

5. The output is placed a couple rows below the data. Make the columns wider so that you can read the entire labels.

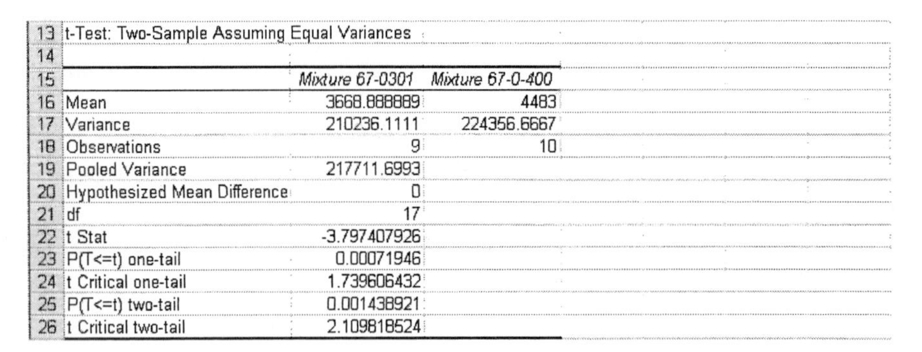

13	t-Test: Two-Sample Assuming Equal Variances		
14			
15		*Mixture 67-0301*	*Mixture 67-0-400*
16	Mean	3668.888889	4483
17	Variance	210236.1111	224356.6667
18	Observations	9	10
19	Pooled Variance	217711.6993	
20	Hypothesized Mean Difference	0	
21	df	17	
22	t Stat	-3.797407926	
23	P(T<=t) one-tail	0.00071946	
24	t Critical one-tail	1.739606432	
25	P(T<=t) two-tail	0.001438921	
26	t Critical two-tail	2.109818524	

6. Next, you will draw box plots of each data set beginning with Mixture 67-0-301. At the top of the screen, click **PHStat**.

7. Select **Descriptive Statistics → Box-and-Whisker Plot**.

8. Complete the Box-and-Whisker Plot dialog box as shown below. Click **OK**.

The Raw Data Cell Range was entered by clicking and dragging over the data range in the worksheet. If you prefer, you can type ***A1:A10****.*

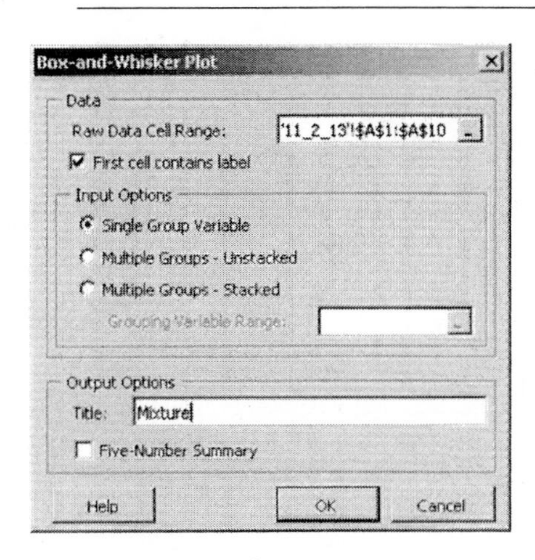

9. The box plot is displayed in a worksheet named BoxWhiskerPlot. Next, you will draw a box plot of the Mixture 67-0-400 data. Return to the worksheet containing the data by clicking on the **11_2_13** sheet tab at the bottom of the screen. At the top of the screen, click **PHStat**.

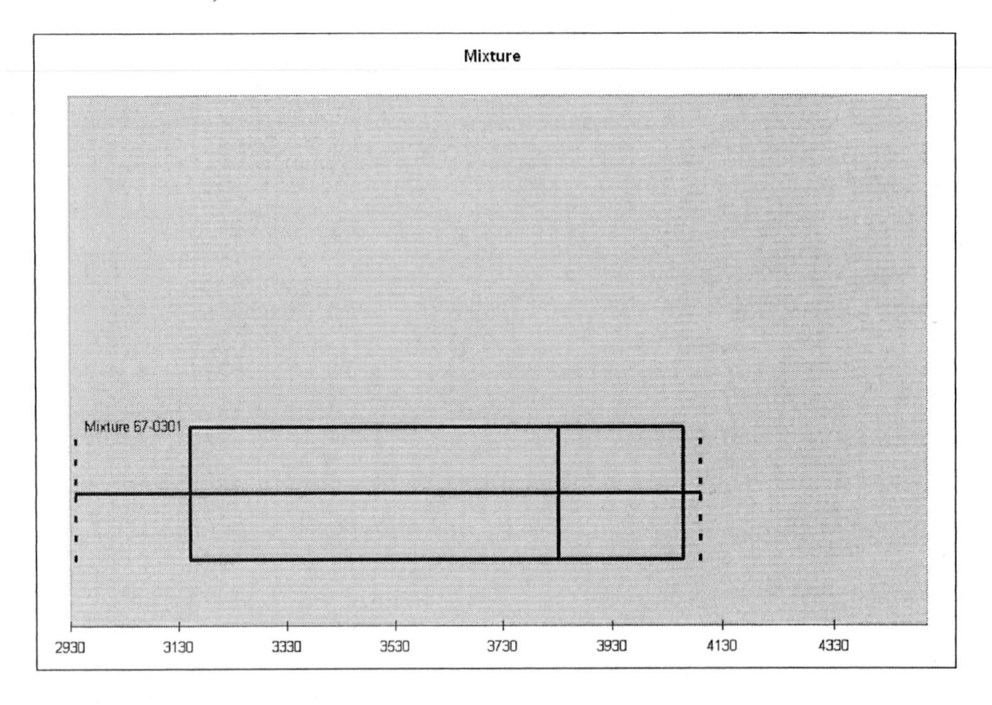

10. Select **Descriptive Statistics → Box-and-Whisker Plot**.

11. Complete the Box-and-Whisker Plot dialog box as shown below. Click **OK**.

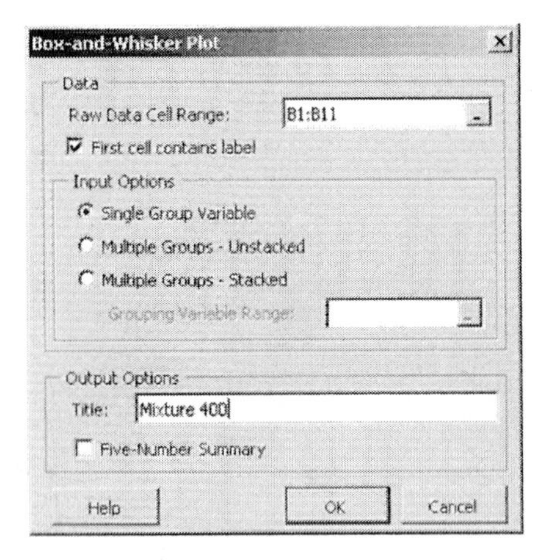

The output is displayed in a worksheet named BoxWhiskerPlot2.

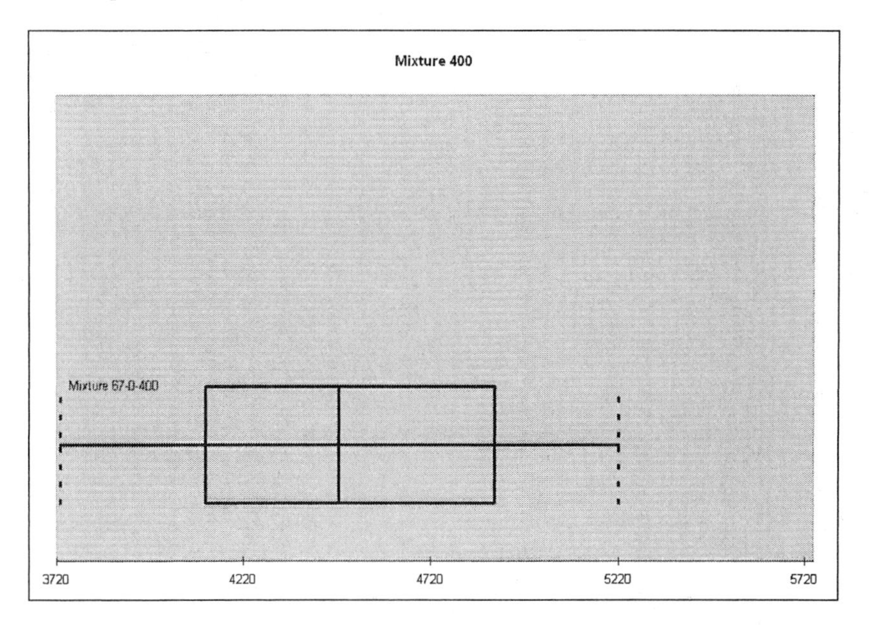

Section 11.3 Inference about Two Population Proportions

▶ Problem 13 (pg. 610) | Testing a Hypothesis Regarding Cholesterol

If the PHStat add-in has not been loaded, you will need to load it before continuing. Follow the instructions in Section GS 8.2.

1. Open a new Excel worksheet.
2. At the top of the screen, click **PHStat**.
3. Select **Two-Sample Tests → Z Test for Differences in Two Proportions**.

4. Complete the Z Test for the Difference in Two Proportions dialog box as shown below. Click **OK**.

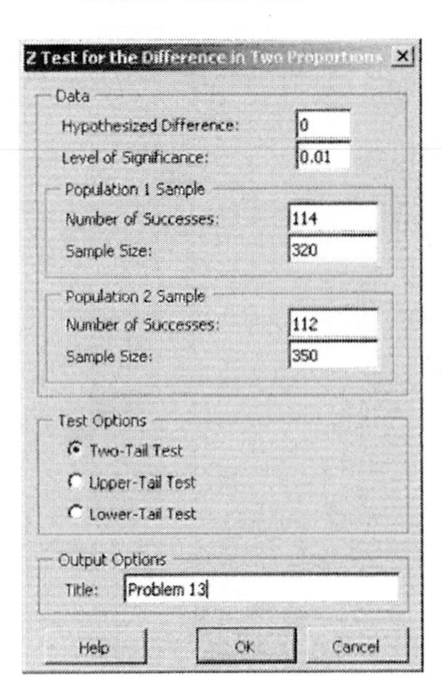

The output is displayed in a worksheet named Hypothesis.

	A	B
1	Problem 13	
2		
3	Data	
4	Hypothesized Difference	0
5	Level of Significance	0.01
6	Group 1	
7	Number of Successes	114
8	Sample Size	320
9	Group 2	
10	Number of Successes	112
11	Sample Size	350
12		
13	Intermediate Calculations	
14	Group 1 Proportion	0.35625
15	Group 2 Proportion	0.32
16	Difference in Two Proportions	0.03625
17	Average Proportion	0.337313433
18	Z Test Statistic	0.991308246
19		
20	Two-Tail Test	
21	Lower Critical Value	-2.575829304
22	Upper Critical Value	2.575829304
23	p-Value	0.321535088
24	Do not reject the null hypothesis	

◀

Section 11.4 Inference about Two Population Standard Deviations

► Problem 21 (pg. 621)

Testing a Claim Regarding Wait Time Standard Deviations

1. Open worksheet "11_4_21" in the Chapter 10 folder. The first few rows are shown below.

	A	B	C	D	E	F	G
1	Single Line	Multiple Lines					
2	1.2	1.1					
3	1.9	3.8					
4	2.1	4.3					

2. At the top of the screen, click **Tools** and select **Data Analysis**.

If Data Analysis does not appear as a choice in the Tools menu, you will need to load the Microsoft Excel Analysis ToolPak add-in. Follow the procedure in Section GS 8.1 before continuing.

---

3. Select **F-Test Two-Sample for Variances** and click **OK**.

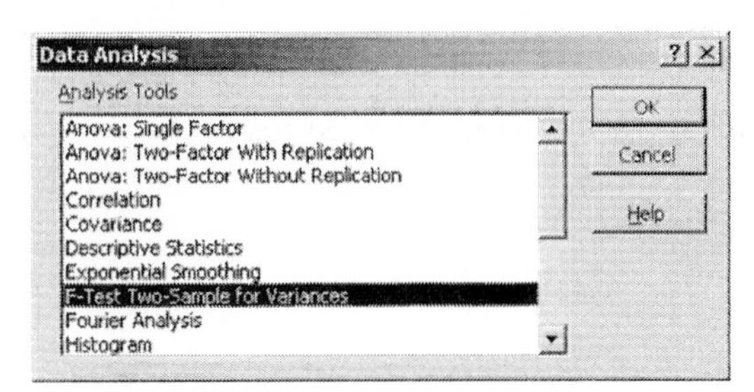

4. Complete the F-Test Two-Sample for Variances dialog box as shown below. Click **OK**.

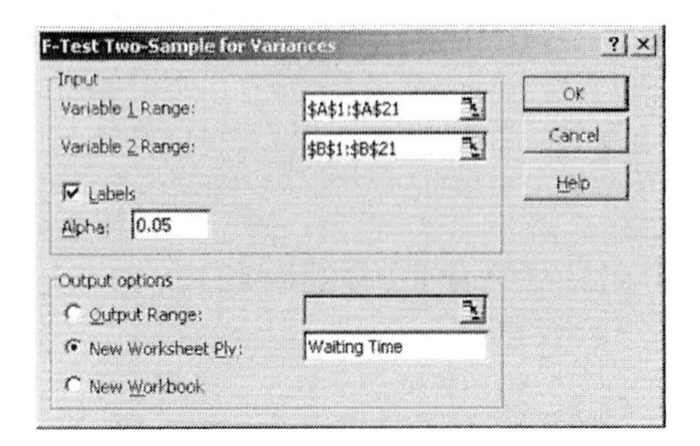

The output is displayed in a worksheet named Waiting Time. Make the columns wider so that you can read the entire labels.

	A	B	C	D	E	F	G
1	F-Test Two-Sample for Variances						
2							
3		*Single Line*	*Multiple Lines*				
4	Mean	2.255	2.475				
5	Variance	0.329973684	1.025131579				
6	Observations	20	20				
7	df	19	19				
8	F	0.321884225					
9	P(F<=f) one-tail	0.008700046					
10	F Critical one-tail	0.461200855					

◀

CHAPTER 12

Inference on Categorical Data

Section 12.2 Contingency Tables and Association

▶ Problem 5 (pg. 648) Constructing a Frequency Marginal Distribution and Conditional Distribution

1. Open a new Excel worksheet and enter the data shown below. Begin by calculating the column totals. Click in cell **B4** where you will place the sum of the LT 18 column.

	A	B	C	D	E	F	G
1	Poverty Level	LT 18	18-44	45-64	GE 65	Row Total	
2	Below	7571	3693	1718	1025		
3	Above	9694	4745	2282	2258		
4	Column Total						

2. At the top of the screen, click the AutoSum button Σ. You will see =SUM(B2:B3) in cell B4. Press [**Enter**].

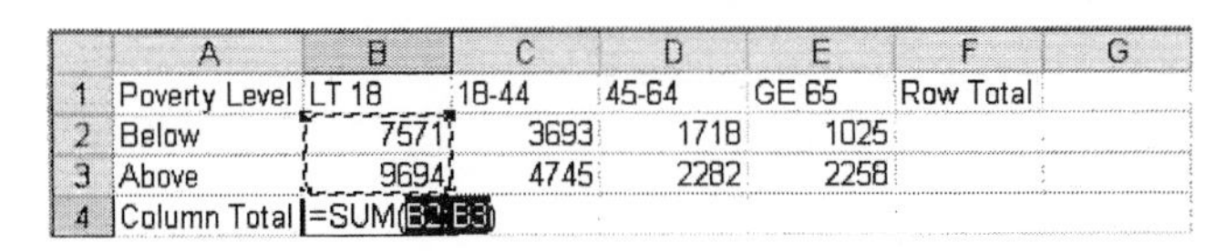

	A	B	C	D	E	F	G
1	Poverty Level	LT 18	18-44	45-64	GE 65	Row Total	
2	Below	7571	3693	1718	1025		
3	Above	9694	4745	2282	2258		
4	Column Total	=SUM(B2:B3)					

3. Copy the SUM function in cell B4 to cells C4 through E4. Then click in cell **F2** where you will place the sum of the Below row.

	A	B	C	D	E	F	G
1	Poverty Level	LT 18	18-44	45-64	GE 65	Row Total	
2	Below	7571	3693	1718	1025		
3	Above	9694	4745	2282	2258		
4	Column Total	17265	8438	4000	3283		

4. At the top of the screen, click the AutoSum button Σ. You will see =SUM(B2:E2). Press [**Enter**].

	A	B	C	D	E	F	G	H
1	Poverty Level	LT 18	18-44	45-64	GE 65	Row Total		
2	Below	7571	3693	1718	1025	=SUM(B2:E2)		
3	Above	9694	4745	2282	2258	SUM(number1, [number2], ...)		
4	Column Total	17265	8438	4000	3283			

5. Copy the SUM function in cell F2 to cells F3 through F4. Then click in cell **B5** where you will place the relative frequency for the LT 18 column.

	A	B	C	D	E	F	G
1	Poverty Level	LT 18	18-44	45-64	GE 65	Row Total	
2	Below	7571	3693	1718	1025	14007	
3	Above	9694	4745	2282	2258	18979	
4	Column Total	17265	8438	4000	3283	32986	
5							

6. Key in the formula shown below to calculate the relative marginal frequency. Note that the dollar signs are necessary here. You want to make the F4 reference absolute so that it will not change when it is copied. Press [**Enter**].

	A	B	C	D	E	F	G
1	Poverty Level	LT 18	18-44	45-64	GE 65	Row Total	
2	Below	7571	3693	1718	1025	14007	
3	Above	9694	4745	2282	2258	18979	
4	Column Total	17265	8438	4000	3283	32986	
5		=B4/$F$4					

7. Copy the formula in cell B5 to cells C5 through E5. Then click in cell **G2** where you will place the relative frequency for the Below row.

	A	B	C	D	E	F	G
1	Poverty Level	LT 18	18-44	45-64	GE 65	Row Total	
2	Below	7571	3693	1718	1025	14007	
3	Above	9694	4745	2282	2258	18979	
4	Column Total	17265	8438	4000	3283	32986	
5		0.523404	0.255805	0.121264	0.099527		

8. Key in the formula shown below to calculate the relative marginal frequency. Press [**Enter**].

	A	B	C	D	E	F	G
1	Poverty Level	LT 18	18-44	45-64	GE 65	Row Total	
2	Below	7571	3693	1718	1025	14007	=F2/$F$4
3	Above	9694	4745	2282	2258	18979	
4	Column Total	17265	8438	4000	3283	32986	
5		0.523404	0.255805	0.121264	0.099527		

9. Copy the formula in cell G2 to cell G3.

	A	B	C	D	E	F	G
1	Poverty Level	LT 18	18-44	45-64	GE 65	Row Total	
2	Below	8550	4356	1520	958	15384	0.535879
3	Above	5884	3741	1756	1943	13324	0.464121
4	Column Total	14434	8097	3276	2901	28708	
5		0.502787	0.282047	0.114115	0.101052		

10. You now will construct a conditional distribution by poverty level. Copy cells **A1:E3** to an area two rows below the previously constructed table.

	A	B	C	D	E	F	G
1	Poverty Level	LT 18	18-44	45-64	GE 65	Row Total	
2	Below	7571	3693	1718	1025	14007	0.424635
3	Above	9694	4745	2282	2258	18979	0.575365
4	Column Total	17265	8438	4000	3283	32986	
5		0.523404	0.255805	0.121264	0.099527		
6							
7							
8	Poverty Level	LT 18	18-44	45-64	GE 65		
9	Below	7571	3693	1718	1025		
10	Above	9694	4745	2282	2258		

11. Begin with the conditional probabilities in the Below row. Click in cell **B9** where you will place the LT 18 probability. Key in the formula shown below. Press [**Enter**].

8	Poverty Level	LT 18	18-44	45-64	GE 65
9	Below	=B2/$F$2	3693	1718	1025
10	Above	9694	4745	2282	2258

12. Copy the formula in cell B9 to cells C9 through E9. Then click in cell **B10** where you will enter a formula to calculate the conditional probability for LT 18 in the Above row.

8	Poverty Level	LT 18	18-44	45-64	GE 65
9	Below	0.540515	0.263654	0.122653	0.073178
10	Above	9694	4745	2282	2258
11					

13. Key in the formula shown below. Press [**Enter**].

8	Poverty Level	LT 18	18-44	45-64	GE 65
9	Below	0.540515	0.263654	0.122653	0.073178
10	Above	=B3/$F$3	4745	2282	2258

14. Copy the formula in cell B10 to cells C10 through E10. The completed conditional probability distribution is shown below.

8	Poverty Level	LT 18	18-44	45-64	GE 65
9	Below	0.540515	0.263654	0.122653	0.073178
10	Above	0.510775	0.250013	0.120238	0.118974

15. You will now draw a bar graph of this conditional distribution. At the top of the screen, click **Insert** and select **Chart**.

16. Under Chart type, select **Column**. Under Chart sub-type, select the leftmost diagram in the top row. Click **Next>**.

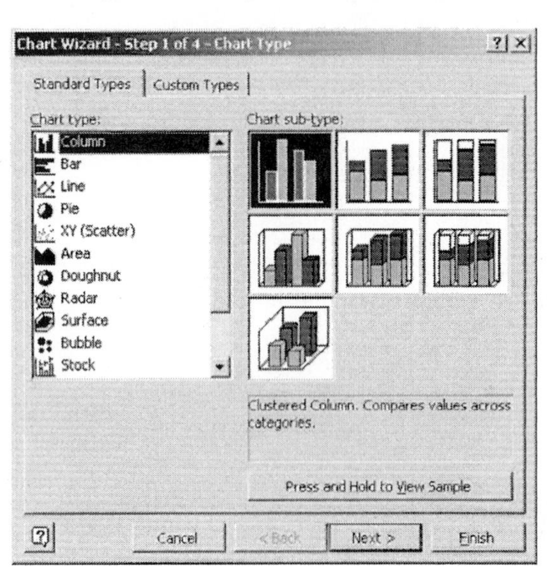

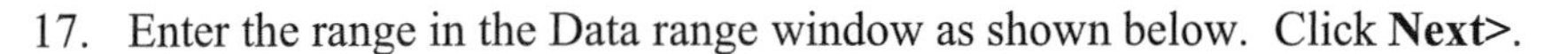

17. Enter the range in the Data range window as shown below. Click **Next>**.

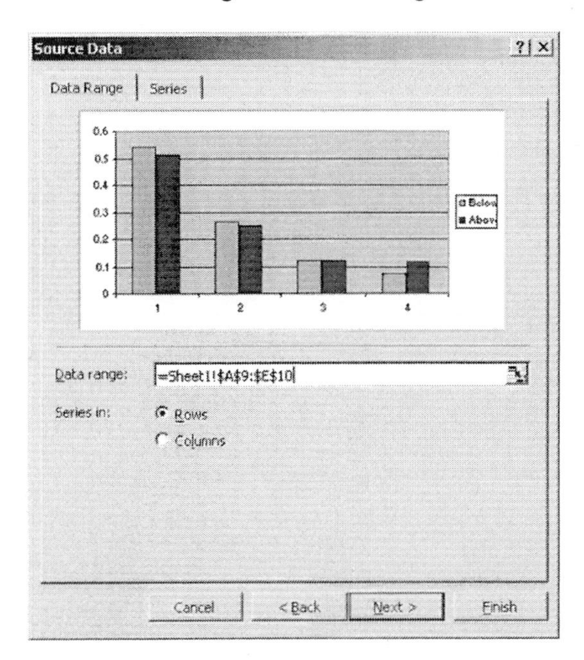

18. Click the **Titles** tab at the top of the Chart Options dialog box. For the Chart title, enter **Conditional Distribution by Poverty Level**. For the Category (X) axis, enter **Age**. For the Value (Y) axis, enter **Relative Frequency**. Click **Next>**.

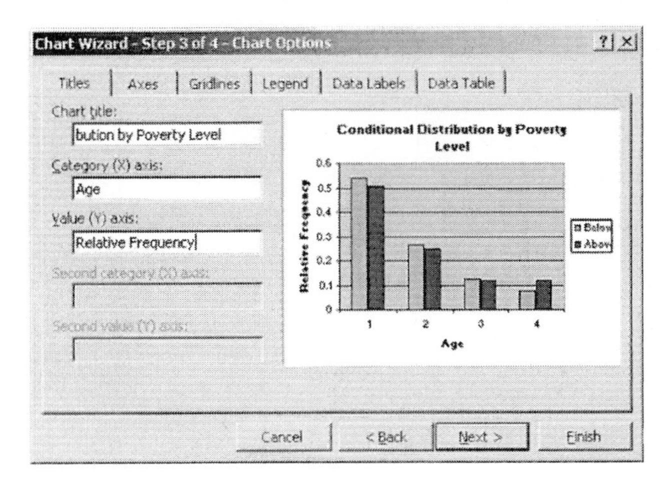

19. The Chart Location dialog box presents two options for placement of the chart. Select **As new sheet**. Click **Finish**.

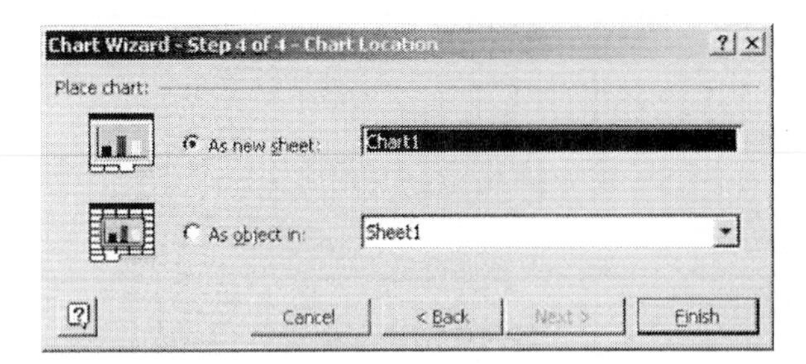

The completed chart is shown below.

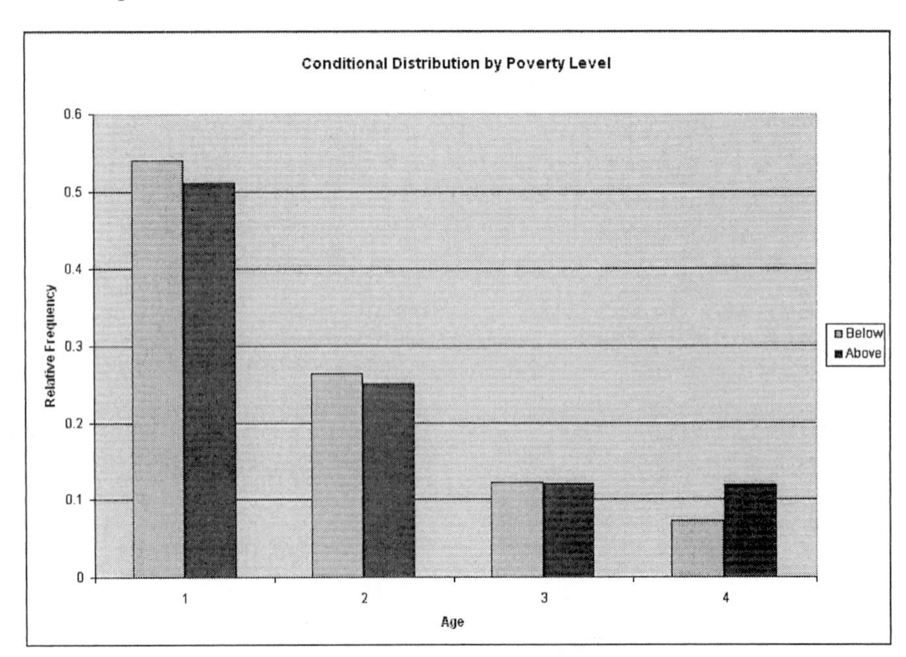

Section 12.3 Tests for Independence and the Homogeneity of Proportions

▶ Problem 3 (pg. 662) — Computing the Value of the Chi-Square Test

If the PHStat add-in has not been loaded, you will need to load it before continuing. Follow the instructions in Section GS 8.2.

1. Open a new Excel worksheet.

2. At the top of the screen, click **PHStat**. Select **Multiple-Sample Tests → Chi-Square Test**.

3. Complete the Chi-Square Test dialog box as shown below. Click **OK**.

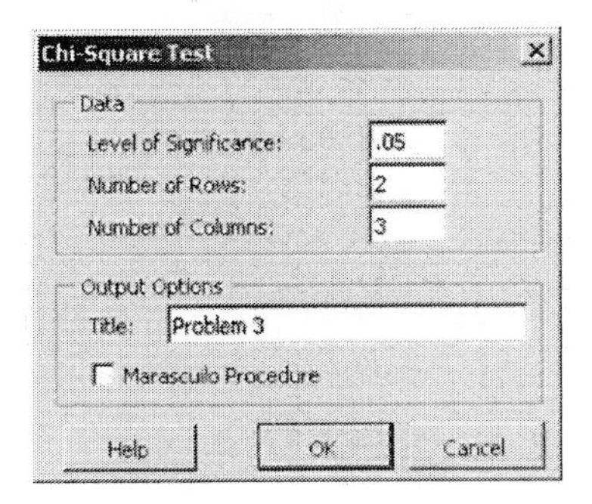

4. Enter observed frequencies in the top section of the worksheet as shown below.

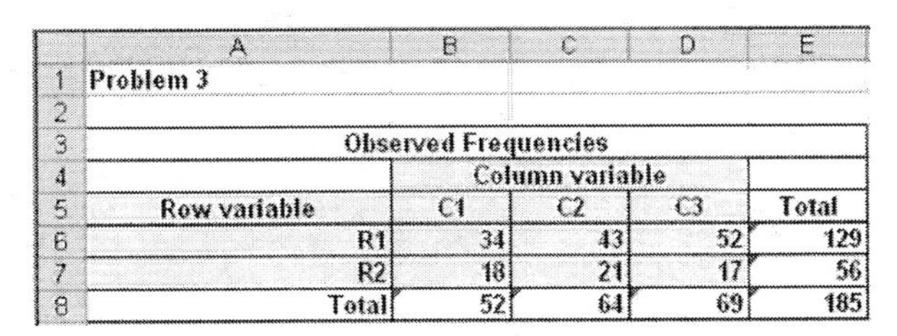

	A	B	C	D	E
1	Problem 3				
2					
3	Observed Frequencies				
4		Column variable			
5	Row variable	C1	C2	C3	Total
6	R1	34	43	52	129
7	R2	18	21	17	56
8	Total	52	64	69	185

5. Scroll down to see the results of the chi-square test.

	Results	
23	Results	
24	Critical Value	5.991465
25	Chi-Square Test Statistic	1.698215
26	*p*-Value	0.427797
27	Do not reject the null hypothesis	
28		
29	*Expected frequency assumption*	
30	*is met.*	

Comparing Three or More Means

CHAPTER 13

Section 13.1 Comparing Three or More Means (One-Way Analysis of Variance)

► Problem 11 (pg. 686) **Testing Whether the Mean Numbers of Plants Are Equal**

1. Open worksheet "13_1_11" in the Chapter 13 folder.

	A	B	C	D	E	F	G
1	Plot Type	C2	C3	C4	C5	C6	C7
2	Sludge Plo	25	27	33	30	28	27
3	Spring Dis	32	30	33	35	34	34
4	No Till	30	26	29	32	25	29

2. At the top of the screen, click **Tools** and select **Data Analysis**.

If Data Analysis does not appear as a choice in the Tools menu, you will need to load the Microsoft Excel Analysis ToolPak add-in. Follow the procedure in Section GS 8.1 before continuing.

3. Select **Anova: Single Factor** and click **OK**.

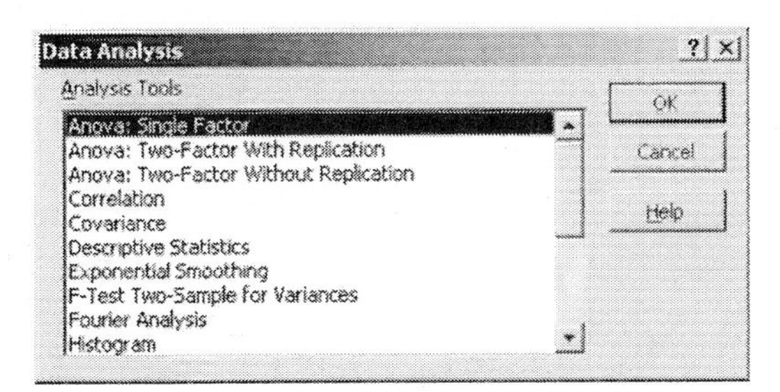

4. Complete the Anova: Single Factor dialog box as shown below. Be sure to select Grouped By Rows. Click **OK**.

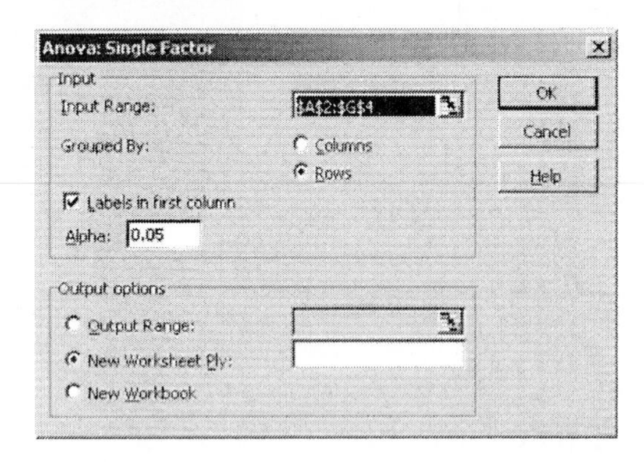

The output is displayed in a new worksheet.

	A	B	C	D	E	F	G
1	Anova: Single Factor						
2							
3	SUMMARY						
4	*Groups*	*Count*	*Sum*	*Average*	*Variance*		
5	Sludge Plot	6	170	28.33333	7.866667		
6	Spring Disk	6	198	33	3.2		
7	No Till	6	171	28.5	6.7		
8							
9							
10	ANOVA						
11	*Source of Variation*	*SS*	*df*	*MS*	*F*	*P-value*	*F crit*
12	Between Groups	84.11111	2	42.05556	7.101313	0.006761	3.68232
13	Within Groups	88.83333	15	5.922222			
14							
15	Total	172.9444	17				

◀

▶ Problem 17 (pg. 689) Testing Whether the Chest Compression Means Are Equal

1. Open worksheet "13_1_17" in the Chapter 13 folder.

2. When using Excel to carry out a one-way analysis of variance, the columns of data must be right next to each other. Copy the Chest Compression data to an area in the worksheet that is two rows below the complete data set. Also copy the labels to row 10 as shown below.

	A	B	C	D	E	F	G
1	Large Fan	Chest Con	Passenge	Chest Con	Midsize U	Compression	
2	Hyundai X	33	Toyota Si	29	Honda Pil	31	
3	Ford Taur	28	Honda Od	28	Toyota 4F	36	
4	Buick LeS	28	Ford Free	27	Mitsubishi	35	
5	Chevrolet	26	Mazda MF	30	Nissan Mu	29	
6	Chrysler 3	34	Chevrolet	26	Ford Explc	29	
7	Pontiac Gr	34	Nissan Qu	33	Jeep Libe	36	
8	Toyota Av	31	Kia Sedon	21	Buick Ren	29	
9							
10	Large Fan	Passenge	Midsize Utility Vehicles				
11	33	29	31				
12	28	28	36				
13	28	27	35				
14	26	30	29				
15	34	26	29				
16	34	33	36				
17	31	21	29				

3. At the top of the screen, click **Tools** and select **Data Analysis.**

If Data Analysis does not appear as a choice in the Tools menu, you will need to load the Microsoft Excel Analysis ToolPak add-in. Follow the procedure in Section GS 8.1 before continuing.

4. Select **Anova: Single Factor** and click **OK**.

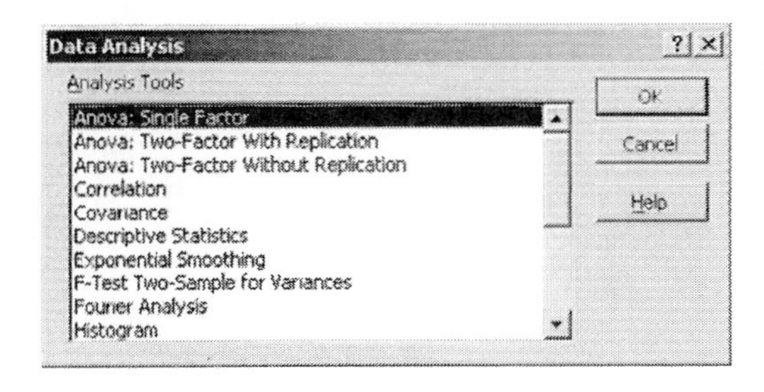

5. Complete the Anova: Single Factor dialog box as shown below. Click **OK**.

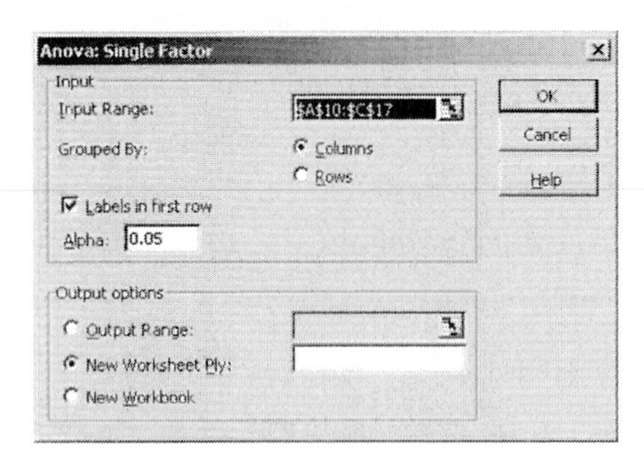

The output is displayed in a new worksheet.

	A	B	C	D	E	F	G
1	Anova: Single Factor						
2							
3	SUMMARY						
4	*Groups*	*Count*	*Sum*	*Average*	*Variance*		
5	Large Family Cars	7	214	30.57143	10.61905		
6	Passenger Vans	7	194	27.71429	13.90476		
7	Midsize Utility Vehicles	7	225	32.14286	11.47619		
8							
9							
10	ANOVA						
11	*Source of Variation*	*SS*	*df*	*MS*	*F*	*P-value*	*F crit*
12	Between Groups	70.57143	2	35.28571	2.940476	0.078521	3.554557
13	Within Groups	216	18	12			
14							
15	Total	286.5714	20				

◀

Section 13.4 Two-Way Analysis of Variance

► Problem 19 (pg. 725) **Testing Whether the Strength and Slump Means Are Equal**

1. Open worksheet 13_4_19 in the Chapter 13 folder. The first two lines are shown below.

	A	B	C	D	E	F	G
1	C1	Mixture 67-	Mixture 67-	Mixture 67-0-353			
2	3.75	3960	4815	4595			

2. At the top of the screen, click **Tools** and select **Data Analysis**.

If Data Analysis does not appear as a choice in the Tools menu, you will need to load the Microsoft Excel Analysis ToolPak add-in. Follow the procedure in Section GS 8.1 before continuing.

3. Select **Anova: Two-Factor With Replication** and click **OK**.

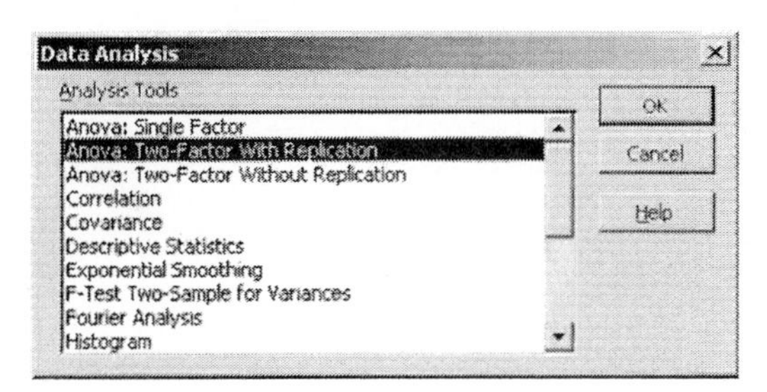

Note that this Excel procedure requires that n's are equal. The data for this problem meet this requirement because there are three observations in each level of the slump variable.

4. Complete the Anova dialog box as shown below. Click **OK**.

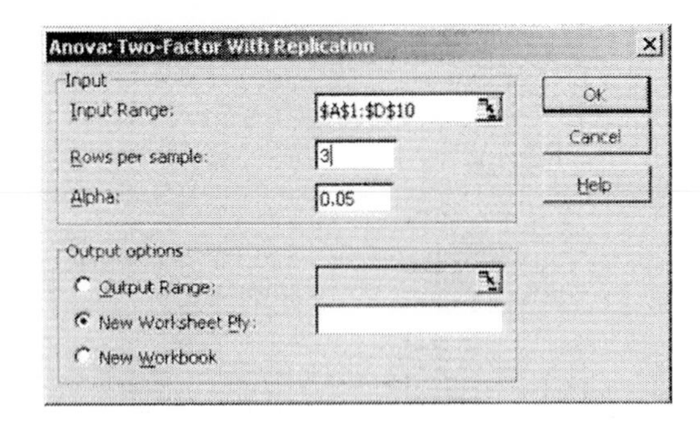

You will have to scroll down to see all of the output. The top of the worksheet displays summary for each slump level and for the total across levels.

	A	B	C	D	E	F	G
1	Anova: Two-Factor With Replication						
2							
3	SUMMARY	Mixture 67-	Mixture 67-	Mixture 67-	Total		
4	3.75						
5	Count	3	3	3	9		
6	Sum	11410	13595	13325	38330		
7	Average	3803.333	4531.667	4441.667	4258.889		
8	Variance	96808.33	102233.3	66033.33	184523.6		
9							
10	4						
11	Count	3	3	3	9		
12	Sum	11135	12790	11640	35565		
13	Average	3711.667	4263.333	3880	3951.667		
14	Variance	88508.33	62258.33	14275	101212.5		
15							
16	5						
17	Count	3	3	3	9		
18	Sum	10420	13110	10925	34455		
19	Average	3473.333	4370	3641.667	3828.333		
20	Variance	55833.33	97975	71458.33	226668.8		
21							
22	Total						
23	Count	9	9	9			
24	Sum	32965	39495	35890			
25	Average	3662.778	4388.333	3987.778			
26	Variance	82050.69	79306.25	164475.7			

The analysis of variance table is displayed in the bottom portion of the worksheet.

29	ANOVA						
30	*Source of Variation*	*SS*	*df*	*MS*	*F*	*P-value*	*F crit*
31	Sample	884924.1	2	442462	6.076075	0.009629	3.554557
32	Columns	2377502	2	1188751	16.32443	9.04E-05	3.554557
33	Interaction	410970.4	4	102742.6	1.410905	0.270456	2.927744
34	Within	1310767	18	72820.37			
35							
36	Total	4984163	26				

Inference on the Least-Squares Regression Model and Multiple Regression

Section 14.1 Testing the Significance of the Least-Squares Regression Model

▶ Problem 13 (pg. 748)

Finding the Least Squares Regression Equation

1. Open worksheet "14_1_13" in the Chapter 14 folder. The first few lines are shown below.

	A	B	C	D	E	F	G
1	Height (inc	Head Circumference (inches)					
2	27.75	17.5					
3	24.5	17.1					

2. At the top of the screen, click **Tools** and select **Data Analysis**.

If Data Analysis does not appear as a choice in the Tools menu, you will need to load the Microsoft Excel Analysis ToolPak add-in. Follow the procedure in Section GS 8.1 before continuing.

3. In the Data Analysis dialog box, select **Regression** and click **OK**.

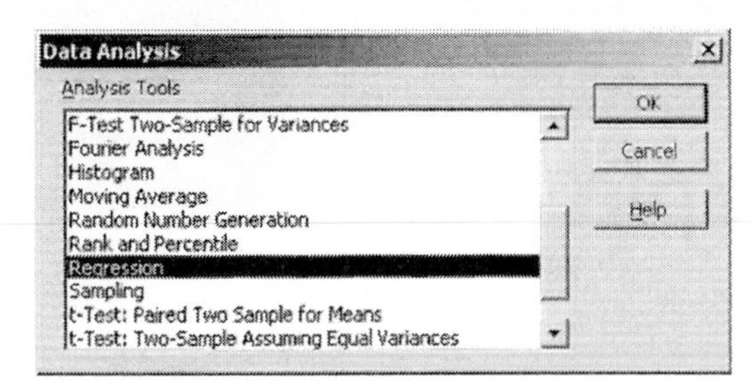

4. Complete the Regression dialog box as shown below. Click **OK**.

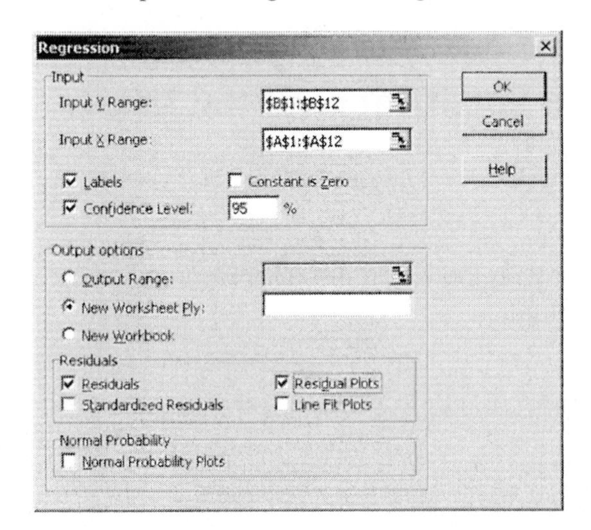

Because you requested many output options, you will need to scroll down and to the right to view all the output. The correlation is 0.9112. The intercept is 12.4932. The slope is 0.1827. The standard error of the slope is 0.0276. The F test is statistically significant with p = .000095. The 95% confidence interval about the slope is 0.1204 to 0.2451.

	A	B	C	D	E	F	G
1	SUMMARY OUTPUT						
2							
3	*Regression Statistics*						
4	Multiple R	0.911073					
5	R Square	0.830054					
6	Adjusted F	0.811171					
7	Standard E	0.095384					
8	Observatio	11					
9							
10	ANOVA						
11		*df*	*SS*	*MS*	*F*	*ignificance F*	
12	Regressior	1	0.399935	0.399935	43.95785	9.59E-05	
13	Residual	9	0.081883	0.009098			
14	Total	10	0.481818				
15							
16		*Coefficients*	*tandard Err*	*t Stat*	*P-value*	*Lower 95%*	*Upper 95%*
17	Intercept	12.49317	0.729685	17.12132	3.56E-08	10.84251	14.14383
18	Height (inc	0.182732	0.027561	6.630072	9.59E-05	0.120385	0.24508

The residuals are shown at the bottom of the worksheet.

	A	B	C
22	RESIDUAL OUTPUT		
23			
24	*Observation*	*d Circumfe*	*Residuals*
25	1	17.56399	-0.06399
26	2	16.97011	0.129886
27	3	17.15285	-0.05285
28	4	17.24421	0.055787
29	5	17.06148	-0.16148
30	6	17.56399	0.036006
31	7	17.33558	-0.03558
32	8	17.42694	0.073055
33	9	17.38126	-0.08126
34	10	17.38126	0.118738
35	11	17.51831	-0.01831

The residual plot is shown at the right of the worksheet. You will want to make this plot larger.

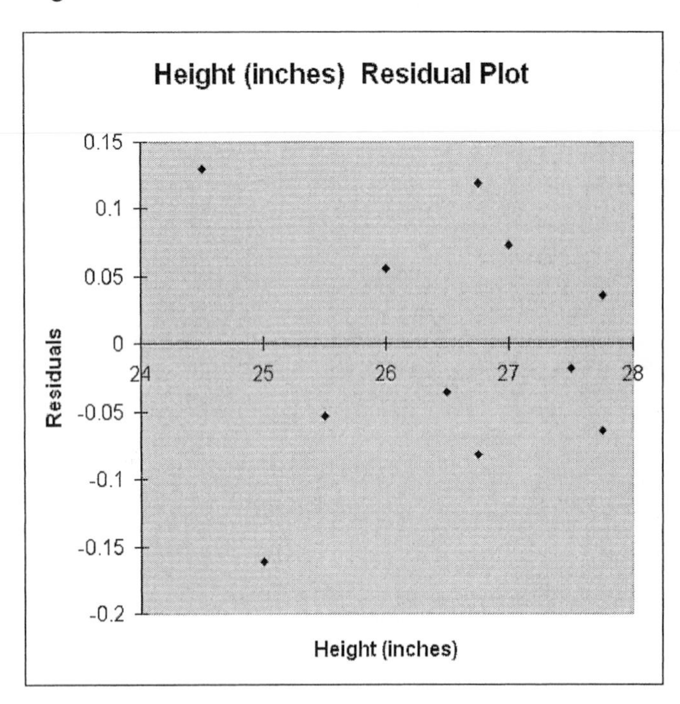

◀

▶ Problem 17 (pg. 749)

Testing the Claim That a Linear Relation Exists

1. Open worksheet "14_1_17" in the Chapter 14 folder. The first few lines are shown below.

	A	B	C	D	E	F	G
1	Month	Rate of Re	Rate of Return in United Tech				
2	Aug-04	0.23	1.21				
3	Sept-04	0.94	-0.57				

2. At the top of the screen, click **Tools** and select **Data Analysis**.

If Data Analysis does not appear as a choice in the Tools menu, you will need to load the Microsoft Excel Analysis ToolPak add-in. Follow the procedure in Section GS 8.1 before continuing.

---

3. In the Data Analysis dialog box, select **Regression** and click **OK**.

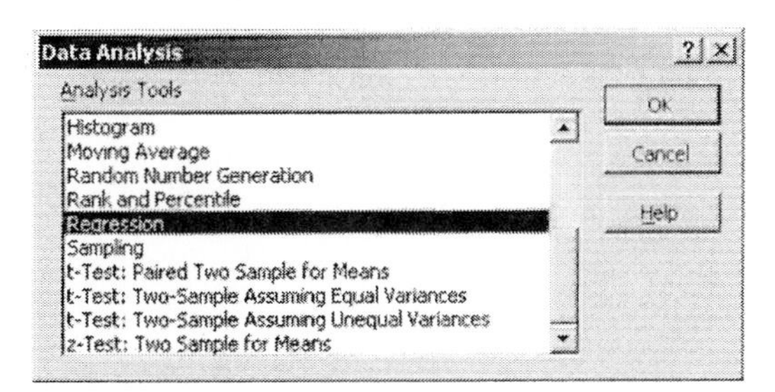

4. Complete the Regression dialog box as shown below. Click **OK**.

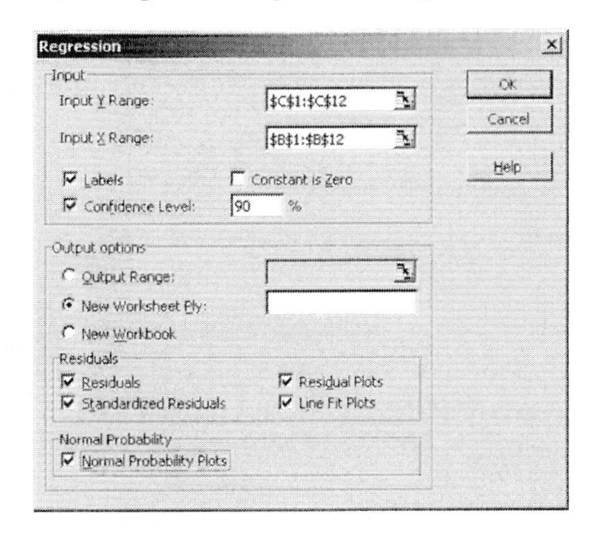

Because you requested many output options, you will need to scroll down and to the right to view all the output. The intercept is 0.5975. The slope is 0.9764. The standard error of the slope is 0.3379. The lower bound of the 90% confidence interval about the slope is 0.3570. The upper bound is 1.5959.

16		Coefficients	Standard Error	t Stat	P-value	Lower 95%	Upper 95%	ower 90.0%	Jpper 90.0%
17	Intercept	0.5974944	0.76130489	0.784829	0.452716	-1.1247	2.319686	-0.798063	1.9930522
18	Rate of Return of S&P	0.97644807	0.337926247	2.88953	0.017898	0.212006	1.74089	0.356991	1.595905

The correlation (Multiple R) is 0.6937. The coefficient of determination (R Square) is 0.4812. The standard error is 2.3518.

1	SUMMARY OUTPUT	
2		
3	Regression Statistics	
4	Multiple R	0.69372156
5	R Square	0.4812496
6	Adjusted R Square	0.42361067
7	Standard Error	2.35176761
8	Observations	11

The p-value for the F test is 0.0179. The test is statistically significant with α set at 0.10.

10	ANOVA					
11		df	SS	MS	F	Significance F
12	Regression	1	46.1788657	46.1788657	8.34938432	0.017897963
13	Residual	9	49.777298	5.53081088		
14	Total	10	95.9561636			

◀

Section 14.2 Confidence and Prediction Intervals

► Problem 3 (pg. 757)

Constructing a 95% Confidence Interval About the Mean Value of y

If the PHStat add-in has not been loaded, you will need to load it before continuing. Follow the instructions in in Section GS 8.2.

1. Open a new Excel worksheet and enter the X and Y values shown below.

	A	B	C	D	E	F	G
1	X	Y					
2	3	4					
3	4	6					
4	5	7					
5	7	12					
6	8	14					

2. At the top of the screen, click **PHStat**.

3. Select **Regression** → **Simple Linear Regression**.

4. Complete the Simple Linear Regression dialog box as shown below. Click **OK**.

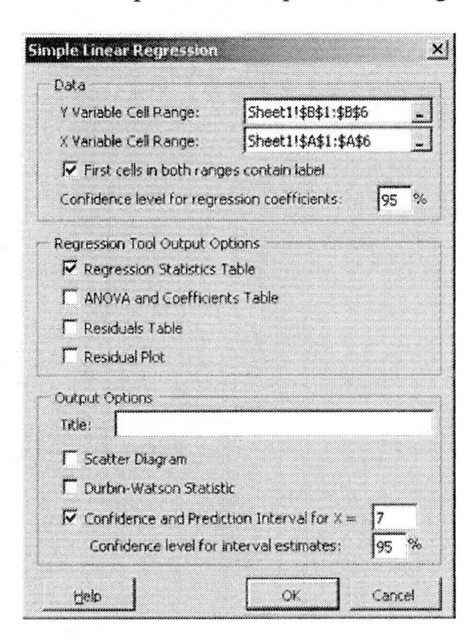

The predicted mean value of y is 11.8372. The 95% confidence interval about the mean value is 10.8722 to 12.8022.

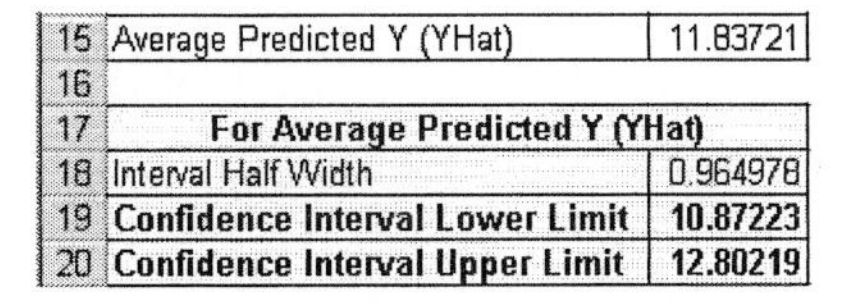

15	Average Predicted Y (YHat)	11.83721
16		
17	**For Average Predicted Y (YHat)**	
18	Interval Half Width	0.964978
19	**Confidence Interval Lower Limit**	**10.87223**
20	**Confidence Interval Upper Limit**	**12.80219**

The 95% prediction interval about the value of y for $x = 7$ is 9.9397 to 13.7347.

22	For Individual Response Y	
23	Interval Half Width	1.897518
24	**Prediction Interval Lower Limit**	**9.939691**
25	**Prediction Interval Upper Limit**	**13.73473**

Section 14.3 Multiple Regression

Building a Multiple Regression Model

1. Open worksheet "14_3_19" in the Chapter 14 folder. The first few lines are shown below.

	A	B	C	D	E	F	G
1	Slump (ind	7-day PSI	28-day PSI				
2	4.5	2330	4025				
3	4.25	2640	4535				

2. You will begin by constructing a correlation matrix. At the top of the screen, click **Tools** and select **Data Analysis**.

If Data Analysis does not appear as a choice in the Tools menu, you will need to load the Microsoft Excel Analysis ToolPak add-in. Follow the procedure in Section GS 8.1 before continuing.

3. Select **Correlation** and click **OK**.

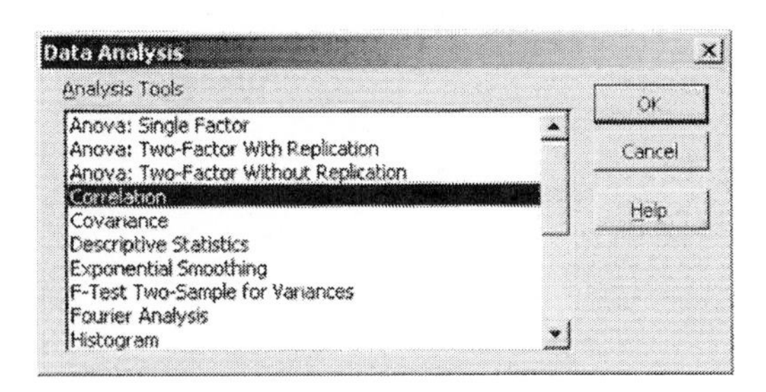

4. Complete the Correlation dialog box as shown below. Click **OK**.

The correlation matrix is displayed in a new worksheet.

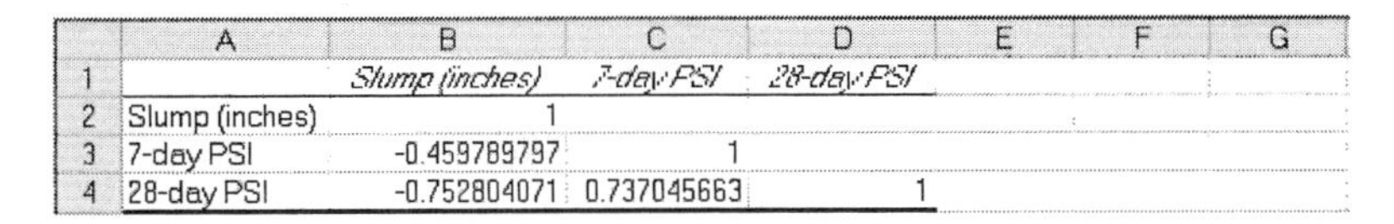

	A	B	C	D	E	F	G
1		*Slump (inches)*	*7-day PSI*	*28-day PSI*			
2	Slump (inches)	1					
3	7-day PSI	-0.459789797	1				
4	28-day PSI	-0.752804071	0.737045663	1			

5. Now you will build the multiple regression model. Return to sheet 1 where the data are displayed by clicking on the **sheet1** tab at the bottom of the screen.

6. At the top of the screen, click **Tools** and select **Data Analysis**.

7. In the Data Analysis dialog box, select **Regression** and click **OK**.

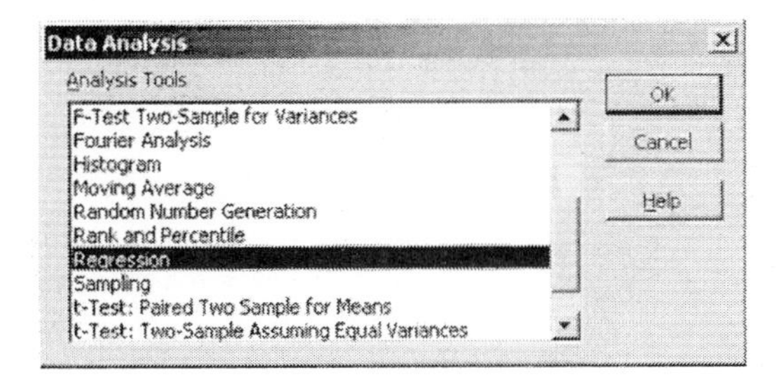

8. Complete the Regression dialog box as shown below. Click **OK**.

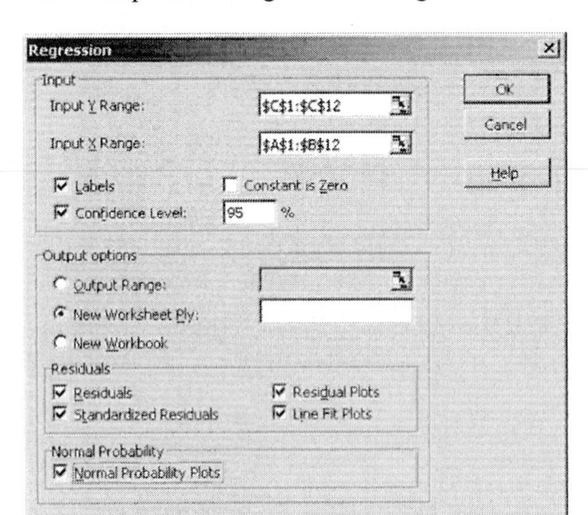

Because you requested many output options, you will need to scroll down and to the right to view all the output. The regression coefficients are displayed in column labeled *Coefficients*. The regression equation is $\hat{y} = 3890.55 - 295.93x_1 + 0.55x_2$.

16		*Coefficients*
17	Intercept	3890.5497
18	Slump (inches)	-295.925512
19	7-day PSI	0.55228367

R^2 and adjusted R^2 are displayed in a section labeled *Regression Statistics*. R^2 is equal to 0.7605 and adjusted R^2 is equal to 0.7006.

3	*Regression Statistics*	
4	Multiple R	0.87206317
5	R Square	0.76049418
6	Adjusted R Square	0.70061772
7	Standard Error	276.504767
8	Observations	11

To test hypotheses regarding β_1 and β_2, refer to the t Stat and p-value output to the right of the coefficients. β_1 is the slope coefficient for Slump. The p-value is less than 0.05, so the t test is statistically significant. β_2 is the slope coefficient for 7-day PSI. The p-value is less than 0.05, so that t test is also statistically significant.

16		Coefficients	Standard Error	t Stat	P-value
17	Intercept	3890.5497	828.7504305	4.694477	0.001553
18	Slump (inches)	-295.925512	109.8519319	-2.69386	0.027333
19	7-day PSI	0.55228367	0.217080363	2.544144	0.034488

9. To construct the 95% confidence and prediction intervals requested in part j, you will be using a procedure available in PHStat. Return to the sheet where the data are displayed by clicking on the **Sheet1** tab at the bottom of the screen.

If the PHStat add-in has not been loaded, you will need to load it before continuing. Follow the instructions in Section GS 8.2.

---

10. At the top of the screen, click **PHStat**.

11. Select **Regression → Multiple Regression**.

12. Complete the Multiple Regression dialog box as shown below. Click **OK**.

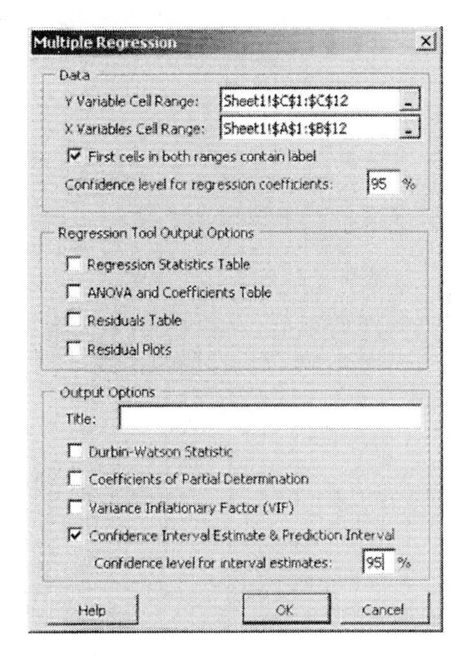

13. Type in **3.5** for slump and press [**Enter**]. Type in **2450** for 7-day psi and press [**Enter**]. The confidence interval limits are displayed at the bottom of the output.

	A	B	C	D	E	F
1	**Confidence and Prediction Estimate Intervals**					
2						
3	**Data**					
4	**Confidence Level**	**95%**				
5		1				
6	**Slump (inches) given value**	**3.5**				
7	**7-day PSI given value**	**2450**				
8						
9	X'X	11				
10		43				
11		27040				
12						
13	Inverse of X'X	8.983432				
14		-0.96953				
15		-0.00208				
16						
17	X'G times Inverse of X'X	0.504519	-0.06574	-6.4E-05		
18						
19	[X'G times Inverse of X'X] times XG	0.118325				
20	t Statistic	2.306004				
21	**Predicted Y (YHat)**	**4207.91**				
22						
23	**For Average Predicted Y (YHat)**					
24	**Interval Half Width**	**219.332**				
25	**Confidence Interval Lower Limit**	**3988.57**				
26	**Confidence Interval Upper Limit**	**4427.24**				
27						
28	**For Individual Response Y**					
29	**Interval Half Width**	**674.29**				
30	**Prediction Interval Lower Limit**	**3533.62**				
31	**Prediction Interval Upper Limit**	**4882.2**				

PHStat2 User Note:

Enter the values for the given X's in the cell range B6:B7.
(You can interactively change these values at any time.)

To delete this note:
Select this note and then select Edit | Cut.

Nonparametric Statistics

CHAPTER 15

Section 15.4 Inferences about the Difference between Two Medians: Dependent Samples

▶ Problem 13 (pg. 15-24) — Testing the Claim That the Median Reaction Times Are the Same

If the PHStat add-in has not been loaded, you will need to load it before continuing. Follow the instructions in Section GS 8.2.

1. Enter the reaction time data as shown below.

	A	B	C	D	E	F	G
1	Blue	Red					
2	0.582	0.408					
3	0.481	0.407					
4	0.841	0.542					
5	0.267	0.402					
6	0.685	0.456					
7	0.45	0.533					

2. At the top of the screen, click **PHStat**. Select Two-Sample Tests → **Wilcoxon Rank Sum Test**.

3. Complete the Wilcoxon Rank Sum Test dialog box as shown below. **Click OK**.

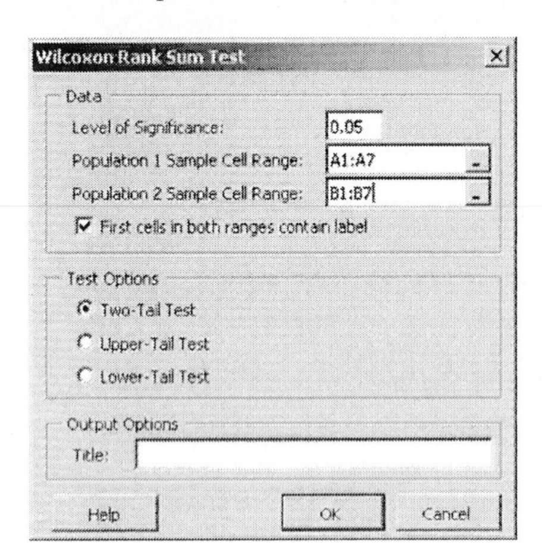

The test results are displayed in a worksheet named Calculations.

	A	B
4	**Level of Significance**	**0.05**
5		
6	Population 1 Sample	
7	Sample Size	6
8	Sum of Ranks	46
9	Population 2 Sample	
10	Sample Size	6
11	Sum of Ranks	32
12		
13	Intermediate Calculations	
14	Total Sample Size n	12
15	*T*1 Test Statistic	46
16	*T*1 Mean	39
17	Standard Error of *T*1	6.244998
18	**Z Test Statistic**	**1.120897**
19		
20	**Two-Tail Test**	
21	**Lower Critical Value**	**-1.95996**
22	**Upper Critical Value**	**1.959964**
23	***p*-Value**	**0.262332**
24	**Do not reject the null hypothesis**	

The data sorted by rank are displayed in a worksheet named Sorted.

	A	B	C
1	Sample	Value	Rank
2	Blue	0.267	1
3	Red	0.402	2
4	Red	0.407	3
5	Red	0.408	4
6	Blue	0.45	5
7	Red	0.456	6
8	Blue	0.481	7
9	Red	0.533	8
10	Red	0.542	9
11	Blue	0.582	10
12	Blue	0.685	11
13	Blue	0.841	12

◀

Section 15.7 Kruskal-Wallis Test of One-Way Analysis of Variance

▶ Problem 9 (pg. 15-49) — Testing the Claim That the Distribution for Each Stimulus Is the Same, $\alpha = 0.01$

If the PHStat add-in has not been loaded, you will need to load it before continuing. Follow the instructions in Section GS 8.2. ***Note that PHStat's Kruskal-Wallis procedure can handle a maximum of four groups. If your problem has more than four groups, the Kruskal-Wallis output will not be accurate.***

4. Open worksheet "15_7_9" in the Chapter 15 folder. The first few lines are shown below.

	A	B	C	D	E	F	G
1	Simple	Go/No Go	Choice				
2	0.43	0.588	0.561				
3	0.498	0.375	0.498				

5. At the top of the screen, click **PHStat**. Select **Multiple-Sample Tests → Kruskal-Wallis Rank Test.**

6. Complete the Kruskal-Wallis Rank Test dialog box as shown below. Click **OK**.

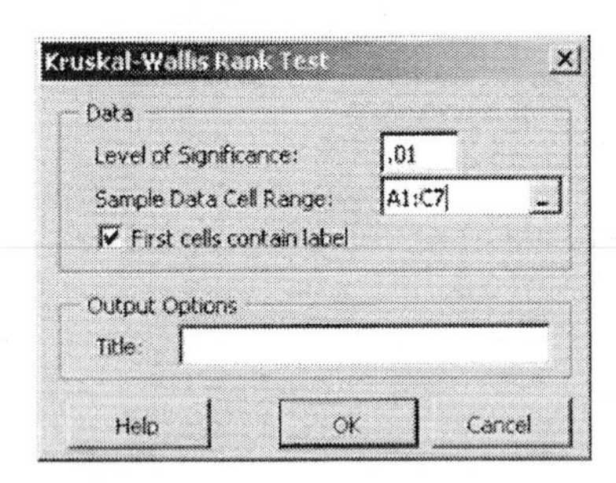

The output is displayed in a worksheet named Kruskal.

	A	B	C	D	E	F	G
1	**Kruskal-Wallis Rank Test for Differences in Medians**						
2							
3	**Data**						
4	**Level of Significance**	**0.01**		Group	Sample Size	Sum of Ranks	Mean Ranks
5				1	6	37.5	6.25
6	Intermediate Calculations			2	6	54	9
7	Sum of Squared Ranks/Sample Size	1773.75		3	6	79.5	13.25
8	Sum of Sample Sizes	18					
9	Number of Groups	3					
10							
11	**Test Result**						
12	**H Test Statistic**	5.236842					
13	**Critical Value**	9.21034					
14	**p-Value**	0.072918					
15	**Do not reject the null hypothesis**						

SPSS Manual

Brent Timothy

Statistics

Informed Decisions Using Data

SECOND EDITION

Michael Sullivan, III

Table of Contents

Getting Started with SPSS

►Opening SPSS

When you first open SPSS, the first screen you should see is the "What would you like to do?" window. This is asking for how you would like to enter the data.

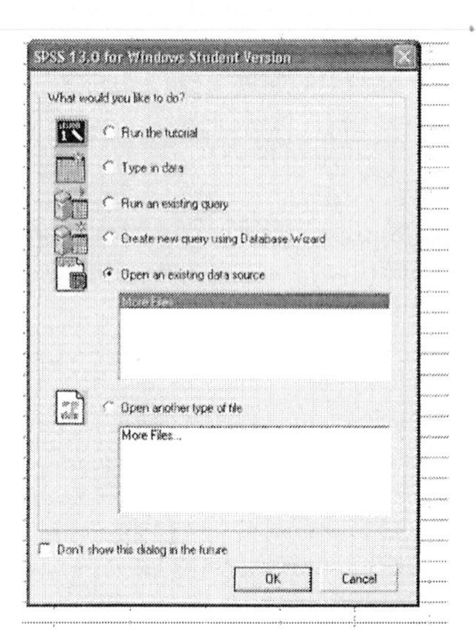

The first option, **Run the tutorial** will take you to a SPSS tutorial. This is a nice little guide on the basics of SPSS. It will walk you through entering data, doing basic analysis, and working with SPSS output. It also gives an introduction to using the SPSS help function.

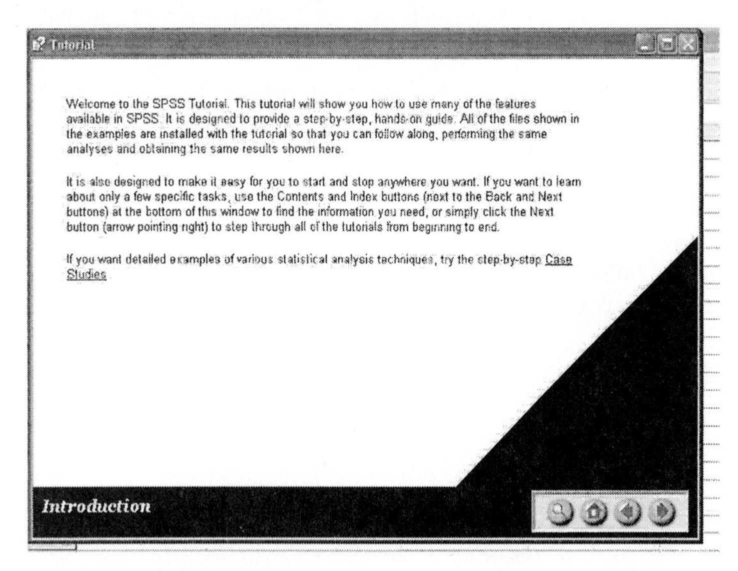

The next option is to **Type in Data**. If you have data from a book, or your own research, or some other place that is not currently on the computer, this is the option you would want to choose.

Run an Existing Query will help you bring data in from another database, which you have used before. When you run a query into an Excel sheet, or another type of database, you can save the commands so that you don't have to start from scratch each time you want to read in the information. This is very useful when you are doing the same type of analysis over and over again on data that gets updated regularly.

Create a new query using Database Wizard walks you step by step in reading in data that exists on the computer in a non-SPSS format.

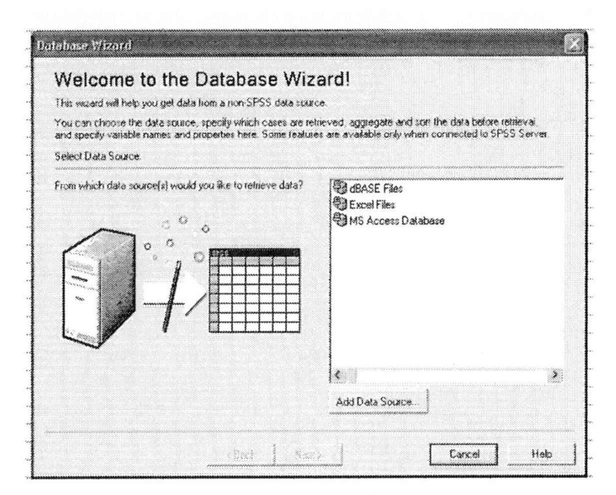

These formats include dBase Files, Excel Files, MS Access Databases, and any other database that you have a database (ODBC) driver for. These can be added through the Add Data Source button. This is an easy way to change from Excel format into SPSS.

Open an existing data source, which is an SPSS data set. The recently used data sets will appear in the box for easy access to the data sets you were last utilizing.

Open another type of file will allow you to open SPSS scripts, which can be useful for randomly created data sets, or for specific programs you have saved.

All of these options are available in SPSS through the file menu, and you can ask to not see this menu by clicking the Do not show this dialog in the future box. This would take you directly into SPSS on future executions.

►Using SPSS Files

SPSS begins with two views, the Data View, and the Variable View. Data View is the default start window, and looks like a spreadsheet. Each column is a variable, and each row is an observation. SPSS is set up to accept survey data, and so it expects data for every row for each column with data.

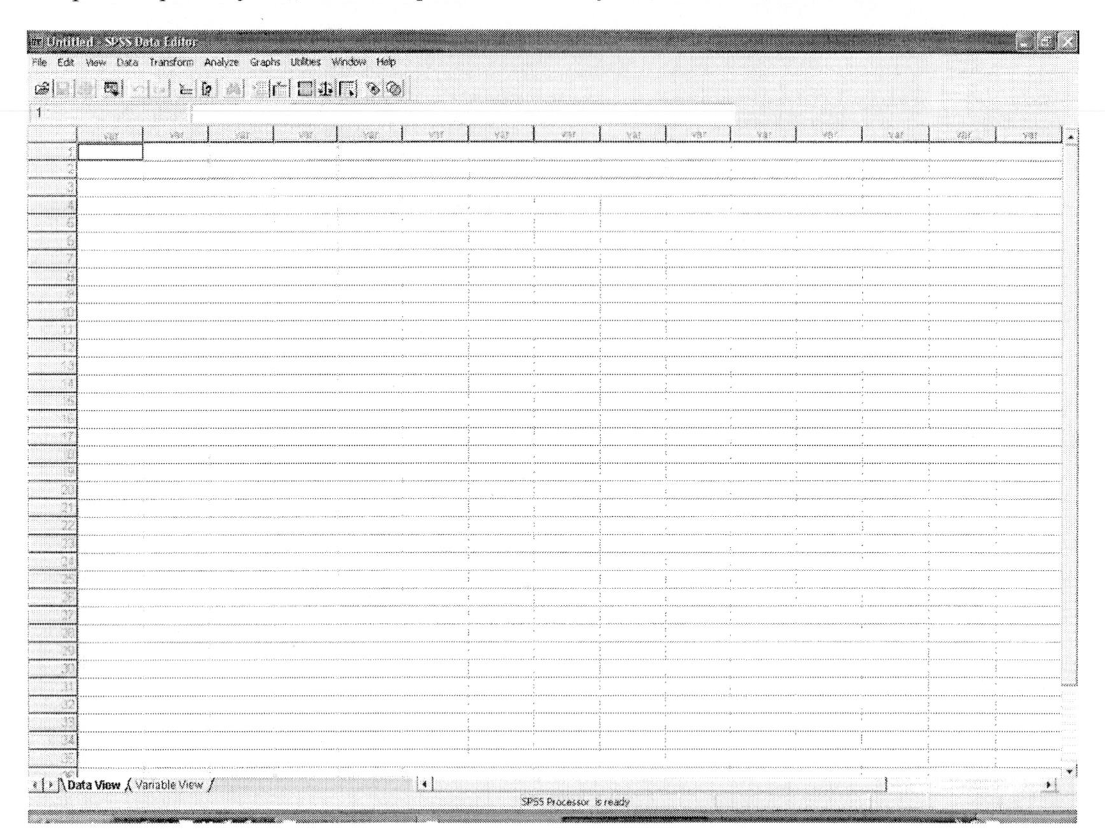

The Variable View window is accessed by clicking on the Variable View tab at the bottom of the Data View window. This is where you can add or change variable names, types, how they are viewed, etc.

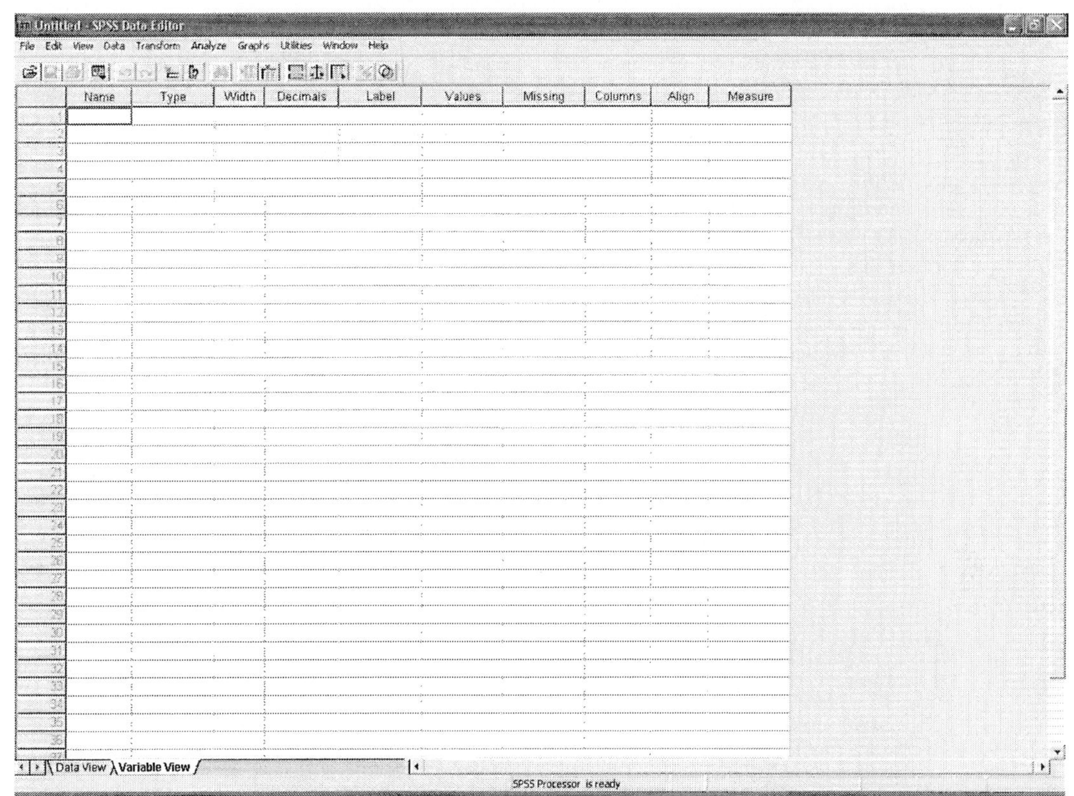

Name is the name of the variable. It cannot be more than 64 characters long, must start with a letter, and cannot use any other symbols than a period, @, #, _, and $. No spaces are allowed, and the name cannot end in a period. While this may seem constricting, this is just a variable name. **Labels** can be used to give better detail on what is contained in the variable. Spaces are allowed in labels, but labels cannot be longer than 256 characters long.

The **Type** of variable can be selected from Numeric (example: 10000.001), Comma (example: 10,000.001), Dot (European convention, example: 10.000,001), Scientific Notation (example: 1.00E+004), Date (example: 01-June-2005), Dollar (example: $10,000), Custom Currency, and string (text, example: Male) .

Width is the width of the variable (how many characters). This is most useful when changing String, but can be used for long numbers as well.

The **Decimals** column gives how many decimal places to show. This does not change the values, just how they are seen in the Data View sheet.

Values allow you to give labels to specific values, such as 0-male 1-female. To view the label instead of the number, go to **View**, and check the **Value Labels**.

Missing allows you to specify what characters, numbers, etc. denote a missing value.

Columns dictate how wide the column in the Data View sheet will be.

Align allows you to adjust the alignment of the values in the Data View Sheet (Right, Left, Center).

Measure is what type of variable you have. The choices are: Nominal (used for string, but can be used for numbers where the numbers just represent names), Ordinal (where order counts, but distance either varies or cannot be measured), and Scale (the normal numerical line). The type of analysis possible depends on which you chose.

►Entering Data into the Data View

Although it is not necessary to define the variables before typing in data, it is suggested to do so. Define the variables in the Variable View window, providing at least a name, and what Type of variable you wish to use. If you define a variable to be numerical, only numbers will be allowed in that column. If you define a variable to be a string, then you can type any text into the cells, provided they do not go past the length defined.

If you begin by typing in a cell in the Data View window instead, an automatic name will be created (usually VAR0001 if it is the first variable) and what you type will determine the type of variable. Putting in any numerical value will define the variable as numerical, and no value in that column will be accepted which is non-numerical. If you start by typing in text, the variable will be defined as a string, and the length will be set to be the length of the text you type in. (Example: Typing FOX into the first cell will define the variable to be a string, with a width of 3 characters. Typing in GOOSE into the next cell will give you GOO, as only 3 characters will be allowed). These definitions can still be changed in the Variable View window.

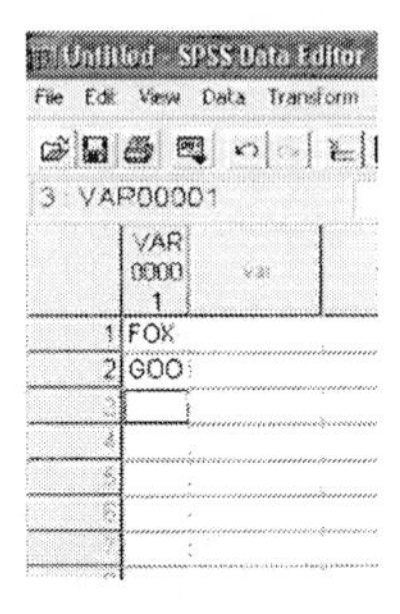

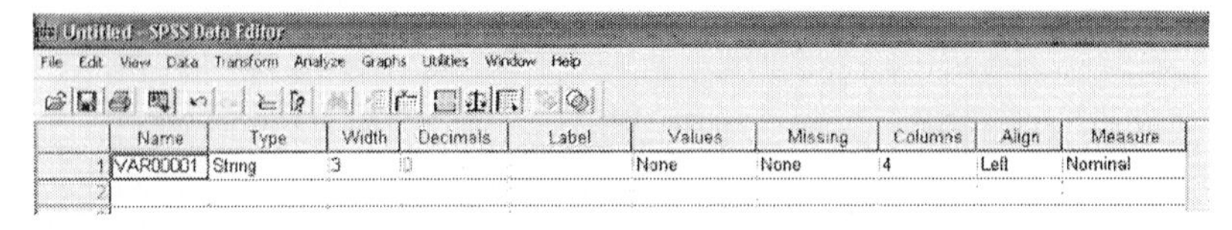

►Working with data

Before entering in data values, it is best to consider what type of variables you are looking at, and what type of analysis you want to run on the data. There are three major types of data to worry about.

-Nominal data: nominal data have no ordering issues, and can easily be entered as seen. For regression, or higher analysis, dummy variables should be used, but these can be created later.

-Ordinal data: ordinal data have a natural ordering to it, and so, the ordering needs to be preserved. This can be accomplished by using numbers to represent the data, in order, and then attaching labels for the numbers. For example, a survey asks "What year of school are you in?" with responses ranging from high school to graduate school. In SPSS, we create a new variable, schoolyear, which will be numeric, no decimals, and a Measure of ordinal. This is all done in the Variable View page.

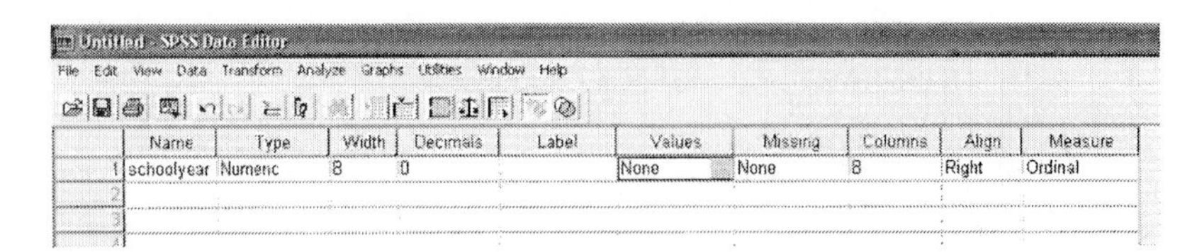

Clicking on the Values box will show a … button, and clicking on the … button will open the Value Labels dialog window.

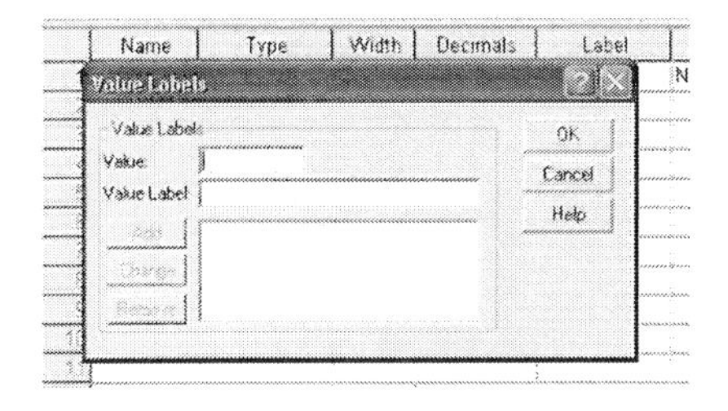

In the Value Labels dialog window, the first box is the Value of the variable, as you will be typing it into SPSS. We will use 1 for high school, so we type a 1 in the Value box, and High School in the Value Label box.

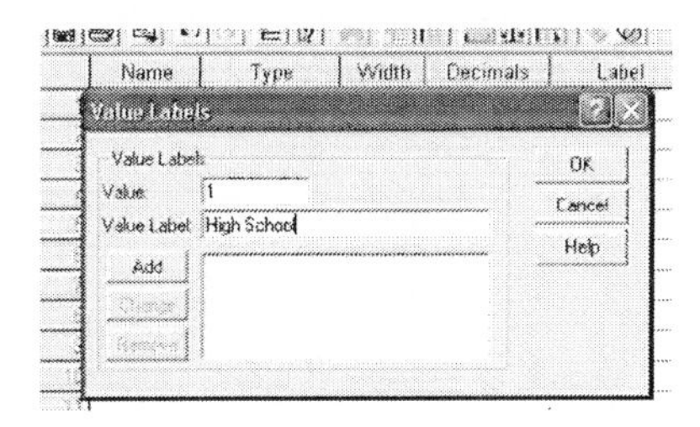

Once values are in both boxes, the Add button is available. Press Add, and any 1 in the column will be recognized by SPSS as High School.

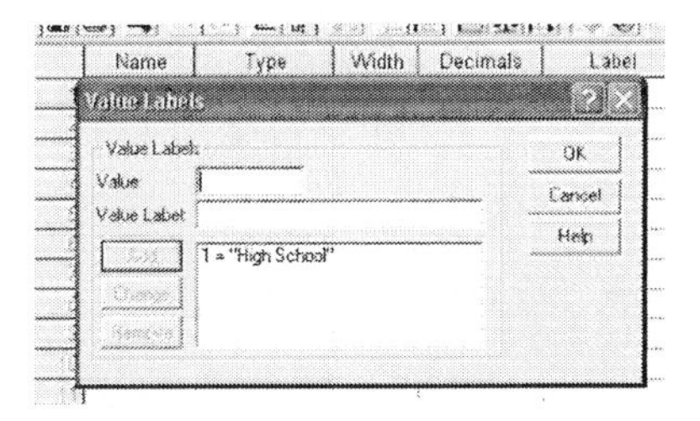

Once you add in all the possible values, push OK. If a value needs to be changed, you can highlight the value, which will put the information back into the Value and Value Label boxes, make the appropriate changes, and push the Change button. You can remove labels by highlighting them, and pressing the Remove button.

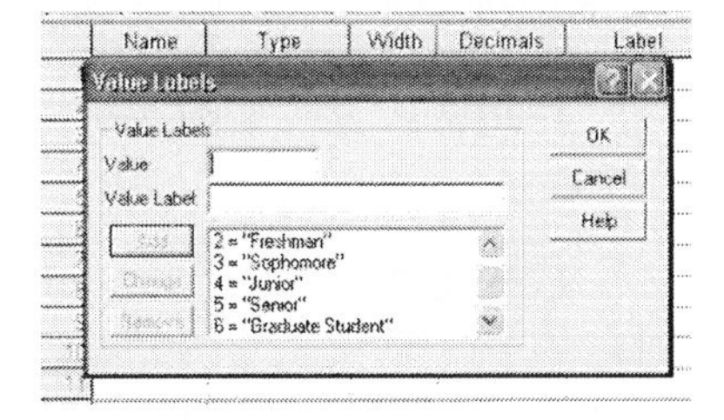

When labels are being used, you will see text in the Labels box, starting with the first value.

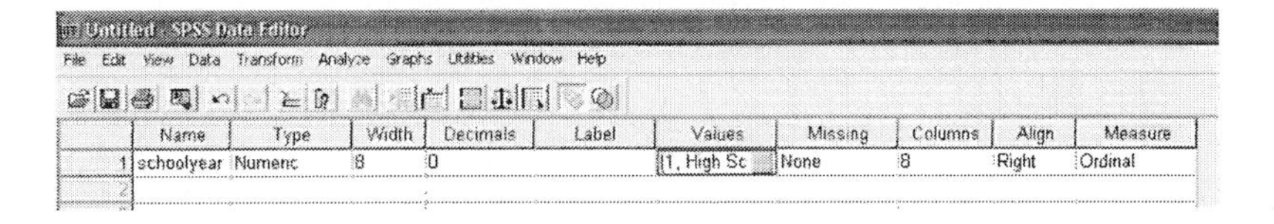

We can then switch to the Data View sheet, and enter in the information.

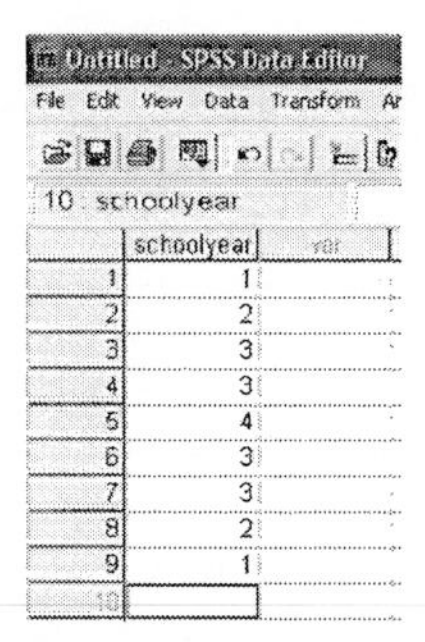

You will see the values as the numbers that you enter, but you can change that, by going to the View menu, and checking the Value Labels option. This will let you see the labels instead of the actual values. You will still type in a 1, 2, etc. but you will see the full names. This is also useful for any nominal values that you don't want to have to type multiple times. It is much easier to type in small numbers than long strings.

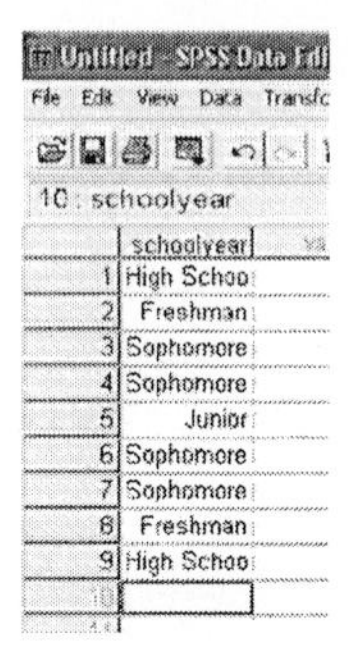

Although the analysis uses the numbers, thus preserving order, the output will print the labels.

-Numerical data: Numbers can be typed in directly, but be aware that you may not see the full number. You can change the number of characters shown in the Variable View sheet, as well as the number of decimal places shown. Reducing the number of decimal places does not change the value of the number. SPSS does not round the actual value, only the value that you see.

►Computations with data

There are times when you will want to perform some type of calculation using your data, such as squaring values, or finding probabilities, or creating a new variable with random values. SPSS has a calculator for such transformations. When you go to **Transform → Compute** you can make many types of changes to your data.

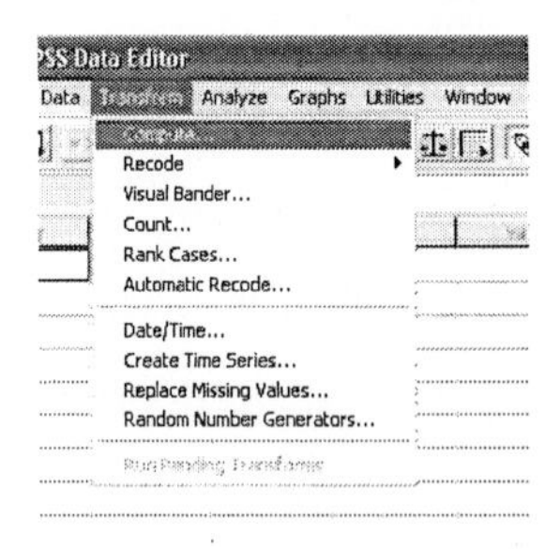

This will open up the Compute Variable window. You must already have data before this will open. The Compute Variable window contains five main boxes. The first is the Target Variable. This is a name for the new variable which you are creating. You can overwrite an existing variable if you so desire, but SPSS will give a warning before allowing you to do so.

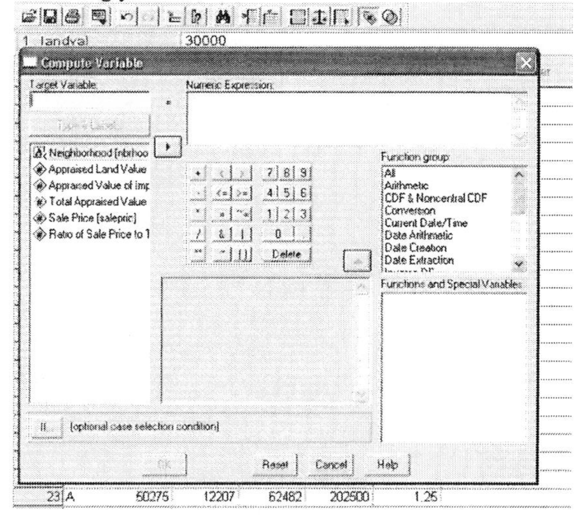

The window below the Target Variable is a list of the current variables. All of these can be used as part of the function, and the list lets you come up with a name that will be different from what already exists. The Numeric Expression box is where you will put in the function you want to use. This could be as simple as 1+1, which would create a new variable, with the same number of rows as the existing data, all with the value of 2. The next box is the Function group box. This is the general category for mathematical or statistical functions that are available to use. For example, highlighting Arithmetic brings up a list of arithmetic operators in the Functions and Special Variables box.

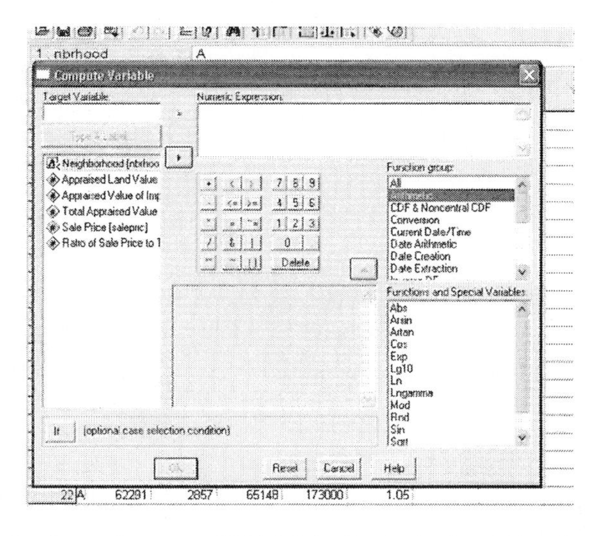

So, to calculate absolute values, we will highlight Abs in the list of Functions and Special Values, and an explanation of the function will appear in the box next to the function list.

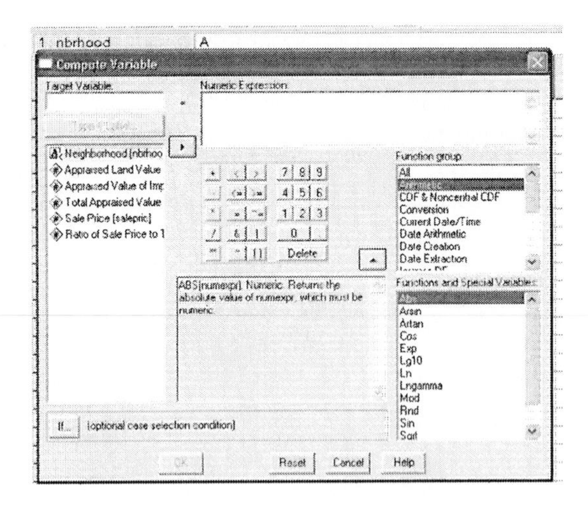

To use the Abs function, we will need a numerical expression to put into the parentheses. This value could be any of the variables or a function of the variables. So, we could use ABS(totval-landval) to get a new variable which is the absolute difference between total appraised value and appraised land value.

Another way to change the data is to recode the information. To recode, go to **Transform → Recode**, and select **Into Same Variables**… which overwrites an existing variable, or **Into Different Variables**… which creates a new variable. It is generally preferred to create a new variable in case of mistakes unless memory becomes a problem.

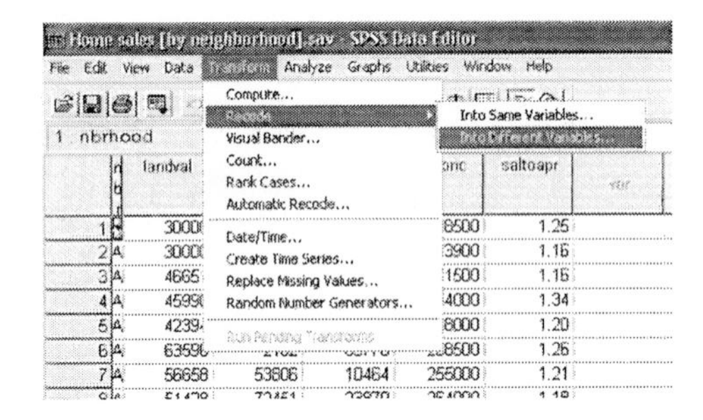

Selecting the different variables option will open up a Recode into Different Variables window. You select which variable you want to recode, and a name for the new variable. To create the variable, you must push the change button, which takes the new name to the right of the → in the Numeric Variable → Output Variable box.

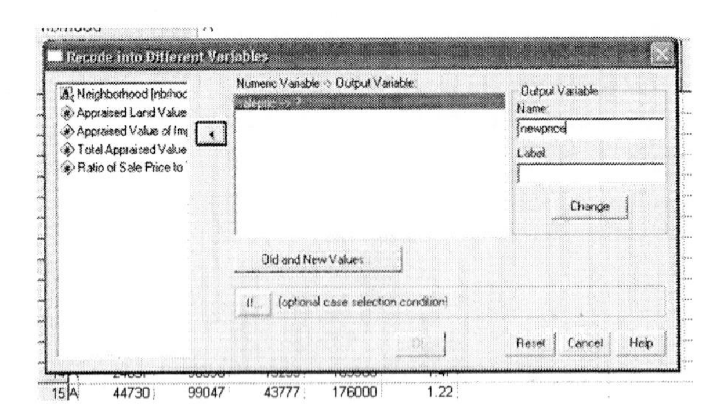

Once you have a name for the variable, click on Old and New Values. This will open the Recode into Different Variables: Old and New Values window. This is where you tell SPSS what changes you would like to make to the data.

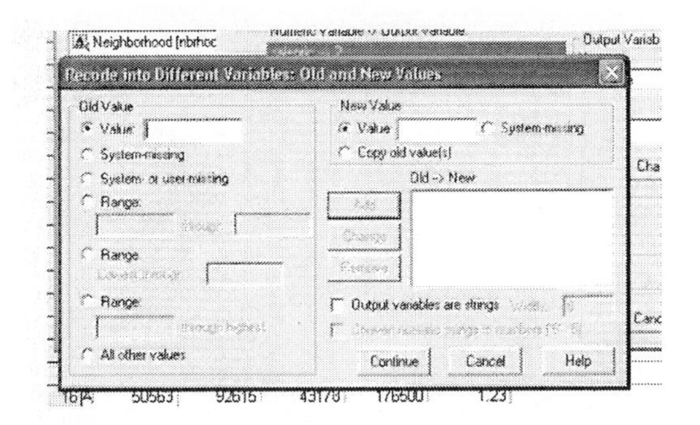

In the Old Value list, you let SPSS know what the old value was. If you have a short list, you can use the Value box. Type in the old value, put the new value in the New value box, and click Add. This is useful when you want to switch order, for example, you receive data that has a 5 point scale, where 1 is high and 5 is low, and you want 5 to be high and 1 to be low. You can then put 1 for the old value, and 5 for the new value, etc. If you have a list of numbers, you can use the Range values. This can be used to categorize numerical data, for example heights from height in inches to short, medium, tall. Once you have completed the full list, push the Continue button, and then you can push OK. You will now have a new, recoded variable. The Old and New Values works similarly for recoding into the same variable, but you don't have a window in which to name the variable.

►Opening Saved Data Sheets

To open a saved SPSS data sheet, click on **File → Open → Data …**

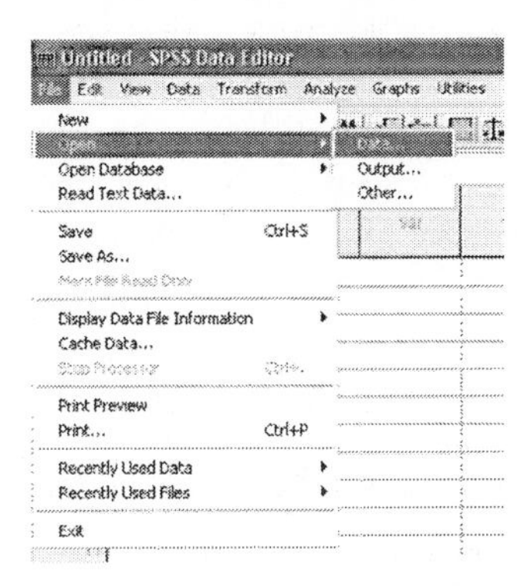

The following screen will appear.

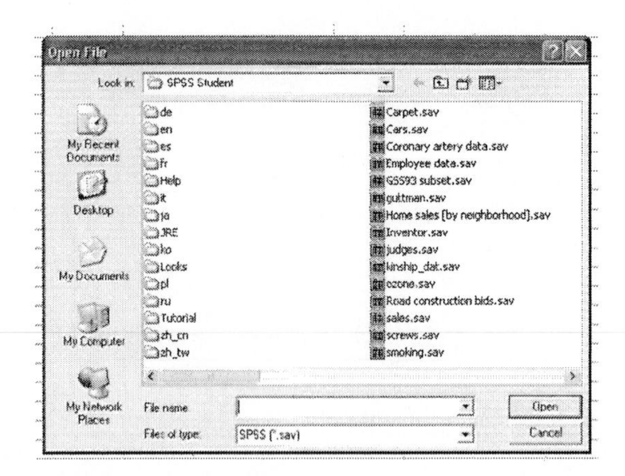

First, you must find where your data files are stored. The default location for SPSS to begin looking is in the data sets that came with the program. You can change the disk location by using the Look in: box. It is best to keep all of your data sets in a location that is easy to find. SPSS data sets will appear as little blue data boxes with SPSS in red above them. They also have the extension .sav following the name. No other data type can be read in as an SPSS data set.

►Importing Data from Excel

As most of the data come in three formats on the disk supplied with ASCII files, Excel files, and MINITAB files, but not SPSS files, it will be advantageous to import from one of these formats rather than recreating the data sets. You will need to save the files from the disk onto your hard drive, as they are not accessible by browsing the CD. Excel files are easier to import from than ASCII, as they follow specific conventions. To import from an excel file, go to **File → Open Database → New Query …**

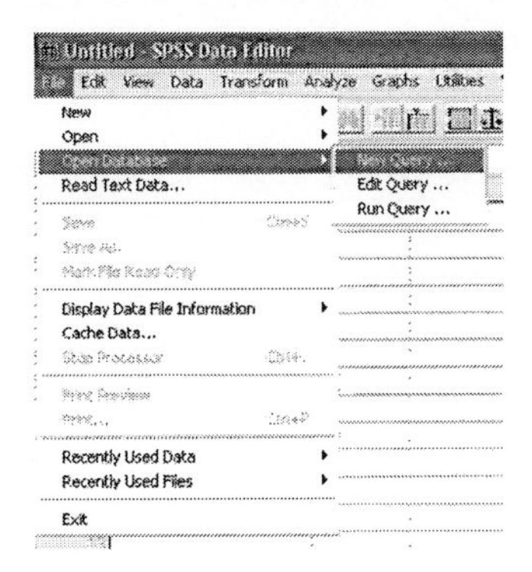

This will open up a Database Wizard window.

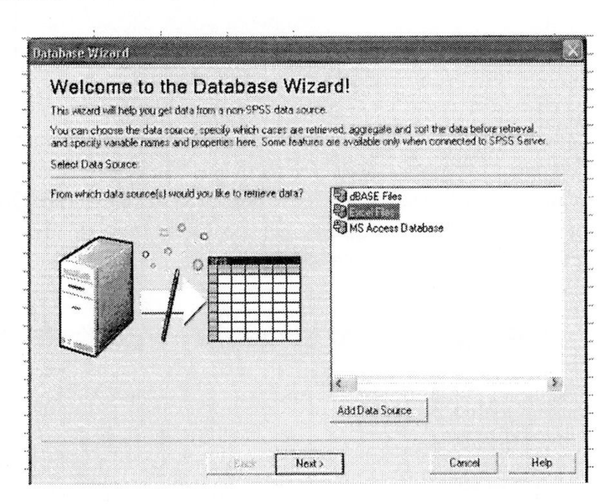

Highlight Excel Files, and push the **Next>** button. This will give you a dialog window.

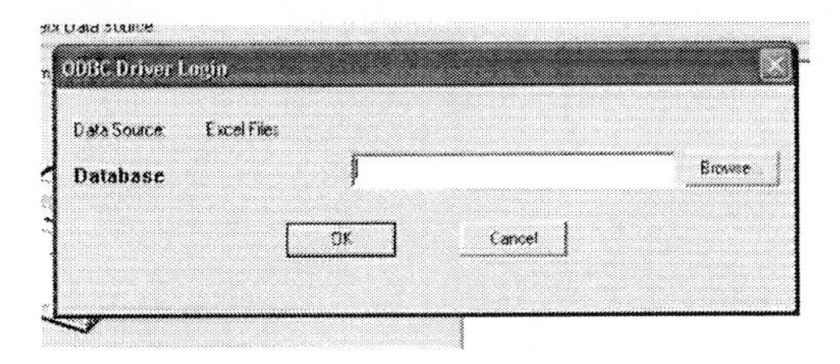

Use the Browse… button to find the file that you wish to open. The Excel file cannot be open in Excel when importing the data, as a sharing violation will occur.

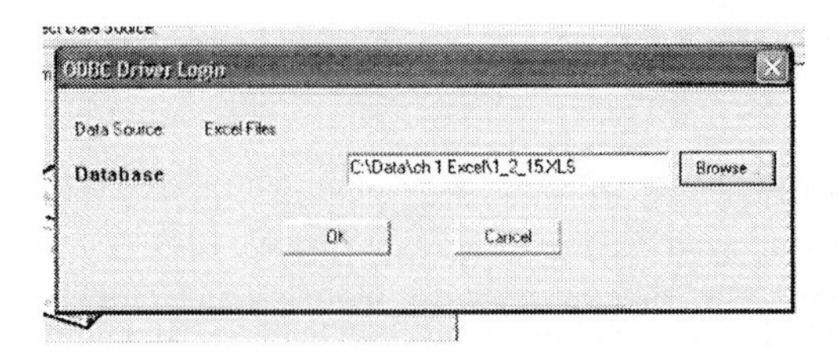

The next screen is the Select Data screen. You may select all of a data set, or portions of the file. If Excel has more than one sheet available, this is also where you would specify which sheet contains the data that you want to use.

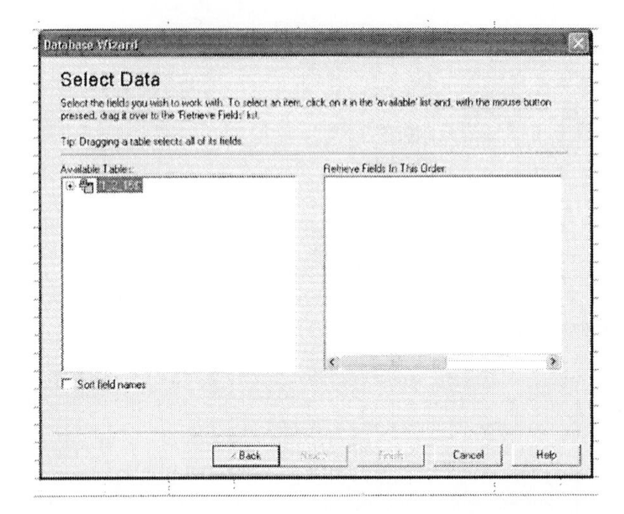

Drag the sheet over from the Available Tables box to the Retrieve Fields in This Order box. When you click on the available table, a hand will appear. Keep the mouse button pressed until the hand is over the Retrieve box. This will give a list of the variables in the data set.

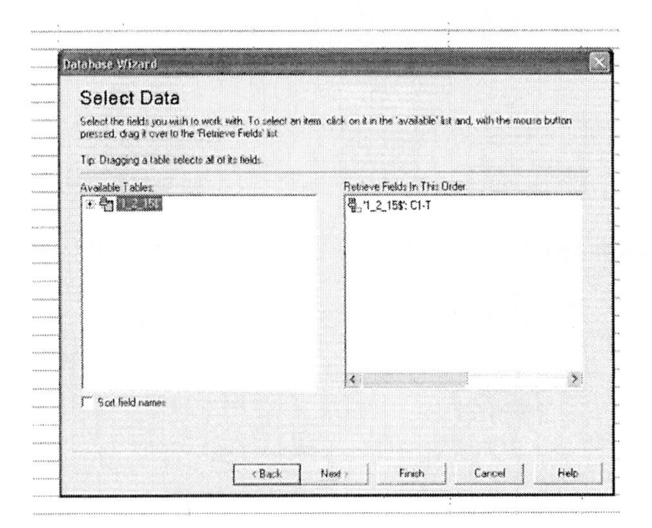

You may change the order of the variables, or remove variables at this point. When you are ready, push Next> or Finish.

Pushing Next> will take you to a window where you can limit the number of observations to bring in, including the possibility of bringing in a random sample of the observations.

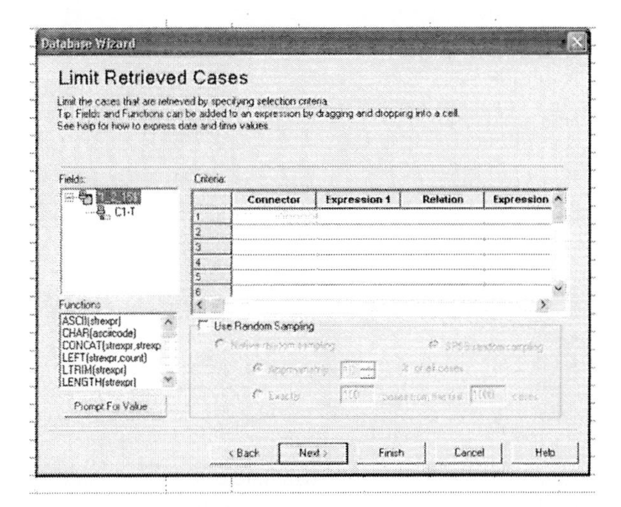

Pushing Next> again will take you to the Define Variables window. Here you can change the variable type, or length.

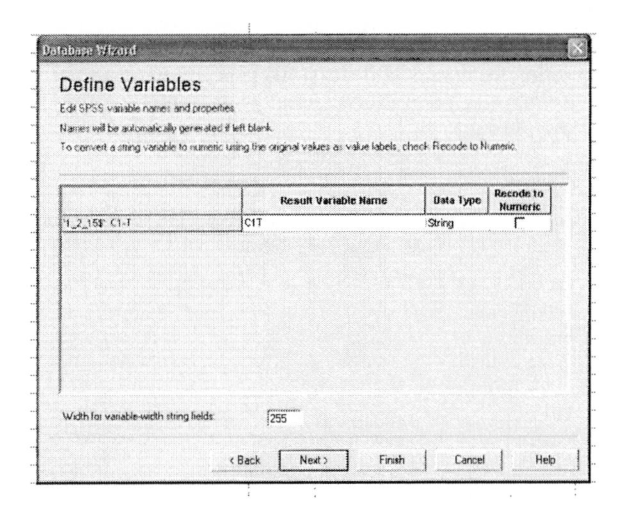

Pushing Next> again will take you to the Results window. Here you have the option of saving the query that you just ran. This is useful if you plan on opening the same Excel file many times. It is easier to save the data as an SPSS file than to re-import the data over and over again, but in business practice, there are times when you will want to open the same file numerous times, after updating the information in the file. When there are a large number of observations available in the data set, it is easier to save the query and just re-run it periodically than to re-make the query each time.

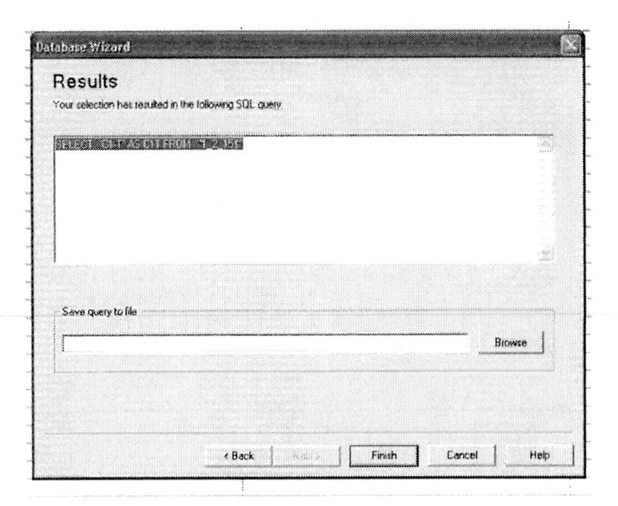

Pushing the Finish button at any time will bring the data into SPSS. The name of the column will be obtained from the first row in Excel, but will be changed to accommodate the SPSS naming conventions.

►Importing Data from ASCII

There are times when importing from an ASCII (text) file is necessary, especially when retrieving data from web sites. To read in a text file, go to **File → Read Text Data…**

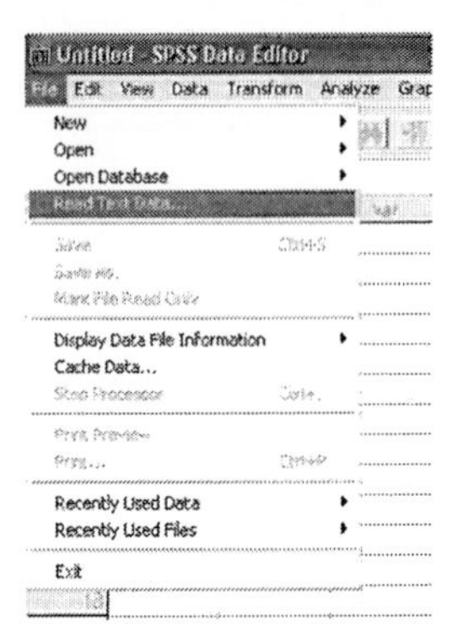

This will take you to an Open File window, where you will browse for the file that you want to import the data from.

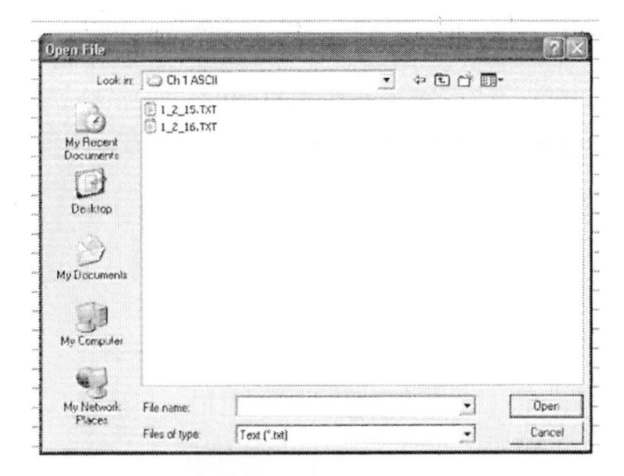

Once you select which file you want to use, the text import wizard will open.

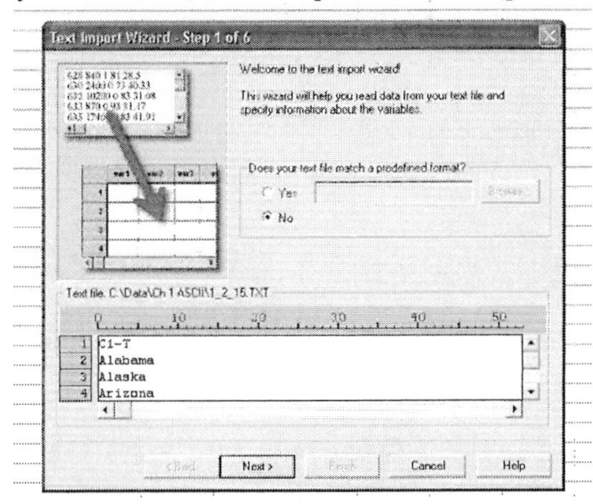

The first step is to see if the data is the correct one, you will see the first couple lines of the file in the bottom box. If you are going to read in the same document over and over, you can save the format, and use that in the predefined format box.

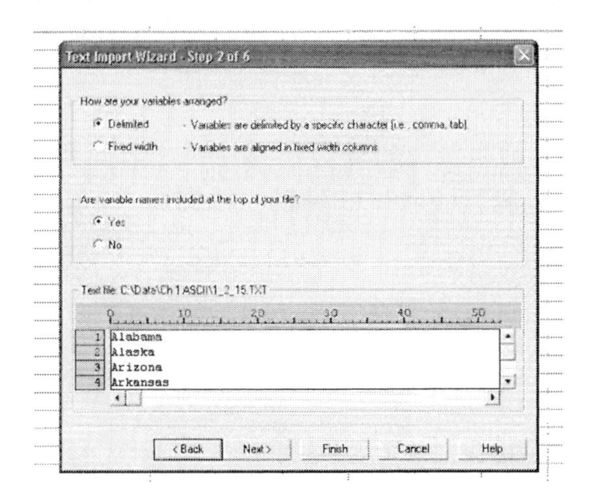

Step 2 is to tell SPSS how the data looks. If you are using a file which is Delimited (using the same character to distinguish between variables) then use the Delimited button. If the variables are EXACTLY in line, you can use the Fixed width button. Files can be delimited with a space, but this means that spaces cannot be inside strings. If the data is not exactly lined up, the fixed width will not work very well. Also, define whether the first row is the variable name, or part of the data.

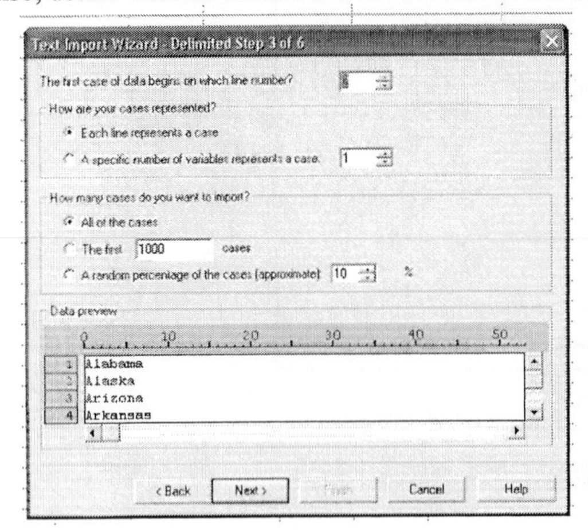

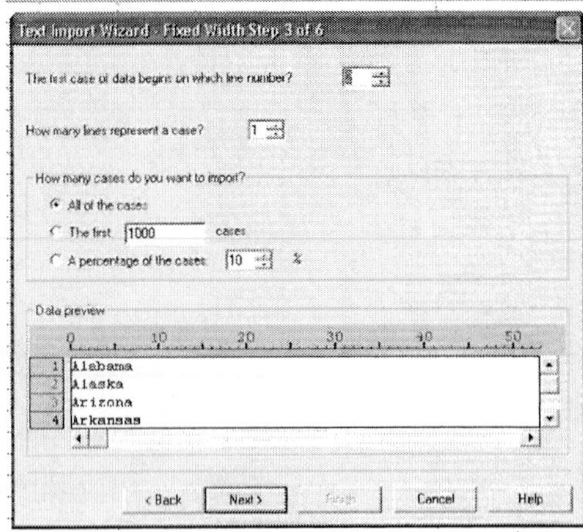

Step 3 is either to tell SPSS where the data starts. If the data starts on any row other than the second, you can change that information here as well. If you don't want to read in the whole data set, you can limit the number of cases.

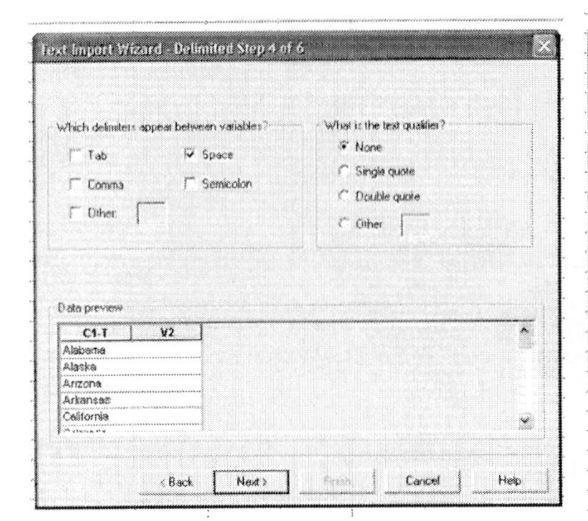

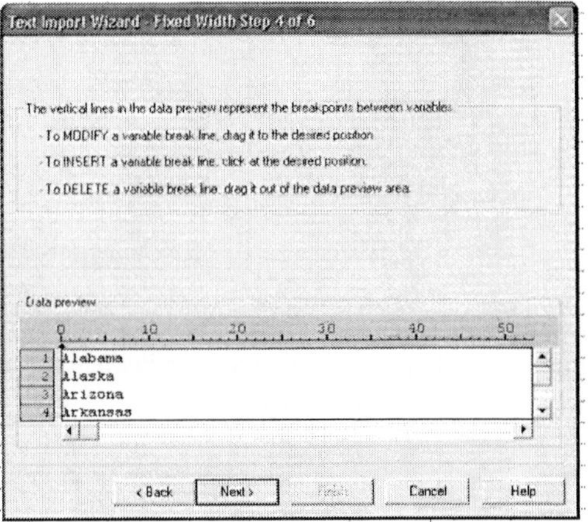

Step 4 is to define the delimiter or to define the widths. SPSS tries to define a delimiter based on general conventions, but these can be changed. The common delimiters are listed (Tab, Space, etc.) but a box is available for different delimiters. With the fixed width, you will put in a line delineating the variables.

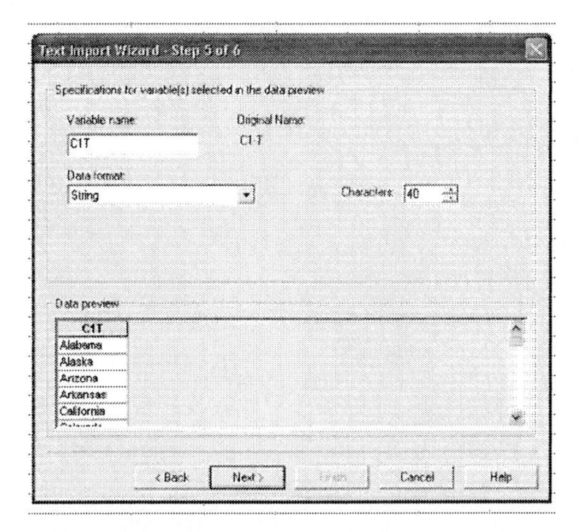

Step 5 is to specify a name. You will get an error message if the name in the first line (if you specified that the name was at the top of the file) does not satisfy the naming restrictions. SPSS will take out all illegal characters, but you have the possibility of renaming the variable altogether as well. Also, you can specify what type of data the variable should be. The default is numeric, so if there are any characters, such as commas or dollar signs, this will need to be changed.

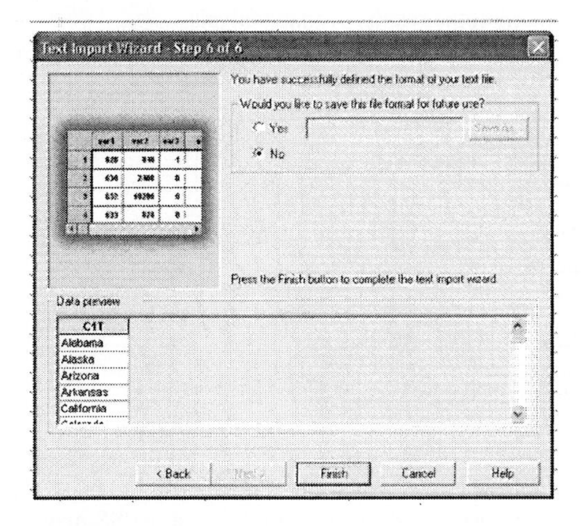

Step 6 allows you to save the format for future use.
Pressing Finish at any time will read in the data.

►Using the Output Viewer

After running an analysis in SPSS, you will usually see the results in a new window, called the Output Viewer window.

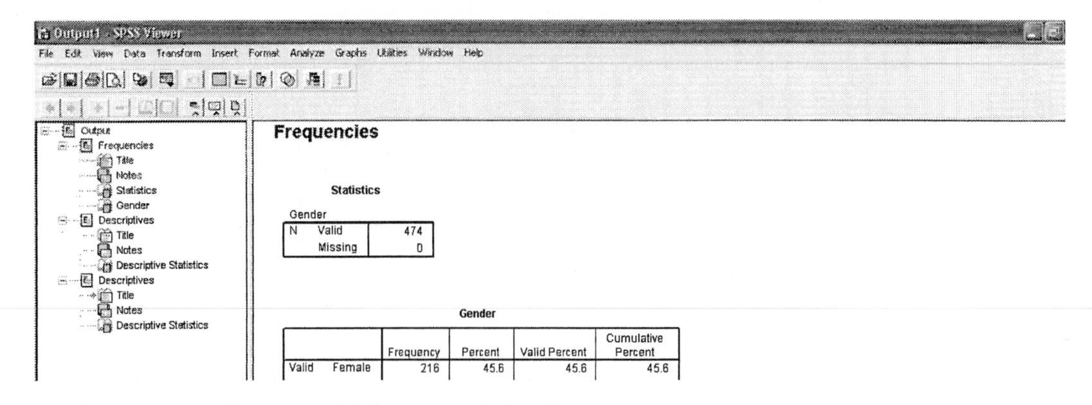

On the left hand side of the window is a list of all the analyses that have been run, and the pieces of the output that are available. Any of these can be removed, or, if the list is getting too long, you can push the – sign to the left of the analysis and it will hide, but not delete, the output. To show the analysis again, push the + button to the left of the closed output. You can continue to run analyses from the Output window. The same options appear in the output window as in the data view window, so you do not need to return to the data to keep going.

While in the output window, if you click on any piece of output with the right mouse button, the option will appear to use the Results Coach. This will take you to the tutorial for that type of analysis, and will walk you step by step on how to interpret the results of that analysis.

Gender

		Frequency	Percent	Valid Percent	Cumulative Percent
Valid	Female	216	45.6	45.6	45.6
	Male	258	54.4	54.4	100.0
	Total	474	100.0		

What's This?
Cut
Copy
Copy objects
Paste After
Export...
Results Coach
Case Studies
SPSS Pivot Table Object

Descriptives

Descriptive Stat

	N	Minimum			Std. Deviation
Educational Level (years)	474	8			2.885
Valid N (listwise)	474				

You can also copy the table from SPSS into Excel, or into Word for better formatting and printing options, but this should be done one table at a time, as problems arise trying to copy a whole page. You will want to look at the print preview when printing directly from SPSS, as SPSS may put page breaks at interesting times, and some output will run off of the page. Watching the print preview can save paper.

Chapter 1. Data Collection

SPSS is a data analysis tool, not a data creation tool. Because of this, it is not as powerful as other programs in generating data. It is easier to generate the data in Excel, or another tool, and import it into SPSS.

Section 1.1 Introduction to the Practice of Statistics

SPSS is particular about the type of data that you collect. Qualitative variables are broken up into two cases, Nominal, and Ordinal (see problem 57). Nominal comes from name, and represent qualitative data which have no specific order. Ordinal data on the other hand have an implied or specified order to them. Example:

a) Gender -Nominal, there is no order implied in being male or female.

b) Zip Code -Nominal, although the first digit can be used to specify a region in the US, 0 being east and 9 being west, the other digits relate to a city, but have no real order.

c) Class Standing -Ordinal, Freshmen have less credit hours than Sophomores, who have less credit hours than Juniors, who have less credit hours than Seniors, so an order is implied. You move from being a Freshman to being a Sophomore to being a Junior to being a Senior.

d) Leikert Scale -a scale from 1 to 5, or 1 to 7 etc. measuring agree to disagree, or like to dislike, or other opinions. Although there is an order specified, the distance between the values changes from view to view. In other words, a 5 doesn't dislike something twice as much as a 4, who dislikes the thing twice as much as a 3, etc.

SPSS does not break Quantitative variables into discrete and continuous. Both are given the title Scale. SPSS will identify text values as Qualitative variables, but must be told whether numbers are Nominal, Ordinal, or Quantitative. This is done in the Variable View window.

Section 1.2 Observational Studies, Experiments, and Simple Random Sampling

SPSS is a data analysis tool. It assumes you already have some data to analyze. Unlike some other software packages, it is difficult to create data from scratch in SPSS. SPSS will, however, take random samples.

►Obtaining a Simple Random Sample from a list

Example 2, page 17.

To select five clients out of the list of thirty, first we type all thirty clients into a blank SPSS sheet.

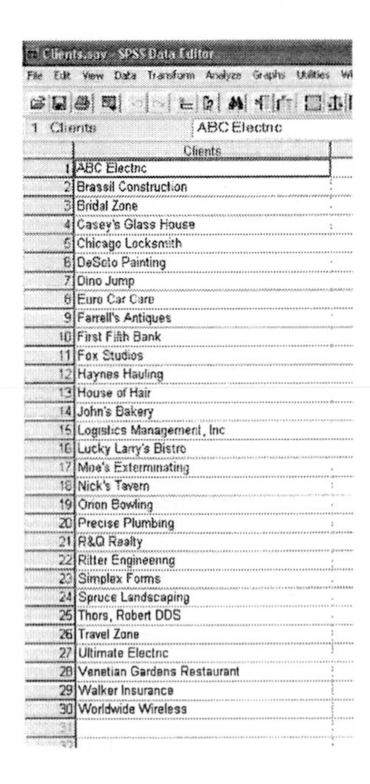

We then go to **Data → Select Cases** to take a sample.

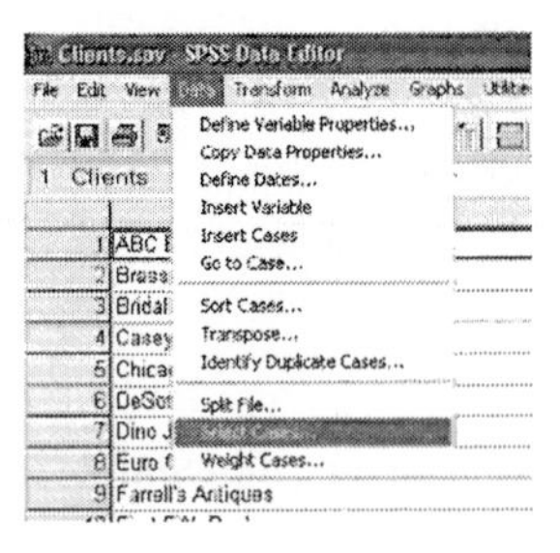

There are different types of selections we can take from SPSS. We can select specific values, or in this case, we can take a random sample of cases. In the Select Cases window, **select Random sample of cases**, and push the **Sample...** button. (The Sample... button will not be available until the Random sample of cases has been selected.)

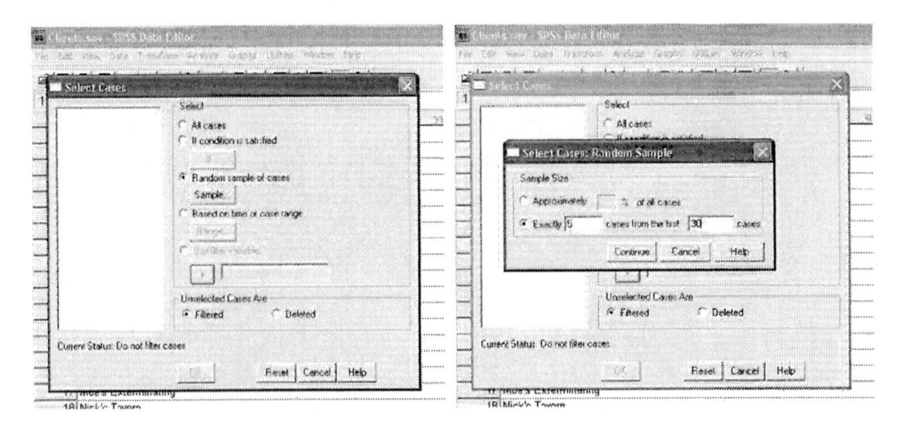

Since the exact sample size is known, choose the Exactly row, and type in the sample size wanted (5) and the population size (30), and push continue. This will return you to the Select Cases window, where you can now push OK.

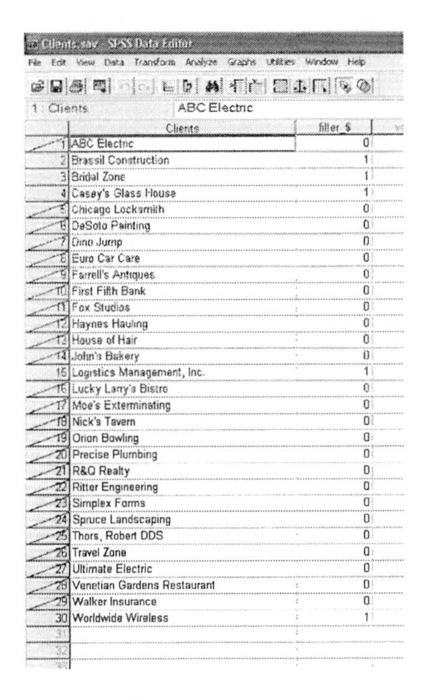

	Clients	filter_$
1	ABC Electric	0
2	Brassil Construction	1
3	Bridal Zone	1
4	Casey's Glass House	1
5	Chicago Locksmith	0
6	DeSoto Painting	0
7	Dino Jump	0
8	Euro Car Care	0
9	Farrell's Antiques	0
10	First Fifth Bank	0
11	Fox Studios	0
12	Haynes Hauling	0
13	House of Hair	0
14	John's Bakery	0
15	Logistics Management, Inc.	1
16	Lucky Larry's Bistro	0
17	Moe's Exterminating	0
18	Nick's Tavern	0
19	Orion Bowling	0
20	Precise Plumbing	0
21	R&Q Realty	0
22	Ritter Engineering	0
23	Simplex Forms	0
24	Spruce Landscaping	0
25	Thors, Robert DDS	0
26	Travel Zone	0
27	Ultimate Electric	0
28	Venetian Gardens Restaurant	0
29	Walker Insurance	0
30	Worldwide Wireless	1

A new column has been added to the data, called filter_$. This has a 1 if the row is in the sample, and a 0 if it is not in the sample. The rows that have not been selected also have a line through the observation number. If the Deleted option was chosen in the Select Cases window, these values would no longer be in the data set. Although all values can be seen, only those with a filter_$ value of 1 will be used for any analysis. In this example, the clients to be surveyed are Brassil Construction, Bridal Zone, Casey's Glass House, Logistics Management, Inc., and Worldwide Wireless.

If another sample is desired, going back through the steps will select another sample, without having to clear the information from the first.

To change the random seed (starting location in the random number table), go to **Transform → Random Number Generators...**

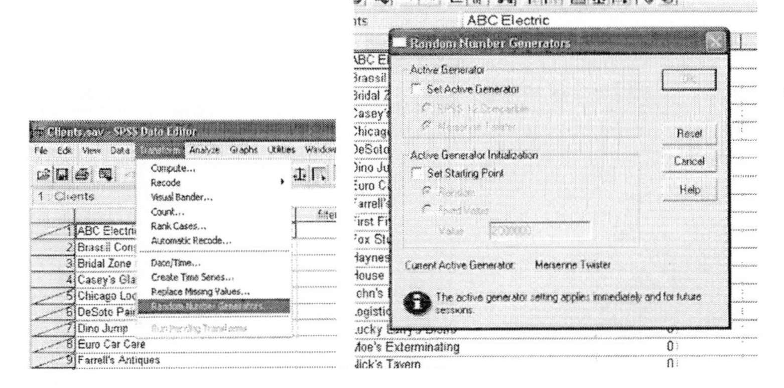

The Random Number Generators window has two options you can change. The first is the Active Generator. If you want to reproduce results as done in SPSS version 12, the SPSS 12 Compatible should be selected. Otherwise, the Mersenne Twister should be selected, as it is a more reliable randomization tool. Under Active Generator Initialization, you can select the starting point. The default is Random, which is generally based on the computer clock, down to a fraction of a second, so every person who takes a sample will get a different starting point. If you want more replicable results, a fixed value can be used, which will ensure the same starting location, and thus the same sample, each time.

►Obtaining a Simple Random Sample without a list in SPSS

There are times that you do not want to enter a list into SPSS, due to length or other constraints. SPSS can still help chose the sample, although you must have a numbered list to choose from. SPSS has a random number generating tool, but data must exist before using it. So, the first step is to decide on the sample size. Similar to Excel, or Minitab, we will have to create a few extra observations in case of repeats.
In a blank SPSS Data sheet, create a column of integers, from 1 to n. You will want to add a few more after this in case of repeated values.

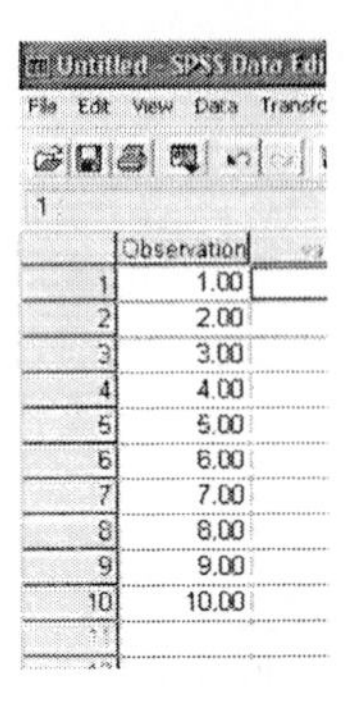

Here we want a sample of five from the thirty, but to be safe, we added five extra in case of repeats. In the Variable View window, create a new variable, which is numerical with no decimal places.

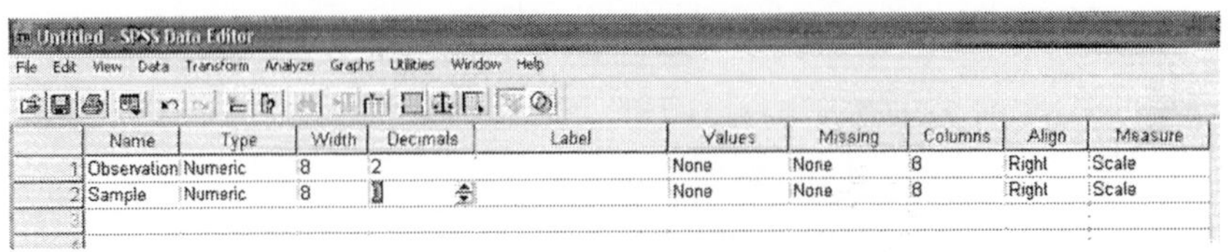

Go to **Transform → Compute...**

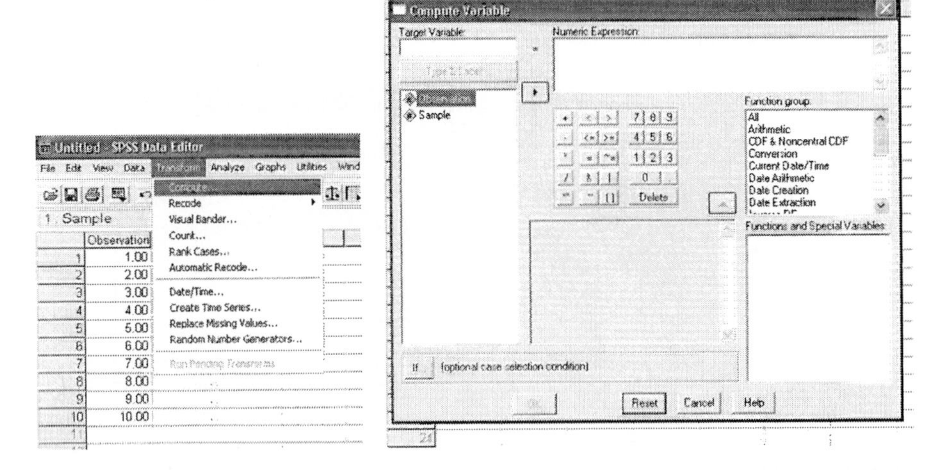

There are two windows that we must fill. The first is the Target Variable, for which we want to use the name of the blank column we just created.. In the Function Group box, find Random Numbers

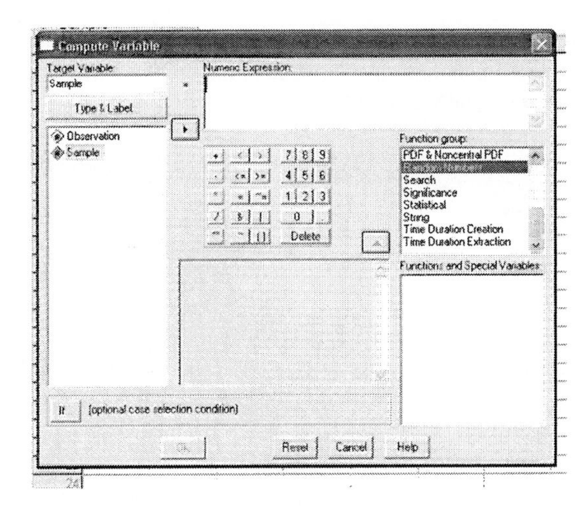

Clicking on Random Numbers will bring up a list of possible random variables. Find RV.Uniform. RV.Uniform will give every value an equal chance, which is what we want for a Simple Random Sample.

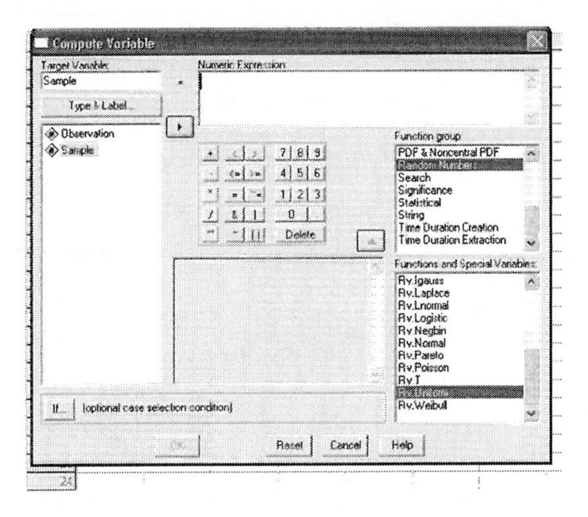

Click on the RV.Uniform until it appears in the Numeric Expression box.

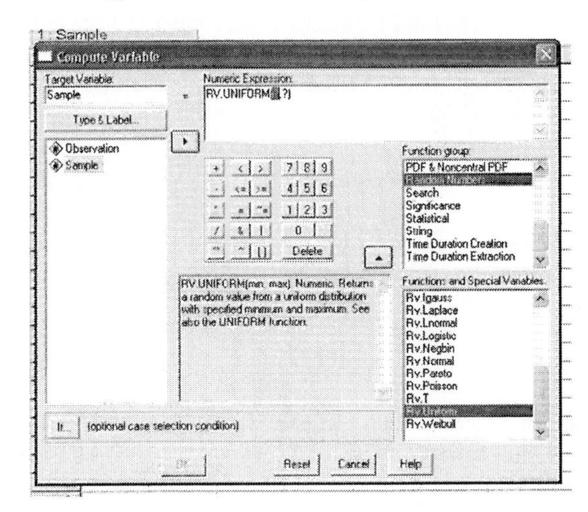

The question marks in the Numeric Expression are values that you must provide. From the explanation in the box, we can see that we need a minimum value, and a maximum value. Since we are selecting values from 1 to 30, these are the values that we enter.

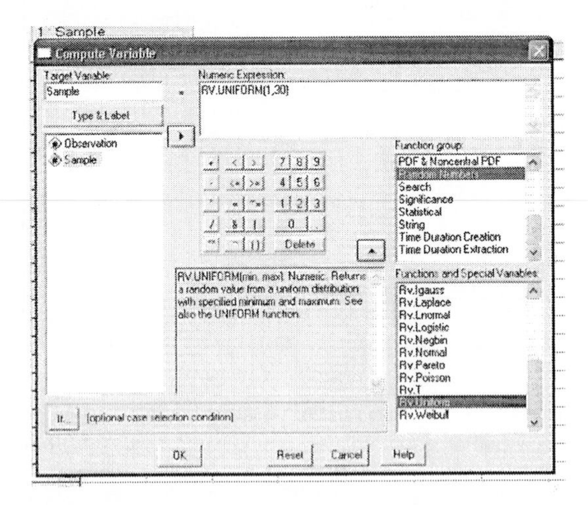

When the Numeric Expression looks correct, push OK. You will be asked if you want to change the existing variable. Since the existing variable Sample is currently empty, we can safely push OK. When you get this warning, you do want to make sure that you are not going to overwrite data that already exists. We will now see the values that have been selected in the new column.

Untitled - SPSS Data Editor

File Edit View Data Transform Anal

1 : Observation 1

	Observation	Sample
1	1.00	21
2	2.00	4
3	3.00	13
4	4.00	4
5	5.00	22
6	6.00	5
7	7.00	25
8	8.00	14
9	9.00	5
10	10.00	19
11		

Our sample in this example would be the 21st, 4th, 13th, 22nd, and 5th values in the list. Note that 4 was listed twice, so we ignore the second occurrence, and use the sixth value in the list. We can ignore the other values, as we only want the first five unique values.

Section 1.3 Other Effective Types of Sampling

SPSS only takes Simple Random Samples. Other types of sampling can be achieved through other methods, but are much harder to do. This is a result of SPSS being an analysis tool, where it is assumed that the data is already from a sample.

Obtaining a Stratified Sample:

To obtain a Stratified Sample, you can create a Simple Random Sample for each stratum. It must be remembered that SPSS will only create a value for a row that is already in use by another variable.

Obtaining a Systematic Random Sample:

SPSS can help create a Systematic Random Sample by helping you choose the first value from the first k observations. You will then choose every k^{th} individual after the randomly selected starting point.

Obtaining a Cluster Sample:

To obtain a Cluster Sample, you can use SPSS to help take a Simple Random Sample of the clusters.

Chapter 2. Organizing and Summarizing Data

Section 2.1 Organizing Qualitative Data

There are two different ways to enter data into SPSS, in raw format, or summarized information. Raw format is the actual data, in a full list, each row representing an appearance of the observation. Summarized format is where one column is the value of the observation, and another column represents how many times that observation was observed.

Example 1, page 61

To construct a frequency distribution and/or a frequency bar graph in SPSS, you will first need to enter the data into a Data View sheet. For example 3, we will use the information as presented in Table 1 on page 61. This is in raw format. Each row represents an individual observation.

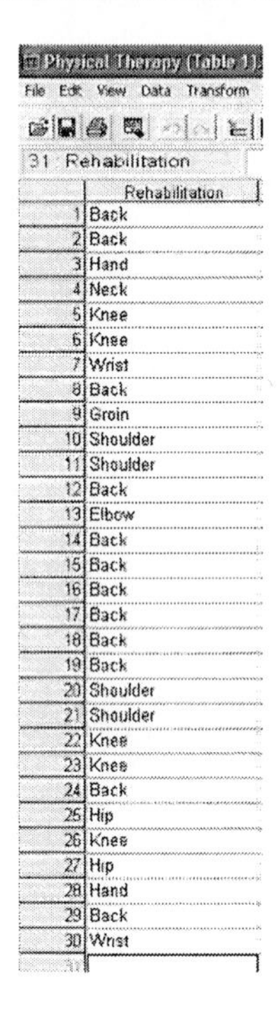

►Creating a Frequency Distribution

To create a frequency distribution, we select **Analyze → Descriptive Statistics → Frequencies…**

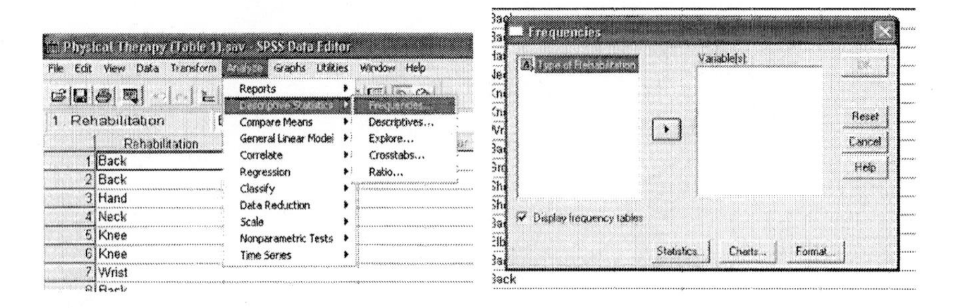

Highlight the variable that you want the frequency distribution of, in this case Type of Rehabilitation, and push the ▶ button. This will take the variable you chose into the Variable(s) list. Make sure that the Display frequency tables box is selected. Push OK. The output will appear a new window called the Output Viewer.

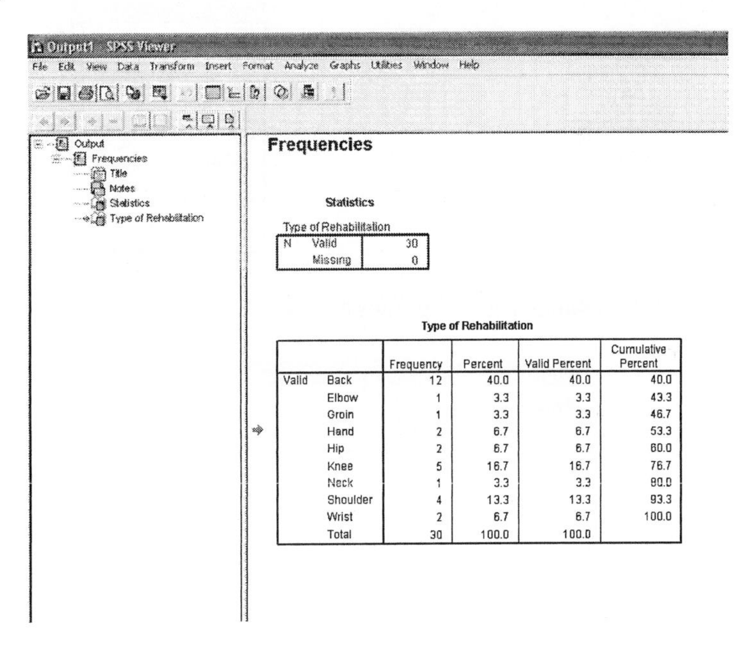

Type of Rehabilitation

N	Valid	30
	Missing	0

Type of Rehabilitation

		Frequency	Percent	Valid Percent	Cumulative Percent
Valid	Back	12	40.0	40.0	40.0
	Elbow	1	3.3	3.3	43.3
	Groin	1	3.3	3.3	46.7
	Hand	2	6.7	6.7	53.3
	Hip	2	6.7	6.7	60.0
	Knee	5	16.7	16.7	76.7
	Neck	1	3.3	3.3	80.0
	Shoulder	4	13.3	13.3	93.3
	Wrist	2	6.7	6.7	100.0
	Total	30	100.0	100.0	

Included in the table are the Frequencies, Percent (Relative Frequency, but will take n to be the sample size including missing values), Valid Percent (which is the Relative Frequency without missing values) and Cumulative Percent (Cumulative Relative Frequency, based on the Valid Percent).
The order that SPSS uses by default is the alphabetical order. To change this, while in the Frequencies window (before pushing OK), select Format. This will open the Format window.

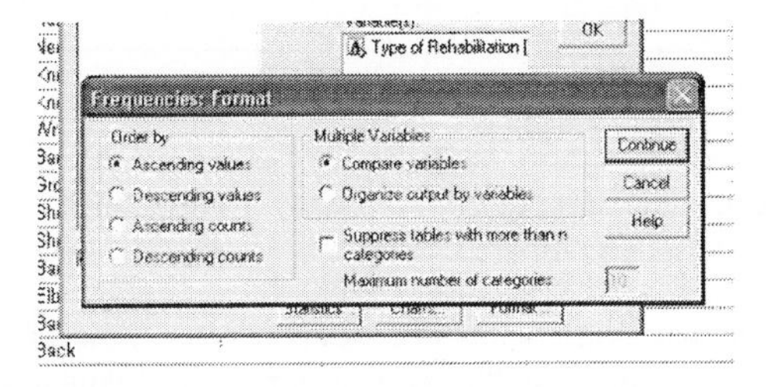

Here you can choose to order the variables by Ascending values (alphabetical), Descending values, Ascending counts, and Descending counts. You can also choose to not show values that appear infrequently.

▶Creating a Frequency and Relative Frequency Bar Graph

Using the same data as for a frequency table, select **Graphs → Bar**

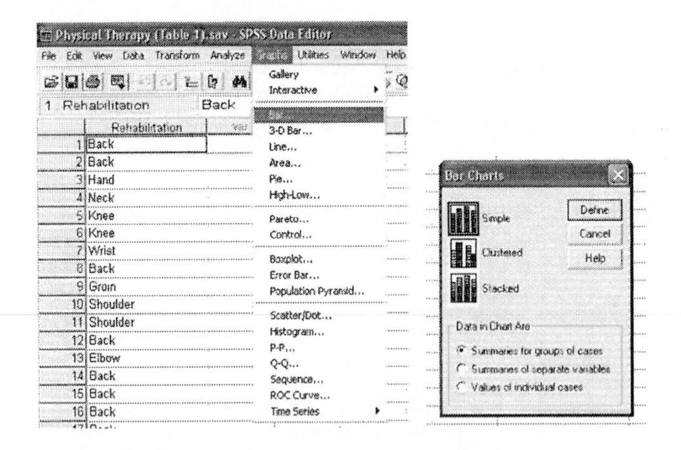

In the box, select **Simple**, and **Summaries for groups of cases**

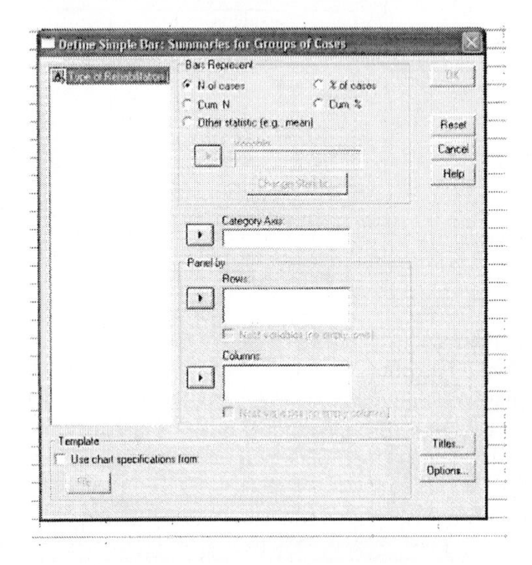

Highlight Type of Rehabilitation, and push the ► key by Category Axis. This selects the category for the bottom of the graph. You can panel, or split the bar plot by placing other variables in the Rows, and Columns boxes. This can be used to compare frequencies among different groups.

For a Frequency Distribution, make sure that **N of cases** is selected. Select OK. For a Relative Frequency Distribution, select % of Cases. Cum. N will provide a cumulative frequency plot, and Cum. % will provide a cumulative relative frequency plot. You can use the Titles button to add titles to the plot.

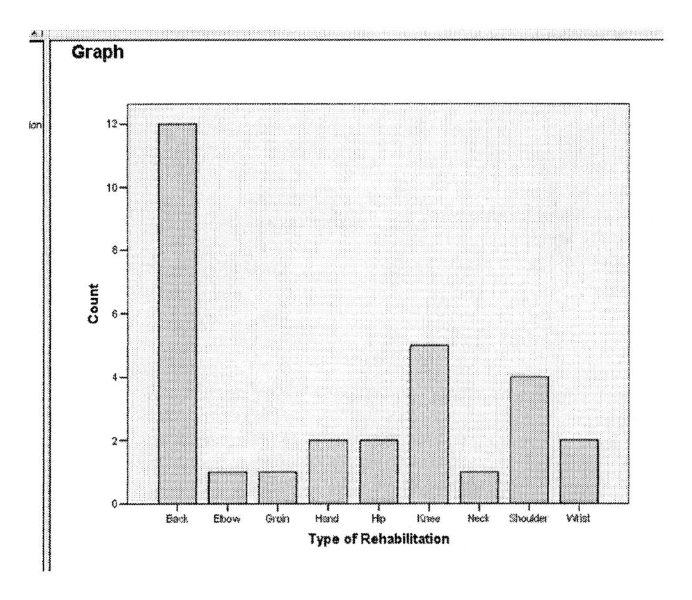

Note that SPSS puts values into Alphabetical Order, so although values are the same, the graph may not look exactly like you would expect. This is ok, as order doesn't matter in a bar graph.

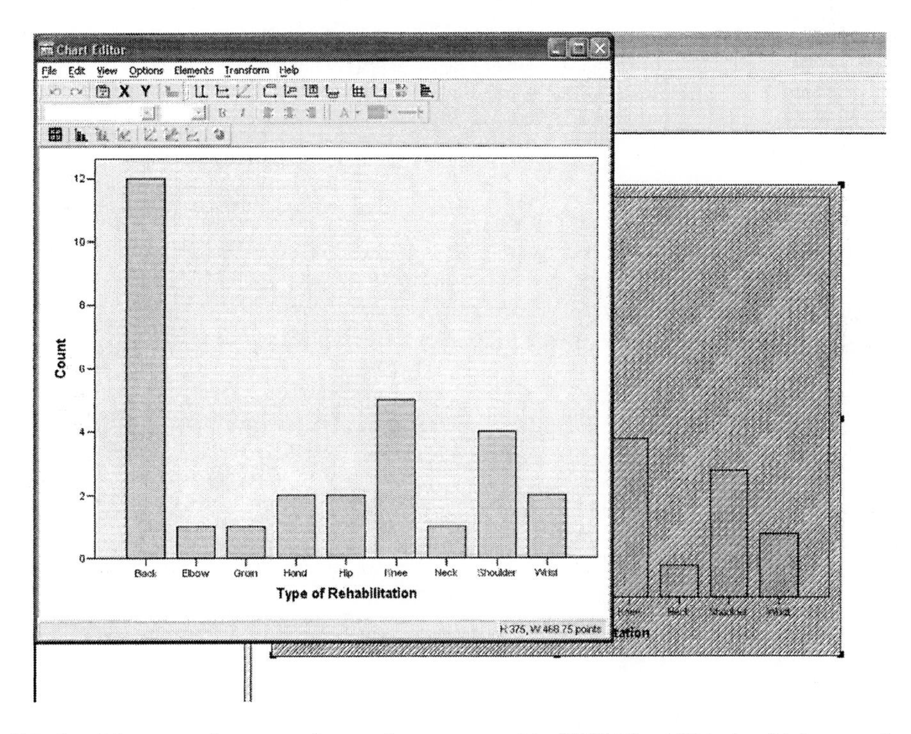

If you click the right mouse button on the graph, you can get to SPSS Chart Object, which opens the graph editor. Here you can change the order of the bars in the **Edit → Properties** menu, as well as changing bar width, changing color, etc.

Example 5, page 65

Sometimes data is already summarized. When this occurs, SPSS must know where the summarized information lies. This form of data entry is useful for large amounts of information, and can save on many hours of typing, as each value only needs to be entered once, with another column describing the actual frequency of observations.
To construct a frequency distribution: Type in the data (do not include totals)

	Educational_Attainment	yr1990	yr2003
1	Less than 9th grade	16,502	12,276
2	9th-12th grade, no diploma	22,842	16,323
3	High school diploma	47,643	59,292
4	Some college, no degree	29,780	31,762
5	Associates degree	9,792	15,147
6	Bachelor's degree	20,833	33,213
7	Graduate/professional degree	11,478	17,169

Select Data → Weight Cases

Adding weights allows SPSS to "see" the data a different number of times than what appear in data sheet. Although the actual value was only typed one time, the computer will "see" it as many times as the value actually occurred in the sample. We must select which variable to weight cases by, but only one variable may be selected at a time. You cannot have SPSS see two different sets of data at once.

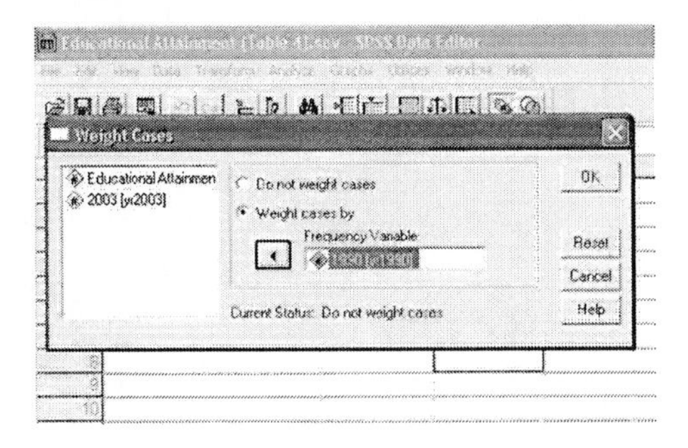

Select OK. The methods above can now be used to find a Frequency distribution.

Frequencies

Statistics

Educational Attainment

N	Valid	158870
	Missing	0

Educational Attainment

		Frequency	Percent	Valid Percent	Cumulative Percent
Valid	Less than 9th grade	16502	10.4	10.4	10.4
	9th-12th grade, no diploma	22842	14.4	14.4	24.8
	High school diploma	47643	30.0	30.0	54.8
	Some college, no degree	29780	18.7	18.7	73.5
	Associates degree	9792	6.2	6.2	79.7
	Bachelor's degree	20833	13.1	13.1	92.8
	Graduate/professional degree	11478	7.2	7.2	100.0
	Total	158870	100.0	100.0	

The frequencies should match the original data, and relative frequencies (percent), valid percent (percent when there are missing values), and cumulative relative frequencies (cumulative percent) are given. Note again that the values will be in alphabetical order. If the data is recorded as nominal, SPSS will put these values into either alphabetical order, or ordered by counts, so when typing in the values, it is usually best to include an ordering value, or to use labels. Using numbers for the data and attaching a value to the number will preserve the order without having to have the numbers out in front of the text.

If values are too large to be seen in the window, SPSS will use scientific notation for them. You can click on the table, and enlarge the width of the columns so that the values can be seen well.

►Creating a Side-by-Side Frequency Bar Graph

We can use the same steps as above to create a regular Bar Graph for an individual set of data. However, this is not the case for a Side-by-Side Bar Graph. We cannot use the weighting method, as we can only weigh data by one set of values at a time. So, to get a side-by-side plot, we must first take off any weights we have used. This is accomplished by going to the weight cases window and clicking on Do not weigh cases. As above, to make sure the order of the bar graph is in the order that we want, we must use variable labels. You can view the data values still by using **View** and checking **Value Labels**.

As in the regular bar graph, select **Graphs → Bar...** but, unlike regular bar graphs, select Clustered, for side-by-side, and Summaries of separate variables. This utilizes the summarized information instead of looking for raw data.

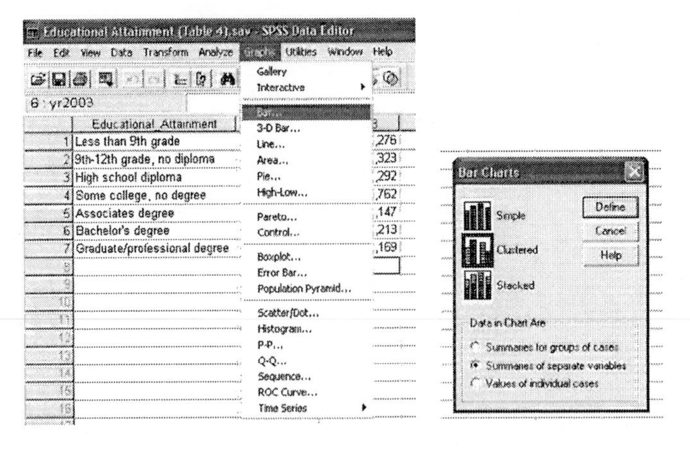

Select Define, and select the columns with the counts for the Bars Represent box, and the values as the Category Axis. The bars must represent a number, and there is a list of functions that can be used. Mean works for the type of data we have, since the mean of a single value is the value itself. Other functions, such as the median, mode, number of cases, sum, standard deviation, variance, minimum, maximum, and cumulative sum can be used. Number of cases sounds like the one that should be used, but would require raw data.

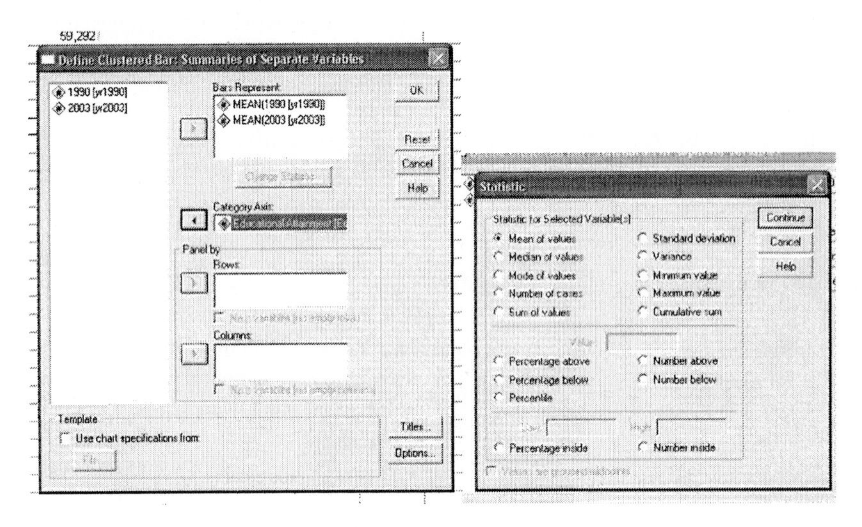

Push OK and the side-by-side frequency graph will be in the output window.

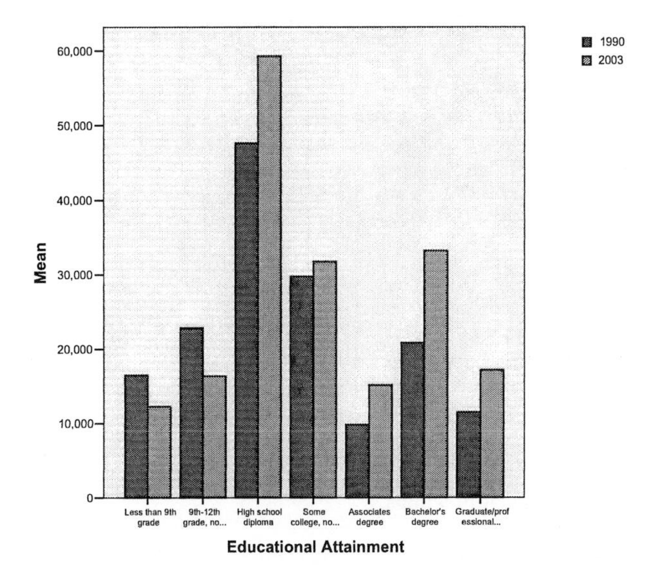

We can use the chart editor (right click on the graph to get to chart editor) to change the label on the y-axis to Frequency instead of Mean. You can also add a title, move the legend, change colors, and make other alterations that may seem necessary.

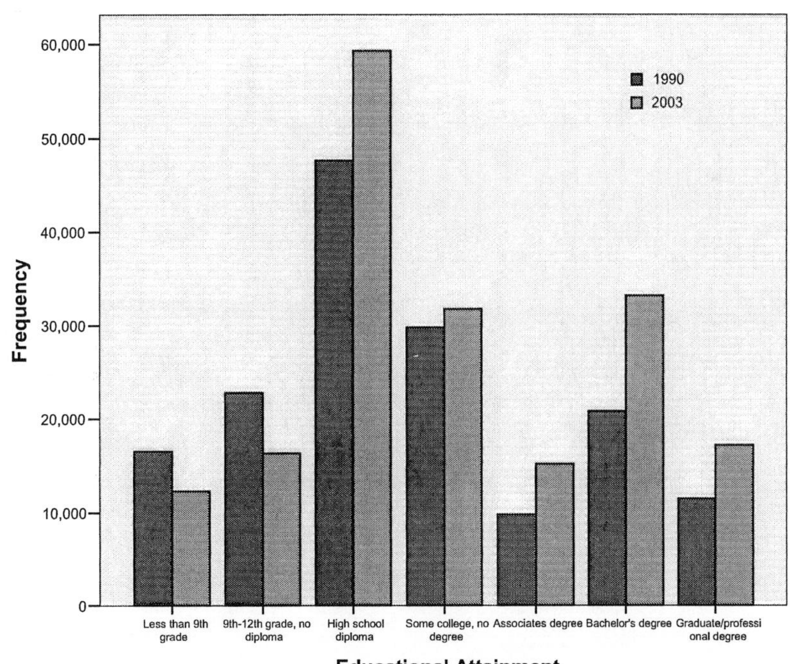

Of course, with side by side bar plots, it is better to use relative frequency rather than actual frequency, so better comparisons may be done. To do this, we will use the SPSS calculator to compute the relative frequencies. Go to **Transform → Compute…**

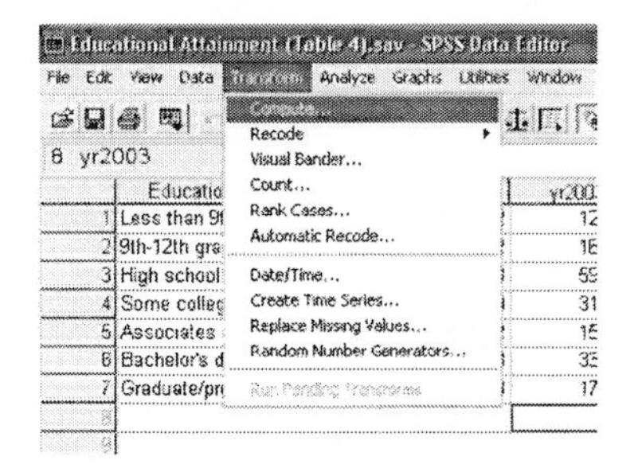

In the calculator, there are four boxes. The Target Variable is the name for the new column you are going to create, such as pct1990. This name must follow SPSS naming conventions. The second box is the Numeric Expression box. This is where the equation will go. Under the Target Variable box is the current variable box. This provides a list of all the variables currently being used, so you can use them in an equation, and also make sure you don't copy over an existing variable. The fourth box is the function box. This is a list of functions available in SPSS. Also provided is a calculator keypad, although the keyboard can provide all necessary mathematical functions.

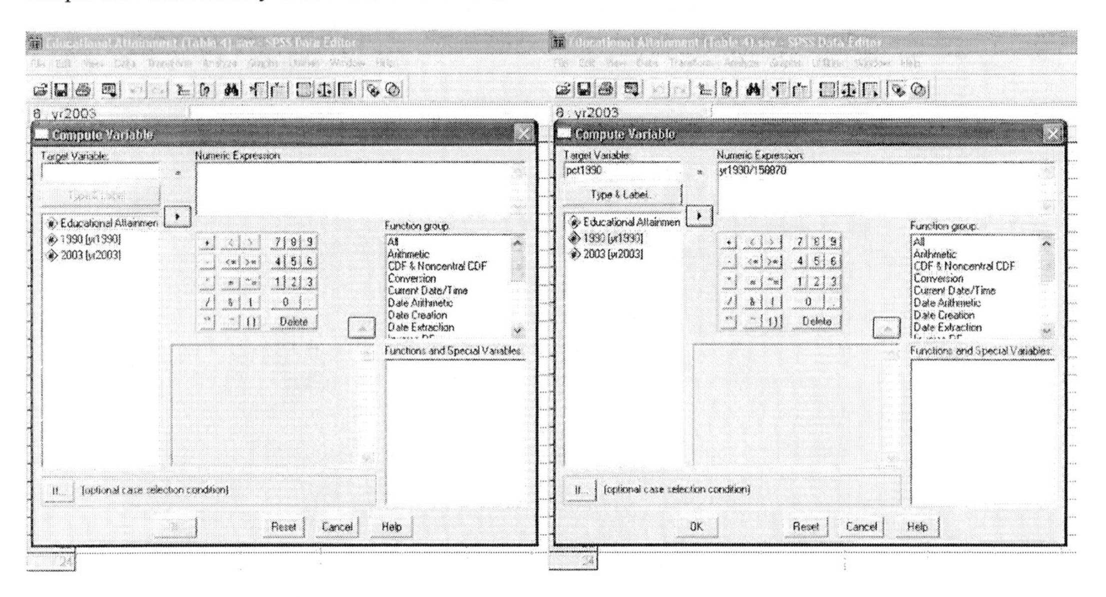

Type the function into the Numeric Expression box, you can either type in the variable name that you want to use, or click on the variable name in the list and press the ► key. This will perform the same calculation for each cell in the variable selected. Doing this for each year will provide us with two new columns in the Data View sheet, the relative frequencies for each year.

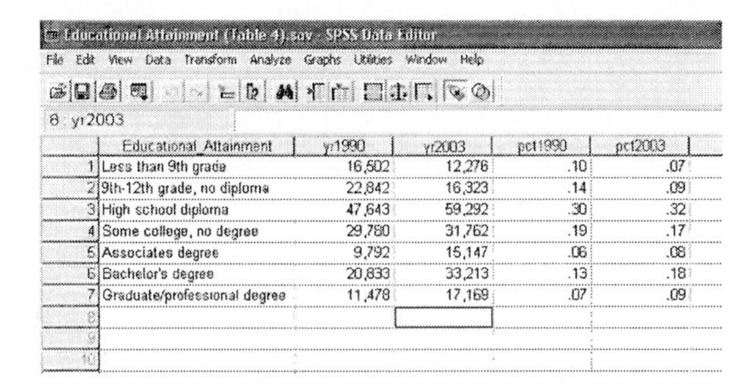

	Educational_Attainment	yr1990	yr2003	pct1990	pct2003
1	Less than 9th grade	16,502	12,276	.10	.07
2	9th-12th grade, no diploma	22,842	16,323	.14	.09
3	High school diploma	47,643	59,292	.30	.32
4	Some college, no degree	29,780	31,762	.19	.17
5	Associates degree	9,792	15,147	.06	.08
6	Bachelor's degree	20,833	33,213	.13	.18
7	Graduate/professional degree	11,478	17,169	.07	.09

We can then redo the bar plot, using the new variables instead of the counts.

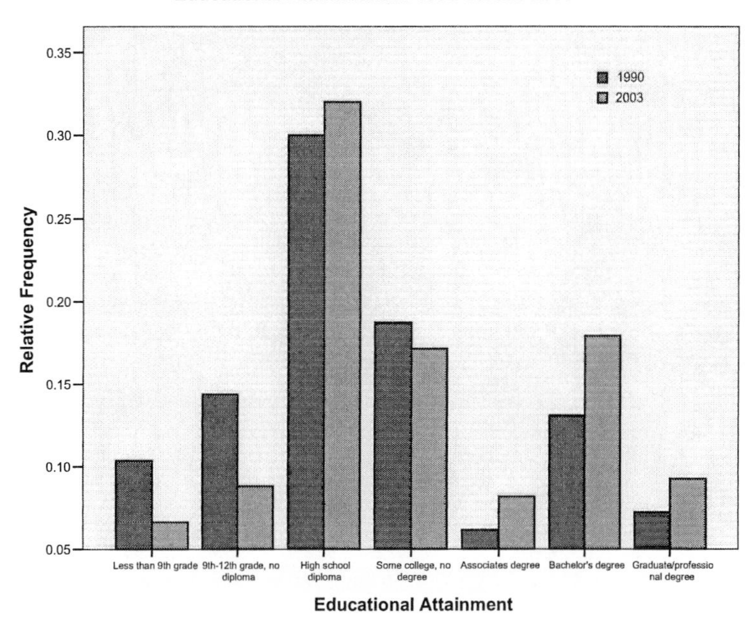

►Creating a Pie Chart

Example 6, page 66

Using the same data as above, the methods for producing a pie chart are similar to producing a frequency table. We follow all the steps of producing a frequency table until we get to the table itself. Instead of creating a table, we go to **Graph → Pie...**,

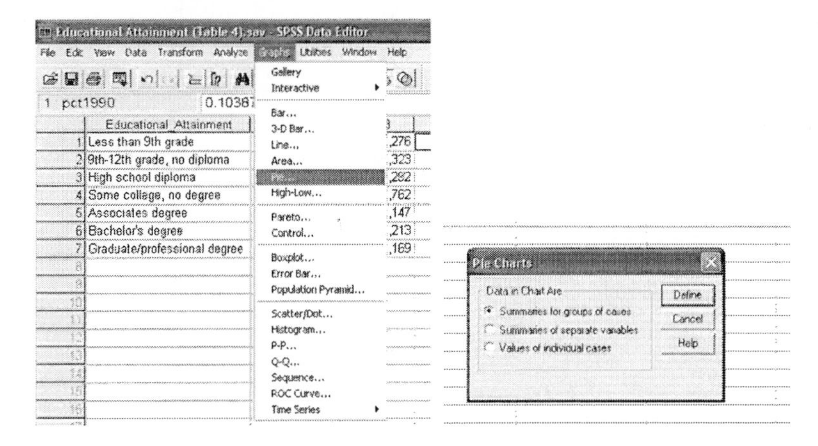

For the raw data, or weighted summarized data, use the Summaries for groups of cases, and push Define. This will open the Summaries for Groups of Cases window. We then select the variable as the Define Slices by, and if we want to create a pie chart based on N(umber) of cases, % of cases, or the sum of another variable.

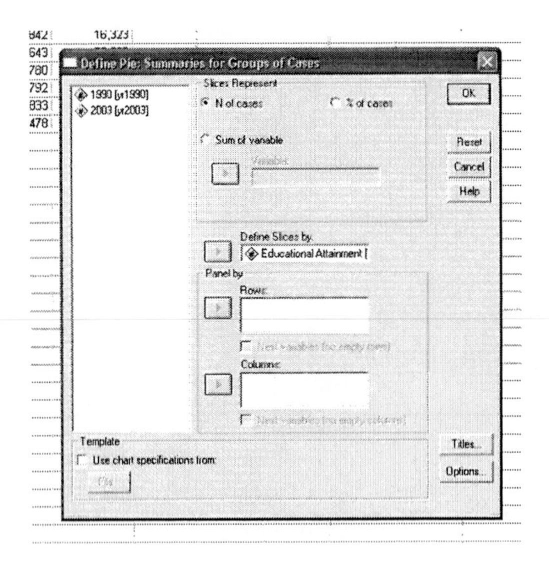

For un-weighted summarized data, use the Values of individual cases.

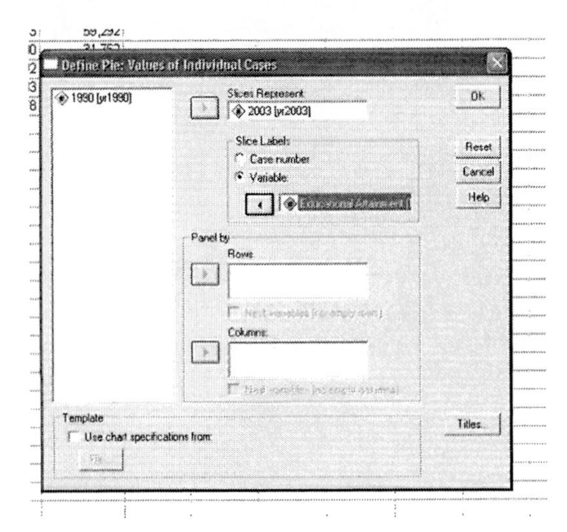

Select the counts as the Slices Represent, and the variable categories as the Variable in the Slice Labels, and push OK.

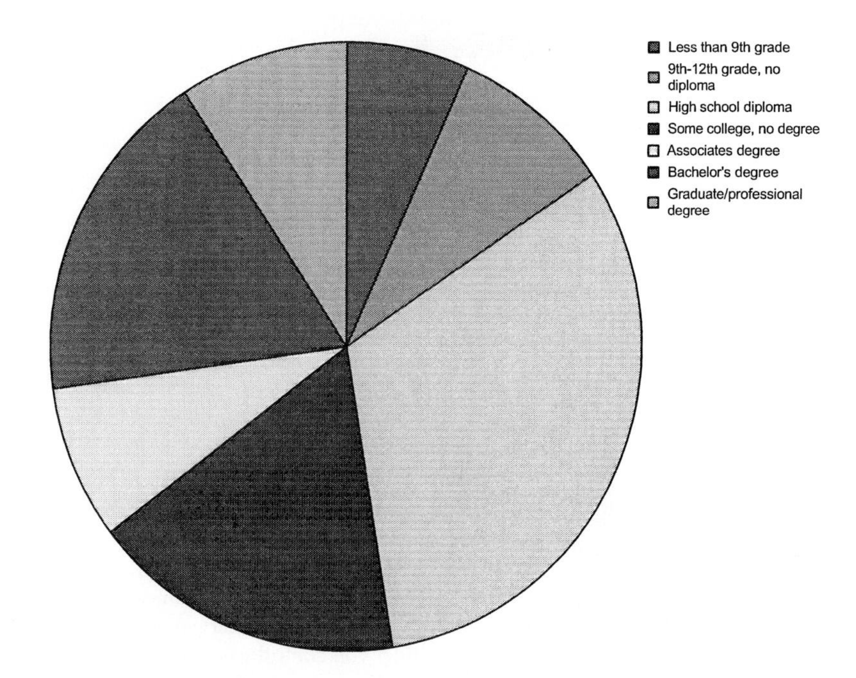

The colors and order can be changed by going to the SPSS chart object (right mouse button), and the percent labels can be added there as well. To add the labels, go to the SPSS chart object, and go to **Elements →Show Data Labels**

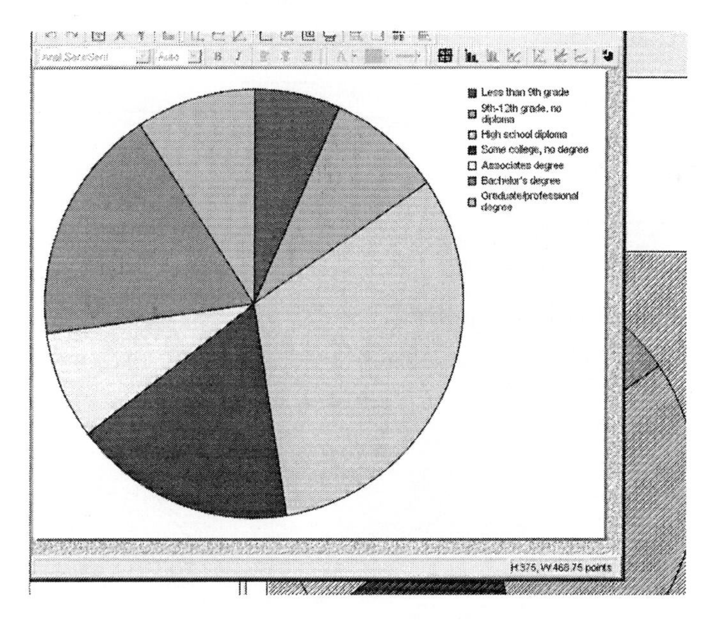

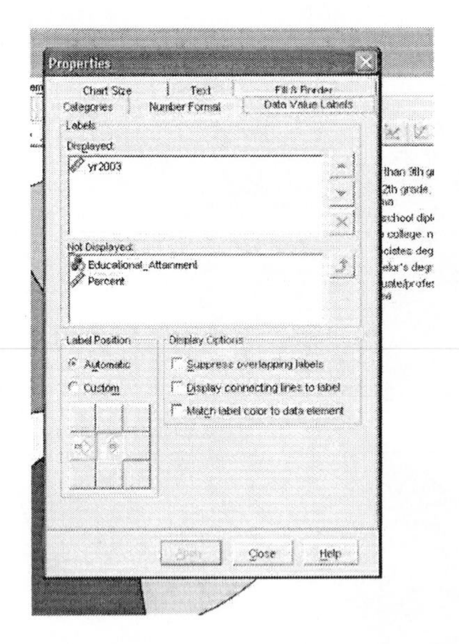

In the Properties window, select Percent and hit the green arrow button. The values in the Displayed box will be displayed on the graph. The ruler by the variable represents a numerical value, while the three colored balls represent a string. The Label Position will let you control where on the plot the value will appear (inside the wedge or outside the wedge). Other options may be played with as well.

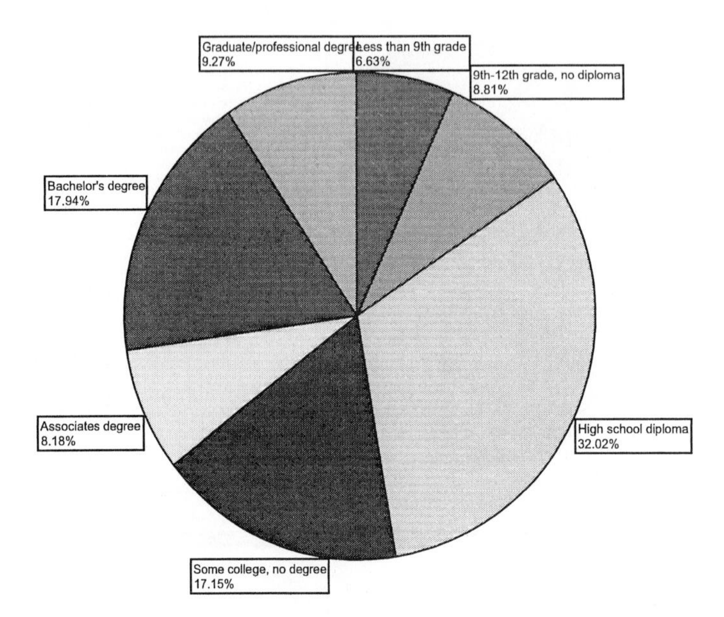

Section 2.2 Organizing Quantitative Data: The Popular Displays

►Creating a Histogram

Example 2, Page 78

To create a histogram, first, we enter the data values.

	Arrivals
1	7
2	6
3	6
4	6
5	4
6	5
7	6
8	6
9	11
10	4
11	2
12	7
13	1
14	2
15	4
16	6
17	5
18	5
19	3
20	7
21	2
22	2
23	9
24	7
25	5
26	6
27	2
28	6
29	5
30	7
31	6
32	8
33	2
34	6
35	5
36	4
37	6
38	9
39	8
40	5

Once all 40 values have been entered, we are ready to create the histogram. Go to **Graphs → Histogram…**

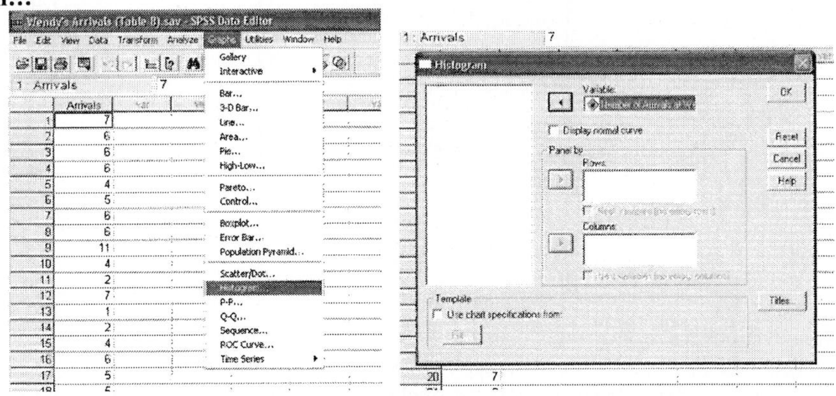

This will open up the Histogram window. Highlight the variable you are using to create the histogram and push the ► button next to the Variable box. You can add titles by clicking on the Titles… button. This allows for a two line title, a subtitle, and two lines of footnotes. Once you are ready, push OK. The histogram will appear in the Output window.

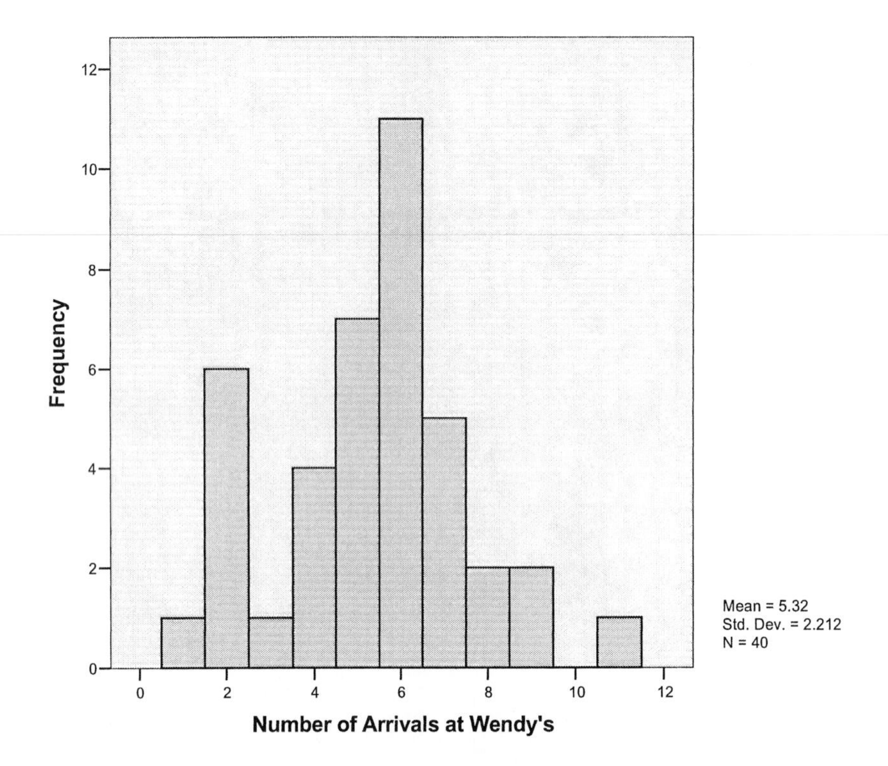

You can change the colors, and how many values are displayed on the horizontal axis by opening the SPSS Chart Object. Using the right mouse button, click on the histogram to open the chart object. In the chart editor window, click on the histogram, and select **Edit → Properties**. This will open the Histogram Options box.

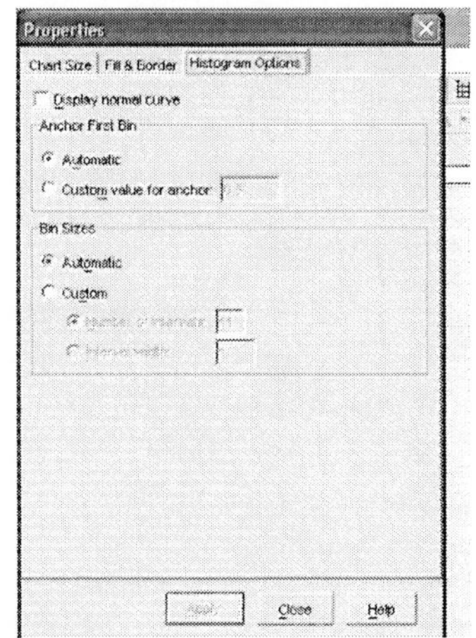

The histogram Options will allow you to change where the lowest point on the graph will be (but you cannot set it at a higher point than already is set by SPSS, and once you change it to a lower value, it cannot be increased again). You can change the number of classes, as well as the class width as well. To change colors, click on the Fill & Border tab.

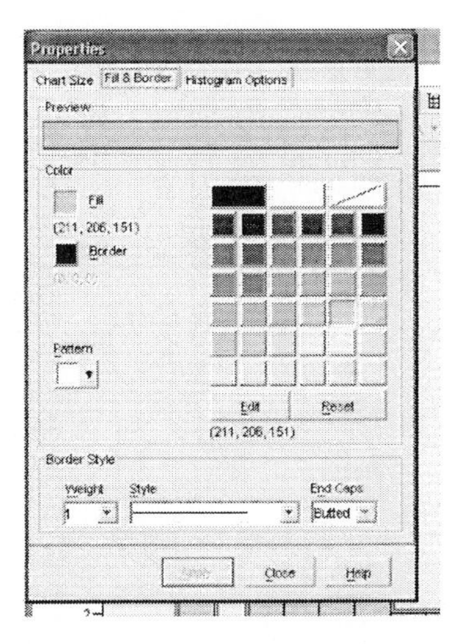

Since you clicked on the histogram before choosing the preferences, the color that appears is the color for the bars of the histogram. To change the color of the background, click on the background and select preferences. To change the numbers on the horizontal axis, click on the numbers on the current axis. Make sure that only the numbers are highlighted. Then go to preferences.

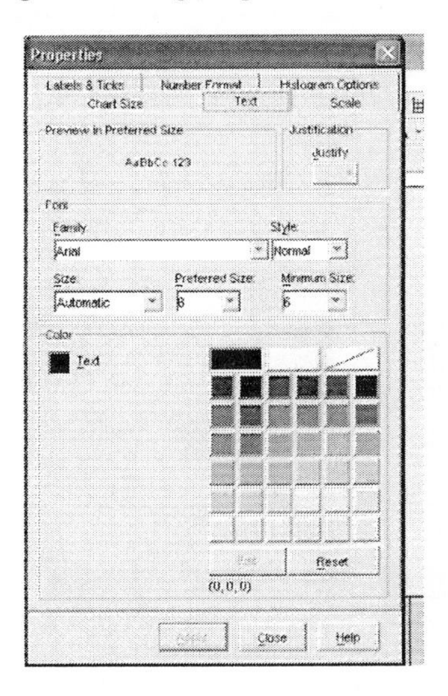

By default, you should enter the Text properties. Here you can change the font, style, and size for the numbers. Clicking on the Scale tab will take you to where you can change the numbers themselves.

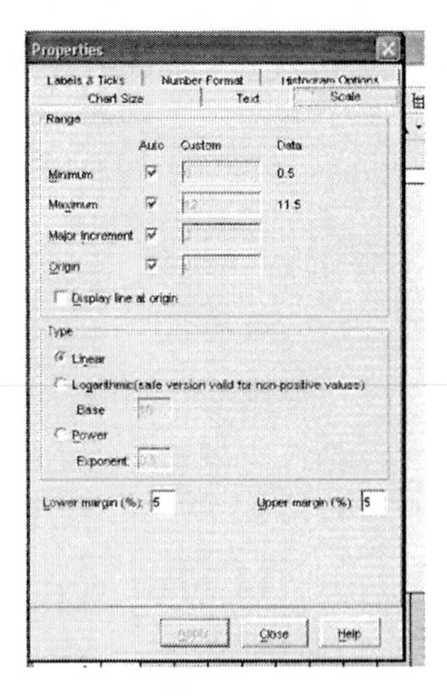

By changing the Major Increment, you change how many values will be shown. For example, to show all values in the Wendy's histogram, change the Major Increment to 1 instead of 2.

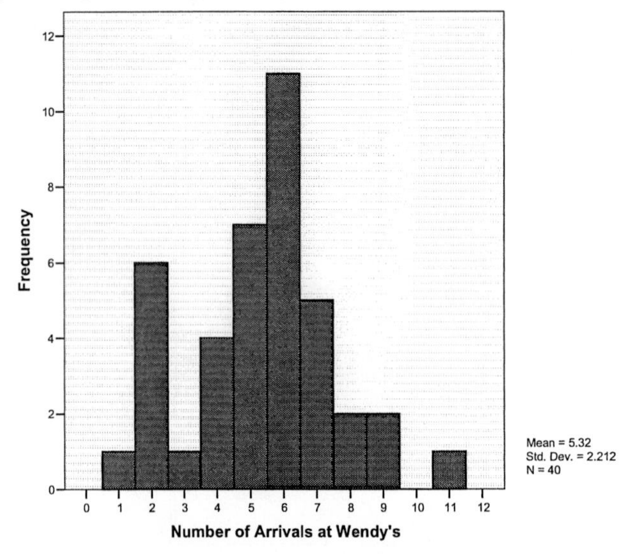

So, by changing the color, and the numbers on the horizontal axis, we can create a histogram that looks like the one on page 79. To change this into a relative frequency histogram, again go to the SPSS Chart Object. Click on the vertical axis (the numbers on the vertical (Frequency) axis should be the only things highlighted) and go to properties. Click on the Number Format Tab.

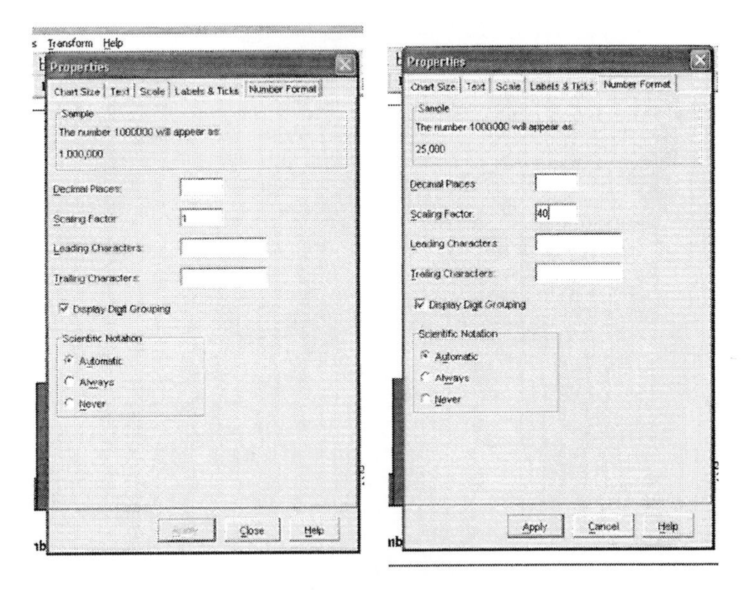

Where you see the scaling factor of 1, change this to the sample size. If you are unsure of the sample size, it appears to the right of the histogram, in this example, 40. This will change the scale to be divided by 40, which is the relative frequency. You will also need to change the label. After pushing Apply and close, highlight the Frequency label until it opens into a text box (it will change to horizontal to make typing easier), and type in what you want the label to be.

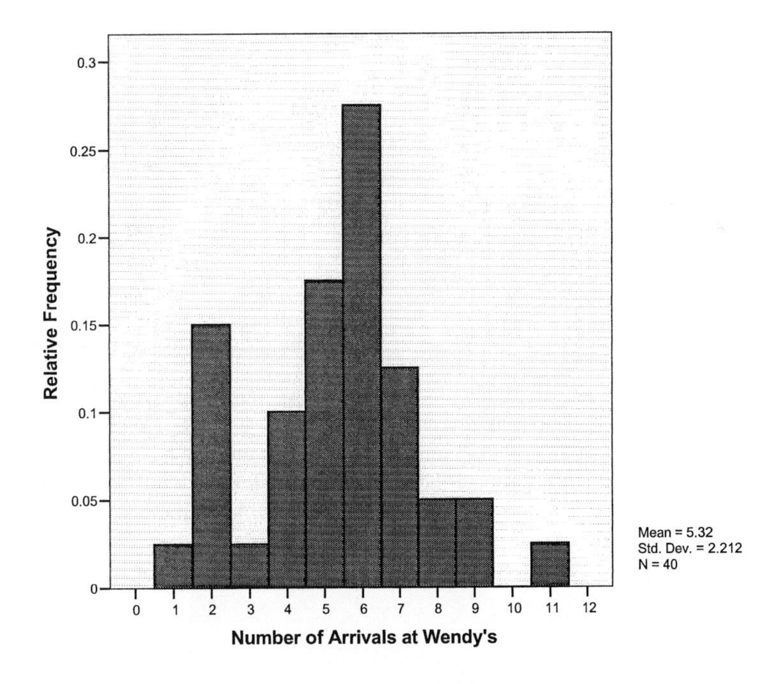

►Creating a Stem-and-Leaf Plot

Example 6, Page 82

To create a stem-and-leaf plot, first, we enter the data values.

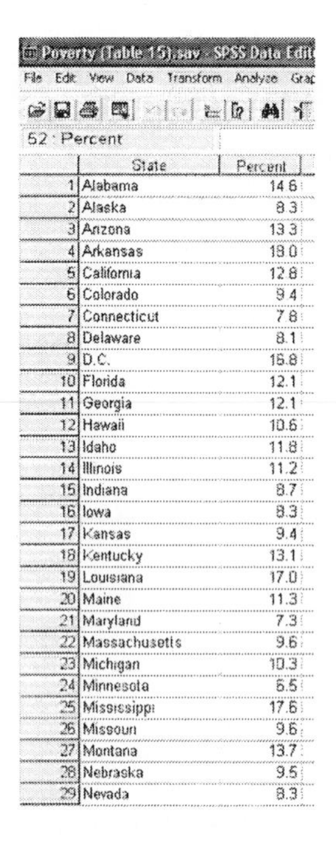

52 : Percent

	State	Percent
1	Alabama	14.6
2	Alaska	8.3
3	Arizona	13.3
4	Arkansas	18.0
5	California	12.8
6	Colorado	9.4
7	Connecticut	7.8
8	Delaware	8.1
9	D.C.	16.8
10	Florida	12.1
11	Georgia	12.1
12	Hawaii	10.6
13	Idaho	11.8
14	Illinois	11.2
15	Indiana	8.7
16	Iowa	8.3
17	Kansas	9.4
18	Kentucky	13.1
19	Louisiana	17.0
20	Maine	11.3
21	Maryland	7.3
22	Massachusetts	9.6
23	Michigan	10.3
24	Minnesota	6.5
25	Mississippi	17.6
26	Missouri	9.6
27	Montana	13.7
28	Nebraska	9.5
29	Nevada	8.3

Once the data have been entered, go to **Analyze → Descriptive Statistics → Explore…**.

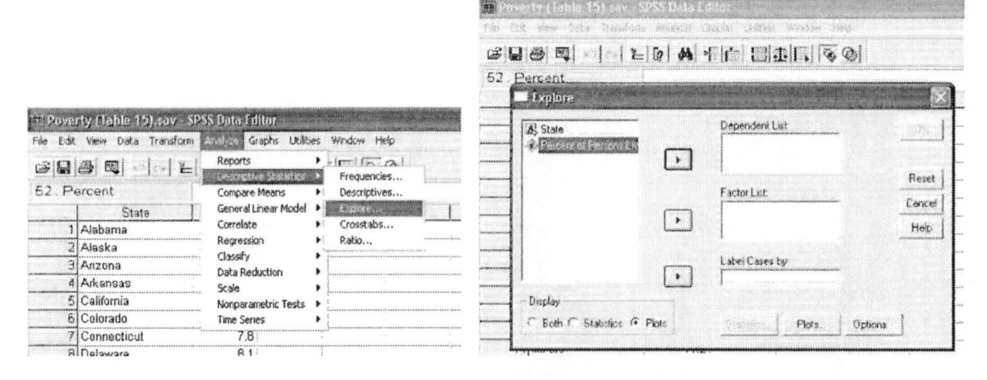

In the Explore window, highlight the value you wish to use, which must be numerical, and click on the ▶ button by the Dependent List box. Clicking on the Plots button, you will see that stem-and-leaf is a default plot for the data exploration. A box-plot is also default, but you can remove the box-plot by checking the **none** radio button under box-plots. To only get the stem-and-leaf plot, click on the Plots button. Statistics provides some summary descriptive statistics, and the both will provide both the statistics and the plots.

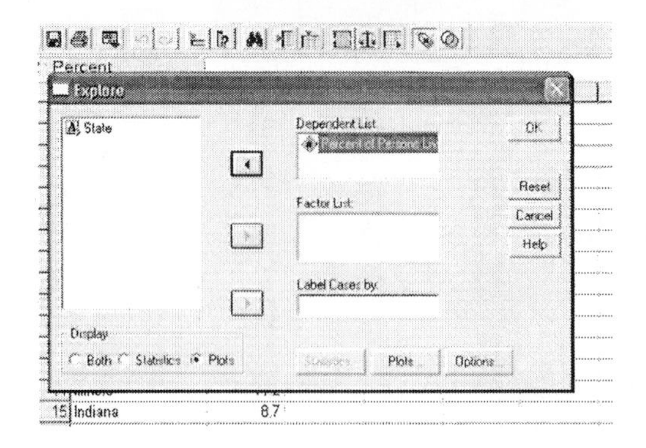

Once you are ready, push OK. The stem-and-leaf plot will appear in the output window.

Percent of Persons Living in Poverty, 2002 Stem-and-Leaf Plot

Frequency	Stem &	Leaf
1.00	0 .	5
4.00	0 .	6777
16.00	0 .	8888888999999999
11.00	1 .	00000011111
8.00	1 .	22233333
5.00	1 .	44445
5.00	1 .	66777
1.00	1 .	8

Stem width: 10.0
Each leaf: 1 case(s)

Unfortunately, SPSS does not have a method to change the stem-and-leaf plot. What the program decides to use is what you get. In the SPSS output, the first column is the frequency of leaves for each stem. This is to help show where the median would be. It is not cumulative. The next column is the Stem. SPSS has chosen to use a stem width of 10, so each stem represents the tens place. To the right of the stem is the leaf. Since the stem is in units of 10, the leaves are in units of 1. In this example, SPSS chose a split stem, which means each stem value appears more than once, to make the plot easier to read. So, the first data point in this example has 0 tens, and one 5, so it would be 5. The sixes and sevens appear in the next stem, the eights and nines in the next, etc.

►Creating a Dot Plot

Example 9, page 86

To create a dot plot, we first enter the data, which in this example is the same as in example 1. Once we have the data, go to **Graphs → Scatter/Dot…**. From the five choices for type of graph, select Simple Dot.

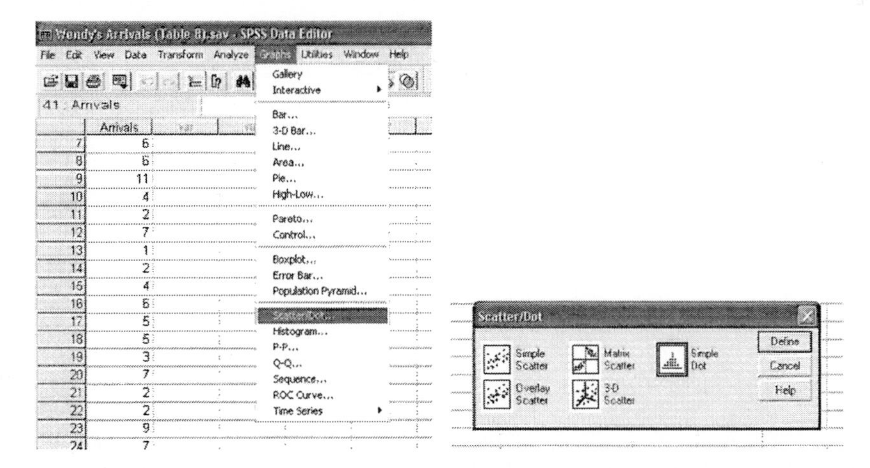

Select the variable that you want to make the dot plot of and click on the ► button to place it in the X-Axis Variable. Then, push OK.

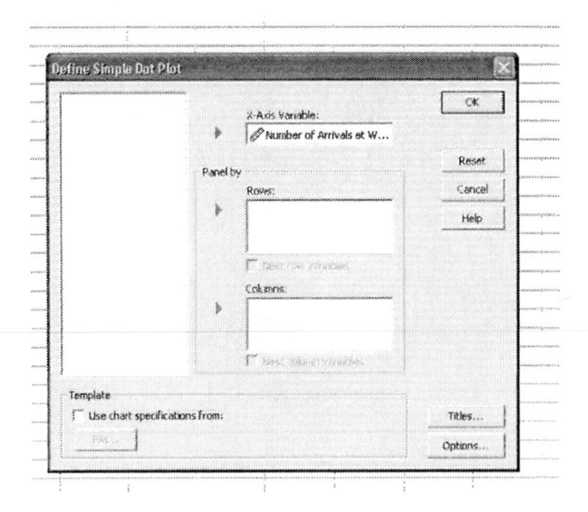

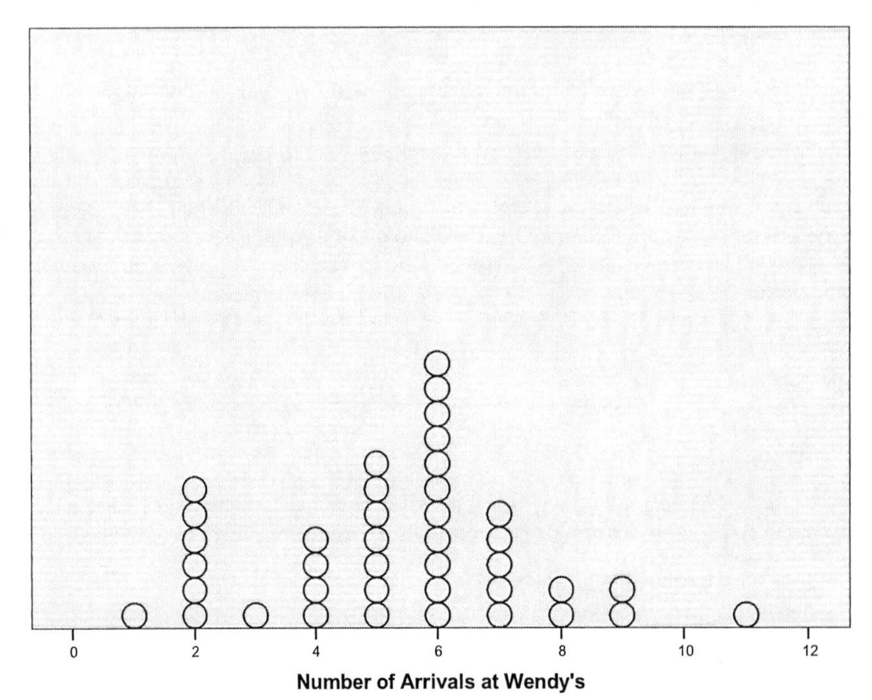

You will have a dot plot of the data. Using the chart editor, you can change the size and shape of the dots, as well as color, labels, titles, etc.

►Creating a Back-to-Back Histogram

Problem 44, page 96

Creating a back-to-back histogram is new to SPSS version 13. If you are using a previous version of SPSS, you cannot use these methods. As will often be the case with SPSS, the data must follow a specific format. This is due to the survey analysis nature of SPSS. In this case, we need to create two columns. One column will be the value of interest, in this case homerun distance. The second column will be the player's name.

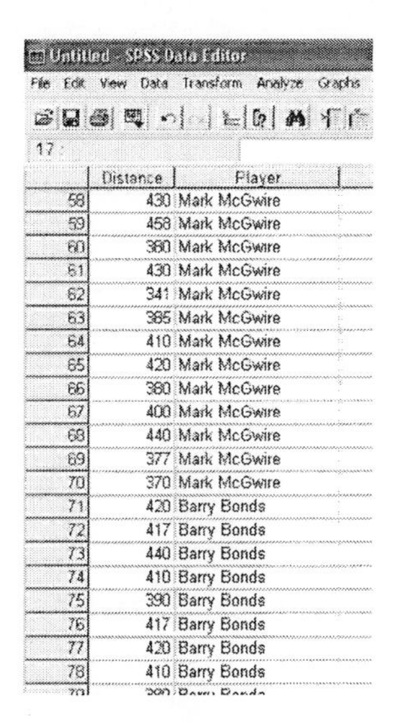

Once the data have been entered, go to **Graphs → Population Pyramid…**

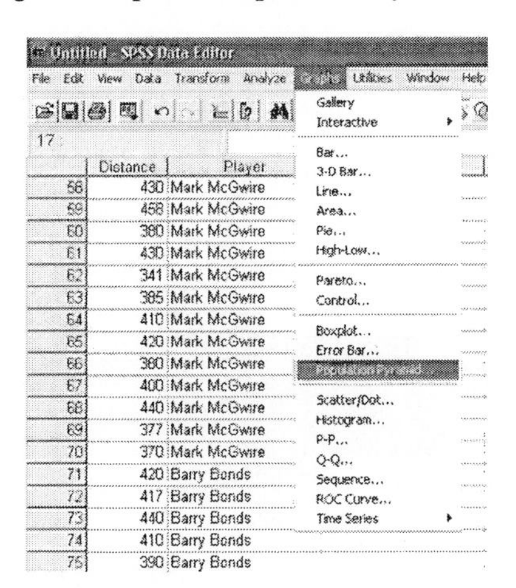

The population pyramid is the back-to-back histogram. Once you get to the Define Population Pyramid window, highlight the category (player) and the ▶ button by the **Show Distribution Over:** box. Highlight the grouping variable (homerun distance) and the ▶ button by the **Split by**: box. Once the variables are in the correct places, push OK.

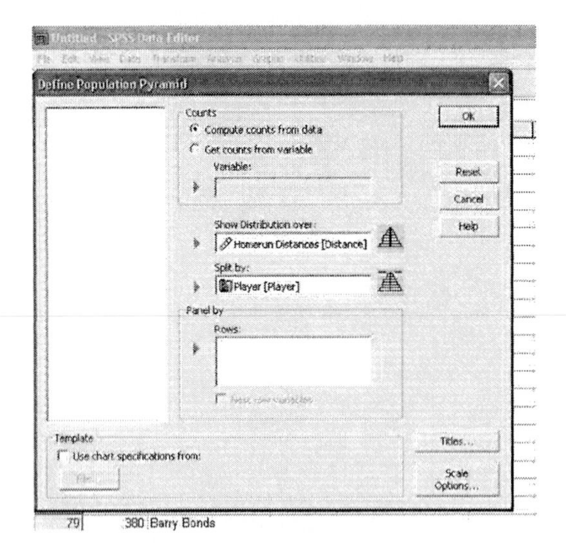

The output window will now contain the back-to-back frequency histogram.

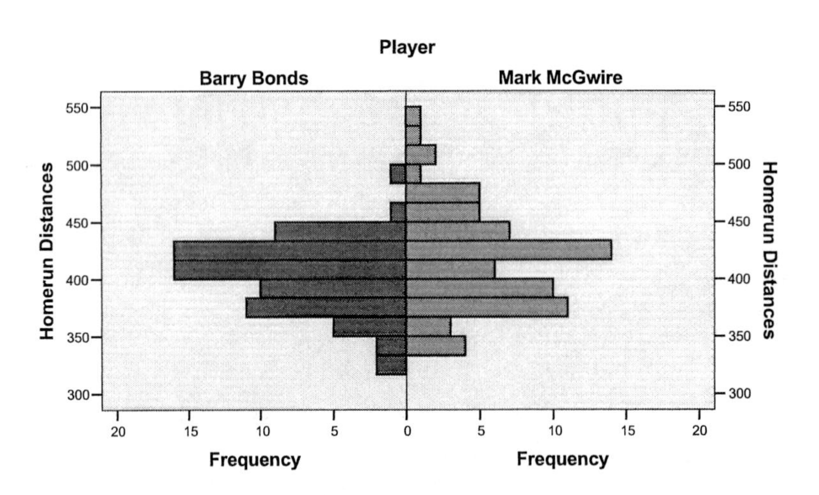

To change this to a relative frequency histogram, go to the Chart Editor, highlight the numbers on the horizontal axis (the frequency), and go to properties. Click on the Number Format tab, and change the Scaling Factor to the larger of the two sample sizes, in this case you can use 73.

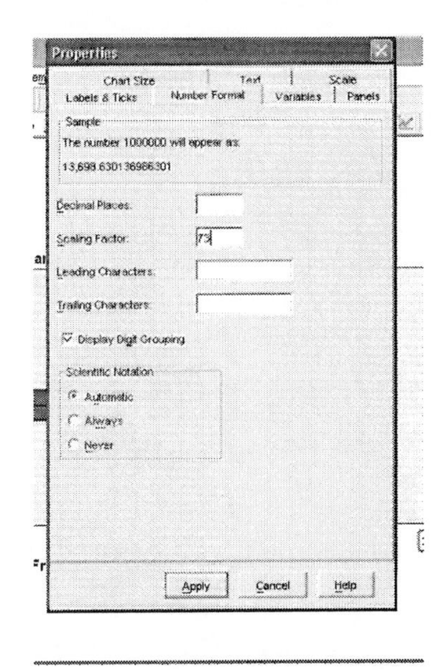

This will change the frequency histogram into a relative frequency histogram. You will also want to change the label from sum to percent. Changing the label for one group will change the label for both.

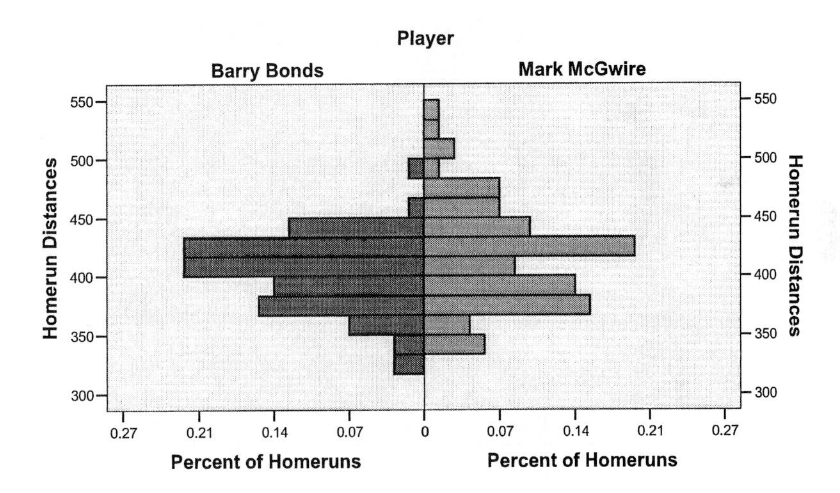

Section 2.3 Additional Displays of Quantitative Data

►Creating a Frequency Polygon

Example 1, Page 97

To create a frequency polygon, first we enter the data. We will create two columns for the classes, and one column for the frequency. Creating two columns for the classes will allow SPSS to do the work for us.

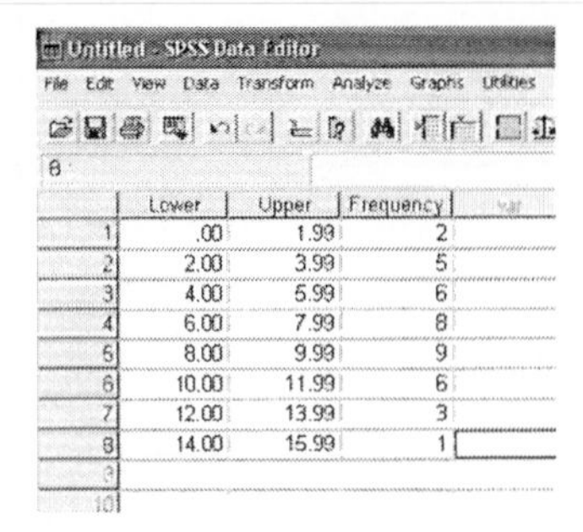

	Lower	Upper	Frequency
1	.00	1.99	2
2	2.00	3.99	5
3	4.00	5.99	6
4	6.00	7.99	8
5	8.00	9.99	9
6	10.00	11.99	6
7	12.00	13.99	3
8	14.00	15.99	1

Once the data have been entered, we go to **Transform** → **Compute** to calculate the midpoints. We type the name for the new variable, midpoint, in the Target Variable box. In the Numeric Expression box, we type in the formula. We just need to remember to use parentheses for the addition, so that the numbers are added before dividing by 2.

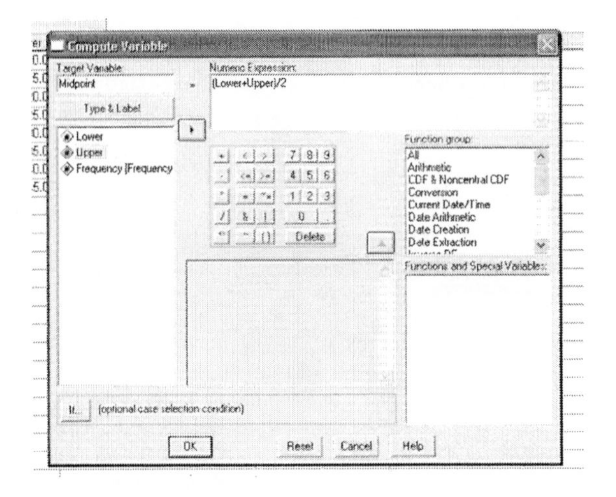

This creates the column of midpoints for us. Similarly, we can use the **Transform** → **Compute** to calculate the relative frequency.

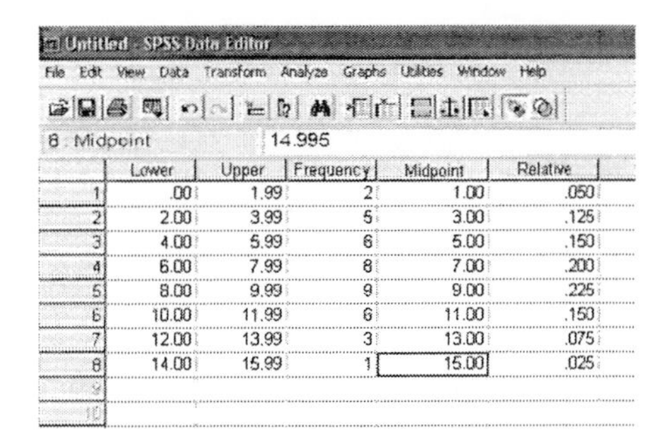

	Lower	Upper	Frequency	Midpoint	Relative
1	.00	1.99	2	1.00	.050
2	2.00	3.99	5	3.00	.125
3	4.00	5.99	6	5.00	.150
4	6.00	7.99	8	7.00	.200
5	8.00	9.99	9	9.00	.225
6	10.00	11.99	6	11.00	.150
7	12.00	13.99	3	13.00	.075
8	14.00	15.99	1	15.00	.025

Once the data have been entered, we go to **Graphs → Scatter…**

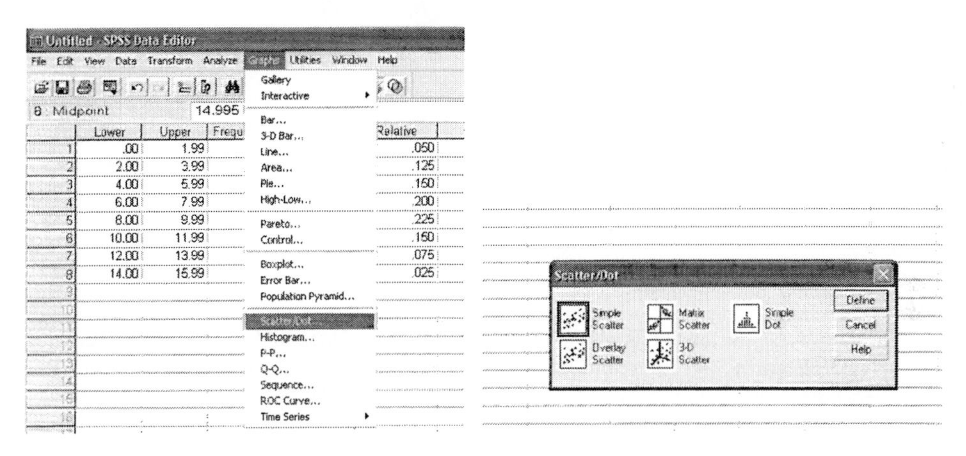

Select Simple Scatter and push the Define button. In the Simple Scatterplot window, select the frequency, or relative frequency for the Y-Axis, depending on whether you want a frequency polygon or a relative frequency polygon. Select the Midpoints as the X—Axis.

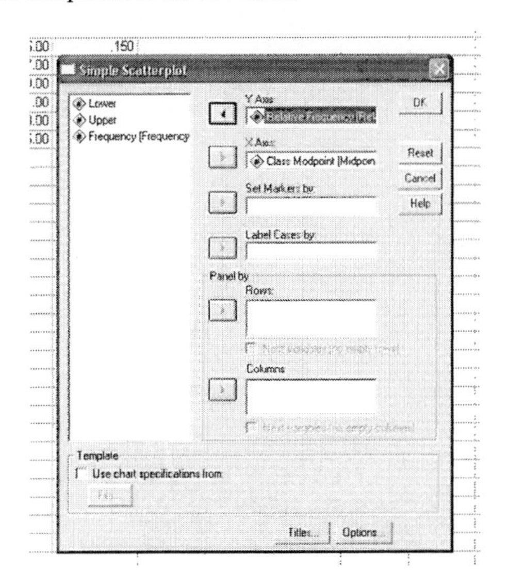

Once the variables have been selected, press OK.

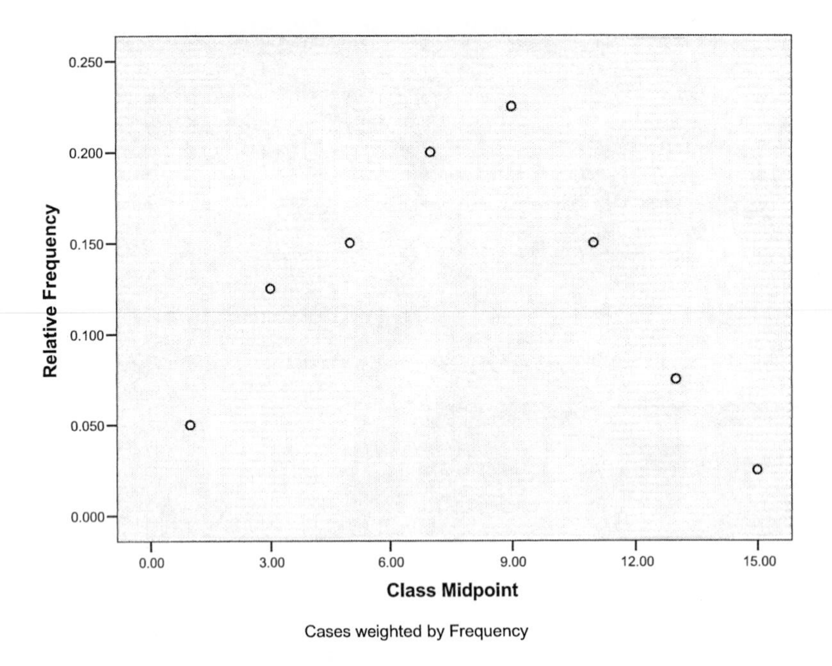

This provides the "dots" for the polygon. Go to the Chart Editor. Once in the Chart Editor, click on any of the dots, and go to **Elements → Interpolation Line**. The line is automatically added, so push close, and close the Chart Editor. This connects the dots, and provides your (relative) frequency polygon.

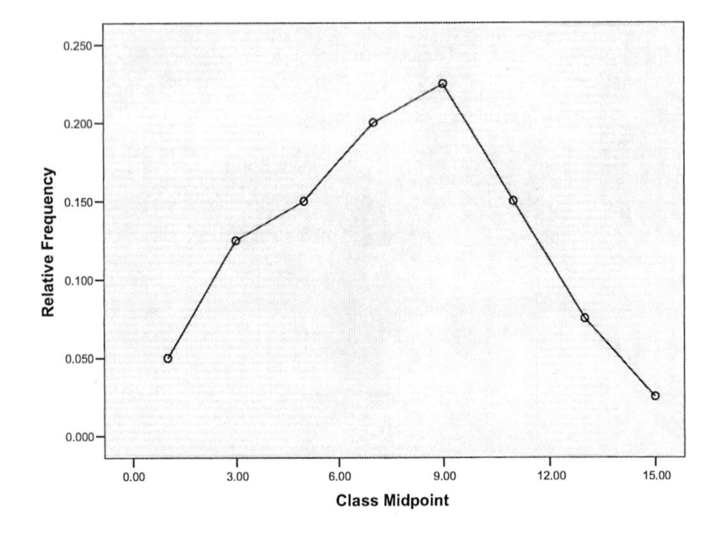

Ogives can be created in the same way, using cumulative frequencies instead of frequencies.

►Creating a Time Series Plot

Example 1, Page 100

To create a time series plot, enter the data into SPSS, one column for the dates, and one column for the values.

Cisco Closing Price (Table 19).s

File Edit View Data Transform Anal

25 : Closing

	Date	Closing
1	MAR 03	12.98
2	APR 03	15.00
3	MAY 03	16.41
4	JUN 03	16.79
5	JUL 03	19.49
6	AUG 03	19.14
7	SEP 03	19.59
8	OCT 03	20.93
9	NOV 03	22.70
10	DEC 03	24.23
11	JAN 04	25.71
12	FEB 04	23.16
13	MAR 04	23.57
14	APR 04	20.91
15	MAY 04	22.37
16	JUN 04	23.70
17	JUL 04	20.92
18	AUG 04	18.76
19	SEP 04	18.10
20	OCT 04	19.21
21	NOV 04	18.75
22	DEC 04	19.32
23	JAN 05	18.04
24	FEB 05	17.42

Once the data have been entered, go to **Graphs** → **Sequence…**. Highlight the date variable and click the ► button next to the Time Axis Labels: box. Highlight the values variable, and click the ► button next to the Variable: box.

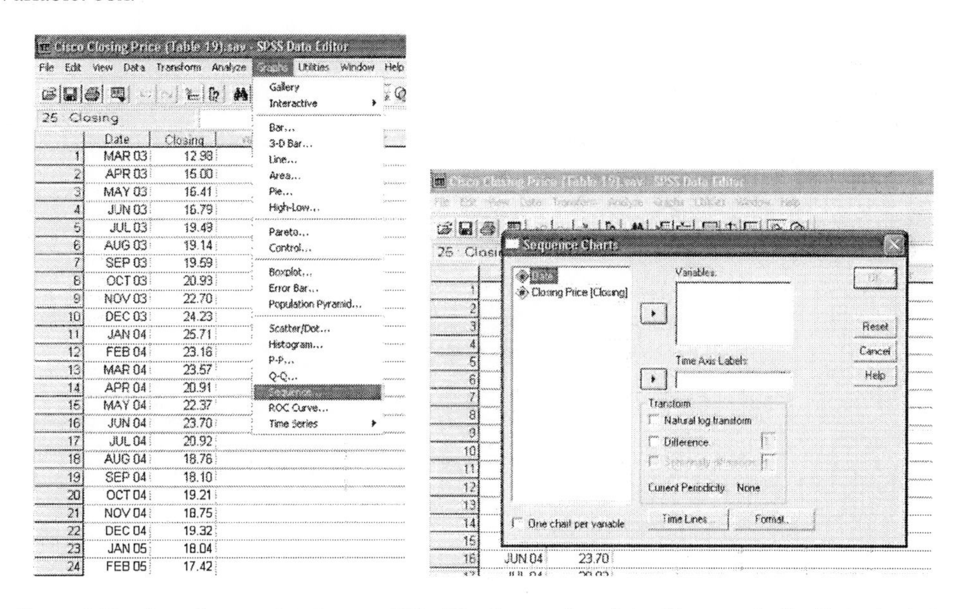

Once the variables have been entered, press OK. The time series plot will appear in the Output window.

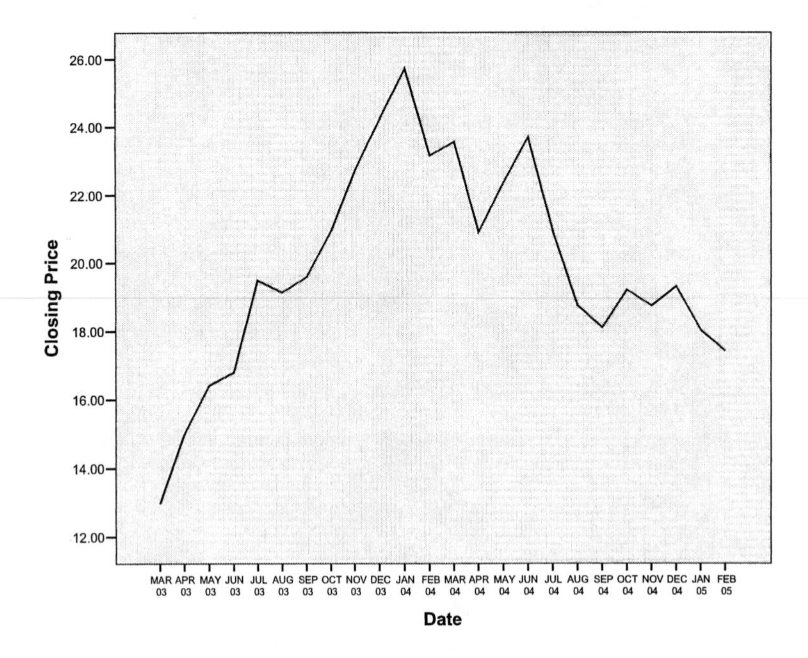

Using the Chart Editor, you can change the scale, add a title, etc.

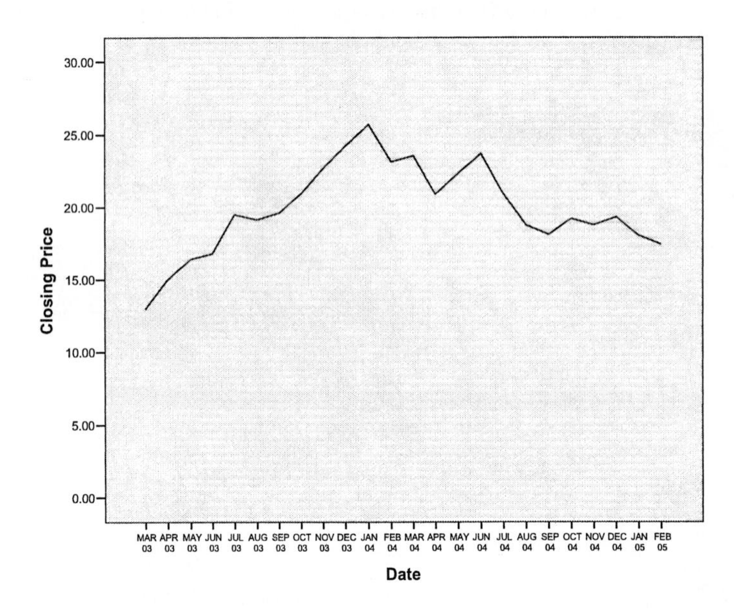

Chapter 3. Numerically Summarizing Data

Finding descriptive statistics is very easy to do in SPSS, but there are many possible ways to do so. SPSS is an analysis tool, not a scholastic tool, and as such, populations are not available in SPSS. SPSS automatically assumes that the information comes from a sample, as most real data does. If the data does represent a population, some adjustments to the SPSS values must be made.

► Finding the basics: mean, standard deviation, percentiles, etc.

Example 1, Page 122

The basic descriptive statistics are the mean and standard deviation. These are very easy to obtain in SPSS. Once the data have been entered, go to **Analyze → Descriptive Statistics → Descriptives…**.

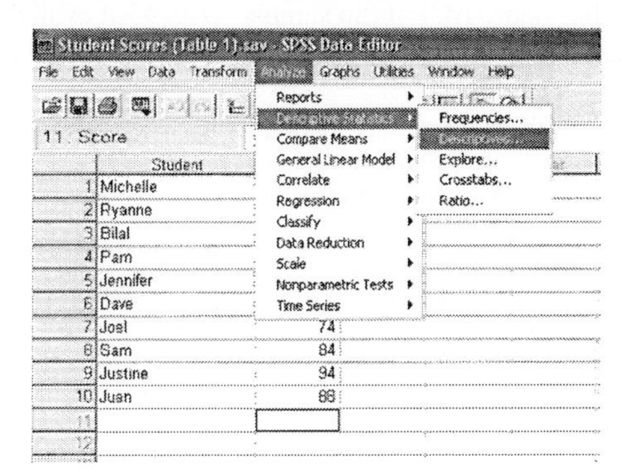

Once in the Descriptives window, highlight the variable(s) you want the mean and standard deviation of, and push the ► button by the Variable(s) window.

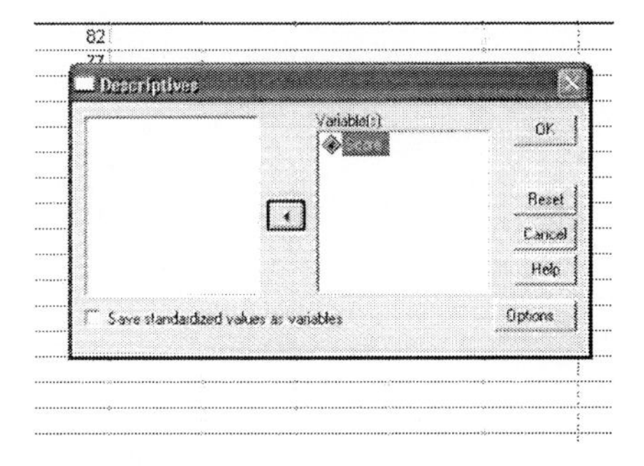

Pushing the Options button will allow you to select which descriptive statistics you want to use.

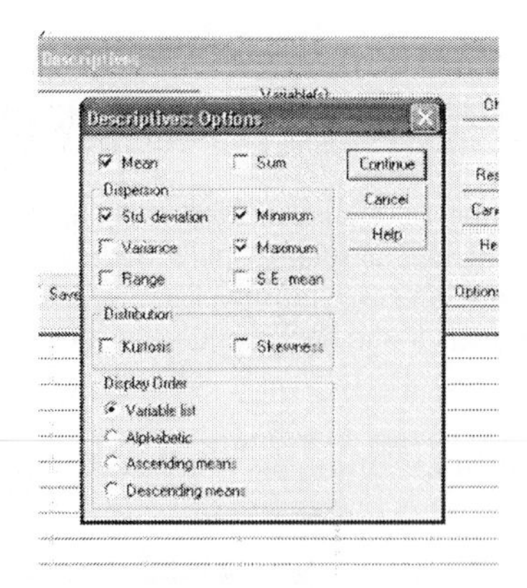

The mean, standard deviation, minimum, and maximum are default values. You can also select the Sum, or total of all the values, the Variance, Range, and standard error of the mean (S.E. mean), which is the standard deviation divided by the square root of the sample size. Also available are the Kurtosis, which measures peakedness, or the amount to which the data values cluster around the center, and the skewness, which measures how asymmetric the values are.
Once you have selected your variable(s), push OK. The results will appear in the output window.

Descriptive Statistics

	N	Minimum	Maximum	Mean	Std. Deviation
Score	10	62	94	79.00	10.349
Valid N (listwise)	10				

If the list of possible statistics in descriptives is not satisfying, there are more options available by using **Analyze → Descriptive Statistics → Frequencies…**. Once in the Frequencies window, highlight the variable(s) you want the mean and standard deviation of, and push the ▶ button by the Variable(s) window.

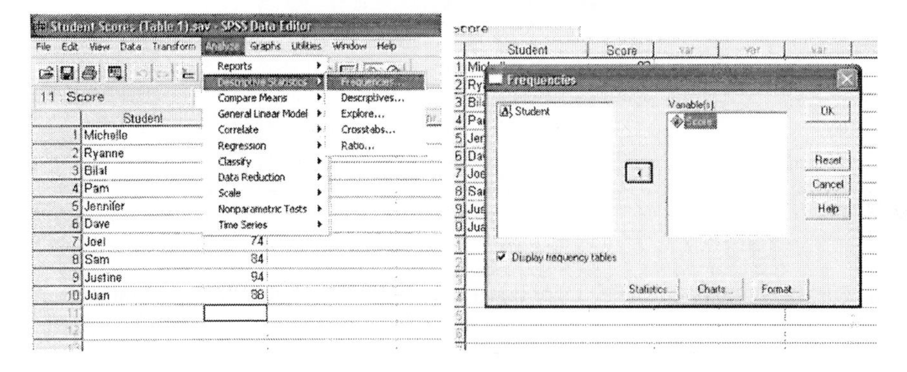

You do not usually want to see a frequency table of numerical data, as this is not a grouped frequency, but actual frequency. Instead you want to select some statistics. If you uncheck the Display frequency tables before selecting some statistics, you will get a warning, but it is only to remind you that without selecting something, there will be no output. Click on the Statistics button.

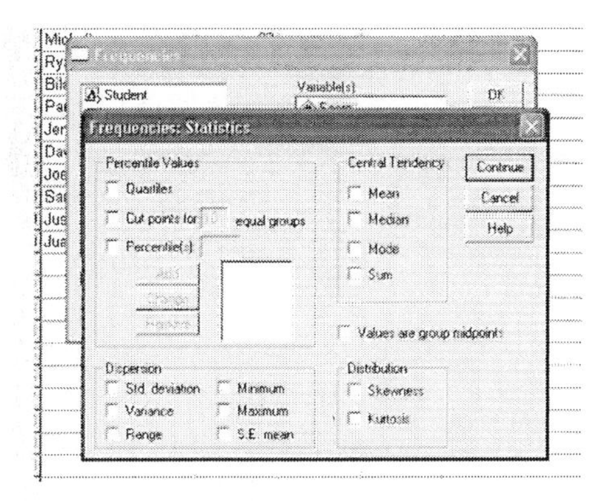

Once in the Frequencies: Statistics window, you can select which statistics you want. These are grouped into Central Tendency (Section 1) on the top right, Dispersion (Section 2) on the bottom left, Percentile Values (Section 3) on the top left, and Distribution (not in book) on the bottom right.
For measures of Central Tendency, you can select the mean, median, mode (if one exists), and the sum. From dispersion, you can select standard deviation, variance, range, maximum, minimum, and standard error of the mean (standard deviation divided by the square root of n). Again, it must be remembered that these are the sample values, not population values.
From the Percentiles, you can select the quartiles, values to cut the data into equal groups (using both quartiles and cut points for 4 equal groups will be redundant, and SPSS only prints each value once), or specific percentile values. For percentiles, check the box to the left of Percentile(s), and type the value in the box to the right of Percentile(s), and push the Add button. You can ask for as many percentiles as you wish. Again, asking for the quartiles, along with the 25th, 50th, and 75th percentiles is redundant, and will only be printed once.

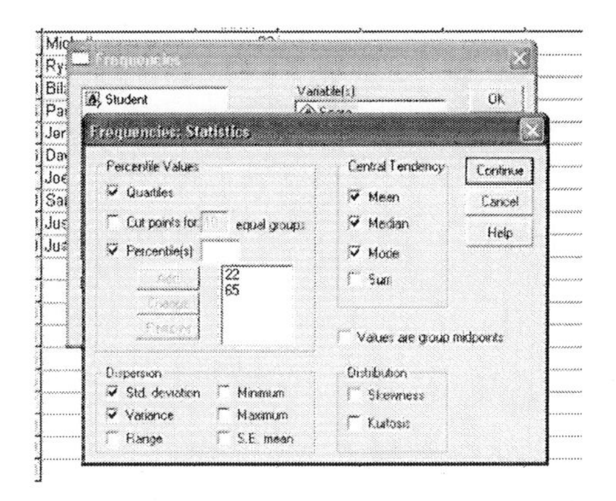

Once you have selected which statistics you want, push Continue, make sure the Display Frequency Tables is unchecked, and push OK. The results will appear in the output window.

Statistics

Score

N	Valid	10
	Missing	0
Mean		79.00
Median		79.50
Mode		62[a]
Std. Deviation		10.349
Variance		107.111
Percentiles	22	69.26
	25	70.25
	50	79.50
	65	84.60
	75	88.50

a. Multiple modes exist. The smallest value is shown

Note that SPSS will attempt to find a mode, but that does not mean that the mode actually exists.

Other calculations, such as the population variance, population standard deviation, inter-quartile range, and z-scores, must be found by hand. SPSS does not calculate these directly.

Another method to obtain summary statistics is **Analyze → Descriptive Statistics → Explore**.

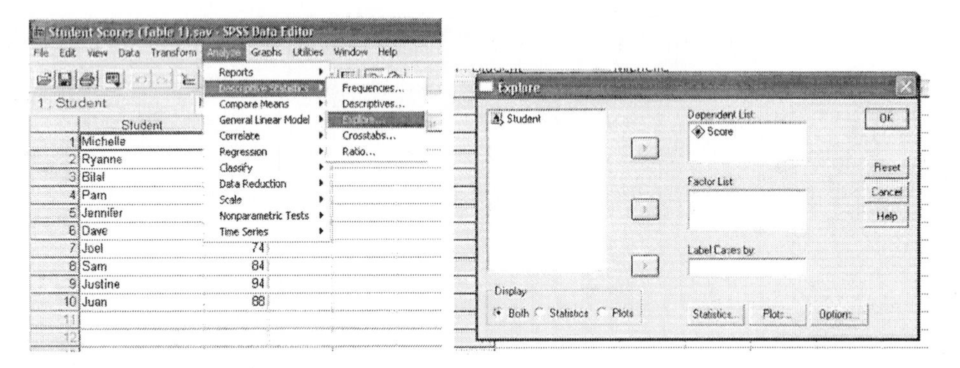

Choose the variable of interest and put it in the Dependent List. To select which statistics you want displayed, click on the **Statistics** button.

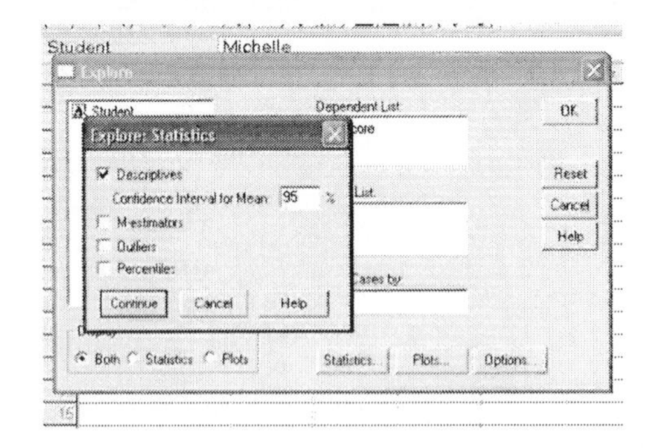

Descriptive will provide the mean, median, mode, trimmed mean (which is the mean after dropping out the points which make the outer 5% of the data), the standard error (which is the standard deviation divided by the square root of the sample size, and will be used in later chapters), variance, standard deviation, minimum, maximum, range, IQR, skewness (a measure of how skewed the distribution is, which is not used in this book), skewness standard error, kurtosis (which is a measure of how much of the data is in the center, also not used in the book), and kurtosis standard error. You will also be provided with a Confidence interval for the Mean, which will be addressed in chapter 8. M-estimators are better estimators of the center when there are extreme values, but are not addressed in this book. Outliers will provide the 5 smallest and 5 largest observations so you can check if they are outliers. Percentiles provides the 5th, 10th, 25th, 50th, 75th, 90th, and 95th percentiles.

Clicking on the Plots option will provide the options for creating a Boxplot, Stem-and-Leaf plot, and Histogram. You can choose to have only statistics, only plots, or both in the output.

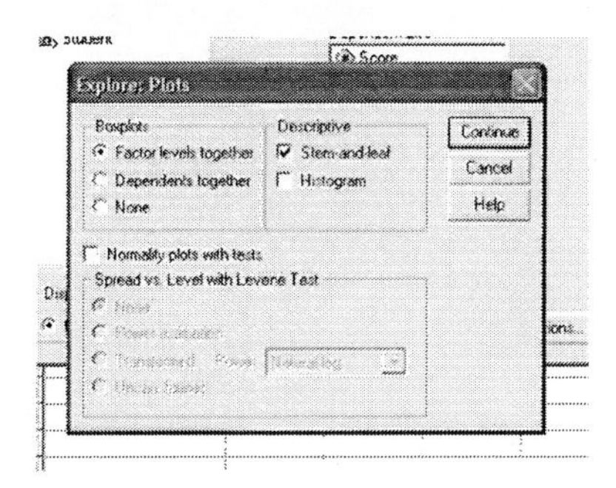

The output will look similar to the below, using the default options.

Descriptives

			Statistic	Std. Error
Score	Mean		79.00	3.273
	95% Confidence Interval for Mean	Lower Bound	71.60	
		Upper Bound	86.40	
	5% Trimmed Mean		79.11	
	Median		79.50	
	Variance		107.111	
	Std. Deviation		10.349	
	Minimum		62	
	Maximum		94	
	Range		32	
	Interquartile Range		18	
	Skewness		-.163	.687
	Kurtosis		-1.002	1.334

►Finding measures based on Grouped data

Example 1, Page 158

When entering the data, it makes things a little easier to enter the lower class limit and upper class limit as separate variables. This will allow you to make calculations in SPSS rather than by hand.

Rate of Return (Table 15).sav - SPSS Data Editor

File Edit View Data Transform Analyze Graphs Utilities

9 : Frequency

	Lower	Upper	Frequency
1	.00	1.99	2
2	2.00	3.99	5
3	4.00	5.99	6
4	6.00	7.99	8
5	8.00	9.99	9
6	10.00	11.99	6
7	12.00	13.99	3
8	14.00	15.99	1
9			
10			

Once the data have been entered, we must find the class midpoints. We can do that by going to **Transform → Compute…**.

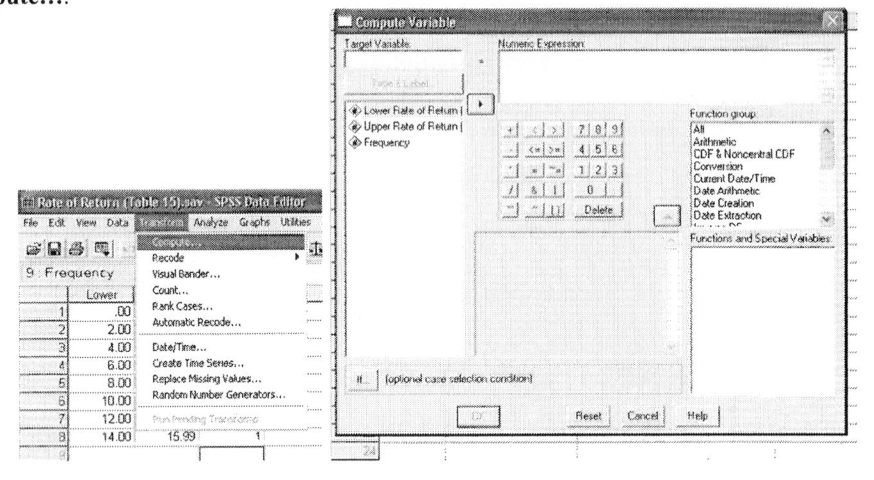

In the Target Variable box, create a name for the midpoint. In the Numeric Expression box, we will create the equation (Upper + Lower)/2, which is why it is easier to have these as separate variables. This will not work if you create one text variable to define the classes.

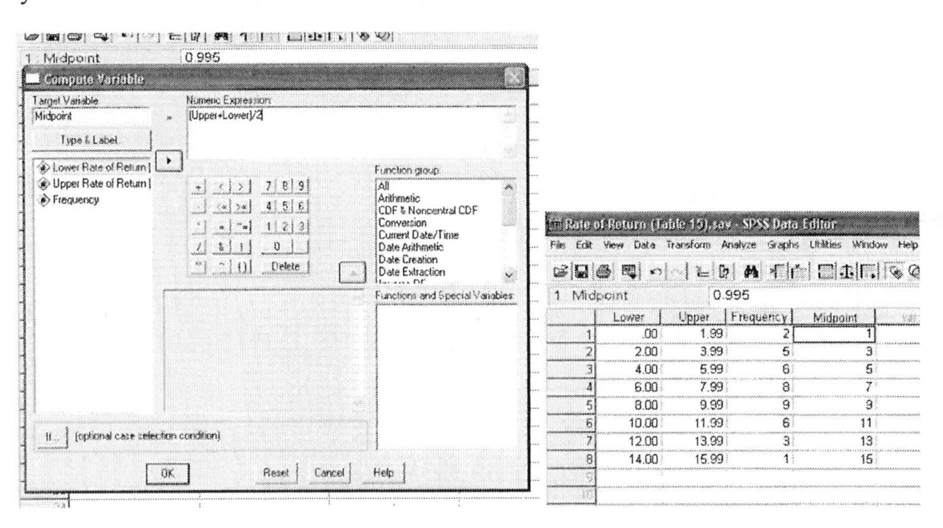

	Lower	Upper	Frequency	Midpoint
1	.00	1.99	2	1
2	2.00	3.99	5	3
3	4.00	5.99	6	5
4	6.00	7.99	8	7
5	8.00	9.99	9	9
6	10.00	11.99	6	11
7	12.00	13.99	3	13
8	14.00	15.99	1	15

When the equation looks correct, push OK. You will now have a variable of frequencies, and a variable of midpoints. **Go to Data → Weight Cases…**. We want to weigh the cases by the frequency. This tells SPSS to see the midpoints as many times as we have a frequency instead of once.

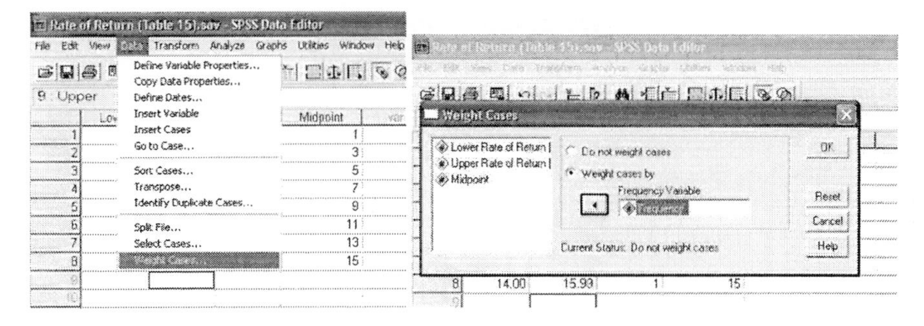

Now, we can use the same procedures as for ungrouped data, but we must remember that these are only approximate values.

Descriptives

			Statistic	Std. Error
Midpoint	Mean		7.5950	.54632
	95% Confidence Interval for Mean	Lower Bound	6.4900	
		Upper Bound	8.7000	
	5% Trimmed Mean		7.6061	
	Median		6.9950	
	Variance		11.938	
	Std. Deviation		3.45521	
	Minimum		1.00	
	Maximum		15.00	
	Range		14.00	
	Interquartile Range		5.50	
	Skewness		-.020	.374
	Kurtosis		-.581	.733

This is now the weighted mean, median, etc. Percentiles should refer to the group rather than a specific number, as they are referring to midpoints, not actual values.

►Finding a z-score

Example 1, Page 166

Calculating a single z-score in SPSS is the same as using a calculator. The only real advantage of using SPSS is when you have a list of values to calculate z-scores for, all with the same mean and standard deviation. As SPSS does not calculate population means and variances, these must be done by hand. To calculate the z-score, type in the x value into a cell, then go to **Transform → Compute**, which is the SPSS calculator. You must have at least one value in a cell before being able to use the SPSS calculator.

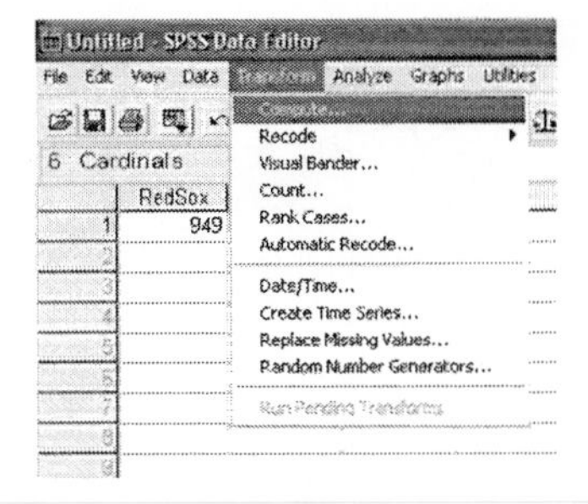

In the calculator, you create a new variable, so you need to name the variable. This can be any name under the normal SPSS naming conventions (so z-score will not work, as it has a -, but zscore will be fine). This name goes in the Target Variable box. You will type in the z-score equation in the Numeric Expression box. You can use the value(s) of a current variable by typing the name of the variable, or selecting the variable from the current variable list and clicking the ► button. You must watch the order of operations. Use parentheses to ensure that the order in which SPSS will do the math is what you expect.

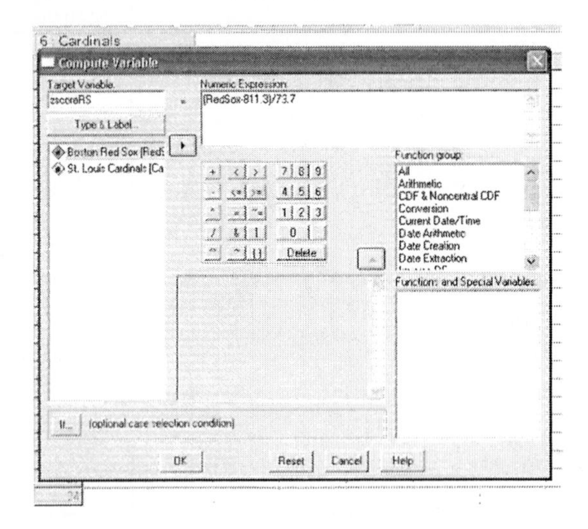

Once the formula has been correctly entered, push OK. This will create a new column in the Data View window, with the value of the z-score.

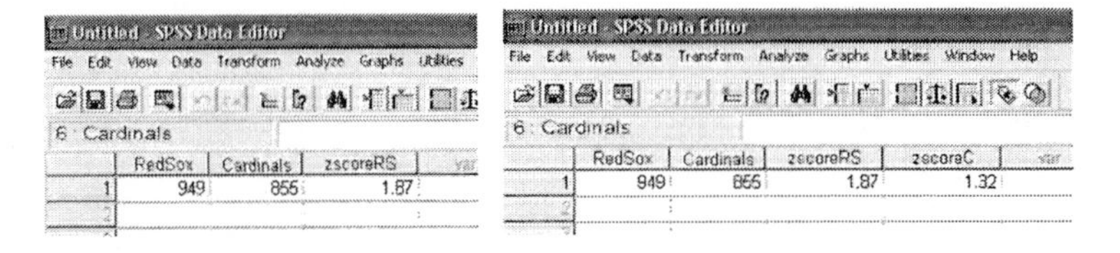

We can then compare the z-values. The Boston Red Sox had a z-score of 1.87, or a score 1.87 standard deviations above the mean, while the St. Louis Cardinals had a z-score of 1.32, or 1.32 standard deviations above the mean, so the Red Sox have a relatively higher score than the Cardinals. The advantage of using SPSS will come when calculating the z-scores for a list of values, such as calculating the z-values for all the teams in the American League, or National League.

►Creating a box-plot

Example 1, Page 176

Once the data have been entered, go to **Graphs → Boxplot…**.

	FinishingTime
1	19.95
2	23.25
3	23.32
4	25.55
5	25.83
6	26.28
7	28.58
8	28.72
9	30.18
10	30.35
11	30.95
12	32.13
13	33.23
14	33.53
15	36.68
16	37.05
17	37.43
18	41.42
19	42.47
20	49.17
21	64.63

In the Boxplot window, you will see two choices for type of box-plot. The first is the simple box-plot. This is the one that you will use most often. The simple box-plot is for a single, or for grouped data. The clustered box-plot is used when looking at groups within a cluster. An example could be box-plots for males and females within each class (freshman, sophomore, junior, senior). After selecting Simple, the data can be presented in one of two ways. **Summaries for groups of cases** is used when you have all the values for all groups in one column, and which group they belong to in a separate column. **Summaries of separate variables** is used when the data are in separate columns. In this example, since we don't have a group, we will use summaries of separate variables.

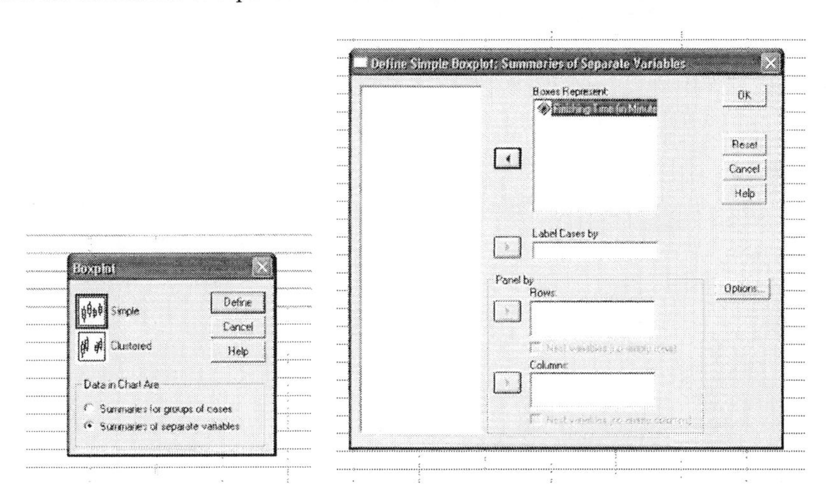

Once in the Define Simple Boxplot: Summaries of Separate Variables window, select the variable of interest, and push the ▶ button next to the Boxes Represent box. As we don't have a grouping variable, which would go in the Label Cases by: box we just push OK.

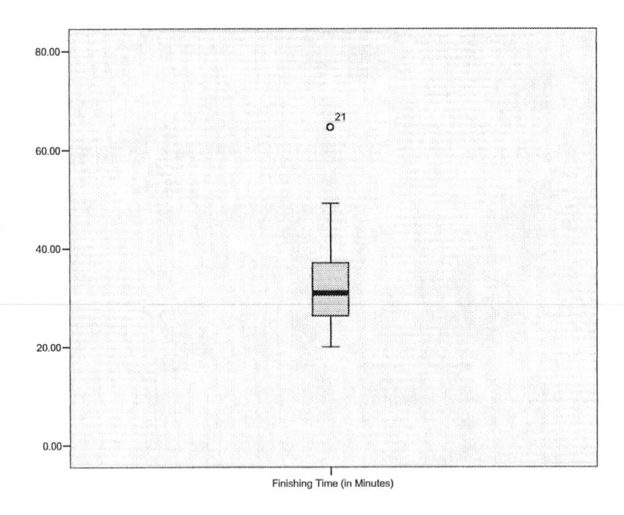

The box-plot will appear in the output window. SPSS creates the box-plot by identifying the 1st, 2nd, and 3rd Quartiles, and then the maximum and minimum values inside the lower and upper fences. The median is marked by the solid black line inside the box. Any values that lie outside the fences are marked as outliers. SPSS uses two sets of fences. If a value falls outside 1.5*IQR but inside 3*IQR then an O is used to label the observation as an outlier. If a value falls outside 3*IQR the outlier is marked as an extreme outlier, and an asterisk is used. The observation number is printed by the outlier label so you can check to make sure that observation was entered correctly. Most people don't like to read box-plots from bottom-to-top, as is presented by SPSS. This can be changed by going to the SPSS Chart Object editor. Once in the editor, there is a button on the far right that looks like a histogram on its side. Clicking on this button will transpose the x and y axes, so that the box-plot will be easier to read. The same option appears under **Options → Transpose Chart**.

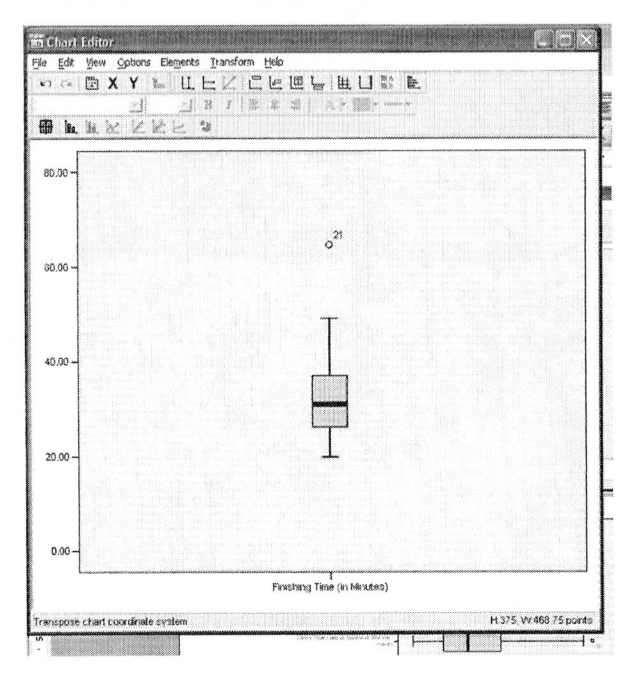

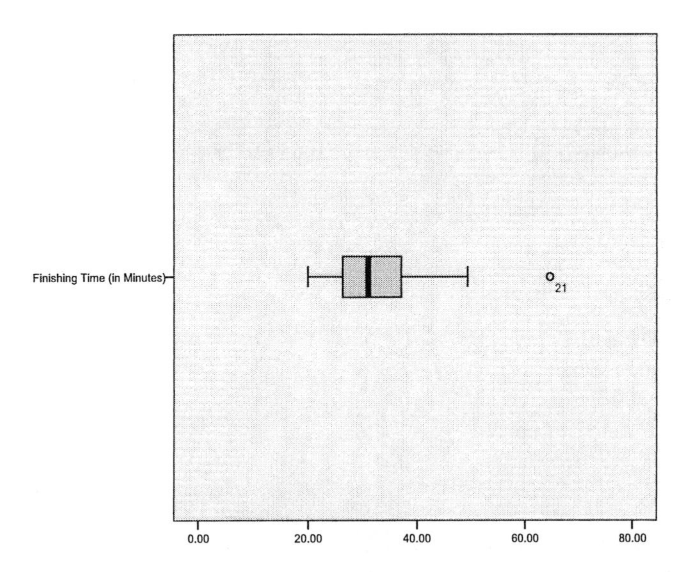

To conserve space, the output box can also be resized, although it may take some practice to get it to come out how you want it.

Chapter 4. Describing the Relation between Two Variables

Similar to descriptive statistics, there are various ways to calculate the relationship between variables. Only the simplest methods will be shown here.

Section 4.1 Scatter Diagrams and Correlation

►Creating a Scatter diagram

Example 1, Page 195

After typing the data into SPSS, go to **Graphs → Scatter/Dot...** and select Simple Scatter.

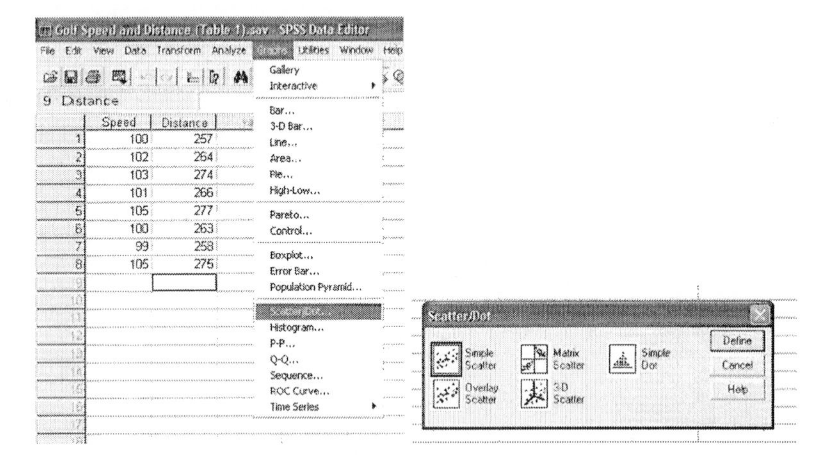

Selecting simple scatter will open the Simple Scatterplot window. Here you are asked to define the Y-axis, or the response variable, and the X-axis, or the predictor variable. You can also Set **Markers by:** which allows you to set different markers (color or shape) for a group (such as males and females). The **Label Cases by** can be used to identify specific points, if a categorical variable exists.

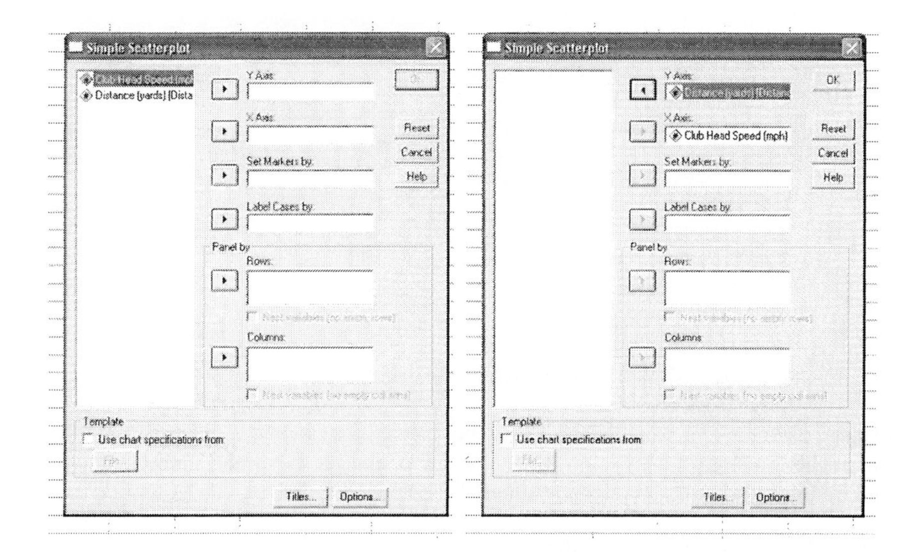

Once you have selected the x-axis and y-axis values, you can create the graph. To add labels, you must select a variable to label cases by, and go to options, and select **Display Chart with case labels**. Once you are ready, push OK. The scatter plot will appear in the output window.

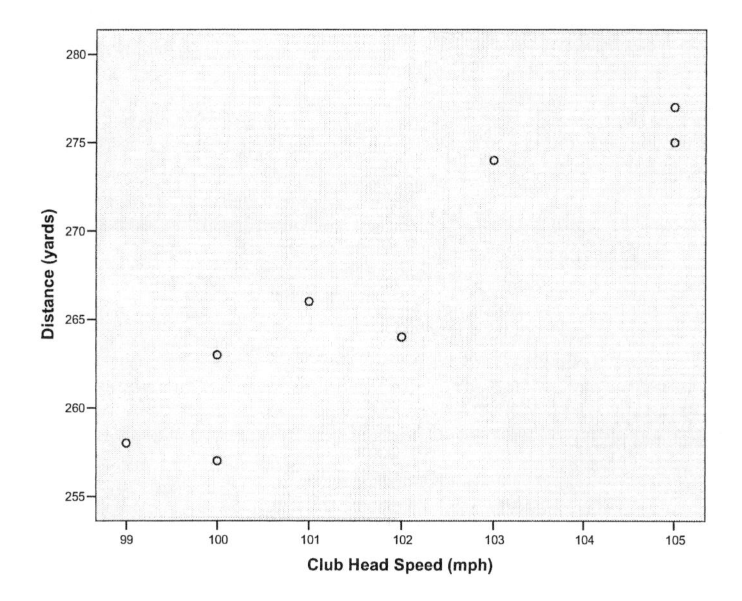

►Finding Correlation

Example 2, Page 200

After entering the data into SPSS, go to **Analyze → Correlate → Bivariate...** (two variables)

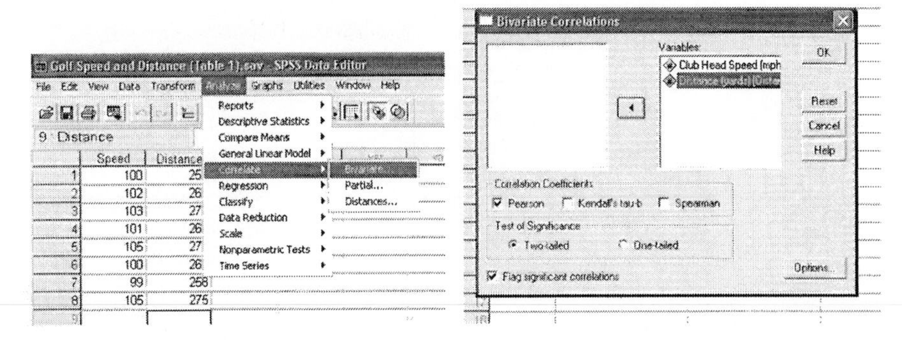

Highlight the variables you wish to find the correlation between and push the ▶ button.

Make sure that Pearson is selected. Spearman may be selected for chapter 13 (Non-Parametric Correlation). Other options (in the Options button) allow you to get the means, and standard deviations, as well as the covariance and cross product deviations.

When you have the options you want, push OK

Correlations

		Club Head Speed (mph)	Distance (yards)
Club Head Speed (mph)	Pearson Correlation	1	.939**
	Sig. (2-tailed)		.001
	N	8	8
Distance (yards)	Pearson Correlation	.939**	1
	Sig. (2-tailed)	.001	
	N	8	8

**. Correlation is significant at the 0.01 level (2-tailed).

The correlation will appear in the Output Viewer Screen, in this case 0.939. The correlation is provided for all combinations of variables (you can have more than two at a time). The correlation of something with itself is always 1, which appear in the boxes relating the variables to themselves. The other values, Sig. (2-tailed) and the asterisks will not be used until after chapter 9, when hypothesis testing is discussed.

Section 4.2 Least Squares Regression

▶Fitting a line in a Scatter diagram

Example 1, Page 213

After creating the scatter plot (see Creating a Scatter diagram), go to the chart editor and click on any point in the graph, so that the points are highlighted.

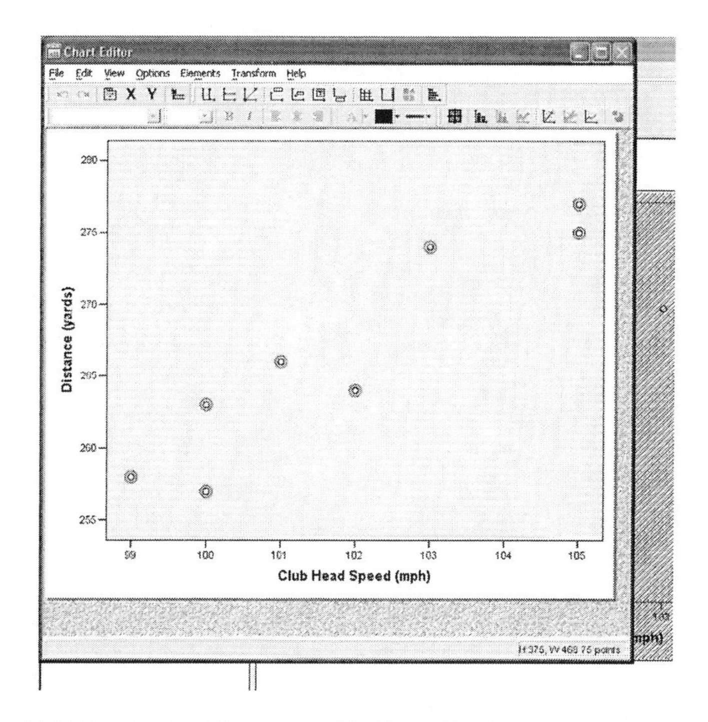

Once the points are highlighted, select Elements → Fit Line at Total

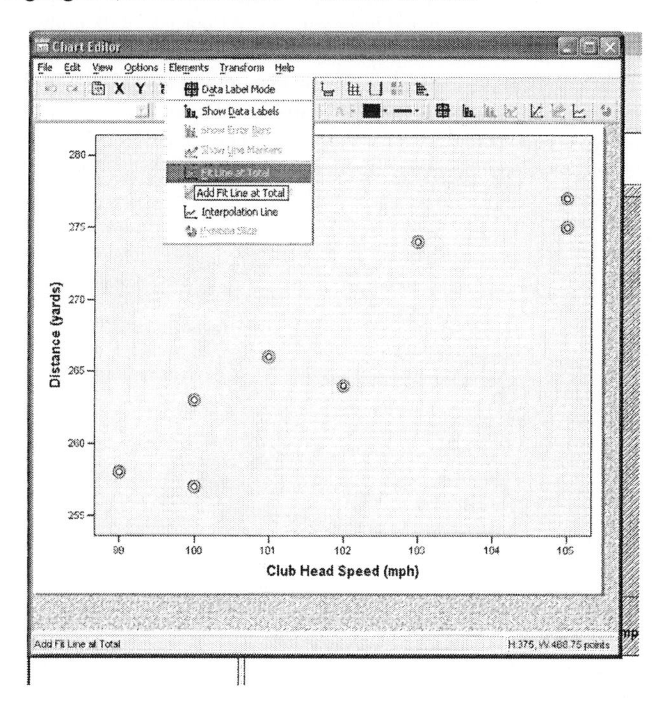

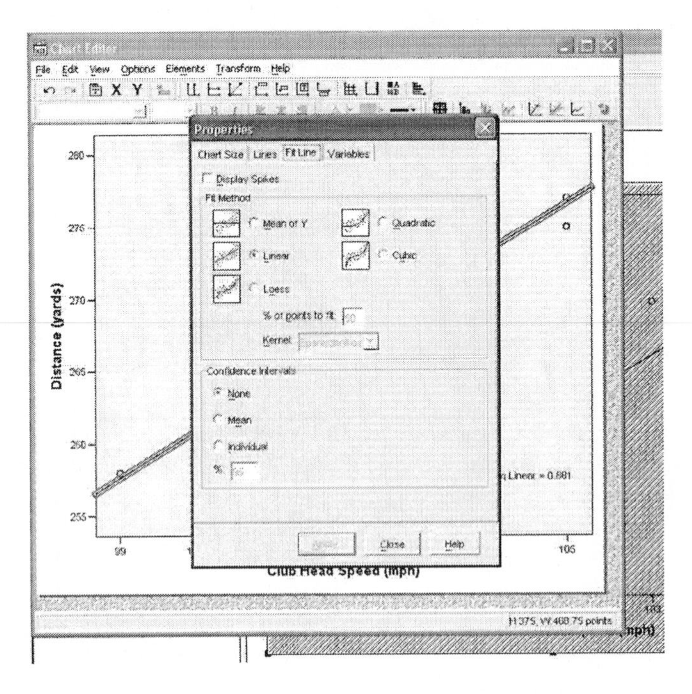

Select Linear (which is the default). This will automatically add in the Least Squares Regression line to the plot. Mean of Y will create a horizontal line at the mean of the response variable. Loess will create a line that will draw a fit line using iterative weighted least squares. At least 13 data points are needed. This method fits a specified percentage of the data points, with the default being 50%. In addition to changing the percentage, you can select a specific kernel function. The default kernel (probability function) works well for most data. Quadratic will fit a quadratic curve to the data, and Cubic will fit a cube function to the data. These are beyond what you should need for this class. The Linear option is all you really need at this point. When you have selected the option you want, push Close.

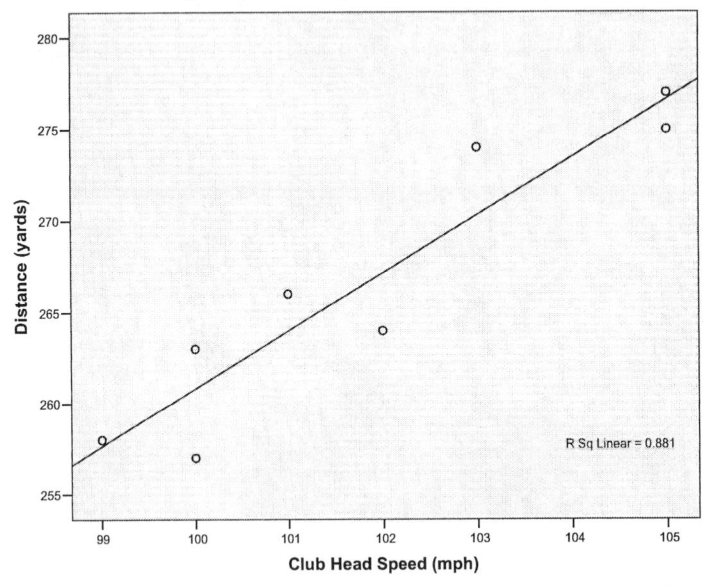

►Finding the Least-Squares Regression Equation

Example 2, Page 216

Once the data have been entered into SPSS, go to **Analyze → Regression → Linear…**

This will open the linear regression window. Select the response variable, and click the ▶ button by the Dependent box. Select the predictor variable and click the ▶ button by the Independent(s) box.

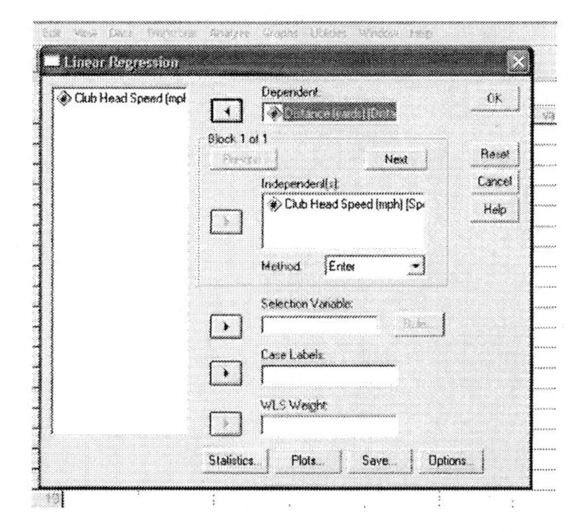

Although there are many other things we can, and will, do while in this window, for now we will just proceed by clicking OK. This will provide the following output.

Variables Entered/Removed[b]

Model	Variables Entered	Variables Removed	Method
1	Club Head Speed (mph)[a]	.	Enter

a. All requested variables entered.

b. Dependent Variable: Distance (yards)

This first box is only for other types of regression, and we will ignore it.

Model Summary

Model	R	R Square	Adjusted R Square	Std. Error of the Estimate
1	.939[a]	.881	.861	2.883

a. Predictors: (Constant), Club Head Speed (mph)

This table provides the correlation coefficient, similar to what was provided by the **Analyze → Correlate → Bivariate...** command used previously. If you plan on obtaining both the correlation and the regression equation, only the **Analyze → Regression → Linear...** command needs to be used. The other values will be discussed later.

ANOVA[b]

Model		Sum of Squares	df	Mean Square	F	Sig.
1	Regression	369.642	1	369.642	44.484	.001[a]
	Residual	49.858	6	8.310		
	Total	419.500	7			

a. Predictors: (Constant), Club Head Speed (mph)

b. Dependent Variable: Distance (yards)

The third table provides information that will not be discussed until chapter 12, but provides some important information for later.

Coefficients[a]

Model		Unstandardized Coefficients		Standardized Coefficients	t	Sig.
		B	Std. Error	Beta		
1	(Constant)	-55.797	48.371		-1.154	.293
	Club Head Speed (mph)	3.166	.475	.939	6.670	.001

a. Dependent Variable: Distance (yards)

The last table is the one of most interest for this section. The first column, under the Unstandardized Coefficients, is the B values. The first of these is b_0, or the intercept, in this example -55.797. The second, next to the predictor variable, is the slope, in this example 3.166. So our regression equation would be Distance (yards) = -55.797 + 3.166 * Club Head Speed (mph). The other values in the table will be used in later chapters.

Section 4.3 Diagnostics on the Least-Squares Regression Line

►Coefficient of Determination

Example 1, Page 228

Using the information in the regression, the R^2 value is given in the second table, right next to the correlation coefficient.

Model Summary

Model	R	R Square	Adjusted R Square	Std. Error of the Estimate
1	.939[a]	.881	.861	2.883

a. Predictors: (Constant), Club Head Speed (mph)

So, in this example, R^2 is 0.881, or 88.1%. The adjusted R^2 value is used in more complicated models.

►Producing a Residual Plot

Example 2, Page 229

After entering the data into SPSS, go to **Analyze → Regression → Linear...**, select the dependent and independent variables, and push the Save button.

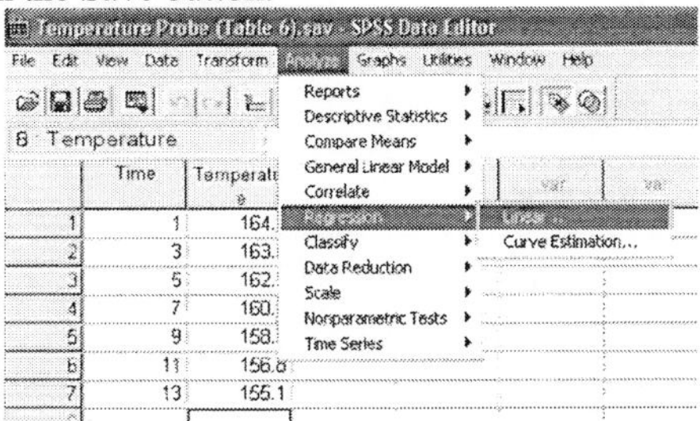

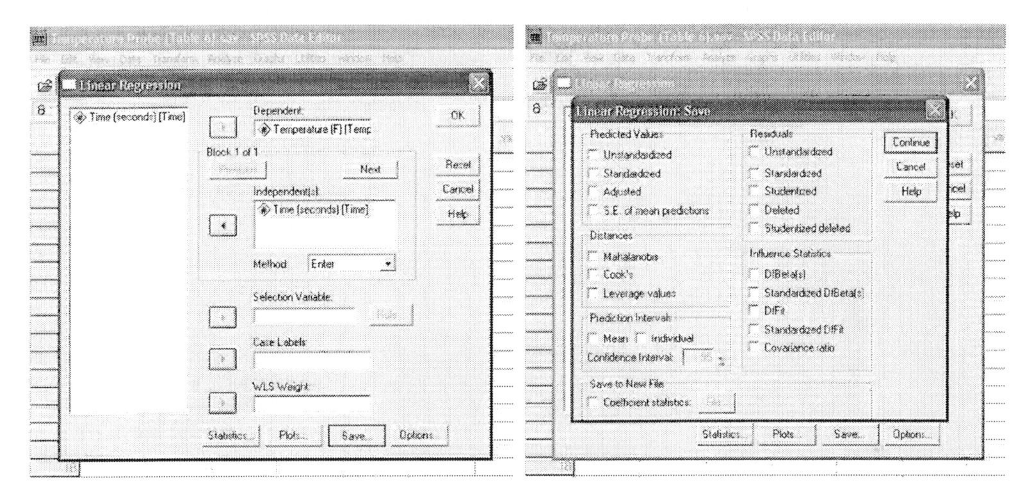

In the Linear Regression: Save menu, you will see Residuals in the top Right section of the window. Select Unstandardized. This will not put the residuals into the output window, but will create a new variable, RES_1, which are the residuals at each point. After selecting Unstandardized, press continue, and OK. You will then go to Graph → Scatter/Dot, (you do not have to switch back to the data window to do this, all menus are available in the output window as well) and select simple scatter. Select the new variable, RES_1 as the Y-Axis, and the predictor variable as the X-Axis. This will create the Residual plot.

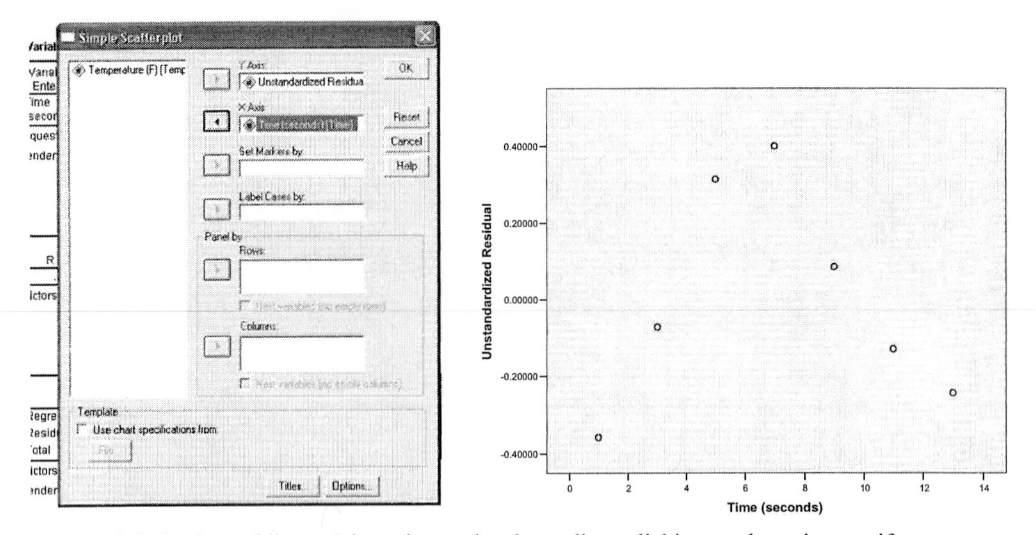

You can add the horizontal line at 0 by going to the chart editor, clicking on the points, as if you were going to add the least squares line, Go to **Elements → Fit Line at Total** and select Mean of Y. The mean of the residuals is 0, so it will create a horizontal line at 0.

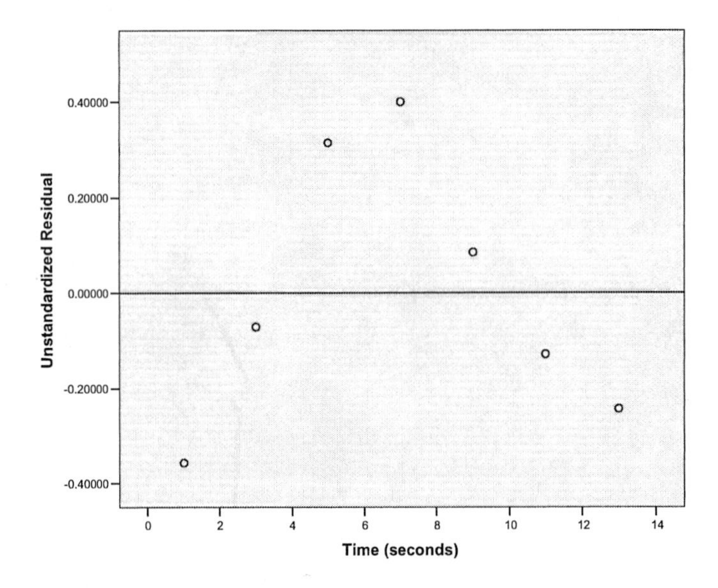

SPSS does have its own plots that it produces, under the plots button in the regression window, which are similar in function, but are different than the ones provided in the book.

Section 4.4 Nonlinear Regression: Transformations (on CD)

►Non-linear transformations

Example 4 (4-4)

There are two ways to find regression equations for non-linear transformations in SPSS. The first is to create a new variable and using the simple linear regression methods already produced. To do this, enter the data into SPSS. Then go to **Transform → Compute...** to create the new variable. Type a new variable name in the Target Variable box, and the transformation in the Numeric Expression box.

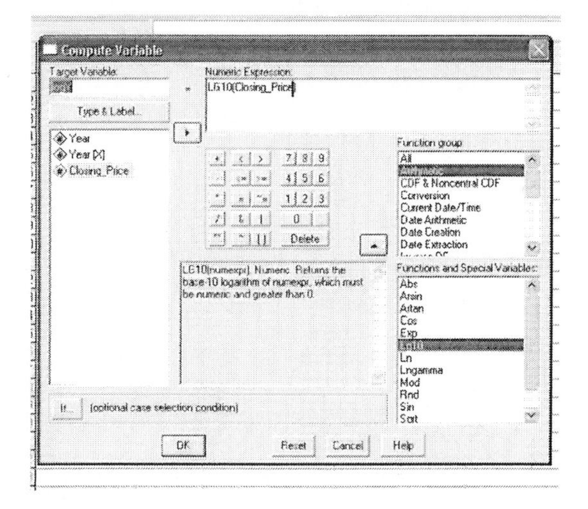

In this case, the transformation is the LG10, or $\log_{10}$ transformation. Other transformations are possible, such as using the natural log, ln. These are listed under Arithmetic functions. Once the variable has been created, a regular simple linear regression can be used, using the new variable as the response, and the regular x as the predictor.

Coefficients[a]

Model		Unstandardized Coefficients		Standardized Coefficients	t	Sig.
		B	Std. Error	Beta		
1	(Constant)	.211	.068		3.084	.009
	Year	.116	.008	.974	15.445	.000

a. Dependent Variable: logY

SPSS knows that sometimes transformations must be used. They have some transformations built in. Go to **Analyze → Regression → Curve Estimation**.

Harley Davidson (Table 7).sav - SPSS Data Editor

File Edit View Data Transform Analyze Graphs Utilities Window Help

Analyze menu: Reports; Descriptive Statistics; Compare Means; General Linear Model; Correlate; Regression (Linear..., Curve Estimation...); Classify; Data Reduction; Scale; Nonparametric Tests; Time Series

11 : logY

	Year			logY	var
1	1990				
2	1991				
3	1992			.66	
4	1993			.73	
5	1994			.83	
6	1995	6.00	7.0326	.85	
7	1996	7.00	11.5585	1.06	
8	1997	8.00	13.4799	1.13	
9	1998	9.00	23.5424	1.37	
10	1999	10.00	31.9342	1.50	
11	2000	11.00	39.7277	1.60	
12	2001	12.00	54.3100	1.73	
13	2002	13.00	46.2000	1.66	
14	2003	14.00	47.5300	1.68	
15	2004	15.00	60.7500	1.78	

This will open the Curve Estimation window. Here you can select from a variety of possible transformations. The advantage of this is, when you are unsure of which transformation to use, you can select more than one at a time, and compare the results to see which is best.

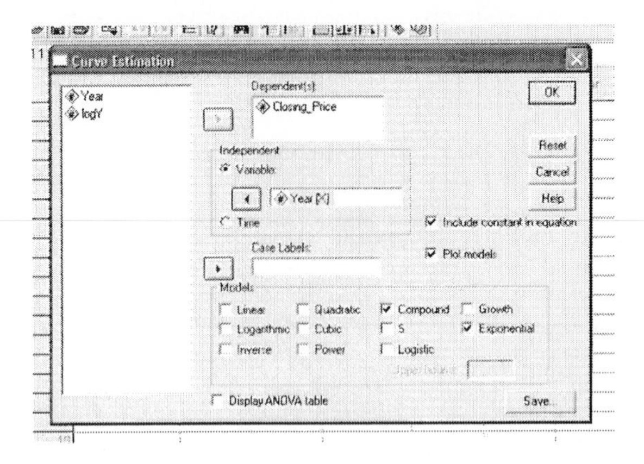

Here, the method that was used was the Compound, which uses $\log_{10}$. Also possible is the exponential, which uses the natural log. Here we can compare the two directly to each other. Using the right mouse button on any of the models will provide you with information on what the model looks like. For example, using the right mouse button on the Compound will provide the following information:

Model whose equation is Y=b0*(b1**t) or ln(Y)=ln(b0)+(ln(b1)*t),

which tells us that the equation will be $Y = b_0 b_1^x$ which is the equation we used in the example. The output provided is

Model Summary and Parameter Estimates

Dependent Variable: Closing_Price

	Model Summary					Parameter Estimates	
Equation	R Square	F	df1	df2	Sig.	Constant	b1
Compound	.948	238.536	1	13	.000	1.624	1.306
Exponential	.948	238.536	1	13	.000	1.624	.267

The independent variable is Year.

Here we can compare, using R^2 values, the different equations. In this case, both models are the same, although the models look slightly different. (In the Compound model, we are using $1.624 * (1.306^x)$ and in the Exponential model we are using $1.624 * e^{.0.267*x}$) Using the curve estimation also provides a graph, in which we can see that both lines are the same.

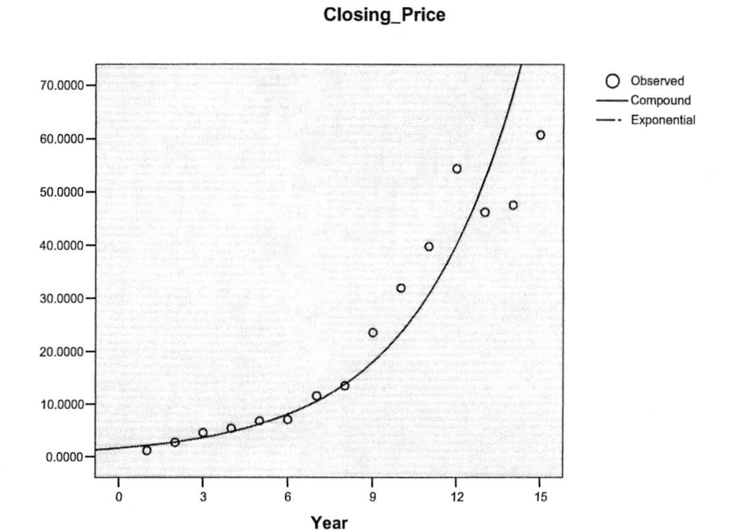

Similarly, we could fit power models, or other models using this same method, without having to change the original data.

Chapter 5. Probability

SPSS is not a data creation tool, and there are some difficulties in creating data sets in SPSS. SPSS is a data analysis tool, and should be used as such. Despite this, once you have a data set, you can create new variables that can be used with this section.

Section 5.1 Probability Rules

►Simulating Probabilities

Example 8, Page 259

To create 100 random numbers, you must first create a data set with 100 observations. SPSS will create new variables, with the same sample size as already exists, but will not create more observations than are already in the data set. To create the original sample of 100, you can type in the values 1 to 100 by hand, or use another program, such as Excel, where data generation is easier. After creating the data set with 100 observations, go to **Transform → Compute...**

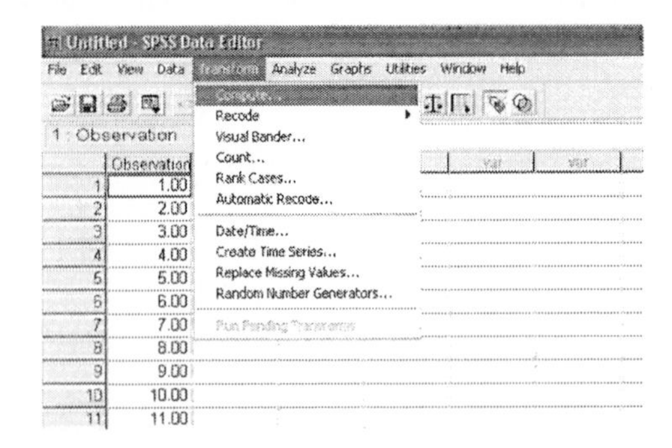

Once in the Compute Variable window, you can choose a name for the new variable you wish to create. This name must follow SPSS naming conventions (no strange characters or spaces).

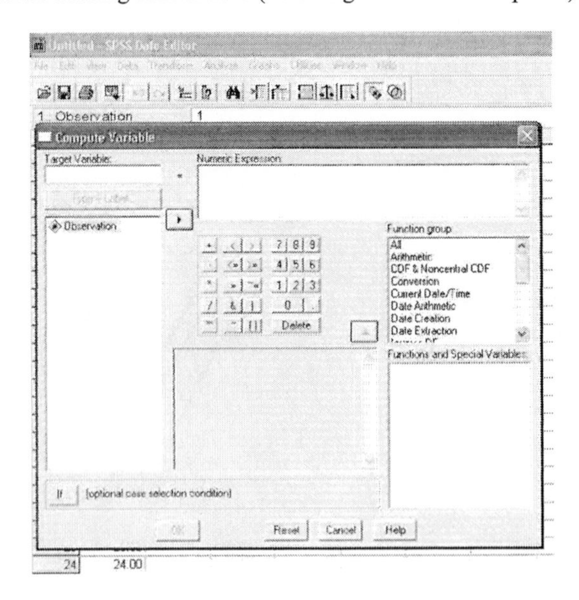

The name of the variable goes in the Target Variable box. In the Function group box, find and select Random Numbers.

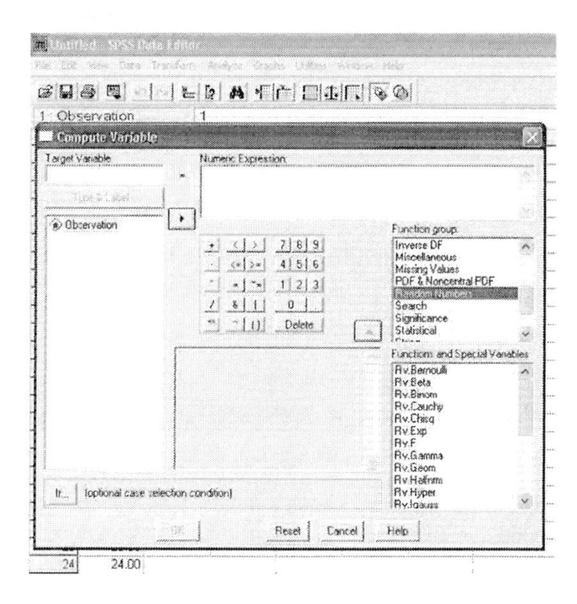

This provides a list of all of the random number distributions possible. We will discuss more about what distributions are in the next chapter. To choose a value of 0 or 1, select Rv.Bernouli. In the box to the left of the Functions and Special Variables box will be a description of what the function does.

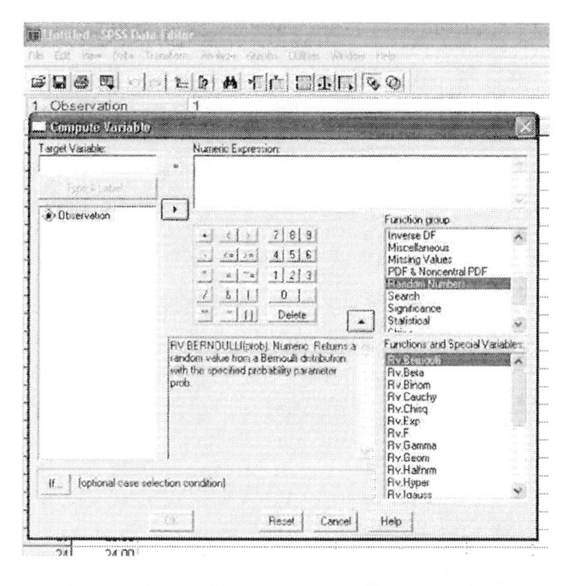

In this case, Bernoulli will provide a value of either 0 or 1 (a Bernoulli trial) randomly.

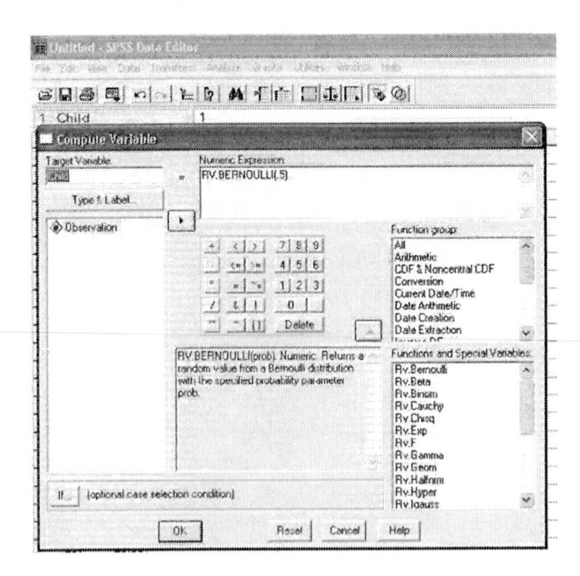

The random variable requires an expected probability (prob), which for this example we will select to be .5 (half girls and half boys). Any value between 0 and 1 can be selected as the prob value. Once you have selected a name for the variable, and the function you wish to use, push OK. This will add a new column to the data window.

Untitled - SPSS Data Editor

File Edit View Data Transform Analyze

1 : Child 1

	Observation	Child
1	1.00	1.00
2	2.00	.00
3	3.00	1.00
4	4.00	.00
5	5.00	.00
6	6.00	.00
7	7.00	.00
8	8.00	.00
9	9.00	1.00
10	10.00	.00
11	11.00	1.00
12	12.00	1.00
13	13.00	.00
14	14.00	1.00
15	15.00	.00
16	16.00	1.00
17	17.00	.00
18	18.00	.00
19	19.00	1.00
20	20.00	1.00
21	21.00	.00
22	22.00	.00
23	23.00	.00
24	24.00	.00

You may go to the variable view to define 0 to be boys and 1 to be girls, as in the example. To get the count, we use the frequency methods discussed in chapter 2.

Child

		Frequency	Percent	Valid Percent	Cumulative Percent
Valid	.00	54	54.0	54.0	54.0
	1.00	46	46.0	46.0	100.0
	Total	100	100.0	100.0	

So, in this example, we had 54 boys and 46 girls out of the 100 children.

To set a seed value, or starting place in the random number table, go to **Transform → Random Number Generators…**

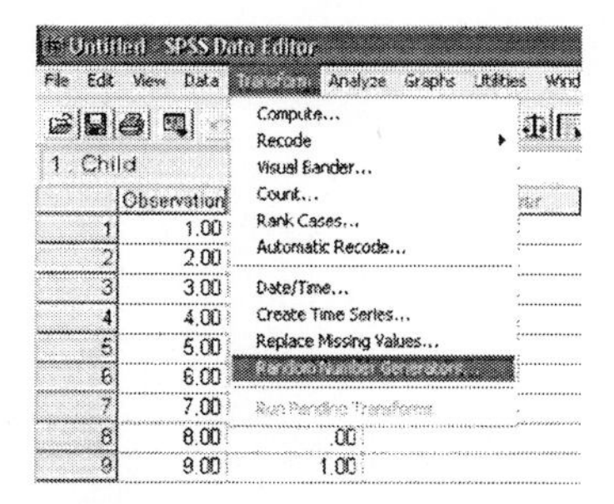

In the Random Number Generators window, you can select Set Starting Point, and Fixed Value, and type in the seed value. This will make it so every time you open SPSS, the starting point in the random number table will be the same. Using Random will make it so the starting point changes every time you open SPSS. Using a set value is useful if you want everybody to get the exact same results, while using the Random option is better for actual simulation.

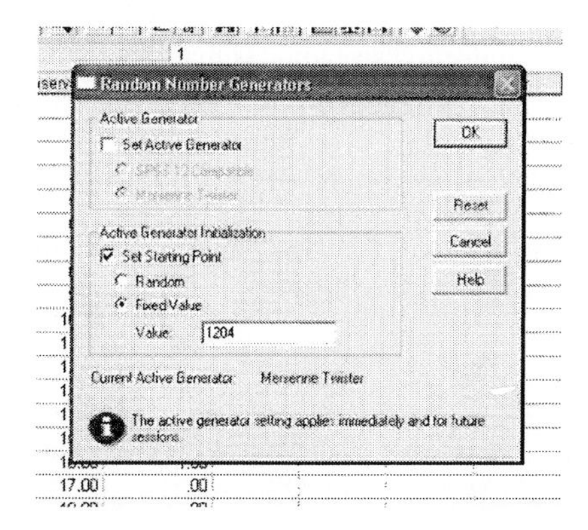

Technology Step-by-step Page 265

To create a simulated die roll, follow the steps above, except instead of choosing Bernoulli, we will use a uniform distribution, which means all values are equal. Again, the sample size will be the same as how many samples are currently in the data set, so we will create 100 tosses, as we already have 100 observations.

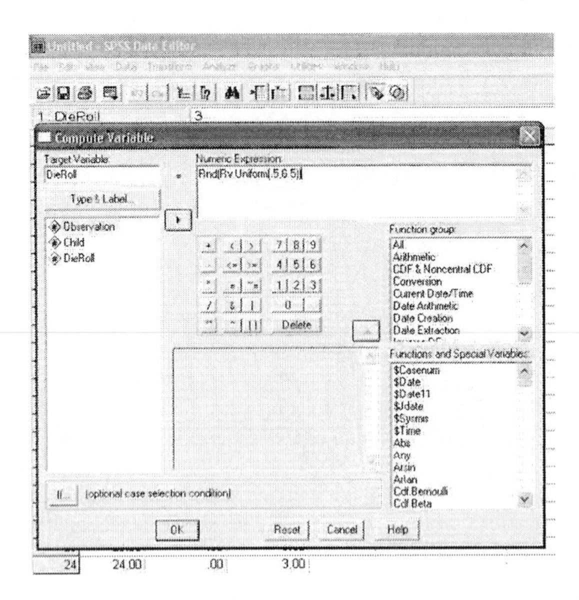

One of the problems with using the uniform distribution is that it is a continuous instead of discrete uniform distribution. Because of this, instead of selecting numbers between 1 and 6, as we would think, we will select between .5 and 6.5. This gives 1 and 6 equal chances instead of giving them only ½ of a chance each. We then round the random numbers to the nearest whole number. (This is the Rnd function).

This provides us with a new column of values between 1 and 6. Again using the frequency procedures from chapter 2, we can count the number of times we roll each number.

DieRoll

		Frequency	Percent	Valid Percent	Cumulative Percent
Valid	1.00	14	14.0	14.0	14.0
	2.00	18	18.0	18.0	32.0
	3.00	17	17.0	17.0	49.0
	4.00	18	18.0	18.0	67.0
	5.00	14	14.0	14.0	81.0
	6.00	19	19.0	19.0	100.0
	Total	100	100.0	100.0	

Chapter 6. Discrete Probability Distributions

Many of the methods required for this chapter are found in previous chapters, including frequency distributions, and histograms.

Section 6.1 Expected Values

You can find the expected value of a distribution in SPSS, but you cannot find the variance. This is because SPSS is still assuming that the data comes from a sample, not a population. To find the expected value, enter the values of x and the probabilities. We then follow the methods of finding a weighted mean (from chapter 3, section 4). We weight the data by the probability, and then find the descriptive statistics on x.

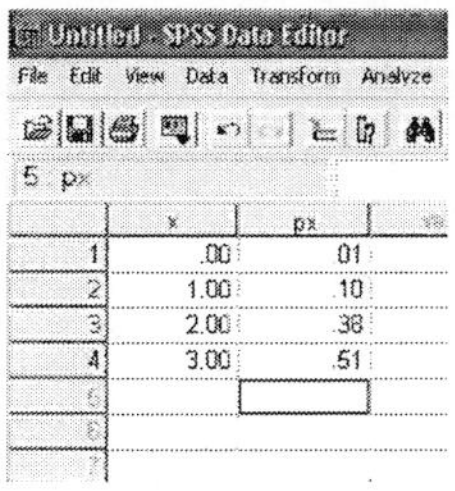

Untitled - SPSS Data Editor

File Edit View Data Transform Analyze

5 : px

	x	px
1	.00	.01
2	1.00	.10
3	2.00	.38
4	3.00	.51

Descriptive Statistics

	N	Minimum	Maximum	Mean	Std. Deviation
x	1	.00	3.00	2.3900	.
Valid N (listwise)	1				

The standard deviation is missing, because n-1 is 0, so cannot be calculated. Even if you changed the weights to be whole numbers (say by multiplying by 100), the standard deviation would be calculated using n-1 instead of n, and would provide the wrong value. This value may be adjusted for, however. To get the population variance from the sample variance, use $\frac{N-1}{N}s^2$.

Section 6.2 The Binomial Probability Distribution

▶ Binomial Distribution

Example 2, Page 330

To create a Binomial Probability Distribution, create a variable with the possible values of x, in this example, 0, 1, 2, 3, and 4. Once the variable has been defined, go to **Transform → Compute...**

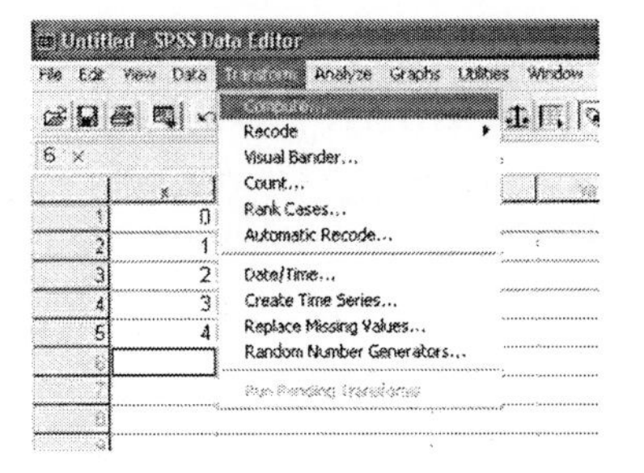

While in the compute variable window, create a name for the probabilities in the Target Variable box. In the Function Group box, select PDF and Noncentral PDF. PDF is the Probability Distribution Function. In the Functions and Special Variables, select Pdf.Binom.

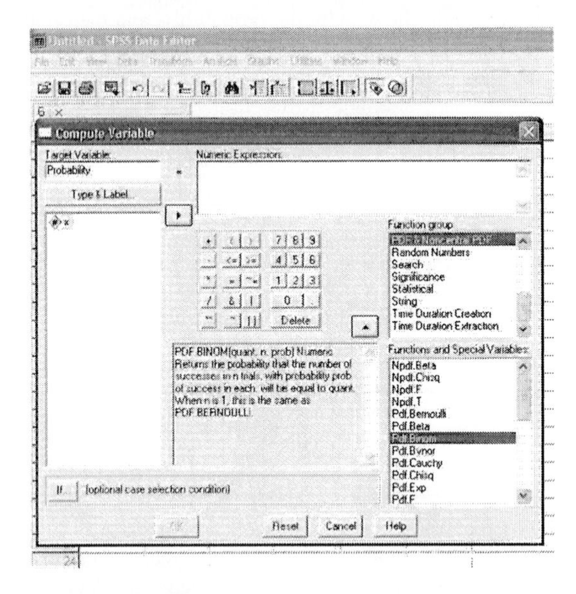

For this function to work, you need three values. The first is x, which is the variable we created earlier. The second is the number of trials, n, and the third is the probability of a success, or p. So, we will select Pdf.Binom, and replace the question marks that appear with x, 4, and .06. When selecting the name of the variable we wish to use, we can either type the name (if it is short), or highlight the name, and push the ▶ button next to the Numeric Expression box.

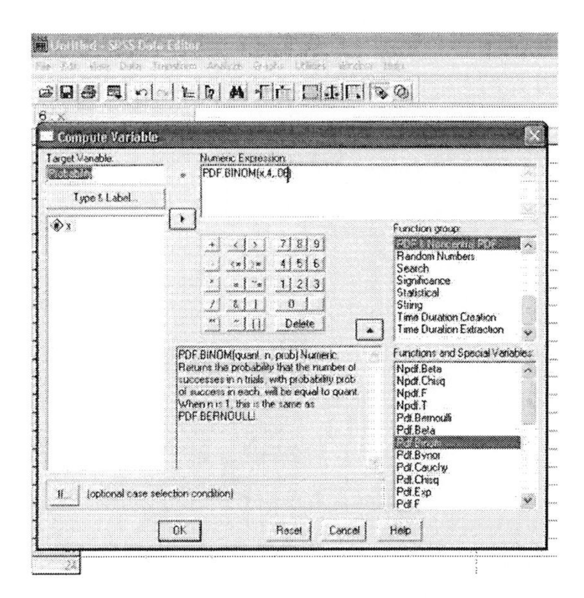

Once the name and expression have been completed, push the OK button. This will create a new column in the data view window with the probabilities of observing each x value. (You may need to change the number of decimal places shown in the variable view window to see some of the values).

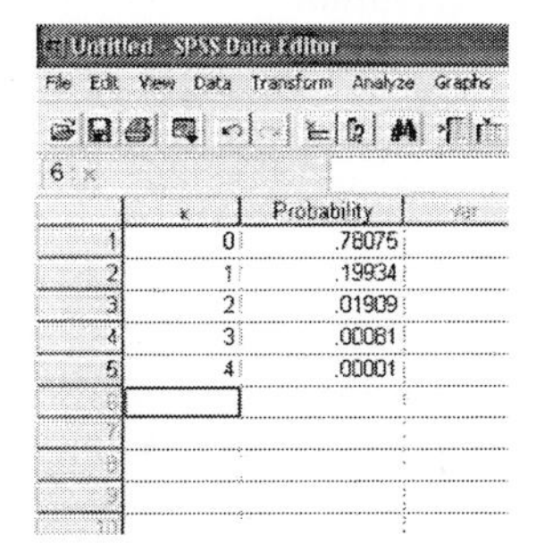

The probabilities provided are the probabilities of being equal to x for the pdf. We can also find the Cumulative Distribution Function by changing the P in PDF to C in the Numeric Expression box. This is the probability of getting a value less than or equal to x.

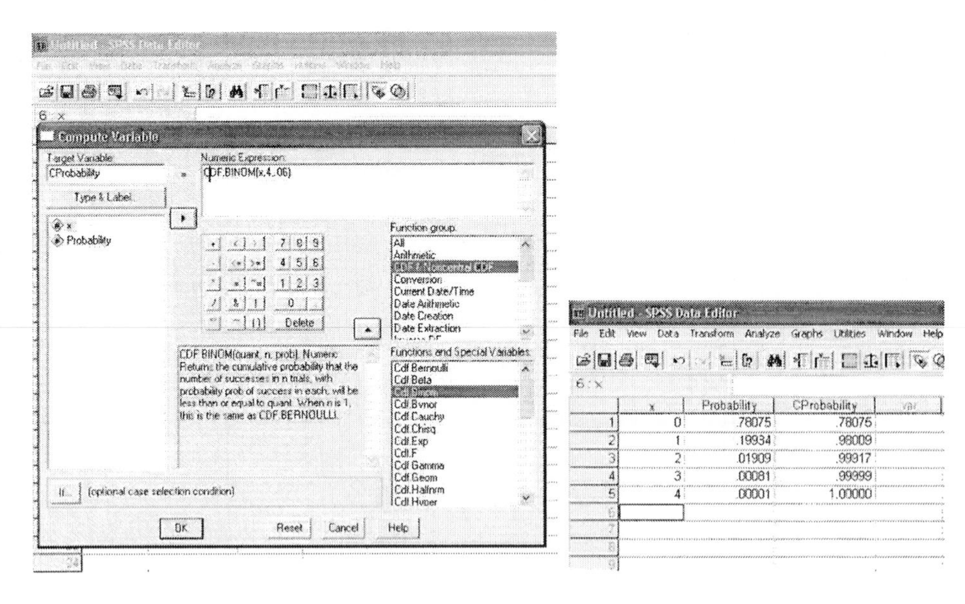

To create a Binomial Probability Histogram, follow the methods used in chapter 2.

Section 6.3 The Poisson Probability Distribution

►Poisson Distribution

Example 2, Page 346

To create a Poisson Probability Distribution, create a variable with the possible values of x. This is a little more difficult than for the Binomial distribution, as the values of x can go on to infinity. Choose appropriate x values for the problem. For example, to find the probability that exactly 6 cars arrive between 12 noon and 12:05 P.M., one of the x values needs to be 6. To get the probability of fewer than 6, we need an x value of 5 (since the cumulative probability up to 5 is the probability of less than 6). To get the probability of at least 6 cars arriving, you need to have a value of 5, and use the complement rule. We will use the values 0 through 6 for this example, so that the values can be related to the ones in the book. Once the variable has been defined, go to **Transform → Compute…**

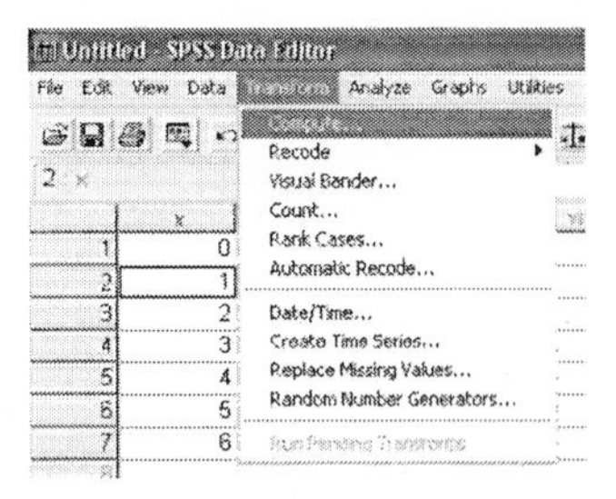

While in the compute variable window, create a name for the probabilities in the Target Variable box. In the Function Group box, select PDF and Noncentral PDF. PDF is the Probability Distribution Function. In the Functions and Special Variables, select Pdf.Poisson.

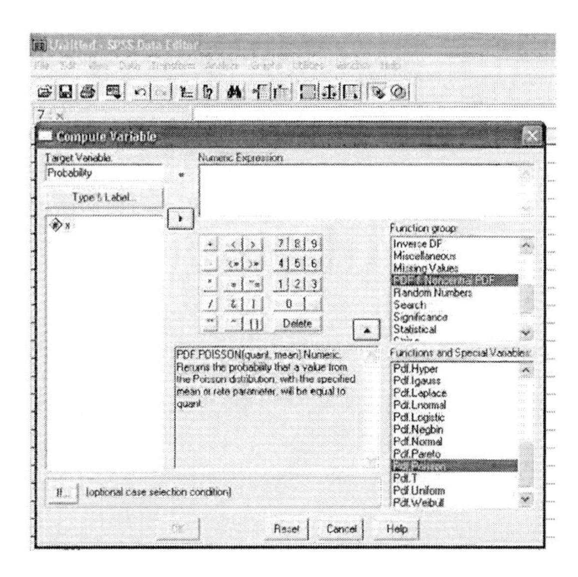

For this function to work, you need two values. The first is x, which is the variable we created earlier. The second is the mean, or the average rate, which is λt, or 10 for this example. So, we will select Pdf.Poisson, and replace the question marks that appear with x, and 2. When selecting the name of the variable we wish to use, we can either type the name (if it is short), or highlight the name, and push the ▶ button next to the Numeric Expression box.

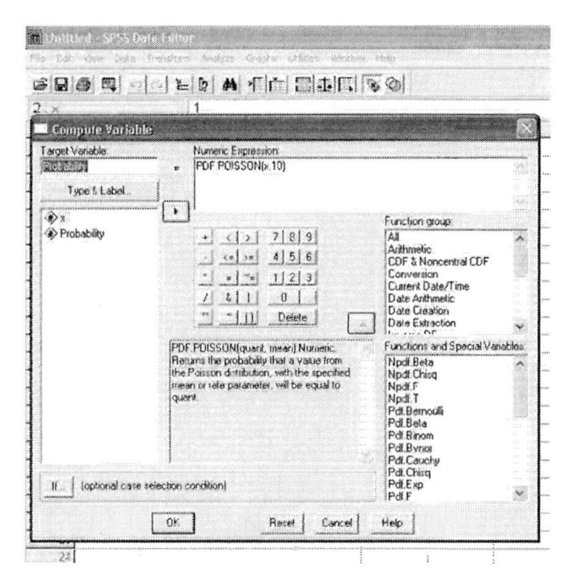

Once the name and expression have been completed, push the OK button. This will create a new column in the data view window with the probabilities of observing each x value. (You may need to change the number of decimal places shown in the variable view window to see some of the values).

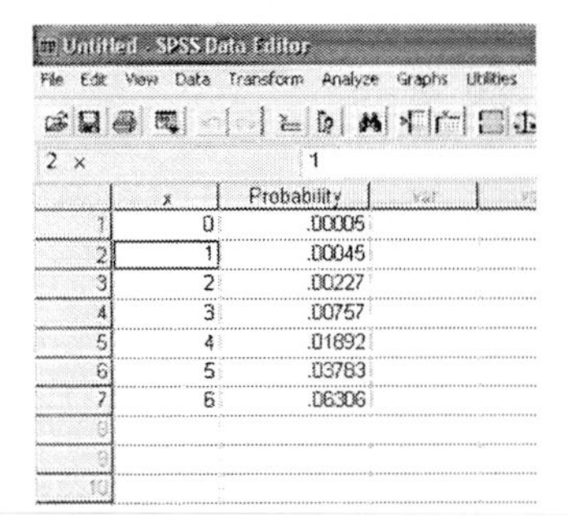

The probabilities provided are the probabilities of being equal to x for the pdf. We can also find the Cumulative Distribution Function by changing the P in PDF to C in the Numeric Expression box. This is the probability of getting a value less than or equal to x.

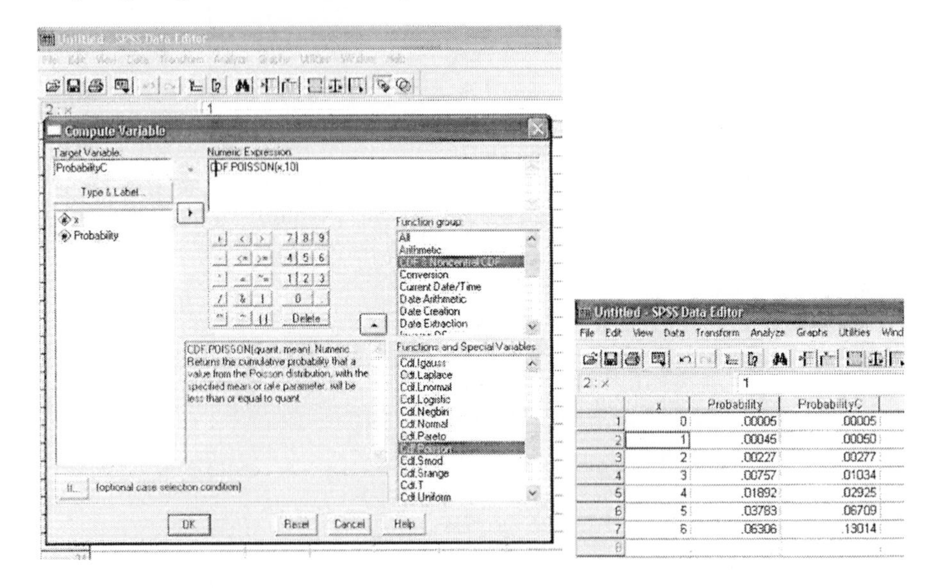

To find the complement, we take the same function, and subtract it from 1.

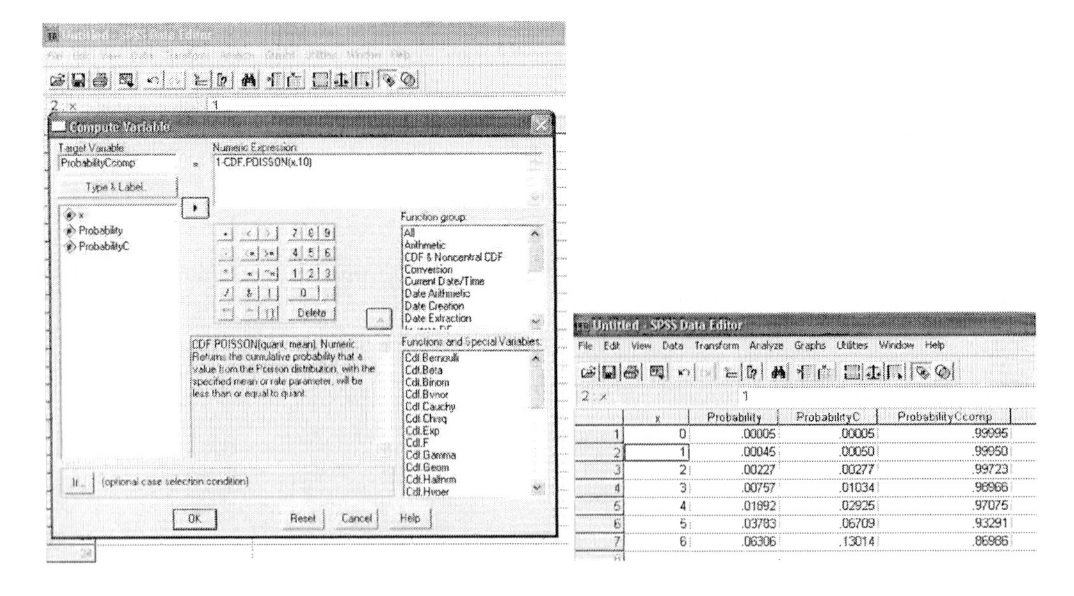

So, using the tables we have just created, the probability of having exactly 6 cars arrive between 12 noon and 12:05 P.M. is .06306. The probability of less than 6 cars arriving between 12 noon and 12:05 P.M. is .06709, which is the cumulative probability where x=5. The probability that at least 6 cars arrive between 12:00 noon and 12:05 P.M. is .93291, which is the complement of the cumulative probability when x=5.

Chapter 7. The Normal Probability Distribution

Section 7.1 Properties of the Normal Distribution

▶ Uniform Distribution

Example 2, Page 362

To find the probability between two values in a continuous distribution, we must use the cumulative distribution functions. In SPSS, type the two values that you want to find the area between as a new variable. Once the variable has been defined, go to **Transform → Compute...**

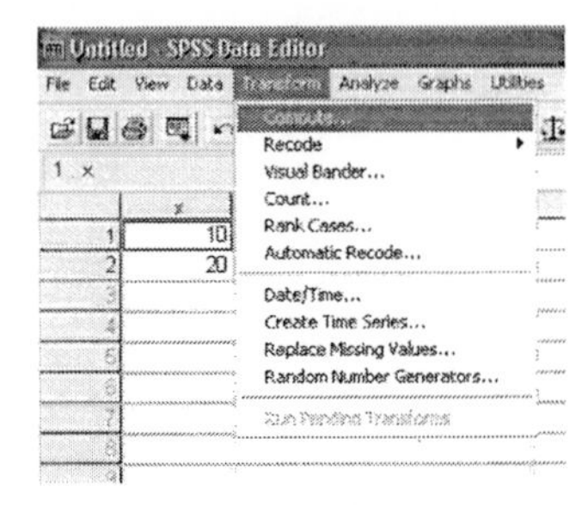

While in the compute variable window, create a name for the probabilities in the Target Variable box. In the Function Group box, select CDF and Noncentral CDF. CDF is the Cumulative Distribution Function. In the Functions and Special Variables, select Cdf.Uniform.

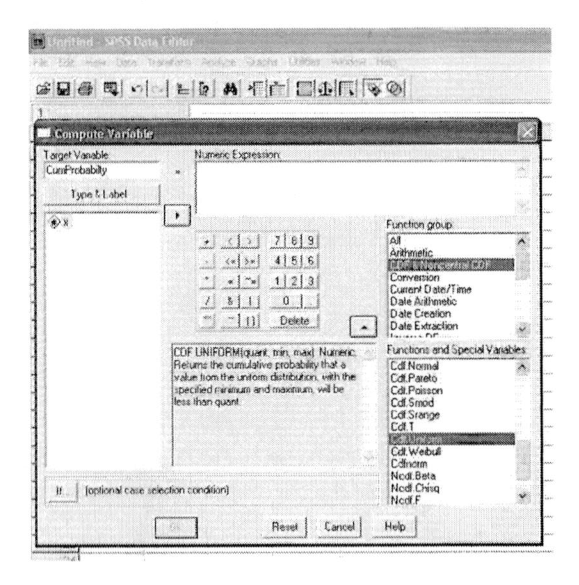

For this function to work, you need three values. The first is x, which is the variable we created earlier. The second is the minimum value for the distribution, and the third is the maximum value for the distribution. These are the maximum and minimum values over the whole distribution, not the values of x. So, we will select Cdf.Uniform, and replace the question marks that appear with x, 0, and 30. When selecting the name of the variable we wish to use, we can either type the name (if it is short), or highlight the name, and push the ► button next to the Numeric Expression box.

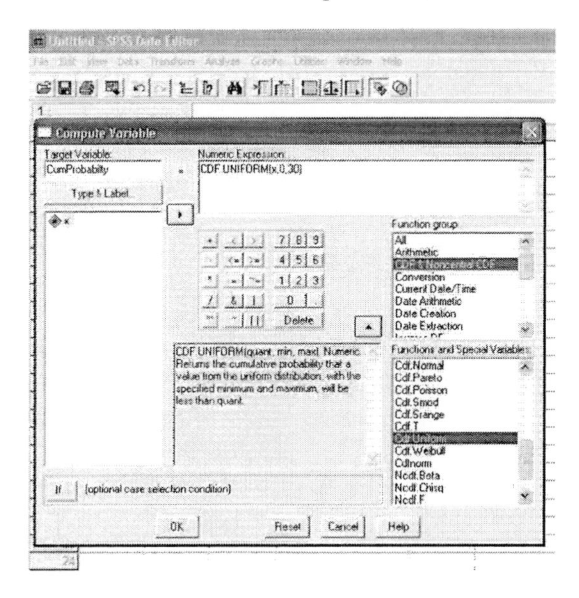

Once the name and expression have been completed, push the OK button. This will create a new column in the data view window with the probabilities of observing less than or equal to each x value. (You may need to change the number of decimal places shown in the variable view window to see some of the values).

To get the area between the two, you will need to find the difference. .666667-.33333=.33333 or 1/3. Using SPSS is only an advantage in this case when there are many x values that you will use.
To find the probability between two points, you can start with typing in one value in the first column. This value is irrelevant, as it serves only to define the sample size. . Once the variable has been defined, go to **Transform → Compute...**

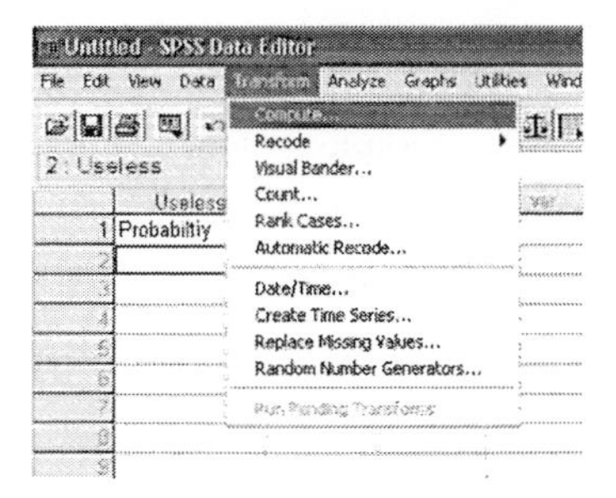

While in the compute variable window, create a name for the probabilities in the Target Variable box. In the Function Group box, select CDF and Noncentral CDF. CDF is the Cumulative Distribution Function. In the Functions and Special Variables, select Cdf.Uniform. Instead of using a variable, x, for the first value, we type in the larger of the two x values. We then subtract the Cdf.Uniform of the smaller of the two x values from the first.

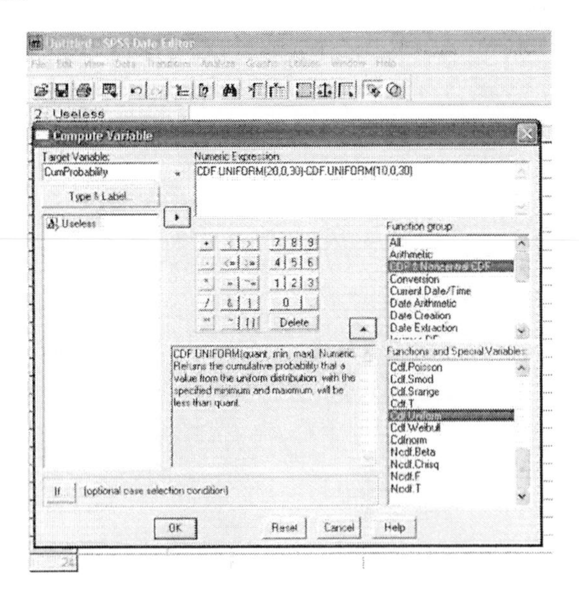

Once the name and expression have been completed, push the OK button. This will create a new column in the data view window with the probabilities of observing an x value between 10 and 20. (You may need to change the number of decimal places shown in the variable view window to see some of the values).

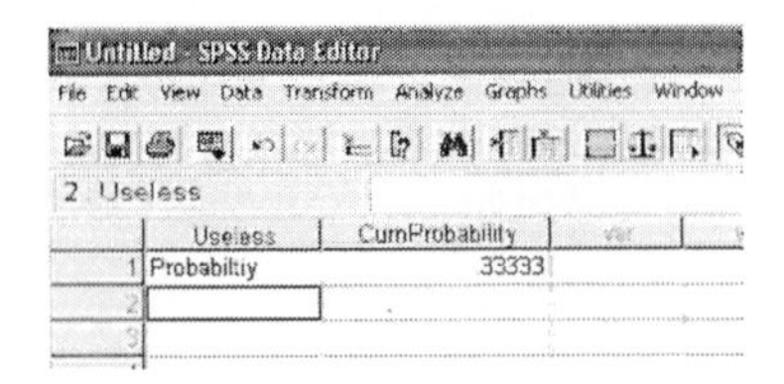

If you don't create the dummy variable first, you will get an error, as SPSS will not know how many answers to create. If you have more than one value, the answer will be repeated as many times as you have values in the data set.

►Adding a Normal Curve to a Histogram

Problem 37, Page 373

To create a histogram, see the information presented in chapter 2. Once all of the values have been entered, we are ready to create the histogram. Go to **Graphs → Histogram…**

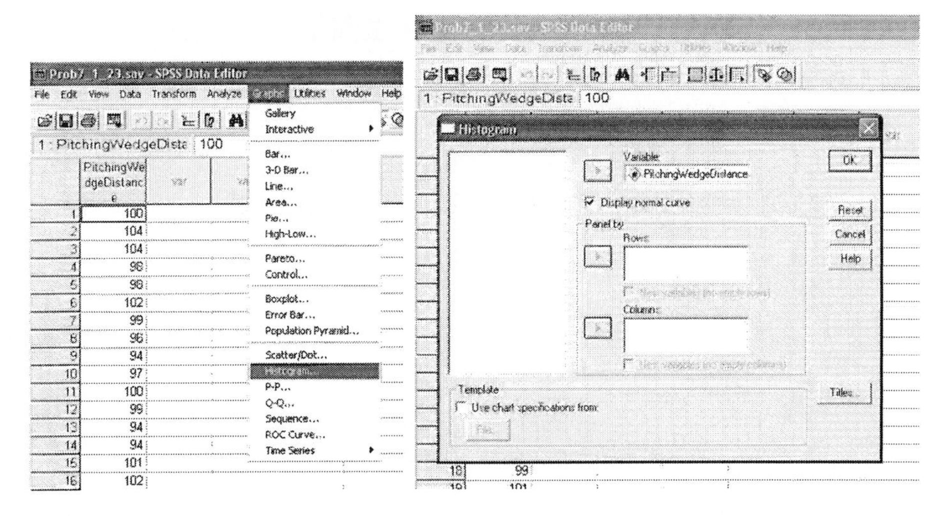

To add the Normal Curve, check the Display normal curve box underneath the Variable box. Then push OK. The histogram will be provided in the Output window, with the normal curve added.

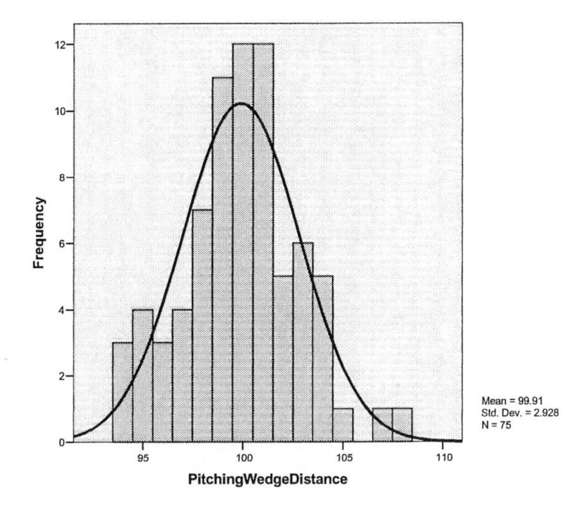

If you forget to click the Display normal curve when creating the histogram, the curve may be added to the histogram through the SPSS Chart Editor. While in the editor, click on the histogram and go to **Edit → Properties**. Here you will be given the option of adding the normal curve. Again, all you need to do is check the Display normal curve box.

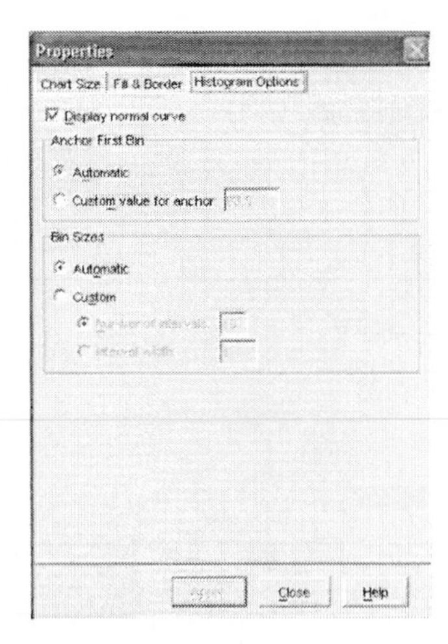

Section 7.2 The Standard Normal Distribution

► Finding the Area under the Standard Normal Curve

Rather than going into detail of how to find the area under the Standard Normal Curve, we will show how to find the area under any Normal curve. To find areas under the Standard Normal, use the same methods provided below with a mean of 0 and a standard deviation of 1.

Section 7.3 Applications of the Normal Distribution

► Finding the Area under a Normal Curve

► *Area less than or equal to a value*:
Example 1, Page 387

To find the proportion of three-year-old females who have a height less than 35 inches, we create a new variable with the value 35. Once the variable has been defined, go to **Transform → Compute...**

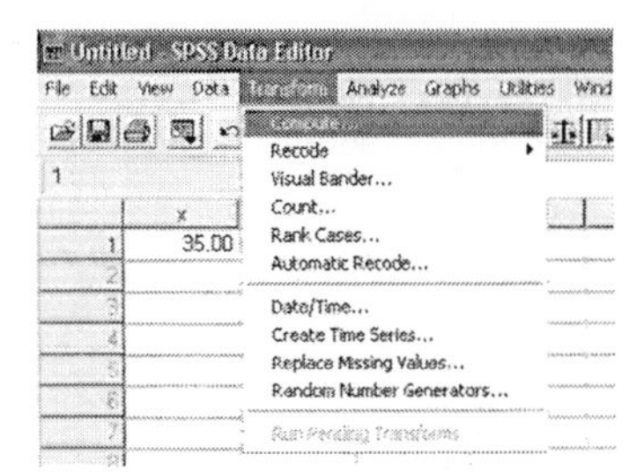

While in the compute variable window, create a name for the probabilities in the Target Variable box. In the Function Group box, select CDF and Noncentral CDF. CDF is the Cumulative Distribution Function. In the Functions and Special Variables, select CDF.NORMAL.

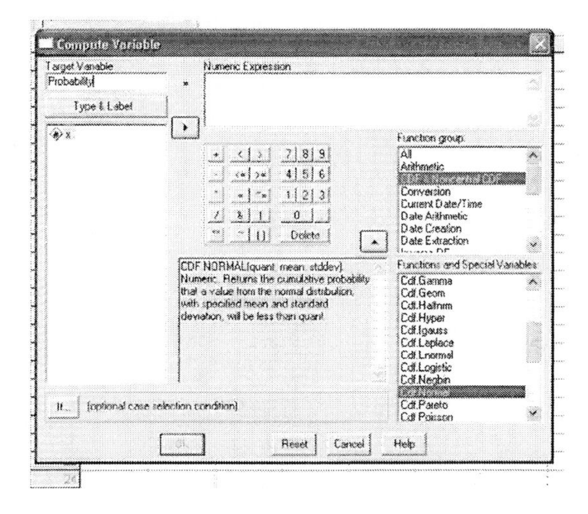

For this function to work, you need three values. The first is x, which is the variable we created earlier. The second is the mean of the Normal distribution, and the third is the standard deviation of the Normal distribution. So, we will select CDF.NORMAL, and replace the question marks that appear with x, 38.72, and 3.17. When selecting the name of the variable we wish to use, we can either type the name (if it is short), or highlight the name, and push the ► button next to the Numeric Expression box.

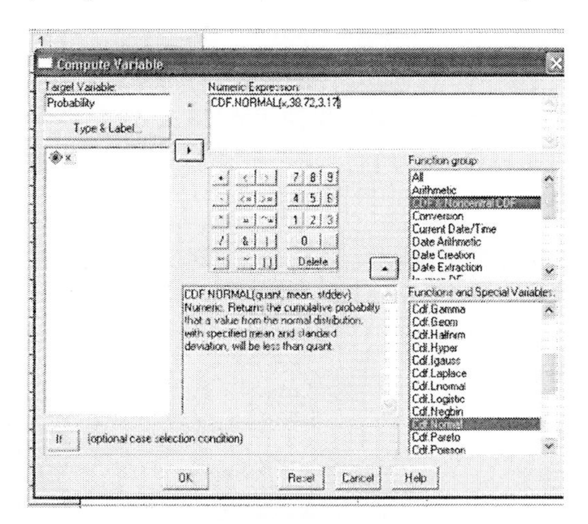

Once the name and expression have been completed, push the OK button. This will create a new column in the data view window with the probabilities of observing an x value less than or equal to 35 inches. (You may need to change the number of decimal places shown in the variable view window to see some of the values).

Note: SPSS answers may vary slightly from book answers due to the book rounding to 2 decimal places for the z-score. SPSS does not round in its calculations.

Untitled - SPSS Data Editor

File Edit View Data Transform Analyze Graphs

1 : Probability 0.1202973623

	x	Probability	var
1	35.00	.12030	
2			
3			
4			

So, 12.03% of three-year-old females have a height less than 35 inches.

► *Area greater than or equal to a value*:

To find the area greater than or equal to a value, use 1-CDF.NORMAL as the equation. The CDF is always the area to the left, so we can use the complement rule to find the area to the right.

► *Area between two values*:

Example 3, Page 389

To find the area between two values, you can start with typing in one value in the first column. This value is irrelevant, as it serves only to define the sample size. . Once the variable has been defined, go to **Transform → Compute...**

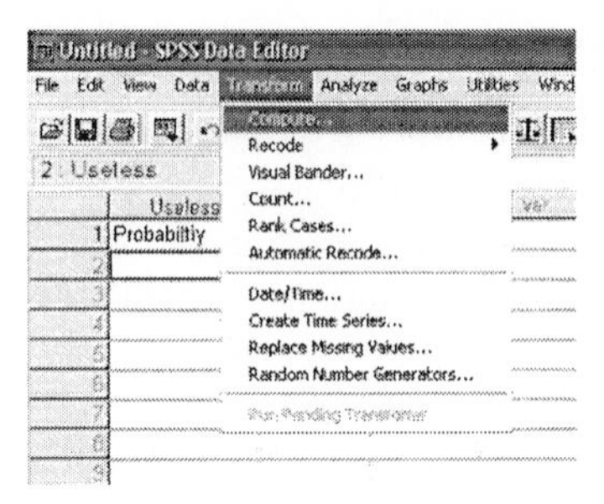

While in the compute variable window, create a name for the probabilities in the Target Variable box. In the Function Group box, select CDF and Noncentral CDF. CDF is the Cumulative Distribution Function. In the Functions and Special Variables, select CDF.NORMAL. Instead of using a variable, x, for the first value, we type in the larger of the two x values. We then subtract the CDF.NORMAL of the smaller of the two x values from the first.

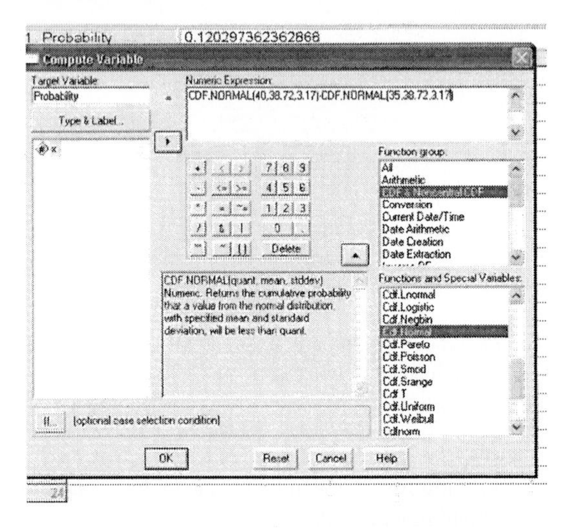

Once the name and expression have been completed, push the OK button. This will create a new column in the data view window with the probabilities of observing an x value between 35 and 40. (You may need to change the number of decimal places shown in the variable view window to see some of the values).

If you don't create the dummy variable first, you will get an error, as SPSS will not know how many answers to create. If you have more than one value, the answer will be repeated as many times as you have values in the data set.

The probability of finding a three-year-old female between 35 inches and 40 inches is .53652.

►Finding Values of Normal Random Variables

► *Area less than or equal to a value*:

Example 4, Page 390

When looking for a specific value of x under the Normal Distribution, SPSS must look for values using a cumulative distribution. So, probabilities must be areas to the left of x. If the information does not come in this way, it must be changed to fit how the computer can use it.

In this example, we are looking for the 20th percentile, which is the point at which 20% of the information is less than or equal to x, or the cumulative probability is .20. To find the height of three-year-old females which marks the 20th percentile, we create a new variable with the value .20. Once the variable has been defined, go to **Transform → Compute…**

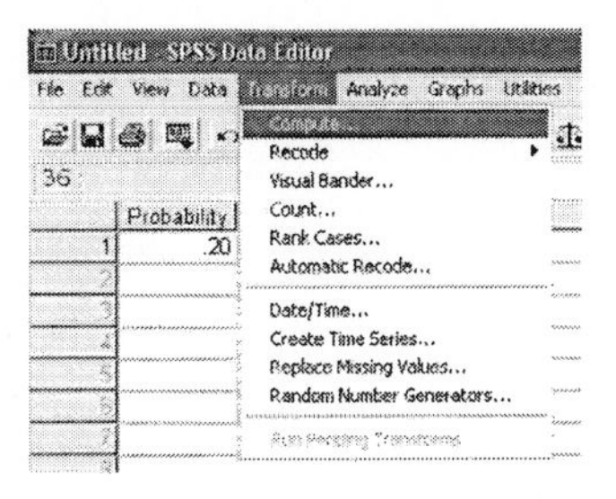

While in the compute variable window, create a name for the probabilities in the Target Variable box. In the Function Group box, select Inverse DF. Inverse DF is the Inverse Distribution Function. In the Functions and Special Variables, select IDF.NORMAL.

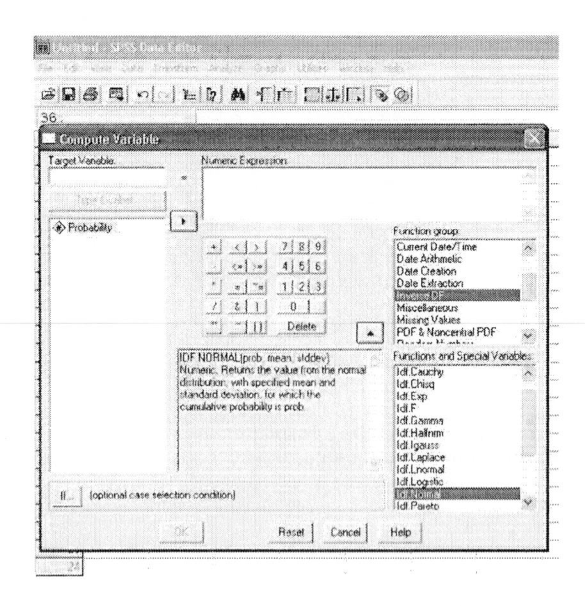

For this function to work, you need three values. The first is the probability to the left, which is the variable we created earlier. This only works with the cumulative probability. The second is the mean of the Normal distribution, and the third is the standard deviation of the Normal distribution. So, we will select IDF.NORMAL, and replace the question marks that appear with probability, 38.72, and 3.17. When selecting the name of the variable we wish to use, we can either type the name (if it is short), or highlight the name, and push the ► button next to the Numeric Expression box.

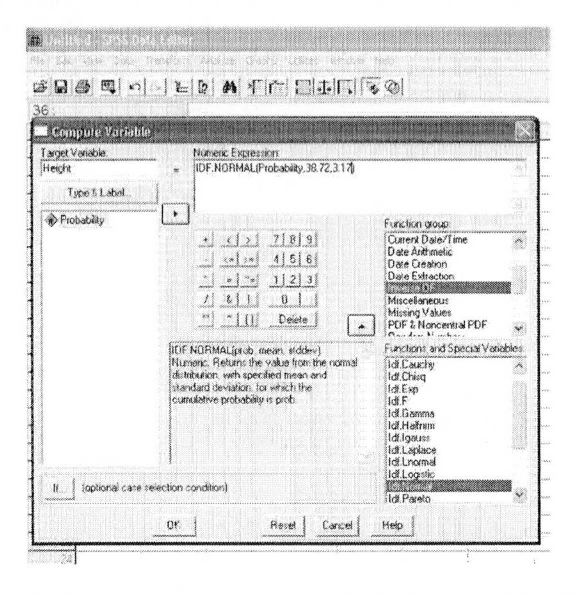

Once the name and expression have been completed, push the OK button. This will create a new column in the data view window with the x value. (You may need to change the number of decimal places shown in the variable view window to see some of the values).

The height of a three-year-old female that separates the top 80% from the bottom 20% is 36.05.

► *Area more than or equal to a value*:

Example 6, Page 391

To get the x value which marks where 1% of the information is to the right of that value, we have to use the complement rule. We know that if 1% is to the right, then 99% is to the left, and that is the information we give to SPSS. We can then follow the steps for area to the left.

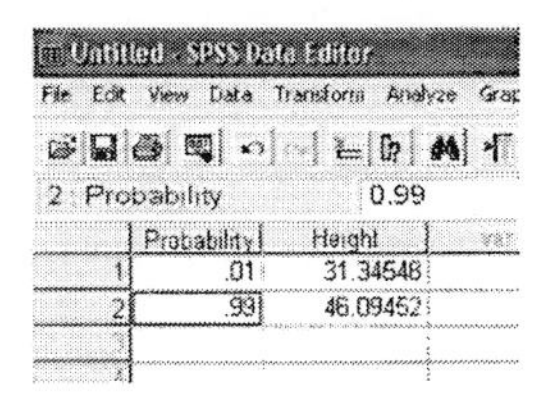

There are two advantages of using SPSS in these situations. The first is that we don't need to find the closest value in the table, as SPSS has the complete table to work from. In other words, we don't have to worry about rounding for the z-scores. The second is we can look up the values of more than one probability at a time, without adding work.

Section 7.4 Assessing Normality

►Creating a Normal Probability or Q-Q Plot

Example 1, Page 397

Once we have entered the data, we go to **Graphs → Q-Q…** (Q-Q stands for Quantile - Quantile).

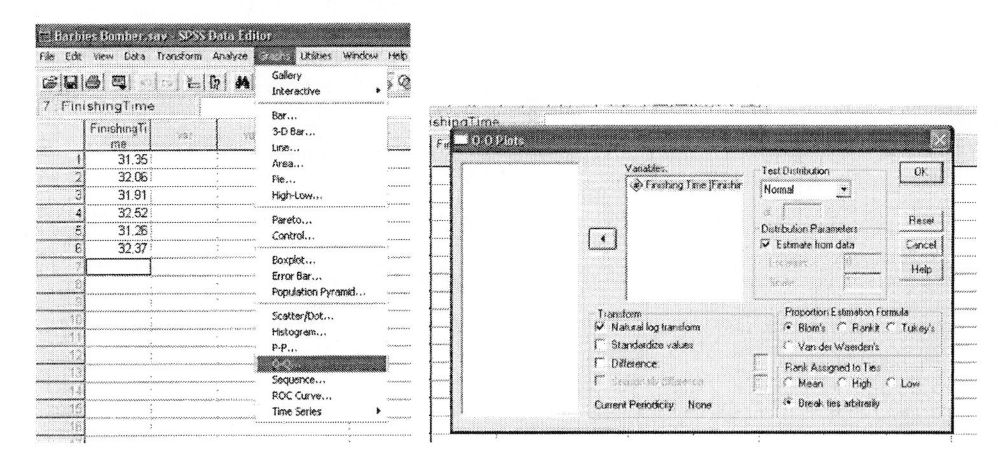

Highlight the variable of interest, and push the ▶ button. Make sure that the Test Distribution is Normal (if you want to test if the distribution is other than normal, there are other choices possible).

The Blom's Proportion Estimation Formula is the one used in the book, $f_i = \dfrac{i - 0.375}{n + .25}$.

Rankit uses $f_i = \dfrac{i - 0.5}{n}$, Tukey's uses $f_i = \dfrac{i - 0.333}{n + .333}$, and Van der Waerden's uses $f_i = \dfrac{i}{n+1}$. We will just use Blom's Proportion Estimation Formula, as on page 397. Also, to make the Q-Q plot look like the ones in the book, select the Break ties arbitrarily under the Rank Assigned to Ties options. The default is to give values that have the same rank the mean of the values, where breaking ties arbitrarily will provide two different values for the same observed value, based on the different indexes.
Push OK

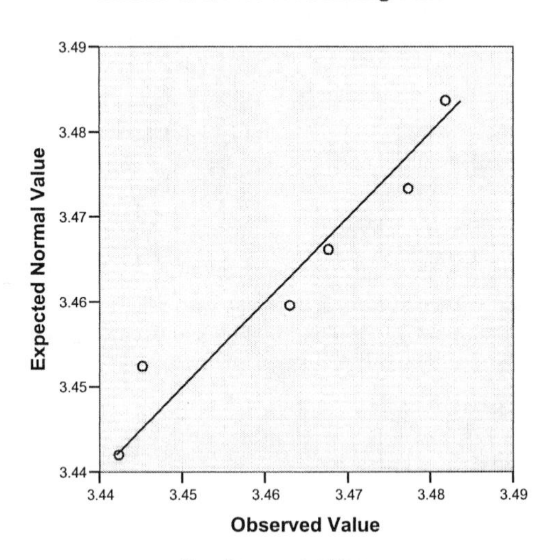

The plot will appear in the Output window, with a de-trended plot. We will work with the Q-Q plot, as in the book. SPSS does not put the interval curves, as Minitab does. Instead, we must use our best judgment as to how close the points are to the line. Some general guidelines are as follows:

Quantile - Quantile Plot Diagnostics

Description of Point Pattern	**Possible Interpretation**
All but a few points fall on a line	Outliers in the data
Left end of pattern is below the line; right end of pattern is above the line	Long tails at both ends of the data distribution
Left end of pattern is above the line; right end of pattern is below the line	Short tails at both ends of the distribution
Curved pattern	Data distribution is skewed
Staircase pattern (plateaus and gaps)	Data have been rounded or are discrete

Here is the graph using same method for example 3, showing a curved pattern, indicating a skewed distribution.

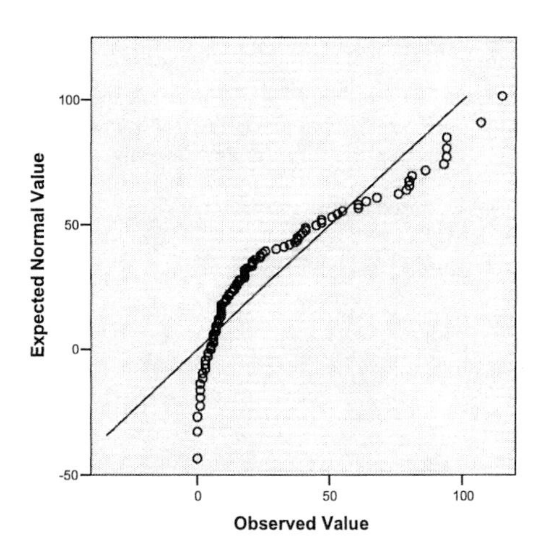

Section 7.5 The Normal Approximation to the Binomial Probability Distribution

►Normal Approximation to the Binomial

You can use SPSS to get the actual probability for the Binomial, so that the Normal Approximation is no longer necessary. It was more useful before computers. But, the approximation can still be done using the same methods as in section 7.3, but you must remember to make the continuity corrections.

Chapter 8. Sampling Distributions

Section 8.1 Distribution of the Sample Mean

► Sampling Distribution Simulation - Mean

Example 2, Page 420

Similar to Chapter 5, section 1, to create 100 random numbers, you must first create a data set with 100 observations. SPSS will create new variables, with the same sample size as already exists, but will not create more observations than are already in the data set. To create the original sample of 100, you can type in the values 1 to 100 by hand, or use another program, such as Excel, where data generation is easier. After creating the data set with 100 observations, go to **Transform → Compute...**

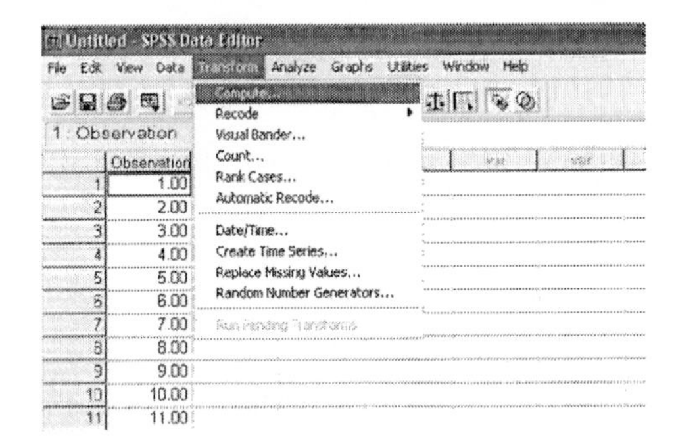

Once in the Compute Variable window, you can choose a name for the new variable you wish to create. This name must follow SPSS naming conventions (no strange characters or spaces). You will be creating five variables in the same way, so it may be easiest to add a number to the end of the name, to keep the sample straight. In the Function Group box, select Random Numbers. In the Functions and Special Variables box, select Rv.Normal. This will create a random observation from a Normal distribution. There are two pieces of information that this requires, which are the mean and standard deviation of the distribution. We replace the two question marks with 38.72 and 3.17, for a mean of 38.72 and standard deviation of 3.17.

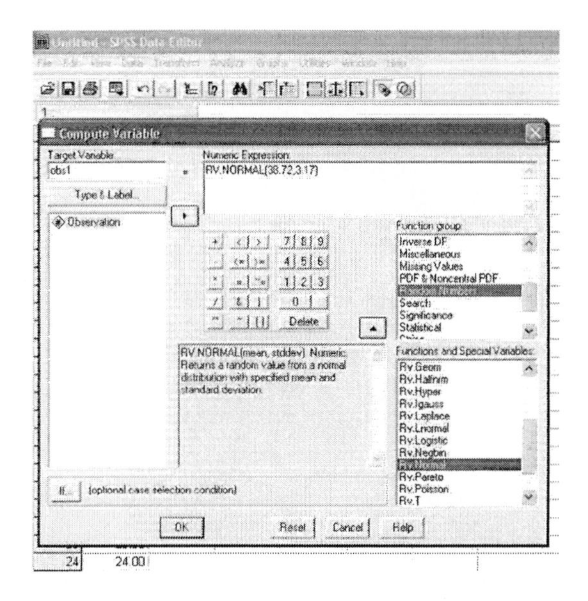

Once the equation and target variable name is correct, push OK. This will create a new variable of random Normal values. To create the other four observations, go back to **Transform → Compute...** and repeat the process. The equation will still be there, so all you have to do is change the variable name.

Untitled - SPSS Data Editor

File Edit View Data Transform Analyze Graphs Utilities Window Help

1 : obs1 42.5362990418493

	Observation	obs1	obs2	obs3	obs4	obs5
1	1.00	42.54	38.53	42.17	35.63	35.70
2	2.00	37.20	33.31	39.92	36.58	32.53
3	3.00	42.01	37.60	31.44	34.40	38.06
4	4.00	38.10	38.80	34.00	40.90	39.94
5	5.00	40.09	40.40	37.86	41.91	43.66
6	6.00	43.34	36.87	41.01	36.55	35.83
7	7.00	39.51	44.54	39.46	33.12	39.39
8	8.00	43.11	37.23	36.59	34.94	38.92
9	9.00	39.64	32.46	33.66	40.33	36.43
10	10.00	40.86	41.98	40.14	40.53	38.00
11	11.00	45.95	39.76	35.64	43.14	40.87
12	12.00	39.92	35.94	43.47	40.81	39.32
13	13.00	41.36	37.32	44.39	41.25	35.70
14	14.00	36.06	35.02	39.84	39.74	33.50
15	15.00	33.26	44.44	36.19	38.06	44.49
16	16.00	33.66	40.58	41.21	37.36	39.49

Once you have all five observations, return to **Transform → Compute...**, but this time, we will be changing both the name and equation. Choose a name to represent the mean of the five observations. After clearing the equation in the Numeric Expression box, select Statistical in the Function Group box. From the Functions and Special Variables box, select Mean. The mean requires the data to find the mean of, so type in the name of all five variables that we just created, separating each name with a comma.

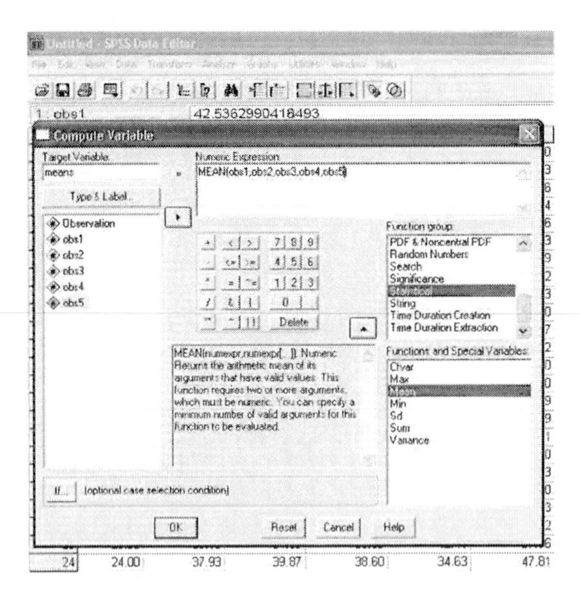

When the equation looks correct, and the name is as you want it, click OK. We now have a variable of sample means.

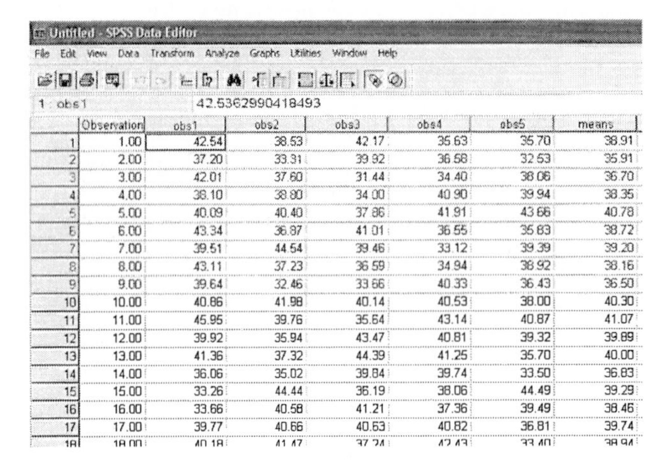
Untitled - SPSS Data Editor

File Edit View Data Transform Analyze Graphs Utilities Window Help

1 : obs1 42.5362990418493

	Observation	obs1	obs2	obs3	obs4	obs5	means
1	1.00	42.54	38.53	42.17	35.63	35.70	38.91
2	2.00	37.20	33.31	39.92	36.58	32.53	35.91
3	3.00	42.01	37.60	31.44	34.40	38.06	36.70
4	4.00	38.10	38.80	34.00	40.90	39.94	38.35
5	5.00	40.09	40.40	37.86	41.91	43.66	40.78
6	6.00	43.34	36.87	41.01	36.55	35.83	38.72
7	7.00	39.51	44.54	39.46	33.12	39.39	39.20
8	8.00	43.11	37.23	36.59	34.94	38.92	38.16
9	9.00	39.64	32.46	33.66	40.33	36.43	36.50
10	10.00	40.86	41.98	40.14	40.53	38.00	40.30
11	11.00	45.95	39.76	35.64	43.14	40.87	41.07
12	12.00	39.92	35.94	43.47	40.81	39.32	39.89
13	13.00	41.36	37.32	44.39	41.25	35.70	40.00
14	14.00	36.06	35.02	39.84	39.74	33.50	36.83
15	15.00	33.26	44.44	36.19	38.06	44.49	39.29
16	16.00	33.66	40.58	41.21	37.36	39.49	38.46
17	17.00	39.77	40.66	40.63	40.82	36.81	39.74
18	18.00	40.18	41.47	37.74	42.43	33.40	38.94

Now, we can use the methods in chapter 2, section 2 to create a histogram of the means.

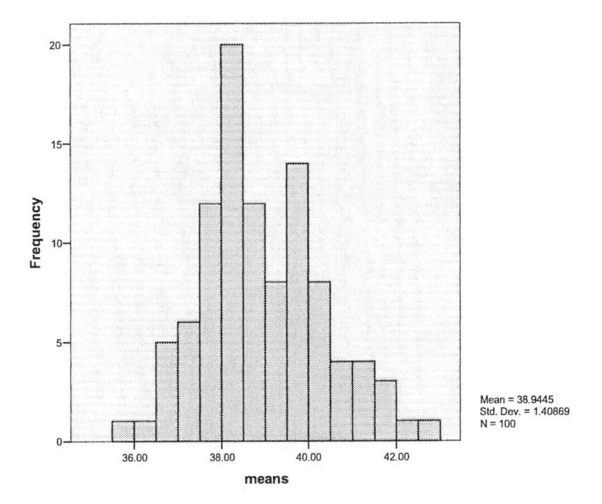

To sample from a non-Normal distribution, you can select from any of the other random number functions provided by SPSS, and repeat the same process.

►Applying the Central Limit Theorem

Example 6, Page 429

To find probabilities applying the Central Limit Theorem, we can use the same methods as in chapter 7, except now we must calculate the correct standard error first. So, where we selected Cdf.Normal, and replaced the question marks that appear with x, 2716, and 72.8 for a single observation, we will now use x, 2716, and 72.8/sqrt(35) , (SPSS uses sqrt(value) to calculate square roots). SPSS will accept an equation for the standard error, so we don't have to worry about rounding. SPSS does not care if the variable x contains single observations or sample means, as it assumes that you know which you wish to use, and that you will adjust the standard error appropriately. The same goes for the Idf.Normal. As long as you know if you are putting in the standard deviation or standard error, SPSS will give you the correct values.

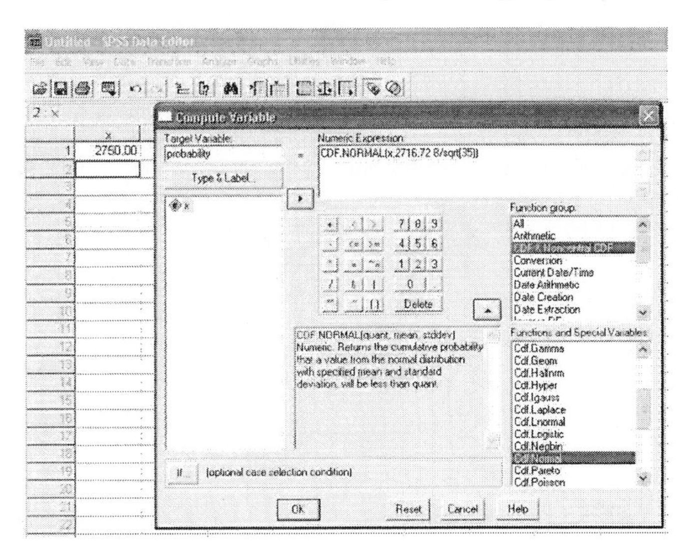

Section 8.2 Distribution of the Sample Proportion

►Sampling Distribution Simulation - Proportion

Example 2, Page 435

Similar to Chapter 5, and the sampling distribution of means, to create 100 random numbers, you must first create a data set with 100 observations. SPSS will create new variables, with the same sample size as already exists, but will not create more observations than are already in the data set. To create the original sample of 100, you can type in the values 1 to 100 by hand, or use another program, such as Excel, where data generation is easier. After creating the data set with 100 observations, go to **Transform → Compute…**

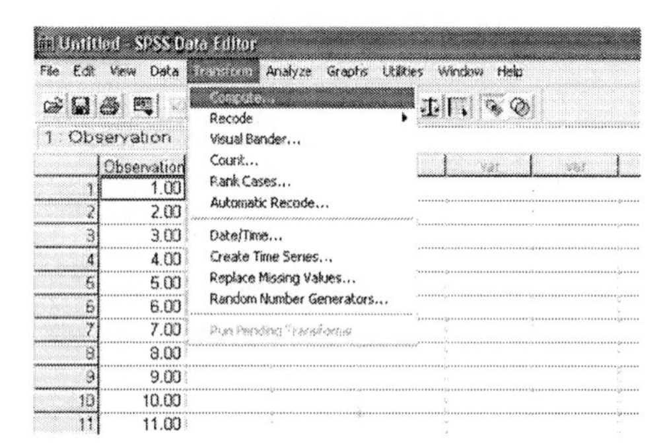

Once in the Compute Variable window, you can choose a name for the new variable you wish to create. This name must follow SPSS naming conventions (no strange characters or spaces). In the Function Group box, select Random Numbers. In the Functions and Special Variables box, select Rv.Binom. This will create a random observation from a Binomial distribution. There are two pieces of information that this requires, which are the sample size and the probability of a success. We replace the two question marks with 10 (40, 80) and 0.17. This will give a binomial value, which will be a number between 1 and 10. Because we are looking for proportions, we will divide the random variable by the sample size (10).

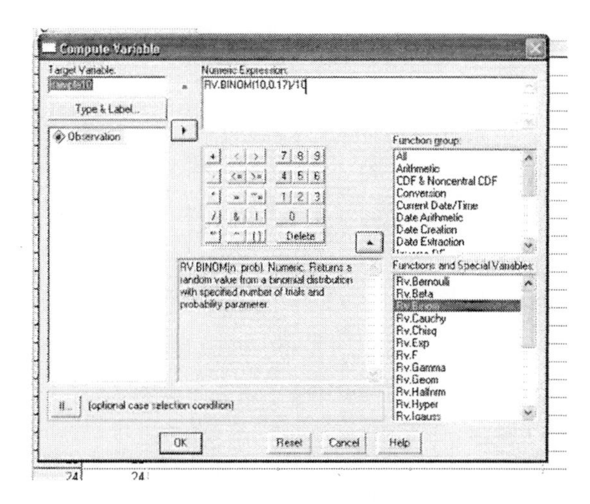

Once the equation and target variable name is correct, push OK. This will create a new variable of random proportions based on the Binomial distribution.

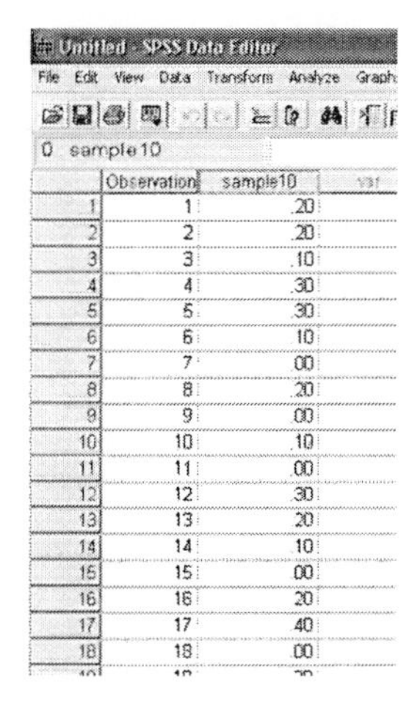

	Observation	sample10
1	1	.20
2	2	.20
3	3	.10
4	4	.30
5	5	.30
6	6	.10
7	7	.00
8	8	.20
9	9	.00
10	10	.10
11	11	.00
12	12	.30
13	13	.20
14	14	.10
15	15	.00
16	16	.20
17	17	.40
18	18	.00

Now, we can use the methods in chapter 2, section 2 to create a histogram of the means.

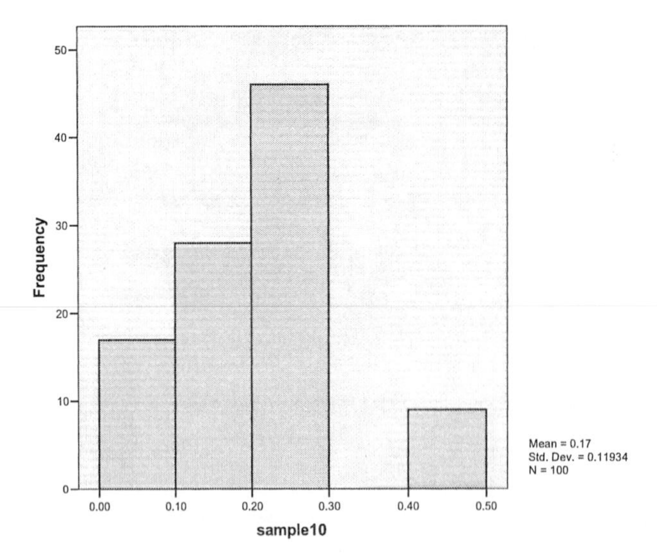

To obtain the histograms for sample of size 40 and 80, repeat the process, changing the sample size in the equation.

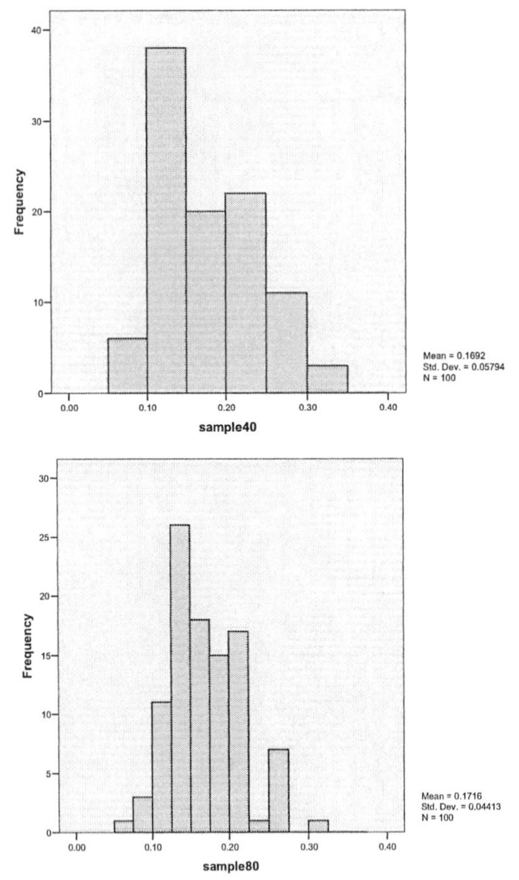

To sample from other non-Normal distribution, you can select from any of the other random number functions provided by SPSS, and repeat the same process.

Chapter 9. Estimating the Value of a Parameter

SPSS assumes that all data come from a sample. Because of this, all packaged methods assume σ to be unknown. Methods where σ is known must be done, at least in part, by hand.

Section 9.1 The Logic in Constructing Confidence Intervals about a Population Mean

The point estimate, or sample mean, can be obtained through the methods covered in chapter 3.

►Constructing 20 95% Confidence Intervals Based on 20 Samples

Example 2, Page 450

We can create samples using the method provided in chapter 8. Once we have the sample means in our data set, got to **Transform → Compute...**

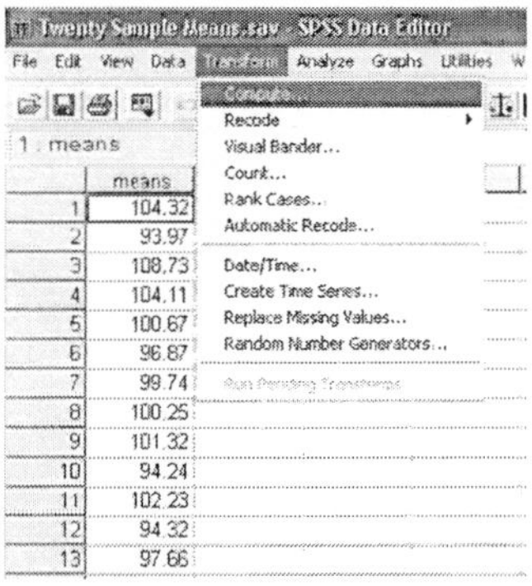

We will have to create two values, one for the lower confidence limit, and one for the upper confidence limit. Create a name for the lower limit. There are two ways we can now create the confidence interval. We can type in the full equation after looking up the z-value on the table, which would be means – 1.96*16/sqrt(15) and means+1.96*16/sqrt(15), or, we can let SPSS find the z-value for us, using the IDF, or Inverse Distribution Function. IDF.NORMAL finds the value of z (when we use a mean of 0 and standard deviation of 1) where the area **to the left** of z is the probability we provide. In the case of a 95% confidence interval, the area in the tails is .05, so the area to the left for the lower limit is .025, or $\alpha/2$. The area to the left for the upper limit will be 1-$\alpha/2$, or 1-.025, which is .975. Here we have means + IDF.NORMAL(0.025,0,1)*16/sqrt(15), and means + IDF.NORMAL(0.975,0,1)*16/sqrt(15). Note that we have to use the area to the left, not the confidence level. We can also use the IDF for the full equation. Knowing that the IDF.NORMAL uses the probability to the left, a mean, and a standard error, we can use IDF.NORMAL(0.025, means, 16/sqrt(15)) as our equation, and receive the same values. Also, in both cases we are **adding** the IDF.NORMAL value, since the z-value at .025 will be negative. If we subtract the IDF.NORMAL(0.025,0,1) it will provide the same value as adding the IDF.NORMAL(0.975,0,1). An advantage of allowing SPSS to find the values is that SPSS won't round the values, as we have to when using the table.

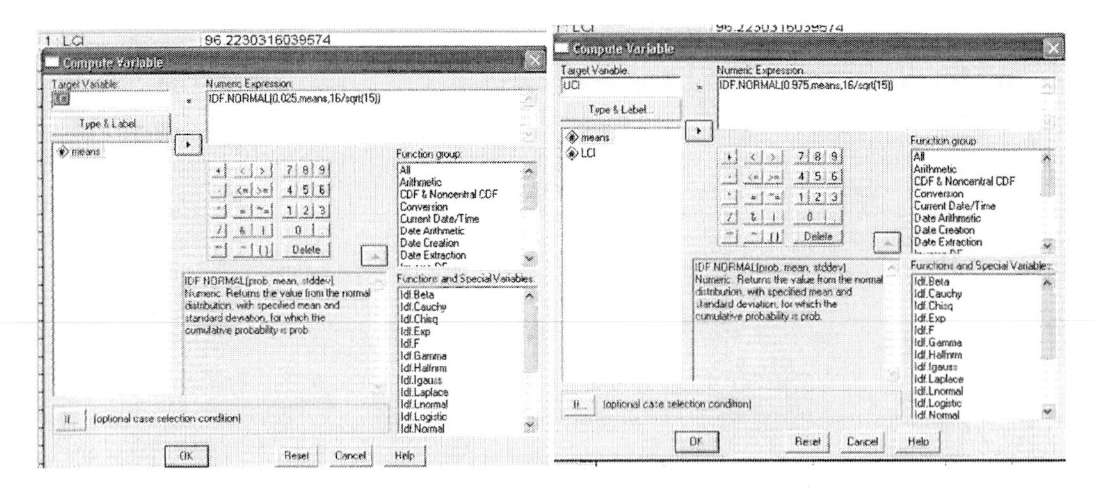

When you have the equation as you want it, push OK. Do this again for the upper limit, changing the probability value. This will provide the lower and upper limits for the confidence interval.

Twenty Sample Means.sav - SPSS Data Editor

File Edit View Data Transform Analyze Graphs Utilities

1 : LCI 96.2230316039574

	means	LCI	UCI
1	104.32	96.22	112.42
2	93.97	85.87	102.07
3	108.73	100.63	116.83
4	104.11	96.01	112.21
5	100.67	92.57	108.77
6	96.87	88.77	104.97
7	99.74	91.64	107.84
8	100.25	92.15	108.35
9	101.32	93.22	109.42
10	94.24	86.14	102.34
11	102.23	94.13	110.33
12	94.32	86.22	102.42
13	97.66	89.56	105.76
14	101.44	93.34	109.54
15	98.19	90.09	106.29
16	107.15	99.05	115.25
17	100.38	92.28	108.48
18	95.89	87.79	103.99
19	104.43	96.33	112.53
20	102.28	94.18	110.38
21			

SPSS does not work directly in computing confidence intervals using the standard normal table, as SPSS assumes that the population standard deviation is not known. SPSS can provide the point estimate from a data set, but the interval must be calculated by hand.

Example 4, Page 455

We follow the methods in chapter 3 to obtain the following information.

Descriptive Statistics

	N	Minimum	Maximum	Mean	Std. Deviation
Speed	12	43.9	70.3	59.592	7.0171
Valid N (listwise)	12				

We can also use SPSS to create the box-plot, and Normal Probability Plot to check the data. We will note that the sample standard deviation is not the same as the population standard deviation. SPSS does not know the population standard deviation. To obtain the confidence interval we create a new column which contains the value of the point estimate.

Highway Speeds.sav - SPSS Data Editor

1 : PointEstimate 59.59166

	Speed	PointEstimate
1	57.4	59.59
2	56.1	.
3	70.3	.
4	65.6	.
5	43.9	.
6	58.6	.
7	66.1	.
8	57.3	.
9	62.2	.
10	60.4	.
11	64.5	.
12	52.7	.

We can then follow the same steps as example 2 to create the confidence interval.

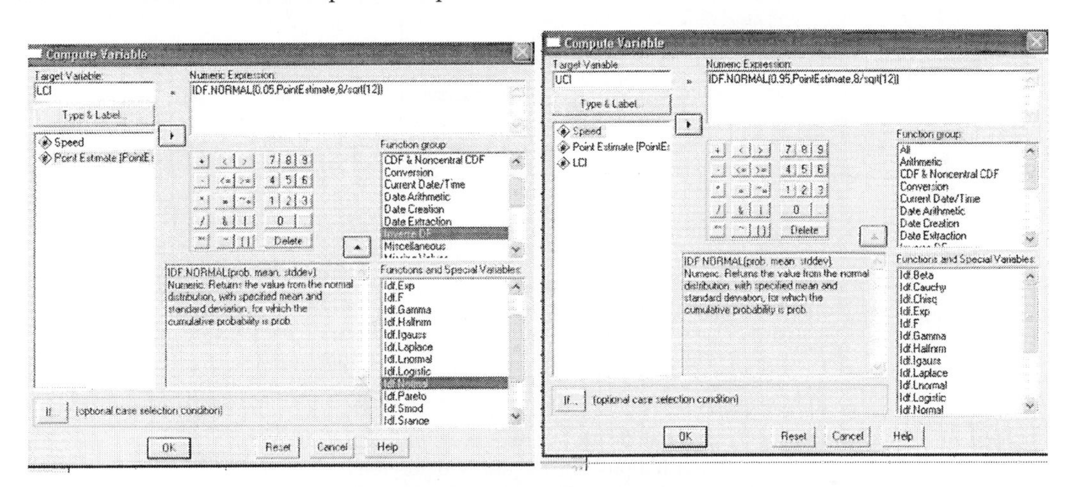

Highway Speeds.sav - SPSS Data Editor

0 : LCI

	Speed	PointEstimate	LCI	UCI
1	57.4	59.59	55.7930	63.3903
2	56.1	.	.	.
3	70.3	.	.	.
4	65.6	.	.	.
5	43.9	.	.	.
6	58.6	.	.	.
7	66.1	.	.	.
8	57.3	.	.	.
9	62.2	.	.	.
10	60.4	.	.	.
11	64.5	.	.	.
12	52.7	.	.	.

So, we are 90% confident that the mean speed of all cars traveling on the highway outside the subdivision is between 55.79 and 63.39 miles per hour.

Section 9.2 Confidence Intervals about a Population in Practice

►Finding t-values

Example 2, page 468
We can use the IDF.T to find t-values. We must remember that IDF is always looking for area to the left, so if we want to find the t-value where the area under the t-distribution to the right of the t-value is .10, assuming 15 degrees of freedom, we can use IDF.T(0.90,15). The IDF.T requires the area to the left, and the degrees of freedom. Unlike the IDF.NORMAL, you cannot get back to an x-value directly by using the IDF.T.

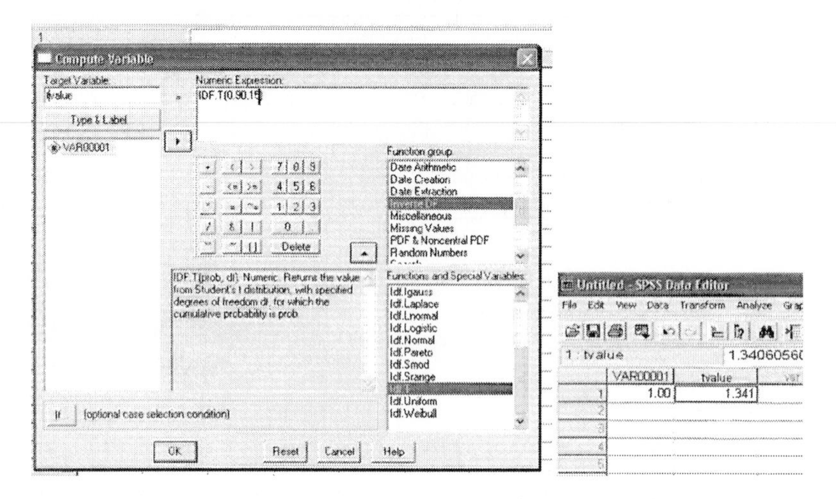

The value of $t_{0.10}$ with 15 degrees of freedom is 1.341.

►Constructing Confidence Intervals when σ is unknown

Example 3, page 470

First, we must enter the data. We can then do all the steps of a confidence interval.
Step 1. To create the box-plot and normal probability plot, see previous sections.
Step 2-Step 4. These are done in the computer. Go to **Analyze → Compare Means → One-Sample T Test…**

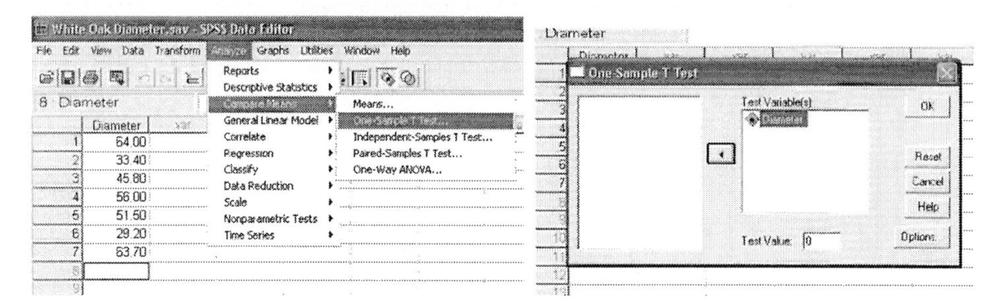

Highlight the variable of interest, and push the ► button. We will ignore the Test Value box for now. For the confidence interval, we want to leave this at 0.

To chose the level of confidence, click on the Options button.

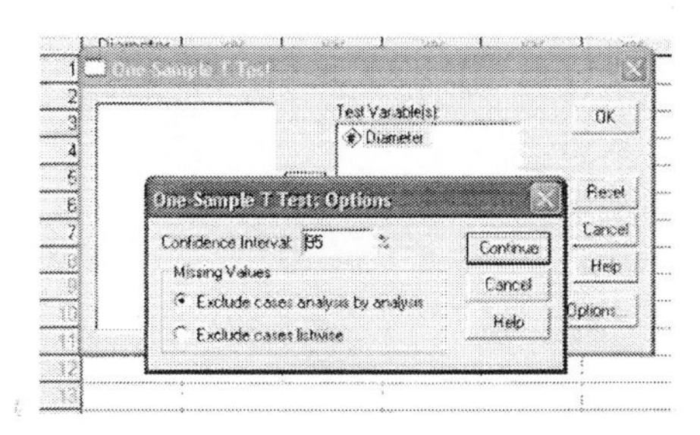

Type the level of confidence in the box. Note-this is **not** in decimal form. When you have chosen the correct level of confidence, click Continue, and OK. The output window will have the following:

One-Sample Statistics

	N	Mean	Std. Deviation	Std. Error Mean
Diameter	7	49.0857	13.79570	5.21429

This provides the sample size, sample mean, sample standard deviation, and standard error, which is the sample standard deviation over the square root of the sample size. If you want to perform a confidence interval by hand, these are all the values you will need, except the table value.

One-Sample Test

	Test Value = 0					
					95% Confidence Interval of the Difference	
	t	df	Sig. (2-tailed)	Mean Difference	Lower	Upper
Diameter	9.414	6	.000	49.08571	36.3268	61.8446

The confidence interval is provided in the **last two boxes**. For now, we will ignore the first four boxes. NOTE- the t value given in the first box is not the t-value used for the confidence interval. The df is the correct degrees of freedom, or sample size -1.

Step 5. Interpretation: We are 95% confident that the mean diameter from the base of mature white oak trees is between 36.33 and 61.84 centimeters

Section 9.3 Confidence Intervals about a Population Proportion

►Constructing Confidence Intervals about a Population Proportion

Oddly enough for a package originally designed for the Social Sciences, SPSS does not work directly with one or two proportions. There is very little advantage to using SPSS over a calculator unless you are working with multiple proportions at the same time, or you are worried about rounding.

Example 2, Page 480

Step 1: Compute the value of $\hat{p}$

In SPSS, create one variable for the value x, and another for the value n. Go to **Transform** → **Compute…**, create a new variable, called phat, by putting the name phat in the Target Variable box, and the equation x/n in the Numeric Expression box.

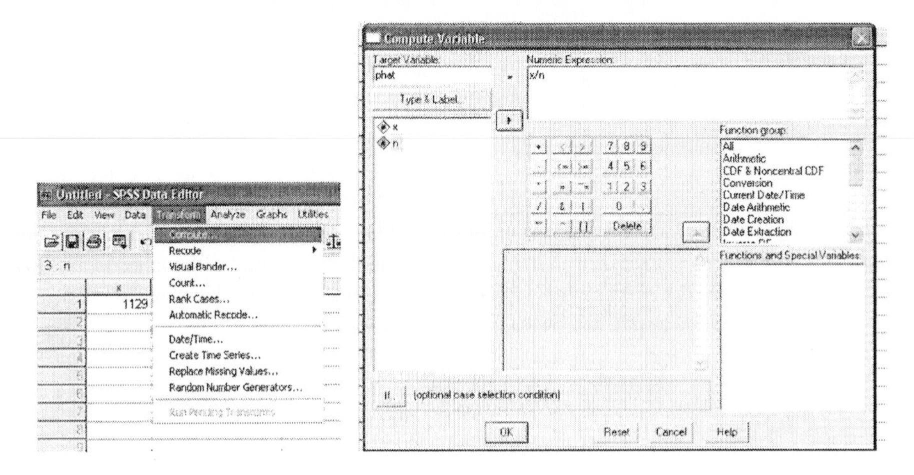

Remember that SPSS will not accept a name with odd characters, so p-hat is not accepted.

	x	n	phat
1	1129	1505	750166

The point estimate will appear in the Data View sheet, although you may need to change the number of decimal places, using the Variable View sheet to see it better. In this example, $\hat{p}$ is 0.7502, or, rounding to 4 decimal places, 0.0750. SPSS does not round.

Step 2: We can check this using the **Transform** → **Compute…** as well. Create a new variable, check, by using the equation n*phat*(1-phat) in the numeric expression box.

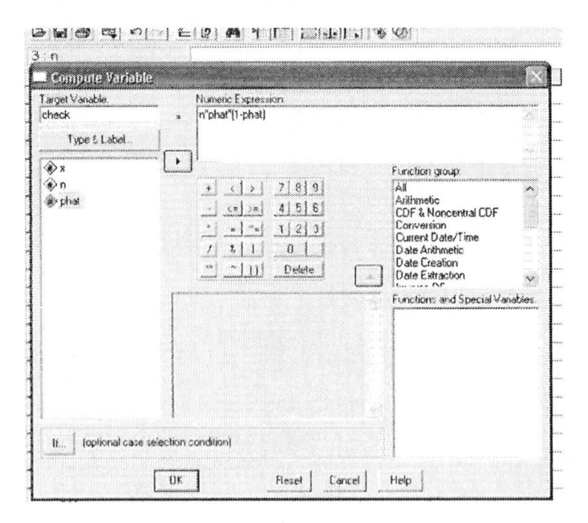

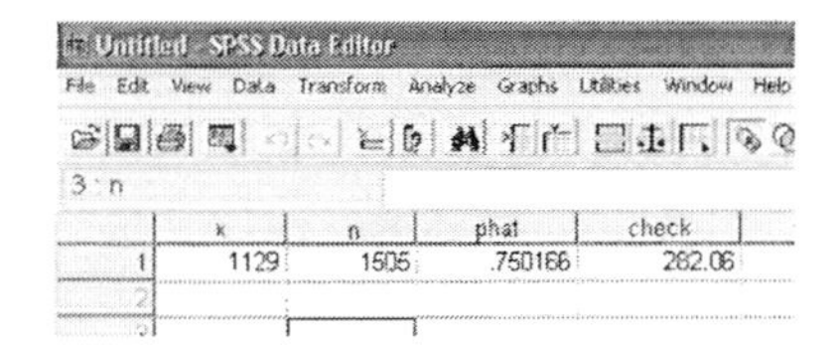

As the value is 282.06, which is greater than 10, we can proceed to construct the confidence interval. One advantage of using the computer is that you don't have to worry about rounding, but this may present slight differences from what you see in the book.

Step 3-4: This is similar to constructing a confidence interval with σ known. We will create two new variables, one for the lower bound, and one for the upper bound. Returning to **Transform → Compute...**, we will start with the lower bound, so we create a name, and in the Numeric Expression box, type the equation. We can again either use a table value, or use the SPSS tables. For a 95% confidence interval, the areas to the left are 0.025 and 0.975, leaving 0.95 in the middle. So, our equation for the lower bound will be phat + IDF.NORMAL(.025,0,1)*sqrt(phat*(1-phat)/n), and the equation for the upper bound will be phat + IDF.NORMAL (.975,0,1)*sqrt(phat*(1-phat)/n), or, similar to means, we can use IDF.NORMAL(probability to the left, point estimate, standard error).

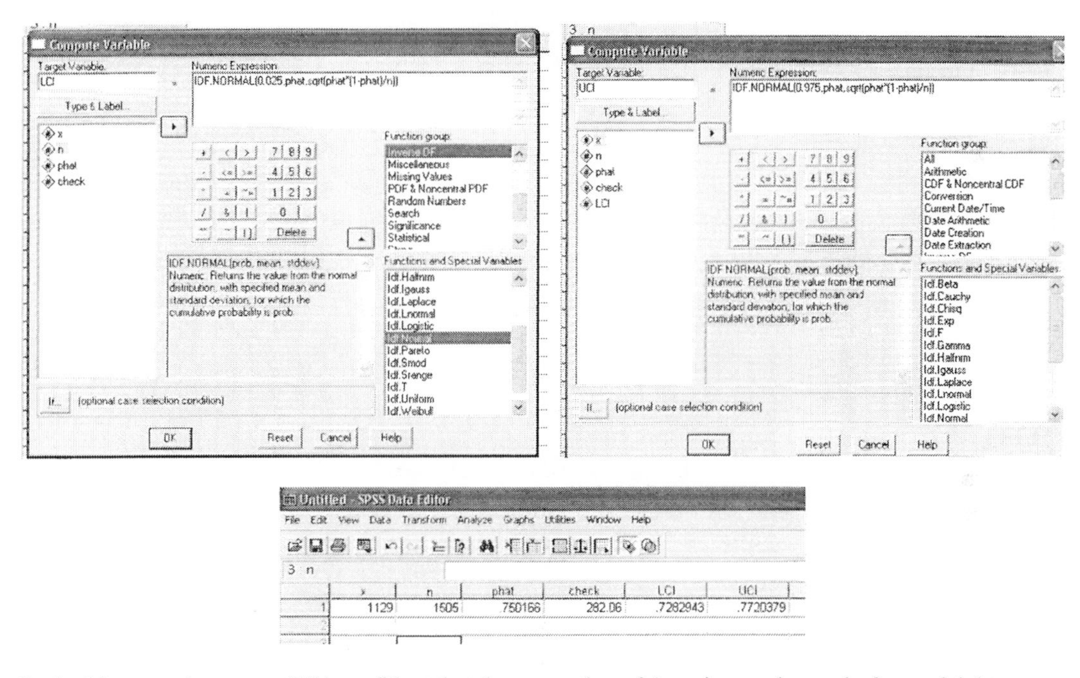

So, in this example, we are 95% confident that the proportion of Americans who are in favor of tighter enforcement of government rules on TV content during hours when children are most likely to be watching is between .728 and .772.

Section 9.4 Confidence Intervals about a Population Standard Deviation

SPSS does not have a method to directly create confidence intervals for standard deviations. You can use the IDF.CHISQ in **Transform → Compute...** to find the critical values for the Chi-Square distribution, but the rest must be done either by hand, or similar to the method for proportions.

►Finding Critical values for the Chi-Square Distribution

Example 1, Page 488

Before calculating the critical values, you must have a variable already existing. To find the critical values, go to **Transform → Compute…**. Create a name for the new variable, such as critical1. In the Function Group, select Inverse DF, and from the Functions and Special Variables, select IDF.CHISQ.

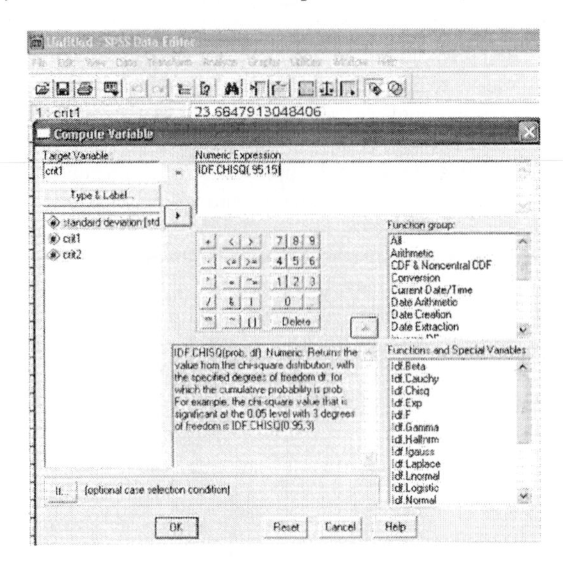

The Inverse Chi-Square requires two pieces of information. The first is the probability to the left. This will be opposite of the tables in the book, which are areas to the right. So, for $\chi^2_{.95}$ in the book, we will look up the value with an area of .05 to the left of the value, and for $\chi^2_{.05}$, we will look up the value with an area of .95 to the left. The second piece of information is the degrees of freedom. So, for $\chi^2_{.95}$, we will use the equation IDF.CHISQ(.05,15), and for $\chi^2_{.05}$, we will use the equation IDF.CHISQ(.95,15).

Untitled - SPSS Data Editor

File Edit View Data Transform Analyze Graphs Utilities W

2 : crit1

	stddev	crit1	crit2
1	4522.00	24.99579	7.26094
2			

So, we get $\chi^2_{.95}$=7.26094, and $\chi^2_{.05}$=24.99579.

►Constructing Confidence Intervals about a Population Standard Deviation

Example 2, Page 490

Calculating a confidence interval for the standard deviation in SPSS is similar to finding a confidence interval for μ when σ is known, or finding the confidence interval for a proportion. First, we can find the sample standard deviation.

Descriptive Statistics

	N	Minimum	Maximum	Mean	Std. Deviation
Price of 3-year old Chevy	12	$35,950	$43,995	$40,012.42	$2,615.187
Valid N (listwise)	12				

We can then copy the value of the standard deviation into a new column.

Untitled - SPSS Data Editor

File Edit View Data Transform Anal

1 : stddev 2615

	Price	stddev
1	$41,844	2615.19
2	$41,500	
3	$39,995	
4	$36,995	
5	$40,990	
6	$37,995	
7	$41,995	
8	$38,900	
9	$42,995	
10	$36,995	
11	$43,995	
12	$35,950	
13		

We then go to **Transform → Compute….** Create a name for the lower bound in the Target Variable box, and in the Numeric Expression box, use the equation (15-1)*stddev**2/ IDF.CHISQ (.95,14). The double asterisk represents a power, so stddev**2 is stddev^2. Remember that the tables in SPSS are to the left, where the ones in the book are to the right, so where the book has $\chi^2_{\alpha/2}$, SPSS wants $\chi^2_{1-\alpha/2}$.

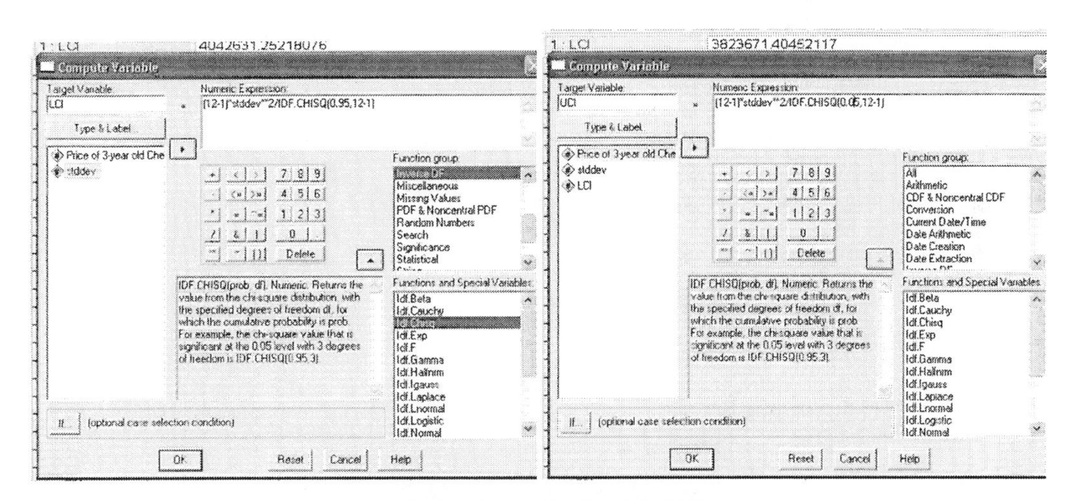

Untitled - SPSS Data Editor

File Edit View Data Transform Analyze Graphs Utilities Window Help

1 : LCI 3823671.40452117

	Price	stddev	LCI	UCI
1	$41,844	2615.19	3823671.40	16444663.3
2	$41,500			
3	$39,995			
4	$36,995			
5	$40,990			

We can then use the **Transform → Compute…** to take the square root of these values, to get the standard deviations.

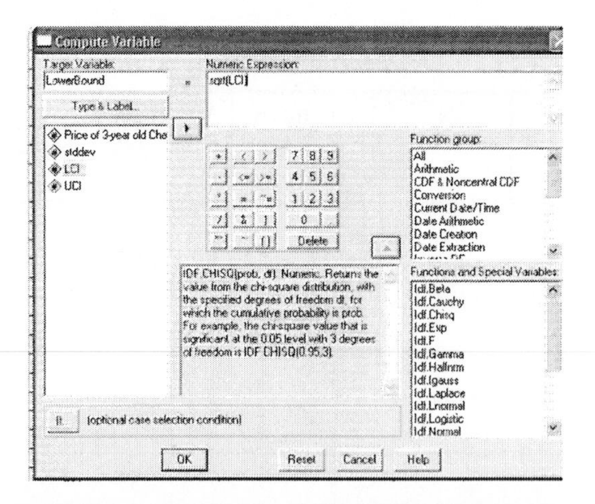

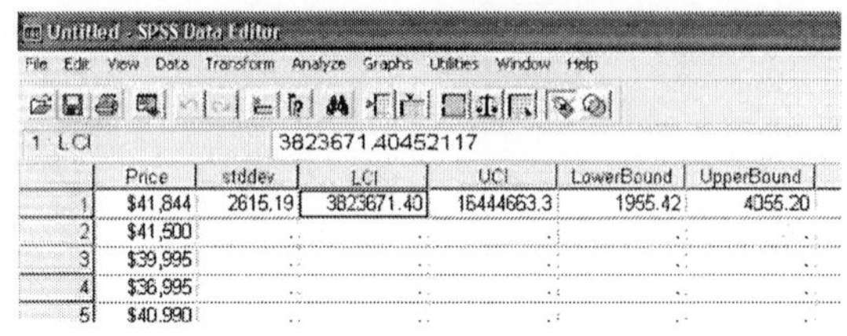

	Price	stddev	LCI	UCI	LowerBound	UpperBound
1	$41,844	2615.19	3823671.40	16444663.3	1955.42	4055.20
2	$41,500	.	.	.	.	.
3	$39,995	.	.	.	.	.
4	$36,995	.	.	.	.	.
5	$40,990	.	.	.	.	.

So, we are 90% confident that the population standard deviation of the price of a three year old Chevy Corvette is between $1,955.42 and $4,055.20.

Chapter 10. Testing Claims Regarding a Parameter

SPSS assumes that all data come from a sample. Because of this, all packaged methods assume σ to be unknown. Methods where σ is known must be done, at least in part, by hand.

Section 10.2 A Model for Testing Claims about a Population Mean

►Testing a hypothesis about μ, σ known

Example 5, Page 524

Since SPSS assumes that σ is unknown, the only pieces of the hypothesis test that SPSS will provide are the sample mean and the p-value. First, we must enter the data. Once the data have been entered, go to **Analyze → Descriptive Statistics → Descriptives...**, select the variable that you want the mean of, and push OK. The sample mean will be in the output window.

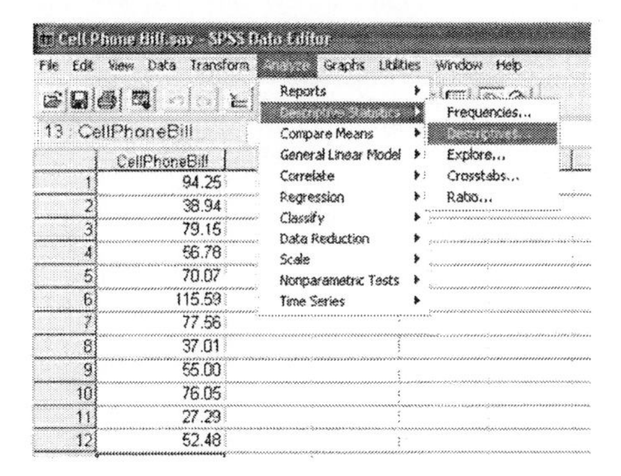

Descriptive Statistics

	N	Minimum	Maximum	Mean	Std. Deviation
CellPhoneBill	12	27.29	115.59	65.0142	25.45873
Valid N (listwise)	12				

If you click on the cell that contains the mean, until the mean is highlighted, you can copy the value of the sample mean into the clipboard, 65.01416666667. This will prevent rounding problems. You can then paste this into a new column. To calculate the z-value, go to **Transform → Compute...**. Once in the Compute Variable window, create a name for the z-value, and use the copied sample mean value in the z-equation.

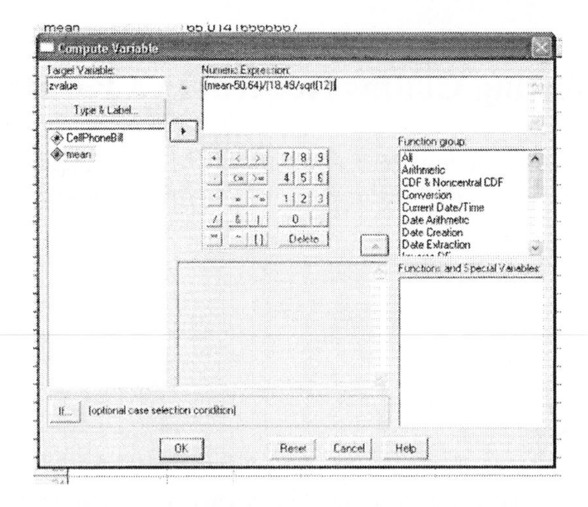

Once the equation is correct, push OK. Remember to get the parentheses in the correct places. Putting too many parentheses is better than not having enough to make sure the order of operations for the computer is what you believe they should be. This will create a new variable, with the value of z, 2.69.

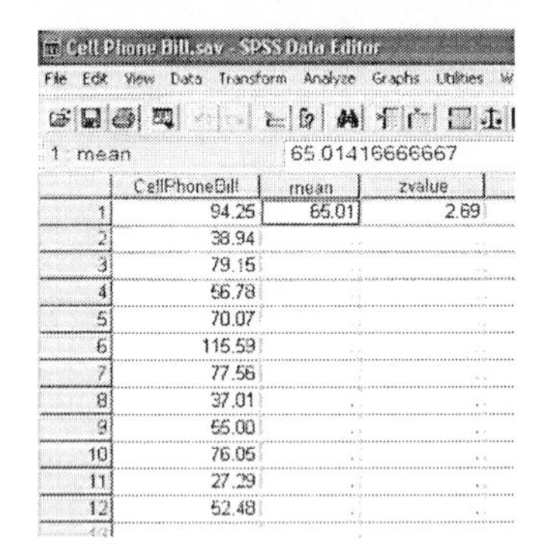

The test statistic is 2.69, or the sample mean is 2.69 standard errors above the hypothesized mean. To get the two-sided p-value, we will again use **Transform → Compute…**. Clear the z equation in the Compute Variable window, and click on CDF and Noncentral CDF in the Function Group box. In the Functions and Special Variables, select CDF.NORMAL. It must be remembered that this is the area to the left of the z-value. Since the CDF is the area to the left of z, and we want the tail areas, we will use 1-CDF.NORMAL, with a mean of 0 and standard deviation of 1, and since this is a two tailed test, we will multiply this value by 2.

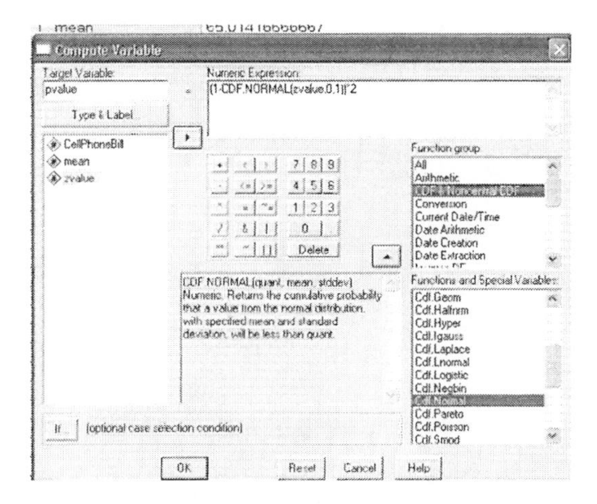

Once the equation is done, push OK. The p-value will be a new variable in the data set. You may need to go to Variable View to change the number of decimal places to see the actual p-value. The default is to show only two decimal places.

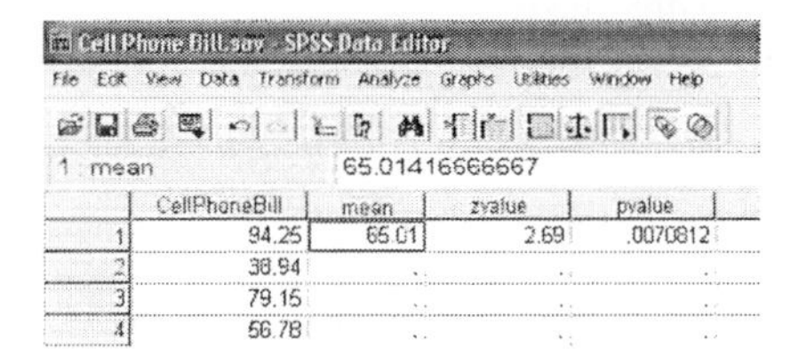

This value may be slightly different from what you would get by hand, as the z-value is not rounded to two decimal places. We get a p-value of .0071 which is less than .05, so we reject the null hypothesis.

If you don't need to know the z-value, and just want to calculate the p-value, you can skip the z-value calculation. We can calculate the tail probabilities directly for the sample mean.

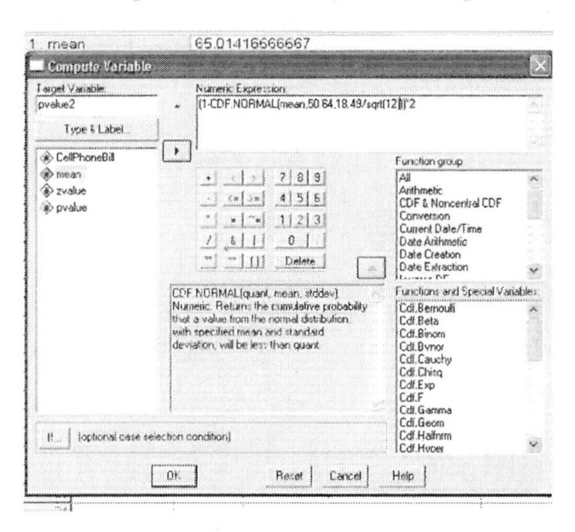

Use the CDF.NORMAL, with the value of the sample mean as the first value, the hypothesized mean as the second value, and the standard deviation over the square root of the sample size as the third value. Here you will have to see if the sample mean is larger or smaller than the hypothesized mean in order to see if you want the CDF or 1-CDF to get the tail values.

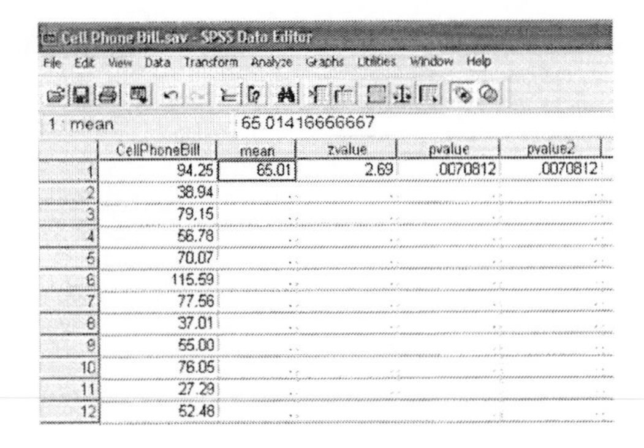

The p-value is the same, whether you do the test step-by-step, or whether you calculate the p-value directly.

Section 10.3 Testing Claims about a Population Mean in Practice

►Testing a hypothesis about μ, σ unknown

Example 3, Page 535

First, we must enter the data.

Normality Check

To create the box-plot and normal probability plot, see previous sections.

Step 1: Based on the quality control engineer's claim that the "fun size" Snickers bars **do not** weigh an average of 20.1 grams, we get

H_0: $\mu = 20.1$ -status quo (what M&M-Mars states on the bar)

H_1: $\mu \neq 20.1$ -quality control engineer's claim

Step 2: α=.01. If the p-value is less than .01 we will reject the null hypothesis. This is similar to finding a critical region, but since the computer calculates the p-value for us, there is no necessity in looking up values on the t-table.

Step 3-4: done in SPSS

Go to Compare **Means → One-Sample T Test…**

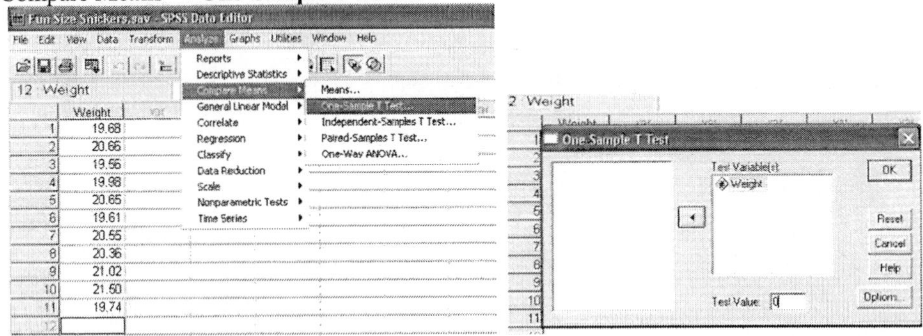

Highlight the variable of interest, and push the ► button.

In the Test Value box, type in the value from the **Null Hypothesis**. ($\mu = 20.1$)

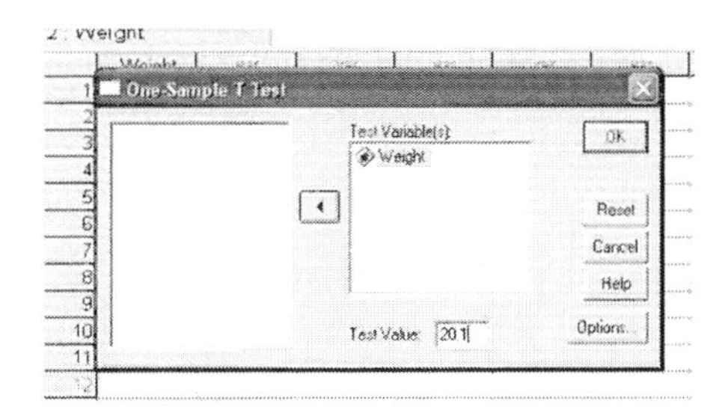

Click OK. The output window will have the following:

One-Sample Statistics

	N	Mean	Std. Deviation	Std. Error Mean
Weight	11	20.3009	.64037	.19308

This provides the sample size, sample mean, sample standard deviation, and standard error, which is the sample standard deviation over the square root of the sample size. If you want to perform a test by hand, these are all the values you will need.

One-Sample Test

	Test Value = 20.1					
					95% Confidence Interval of the Difference	
	t	df	Sig. (2-tailed)	Mean Difference	Lower	Upper
Weight	1.041	10	.323	.20091	-.2293	.6311

Note that the hypothesized mean appears above the test. In creating confidence intervals, this value should be 0. The value of t in the first box is the **Test Statistic** (calculated in Step 3). The sample mean is 1.041 standard errors (standard deviation of the sampling distribution) above the hypothesized mean.
The degrees of freedom in the second box are the degrees of freedom for the test.
The p-value is .323 which is larger than α (.323>.01) so we fail to reject the null hypothesis.
The confidence interval provided is the confidence interval for the difference between the sample mean and the hypothesized mean. We are 95% confident that the sample mean is 0.2293 below to 0.6311 grams above the hypothesized mean. Since 0 is in this interval, we will fail to reject the null hypothesis.

Note: SPSS always provides the p-value for a two-tailed test. If we want a 1-sided test, we can divide this in half, but we need to pay attention to the direction of the test and the sign of the test statistic. (0.323/2=0.1615), so p-value for a test with H_1: $\mu>20.1$, the p-value would be 0.1615. For a test with H_1: $\mu<20.1$, the p-value would be 1-0.1615, or 0.8385.

Step 5: We do not have sufficient evidence to show that the mean "fun size" Snickers bar weighs something other than 20.1 grams at the α=.01 level of significance.

Interpreting the result in step 4 is more important than getting the result. If others cannot understand the results then we have wasted the time in performing the test. Make sure that the statement declares what was measured (not just that the mean has(n't) changed) and the level of significance that was used, or the p-value, so that future researchers can verify the results.

We can also use SPSS to get the t-value to use for the critical region; we use **Transform → Compute…**, similar to finding the critical values for a confidence interval. Create a name for the column, t, and use IDF.T(p,df), where p is the area to the left, (so for an upper tail test, use 1- α, for a lower tail test, use α, and for a two tailed test, use α/2 and 1- α/2) and the correct degrees of freedom.

Section 10.4 Testing Claims about a Population Proportion

►Testing a Hypothesis about a Population Proportion

SPSS does not work directly with proportions. There is very little advantage to using SPSS over a calculator unless you are working with multiple proportions at the same time.

Example 1, Page 546

Step 1: The claim is that the proportion of adult Americans who thought that the death penalty was morally acceptable has increased since 2004.

H_0: p = 0.65
H_1: p > 0.65

Step 2: α=0.05.

Step 3: Compute the value of $\hat{p}$. In SPSS, create one variable for the value x, and another for the value n. Go to **Transform → Compute...**, create a new variable, called phat, by putting the name phat in the Target Variable box, and the equation x/n in the Numeric Expression box.

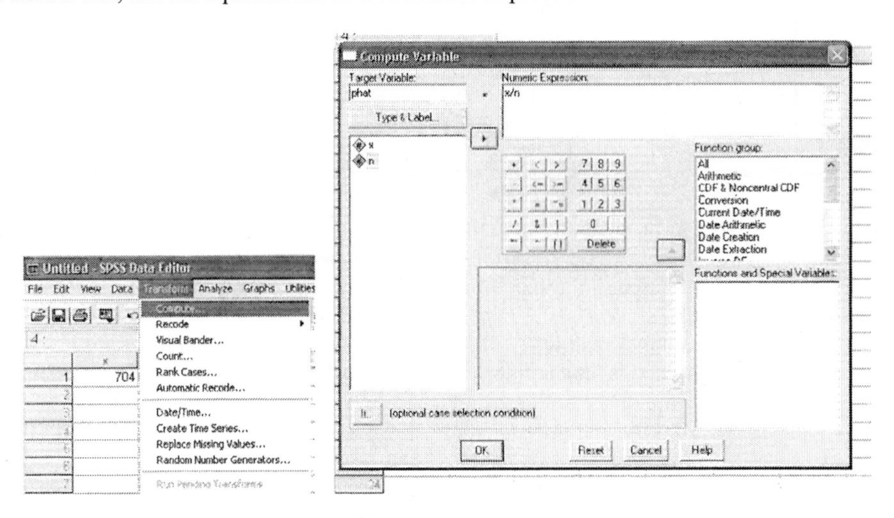

When the name and equation are ready, push OK.

Untitled - SPSS Data Editor

1 : phat 0.70049751243

	x	n	phat
1	704	1005	.7004975
2			
3			

The point estimate will appear in the Data View sheet, although you may need to change the number of decimal places, using the Variable View sheet to see it better. In this example, $\hat{p}$ is 0.7004975, or 0.70.

We can check the normality assumption using the **Transform → Compute...** as well. Create a new variable, check, by using the equation n*phat*(1-phat) in the numeric expression box.

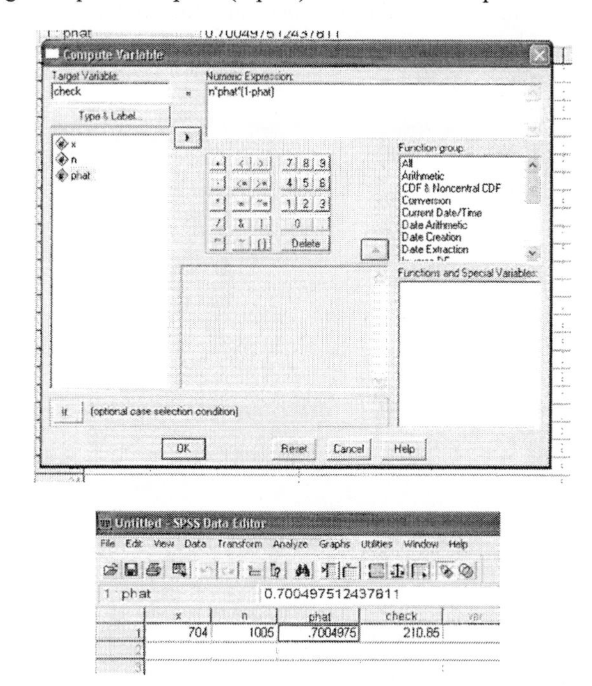

As the value is 210.85, which is greater than 10, we can proceed with the test. One advantage of using the computer is that you don't have to worry about rounding, but this may present slight differences from what you see in the book.

Returning to **Transform → Compute...**, we will calculate the z-value, similar to the test for μ with σ unknown. Type in the Z-equation, z=(phat-p_0)/sqrt(p_0*(1-p_0)/n), substituting the value of p_0 into the equation.

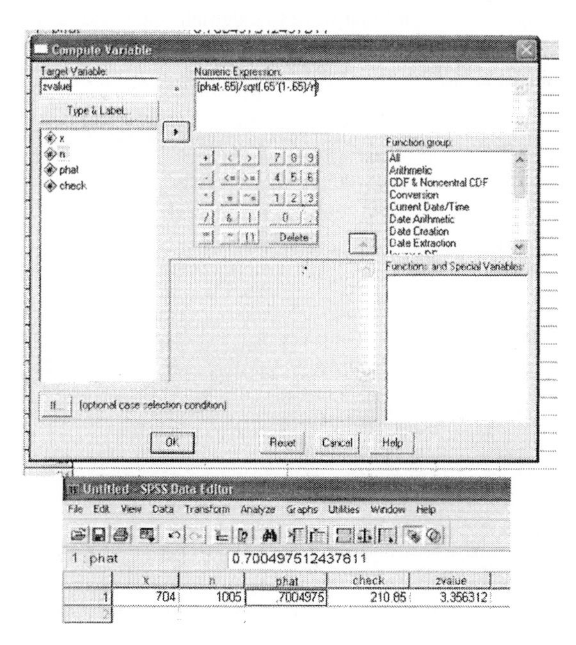

So, in this example, z = 3.36.

Step 5: We can then use SPSS to calculate the p-value. To get the one-sided p-value, we will again use **Transform → Compute….** This is similar to what was done in chapter 7. Clear the z equation in the Compute Variable window, and click on CDF and Noncentral CDF in the Function Group box. In the Functions and Special Variables, select CDF.NORMAL. It must be remembered that this is the area to the left of the z-value.

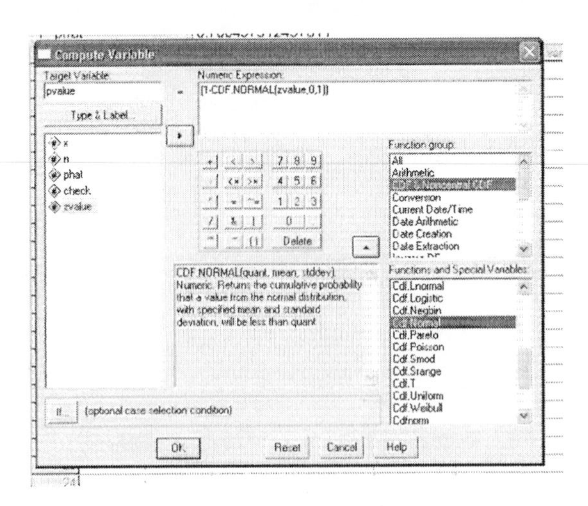

Once the equation is done, push OK. The p-value will be a new variable in the data set. Again, it will appear as many times as you had variables, and you may need to go to Variable View to change the number of decimal places.

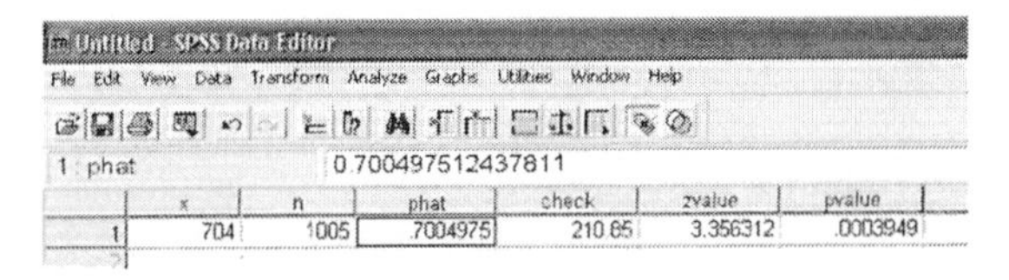

If you don't need to know the z-value, and just want to calculate the p-value, you can skip the z-value calculation. We can calculate the tail probabilities directly for the sample proportion, we just need to make sure to use the correct standard error.

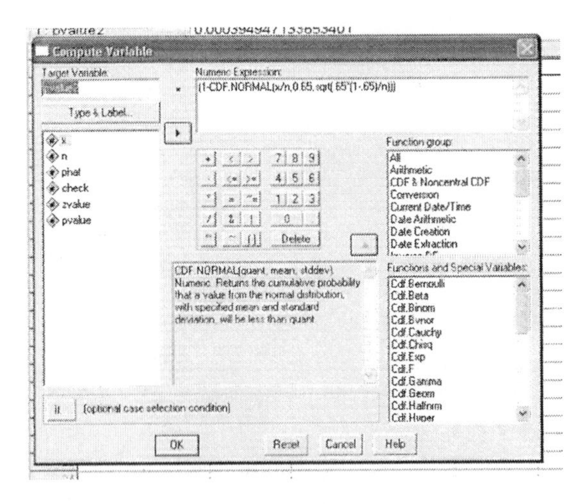

Use the CDF.NORMAL, with the value of the x/n as the first value, the hypothesized proportion as the second value, and the standard error equation as the third value. Here you will have to see if the sample proportion is larger or smaller than the hypothesized mean in order to see if you want the CDF or 1-CDF to get the tail values.

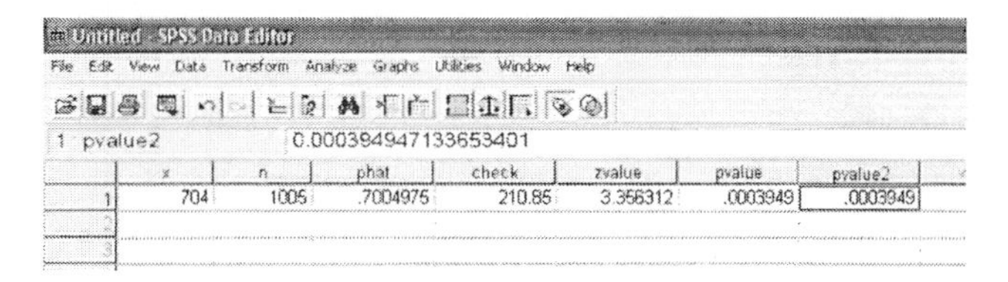

Untitled - SPSS Data Editor

File Edit View Data Transform Analyze Graphs Utilities Window Help

1 : pvalue2 0.000394947133653401

	x	n	phat	check	zvalue	pvalue	pvalue2
1	704	1005	.7004975	210.85	3.356312	.0003949	.0003949
2							
3							

The p-value is the same, whether you do the test step-by-step, or whether you calculate the p-value directly.

Section 10.5 Testing a Claim about a Population Standard Deviation

SPSS does not have a method to test standard deviations. You can use the CDF.CHISQ in **Transform → Compute…** to find the p-values for the Chi-Square distribution, but the rest must be done either by hand, or similar to the method for proportions.

►Testing a Hypothesis about σ

Example 1, Page 555

Calculating a test statistic for the standard deviation in SPSS is similar to finding a test statistic for μ when σ is unknown, or finding the test statistic for a proportion. First, we create a variable with the sample standard deviation. This can be done by hand, or we can use **Analyze → Descriptive Statistics → Descriptives…** to calculate s, and copy it into a cell on the data page.

Descriptive Statistics

	N	Minimum	Maximum	Mean	Std. Deviation
Weight	11	19.56	21.50	20.3009	.64037
Valid N (listwise)	11				

Fun Size Snickers.sav - SPSS Data Editor

File Edit View Data Transform Analyze Graphs Utilities

	Weight	stddev	var	var
1	19.68	.640366		
2	20.66			
3	19.56			
4	19.98			
5	20.65			
6	19.61			
7	20.55			
8	20.36			
9	21.02			
10	21.50			
11	19.74			
12				

We then go to **Transform → Compute….** Create a name for the test statistic in the Target Variable box, and in the Numeric Expression box, use the equation (11-1)*s**2/hypothesized value**2. The double asterisk represents a power, so s**2 is s^2.

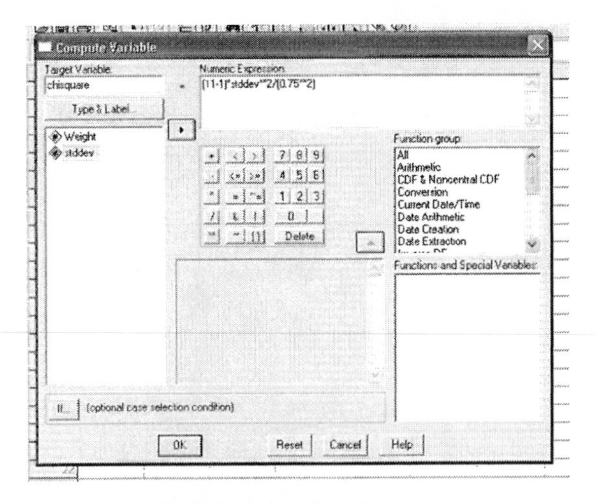

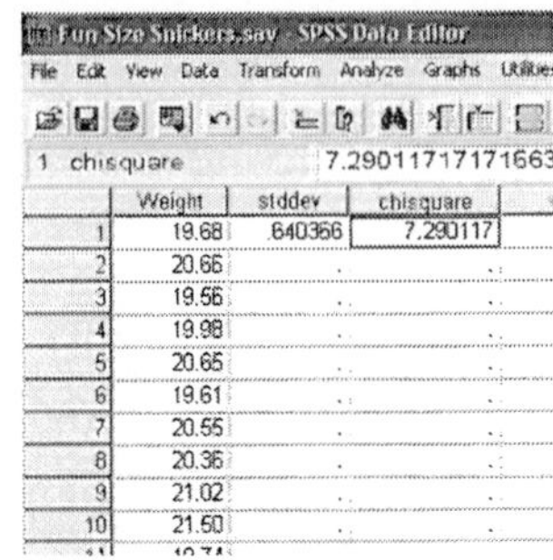

We can then use CDF.CHISQUARE in **Transform → Compute…** to find the p-value.

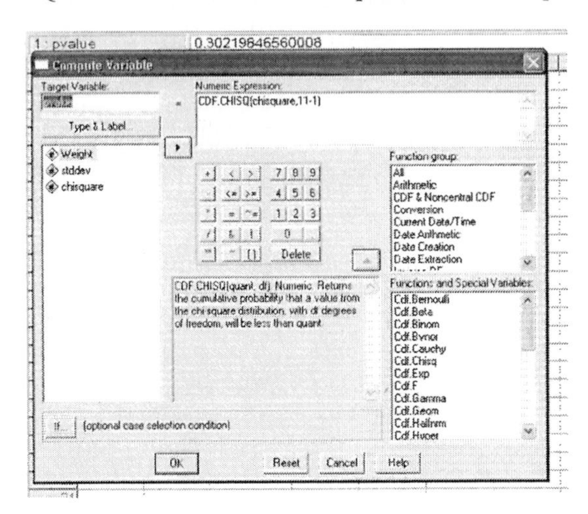

Here, we must remember that the CDF is the area to the left. The CDF.CHISQ requires a chi-square value, and the degrees of freedom for the test. You cannot calculate the p-value without the chi-square value as you could with the normal distribution.

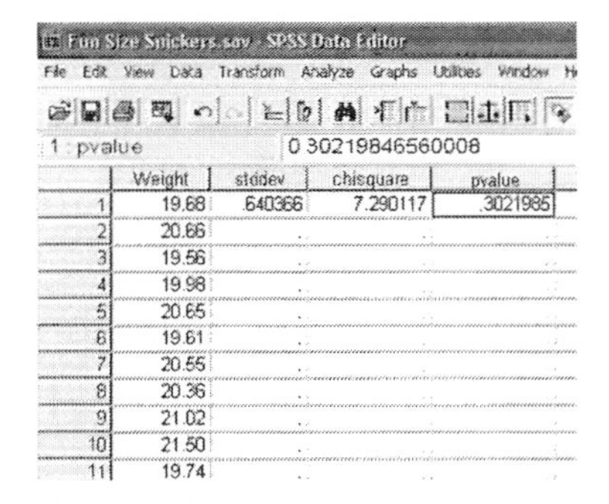

The p-value is .3022 which is larger than .05, so we fail to reject the null hypothesis. There is not sufficient evidence to support the claim that the standard deviation is less than 0.75 miles at the α=.05 level of significance.

Section 10.7 The Probability of a Type II Error and the Power of the Test

►The probability of a type II error of the test

Finding the probability of a type II error in SPSS is similar to finding the p-value for the test, we just change the values to be those for the rejection region under the true distribution instead of the hypothesized distribution.

Example 1, Page 563

We had the following problem: H_0: μ=12,200, and H_1: μ>12,200. One of the problems with computing the probability of a type II error in practice is the fact that there are an infinite number of values under the alternate hypothesis. We had n=35, σ=3,800, and α=0.1. Based on this information, we know that we will reject the null hypothesis for any sample mean larger than $\mu_0 + z_{0.1} * \sigma/\sqrt{n}$, we can find this using SPSS.

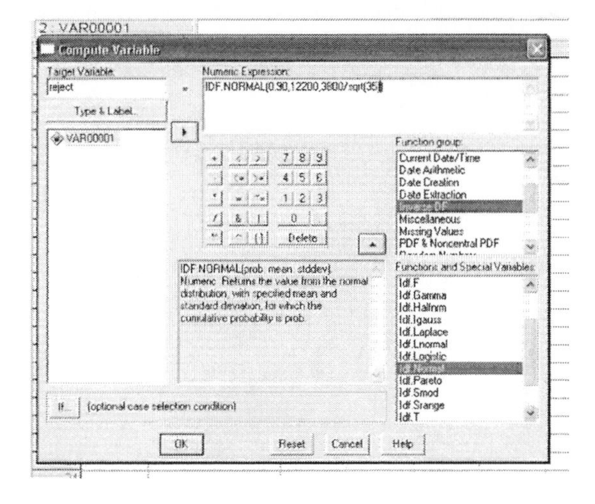

Using the IDF.NORMAL provides the value under the Normal curve for which the area to the left is 0.90 (which means the area to the right is α).

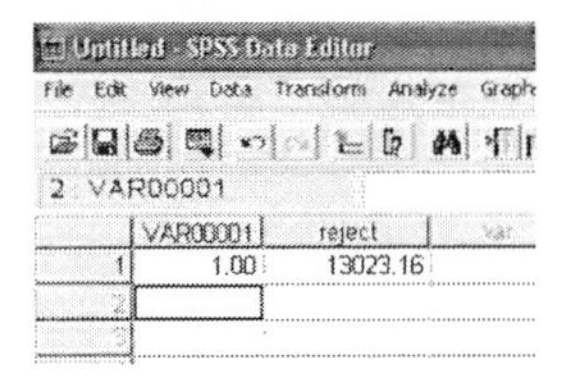

So, we will reject the null hypothesis for any sample mean larger than 13,023.16. Any sample mean less than this will lead us to fail to reject the null hypothesis. The probability of making a type I error is already known to us, but now we want to calculate the probability of a type II error. To calculate this, we are assuming that the true mean is 12,500. Since we had α be the area to the right of 13,023.16 under the Normal curve with the hypothesized mean, the probability of a type II error will be the area to the left (opposite direction of the type I error) of the same point, 13,023.16, with the true mean. So, now we will use the CDF.NORMAL to calculate the probability.

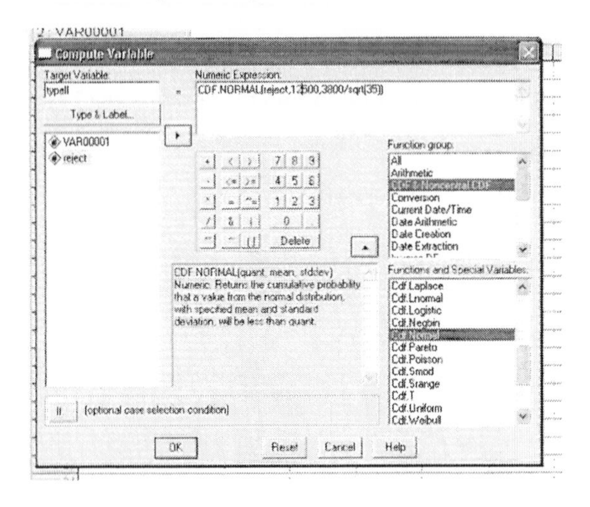

This should look the same as finding a p-value, except that we are using the critical value of the rejection region for x, and the new true mean instead of the hypothesized mean as the center.

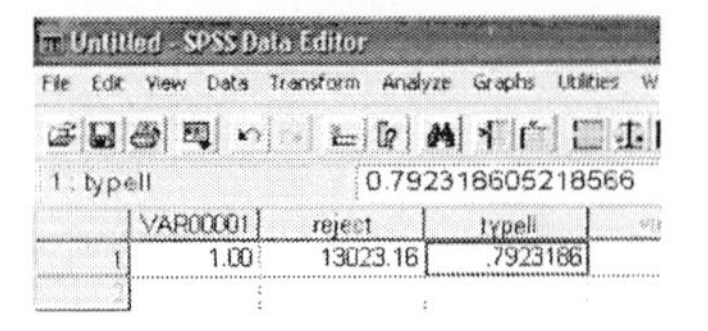

There is a probability of 0.7923 of making a type II error if the true mean is 12,500 for a test with α=0.10.

Chapter 11. Inferences on Two Samples

SPSS assumes that all data come from a sample. Because of this, all packaged methods assume σ to be unknown. Methods where σ is known must be done, at least in part, by hand.

Section 11.1 Inference about Two Means: Dependent Samples

►Inferences about Two Means: Dependent Samples

Example 2, Page 577

First, we type in the data. For dependent samples, each value is it's own variable, and it is paired by the observation.

Meter Stick.sav - SPSS Data Editor

File Edit View Data Transform Analyze Graphs Utilities

1 : Student 1

	Student	xi	yi
1	1	.177	.179
2	2	.210	.202
3	3	.186	.208
4	4	.189	.184
5	5	.198	.215
6	6	.194	.193
7	7	.160	.194
8	8	.163	.160
9	9	.166	.209
10	10	.152	.164
11	11	.190	.210
12	12	.172	.197
13			

Normality Check

Use **Transform → Compute...** to calculate the differences. In the equation box, take the first variable minus the second variable. These are not needed for the test, but are necessary to check the normality of the differences. To create the box-plot and normal probability plot, see previous sections.

Step 1: Based on Professor Neill's claim that reaction time in the dominant hand is **less** than the reaction time in the non-dominant hand, we use

H_0: $\mu_d = 0$ -status quo (no difference in hands)

H_1: $\mu_d < 0$ -Professor's claim

assuming that the difference is Dominant-Non-dominant. If we took Non-dominant-Dominant we would expect the mean of the differences to be positive. This is important to be aware of because SPSS **always** takes the first column minus the second column. The test will be the same either way, only the sign on the test statistic will change.

Step 2: α=0.05

Step 3-4: done in SPSS

Go to Compare **Means → Paired-Samples T Test...**

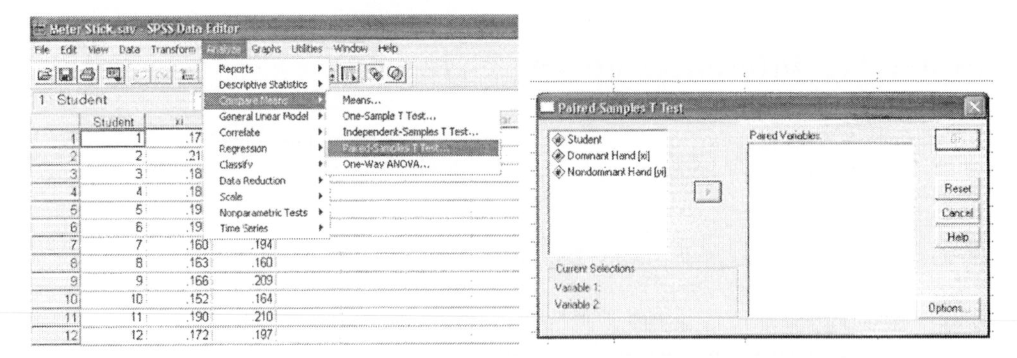

Highlight the variables of interest, and push the ▶ button. Note that the order in which you select the variables does not matter. Use the shift key when highlighting the variables if they are next to each other, or the Ctrl key if they are not next to each other so that both variables can be highlighted.

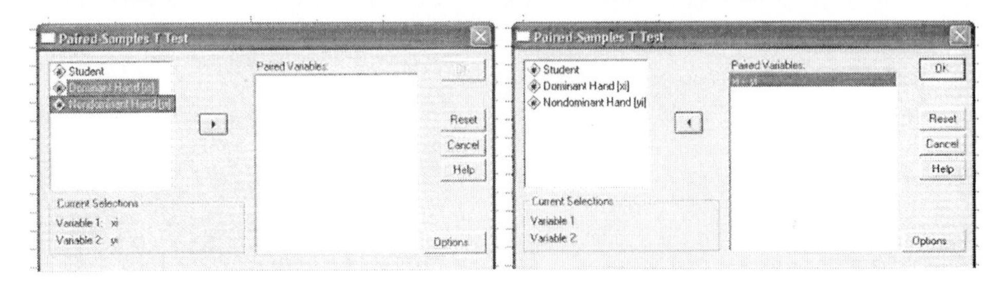

Using the Options button will allow you to set the confidence level for a confidence interval on the mean difference.
Click Continue, and OK. The output window will have the following:

Paired Samples Statistics

		Mean	N	Std. Deviation	Std. Error Mean
Pair 1	Dominant Hand	.17975	12	.017525	.005059
	Nondominant Hand	.19292	12	.017987	.005192

This provides the sample size, sample means, sample standard deviations, and standard errors, which is the sample standard deviation over the square root of the sample size. Note that the label appears as the name, not the variable name. This way, you can keep the variable names simple, and still remember which value is which in the output.

Paired Samples Correlations

		N	Correlation	Sig.
Pair 1	Dominant Hand & Nondominant Hand	12	.572	.052

This provides a correlation coefficient for the Paired samples. SPSS tests this, seeing how the correlation should be fairly high (the p-value small), but it is the method of sampling used that should determine whether or not the paired test should be performed.

Paired Samples Test

		Paired Differences					t	df	Sig. (2-tailed)
					95% Confidence Interval of the Difference				
		Mean	Std. Deviation	Std. Error Mean	Lower	Upper			
Pair 1	Dominant Hand - Nondominant Hand	-.01317	.016431	.004743	-.023606	-.002727	-2.776	11	.018

The mean of the differences, standard deviation of the differences, and standard error of the differences are provided here. Here you can calculate the test statistic by hand if you want to compare the by hand methods to SPSS.

The value of t in the seventh box is the **Test Statistic** (calculated in Step 3). The sample mean difference is 2.776 standard errors (standard deviation of the sampling distribution) below the hypothesized mean difference of 0.

The degrees of freedom, in the eighth box, are the degrees of freedom for the test, which is the number of pairs minus one. The Sig. (2-tailed) is the p-value for a two sided test. Since we want a 1-sided test, we divide this in half. (.018/2=.009) so the p-value is .009 which is smaller than α (.009<.05) so we reject the null hypothesis.

Note: The p-value calculated is **always** the tail values for a two-tailed test. Make sure that the p-value corresponds to the hypotheses that you created in step 1. If you are expecting a positive t-value (H_1: $\mu >$ 142.8) and the sample yields a negative t-value (sample mean was less than the hypothesized mean instead of larger), then the p-value would be one minus the p-value calculated above. P(t>Test Statistic)=1-P(t<Test Statistic). Also remember, p-values MUST be POSITIVE!!! Just because the test statistic is negative does not make the p-value negative.

Also provided is the confidence interval, as in example 4. We are 95% confident that the mean difference of reaction time between dominant and non-dominant hands is between -0.023606 and -0.002727, or the average reaction time for the dominant hand is 0.0027 to 0.0236 seconds faster than the reaction time for the non-dominant hand.

Note: The confidence interval does not have to agree with the results of the test, unless you are performing a two tailed test with the same level of confidence. One tailed tests can provide different results, as you are only looking at one tail instead of both.

Step 5: We have sufficient evidence to show that the reaction time for the dominant hand is less than the reaction time for non-dominant hand at the α=.05 level of significance.
We have sufficient evidence to show that the reaction time for the dominant hand is less than the reaction time for non-dominant hand (p-value=.009).

Interpreting the result in step 4 is more important than getting the result. If others cannot understand the results then we have wasted the time in performing the test. Make sure that the statement declares what was measured (not just that the mean has(n't) changed) and the level of significance that was used, or the p-value, so that future researchers can verify the results.

Section 11.2 Inference about Two Means: Independent Samples

►Inferences about Two Means: Independent Samples

Example 1, Page 590

First we must type in the data. This is a little different from the paired samples, in that we can no longer have two variables to contain the values. Since SPSS is built around each observation being the same person, we must create two variables, one for the value that we want to use, and one for which group the value belongs to. The group should be a number (1 and 2) which can be labeled in the Variable View screen.

Spacelab Rats.sav - SPSS Data Editor

1 : Mass 8.59

	Mass	Group
1	8.59	Flight
2	6.87	Flight
3	7.00	Flight
4	6.39	Flight
5	7.43	Flight
6	9.79	Flight
7	9.30	Flight
8	8.64	Flight
9	7.89	Flight
10	8.80	Flight
11	7.54	Flight
12	7.21	Flight
13	6.85	Flight
14	8.03	Flight
15	8.65	Control
16	7.62	Control
17	7.33	Control
18	7.14	Control
19	8.40	Control
20	8.55	Control
21	9.88	Control
22	6.99	Control
23	7.44	Control
24	8.58	Control
25	9.14	Control
26	9.66	Control
27	8.70	Control
28	9.94	Control

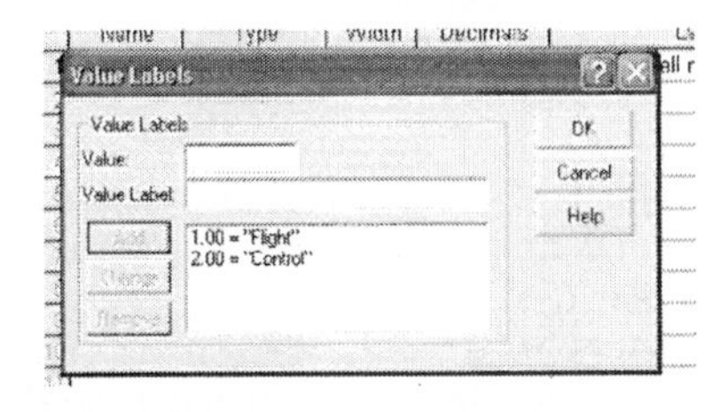

This allows for longer group descriptions as well as keeping SPSS happier.

Normality Check

To create the box-plot and normal probability plot, see previous sections. Each group should be independently normally distributed. In order to create the normal probability plots, we will have to separate out the groups. To do this, go to **Data → Select Cases**. This allows you to look at each group separately.

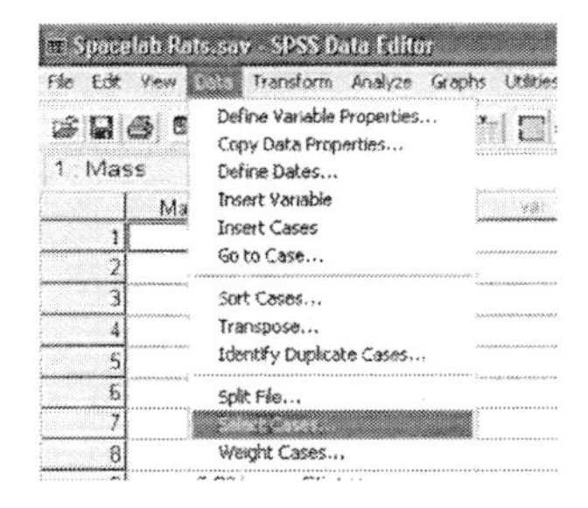

In the Select Cases menu, select If condition is satisfied, and click on the **If...** button.

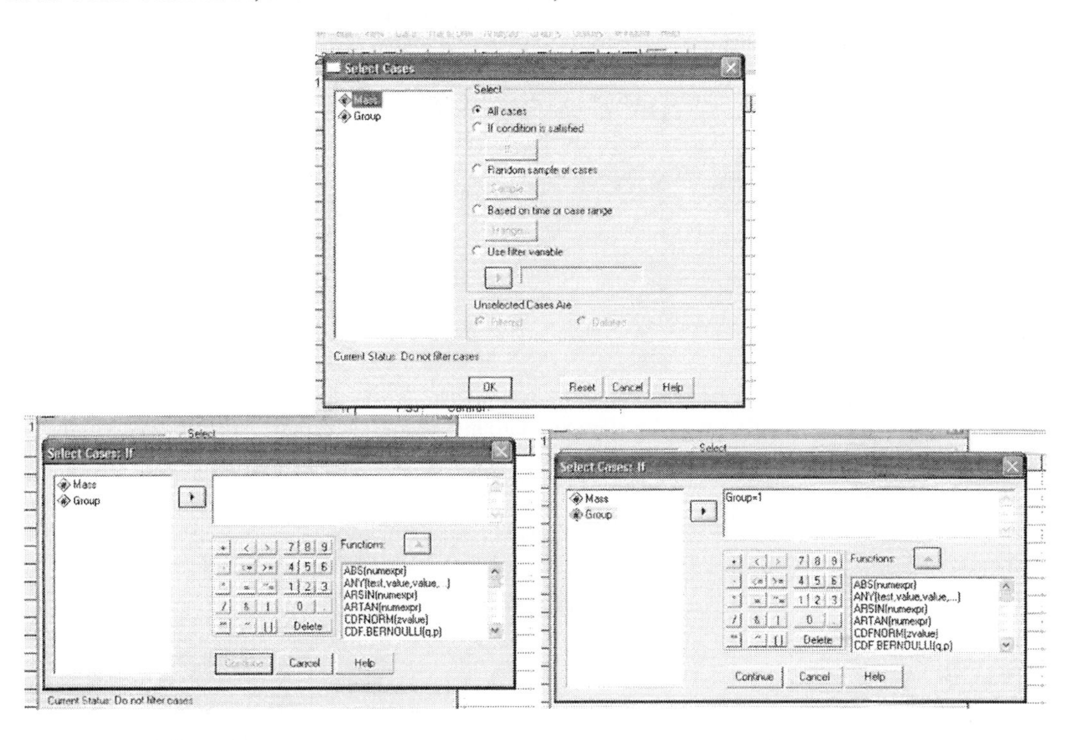

You can now define which group, by name or number, you want to look at. You must remember to return to All cases before continuing the analysis.

Step 1: Based on the claim that flight animals have **different** red blood cell mass from the control animals, we use

H_0: μ_1-μ_2 = 0 -status quo (μ_1=μ_2, no difference in the means)

H_1: μ_1-$\mu_2 \neq 0$ -claim-there is a difference in the means

Note that it doesn't matter which is the first group, and which is the second, as long as you keep in mind which is which and use the appropriate relation (<. >) for the difference. Switching groups will only switch the sign of the test statistic.

Step 2: α=0.05.

Step 3-4: done in SPSS

Go to **Compare Means → Independent-Samples T Test...**

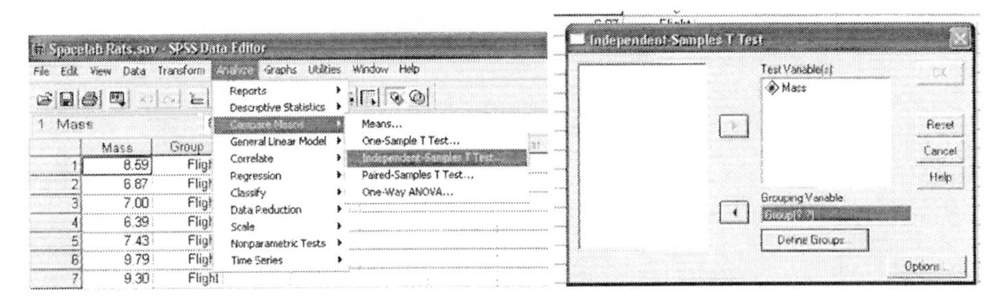

Select the column which contains the values, and put that in the Test Variable(s) box. Select the column which contains the group value and put that in the Grouping Variable box. You then have to let SPSS know what groups to use, by clicking on the Define Groups button.

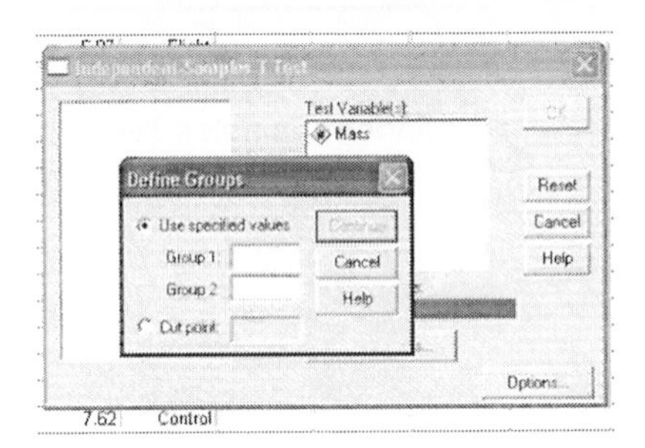

Type in the name or number for the two groups, (this is another reason short numbers are better than long names). SPSS will take the difference to be Group 1-Group 2, so use the group that corresponds to the direction you want the difference to be taken.

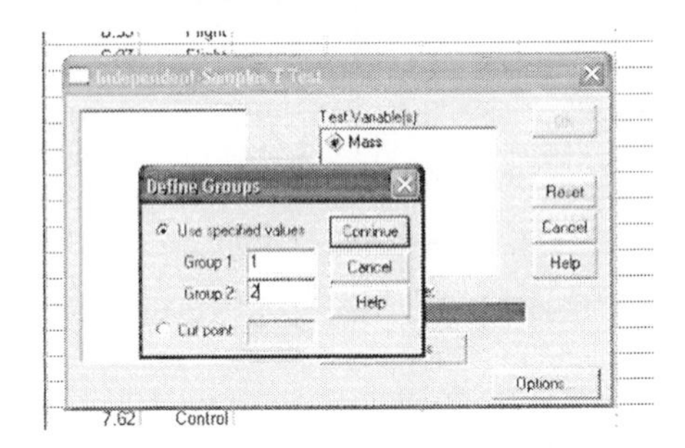

Using the Options button will allow you to set the confidence level for a confidence interval on the difference in the means.
Click Continue, and OK. The output window will have the following:

Group Statistics

	Group	N	Mean	Std. Deviation	Std. Error Mean
Mass	Flight	14	7.8807	1.01745	.27193
	Control	14	8.4300	1.00547	.26872

This provides the sample size, sample means, sample standard deviations, and standard errors for each group.

Independent Samples Test

		Levene's Test for Equality of Variances		t-test for Equality of Means		
		F	Sig.	t	df	Sig. (2-tailed)
Mass	Equal variances assumed	.023	.882	-1.437	26	.163
	Equal variances not assumed			-1.437	25.996	.163

Independent Samples Test

		t-test for Equality of Means			
				95% Confidence Interval of the Difference	
		Mean Difference	Std. Error Difference	Lower	Upper
Mass	Equal variances assumed	-.54929	.38230	-1.33512	.23655
	Equal variances not assumed	-.54929	.38230	-1.33513	.23655

The difference in the means, standard deviation of the differences, and standard error of the differences are provided here. Here you can calculate the test statistic by hand.

SPSS provides the output of two separate tests, one when equal variances are assumed (the pooled t-test), and one where equal variances are not assumed (known as Welch's t-test). The Levene test for equality of variances is a robust test, which allows you to better select which test you wish to use, and as the tests are both presented, you can see what differences there may be.

We want to use the Welch's t-test, which is the second row. So we have a t-value of -1.437, with 25.996 degrees of freedom, which is found using the formula on page 594, and the p-value for a two-tailed test of 0.163.

Note: The p-value calculated is the tail values for a two-tailed test. Make sure that the p-value corresponds to the hypotheses that you created in step 1. If you are expecting a positive t-value (H_1: $\mu_1 > \mu_2$) and the sample yields a negative t-value (sample mean was less than the hypothesized mean instead of larger), then the p-value would be one minus the p-value calculated above. P(t>Test Statistic)=1-P(t<Test Statistic). Also remember, p-values MUST be POSITIVE!!! Just because the test statistic is negative does not make the p-value negative.

Also provided is the confidence interval. We are 95% confident that the difference in the means between flight and control rats is between -1.3351 and 0.2365, or the average for rats in flight is 1.335 milliliters below to 0.237 milliliters above the average for control rats.

Note: The confidence interval does not have to agree with the results of the test, unless you are performing a two tailed test with the same level of confidence. One tailed tests can provide different results, as you are only looking at one tail instead of both.

Step 5: We do not have sufficient evidence to show that red blood cell mass in flight animals is different from red blood cell mass in control animals at the α=.05 level of significance.
We do not have sufficient evidence to show that red blood cell mass in flight animals is different from red blood cell mass in control animals (p-value=.1627).

Section 11.3 Inference about Two Population Proportions

►Inferences about Two Population Proportions

As we have seen before, SPSS does not work directly with proportions. We can use the SPSS calculator, and the SPSS z-table for the test, but there is little advantage to using SPSS for a proportion test over a calculator.

Example 1, Page 604

Step 1: The claim is that the proportion of Americans 18 years old or older who believe that men are more aggressive than women is less than 0.74.

H_0: $p_1 = p_2$
H_1: $p_1 > p_2$

Step 2: We will use p-values instead of critical values.
Step 3: Compute the values of $\hat{p}$. In SPSS, create two variables for the value x, and another two for the values of n.

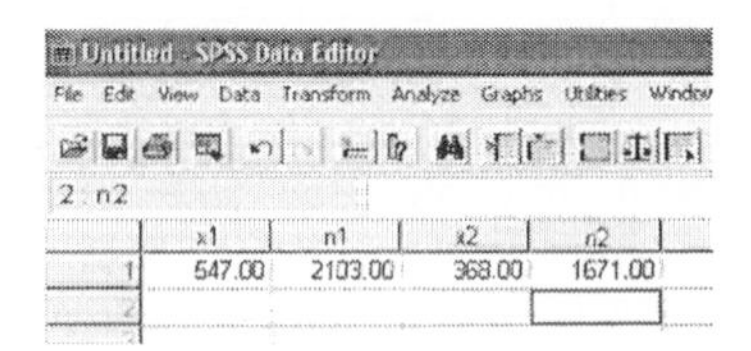

Go to **Transform → Compute...**, create a new variable, called phat1, by putting the name phat1 in the Target Variable box, and the equation x1/n1 in the Numeric Expression box. Similarly create phat2, and phat, which is the sum of the x values divided by the sum of the n values.

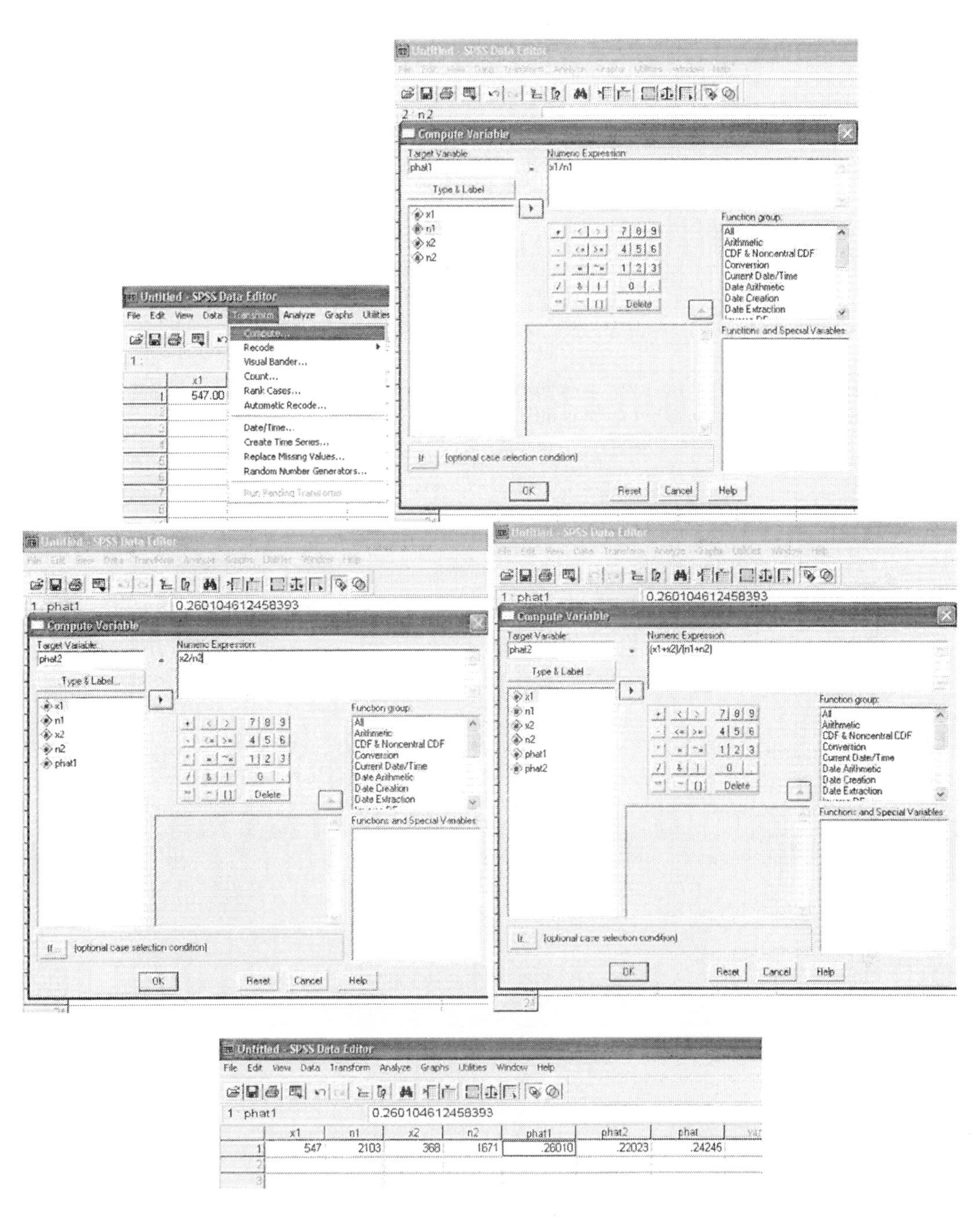

We can check the sample sizes, using phat1 and phat2, similar to chapter 9.
Returning to **Transform → Compute…**, we will calculate the z-value, similar to the test for μ with σ unknown. Type in the Z-equation, z=(phat1-phat2)/sqrt(phat*(1-phat)*(1/n1+1/n2)).

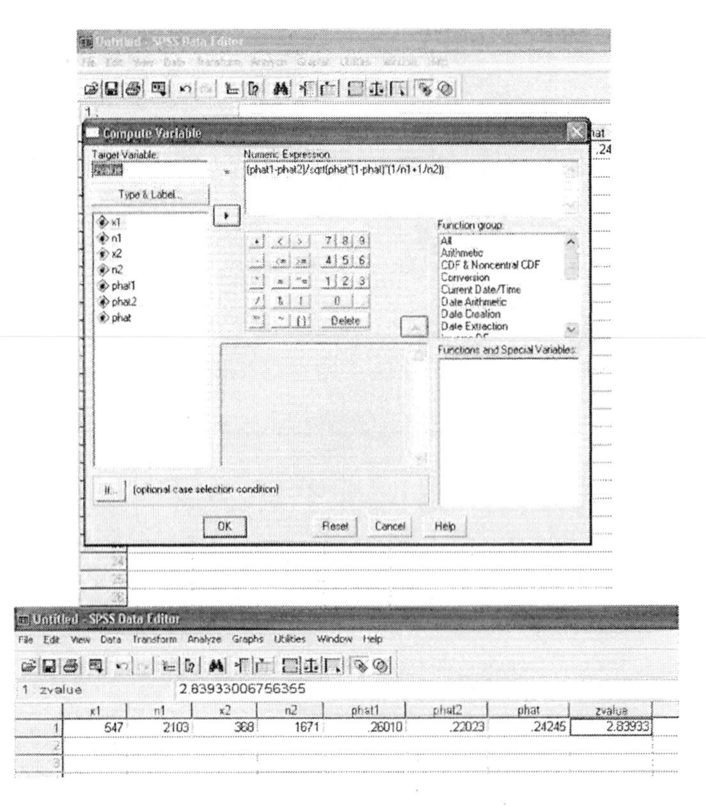

So, in this example, z=2.84.

Step 5: We can then use SPSS to calculate the p-value. To get the one-sided p-value, we will again use **Transform → Compute…**. This is similar to what was done in chapter 7. Clear the z equation in the Compute Variable window, and click on CDF and Noncentral CDF in the Function Group box. In the Functions and Special Variables, select CDF.NORMAL. It must be remembered that this is the area to the left of the z-value.

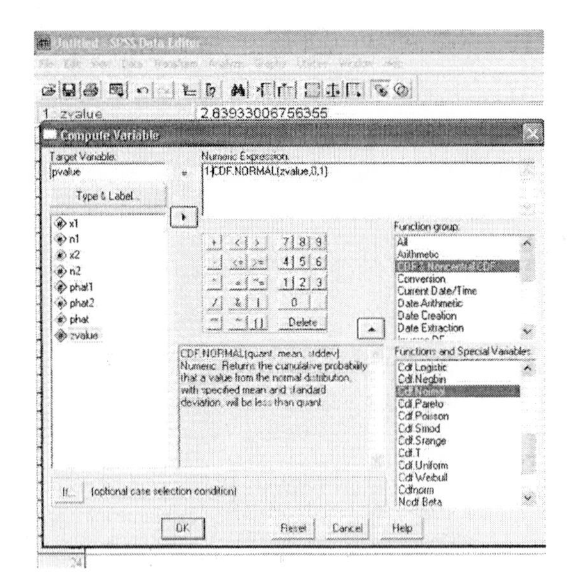

Once the equation is done, push OK. The p-value will be a new variable in the data set. Again, it will appear as many times as you had variables, and you may need to go to Variable View to change the number of decimal places.

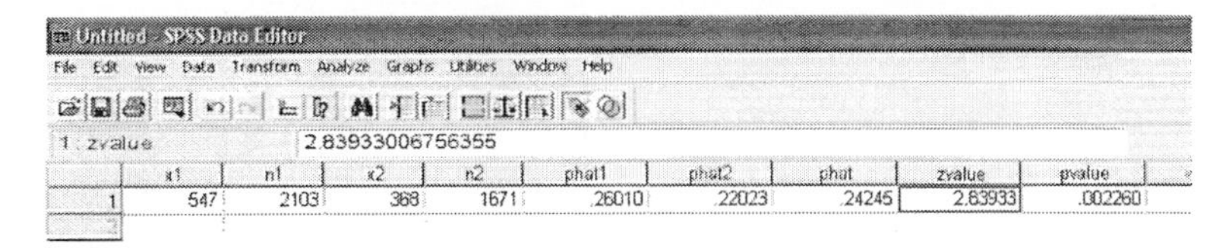

	x1	n1	x2	n2	phat1	phat2	phat	zvalue	pvalue
1	547	2103	368	1671	.26010	.22023	.24245	2.83933	.002260

If you don't need to know the z-value, and just want to calculate the p-value, you can skip the z-value calculation. We can calculate the tail probabilities, similar to chapter 7, directly for the sample mean.

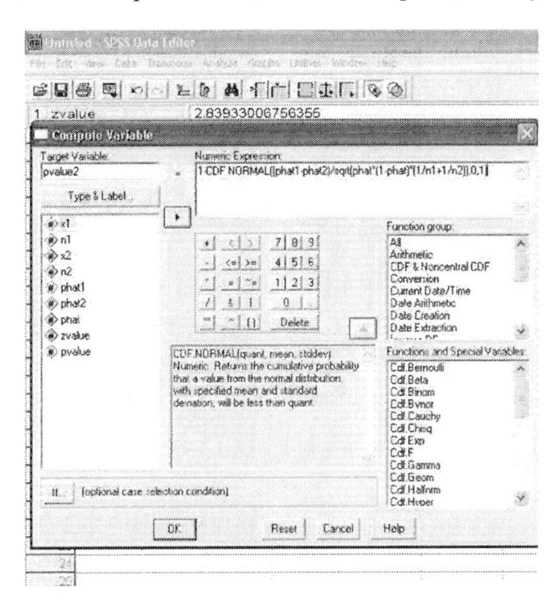

Use the CDF.NORMAL, with the value of the phat as the first value, the hypothesized proportion as the second value, and the standard error equation as the third value. Here you will have to see if the sample proportion is larger or smaller than the hypothesized mean in order to see if you want the CDF or 1-CDF to get the tail values.

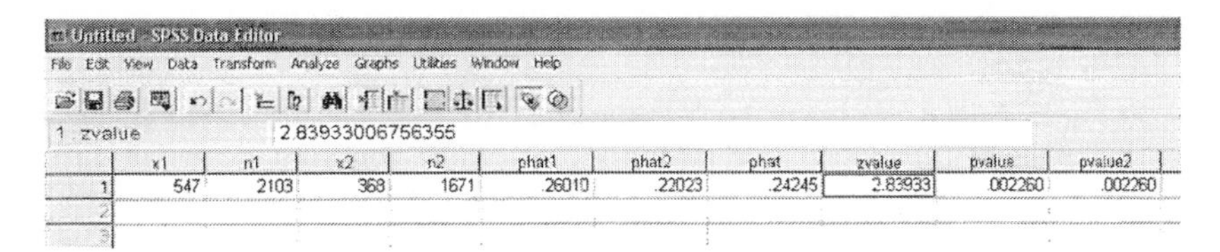

	x1	n1	x2	n2	phat1	phat2	phat	zvalue	pvalue	pvalue2
1	547	2103	368	1671	.26010	.22023	.24245	2.83933	.002260	.002260

The p-value is the same, whether you do the test step-by-step, or whether you calculate the p-value directly. Since the p-value is less than 0.05, we reject the null hypothesis. There is enough evidence to show that the proportion of individuals 12 years and older taking 200 mcg of Nasonex who experience headaches is greater than the proportion of individuals 12 and older taking a placebo who experience headaches.

Confidence intervals can be similarly produced, using the IDF.NORMAL function, as in chapter 7. For a 95% confidence interval we would create a lower bound using the area to the left as .025 and the upper bound using the area to the left as .975.

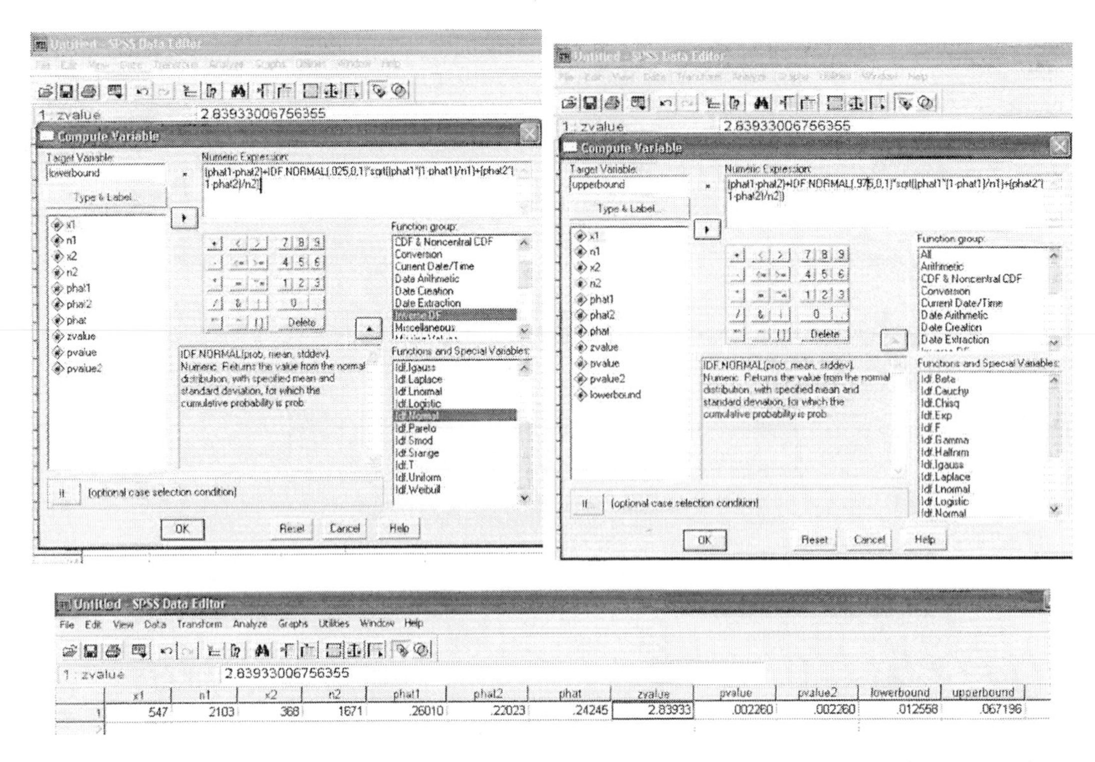

So we are 95% confident that the proportion of individuals 12 and older taking 200 mcg of Nasonex who claim a headache as a side effect is 1.3% to 6.7% more than individuals 12 and older taking a placebo.

Section 11.4 Inference about Two Population Standard Deviations

Instead of using the F-test, which is not robust, SPSS uses the Levene test, which is robust. This is calculated at the same time as performing a two-sample t-test for independent samples.

► Finding Critical Values for the F-Distribution

Example 1, page 614

To find critical values under the F-distribution, we turn to the SPSS calculator. Go to **Transform → Compute**. We find the function group Inverse DF and find IDF.F. The F distribution requires three values, which are the probability to the left (which is complement of what you see in the book), and the degrees of freedom for the numerator and denominator. So, for a right-tailed test with α=0.05, degrees of freedom in the numerator =10, and degrees of freedom in the denominator =7, we would have 0.95 as the probability to the left, and would use IDF.F(0.95,10,7).

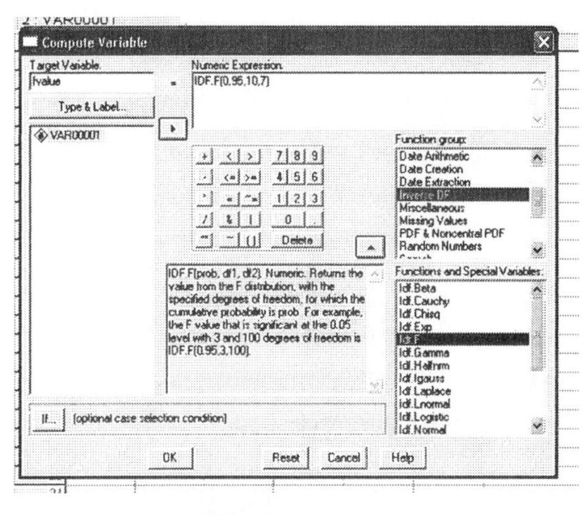

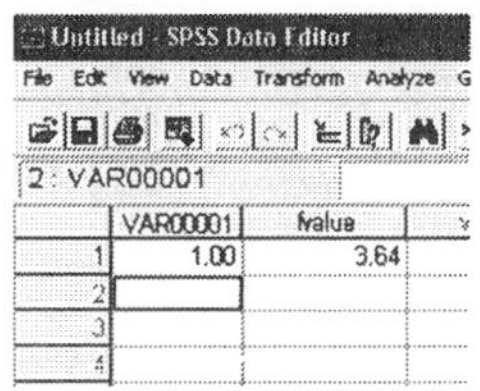

So, for this test, we would have a critical F value of 3.64.

►Inferences about Two Population Standard Deviations

Example 2, page 616

We enter the data as for an independent sample t-test in Section 11.2. It must be remembered that all values are in one column, and which group they belong to in another column.

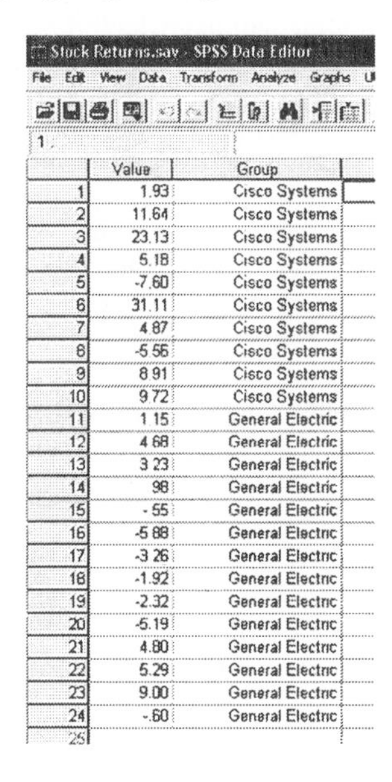

Stock Returns.sav - SPSS Data Editor

	Value	Group
1	1.93	Cisco Systems
2	11.64	Cisco Systems
3	23.13	Cisco Systems
4	5.18	Cisco Systems
5	-7.60	Cisco Systems
6	31.11	Cisco Systems
7	4.87	Cisco Systems
8	-5.56	Cisco Systems
9	8.91	Cisco Systems
10	9.72	Cisco Systems
11	1.15	General Electric
12	4.68	General Electric
13	3.23	General Electric
14	.98	General Electric
15	-.55	General Electric
16	-5.88	General Electric
17	-3.26	General Electric
18	-1.92	General Electric
19	-2.32	General Electric
20	-5.19	General Electric
21	4.80	General Electric
22	5.29	General Electric
23	9.00	General Electric
24	-.60	General Electric
25		

Step 1: Based on the claim that Cisco systems is more volatile than General Electric stock

H_0: $\sigma_1 = \sigma_2$ -status quo (no difference in the variation)

H_1: $\sigma_1 > \sigma_2$ -claim- Cisco systems is more volatile

Note that it doesn't matter which is the first group, and which is the second, as long as you keep in mind which is which and use the appropriate relation (<. >) for the difference.

Step 2: We will use the p-value instead of a critical value.

Step 3-4: done in SPSS

Go to **Compare Means → Independent-Samples T Test...**

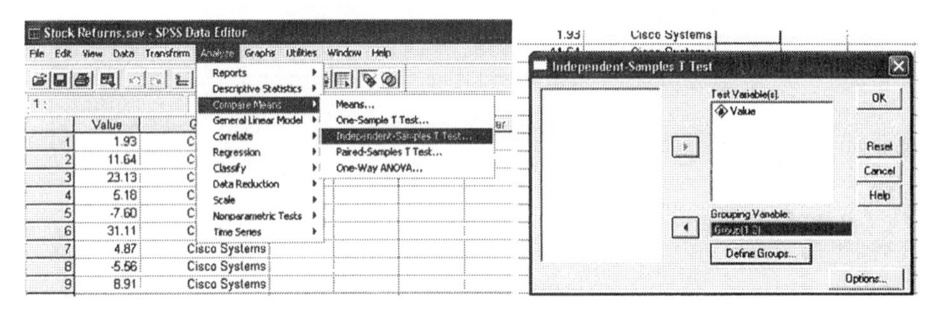

Select the column which contains the values, and put that in the Test Variable(s) box. Select the column which contains the group value and put that in the Grouping Variable box. You then have to let SPSS know what groups to use, by clicking on the Define Groups button.

Type in the name or number for the two groups, (this is another reason short numbers are better than long names). SPSS will take the difference to be Group 1-Group 2, so use the group that corresponds to the direction you want the difference to be taken.

Using the Options button will allow you to set the confidence level for a confidence interval on the difference in the means.
Click Continue, and OK. The output window will have the following:

Group Statistics

	Group	N	Mean	Std. Deviation	Std. Error Mean
Value	Cisco Systems	10	8.3330	11.83565	3.74276
	General Electric	14	.6721	4.31622	1.15356

This provides the sample size, sample means, sample standard deviations, and standard errors for each group.

Independent Samples Test

		Levene's Test for Equality of Variances	
		F	Sig.
Value	Equal variances assumed	5.536	.028

So, with an F-value of 5.536, and a p-value of 0.028, we would reject the null hypothesis. There is enough evidence to show that Cisco systems is more volatile than General Electric stock.

This is a different test than provided in the book. To perform the non-robust F test, we need to do some work by hand. We can copy the standard deviation values from the group statistics and put them into two separate columns.

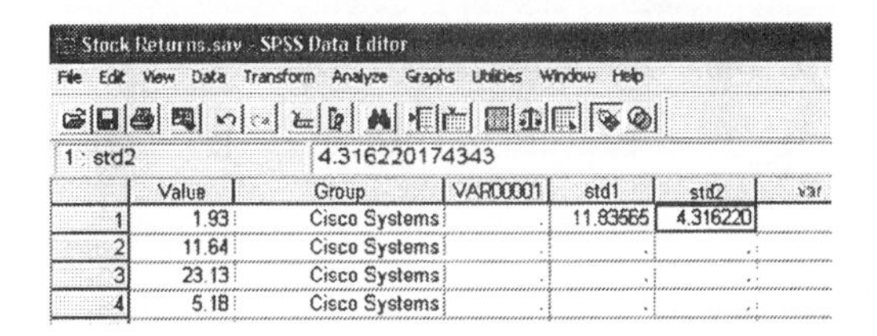

We then turn to **Transform → Compute** to calculate F for us. We can use **2 to square the standard deviations.

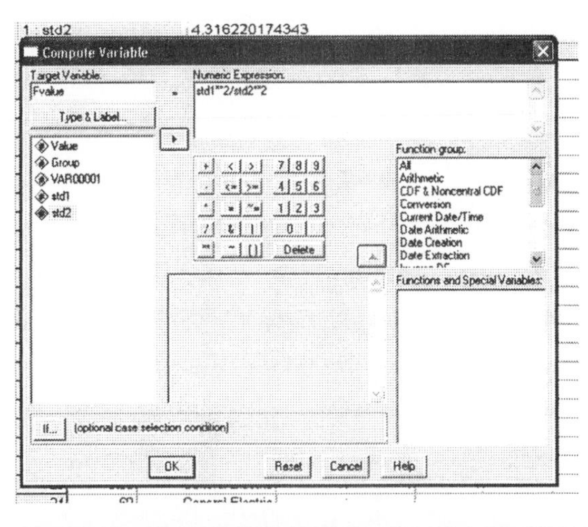

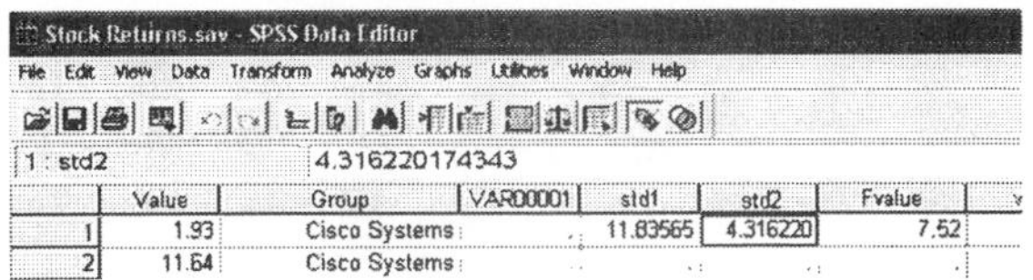

Now we need to calculate the p-value for the test. We again turn to **Transform → Compute**, and use the function group CDF & Noncentral CDF. We find the CDF.F which requires the F-value, and the numerator and denominator degrees of freedom. Because we want the right tail value, we have to use the complement rule to find the p-vlaue.

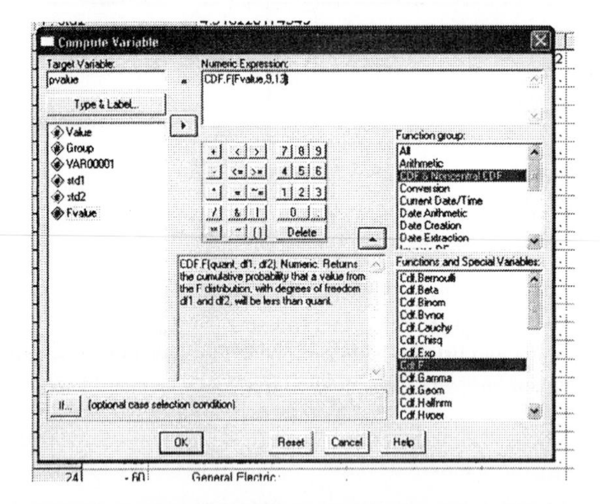

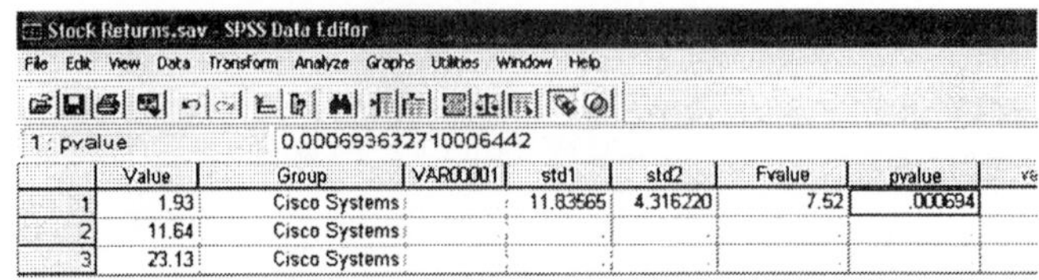

	Value	Group	VAR00001	std1	std2	Fvalue	pvalue
1	1.93	Cisco Systems		11.83565	4.316220	7.52	.000694
2	11.64	Cisco Systems					
3	23.13	Cisco Systems					

We get a p-value of 0.0007 which is less than 0.05, so we reject the null hypothesis.

Chapter 12. Inference on Categorical Data

Section 12.1 Goodness of Fit Test

▶Goodness-of-Fit

Example 2, Page 634

First we must type in the data. It is very important to know the exact order of the data, so it is easiest to type the categories as numbers and use the label options in the Variable View page. We need to create one variable with the category, and one with the observed frequency.

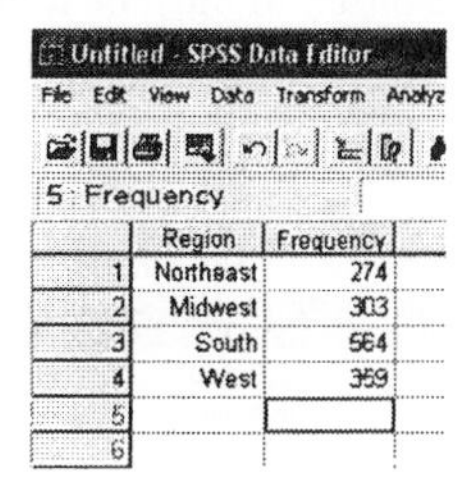

Once the data have been entered, we go to Data → **Weight Cases…** to tell SPSS how many times each category was observed. Weigh the cases by frequency.

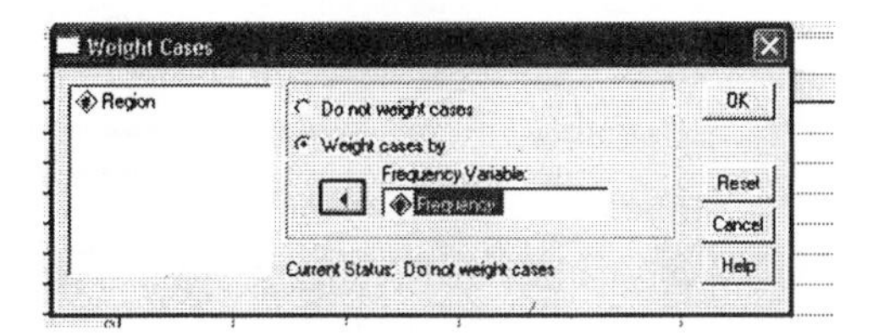

Once SPSS can see the data the way we want it to, go to **Analyze → Nonparametric Tests → Chi-Square….** Note that this Chi-Square test only works for the goodness-of-fit test. We will use another method for table data.

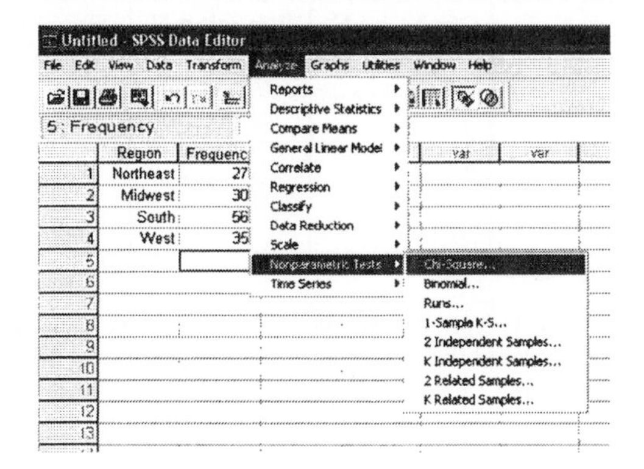

Once in the Chi-Square test window, select the category as the Test Variable List.

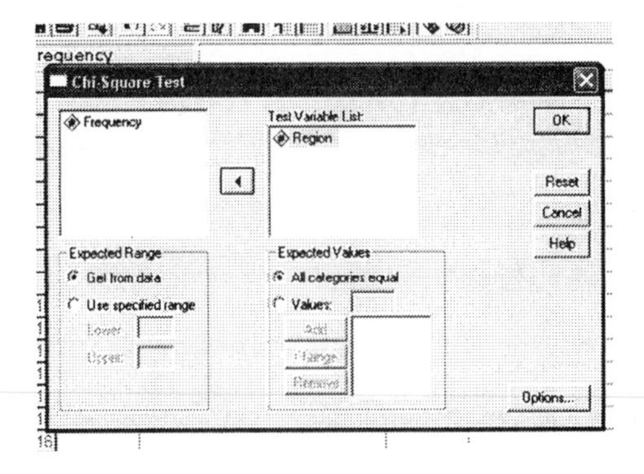

We now need to let SPSS know what the null hypothesis is. This is done in the Expected Values section. If the null hypothesis is that all categories are equal, we can use the default. In this case, the proportions are different, depending on the area. Northeast is 19.0%, Midwest is 22.9%, South is 35.6% and West is 22.5%. This is where we need to know the order. It must be remembered that SPSS likes to reorder things alphabetically, so if you used the actual names, you will have to figure out the alphabetical order. Click on the Values button. In order, from first to last, type in the hypothesized proportion, clicking the Add button between each value.

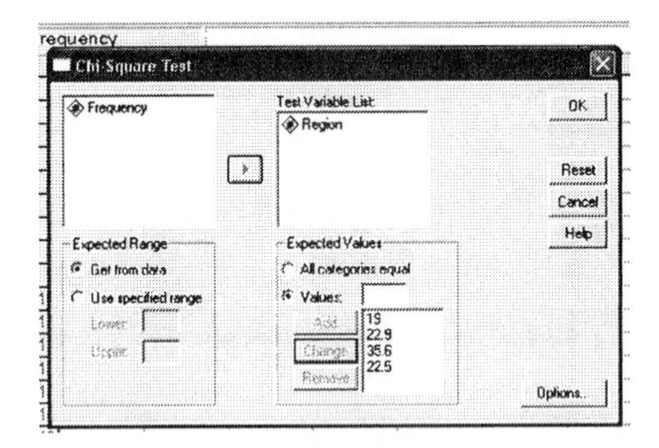

If you get these out of order, you will get different expected values, and, thus, a different chi-square value.

Once you have the values you want, in the correct order, press OK. In the output window, you will see the following:

Region

	Observed N	Expected N	Residual
Northeast	274	285.0	-11.0
Midwest	303	343.5	-40.5
South	564	534.0	30.0
West	359	337.5	21.5
Total	1500		

This is the summary table. This provides the observed value, expected value, and residual value, which is the difference between the observed and the expected. This is a good way to check yourself and make sure that the order you typed the values in was the order you expected.

Test Statistics

	Region
Chi-Square[a]	8.255
df	3
Asymp. Sig.	.041

a. 0 cells (.0%) have expected frequencies less than 5. The minimum expected cell frequency is 285.0.

The second table contains the test information. We get a chi-square value of 8.255, with 3 degrees of freedom, and a p-value of .041. Since .041 is less than .05, we reject the null hypothesis. There is sufficient evidence to show that the proportions have changed since 2000. Note that SPSS also provides the information needed to check for Normality. 0% cells have expected frequencies less than 5, so we have less than 20%, and the minimum expected cell frequency is 285 which is greater than 1, so the conditions have been met.

Although this method will provide the correct p-value for a single sample test on proportions (see problems 23 and 24), it will not provide confidence intervals.

Section 12.3 Tests for Independence and Homogeneity of Proportions

►Contingency Tables and Independence Tests

Example 2, Page 655

Similar to the independent sample t-test in chapter 10, we have to enter the data into SPSS in a very specific manner. We will need to create three variables, one with the value of category 1, one with the value of category 2, and one with the observed frequency.

Blood Type.sav - SPSS Data Editor

	BloodType	RhStatus	Frequency
1	A	Rh+	176.00
2	B	Rh+	28.00
3	AB	Rh+	22.00
4	O	Rh+	198.00
5	A	Rh-	30.00
6	B	Rh-	12.00
7	AB	Rh-	4.00
8	O	Rh-	30.00
9			

To make the table look the way you expect, use numbers and labels, otherwise, SPSS will rearrange the table in alphabetical order. Once the data have been entered, go to **Data → Weight Cases**.

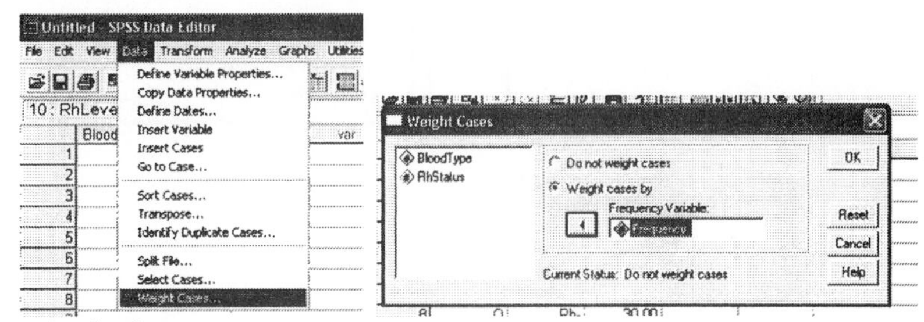

Weigh the cases by the frequency.

Step 1: H_0: Blood Type and Rh-status are Independent
OR H_0: Proportion of Rh^+ and A is $P(Rh^+ \text{ and } A)=P(Rh^+)*P(A)$; Proportion of Rh^+ and B is $P(Rh^+ \text{ and } B)=P(Rh^+)P(B)$; etc.

Note: although we write the hypothesis in words, the test is based on the probability definition of independence. This comes in how we calculate the expected values. Unlike the Goodness of Fit test, we don't need to worry about writing these out, as the computer can calculate them for us.

The alternate hypothesis would be:

H_1: Blood Type and Rh-status are Dependent
OR H_1: at least one proportion is different.

Step 4: we will use the p-value instead of a critical value.

Step 2, 3, 5: done in SPSS

Go to **Analyze → Descriptive Statistics → Crosstabs...**

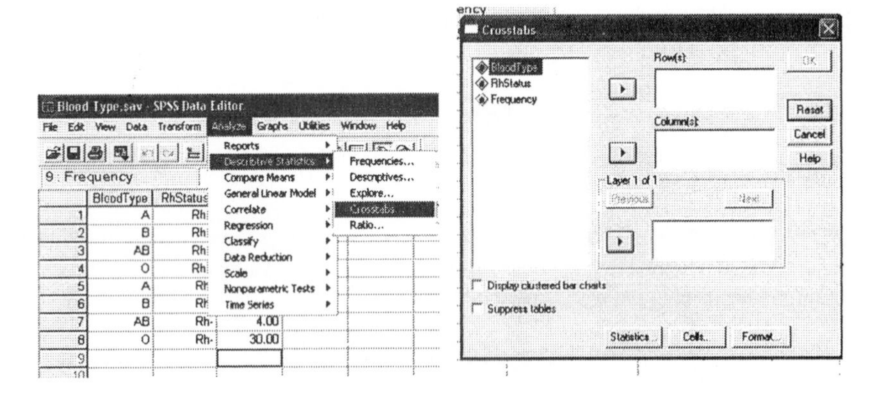

Place the Row variable in the Row(s) box, and the Column variable in the Column(s) box. Then click on the Statistics box.

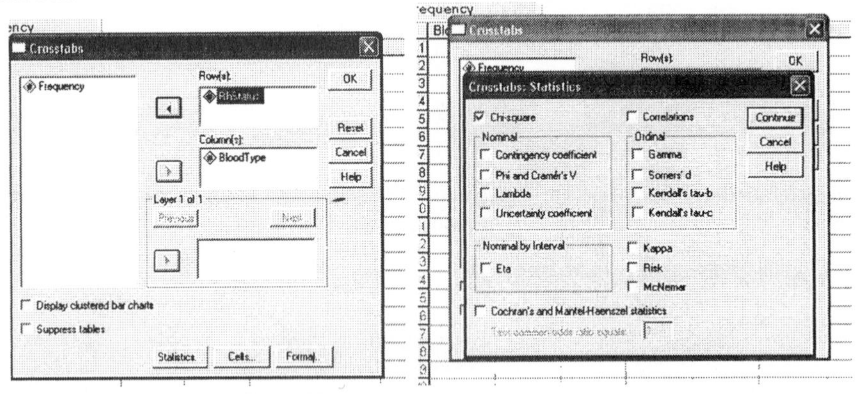

Check the Chi-square box and push Continue.

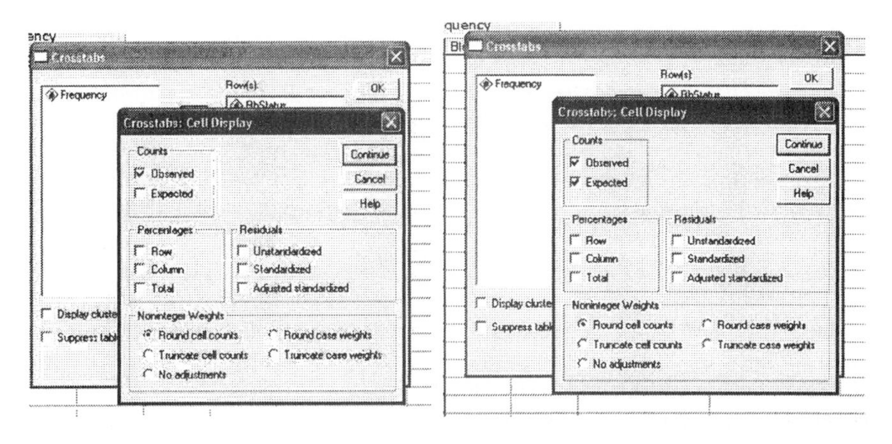

Pushing the Cells button will bring up the Cell Display window. If you want SPSS to calculate the Expected Values for you, check the Expected box. SPSS can also give the proportions for each cell, as well as the residuals. Residuals are useful for checking which cell contributes the most to a rejected null hypothesis, and the standardized is the easiest to interpret for this.
Push Continue, and OK

The output will provide the following tables:

RhStatus * BloodType Crosstabulation

			BloodType				
			A	B	AB	O	Total
RhStatus	Rh+	Count	176	28	22	198	424
		Expected Count	174.7	33.9	22.0	193.3	424.0
	Rh-	Count	30	12	4	30	76
		Expected Count	31.3	6.1	4.0	34.7	76.0
Total		Count	206	40	26	228	500
		Expected Count	206.0	40.0	26.0	228.0	500.0

Here we can see the marginal frequencies, as well as the observed and expected frequencies for each cell.

Chi-Square Tests

	Value	df	Asymp. Sig. (2-sided)
Pearson Chi-Square	7.601[a]	3	.055
Likelihood Ratio	6.410	3	.093
Linear-by-Linear Association	.494	1	.482
N of Valid Cases	500		

a. 1 cells (12.5%) have expected count less than 5. The minimum expected count is 3.95.

The test statistic (7.601) is provided in the next box, as well as the check for normality, or Step 3, SPSS will verify that no more than 20% of the cells are less than 5 and that all the expected values are greater than or equal to 1. If either of these is not met we should not perform the test.

Step 6. Because the p-value (.055) is not less than α (.05), we fail to reject the null hypothesis.
Step 7. There is insufficient evidence at the α=.05 level of significance to show that blood type and Rh-status are dependent.

It is difficult based on a global test, such as the Chi-Square test, or ANOVA test, to show where the differences really are once a difference has been found. In the case of ANOVA, post-hoc tests can be run to see where the differences are. In the case of Chi-Square, the cell with the largest individual chi-square value is seen to be the one contributing the most to the rejection, and is thus the one that causes the difference. All other differences may not be significant, and should be tested separately. To find the largest contributor to the Chi-square value, find the cell with the largest absolute standardized residual.

Chapter 13. Comparing Three or More Means

Section 13.1 Comparing Three or More Means (One-way Analysis of Variance)

►One-Way Analysis of Variance

Example 1, Page 679
First, we must type in the data. This is done similar to the independent sample t-test in chapter 11, with the value of interest in one column, and the group in another. There can now be more than two groups. The groups must be numerical, but you can use labels, which will help in the output.

Blood Glucose.sav - SPSS Data E

File Edit View Data Transform Anal

	Glucose	Group
1	288.1	Control
2	296.8	Control
3	267.8	Control
4	256.7	Control
5	292.1	Control
6	282.9	Control
7	260.3	Control
8	283.8	Control
9	229.1	Fenugreek
10	240.7	Fenugreek
11	239.4	Fenugreek
12	207.7	Fenugreek
13	225.7	Fenugreek
14	230.8	Fenugreek
15	206.6	Fenugreek
16	213.3	Fenugreek
17	177.4	Garlic
18	202.2	Garlic
19	163.1	Garlic
20	184.7	Garlic
21	197.9	Garlic
22	164.6	Garlic
23	193.9	Garlic
24	158.1	Garlic
25	298.7	Onion
26	258.3	Onion
27	286.8	Onion
28	244.0	Onion
29	267.1	Onion
30	297.1	Onion
31	249.9	Onion
32	265.1	Onion
33		

The normality must be tested on each group separately. To do this, you can use the **Data → Select Cases** option.

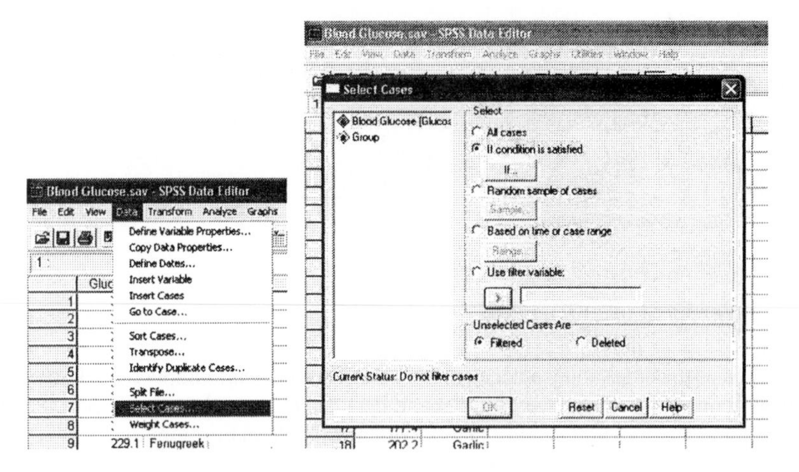

In the Select Cases window, you can select If condition is satisfied, click on the **if** button, and create a condition that separates out a specific group to test the normality on, for example, Group=1 for the first group.

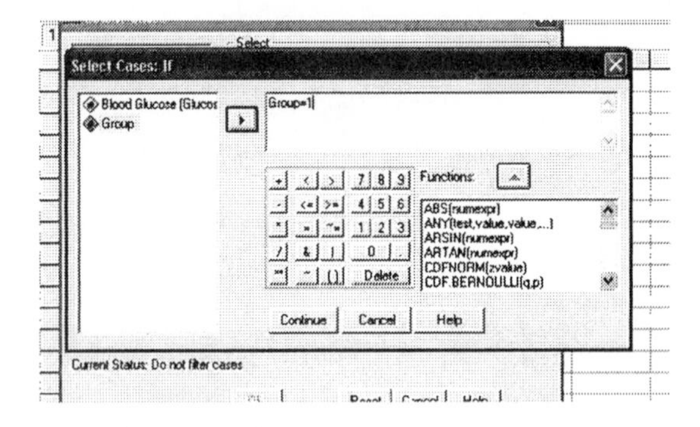

Once you have selected the group, you can test for normality the same as in earlier chapters. After checking the normality of each group, you must remember to go back and select all cases before going to the Analysis of Variance.

The claim is that the mean blood glucose levels are the same for all diets.

H_0: $\mu_{\text{Control}} = \mu_{\text{Fenugreek}} = \mu_{\text{Garlic}} = \mu_{\text{Onion}}$

OR H_0: mean blood glucose level is independent of diet

The second statement shows that the test is similar to the chi-square tests, but because we are looking at means, we no longer have the restrictions that everything adds up to 100%. Note: if there were only two means, this would be the same as the independent samples t-test hypothesis from chapter 11.

The alternate hypothesis would be:

H_1: at least one mean is different

We set α=0.05.

Go to **Analyze → Compare Means → One-Way ANOVA…**

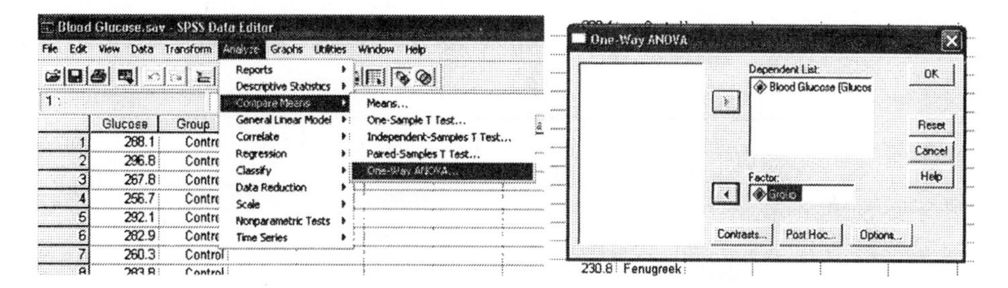

Use the column with the values for the Dependent list, and the variable containing the groups as the factor. This should remind you of the method used in chapter 11, for the independent t-test, except now we do not have to restrict to two groups, so we don't need to define which two groups to use. Also, since this is a two tailed test, the order in which the groups are taken doesn't matter, as it does in the independent samples t-test.

Click on the Options box

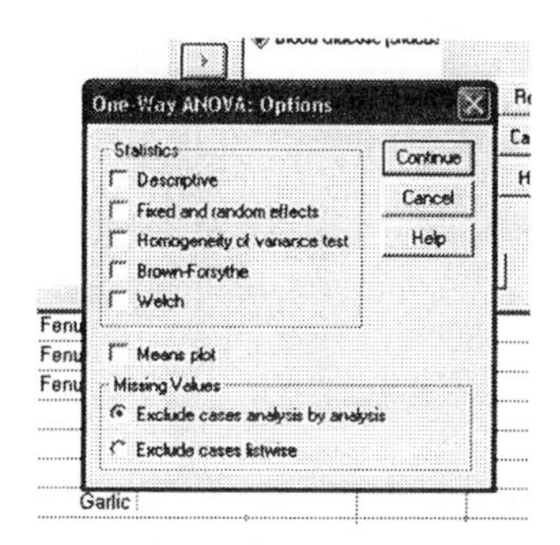

Checking Descriptive will provide you with descriptive statistics for each group. The Fixed and random effects will provide the different results based on whether the effects (groups) are fixed or random. This is for a design of experiments class, and is beyond the scope of this class. The Homogeneity of Variances test will provide a check to see that the assumption of equal variances is met. This is the Levene's test, as used in chapter 11. The Brown-Forsythe, and Welch tests are better tests than the F test if the assumption of equal variances doesn't hold. They also tests the equality of means. The Means Plot provides a graphical way to look at the means to see where a difference occurs, if one exists.

Once you have the options you want to use, click continue.

Clicking on the Contrasts button will open up a contrasts option window. This is also for a design of experiments class, and is beyond what we will discuss here.

Clicking the Post-Hoc Test button provides the possible Post-Hoc (after this) tests. These are tests to show, if a difference was found, where the differences are. We will discuss these later.

Once you have selected the options you want to try, push continue and OK.

The output (using only default values) will provide the following table.

ANOVA

Blood Glucose

	Sum of Squares	df	Mean Square	F	Sig.
Between Groups	50090.691	3	16696.897	58.209	.000
Within Groups	8031.616	28	286.843		
Total	58122.307	31			

Here we see a similar table to that provided by MINITAB on page 681.

Adding the descriptive statistics provides the following table

Descriptives

Blood Glucose

	N	Mean	Std. Deviation	Std. Error	95% Confidence Interval for Mean		Minimum	Maximum
					Lower Bound	Upper Bound		
Control	8	278.563	15.0257	5.3124	266.001	291.124	256.7	296.8
Fenugreek	8	224.163	13.4903	4.7695	212.884	235.441	206.6	240.7
Garlic	8	180.238	17.0597	6.0315	165.975	194.500	158.1	202.2
Onion	8	271.000	21.1797	7.4882	253.293	288.707	244.0	299.7
Total	32	238.491	43.3003	7.6545	222.879	254.102	158.1	299.7

Based on the fact that we reject the null hypothesis, so we know that at least two means are different, we can use this table to find where a difference may be. In this example, the most extreme means are the Control, with a mean of 278.56 and the Garlic, with a mean of 180.24. Because we reject the null hypothesis, we can show that these two are different. We cannot, however, look at the other values without some further analysis.

Section 13.2 Post-Hoc Tests on One-Way Analysis of Variance

►Performing Post-Hoc Tests

We use post-hoc tests after finding a difference in the means to show us where those means might be. When running the one-way ANOVA test, you can select post-hoc tests by clicking on the post-hoc button.

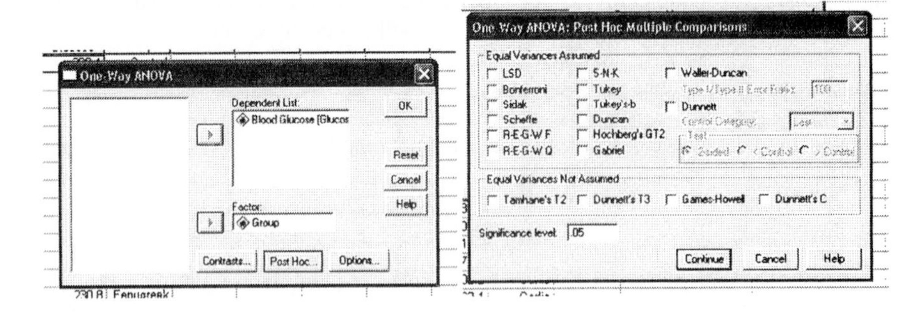

This provides a list of different post-hoc tests. Each test has its own way of conserving the α value that we have selected, but some are more conservative (will find less differences) than others. You can experiment with the different tests to see the differences, but we will present the results of the Tukey test.

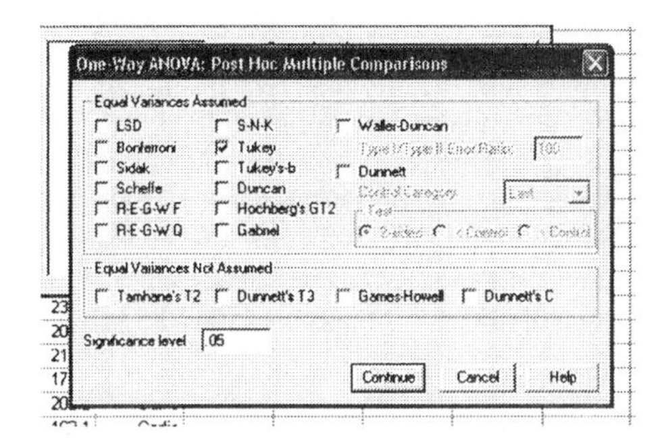

We select the Tukey test, and the Significance level (α). This will add the following tables to the output.

Multiple Comparisons

Dependent Variable: Blood Glucose
Tukey HSD

(I) Group	(J) Group	Mean Difference (I-J)	Std. Error	Sig.	95% Confidence Interval	
					Lower Bound	Upper Bound
Control	Fenugreek	54.4000*	8.4682	.000	31.279	77.521
	Garlic	98.3250*	8.4682	.000	75.204	121.446
	Onion	7.5625	8.4682	.808	-15.558	30.683
Fenugreek	Control	-54.4000*	8.4682	.000	-77.521	-31.279
	Garlic	43.9250*	8.4682	.000	20.804	67.046
	Onion	-46.8375*	8.4682	.000	-69.958	-23.717
Garlic	Control	-98.3250*	8.4682	.000	-121.446	-75.204
	Fenugreek	-43.9250*	8.4682	.000	-67.046	-20.804
	Onion	-90.7625*	8.4682	.000	-113.883	-67.642
Onion	Control	-7.5625	8.4682	.808	-30.683	15.558
	Fenugreek	46.8375*	8.4682	.000	23.717	69.958
	Garlic	90.7625*	8.4682	.000	67.642	113.883

*. The mean difference is significant at the .05 level.

Blood Glucose

Tukey HSD[a]

Group	N	Subset for alpha = .05		
		1	2	3
Garlic	8	180.238		
Fenugreek	8		224.163	
Onion	8			271.000
Control	8			278.563
Sig.		1.000	1.000	.808

Means for groups in homogeneous subsets are displayed.

a. Uses Harmonic Mean Sample Size = 8.000.

The first table provides the actual difference in the means, the p-value (Sig.) for the Tukey test, and a confidence interval for the difference. Each mean that is significantly different at α is marked by an *. We can see in this example that Control is significantly different from Fenugreek and Garlic, but not significantly different from Onion. We can also use the confidence intervals to show that Control is 31.279 to 77.521 mg/dl higher than Fenugreek. This is because when we look at the row where Control comes first, the difference is positive, and the upper and lower bounds of the confidence interval are both positive. If we look at the row with Fenugreek in the first column, then when we look at control, the mean difference is negative, as are the upper and lower bounds of the confidence interval. That tells us that Fenugreek had a lower mean than Control. The standard error column provides the value for

$$\sqrt{MSE\left(\frac{1}{n_1}+\frac{1}{n_2}\right)}$$

which is similar to what we calculate by hand. SPSS uses a table that incorporates the constant $\sqrt{2}$ in the denominator.

Although the first table provides the most information, for most students, the second table is the easier table to follow. Here we place the diets into separate groups, where we can see a difference between groups but not within groups. Garlic is in its own group, and has the lowest mean. This means that Onion is significantly lower than the other diets. Fenugreek is also different than the other diets. It has a higher mean than Garlic, but a lower mean than Onion or Control. Since Onion and Control are in the same group, we cannot show a significant difference between them.

Section 13.3 The Randomized Complete Block Design

►Performing an ANOVA for random block designs

Example 2 page 704

For a two-way ANOVA, we need to enter the data similarly to a one-way ANOVA, but we will have an extra column. So, we need one column for the response variable, one column for the first treatment, and a second column for the second treatment, or the block.

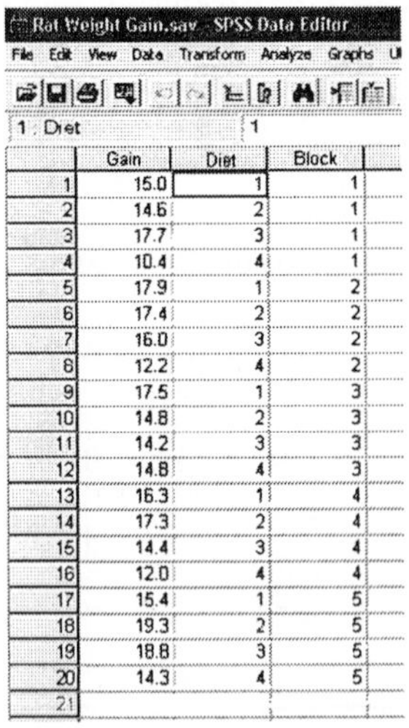

Rat Weight Gain.sav - SPSS Data Editor

File Edit View Data Transform Analyze Graphs U

1 : Diet 1

	Gain	Diet	Block
1	15.0	1	1
2	14.6	2	1
3	17.7	3	1
4	10.4	4	1
5	17.9	1	2
6	17.4	2	2
7	16.0	3	2
8	12.2	4	2
9	17.5	1	3
10	14.8	2	3
11	14.2	3	3
12	14.8	4	3
13	16.3	1	4
14	17.3	2	4
15	14.4	3	4
16	12.0	4	4
17	15.4	1	5
18	19.3	2	5
19	18.8	3	5
20	14.3	4	5
21			

To run the two-way ANOVA, we go to **Analyze → General Linear Model → Univariate.**

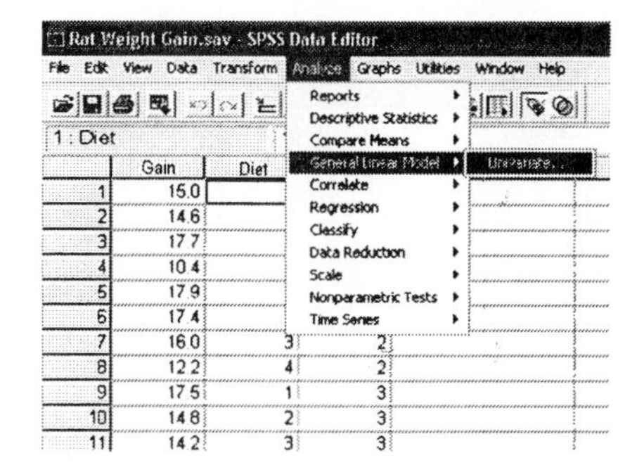

The General Linear Model can be used for more complicated models. In this example, we are going to treat both treatments as fixed.

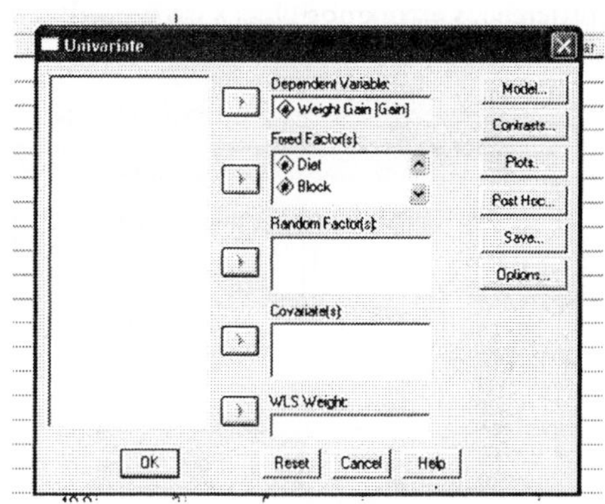

We put the response variable into the Dependent Variable box. We place both the treatment of interest, and the blocking variable into the Fixed Factor(s) box. We then have to define what the model looks like. To do this, click on the **Model** button. We are not performing a full factorial design, so we need to make a custom model.

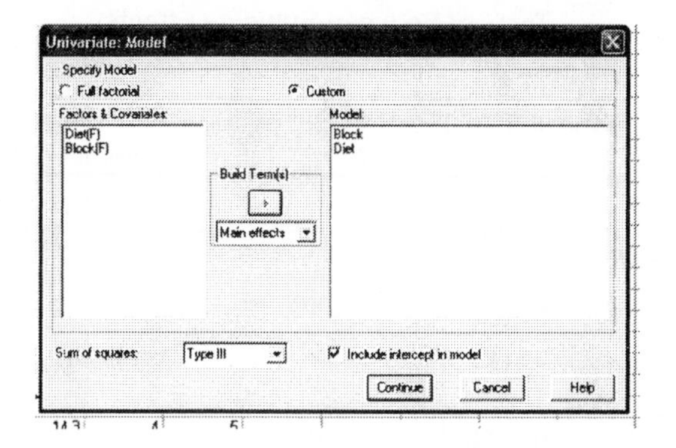

Select each factor and put them in the model, and select only Main effects under the Build Term(s). This will provide the ANOVA that we want. You can either keep the intercept in the model, or take it out. Neither method affects the portion of the ANOVA table that we want to look at.
Back in the main window, we can also add a post-hoc test, by selecting the post-hoc button.

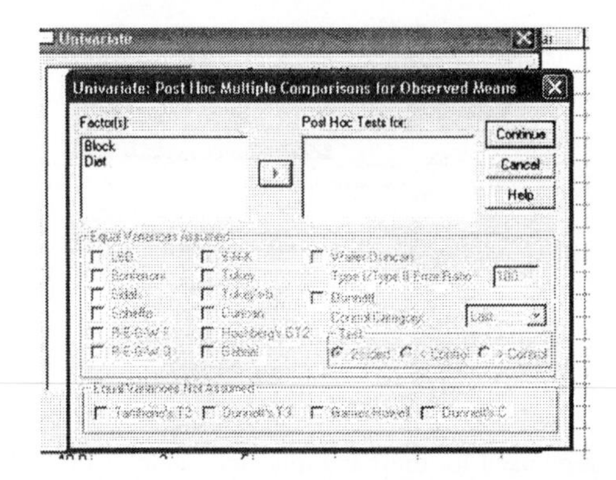

We are not really interested in differences among the blocks, so we will select Diet. Once we have selected a variable, the different post-hoc tests become available. We will use the Tukey.

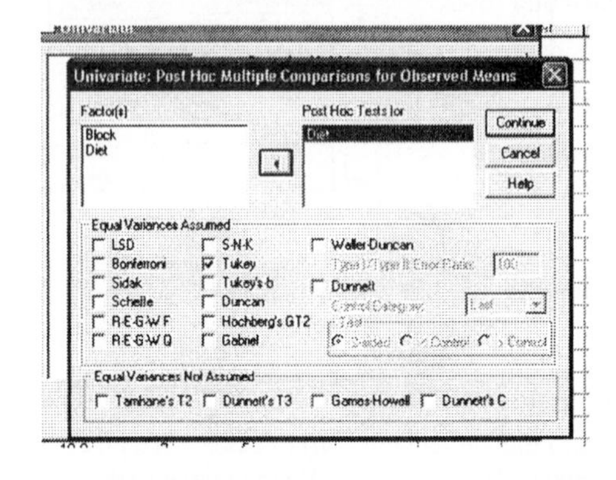

The results of the test should look similar to the following:

Tests of Between-Subjects Effects

Dependent Variable: Weight Gain

Source	Type III Sum of Squares	df	Mean Square	F	Sig.
Corrected Model	66.583[a]	7	9.512	3.117	.041
Intercept	4814.305	1	4814.305	1577.469	.000
Block	14.713	4	3.678	1.205	.359
Diet	51.870	3	17.290	5.665	.012
Error	36.623	12	3.052		
Total	4917.510	20			
Corrected Total	103.206	19			

a. R Squared = .645 (Adjusted R Squared = .438)

This is the ANOVA with the intercept included. We can ignore the first two rows. The rows with Block, Diet, Error, and Corrected Total are all we actually require, and are the same as provided by MINITAB on page 705. If you opted to not have the Intercept, the output will look like

Tests of Between-Subjects Effects

Dependent Variable: Weight Gain

Source	Type III Sum of Squares	df	Mean Square	F	Sig.
Model	4880.887[a]	8	610.111	199.911	.000
Block	14.713	4	3.678	1.205	.359
Diet	51.870	3	17.290	5.665	.012
Error	36.623	12	3.052		
Total	4917.510	20			

a. R Squared = .993 (Adjusted R Squared = .988)

The output is the same, except the values in the intercept line have now been added to the Model line. We will still ignore the Model and Total rows.

We are not interested in differences in the blocking variable, so we look only at the diet. Diet has a p-value of 0.012, which is less than 0.05, so we reject the null hypothesis. This means we can show a difference in the means of the diets, so we should look at the tukey test to see where those differences are. The output for the Tukey test are similar to those in the previous section.

Multiple Comparisons

Dependent Variable: Weight Gain

Tukey HSD

(I) Diet	(J) Diet	Mean Difference (I-J)	Std. Error	Sig.	95% Confidence Interval Lower Bound	Upper Bound
1	2	-.260	1.1049	.995	-3.540	3.020
	3	.200	1.1049	.998	-3.080	3.480
	4	3.680*	1.1049	.027	.400	6.960
2	1	.260	1.1049	.995	-3.020	3.540
	3	.460	1.1049	.975	-2.820	3.740
	4	3.940*	1.1049	.018	.660	7.220
3	1	-.200	1.1049	.998	-3.480	3.080
	2	-.460	1.1049	.975	-3.740	2.820
	4	3.480*	1.1049	.037	.200	6.760
4	1	-3.680*	1.1049	.027	-6.960	-.400
	2	-3.940*	1.1049	.018	-7.220	-.660
	3	-3.480*	1.1049	.037	-6.760	-.200

Based on observed means.

*. The mean difference is significant at the .05 level.

Weight Gain

Tukey HSD[a,b]

Diet	N	Subset	
		1	2
4	5	12.740	
3	5		16.220
1	5		16.420
2	5		16.680
Sig.		1.000	.975

Means for groups in homogeneous subsets are displayed.
Based on Type III Sum of Squares
The error term is Mean Square(Error) = 3.052.

a. Uses Harmonic Mean Sample Size = 5.000.

b. Alpha = .05.

We can show that Diet 4 is different from the other three diets, but we cannot show a difference in the other three diets at all. Diet 4 results in the least amount of weight gain.

Section 13.4 Two-way Analysis of Variance

►Performing a two-way ANOVA

Example 3 page 716

Performing a two-way ANOVA is similar to the random block design, but now we are interested in both treatments. We put the data into SPSS like we do for the random block design.

HDL Cholesterol.sav - SPSS Data Editor

File Edit View Data Transform Analyze Graphs U

15 :

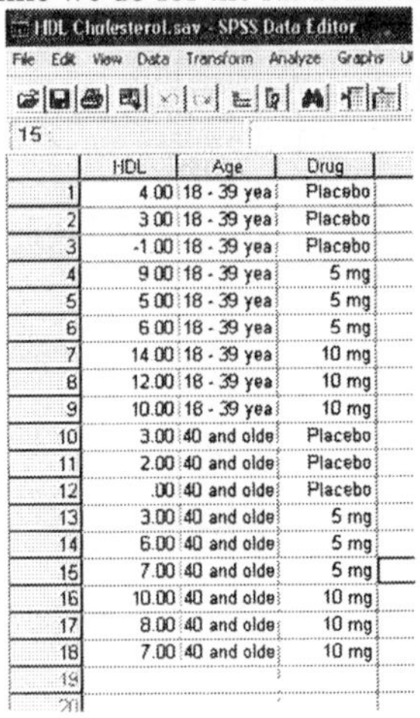

	HDL	Age	Drug
1	4.00	18 - 39 yea	Placebo
2	3.00	18 - 39 yea	Placebo
3	-1.00	18 - 39 yea	Placebo
4	9.00	18 - 39 yea	5 mg
5	5.00	18 - 39 yea	5 mg
6	6.00	18 - 39 yea	5 mg
7	14.00	18 - 39 yea	10 mg
8	12.00	18 - 39 yea	10 mg
9	10.00	18 - 39 yea	10 mg
10	3.00	40 and olde	Placebo
11	2.00	40 and olde	Placebo
12	.00	40 and olde	Placebo
13	3.00	40 and olde	5 mg
14	6.00	40 and olde	5 mg
15	7.00	40 and olde	5 mg
16	10.00	40 and olde	10 mg
17	8.00	40 and olde	10 mg
18	7.00	40 and olde	10 mg
19			
20			

Once the data have been entered, we go to **Analyze → General Linear Model → Univariate**. We enter the variables similarly to the random block design.

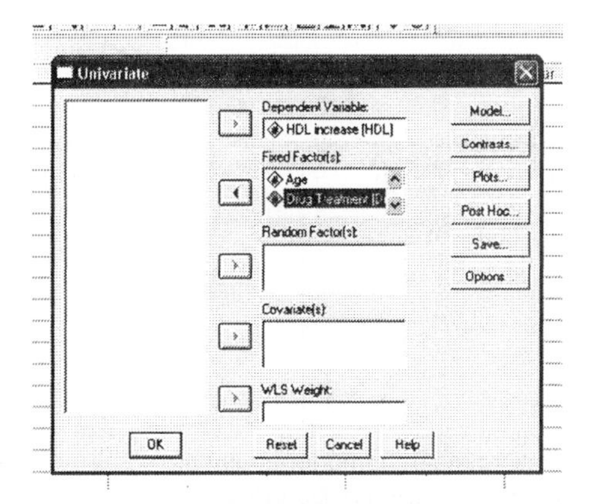

Now comes the difference between the random block design and the two-way ANOVA. When we go to the Model menu, we want the full factorial. We do not have to use a custom design. The full factorial design will include both variables as well as the interaction between the two variables. We may want an interaction plot, so we will click on the Plots menu.

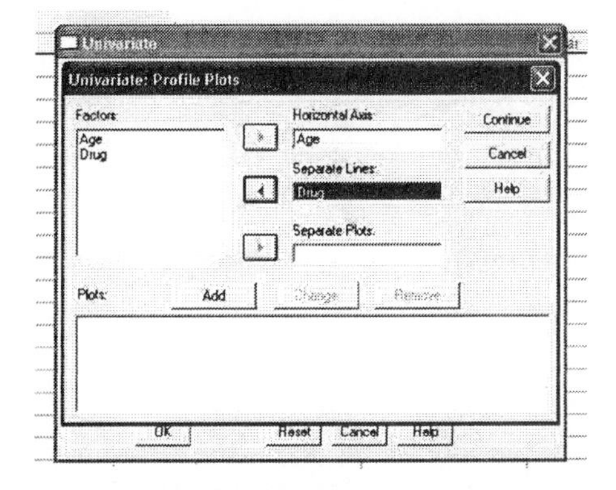

We will most likely want both interaction plots, to make interactions easier to see, if an interaction exists.

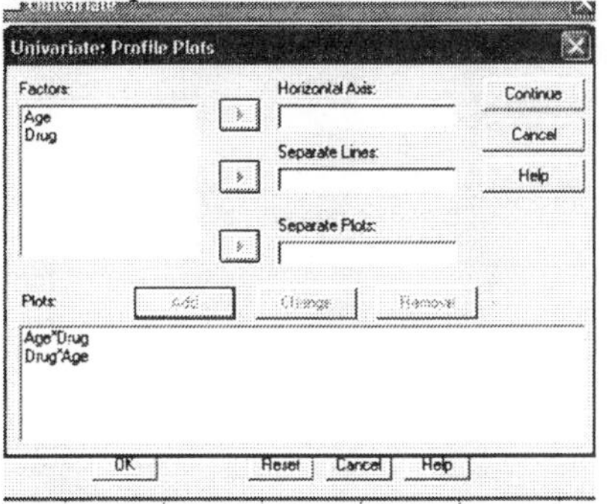

We can then continue to the output.

Tests of Between-Subjects Effects

Dependent Variable: HDL increase

Source	Type III Sum of Squares	df	Mean Square	F	Sig.
Corrected Model	231.333[a]	5	46.267	11.408	.000
Intercept	648.000	1	648.000	159.781	.000
Age	14.222	1	14.222	3.507	.086
Drug	208.333	2	104.167	25.685	.000
Age * Drug	8.778	2	4.389	1.082	.370
Error	48.667	12	4.056		
Total	928.000	18			
Corrected Total	280.000	17			

a. R Squared = .826 (Adjusted R Squared = .754)

Again, we can ignore the Corrected Model, Intercept, and Total rows. (If you didn't include the intercept, ignore the Model and Total rows). We do not see a significant interaction in this model, but there is a significant difference in the means for the different drugs. We could do a post-hoc test to see where the differences were. We can still look at the interaction plots, although they aren't significant.

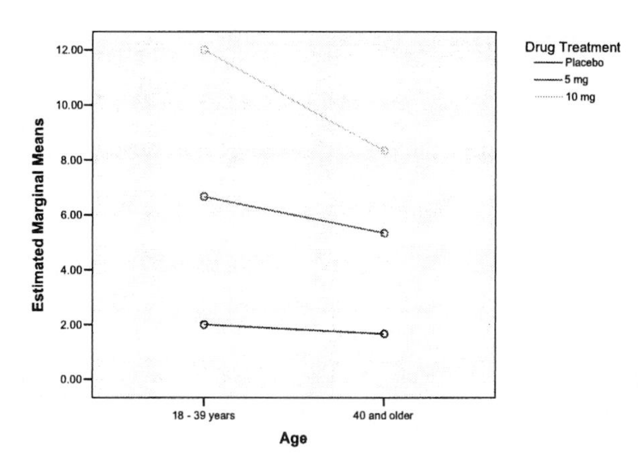

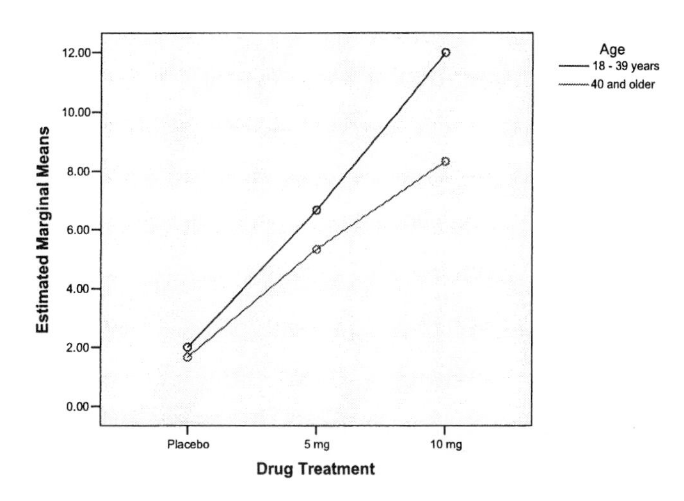

If there was an interaction, we would see a crossing, or non-parallelism in the lines that were outside the error of the model. We cannot actually interpret these plots, because there was no significant interaction.

Chapter 14. Inference on the Least-Squares Regression Model and Multiple Regression

Section 14.1 Testing the Significance of the Least-Squares Regression Model

►Testing the Least-Squares Regression Model

Example 1, Page 737

Much of what we will do for regression in chapter 12 was already done in chapter 4. The only difference is that we will now look at the tests as well as the coefficients. For this test, the data is entered in two columns, one with the predictor variable and the next with the response variable. Both must be numerical.

The claim is that the distribution of residents is the same today as it was in 1999. The null hypothesis is:

H_0: There is no linear relation between Age and Total Cholesterol

OR H_0: $\rho=0$

OR H_0: $\beta_1=0$

All three hypotheses mean the same thing. The first is the hypothesis in words, the second using correlation, and the third using the slope. The second and third are useful in calculation, while the first is useful in interpretation.

The alternate hypothesis would be:

H_1: There is a linear relation between Age and Total Cholesterol

The alternate hypotheses can be one or two tailed by specifying a positive or negative linear relation. For this example we will use

H_1: $\rho\neq0$

OR H_1: $\beta_1\neq0$ or the two-tailed test.

Go to **Analyze → Regression → Linear…**

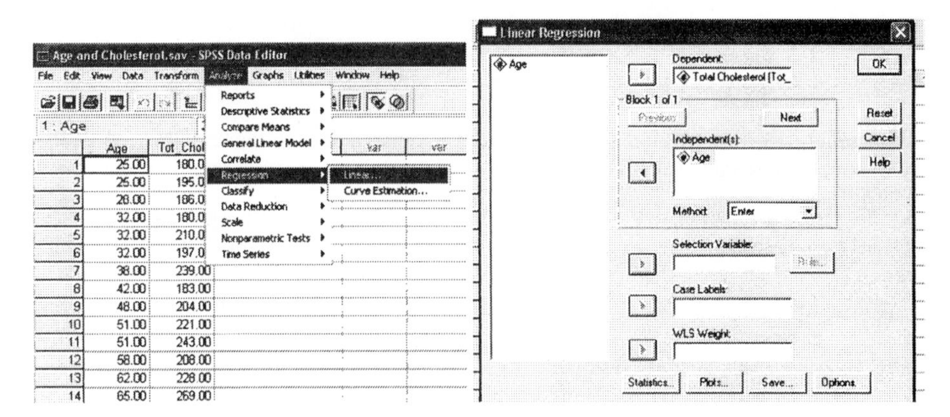

Use the Dependent, or Y, or response variable in the Dependent box. The Independent, or X, or predictor variable goes in the Independent(s) box. When you have selected the variables the way you want, push OK.

The output will provide the following table.

Model Summary

Model	R	R Square	Adjusted R Square	Std. Error of the Estimate
1	.718[a]	.515	.475	19.48054

a. Predictors: (Constant), Age

This provides the correlation, coefficient of determination, coefficient of determination adjusted for number of predictor variables (which is of more use later, when doing more than simple linear regression), and the **standard error of the estimate**, s_e, similar to example 2, page 740.

ANOVA[b]

Model		Sum of Squares	df	Mean Square	F	Sig.
1	Regression	4840.462	1	4840.462	12.755	.004[a]
	Residual	4553.895	12	379.491		
	Total	9394.357	13			

a. Predictors: (Constant), Age

b. Dependent Variable: Total Cholesterol

The ANOVA table provides values used in both the hypothesis test. This is the test that at least one predictor variable has a linear relationship with the response. With a simple linear regression, the F-value is t^2, and the p-values are the same. The Mean Square for the Residual Row is s_e^2, or the variance of the estimate.

Coefficients[a]

Model		Unstandardized Coefficients		Standardized Coefficients	t	Sig.
		B	Std. Error	Beta		
1	(Constant)	151.354	17.284		8.757	.000
	Age	1.399	.392	.718	3.571	.004

a. Dependent Variable: Total Cholesterol

For the Coefficients table, the (Constant) row relates to the Intercept, and the Age row relates to the slope. The B column is the estimated coefficients; the intercept and the slope. In this example, the regression line is Total Cholesterol=151.354+1.399*Age. The standard error column is the standard error for that estimate, so the standard error for the slope is .392. The t column is the t-test testing whether the population coefficient is 0. We don't usually worry about the intercept being 0, unless x=0 is part of the data, since we don't worry about values outside the scope of the sample. The significance (Sig.) value is the p-value for the test.

Because the p-value (.004) is less than α (.05), we reject the null hypothesis (See example 5 page 744). There is sufficient evidence at the α=.05 level of significance to show that there is a linear relationship between Age and Total Cholesterol.

It must be remembered that relation is NOT the same as causation. Just because we can show a relationship, even if the correlation is 1 or -1, that does not mean we know what causes the values of either variable.

To create a confidence interval for the slope (see Example 7, page 747), when in the **Linear Regression** window, click on the **Statistics** button, and check the Confidence Intervals option. This will create a 95% confidence interval for the intercept and slope. You cannot change the level of confidence. For a different confidence level, you will have to do it by hand.

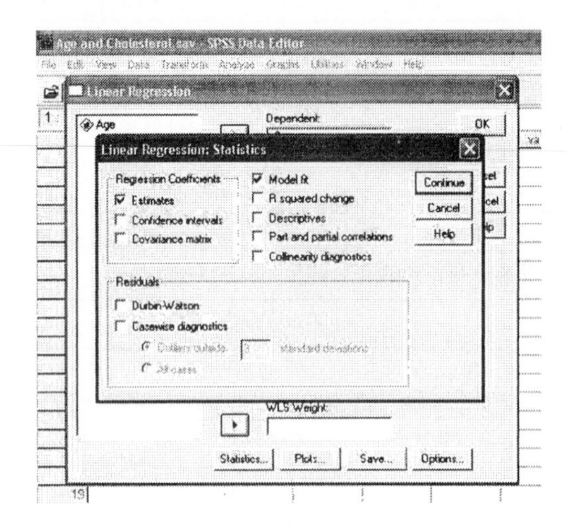

Coefficients[a]

Model		Unstandardized Coefficients		Standardized Coefficients	t	Sig.	95% Confidence Interval for B	
		B	Std. Error	Beta			Lower Bound	Upper Bound
1	(Constant)	151.354	17.284		8.757	.000	113.696	189.012
	Age	1.399	.392	.718	3.571	.004	.546	2.253

a. Dependent Variable: Total Cholesterol

We are 95% confident that the slope is between 0.546 and 2.253. Again, we normally do not worry about the intercept, unless it is in the scope of the data.

▶Verifying that the Residuals are Normally Distributed

Example 4, page 742

To verify that the residuals are normally distributed, we go to the plots menu in the linear regression window.

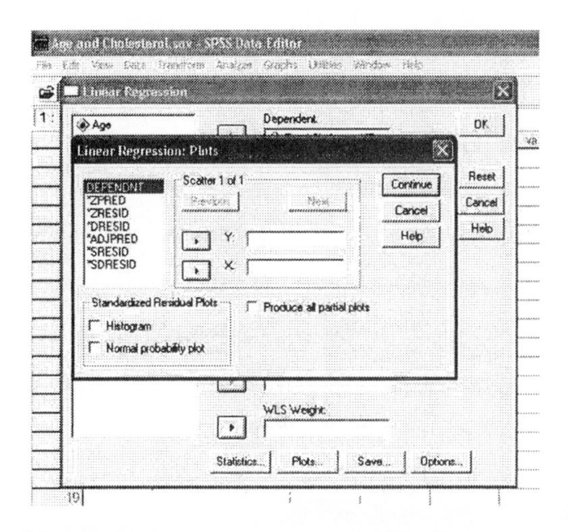

In the bottom left hand corner, we see the Standardized Residual Plots. We can get a Normal probability plot by checking the box. Also, if wanted, we could also get a histogram of the residuals.

Normal P-P Plot of Regression Standardized Residual

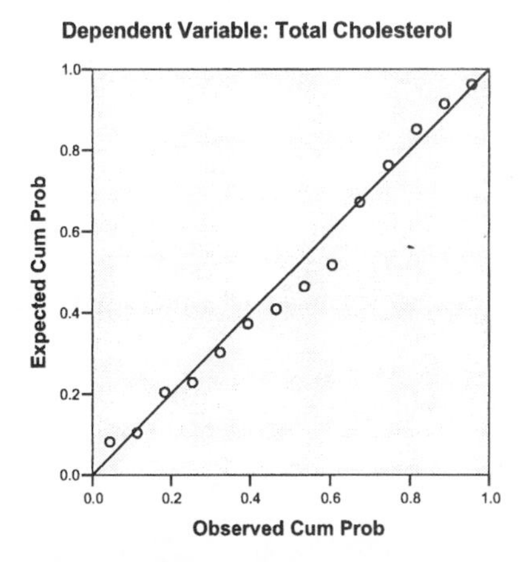

Section 14.2 Confidence and Prediction Intervals

►Creating Confidence and Prediction Intervals

Example 1, Page 754

To obtain a prediction or confidence interval, add the value of x that you want to predict for (if it is not already a part of the list), leaving a missing value for the y.

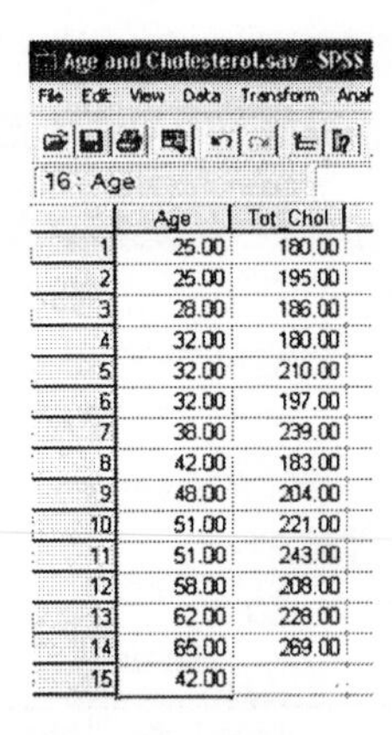

	Age	Tot_Chol
1	25.00	180.00
2	25.00	195.00
3	28.00	186.00
4	32.00	180.00
5	32.00	210.00
6	32.00	197.00
7	38.00	239.00
8	42.00	183.00
9	48.00	204.00
10	51.00	221.00
11	51.00	243.00
12	58.00	208.00
13	62.00	228.00
14	65.00	269.00
15	42.00	.

Go to **Analyze → Regression → Linear…**

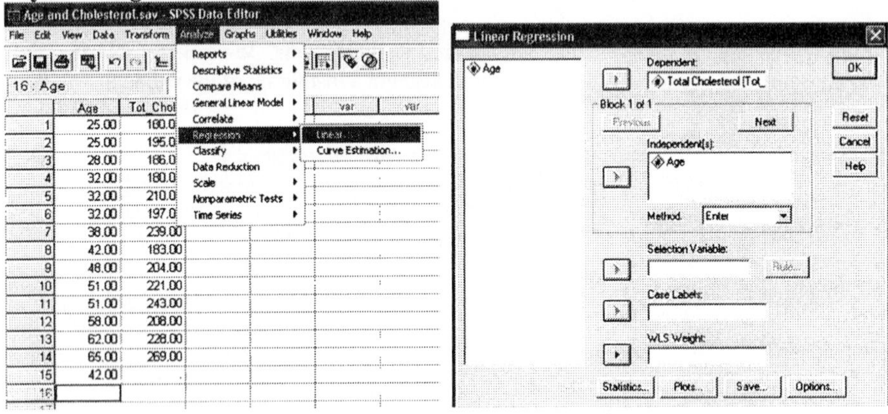

Use the Dependent, or Y, or response variable in the Dependent box. The Independent, or X, or predictor variable goes in the Independent(s) box.

Before going to OK, push the **Save** Button.

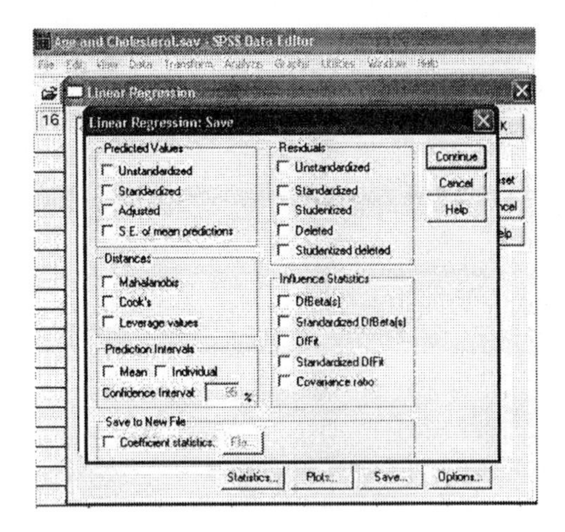

Select the options Unstandardized Predicted Value, which will give you $\hat{y}$, the predicted value for each x value, and click the Mean (confidence interval) and Individual (prediction interval) boxes in the Prediction Intervals section, and select the level of confidence.

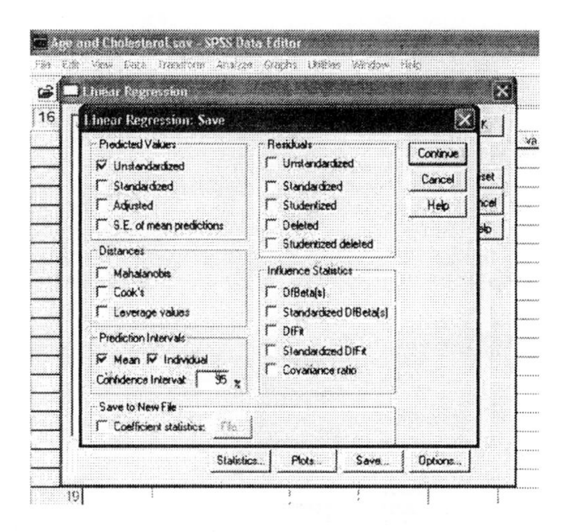

Push Continue, and OK

The output will look the same, with only one table added, which is a table of residual statistics. The predicted values, and intervals are back in the data table.

Age and Cholesterol.sav - SPSS Data Editor

16 : Age

	Age	Tot_Chol	PRE_1	LMCI_1	UMCI_1	LICI_1	UICI_1
1	25.00	180.00	186.33026	167.86434	204.79619	140.04289	232.61763
2	25.00	195.00	186.33026	167.86434	204.79619	140.04289	232.61763
3	28.00	186.00	190.52745	174.00693	207.04798	144.98124	236.07367
4	32.00	180.00	196.12371	181.89082	210.35660	151.35647	240.89095
5	32.00	210.00	196.12371	181.89082	210.35660	151.35647	240.89095
6	32.00	197.00	196.12371	181.89082	210.35660	151.35647	240.89095
7	38.00	239.00	204.51810	192.65400	216.38219	160.44671	248.58948
8	42.00	183.00	210.11435	198.77044	221.45827	166.18014	254.04856
9	48.00	204.00	218.50874	206.08754	230.92993	174.28412	262.73335
10	51.00	221.00	222.70593	209.04005	236.37181	178.11572	267.29614
11	51.00	243.00	222.70593	209.04005	236.37181	178.11572	267.29614
12	58.00	208.00	232.49938	214.79301	250.20575	186.50975	278.48901
13	62.00	228.00	238.09564	217.65051	258.54077	190.98371	285.20756
14	65.00	269.00	242.29283	219.67275	264.91290	194.19711	290.38855
15	42.00	.	210.11435	198.77044	221.45827	166.18014	254.04856
16							
17							

Here, we have 5 new variables. The first, *Pre_1* is the predicted value, so the predicted value when age is 42 is 210.11435. The next two, *LMCI_1* and *UMCI_1* are the lower and upper (L and U) limits for the **Mean** Confidence Interval (MCI). This is the 95% confidence interval for the mean of a group of people all aged 42. *LICI_1*, and *UICI_1* are the lower and upper limits for the **Individual** Confidence Interval (ICI), or the prediction interval. This is the 95% prediction interval for an individual aged 42.
We are 95% confident that the mean total cholesterol of all 42-year-old females is between 198.77044 and 221.45827. (198.77 and 221.46).
We are 95% confident that the total cholesterol for a randomly selected 42-year-old female is between 166.18014 and 254.04856 (166.18 and 254.05).

You can add as many values as is necessary to predict for, just keep in mind that they should be in the scope of the data.

Section 14.3 Multiple Linear Regression

►Obtaining the Correlation Matrix

Example 1, Page 760

To find the correlation matrix, we follow many of the same steps as in chapter 4. First, we enter the data into an SPSS worksheet.

	Age	Fat	Tot_Chol
1	25	19	180
2	25	28	195
3	28	19	186
4	32	16	180
5	32	24	210
6	32	20	197
7	38	31	239
8	42	20	183
9	48	26	204
10	51	24	221
11	51	32	243
12	58	21	208
13	62	21	228
14	65	30	269

Once the data have been entered, we go to **Analyze → Correlate → Bivariate**.

Select all the variables that you want in the correlation matrix.

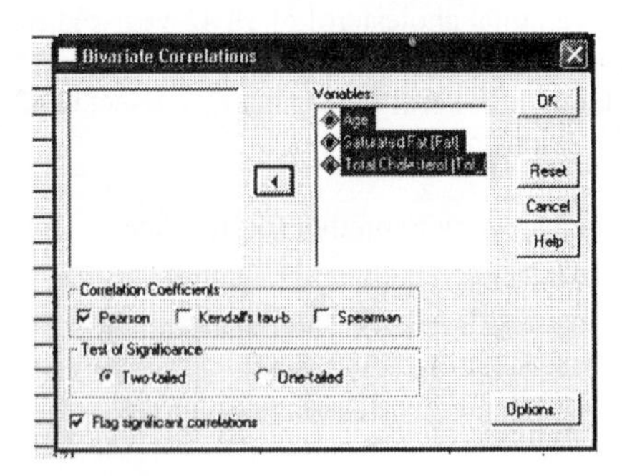

We are using Pearson Correlations, so make sure the Pearson box is checked. Push OK, and the output will provide you with the following table:

Correlations

		Age	Saturated Fat	Total Cholesterol
Age	Pearson Correlation	1	.324	.718**
	Sig. (2-tailed)		.258	.004
	N	14	14	14
Saturated Fat	Pearson Correlation	.324	1	.778**
	Sig. (2-tailed)	.258		.001
	N	14	14	14
Total Cholesterol	Pearson Correlation	.718**	.778**	1
	Sig. (2-tailed)	.004	.001	
	N	14	14	14

**. Correlation is significant at the 0.01 level (2-tailed).

The first value in each cell is the correlation coefficient. So the correlation between Saturated Fat and Age is 0.324, which has a p-value for the test of being 0 of 0.258, so we cannot show that age and saturated fat is correlated. The values with the ** by them are significant at α=0.01, and should be not be used together in the multiple regression for the same response. The response should be significantly correlated with the predictors in order to have a significant model. Since age and saturated fat are not significantly correlated, we do not have to worry about multicollinearity.

►Obtaining the Multiple Regression Equation

Example 2, Page 761

To find the multiple regression equation, we go to Analyze → Regression → Linear, just like in the simple linear regression. In the simple linear regression, we only picked one predictor variable, now we want to select both predictor variables.

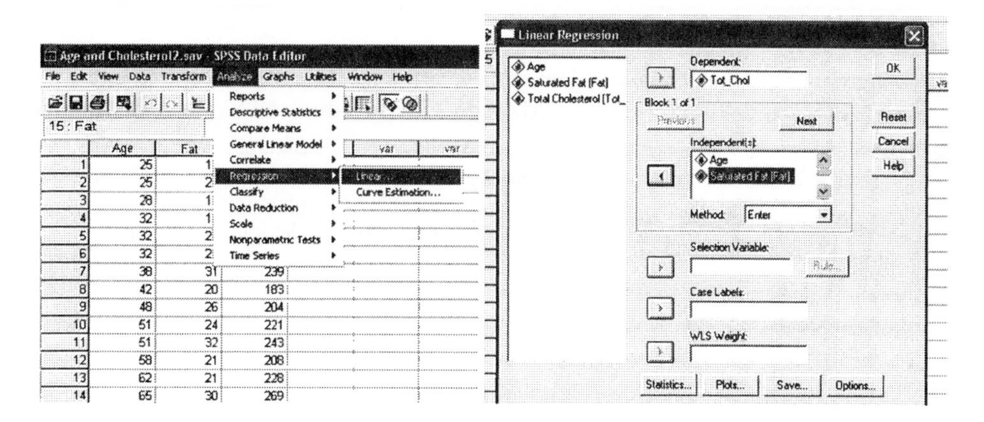

We can follow the same steps as in the simple linear regression for plots, confidence intervals, etc. The output should be as follows:

Model Summary

Model	R	R Square	Adjusted R Square	Std. Error of the Estimate
1	.921[a]	.847	.820	11.418

a. Predictors: (Constant), Saturated Fat, Age

This provides us with the model correlation, the model R^2 value, and the R^2 value after adjusting for having two predictors and only 14 observations. The adjusted R^2 value will penalize the model for having too many predictors, so that you don't get a model that just connects the dots with little or no meaning. (See example 4, page 765) We also get the standard error for the model.

ANOVA[b]

Model		Sum of Squares	df	Mean Square	F	Sig.
1	Regression	7960.297	2	3980.148	30.530	.000[a]
	Residual	1434.060	11	130.369		
	Total	9394.357	13			

a. Predictors: (Constant), Saturated Fat, Age

b. Dependent Variable: Total Cholesterol

The ANOVA table tests whether there is something significant in the model or not. This is a global test of the full model, or the F-test for Lack of Fit (see example 5, page 766). If the ANOVA test is not significant, we can throw out all of the predictor variables as being insignificant.

Coefficients[a]

Model		Unstandardized Coefficients		Standardized Coefficients	t	Sig.
		B	Std. Error	Beta		
1	(Constant)	90.842	15.989		5.682	.000
	Age	1.014	.243	.520	4.179	.002
	Saturated Fat	3.244	.663	.609	4.892	.000

a. Dependent Variable: Total Cholesterol

If we do not throw out the whole model, we look at the Coefficients table to tell us what the model would be. Here, we have the model $\hat{y} = 90.842 + 1.014\text{Age} + 3.244\text{Saturated Fat}$. We can also test each coefficient using the t-tests. These test whether the coefficient is significant, provided the other coefficient is still in the model. In this example, both t-tests have p-values less than .05, so both coefficients are significant at the α=0.05 level. We still do not need to worry about the intercept unless 0 is a valid point for both Age and Saturated Fat. We need to stay within the scope of the model with both variables.

►Obtaining the Residual Plots

Example 2, Page 761

To obtain the residual plots, we must first find the residuals. While you are in the linear regression window, click on the **Save** button. In the save menu, you can obtain the predicted values, as we did with simple linear regression, and in the right hand column, you can obtain the residual values.

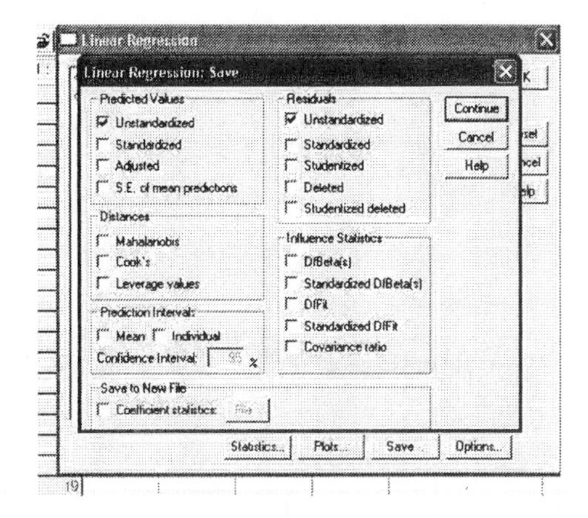

There are different types of residuals. The first on the list is the Unstandardized residuals. These are the residuals that you would calculate by hand, or the observed value-the expected value. The Standardized residuals are the residuals divided by an estimate of the standard error of the residuals. Since the standardized residuals are similar to z-values, we can look for outliers by looking for standardized values larger than 3, or less than -3. The Studentized residuals are similar to the standardized residuals, except the residual is divided by a standard error that is based on the location, similar to the confidence and prediction intervals. These three are common for residual plots.

The deleted and studentized deleted residuals are used for finding influential points. The Deleted residual is the residual value for a model that is made without the observation. These help to find influential points by omitting the value from the model, and seeing how well the observation fits with the predicted values for the rest of the observations.

We will select unstandardized residuals to match what we would get by hand. We also want the unstandardized predicted (fitted) values for use in the residual plots. Using the save menu does not add anything to the output window. It adds values to the data set. In the data, you will see a new column, RES_1.

Age and Cholesterol2.sav - SPSS Data Editor

File Edit View Data Transform Analyze Graphs Utilities Window Help

1 : PRE_1 177.837691353626

	Age	Fat	Tot_Chol	PRE_1	RES_1
1	25	19	180	177.83769	2.16231
2	25	28	195	207.03608	-12.03608
3	28	19	186	180.88031	5.11969
4	32	16	180	175.20433	4.79567
5	32	24	210	201.15846	8.84154
6	32	20	197	188.18139	8.81861
7	38	31	239	229.95355	9.04645
8	42	20	183	198.32345	-15.32345
9	48	26	204	223.87427	-19.87427
10	51	24	221	220.42836	.57164
11	51	32	243	246.38248	-3.38248
12	58	21	208	217.79500	-9.79500
13	62	21	228	221.85182	6.14818
14	65	30	269	254.09282	14.90718
15					

These are the residuals. We can now create plots. To create the plots, go to **Graphs → Scatter/Dot**, and select Simple Scatter.

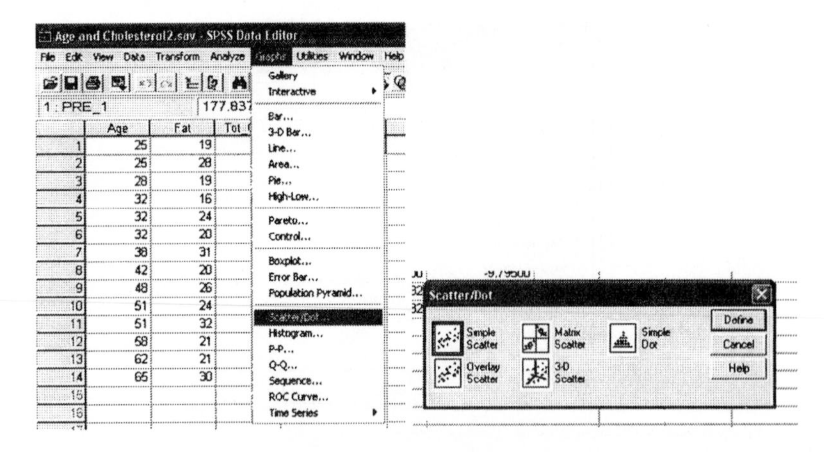

We now need to choose which plot we want to look at first. We will first look at the Residuals versus the fitted values plot.

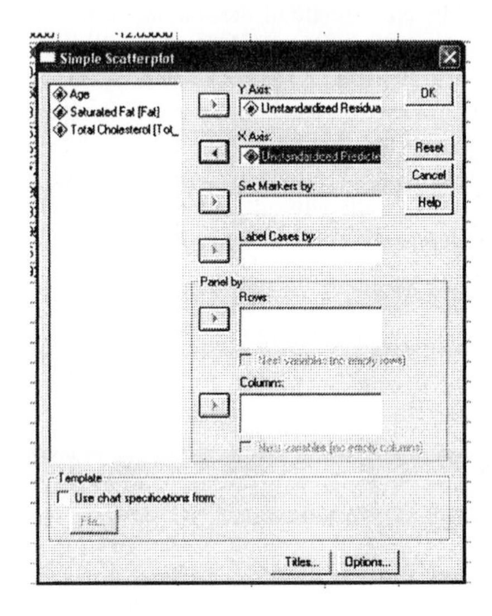

We select the residuals as the Y-Axis, and the predicted (fitted) values for the X-Axis. We then push OK.

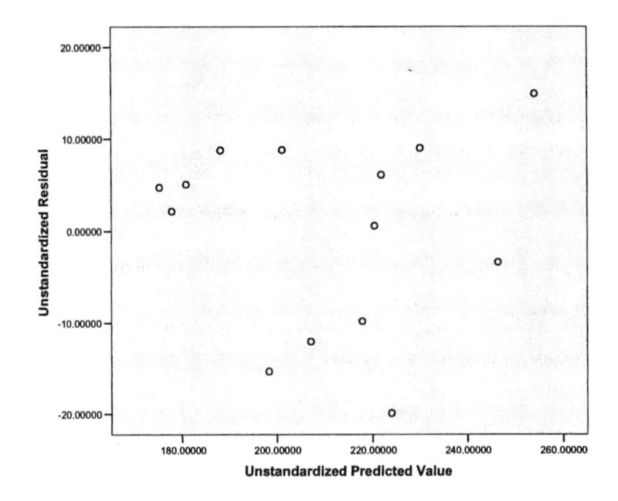

This provides a scatter plot, but we need to add the reference line at 0. To do this, go to the chart editor. In the chart editor, find **Options → Y Axis Reference Line**.

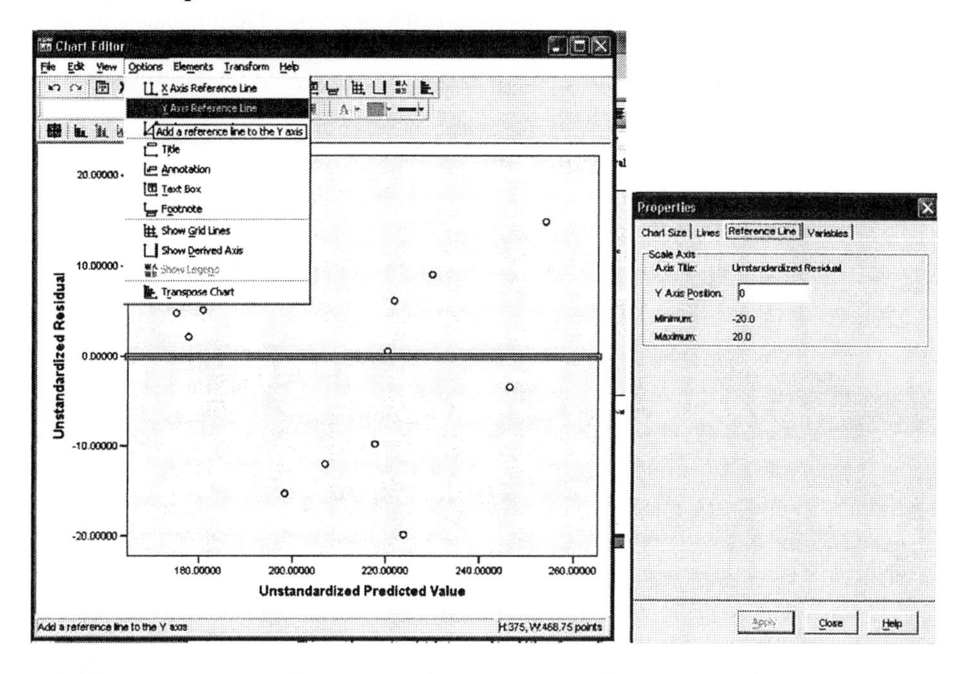

Since we want a line at the mean (0), we put 0 in the Y Axis Position, and push the **Close** button. We can then close the chart editor, and we have a residual plot.

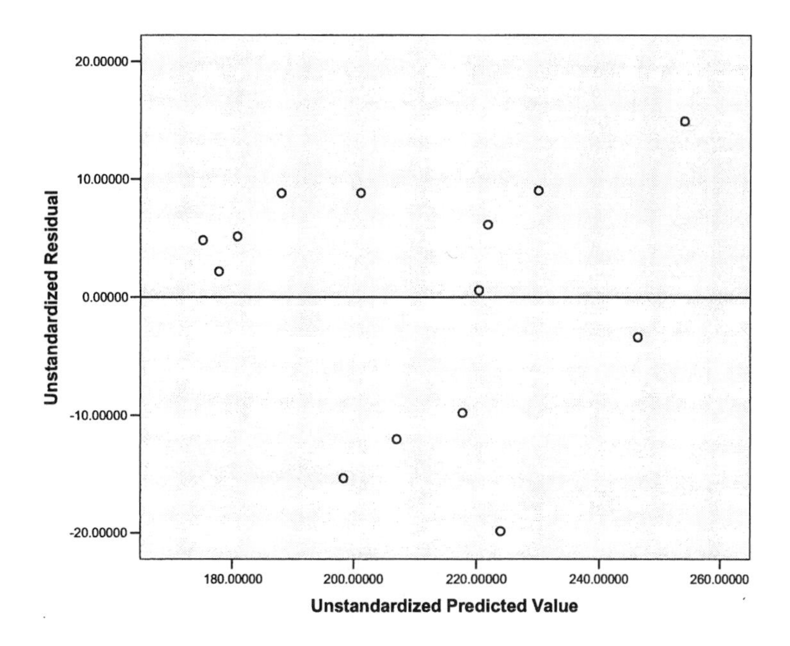

We can create the other residual plots in the same way, selecting different X-Axes.

►Creating Confidence and Prediction Intervals

Obtaining a confidence and prediction interval is similar to the method used for the simple linear regression, except now we have to have two x values for each prediction.
Example 7, Page 768

To obtain a prediction or confidence interval, add the values of x that you want to predict for (if it is not already a part of the list), leaving a missing value for the y.

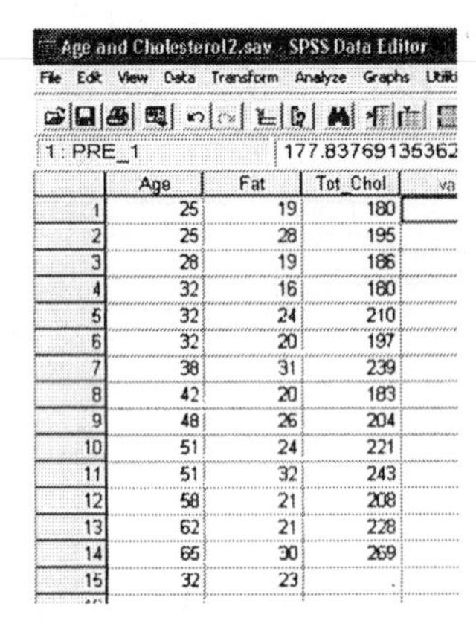
Age and Cholesterol2.sav - SPSS Data Editor

1 : PRE_1 177.83769135362

	Age	Fat	Tot_Chol
1	25	19	180
2	25	28	195
3	28	19	186
4	32	16	180
5	32	24	210
6	32	20	197
7	38	31	239
8	42	20	183
9	48	26	204
10	51	24	221
11	51	32	243
12	58	21	208
13	62	21	228
14	65	30	269
15	32	23	.

Go to **Analyze → Regression → Linear…**

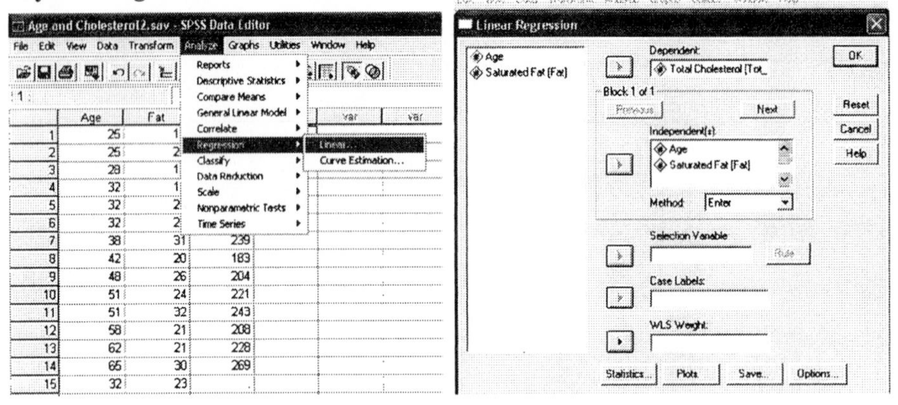

Put in the model that you want to use for the estimates.

Before going to OK, push the **Save** Button.

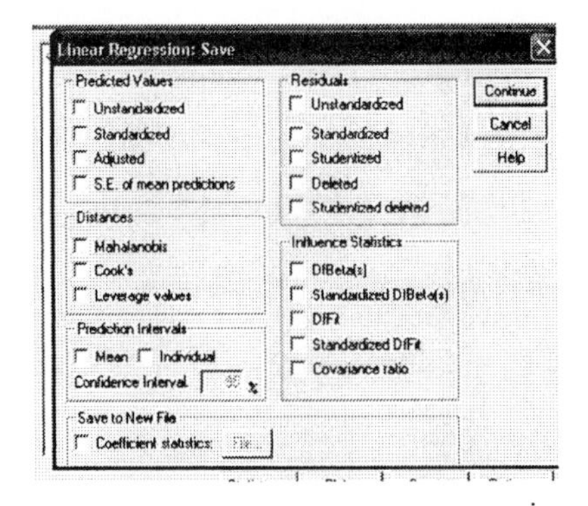

Select the options Unstandardized Predicted Value, which will give you $\hat{y}$, the predicted value for each x value, and click the Mean (confidence interval) and Individual (prediction interval) boxes in the Prediction Intervals section, and select the level of confidence.

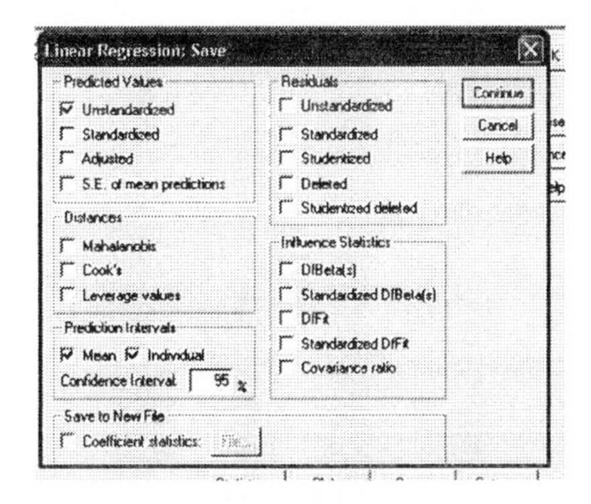

Push Continue, and OK

The output will look the same, with only one table added, which is a table of residual statistics. The predicted values, and intervals are back in the data table.

Age and Cholesterol2.sav - SPSS Data Editor

File Edit View Data Transform Analyze Graphs Utilities Window Help

1 : PRE_1 177.837691353626

	Age	Fat	Tot_Chol	PRE_1	LMCI_1	UMCI_1	LICI_1	UICI_1
1	25	19	180	177.83769	166.25585	189.41953	150.16657	205.50881
2	25	28	195	207.03608	192.67198	221.40018	178.08994	235.98222
3	28	19	186	180.88031	170.17898	191.58163	153.56602	208.19459
4	32	16	180	175.20433	162.57090	187.83776	147.07685	203.33182
5	32	24	210	201.15846	192.43223	209.88468	174.55585	227.76106
6	32	20	197	188.18139	179.02798	197.33481	161.43561	214.92717
7	38	31	239	229.95355	216.52561	243.38148	201.46037	258.44672
8	42	20	183	198.32345	189.76452	206.88237	171.77525	224.87165
9	48	26	204	223.87427	216.13380	231.61474	197.57852	250.17002
10	51	24	221	220.42836	212.27237	228.58434	194.00731	246.84941
11	51	32	243	246.38248	233.00531	259.75965	217.91319	274.85177
12	58	21	208	217.79500	205.39836	230.19164	189.77307	245.81692
13	62	21	228	221.85182	207.71142	235.99221	193.01603	250.68760
14	65	30	269	254.09282	239.68592	268.49972	225.12542	283.06023
15	32	23	.	197.91419	189.44870	206.37968	171.39596	224.43242

Here, we have 5 new variables. The first, *Pre_1* is the predicted value, so the predicted value when age is 32 and saturated fat is 23 grams is 197.91419. The next two, *LMCI_1* and *UMCI_1* are the lower and upper (L and U) limits for the **Mean** Confidence Interval (MCI). This is the 95% confidence interval for the mean of a group of people all aged 42. *LICI_1*, and *UICI_1* are the lower and upper limits for the **Individual** Confidence Interval (ICI), or the prediction interval. This is the 95% prediction interval for an individual aged 42.

We are 95% confident that the mean total cholesterol of all 32-year-old females who consume 23 grams of saturated fat daily is between 189.44870 and 206.37968 (189.45, 206.38).

We are 95% confident that the total cholesterol for a randomly selected 32-year-old females who consume 23 grams of saturated fat daily is between 171.39596 and 224.4342 (171.40 and 224.43).

You can add as many values as is necessary to predict for, just keep in mind that they should be in the scope of the data.

Chapter 15. Nonparametric Statistics

Section 15.2 Runs Test for Randomness

►Runs Test for Randomness

Example 3, Page 15-7

Enter the data in to SPSS. The values must be numerical, but labels can be attached. For this example, 0 was used to represent males, and 1 was used to represent females.

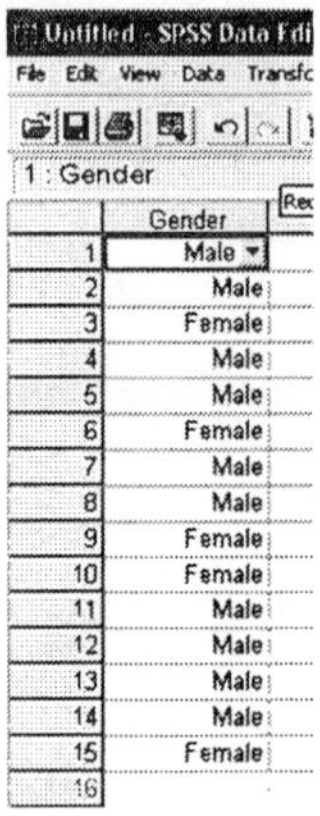

Once the data have been entered, go to **Analyze → Nonparametric Tests → Runs...**

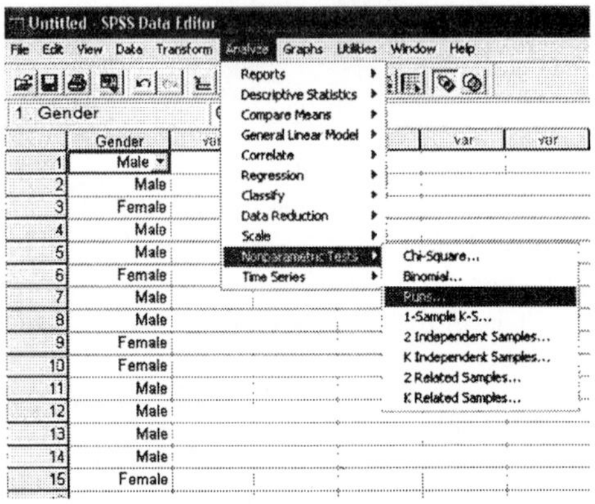

In the Runs Test window, select the variable you wish to run the test on, and click on the ► button. There are four options you can use to separate the different groups. Observations with value less than the cut point are placed in one group, while observations with value higher than the cut point are placed in the second group. Two options are the median and mean. One problem with these choices, is that the median is most likely a part of one of the two groups. The median, and mean should be used when looking at more continuous values. The mode will be part of a group, and can also provide different answers. The easiest to use is the Custom. Here, type in a value that is between the two groups (this is why the values must be numerical). For this example, we use the value of 0.5, since that is between 0 and 1.

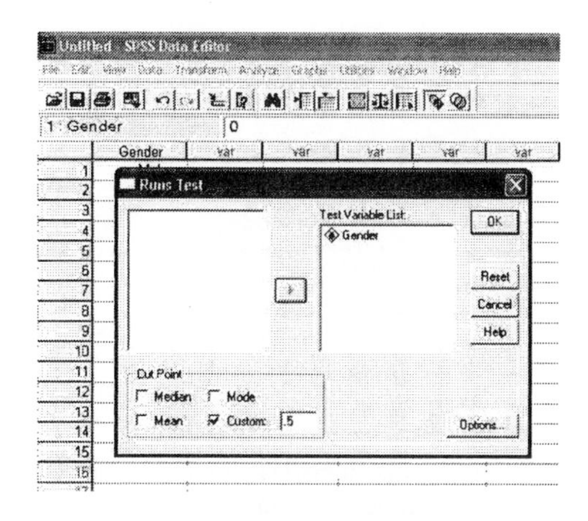

Once the variables and cut point are correct, push OK. The following will appear in the output window.

Runs Test

	Gender
Test Value[a]	.5000
Total Cases	15
Number of Runs	8
Z	.000
Asymp. Sig. (2-tailed)	1.000

a. User-specified.

Here, we see that we had a total of 15 observations, with 8 runs. We can use this in the small sample case, and look up 8 in the table. Because 8 is not less than or equal to the lower critical value, 3, and 8 is not greater than or equal to the upper critical value, 12, we cannot reject the null hypothesis. The p-value (Asymp. Sig. (2-tailed)) should only be used in large sample cases.

Example 4, page 15-7

Example 4 is a large sample case, so we can see how the z-score of SPSS correlates with what we would find by hand. First, we enter the data. This can be done by using the Ps and Ns as in the book, but in SPSS we can work with the original values, so there is no need to convert them.

Standard and Poors Index.sav - SPSS Data E

File Edit View Data Transform Analyze Graphs

43 : Return

	Date	Return	var
1	JAN 2002	-2.12	
2	FEB 2002	2.06	
3	MAR 2002	3.67	
4	APR 2002	-6.14	
5	MAY 2002	-.91	
6	JUN 2002	-7.25	
7	JUL 2002	-7.90	
8	AUG 2002	.49	
9	SEP 2002	-11.00	
10	OCT 2002	8.64	
11	NOV 2002	5.71	
12	DEC 2002	-6.03	
13	JAN 2003	-2.74	
14	FEB 2003	-1.70	
15	MAR 2003	.84	
16	APR 2003	8.10	
17	MAY 2003	5.09	
18	JUN 2003	1.13	
19	JUL 2003	1.62	
20	AUG 2003	1.79	
21	SEP 2003	-1.19	
22	OCT 2003	5.50	
23	NOV 2003	.71	
24	DEC 2003	5.08	
25	JAN 2004	1.73	
26	FEB 2004	1.22	
27	MAR 2004	-1.64	
28	APR 2004	-1.68	
29	MAY 2004	1.21	
30	JUN 2004	1.80	
31	JUL 2004	-3.43	
32	AUG 2004	.23	
33	SEP 2004	.94	
34	OCT 2004	1.40	
35	NOV 2004	3.86	
36	DEC 2004	3.25	
37	JAN 2005	-2.53	
38	FEB 2005	1.89	
39	MAR 2005	-1.91	
40	APR 2005	-2.01	
41	MAY 2005	3.00	
42	JUN 2005	.90	
43			

Once the data have been entered, go to **Analyze → Nonparametric Tests → Runs....** We follow the same steps as for a small data set. We can set the cut point to 0, so that positive values would be in one group, and negative values in the second group. The output is then

Runs Test

	Return
Test Value[a]	.0000
Total Cases	42
Number of Runs	18
Z	-.889
Asymp. Sig. (2-tailed)	.374

a. User-specified.

We see 42 observations, with 18 runs. We get a z-value of -0.889, which is different from what we get by hand. SPSS uses a correction for sample sizes less than 50, of $z = \begin{cases} (R-\mu_r+0.5)/\sigma, \text{if R - } \mu \leq 0.5 \\ (R-\mu_r-0.5)/\sigma, \text{if R - } \mu \geq 0.5 \end{cases}$. We get a two-tailed p-value of 0.374, which is larger than 0.05, so we fail to reject the null hypothesis.

Section 15.3 Inferences about Measures of Central Tendency

►One-Sample Sign Test

Example 1, Page 15-13

Type the data into SPSS.

	Debt
1	6000
2	0
3	200
4	0
5	400
6	1060
7	0
8	1200
9	200
10	250
11	250
12	580
13	1000
14	0
15	0
16	200
17	400
18	800
19	700
20	1000

Once the data have been entered, go to **Analyze → Nonparametric Tests → Binomial...**.

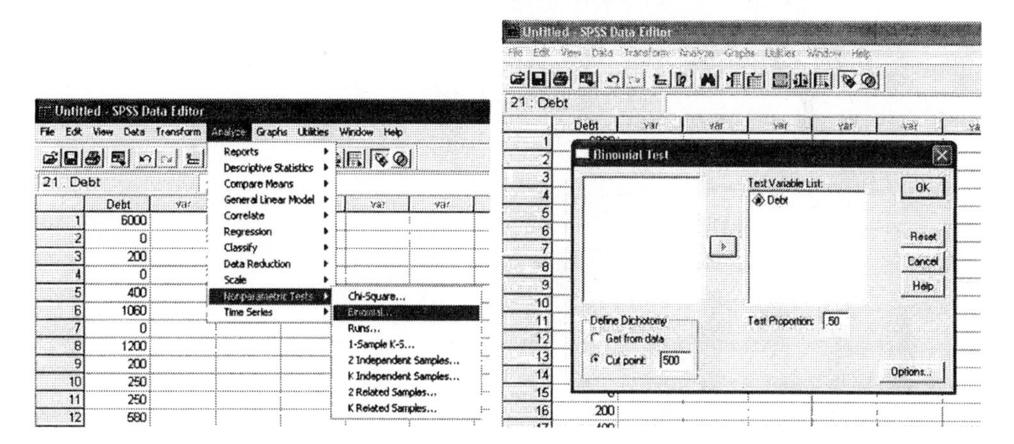

Select the variable of interest as the Test Variable List, and your hypothesized median as the Cut Point. Keep the Test Proportion at .50 (the median). When the values have been entered, push OK.

Binomial Test

		Category	N	Observed Prop.	Test Prop.	Exact Sig. (2-tailed)
Debt	Group 1	<= 500	12	.60	.50	.503
	Group 2	> 500	8	.40		
	Total		20	1.00		

The output shows how many observations were less than the hypothesized median (12), and how many were above the hypothesized median (8). It also provides the p-value. We do not, in this case, have enough evidence to show that the median is not 500.

One note: SPSS does not take out ties. It groups the less than or equal to in the same group instead of removing those values which are equal to the hypothesized median.

Section 15.4 Inferences about the Difference between Two Measures of Central Tendency: Dependent Samples

▶ Wilcoxon Matched-Pairs Signed Rank Test

Example 1, Page 15-21

We start by typing in the data. This is similar to the paired t-test. We create two variables.

Stock Monthly Volume.sav - SPSS Data Editor

File Edit View Data Transform Analyze Graphs Utilities Wi

15 : JDS

	Day	Nortel	JDS
1	1.00	11.50	26.00
2	2.00	14.10	26.20
3	3.00	19.30	24.60
4	4.00	35.00	30.60
5	5.00	15.90	37.50
6	6.00	21.70	36.00
7	7.00	11.70	25.90
8	8.00	17.10	16.90
9	9.00	27.30	50.20
10	10.00	13.80	33.10
11	11.00	43.20	99.90
12	12.00	11.20	26.10
13	13.00	34.20	43.80
14	14.00	26.70	63.60
15			
16			

Once the data have been entered, go to **Analyze → Nonparametric Tests → 2 Related Samples….**

Similar to the dependent sample t-test, select the two variables. Again, the test will be run in the order of the columns.

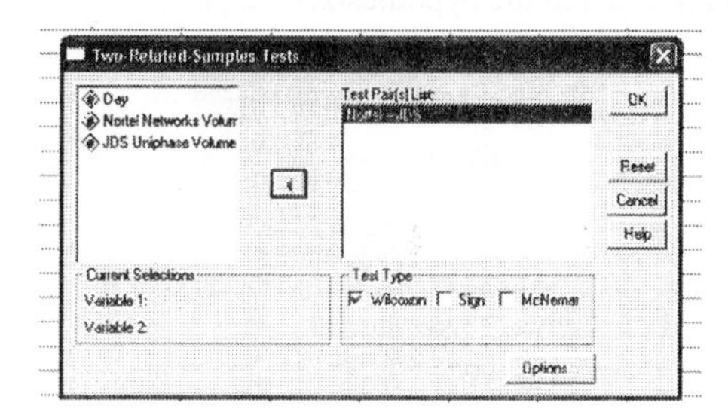

Make sure that the Wilcoxon option is checked, and press OK.

Ranks

		N	Mean Rank	Sum of Ranks
JDS Uniphase Volume - Nortel Networks Volume	Negative Ranks	2[a]	1.50	3.00
	Positive Ranks	12[b]	8.50	102.00
	Ties	0[c]		
	Total	14		

a. JDS Uniphase Volume < Nortel Networks Volume

b. JDS Uniphase Volume > Nortel Networks Volume

c. JDS Uniphase Volume = Nortel Networks Volume

Test Statistics[b]

	JDS Uniphase Volume - Nortel Networks Volume
Z	-3.107[a]
Asymp. Sig. (2-tailed)	.002

a. Based on negative ranks.

b. Wilcoxon Signed Ranks Test

The output provides the mean (1.50) and sum (2) of the negative ranks, the mean (8.50) and sum (12) of the positive ranks, how many ties (0) there were, and the p-value for a two tailed test (0.002). We can get the p-value for the one-tailed test by dividing this by 2.

Section 15.5 Inferences about the Difference between Two Measures of Central Tendency: Independent Samples

►Mann-Whitney Test

Example 1, Page 15-30

We start by typing in the data. This is similar to the independent t-test. We create two variables, one for the values and one for the group.

Untitled - SPSS Data Editor

File Edit View Data Transform Analyze

23 : Group

	titers	Group
1	640.00	III
2	160.00	III
3	1280.00	III
4	320.00	III
5	80.00	III
6	640.00	III
7	640.00	III
8	160.00	III
9	1280.00	III
10	640.00	III
11	160.00	III
12	10.00	Healthy
13	320.00	Healthy
14	160.00	Healthy
15	160.00	Healthy
16	320.00	Healthy
17	320.00	Healthy
18	10.00	Healthy
19	320.00	Healthy
20	320.00	Healthy
21	80.00	Healthy
22	640.00	Healthy

Once the data have been entered, go to **Analyze → Nonparametric Tests → 2 Independent Samples….**

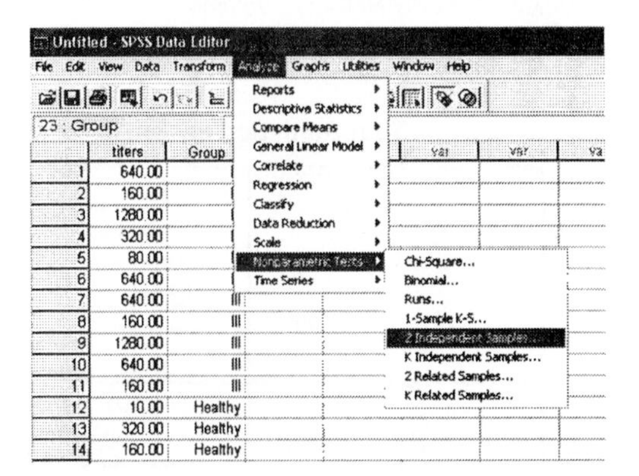

In the Two-Independent Samples Test window, we follow the basic steps of the independent t-test. We select the variable with the values for the Test Variable List, and the variable defining the group as the Grouping Variable. We must also define what the two groups are.

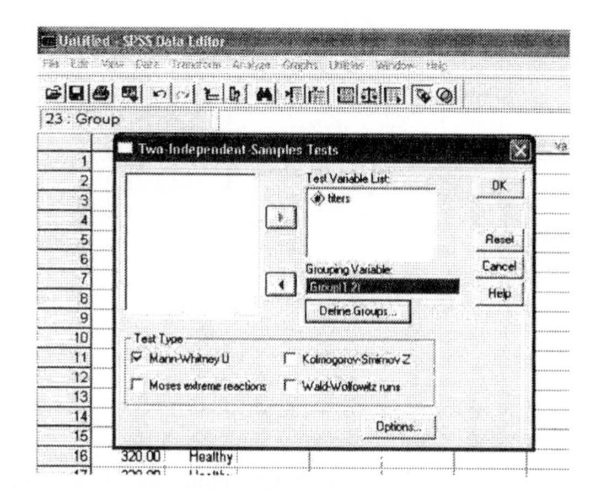

Make sure that Mann-Whitney U is checked, and push OK.

Ranks

	Group	N	Mean Rank	Sum of Ranks
titers	Ill	11	13.82	152.00
	Healthy	11	9.18	101.00
	Total	22		

Test Statistics[b]

	titers
Mann-Whitney U	35.000
Wilcoxon W	101.000
Z	-1.713
Asymp. Sig. (2-tailed)	.087
Exact Sig. [2*(1-tailed Sig.)]	.101[a]

a. Not corrected for ties.

b. Grouping Variable: Group

The output provides the mean and sum of the ranks for each group, along with the Mann-Whitney U statistic, and p-values. The p-value is for the two-tailed test, so for a one-tailed test we divide this in half, or 0.0507 in this example. Since the p-value is smaller than .1, we reject the null hypothesis.

Example 2, Page 15-32

Following the same steps as Example 1, we get the following output:

Ranks

	State	N	Mean Rank	Sum of Ranks
pH	Texas	22	16.09	354.00
	Montana	20	27.45	549.00
	Total	42		

Test Statistics[a]

	pH
Mann-Whitney U	101.000
Wilcoxon W	354.000
Z	-2.997
Asymp. Sig. (2-tailed)	.003

a. Grouping Variable: State

This shows the test statistic of 101 for the Mann-Whitney test, as well as the z-value of -2.997. Note that the Exact Sig. value (Exact Significance is based on the Mann-Whitney table) is no longer part of the output. Only the Asymptotic p-value, for the z-test is available.

Section 15.6 Spearman's Rank-Correlation Test

►Spearman's Rank-Correlation Test

Example 1, Page 15-39

This test follows the steps for the parametric (Pearson Correlation) test almost exactly in SPSS. We begin by typing in the data. After entering the data into SPSS, go to **Analyze → Correlate → Bivariate...** (two variables)

Highlight the variables you wish to find the correlation between and push the ► button.

Make sure that Spearman is selected. Pearson may be selected for the parametric correlation tests.

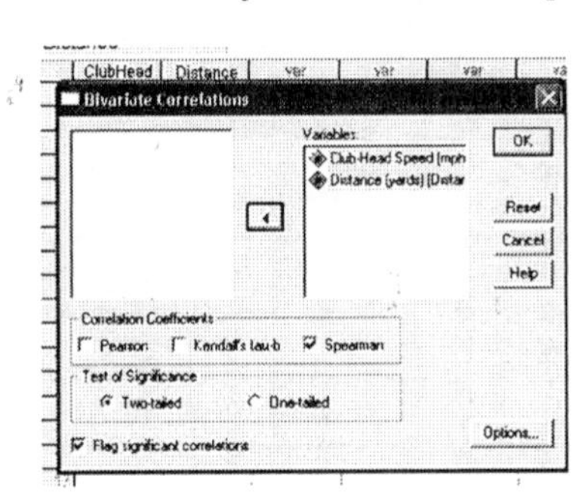

Push OK

Correlations

			Club-Head Speed (mph)	Distance (yards)
Spearman's rho	Club-Head Speed (mph)	Correlation Coefficient	1.000	.928**
		Sig. (2-tailed)	.	.001
		N	8	8
	Distance (yards)	Correlation Coefficient	.928**	1.000
		Sig. (2-tailed)	.001	.
		N	8	8

**. Correlation is significant at the 0.01 level (2-tailed).

Here we have a correlation value of 0.928, with a p-value of 0.001, which shows that there is significant evidence to support the claim that GDP and Life Expectancy are associated.

Section 15.7 Kruskal-Wallis Test of One-Way Analysis of Variance

►Kruskal-Wallis One-Way Analysis of Variance Test

Example 1, Page 15-45

First, we must type in the data. This is done similar to the regular ANOVA test, or the Mann-Whitney test, with the value of interest in one column, and the group in another. The groups **must** be numerical, but you can use labels, which will help in the output.

HDL Cholesterol.sav - SPSS Data Editor

File Edit View Data Transform Analyze Graphs

37 : Group

	HDL	Group
1	54	20 to 29 years old
2	43	20 to 29 years old
3	38	20 to 29 years old
4	30	20 to 29 years old
5	61	20 to 29 years old
6	53	20 to 29 years old
7	35	20 to 29 years old
8	34	20 to 29 years old
9	39	20 to 29 years old
10	46	20 to 29 years old
11	50	20 to 29 years old
12	35	20 to 29 years old
13	61	40 to 49 years old
14	41	40 to 49 years old
15	44	40 to 49 years old
16	47	40 to 49 years old
17	33	40 to 49 years old
18	29	40 to 49 years old
19	59	40 to 49 years old
20	35	40 to 49 years old
21	34	40 to 49 years old
22	74	40 to 49 years old
23	50	40 to 49 years old
24	65	40 to 49 years old
25	44	60 to 69 years old
26	65	60 to 69 years old
27	62	60 to 69 years old
28	53	60 to 69 years old
29	51	60 to 69 years old
30	49	60 to 69 years old
31	49	60 to 69 years old
32	42	60 to 69 years old
33	35	60 to 69 years old
34	44	60 to 69 years old
35	37	60 to 69 years old
36	38	60 to 69 years old

Once the data have been entered, go to **Analyze → Correlate → K Independent Samples...**

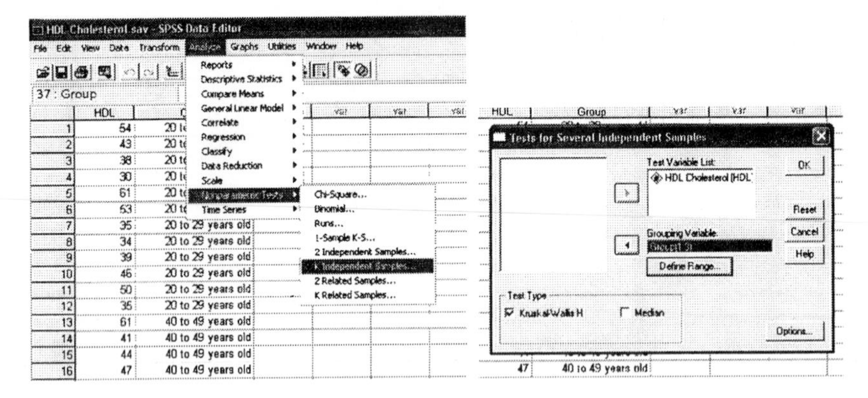

Similar to the one-way ANOVA, we select the values as the Test Variable List, and the variable defining the groups as the Grouping Variable. Unlike the one-way ANOVA, we must define which groups we want to use, in a range.

Ranks

	Age Group	N	Mean Rank
HDL Cholesterol	20 to 29 years old	12	16.21
	40 to 49 years old	12	18.79
	60 to 69 years old	12	20.50
	Total	36	

Test Statistics[a,b]

	HDL Cholesterol
Chi-Square	1.012
df	2
Asymp. Sig.	.603

a. Kruskal Wallis Test

b. Grouping Variable: Age Group

The output provides the mean rank for each group, the Kruskal-Wallis test statistic, and the p-value (Asymp. Sig). Because the p-value is not less than 0.05, we fail to reject the null hypothesis. There is not sufficient evidence to show that the distributions of HDL cholesterol for the three age groups are different.

PHStat 2.5 with Data Files for use with the Technology Manual to accompany Statistics: Informed Decisions Using Data, 2e
Michael Sullivan, III
ISBN 0-13-227556-2
CD License Agreement

Pearson Prentice Hall
Pearson Education, Inc.
Upper Saddle River, NJ 07458

READ THIS LICENSE CAREFULLY BEFORE OPENING THIS PACKAGE. BY OPENING THIS PACKAGE, YOU ARE AGREEING TO THE TERMS AND CONDITIONS OF THIS LICENSE. IF YOU DO NOT AGREE, DO NOT OPEN THE PACKAGE. PROMPTLY RETURN THE UNOPENED PACKAGE AND ALL ACCOMPANYING ITEMS TO THE PLACE YOU OBTAINED THEM. THESE TERMS APPLY TO ALL LICENSED SOFTWARE ON THE DISK EXCEPT THAT THE TERMS FOR USE OF ANY SHAREWARE OR FREEWARE ON THE DISKETTES ARE AS SET FORTH IN THE ELECTRONIC LICENSE LOCATED ON THE DISK:

1. **GRANT OF LICENSE and OWNERSHIP:** The enclosed CD-ROM ("Software") is licensed, not sold, to you by Pearson Education, Inc. publishing as Pearson Prentice Hall ("We" or the "Company") in consideration of your adoption of the accompanying Company textbooks and/or other materials, and your agreement to these terms. You own only the disk(s) but we and/or our licensors own the Software itself. This license allows instructors and students enrolled in the course using the Company textbook that accompanies this Software (the "Course") to use and display the enclosed copy of the Software on up to one computer of an educational institution, for academic use only, so long as you comply with the terms of this Agreement. You may make one copy for back up only. We reserve any rights not granted to you.
2. **USE RESTRICTIONS:** You may not sell or license copies of the Software or the Documentation to others. You may not transfer, distribute or make available the Software or the Documentation, except to instructors and students in your school who are users of the adopted Company textbook that accompanies this Software in connection with the course for which the textbook was adopted. You may not reverse engineer, disassemble, decompile, modify, adapt, translate or create derivative works based on the Software or the Documentation. You may be held legally responsible for any copying or copyright infringement that is caused by your failure to abide by the terms of these restrictions.
3. **TERMINATION:** This license is effective until terminated. This license will terminate automatically without notice from the Company if you fail to comply with any provisions or limitations of this license. Upon termination, you shall destroy the Documentation and all copies of the Software. All provisions of this Agreement as to limitation and disclaimer of warranties, limitation of liability, remedies or damages, and our ownership rights shall survive termination.
4. **DISCLAIMER OF WARRANTY: THE COMPANY AND ITS LICENSORS MAKE NO WARRANTIES ABOUT THE SOFTWARE, WHICH IS PROVIDED "AS-IS." IF THE DISK IS DEFECTIVE IN MATERIALS OR WORKMANSHIP, YOUR ONLY REMEDY IS TO RETURN IT TO THE COMPANY WITHIN 30 DAYS FOR REPLACEMENT UNLESS THE COMPANY DETERMINES IN GOOD FAITH THAT THE DISK HAS BEEN MISUSED OR IMPROPERLY INSTALLED, REPAIRED, ALTERED OR DAMAGED. THE COMPANY DISCLAIMS ALL WARRANTIES, EXPRESS OR IMPLIED, INCLUDING WITHOUT LIMITATION, THE IMPLIED WARRANTIES OF MERCHANTABILITY AND FITNESS FOR A PARTICULAR PURPOSE. THE COMPANY DOES NOT WARRANT, GUARANTEE OR MAKE ANY REPRESENTATION REGARDING THE ACCURACY, RELIABILITY, CURRENTNESS, USE, OR RESULTS OF USE, OF THE SOFTWARE.**
5. **LIMITATION OF REMEDIES AND DAMAGES: IN NO EVENT, SHALL THE COMPANY OR ITS EMPLOYEES, AGENTS, LICENSORS OR CONTRACTORS BE LIABLE FOR ANY INCIDENTAL, INDIRECT, SPECIAL OR CONSEQUENTIAL DAMAGES ARISING OUT OF OR IN CONNECTION WITH THIS LICENSE OR THE SOFTWARE, INCLUDING, WITHOUT LIMITATION, LOSS OF USE, LOSS OF DATA, LOSS OF INCOME OR PROFIT, OR OTHER LOSSES SUSTAINED AS A RESULT OF INJURY TO ANY PERSON, OR LOSS OF OR DAMAGE TO PROPERTY, OR CLAIMS OF THIRD PARTIES, EVEN IF THE COMPANY OR AN AUTHORIZED REPRESENTATIVE OF THE COMPANY HAS BEEN ADVISED OF THE POSSIBILITY OF SUCH DAMAGES.** SOME JURISDICTIONS DO NOT ALLOW THE LIMITATION OF DAMAGES IN CERTAIN CIRCUMSTANCES, SO THE ABOVE LIMITATIONS MAY NOT ALWAYS APPLY.
6. **GENERAL:** THIS AGREEMENT SHALL BE CONSTRUED IN ACCORDANCE WITH THE LAWS OF THE UNITED STATES OF AMERICA AND THE STATE OF NEW YORK, APPLICABLE TO CONTRACTS MADE IN NEW YORK, EXCLUDING THE STATE'S LAWS AND POLICIES ON CONFLICTS OF LAW, AND SHALL BENEFIT THE COMPANY, ITS AFFILIATES AND ASSIGNEES. This Agreement is the complete and exclusive statement of the agreement between you and the Company and supersedes all proposals, prior agreements, oral or written, and any other communications between you and the company or any of its representatives relating to the subject matter. If you are a U.S. Government user, this Software is licensed with "restricted rights" as set forth in subparagraphs (a)-(d) of the Commercial Computer-Restricted Rights clause at FAR 52.227-19 or in subparagraphs (c)(1)(ii) of the Rights in Technical Data and Computer Software clause at DFARS 252.227-7013, and similar clauses, as applicable. Should you have any questions concerning this agreement or if you wish to contact the Company for any reason, please contact in writing: Pearson Education, Inc., One Lake Street, Upper Saddle River, New Jersey 07458 "AS IS" LICENSE

****
SYSTEM REQUIREMENTS
****
*Microsoft Windows, Windows 98, Windows NT, Windows 2000, Windows ME, or Windows XP
*In addition to the minimum processor requirements for the operating system your computer is running, this CD requires a Pentium II, 200 MHz or higher processor
*In addition to the RAM required by the operating system your computer is running, this CD requires 64 MB RAM for Windows 98, Windows NT 4.0, Windows 2000, Windows ME, and Windows XP
*Macintosh OS 9.x or 10.x
*In addition to the minimum processor requirements for the operating system your computer is running, this CD requires a PowerPC G3 233 MHz or better
*In addition to the RAM required by the operating system your computer is running, this CD requires 64 MB RAM
*Microsoft Excel 97, 2000, 2002, or 2003 (Excel 97 use must apply the SR-2 or a later free update from Microsoft in order to use PHStat 2.5. Excel 2000 and 2002 must have the macro security level set to Medium) for Windows; MINITAB Version 14 or 12; JMP version 5.1; SPSS versions 13.0, 12.0, 11.0; or other statistics software.
*PHStat will not work on the Macintosh
*Microsoft Excel Data Analysis ToolPak and Analysis ToolPak VBA installed (supplied on the Microsoft Office/Excel program CD)
*CD-ROM or DVD-ROM drive; Mouse and keyboard; Color monitor (256 or more colors and screen resolution settings set to 800 by 600 pixels or 1024 by 748 pixels)
*Internet browser for Windows (Netscape 4.x, 6.x or 7.x, or Internet Explorer 5.x or 6.x) and Internet connection suggested but not required
*Internet browser for Macintosh (Internet Explorer 5.x or Safari 1.x) and Internet connection suggested but not required
*This CD-ROM is intended for stand-alone use only. It is not meant for use on a network.
****
CD-ROM CONTENTS
****
--Data Files : Included within that folder are:
--JMP Data Files (Folder: JMP_Sullivan_Data)
JMP is required to view and use these files. Information about JMP can be found on the internet at http://www.jmp.com
--SPSS Data Files (Folder: SPSS_Sullivan_Data)
SPSS is required to view and use these files. Information about SPSS can be found on the internet at http://www.spss.com
--MINITAB Data Files (Folder: minitab_Sullivan_Data)
MINITAB is required to view and use these files. Information about MINITAB can be found on the internet at http://www.minitab.com/support/index.htm.
--Data Files (Folder: ASCII_Sullivan_Data)
For use with SPSS or other statistics software.
--Excel Files (Folder: Excel_Sullivan_Data)
Excel is required to view and use these files. Information about Excel can be found on the Internet at http://office.microsoft.com/en-us/default.aspx
--TI-83/84 Files (Folder: TI-8x_Sullivan_Data)
TI-83/84 calculator is required http://education.ti.com/us/product/main.html.
--PHStat 2.5
Prentice Hall's PHStat statistical add-in system enhances Microsoft Excel to better support learning in an introductory statistics course. http://www.prenhall.com/phstat/
--Readme.txt
****
TECHNICAL SUPPORT
****
If you continue to experience difficulties, call 1 (800) 677-6337, 8 am to 8 pm Monday through Friday and 5 pm to 12 am Sunday (all times Eastern) or visit Prentice Hall's Technical Support Web site at http://247.prenhall.com/mediaform.
Our technical staff will need to know certain things about your system in order to help us solve your problems more quickly and efficiently. If possible, please be at your computer when you call for support. You should have the following information ready:

- Textbook ISBN
- CD-Rom/Diskette ISBN
- Corresponding product and title
- Computer make and model
- Operating System (Windows or Macintosh) and Version
- RAM available
- Hard disk space available
- Sound card? Yes or No
- Printer make and model
- Network connection
- Detailed description of the problem, including the exact wording of any error messages.

NOTE: Pearson does not support and/or assist with the following:
- 3d-party software (i.e. Microsoft including Microsoft Office Suite, Apple, Borland, etc.)
- Homework assistance
- Textbooks and CD-ROMs purchased used are not supported and are non-replaceable.

For assistance with third-party software, please visit:
JMP Support (for JMP): http://www.jmp.com/support/techsup/index.shtml
MINITAB Support: http://www.minitab.com/support/index.htm
SPSS Support: http://www.spss.com/tech/spssdefault.htm
TI-8x Support: http://education.ti.com/us/support/main.html
Excel Support: http://support.microsoft.com/
PHStat Support: http://www.prenhall.com/phstat/phstat2/phstat2(main).htm
----